PRAISE FOR *THE DECEIVERS*

"Thad Holt has written a brilliant account of Allied military deceptions in World War II. It will become a standard work on military intelligence tactics. And best of all, it reads like a novel!"
> — Joseph S. Nye, Jr., Dean of the Kennedy School of Government, Harvard University

"Mr. Holt's history of Allied deception in World War II is definitive. He has trawled through all the documentation, interviewed all the survivors, and put together a history as comprehensive as it is readable and entertaining. It is an astonishing achievement, and no library of the war can afford to be without it."
> — Professor Sir Michael Howard, author of *British Intelligence in the Second World War: Volume 5, Strategic Deception*

"Just when you have convinced yourself that you have long ago imbibed the last word on the secret side of World War II, along comes Thaddeus Holt and his remarkable study. Superbly researched and full of fresh revelations, *The Deceivers* is not only immaculately written but wonderfully readable."
> — Robert Cowley, founding editor of *MHQ: The Quarterly Journal of Military History*

"A highly professional yet entertaining analysis of the dirty tricks ingeniously dreamed up by unscrupulous Allied intelligence personnel in World War II to defeat the enemy. Easily the best book yet written, or ever likely to be, on the subject, drawing on the most recent documents declassified on both sides of the Atlantic."
> — Nigel West, author of *M15*, *M16* and *Seven Spies Who Changed the World*

"A truly wonderful book! Deeply researched and written with authority and verve, it tells the full story of Allied deception during World War II. *The Deceivers* not only recounts every major operation, it describes in detail how each operation affected the enemy. It will be essential for the bookshelf of every serious student of World War II."
> — Ernest May, Charles Warren Professor of History, Harvard University

Brigadier Dudley W. Clarke, C.B., C.B.E.
The Master of the Game

THE DECEIVERS

ALLIED MILITARY DECEPTION
IN THE SECOND WORLD WAR

THADDEUS HOLT

SKYHORSE PUBLISHING

www.skyhorsepublishing.com

10 9 8 7 6 5 4 3 2 1

Library of Congress Cataloging-in-Publication Data
Holt, Thaddeus.
  The deceivers : Allied military deception in the Second World War / Thaddeus Holt.
    p. cm.
  Originally published: New York : Scribner, c2004.
  ISBN-13: 978-1-60239-142-0 (alk. paper)
  ISBN-10: 1-60239-142-4 (alk. paper)
  1. World War, 1939-1945—Deception—Great Britain. 2. World War, 1939-1945-Deception-United States. 3. Deception (Military science) I. Title.

D744.H64 2007
940.54'86-dc22
                          2007015369

Printed in India

For W.
Thank you

CONTENTS

CONTENTS

ILLUSTRATIONS

PREFACE

Virgil put it with Roman bluntness and economy. *Dolus an virtus, quis in hoste requirat?* cried Aeneas's comrade as they fought their way out of burning Troy disguised in Greek armor; which may be loosely translated: It won't matter to the enemy whether you beat him by guile or by valor.

The Western Allies in the Second World War beat their enemies by valor in full measure. But that valor was aided by guile on a level never before seen; the most systematic and skillful deception ever practiced in warfare. This is the story of that guile. In particular, it is the story of the people who practiced it; of the cadre of officers, British and American, who developed and applied that system and that skill. Some were regular soldiers and sailors and security men, but the men of guile, like the men of valor, were mostly plain civilians who had left their peaceful occupations for a time to help their country. They were a cross-section of free citizens: old Etonians and plain Midwesterners, a City of London financier, an international banker, a railroad executive, a gentleman wine-merchant and thriller-writer, a construction contractor, an art dealer, a sociology professor, a film star, a colonial administrator, one of New York's leading journalists, one of the pioneers of stereophonic sound, a world traveler and adventurer; businessmen and schoolmasters and lawyers and Oxford dons. This is their tale.

That tale, particularly its American side, has never been fully told. During the war and for years afterwards the very fact that the Allies engaged in strategic deception at all was a closely-held secret. Portions of the story came out over the years, often in distorted form, often sensationalized. Balance and accuracy were restored, at least for the British side, when Professor Sir Michael Howard's British official history was published in 1990. But only in recent years have important original documents been declassified. And the present author is the first nonofficial historian to have had access to the surviving records of the American Joint Security Control.

Mistakes are inevitable in any work of this length. They should be charged to the author and not to any of the scores of people who so kindly helped him, and without whose aid it could not have been written.

First and foremost are those who shared with the author their first-hand memories, ranging from long interviews, in person or by telephone, to brief notes. They include Mrs. Clarence Beck, Lady Jane Bethell, Desmond Bristow, Anita Burris-Meyer, Alice Crichton, Kenneth M. Crosby, Philip C. Curtis, Arne Ekstrom, Diana Eldredge, Douglas Fairbanks, Mark Felt, Richard D. Fletcher, Stuart Giles, Carl Goldbranson, Jr., Rex Hamer, Christopher Harmer, Sir John Harvey-Jones, Erwin C. Lochmueller, Robert A. Maheu, Lady Maynard, Colonel John R. Nickel, John B. Oakes, Captain Wyman Packard, General Robert Porter, The Countess of Ranfurly, Canon David Strangeways, Father Eugene J. Sweeney, Will Hill Tankersley, Sir Peter Thorne, W. Ray Wannall, Colonel H. N. H. Wild, and Sir Edgar Williams.

Family members of participants who generously made personal papers and other sources of information available include Natalie Baumer, Julian Bevan, Nicolas Ekstrom, James W. Eldredge, Nicholas P. V. Fleming, Carl Goldbranson, Jr., Donna Goldbranson, Ann Greenwood, Kate Grimond, Richard Kehm, Eleanor Lanahan, Wayne Lawrence, Lady Lowther, Mrs. Eldon Schrup, Harriet Harris Stroup, and Susan Train.

Others who helped in identifying and gaining access to participants and first-hand witnesses, family members, and relevant books and papers include The Earl Alexander of Tunis, Elizabeth Bancroft, Stuyvesant Bearns, George Blow, Christopher Buchanan, Carleton Coon, Katya Doolittle Coon, The Earl of Erne, Mikell Grafton, Duff Hart-Davis, Malcolm McLean, Ernest May, Joseph McPhillips, James Parton, Nell Pillard, and John Train.

Many authorities in relevant fields generously shared their knowledge, and in some cases gave of their time to read portions of the draft manuscript: Rupert Allason, Professor Carl Boyd, Thomas Buell, Professor Matthew J. Bruccoli, Professor John P. Campbell, Brice M. Clagett, Thomas M. Constant, John Costello, Robert Cowley, Carlo W. d'Este, Douglas W. Dearth, Edward J. Drea, John B. Dwyer, Keith Eiler, Colonel the Hon. P. N. Trustram Eve, Professor M. R. D. Foot, Katherine L. Herbig, Professor Sir Michael Howard, Thomas M. Huber, David Irving, David Kahn, Ilana Kass, Professor MacGregor Knox, Arnold Kranisch, Thomas J. Macpeak, Mel and Jim Meehan, Timothy J. Naftali, George Nash, Galen Roger Perras, Professor Carol M. Petillo, Walter Pforzheimer, Thomas Powers, Professor Richard Gidd Powers, Professor Donal J. Sexton, William Spahr, Robbie Stamp, Brian Sullivan, Thomas Troy, John H. Waller, C. M. Woods, and Martin Young. Robert C. Thweatt read the entire draft manuscript and contributed many valuable comments.

Very special thanks are owed to Brigadier General David A. Armstrong, Director for Joint History, Office of the Chairman, Joint Chiefs of Staff, who located the surviving files of Joint Security Control in some unlikely place in the bowels of the Pentagon and made arrangements for the author to examine them; to Lieutenant General Theodore G. Stroup, Jr., through whose good offices the author was first put in touch with General Armstrong; to Commander Moira Wurzel and Will Kammer of the Documents Division, Joint Secretariat, and their colleagues for their help, hospitality, and patience during the long process of reviewing those files; and to Anthony Passarella, Director for Freedom of Information and Security Review in the Office of the Secretary of Defense, for expediting the otherwise interminable process of review and declassification. Liles Creighton gave the author access to other important documents in Pentagon files and arranged for their declassification.

John Taylor of the National Archives in Washington (now in College Park, Maryland), gave to the author freely and abundantly of his unrivaled knowledge of American records of the Second World War, as he has done for other researchers beyond number. He has been justly characterized as an American living national treasure; his contribution to our lasting knowledge of that era can never be measured.

Others at public and private institutions who generously aided the author's quest include Dixie Dysart, Lieutenant Sergio Lopez, and Warren Trest of the Air Force Historical Agency, Maxwell AFB, Alabama; Robert Lane of the Air University Library, Maxwell AFB, Alabama; Peter Geissler of the Auswärtiges Amt, Bonn; Howard Gotlieb and Charles Niles of the Boston University Library, Boston; Pat Andrews of the Cabinet Office, London; Neil Cooke and Phil Reed of the Cabinet War Rooms, London; Louise King and Katherine Thomson of the Churchill

College Archives Centre, Cambridge; Dwight Strandberg and Herbert Pankratz of the Eisenhower Library, Abilene, Kansas; Jane Barton, Susan Diemert, and Patty Gipson of the Fairhope Public Library, Fairhope, Alabama; Rex Tomb of the Federal Bureau of Investigation, Washington; Glenn Cook of the George C. Marshall Papers, Lexington, Virginia; Barbara Heuer of the Georgia Historical Society, Savannah; Lena Daniels, Carol Leadenham, and Catherine Reynolds of the Hoover Institution Library, Palo Alto, California; Simon Robbins and Stephen Walton of the Imperial War Museum, London; Wil Mahoney of the National Archives, Washington and College Park, Maryland; Bernard Cavalcante, Kathy Lloyd, and Eleanor Nargele of the Naval Operational Archives, Washington; Ralph Bryan, Ian Malcolm, and Gerry Toop of the Public Record Office, Kew; Brigadier General John W. Mountcastle of the U.S. Army Center for Military History, Washington; Ted Riddick of the U.S. Army Command and General Staff School Library, Fort Leavenworth, Kansas; Thomas B. Proffitt of the U.S. Army Institute of Heraldry, Alexandria, Virginia; Louise Arnold Friend, John Slonaker, and Richard Sommers of the U.S. Army Military History Institute, Carlisle Barracks, Pennsylvania; Alice Creighton, Susan Dean, and Gary LaValley of the U.S. Naval Academy Special Collections and Archives, Annapolis, Maryland; and Paul Stilwell of the U.S. Naval Institute, Annapolis, Maryland. Jamie Bronstein and William Scaring searched specific files for the author at the Hoover Institution and Boston University libraries, respectively. Professor Brian P. Farrell provided good company during a visit to the Public Record Office, and helped clear up some research points for the author.

The author was blessed with two remarkable editors in succession, Marc Jaffe and Lisa Drew. They were the editors that any author dreams of: loyal, patient, astute, understanding, and overcoming unexpected obstacles with steady firmness.

Last, and very far from least, special thanks to Phyllis Westberg of Harold Ober Associates for her help and her patience.

T. H.

Point Clear, Alabama

January 2004

LIST OF ABBREVIATIONS

AAI	Allied Armies in Italy. The term used from March to December 1944 for the army group consisting of the British Eighth and American Fifth Armies; known as "Allied Central Mediterranean Force" from January to March 1944, and as 15th Army Group at all other times.
Abwehr	Nominally the security element of the German Amt Auslandsnachrichten und Abwehr, or Office of Foreign and Counter Intelligence, of OKW. In fact, the German secret intelligence service.
AFHQ	(Pronounced "Affkew.") Allied Force Headquarters. Headquarters for the Mediterranean theater of operations.
ATS	Auxiliary Territorial Service. The British Army's women's organization.
BCRA	Bureau Centrale de Renseignements et d'Action. General de Gaulle's Free French intelligence service, merged in November 1943 with Paul Paillole's DSM to form the DGSS (later renamed the DGER).
BSC	British Security Coordination. The office of MI6 in the United States.
CBI	The American China-Burma-India Theater.
CEA	Controlled Enemy Agent. The OSS term for what is usually called a "double agent."
CIGS	Chief of the Imperial General Staff. The chief of staff of the British Army.
CINCAFPAC	Commander in Chief, Army Forces Pacific. MacArthur's command from April 3, 1945, in addition to his role as CINCSWPA; embracing all Army ground and air forces in the Pacific.
CINCPAC	Commander in Chief, Pacific. Nimitz's command from April 3, 1945, in addition to his role as CINCPOA; embracing all Navy forces in the Pacific.
CINCPOA	Commander in Chief, Pacific Ocean Areas. Admiral Nimitz's overall theater in the Pacific.

COMINCH   Commander in Chief, United States Fleet. The title held by Admiral Ernest J. King in his capacity as overall commander of the Fleet; also refers to the separate staff that served him in that capacity. See CNO.

CMBE   Combined Bureau Middle East. A Middle Eastern offshoot of the cryptographic activity headquartered at Bletchley Park in England.

CNO   Chief of Naval Operations. The title held by Admiral Ernest J. King in his capacity as head of supporting activities for the Fleet; also refers to the separate staff that served him in that capacity. See COMINCH.

COMNORPAC   Commander, North Pacific. The United States Navy command for Alaskan waters; subsidiary to CINCPOA.

Deuxième Bureau   The traditional name for the French intelligence and counterintelligence service. The name was officially discontinued shortly after the beginning of the war but continued to be used as a shorthand compendious term for the various successor organizations.

DGER   Direction Générale des Études et Recherches. The former DGSS, renamed.

DGSS   Direction Générale des Services Spéciaux. The intelligence service formed in November 1943 by the merger of BCRA and Paul Paillole's DSM; commonly still referred to as the DSM. Later renamed the DGER.

DIB   Delhi Intelligence Bureau. The Intelligence Bureau of the Home Department of the Government of India. "DIB" referred to the organization itself; "*the* DIB" referred to the Director of DIB.

DSDOC   Direction Générale des Services de Documentation. The counterespionage section of the DGER.

DSM   Direction des Services de Renseignements et des Services de Securité Militaire. The French intelligence and counterespionage service; formerly the Deuxième Bureau, renamed after the beginning of the war.

DSO   Defense Security Officer. An officer in charge of MI5 affairs at an outstation. Also Distinguished Service Order, a high British decoration.

ETOUSA   European Theater of Operations, United States Army.

FBI   Federal Bureau of Investigation.

FFI   Forces Françaises de l'Intérieur, French Forces of the Interior, the army of the French Resistance.

FHW   Fremde Heere West. The intelligence office of OKH responsible for keeping up with American and British forces.

FOPS   Future Operational Planning Section. A section of the British Joint Planners whose main business was to plan for the return of British arms to the Continent of Europe.

GC&CS   Government Code and Cipher School. The British cryptographic organization.

ISLD   Inter-Service Liaison Department. A cover name for MI6 in the Middle East and India.

ISOS   Intelligence Service, Oliver Strachey. The term for ULTRA Abwehr decrypts.

ISSB   Inter-Service Security Board. A British War Office committee of representatives from the three fighting services and from MI5 and MI6, in charge of overall security.

MI5   The term commonly used for the Security Service, the British counterintelligence service for Britain and British territory overseas.

MI6   The term commonly used for the Secret Intelligence Service, the British foreign intelligence and counter-intelligence service.

OKH   Oberkommando des Heeres. The High Command of the German Army.

OKL   Oberkommando der Luftwaffe. The High Command of the German Luftwaffe.

OKM   Oberkommando der Marine. The High Command of the German Navy.

OKW   Oberkommando der Wehrmacht. The High Command of the German armed forces.

OPD   Operations Division, War Department General Staff. The Washington planning and operational command post for the United States Army.

OSS   Office of Strategic Services. The American secret intelligence and covert action service.

OSS/X-2   The security element of OSS.

OVRA   The secret police and security service of the Italian Fascist Party. The name is variously said to be an acronym for Organizzazione per la Vigilanza e la Repressione dell'Antifascismo, or for Opera Voluntario de Repressione Antifascista, or for Organo di Vigilanza dei Reati Antistatali; or to be a meaningless name adopted by Mussolini to suggest mystery and ubiquity, along the line of the Soviet OGPU.

PPF   Parti Populaire Français, a French fascist and collaborationist political party.

RSHA   Reichssicherheitshauptamt. The security service responsible to Heinrich Himmler, which eventually took over the Abwehr.

SAC   Special Agent in Charge. The FBI special agent in charge of a field office.

SACSEA   Supreme Allied Commander, Southeast Asia (Admiral Lord Louis Mountbatten).

SCI   Special Counterintelligence. In British usage, refers to a special detachment from MI5 and MI6, responsible on the Continent for double agent operation. In American usage, refers to a detachment from OSS X-2 responsible for ULTRA support to counterintelligence, and secondarily for double agent operation.

SCU   Special Case Unit. A sub-unit of an American SCI, responsible for double agent operation.

SD   The Sicherheitsdienst; the Nazi Party intelligence service, controlled by the RSHA.

SEAC   Southeast Asia Command. The Allied command covering Burma, Ceylon, Sumatra, and Malaya, operating informally in Thailand and French Indochina as well.

Section V   The security element of MI6.

SHAEF   Supreme Headquarters, Allied Expeditionary Force. Eisenhower's headquarters for the invasion and liberation of Western Europe.

SIM   Servizio de Informazione Militare. The intelligence agency of the Italian Army.

SIME   Security Intelligence, Middle East. The coordinating body for security matters in the British Middle East Command.

SIS   (a) The Secret Intelligence Service, the British foreign intelligence and counterintelligence service better known as MI6; (b) the Special Intelligence Service of the FBI, the American overseas intelligence service in the Western Hemisphere; (c) the Signal Intelligence Service, the communications intelligence service of the United States Army; (d) Servizio Informazioni Segrete, the intelligence service of the Italian Navy.

SOE   Special Operations Executive. The British covert action service for assistance to anti-Axis underground and partisan activity.

SOG   Seat of Government. The FBI term for its head office in Washington.

SWPA   Southwest Pacific Area. General MacArthur's theater.

TR   Travaux Ruraux (actually Tous Renseignements). The double agent element of the clandestine DSM organized by Paul Paillole after the Vichy government abolished the official DSM.

Ultra   Generic term for decrypted Axis communications.

W/T   Wireless telegraphy. Morse code radio signaling.

WAAC   Women's Auxiliary Army Corps. The United States Army women's organization.

WAAF   Women's Auxiliary Air Force. The Royal Air Force women's organization.

WAVES   "Wave." Women Accepted for Volunteer Emergency

Service. The United States Navy women's organization.

WRN "Wren." Women's Royal Navy. The Royal Navy women's organization.

Prologue, 1862–1940

June 1862. For two months Stonewall Jackson has marched and counter-marched his little Confederate army in a bewildering choreography up and down the Shenandoah Valley of Virginia, striking where least expected and disappearing again, leaving four different Union commanders wondering what had hit them. Now he has slipped his army across the Blue Ridge to join Lee's main body for a surprise attack upon McClellan's host bearing down on Richmond. If the Yankees should suspect even for a moment that this is happening, the telegraph will flash the word to Washington and thence to McClellan. So they must be made to act on the belief that Jackson is headed down the Valley towards the Potomac in pursuit of retreating Federals.

To this end Jackson has directed his engineers to perform a new topographical survey of the Valley, as if he were planning a further campaign there. He has ordered rumors spread of an impending advance to the Potomac. He has sent cavalry to follow the enemy retreat, and the troopers themselves have no idea where their infantry is. His outpost lines and cavalry screen are airtight. His officers have been told nothing. His men have no notion what is afoot; they have been instructed to answer all questions with "I don't know," and have been forbidden even to ask the names of villages they pass through. He himself is riding ahead to Richmond incognito. And in a few days his men will pour yelling out of the woods against McClellan's right wing. "Always mystify, mislead, and surprise the enemy," Jackson said once to one of his generals. He is a master of that game.

Mystify, mislead, and surprise. In principle this is nothing new. Deception of one sort or another has been practiced in war from the dawn of time. Joshua overthrew the men of Ai by the ruse of a feigned retreat. "Warfare is the *Tao* of deception," wrote Sun Tzu, the Chinese sage of war, in the fourth century B.C. Thrasybulus of Miletus two hundred years before Sun Tzu, Leonidas and Themistocles in the Persian Wars, Belisarius nine centuries later, all supposedly tricked their enemies into thinking their forces were stronger than they really were. Feudal Japan knew the *kagemusha,* a man who pretended to be the warlord, dressed in his armor, to decoy the enemy away from the real commander. In 1704 Marlborough deceived all Europe into believing that his march from the Netherlands to Bavaria, leading up to the victory of Blenheim, was really aimed at Alsace. Frederick the Great before the battle of Hohenfriedberg in 1745, Napoleon in the campaign of Ulm in 1805, used carefully planned deceptions to disguise their movements. But just as Jackson's war is the first in which the railroad and the telegraph have enabled movement with unprecedented speed and instantaneous communication over long distances, so Jackson is the first to adjust his stratagems to modern technology. With instant communication, secrecy is more essential than ever before: so Jackson keeps newspaper correspondents out of his camp, and not only do his men never know where they are going (or, often, where they are); neither does his closest staff. A tight cavalry screen always masks his movements. Secret marches, often by the worst road available; systematic and industrious spreading of false rumors; sending seeming deserters across the lines primed with false information; feigned retreats—he uses them all, with the sure touch of a master.

Fast forward now to 1900.

Colonel G. F. R. Henderson is a distinguished military historian and scholar, who since 1892 has been Professor of Military Art and History at the British Staff College, where, as *The Times* will say in his obituary a few years hence, "he exercised by his lectures and his personality an influence upon the younger generation of the officers of the British army for which it would be difficult to find a parallel nearer home than that of Moltke in Prussia." Henderson is the closest of all students of Stonewall Jackson. His two-volume biography of the Confederate genius, published in 1898, is (and a century later will still be) one of the masterpieces of Civil War studies. To research it he visited Virginia in the early 1880s and tramped over Jackson's battlefields, and then conducted an extensive correspondence with Jackson's surviving officers. In it he has been the first to emphasize and analyze Jackson's systematic mystifying and misleading of the enemy. The greatest general, says Henderson, is "he who compels his adversary to make the most mistakes," whose imagination can produce "stratagems which bring mistakes about;" and in this respect he compares Jackson to Wellington—"Both were masters of ruse and stratagem"—and contrasts him with Grant, who had "no mystery about his operations" and "no skill in deceiving his adversary."

When Field-Marshal Lord Roberts is sent out to retrieve the initial British disasters in the South African War, he brings Henderson with him as his chief of intelligence. Roberts's first task is to lift the siege of Kimberley, well behind the main Boer line. His basic plan for this involves a traditional feint. His cavalry will demonstrate conspicuously against the Boers' right; then the bulk of the force will slip away and swing around the Boer left flank and make for Kimberley. Henderson's job is to keep the Boers' attention focused on their right and prevent their catching on to Roberts's real plan till too late.

Henderson is overjoyed at this opportunity to try his own hand at the skill he so much admires in Jackson. As his assistant will record years later, "Henderson, always an ardent advocate for mystifying and misleading the enemy . . . reveled in the deceits he practised." He sends out fictitious orders in clear, and then cancels them in cipher. He circulates false orders directing concentration opposite the Boers' right. He gives "confidential" tips to people he knows will divulge them. He gives a London newspaper correspondent a particularly juicy piece of misinformation with a stern injunction to keep it to himself; it promptly appears in the London papers (which evokes a sharp warning from the War Office about the indiscretion of someone on the staff). His intelligence officers and agents continually reconnoiter the enemy's right. Information is sought about availability of water and good campsites along the route that Roberts does not intend to follow. Telegrams in an easily-broken cipher are allowed to fall into enemy hands. And in due course, with the Boers' attention thus focused on their right, Roberts's cavalry swings round their left, rides for Kimberley, and lifts the siege.

Riding with Roberts's cavalry is a thirty-eight-year-old officer named Edmund Allenby. He observes, and remembers.

Fast forward again now, to 1917.

To protect the Suez Canal against Germany's Turkish allies operating from Palestine, the British have advanced an army from British-occupied

Egypt across the Sinai. With the failure of a British attack on Gaza in March, this force has bogged down at the borders of Palestine along the Gaza-Beersheba line. A second attack on Gaza in April fails too. In June the British government sends out that same Allenby, now a full general, to take over the command, with orders to mount an all-out effort to conquer Palestine.

Allenby decides to hit the Turkish left and roll their line up from Beersheba rather than making yet another direct attack on their right at Gaza. His intelligence (presided over by Colonel Richard Meinertzhagen, an exceptionally energetic and resourceful officer) is highly efficient, including decipherment of Turkish radio messages, aerial reconnaissance, and an effective spy network; from it Allenby knows that the Turks expect the main blow to fall at Gaza once again and are also concerned over a possible British landing on the Mediterranean coast in their rear. Harking back to Roberts and Henderson before Kimberley, he orders steps taken to reinforce both of these concerns. To emphasize a possible landing in the Turkish rear, British forces in Cyprus build new camps, the ports of Cyprus bustle with preparations, local dealers show interest in buying large quantities of supplies, wireless traffic is stepped up, the Royal Navy makes shows of activity north of Gaza. On the main front, extensive false information is planted in radio messages in a cipher whose key has been deliberately leaked to the Turks; rumors confirming the deception plan are spread; patrol and artillery activity reinforce the appearance of an attack at Gaza with a mere demonstration at Beersheba; shift of forces to the British right is conducted in slow stages only at night.

Most memorable is the famous "Haversack Ruse." Meinertzhagen himself rides out and allows the Turks to fire on him. He gallops away, pretending to be wounded, dropping a bloodstained haversack in his flight. When the Turks open it they find documents tending to confirm the deception plan, plus money and personal items. The documents are fakes; the other items were added to lend authenticity. The loss of valuable papers is reported in radio messages, patrols are sent out to search for the compromising haversack, a sandwich wrapped in a daily order dealing with its loss is left behind by a patrol near enemy lines for the Turks to find.

The Turks remain focused on their right. The Third Battle of Gaza opens on October 31 with the primary assault against their left. It is an unqualified victory for Allenby.

The Turks fall back and Allenby slowly advances into Palestine, taking Jerusalem in December and setting up a line running from north of Jerusalem to the Mediterranean. He has to send reinforcements to France to help stem Ludendorff's spring offensive and his front is relatively quiet for much of 1918. By September he is ready to attack again. This time his main thrust will be on the coastal plain on the Turkish right, so his goal is to make the Turks focus on the Jordan Valley on their left; moreover, this time he hopes to conceal from the enemy not merely the direction of his attack but the very fact that he means to make one.

To accomplish this, Allenby implements an even more elaborate deception than before. He unleashes Lawrence of Arabia and his Arabs in the Transjordan, attracting Turkish attention well to the left of their main line. Lawrence's Arabs spread the word throughout the Transjordan that the British are soon coming and will need to buy fodder and sheep. Allenby moves his units towards his left at night and they lie under cover during the day. In the area where the main attacking force will be concentrated, camps are built and occupied by skeleton units long before the scheduled date of the offensive, to accustom the Turks to the presence there of large forces with no aggressive intent. Vacated camps are occupied by limited-service troops who keep up a regular camp routine, and no less than fifteen thousand dummy wooden horses are set out in vacated cavalry camps for the benefit of enemy aerial reconnaissance. Mules dragging wooden sledges and tree limbs raise huge dust clouds in areas where no real activity is taking place. At key river crossings, bridges are repeatedly built and dismantled so that the actual assault crossing will seem to be just another training session. Every day two battalions march to the Jordan Valley; they go back in trucks every night, and the same two battalions march to the Jordan again the next day. Rumors are launched that Allenby's headquarters will be shifted to Jerusalem opposite the Turkish left; billets are marked and a hotel is requisitioned; the doors of its rooms are labeled with the names of headquarters departments and special telephone lines are conspicuously installed.

The net result is a four-to-one British superiority in the crucial sector. At dawn September 19, Allenby launches the Battle of Megiddo. He smashes through the Turkish right. His cavalry pours through the gap and swings east, and the Turkish army is trapped. Allenby drives on to Damascus and Aleppo, and by the end of October Turkey is out of the war.

Allenby has taken Henderson one step further, by institutionalizing and sytematizing deception on the basis of an orderly assessment of the situation and integrating it with his operational planning. "Deceptions which for ordinary generals were just witty hors d'oeuvres before battle, had become for Allenby a main point of strategy," Lawrence will write in later years. In effect, Allenby has updated the methods of Stonewall Jackson as carried forward by Henderson, applying to them the elaborate staff procedures of the twentieth century.

Attached to Allenby's staff, and subsequently on the staff of one of his corps, is a thirty-five-year-old officer named Archibald Wavell. He observes, and remembers.

Fast forward once more, to November of 1940.

General Sir Archibald Wavell is now British Commander-in-Chief, Middle East, with headquarters in Cairo. He commands a vast theater extending from Iran on the east to Libya on the west, and south to East Africa. Since the fall of France in June, Britain has stood alone. The Blitz is at its height, and London writhes under the bombs of the Luftwaffe; the RAF has won the Battle of Britain, but nobody can be sure that Hitler has given up the thought of invasion. Only Wavell is in direct contact with Axis forces on the ground and in a position to give the British people a tangible victory. His immediate adversary is not Germany but Italy, which Mussolini brought into the war shortly before France collapsed. An Italian army under Marshal Rodolfo Graziani, based in the Italian colony of Libya, on the western border of Egypt, has established itself at Sidi Barrani on Egyptian territory, while forces operating from Italian East Africa under the Duke of Aosta have occupied British Somaliland. Wavell plans to attack Graziani first, in early December, followed by an offensive in East Africa. In Allenby style, his plan of attack on Graziani includes deception of the enemy, by spreading word through known Axis sources of information in Cairo (including the Japanese consulate) that his force is being weakened owing to the detachment of troops to be sent to Greece to aid in repelling the invasion of Greece which Mussolini launched in October; by taking administrative measures and broadcasting dummy radio traffic that tend to confirm a sizable withdrawal from his front; and by dropping hints to the press that maneuvers and training exercises are going on in the desert. To show enemy agents that he has nothing afoot, on the eve of the attack he ostentatiously attends the afternoon races with Lady Wavell and their daughters, and that evening he entertains his senior officers at a party, looking carefree and relaxed.

As he presides over the planning and implementation of this process, Wavell decides that Allenby's approach should be taken one ultimate further step. Not only should deception be institutionalized, but it should be entrusted to a permanent specialized staff element dedicated solely to that function. On November 13 he advises London by personal signal that he intends to form "a special section of Intelligence for Deception of the enemy," and requests that there be assigned to that responsibility an officer, now a lieutenant-colonel, who had served under him in Palestine in the 1930s, and in whom he had, in his own words, "recognized an original, unorthodox outlook on soldiering," coupled with "originality, ingenuity, and [a] somewhat impish sense of humor."

On December 19 that officer reports to Wavell for duty. In an absolutely true sense the fourth-generation heir of Stonewall Jackson in the direct line, he will prove to be not merely a worthy successor to Jackson but the master of the game, the man who perfects the art of military deception in its modern form and is ultimately responsible for the greatest military deceptions in history. His name is Dudley Clarke.

CHAPTER 1

The Master of the Game

Lieutenant-Colonel Dudley Wrangel Clarke was forty-one that late 1940, having been born in Johannesburg on April 27, 1899. He was the eldest child of a venturesome Yorkshireman named Ernest Clarke, who in his late twenties grew bored after nine years in the family shipping business at Hull and went out to South Africa. There he knocked about a bit and at the end of 1895 found himself carrying dispatches in the abortive plot to take over the Transvaal for the British Empire known as the "Jameson Raid." He did not go to jail for his part in the Raid, but some friends were not so lucky. Visiting one of these, Clarke met his friend's cellmate, who was the head of a gold mining company, and he offered Clarke a job. Clarke took it, and did well with the company. He married Madeline Gardiner, a bank manager's daughter who had been brought out to South Africa in her infancy. Her father was Irish and her mother Austrian. Ernest and Madeline had two children born in South Africa, Dudley and Dorothy, known as Dollie.

After the Boer War the company sent Ernest to its London office. He bought a house at Watford, in the northwestern exurbs of London (largely, he claimed, from his winnings in the ship's pool on the daily run, having bought up all the low numbers on a day when he had ascertained that the ship had stopped in the night). After several years two more children were born to the Clarkes: Tom, known outside the family as Tibby, and Sybil. Clarke continued to prosper with his company, and thereafter on his own as an investment advisor. In 1910 he retired first to Frinton-on-Sea, and subsequently to Oxted in Surrey. Soon after the outbreak of the First World War he was largely responsible for founding, and financing, the Motor Ambulance Brigade. For this he was knighted and became Sir Ernest.

The children seem all to have inherited English ingenuity, Irish charm, and Austrian *Gemütlichkeit,* plus a generous share of brains. There appear to have been some genes on the Clarke side in particular that Dudley shared with their Uncle Sidney, Sir Ernest's eldest brother; besides being a barrister who wrote a legal advice column for the *News of the World,* and the sometime author of *Old Moore's Almanack,* Uncle Sidney was for years chairman of the Magic Circle, the magicians' society of England, and could always be relied upon to find a half-crown in a nephew's ear, or, as happened once to Tom, a ten-shilling note in his boiled egg. The boys were sent to Charterhouse, one of the major English public schools, and Tom went on to Cambridge. Dudley and Dollie never married; the other two did. Dudley went into the army and Dollie became a journalist and author. Tom became an author, editor, and screenwriter; he wrote the scripts for the classic Ealing Studios comedies *The Titfield Thunderbolt, The Lavender Hill Mob,* and *Passport to Pimlico,* among other films. Sybil married a Canadian and moved to Toronto.

In May of 1916, just turned seventeen, Dudley Clarke entered the Royal Military Academy, Woolwich, where engineer and artillery officers were trained. Commissioned in November in the Royal Artillery, he tried to get sent to France; but in the land forces you had to be nineteen to serve overseas. So he transferred to the Royal Flying Corps and was a pilot in Egypt for the rest of the war. In 1919 he returned to the artillery, but he proudly wore his RFC wings for the rest of his army career (fighting a running battle for a number of years with War Office bureaucrats who said that it was improper to do so).

He was next posted to Mesopotamia for three years. Then in 1922 he took a long leave to travel in Europe, with a detour to carry out a minor special mission for the British commander in Constantinople during the troubles when Mustafa Kemal drove out the Greeks and ousted the Sultan. In 1924, again on leave from the army, he covered the Riff rebellion in Morocco for the London *Morning Post.* From 1926, he spent four years in Sussex attached to the Territorial Army (the British equivalent of the National Guard in the United States), and in 1930 volunteered for the Transjordan Frontier Force in present-day Jordan. In 1933–34 he attended the Staff College at Camberley. There he came under the favorable notice of the commandant, General Sir John Dill. As part of the course he visited Italy and Germany and got to know a number of German officers. He was in Nuremberg at the time of the great Party Rally of 1934 memorialized in Leni Riefenstahl's *Triumph of the Will,* and was in the Saarland in 1935 when the plebiscite returning it to Germany was held.

After Camberley he served in coastal defense at Aden, and in 1936 was transferred to Palestine. There he was for practical purposes the chief of the operations staff for General Dill, now the commander in Palestine, and for Dill's successor Wavell. There, too, from a period of time when he was on an Arab death squad's hit list, he picked up the habit of never accepting a table in a restaurant that was not against a wall.

From Palestine, now a major, he was assigned to the War Office in London. At Eastertide 1939, he paid a visit to his old friend Kenneth Strong, the assistant military attaché in Berlin (who was later to be Eisenhower's chief of intelligence), and when there he made the acquaintance of several German intelligence officers, of whom, he wrote after the war, "two at least were to cross my path again before very long—on the other side of the fence." When war broke out at the beginning of September he was promoted to lieutenant-colonel.

In the first year of the war he performed a succession of interesting tasks. He was sent to Africa to reconnoiter an overland route from Mombasa, the port of Kenya, to Egypt, by which the Middle East could be supplied if Italy should enter the war and succeed in closing both the Mediterranean and the Red Sea. (This meant working with Wavell once more. "Would you like to come back to my staff again?" asked Wavell as Clarke departed. "I knew it was as good as fixed," wrote Clarke later.) He went twice to Norway during that ill-fated campaign in the spring of 1940. He worked on an abortive effort to hold Calais during the collapse on the Western Front in May. He was sent on a secret mission to Eire, involving contingency planning in case the Germans should make a sudden descent on that country.

When later in that May Sir John Dill became Chief of the Imperial General Staff, he made Clarke one of his military assistants. Clarke was at Dill's side during the Dunkirk evacuation; thereafter he was deeply involved in the original formation and organization of the Commandos. (He took credit for giving the Commandos their name, after the legendary fast, hard-hitting Boer units of his South African childhood.) Among those he worked with was the actor David Niven, who had left Hollywood and rejoined the Army when war broke out. Clarke had known Niven as a young regular officer before he turned to acting; his brother Henri, known as "Max," would be one of Clarke's officers before the war was out. Clarke went personally on the first Commando raid, on the French coast barely three weeks after Dunkirk, and had an ear shot nearly off. For the next five months, during the memorably beautiful summer of the Battle of Britain and on into the autumn of the Blitz, he was immersed in Commando affairs.

Then in November came the summons from Wavell. It had indeed been "as good as fixed."

Nobody who knew Dudley Clarke ever forgot him. He was fair-haired and blue-eyed, with a twinkling countenance, all rounded with no planes or corners—"a face like a sort of merry-eyed potato," recalled Sir Edgar Williams, Montgomery's chief of intelligence—which reminded his secretary of a guinea pig; "a sharp little man with bright, quick eyes," said Malcolm Muggeridge. Everyone remembered the merry eyes, which he had a habit of blinking incessantly. He was a small and compact man with a "gently booming" voice, always very neat, hair always smoothed back, almost always with a pipe or a long cigarette holder. "A man of few words, as soon as he looked at you and spoke laconically, but very courteously, one became conscious of a sphinxlike quality of sardonic humor and absorbent watchfulness," who could "seem like someone's not at all dear old butler . . . confronted with a nervous and uncertain guest," wrote David Mure, one of his officers and one of his greatest admirers; "beneath his bland rather old world exterior it was impossible to guess what he was thinking, and what he said nearly always came as a surprise." The Countess of Ranfurly, General "Jumbo" Wilson's secretary, remembered him as "brilliantly clever and imaginative and always on the edge of laughter," and Dennis Wheatley remembered his "quiet chuckle which used to make his shoulders shake slightly." "An excellent raconteur and great company in a party," said Wheatley, no mean judge of such matters, "but with a strange quietness about his movements and an uncanny habit of suddenly appearing in a room without anyone having noticed him enter it"; "a truly legendary figure" who "got a great deal of fun out of intrigue of every kind." Wheatley observed that Clarke even affixed his insignia of rank to his uniform with snaps so they could be quickly removed to transform him into an inconspicuous junior officer.

Clarke enjoyed tilting at bureaucratic windmills. He had been born during the siege of Ladysmith by the Boers, and he engaged for years in a cheerful struggle with the War Office, claiming to be entitled to the Boer

War campaign medal even though he had been a newborn infant at the time—because, he said, he had been fed on garrison rations, and an order had decreed that all persons on the ration strength of the garrison were entitled to the medal. But, unlike such other military gadflies as his colleague Orde Wingate, Clarke never went out of his way "to flout, disagree or to insult Authority; he merely smiled sweetly, got his own way, or went in another direction."

Among his many talents, Clarke was a proficient horseman who at the Empire Exhibition Rodeo at Wembley in 1924 was the only non-cowboy to last for thirty seconds on a bucking bronco. And he was a dedicated moviegoer; some of his best ideas came to him by inspiration as he sat alone in a darkened cinema watching the images flicker across the screen.

"He did not suffer fools gladly," said his old friend and deputy Noel Wild, "but he was always scrupulously fair. If you made a mistake you were either fired or forgiven and there it ended. He could be difficult but always good tempered, and never moody. He enjoyed an argument. It was always tempered by wit and good humor, and if ever it looked likely to become heated, he would quickly wind it up with some outrageously unexpected remark, and that was the end."

Clarke was a lifelong bachelor. He detested children (or affected to do so, in a W. C. Fields sort of way). He was a thoughtful friend and liked giving imaginative little gifts; he had a feminine sensibility, you might say, though there was nothing effeminate about him. He had been unlucky in love, with at least two experiences that might well leave any man wary. In Wiesbaden in 1922 during his tour of Europe he had fallen under the spell of an attractive Russian named Nina, daughter of a court official of the Tsar who had been killed in the revolution. "She had a distinctive Slav beauty, with fine, high cheekbones and sparkling dark eyes," remembered Clarke many years later, "but she also had a very un-Slav-like sense of fun. I fell for her rather heavily—so heavily that it led me into undertaking the most foolish mission of my life." That mission was to take an envelope containing a letter and some money to a Russian friend of Nina's in Bulgaria in violation of currency regulations. He did so at the cost of some hair-raising scrapes and much of his own money, but he never saw Nina again; evidently all he had done was to enable her Russian sweetheart to join her. Then during his years in Sussex with the Territorials in the late 1920s he had had what he called "a romance which had meant everything in the world to me"; but the lady turned down his proposal, and married an officer in the Guards. ("She made the most wonderful treacle tarts," he said wistfully once in Cairo.) He seems not to have been willing to risk another such disappointment; nevertheless, all his life he greatly enjoyed the company of beautiful and elegant women and made every effort to have them around. "Dudley's Duchesses," his friends called them.

Indeed, he liked the good life in general: good food, good wine, select company, living in the right place, moving in the right circles. It was characteristic that when posted to the War Office in 1939 he found himself a cozy little flat on the top floor of a centuries-old house in Mayfair. It was characteristic that he made so many friends in the 11th Hussars—"Prince Albert's Own," one of the snootiest regiments of the army—that he, a mere artilleryman, was asked to write the history of the 11th's service in the Second World War. It was characteristic that in Cairo he took up residence at Shepheard's Hotel. He was, it must be confessed, a bit class-conscious and a bit of a lion-hunter, and as the war progressed he would attract to his "A" Force sprigs of the Establishment and people of conspicuous accomplishments, and less exalted members of the staff sometimes felt themselves somewhat on the outside looking in.

But these were venial faults by comparison with the phenomenal talent that he brought to his work. "He had such a fantastically quick brain although it was not quite like anybody else's," recalled Oliver Thynne, one of his officers; "the most all-containing brain of anyone I ever met." "At any time, as well as complete deception orders of battle and battle plans for say two particular situations . . . which were worked out in all detail on paper, there would be another six embryo plans in his mind which could be translated to paper in 24 hours. I have never known a brain so full of stuff! And he never forgot anything in transmitting it to paper, either!" "He was certainly the most unusual Intelligence officer of his time, very likely of all time," wrote Mure; "his mind worked differently from anyone else's and far quicker; he looked out on the world through the eyes of his opponents."

The London Blitz was building to its climax that 5th of December 1940 when Dudley Clarke took his leave of his Mayfair flat; a permanent leave, as things turned out, for the house would collapse in the last big raid on London five months later. To get from England to Egypt in those days took a week of flying time, to avoid Vichy French territory. Clarke flew therefore by way of neutral Portugal, the Canary Islands, and Bathurst in Gambia, to Freetown in Sierra Leone; and from there to Lagos in Nigeria—apparently in the guise of an American war correspondent—where he met with Free French officers from Chad to discuss the possibility of long-range operations across the Sahara against the Italians in Libya ("a tang of 'Beau Geste,' " said Clarke of these talks).

From Lagos to Kano in Nigeria; thence on past the Free French outpost at Fort Lamy and to El Fasher ("an evening flight low down, with ostrich galloping below in the sunset"), on to Khartoum and down the Nile Valley to Wadi Halfa. The next day to Luxor, and finally to Cairo on the afternoon of December 18, where he was met by an old friend from the Palestine days, Lieutenant-Colonel Tony Simonds, who was startled to see Clarke step off the plane in his "American war correspondent" costume—a loud pair of black-and-white plus-fours. He was to report to Wavell the next morning.

General Sir Archibald Wavell was a remarkable man. "Behind an inarticulate and ruggedly orthodox exterior," Sir Michael Howard has written, "Wavell concealed one of the most fertile minds ever possessed by a British officer." The inarticulateness was legendary, a reserve that discomfited many, with long periods of silence that baffled many more.

He could recite poetry by the hour; he was a Latin scholar; he was a clear thinker and a deep one, and a highly creative and imaginative strategist; he had a knack for spotting men of special, out-of-the-ordinary talents. Wavell and Churchill never got on—which might be thought curious, since they were alike in many ways, but to one of Churchill's flamboyant, outgoing, and articulate temperament, the reserve and silences were frustrating and ultimately exasperating. That was a pity, for it led Churchill eventually to do Wavell great injustice.

From his Cairo headquarters Wavell presided over a vast and complex theater. To the north the British mandates of Palestine and Transjordan abutted the French mandate of Syria (including Lebanon), while British-ruled Cyprus lay off the Syrian-Turkish coast. Farther east lay Iraq, likewise a British mandate until 1932 and then at least nominally an independent kingdom. East of Iraq lay the independent kingdom of Iran (then often still called Persia), abutting to the east upon British India; the latter was a wholly separate theater. North of Syria and Iraq lay neutral Turkey. North of Iran and Turkey lay the Soviet Union. Iran was important not only because of its oil—for a generation the main source of fuel for the Royal Navy—but as a back-door supply route for the Soviet Union. To the south, Saudi Arabia and the scattered sheikdoms were drowsy backwaters in those pre-oil days.

South of Egypt lay the vast Sudan, nominally an Anglo-Egyptian condominium but under British control for all practical purposes. South of the Sudan was Italian East Africa, composed of the long-established Italian colonies of Eritrea and Italian Somaliland, plus Ethiopia, which Mussolini had conquered in 1936; the Duke of Aosta was the Italian commander-in-chief in that region. On the north side of the outlet from the Red Sea to the Indian Ocean lay British-ruled Aden. On its south side was the little province of French Somaliland, Vichy-controlled and closely blockaded; along the coast to the east lay British Somaliland, which to Churchill's annoyance had been occupied by the Duke of Aosta. South of Italian East Africa were the British territories of Kenya and Tanganyika (present-day Tanzania).

Most important of all, west of Egypt was Libya, ruled by Italy since just before the First World War, and hence the logical Axis base for a drive against Suez.

Axis intrigue was alive in the northern tier of neutrals. Turkey was a hotbed of espionage. Syria and Lebanon were in Vichy French hands, and there was a disturbing amount of German interest in them as possible bases for air attacks on the Middle East. The King of Iraq and the Shah of Iran owed their thrones to Britain, but there were influential pro-Axis elements in both countries.

For the British, the focus of this huge theater was the Suez Canal, the most vulnerable link in the British "lifeline" from England past British-held Gibraltar and Malta, through the Canal, down the Red Sea, past Aden and out into the Indian Ocean and on to British India, Burma, Malaya, Singapore, and Hong Kong. So Cairo, capital and metropolis of Egypt, only some eighty miles from the canal as the crow flies, was the logical place for the headquarters and nerve center of the theater.

* * *

Though it was a seat of war and the headquarters and base of the British effort in the Middle East, Egypt was not part of the British Empire but a nominally independent monarchy under the young King Farouk. By treaty, the British preserved the right to defend Egypt and the route to India, and the administration of the Sudan; and some key civil service posts were held by Britons. Cairo was the British Middle Eastern Theater headquarters, Alexandria was the British naval base for the eastern Mediterranean, and the Suez Canal was heavily garrisoned by British troops.

Nevertheless, Egypt was not only legally independent but legally neutral. Though German nationals had been rounded up and interned, rail and air transport had been put under British control, and censorship instituted, there was substantial Axis influence on Farouk's entourage, and indeed sympathy for the Axis was widespread, promoted by an efficient propaganda machine that was especially skillful at spreading rumors. Cairo was believed to be aswarm with enemy agents, though in fact British security was tight and effective. The country was practically untouched by the hardships and shortages that war had brought to Europe. The stores bulged with goods; as long and tempting as ever were the bills of fare and wine lists at the great hotels like Shepheard's and the Continental, the fashionable cafés and restaurants like Groppi's, the posh private establishments like the Muhammad Ali Club.

Cairo was then a city of half a million, at once exotic and cosmopolitan. In the jammed streets the senses were assailed by all the sights, sounds, and smells of the Middle East, the air filled with the honking of horns, the cries of vendors, a babel of Arabic, Greek, French, English. Moslems, Jews, Coptic Christians, Syrian Christians, Egyptians, Englishmen, Frenchmen, Italians, Greeks, Maltese, Cypriots, Turks, thronged the bazaars and cafés. And everywhere the British Army, its trucks and staff cars threading impatiently through the crowds, its boisterous men crowding the Greek bars and taverns, its officers drinking at the Turf Club or at the Long Bar in Shepheard's, or watching (or playing) polo or cricket at the Sporting Club on the island of Gezira in the Nile.

The main home of the British Army in Cairo was a huge fortress complex on the southeastern edge of the old city called the Citadel of Muhammad Ali. A smaller detachment resided in the Kasr-el-Nil Barracks on the banks of the Nile opposite Gezira. The headquarters of Middle East Command—GHQ Middle East, usually called just GHQ—was in a modern apartment building called Grey Pillars, half a mile south of Kasr-el-Nil Barracks in a residential area of trees and winding streets known as Garden City. As the war wore on, GHQ would outgrow Grey Pillars and eventually a substantial part of Garden City would be transformed into a huge headquarters compound. When Clarke arrived at the end of 1940, Grey Pillars was already stretched at the seams, a maze of bedrooms and sitting rooms partitioned into cubicles, offices set up in converted bathrooms and kitchens, a busy hive of sweating, khaki-clad men (and, as time went on, more and more women) with much going and coming and jangling of telephones, reminding one war correspondent of a busy department store trying to cope with a flood of business during alterations; two waves of this each day, for Cairo still observed the ancient custom of the siesta, and there was a long break for luncheon and into the afternoon. The men facing death at the front referred to the paper-pushers back at Grey Pillars by such scornful appellations as the Gabardine Swine, or Groppi's Light Horse, or the Short Range Shepheard's Group (a contribution by Michael Crichton, whom we will meet later, for before becoming a Gabardine Swine himself he had been involved with the legendary force called the Long Range Desert Group).

So on the morning of December 19, 1940, Dudley Clarke found himself in Grey Pillars, standing once again in front of General Wavell to report for a new assignment. He found that his job would be "Personal Intelligence Officer (Special Duties) to the Commander-in-Chief." It entailed not only the planning and conduct of deception activities, but the organization and operation of a Middle Eastern equivalent of MI9, the element of Military Intelligence at the War Office responsible for assisting British soldiers to evade capture and for securing information from prisoners of war in enemy hands and assisting them to escape. So it was quite a job that had been "as good as fixed." *

For the first few months Clarke was a one-man show. His office was a small converted bathroom in Grey Pillars. His clerical staff consisted of

---

* Clarke ran MI9 or escape and evasion activities for the Middle East, and subsequently the whole Mediterranean, from January 5, 1941, until August of 1944. Being less secret than deception, it served throughout the war as a cover for Clarke's deception work. Otherwise there was little or no connection between them and this book will accordingly touch only tangentially on Clarke's escape and evasion activity—which was itself quite remarkable.

the part-time services of Wavell's private secretary and of the personal assistant to Wavell's chief of staff. He himself took up residence in a room at Shepheard's Hotel (where the rice pudding was excellent, he said). And it was a tightly held one-man show. Forty years later, in correspondence with one of Clarke's former officers Wavell's regular chief of intelligence still had no clear understanding of what Clarke had been up to during those early months.

A very early order of business for Clarke would have been to get acquainted with Lieutenant-Colonel Raymund Maunsell and with his organization, Security Intelligence Middle East, or SIME, for SIME was to be the primary agency for passing deceptive information to the enemy in that theater.

SIME was an element of Wavell's staff set up in December 1939 to coordinate competing bureaucratic interests in counterespionage and related matters for the theater. Under British practice, counterespionage and counter-sabotage within the British Empire (which for this purpose included such British-controlled territory as Egypt and Palestine) was the responsibility of the Security Service, known usually as MI5, while elsewhere it was the responsibility of the Secret Intelligence Service, called the SIS or MI6. (India had separate arrangements.) This division of authority led to turf wars and failures of coordination. SIME was designed to obviate such problems. Its charter included watching the activities of hostile agents in the Middle East and coordinating measures to counter them, maintaining liaison with the Indian government and military with respect to activity targeted on India, reporting periodically on hostile action and counteraction, and expanding the security intelligence organization in the theater.

Maunsell was a thirty-six-year-old tank officer with ten years' Middle East experience who had since 1935 been the MI5 chief in Cairo. SIME was his own brainchild. To keep it effective he fended off other staff elements until its independence was finally confirmed in 1942; at which point SIME acquired its own headquarters building just outside the Grey Pillars compound, zealously guarded by an intelligent and loyal Egyptian named Abdu, who had no legs and got about on a small platform with roller-skate wheels. Maunsell ran SIME on loose and informal lines, with little attention to military proprieties. First names were encouraged, and his senior people called him "R. J."; informal communication was encouraged without worrying about the chain of command.

Even before SIME was formed, Maunsell had built an efficient organization with slender resources. He had on his unofficial payroll certain Egyptian policemen and concierges of Cairo apartment buildings. He had penetrated the Muslim Brotherhood by means of a Welshman posing as a Syrian. He had recruited the assistance of a volunteer organization of Sephardic Jews who operated a nationwide organization of agents, rumormongers, and useful underworld mouthpieces. He had established an efficient censorship bureau. He had deeply penetrated the Spanish consulate, which was a major source of Axis information, and he kept on Japanese consular and diplomatic officials a close and sometimes humorous eye.* He had established a close working relationship with the Turkish secret police. By the time Clarke arrived in the theater, SIME had outstations in Aden and Palestine as well as Cairo, together with a newly established acting MI5 officer in Istanbul, the redoubtable Commander Wolfson (of whom more hereafter). This last reflected a valuable arrangement made by Wolfson with the Turkish secret service to set up an "Anglo-Turkish Security Bureau," which among other things gathered details on everyone who traveled between the Balkans and Turkey.

SIME had already had some success in passing deceptive information to the Axis, both in connection with Wavell's December offensive against Graziani and in connection with cover plans directed from London. These were simple and relatively unsophisticated activities, but SIME would soon prove to be an invaluable tool for Clarke's more complex work.

There were other resources to supplement those of SIME. The Field Security branch of the Cairo Military Police cooperated with Maunsell and could be useful in catching Axis agents who might be turned against their masters. Its commander was Major A. E. W. Sansom, an Englishman born and raised in Egypt and fluent in Arabic, French, Italian, and Greek, who had worked in his family's Cairo insurance business till joining the Army in 1940. The regular Cairo police were headed by an Englishman, the suave and elegant Sir Thomas Russell Pasha, who on three

---

* In 1939 a Japanese military attaché had conducted an ingenious reconnaissance of Syria and Palestine, motoring through them with a motion picture camera mounted on his dashboard to film the roads. When the attaché passed back through Egypt, Maunsell managed to have the film in his camera replaced by a particularly vile Egyptian pornographic movie.

or four afternoons a week would show the flag of law and order by riding about the city on his white horse, clad in black uniform and red tarboosh. And the Foreign Office had a body of 350 Egyptians on its payroll to spread pro-Allied rumors, including fortune-tellers and holy men who were paid to predict Allied success.

When Clarke arrived in Cairo, Wavell had driven Graziani out of Egypt and a British drive into Libya would soon begin. For the moment, Wavell was putting the finishing touches on a plan to attack the Duke of Aosta, conquer Italian East Africa, and drive the Italian invaders out of British Somaliland. His opening moves would involve an advance into Eritrea by forces from the Sudan, to be reinforced by a division (the 4th Indian Division) which he hoped to shift in secrecy from Egypt to the Sudanese front.

A couple of days before Clarke's arrival, Wavell had signed off on an associated deception plan of his own design. Called CAMILLA, it was intended to make the Duke of Aosta believe that Wavell's attack would come not from the north but from the east, opening with the recapture of British Somaliland by a force based in Aden consisting of Indian troops moved from Egypt and South African troops moved from Kenya. (Among other things, this would account for the departure of the 4th Indian Division from Egypt if Italian agents were to notice it.) It was Clarke's first assignment.

Clarke designed an implementing plan for CAMILLA, no doubt in conjunction with Maunsell, which Wavell approved three days after Christmas. For the next four weeks Clarke was busy night and day. British Somaliland was raided by sea and air. Maps and pamphlets about Somaliland were issued to the troops. Administrative activity suggesting that great things were afoot was undertaken conspicuously at Aden. Radio links between Aden, Nairobi, Pretoria, and Delhi on the one hand, and Cairo and Khartoum on the other, were filled with dummy messages. Seeming security leaks were sprung in Egypt, Aden, India, and South Africa, rumors were spread in Egypt and among the troops involved, phony information was planted on the Japanese consul in Port Said, and various indiscreet unenciphered private telegrams were dispatched.

CAMILLA no doubt contributed to the surprise that Wavell's main offensive achieved when it opened in January. But from Wavell's perspective, CAMILLA appeared to have worked *too* well. The Duke of Aosta was apparently indeed convinced that Wavell meant to begin with an attack on British Somaliland. But he did not shift his troops away from the Sudan front to meet it. Instead, apparently concluding that the assault would be too strong to resist, he moved troops out of British Somaliland instead of reinforcing it—thus putting his forces in a better position to oppose Wavell's real plan of campaign, and requiring Wavell to draw on his own limited forces to reoccupy British Somaliland.

CAMILLA was as important as any operation that Clarke ever undertook, even though by later standards it was small and unsophisticated. It gave him his first practice in the mechanics of deceiving, and that was important enough; but of bedrock importance was the fundamental lesson that what you must focus on is not what you want the enemy to *think* but what you want him to *do*—that misinforming him does you no good if as a result of his misinformation he takes undesirable action. That lesson Clarke never forgot. From then on he regarded it as the fundamental axiom of deception, and he never tired of repeating it.

In all likelihood CAMILLA also brought home to Clarke how much it helps in deceiving your adversary if you know his real plans and fears, for the British were intercepting and deciphering perhaps 90 percent of Italian radio messages in East Africa. Perhaps with this lesson in mind, by the time CAMILLA ended in late January Clarke had had up and running for ten days the first operation entirely of his own creation, called ABEAM, in support of Wavell's new drive into Libya. It was known that the Italians were concerned that the British might land airborne forces in their rear. Wavell in fact had no such troops; but Clarke set out to convey the notion that a British unit called "First Special Air Service Brigade" was training secretly in the Transjordan desert. Officers wearing parachutist insignia and armbands of the staff of the "1st Airborne Division" showed up in various places in the Middle East. Two soldiers were brought into the secret, and circulated in Egypt dressed as paratroopers and describing themselves as convalescents from the "SAS Brigade." Pictures of parachutists in training (including one Ethiopian) were printed in a Cairo illustrated paper. Troops of the Arab Legion cordoned off a "prohibited area" in Transjordan to encourage speculation that this was the airborne training area. RAF pilots were warned to watch out for towed gliders in flight. Rumors and "leaks" were spread—the Jewish Agency in Palestine helped with this—and documents identifying "1-SAS Brigade" were planted in Egypt and Palestine; one was planted on a Japanese consular official traveling to Turkey. Construction was started on dummy gliders for the benefit of Axis reconnaissance.

Active in one form or another for six months, ABEAM had no specific short-run goal. Its only purpose was the long-term one of persuading the Axis that there was a British airborne brigade in the Middle East. In this it had some success, as captured documents subsequently confirmed. The direct value of ABEAM was that Axis generals had to take these imaginary airborne forces into account in their plans, and that future deception operations could build on their perceived existence.

The indirect value was incalculably greater. ABEAM was Clarke's first tentative foray into the long-term order of battle deception that was to be his masterwork: building up in the enemy's mind, slowly and over time, an inflated notion of your strength—or, more precisely, to force him to take into account in his planning the possibility that your strength may be greater than it is.*

Clarke's next idea was an abortive one called the K-SHELL PLAN. It consisted of spreading a rumor that the British had a new artillery shell of Australian invention which worked by producing a titanic concussion rather than by the usual fragmentation effect. This plan was operated for the last week in January and then dropped because the press picked up the story and it could not be carried further without either lying to reporters or taking them into the deception secret, neither course being deemed acceptable. But from even this trivial fiasco, Clarke drew a valuable lesson that became a second axiom of deception: Do not try to run a deception plan with no clear object just because the means are available to do it.

Immediately thereafter he started his second modest order of battle deception, called simply the 10TH ARMORED DIVISION PLAN. Wavell

---

* Another result of ABEAM was that when in August 1941 the legendary Major David Stirling formed a small parachute-qualified unit for his "Long Range Desert Group" he dubbed it "1-SAS Brigade" as a cover and to help the ABEAM deception. The prototype of special forces the world over, its direct descendant is the Special Air Service of today. So Dudley Clarke could claim to have christened both of the British elite special formations, the Commandos and the SAS.

had at that time no prospect of armored reinforcement until the scheduled arrival of a single tank battalion nearly three months in the future, but a motorized brigade of the Indian Army had joined him in January, and it was represented as the advance unit of an imaginary British "10th Armored Division" by methods like those being used to put over ABEAM.

Then Clarke was put out of action for six weeks from February 5 with an attack of jaundice. His activities came to a halt, though Maunsell visited him almost every day, and ensured that misleading information continued to be "percolated in the direction of the enemy," in Clarke's phrase; and between them they managed to cook up two new plans of which nothing came, called ABAFT (to cover a projected landing in Eritrea that never took place), and WAR OF NERVES PLAN, a general rumor-spreading operation.

By the time Clarke returned from sick leave in late March, Wavell had driven the Italians from Cyrenaica (the eastern half of Libya), and advance British patrols had reached El Agheila at the southern end of the Gulf of Sirte; while in Europe the Germans had entered Bulgaria and threatened Greece, which was already at war with Italy. Wavell sent a force to the aid of the Greeks, and earmarked a division at Cairo for an operation (Operation CORDITE) to seize Rhodes and the rest of the Italian-held Dodecanese Islands, off the Turkish coast. For this, Clarke devised a project called simply the CORDITE COVER PLAN. Its object was to cause the Italians to divert forces from Rhodes to the weakly-held island of Scarpanto (now better known by its Greek name, "Karpathos"), halfway between Rhodes and Crete, and on Rhodes itself to shift forces from the north side of the island, where the actual landing would take place, to the south side. The "story" to be palmed off on the enemy was that Scarpanto was the real objective (this had actually been considered), with a simultaneous diversionary raid on an airfield at the south end of Rhodes, to be carried out by the notional* "1-SAS Air Brigade," which would move from Transjordan to Crete for that purpose; D-Day was to be

---

* "Notional," a term much used in British philosophical writing meaning conceptual or imaginary, was a standard term in deception in the Second World War and will be frequently met with in these pages. Typists and communications personnel had a way of correcting it to "national," and late in the war the Americans made a systematic effort to use "fictional" instead. ("Fictitious" would perhaps have been more correct.)

one week later than the actual D-Day for the Rhodes landing. Since Axis intelligence would be alerted when the division earmarked for the operation left Egypt, to account for the delay the story would be put out that it planned to stop off at Cyprus to conduct a dress rehearsal of the operation.

The plan was never put into effect; CORDITE was canceled when at the end of March German troops under General Erwin Rommel arrived on the North African scene and quickly threw the British back into Egypt. But it was a valuable exercise, for it brought out four elements of the successful deception plans of the future.

First, it showed the value of starting a deception with a thought-out "story," a scenario around which deceptive activity would consistently be built.

Second, it demonstrated the utility of building the "story" on an alternate plan of action that had actually been considered and discarded.

Third, it introduced the factor of timing—persuading the enemy that the main operation was to come later than the real one.

Finally, it introduced the concept of persuading the enemy that the real operation was merely a feint.*

When the CORDITE COVER PLAN was suspended, Wavell told Clarke to get cracking instead on a plan aimed at making Rommel waste resources to cover his rear. The resulting plan—called simply PLAN A-R (meaning "anti-Rommel")—was yet another of the early schemes that had no perceptible result but forwarded Clarke's education, notably by affording the first actual employment of notional forces and the first use of dummy equipment.

Its "story" was that the British and Free French planned a multi-pronged attack on Rommel's line of communications reaching back to his main supply base at Tripoli. A landing would be made somewhere on the coast of the Gulf of Sirte to cut Rommel off from Tripoli. (Since enemy agents were presumed to know that there had been activity among the troops that would have conducted CORDITE, including a force of Commandos that had recently reached Port Said from England

* For CORDITE, Wavell also gave Clarke a nondeception assignment, to procure intelligence about Rhodes. This produced two extraordinary operations, called DOLPHIN, involving sailing a small ketch through the Dodecanese, and BYNG BOYS, involving the secret establishment of a Turkish-manned post on a rocky crag on the Turkish mainland, observing Rhodes from sixteen miles away through a special gigantic telescope borrowed from the RAF.

in special assault transports, these were supposed to be the units earmarked for this operation.) Newfangled American air-conditioned tanks were being brought in so as to render feasible an attack on Tripoli in the hot months—it being believed that the Germans thought tanks could not be operated for prolonged periods at that season in the desert. The notional 1-SAS Brigade, with gliders, was moving from Transjordan to Egypt for an airborne attack against Rommel's line of communications. The notional 10th Armored Division, reinforced by tanks being brought on transporters overland from Nigeria via Fort Lamy and the Sudan, was moving into Libya to attack his line of communications from the desert flank. Free French troops would make long-range raids on Rommel's communications, striking across the Sahara from French Equatorial Africa. In mid-April, to discourage Rommel from trying to swing round the British flank, Wavell added to the "story" the proposition that in 1940 extensive minefields had been laid in the desert south of Mersa Matruh, some of which could be set off electrically by remote control.

Throughout April and May Clarke and SIME worked at PLAN A-R with every means at their disposal. Elaborate rumor campaigns, "leakages," and "indiscretions" were conducted in Egypt, Istanbul, Athens, Palestine, Free French headquarters in West Africa, and Lisbon. The special assault transports put out from Port Said and the Japanese consul there scrambled off to tell his superiors in Cairo about it. Exaggerated stories were published in the press about the arrival of an American team to train British crews in the use of new American tanks, as were photographs of a new model tank transporter against a desert background (it was in fact the only such vehicle in Africa).

Dummy gliders were displayed, and dummy parachutists simulating airborne training were dropped, at a disused airfield near Cairo. The British Desert Air Force being dangerously depleted, some sixteen dummy gliders were converted to dummy bombers (subsequently christened MECCANOS), and beginning in mid-May were displayed at another disused airfield—where they were duly attacked by the Luftwaffe on several occasions.

This busy time saw the christening of Clarke's outfit by the name under which it has gone down in history, together with the setting up of its own home and an expansion of staff.

When CORDITE was being planned, the unit slated for it, the 6th Division, had set up headquarters in a requisitioned house in the Sharia Kasr-el-Nil, the boulevard that ran straight from the center of Cairo to the Kasr-el-Nil Barracks on the riverbank. Clarke, meanwhile, was working not only on the CORDITE COVER PLAN and PLAN A-R but also on DOLPHIN and BYNG BOYS. These last required consultation with local inhabitants and others who were knowledgeable about Rhodes. In Cairo such consultation could not long be kept a secret, and it would not do for the 6th Division to be associated with Rhodes. So Clarke could not very well set up shop in division headquarters.

But for him to set up as an element of an airborne unit would fit nicely with the story that an airborne diversion would be launched against Rhodes. Headquarters of 1-SAS Brigade was still supposed to be in Transjordan, so it would fit the story for Clarke's office to be just a subsidiary "Advanced Headquarters." And it might be too obvious to label it as "Advanced Headquarters 1-SAS Brigade," since that unit itself was supposed to be secret. "Advanced Headquarters Airborne Force"? Better, but still too obvious. What about "Advanced Headquarters 'A' Force"? Give the curious the satisfaction of thinking they had divined that "A" stood for "Airborne." Excellent. So "Adv HQ 'A' Force" it became—officially, on March 28—and as "A" Force Dudley Clarke's great pioneering deception organization is known to history.

Next, offices were needed. The converted bathroom at Grey Pillars was now outgrown and would not do for receiving visitors in any case. Clarke needed to be reasonably close to 6th Division headquarters and not too far from GHQ. At No. 6 Sharia Kasr-el-Nil, across the street from 6th Division headquarters, was an apartment building of questionable reputation—it housed a fashionable brothel—but otherwise eminently suitable. Clarke took over two apartments and "A" Force moved in on April 8, 1941. It would remain there till the end of the European war in May 1945. No. 6 was a pleasant building with a quiet and private courtyard in the rear, invisible to anyone except a watcher on a rooftop. Clarke gallantly permitted the ladies to continue their business on an upper floor, and they and "A" Force's staff would exchange cordial greetings when they met.

Next, manpower. Clarke was stretched too thin, managing single-handed three or four deception operations, two sensitive intelligence ventures, and MI9 affairs. So he persuaded the army bureaucracy to put "A" Force officially on the rolls, with a "War Establishment" of three officers plus a junior officer as personal assistant, ten enlisted men, a car, and four trucks. By mid-April he had his officers on board. They were a Major Victor Jones, a tank officer, on the deception side, a Captain Ogilvie-Grant of the Scots Guards on the MI9 side, and an efficient Miss Hopkins, an ATS* subaltern, as personal assistant. In addition, he had the help of two extraordinary specialists, Major Jasper Maskelyne and Major E. Titterington.

Major Jones was a regular officer, small and florid and charming, with good social contacts in Cairo, about which he was inclined to talk at length. He had already been active in the business of running dummy vehicles; in August he was promoted to lieutenant-colonel and formally put in charge of all dummy tank operations. Wavell himself had had a small unit operating crude folding wood-and-canvas dummy tanks and trucks in his early COMPASS offensive. Jones took up where these had left off. During the spring the Royal Engineers worked out an improved model of folding dummy tank. By late April they were turning them out at the rate of two a day. From April through June, Jones raised, trained, and equipped three "Royal Tank Regiments" of dummies, and a fourth was formed in November. All three saw good service in 1941, and equipment was steadily improved. In the autumn an improved new model self-propelled dummy tank made of steel piping covered with painted canvas and mounted on a truck chassis was introduced, together with equipment for both simulating and erasing the tracks left by tanks in the sand; plus a device called a SUNSHIELD, a painted canvas cover which when placed over a tank made it seem to be a truck. "By combining Dummy Tanks, Track Markers and Erasers, with 'SUNSHIELDS,' " recorded Clarke, "it would be possible to ring the changes between real and dummy tanks with bewildering rapidity."

The dummies and the associated equipment like the tank track markers and erasers were largely designed by, and manufactured under the supervision of, one of the world's great stage magicians, Major Jasper

---

* The Auxiliary Territorial Service or "ATS" was the British women's army corps, corresponding to the American WAACs. There were also a Women's Royal Navy (known as "Wrens"), corresponding to the American WAVES, and a WAAF, or Women's Auxiliary Air Force.

Maskelyne. Maskelyne was his generation's representative of a remarkable family who for generations had been England's leading illusionists (with a sideline in exposing the tricks of fake spiritualist mediums). Now he and his "Crazy Gang," ensconced in a cordoned-off area they called "Magic Valley," had taken up this more serious game of illusion. Maskelyne applied his skills both to deception and, even more, to designing devices to aid prisoners of war to escape, like hacksaws concealed in boot heels.

Later in the year, "A" Force acquired a different sort of expertise. In Maunsell's censorship bureau was a special section under a certain Professor Titterington, who was officially employed by the Egyptian government to protect the king against assassination by poison and was therefore known to his friends as Titters the Taster. Its relatively overt function was to test correspondence for secret inks and the like. A more closely guarded function was to prepare fake documents, for which purpose Titterington had acquired the services of a talented Polish forger and a few other experts, together with some special equipment. "A" Force called on Titterington originally to prepare forged identity cards and travel papers in connection with its MI9 work, but there soon arose a need for fake Axis documents in connection with deception. Demand for these services increased, and finally in October 1941 a new army unit was formed—originally called "Printing Section (Type X)" and at the end of March 1942, renamed " 'A' Force Technical Unit"—with Titters the Taster, now Major Titterington, as its commander.

"A" Force Technical Unit functioned until December 1944, on behalf of all secret organizations in the theater. It accumulated from India and the whole Middle East a stock of more than twelve hundred different types of paper, built up an enormous "library" of enemy documents, forms, stamps, and specimen signatures, kept up with German and Italian travel regulations, and maintained a meticulous card-index of the movements and locations of Axis officials who were authorized to sign travel documents. It could produce almost anything within its scope. It detected enemy forgeries and secret ink messages and reconstructed shredded or burned enemy documents. It compounded secret inks and marked carrier pigeons with secret messages, and at least once it dyed a man brown to simulate an Arab. It forged documents, passports, orders, letters, signatures. In its last two years of operation alone it counterfeited 86 different kinds of revenue stamps, 99 different embossed stamps, 888 metal stamps, 496 rubber stamps, 3,210 identity cards, and 3,418 permits; and forged 3,149 signatures.

With Jones and Ogilvie-Grant on board, Clarke could afford an absence from Cairo. On April 26 he left on the first of his incognito wartime journeys, a four-week sojourn in Turkey to build up his channels through which misinformation could be passed, and to set up MI9 escape routes from the Balkans. His false identity for this trip is unknown, but evidently he passed as something other than a British subject. By Egyptian airline and Turkish train he made his way to Ankara. There he met with the British ambassador and the military attaché, Brigadier Allan Arnold. Thence he took the night train for Istanbul, and on the morning of April 29 checked in at the Park Hotel, next door to the German embassy and a favorite haunt of Germans and Italians. That evening he kept "an unobtrusive rendezvous" with Commander Vladimir Wolfson of the Royal Navy, whose official title was assistant naval attaché.

Both Wolfson and Arnold were important figures in Middle Eastern wartime intrigue. Wolfson was a multilingual reserve officer of Russian Jewish origin, a native of Odessa, who is said to have served in the Russian Imperial and United States navies; Clarke's meeting with him began what Clarke called "a long and profitable partnership for both Deception and MI9 matters in Turkey which was to last for the rest of the war." Arnold headed an "Inter-Service Balkan Intelligence Center" located at Istanbul that collated information from the Balkans; in what Clarke called "an equally long-lasting and no less profitable" liaison, he undertook general supervision of all "A" Force affairs in Turkey.

Clarke stayed in Istanbul till May 16. During that time he and Wolfson set up nine channels in Istanbul and one in Ankara for passing false information to the Axis. There were Swedes, Greeks, Russians, Hungarians, Iraqis, and Turks; a banker, a rug merchant, a journalist, a diplomat, a stenographer. During that fortnight much of the story of PLAN A-R was transmitted through these channels. Over time these and other Turkish channels became and remained the main conduits for getting false information quickly to the Axis in the Middle East. He was back in Cairo on May 21, after an adventurous trip through Syria, where tensions were rising, for London had directed Wavell to prepare to wrest the province from the Vichy French.

Wavell had a new assignment for him. Preparations for the Syrian operation—it was codenamed EXPORTER—would be pretty obvious, what with RAF activity, troop movements in Palestine, and a very conspicuous visit of De Gaulle to Cairo. Clarke's job was to lull the French into thinking that nevertheless it might not happen. In four days Clarke had EXPORTER COVER PLAN under way. Its "story" was that De Gaulle was in town to mediate a quarrel between Wavell and General Catroux, the courtly and dapper commander of the Free French troops in the Middle East, over the latter's failure to persuade Wavell to invade Syria. On June 6, two days before the scheduled D-Day for the Syrian operation, De Gaulle would notionally leave Cairo, having failed to persuade Wavell to act.

The customary rumors and seeming "leaks" were put into motion. De Gaulle's cooperation was secured, and early in June Shepheard's Hotel was notified that De Gaulle meant to check out sooner than expected. Baggage was packed. A special plane was laid on. The Governor-General in Khartoum was warned to expect the General's party for the night of the 6th on their way back to Fort Lamy. Clarke flew to Jerusalem on June 4 and arranged to infiltrate an Arab agent across the border in the early hours of the 6th with the story that De Gaulle was flying to Khartoum and that the invasion had been called off.

EXPORTER COVER PLAN seemed to work in some measure. It transpired later that French intelligence had absorbed the main story in time. Perhaps at least partly as a result, resistance to the opening moves was light, though the Syrian campaign saw hard fighting before it was over.

The operation underscored how useful it could be for a deception "story" to have a core of truth. For it was indeed true that Catroux had pressed Wavell to act more swiftly than Wavell thought prudent, and Wavell had resisted Catroux's pressure to the point of offering his resignation.

No rest for the weary. Clarke was still in Jerusalem on June 7, when he got an urgent message from Wavell with yet another assignment. Greece and Crete had fallen, and there was concern that Cyprus might be next on the German list. Clarke was to devise a plan calculated to delay any such attack for two weeks while reinforcements were sent to the island. So Clarke prepared what was simply called the CYPRUS DEFENSE PLAN. More sophisticated than ABEAM or the 10TH ARMORED DIVISION PLAN, it proved to be the real beginning of long-term order of battle deception.

Clarke set out to represent to the Axis the four thousand troops actually on Cyprus as more like twenty thousand—a full division plus local
forces. On June 13, orders went out to rename the command on Cyprus
for all purposes as the "7th Division" and for the brigadier commanding
to assume the rank of major-general. Fake headquarters were opened on
Cyprus, buildings were requisitioned, phony signs posted. The customary rumors were set afloat in Egypt and Palestine. Fake orders were circulated. Fake radiograms and civil telegrams flew back and forth between
Cyprus and the mainland. A squadron of dummy tanks went out to the
island. In the first application of a variant of Meinertzhagen's Haversack
Ruse, a phony defense plan of Cyprus complete with maps and order of
battle was planted upon a Cairo woman known to be in touch both with
Japanese intelligence and with a female German agent.

While the Germans had in fact never intended a Cyprus operation,
the plan successfully fixed in their minds the presence there of one to two
divisions, totaling over twenty thousand men—as was shown by documents captured in early July, confirmed by a report from Arnold that the
German military attaché in Ankara had said that Cyprus was more
strongly defended than Crete had been; and confirmed most dramatically by the capture, six months later, of an Italian intelligence bulletin
containing a map of Cyprus with all the chief features of the haversack
ruse documents plus further embellishment, together with a troop estimate of some thirty thousand men.

By the middle of July, Cyprus had been reinforced by the genuine
50th Division. This led to a momentous decision: The 7th Division fiction was continued, and to account for the presence of two divisions a
phony corps headquarters on Cyprus (initially called XVIII Corps, subsequently XXV Corps) was added. This fake order of battle was to be
continued, under successive names, for the next three years. Captured
documents and, subsequently, ULTRA intercepts, showed throughout the
period that the Axis accepted it without question. It was around this nucleus that Clarke's great accomplishment in strategic deception, the
long-term bogus order of battle, was formed.

In August Clarke installed on Cyprus a permanent representative of
"A" Force, Lieutenant Philip Druiff, to monitor this activity. In various
capacities he would be a fixture of "A" Force for the next three years.

*   *   *

Before leaving for Palestine Clarke had learned of BATTLEAXE, an offensive against Rommel planned for mid-June—much against Wavell's better judgment, for he well knew that it was hopelessly premature, but Churchill forced it on him. Cover and deception for it was cobbled together prematurely like BATTLEAXE itself, and was no more successful. All that could be done was to put out a rumor, after the invasion of Syria had started, that the British were nervous about the safety of the Western Desert during the Syrian campaign, and were sending a reserve tank unit to guard their forward supply dumps there. It was hoped that this might at least afford an explanation for the forward movement of armor in preparation for BATTLEAXE. London helped in getting the story across; indeed, it found its way into the *New York Post* two days before BATTLEAXE opened. But by that time the Germans were well aware that a real offensive was imminent.

BATTLEAXE was launched on June 15. In two days it failed utterly, as Wavell had known it would. In an act of gross injustice, Churchill fired Wavell, sending him to be Commander-in-Chief in India and bringing General Sir Claude Auchinleck, the previous C-in-C India, to take over the Middle East.

Wavell turned over Middle East Command to Auchinleck on July 5, 1941. Just before his departure, he decreed an organizational change. Tactical deception by methods other than camouflage or dummy units was assigned to a new subsection of the theater general staff, called GSI(d). The theory was that GSI(d) would give field commanders "machinery for operating tactical deception plans in the field on the pattern of that afforded on a strategic level by the original 'A' Force organization." Jones managed Advanced HQ "A" Force; GSI(d) was under Major A. D. Wintle of The Royal Dragoons, with Lieutenant R. A. Bromley-Davenport as his intelligence officer. Clarke was over both.

This experiment did not last long. As things turned out, Wintle's GSI(d) had to concentrate mainly on defensive deception employing fifth-column activities behind enemy lines, and most of this was focused on Cyprus. In mid-July, it came up with a project called COPPERS SCHEME. More "dirty tricks" and psychological warfare than deception, COPPERS SCHEME sought to sow among the Axis distrust of their own supplies and suspicion of sabotage and disloyalty at home, by doctoring captured German and Italian ammunition so it would explode when fired, bursting the weapon and injuring its user. Moreover, in each doctored cartridge case was placed a slip of paper bearing an anti-Hitler or anti-Mussolini message and perhaps a "V" sign or a hammer and sickle. In the autumn these were distributed to Eighth Army, and a supply of them was taken to the Tobruk garrison for scattering in No Man's Land by nightly patrols. Some were infiltrated into Europe by surreptitious means. No firm evidence of the result of this scheme was ever forthcoming.

Wintle was a truly extraordinary character. Behind a Blimpish façade, complete with monocle, was a man of resolution, courage, and principle; unfortunately, Clarke was to have his services only until September, when he left on a mission to rescue British prisoners in France. Bromley-Davenport—a less unconventional officer, but nobody wholly conventional could have lasted long with Clarke—was to remain with "A" Force throughout the war. With Wintle's departure, the GSI(d) experiment came to an end and the former organization was resumed.

Auchinleck—the soldiers called him "the Auk," and the sobriquet has clung—proved to be as supportive of Clarke's work as Wavell had been. His assignment from Churchill was to attack the Germans and Italians, raise the siege of Tobruk, and advance into Libya. He had first to rebuild the army of the Western Desert (soon to bear the ever-glorious name, Eighth Army) after the diversions of Greece and Syria and the losses of the BATTLEAXE disaster; so his offensive—codenamed CRUSADER—could not be launched before November.

He had also to be concerned about his distant rear. A fortnight before he took up his new duties, the Germans had invaded the Soviet Union. Throughout that summer the betting was that the USSR would collapse. German occupation of the northern and eastern shores of the Black Sea and an advance through the Ukraine and the Caucasus could produce a vast pincers movement against the whole Middle East, with irresistible pressure on Turkey and on Iran. So Clarke got three assignments from Auchinleck. The main one was to prevent Rommel from mounting spoiling attacks, and force him into premature defensive preparations, by making him prepare for a British offensive to begin as early as August. Clarke was also to do what he could to give the impression that the Turkish border was strongly held notwithstanding that the armored units that had conquered Syria were being brought back to Egypt, and to cover preparations for the occupation of Iran in late August.

Two minor operations, dubbed EUPHRATES PLAN and IRAN COVER PLAN, carried out these latter two assignments. The EUPHRATES PLAN "story" was that Canadian armored units equipped with the new American M-3 Grant tanks had arrived at Suez direct from Canada, moved by rail to Palestine, and gone into a tightly-guarded training area in the desert around Palmyra. The usual leaks and rumors were circulated. Palmyra was cordoned off into a prohibited area, with access only by way of specially printed passes. Soldiers wearing Canadian insignia appeared in various towns. Money-changers in Jerusalem and Haifa received a sudden influx of Canadian dollars. Passes and identity cards were "lost" in useful places. The pro-Vichy security forces in Beirut were asked to look out for an (imaginary) French-Canadian deserter who might try to escape to France by stowing away on a Vichyite repatriation ship. Dummy tanks were deployed, and trainloads of other dummy tanks loaded on flatcars under tarpaulins were routed through populous areas of Palestine.

The IRAN COVER PLAN was even simpler. In preparation for the invasion and occupation of Iran, supply dumps were erected on the desert route between Palestine and Iraq. Rumors and leaks portrayed these, as well as increased shipping activity in the Persian Gulf, as preparations for a division to move from India to disembark at Basra and move overland to Palestine.

The deception plan targeted on Rommel, called COLLECT, was of a wholly different order of magnitude, and introduced several important new techniques. One was the "postponing" of a notional event in a plausible way, without raising suspicions in the enemy's mind. Another was the risky art of the "double bluff"—making the real operation the "story" in the hope that the enemy will believe it to be a deception or other misinformation. And COLLECT brought on to the scene in the Middle East what proved to be the most powerful of all tools of deception, the controlled enemy agent or "double agent" reporting at long range to Axis intelligence.

COLLECT was to be a multistage operation. First, word was to be spread that D-Day for Auchinleck's attack would be August 9, the earliest date at which troops that had been engaged in the Syrian operation could conceivably have reached their new concentration areas in Egypt.

Then at some time around August 3 the story would be put out that this offensive had been postponed to a new date. This process would be repeated, and it was hoped that successive cries of "wolf" would lull Rommel into "a sense of apathy and false security" by the time the real CRUSADER offensive was ready, so that he would disregard information that the real operation was imminent.

Once more, the usual rumors and leaks were set afoot. In addition, British forces themselves were deceived into thinking that an early offensive was being planned, by issuing various orders with this end in view. By early August, Maunsell's people were reporting that rumors were circulating that a British offensive in the Western Desert was under preparation and probably imminent. On August 5, word was passed that D-Day was postponed to August 30. It was subsequently "postponed" again to September 15.*

With COLLECT apparently running satisfactorily, Clarke left Egypt for Lisbon on August 17, an absence that was unexpectedly to last for three months. Jones ran COLLECT while he was away, with the aid of Maunsell and Brigadier Shearer, the director of military intelligence for the Middle East theater (and sometime managing director of Fortnum & Mason). On their watch there opened up the first major double agent channel for the eastern Mediterranean. Called early in its history by several different names, it finally received the permanent appellation of CHEESE. It was to be one of the four *hors de classe* Allied double agent channels of the war.

The original CHEESE was an Italian Jew (apparently technically a British national) named Renato Levi, about thirty-five years old, of medium height and slight build, with a fresh complexion and dark slicked-back hair, who had lived in India, Switzerland, Italy, and Australia, and spoke English reasonably well. Bright, resourceful, and well-traveled, but also easy-going and lazy, he had been recruited by MI6 before the war, and in December of 1939—when Italy was still neutral and he was living in Genoa—he reported that the Abwehr (the German military intelligence

---

* As part of COLLECT a portion of the old Plan A-R, referred to as TRIPOLI PLAN, was kept afloat. While Auchinleck did not want to suggest any specific objective for the notional COLLECT offensive, he did agree to continuing PLAN A-R's notional threat to Tripoli, hoping that Rommel would be induced to hold reserves back at his base there, eight hundred miles from the front. So various measures designed to increase German nervousness about Tripoli were

organization) had approached him to work for them. The British told him to accept. He acted as an Abwehr agent in France—keeping in touch with French intelligence, the Deuxième Bureau—until the French capitulation in June of 1940. The Abwehr then offered to share his services with the Italian military intelligence agency, the Servizio de Informazione Militare, known as SIM. SIM dispatched him to Cairo in October 1940 in company with a radio operator. A transmitter was to be made available to them in Cairo by way of the Hungarian diplomatic bag, together with trustworthy contacts. They were to transmit military information to a station at Bari in the south of Italy.

When Levi and his operator reached Istanbul en route for Egypt the Turkish authorities arrested them for dealing in counterfeit money. MI6 arranged their release. The radio operator lost his nerve and went home. MI6 got Levi to Cairo in February 1941. There he passed under SIME's control, being initially in charge of Lieutenant-Colonel William Kenyon-Jones, Maunsell's deputy. He was installed in a pension in Cairo under a false name. But nothing happened. Levi himself proved to be mainly interested in women and bar-hopping, while neither the promised transmitter nor the promised trustworthy contacts were forthcoming; the only names with which SIM had supplied him proved to be unsuitable low-level people. SIME thus had on their hands a potentially valuable double agent with no way to report back to his control. He needed a transmitter and an operator, and when he got them he would need a credible story to account for them both. Meanwhile, Kenyon-Jones—the only British officer actually in touch with Levi—was beginning to get tired of his charge. At this stage very little importance was being attached to him.

Kenyon-Jones decided to perk the situation up. He suggested to Levi that the British produce the transmitter and that Levi would have to convince the Italians that he had managed to do this through his Cairo contacts. Levi was enthusiastic. Kenyon-Jones then went to see the signals section of GHQ. They were not receptive. Their job, they told him, was

---

undertaken. Among other things, a radio message in an insecure cipher was sent to the British consul at Tangier, asking him to find an imaginary British merchant seaman who was believed to be in hospital at Tangier after escaping from an Italian POW camp at Tripoli, and put to him several questions designed to suggest that information was being gathered for use in planning an amphibious assault on Tripoli.

not to help the Axis to get at British communications but to keep them from doing so. But eventually Kenyon-Jones found a non-com of the signals department who understood the situation. He put together a transmitter from parts that could plausibly have been obtained in Cairo before the war, and undertook to act as Levi's radio operator. The story was then concocted for Levi that he had by happy chance been able to buy a transmitter from an Italian living in Cairo, and had installed it in a flat he had rented. He had then managed to befriend one "Paul Nicossof," supposedly a Syrian of Slavic extraction (wholly imaginary, of course), who could send Morse and had undertaken to act as radio operator. SIME then dispatched Levi back to Italy to tell this story, furnish his control with a cipher provided by SIME, and collect additional money. He would tell his control that, beginning in late May, NICOSSOF would come on the air every Monday and Thursday till Levi's return to Cairo.

In April, Levi took ship at Haifa—and was not heard of again for three years. It transpired later that back in Italy he had been tried, convicted, and in November jailed, for black-market activity. The Germans had thought highly of him. The Italians had been suspicious. Jailing him may have been SIM's way of resolving this divergence of views in their favor. Levi was a brave man; though repeatedly interrogated, he never broke.

Meanwhile, back in Cairo NICOSSOF duly went on the air towards the end of May. After some weeks of effort and fiddling with different frequencies—word had to be sent back to Levi through Turkey what frequencies to change to—he finally achieved contact with Bari on July 14. (From the ineptness with which Bari handled the traffic, it appeared that the notoriously incompetent Athens station of the Abwehr was running the case and SIM had washed its hands of it.)

So began the great channel known as CHEESE. From that date till February 10, 1945, when CHEESE went off the air, it transmitted no less than 432 messages to the Germans and Italians, at least half of which was deceptive material.

For the rest of the summer of 1941 and into the autumn, NICOSSOF transmitted "chickenfeed," as true but unimportant information to build up an agent's credibility is called. At some point Major James Robertson took over from Kenyon-Jones as case officer; Captain Evan John Simpson—a man of imagination, who in civil life had been a novelist—worked under Robertson and composed NICOSSOF's messages.

In due course NICOSSOF acquired a notional source described to his handlers as a promising contact at GHQ. Through the latter part of the summer, he transmitted information supposedly from this contact, supporting the basic COLLECT "story" that a British offensive was imminent. By this time, the British were reading the Abwehr hand cipher with some regularity. It was clear from these decrypts that even though the Italians might be suspicious of CHEESE, the Germans were not; and that in particular they thought well of his supposed new source.

This nerved Maunsell and Shearer to use the CHEESE channel to convey the definitive "postponement" of Auchinleck's offensive by a double bluff. The regular rumor mill would continue to say that it was imminent. CHEESE, however, would produce supposedly more reliable information from his new contact to the effect that the offensive was postponed till the end of the year, leading the Axis to regard as efforts at deception not only the existing rumors but the actual preparations for CRUSADER. On October 6, CHEESE identified his contact as "PIET," a disaffected South African confidential clerk to the chief of the operations staff at GHQ Middle East, a habitual complainer dogged by money troubles and problems with women, who had "hot information" that he might be willing to sell. For two weeks, CHEESE's reports teased Bari with prospects of exciting disclosures by PIET, whose financial difficulties NICOSSOF had supposedly eased, and who supposedly grew talkative under NICOSSOF's generous hospitality.

On the night of October 20, NICOSSOF radioed to his control that PIET, desperate for money, had brought in important information, with the promise of more if he were paid for it. Wavell, PIET reported, had just visited GHQ, having come from Tiflis in Soviet Georgia. Four divisions, one armored and three infantry, were to be detached from Eighth Army and sent to help the Soviets defend the Caucasus; Wavell was now on the way to Iraq to prepare to receive these troops. NICOSSOF reported that he had paid PIET forty pounds Egyptian for this information out of his own pocket, and told his control that money had to be sent at once to get further information from PIET. On October 27, NICOSSOF reported the crucial news from PIET that because his army had been weakened to help Wavell, Auchinleck could not open his offensive till after Christmas.

Messages from CHEESE's control radiated gratification over this coup.

NICOSSOF prudently eased up on PIET, and for the next few weeks CHEESE sent only commonplace chickenfeed. Auchinleck launched his CRUSADER offensive on November 18. It achieved total surprise. By the end of the year Eighth Army was back at El Agheila, whence Wavell had been driven the previous spring.

The tone of CHEESE's control grew perceptibly colder. Shearer and Maunsell—rejoined now by Clarke, who got back to Egypt the very day CRUSADER opened—took it for granted that CHEESE was irretrievably blown; a sacrifice they had assumed from the outset and had been willing to make. They wrote CHEESE off. Simpson disagreed. The Germans, he argued, had no reason to suspect NICOSSOF himself. Maybe he had been deceived by PIET. Maybe PIET himself had been deceived by his boss, who might have recognized his clerk as a security risk. Simpson thought CHEESE could be nursed back into the good graces of the Axis. Though they thought the case was hopeless, Clarke, Maunsell, Shearer, and their colleagues were willing to let him try.*

Clarke had been away for most of COLLECT. He left Egypt on August 17, it will be recalled, for a visit to Lisbon. His purpose was to open channels to the Axis in that neutral capital, as insurance against the possibility that Soviet resistance might collapse and Turkey be drawn into the war, shutting off his channels in Istanbul and Ankara. He once again traveled by the familiar route via Khartoum, the Sahara, and West Africa, reaching Lisbon on August 22 in the guise of a civilian government employee (probably named "Derek William Carter"; this was certainly his incognito later in the war). He remained in Portugal for a month, with gratify-

---

* According to unofficial accounts, all of which trace ultimately to Brigadier Shearer many years after the war, two German agents besides Levi entered into the CHEESE story. One was supposedly an agent codenamed the GAULEITER OF MANNHEIM because it transpired that he had a Nazi Party past, who parachuted into Palestine claiming to be a Jewish refugee. Notionally becoming a waiter in a British officers' mess, he allegedly became a source for CHEESE up through the CRUSADER offensive. This perhaps refers to one Ernst Paul Fackenheim, a Jew working for the Abwehr, who parachuted into Palestine in October 1941, intending to be a triple agent. The other was supposedly an agent working back to the Abwehr in Athens, named Klein (codenamed STEPHAN), who turned himself in together with a radio transmitter. Supposedly Warrant Officer Ellis operated the LAMBERT transmitter back to Bari and Sergeant Shears operated the STEPHAN transmitter back to Athens. Whatever the facts, there is no doubt that the CHEESE channel in its established long-running form consisted solely of NICOSSOF, impersonated on the air by Sergeant Shears.

ing results—at least over the short run. "As a spy center," he later recorded, "Lisbon turned out to be more prolific even than Istanbul." No less than sixteen separate channels to the Germans were opened up. They were a colorful set, as Clarke recorded; "Germans, Portuguese, Spaniards, international Americans, French and Swiss, of both sexes and mostly of doubtful occupation." But, though they were used to pass the COLLECT story, they proved disappointing in the long run. After a few months, they faded away, and channels in Portugal were not again seriously worked by "A" Force till the spring of 1944.

Meanwhile, back in England a feeling was germinating that "A" Force was learning lessons that could be applied on an even broader scale than the Middle East. It seemed sensible to take advantage of Clarke's presence in Western Europe to get a report from him at first hand. So, after a month in Lisbon and Estoril, Clarke was ready to return to Egypt, when he was suddenly ordered to visit London.

From the beginning of the war, such deception as existed was overseen at the War Office by a committee of representatives from the three fighting services and from MI5 and MI6, called the Inter-Service Security Board, or "ISSB." The ISSB had played some role in Clarke's work. It reviewed "A" Force's deception plans once Wavell or Auchinleck had approved them, and warned of any risk of their clashing with activities elsewhere. From time to time "A" Force had asked it to arrange for selected items of misinformation to be passed to the Axis through double agents operated from England by MI5. But the ISSB's focus was primarily on security. Such deception as it planned was in the nature of cover for convoys or troop movements from Britain, "aimed," as Clarke put it, "more at concealing our own intentions than at persuading the enemy to make a calculated false move of which we could take active operational advantage." There was a growing feeling that something more than this could be done and that "A" Force might in some respects be the model for a central deception entity.

Clarke reached London on September 18. There he was asked to write and circulate a paper on the experience of Middle East Command in strategic deception. On October 2 he attended a Twenty Committee meeting and met with the Joint Intelligence Subcommittee and the Joint Planning Staff at the War Cabinet Office. They liked what they read and heard. The Chiefs of Staff themselves met with Clarke on October 7. On

October 8, the Joint Planning Staff recommended to the Chiefs that a section be set up in the planning organization of the War Cabinet Office to control deception world wide, analogous to "A" Force in the Middle East. Like "A" Force, it should function as an intrinsic element of the operational planning staffs. Like "A" Force, it should not be a committee but a commander with a staff to assist him, whom they called the "Controlling Officer." Like "A" Force, it should both plan cover and deception operations and supervise their execution, calling on the intelligence, security, psychological warfare, and military operational authorities as needed. And again like "A" Force, its success would depend directly on the personality of the Controlling Officer, who must be a man of "considerable ingenuity and imagination, with an aptitude for improvisation, plenty of initiative, and a good military background." The Chiefs met with Clarke again on October 9, and promptly approved the proposal.

Admiral of the Fleet Sir Dudley Pound, the First Sea Lord, asked Clarke if he would be interested in taking the job of Controlling Officer. He responded that he was a staff officer of Sir Claude Auchinleck, "who alone was conducting active operations at the time, and I felt sure my place was to stay with him. I even used the simile that you can't pinch a man's butler when he has only been lent to you for the night."

On October 11, the Chiefs saw Colonel Oliver Stanley, the head of the Future Operational Planning Section of the Joint Planning Staff, and asked him to undertake the controlling authority of the new organization and to consider what staff he would require and whence they should come. For the time being at least, "A" Force would communicate with the Chiefs of Staff not through Stanley but through the ISSB as hitherto.

Meanwhile, on September 25 an officer had arrived from the Middle East on other business but bearing an oral message for Clarke from Shearer describing the planned double-bluff through CHEESE to implement the last "postponement" in COLLECT. Clarke and the intelligence people in London concluded that this scheme was so delicate that all that should be tried in aid of it at the Western European end was to pass word through the new channels in Portugal the simple fact that Auchinleck's offensive could not start before Christmas.

Clarke left London as soon as Stanley had been appointed, and reached Lisbon on October 12, apparently traveling in the guise of a journalist. Within a few days he had succeeded in filtering through to the

German Embassy word that there would be no offensive in Libya before Christmas. Clarke moved on to Madrid, intending to pass confirming evidence to the German military attaché; and here his fun and games as an undercover agent came to an inglorious and comical end. The details of his Madrid "caper" are obscure, but evidently it involved his disguising himself as a woman, for on or about October 20, the Spanish authorities arrested him, dressed in female garments. (They put it out that they had arrested "Wrangal Craker," correspondent of *The Times,* in drag.) He was soon released, and proceeded to Gibraltar. There he was ordered back to London to explain. "I am afraid that after his stay in Lisbon as a bogus journalist [Clarke] has got rather over-confident about his powers as [a secret service] agent," said Guy Liddell, head of the counterespionage division of MI5, who had met Clarke twice in London and had been favorably impressed. "It would be much better if these people confined themselves to their proper jobs." Some seem to have wondered if Clarke had gone off the deep end.

Clarke set out for England by sea, but his ship was torpedoed and he was in Gibraltar again by October 30. Auchinleck urgently wanted him back, and Clarke cabled London to ask whether he should still return to the capital or go on to the Middle East. Sir John Dill sent a memorandum to Churchill himself, suggesting that Field-Marshal Lord Gort, who was at that time Governor of Gibraltar, be directed to interview Clarke, get an "explanation of his doings," and "if he considers his story is reasonable and that he is sound in mind and body, to send him on to Middle East by first possible aircraft as he is urgently required there." Churchill agreed. So Gort interviewed Clarke, and concluded in a long report that Clarke was "mentally stable," and that "we can reasonably expect that this escapade and its consequences will have given him a sufficient shock to make him more prudent in the immediate future."

Gort's report, Dill told Churchill, clearly showed that Clarke had simply undertaken "a foolhardy and misjudged action with a definite purpose, for which he had rehearsed his part beforehand. As a result he gravely risked undoing some of the excellent work already done in the U.K. and en route there. . . . I feel we can safely leave it to General Auchinleck to deal with him." (The "definite purpose," Dill explained in response to Churchill's further question, was that Clarke "had worked up contact with certain German or German-controlled elements, with a view, later, to their providing a channel for the dissemination of false information which was designed to provide a 'cover' for British operations (in the Middle East).")

The Soviet spy Kim Philby heard about the matter and reported it to his control, noting that the case was "being shrouded in the greatest secrecy." In the long run it would perhaps have been better if the affair had not been so shrouded. For the cross-dressing aspect of this "escapade"—which obviously had no more to it than *Charley's Aunt*, or d'Artagnan's escape from Milady—was whispered around, and enabled the ill-disposed to spread rumors as to Clarke's personal orientation that those who knew him well were vigorously, sometimes angrily, denying even after his death. But this comical affair plainly did Clarke's career no harm; and it seems, as Gort had anticipated, to have pretty thoroughly cured Clarke of trying to act as his own secret agent.

There was a further delay getting passage on one of the Catalina flying boats that periodically ran the Axis gauntlet to Egypt via Malta. Then bad weather held Clarke up on Malta. He finally reached Egypt rather dramatically on November 18, just as the CRUSADER offensive was opening. "As the Catalina flew through the night along the coastline of Cyrenaica," he wrote, "bombs could be seen exploding around Derna, where the RAF were dealing with the German airfields while the Eighth Army's tanks moved forward in the attack."

After Clarke's return, the rest of "A" Force's first formative year saw a new tactical deception device; three more deception plans, one of them an instructive failure; a nearly ruinous bureaucratic power-grab; and two external developments of enormous significance. The external developments were the breaking of the Abwehr's machine cipher and the entry into the war of the United States and Japan; of these, more later. The tactical device was sonic deception. The plans were a tactical ruse called GRIPFIX, a strategic followup to COLLECT called ADVOCATE, and an operational deception called BASTION. The power-grab had a happy ending after a near-disaster.

First, sonic deception. When Rommel fell back after the CRUSADER attack he left behind a garrison holding the strategic Halfaya Pass under a certain Major Bach, a former Lutheran pastor who had exchanged his cassock for a uniform and become one of Rommel's most formidable commanders. "A" Force was involved in two efforts to dislodge him. The first was to support an attack on Bach's position on Christmas Day by leading him to expect an attack on a different part of his line from the actual target area. Bromley-Davenport hastily arranged for an Egyptian film company to record sound tracks of tank movements. These were run up to the front lines, and on Christmas Eve night they were played through propaganda loudspeakers opposite the point where it was hoped to make Bach expect an attack. The assault the next day failed, but sonic deception was added to "A" Force's arsenal. Experimentation eventually produced an efficient method of projecting recorded sound from a specially equipped armored car. The first of these was to go into service in mid-1942.

The tactical ruse, GRIPFIX, was a scheme to procure Bach's surrender by dropping into his perimeter a forged letter purportedly from Rommel himself giving Bach permission to make an honorable capitulation. The letter was prepared with much care by Major Titterington's forgers, written on captured German stationery, stamped with classification markings made with captured German rubber stamps, sealed with a German Army seal specially forged for this project, and dropped in captured German containers on the night of December 29. Unfortunately, the stationery used was obsolete ("A" Force knew this, but decided to take a chance, reasoning that a commander with supply difficulties such as Rommel labored under would frugally use up his old stocks), and this raised suspicion in Bach's mind. Still more unfortunately, the day after the drop Bach got a message from Hitler himself congratulating the defenders of Halfaya and urging them to hold on as long as they could. So GRIPFIX came to nothing, and Bach did not give up till January 17.

ADVOCATE (known for its first few days as "XMAS PLAN II") was the deception scheme for Auchinleck's plan (codenamed ACROBAT) to follow up CRUSADER by clearing the Axis from the rest of Libya. Its "story" was that, having driven the enemy from Egypt and Cyrenaica, rather than advancing farther the British would establish a strong defensive line at El Agheila and shift major forces back to Syria, Iraq, and Iran so as to back up Turkey and support the Soviets. Implementation of ADVOCATE was to be by the usual leakages, by sundry order of battle deception measures to simulate transfer of units to the northern frontiers of the theater, and by encouraging the press to belittle the advantages of advancing farther in North Africa and to suggest that the British wished to avoid giving the

Axis an excuse to intervene in Tunisia and transfer to that province their base at Tripoli.

ADVOCATE had an embarrassing history, reflecting the lack of centralized control of deception notwithstanding the appointment of Colonel Stanley. The Japanese struck Pearl Harbor just as it was getting under way. Unknown to Cairo, the whole ACROBAT offensive was thereby cast in doubt, since troops might have to be rushed from Auchinleck's command to reinforce the Far East. Moreover, again unknown to Cairo, an invasion of Algeria and Tunisia was actively under discussion between Roosevelt and Churchill and their staffs in Washington; the last thing London needed was a press campaign drawing attention to French North Africa.

London tried to stop ADVOCATE, but it was already under way. The Middle East was awash with tales of the strength of British defenses in Libya, bogus telegrams had been passed between Egypt and India indicating notional movements of troops from Libya to Iraq, and the press had been vigorously steered in the desired direction: Reuters had filed a story proclaiming that "the occupation of this arid land is unimportant" and that with the threat to Egypt removed, "large Allied forces are thus set free for any campaigns the Allies may decide on for the New Year." Eighth Army put on a wireless deception operation suggesting that an armored brigade was being sent back to the Delta. New channels to the Italians that had just been opened in Malta were put to use. Special press conferences in Cairo and Delhi on January 14 and 15 produced a number of stories in Egyptian and Indian newspapers emphasizing the strength of British defenses in Libya, while reports from Syria and Turkey confirmed that the "story" had taken root there. (Yet back in London the press was being encouraged to talk up the strength of the reinforcements that were being sent from Libya to Malaya. "There is so much deception and counterdeception going on that it will not be surprising if our troops land up in the Arctic dressed in crepe-de-chine," said Guy Liddell of MI5.)

Then on January 21, 1942, in a surprise attack at El Agheila, Rommel struck back in a counteroffensive that would soon return him to Egypt. ACROBAT went into the dustbin and with it ADVOCATE's stories about the strength of British defenses in Libya. There was considerable embarrassment back in London. Stanley decreed that thenceforth advance London approval would be necessary for any theater deception plan that related to a neutral country, might involve repercussions outside the theater, or involved the use of the press or other overt propaganda on a large scale. On February 11, the Chiefs of Staff went further, directing that all future major deception plans must be referred to them for approval, and that directives to military Directors of Public Relations were to be prepared by the Joint Planning Staff in consultation with Stanley and the Foreign Office. A no doubt chastened Clarke received a friendly word of wisdom from the ISSB, warning that use of the press for deception was "a very ticklish matter, so much so indeed, that its use at all for deception purposes is under consideration at the highest level."

Clarke's difficulties were compounded by organizational problems. During his unexpectedly long absence, and with the departure of Wintle, Lieutenant-Colonel Ralph Bagnold, who had distinguished himself in the organization and early exploits of the Long Range Desert Group and was currently "Inspector of Desert Troops" at GHQ, had been promoted to full colonel and had been made "Chief Deception Officer" in the operations division of GHQ, taking over the tactical side of deception and leaving "A" Force in charge of the strategic side (though nobody knew where the line between them could be drawn).

Clarke got back in November to find that this had been in place for nearly a month. He was dismayed. Rightly or wrongly, but understandably, he saw this as a power-grab by Bagnold. Not only did Bagnold know nothing about what he was supposed to do, but the publicity given to his role was in stark contrast to, as Clarke put it, "the closely guarded secrecy with which 'A' Force had tried to cloak its very existence, let alone its methods for which complete secrecy was the sole safeguard." It soon transpired that Bagnold expected 'A' Force to train his officers and run all radio deception. Nothing seemed to slow him down. In mid-January he took over complete control of all camouflage and dummies and the units operating them. At the end of January he procured a decree from GHQ that his operation would now plan and coordinate all forms of deception at headquarters of units down to brigade level and announcing a three-week course for senior deception officers to start in February, to be followed by a five-week course for juniors. He had of course no one to teach these courses.

It took Erwin Rommel to put an end to this, with his offensive of late January. At eight o'clock the night of Sunday, February 1, the Auk, having just flown back to Cairo from a visit to Eighth Army headquarters in the desert, summoned Clarke and told him to put together a crash deception operation to help slow down Rommel. Clarke flew out from Cairo at dawn the next morning, leaving behind "a frank memorandum to the Directors of both Operations and Intelligence at GHQ, bewailing the unfortunate position in which he now found himself, and begging for the whole question of deception machinery in the Command to be reviewed completely afresh during his absence."

Clarke landed near Eighth Army headquarters at Gazala just before noon in a blinding sandstorm. Vehicles were flowing back along the coast road in a dense stream. Headquarters was busily packing up to move back to the Egyptian frontier. Rommel had already driven forward two hundred miles in ten days; the Eighth Army had only two shaken divisions to hold a seventy-five-mile front, and desperately needed a few days' breathing space to ready a defensive line covering Tobruk and bring up additional divisions from the Nile Delta.

On the spot, Clarke roughed out a plan, which received the codename BASTION. Over the next three days he reconnoitered the terrain and put in urgent calls to Jones, Maunsell, and others. On February 5 the plan was complete and the Eighth Army and RAF commanders signed off on it. Its formal object was to induce Rommel to make no further advance before the first week in March. It proposed to do this by making the German believe that he was being led into a trap. Its "story" was that the British were waiting for Rommel to overcommit himself against the British defensive line at Gazala, at which time strong tank reinforcements on the British right in the Tobruk area and the British left deep in the desert would strike to cut him off from his base. Tobruk would be the key advance base for the counterattack. Clarke contemplated that the "story" would begin to reach Rommel's staff on February 15, reach its full effect on February 20, and begin to peter out during the first week of March. The keys to the plan were the creation of phony tank reserves on each flank together with simulated strengthening of the Gazala line, plus evidence that the British were in Tobruk to stay.

Clarke had already brought Jones up to Eighth Army and told him to put every dummy tank he could scrape up into Tobruk. Jones's broader assignment was to simulate a powerful force of three hundred tanks. There were not that many dummies in the whole Middle East, so Clarke and

Jones borrowed a trick from Rommel himself. The RAF had learned to watch Arab encampments for tank tracks, for the Germans liked to hide tanks under Bedouin tents. Presumably the Luftwaffe would be alert for the same evidence. Cairo was able to furnish remarkably quickly a supply of unserviceable canvas that would pass for Arab tenting. Overnight on the night of February 15, Jones put up some 150 tents deep in the desert behind the left wing of the army (four of them with dummy tanks peeping out as if carelessly placed), surrounded by "tank tracks" created by one of Maskelyne's devices and suitably augmented by campfires and other signs of humanity together with fake radio traffic. Supply convoys were routed past the "camp" as if dropping off materiel for the tankers hidden there, and aerial reconnaissance was increased over the southern flank as if in preparation for an attack. After a few days and several gratifying attacks by the Luftwaffe, the "camp" was changed so as to seem to have been abandoned, and the dummies were moved to a nearby wadi.

Meanwhile, Jones stationed fifty-two dummy tanks at a former airfield in Tobruk itself, where they too were visited by the Luftwaffe, and used fifty-two more to stiffen a position on the front line. Additional dummies were used to augment the apparent strength of the main line of resistance together with fake minefields, guns, and trenches prepared by the camouflage units. Steps were taken to counter the widespread feeling in Tobruk that—unlike the last German advance, when the town had held out for an eight-month siege—this time it would be abandoned to Rommel.

At a more rarified level, stories were planted through the established channels in Portugal and Turkey; and SIME, to quote Clarke, "managed a neat 'leak' to a party of disgruntled Vichy diplomats as they were being transported from Cairo to France through Istanbul." (CHEESE was still under a cloud and of no use for BASTION.)

Rommel did in fact stop before Eighth Army's Gazala line and did not resume his offensive till late May. "A" Force liked to think that BASTION contributed to this. As so often in deception, there was no way to prove it. But BASTION was important in a different way. "We learnt more *LESSONS* from it than from almost any other Plan," wrote Clarke after the war, "and it helped us to evolve three important Principles." Indeed, it "led to a good deal of thought on the Theory and Practice of Deception in general. Several papers on the subject were prepared about this time, and many of the theories upon which Deception was conducted later evolved during this period."

* * *

When Clarke returned to Cairo on February 15, he found that the memorandum he had left behind on departing for the front had borne fruit. Apprised at last of the Bagnold problem, on February 12 Auchinleck had decreed that for the future "A" Force was to run both tactical and strategic deception through the Operations (not the Intelligence) branch of GHQ, with deception officers below corps level eliminated.* Colonel Bagnold vanished, as far as "A" Force was concerned. (He went on to be a brigadier and deputy chief signal officer for the theater, and to have a distinguished scientific career after the war.) In late March, after BASTION was over and there had been time to think about housekeeping matters, Clarke was promoted to full colonel and "A" Force was expanded and reorganized.

This new charter was in effect Clarke's doctoral degree in the art of deception, as the red tabs of a colonel on his tunic were its visible sign. In a little more than a year after he had first stood before Wavell to get his new assignment, he had learned the game by trial and error; sometimes largely error, but those are the most effective lessons. "The organ was now built," he wrote long afterwards; "its stops were ready for us to pull at will, and all we had to do now was write the music and gain a little more practice in playing it." With the intense experience of BASTION to pull it all together, he had consciously evolved a coherent theory and practice of deception.

Specifically, he had learned:

First, the CAMILLA principle: The object of a deception is not to induce the enemy commander to *think* something, but to induce him to *do* something: To act as you want him to act. And its corollaries: Your *target* is the mind of the enemy commander. You must judge what estimate of the situation given to him by his intelligence services will induce him to act as you wish. Your *customers* are the enemy intelligence services. You need to know how they operate, and what information given to them will induce them to give their commander the estimate of the situation that will cause him to act as you want him to act.

Second, the K-SHELL principle: Never conduct a deception with no clear object simply because you can do so.

---

* Within a few months it was recognized that corps did not need them, and by the time the first army group was organized a year later in Tunisia it was recognized that they were not needed at army headquarters level either.

Third, a proper deception plan must have time to work. Only a quick and simple tactical deception can be expected to work on short notice. A major operational deception may take weeks to percolate through the enemy system; a large-scale strategic one may take months.

He had learned, too, some valuable devices as components of deception plans, such as that the most effective deception is one that confirms what the enemy already wants to believe; how a "story" can be based on an operation that had been considered and rejected; leading the enemy to believe that an operation was scheduled for earlier or later than a true operation; notional "postponements" of operations. Most important of all, ABEAM, 10TH ARMORED DIVISION, PLAN A-R, and CYPRUS DEFENSE PLAN had opened up the possibilities of long-term order of battle deception—building up over time the belief that your forces are stronger than they really are, thus making all sorts of fictitious operations seem plausible to the enemy. /

The Bagnold affair had confirmed views he already held about deception organization. Deception mingles both operations and intelligence. Operations, because a deception must be related to your side's actual plans; intelligence, because the deceiver needs to know what the other side believes and knows and how the other side's intelligence works. Ideally, the deceiver should report directly to the commander. If that is not possible, it is probably better to be part of the operations side of the staff than the intelligence side.

He had seen how intelligence staffs build up a picture from little details, and had learned how to feed little details to enemy intelligence to encourage them to build up the desired picture. He would soon develop the technique of constructing a "story," tabulating elements that would contribute to the "story," and assigning them to different channels on a regular and systematic basis.

He had gained experience with the tools of the trade; with dummy vehicles and aircraft on the tactical level, planted rumors and deceptive administrative measures, and, most of all, through CHEESE he had seen the power of a properly managed double agent channel. And he had learned how to orchestrate all these instruments to achieve the desired effect.

He was ready to begin his first masterpiece, to be called CASCADE. He had become the master of the game.

CHAPTER 2

The Art of Deception

With Dudley Clarke's education substantially complete (it would never be complete; he learned something from every experience), let us step back and examine the theory and practice of deception as fully developed by the Western Allies—which is almost the same thing as saying by Dudley Clarke.

It falls naturally into four parts: The basic principles of deception; the structure of a deception organization; the planning and execution of a deception operation; and the tools of the trade.

First, the basic principles.

They were British in their inception and British in almost every aspect of their development, so it is but fitting to start with a paradigm from the very beginning of the Matter of Britain: the tale, as told five hundred years ago by Sir Thomas Malory, of Uther Pendragon, King of All England, and the fair Igraine, wife of the Duke of Tintagil.

Uther fell in love with Igraine and made war on the Duke. The Duke was slain in battle with Uther's men. Before Igraine could learn of this, Uther arranged with the wizard Merlin to take by magic the likeness of the Duke, rode to the Duke's castle, and entered Igraine's bedchamber. "So after the deth of the duke kyng Uther lay with Igrayne, more than thre houres after his deth, and begat on her that nyght Arthur"; and departed ere daybreak. Uther subsequently married the fair Igraine and made her his queen. When "quene Igrayne waxid daily gretter and gretter," Uther asked whose child it was.

> "Syre," saide she, "I shalle telle you the trouthe. The same nyghte that my lord was ded, the hour of his deth as his knyghtes record, ther came into my castel of Tyntigaill a man lyke my lord in speche and in countenaunce . . . and soo I went unto bed with hym as I ought to do with my

lord; and the same nyght, as I shal ansuer unto God, this child was be-
goten upon me."

Uther told her what had happened. "Thenne the quene made grete joy
whan she knewe who was the fader of her child." The child, of course,
grew up to be King Arthur, of Camelot and the Round Table.

All the fundamental principles of deception are in this ancient tale. In
fact, of *cover and deception.* These two concepts, so often mentioned together in one phrase, are strictly speaking not the same. In military
usage, *cover,* as defined officially in 1945 by Joint Security Control, the
American central deception staff in Washington, consists of "Planned
measures for disguising or concealing an operation against an objective,
such measures being directed against an enemy, and as desired against
friendly forces." *Deception,* as similarly defined, consists of "Planned
measures for revealing or conveying to the enemy true information (or
false information which could be evaluated as true) regarding our strategic plans, strength, dispositions, operations or tactics, with the purpose
of causing him to reach false estimates and to act thereon." Cover basically implies making your opponent behave on the belief that something
true is something false. Deception basically implies making him behave
on the belief that something false is something true. Cover conceals
truth; deception conveys falsehood. Cover induces nonaction; deception
induces action. If the Duke's sentinels saw Uther Pendragon go to
Igraine's chamber but did nothing because they thought he was the
Duke, that was cover. When Igraine "went unto bed with hym as I ought
to do with my lord" because she thought he was the Duke, that was deception.

(Actually, this distinction—which was not made by the experienced
British and American deceivers of the Second World War, and only entered the vocabulary in 1945 at the insistence of the United States
Navy—is chopping the semantics pretty fine. There is not that bright a
line between the two. Nonaction is a form of action; the decision to do
nothing is still a decision, as Secretary of Defense Robert McNamara
used to say.)

First and most important, the object was to induce Igraine to *do something.* The objective of military deception, said Joint Security Control,
"is to cause the enemy to make false estimates and mistakes in his mili-

tary decisions and consequent actions, thereby contributing to the ac-complishment of tasks in our over-all military mission." It did not mat-ter to Uther how thoroughly he might fool her if she did not *act* on her misapprehension as Uther wanted her to. (To change the example, it would have done Jacob no good to have Isaac simply *think* he was Esau; the object was to induce his father to *act* on that belief and give him the paternal blessing.) That was the lesson of CAMILLA, when the Duke of Aosta evidently believed what Wavell wanted him to believe but did not act on it in a way that did Wavell any good. And it is the first and great commandment of military deception: *Your goal is not to make the enemy* think *something; it is to make him* do *something.** Breach of this com-mandment underlies many deception failures.

Some people understand this as it were by instinct. (Mao Tse-tung was one. "It is often possible," he said, "by adopting all kinds of measures of deception to drive the enemy into making erroneous judgments and tak-ing erroneous actions, thus depriving him of his superiority and initia-tive.") But it is surprising how many thoroughly intelligent people in the Second World War were never able to get this into their heads. They con-fused deception with psychological warfare. But psychological warfare does not try to make the enemy *do* anything specific. It seeks to affect his frame of mind—to feel that the war is hopeless, that his leaders are in-competent, that his cause is unjust, that the men and weapons opposing him are irresistible.

Both American and British deceivers had to deal with this misappre-hension. As an example, General "Freddie" Morgan, the chief advance planner of OVERLORD, the Normandy invasion, knew about Dudley Clarke's work and should have understood the basics. Yet as late as 1943 he suggested to Colonel John Bevan, chief of the London Controlling Section, the British central deception staff, that deception focus on telling the Germans that Allied power was irresistible. Bevan gently ex-plained that taking this line with the Germans was a matter for the Polit-ical Warfare Executive, and that he, Bevan, had "a very different task." "My channels lead to the German General Staff, or I hope they do, and

---

* Even less is it your goal merely to confuse him (what the French call *intoxication*). Rarely do you have any sure sense of how he will react if he is merely confused. If Igraine had been merely confused as to who was coming into her bedroom she might have stabbed first and asked questions afterward.

my job is to deceive them and thereby induce them to DO something or NOT DO something which will assist our real operations." The German General Staff, he pointed out, was not unsophisticated; it would not be safe to exaggerate Allied strength to them by more than, say, 20 percent; this would either be so implausible as to discredit the channels through which the Germans were fed misinformation, or alternatively "they might reinforce France to such an extent that OVERLORD would prove an impossible proposition."

Moreover, you want the enemy not only to *do*, but to do something *specific*. Usually what you want him to do is deploy forces in a specific manner to meet a specific imaginary threat, or otherwise act in a way that will enhance the strength of your force as compared to his. When you are on the defensive with inferior forces, as was so often true in the Western Desert in 1941–42, the function of deception is to make the enemy worry that you are stronger than you really are, so as to misdirect, delay, or deter his attack. But even if you are on the offensive with superior forces, it can enhance your superiority where it counts by the "stretch" effect—inducing him to stretch his forces so as to cover every seemingly threatened point. This was the principle employed in BODYGUARD, the overall deception for the invasion of the Continent, to keep German forces stretched from Norway to Greece and away from Normandy; it was the principle employed in the WEDLOCK series of deceptions in the Pacific, to keep Japanese forces stretched to the Kuriles and away from the Marianas and the Philippines. Often called a "force multiplier," deception is more accurately a force *ratio* multiplier; it enhances your (actual or perceived) relative strength at the decisive point.

Bear in mind—and this is a vitally important point, often overlooked—that *it is not necessary to make the enemy actually* believe *in the false state of affairs that you want to project. It is enough if you can make him so* concerned *over its likelihood that he feels that he must provide for it.*

Moreover, you want to be sure that what the enemy does is something that will help you, not harm you. That was the other half of the CAMILLA lesson. When the Duke of Aosta acted on the belief that Wavell meant to attack British Somaliland, he acted in such a way as to put Wavell in a worse position than before. Sometimes it requires a deft touch to ensure that the enemy's reaction is the right one; to induce him, for instance, to tie up forces to guard against an imaginary threat without leading him to try to eliminate the threat with an attack, as was the challenge faced by the American deceivers on the Franco-Italian border in 1944–45. Even at the peak of their skill and success the Allied deceivers could not always predict accurately how the enemy would react. Thus, for FORTITUDE NORTH, the notional invasion of Norway in the spring of 1944, the deceivers originally planned to persuade the Germans that a force to land in southern Norway was being gathered at ports on the east coast of Scotland. But the logisticians insisted that this would not be plausible because the capacity of those ports was inadequate. So the forces for the operation were notionally concentrated in the Clyde on the west coast instead. But when the Germans learned of this concentration they assumed that the forces in question must be destined not for Norway at all but for France.

Deception is aimed at the enemy commander, by way of his intelligence people. "The target of our cover and deception efforts," in the words of Joint Security Control, "is the enemy's command decision, as influenced by his strategic and combat intelligence. The target is attacked by causing the enemy to acquire, through his sources of intelligence, a mixture of true and false information in accordance with a logical, carefully-timed plan." To gauge the enemy commander's reactions you must *know* him; know how his mind works, know what makes him tick, know what kind of stimulus will evoke what kind of response. Moreover, you need to know how individual circumstances may affect his decision, and what those circumstances are. Uther Pendragon would have been a disappointed man if Igraine had proved to be uninterested in the Duke that evening because of an earlier spat. He would have been an unhappy man if she had proved to be a warrior-woman like Brunhild in the *Nibelungenlied,* who trussed up her husband and hung him on a nail for the night. But he correctly judged that she was a mild and dutiful wife who would do "as I ought to do with my lord."

So you need to put yourself in the enemy commander's mind as closely as you can; as Clarke recognized early on, when he concocted ABEAM after learning that the Italians were concerned that the British might land airborne forces in their rear. If you know him as well as you should know him before you try to deceive him, you will learn that he will swallow some astonishing things, and will be oblivious to other things that you would think would infallibly provoke him to action. In the German war, especially after the end of 1941, strategic deception entailed trying to enter the erratic mind of Adolf Hitler. In the words of the official report written after the war by Sir Ronald Wingate, second in command of the London Controlling Section:

> Psychologically the deceivers in London had to endeavor to put themselves in the position not of a German Supreme Headquarters Staff on the model of the British Chiefs of Staff organization, but of a small coterie of men under the direct personal control of a monomaniac believing that his intuition could solve any political or military problem. . . . The risk was taken that Hitler's world strategy complex and the anxieties caused by it would outweigh with him the probability, and it should have been the certainty, that the Allies had not the resources to undertake many of the operations against which he felt compelled to insure himself.

One prime example of this was Hitler's endless fretting over Norway and the Balkans. There the Allied deceivers found they could keep crying wolf. Over and over, notional operations would be mounted against those areas, only to be stood down on one pretext or another; yet the Germans seem never to have smelled a rat. This was because from Hitler's point of view, Norway and the Balkans were vital to Germany's security: Norway because of its proximity to northern Germany, the Balkans because they lay behind the front in the southern USSR. Norway and the Balkans were always in jeopardy if the Allies had sufficient resources; and the deceivers successfully set out to convince the Germans that the Allies did indeed have those resources.

Another example was German hypersensitivity to the idea that the major neutrals—Turkey, Sweden, Spain, Portugal—might join the Allied side. In fact, nothing whatever was likely to induce these countries to endanger their neutrality. The Foreign Office pointed this out when the deceivers proposed to utilize this threat in connection with BODYGUARD. The deceivers explained patiently that what mattered was not what the neutral countries intended to do, but what the Germans might believe they intended to do.

As a third example, the Germans could be made to guard against amphibious operations that were, if not impossible, at least too difficult to be plausible to a knowledgeable person. For all the brilliance of their staff work, they were abysmally ignorant of amphibious warfare. They knew nothing about landing craft requirements, what it takes to put a division ashore and maintain it across the beach, how to mobilize and deploy naval fire support; such operations were totally outside their experience, and they had no sense whatever of the limiting factors that might make them unfeasible.

A similar failing—and a more surprising one, in view of German pioneering in tactical airpower and close air support—was relative neglect of the need for fighter cover as a limiting factor on Allied operations; a consideration the Allies, giving the Germans too much credit, often perceived as limiting their choice of deception "stories."

Worries are magnified when you lose the initiative. It makes fear of the unknown more potent—"a far more potent factor in influencing strategic plans than deductions from restricted observation," as Sir Ronald Wingate put it. One who has lost the sense of being in control of things is more likely to take counsel of his fears. If you play to your opponent's uncertainties and speculations, he will begin to find confirmation on his own. During the last three years of the war the German high command focused constantly on their own vulnerability, tending to assume that if the Allies needed to do something they would do it, without ever thinking through whether the operation would be difficult or even impossible for them. To quote Wingate again:

> If these two conditions exist, namely apprehension and plausibility, then deception can turn what was a vague fear in the enemy's mind into a certainty and to such a degree can this be achieved that the enemy may go out of his way to try and find evidence to confirm his pre-conceived ideas in the face even of negative evidence to support them. For instance if Germany feared an Allied assault on Norway, she would assume that Allied resources and troops were available, even though there was no direct observational intelligence on the subject. Then anything she picked up by logging the simulated W/T traffic, or through planted reports, would confirm her fears. But the reverse would not be true; namely, that because she could not locate exactly divisions in Scotland, or calculated that a great deal of Allied shipping was in the Mediterranean, she would write off the possibility of a Norwegian assault. . . . It was assumed in London that it was certain that the Germans would rule out all this as operationally or politically impracticable. Yet they did not.

* * *

That your goal is to induce the enemy to *do* something does not mean he has to do it right away. Building up a pattern in his mind that will be the basis for future specific actions is a perfectly legitimate goal. This was the basis of Dudley Clarke's masterpieces, the false orders of battle that he built up in German minds, patiently and bit by bit over a long time, which not only served to tie down great numbers of troops that the Germans sorely needed on the active fronts, but were the backdrop that made individual deceptions plausible. Beginning with the tiny seed of ABEAM and the 10TH ARMORED DIVISION PLAN, on through the actual employment of notional units in PLAN A-R, and the almost fortuitous discovery of long-term possibilities in connection with the CYPRUS DEFENSE PLAN, order of battle deception grew under Clarke's hand to such a point that by the spring of 1944 the Germans believed there were fourteen Allied divisions in Egypt and Libya when there were really only three, none of them battleworthy.

The enemy commander is the deceiver's *target,* but he is not the deceiver's *customer.* The customer is the enemy intelligence service, who gather and process information and present it to their commander. The deceiver must know all he can about them, so as to judge what data fed to them will cause them to make the appreciation of the situation that he wants their commander to receive. How much faith does their commander put in them? What sources of intelligence do they rely on and what do they mistrust? What technical methods of gathering data have they got? Are they sophisticated or gullible? Are they cautious or bold? Do they know how to build an appreciation from bits and pieces of data, or do you have to rub their noses in your story? Will they admit mistakes, or cover them up? Are there rivals within the organization, or rival organizations, that can be played off against each other?

In general, in the Second World War the Allies overrated the efficiency of German intelligence. For one thing, they did not realize until postwar interrogations just how fragmented and disorganized it was at the top level, the level of the Oberkommando der Wehrmacht (OKW) and Hitler himself. There was no central organization comparable to the Joint Intelligence Committees of the British and American Chiefs of Staff to collate and integrate information from all sources. Items of unevaluated raw data reached the top commanders and Hitler himself almost at random, depending on who had personal access at the moment. As a result, the risk that a deception story concerning Allied preparations would be contradicted by an operational report showing no observation of such preparations was much smaller than the deceivers realized.

On a more operational level, in the latter years of the war the Allies overrated the efficiency of German radio interception and associated direction-finding and traffic analysis (an activity that in the Second World War was commonly called by the British term " 'Y' Service"). In the opening years and in the Western Desert the German "Y" Service had been highly efficient. By 1943 it had fallen off badly, at least in the West. But the Allied deceivers did not realize this; accordingly, they were unnecessarily cautious in some respects, cutting back on plans because the available personnel and equipment to simulate radio traffic was deemed inadequate, or alternatively providing a level of such traffic that proved to have been wasted.

The means of intelligence varies with the situation and the means of deception have to be adjusted accordingly. When armies are in contact, straight observational intelligence has priority. At a distance, non-observational intelligence perforce takes the fore. For four years, from Dunkirk to D-Day, there was no contact between the armies of Germany and the Western Allies except in the Mediterranean, and from 1942, little German aerial reconnaissance of the United Kingdom was possible. As a result, for intelligence about Britain and the United States the Germans had almost nothing but their agents to rely on, and these were all or almost all controlled by the Allies. There was little or no need to reinsure their reports with devices such as dummy tanks and landing craft, or fake insignia on vehicles and uniforms; whereas in the Mediterranean and the Middle East, such devices played a leading role. (On the negative side, by the same token active deception of observational intelligence on the Channel front was not possible either. In 1943 the Germans failed entirely to observe the elaborate feint invasion that culminated Operation COCKADE, though convinced at the same time from agents' reports that an attack was imminent.)

Similarly, the geographical features of the Pacific Ocean Areas theater—small, widely separated land masses and large areas of water—

combined with American command of the sea and air, meant that in the later stages of the war the Japanese in that theater were forced to rely on intelligence reports from Berlin and neutral capitals, plus radio traffic analysis and inferences from American sea and air activity.

For successful deception you must have a firm plan for what you really want to do; you must have adequate security; and you must have time.

The firm plan is an absolute prerequisite. "It is impossible, or at least highly dangerous," said Peter Fleming, "to attempt to tell a lie until you know what the truth is." He knew this best of all the deceivers in the Second World War, for until the very end of the war there were hardly any firm plans on his front for anything, and it became a standing joke at theater headquarters that deception plans turned out to be real operational plans (which were then discarded in their turn). The same problem nagged the American planners well into 1945 in their efforts to formulate some sort of overall deception policy for the war with Japan.

Nevertheless, there are occasions when your hand is forced, and you have to operate a deception plan even when the real plan is not yet firm. "[O]nce you have started on the road to deception you cannot halt," said Sir Ronald Wingate, adding:

> In real operations you may perforce have to stop and regroup and consider alternative courses of action for the future. But whatever you do the enemy is the whole time endeavoring to appreciate your war potential, your dispositions and your intentions. He is being fed continuously by his own sources of information and by your sources of information to him. The deceivers cannot stop; they must continue to feed the enemy and to feed him plausibly. To do this the deceivers must have a background. If because of the lack of a firm real policy, the deceptive background is blurred and vague, well then so much the worse. Deception policy has then the very difficult task of avoiding compromise to one or several of a number of possible real operations in the future. It, therefore, becomes less positive and will almost certainly be less effective; but it must be there.

As for good security, obviously there can be no deception if the enemy knows the true state of affairs; deception takes advantage of the fact that

his information is imperfect and fills out the imperfections in such a way as to give him a false overall picture. Nor, obviously, can there be decep-
/tion if the enemy can discern that the deception operation is a fake.* In the Second World War—and for some years thereafter—the very fact that the Allies were engaged in deception *at all* at the strategic and higher operational level was a secret almost as closely held as ULTRA or the Manhattan Project; held almost as closely vis-à-vis Allied personnel who were not in the know as it was against the Axis itself.

Methods of compartmenting special information of this exceptionally sensitive nature grew more sophisticated as the war progressed. It was not until 1944 that the now-familiar three-tiered system of CONFIDENTIAL, SECRET, and TOP SECRET was formally introduced, though a two-tiered system was followed from the outset. The super-secret ULTRA material derived from breaking enemy codes and ciphers developed its own communications channels and system of special handling, through "SCUs" and "SLUs"—Special Communications Units and Special Liaison Units. Such material was not distributed to the Middle East till March 1941 and the SCU/SLU communications system was not established in Cairo till August of that year. Known sometimes as "the Link," by the latter part of the war the SCU/SLU system was also used for especially sensitive eyes-only communications such as those between the deception units on the Continent in 1944–45.†

But when Dudley Clarke left London in December 1940, there was no system ready to hand for compartmenting deception and its associated communications from people who did not need to know about it. So he arranged with the ISSB a rough-and-ready code to conceal the fact

* On the other hand, if the enemy knows you do deception he may conclude that your real operation is a fake. Rear Admiral Robert A. Theobald, commander of American naval forces in the Aleutians in 1942, was warned by his chief, Admiral Nimitz, that "The Japanese are adept at the practice of deception." Navy intelligence then alerted Theobald that the Japanese would merely bomb Dutch Harbor but actually seize Attu and Kiska. He was not privy to the fact that Japanese codes had been broken; he assumed therefore that the intelligence warning was based on analysis of Japanese radio traffic; assumed that the traffic was deceptive and that the Japanese would really try to seize Dutch Harbor; positioned himself accordingly—and was caught flatfooted when the Japanese did exactly what Navy intelligence had told him they would do.

† Used also for eyes-only communications among high-ranking officers and officials, this communications system and its postwar successors came to be called the "back channel," a term discovered and misunderstood by the media in the 1970s, and misused ever since.

that a message dealt with deception. Messages to or from Clarke in Cairo dealing with deception were to be addressed to, or come from, "COLONEL CROFT CONSTABLE," "MAYHEW," or "U.S.R.C." (these being the initials of an imaginary "United Service Research Committee"). A message so addressed or signed, asking or directing that something be *"countered,"* would in reality be asking or directing that it be *encouraged.* Thus, a message from COLONEL CROFT CONSTABLE asking ISSB: "Please counter leakages regarding early sailing of Royal Dragoons for Middle East" meant that "A" Force wanted the Axis to believe that the Royal Dragoons would soon sail for the Middle East. As the war went on, "MAYHEW" was changed to "GALVESTON," and later still to "DOWAGER," and "CROFT CONSTABLE" was changed to "TALENTS," but the basic system—it was christened the Counteraction Communications Procedure ("C.C. Procedure" for short)—remained in place throughout the war.

When Clarke visited London in October 1941, the system was enlarged by adding a special list of codenames* reserved exclusively for deception operations, called the "C.C. list." Those who were in the know could thereby recognize a deception operation on its face. Thus, if PARAFFIN was on the C.C. list, a message reading "Codename PARAFFIN has been allotted to the move of 50 Armored Division from Suez to Socotra" would tell a deceiver at once that a deception plan was being operated to make the Axis believe that this move had been ordered, while the uninitiated would be none the wiser.

A third prerequisite for successful deception—time—is too often overlooked. It is idle for a commander to ask his deceivers for a quick fix, except perhaps on the very lowest tactical level. "Tactical Deception Plans," wrote Dudley Clarke, "unlike Strategic ones, must usually be designed for short-term fruition, since the real tactical plans of an Army will inevitably fluctuate." But on the strategic and higher operational levels, the more time the deceiver has the better. Decisions, even those made

---

* For security, codenames were assigned throughout the war to people, places, things, and activities. As for activities, it should be noted that, strictly speaking, "Operation" should properly be used for an actual operation actually embarked upon, "Exercise" should be used for a practice activity, and everything else, including deception schemes, should be called a "Plan." Operation OVERLORD; Exercise FABIUS; Plan BODYGUARD. But practical usage plays fast and loose with these distinctions and so does this book.

on impulse by an "intuitive" decision maker like Hitler, are reached against a mosaic of information, belief, and experience, the cumulative effect of assimilating over time a long series of events and scraps of data. The deceiver's task is to build up that mosaic, a bit here, a bit there. That takes time. And the higher the strategic level, the more time it takes. That was the secret of the false order of battle built up over two years by "A" Force, to the incalculably great profit of the Allied cause. Had the war in the Far East not ended abruptly, the false order of battle built up over three years by Peter Fleming might have proved equally valuable in Southeast Asia.

The first and great commandment, once again, is the CAMILLA commandment that you don't want just to make the enemy *think,* you want to make him *do* something that will help you. And the second is like unto it: Never—never *ever*—mount a deception operation with no clear object just because you have the means of doing so. This we might call the K-SHELL commandment, for it was the lesson Dudley Clarke learned from the abortive K-SHELL PLAN, which had no real goal and nearly ran into trouble. Deception is a weapon with many sharp edges. A deception plan has to have a clear purpose, and it has to be integrated with everything else you are doing. If Uther Pendragon had not had the specific object of inducing the fair Igraine to go "unto bed with hym," the only point of his seeming to be the Duke would have been to gratify his ego by demonstrating to his own satisfaction that he had on his side a magician who could perform such wonders. And this could have worked against his interests: It might, for example, have led the Duke's vassals to disbelieve reports of their master's death and fight Uther to the end.

The temptation to make this mistake can be very great. It was forced even upon the mature and experienced "A" Force by General "Jumbo" Wilson, the Allied commander in the Mediterranean, as late as the autumn of 1944, when he insisted that a deception plan called UNDERCUT, designed to cover his preparations for occupying Athens, disguise them as preparations to land in Istria and on Rhodes. But it became evident that the Germans were not interested in making a fighting withdrawal from Greece; they just wanted to get out of there. There was thus no need for UNDERCUT. Wilson nevertheless insisted that it proceed, so as to keep his plans a secret in Greece itself, because of the complicated political situation; though Clarke pointed out that he had no machinery for deceiving Greeks and that an attempt to operate a deception plan under those circumstances would be useless and perhaps positively harmful. And it had that result: Greek forces involved in the operation protested at the thought they were being sent anywhere but Greece, and the fact that Greek resistance leaders were brought over to Italy to meet the commander of the expedition made the whole thing absurd. So it was canceled. UNDERCUT taught once again "the lesson that Deception is in no way an alternative to Secrecy," to quote Dudley Clarke, "and that real harm may be done if an attempt is made to use the delicate machinery of Deception when nothing more than Secrecy is required. 'UNDERCUT' was a fair illustration of the consequences of demanding a Deception Plan largely because a Deception Staff was at hand with little else to do."

There is a long-term lesson here. Any mechanical requirement such as "every operational plan must be accompanied by a deception annex" is dangerous in at least two ways. It runs the risk of violating the K-SHELL commandment and producing a deception plan without any demonstrable need or suitable object. And it runs the risk of having a deception plan concocted by people who do not really know what they are doing—which can easily be worse than no plan at all. This almost happened with the Normandy invasion, until the clumsy APPENDIX Y to the original cross-Channel assault plan was transformed by David Strangeways into the brilliantly successful FORTITUDE SOUTH. (Yet even Eisenhower, of all American commanders the one most appreciative of deception and what it could do, decreed when Chief of Staff after the war that all operational plans should have a deception annex.)

Turning now to the structure of a deception organization.

Deception is a hybrid, part operations (including plans), and part intelligence. This troubles the bureaucratic mind, because staff organizations always put the two in separate compartments. With Wavell, Clarke started out on the intelligence side. His early deceptions were implemented entirely through Maunsell's counterintelligence system, so the operations people hardly noticed. They began to notice when dummies and administrative measures began to play a more important role. This lay at the root of Clarke's troubles in the Bagnold affair. It was a matter, said Clarke, of "both sides first tugging at the body, and then trying for a compromise by dividing it in half."

The compromise failed, and from it emerged the first lesson. Deception will pay its best dividends when both the planning and the implementation of plans by all methods is made the responsibility of one controlling mind. Furthermore the same mind must exercise an overall control over ALL plans operated within the theater, both tactical and strategic. There is no clear dividing line between the two, and any attempt to define the point at which they merge in order to divide responsibilities between two separate authorities will be fatal to success. As a rule it will also, incidentally, prejudice the security of highly secret machinery.

The question still remains, however: Do you assign this unified function to the intelligence part of your staff, or to the operations part, or let it stand alone? In Clarke's view "the real answer is that it must fall somewhere between intelligence and operations, so that where it is possible it will usually function best under the direct instructions of the Chief of the General Staff, to whom of course the Operations and Intelligence are both answerable. But, whatever the Staff arrangements, successful deception can only result from the joint action of 'O' and 'I' in close partnership, and when the Deception Staff enjoys the full and complete confidence of both." After the Bagnold affair was cleared up, "A" Force still found itself permanently assigned to Operations at GHQ Middle East. Clarke was able to live with that, though his old colleagues in Intelligence took it rather harder, for, as Clarke put it, they "had 'invented' Deception: through their initiative the fractious infant had entered the world and they had nurtured it through a stormy childhood." But, he conceded, it was "good sense" to assign it to the operations staff. "They are the real users, they must dictate the object, they will direct its tempo and decide when it should be replaced by a new one." (Meanwhile, "A" Force's field representatives matched Clarke's ideal pattern, as independent and coequal general staff elements of the forces to which they were assigned.)

Operations people were generally skeptical of deception till they saw its successes. The visible results especially of order of battle deception embodied in captured enemy maps "sometimes appeared as Black Magic" to the operating staffs, said Clarke, "and they never failed to secure us an even greater degree of co-operation."

The United States Army found itself less able than the British to adapt its staff system to the needs of deception. That system was rigid to the point of ossification. There were always at least four basic staff elements: G-1 (Personnel), G-2 (Intelligence), G-3 (Operations, including Plans), and G-4 (Supply), each headed by an assistant chief of staff; and in a large headquarters each of these would have a number of subsections for particular purposes. The assistant chiefs of staff did not deal directly with the commander; they dealt with his chief of staff, who in turn reported to the commander. Thus, in contrast with the direct access to Wavell and Auchinleck that Clarke had had at GHQ Cairo, at SHAEF, Eisenhower's headquarters on the Continent, Noel Wild, as head of "Ops (B)," the cover and deception section of G-3, had to get plans approved by the chief of the operations subsection, Brigadier General A. S. Nevins, who in turn had to get the approval of the G-3 himself, Major General H. R. Bull, who in turn had to get the approval of Eisenhower's chief of staff, Lieutenant General Bedell Smith, who in turn was the only one with an unquestionable right of access to Ike. And the United States Navy, although its staff system was less rigid than that of the Army, found it equally hard to accept Clarke's ideal. When in early 1945 a formal deception organization was finally established on the staff of Admiral King as Commander in Chief, United States Fleet ("COMINCH"), the recommendation that it report directly to the chief of staff was rejected and it was placed in the intelligence division instead.

The British Army staff system was more flexible than the American. The staff was divided into two parts, general ("G") and administrative ("Adm"). (In the largest units there was a third part, a military secretariat ("MS").) The work of each was coordinated by a senior officer, but the officers in charge of operations, intelligence, and so on had the right of access to the commander. There was no chief of staff in the American sense, although at a theater or army group headquarters the senior "G" officer bore the title "Chief of the General Staff"; at lower headquarters he was just called the "Brigadier General Staff" (at an army headquarters sometimes "Major-General General Staff"), usually abbreviated "BGS" (or "MGGS"). Within "G" there were officers for plans and operations ("G(O)"), intelligence ("G(I)"), and so on. General staff officer slots were referred to by their grade: GSO1 (colonels and lieutenant-colonels), GSO2 (majors), GSO3 (captains).

Under the flexible British system, deception could be fitted into Clarke's ideal mold. Thus, David Strangeways, chief of deception for Alexander's and later for Montgomery's army groups as well as commander of their tactical deception forces, dealt directly with Alexander and Montgomery (or with Monty's Chief of General Staff, "Freddie" de Guingand). Without this immediate access and centralized authority, FORTITUDE SOUTH would never have happened in 1944; and without FORTITUDE SOUTH it is a moot question whether the Normandy invasion would have succeeded.

Less consistently happy were the experiences of the deceivers on staffs with American-style organizations. Those with "AFHQ," Mediterranean theater headquarters, and with General Jacob M. Devers's 6th Army Group, fared well. Both were offshoots of "A" Force. AFHQ was the first inter-Allied supreme theater headquarters, set up under Eisenhower in Algiers after the landings in North Africa and structured on the American model. As will be seen, a branch of "A" Force was opened to serve the new theater. There was apparently never any question of its getting a direct link to the top; it would be in G-2 or G-3. It would appear that G-2 was the first choice of Bedell Smith, Ike's chief of staff; very possibly because the nascent deception control back in Washington was a G-2 function. But AFHQ's (British) G-2 was Kenneth Strong; he was an old friend of Clarke's but Clarke evidently objected to serving under him. So the new Advanced HQ "A" Force was placed in AFHQ's G-3, where it functioned quite well. It was then inevitable that when Devers's 6th Army Group was formed out of Mediterranean forces for the southern France invasion his deception section, known as "No. 2 Tac HQ 'A' Force," would be situated in his G-3. After a rocky start, it accomplished more than any other deception outfit on the Continent.

The story was different at SHAEF and at General Omar Bradley's 12th Army Group headquarters. SHAEF deception, buried in G-3 as "Ops (B)," accomplished relatively little. Bradley's deception section, known as Special Plans Branch of the operations section of his G-3, was able to accomplish little more. Bradley himself was hardly aware of its existence.

The great trouble with the American staff system was that it smothered deception, which of all activities needs to be imaginative and flexible, under layers of bureaucracy and bureaucratic coordination. Long afterwards, David Strangeways recalled how Billy Harris, his opposite number with Bradley's army group, "used to open his heart to me a bit when just after the initial activities in D-Day he set up his own branch on the American side, and he found that whereas I was able to go to anybody and do anything, he had to go through the awful, ghastly staff procedures they had."

The more flexible British staff organization was rendered even more flexible by the looser British approach to rank. Thus, Strangeways, a mere lieutenant colonel, had Montgomery's full backing and, as he put it, "I would make a signal to a corps saying do so-and-so and they did it. And I submitted that to nobody. And [the Americans] can't understand that. But I was serving Montgomery and I did everything in his name. Period."*

Keeping deception a separate element responsible directly to the top helps not only flexibility but coordination. Everything has to work together. If a double agent in England reports that the (imaginary) U.S. 21st Airborne Division is at Fulbeck in Lincolnshire and is part of the (imaginary) U.S. Fourteenth Army, whose headquarters are at Little Waltham in Essex, radio traffic suitable to an airborne division should emanate from Fulbeck and should communicate with an apparent army headquarters at Little Waltham. If a double agent and radio traffic suggest that the (imaginary) 3d South African Division is at Cairo, and it is believed that there are uncontrolled enemy agents in Cairo, soldiers and vehicles bearing the insignia of the 3d South African Division should be seen in the area. And the deceiver's standing nightmare is that somebody on his own side who is not in the know will do something that blows a deception scheme. One minor example: When the Americans were first getting into the deception business in 1942, an imaginary "10th (Light) Division" was allocated for use in Middle East order of battle deception.

---

* Another British flexibility that astonished Americans was the ease with which they conferred temporary or local rank. When Second Lieutenant Arne Ekstrom was recruited to "A" Force in Algiers in early 1943, Dudley Clarke "went down to tell Eisenhower that they had found the officer they liked and they wanted me made a major. And his answer was that in the American Army when we need a major we send a major." (Ekstrom was promoted to first lieutenant, however, and eventually made captain.)

Deception was being kept so super-secret that nobody told Army Ground Forces, who duly activated a genuine 10th (Light) Division; fortunately it was not too late to deactivate the notional one, and AGF was thereafter brought into the deception picture. (Even so, in late April 1945, as part of a deception plan, enemy intelligence was told through a double agent that men wearing the insignia of the 10th Light Division had been seen in New York.)

The only sure way to cover all these possibilities is for all the strings to be held in one hand. This is as true on the national or world level as it is on the army group or theater level. If you are trying to convince the Japanese that, say, the 119th Division is set to invade the Kuriles, you do not want somebody else to try to convince the Germans that it is training in Algeria for a landing in southern France. Sir Ronald Wingate again:

> In this vast world complex some quite unsuspected and innocent action by one of the Allies in some particular sphere might seriously compromise an elaborate story being slowly and systematically fed to the enemy with the object of securing surprise or enemy dispositions favorable to ourselves in a vital forthcoming military operation. It may be truly said of the deception staff that they had to know everything and to be prepared to advise and suggest action in the most unlikely spheres if by so doing they could prevent compromising situations arising or, what was better, create a situation which might be interpreted by the enemy in a manner favorable to our real plans.

The London Controlling Section and Joint Security Control were supposed to maintain continuous liaison with each other for just this purpose. At least once an effort was made to coordinate deception with the Soviets as well, when the British Colonel Bevan and the American Lieutenant Colonel Baumer flew to Moscow early in 1944 to secure their adherence to BODYGUARD.

Yet control can be *too* centralized. When arrangements for the Mediterranean theater were being debated in Algiers in March of 1943, there was a school of thought that held that the theater deception organization should be a branch of the central London Controlling Section. But it was concluded, and wisely so, that this could cause friction and resentment; the theater deception staff must be, and be seen to be, part of the theater commander's staff and responsible to him. (As will be seen, this was a distinction that Colonel Ormonde Hunter, for some time the American deceiver in India, could never grasp; to the end of the war he was convinced that theater deceivers were a branch of Joint Security Control in Washington.)

Turning now to the planning and execution of a deception operation.

You begin by getting an answer to the fundamental CAMILLA question: What is the *object* of the deception—what is it that you want the enemy commander to *do?* That question has to be answered by your commander and his operations staff in light of their own plans. In 1942 and 1943 in Northern Europe the answer was that we wanted the Germans to keep as many men as possible away from the Soviet and Mediterranean fronts. In 1944 in the same theater it was that we wanted them to hold back forces in readiness to repel a major assault on the Pas-de-Calais. In 1945 in the Pacific it was that we wanted the Japanese to prepare to repel an attack on Formosa or the China coast rather than reinforce Okinawa or Kyushu. At El Alamein it was that the Germans should keep their strength on the right wing.

The commander and his operations staff may well need guidance in this from experienced deceivers, for not all objects are equally desirable or feasible. To begin with, the deceiver has to answer the other fundamental question, the K-SHELL one: Is there a suitable object for a deception plan at all? The criteria are relatively simple. The object should be simple and unitary. It should be something that you are confident that the enemy would do if given a stimulus that you can give him. If you cannot get by both these tests you should not try a deception plan. And any object you decide on should be phrased very carefully.

The London Controlling Section learned all these lessons the hard way with an unnamed overall policy for deception of Germany and Italy in 1943, and the complex plan called COCKADE that implemented it in Northwestern Europe. The overall object should have been simply to induce the Germans to keep as many forces as possible in Northwestern Europe by threatening a 1943 cross-Channel invasion. The policy was overcomplicated by adding such further objects as bringing the Luftwaffe to battle in the West (without any idea whether the Luftwaffe's strategy might not be, as it in fact was, to husband its resources until *after* an Allied landing); containing the German U-boats in the North Atlantic; preventing German reinforcement of the Mediterranean; and lowering Italian morale and will to resist. Any of these might have been a bonus had they happened as an incidental result of an operation focused on the main object, but including them in the overall policy led to including them in implementing plans, with a consequent diffusion of effort.

Including "lowering Italian morale and will to resist" as a goal for 1943 brought home how important it is that a deception object be properly formulated. Lowering morale is a psychological warfare object, not a deception one; it breaches the CAMILLA commandment: It involves *thinking,* not *doing.* A proper deception object might have been something like "to endeavor to cause an early surrender or collapse of Italian forces." Phrasing it the way it was phrased triggered the turf-protecting instincts of the psychological warfare agencies, and the deceivers "got themselves involved in an endless academic discussion as to the charters and prerogatives of various Departments," as Wingate put it; wasting time and energy that could have been put to more productive use.

Having settled on your object with the operations side of your organization, the next question is for the intelligence side: What perception of the situation by the enemy commander and his intelligence staff is most likely to make him decide to do what you want him to do? What story, in short, do you want to tell him? (And "story" is indeed the term used in the trade.)

Every situation and every object calls of course for a different "story," but a few general propositions can be safely made. It has to be as plausible as possible, of course; and as has already been observed, this means plausible to the enemy, not necessarily to you. And it helps to play to his fears and try to reinforce what he already believes. This leads in turn to the proposition that it helps to have the greatest possible kernel of truth in the "story." *

One device that is often useful in this regard is to base the "story" on a

---

* Deceivers have pointed out more than once that from their perspective, Churchill's oft-quoted remark to Stalin that in war, "truth deserves a bodyguard of lies," had the wrong end of the stick. For deceivers, lies are so precious that they should be attended by a bodyguard of truth. "A lie that is half a truth is a harder matter to fight," said Tennyson.

real plan that was actually considered and discarded; for an operation that seemed to your own planners sufficiently plausible to be seriously considered, is likely to seem plausible to your opponent too. This was the basis, it will be recalled, for Dudley Clarke's early CORDITE COVER PLAN. Again, the original plans for the invasion of Sicily called for landings in the western part of the island as well as the eastern; the western landing was later deleted but was the basis for DERRICK, the deception plan for the invasion. WEDLOCK, the bogus threat to the Kuriles in 1944, followed on earlier discarded plans for an advance along that route. BLUE-BIRD, the deception plan for the invasion of Okinawa, was designed to make the Japanese defend against an invasion of Formosa, which had in fact been considered and rejected by the Americans. The first phase of PASTEL TWO, the deception plan for the invasion of the Japanese homeland, was built around the "story" of a landing in the Shanghai area of China for which contingency plans had actually been made.

By the same token, it helps to build as many real events into your "story" as you can; such as Clarke's use of de Gaulle's visit to Cairo for his EXPORTER COVER PLAN. Genuine political troubles among the Greek troops in the Mediterranean theater in 1944 underpinned the "story" of VENDETTA, the false landing in the south of France at the same time as the Normandy invasion, furnishing a plausible reason for the Allies to have abandoned an (imaginary) invasion of the Balkans and shifted their focus to the Western Mediterranean. And one of the strengths of FORTI-TUDE SOUTH was that there really was a "1st United States Army Group" or "FUSAG."

But you do not want too much truth. The temptation often arises to try a "double bluff"—which, as already noted, means making the real operation the "story" in the hope that the enemy will believe it to be a deception or other misinformation. A double bluff might do in rare cases for a single element of a broader plan—"A" Force once disguised real Spitfires as fakes by adding wooden struts that made them look like dummies, and the last move in COLLECT involved a double bluff—but on any larger scale such a temptation should be firmly resisted. It is hard to sell, and it runs the risk of exposing the methods of implementing a deception plan. "We should never resort to it unless in absolute despair," said Sir Ronald Wingate. Unfortunately, a double bluff can be forced on you when the commander decides to change his real operation to match the deceptive "story," as happened to Clarke in Italy in August of 1944 when at the last minute Alexander shifted his attack from his center to his right.

A "story" can, of course, come true wholly or partially for reasons beyond your control. This happened to some degree with ADVOCATE, when the Far Eastern war opened up, Rommel launched his attack, and Auchinleck found himself on the defensive in reality. It happened in a minor but amusing way in connection with the EUPHRATES PLAN, in which, as will be recalled, in support of the "story" that Canadian tank units were gathering at Palmyra, pro-Vichy security forces in Beirut were asked to look out for an (imaginary) French-Canadian deserter who might try to escape to France by stowing away on a Vichyite repatriation ship. To the chagrin of "A" Force, the French announced that they had captured the wanted man, who proved to be a French-speaking British soldier brought up in Belgium who had made a daring escape from the Germans in Greece. "Poor man," said Clarke, "we did the best we could for him, but were quite unable to explain the reason for his inhospitable reception at the end of a truly gallant escape from Enemy territory."

Normally, however, your "story" will *not* come true; and no aspect of crafting a "story" is more important than leaving enemy intelligence a "get out" or "breakoff," an excuse why it failed to come true that will not leave him suspicious—and, what is just as important, will not weaken the enemy decision maker's faith in his intelligence staff. (You and the enemy intelligence officer share the same interest on this latter point, for he has every incentive to believe the breakoff and sell it to his boss.) Many deception "stories" involve timing, on either the COLLECT pattern of telling the enemy that you plan to act sooner than you mean to, so as to keep him off balance, or the CORDITE COVER PLAN pattern of telling him you plan to act later than you mean to, so as to surprise him. The latter type often involves a change in apparent objectives once the enemy can be assumed to have perceived that an operation is imminent; thus, PASTEL TWO, the deception plan for the invasion of Kyushu in 1945, entailed a story that the Americans planned to land on the China coast up to the point where the Japanese would clearly perceive that Japan itself was the target, followed by a switch to show Shikoku rather than Kyushu as the objective. Or it may involve keeping the notional objective alive after the real operation has been lunched, combined with a story that the real operation is just a feint, on the pattern again of the CORDITE COVER PLAN. All these obviously have to have built-in breakoffs.

Sometimes a breakoff can be worked merely by fading away the evidence being fed to support the old story and fading in the support for the new one. For example, for the first phase of BODYGUARD, the "story" was that invasion would not come till late in the year if at all, because the Allies were relying on POINTBLANK, the bomber offensive, to bring Germany to her knees. Tidbits of intelligence pointing in that direction simply faded away as preparations for the actual invasion became unmistakable. This has to be done deftly, of course. The simplest and commonest breakoff is simply an asserted change of plan. This may be tied to some objective event, like the unrest among the Greek troops in the Mediterranean already referred to, or to a fictitious event, like the alleged demotion of Patton from command of FUSAG in 1944. More often than not, it can simply be justified as a shift in strategy; these happen often enough in real operations. For example, in the spring of 1945 Operation BLUEBIRD had been focused on distracting Japanese attention from Okinawa by threatening action against Formosa and the Chinese coast. Once Okinawa was actually invaded, the breakoff "story" was that Formosa had been postponed because the failure of the European war to end by the close of 1944, as had been expected, plus prolonged Japanese resistance in the Philippines, meant that seizure of Iwo and Okinawa to protect his right had been the only operation Nimitz could mount at that time.

Working a breakoff ties in with the way a "story" is fed to the enemy in bits and pieces from which he will deduce what he is supposed to deduce. "You don't take a great big silver salver and give it to him on that," said David Strangeways. "He's got to make the story up himself. Then if the story goes wrong he blames himself, not you." He works the breakoff upon himself; realizes that the word in the crossword is "strike" rather than "stripe," decides that the inkblot looks more like a whale than like a weasel after all.

A special breakoff problem is disposing of a notional unit once it has served its purpose. This can be harder than it sounds. It can risk compromise of secret channels, and sometimes the enemy just does not want to let go; it was especially hard to induce German intelligence officers to take a unit off their order of battle rosters once it got there. A big challenge of this sort came after FORTITUDE SOUTH, when it was necessary to liquidate a large block of fictitious American units, including an army group, an army, three corps, and seven divisions. A variety of devices was used; units were notionally disbanded, broken up for replacements, merged, or just never heard of again. An even bigger challenge was "A" Force's liquidation, during the last months of the war, of the huge notional order of battle—two armies, seven corps, twenty-seven divisions—that it had built up in the Mediterranean

Once you, the deceiver, have worked out with your commander and his planning and operations people what they want the enemy commander to do, and worked out with your intelligence people what perception of the situation will cause him to do it, it is up to you to decide what information you can palm off on the enemy's intelligence organization that will cause them to give him that perception, and to organize the means for passing that information to them. You have to select among the available tools, work out how they are to be used, and supervise their use. Dudley Clarke used to compare the deceiver's work to orchestrating and conducting a symphony, and to producing and directing a play.

Under the fully-developed system in place in the latter part of the war, the Combined Chiefs of Staff would adopt a broad policy governing deception for a campaigning season (usually for a year or so), one for the war against the European Axis, for which the British had the primary responsibility, and one for the war with Japan, with primary responsibility on the Americans. This was high-level stuff, and there were only a handful of these in the whole war. Within these overall policies, deception plans were prepared at theater or army group level, coordinated in London and Washington by the London Controlling Section and Joint Security Control, and, if important enough, passed upon by the appropriate Chiefs of Staff.

A correctly prepared deception plan was a carefully drawn document. If properly drafted it would first state the object in very specific terms; discuss the factors bearing on the problem; state the "story" resulting from consideration of the object and the relevant factors; and describe the proposed implementation in general terms, leaving specific implementing plans to be drawn up separately. In turn, a properly drawn implementation plan would have a detailed timetable for each piece of information that the enemy was to receive. There would be a meticulous scheduling of the real and bogus units that were supposedly earmarked for the operation, their physical locations according to a detailed timetable, and a plan for wireless simulation of the communications of such units at those locations. If the display of dummies was involved, there would be a schedule for what dummies should be situated where and when. There would be a detailed plan for passing information by controlled enemy agents, channels to the enemy in neutral capitals, and the like. All these would be carefully integrated and continuously updated.

It is this phase that calls forth the very special skills of the deceiver. He cannot, as Merlin could, wave a magic wand and cause Uther Pendragon to take the likeness of the Duke of Tintagil. He has instead to play up to the methods of the enemy's intelligence. An intelligence picture is built up slowly, like a mosaic, from bits and pieces of information. "All the business of war, and indeed all the business of life," said the Duke of Wellington, "is to find out what you don't know by what you do; that's what I called 'guessing what was at the other side of the hill.'" Dudley

The Jigsaw Poster

(Office of War Information)

Clarke kept on his office wall a "careless talk" poster showing a swastikaed hand fitting together a jigsaw puzzle with the caption "Bits of careless talk are pieced together by the enemy." (Actually, the common "mosaic" and even commoner "jigsaw" analogies are inexact. It is a process of connecting dots, discerning a pattern or *Gestalt,* having more in common with crossword puzzles—or, indeed, with reading Rorschach blots—than it has with jigsaw puzzles.) An intelligence officer applies much the methods of Sherlock Holmes, as when in *The Sign of Four* he deduced from the dents and scratches on Watson's watch that Watson's elder brother had been a man of untidy and careless habits who was left with good prospects but threw away his chances, lived for some time in poverty with occasional short intervals of prosperity, and finally took to drink and died. Similarly, an intelligence officer deduces the enemy's capabilities and intentions and other relevant information from a rumor here, an observation there, a captured document somewhere else.

Ingenuity and imagination are essential. In 1944, for example, the Americans realized that it was not necessary to sacrifice brave men and women of the French Resistance to ascertain how effectively Allied bombing was interdicting rail communications south of Paris; all that was necessary was to watch through neutral sources the price of oranges at Les Halles. When the lines to Spain were cut, the price went up. When the price went back down, it was time to bomb out some bridges again.

Once again, knowledge of the opposition's skills and modus operandi is required. The art of implementing a deception consists in knowing the enemy's methods, breaking your story into bits and pieces, and feeding him those bits and pieces through selected channels and according to a precise timetable, designed to lead him to draw the desired conclusion for himself. Some pieces may be very significant; others may be "merely corroborative detail, to add artistic verisimilitude to an otherwise bald and unconvincing narrative," to quote from *The Mikado.* A German deceiver who was so astute as to judge that the Allies might draw conclusions from the price of oranges in Paris, might see to it that false prices were printed in the newspapers.* Someone who knew Holmes's methods

---

* Or if he had a big enough budget he might spend whatever it took to manipulate the price of oranges. Such things have been done. When in 1943 Plan BARCLAY set out to worry the Germans about a possible invasion of Greece, the Chief Paymaster of GHQ Middle East started a heavy buying of Greek drachmas.

and wanted him to believe that Watson had a dissolute elder brother might have successfully palmed off on him a specially-prepared watch. Someone who did *not* know Holmes's methods and tried to deceive him might come out far worse than before (as Jonas Oldacre found in *The Adventure of the Norwood Builder,* when he tried to frame an innocent man with a fake thumbprint).

This calls for a great deal of meticulous paperwork. In the Mediterranean and Middle East, beginning in early 1942, when permanent elements of "A" Force became attached to field armies (and subsequently "A" Force "outstations" were opened in places as far afield as Nairobi), Dudley Clarke operated a "Strategic Addendum" system for control of deception across the whole theater. The Strategic Addendum was a document, normally written by Clarke himself, that was issued as a closely-held appendix to the more generally distributed newsletter called " 'A' Force Instructions," and distributed at least once a month to tactical deception units and outstations with copies to London and India. It was composed of three parts. Part One set forth in a "General" paragraph the current deception policy (stated as if it were a plan for actual operations). A "Moves" paragraph then gave every important actual troop movement, in progress or completed. The final paragraph gave news of the progress of every deception plan, whether under way or being planned, including any run out of London or India that might have effects in the Mediterranean. Part Two provided true items of news that could be used as chickenfeed to build up the credibility of "special means" channels. Most significant was Part Three. Written in the double-talk of "C.C. Procedure," it consisted of misleading information in the form of the picture that it was desired that the Axis build up for itself. To quote Clarke:

> In it there appeared the "Story" of each plan in operation, broken down into single sentences and transposed in such a manner as to form the meaningless pieces of a jigsaw puzzle which would show the coherent picture only when fitted together in some center of the enemy's Intelligence. By means of Part Three we were able to avoid the distribution of complete Plans outside the Main Headquarters, and thereby avoid a serious risk to secrecy. At the end of Part Three "Notes" would usually be added to indicate the more important items, to allocate particular items to particular

channels on occasion, and to show which were to be circulated in the form of "rumors" rather than as deliberate "leakages" to the hostile Secret Services. In this the codename "ENTWHISTLE" was used to indicate SIME's organization for the circulation of rumors.

The key instrument enabling Clarke, in his words, "to control both the thoughts and the actions of a variety of individual officers working upon the same theme in isolated stations over a wide geographical area," conveying to them "the essentials of the wide overall picture which he alone can see," the Strategic Addendum grew with the growth of deception in the Mediterranean from a modest fifteen-paragraph document at the outset until by 1944 more than one hundred paragraphs was a not uncommon length.

To borrow Clarke's favorite metaphor, once the deceiver had written his play and composed his scenario, the "Lighting, Scenery, Costumes and Property"—not to mention actors—available for staging his production were many and varied. Enemy intelligence derived information through many channels—"sensors" is a currently fashionable word, though not one that was used in the Second World War—and the deceiver had to be able to convey information through all of them. They can be roughly broken down into eight categories. There were the human physical senses of sight, hearing, and occasionally smell. There was radar. There were intercepted radio transmissions. There were actual troop movements and other operations. There were administrative measures taken by the deceiver's forces of which the enemy might become aware. There were captured documents and other tangible things, and prisoner interrogations. There was rumor, ranging from the gossip of the Middle Eastern souks to the cocktail-party chatter of diplomats in neutral capitals. And there were the reports of spies.

Visual deception devices were designed primarily with enemy airborne photoreconnaissance in mind. Such missions were normally flown fairly high, for safety against antiaircraft fire and for maximum coverage; so dummy equipment was designed to deceive vertical photography at eight thousand feet with Axis equipment. This was adequate as well for most ground observation, which is not usually very accurate—as witness the wooden "Quaker guns" used in the American Civil War.* (Low-level oblique photography remained a concern.)

First and most familiar of visual deception devices was ordinary camouflage. While often used in connection with formal deception plans, it was even more often used simply as a matter of routine. More directly related to active deception was the Allied deceivers' extraordinary range of dummies and similar devices. Foremost among these was a numerous and varied stock of dummy vehicles and other equipment. As the war progressed, dummies became available for every major type of tank, both British and American, for several types of trucks, and for a variety of field and antiaircraft artillery. Folding models were made of wood or, later, metal tubing, and fabric. Collapsible models were subsequently developed, made of metal tube posts supported by guys, with fabric covering; the Americans made a great further improvement by developing inflatable models made of rubber tubing with fabric covering. Collapsible and inflatable models had the advantage of greater compactness—twenty-five to thirty pieces of equipment could be loaded into one three-ton truck.

Lightness and portability could cause mishaps that, in Clarke's words, made for "a distressing struggle for Sense of Security over Sense of Humor." Railway engine drivers who thought they were pulling a train of flatcars loaded with tanks and weighing many tons were often astonished to find that the train shot out of the station like a jackrabbit; this happened when the bogus Canadian tanks were being hauled through Palestine in the EUPHRATES COVER PLAN, and was compounded when a "tank" blew off the train in a high wind. Inflatable landing craft had to be watched lest they be blown on shore; and once a shipment of dummy landing craft being brought ostentatiously past the German coast-watchers at Gibraltar blew off of the deck of the ship carrying them.

Already mentioned has been a contraption that proved invaluable in the desert and contributed greatly to the victory at El Alamein, called a SUNSHIELD in the Middle East and a HOUSEBOAT in Britain. This was a canvas cover that could be fitted over a tank by four men in five minutes

---

* And, in the world of fiction, the dead bodies propped up on the battlements of Fort Zinderneuf by Sergeant Lejaune in *Beau Geste*.

to make it appear to be a truck. A similar contraption called a CANNIBAL could turn a field gun and its trailer into a truck. Associated with vehicular dummies were track-making and erasing equipment and "mess-making equipment" that could spread bleach and other powders on grassland or plowed fields to represent general scarring from the passage of numerous men and vehicles.

Dummy equipment was not limited to vehicles. There were imitation Bailey bridges, which could be used as decoys to draw enemy fire away from real bridges, or to give the impression that a major river crossing was imminent at a place where no crossing was intended. There were imitation air observation posts, which could be stationed near a dummy headquarters to give the impression of an important battle headquarters. There were smoke generators that could simulate the presence of cooking fires in the woods or in ruined buildings. There were flash simulators that gave the appearance of artillery fire.

Dummy parachutists, sometimes called by the British codename PARAGONS (or SAINTS), were composed of a plain burlap body and legs attached to a small cotton parachute. Douglas Fairbanks invented an improved model that self-destructed after landing, to leave as little evidence as possible behind; "A" Force had a self-destroying model as early as 1942. The United States Navy's version as of the last months of the war, called PD PACKS, weighed forty pounds, were self-inflating and self-destroying, and simulated rifle fire commencing three minutes after jettisoning. When both genuine and diversionary dummy airborne drops were made, non–self-destroying dummies might be mixed in with the real parachutists as well, so that when the enemy found the dummies for the diversionary drop he could not be sure they had not been intermixed with a real one. Dummies could be booby-trapped, a practice denounced by Radio Berlin as a deception which "could only have been conceived in the sinister Anglo-Saxon mind." Peter Fleming invented a device called a PINTAIL for use with dummy paratroops. A PINTAIL landed upright with a spike stuck into the ground and fired off a Verey light as if an officer were signaling to his men; PINTAILS were generally mixed in with dummy parachutists in a ratio of five to one.

First cousin to dummy paratroops were GOONEY BIRDS, inflatable dummies of assault troops, weighted at the bottom, which could be dropped into the water to simulate infantry going ashore; either wading or swimming could be simulated, depending on the degree of inflation.

Dummy aircraft were also available. The demands of realistic appearance, strength in windy weather, and portability made it easier to imitate fighters and small naval aviation bombers than larger planes or gliders (though Clarke had some success with dummy gliders). A trained crew of six could easily erect an entire squadron of bogus Spitfires or Mustangs in a single night. The U.S. Navy had standard plans for dummy carrier-based planes of all kinds.

Another air-related deception activity was display lighting. Known to the British as "Q" Sets, dummy lighting packages could be used to simulate airstrips, ports, railroad centers, dumps, large or small camps, a convoy moving along a road, or a motor park with the vehicles' side lights on. The RAF early on developed artificial or decoy landing strips to a fine art. This art, and the associated camouflage of real airfields, was known as "C.T.D." for Colonel Turner's Department, drawing its name from Colonel Sir John Turner, the master of the art.

The Allied navies used visual deception extensively. Camouflage of various kinds was well established. Tankers on the dangerous run to Murmansk were disguised as ships with their engines amidships. There were flame and smoke floats that could be dropped from airplanes to simulate a burning motor torpedo boat. There was a WATER SNOWFLAKE, a float that could be dropped from a ship or plane and would release an illuminating rocket (called a SNOWFLAKE) at any predetermined time up to six hours, to decoy U-boats away from a convoy. There were also floating PINTAILS, called by the British AQUASKITS and AQUATAILS. Drone boats, containing explosives which when blown cleared underwater obstacles from a beach, could also be used on the flanks of a landing to simulate an assault boat disaster.

Over the long run, the most useful seaborne visual deception devices were dummy landing craft. They were moored in great numbers in ports from which it was desired to make the enemy believe that an amphibious assault would be mounted. "A" Force was the pioneer in these devices. The first model was called DRYBOB;* it did not float but rested on the

---

* Plainly christened by an Old Etonian (perhaps Noel Wild). At Eton a boy whose summer sport is cricket is called a "drybob," while one whose summer sport is rowing is a "wetbob."

beach or on stilts in the water. Later came an inflatable floating model, called WETBOB, a dummy LCM (Landing Craft, Mechanized) called MIDDLEBOB, and BIGBOB, an inflatable dummy LCT (Landing Craft, Tank) nearly two hundred feet long. (There were also rigid WETBOBS and BIGBOBS as well as inflatable ones.) In the Middle East there were locally-produced inflatable landing craft of various sizes known as BAGPIPES. There was briefly in Egypt a dummy submarine known as SLY-BOB, about which more later.

Early in the war, three British merchant ships, referred to as "Fleet Tenders," were disguised as two battleships and a carrier, and stationed at Rosyth in the Firth of Forth in the hope of luring German air attacks to that place (and incidentally divert them from the main base at Scapa Flow); and special copies of *The Times* were printed containing doctored photographs showing battleships in the Firth of Forth. (This deception had no provable result.) Again, the old British battleship *Centurion*, disarmed as a result of the Washington Naval Treaty, was fitted out as an imitation of the new battleship *Anson* and went out to India and back, drawing heavy attacks while part of a Malta convoy. The U.S. Navy worked out systematic methods of disguising relatively low-level ships as more powerful craft. A frigate could be converted into a *Colorado* class battleship, a minesweeper into a *Northampton* class cruiser, a subchaser into an escort carrier; and from such conversions could be formed a "Swiss Navy"—an amphibious diversion or a decoy task force—though it is not clear that this was ever done on any substantial scale.

The U.S. Navy also produced a kit called a FILBERT, which could imitate either of the two main types of standard prefabricated Navy shore installations (known respectively as ACORN and CUB). A CUB FILBERT, for example, had dummy huts, utility buildings, fuel tanks, piers, wharves, ramps, pontoon bridges and barges, magazines, oil drums, trucks, picket boats and so on and on, all intermixed with real equipment. A real CUB base took up 55,000 tons of shipping space and required 4000 officers and men. A CUB FILBERT took up 900 tons of shipping space and required 150 officers and men, or one-fourth of a Seabee battalion, plus a "Beach Jumper" visual deception team.

In the category of visual deception should be placed the insignia of bogus units. Unit insignia proliferated in both the British and American armies. In the United States Army they appeared most commonly in the form of embroidered cloth patches sewn on the left arm of the uniform, whence the Americans commonly called them "shoulder patches," or just "patches." In the British Army, the official term was "formation badges," and they were more widely displayed than in the American service, being not only worn on the right sleeve (or, in some theaters, on the shoulder strap or the turned-up hat brim), but stenciled on all vehicles, used on signposts to mark the routes allotted to formations during operations, painted on billets and on stores and captured equipment, and displayed at recreational facilities.

Observers on the scene were thus bound to see the insignia of a unit, British or American, whenever it was present. Bogus insignia accordingly became necessary as soon as order of battle deception was introduced; every notional formation had to have a notional sign. In Britain and the United States, where there were believed to be no uncontrolled enemy agents, visible role-playing was unnecessary; descriptions of insignia of bogus units were simply incorporated in controlled agents' messages and other special means communications. In the Middle East and the Mediterranean, where there could be no assurance that live enemy agents were not lurking about, a measure of role-playing was called for. At this the British excelled, while the Americans were too much hobbled by bureaucratic procedures to join in.

When the deceivers wished to activate a notional British division, the War Office would generally assign for notional purposes a division that had served in the First World War and had not been activated since. It would have the same component units—"8th Somerset Light Infantry," or whatever—as its First War avatar, and its 1914–18 division sign would normally be employed.* Once a bogus division was activated, the British had far more flexibility in displaying it on the ground than did the Americans. A combat brigade, garrison battalion, or base area—or a combination of all three—would be selected to play the role. An administrative order would be prepared with special distribution "activating" the division, and directing that no action be taken with respect to disbanding the unit playing the role or in any other respect other than to

---

* British deceivers themselves sometimes concocted the insignia for their notional formations, for in the British service, the choice of formation signs was traditionally left to the formations themselves.

change the designation on working records; the true situation would be explained in a classified letter sent to a much smaller distribution list. The bogus division thenceforth would appear on order of battle lists; mail would be addressed to it; the commander of the role-playing unit— in reality a colonel or brigadier—would be authorized to wear a major-general's insignia, a major-general's pennant would be flown on his car, and members of his staff would be correspondingly upgraded to appropriate notional ranks and authorized to wear the corresponding insignia (but not, be it noted, to draw the corresponding pay and allowances). The bogus divisional insignia would be worn by a large proportion of the personnel of the role-playing unit, and it would be displayed on its vehicles and along roads and outside headquarters.

American procedures were far too rigid for this. The United States Army had no, or next to no, system of "local ranks" for officers, nor could units be activated on the authority of a theater of operations; that had to come from the War Department. "In order factually to devise a similar system in U.S. Forces," wrote Lieutenant Colonel George Train of "A" Force to an American general when the latter wondered why something similar could not be done by American forces, "so much explanation and so many changes in custom and procedure would be required as to 'blow' the scheme pretty thoroughly before it could be put into effect." American bogus units were drawn from a master list prepared in Washington by Army Ground Forces, with insignia designed by the same office that designed insignia for genuine units; and real generals were designated as their notional commanders when required. There was little or no role-playing on the ground comparable to that engaged in by British forces.*

By contrast with visual deception, sonic deception—which carried the overall codename HEATER in the American service, and POPLIN in the British—could be of considerable tactical and operational use but had little strategic value. Its chief form was the projection of recorded noises such as the sound of tanks or other vehicles moving forward, bedding down, or starting up in landing craft; engineers building a bridge; the engines of seaborne craft of various kinds; or the loading of assault craft.

* See Appendix III for details of bogus units and their insignia.

Magnetic tape had not yet been invented; the British recorded generally on sound film, the Americans on magnetic wire (a novelty in that era, and the forerunner of magnetic tape). The British system had greater fidelity and the American system had greater volume.

POPLIN or HEATER equipment could be carried on land by light scout cars or armored cars, and in small craft at sea. Light models could be dropped with paratroops and used to project the voices of soldiers in battle. The American Lieutenant Commander Harold Burris-Meyer developed POLLY, an airplane fitted with HEATER (AIR HEATER) as a public-address system which could be used at two thousand to five thousand feet to call on enemy garrisons to surrender, talk to civilian populations, or direct movements. GAS HEATER projected motor sounds by direct horn amplification of a gasoline motor exhaust. WATER HEATER, also called CANARY HEAD, was a torpedo that proceeded to a selected point, sat until a scheduled time, and stuck up its head to provide a sound program to fit the situation. The same device could be used to send radio transmissions and was then called DUCK HEAD. Another American device called a BUNSEN BURNER consisted of a loudspeaker and radio receiver that could be dropped by parachute and operate for up to four hours, sending out sounds or voice messages transmitted to it from a remote location and then destroying itself.

High-frequency sound waves are rapidly absorbed in air, and low frequencies tend to be masked. Weather conditions may distort sound as well. A direct sound reproduction may thus lose its characteristics at a distance, and it was therefore often necessary to synthesize the characteristic elements of a sound rather than attempt actual reproduction, and necessary too to check the result from a vantage point similar to that of the enemy. All this took special skills and training. The American Beach Jumper teams carried the art of sonic deception to its highest proficiency, but the British Light Scout Car units were not far behind.

Not all sonic deception was electronic. There were paper noise bombs that produced the sound and flash of an explosion but no lethal fragmentation. There were rifle fire and grenade simulators, called by the British PARAFEXES, that could be dropped with dummy paratroopers; as with PINTAILS, a five-to-one ratio was standard. The corresponding devices used by the British in Burma in 1943–45 to simulate small arms fire were known in honor of their inventor, Major André Bicat, as BICAT SAUSAGES, later called BICAT STRIPS and BICAT CRACKERS; each had fifty crackers to a fuse length. With a judicious mixture of dummy parachutists, PINTAILS, fire simulators, chemicals reproducing the smell of cordite, and POPLIN or HEATER recordings of soldiers' voices (perhaps with a BUNSEN BURNER), a single sortie by one large airplane could lay down equipment that could simulate a platoon-level firefight lasting for from one to six hours. As part of Operation TITANIC, such a bogus battle played a key role in diverting German troops at D-Day.

The deceivers even put the sense of smell to use. To corroborate the evidence of the enemy's ears that there were tanks or landing craft somewhere out in the darkness or behind a smoke screen, there were devices that produced diesel fumes smelling like those emitted by armored vehicles or LCTs. There were canisters of chemicals that could produce the smell of cordite. And if gas warfare had ever erupted, the British at least were prepared on a few months' notice to produce chemicals and equipment that would emit bogus smells causing the enemy to assume the use of gas; in particular, they could produce a substance with all the attributes of mustard gas except that of burning the skin, including its effect on gas detectors.

The Second World War saw the advent of a new sensor: radar. And it saw too the first forms of radar jamming and spoofing, the forerunners of the complex electronic countermeasures (ECM) of today.

Jamming of radar involved the same principle as the jamming of foreign propaganda broadcasts: Transmitting on the same frequency a signal that blanketed the hostile signal and rendered it useless. A simpler approach was a "barrage" jammer, which jammed continuously over a band of frequencies; later, scanning devices were developed that would search a band of frequencies and automatically stop and jam any signal found within the band. Devices to perform these functions were built in bewildering variety. CARPET I (and improvements called CARPET II and CARPET III) provided barrage jamming over a selected part of the band of frequencies used by the Germans for ground control and coast-watching. AIRBORNE GROCER was a barrage jammer aimed at German air intercept radar. MANDREL was another barrage jammer used against German ground control.

Electronic devices for spoofing or blanketing enemy radar included MOONSHINE, which produced multiple images all on the same bearing, making one aircraft seem to be many; RUG, a low-power jammer aimed at German coast-watching radar; PIMPERNEL, an improved automatic self-monitoring RUG; and PETER, a device to falsify the bearing of enemy naval fire-control radar. Mechanical devices included WINDOW, bundles of strips of paper backed by aluminum foil, which when dumped from planes produced a cloud of spurious echoes; WINDOW properly employed could also simulate the radar image of a ship of desired size. A ship of almost any size could be simulated by a balloon carrying a "corner reflector" (intersecting disks of reflecting fabric) or trailing metallic streamers. A similar device floated on a buoy could simulate a submarine (the Germans had a decoy of this type, called NARCISSUS). A corner reflector could be mounted in a small ship to make it appear on radar to be a large one, or a number of streamer balloons and corner reflectors could be secured to sea anchors, anchors, and rafts to simulate an entire convoy or task force. ANGELS, strips of foil 13 feet by 2 inches suspended from balloons tethered to small boats, simulated a carrier on radar; CHICKENS were wire cages that showed up as cruisers on radar.

More comprehensible, and more interesting, are the techniques for radio deception employed during the Second World War.

Radio communication was of two types: wireless telegraphy (for which the British abbreviation "W/T" is convenient), *i.e.,* dot-and-dash Morse code,* and voice radio (radiotelephony or "R/T"). Voice radio was relatively less used than W/T, and largely on a tactical level. It was insecure; only on major top-level circuits was it feasible to maintain "scramblers," whereas W/T lent itself to state-of-the-art encryption.

Intelligence can be derived from W/T in two ways.† The obvious way is by breaking the adversary's encryption system and reading the actual

---

* There was also radioteletype. It did not use Morse but a different system of signals to indicate individual letters (the predecessors of the "bytes" of latter-day computers), and the signals were not manually formed; but for the present discussion it can be regarded for many purposes as a type of W/T.

† Though phrased in the present tense, much—not all—of the following discussion is obsolete today. In the computer era, the communications and cryptology of 1939–45 have joined the horse cavalry and the sailing frigate.

text. Less obvious is "traffic analysis." In skillful hands, an astonishing amount of information can be derived from the enemy's messages even without being able to read a word of them. An individual transmitter has recognizable characteristics, and each individual Morse operator has a unique way of sending, a fist that is as recognizable to the practiced ear as is individual handwriting to the eye.* By direction-finding you can ascertain where each transmitter is located. You can determine who talks to whom and derive thereby a picture of the opposing command structure; determine, for example, that a station in Wentworth with the characteristics of an army group headquarters talks to a station at Little Waltham with the characteristics of an army headquarters, which in turn talks to stations at Bury St. Edmunds and Chelmsford with the characteristics of corps headquarters, and so on and on till you have reconstructed the whole communications net and have a pretty clear picture of the entire order of battle, where the units are located, and even their state of training, even though you know the numbers of no units and understand not one word of what they say to each other.

The characteristics and volume of traffic are significant. From designators that prefix the message it is reasonably easy to determine its level of priority (in the American service: Urgent, Operational Priority, Priority, Routine, Deferred), its level of classification (Top Secret, Secret, Confidential, Restricted, Unclassified), the cryptosystem being used, its originator, its addressee or addressees, and its transmission time. Long messages tend to be about administrative matters or intelligence situation reports. Traffic commonly mounts to a peak prior to an operation and then falls off, perhaps with a period of radio silence. A sudden flurry of highly classified high-priority messages is a pretty clear sign that something big is about to happen. And by identifying the participants you may well be able to judge what it is and where it will take place. In the early part of the air war against Germany, the Germans were almost in-

---

* At the Battle of Midway, Nimitz's specialists at Pearl Harbor identified the Japanese Admiral Nagumo's flagship, the carrier *Akagi*, by the fist of her radio operator, who, as one of them commented, "hits the key like he's kicking it with his foot." Some time later, they recognized the fist of the chief radioman of the cruiser *Nagara* using Nagumo's call sign—from which they correctly deduced that the carrier had been damaged or sunk and Nagumo had shifted his flag to the cruiser.

variably able to deduce the strength, direction, and timing of a raid from signals intelligence alone. Depending upon propagation characteristics, this can be done from a long way off. The Japanese intercept facilities at Rabaul, a thousand miles south of the American B-29 bases in the Marianas, were frequently able to warn Tokyo when a raid upon Japan was being launched.

Radio deception is conventionally divided into two categories: *imitative* and *manipulative*. *Imitative* deception is the use of the enemy's own communications channels to simulate his own operations in order to confuse or deceive him. *Manipulative* deception is the use of your own channels to confuse or deceive him, and in particular to defeat or deceive his traffic analysis. You defeat his traffic analysis through "defensive manipulative deception" by "leveling" your traffic—padding it with fake messages, thus smoothing out its peaks and valleys. Or you thrust into it dummy traffic that provides false peaks, or random periods of radio silence that will lull his suspicions if radio silence is enforced when a real operation begins. You deceive his traffic analysis through "offensive manipulative deception" by sending out fake traffic, often in association with an enforced silence or reduction of real traffic, that causes him to believe that real or imaginary units are in particular locations or imaginary operations are under way. Or you distort what are called "directional trends," by building up activity in channels other than those that would normally build up when a particular activity is taking place, while at the same time checking the increase on the important channels by rerouting or other means.

For fake traffic your dummy messages have to be absolutely indistinguishable from real messages to any person who has not broken your codes and ciphers. A skilled enemy cryptographer will quickly distinguish random gibberish from the real thing. Moreover, the receiving code clerk cannot be sure that it is not a real message containing some transmission error rendering it indecipherable. The ideal message is a comprehensible text in plain English, containing some agreed-upon codeword or other characteristic that signifies to the recipient that it is a fake. This too is risky—the recipient may miss the fact that it is a dummy—and it requires an immense amount of work to compose the volume of messages normally required. The standard method of creating a random text—until IBM produced, late in the war, a machine that generated random strings of words—was to pick the first word on each line, or every fourth or fifth word, of a newspaper article.

Moreover, to fool enemy traffic analysis the dummy traffic has to match all the characteristics of the real traffic of the station whose traffic you are padding, or of a formation (say, an armored division) that you are pretending to be, or of some activity, like the sailing of a convoy, that you are simulating. Over any short period of time—two hours is the customary unit—it has to correspond with the real traffic for the same period (or with the typical traffic of, say, an armored division for the same period, or of a convoy putting out from New York) with respect to length of messages, priority of messages, security classification of messages, and originators and addressees. If, for example, you are preparing dummy traffic for the War Department's home station WAR, and 4 percent of the real messages sent by WAR between 10 A.M. and noon are Routine priority "book" (*i.e.,* multiple addressee) messages relayed from New York Port of Embarkation to European Theater of Operations with informational copies to AFHQ in Algiers, classified Confidential and from forty to fifty groups* in length, then 4 percent of the dummy messages sent in that period should share those features. (If, as is often the case, the elapsed time between transmission and filing of a message can be determined without deciphering the text, your dummies must fit that profile too.)† Finally, your W/T operators should never be told which traffic they are sending is real and which is dummy. It should all look the same to them; you do not want to risk their treating the dummy messages in some different way that will be perceptible to the enemy.

Dummy traffic on a voice radio circuit presents an entirely different set of problems. It is less concerned with traffic analysis than with imitative content; making the enemy believe that a particular tactical operation is under way, or helping create the image of a bogus unit. Deceptive voice transmissions are commonly done from a written script, recorded,

---

* W/T traffic is commonly sent in groups of five letters or numbers, regardless of the word divisions of the clear text: RRUTE UIFII UERIE CTTRN, perhaps, or perhaps 31415 92718 28182.

† You have to watch for other peculiarities as well. If the originating operator sends his opening designators in the form "Able Baker Six Five" rather than "AB 65," you have to imitate that too.

and played over the appropriate transmitter at the appropriate time and place. Voice radio deception in the Second World War was usually on a tactical or low operational level.

Any particular station has distinct "fingerprints," and these can be changed by varying power between transmissions, adjusting the bias, filaments and antenna input, changing the filtering characteristics of the power supply, modifying frequencies slightly, and, for W/T, changing operators and the hands they operate the key with (a right-handed operator acquires a wholly different fist when he sends with his left hand). A single transmitter can simulate an entire net if these variations are consistently maintained. The air forces, British and American, operated "spoof vans." A spoof van was a truck carrying a transmitter, which simulated an aircraft unit. The truck was driven around the circumference of a landing field, the transmitter was shaken by a special vibrator that gave the noise or "note" characteristic of aircraft transmitters, the antenna was tilted to cause fading, and the W/T operators rotated, all giving the effect of several aircraft in operation. The spoof van would then be driven some distance away and operations resumed, simulating the movement of an aircraft unit. United States Navy Beach Jumper teams were comparably equipped to simulate the traffic of a substantial naval task force. Such dummy operations might simulate a notional formation or act as a decoy to draw enemy jamming to itself and away from the genuine unit's communications.

Imitative deception is only of tactical use. It may take the simple form of issuing fake instructions in the enemy's language on a voice frequency used by him. More sophisticated imitative deceptions can be practiced if you are familiar with his operating procedure and authenticator systems and can imitate his equipment and operator idiosyncrasies. Knowledge of his cryptosystem may be necessary too. Imitative deception at this level is exceedingly difficult and requires not only intimate knowledge of the enemy's methods but highly skilled personnel of your own. A somewhat easier variant is "nuisance imitative deception." If you have deduced the enemy's procedure signals you can enter his net and snarl his communications by constant requests for repeats, confirmations, signal strength reports, and the like, without ever employing actual fake messages.

Closely related to this sort of activity, of course, is straight jamming of enemy channels. A mild form of jamming not requiring special equipment is simply to sit on the frequency interspacing dots and dashes, or repeating everything that is intercepted. Special jamming equipment in the Second World War included devices for jamming enemy fighter control, and sometimes paratroop communications, known as airborne CIGAR, JOSTLE II, and JOSTLE IV. And TINSEL enabled use of the existing communications transmitter in a bomber to jam enemy fighter control voice radio.

A sophisticated intelligence establishment constantly performs traffic analysis on the communications of its own side, both for communications security, so as to judge what hostile traffic analysis might be able to determine and take defensive measures, and to judge the effectiveness of its offensive deception. It is best if the analysts themselves have no more knowledge about actual plans than would enemy intelligence. As one example, during 1945 the Wahiawa naval communications station in Hawaii regularly reported on the conclusions it drew from American traffic about the progress and plans of American forces in the Pacific.

Radio deception is such an interesting and indeed amusing challenge that in the Second World War the temptation to overdo it had constantly to be warned against, and control over it was kept tightly in the grasp of the theater commanders and ultimately of Washington and London.

Actual troop movements and other operations are of course a legitimate component of deception schemes, to the extent that operational commanders can be persuaded to go along with them. The pattern of bombing in France prior to OVERLORD, for example, was carefully tailored to give equal weight to attacks north of the Seine in the supposed objective area of the deception operation FORTITUDE SOUTH; this was in fact not a diversion of effort, for raids on either side of the river furthered the real objective, which was to interdict German movement from one side to the other. Air raids and shore bombardments furthered the "stories" of WEDLOCK and BLUEBIRD, the notional threats to the Kuriles and Formosa in 1944–45. Indeed, there was hardly a major deception plan in which actual operations did not have a role to play.

All the other miscellaneous methods of deceiving enemy intelligence—calculated indiscretions, controlled leakage, manipulative administrative activity—were lumped together early in the war as "special means." Later on, that term was generally limited to the reports of controlled enemy agents, with other methods referred to as "related means."

Variants of Meinertzhagen's Haversack Ruse—leaving false documents or other tangible things where the enemy would find them (or, indeed, selling them to the enemy)—were tried a number of times in the Second World War. Rarely was it possible to show any results from it. The outstanding exception was MINCEMEAT, the famous "Man Who Never Was," whose story will be told in due course. Another that may have been successful was a fake map, known as the "false going map" showing a dangerous stretch of desert as "fair going," which was planted on the Germans before the Battle of Alam Halfa in 1942.

The old ruse of priming an apparent deserter with false information does not appear to have been used by the Allies in the Second World War. A related practice, however, was to give your own troops misleading information about their objective or destination, which might reach the ears of the enemy.

Rumormongering as a channel for deception was not limited to the souks and dives frequented by most of Maunsell's spokesmen. Thus, as part of TREATMENT, the strategic deception plan covering the Alamein offensive, the heads of all British departments in Cairo, including the British embassy, were briefed to speak disparagingly of the new American tanks, to plan social engagements and holiday trips for the end of October, and to base arrangements on a speculation that things would be quiet in the desert until after the full moon; while one or two individuals were asked to drop deliberately indiscreet hints. Perhaps no other area was so fertile a field for rumor as the Middle East, but gossip was planted from time to time in neutral capitals as far afield as Buenos Aires, and topics to be dropped into conversation were regularly supplied to Allied diplomats.

Little use was made of the Allied press for deception after the painful experience of ADVOCATE. Clarke again had an unfortunate experience later in 1942. After that, both the British and the Americans forbade as a matter of policy the use of the Allied press in implementing deception. The most that was permitted was to encourage them to write perfectly legitimate stories that would tend to support a deception plan—as was done in connection with WEDLOCK in the Pacific in 1944, when the

American press was encouraged to discuss the advantages of the short northern great circle route to Tokyo and to speculate about what was going on in Alaska. (Of course, if a journalist made a useful mistake on his own, nobody saw any need to correct it.)

Deceiving the Axis and neutral press was another matter. Commander Wolfson opened a channel through which he was able to contribute "forecasts" of military operations to "a certain Central European newspaper in the German language" (probably the *Pester Lloyd*). Its unwitting columns were used to further TREATMENT, the strategic deception plan for the El Alamein offensive, BARCLAY, the overall Mediterranean deception plan for 1943, and OAKFIELD, the deception plan for the Anzio landing, among other operations. BARCLAY was likewise furthered by sending an innocent Turkish newsman on a wild goose chase.

Administrative measures taken by the deceiver's forces of which the enemy might become aware were often an important component of deception operations. These took many forms. Closing the borders and shutting off communications with an adjacent territory was a surefire way to call attention to a supposed imminent operation. Conversely, administrative measures might be invoked to lull observers into thinking no action was planned when in fact an operation was imminent. Thus, a significant part of FABRIC, the deception plan for Auchinleck's aborted spring 1942 offensive in the Western Desert, involved opening of a leave period, hiring of accommodations for officers on leave as far away as Palestine, a noticeable increase in entertainment arrangements, and encouragement to war correspondents and senior officers to plan tours away from the desert and freely to accept social invitations from the cosmopolitan high society of Cairo.

The device of "setting the stage" was often used. More than once, in order to lull suspicions conspicuous arrangements were made for an important general to travel from the theater—notionally—when in fact an offensive was about to begin. Or setting the stage might involve the actual (or apparent) presence of a leading figure at a place to which attention was to be drawn.

A special type of "administrative measure" is a misleading diplomatic approach to a neutral government, such as asking for special facilities that would be useful in a (bogus) operation.

Almost anything that exuberant imagination might devise was fair game, with one exception. Notwithstanding canards disseminated after the war under Communist auspices, nothing so cold-blooded was done as to sacrifice a genuine resistance group by allowing it actually to rise up in aid of a deception plan. Quite the contrary: When COCKADE, the threatened invasion of France in 1943, was being mounted, the Resistance was told over the BBC that they should hold their hand till they received a direct order from London to rise up; and just before the notional D-Day, leaflets were dropped telling them that the forthcoming activities were only a rehearsal.

Last, but most important of all in the war with Germany, there were the reports of spies. How this channel of information was manipulated—"special means" in the strictest sense—is the most important story of all, told in Chapter 4.

One last fundamental question is often asked: All right, so you can do deception. It's a lot of trouble. No doubt it's fun. Dudley Clarke may have learned a lot that first year about putting on a show, but he could point to very little unambiguous evidence that his productions had played to any audience at all. Granted, once in a blue moon, maybe— once in a *very* blue moon—you can turn up proof positive that a specific deception produced a specific result. Setting those rare cases aside, are you doing anything more than playing games and having a fine time? How do you know it's not wasted effort? How do you know whether you have done any good at all? Even if your opponent did what you wanted him to do, how do you know that he did it because of your deception? (Was Igraine just too sleepy or amorous to care?)

The answer is that more often than not, you don't know. But that does not mean your efforts are wasted. Deception is like insurance. If your house did not burn down last year, you "wasted" the money you spent on premiums. But you don't know that until the year is out. To change the simile, it is like advertising. "I know half the money I spend on advertising is wasted," the department store magnate John Wanamaker is supposed to have said, "but I can never find out which half."

So it is with deception. Even if you cannot prove that a particular action by the enemy resulted from a particular deception, it is always relevant to ask what he might have done if he had been absolutely certain that you had *not* had the capabilities or intentions that deception suggested. This is particularly true for order of battle deception. What would the Germans have done if they could have been certain that the Allies did *not* have the two armies, seven corps, and twenty-seven divisions in the Eastern Mediterranean that "A" Force pretended they had? What would the Japanese have done with the seventy thousand men that sat idle in the Kuriles on constant alert if they had been certain that the United States and Canada would *not* move against their northern outposts?

If you win the war, of course, you may be able to interrogate the enemy generals and find out what information they acted on (just as Uther Pendragon's interrogation of Igraine after their marriage confirmed that it was his deception that had achieved his object). Such debriefing was done extensively with the Germans, much less so with the Japanese; Fleming managed to interview a number of Japanese commanders in Southeast Asia after V-J Day, but MacArthur refused to allow the American deceivers to come to occupied Japan to interview their former opponents. Even during hostilities there is feedback from which success can be gauged. Fairly regularly in Europe and the Mediterranean the Allies captured Axis intelligence maps that showed gratifying success with order of battle deception.

But the most powerful source of feedback in the Second World War was the breaking of Axis ciphers known since the war by the umbrella term ULTRA, and during the war most commonly referred to as "most secret sources." "Most secret sources" and "special means" were the two most powerful tools of the trade and were the keys to Allied success with deception. They deserve a chapter to themselves. But first, it will be useful to get better acquainted with the Axis intelligence services, the customers for Allied deception.

CHAPTER 3

The Customers

Of the three Axis intelligence services, the elaborate but poorly managed German system was the most successfully deceived. Deception efforts against the Japanese fell upon stony ground, for their even more elaborate system was too incompetent to understand what was being told them, and stood too low in the estimation of the decisionmakers for it to have done much good if they had. The Italians were by far the most competent, but Italy was not a prime target of major strategic deception efforts.

The mind of Adolf Hitler, the real target of Allied deception against Germany, was largely reached through the Supreme Command of the Defense Force, the Oberkommando der Wehrmacht, or OKW, of which Field-Marshal Wilhelm Keitel was chief. Under Keitel and Hitler, operations were directed by the Defense Force Operations Staff or Wehrmachtführungsstab, headed by Colonel-General Alfred Jodl. For intelligence about the Western Allies, the Wehrmachtführungsstab relied chiefly on the intelligence staff of the High Command of the Army, the Oberkommando des Heeres, or OKH; and specifically on a branch of OKH known as Foreign Armies West, Fremde Heere West, or FHW.

FHW's original head was Colonel Ulrich Liss, a bachelor artilleryman, an outstanding horseman, and a lover of England and the English. He staffed FHW with reserve officers, especially preferring journalists and businessmen who had lived in other countries, though the key positions were reserved for regular General Staff officers. Gradually he built up an organization of which he was proud. From its headquarters in a double-level A-frame bunker at OKH field headquarters at Zossen, twenty miles south of Berlin, it performed almost flawlessly in the campaign against the Low Countries and France in 1940. Thereafter Liss broadened FHW's work to include assessments of enemy intentions as well as capabilities. (But he failed to predict the Allied landing in North Africa.) In the first part of 1943, Liss left FHW for a field command on the Eastern Front, where he became a general, received command of a division, and was wounded and captured by the Red Army in January of 1945.

Liss's successor at FHW as of March 1, 1943, was Lieutenant Colonel Alexis Baron von Roenne. Roenne was a Baltic German, a scion of the Teutonic gentry who more or less ruled Latvia in Tsarist days. Frail of build, stiff and precise in manner, peering through gold-rimmed spectacles, sometimes rude to his subordinates, meticulous and highly intelligent, he was "an intellectual but aloof person, impossible to make friends with," according to his operations officer. Until early 1942 he had worked for Liss in FHW—he had been in charge of the crucial French order of battle during the 1940 campaign—and then was transferred to Fremde Heere Ost, or FHO, the equivalent of FHW for the Soviet front, where his fluent Russian could be put to use. There his performance so impressed FHO's chief, the legendary Reinhard Gehlen, that he made Roenne his chief of staff, and then recommended him for FHW when that post opened up.

Under Roenne, FHW issued a daily Lagebericht West, or Situation Report West, a two- or three-page discussion of enemy activity together with a listing of newly discovered or reconfirmed British and American units. Every so often FHW would issue in addition a one- or two-page Kurze Feindbeurteilung West, or Brief Enemy Estimate West, furnishing a longer-range overview and sometimes assessing Allied intentions. Every two weeks would be issued a comprehensive five- or six-page survey of the British and American armies. These were furnished not only to OKW but to major commands in the field. The aim of Allied deception, therefore, was to have information included in these documents.

Central to all this was FHW's estimate of the Allied order of battle. This it consistently overstated from the latter part of 1943 onwards. It resolved all doubts in favor of accepting the existence of a unit, however shaky the evidence; and once part of a unit was accepted the whole one was. This inflation was deliberate on Roenne's part. His motivation is still uncertain, but it is probably not far to seek. One of his subordinates claimed that Roenne knew that the Wehrmachtführungsstab slashed order of battle statistics, and decided to correct for that in advance.

Moreover, if OKW and the Führer did not allocate enough forces to defend the West, drawing off forces into the insatiable maw of Russia, he would be blamed. So he played it safe. (Intelligence people in all wars tend by nature to inflate order of battle figures for the same reason; savvy commanders know this and make allowances for it.) Conceivably yet another motivation was at work. Like many others of his background, Roenne—a profoundly devout Christian—despised Hitler and the Nazis, yet feared the irruption of Bolshevism into Central Europe. He would not have been the only officer to fantasize that if Germany should give in to superior force in the West the Allies would help hold back the Soviets; and inflating Allied strength was a means to that end.

"Deception can never be effective either in love or in war," Sir Michael Howard has well said, "unless there is a certain willingness to be deceived." And whatever Roenne's exact motivation, this willingness he had in full measure. Every single phony unit palmed off on the Germans by the Allied deceivers was accepted by FHW.

Roenne did not participate in the conspiracy that culminated in the attempt on Hitler's life in July 1944, but he was close enough to the resistance movement to be shot in its aftermath. His successor was a certain Colonel Willi Bürkelein, who had been in FHW in 1939–40 before serving at the front, where he had been severely wounded. By that time the German and Allied armies were in full contact on the Western Front, conventional direct sources of intelligence had come again to the fore, and the great days of Allied order of battle deception were over. So Bürkelein plays only a small role in the history of deception.

FHW employed input from all the usual intelligence sources, such as operational data (reconnaissance reports, prisoner interrogation, captured documents), signals intelligence, and the reports of spies. At gathering operational data the German Army was as good as any and better than most. Communications intelligence, including cryptography, was gathered by nine different agencies. Though these never produced a coup comparable to the massive Allied breaking of the German and Japanese machine ciphers, they scored a number of successes. In the field, the German "Y" service was highly efficient and German tactical cryptanalysis was fully competent.

But, as has already been discussed, for a good part of the war the Germans were in only limited contact with the Allies on land and had little access to these relatively reliable sources of raw intelligence. So FHW had to rely extensively on input from human espionage agents. Like all good intelligence evaluators, FHW was skeptical of these unless their reports were clearly corroborated by more tangible evidence. Natural skepticism was reinforced in this instance by lively mistrust of their chief source, an element of the Wehrmacht known as the Abwehr.

Such mistrust was justified. The Abwehr, as Maunsell wrote after the war, "was a thoroughly corrupt organization, careless, dilettante and foreign to exact administration which is essential to good intelligence work."

The Abwehr began as an underground organization designed to evade the Treaty of Versailles, which forbade Germany to engage in espionage or other covert "offensive" intelligence-gathering, but permitted "defensive" counterespionage. So covert intelligence activity, under the control of the High Command, was revived in the guise of the Amt Auslandsnachrichten und Abwehr, which may be translated as Office of Foreign and Counter Intelligence. The "Foreign Intelligence" element coordinated the legitimate overt information-gathering of military and naval attachés. The Abwehr proper or "Counter" element, nominally the legitimate security service, was also the illegitimate secret intelligence service. Rooted not in the parvenu soil of the Nazified new Germany but in the old imperial military, its officers came from the pre-1919 armed services, both regular and reserve, and from the regular officer corps that remained standoffish to Hitler and his minions. A small and close organization in its early years, with the mid-1930s rearming of Germany it was substantially enlarged, beginning when Admiral Wilhelm Canaris took charge in January of 1935.

Canaris, born in 1888, had reached the end of a reasonably successful navy career when he was appointed to the Abwehr quite by chance. As a young officer he had served in von Spee's Pacific squadron, had fought at Coronel and the Falklands in 1914, had escaped from internment in neutral Chile, had spent time in Spain as a secret agent gathering naval intelligence under civilian cover, and had ended the war commanding a U-boat in the Adriatic. Known as a royalist and a reactionary and considered to have a taste for intrigue, in the tumultuous years after the end of the Great War he had been a member of the court-martial that acquitted the killers of Karl Liebknecht and Rosa Luxemburg in the Communist uprising of 1919, and had supported the abortive Kapp Putsch. Then he returned to active duty, serving largely in the Ministry of Defense. In 1924 he visited Japan in connection with a clandestine submarine construction project, and in 1925 he went to Spain in connection with a similar project and to establish a German intelligence network.

Canaris spoke Spanish and had many links with Spain. With the coming of the Spanish Civil War he established friendly personal relationships with Franco himself as well as with several of his key generals, not least Martinez Campos, Franco's intelligence chief.

Since the Abwehr was an element of the OKW, Canaris reported directly to Keitel, and through Keitel to Hitler. The Abwehr's Berlin headquarters, in a block of houses on the Tirpitzufer, a shady street on the Landwehr Canal, were hard by the offices of OKW in the Bendlerstrasse; but in April 1943 the increasing tempo of Allied bombing forced a move out to OKH headquarters at Zossen.

Under Canaris the wartime Abwehr was divided into three main sections or Abteilungen—"Abts" for short—plus an administrative Abt, each containing representation from the three fighting services. Abt I was responsible for espionage. As of the outbreak of war in 1939, its head, and also Canaris's deputy chief, was a cheerful Westphalian who was an old friend of the Admiral's, Colonel Hans Pieckenbrock by name. Independently wealthy, lazy, inefficient, and humorous, Pieckenbrock appeared to his colleagues to be a technocrat, albeit an inept technocrat, and seemingly aloof from politics. In March 1943 he moved on to command troops on the Eastern Front and was replaced by an officer from FHW, Colonel Georg Hansen. Energetic, efficient, and a bit of a roughneck as compared with his elegant predecessor, Hansen did much to straighten out the mess Pieckenbrock had left. Abt II, under Colonel Erwin Lahousen, an Austrian of an old military family, dealt in sabotage and subversion. Abt III, under Colonel Egbert von Bentivegni, an ambitious Prussian and apolitical technocrat, was responsible for security and counterespionage. Abt Z (*i.e.,* Zentral) was the administrative department, headed by the energetic Colonel Hans Oster.

Only certain elements of these four Abteilungen were to be important to Allied deception. Abt I's spies,* captured and doubled against their employer, were to become principal channels of Allied deception. The same would be true to a much lesser degree of Abt II's saboteurs; and an Abt II officer would play a major role in the SILVER channel run out of India. The subsection of Abt III concerned with counterespionage, denominated IIIF, was to prove enormously successful against British clandestine operations but would play little role in deception.

Abwehr headquarters concerned itself only with planning and organization. Execution was delegated to subordinate stations, called Abwehrstellen ("Asts" for short); some twenty-three throughout the Reich, plus additional Asts in the occupied territories, numbering eventually fifteen. Important Asts frequently had subordinate stations of their own, known as Nebenstellen or "Nests." There were also smaller, specialized offices, called Aussenstellen or outstations.

The relatively overt Abwehr presence in a neutral country, which usually operated under diplomatic (or sometimes commercial) cover, was called a War Organization—Kriegsorganisation, or KO. A KO was not normally directed against the host country but was used merely as a base for operations. At the height of the war the most significant were in Turkey, Spain, Portugal, Switzerland, and Sweden; the small Afghanistan station watched India, while a KO at Shanghai liaised with the Japanese. Important KOs would have branch offices corresponding to the Nests of Asts.

Asts, Nests, and KOs were organized on substantially the same basis as the Tirpitzufer headquarters, with representation from Abts I, II, and III, and service representation within each Abt group.

Abt II was the parent of special forces elements called Brandenburg units, eventually the "Brandenburg Division," which carried out special operations in Belgium and France in 1940 and in Yugoslavia in 1941. It was used for such special purposes as anti-partisan operations in Yugoslavia, and tracking down Allied POWs released in Italy after the Italian armistice. By 1942 its connection with the Abwehr had largely faded away.

The Abwehr was very badly run. The Asts operated independently

---

* Individual spies were called V-Männer, short for Vertrauensmänner, or trustworthy men.

with no central direction or coordination. Each Ast—indeed, even each Nest—was free to recruit and train its own V-Männer and to send them, and initiate operations, anywhere. Two Asts (or more) could very easily have V-Männer in the same location without either knowing about the other's. As a practical matter, Ast Hamburg concentrated on Britain and the United States, but there was nothing to keep other Asts or KOs from mounting missions there, and they often did.

Corruption compounded inefficiency. Officers in Asts and Nests and KOs recruited, trained, and sent out their V-Männer according to no clear plan and without much regard for their talents or their loyalty, and behaved much as if they were keeping score in some sort of competitive game by the number of agents they could claim were reporting to them; "the whole organization was permeated with 'pins-in-the-map syndrome,'" said Maunsell after the war. This rendered them easy prey for deception by the Allies, by the agents themselves, or by both, for they were reluctant to remove a pin from the map, and had every incentive to defend an agent's reliability against any doubts or criticisms by rivals or by Berlin. As the American deception officer Arne Ekstrom put it after the war: "It is probably fair to state that on such occasions each agent benefited by the confidence and vanity of his German case officer, just as in the case of an adulterous marriage the deceived husband is slow and unwilling to perceive his horns. This, to pursue the analogy further, is of uncontested advantage to the wife; in fact it may preserve the marriage."

Especially in the neutral capitals, free of wartime austerity and far from Berlin, some officers yielded to a more tangible corruption, skimming the money entrusted to them for payment of their agents, padding expense accounts, and speculating on the local currency market with Abwehr funds.

Inefficiency and corruption were further compounded by disloyalty. Canaris himself grew increasingly hostile to the Hitler regime and increasingly defeatist. Agonizing over his disaffection and over Germany's slide towards the abyss (without doing anything concrete about it), he paid less and less attention to Abwehr business. He consistently advised his friend Franco to stay out of a war that Germany was certain to lose. He was undoubtedly aware of, though not a participant in, the July 1944 plot on the Führer's life. Oster, the head of Abt Z, was also a ringleader in anti-Nazi resistance, as was a close associate of his, Hans von Dohnanyi by name. Pieckenbrock, the cheerful and seemingly apolitical bumbler, was in fact an anti-Nazi. Hansen, who replaced him, was even more so, and was actively engaged in the July 1944 conspiracy.

Despite its weaknesses, the Abwehr performed well in the campaigns in Poland, Norway, the Low Countries, France, and Yugoslavia. Working with the Gestapo it broke the Soviet "Rote Kapelle" spy ring, penetrated major Resistance networks in France, seriously damaged British clandestine operations in Belgium, and controlled and doubled back those in Holland.

A considerable role would be played in Allied deception by the substantial facilities which the Abwehr set up in Spain. The main KO in Madrid, staffed by some twenty officers, ran substations in every port of any importance. The Algeciras station, working with the Tetuan station (which controlled substations in Tangier, Ceuta, and Melilla), watched shipping at Gibraltar. The San Sebastián station organized and monitored frontier crossings. At Barcelona was a training school for agents. Bilbao was a center for courier traffic via Spanish merchant ships. Closely related was the operation in Lisbon, with substations in the Canaries, the Azores, Portuguese Guinea, and Mozambique.

Beginning in a small way in 1939, and more actively from 1941, the Abwehr built an espionage organization in Latin America. This seems to have been driven at least in part by the technical fact that at the current stage of the sunspot cycle, east-west radio communication across the Atlantic was unreliable by comparison with north-south transmission; it made sense, therefore, to set up W/T facilities in Latin America to which V-Männer in the United States could send their messages for relay to Europe. Thus, the V-Männer known to the Allies as TRICYCLE and RUDLOFF were directed to set up radio communication from the United States to Rio and thence to Lisbon, and to Montevideo and thence to Hamburg, respectively.

Against Britain and the United States the Abwehr did less well than it had on the Continent. In the United States it scored a coup when it obtained the plans for the super-secret Norden bombsight. But it never recovered from the FBI's blowing open of the Fritz Duquesne spy ring in the Sebold case of 1941, plus subsequent less dramatic cases. Thereafter every, or nearly every, agent it sent to the United States was caught and sometimes doubled.

In England it began the war with but half a dozen agents, five of whom were soon rounded up while the sixth was a double agent working for the British under the codename Snow. With the unexpected collapse of France and the possibility of an invasion of England, in September 1940 the Abwehr mounted a hasty espionage offensive against the United Kingdom, scraping up and sending in by U-boat and parachute every would-be spy it could lay hands on, many of them totally unsuited for the work and most of them inadequately trained. Every one of these, plus all those sent into Britain in subsequent years, was captured and many were doubled back against their German employers.

Mussolini's abrupt entry into the war in 1940 caught the Abwehr with no Middle Eastern agent network in place. It hastily set up a KO in Ankara with Nests at Istanbul and Tehran. But by the end of 1941, with British forces in Iraq and Iran and the Vichy French cleared from Syria and Lebanon, SIME with the close cooperation of the Turkish secret police had the Middle East largely sealed off. The Abwehr did its best; by 1942 they could report that they had half a dozen W/T agents in the Middle East. They did not know, of course, that all of them were under British control.

There was in Germany a rival intelligence and counterespionage service separate from the military services, which finally absorbed the Abwehr. This was the Reich Security Head Office, the Reichssicherheitshauptamt, or RSHA. Ruled by Reinhard Heydrich from 1939 until he was assassinated by Czech patriots in 1942, and thereafter by Ernst Kaltenbrunner, the RSHA was responsible ultimately to Heinrich Himmler, the Reichsführer SS, in his capacity as chief of the German police. The RSHA controlled both the Nazi Party intelligence service, known as the Security Service, the Sicherheitsdienst or SD; and the Security Police, the Sicherheitspolizei or Sipo—of which the Gestapo was one element.

Among other things, the RSHA engaged in domestic and foreign political intelligence activity, and overlaps and conflicts with Abwehr responsibilities were clearly possible. In 1936 Canaris and Heydrich entered into an agreement, called in the Abwehr the Ten Commandments, aimed at minimizing these. In general, the Abwehr would focus on military intelligence, the SD on political intelligence; each would pass on any information it might come across bearing on the other's field. The Abwehr would be responsible for counterespionage (though it had no arresting authority; that belonged to the Gestapo), and a political investigation by the RSHA that turned up an espionage component would be referred to it, while in turn it would refer to the Gestapo any spying case that proved to have a political component.

As the war wore on and Germany came to occupy much of Europe, the power of the Gestapo expanded and pressed against that of the Abwehr's Abt IIIF. And from 1942 on, the RSHA's foreign political intelligence arm steadily expanded its presence abroad under the leadership of a gifted protégé of Himmler's and Heydrich's named Walter Schellenberg. Schellenberg had performed with distinction various jobs for Heydrich in the prewar years, and continued to distinguish himself in wartime. He perceived the weakness of the Abwehr and was ambitious to knit together a unified intelligence service under his control. The Ten Commandments began to crack under the strain.

At the end of 1942, there began a succession of intelligence failures by the Abwehr that, together with the disaffection in its upper echelons, eventually brought it down. It failed to provide warning of the El Alamein offensive in October 1942 or of the North Africa landings the next month. It failed to warn of the invasion of Sicily in July 1943. It misjudged the position of Badoglio after the fall of Mussolini. Hansen replaced Pieckenbrock, a certain Colonel von Freytag-Loringhoven replaced Lahousen, and Oster was suspended on a corruption charge. The Abwehr failed again to warn of the Anzio landing in January 1944. Then the next month a high Abwehr officer and his wife defected to the Allies, soon followed by others. At about the same time, Canaris was perceived to have lost his last unique value to the Reich when he was unable to dissuade his Spanish friends from yielding to British pressure to restrict Abwehr activities in their country.

In February 1944, Canaris was fired, Hansen was put in charge on an interim basis, and Hitler decreed the unification of the intelligence services. This had barely gotten under way when the July attempt on Hitler's life took place. Dissident Abwehr figures were rounded up. Hansen and others were shot, Freytag-Loringhoven killed himself, Canaris and Oster went to a concentration camp, where they would be killed in the last days of the war. Schellenberg took over completely.

He tried to introduce the centralized control and planning that had been so conspicuously absent. Deadwood was cleared out. The whole roster of V-Männer was reviewed and efforts were made to cull out those who were unproductive or seemed on examination to have been doubled by the Allies. A new cipher system was introduced in October 1944, with disastrous effect on Allied ability to read German intelligence traffic.

But it was too little and too late. Nazi loyalty and efficiency could not make up for lack of experience, nor for the desperate situation of Germany during the last year of the war. No significant new sources of intelligence directed against the British and Americans were opened, German staybehind agents in France, Belgium, and Italy were swept up by the Allies, and nothing was done to prepare resistance to Allied occupation.

Any Western stereotypes about the sly and subtle Oriental were soon laid to rest as far as the Japanese intelligence services were concerned. They were uncoordinated, unsophisticated, and inept.

For strategic deception there was no target comparable to Hitler in the war with Japan, for clear control and lines of authority were missing even at the top. For the last year of the war, after the fall of the Tojo ministry in August 1944, domestic and foreign policies were supposedly determined by something called the Saiko Senso Shido Kaigi, or Supreme Council for the Direction of the War, composed of the Prime Minister, the War, Navy, and Foreign Ministers, and the Chiefs of the Army and Navy General Staffs. But it had no responsibility for military and naval strategy. That was the responsibility of the Dai Honei or Imperial General Staff.

The latter, however, was not a real general staff; nor was it a joint service body like the German OKW, the British Chiefs of Staff Committee, or even the American Joint Chiefs of Staff organization. It had no chief of staff and was composed of the War and Navy Ministers, the Army and Navy Chiefs of Staff, and their retinues. It was supposedly aided by an advisory body of "wise men," the Gunji Sangi In, or Supreme War Council, composed of the Army and Navy Ministers and Chiefs of Staff, the field marshals and admirals of the fleet, and designated distinguished high officers, many of them retired. As the war progressed, the Imperial General Staff declined both in authority and in cohesion, with the Navy often going one way and the Army another, submitting differences to the Emperor for his decision.

The Imperial General Staff had no intelligence organization of its own. Intelligence was gathered, assessed, and distributed independently by the Army and the Navy (the air force was part of the Army), as well as by two civilian departments, the Foreign Ministry and the Greater East Asia Ministry.

The Rikugun Sambo, or Army General Staff, had a Department II responsible for intelligence, which collated information and made recommendations to Department I (Plans and Operations). There was also a small air intelligence section in the air element of the Army General Staff. Department I had far more prestige than Department II and attracted stronger officers, and it might or might not pay attention to what Department II told it. (The chief of Army intelligence told the Americans after the war that "there was a feeling on the part of the general army officers that intelligence was not necessary.") Especially in the later stages of the war, there was a tendency to ignore unpleasant news. Army officers were unwilling to admit that they were losing the war, and any information that tended to indicate that they were was considered untrue.

The Kaigun Gunrei Bu, or Naval General Staff, had a Department III for intelligence; it too was overshadowed by the operations department (Department I). There was no systematic training of intelligence officers. Intelligence officers were permitted to attend planning meetings only in the early stages of a plan, and final estimates of Allied capabilities and intentions were made in the operations department, not the intelligence department. The operations department often distributed intelligence reports. There was no centralized or coordinated flow of information out of the Naval General Staff. The Navy Ministry engaged in some intelligence gathering as well.

The Gaimu Sho, or Foreign Affairs Ministry, was extensively engaged in intelligence collection, primarily of course political and economic. In November 1942, a separate ministry, the Dai Toa Sho, or Greater East Asia Ministry, was established to deal with all matters, other than purely diplomatic ones, relating to political, economic, and cultural affairs in the Far East. It took over wholesale the field organization of the Foreign Ministry in the Far East. Although the Foreign Minister served concurrently as Greater East Asia Minister, the new ministry was dominated by the Army; it was well staffed with intelligence personnel and engaged in intelligence collection, censorship, counterintelligence, and propaganda. In general, after the Greater East Asia Ministry was formed the Foreign Ministry concentrated on Soviet intelligence while the former dealt with the rest of the Far East.

The Japanese stressed espionage, sabotage, and fifth-column techniques, and espionage in particular; to an extent that Western intelligence agencies would not have regarded as good practice. Spy reports made up a large proportion of the intelligence reports sent to Tokyo. Almost every diplomatic post had its spies, sometimes elaborate networks of them. Particularly noteworthy was a redoubtable espionage network run by the consul at Canton, one Tonegi, known as the Tonegi Kikan and also called "Southwest Activities," one of whose agents may have been the channel for the British deception known as PURPLE WHALES.

A feature of Japanese practice foreign to Western thinking was the commingling of intelligence and propaganda; even military intelligence summaries often had a "cheerleading" tone. Propaganda agencies frequently performed intelligence functions; the Domei News Agency and the chief Japanese newspaper offices were exceedingly active fronts for intelligence work. This commingling probably contributed to the frequent inclusion of rumors and general fantasies in Japanese intelligence reports and to the weakness of intelligence evaluation by the diplomatic services.

In cryptanalysis, the Japanese had considerable success against Chinese codes and ciphers of all types (thus making Chinese attachés possible unwitting channels for Allied deceptive information), but only limited success against American and British cryptosystems, mainly low-grade aviation and weather report systems. Much of Japanese communications intelligence was concentrated on traffic analysis, at which they were quite proficient.

Both the Army and the Navy maintained special operations units for espionage, sabotage, and other irregular activities, called in the Army Tokumu Kikan, Special Service Organizations, or Renraku Bu, Liaison Departments; and in the Navy, Tokumu Bu or Special Service Departments.

Counterintelligence was the special province of the Kempei Tai, literally "Military Police Units," though this organization had little in common with the "MPs" familiar to Western armies, and was more nearly comparable to the Gestapo and other elements of the RHSA. Responsible to the Ministry of War rather than to the Army General Staff, the Kempei was charged with security and counterintelligence in combat and occupied areas, and shared this duty with the regular civil police in the homeland. Acting often in conjunction with Tokumu units, it suppressed espionage, sabotage, and subversion, conducted "pacification" of occupied populations, promoted an all-out war effort, gathered intelligence, and coordinated civil and military security measures. Though its personnel were generally of low caliber, the Kempei carried out its duties with ruthless and brutal effectiveness. Kempei Tai and Tokumu Kikan did a profitable business running opium dens in Manchukuo.

Japanese intelligence in the United States, active before Pearl Harbor, collapsed immediately thereafter. With the aggressive action of the FBI's Special Intelligence Service and the cooperation of local governments, so did the activity in Latin America that had been meant to take up the slack. By 1943, apart from abortive efforts to penetrate the United States through the Spanish freelancer Alcázar de Velasco, there was little Japanese activity anywhere in the Western Hemisphere save in Argentina, and that was based largely on American overt publications and Buenos Aires gossip. Thereafter, Japanese intelligence focused on the gathering of information by military and naval attachés in the neutral and pro-Axis capitals of Europe, plus what could be obtained from liaison with the Germans.

At Stockholm was located the one Japanese intelligence officer of the entire war who was clearly of the first rank. Makoto Onodera became military attaché to Sweden in 1941 at the age of forty-four. At the time of his posting he had made a considerable reputation for himself as a Russian specialist, and enjoyed the confidence of the Army General Staff to such an extent that he was allowed to work pretty much as he saw fit—which was unusual in the Japanese system. He had been attached to the Red Army as an observer, and had held posts in Riga and Warsaw; he spoke Russian fluently and was reputed to admire the Russian people. At the outset, his assignment was to cover the Soviet Union, Scandinavia, and Germany, and to report on new tactical methods and the strategic developments of the war in Europe. Tokyo added special assignments several times thereafter: In February 1941 to get material on the German invasion of Britain, in September 1944 to try to absorb Axis espionage nets as the German conquests were rolled back, in May 1945 to get information on Allied redeployment to the Far East.

Ambitious, energetic, active, and talkative, Onodera traveled in Europe a good deal during the war, visiting Berlin and other capitals. He made good contact with important Swedish Army officers, including the chief of the general staff. The relationships he developed were the most extensive and useful of any Japanese official in Europe. This put him in a position to exchange favors with important fellow Japanese officials and with the intelligence services and general staffs of the Germans, Swedes, Finns, and Hungarians, plus the exiled Latvians and Estonians. He gathered strategic intelligence and participated in counterespionage; he even bought up a supply of ball bearings, that rarest of commodities in those days, for attempted shipment to Japan through the diplomatic pouch. By the end of the war he was directing most or all of Japanese undercover work in northern Europe and trying to cover the whole European theater of operations, operating on a budget of two million yen a year. In the autumn of 1944 he tried to arrange to take over the Abwehr's Oslo Ast in the event of German evacuation of Norway, but this plan was never carried through.

Even Onodera slipped occasionally. He paid good money for bad information to a German confidence man, Karl Heinz Kraemer, and considered taking over Kraemer's imaginary network at the German collapse. Some of his other sources—a certain Colonel Maasing, an Estonian who had contacts with Swedish intelligence and military circles, and a Captain Per Wilhelm Gunnar Grip, of Swedish intelligence—gave him largely inaccurate information. And he was the victim of a British special means channel known as SUNRISE, through whom high-level deception material was passed. But such mishaps could have been experienced by anyone, given the difficult conditions Onodera had to work under towards the end of the war.

Onodera was justly appreciated by friend and foe alike. His own service promoted him to major general. His German allies awarded him a high decoration, the Verdienst Kreuz with Swords. "There can be little doubt," recorded the United States Office of Naval Intelligence after hostilities had ended, that Onodera "was one of the outstanding intelligence officers of the war."

But Onodera was unique. In general, second-rate personnel were assigned to intelligence work. The high command simply did not appreciate the role and value of intelligence in modern warfare. There was no central or high-level coordination; this resulted in overlapping and dissipation of effort, including a tendency in the field to collect information wholesale without regard to its value, its reliability, or its relationship to strategic plans or problems. Espionage was not only overemphasized but contained much worthless or inaccurate information. Reports other than espionage reports were rarely graded as to reliability. Exchange of intelligence among army and navy commands, the armed forces, and the civilian agencies, was haphazard and duplication of effort was not infrequent. So-called intelligence summaries—even including those issued daily and weekly by the Army General Staff and disseminated through the major field commands to the lower echelons—were usually simply accounts of current operations with tributes to the indomitable spirit of the Japanese soldier and exhortations to fight on to victory.

Training was poor; such as there was tended to emphasize counterespionage work. Even so, Japanese security in the field was remarkably bad. Individuals were allowed to keep comprehensive diaries. Documents were permitted to fall into Allied hands in copious amounts. Well into the war, the Japanese did not recognize the possibility that the United States Army might practice radio deception except on a low tactical level; they recognized that the United States Navy might do so but were slow in realizing the extent of the practice. Signal security was appalling, with such elementary blunders as sending messages in cipher and then in clear, or the same message in two different ciphers; and intelligence derived from cryptanalysis was handled no differently from ordinary intelligence.

Order of battle intelligence was particularly weak. They had "only the most rudimentary grasp of the technique involved in assessing their opponent's Order of Battle," said Peter Fleming. They were virtually incapable of assimilating identifications below division level, and "they could not, for instance, be relied upon to identify a division from one of its component brigades even when they knew its composition; and it was a waste of time to give them information about battalions or regiments, since, although they were glad to get it, they were unable to make any deductions from it." "The Japanese have been consistently unable to determine the forces before them before actual contact has been made," said an American study prepared in the last days of the war. "Except for information compiled from combat operations or newspaper reports, the Japanese have been unable to produce a substantially accurate order of battle list of American or British dispositions in any theater."

A Japanese summary of Allied dispositions issued in February 1945 showed thirty-seven United States divisions in the Pacific—thirty-one of them identified by number—when the actual total was twenty-seven; moreover, three of these were nonexistent; only nineteen were shown in their correct locations, all but two of which had been identified in press releases; four were shown as being in the last location identified in press releases, when in fact they were no longer there. In India-Burma, the Japanese identified as many as forty-eight Allied divisions when the total was eighteen; in June 1944, a summary identified thirteen British (as opposed to Indian, African, or Chinese) divisions in Burma and India when in fact there was only one. Most of these incorrect identifications came from espionage reports, mainly under Allied control.

Similarly, predictions of Allied attacks were often wide of the mark and sometimes covered so many alternatives as to be useless. There were continual predictions of attacks in the Kuriles and against southern Burma, the Chinese coast, Malaya, Sumatra, and Rabaul, none of which ever transpired.

The most continuous and intimate contact with Japanese intelligence for much of the war was conducted by Peter Fleming, head of D Division, the deception organization for India-Burma. Fleming rated them low from the outset and never found reason to change this opinion.

He early found that subtlety was wasted on them. They "could in no circumstances be relied upon to make the required deductions from even the most obvious hints." Instead, "crudity and boldness paid." You had to lay your misinformation on with a trowel, and when you did, you would find that "the credulity of the Japanese Intelligence Staff left nothing to be desired, and they were well prepared to swallow the most outrageous and implausible fabrications"; indeed, Fleming found after the war that "it was clear that they had swallowed virtually all the misleading information passed to them by us." But this was of little avail if the Japanese did not know what to do with information once they had it. "They are rather like magpies as regards collection of information," he had said in early 1943, and postwar investigation abundantly confirmed this characterization.

While . . . there was never any reason to fear scepticism with regard to the contents of misleading messages transmitted to the Japanese, there was very little guarantee that these messages would be interpreted correctly by Intelligence, let alone acted on by Operations staff. There is abundant evidence to prove that the Japanese, generally for reasons which remain obscure, garbled or distorted a high proportion of the misleading information which they received. This tendency, it is thought, was due in part to the weakness of their cipher staffs and translators; but it cannot all be explained on similar semi-technical grounds and in many cases must be ascribed to the folly and ignorance of the average Japanese Intelligence Officer.

Folly and ignorance showed perhaps most conspicuously in connection with order of battle deception. D Division, aided by the Japanese incompetence in this regard already described, successfully imposed a huge fake order of battle upon the Japanese, but here again crudity paid, and what Fleming called "the low standard of training and thick wits" of Japanese intelligence stood in the way of deft artistic accomplishments such as Clarke was able to impose upon the Germans.

This folly and ignorance extended to tradecraft in espionage, as to both selecting agents and operating them. The men recruited to operate behind the Allied lines, whether by the Japanese or by the entourage of Subhas Chandra Bose, the leading Indian collaborator with the Japanese, were unintelligent and unreliable more often than not. Agents were inadequately trained, and their controls would send them instructions they had never been taught and could not understand.

Even competent agents would have been frustrated by the inept way they were handled. The Japanese showed no interest in them as people, and seemed to have no comprehension of the conditions under which an agent works. They never once in the whole war tried to supply an agent with parts for their radios or with other equipment—or with money. An agent who did not appear to be supplying high-grade information might be simply abandoned. Agents were never told when their control would shift location. The tasks given to agents were often irrelevant or impossible. The questions and orders they were sent were often wholly impractical. Before reaching the agent in the field, questions and orders might pass through as many as five layers of handling, from military headquarters to the appropriate office of the Hikari Kikan* to a representative of Bose to Japanese and/or Indian cipher personnel to Japanese and/or Indian wireless control and eventually to the agent; confusion and misunderstanding could be introduced at each level; and answers and reports had to run the same multinational gauntlet in reverse. Sometimes the intelligence organization running an agent would be at a different location from that of the control radio station communicating with him.

Clandestine radio communications were operated with what Fleming called "scarcely credible inefficiency." The Japanese would send messages using the wrong frequency, the wrong call sign, the wrong cipher. They would pick frequencies that were unsuitable for the time of day or the distance involved, or that were drowned out by powerful regular broadcasting stations. Atmospherics in that part of the world make powerful, reliable equipment particularly important, but they provided their agents with poor-quality transmitters, difficult to maintain and with a range of less then four hundred miles, so that special antennas had to be strung to make them usable at all.

The Japanese had difficulty understanding the rudimentary technique of two-way communication, where both the agent and his control transmit at each scheduled time, usually two or three days a week. Of the eight parties of agents that were in radio communication with the Japanese, four operated on a one-way communication system whereby the agent would transmit at prearranged dates and times—three to five a month, usually—and his control would respond at other prearranged dates and times. This introduced inordinate delays. It often took months to establish contact. The standard of operation at the control end was poor, clearly unsupervised by a knowledgeable intelligence or communications

---

* The Hikari Kikan, or Enlightenment Agency, was the Japanese organization responsible for gathering intelligence on British dispositions in India, spreading propaganda in the subcontinent, and supporting Bose and his Indian National Army.

officer. Use of the wrong call sign was not unknown, and at least once an agent was sent a message in another agent's cipher. Control could usually encipher well enough, but seemed to have trouble distinguishing some letters in the roman alphabet from others. Valuable communication appointments were sometimes wasted by agents' being asked the same questions over and over. Sometimes agents that should have been very productive were wholly wasted through their control's incompetence.

Japanese ineptitude in handling their own agents was matched by ineptitude in playing back channels of their own that they hoped the Allies would not recognize as having been doubled. The style of purported agents' messages would vary, the imaginary agent would sometimes act with wholly implausible swiftness, references would never be made to money problems or other troubles that bedevil real agents.

A final, and insuperable, stumbling block to the practice of deception on the Japanese Army was that it paid little attention to its intelligence anyway. In part this reflected the samurai attitude of the senior officers, who were reluctant to let their plans be affected by threats from a despised foreign enemy. (Perhaps the most noteworthy example of this was the Japanese offensive into India in March 1944, with three divisions and negligible logistical support, when their intelligence staffs believed that the Allies could muster nearly sixty divisions against them.) In part, at least in Southeast Asia, it reflected overall intellectual shortcomings and poor training of commanders and staffs at every level, "which," as Fleming put it, "resulted in the arbitrary distribution of alarmist intelligence appreciations conjuring up so many bogeys that the particular apparition sponsored by D Division became merely one among many."

In sharp contrast to Japanese buffoonery was the skill of the Italians. Theirs was the ablest Axis secret service on the technical level; reflecting, perhaps, an aptitude that went back at least to the age of Machiavelli and Cesare Borgia. But the Italians sometimes did little better than their allies in reaching integrated judgments and assessments on a strategic plane.

Each of the three Italian armed services had its own intelligence branch. The Army intelligence corps was the Servizio de Informazione Militare, or SIM. The Navy's intelligence arm was the Servizio Informazioni Segrete, or SIS. That of the Air Force was the Servizio Informazioni Aeronautiche, or SIA. There was no effective coordination among them.

The naval SIS was understaffed, and concentrated on cryptography and tactical intelligence, focusing on France and Yugoslavia; it routinely read the traffic of those navies. It seems to have maintained no regular coordination with SIM's cryptographic section, which had some success against British naval traffic. Its assessment skills were limited, reflecting its primary attention to tactical intelligence. Its chief till May 1941 was Admiral Giuseppe Lombardi, and thereafter Admiral Franco Maugeri. Air intelligence, the SIA, was even weaker, spending its limited resources on foreign technical developments and protection of air bases against foreign espionage. Its assessment skills were likewise poor; in 1940 it substantially overestimated French air strength. Its chief was General Virgilio Scagliotti.

SIM was infinitely the most important of the three services. Administratively it was a department of the Ministry of War, reporting to a Deputy Chief of Staff—during the important period of the Second World War this was General Mario Roatta, who had himself headed SIM in 1934–37—for its technical and intelligence functions. SIM's successive chiefs during the 1939–43 period until Italy dropped out of the Axis were Roatta's successor Colonel Donato Tripiccione from July 1937 to August 1939, General Giacomo Carboni from November 1939 to September 1940, and Colonel, later General, Cesare Amè from then until August 18, 1943; Amè had since January 1, 1940, been vice-chief under Carboni, with all the technical and operating responsibilities.

Giacomo Carboni was a dapper man with a pencil-thin mustache and a jaunty air, who entertained a high opinion of himself that seems to have been shared by Marshal Badoglio, the Chief of the General Staff, but not necessarily by many others, who regarded him as a careerist who showered praise on Mussolini and cozied up to the dictator's son-in-law and foreign minister Count Ciano until the wind changed and he switched to anti-Fascism. Carboni had seen service with SIM in the course of his career, had become a general at a remarkably young age in 1937, and took over SIM in 1939. To hear him tell it, he found a lazy office that did little more than clip newspapers, and turned it within a few months into the best intelligence service in Europe.

If it was the best in Europe, that was hardly Carboni's doing, and you would assuredly not have known it from the product Carboni delivered in connection with Italy's entry into the war in June 1940. After a visit to Germany early that year, Carboni reported that the country was dejected and appeared bent on self-destruction. Mussolini was displeased. As the Duce approached a decision for war, Carboni furnished estimates of the French and British orders of battle in France and Africa that were grotesquely inflated beyond anything even Roenne in Germany would have dreamed of. He was not alone in this; the SIA wildly inflated Allied airpower as well. These intelligence estimates, combined with other factors such as the reluctance of the King, the Navy, and Badoglio to get into the war, produced near-total inactivity of the Italian forces when they first entered the war, astonishing the German and Japanese attachés in Rome. Carboni later explained the intelligence exaggerations as having been designed to deter Mussolini from going to war: as candid an admission of manipulating intelligence for ulterior reasons as history affords. It seems at least possible that Carboni was ingratiating himself with Badoglio, who did not think the Army and Air Force would be ready for war for years to come.

Carboni was removed from SIM on September 20, 1940, and sent to command troops. He was succeeded by his vice-chief, forty-eight-year-old Colonel (later General) Cesare Amè. Amè was cut from different cloth. Tall, slender, balding, he was a professional with years of experience in SIM. In the interwar years he had carried out missions in North Africa and Corsica, and served in Vienna and Budapest, with extensive travels in Central Europe, the Balkans, and Turkey, often under cover as a functionary of the Italian government tourist bureau. At the end of 1939 he had been plucked from duty with troops to take over as Carboni's vice-chief. Nine months later he found himself Carboni's successor. Badoglio evidently did not like the appointment, and treated Amè rudely; again suggesting that Badoglio might have colluded with Carboni in presenting the latter's absurd assessments to Mussolini in June.

Three weeks after Amè took over SIM, Mussolini decided upon the invasion of Greece—again over Badoglio's objection that far more divisions would be needed than were available. The Marshal was right, of course. The Greek adventure was an embarrassing fiasco from the outset. Badoglio resigned at the end of November. Mussolini's decision to attack

Greece was motivated by pique at the Germans, not by reason, and SIM may not have been asked for an assessment beforehand. (SIM had, however, prepared a report that profoundly misjudged the fighting qualities of the Greek people.)

The SIM that Amè ran for the next three years may or may not have been the best secret service in Europe, but it certainly excelled by far any other on the Continent outside the USSR. It was modest in size—as of Italy's entrance into the war it numbered only 150 officers and 600 enlisted men, and as of the middle of 1941 its strength was a bare 1,000—with its focus on operational activity rather than assessment. Its principal operating sections dealt respectively in espionage, counterespionage and "special services" (including sabotage and assassination), and cryptography; there was also a section that prepared assessments.

Of these, the most effective was the counterespionage and special services section. It had chalked up some notable accomplishments during the 1930s, and especially during the Roatta era. It coordinated Italian support of Macedonian and Croatian separatist movements to weaken Yugoslavia, including arranging the assassination of the King of Yugoslavia in 1934. During the period of the Spanish Civil War, to distract French and British attention from Spain it carried out a campaign of sabotage and murder in France with the aid of the French fascist Cagoulards, and supplied weapons and ammunition to Arab insurgents in Palestine. During the conquest of Ethiopia it participated in subversion of chieftains who should have been loyal to Haile Selassie. In connection with the Ethiopian campaign it is credited with at least one masterpiece of covert action. World public opinion had condemned Italian use of poison gas against the Ethiopians. Ascertaining that a set of horrifying photographs of Ethiopian victims of Italian mustard gas was being sent to a London newspaper, SIM managed to intercept the package and substitute equally horrifying photographs of leprosy patients. When the photographs were published in London, the Italian authorities were able to discredit them as a hoax.

For a brief period in 1940–41 this section was detached from SIM and became in effect a fourth service operating autonomously under the Ministry of War as "CSMSS" (Controspionaggio Militare e Servizi Speziali). This did not work well—it reflected bureaucratic empire-building rather than operational needs—and in January 1941, it was restored to SIM, where it was known as "Sezione Bonsignore." (The corresponding "offensive" or intelligence-gathering section was known as "Sezione Calderini.") In 1942, the counterespionage sections of the SIS and the SIA were merged into the Sezione Bonsignore and it became a unified counterespionage entity for the armed services.

From January 1942, the Servizio Bonsignore was under the command of the then thirty-five-year-old Lieutenant Colonel Giulio Fettarappa-Sandri. Fettarappa-Sandri brought to *controspionaggio*—"CS"—a clear philosophy that included centralized control, flexibility and trust in the initiative of the men in charge, alertness to inflexibility and inertia in the opposing services (especially the tendency of the British and French to repeat a successful modus operandi), and—remarkably, perhaps—the view that "one of the secrets for the success of CS activities is that all activities should be carried out in a gentlemanly fashion." "It is unnecessary and sometimes harmful," said Fettarappa-Sandri, "to adopt blackmailing or some other similar system. This is true for example of the French CS system."

And he got results. SIM found and neutralized or controlled most of the agents sent into Italy by the British, including the only Italian radio operator infiltrated by them into northern Italy—whom SIM ran without his ever realizing he was under control.

SIM's cryptographic section concentrated on military and diplomatic traffic. Gathering its raw material from a network of monitoring stations centered at Forte Bravetta near Rome, it distributed a daily bulletin of decrypts of French, British, Yugoslavian, Greek, and Turkish messages. Not only was it remarkably successful in breaking codes and ciphers of the major powers, but in aid of its work SIM conducted during the 1930s and into the wartime era an especially productive campaign of burglarizing foreign embassies to steal their code books. In late 1935, SIM's ability to read British naval traffic had helped Mussolini call the British bluff over the Ethiopian crisis. Related to SIM's cryptographic work was its program of opening and reading outgoing mail to foreign addresses; by a curious division of labor, inbound foreign mail was read not by SIM but by the police.

At direct espionage SIM lagged behind the times. It had agents abroad, but as of Italy's entrance into the war it had none in such crucial countries as Britain, the USSR, the United States, or Germany. Not until late 1940 did it begin to set up a modern network of secret agents equipped with radio communication; the operation that became Clarke's CHEESE channel in Egypt reflected its opening efforts in that regard.

Relations between SIM and German intelligence were outwardly cordial. Roatta and Canaris had worked well together. Amè and Canaris got along well, and Canaris shared with his Italian colleague some of his doubts about the regime and the war; notably on a visit to Rome in May of 1943, when in the safety of the vast space of St. Peter's the two exchanged their pessimistic views, and again in Venice in early August of 1943, soon after Mussolini had been deposed and Badoglio summoned to head the government (on July 25), when Canaris told Amè that "we also hope that our July 25 will come quickly," and in the course of a long walk on the Lido warned Amè about German designs on Italy. Amè, unlike Canaris, kept his distance from opposition groups; he served Italy, whether her prime minister was Mussolini or Badoglio.

Aside from this high-level cordiality SIM prudently kept the Germans at arm's length. From SIM's perspective they could not be trusted to keep their agreements; and there was no little antagonism between the two countries' services at the working level. Though they had undertaken not to open clandestine services in Italy, the Germans continually tried to do so, their efforts being continually opposed by SIM. Both at home and abroad SIM treated the Germans like any other foreign power, keeping tabs on their intelligence activities and maintaining dossiers on their personnel "as for any other enemy service," to quote Fettarappa-Sandri after the war. Nor were the Germans apprised of any of the doubling of Allied agents which SIM conducted with great success.

This independence made it all the easier, of course, for SIM's facilities to be put at Allied disposal after Italy changed sides in 1943. So did the fact that SIM was likewise independent of the Fascist Party and of its secret police and security service, the OVRA. Like their opposite numbers in Germany, SIM and the OVRA were in some measure rivals and competitors. OVRA managed to double certain British agents run out of Switzerland, unknown to SIM; and at least once SIM and OVRA found they were both pursuing the same quarry. OVRA, of course, could not emulate Schellenberg's success in taking over the rival service. Mussolini did not enjoy the absolute power over both party and state that Hitler did. The king, after all, remained on the throne, and the armed forces, of which SIM was a part, remained ultimately loyal to him.

A few days after Amè's Venice meeting with Canaris, Badoglio fired him and put his old protégé Carboni back in charge of SIM. Carboni, who had been with the Italian occupying force in Corsica, had been recalled to the mainland in March. At the time of Mussolini's overthrow he was put in command of a motorized unit that was supposed to defend Rome. With the sacking of Amè, to this duty was added a designation as "commissioner extraordinary" in charge of SIM. A few days after that, Italy surrendered.

This was in effect the end of the SIM that Amè had led so well. On October 13, Italy declared war on Germany. SIM began to reconstitute itself to assist its new allies—in a few cases, such as the SIM station in Turkey, handing over lists of its agents to the British.

SIM had its weaknesses. It did not view its agents' reports with a sufficiently skeptical eye. And, quite apart from Carboni's absurdities, its assessments were sometimes poor—though Montgomery's chief intelligence officer felt that they "made far more intelligent deductions from the information they received than did the Germans." But Dudley Clarke always regarded SIM as worthy of his steel. It was SIM, it will be recalled, who suspected the bona fides of CHEESE from an early date while the Abwehr swallowed it whole. SIM and "A" Force, as will be seen, played a tricky game of mirrors with each other in the CAPRICORN and LILOU cases in early 1943. CAPRICORN, said Clarke later, "was one of several cases which went to confirm a lasting impression in 'A' Force that the Italian was a more difficult man to deceive than his German ally," and LILOU "showed us in a marked degree the skill and ingenuity with which the wily Italian could play us at our own game. We felt fortunate indeed that most of our other channels worked to the ingenuous, and frequently dishonest, representatives of the Third Reich."

CHAPTER 4

Most Secret Sources and
Special Means

Colonel John Bevan, who managed British deception from London, used to say that there were three essentials to strategic deception: good plans, double agents, and codebreaking. While Dudley Clarke was feeling his way with the first of these essentials, back in England the other two were under way.

The key to both was security. The success of Allied strategic deception in the Second World War, especially against Germany, was made possible in the first place by security from Axis penetration. In Great Britain such security was absolute, and in North America it was substantially if not entirely so. For intelligence from the heartland of their enemies, the Axis had only the reports of their clandestine agents; and every one of them in Britain, and probably every one in North America, was under Allied control. While security in India, the Middle East, the Mediterranean, and, later, in France, was not so total, it was nearly so. This security was in turn made possible by Allied breaking of German and Japanese codes and ciphers; which also enabled the deceivers to read enemy fears and intentions and to monitor the effects of their deception. Now commonly called ULTRA, during the war intelligence from such sources was more usually referred to, at least in the British sphere, as "Most Secret Sources," and sometimes as "special intelligence." (In the Middle East it was sometimes called TRIANGLE.)

The story of how the seemingly unbreakable machine ciphers (*i.e.,* ciphers generated by special machines, as contrasted with traditional "hand ciphers" using pencil and paper) of both the Germans and the Japanese were cracked has often been told—best and most accurately by David Kahn in his masterly works called *The Codebreakers* and *Seizing the Enigma,* and by Edward J. Drea in *MacArthur's Ultra*—and will not be repeated in detail here. In summary, the initial breakthroughs with respect to the "Enigma" machine used by the Germans were made in the 1930s by Polish mathematicians, with an assist from material given to French intelligence by a German traitor. Early in 1939, with war impending, the Poles shared their work with the French and British. The British cryptographic organization, an element of MI6 known as the Government Code and Cipher School (GC&CS), built on this foundation, and within a comparatively short time a substantial proportion of German Enigma traffic—and traffic encrypted in other systems as well—was being read at GC&CS's facility at Bletchley Park, a country house some fifty miles northwest of London.

Inside the Reich the Abwehr communicated by landline and its messages could not be intercepted. But its branch offices in foreign countries communicated with their agents by radio,* as did Abwehr headquarters with stations in neutral and occupied countries, and all these were subject to interception and furnished grist for Bletchley's mill. The first break into Abwehr hand ciphers was made in March 1940, when messages between Hamburg and a German spy ship off the Norwegian coast were identified from experience gained in operating the wireless set of the double agent called SNOW. This enabled the British to learn in August 1940 of the wave of new German agents soon to be sent to the United Kingdom. Beginning in October 1940, Abwehr decrypts were distributed under the designation "ISOS," standing for Intelligence Section, Oliver Strachey—Strachey being the head of the section at Bletchley concentrating on Abwehr traffic. A major further advance came two months later, in December 1940, when the hand cipher of the main Abwehr group was broken. In addition, from June 1941, a Middle Eastern offshoot of Bletchley called Combined Bureau Middle East was decrypting locally intercepted signals in hand ciphers used by the Abwehr in Turkey and later by the Abwehr and the Sicherheitsdienst in the Balkans, the Aegean, and North Africa.

The great breakthrough against the Abwehr came in December 1941, when a Bletchley team headed by Dillwyn Knox cracked the Enigma system used by the Abwehr for W/T communications between Berlin and

---

* Not by voice radio but by W/T using Morse code.

their main stations in occupied and neutral countries. For internal purposes, Abwehr Enigma decrypts were labeled "ISK," for "Intelligence Service Knox." Traffic between Berlin and the Abwehr stations in the Strait of Gibraltar by way of Madrid used a separate key, labeled "GGG"; this was broken in February 1942. Early that year also, the cipher of the Abwehr's teletype system, called TUNNY, was solved as well; this traffic was labeled ISTUN.

In distribution no distinction was made among these various sources, few consumers knew the difference, and "ISOS" was used generically for all decrypts or translations of German intelligence messages, hand or machine. In some quarters, "ISOS" was called "ice" and people who were cleared to read it were said to be "iced."

In the latter part of 1943, the Abwehr improved its ciphers and its wireless security. For several months, Abwehr hand ciphers were virtually impossible to read. Then Combined Bureau Middle East broke into the new systems in its territory, and in January 1944 the head of its Abwehr section returned to Bletchley to assist in extending this success to other areas.

The Sicherheitsdienst's communications security was consistently better than that of the Abwehr, so that throughout the war there were fewer SD than Abwehr decrypts. Its hand ciphers were more difficult than those of the Abwehr, but Bletchley read them from time to time beginning in early 1941, and at the end of May 1941 it inaugurated a new ISOS series of Sicherheitsdienst decrypts called ISOSICLE. The Sicherheitsdienst did not use the Enigma to any great extent before the middle of 1943; its Berlin-Rome Enigma traffic was read until a new machine was introduced in 1944. Not much SD machine traffic was read thereafter except for some traffic with Turkey on occasion.

In October of 1944, both the Abwehr and the SD changed their cipher procedures, and ISOS was lost except for traffic between Berlin on the one hand and Spain and Portugal on the other. Traffic of their mobile units was recovered in mid-December—just too late to give warning of the Ardennes counteroffensive.

The volume of ISOS decrypts rose steadily, from 70 a day at the end of 1941 to 282 a day at the peak in May of 1944. Together with the decrypted daily situation reports of Fremde Heere West, which enabled close monitoring of order of battle deception, ISOS was the great working-level "Most Secret Sources" tool for the Allied deceivers. It enabled the Allies to identify German agents before they ever set out. It produced a complete understanding of the Abwehr's organization, plans, and state of knowledge. It made it possible to check on the effectiveness of the double agents, and of their loyalty in the case of those like ZIGZAG and TRICYCLE who dealt with the Germans directly.

Not all these advantages accrued as quickly as they should have. GC&CS was an element of MI6, whose Section V controlled the dissemination of ISOS. For the first eighteen months, it shared ISOS information only grudgingly. Not until June 1942, in response to a direct appeal by John Masterman to the chief of MI6, did MI6 consent to share with MI5 and members of the Twenty Committee all ISOS decrypts relating to the double agents.

While British cryptography focused on German communications, the Americans concentrated on Japan. In August of 1940, after massive effort, a team headed by the American genius William Friedman—who back in 1924 had been the first man to break a machine cipher—cracked the Japanese diplomatic cipher in one of the greatest of all (known) cryptographic accomplishments. Without having ever laid eyes on the Japanese code machine, without the benefit of a commercial version or the aid of a treacherous Japanese, they constructed a device that duplicated it. As a result of Friedman's work, throughout the war the Allies read the most secret and sensitive of Japanese diplomatic messages almost as quickly as they read their own. The system received the codename MAGIC.

Though less copious than ISOS or the other products of Bletchley's attack on the German cryptosystems, MAGIC was perhaps the most valuable single window not only upon Japanese but on German concerns, intentions, and beliefs at the very highest level. Japanese service attachès regularly reported to Tokyo on conversations with top German officers. Most extraordinary of all were the regular reports of General Baron Hiroshi Oshima, the Ambassador of His Imperial Japanese Majesty to the Greater German Reich. Oshima had known the Reich from the beginning in 1933—he had been the Japanese military attachè in Berlin for six years before becoming ambassador in 1939—and some people considered him "more Nazi than the Nazis." Hitler himself often received

Oshima for wide-ranging talks about the state of the world, which Oshima dutifully reported in word-for-word detail to Tokyo—and, unwittingly, to Washington and London as well. Even today, Oshima's reports convey an extraordinary sensation of immediacy, as if one were eavesdropping through hidden microphones in the Reich Chancery, Hitler's Wolfschanze headquarters in the forests of East Prussia, and his Eagle's Nest at Berchtesgaden itself. How they must have affected Churchill, Roosevelt, and Eisenhower can only be imagined.

On a less dramatic scale, by the latter part of the war Japanese traffic of many kinds was being extensively intercepted and read. This confirmed, for example, Japanese expectation of an American landing on the China coast in the latter part of the war, enabling American deceivers to play to this concern in BLUEBIRD, the deception plan for the Okinawa operation, and PASTEL, the deception plan for the invasion of Japan itself.

No accomplishment can approach in drama and importance the work of the cryptographers, but the contribution of the British postal censorship to counterespionage, and ultimately therefore to "special means" deception, should not be forgotten. It steadily spread throughout British Commonwealth and British-controlled territory. By 1941 it employed a staff of some twelve thousand. From November 1940, more than a year before Pearl Harbor, it had the cooperation of the United States authorities in channeling substantially all transatlantic mail through the British censorship at Trinidad and Bermuda, with interesting intercepts being shared with the Americans (thereby avoiding U.S. legal prohibitions against opening mail, and enabling the FBI to crack several spy cases). Chemists had to keep up with German progress in formulating ever more sophisticated secret inks and their methods of development, and thousands of letters had to be opened and tested for their presence and for the presence of microdots. Addresses had to be checked against lists of known cover addresses used by the Germans. Though of limited value to security in the United Kingdom itself, throughout the war this activity paid off generously overseas, especially in the United States and Latin America.

The term "special means" was adopted early in the war for all channels other than visual deception and W/T messages by which false information could be fed to the enemy. Eventually, however, the term was usually restricted to a specific category of such means, a real or notional person who is a member of, or in touch with, enemy intelligence; most notably, a "double agent" or "controlled enemy agent" who the enemy thinks is working for him but who is really working for you.*

To use a "controlled enemy agent" you must have an "enemy agent" real or imaginary. If he is real, you have to catch him or otherwise bring him under control. The main Allied organizations responsible for tracking down enemy agents and running double agents in the Second World War were MI5 in the United Kingdom and the Federal Bureau of Investigation in the United States; and overseas, SIME, Section V of MI6, X-2 of OSS, the Indian Police and the Intelligence Bureau of the Home Department of the Government of India (known as "DIB"), the French Deuxième Bureau and its reincarnations and, after Italy joined the Allies, the Italian SIM.

For the British, security against espionage and sabotage in the United Kingdom and the Empire overseas, including Egypt and Palestine, was the primary responsibility of the Security Service, commonly known as MI5. Except for officers called Defense Security Officers (DSOs) stationed in a few crucial points—as of the outbreak of war, these were Gibraltar, Malta, Cairo, Aden, Singapore, and Hong Kong—outside the United Kingdom itself MI5 functioned through liaison with the police or with the colonial governor.

MI5 had started life in 1909 as the home section of a newly-formed entity called the Secret Service Bureau, and in 1916 became part of the Directorate of Military Intelligence (hence the "MI" in its name). In the

---

* "Double agent" was the term largely used by the British in the Second World War, though they disliked putting the term on paper, preferring simply references to "special means." (On the Continent in 1944–45, at least in 21 Army Group, the codeword DESPOT was used.) Some purists, American and French, think that "double agent" should only be applied to agents that actually work for both sides at once. During the Second World War, OSS/X-2 referred to agents who dealt with the opposition by W/T or secret ink as "Controlled Enemy Agents" ("CEAs"), and to agents who communicated with the enemy face to face or through intermediaries as "double agents," on the theory that you could not be certain that the latter were fully "controlled." This terminology appears eventually to have been adopted by SHAEF. All this seems unduly pedantic and this book generally uses the familiar term "double agent."

The French called a penetration agent inserted into the enemy camp a "W," and an enemy penetration agent who was known to French counterespionage a "W2."

interwar years it absorbed some of the functions and personnel of the domestic intelligence side of the Special Branch of Scotland Yard, and became an interdepartmental intelligence service without executive powers but accountable in many respects to the Home Secretary. In the course of the war, MI5 underwent substantial reorganization and enlargement, but it always remained a relatively small organization by comparison with such broad agencies as the American FBI. During the war, a substantial fraction of its key staff consisted of able men recruited from the outside just for the duration: Oxbridge dons, barristers and solicitors, and other talented members of the Establishment.

The key element of MI5 for the story of deception was known as B Division. Originally responsible for investigating all threats to security, from the spring of 1941 B Division concentrated on counterespionage. Its operational sections were grouped in a subdivision known as B1. In turn, the element of B1 that ran double agents and thus came to be essential to strategic deception was called B1A.

When war began in 1939 the Director of MI5 was Major-General Sir Vernon Kell, aged sixty-seven and in poor health, who had held the position since 1909. Soon after Churchill came to power, he shook up MI5. Kell was retired in June of 1940; Sir David Petrie, a sixty-two-year-old retired officer of the Indian Police and sometime director of the Intelligence Bureau of the Government of India, was commissioned to study and make recommendations concerning the organization while the agency functioned under a temporary director. Petrie's recommendations were approved; in April 1941, he himself—described by J. C. Masterman as "a rock of integrity, the type of Scot whose reliability in all conditions was beyond question, with strong and independent judgment, but ready and willing to delegate and to trust," and by another MI5 officer as "one of the best man managers I ever met"—became head of MI5 with the new title of Director General.

Petrie completed his reorganization by August 1941. B Division was headed by Guy Liddell, its former deputy chief, a cultivated man, cordial and humorous, and an accomplished cellist. B1A was headed by T. A. Robertson (known always as "Tar"), a charming man of great courtesy, described by one who knew him as "a big haughty fellow with friendly eyes and an assertive way about him," and by another as "a perfect officer type, who could have been played by Ronald Colman," who was a natural leader with instinctive good judgment in everything connected with his work.

Robertson's chief assistants were J. C. Masterman and J. A. Marriott. Marriott, a quiet, precise nuts-and-bolts man who never missed a detail, had in peacetime been a solicitor in the City of London. Masterman was a fifty-year-old bachelor, a history don from Christ Church, Oxford. He knew the Germans well; he had been caught in Germany when the First World War broke out, and sat out the war as a civilian internee. Between the wars, besides a distinguished career as a tutor at Christ Church, he had been a notable cricket player and a scarcely less notable player of tennis and field hockey. He had struggled to join the Army in 1939, being finally accepted as an intelligence officer in March 1940. After some training and a stint as secretary of an investigating committee, he found himself, quite to his own surprise, assigned to MI5—knowing nothing about secret service work.

Not long before the outbreak of the war, MI5's headquarters were moved from central London to strange quarters in Wormwood Scrubs Prison; a set of its voluminous files and some of its activities were moved to Blenheim Palace a few miles from Oxford. After the Blitz was clearly over the whole department returned to a more convenient location in St. James's.

MI5's link in Canada was the Royal Canadian Mounted Police, known universally as the Mounties (except within the organization itself, where it was always called "The Force.") In conjunction with an MI5 representative, in 1942–43 the Mounties ran one double agent, called WATCHDOG, back to Hamburg, and cooperated with MI5 and the FBI in running GARBO's subagent MOONBEAM.

MI5 had a presence in India, but was overshadowed there by the highly professional and experienced DIB, which played the major role in double agent activity in that theater.

Outside the Empire, counterespionage was the responsibility of Section V of the Secret Intelligence Service, or "SIS." Often erroneously called "the Secret Service," the SIS originated as the foreign section of that same Secret Service Bureau of the Edwardian era whose home section became MI5. On the outbreak of war, the SIS became part of the Military Intelligence Directorate with the designation MI6.* Its chief (known always as "C") throughout the war years was Major-General Sir Stewart Menzies, a well-connected regular officer with a distinguished war record and years of service in MI6.

Counterintelligence had had a low priority for MI6 in the interwar years; Section V had been composed of only two officers till shortly after the Munich crisis of 1938, when a third was added, and its focus had been almost exclusively on Communist penetration. When the war began it had still had only three officers at home and two overseas, and by the summer of 1942 it still had only twelve at home and another twelve overseas.

When ISOS came on the scene Section V had the responsibility for processing it. This had stretched its slender resources to near the breaking point, and in turn contributed to a feeling in MI5 that it was not being kept adequately in the picture by MI6. Eventually there erupted a bureaucratic turf battle over control of counterespionage generally, which ran from March through October 1942 and then subsided with some adjustments on both sides (including the sharing of ISOS already referred to). By 1944 Section V had expanded to some sixty officers at home and another sixty overseas. An able man named Felix Cowgill, a veteran of the Indian Police and DIB and by his own description "a born anti-Communist," headed it from the beginning of 1941 until he was eased out as a result of bureaucratic maneuvering by the Soviet agent Kim Philby in late 1944.

The American parallel to MI5 was the Domestic Intelligence Division of the Federal Bureau of Investigation. But whereas MI5 had a military origin and its officers generally held commissions, the FBI was determinedly civilian (its agents were exempt from military service as holding an "essential occupation," and these sturdy young men were often embarrassed by pointed questions as to why they were not in uniform). Moreover, while MI5 was a security service and nothing else, the FBI was

* Though purists prefer "SIS," "MI6" will be the term used in this book, to avoid confusion with the Special Intelligence Service of the FBI, the Signal Intelligence Service of the United States Army, and the Italian Navy's Servizio Informazioni Segrete, all of which were also called the "SIS."

a plainclothes police force. Counterespionage was only one of its many functions, and it tended to approach such work as a form of law enforcement.

As an institution, the FBI was the lengthened shadow of one man, its Director, J. Edgar Hoover. Hoover was a compactly built man of bulldog mien, a rather puritanical bachelor whose only life was his Bureau. Since being given charge in 1924, as a twenty-nine-year-old government lawyer, of what was then simply called the Bureau of Investigation of the Department of Justice, he had transformed an unprofessional office staffed by political hacks into a force of remarkable effectiveness and power. Its officers—"special agents," they were officially called; "G-Men" in popular parlance—lived in fear of The Director, who ruled them with a stern and priggish paternalism. Until wartime expansion forced a relaxation of entrance qualifications (the number of agents on the rolls doubled during the war), FBI special agents were required to be lawyers or accountants. But the Bureau did not draw heavily from the elite institutions or the American equivalent of the English public-school class. A Harvard G-Man, though not wholly unknown, was a rarity.

Hoover was a legendary in-fighter in the turf wars of Washington. The Bureau's jurisdiction was rarely reduced and not infrequently extended, and he defended it against trespassers with no holds barred. (Not the least of his weapons was the information he had on practically everybody who mattered.) Though nominally a subordinate of the Attorney-General and just one among a number of bureau chiefs of the Department of Justice, he was to a considerable degree a law unto himself; in that era the Bureau not only interpreted the law on wiretapping rather generously, but engaged when it thought necessary in "black bag jobs"— warrantless (and clearly unlawful) burglary of premises to obtain information.

Hoover had an instinct for public relations and self-promotion, and he saw to it that the FBI got credit for every possible success. He was at least equally eager to ensure that the FBI was never blamed for failure; and few crimes were worse in his book than "embarrassing the Bureau." To have let his special agents be commissioned in the Army like many officers of MI5 would have been to allow an intolerable trespass on his territory. Nor did the FBI recruit temporary wartime assistance from academics, professionals, and other talented outsiders, as did MI5, MI6, or the American OSS (which Hoover particularly detested). To have done so would have "embarrassed the Bureau" by implying that the G-Men could not do their job themselves.

The FBI's field work was conducted through field offices—close to one hundred of them, in every substantial city, each under a "special agent in charge," or "SAC." Washington headquarters—which occupied nearly half of the Justice Department building on Pennsylvania Avenue—was known as the Seat of Government, or "SOG." There under the Director's eye were the major divisions; with respect to enemy agents in the United States the Domestic Intelligence Division was the relevant one. The Assistant Director in charge of it was D. Milton Ladd, known as "Mickey," the hard-driving son of a senator from South Dakota, who had been with the Bureau since graduation from the law school of George Washington University in 1928, and was much respected by his subordinates because when Hoover disapproved of something done or left undone by his division, he "took the heat and never passed it down," as one special agent recalled. Under Ladd, the Espionage Section was headed by Harry Kimball and subsequently by Robert H. Cunningham. It had a General Desk for the humdrum items, staffed by junior officers, and a Major Case Desk for the interesting ones, staffed by half a dozen or so more senior men. Ladd reported to Edward Tamm, Assistant to the Director, and Tamm reported to Hoover.

With all his faults, indeed to some extent because of them, Hoover had created one of the world's great police forces for conventional law enforcement and scientific criminal investigation. But at international intelligence and counterintelligence work it was still a novice when the Second World War began. Beginning in 1940, it extended its operations to Latin America, establishing a Special Intelligence Service, or "SIS," which was responsible for United States nonmilitary intelligence coverage in the Western Hemisphere. It learned early to play back double agents, including W/T playback, in order to catch other agents. But at the outset it did not have the kind of sophistication required to run double agents for more delicate penetration and certainly not for deception.

The nearest American equivalent to MI6 was OSS, the Office of Strategic Services, a new organization—it only received its name, and its broad charter, in June of 1942—responsible directly to the Joint Chiefs of Staff and engaged in both intelligence (corresponding to the British

MI6) and covert action (corresponding to the British Special Operations Executive). Its security element, corresponding to, and modeled on, Section V of MI6, was set up as a separate branch in June 1943 and dubbed X-2 (in imitation of "XB," which was a cryptonym for Section V). It was headed by James Murphy, a Washington lawyer who was an old protégé of "Wild Bill" Donovan, the chief of OSS. OSS counterespionage had an inauspicious and largely unsuccessful start in North Africa. Far better was its experience in England. Even before the formal establishment of X-2, in the spring of 1943 a group of OSS officers moved in with Section V at the latter's headquarters (in St. Albans outside London for the first months; then, from July 1943, in Ryder Street, Mayfair). They were given full run of Section V; extensively briefed, given access to almost all intelligence material, and allowed to sit in on all phases of operations.

Jurisdictional rivalry between MI5 and MI6 flared up in early 1944 in preparation for the invasion of France. Should enemy agents captured on the Continent and doubled be run by MI6, because it was foreign territory? Or by MI5, because of its experience in the business, the fact that they would be run in British-controlled territory, and the need for any double agent activity on the Continent to be integrated with B1A's existing system? The solution was to constitute a "Special Counterintelligence" (SCI) unit—denominated "104 SCI Unit" (there was a reserve "106 SCI Unit" as well)—nominally controlled by Section V but staffed by MI5 officers, whose sole function would be to run double agents. Cowgill exacted as a quid pro quo that Section V's protégé, X-2, should participate as well. Accordingly, beginning in March 1944, a succession of American officers were read into the double agent work of B1A. They received informal lectures from Robertson and Masterman, read extensively in the files of actual cases, and went through a final training course jointly with the personnel of the British 104 SCI Unit. Three X-2 units were ultimately formed for work on the Continent, labeled 31 SCI, 54 SCI, and 62 SCI.

For the French, counterespionage was historically a function of the Deuxième Bureau, which soon after the beginning of the war lost its historic name and was rechristened Direction des Services de Renseignements et des Services de Securité Militaire, or "DSM." (But the term "Deuxième Bureau" continued to be used generically.) The deputy chief of its counterespionage branch at the outset of the war was an energetic young St. Cyr graduate, a darkly handsome man looking perhaps more Spanish than French, named Paul Paillole. Paillole—remembered by a colleague as "one of those people who could walk into a crowded room and everyone would notice him; tall, quiet, very good-looking, cool, calm, and collected"—had worked closely with MI5 and Section V till the French surrender. After the fall of France the Vichy government dissolved the DSM, but Paillole and some others continued it underground. Within this clandestine DSM, Paillole organized a subsidiary service focusing particularly on double agents, bearing the cover name "Travaux Ruraux," known usually as "TR." (The initiated knew that "TR" really stood for "Tous Renseignements," or "All Intelligence.") He soon had six "TR" stations in unoccupied France, Spain, and North Africa. His "TR" maintained some liaison with MI6.

After the Germans occupied the whole of France in November 1942 following the TORCH landings, Paillole made his way to North Africa, arriving in January 1943, and put his double agent network at the service of "A" Force for deception. He and his organization made possible the rounding up of German staybehind agents after the surrender of German forces in North Africa in May 1943.

Paillole and the DSM had been associated with General Giraud, who finally lost out to de Gaulle in the struggle for recognition as leader of the fighting French. De Gaulle had his own intelligence service, which loosely merged with Paillole's in November 1943. Paillole continued to function with considerable autonomy. By the beginning of 1944, he had some 150 operatives in his "TR" network in France, with another 100 ready to emerge from hiding on D-Day. When the invasion came, SHAEF conceded to Paillole's DSM responsibility for security in the zone of the interior in France. Paillole's officers worked with the Anglo-American SCI units on the Continent; in the southern part of France, his DSM and X-2 worked closely together; W/T double agents on the Atlantic coast were worked by his DSM, those on the Mediterranean by X-2. But the French were not privy to Most Secret Sources until February of 1945, and they did not participate in the 212 Committee in the north.

Paillole himself fell afoul of French internal politics in November 1944, when he resigned in protest against de Gaulle's bringing of "TR" under direct command of his own organization. In the waning months of the war in Europe, despite some nominal gestures such as allowing French access to Most Secret Sources at last, the close Franco-Anglo-American working relationship that had characterized the Paillole era melted away.

These, then, were the chief Allied spycatchers. But once you have caught a spy or suspected spy, what do you do with him?

You have several ways to go. You can try him and punish him (or her—but most Axis agents were men) in the full glare of publicity. Politicians, and publicity-minded law enforcers such as J. Edgar Hoover, favor this approach. The FBI extracted maximum publicity, for instance, from a German sabotage team caught on Long Island in 1942, whose case was featured in the American papers for months and made its way to the Supreme Court before they died in the electric chair. Soon after the first Abwehr offensive against Britain began in September 1940, Churchill asked why none of the captured spies had been shot, and for the rest of the war MI5 labored under what it regarded as undue limitation on its ability to offer leniency to a spy to induce his cooperation. If there has been publicity about the spy's capture, you may of course have no choice but to prosecute. This happened, for example, to an unfortunate German named Joseph Jakobs, who broke his leg parachuting into England in January 1941 and whose case attracted a good deal of press attention.

But most professional counterintelligence officers prefer to work more quietly. At a minimum, it is likely to be unsettling to your opponents if their man simply vanishes. It is much more beneficial, though, if they think he is hard at work. That may make them think that his target area is covered and that there is no need to send anybody else in. Or they may send their man reinforcements that you can catch, or put him in touch with other agents to be rounded up.

You can reach this result by convincing enemy agents that they should work for you to save their necks. And sometimes, as in the case of CHEESE, they already work for you when the enemy recruits them. Sometimes they report their recruitment and you tell them to go ahead, like TRICYCLE, a Yugoslav who was approached by the Abwehr in Belgrade and reported it to MI6. You may dangle one of your own people before the opposition for recruitment, a process known as "coat-trailing." BRONX, daughter of a Peruvian diplomat stationed in Vichy, who was recruited by the Germans when visiting her father, was probably an example of this.

Many of the German agents run as doubles in the Second World War simply gave themselves up at the first opportunity, either promptly on arrival in Britain by parachute or boat (or, in the case of staybehind agents, as soon as their territory came under Allied control), or at Allied consulates before ever reaching the target country, like PAT J, a Hollander who turned himself in at the Dutch consulate in Madrid, or RUDLOFF, who turned himself in at the American consulate in Montevideo. Agents in the latter category usually claimed that they had signed up with the Abwehr so as to escape to Allied territory. Such a claim had to be evaluated carefully, of course, for it was just the line that a would-be triple agent loyal to the Abwehr would take.

Once you decide to run an agent back, he is assigned a "case officer" (or "case agent" in FBI terminology). The case officer for a double agent is his hand-holder, guru, rabbi, therapist, shoulder to cry on, taskmaster, Dutch uncle, guard, and parole officer, in a foreign country and a strange world. The case officer arranges his agent's new identity, procures his ration books and insurance cards, teaches him the customs of his new home. He has to know details of his agent's background and personal history, has to understand his psychology, has to sorrow with him in his failures and rejoice with him in his successes, watch him for any sign of treachery or instability, and keep him worried, if necessary, about what will happen if he does not deliver. He keeps meticulous records of everything his man does, cross-indexed against what others have done; for any tiny inconsistency can raise a suspicion in the opposition's mind. And he must understand the psychology not only of his agent but of the officer in the opposing service who thinks he is the agent's control; must know what will appeal to him and what will not, how to arouse his sympathy when that is needed, what excuses for nonperformance he will accept.

To make your opponents think their man is hard at work you have to ensure that harmless but plausible information goes back to them from him by whatever type of communication they expect. In rare instances, you can do this by leaving him in place, not letting him suspect you are on to him, and ensuring that he learns only the information you want them to learn. More often you have to find out how he was expected to send information back, and send desirable information over that channel yourself. If he is cooperative, either because he is genuinely on your side or has come over to it to save his neck, he himself can send it. Otherwise you can pretend you are the agent and send it yourself.

If you can establish communication through him with the opposing intelligence service there are many benefits besides making the opposition think their man is hard at work. The information they ask him to get (his "questionnaires," they are called) may give you clues as to enemy intentions and as to what they already know. You may pick up valuable tips as to their cipher systems. Thus, traffic of the early double agent called SNOW helped in the original ISOS breakthrough; after the Abwehr tightened security in 1943 it thoughtfully sent its agent GARBO a cipher based on its new one; and the communications of TREASURE and BRUTUS were useful after the major loss of ISOS in the autumn of 1944. You may learn a lot about their organization and modus operandi. You may get a line on other agents already in place or being sent in. You may be able to pass information to the opposition that will help you to identify or entrap new agents when they are picked up; thus, a number of German agents caught in England in 1940–41 carried ration books and identity cards bearing names and numbers furnished to the Abwehr by the double agent SNOW. You may entrap more than people: Through double agents and with the help of the Turkish police, SIME lured one U-boat attempting to land saboteurs on the Lebanese coast to destruction and another to capture.

Another benefit is waste of the opposition's resources. You may tie up his case officers, code clerks, communications specialists, intelligence analysts, couriers, maybe even aircraft dropping new supplies, all working on a case that in fact does not exist. You can even get the opposition to send you money. The United States Treasury benefited to the tune of $366,125 (at 1940s prices) in confiscated payments to German agents in the Second World War.

These are useful features of the double agent as a "penetration agent." They are standard practice for any sophisticated counterintelligence system. The British were feeling their way along these lines even before the war. The French were ahead of them, in good measure because of the energy of Paul Paillole. Paillole visited England in June 1938 to encourage MI5 to set up a counterintelligence section, and again in May 1939 to lecture officers of the newly formed B Branch and MI6's Section V on the subject of double agents.

If you judge that your own agent has been doubled by the opposition, you can use him for penetration purposes and perhaps even for deception. Knowing that his messages to you are controlled by the opposition, you may be able to deduce from them what the opposition wants you to believe and hence what their real concerns are; and in turn mislead them by the questions you send to your agent. You may even try to bring this situation about on purpose, by sending in an agent with orders to surrender and volunteer to be doubled. (Your biggest concern with an enemy agent who voluntarily turns himself in to you is that he may be designed to be such a triple-cross agent.) Running such schemes is as risky as any other double bluff. Peter Fleming played this game successfully in Southeast Asia, notably with the OATMEAL party in Malaya, which managed to send its security check after capture, and again with BACKHAND, who was designed as a triple-cross agent from the outset and functioned successfully for a time; but these were exceptional cases.

There are various ways for agents to communicate with their masters. They can do it in person, directly or through intermediaries or "cutouts." The elementary example of this is the simple "line crosser," who manages to make his way back and forth through opposing front lines bearing information; this of course is likely to involve only low-level data of local interest. A higher-level agent may in peacetime be able to travel home (or, rather, to his employer's home) to report in person. In wartime he may be able to travel to a neutral country (Portugal was a favorite in the Second World War, followed by Turkey) and there make personal contact.

More common in that war was the use of cooperative neutrals as cutouts and couriers to convey payments, messages, and occasionally equipment and supplies. The Germans had a standing problem getting their agents paid. Couriers were used to take them not only money—at least once, a seaman courier was found by the FBI to have a large-denomination bill hidden in his bridgework—but such compact valuables as diamonds and rare stamps. Couriers brought in messages hidden in fountain pens or spelled out by pinpricks under individual letters in a book or magazine. To and from the United States, Spanish seamen were much employed by the Germans for such work. The ship *Manuel Calvo,* which plied between Bilbao and Philadelphia (inbound ships were diverted to Philadelphia because New York was a military port of embarkation), seems to have been a particular favorite. When she was in port there would be a small parade of FBI agents tailing suspects up and down the waterfront; at least three of her seamen were arrested as couriers or agents by the FBI, and one of her seamen became the double agent known to his immediate handlers as LITTLE JOE, though officially dubbed BROMO.* Before Pearl Harbor, Japanese diplomats were used; the double agent in England called TATE was sometimes paid through a Japanese assistant naval attaché. Until 1944 the Spanish diplomatic service was of use. And in one instance, sixteen thousand dollars was dropped out of an airplane from Buenos Aires to the home of an intermediary in Montevideo who was supposed to forward it to pay the German agent in New York known as RUDLOFF; unfortunately, the man in Montevideo was in touch with the FBI.

For passing information rather than money and supplies, less direct means were far more likely. Chief among these were short-wave radio— not voice radio, but W/T in code or cipher—and written communication by mail to drops in neutral countries, again Spain or Portugal more often than not.

The Abwehr remained remarkably naive in thinking that in a densely populated and spy-conscious country like England an agent would be able to set up a transmitter and antenna without attracting attention. Moreover, it seems not to have smelled a rat from the fact that some agents, notably GARBO, were able to remain on the air for very long periods without being disturbed. It did have the good sense to furnish agents sent to Britain with only low-power sets that would cause minimal interference to neighbors' receivers and would be more difficult for the British to monitor—though they also afforded less reliable communication.

---

* For security, agents are always referred to by codenames. MI5 case officers made up their own, often with a humorous basis. GARBO was so called because he was such a consummate actor, GELATINE because she was a "jolly little thing," TATE because he looked like the music-hall performer Harry Tate. On the American side they seem to have been picked at random. The FBI commonly referred to "cases" rather than to individuals, forming the case name from the first syllable of the individual's name. Thus COCASE referred to the double agent Dieudonné Costes, and MICASE to an agent codenamed MIKE.

Once again, GARBO was an exception. Telling the Germans that he had recruited a radio operator with a powerful transmitter, he sent his messages at 100 watts from a high-grade set. Even this did not raise the Abwehr's suspicions.

Written, as opposed to radio, communication from agents to the Abwehr was normally accomplished by "secret writing" in invisible ink, generally written between the lines of a seemingly innocuous ordinary letter mailed to a cover address in a neutral country (Spain or Portugal more often than not). This technique was historically a German favorite, reflecting their longtime lead over the rest of the world in chemical research. It had been the chief method used by German spies in the First World War. By the time of the Second World War the Germans had a variety of formulas for such ink and its development, some quite sophisticated and involving obscure chemicals. Agents would be given recipes for making their own ink from chemicals and medicines believed to be available from ordinary pharmacies—though the Abwehr evidently did not realize that some of these were poisons that could not be bought in England without signing a book, or that one favorite item, pyramidon, was a painkiller widely used in Europe but not in the United States. Or secret ink might be hidden in a hollow tooth, or impregnated into seemingly ordinary matches carried by the agent.

The Germans invented and employed a third method of communication, generally useful only for sending instructions to an agent, for men in the field would not normally have the equipment for using it to send messages back. This was the "microdot," a photographic reduction of a message to a tiny spot smaller than a pinhead, stuck on a letter or printed sheet where it would be mistaken for a punctuation mark, or hidden under a stamp, and read through a microscope by the recipient. The agent known as RAINBOW was alerted to expect messages by this technique when he was recruited in April 1940, but its first known actual use came in August 1941, in a letter to RAINBOW and a contemporary questionnaire given to TRICYCLE.

Using these channels for double agent work is tricky. For W/T, you have to assume that the other side knows the agent's fist—indeed, that they have probably recorded it for future reference. Imitating another's fist is possible but difficult and risky, like imitating another's handwriting.

When it was decided that the genuine double agent PEASANT should be notionally moved from England to the United States and transmit from there, an FBI radio operator studied recordings of his fist provided by MI5. Similarly, though PAT J was genuinely in New York, the FBI did not trust him to do his own radio transmissions. His sending style was extensively recorded, and three FBI technicians studied the records thoroughly. (He had a peculiarity of making his dots and dashes almost the same length.)

Similar considerations obviously apply with handwritten messages in secret ink; you have to assume that the other side has a record of your man's handwriting, so his letters should be written by him or by a skilled forger. (When GARBO's imaginary subagent MOONBEAM notionally moved to Canada in 1943 to send reports in secret ink, GARBO sent the Abwehr a sample of his handwriting—actually written by Cyril Mills, the MI5 officer who would be running the case in Ottawa.)

More broadly, the text of what the agent sends has to fit the style the opposition has come to expect from him. GARBO, for example, would pen long essays in an inimitably florid and discursive rhetoric. PAT J, a Dutchman communicating in German, had some odd idioms that had to be replicated with care. Similarly, each agent should be given a distinctive personality. MOONBEAM, for example, the imaginary agent who reported from Canada, was built up as a cheapskate who demanded to be paid by the Germans even for the cost of shoveling snow from his front walk. Carefully controlled small errors can add to verisimilitude. For instance, the CHEESE channel out of Cairo, who communicated with the Abwehr in French, informed the Abwehr in 1942 that he had heard that the British were using captured German tanks, *chars allemands*. A fortnight later, CHEESE reported shamefacedly that he had found that what he had heard about was not *German* tanks but *Sherman* tanks.

You are on safer ground if you can get the agent himself to do the sending. But you have to be confident that he will not do something to give away the fact that he is under control. On the most elementary level, one of your own people who understands Morse code has to be next to him whenever he sends a W/T message. But you have to worry about more subtle devices. Though he may do something as rough and ready as distorting his own fist or handwriting in some way, if his masters have any sophistication they will have briefed him with a "security check," a specific means for authenticating messages or for indicating that he is under control. For example, all authentic messages may have to contain a misspelling at a particular location in the text, or have to use certain words at certain places; or he may be directed to include a certain word, abbreviation, or misspelling only if he is under control. If the authentication is missing or the control sign is included, his control should recognize that something is wrong.

As long as you are confident that the opposition will not catch on— from the fist in which radio transmissions are sent, or otherwise—you can run your agent as a "notional" in every sense except that somewhere he does (or did) exist. He may be dead, or safely locked away, or working at some harmless job with no knowledge of what is being done in his name. The case officer prepares the traffic and sends it through a staff radio operator. The safest kind of agent is of course a total notional. You do not have to pay him money, find him a job, console him when he grows homesick or worries about being a traitor. NICOSSOF, the supposed sender of the CHEESE traffic, was a classic example. An equally classic example of the near-notional was PEASANT, whose traffic was sent from Washington by the FBI while he himself remained at large (but under surveillance) in England. Sometimes the real becomes notional. MI5's double agent TREASURE admitted to her case officer that, piqued that her dog had been kept in quarantine and died there, she had not disclosed her security check to MI5, intending to vent her spite by alerting the Germans that she was under control. She had to be discontinued, but an operator continued to send her traffic.

A big part of building up a major double agent is to let him develop a network of conscious subagents and unwitting informers—whose members, of course, are generally notional even if the agent himself is not. This helps convince the other side that he is working hard and delivering the goods, especially if he produces "sources" with access to special information. Here as in so many things, GARBO was the champion in the Second World War. At his peak he notionally controlled a menagerie of no less than twenty-seven witting and unwitting agents and sources. There were airline employees, a pair of Venezuelan brothers, various seamen, employees of the Ministry of Information, miscellaneous British and American servicemen, an Indian poet and his mistress, a Welsh fascist, a leader of an imaginary "Aryan World Order," and sundry others—all wholly fanciful. A more representative example was the FBI's double agent RUDLOFF, who had three main notional subsources—a civilian in the Navy Department in Washington named NEVI, one in the War Department in Washington named WASCH, and an employee of Republic Aircraft Corporation on Long Island called REP, while NEVI in turn had a subsource called OSTEN, and WASCH had a notional officer in the Army Service Forces in San Francisco.

Once your double agent starts to communicate with the other side, he has to have something to communicate. A certain amount of his traffic can be administrative—reports on his living conditions, pleas for money, discussion of communications procedure, and the like. But if his masters are not to write him off, he has to pass information of substance that a person in his position could plausibly obtain. You need a supply of such information; what is called "foodstuff" in the trade.

You can make your foodstuff up, of course, but you are likely soon to be found out (though there were freelance operatives who did just that and got away with it). Moreover, before using your agent to send false information you want to establish his credibility by having him send true, verifiable information that in fact does your side no harm, sometimes called "tonic." Routine tonic is called "chickenfeed" or in British usage sometimes "chickenfood." Chickenfeed, harmless in itself, which is deftly selected can contribute a great deal to deception. For example, news that an American general has arrived in Cairo to take command of all United States forces in the Middle East may be perfectly true, enhancing the agent's reputation, but also unimportant, since there are few American forces in the Middle East; yet the recipient might easily conclude incorrectly from the item that American forces in the Middle East are to be expanded.

You can also build up an agent's credibility by having him warn of some major action too late to do the enemy any good. GARBO warned of the Normandy landings in 1944, just too late to help; and SILVER was primed in 1945 to warn the Japanese of the imminent offensive to retake Singapore, again just too late to do them any good, but the war ended before that card needed to be played. The same device can be used, of course, to bolster the credibility of special means channels who are not strictly double agents. Thus, RUPERT, a German in touch with Japanese intelligence in Buenos Aires, was directed to warn his contact that the Bonins would be assaulted just before the Marines landed on Iwo Jima. And prior to Hiroshima, GUS, a Polish journalist in Buenos Aires whom the Japanese believed to be their man but who was in fact working for the Americans, was directed to warn his Japanese case officer that some surprises in the form of new weapons were in store for the Japanese, which could easily result in their asking the Americans to walk into Japan without fighting.

Good true foodstuff, whether mere chickenfeed or more noteworthy information, has to be supplied by the armed services, civil defense agencies, and other operating entities. And these instinctively shrink from the thought of handing information to the enemy. Typical was the reaction of one military representative during the British debates on the subject in 1942, who grumbled that there was no point in "trying to control the enemy's espionage system by paying the Danegeld of good information." You need to clear with the operating entities information from any source before sending it, including any information you simply make up, to be sure you are not inadvertently giving important secrets away, or saying something that conflicts with something the enemy is bound to learn in the future. Moreover, especially as you build up a stable of controlled agents, you need to make sure that their information is consistent and that one does not say something that will cast suspicion on another. These are problems in bureaucratic coordination and call for bureaucratic solutions. The British, who had more controlled agents and a bigger problem, solved it by a committee system that actually worked. The Americans solved it by direct liaison.

For the first year of the war, before the Abwehr offensive of September 1940, the only double agents being run by MI5 were SNOW and his colleagues G.W., CHARLIE, and BISCUIT, plus an inactive agent called RAINBOW, so the foodstuff problem was not a serious one. It became serious when the Abwehr offensive of September 1940 netted twenty captured spies, four of whom became double agents (codenamed SUMMER, TATE, GANDER, and GIRAFFE), with the prospect of more. MI5 was able to get only the most general guidance from the operating entities for their foodstuff. The Chiefs of Staff simply decreed that information passed should stress the strength of British defenses against invasion.

Then when the invasion scare waned and the Blitz began, the Abwehr began pressing its agents for information about the effects of bombing and the state of the raw material and food supply, so the question arose what slant should be given to information on those subjects. At the end of September or the beginning of October 1940, the Chairman of the Joint Intelligence Subcommittee (which included, among others, the directors of intelligence of the three fighting services, "C" of MI6, and Liddell from MI5), informally constituted the latter five a committee called the "W Board" to supervise the foodstuff of double agents. The membership of the W Board, which was subsequently enlarged by adding the chairman of the Home Defense Executive, was too high-level to exercise day-to-day supervision, so there soon evolved a working-level body subordinate to it; initially called the "U Section," this was soon dubbed the XX Committee or Twenty Committee, from the Roman numeral "XX," *i.e.,* double-cross.*

With Masterman as the chairman—rather to his own embarrassment at the outset, for he was junior in rank if not in years—the Twenty Committee also included representatives of the three service intelligence chiefs and of MI6, GHQ Home Forces, C. T. D., and the Home Defense Executive (the last three over the objection of the Director of Naval Intelligence). It held its first meeting on January 2, 1941, and for the rest of the German war it met every Thursday afternoon at MI5's offices to review the activities of the double agent system, address problems, and decide on proposed new initiatives, being joined when appropriate by the case officers of individual double agents. It worked smoothly and by consensus; only once in its whole history was it necessary to put a question to a vote. The Twenty Committee, it should be made plain, did not "run" the agents. MI5, or in some overseas cases MI6, did that.

As general rules got settled and the Twenty Committee took hold, the need for policy guidance from the W Board declined. It met seven times

* No one seems to have claimed credit for this pleasantry. Perhaps it was partly inspired by the film *The Great Dictator,* released in 1940, in which Charlie Chaplin played a burlesque of Hitler whose party emblem was not the swastika but an XX "Double Cross." The Twenty Committee was sometimes called by its members the "Twenty Club." Masterman, the chairman, was a don at Christ Church, Oxford, and there was once a "Twenty Club" there; perhaps that had something to do with the name.

The name is written both as "XX Committee" and as "Twenty Committee" in the original documents. The present book follows the usage of the official history.

in 1941, and only four times in 1942, twice in 1943, and once each in 1944 and 1945.

Overseas, SIME was itself a joint body; this fact coupled with its close relationship with "A" Force forestalled any need for committee coordination of this kind in the early days. In March 1942 Maunsell set up a "Special Section" of SIME to help in the provision and operation of such channels for "A" Force. As the war widened and double agents were operated out of a number of locations, more coordination was required. In March 1943 Clarke, Maunsell, and the head of MI6 for the Middle East agreed on a charter for the formation of "Thirty Committees" in crucial locations. These consisted of one representative each of "A" Force, MI6, and the security service relevant to the locality (SIME for committees controlled by Cairo; the French Deuxième Bureau for those controlled by Algiers; the Anglo-American counterespionage organization in Italy; and the local authorities in Kenya and South Africa), with participation by army, air, and naval officers as appropriate. Originally, there were seven Thirty Committees, controlled out of Cairo and bearing the numbers "30" (for Cairo itself) through "36." The system soon grew to encompass Forty and Fifty Committees as well, the Forty series being controlled out of Algiers and subsequently from Italy, and the Fifty series out of Nairobi (there was also a Sixty Committee at Lisbon). In all, there were twenty-one such committees, the total fluctuating from time to time.*

This charter solved at one blow all the major problems that had embroiled London in the early days. The Thirty Committees were specifically charged with "establishing and maintaining channels for passing false information to the enemy through the medium of double agents," and "transmitting information to the enemy on the instructions of Commander, 'A' Force, or of his representative," operating under the guidance of the current "party line" in Clarke's "A" Force Instructions and Strate-

---

* The numerical designations changed somewhat over time and the same number was sometimes used more than once as committees shut down and opened up. Cairo (30) controlled Beirut (31), Baghdad (32), Asmara (33), Tehran (34), Istanbul (35), Nairobi (36, later 50), Tripoli (33 and 39), Cyprus (36), Algiers (37 and 40), and Athens (39). The Forty series included, besides Algiers, Oran (41), Casablanca (42), Tunis (43), Gibraltar (42), Naples (44), Bari (42), Rome (45), and Florence (44). Nairobi became a separate headquarters and controlled Cape Town (51). Lisbon was 60.

gic Addendum. In each committee, to "A" Force was assigned policy regarding the use of channels, and the content of all messages. To the relevant security service was assigned administrative maintenance of agents, provision of case officers, development of potential channels, and advice on reliability of channels, all in countries occupied by Allied forces. To MI6 was assigned provision of staff other than case officers, communications and communications security, finance, and responsibility for agents in countries not occupied by Allied forces. And it was made clear that once an agent was designated as a deception agent ("Special Agent" was the term used), "his primary use is to deceive the enemy and not to penetrate their Intelligence Organization."

This system worked beautifully—including the full cooperation of the French. "Altogether," said Dudley Clarke after the war, "the 30-Committees formed one of the happiest elements in the whole task of deception."

Even after the coordinating machinery of the Twenty Committee had been put in place, it was not always easy to get usable foodstuff out of the operating entities in London. The Home Defense Executive and the fighting services gave vague general guidance, but B1A's specific proposals within that guidance were vetoed more often than not. They got enough to keep the agents going, but it was not very exciting stuff; just "a dull sort of bowdlerized Baedeker," complained Commander Ewen Montagu, the representative of the Director of Naval Intelligence on the Twenty Committee.

So things moved along through 1941 and into 1942. The roster of controlled agents in England grew rapidly. The notable Yugoslav double agent TRICYCLE arrived in December 1940, and acquired subagents, BALLOON and GELATINE, within a few months. SNOW acquired a fourth member of his network, CELERY, but his network soon seemed compromised and was terminated (though G.W. continued independently for a while). DRAGONFLY, an Englishman whom the Germans thought they had recruited, was added in March. MUTT and JEFF, two Norwegians, arrived in April 1941; a Belgian, FATHER, a Pole, CARELESS, and a Yugoslav, THE SNARK, arrived in June and July. SWEET WILLIAM and PEPPERMINT, channels through the Spanish embassy, were added in August and December. In 1942, GARBO, WEASEL, CARROT, WASHOUT, JOSEF, BRONX,

BRUTUS, LIPSTICK, and ZIGZAG joined the roster. A few of these were failures, but most continued usable. By June of 1942, Robertson, in a memorandum to the W Board, could assert confidently that MI5 controlled every German agent in the country.

Even with the example of "A" Force's use of CHEESE before its eyes, the Allied side took a remarkably long time to realize that double agents could be used not only for penetration but for deception. Some intelligence officers seem never fully to have grasped this. The French—notwithstanding their early recognition of the value of double agents—seem to have remained fixated upon *"intoxication,"* the use of the agent's traffic to confuse the enemy's picture rather than to help him paint a false one. And the military mind is especially suspicious of double agents. "The German General Staff will not move a single division on an agent's report alone," wrote the Army member of the Twenty Committee when their use for deception was being debated.

All messages by double agents are in one sense deceptive, of course, but they do not necessarily further some operational or strategic deception. Under the Twenty Committee's aegis a few isolated efforts at true deception were made early on. Rumors were passed on about defenses in specific areas of England and about a couple of naval matters. Some phony charts of minefields were passed through TRICYCLE in January 1941 (Plan MACHIAVELLI), and later in 1941 G.W. passed through the Spanish embassy a fake listing of divisional insignia (PLAN A.B.), spurious information about a Malta convoy, and a phony file of documents purporting to report on air raid damage designed to lead the Germans to concentrate their bombing on well-defended air bases rather than on cities (PLAN IV). None of these seemed to have much effect, though examination of German records after the war suggested that the air raid damage file had been taken seriously.

All this was unsystematic, and MI5 seems for some time to have been of two minds about using the double agents for deception. In late July 1941 MI5 reiterated to the W Board that they could and should be so used. But when Dudley Clarke, fresh from "A" Force's experience with CHEESE, suggested to the Twenty Committee during his visit to London in October that they be used to support his deception operations, and soon thereafter Colonel Oliver Stanley was appointed the first Controlling Officer of deception, MI5 (and MI6) grew apprehensive lest things get out of control, as did the naval intelligence authorities. A new MI5 memorandum to the W Board emphasized that while it had always been contemplated that the double agents would be used for deception, that should not jeopardize their fundamental counterespionage role, and emphasized further that MI5 and MI6, not some deception officer, should be the sole judges of how they should be used. Montagu in particular, who was now secretary of the W Board, busied himself lobbying to keep Stanley off of both the W Board and the Twenty Committee, producing decisions by both those bodies that Stanley was "not to be informed of work of double agents, but only that the W Board has channels for misinforming the enemy. The I.S.S.B. is to be the channel through which Colonel Stanley's requirements will be conveyed to the Twenty Committee." (MI5 was disappointed but stayed out of this dispute. In fact, Guy Liddell had already explained to Stanley "exactly what machinery we had for misleading the enemy.")

As will be seen, the only strategic deception attempted by Stanley was HARDBOILED in early 1942, the first of many spurious threatened invasions of Norway. The double agents controlled from London passed misinformation in support of it. Then came a series of events that changed the whole picture. The exceptional double agent GARBO came to England in April 1942, not only opening extraordinary new opportunities for passing deceptive information, but demonstrating that the Germans could be imposed upon to an unexpectedly great extent. In May 1942, John Bevan replaced Stanley. In June came a new directive broadening the Controlling Officer's authority and responsibilities. Twenty Committee members obtained full access to ISOS that same month.

These developments brought things to a head in the summer of that year 1942. At the end of June, Marriott complained that in the Germans' eyes their agents in Britain must seem to be "men who seldom or never say anything untrue but equally, as must be apparent to the Germans, never say anything new." It would be better, he urged, to furnish so much inaccurate information to the Abwehr that their intelligence reports to the High Command "would themselves be misleading and wrong." Montagu concurred; the opportunity was being neglected, he said, "of filling the German intelligence files with a mass of inaccurate information." In July, in the same memorandum to the W Board in which he asserted that the entire German espionage system in Britain was under MI5 control, Robertson pointed out that the Chiefs of Staff had "in MI5 double agents a powerful means of exercising influence over the OKW German High Command" and urged that this be put to better use. No longer, he urged, should the services content themselves with reviewing and vetoing foodstuff proposed by MI5. Rather, there should be an interservice body that could devote its full time to generating foodstuff and otherwise managing the double agents' traffic.

The time had come when such views were broadly acceptable. Bevan now had responsibility for broad deception policy. The services were at last alert to the possibilities of strategic deception. The Americans had entered the war, and the invasion of North Africa was being planned, bringing with it the need for affirmative cover and deception. So the W Board accepted Robertson's proposal in broad outline. With Bevan firmly in charge of deception planning, the Twenty Committee became in effect the interservice body that Robertson had urged, its role now being, in Masterman's words, "not, as it had been in the past, to promote and pass through small plans of our own but to provide channels for deception according to the plans of the Controlling Officer." Unlike his predecessor, Bevan was initiated into ISOS* and made a member of the Twenty Committee; indeed, it was even suggested that he become its chairman, but he wisely declined.

By the end of September 1942, the machinery for strategic deception integrated with "special means" implementation was fully in place at the London level. No longer would finding suitable foodstuff for the double agents be a problem; indeed, MI5 would have its hands full putting through all the material it would be asked to feed to the Germans.

Benefiting from the British experience, the Americans did not have to go through the same learning process. True, the FBI initially thought of double agents only in counterintelligence terms, but it soon learned their use for deception. Nor did it have difficulty getting appropriate foodstuff

---

* Evidently not fully and all at once. According to Ewen Montagu, Bevan did not receive a regular ULTRA service until March 1943, and from September 1942 until that time had to rely on ULTRA "reference sheets" dealing with "reports received by the Germans of our intentions and appreciations on that point by them" which were passed to him by Montagu with "C" 's permission, plus occasional reports direct from "C."

from the services. There was no American Twenty Committee (Hoover was not committee-oriented, to say the least), but the FBI's direct liaison with Joint Security Control, the element of the Joint Chiefs of Staff organization responsible for deception, worked satisfactorily; in part because there were far fewer double agents to be run out of the United States than out of Britain.

Cooperation between MI5 and the FBI ran back to 1937, when MI5 passed to the Bureau information that enabled Hoover's men to arrest a German agent named Rumrich and his confederates. This produced a high-profile trial and commensurate favorable publicity for the FBI. Hoover was eager for more, but the State Department, concerned to preserve American neutrality, forbade it. In 1940, a Canadian named William Stephenson, representing MI6, came to the United States with a wide brief for advancing British interests and furthering Anglo-American collaboration; and he and Hoover between them persuaded Roosevelt to authorize close cooperation between Stephenson and the FBI. In January 1941, Stephenson registered officially with the State Department as a foreign agent under the title "British Security Coordination," referred to as "BSC," with responsibility for coordinating Anglo-American liaison in security matters.

So far, so good. Meanwhile, however, the Bureau had its first major experience running a double agent. In 1939, a naturalized American citizen of German origin named William Sebold had been approached to act as a spy while visiting in Germany. He reported this to the American consulate at Cologne and was told to accept. After training in the tradecraft of espionage, he returned to the United States. The FBI set up a W/T transmitter on Long Island, ostensibly operated by Sebold (code-named TRAMP), and opened communication with Germany. Through this means a substantial German network in the United States, headed by a veteran German agent named Fritz Duquesne, was uncovered over the following months. In June and July 1941, the Bureau struck, rounding up some thirty-three spies for trial, with great attendant publicity. This experience, it would appear, magnified in Hoover's mind the utility of double agents for "sting" operations producing dramatic and high-profile results. His appetite was whetted when leads supplied by British censorship put the Bureau that same summer on the trail of a spy known as JOE K.

Such was the state of affairs when in August 1941 the FBI received its first exposure to a double agent used for sophisticated penetration and deception. This was Dusko Popov, known as Tricycle.

Tricycle was a Yugoslav commercial lawyer of good family, educated in France and Germany and able to move freely at the highest levels of society, who had been approached by the Abwehr in 1940 to work as an agent in England. He had reported this to the British in Belgrade, who encouraged him to continue; the upshot was that he had come to England under MI6's auspices in December 1940. Operating under British control he quickly became one of the Germans' most trusted agents, steadily passing information, recruiting subagents, even traveling to Lisbon twice in early 1941 and meeting with the Abwehr directly. The Germans paid him well; so well, indeed, that he declined any compensation from the British, and lived in London the life of a wealthy and dissolute playboy. As a star German agent, he was selected by the Abwehr after a few months to go to the United States and open an extensive network there (presumably to try to make up for the disastrous losses in the Sebold case). Stopping off in Lisbon to confer about his new assignment, he arrived in New York in the middle of August. He brought with him from Lisbon an extensive questionnaire concealed in several microdots. The FBI developed and read these; among other things, they sought information about American atomic energy research and rather detailed information about installations at Pearl Harbor.*

---

* On which basis, after the war Tricycle himself, or his ghostwriter, suggested that the FBI had failed to prevent the disaster of December 7, 1941, by not passing his Pearl Harbor questionnaire to the military authorities, and Masterman said that from the nature and detail of the questionnaire it was "surely a fair deduction that the questionnaire indicated very clearly that in the event of the United States being at war, Pearl Harbor would be the first point to be attacked, and that plans for this attack had reached an advanced state by August 1941." This charge is unfair both to the Bureau—who in fact did advise the military authorities as to the questionnaire—and to the military authorities themselves. The American forces at Pearl Harbor were concerned over possible sabotage, and the questionnaire (which is printed in full as an appendix to Masterman's book) was at least as consistent with plans for sabotage as with plans for an air strike. Most important, there is absolutely nothing whatever in the questionnaire to suggest either that "Pearl Harbor would be the first point to be attacked" or that "plans for this attack had reached an advanced state by August 1941." Pearl Harbor was of obvious interest to any potential enemy of the United States from May 1940, when the Fleet was first ordered to be based there rather than on the West Coast. Tricycle's questionnaire suggests nothing more.

At this point relations between Hoover and the British went sour. Stephenson's operation had expected that supervision of TRICYCLE's contacts with the Abwehr would be left in its hands, while the FBI would furnish material for replies to his questionnaires. In this they were disappointed. TRICYCLE was on Hoover's turf now, and Hoover took control. The Bureau's only interest in him was as bait to catch more spies with. But TRICYCLE was supposed to start his own network and had brought no leads to existing agents; moreover, at this stage his reports were going by secret ink correspondence, which offered no hope of the kind of quick payoff that Sebold's W/T operation had delivered. Furthermore, TRICYCLE's style was not such as to appeal to Hoover. He had managed to live the high life even in austere wartime London; in peaceful New York he slid readily into an even more sybaritic existence. What had merely amused the worldlings of MI5* was not funny to the prim Hoover; it might "embarrass the Bureau" to entrust too much responsibility to a Continental playboy with a penthouse on the most fashionable stretch of Park Avenue, a country place in Locust Valley, a Chinese houseboy, a luxurious automobile, skiing trips, vacations to Florida with a woman in process of getting a divorce (in violation of the Mann Act, which was taken seriously in those days), and a torrid affair with Simone Simon, the French film star.

For his first month TRICYCLE got no American foodstuff. Thereafter he got enough to fill eight letters back to the Abwehr, but the intelligence in them was pretty thin gruel. The military services, rather than the FBI, were to blame for this. Like their opposite numbers across the Atlantic, they could at this stage see no point in giving any information to the enemy.

Two months after TRICYCLE reached New York, in October 1941, the Germans directed him to go to Rio de Janeiro and make contact with the Abwehr there. He did so. In November the Abwehr representative in Brazil paid him a substantial sum in United States currency and directed him to go back to New York and set up radio communication with Lisbon and Rio, for sending intelligence particularly about convoys, antisubmarine warfare developments, and war production. (By this time, it

---

* By one account, the sobriquet TRICYCLE was supposedly inspired by a fondness for sharing his bed with two ladies at once; though a more prosaic view is that he was given it in view of his having two subagents, GELATINE and BALLOON.

will be recalled, Lend-Lease and Roosevelt's "shoot on sight" order were in full swing and the United States was waging an undeclared war in the Atlantic.)

The prospect of W/T communication with Germany perked up the FBI's interest somewhat. Montagu and Cowgill crossed to the States in early December, arriving just in time for Pearl Harbor, to reassure TRICYCLE and set up a mini–Twenty Committee to work on his foodstuff. The FBI put TRICYCLE notionally on the air in February of 1942, supposedly with a transmitter furnished by a pro-German Croat. But traffic was produced only grudgingly and under prodding from Stephenson, and TRICYCLE himself was allowed no input to it. (The Bureau was probably much more interested in its own major double agent, codenamed RUDLOFF, who had reached the United States in November and was opening radio communication with the Abwehr at the same time as TRICYCLE.) By March 1942, it was evident from ISOS that the flimsiness of TRICYCLE's reporting had rendered the Abwehr suspicious of his bona fides. By summer he had run out of money. He was in debt to the FBI to the tune of $3200, no mean sum in those days, and Hoover had no intention of supporting him further. In August 1942, the FBI told the British that they wanted to give him back and good riddance.

The British had independently concluded that he should be brought back, and had decided to take a chance that he would be able to clear himself with the Abwehr. In October 1942, after careful coaching, TRICYCLE departed for London via Lisbon. In Lisbon he put on a bravura performance with the Abwehr, convincing them that his poor reporting simply reflected the fact that they had not kept him supplied with money. Rehabilitated, he was sent back to London (he could not go back to the States, he said, because his cover assignment with the British Ministry of Information had come to an end) with a new supply of funds and instructions to report on military matters. There, as will be seen, he did useful work till he had to be closed down through no fault of his own in May of 1944.

Another source of friction between the FBI and Stephenson's operation that summer of 1942 was British liaison with the organization that in June became the OSS. Since in OSS were combined the roles of the British Special Operations Executive and his own MI6 and it had nothing to do with MI5, Stephenson saw no reason why he should not deal with it directly. Hoover resented this. Ever jealous of his prerogatives and alert to threats to them real or imagined—especially from the OSS of "Wild Bill" Donovan, who had been his boss in the Justice Department in his early days in charge of the Bureau—Hoover thought that having been Stephenson's original link the Bureau should continue to be his only one; and resented Stephenson's turning over to OSS some functions that rightly belonged to the FBI.

On the British side, MI5 was unhappy that it had been told almost nothing not only about TRICYCLE's American activities but about anything else the FBI was doing. In August 1942, it proposed that Stephenson represent MI5 in the United States and that an MI5 officer be sent across the Atlantic to be a liaison with both the FBI and the Royal Canadian Mounted Police. Hoover was agreeable to this, and in December 1942 sent an FBI man to the United States Embassy in London to assist in maintaining the liaison being established between the Bureau and MI5. After considerable debate in London, it was finally decided, rather to Hoover's disappointment, not to reciprocate by sending an MI5 officer to Washington; liaison would be effected through the attaché in London, with visits to the States from time to time by Cyril Mills, an MI5 officer who had been posted to Ottawa to work with the Mounties at about the same time as the Bureau sent their man to London.

British-FBI relations improved after the low point of mid-1942. The FBI benefited from ISOS information furnished by the British, and leads from the meticulous British censorship continued to help the Bureau find German agents. MI6 had been instrumental in turning the double agent known as PAT J over to the FBI in April of 1942, and MI5 made the double agent PEASANT available in 1944; both were to become significant deception channels from the United States to the Abwehr. In 1943, MOONBEAM (GLOCASE to the FBI), one of GARBO's imaginary subagents, notionally moved from England to Canada and was run by Mills of MI5 from there; with the FBI's cooperation, he acquired a notional subagent in the United States, an imaginary cousin who was a traveling salesman living in Buffalo. In March of 1945, plans were laid for the double agent OUTCAST, who furnished information from England to the Japanese military attaché in Stockholm, to open up a notional source in the United States in the person of an imaginary White Russian exile codenamed JAM.

All this seems to have appeared to the British as a one-way flow of information. Even forty-five years after the end of the war (reflecting, presumably, lasting grievances felt by MI5), the British official history could still complain that about PAT J it knew only that "it is believed that the US authorities made some use of [him] for deception purposes," but that "the authorities in London were by no means well informed about his traffic"; lamenting further that there was "no meeting of minds between the FBI and the British authorities on the subject of deception," and that "the FBI continued to regard double agents primarily as instruments for catching other spies." This last is unwarranted. PAT J became one of the main American deception channels, and especially after the Special Section of Joint Security Control under Newman Smith took charge of American strategic deception and established close relations with the FBI, the Bureau became quite deft at operating such channels.

Surprisingly, in view of the FBI's skill at self-promotion, unlike MI5 it has done little to publicize its role in strategic deception. Perhaps this is due to the fact that nobody had the whole picture of the Bureau's double agent operations other than Hoover himself, Tamm, Ladd, and Cunningham (and perhaps the Director's alter ego Clyde Tolson), coupled with the fact that even they had only a limited knowledge of the deception operations that their activities supported.

When the Special Section of Joint Security Control was fully organized under Colonel Newman Smith in the winter of 1943–44 it became the link between the Bureau and the services, both for general military foodstuff and for deception material. Each double agent operation was supervised by a special agent on the Major Case Desk; Special Agent Mark Felt, for example, handled the PEASANT case, and Special Agent Gil Levy handled the RUDLOFF case. They reported to Cunningham as chief of the Espionage Section; and the normal link between the Bureau and the military services was between Cunningham and Smith, though sometimes the desk officers dealt directly with Smith, just as sometimes officers below Smith at JSC, such as Carl Goldbranson, dealt with the Bureau. Joint Security Control was apprised of the major agents' backgrounds and as to what type of material made useful foodstuff for each, and their questionnaires were passed along by the Bureau; but it was not involved in the day-to-day running of the agents.

If the Bureau needed foodstuff of a particular kind it would ask for it.

For example, when PEASANT was opening up in September 1944, the Bureau gave Joint Security Control his background, noted that he had considerable potential, and asked for some piece of military information of seemingly great importance that he could pass to the Germans right away, because in view of the situation in Europe it would be logical for him to take the position that he would not continue working without being paid, while the Germans would not make any great effort to pay him unless he seemed valuable. On another occasion the Bureau asked for, and got, a list of the bars and such places frequented by sailors in the San Diego area, for use by PAT J. Messages touching on military matters—sometimes a group of them for a backlog—would be submitted to Joint Security Control for clearance. In turn, messages in aid of deception plans would be passed to the Bureau by Joint Security Control; normally with no explanation of the specific deception—nor did the Bureau pry into it.

All the evidence points to a relaxed and comfortable working relationship of mutual respect between the FBI and Joint Security Control, without the need for any formal committee arrangement.

Though double agents are the most romantic-seeming sort of "special means," it can be just as effective to feed misinformation to enemy intelligence through other channels that he trusts—journalists, neutral diplomats, social acquaintances, and the like, who may themselves not realize how they are being used. As has been seen, Maunsell and Clarke had a clutch of these in the Middle East, and British attachés and diplomats had them around the world. When the German war was winding down and new lines direct to the Japanese were needed, the Americans in particular developed a number of such channels, not only in the familiar European neutral capitals but in South America.

Before leaving the subject of special means, one annoying phenomenon should be discussed. This is the confidence man who peddles to the enemy information which he claims to have from his own agents or other sources but which in fact he may have made up out of whole cloth. His "information" may contradict and discredit what you are so carefully trying to do. Aside from GARBO, who started out as such a freelance, several such pests surfaced during the Second World War. The most prominent were a Czech named Paul Fidrmuc, known as OSTRO; an Abwehr officer, Karl Heinz Kraemer; and a Spanish journalist, Angel Alcázar de Velasco.

OSTRO had worked for the Abwehr for some time before settling in Portugal in 1940 as a purported businessman, sending reports back to Berlin through the Abwehr's Lisbon station. He claimed to have agents in many overseas corners of the British Empire and in the United States, as well as four or five in the United Kingdom, who reported to him in secret writing. Through 1942 and 1943 it became increasingly evident from ISOS that at least his United Kingdom reports were not only inaccurate but phony; but it was also evident that the Abwehr regarded him as one of its best sources—one of the Germans' four top agents, according to Himmler's man Walter Schellenberg after the war.

Nervousness over OSTRO increased in 1944 as D-Day in Normandy approached, peaking when in early June he reported that the favored invasion route was a landing on the Cherbourg peninsula. Though alarm over this report subsided when ULTRA showed it had not led the Germans to change their appreciations, OSTRO continued to be a loose cannon. Sometimes he was helpful, as when his reports of the effect on London of the V-1 flying bombs in the summer of 1944, concocted out of whole cloth, helped to prevent the Germans from correcting their aim; sometimes not, as when his report in February 1945 that Canadian troops were being transferred from Italy to the Western Front accidentally hit upon the truth and directly undercut Operation PENKNIFE, designed to conceal this fact. After several efforts to discredit him in German eyes, this last event led to a decision to try to bring him over to the Allies or eliminate him directly, but the war was over before he could be approached.

The second of these loose cannons, Kraemer, was an Abwehr officer attached under diplomatic cover to the German embassy in Stockholm. In the autumn of 1943, Allen Dulles's OSS operation in Switzerland obtained copies of messages from Kraemer to the German Foreign Ministry reporting information about maritime and aviation matters allegedly obtained by agents in England codenamed JOSEPHINE and HEKTOR; and JOSEPHINE and HEKTOR information from Kraemer then began turning up regularly in ISOS decrypts. The JOSEPHINE and HEKTOR reports caused some initial consternation, for they were circumstantial, in some respects they were remarkably accurate, and they appeared to rely on a highly placed source inside the British Air Ministry; but after close investigation MI5 concluded that Kraemer's network was imaginary and that his reports reflected ingenious conjecture, conversations with Swedish officers and the Japanese attachés, some input from Hungarian diplomatic sources, and close study of the press—plus recycling of information in actual Abwehr agent messages, for on occasion Kraemer's reports appeared to confirm misinformation that had been passed through controlled agents.

Like OSTRO, Kraemer created anxiety in the months before the Normandy landings, in particular reporting in February 1944, when the deception line was that the invasion could not come till late summer, that the invasion had been scheduled for June. Like OSTRO, Kraemer accidentally served the Allied cause at a crucial time, when a JOSEPHINE message fortuitously confirmed a GARBO message, resulting in countermanding the move of 1st SS Panzer Division to the Normandy front three days after D-Day. And again like OSTRO, he continued to come up with lucky guesses that might have been disastrous, as when he forecast MARKET-GARDEN, the Arnhem airborne offensive, immediately before it happened.

Kraemer exercised his creativity not only with his superiors in the Abwehr but with the Japanese, who paid him considerable sums for his information. They got from him both his often surprisingly accurate data on the British, and wildly inaccurate American information. Presumably Kraemer was less familiar with American affairs than with British, and his creativity less sure.

The third of these loose cannons, the Spaniard Alcázar de Velasco, unwittingly furnished an important deception channel and at least one actual double agent. Alcázar, a friend of Serrano Suñer, Franco's fiercely Germanophile foreign minister, was stationed in England as Press Attaché at the Spanish embassy early in the war, and then returned to Spain. At this stage of his career he was a genuine German agent with several subagents. The first of these was a journalist named Del Pozo (codenamed POGO by the British), who in the period of the Blitz was an official visitor to Britain, under the auspices of the Spanish Institute of Political Studies and sponsored by the British Council and the British embassy in Madrid. Del Pozo not only sent reports written in secret ink on the back of his regular reports carried back to Spain through the British Ministry of Information's own bag, but he brought money and instructions to SNOW's subagent G.W. He functioned from September 1940 to early 1941. MI5 procured his recall by sending messages through SNOW to the effect that he had attracted too much attention because of dissolute conduct and pro-German statements.

Del Pozo was followed in May 1941 by another journalist and press attaché named Luís Calvo, who dealt with G.W. till his own return to Spain in January 1942; through Calvo, MI5 was able to send important false information back to the Germans through the Spanish diplomatic bag. Alcázar himself revisited England in the late summer of 1941 and met several times with G.W. An assistant press attaché (codename PEP-PERMINT) and an Englishman employed in the press department (code-name SWEET WILLIAM) were also recruited by Alcázar; they too were used by MI5 to send controlled data via the diplomatic bag. The Spanish vice-consul and the military attaché were also involved in espionage in the German interest.

When Abwehr Enigma messages began to be read in December of 1941, they confirmed the activity of Alcázar, Calvo, and others, and suggested that Alcázar controlled a network of nearly two dozen agents in Britain. This brought matters to a head. In February 1942 Calvo was arrested and spilled all. Protests to the Spanish ambassador put an apparent end to the use of his embassy for German espionage—to the distress of the Twenty Committee, for this shut down G.W. and what Masterman called "by far the best channel we ever had for the transmission of documents." After extensive interrogation of Calvo it was concluded that there was nothing to Alcázar's claim to be running an extensive network in Britain.

This setback did not put an end to Alcázar's intrigues or to his inflation of his claims to be a major spymaster, for Pearl Harbor opened up a new market. From June 1942 on, he was selling information to the Japanese ambassador in Madrid, claiming to have a network of ten agents in the United States. The ambassador forwarded them to Tokyo under the description "Toh Reports." Alcázar actually had only three agents in the United States, and he never received information even from them; his reports to the Japanese were pure invention on his part. Moreover, they were all under surveillance through an extensive FBI operation (codenamed SPANIP, for "Spanish/Nipponese") directed against putative collaborative espionage between Spanish and Japanese interests, directed initially by Special Agent P. E. Foxworth, and subsequently by E. J. Connelly, the Assistant Director in New York.

In the spring of 1944, with Allied victory clearly just a matter of time, Franco's government reluctantly accepted Allied demands to expel German agents from Spain; though it dragged its feet in carrying out this undertaking, Alcázar evidently found it prudent to decamp. His Japanese friends hid him out for a time, then arranged in July for the Germans to smuggle him to the Reich through France. His largely imaginary network apparently continued to function.

At least one of Alcázar's real agents, José María Aladren by name, was recruited by the FBI as a double agent in October of 1942. This did not work particularly well, so in March of 1944 he was sent back to Spain to improve his relations with Alcázar or establish his own contacts directly with the Axis intelligence agencies. With Alcázar's departure he took the latter course, and arranged to work for the Germans and the Japanese jointly. In February 1945, he returned to Washington as a correspondent for a Spanish news agency, equipped to send reports in secret ink, by radio, by coded messages in his newspaper stories, and possibly through a crewman on a Spanish ship acting as a courier. His codename as a double agent was ASPIRIN.

There were other loose cannons, smaller fry, who did not pose the risks that OSTRO, Kraemer, and Alcázar did. In Lisbon, in addition to OSTRO, both the Abwehr station itself and an officer attached to it generated fictitious intelligence for Berlin, claiming as sources Portuguese consulates, penetration of the British embassy in Lisbon, and fictitious Swiss agents in Britain. Another such operative in Portugal was a lawyer, Marqués Pacheco, codenamed LUNA, who supplied intelligence to the Japanese; he claimed to get information from agents in Tangier and Gibraltar, from contacts with RAF officers and Pan American and British Overseas Airways people in Lisbon, and from high Portuguese officials whom he had bribed. But the veracity of his material—the Japanese tagged them as "M Intelligence"—was not good, and some reports contained information passed under Allied control.

Another uncontrolled and prolific source, providing information (principally American material) to the Japanese military attaché in Spain, was an agent called "Fu" by the Japanese. This was one Jozsef Fülop, identified variously as a clerk in the Hungarian legation in Lisbon and as Press Attaché in the Hungarian legation in Madrid. Fülop contributed to a network known to the Germans as FULEP, probably run by the Hungarian military attaché in Madrid, which supplied information to Kraemer as well; his own input was American data which he concocted from close reading of the Luce press, aviation magazines, *The Economist,* and newspaper stories.

London was where the uses of Most Secret Sources and "special means" were most thoroughly worked out; and to London we now turn.

CHAPTER 5

London Control

Colonel the Right Honorable Oliver Stanley was forty-five years old when on October 11, 1941, he took over the job as controlling authority for deception that Dudley Clarke had declined. A scion of one of the great aristocratic families of England (his father, the Earl of Derby, was a grandee of whom it was said that he "possessed what Englishmen admire: geniality, generosity, public spirit, great wealth, and successful race-horses"; a previous Earl had been three times Prime Minister under Queen Victoria; the Stanleys had played a role in English history since the Middle Ages), he had always shouldered the obligations that went with the privilege to which he was born. After serving with distinction as a junior officer on the Western Front in 1914–18 he had gone into politics. In the 1920s, along with Harold Macmillan and Robert Boothby, he was one of the "YMCA Conservatives," a group of young men who formed the (comparatively) left wing of the Tories, unafraid of Keynesian economics and a measure of central planning. He served at subcabinet level with the Board of Education and the Home Office—every Briton who has to struggle with passing a road test to get a driving license has him to thank, for he was responsible for the Road Traffic Act (1934)—and less successfully as Minister of Labor.

Regarded as one of the best and most humorous debaters in the House of Commons, Stanley was widely liked and trusted. In Chamberlain's administration he was President of the Board of Trade. In that capacity he was responsible for completing a valuable trade agreement with the United States, and for economic planning for the war. In 1940 he became Secretary of State for War. There he presided over the debacle in Norway, though it could hardly be deemed his fault; and when Churchill came to power in May he removed Stanley from that position. Stanley chose to rejoin the Army rather than accept some other post. He was commissioned a colonel, and made head of the Future Operational Planning Section ("FOPS") of the Joint Planning Staff; the main business of which was to plan for the distant future when British arms would return to the Continent. It was in that capacity that the responsibility of Controlling Officer in charge of deception was laid upon him.

A word here about British organization for the direction of the war at the top.

Churchill was Prime Minister and also Minister of Defense. As Prime Minister he presided over the War Cabinet, a small tight body which was the highest executive authority in the government, and as Minister of Defense he presided over the Defense Committee of the War Cabinet, composed of the Foreign Secretary, the Minister of Production, the three civilian service ministers (the Secretaries of State for War and Air, and the First Lord of the Admiralty), and the three military chiefs (the Chief of the Imperial General Staff for the Army, the First Sea Lord for the Royal Navy, and the Chief of the Air Staff for the Royal Air Force).

In turn, these latter three constituted the Chiefs of Staff Committee, a corporate body that issued unified strategic instructions for military operations. Under it was its own interservice staff, the Joint Planning Staff, as well as such joint entities as the Joint Intelligence Subcommittee and the Inter-Service Security Board or "ISSB," which we have already met. The Joint Planning Staff in turn had three subdivisions: a Strategic Planning Section, an Executive Planning Section (mainly concerned with getting prompt action on planned operations), and Stanley's own FOPS. The work of the Chiefs of Staff Committee and its staff was administratively coordinated by a War Cabinet Secretariat; Churchill's own Chief Staff Officer in his role as Minister of Defense, General Hastings Ismay, universally known as "Pug," acted also as military deputy secretary of the War Cabinet, and in practice as a fourth member of the Chiefs of Staff Committee. Brigadier (subsequently General) Leslie Hollis of the Royal Marines—brisk, stocky, practical, humorous, a lively raconteur, known to all as "Joe"—was secretary of the Chiefs of Staff Committee and Ismay's deputy and alter ego.

Two other organizations reporting to the War Cabinet were to be important to deception work. One was the Political Warfare Executive, responsible for propaganda both overt and "black" (*i.e.,* purportedly coming from sources other than the British themselves). The other was the Special Operations Executive, known always as SOE, responsible for covert action, particularly in support of Resistance movements in Axis-occupied territory.

Stanley prepared a memorandum for his new job that showed a grasp of fundamental principles. Distinguishing between cover and deception (he called them "defensive" and "offensive"), he suggested that minor cover, say for the sailing of a convoy, could continue to be handled by the ISSB; while major cover plans should be prepared along with the real plans and the new organization should have responsibility for preparing it, working out its details, and arranging for its execution. Deception or "offensive cover" plans presented greater difficulties and "this side of the work will have to develop slowly, as experience is gained." In particular, foreshadowing the problems Peter Fleming and others would face, he noted the danger that "the fake operation will suddenly become the actual objective or the fake future strategy turn out to be the real one." His Future Operational Planning Staff would offer good cover for the new organization and provide essential close liaison with the Planners, but the officers engaged in deception should be engaged full-time "and have no part in the ordinary work of the FOPS."

"The chief duty of the Controlling Officer," he said, "will be to maintain close contact with the planning side and to be absolutely conversant with the general run of strategic thought and operational possibilities. It will not be enough for him just to read papers, as they are issued or conclusions, as they are recorded." He should be allowed to listen in when the Chiefs discussed the future with the Planners; attend regular meetings of the Joint Intelligence Committee; attend such Joint Planners meetings as were desirable; periodically attend meetings of the Psychological Warfare Executive; maintain contact with the intelligence services in accordance with instructions from their own directors; and be in touch with SOE through personal contact when required. At the outset, three staff officers, one from each service, should be adequate. They should have "imagination, initiative and a sound administrative background," at GSO1 grade.

The Chiefs gave "general approval" to Stanley's proposals on November 3, and agreed that he should carry out the duties of Controlling Officer.

The first order of business was to find three GSO1s for the new duties. The Navy assigned a certain Commander Halahan, who had had some limited early connection with the Twenty Committee, but he had recently left the ISSB to serve as an instructor at the Turkish Staff College, and the Navy did not want to call him back from that duty till deception should be well under way. (As it turned out, he never did show up; perhaps foreshadowing the troubles Stanley was to have with the Navy.) The Army assigned Lieutenant-Colonel A. F. R. Lumby, known to his friends as "Fritz." Lumby was an elderly and kindly officer, tall, thin, dark, with a quiet voice, a shy manner, a bawdy sense of humor, and a casual attitude to smartness of military dress, who had lost a leg at Loos in 1915. In the interwar years he had been in intelligence in India, and in the later years a member of the Indian Legislative Assembly. He had retired in 1939 but had soon been recalled to organize the Military Intelligence School, where he had been commandant for two and a half years. He came on board as the first officer of the deception section in December.

Stanley had recognized that a suitable RAF officer might not be available, and indeed the Air Ministry told him that it had a war to fight and could not spare a group captain for this kind of thing. But it had a suggestion. Since the opening of hostilities the popular author Dennis Wheatley had been anxious to get into the war. From no official position at all he had written a number of highly imaginative papers about all sorts of aspects of strategy—on resistance to invasion, a project to seize Sardinia, the position of Turkey, Total War, and so on—which had circulated among his friends in high places, had even come to the attention of the King, and had been examined in all seriousness by the Joint Planning Staff on occasion. Sir Louis Grieg, the personal assistant to the Secretary of State for Air, was one of Wheatley's admirers; moreover, Wheatley had come to know well a number of officers connected with the Joint Planning Staff, including Group Captain William Dickson, Director of Plans (Air). One of them in effect said "What about Wheatley?"

Dickson asked Wheatley to lunch to meet Stanley. Wheatley could tell that he was being checked out for an official appointment—propaganda, he assumed. A few days afterwards came a telephone call from Stanley asking Wheatley to his office; where Stanley offered Wheatley the opportunity to work under him on the Joint Planning Staff. Wheatley accepted with delight.

*　　*　　*

Dennis Wheatley was forty-four and one of England's most prolific and best-selling authors. A heavy-faced, shortish man with a full head of slicked-back black hair parted in the middle, a bright red nose that fitted with his justified reputation as a *bon viveur,* and a twinkle of humor in his eyes even when he was trying to look properly grim and military, he was a flamboyant character who enjoyed his success and lived life to the fullest, a generous host and man about town who knew almost everybody worth knowing.

The son of a wine merchant, Wheatley had in his boyhood been a naval cadet after an unhappy school year at Dulwich, but had decided to follow his father's trade rather than the life of a Navy officer. He had spent a year in Germany on the eve of the First World War, and in that war had obtained a commission in the artillery. He served on the Western Front for a year, then was gassed and invalided home. After the war he participated, not very actively, in his father's business, and more actively in living the good life.

The elder Wheatley died in 1927. Dennis inherited the wine business and set out to expand it, but it failed in the economic collapse of 1930. A notable raconteur, he turned that skill to use and began to write. He was an instant success. He cranked out adventure stories, thrillers, and tales of black magic at a great rate (a lifetime output of seventy-five books, or one every seven months in forty-five years as an author), working thirteen hours a day when the fit was on him, leaving it to his secretaries to clean up his hopeless grammar and worse spelling. In 1936, he branched out, producing with a colleague a series of "crime dossiers," facsimiles of actual files dealing with crimes in various settings, from which the reader was invited to deduce the solution; these were best-sellers too.

When war broke out in 1939 Wheatley was too old to be recalled to uniform, but he offered his services to the newly-created Ministry of Information. To his astonishment—and eventual gratitude—he did not even receive the courtesy of a reply. His friends consoled him with the thought that he would be doing his bit by providing entertainment for lonely, bored, and worried people to take their minds off of the war. So he settled in at his comfortable house in St. John's Wood and turned out a series of up-to-the minute spy novels; which required him to keep in close touch with details of current affairs and the strategy of the war, to his later benefit.

Wheatley had four stepchildren by the first marriage of his wife, Joan, to a member of the Scottish brewing family of Younger. His stepdaughter Diana and his stepson Bill had joined MI5 in the first days of the war, while Joan had placed herself and her car at the government's disposal and was a driver for MI5. Late in May 1940, as the Allied front was collapsing in Belgium and the British Expeditionary Force was falling back to Dunkirk, an MI5 officer whom Joan was driving mentioned that he had been told to come up with ideas for resistance to a German invasion but had not been able to think of much. Joan suggested that her husband specialized in original ideas and would love to do something useful. Tell him to see what he can do, said the MI5 officer in effect. Wheatley worked all night and turned out a paper full of ideas bright and otherwise. The MI5 officer thought it was good but said it would move slowly in the bureaucracy. Wheatley suggested that he, Wheatley, send it to a few highly placed friends. They liked it. He was asked to write more think-pieces on various topics.

Thus began fourteen busy months for Wheatley, churning out his thrillers and writing his papers. His work was punctuated by personal tragedy, for in the Blitz in the winter of 1940 his beloved house in St. John's Wood was bombed out. He and Joan took a ground-floor flat in a block called Chatsworth Court, in Earls Court, which had a restaurant in the basement for entertaining; Bill and Diana took flats of their own in the same building. Between May 1940 and August 1941 Wheatley turned out some score of papers. (He also contributed another bit of service by giving a cover job as secretary to Friedle Gaertner, codenamed GELATINE, an Austrian double agent in TRICYCLE's ring.) His friends on the Joint Planning Staff often expressed the wish that they could get him involved in their work on a more official basis. Finally, the need for an RAF officer for Stanley's deception section afforded the opportunity.

Wheatley had, of course, to be commissioned into the RAF, or more precisely the RAFVR or Royal Air Force Volunteer Reserve. Regulations required that he start as a pilot officer, the lowest commissioned grade in the service, equivalent to a second lieutenant in the Army, and that he go through a three-week course for new officers. His highly-placed friends could not exempt him from these requirements. They could, however, rush through his commission and put him at the head of the line. They endorsed his application for a commission on a Wednesday, and on Saturday, clad in a uniform his tailor had worked on all night, he entered the Officers' Intake Course at RAF Depot, Uxbridge. (He thus became one of the few men, perhaps the only man, to have been an officer in all three of the King's fighting services.) On the last day of 1941, after three wasted but amusing weeks of drilling in the course of which he had lost his voice from bellowing orders to his squad, and feeling rather like a new boy on his first day at school, he reported to Stanley for duty.

For the first months Wheatley and Lumby were the only deceivers on board. Their offices were two rooms adjacent to those of the more conventional elements of Stanley's staff, looking out over St. James's Park, three flights up in a large building at the Parliament Square end of Whitehall, known officially as the New Public Offices, that housed a number of ministries; very large rooms, the sometime offices of a cabinet minister, with vast mahogany desks, huge leather armchairs, and "acres of thick pile carpet." Far below, in the basement of the building and under the adjacent street, were the underground Cabinet War Rooms, a rabbit-warren of corridors and small offices where the Cabinet Secretariat and the main body of the Joint Planning Staff lived and worked and to which the central direction of the government itself moved on air-raid nights, with spartan offices and bedroom cubicles for the Prime Minister and other dignitaries, conference rooms, map rooms, and the supposedly secure telephone that connected Churchill directly with Roosevelt.* ("Mr. Rance's room" was the cover name for the premises; Mr. George Rance was the retired warrant officer in charge of the Cabinet War Rooms for the Ministry of Works, who every day set all the clocks by a chronometer that he carried from room to room, and who

---

* Much of the Cabinet War Rooms is now open to the public. (The part in which the offices of the London Controlling Section were ultimately located, the so-called CWR Annexe, now houses the Churchill Museum.) Access is now through an entrance in Great George Street built for the purpose; during wartime, access was from the interior of the New Public Offices building itself. For some now-forgotten reason, during the war the Cabinet War Rooms were often referred to as "Storey's Gate," though that is the name only of a nearby street, from which there was never any access to the War Rooms.

gave all the newcomers a guided tour of the premises.) For the time being, Wheatley's main contact with the War Rooms was the little room used as a mess by members of the Joint Planning Staff when they were not dining at a restaurant or their club.

Wheatley's friends welcomed him warmly; so did Stanley and Lumby. He was given a pile of files to read and was dumbfounded to find that they were the most secret minutes of the War Cabinet, the Defense Committee, and the Chiefs of Staff. He learned that he was to attend the daily meetings of the ISSB. He was introduced to Oliver Stanley's pleasant custom of an informal gathering of the staff for tea every afternoon. He found that there was plenty of opportunity for leading the good life while serving one's country. Major Eddie Combe, for example, secretary of the ISSB and in civil life a wealthy stockbroker, kept a table for six every day at Rules, the smart restaurant in Maiden Lane, where you could lunch with everybody who was anybody in secret war work, after which you might have to have a bit of an afternoon nap.

It was all very pleasant and superficially exciting. The only problem was that there seemed to be nothing to do.

Stanley had no general directive from the Chiefs of Staff. Nobody in Whitehall yet really understood deception. So Lumby and Wheatley got no directions. They could not talk things over with their fellows, for they were forbidden to mention the nature of their responsibility even to other members of the Joint Planning Staff. Lumby had a vast knowledge of intelligence work and understood a good bit about the German mind (his mother had been German), but he was a gentle and kindly soul and not a self-starter. Wheatley, as the new boy and an absurdly junior officer, was not in a position to force action.

Technically, Wheatley's appointment was as secretary to the deception section. This empowered him, despite his lowly rank, to initiate any kind of activity that came within the section's purview. So to fill time he authorized himself to set about writing another of his think-pieces, this one on principles of deception, with a list of every means and method he could think of for passing deceptive information to the enemy. He was reinventing the wheel, of course; but evidently nobody had thought to ask Dudley Clarke to write up something of the sort before he went back to Cairo.

Unknown to Lumby and Wheatley, Stanley was working up a project for them as to which he too could have used Dudley Clarke's advice, for it violated both of the great commandments of deception: It did not propose to induce the Germans to do anything specific, and it was being proposed simply because the means appeared to be available. This was the first of many notional invasions of Norway, which eventually received the codename HARDBOILED.

In December Stanley had suggested to the Chiefs of Staff that a deception be mounted that would at least embarrass the Germans if it succeeded and would not be likely to compromise any future real plan. After some discussion, the Chiefs concluded that a notional attack on the coast of Norway would fit these specifications, since there was no chance at all that Norway would in fact be invaded in the foreseeable future. Stanley suggested a fake attack on Narvik, or in the alternative on Trondheim. The planners thought that the Germans would consider either of these totally implausible as being too far north. So a notional attack on Stavanger, target date around May 1, was fixed upon instead. Such an operation, codenamed DYNAMITE, had actually been studied by the staff.

(In reality, not only did the Germans not think an attack on Narvik was out of the question, but at that very time, with their eastern front bogged down in the snows of Russia and the Soviet winter counteroffensive opening up, Hitler and the OKW were extremely nervous about a British landing in Northern Norway, perhaps with Swedish support. "The fate of the war will be decided in Norway," opined the Führer on January 22.)

The decision filtered down to Lumby and Wheatley in mid-January and they set to work to draft up a plan. They came up at once against an unexpected snag: Their work was considered too secret to be shared with the young women of the typing pool even though they typed the real plans that were actually going to get people killed. So the former head of the Army intelligence school and the best-selling author found themselves pecking out their first deception plan with their own two fingers. Before long, even the authorities who controlled such things saw how absurd this was, and Lumby and Wheatley got their own secretary, a privilege beyond what even Stanley and the Director of Plans himself were allowed. She was, recorded Wheatley, "a pleasant and efficient girl named Joan Eden."

The inexperienced London deceivers of those early days had an exaggerated notion of the Germans' powers of observation. And they were wholly ignorant of the Twenty Committee and the double agent system. So they assumed that a deception plan must be run as a real operation with real forces right up to the last minute, relying on the Germans to get wind of the preparations from their own intelligence and from rumors and press leaks. The Stavanger deception was planned in this way. Genuine forces were to be trained for a Scandinavian operation and brought right up to the point where supplies were ready to be called forward to the ports for launching the operation; at which time it would be postponed, and postponed again, a week at a time. In addition, steps were to be taken for rumors to get about that might reach the ears of the Germans.

When Stanley had signed off on the plan and Miss Eden had finished typing its final form, Lumby explained to Wheatley how to get a codename assigned to it. Wheatley went to see their friends at the ISSB and was invited to pick one from their book of available codewords. He selected HARDBOILED.

"Who was the bloody fool who chose such a silly codeword?" asked Stanley next day. Pilot-Officer Wheatley gently lectured the Colonel and former Secretary for War on the importance of codewords' not bearing any relation to the actual operation.

The next step was a meeting with a cluster of generals and brigadiers, the ISSB, and the Executive Planning Section, to arrange for implementation. It was agreed that the Royal Marine Division was the appropriate lead unit for HARDBOILED, being amphibious by nature and stationed already in Scotland; to which were added an armored brigade group and a Commando. They were to undergo mountain training and be issued clothing and supplies appropriate to a cold climate; officers would be given special training about terrain in Norway; large numbers of maps of the Stavanger region would be printed; Norwegian interpreters would be rounded up; signposts in English and Norwegian would be prepared; and so on.

When Stanley was present at this meeting he was heard courteously. Afterwards Lumby and Wheatley held a more detailed meeting with lower-level officers who did not know that the exercise was not a real operation. That too went well enough. Less well went subsequent meetings, without Stanley's chaperonage, with more senior officers who were in the know. The idea of a deception plan, as opposed to a cover plan, was novel, and they could see no reason to take their men away from their normal training and order quantities of supplies that would never be used. Nevertheless, directives went out to the commanders concerned on January 31. But it was evident as time went on that many ideas for implementation were held up by foot-dragging in various departments; and Lumby and Wheatley were in no position to do anything about it.

February faded into March. Physical implementation of HARDBOILED was suspended because of unexpected conflicting air operations over Norway. Then in mid-March, with the Japanese Navy active in the Indian Ocean after the fall of Singapore, the real units that were supposed to conduct the physical implementation of HARDBOILED were withdrawn and earmarked for an operation to seize bases on Madagascar lest the Vichy government give the Japanese rights to operate from that island. Other units were designated for HARDBOILED, and they were supposed to take part in an exercise in April to lend it credibility, but there was not enough shipping to carry that out. So HARDBOILED never got any physical implementation at all.

Nonphysical measures did go forward. Inquiries were made about a supply of Norwegian currency. Requisition and billeting forms in Norwegian were printed. Inquiries were made for Norwegian interpreters. The general staff of the Norwegian government in exile were asked to update their studies for DYNAMITE. Hints were dropped in chitchat at the Press Club and elsewhere in Fleet Street, in Stock Exchange circles, and at service clubs and merchant marine pools. An important map was reported lost and a search for it was instituted in Scotland and elsewhere on the northeastern coast. Inquiries were set on foot in Portugal, Spain, and Switzerland with respect to contacts and connections in Norway. Most significantly for the future—though Stanley, Lumby, and Wheatley did not know about it—to a limited extent the double agents were enlisted to pass information in support of HARDBOILED to the Germans.

Thus did HARDBOILED peter out; not at all an auspicious beginning for the cause of strategic deception. It did have some detectable results at home. There was a general impression that something was brewing, and the British press quoted articles from the German press about readiness to meet a British invasion of Norway. And there was certainly German concern about Norway, then and later. Reports were current in Stockholm from the end of January that there might be an Allied operation in Norway and that the Germans were closing harbors and strengthening defenses, and in February that the Germans would occupy Sweden to forestall an Allied move on Norway; there were accounts of the Allied threat to Norway in the German press soon after, and in late April and early May the Germans moved some fifty thousand additional men to that country and kept them on alert till mid-May. But whether HARD-BOILED played any part in all this is a moot question.

Having made their plan, Lumby and Wheatley took no part in implementing it, apart from one occasion when Wheatley, dressed in a borrowed group captain's tunic, quizzed two recently arrived Norwegian refugees about possible landing grounds for airplanes in the Stavanger area, in the hopes that this would filter back through the Norwegian exile community. Otherwise they found themselves twiddling their thumbs again. Lumby spent an hour each morning on the crossword in *The Times*. They took long, long lunch hours. Wheatley treated himself, contrary to RAF regulations, to a scarlet lining for his uniform greatcoat and to a couple of swagger sticks covered in blue (with stilettos inside, in case of trouble in the blackout). He started writing papers again to keep busy: one suggesting a Soviet deception threatening an amphibious assault on the mouths of the Danube, another on the situation Britain might face when the war ended.

In April and May the ISSB prepared, and supervised the execution of, the cover plan for Operation IRONCLAD, the invasion of Madagascar. The expedition was to sail from Scotland, pick up supplies in West Africa, proceed around the Cape of Good Hope, re-form at Durban, and proceed to land at Diego Suarez on the north coast of Madagascar. The goal of the cover plan—it received the codename GOLDLEAF-HERITAGE—was to prevent word of the expedition from reaching the French. The original "story" put out was that the expedition was destined eventually for Burma and would stop off at Trincomalee in Ceylon. Later, in mid-April, it was decided that the disposition of the Royal Navy's forces in the Indian Ocean made Trincomalee an implausible objective. A new story was adopted, under which the expedition was destined for the Middle East and an attack on the Dodecanese.

Durban was the point of greatest risk, for by the time the expedition put in at that port, sealed orders would have been opened and the force briefed on the operation, and Axis intelligence was active in nearby Portuguese East Africa. South African authorities were brought into the secret, and Dudley Clarke's help was enlisted. "A" Force sent a supply of maps of the Dodecanese to the local intelligence authorities at Durban along with the necessary instructions, and set to work in Egypt to spread the word that a convoy from Britain was on the way. Preparations were made to receive the ships at Suez and to encamp the troops near Ismailia for a period of rest and final training.

On May 5, the Turkish consul at Rhodes told the British that the Italian commander on Leros had just made a speech warning of a possible British attack based on Egypt, and that there was great activity on both Leros and Rhodes to shore up defenses and mine the bridges and beaches. On that same morning the landing on Madagascar had achieved total surprise. Whether or not the Axis and the Vichyites had connected the threat to the Dodecanese with the convoy at Durban, security both in Britain and at that port had obviously been satisfactory; though the general opinion in Durban was that the expedition must be destined for Madagascar.

Postmortems showed up the weaknesses of what had to be recognized as a "hazy" plan. Even if the Germans and Italians did know of the expedition and did connect it either with the Middle East or Burma, was there any reason they should share that information with the French? Nevertheless, Clarke and London felt justified in writing GOLDLEAF-HERITAGE down as a success. Surprise had been achieved; the tight state of security in the United Kingdom had been demonstrated; it had been shown once again that the enemy in the Mediterranean would swallow stories that were passed to him. Moreover, later in the year GOLDLEAF-HERITAGE would prove to have afforded valuable practice for covering the invasion of French North Africa.

In London, GOLDLEAF-HERITAGE had been an ISSB affair, for cover plans were still the province of that body, with Stanley's section limited to "putting up suggestions to the Chiefs of Staff." But Wheatley's friends on the ISSB had been kind enough to call him and Lumby in for its planning phase. They sketched out the original plan, and Wheatley, as an officer with the lowest possible rank, had the satisfaction of giving the general commanding the expedition his orders with respect to the cover plan. But once again he and Lumby had no role in putting the plan over. A major from ISSB flew out to Cairo—he had been supposed to go to Durban, but bad weather held him up—and worked with "A" Force on local implementation.

Lumby and Wheatley went back to their crossword puzzles and long lunches and rather aimless think-pieces. Wheatley looked at a project for Anglo-American bases on the Continent after the war. He wrote a paper on a psychological warfare project, and one on the situation if Germany should fold by the end of 1942.

March faded away, and April. The rest of Stanley's staff were moved down to the underground facility, leaving Wheatley and Lumby more isolated than ever, without even the daily tea-break to make them feel part of the team. In early May, Wheatley did a paper on deception in the Pacific; pointless, for that was an American monopoly.

One afternoon in that same first week in May, Lumby came joyfully into the office, waving his cap and cane and positively dancing on his one leg, with the glorious news that he had secured an appointment to SOE in West Africa and was getting out of "this deadly deception racket." Though he was to be available for a short time to aid in any transition, he was gone from the office almost at once.*

At about the same time, Wheatley lost such guidance as Stanley had provided. Stanley's wife was terminally ill; for some time he had been absent from the office more often than not; in the latter part of May he went on leave to be with her in her last days. At the same time he asked Churchill to release him from the Army so he could return to politics.

Wheatley was now left entirely alone. He whiled away his time writing more papers. One suggested trying to deceive the Japanese into thinking that the British would attack the Kra Isthmus. One criticized a suggestion by Churchill that a deception scheme be mounted to persuade the Japanese that Madagascar was being turned into a great base to

---

* After a year with SOE in Africa, Lumby was ordered back to London to become a member of the ISSB. At the Lagos airport he exercised a colonel's priority to "bump" from the flight the journalist Malcolm Muggeridge, a wartime officer of MI6, who was returning home from a posting at Lourenço Marques in Mozambique; thereby inadvertently saving Muggeridge's life at the cost of his own, for the plane crashed at Shannon and all on board were killed.

replace Singapore. One was a report on HARDBOILED. One was a comment on the staff's reaction to his psychological warfare proposal. But all this was relative makework. Nothing seemed to be left of the hopes that had followed Dudley Clarke's visit to London seven months before. Nothing seemed to be left of the idea of strategic deception controlled on a world-wide scale for a world-wide war. Just another bright idea that had not worked out and was hardly even remembered by busy people. The situation was made worse, no doubt, by the confusion that had attended ADVOCATE in January.

Above Wheatley's level, Stanley himself had lost heart. His asking Churchill to let him return to politics reflected not merely his personal tragedy but frustration with the position of the Controlling Officer. He had never been given what he had been promised. No naval officer was assigned to him, the only Air Force officer was Wheatley, "a civilian temporarily put into Air Force uniform," as he put it, while on the Army side Lumby was "a soldier with good training and sound background, but completely devoid of the sort of imagination required for this job." The machinery for cooperation which he had envisaged had largely broken down. He was never invited to a Planners meeting except occasionally when a FOPS paper was under consideration. Perhaps most frustrating, he got no cooperation at all from the Navy; specifically, from Rear-Admiral John H. Godfrey, the Director of Naval Intelligence, and his officer for ULTRA and Twenty Committee matters, Lieutenant-Commander Ewen Montagu, whom we have already briefly met.

Godfrey was an able, highly-strung, and sometimes truculent officer, who on his appointment to head naval intelligence in January 1939 found himself in charge of a badly run-down organization that had fallen far off from its high standards of 1914–18. With his personal assistant, the sometime stockbroker and journalist Ian Fleming, younger brother of Peter Fleming (and future creator of James Bond), he had built it up by recruiting a heterogeneous collection of able people from civil life. Godfrey himself was an enthusiast for deception and for imaginative activity generally. As early as November 1939, he had circulated papers on "Deception, Ruses de Guerre, Passing on False Information, Etc.," which contained dozens of ideas for dirty tricks, decoys, and deceptions;

and he was sufficiently directly interested in double agent work to have visited the United States in 1941 in connection with the TRICYCLE affair, accompanied by Fleming. (Godfrey and Fleming visited Washington again in September 1942, during the run-up to the North African invasion.)

Godfrey's subordinate Lieutenant-Commander the Honorable Ewen Montagu was forty-one years old in 1942. A son of the prominent peer Lord Swaythling, in civil life a barrister, Harvard and Cambridge educated, he had entered on active duty in the Navy at the beginning of the war and had worked under Godfrey since November 1940. Godfrey had put him in charge of non-operational ULTRA intelligence, and this had led naturally to the Twenty Committee, where he was the naval representative from its beginning in January 1941 to the very end of the war. If the papers he saw fit to leave behind in the naval archives are any guide, he was also self-satisfied, self-important, and treacherous.

Montagu was particularly nasty about Johnny Bevan behind his back, as will be seen; very likely he was similarly nasty about Stanley. Certainly, as has been seen, Montagu and Godfrey were the prime movers in limiting Stanley's access to information. In Stanley's last memorandum on deception organization before leaving the office, he said: "The chief difficulty is that in order to satisfy the suspicions of [the Director of Naval Intelligence], who regarded me as a most untrustworthy person, I have deliberately refrained from making enquiries as to sources and channels which are being used and from insisting on joining committees on which really I should have been." (Meaning, of course, ULTRA, the double agents, the W Board, and the Twenty Committee.) "Perhaps this objection is merely personal," he added rather sadly, "and will not be continued with my successor."

And then, once more, the instigator of it all, General Wavell, brought deception back to the fore with a bold message direct to Churchill himself.

As a Commander-in-Chief in a major theater, Wavell could communicate directly with the Prime Minister and Minister of Defense. On May 21 he dispatched to Churchill personally a message that laid the foundation for centralized strategic deception. Translated from its original telegraphic style, it said:

I have always had considerable belief in deceiving and disturbing the enemy by false information. The Commander-in-Chief Mideast instituted a special branch of the staff under a selected officer charged with deception of the enemy, and it has had considerable success. I have a similar branch in India and am already involved in several deception plans. These, however, can have local and ephemeral effect only, unless they are part of a general deception plan on a wide scale. This can only be provided from the place where the main strategical policy is decided and the principal intelligence center is located. A coherent and long-term policy of deception must be centered there. I fully appreciate the value of the work done by the ISSB and the Controlling Officer, but I have the impression that the approach is defensive rather than aggressive and confined mainly to cover plans for particular operations. May I suggest for your personal consideration that a policy of bold imaginative deception worked between London, Washington, and the commanders in the field by only officers with special qualifications might show a good dividend, especially in the case of the Japanese?

"A policy of bold imaginative deception." This was exactly the sort of thing to appeal to Churchill's buccaneering side. He circulated Wavell's message to all members of the Defense Committee of the War Cabinet. The Chiefs of Staff took the hint, and referred it to the Joint Planning Staff.

Meanwhile, another event had taken place, equally momentous for the future of Allied strategic deception. On that same May 21 when Wavell sent his message to Churchill, the Chiefs approved as successor to Stanley Lieutenant-Colonel John H. Bevan, who was already designated for Lumby's old job. He joined Wheatley in their cavernous and lonely offices on June 1. He was to become, along with Dudley Clarke, Peter Fleming, and Newman Smith, one of the four preeminent figures of Allied deception in the Second World War.

"Johnny" Bevan, as he was always known, was in civil life a stockbroker and a significant force in the City of London, head of the firm of David A. Bevan & Co. Forty-eight years old in 1942, he was, in Wheatley's words, "a rather frail-looking man of medium build with sleepy, pale-blue eyes and thin fair hair which turned gray from the strain of the remarkable work he accomplished in the three years following his appointment," who walked with a slight slouch, and was much tougher than he looked. "His most notable feature was a very fine forehead, both broad and deep; and one of his greatest assets an extraordinarily attractive smile."

Bevan was a countryman at heart, who liked to get out of the city and get mud on his boots. His hobbies were those of an outdoorsman: birdwatching, golf, gardening at his country place, shooting and fishing on his holidays in the Outer Hebrides (he tied his own flies), making his own detailed topographical maps, carving and painting models of his favorite seabirds in flight; when his children were small he got them enthusiastic over collecting butterflies. Though essentially unmilitary, he was devoted to the Army; "he disliked the other two services and definitely distrusted all civilians," according to Wheatley. He had, or made, friends in high places in the Army, including "Pug" Ismay, "C" of MI6, and his fellow birder Sir Alan Brooke, the Chief of the Imperial General Staff, with whom he dined two or three times a month. He dealt not infrequently with Churchill himself—the American Colonel Bill Baumer re-

(Julian Bevan)

John Bevan

membered going with Bevan to see Churchill one night—for the great man took a personal interest in deception; deception plans were read by him personally and often came back with comments and suggestions scribbled in his own hand; and once, in October 1943, when both Peter Fleming and Dudley Clarke were in London, he made a point of receiving all three together, Bevan, Clarke, and Fleming, in his private quarters.

Bevan was quintessentially Establishment. (He was "so English," recalled Bill Baumer. "In Russia he wouldn't have anything to do with the Russian language.") His firm was a family firm; his father had served as chairman of the Stock Exchange. His wife was Lady Barbara Bingham, a daughter of the Earl of Lucan. (Her sister was married to the future Field-Marshal Alexander.) His school was Eton, where he had had a successful career after a slow start. He had been one of the smallest boys at the school till age fifteen or sixteen, when he suddenly got his height and was at last allowed to wear the traditional tailcoat instead of the short Eton jacket of the younger boys. Then he flourished: elected to Pop, the very exclusive older boys' club; "Keeper of Fives," which is to say captain of the team that played Eton Fives, a type of handball; and a notable cricketer—one of his schoolmates told Dennis Wheatley how "when things were looking pretty bad for his side at cricket, he would shuffle in, about sixth wicket down, knock up 100 and shuffle out again looking rather ashamed of himself."

From Eton he went up to Christ Church, Oxford, in 1913. The outbreak of the First World War cut short his university career. He was commissioned in the Hertfordshire territorial regiment in August 1914, and was sent out to France in November in time for the First Battle of Ypres. Except for two months' sick leave at home, he served throughout the war as an infantry officer on the Western Front, where he earned the Military Cross for gallantry. He was engaged in tactical deception of some kind for some unknown period; "I had great fun with this in the First World War," he wrote, "and to a minor extent in the Second."

In January 1918, Major Bevan, as he had become, was summoned to Versailles as a liaison officer to the Supreme War Council. There he received a very special assignment. He was given all the clerical help he needed and all the intelligence available, and directed to produce an assessment of likely German intentions now that Russia was out of the war. Early in March he delivered to all the top brass a briefing that proved a few weeks later to have called Ludendorff's spring offensive remarkably closely. Winston Churchill, currently Minister of Munitions, was among those who witnessed Bevan's performance and were impressed by it. He sent for the young officer and spent some time with him; and no doubt remembered him a quarter of a century later.

Major Bevan was kept in the Army for a time after the Armistice in some special capacity. When finally demobilized he went into the family business. Wheatley was convinced that he did so only at his father's urging and that he would have liked to remain in the Army. But in that perhaps Wheatley was wrong. Baumer recalled that Bevan would say: "Bless my soul, here I am a businessman and I spend twelve years of my life in the Army since I came out of the class of '14." All but about three of his class were killed, he told Baumer. "We lost our leadership, we lost our brains. I was so discouraged that I went off after the war was over, and just went to Denmark and got lost for a year. I was mentally, physically beat." The firm did indeed send him to Denmark as part of his training. While there he mastered the fiendishly difficult Danish tongue and picked up a few trophies at the Copenhagen Golf Club. Returning to London, he became a partner in the family firm in 1925. He was successful at the business and in the interwar years laid the foundation of what at his retirement in 1974 the chairman of Lloyds Bank called "the very special position you held for so long in the City as a wise elder statesman."

On one of Bill Baumer's visits to the London Controlling Section, Bevan made a decision and started implementing it even before the Combined Chiefs of Staff had authorized the real operation. Baumer asked him about this and asked to whom Bevan was responsible. "To God and history," said Bevan. That sounds priggish, but he was not. He was simply a man of total integrity who was hard-working and conscientious. When he was included in the New Year's honors in 1945 for his work, one of his partners who was in on the secret that his business was deception wrote to him: "You will no doubt recall how in the piping days of D.A. Bevan & Co. I often stated that the trouble with the noble firm, and in particular with its most industrious partner, was that they were too honest. It is indeed a curious world."

As the 1939 war approached, Bevan was responsible for preparing the censorship machinery that would be introduced with the outbreak of

185

hostilities. When war came he went back into uniform as an MI5 officer. After a few months at Wormwood Scrubs he managed to get himself transferred to duty as an active intelligence officer and saw service in the Norwegian campaign.

In Norway he had his first known experience with deception, when he and Peter Fleming contrived a mild haversack ruse, typing up several seeming messages implying that a bridge which the Germans would have to use was mined and covered by a machine gun, and leaving them half-burned where the Germans would find them. After the evacuation of Norway he was posted to duty as the intelligence officer at the headquarters of Western Command.* That was an uninspiring assignment, in which his chief duty was to lecture on security to units of the Home Guard. From that dreary work he was called to take over Lumby's post in May of 1942.

Things like this seldom happened by accident in the British Establishment; they happened because somebody knew somebody. Alexander, who had taken over command in Burma under Wavell shortly before, must have known that his talented brother-in-law was languishing in the backwater of Western Command. Perhaps someone reminded Churchill about the young officer who had impressed him so favorably in 1918. It was then logical that he succeed Stanley—who, as Stanley told Ismay, knew Bevan well personally. He "has very special qualifications for the work," added Stanley, in light of his experience at the beginning of the war and in Norway. Something surely went on behind the scenes. Whatever it was, the result was fortunate for the Allied cause.

Perhaps something similar underlay Wavell's message as well. Peter Fleming knew Dennis Wheatley, and during the period when Fleming was waiting to ship out to India as deception officer he had surely learned something about the aimless drifting of the supposed deception section in London; perhaps Fleming had said something to Wavell that helped inspire the message. (That was a time too when, as will be seen in Chapter 8, an early version of Wavell and Fleming's PURPLE WHALES deception was being rejected in London as too grandiose. Perhaps Wavell's

---

* Britain was divided into territorial "Commands"—Eastern, Southern, Scottish, and so forth, the overall command being called "Home Forces." Western Command included, very roughly, Wales and the western tier of English counties from the Severn to the Scottish border.

emphasis on boldness and imagination reflected some frustration from this.)

The Chiefs directed the Planners to consider Wavell's message and make recommendations. The Planners initially came up with rather unimaginative suggestions that a senior officer and an assistant be assigned deception planning duties within the regular planning organization. No doubt "Joe" Hollis took a hand at this point (three years later, Hollis would refer to himself as "a kind of unofficial godfather (sic) of Bevan's organization"), and their next draft was a sharp departure. Noting for the first time that Stanley's staff (presumably meaning Wheatley) had been consulted, it proposed, in Wavell's language, an office independent of the Planners and authorized to deal with other departments and ministries directly.

Not until this point, it appears, was Stanley himself consulted. (No doubt he had been on leave owing to his wife's illness during this time.) His reading of this draft evoked the memorandum describing the frustrations of his office already mentioned. He said that Bevan had "greater experience and I think a greater aptitude for this kind of work" than Lumby had had; but "I am sure he can never function" unless the breakdown in coordination, particularly with the Navy, could be repaired. He thought the time was not yet ripe for major deception and a corresponding organization. When that time should come, there should be, not an interservice committee, as seemed to be contemplated, but a single "Controlling Officer with whosoever is required as his staff officers," who should continue to serve directly under the Chiefs of Staff and be of sufficiently high rank to deal on terms of equality with the directors of plans and of intelligence. "I had, in fact, asked for a directive on those lines," he noted sadly, "but have never received it." Moreover, "we should never get anywhere unless we have really high grade contact with America, and unless America is bitten by the bug and will set up a Controlling Officer with whom plans can be concerted."

It took more than a fortnight for the fourth and final version to emerge. It took account of many of Stanley's comments but not all. "We are satisfied," it said, "that the present organization under the Controlling Officer is on the right lines." It proposed creation of "The Controlling Section," responsible for global strategic deception, concentrating on broad deception policy and coordination of theater deception plans. Deception specifically aimed at forthcoming operations on the Continent would be the responsibility of the supreme commander's staff. The Americans should be invited to set up a parallel organization in Washington that would maintain close liaison with the Controlling Section.

What really counted was not so much this covering memorandum as the "Draft Directive to Controlling Officer" which was attached (there had been none with the earlier drafts). And in this, Bevan himself had a hand. As Wheatley, who was evidently not privy to these goings-on, remembered it, "two-thirds of the way through June," Bevan said to him:

"Dennis, we are never going to get anywhere like this. We might just as well be on leave for all the good we are doing. No one tells us anything or gives us any orders. We have got to have a directive. And as no one else seems prepared to give us one, we must write one for ourselves."

"Yes, sir," responded Wheatley dutifully. (He was not sure what a "directive" was.)

The directive that emerged became the fundamental charter for top-level British strategic deception for the rest of the war. ("It was cock-eyed," said Bevan long after, "but I would be hard put to it to say exactly how it should be stated.")

On June 20, the Chiefs gave their approval to the final draft of the Planners' memorandum and to the directive, with minor modifications—including the provision that the present deception section should be known in future as the "London Controlling Section." The directive defined the "Controlling Officer" as chief of the "Controlling Section for Deception." (In line with the Chiefs' decision, it was never called anything but the "London Controlling Section," usually shortened to "LCS," with no indication of what was being "controlled.") This was to be a separate office reporting directly to the Chiefs of Staff, no longer part of the Joint Planning Staff. For cover purposes, however, it was to continue to be considered as part of FOPS; moreover, in the initiation and preparation of deception plans the Controlling Officer was to work in close cooperation with the Planners, and submit his plans to the Chiefs through it. He was also to keep in touch with the Joint Intelligence Subcommittee, the Political Warfare Executive, SOE, MI6, and "other Government organizations and departments."

His charter was a sweeping one:

"3. You are to—

"a. Prepare deception plans on a world-wide basis with the object of causing the enemy to waste his military resources.

"b. Co-ordinate deception plans prepared by Commands at home and abroad.

"c. Ensure that 'cover' plans prepared by ISSB fit into the general frame-work of strategic deception.

"d. Watch over the execution by the Service Ministries, Commands and other organizations and departments, of approved deception plans which you have prepared.

"e. Control the support of deception schemes originated by Commanders in Chief, by such means as leakage and propaganda."

And what was most sweeping:

"6. Your work is not limited to strategic deception alone but is to include any matter calculated to mystify or mislead the enemy whenever military advantage may be so gained."

"Mystify or mislead." Stonewall Jackson again. And, appropriately, the directive provided that the Controlling Officer was to arrange for mutual coordination of British and American plans if the United States should set up a similar operation in Washington, with exchange of liaison officers if that should prove necessary.

"And, from that point," recorded Dennis Wheatley, "Johnny Bevan and I went to work."

The first couple of weeks that Bevan was on board passed quietly, as he "read himself in" by studying the files, attending ISSB meetings with Wheatley, and feeling his way in general. There was several weeks' overlap with Stanley, and initially Bevan signed memoranda "for the Controlling Officer"; indeed, he was not formally appointed Controlling Officer till August 5. He took charge of the office in a quiet but nononsense way. He liked things to be tidy and correct. Lumby had loathed files—he once cheerfully told Wheatley that when a file got too bulky he had simply burned it—and such filing as had gone on in the office had been done by him helter-skelter. Bevan summoned Miss Eden and instructed her in the art. The files were properly orderly from that time on.

Again, one day as Bevan was reading through Wheatley's paper on deception, he looked up and said, with some seeming asperity: "Our staff duties are pretty poor, aren't they, Dennis?"

This took Wheatley aback, for he liked to think of himself as the ideal staff officer, helpful, foresighted, self-effacing.

"No, sir," he replied cautiously, "I'm afraid our staff duties aren't quite all they might be."

But it transpired that "staff duties" was British Army lingo for the organization and format of study papers as taught at the staff colleges, with a particular method of numbering the sections and paragraphs, specific sections on the object of the study, factors pro and con, and so on. Bevan rewrote Wheatley's paper in accordance with the rulebook, and with a few additions it became their "deception bible" for the rest of the war.

Then Bevan tried his hand at writing a deception paper himself with Stanley. Building on Bevan's own years in Scandinavia, it suggested measures aimed at bringing Sweden into the war on the Allied side or at least inducing the Germans to take steps to forestall such a move, tying up additional forces in Scandinavia. Nothing came of this idea. For one thing, the Foreign Office was too tender of the sensibilities of neutrals to let anything of the kind be done at this stage of the war.

Johnny Bevan knew the Army. He understood the edgy relationship between line and staff. And he well knew that conventional warriors are just a little dubious about intelligence people and covert operators, and may not ascribe to instructions emanating from such odd folk the unquestioning authority which they do to orders that come through familiar channels.

With these considerations in mind, his directive deliberately gave him no powers of execution. The Controlling Section would plan and coordinate and supervise, but not implement. Bevan's direct access to the Chiefs of Staff, and his exceptionally broad authority, would be of little use if he did not have the confidence of the commanders on the scene. And the way to get and keep that confidence was for them to be assured that deception was part of their operations. Orders for the execution of deception plans would come from the same authorities as ordered execution of actual plans, and they would be implemented by the same commands as implemented actual plans. (As a practical matter, as will be seen, the LCS functioned almost as an implementing theater deception staff in connection with the North African landings mounted from the United Kingdom.)

Along the same line of reasoning, Bevan decided that the Section should leave its majestic offices overlooking the park and move in with the Joint Planning Staff down in the underground Cabinet War Rooms, so as to be seen and accepted as part of the team. That took some persistence, for space down there was at a premium, but he finally got two cubicles in "Mr. Rance's room."

It was a far cry from the lordly splendor upstairs and must have been particularly uncomfortable for an outdoorsman like Bevan. The place was a white-painted labyrinth of brightly lit narrow passages and little rooms, with noisy ventilating pipes painted a cheerful red and electrical conduits in profusion, staffed and guarded by Royal Marines in their dark blue uniforms (Wheatley likened it to the lower deck of a battleship, and the rooms to ship's cabins); you never saw the sun, of course, and it was easy to forget what time of day it was. The smaller of the new rooms just managed to hold Miss Eden and the files while into the larger one—"larger" by comparison only, for it measured perhaps twelve by fourteen feet—were squeezed five small utilitarian oak desks; two for Bevan and Wheatley and three for future expansion. Wheatley talked Mr. Rance into letting him bring down from the old office and crowd into the new one a large armchair for visitors, and brightened up the place a bit by bringing in a Persian rug of his own for the bare linoleum floor and pinning maps on the walls.

Bevan's next step was to request an increase in authorized personnel, for under the new directive he and Wheatley and Miss Eden could not do it alone. Early in August, when planning for the North African invasion was under way, he received a new "establishment" authorizing four GSO2s: one Navy lieutenant-commander, two Army majors, one RAF squadron leader. The latter was Squadron Leader Wheatley; he had moved up one notch and then another, to flying officer and flight lieutenant, as the minimum periods required by regulations expired; the new establishment of an air GSO2 slot gave him another bump up. In theory, one Army major formed an intelligence subsection, while the other three, one from each service, formed an operations subsection. In practice, there was no dividing line except that the operations people were not officially cleared to know about the double agents and ULTRA, though in fact they did know about the former at least.

Meanwhile, however, Bevan had his first assignment. Under consideration that summer of 1942 was an American project, Operation SLEDGE-HAMMER, a landing on the Continent in 1942. Bevan was directed to prepare a cover plan for it. By July 20, when Bevan submitted a memorandum pointing out how complex a task this would be, the Americans had agreed to abandon SLEDGEHAMMER in favor of the invasion of French North Africa, Operation TORCH. The Combined Chiefs of Staff ratified this decision on July 20. Under date of July 27 Bevan was directed to prepare a cover plan for TORCH and a plan for inducing the Germans to keep as large a force as possible in Northwest Europe in the autumn of 1942.

The story of these ventures belongs in a later chapter. Here we will follow the development of the London Controlling Section from its utter impotence in the spring of 1942 to the point where, to quote Dennis Wheatley, during the latter years of the war "the Chiefs of Staff positively ate out of Johnny's hand," and "had Johnny Bevan asked the Chiefs of Staff for a couple of divisions to play with, they would have given them to him without argument."

Personal relationships mattered a great deal in the tight little world of the Establishment. In this, Dennis Wheatley was a great help.

Though Wheatley was a rather famous middle-aged man about town and a practised host who knew any number of important and useful people, he held, initially at least, a grade commonly borne by twenty-year-olds. So he had to tread warily. He punctiliously followed the code of behavior expected of junior officers when he had been in uniform a generation before. Until asked to use names he called all higher-ranking officers "sir" no matter how much younger than he they might be, and when summoned into a superior's office he stood silently at attention until spoken to. This started things off right, and General X or Air Marshal Y or Admiral Z would invite him to sit down and have a cigarette. He would then relax and state his business, speaking as between equals.

When he got up to leave General X, he would say: "I wonder, sir, if you happen to have a day free to lunch with me?" This might surprise General X, who seldom heard such invitations from lowly subalterns, but Wheatley was different, he was more nearly General X's age than was the usual junior officer, General X might well be a fan of his books; so General X overlooked the impertinence and accepted more often than not. Wheatley kept a regular table at the Hungaria, one of the few London restaurants that somehow managed to keep up something approaching prewar standards. At Wheatley's luncheon, General X would find, as Bill Baumer did, that Wheatley had "trotted out a variety of cabinet secretaries and Air Vice Marshals." The next step was often an invitation to dinner at Chatsworth Court, where General X, whose wife was probably in the country, would enjoy the lively company of Joan Wheatley and Diana; and so another friendship would be formed, pleasant in itself and immensely useful to the Controlling Section.

By the time Bevan came on board, Wheatley already had quite an old-boy network going. He was thus able to introduce Bevan on this basis at luncheons and dinners to many officers important to his work. Bevan in turn picked up on this, entertained regularly himself—something Lumby had not been able to afford—and was soon quite at home in the Chiefs of Staff organization on a personal level.

Bevan slowly increased the Controlling Section staff to fit his enlarged establishment.

"Dennis," he declared one day in August, when he and Wheatley were swamped with work on the cover plans for TORCH, "we can't go on like this. We must have help, and at once." He went to see Pug Ismay, and the next day Major Harold Petavel joined the office. In peacetime the manager of a soap factory, Petavel had served with the expeditionary force that was evacuated at Dunkirk, and then had worked for Bevan for nearly two years in Western Command. He was a shy, soft-spoken man whose reserved and solemn manner masked a subtle humor, "with a mass of thick, smooth, black hair, a heavy mustache (later reduced to more elegant proportions on his marriage to the Bishop of Chichester's daughter), and heavy-lensed spectacles," as Wheatley recalled; never without a pipe, switching from one to another when the first got too hot. He was a godsend to Wheatley, who had borne the brunt of Bevan's irritation and frustration over how much they had to do with such few resources. Wheatley and Petavel soon struck a deal: Petavel, experienced in the details of intelligence work and "staff duties" but shy about dealing with senior officers, would handle the office work, while Wheatley attended the conferences and buttonholed generals, admirals, and air marshals as necessary.

Petavel constituted the intelligence subsection. When the offices expanded, he had his own room, lined with "enormous charts marked with cryptic signs" tabulating the double agents through which misinformation was being fed, "sticking pins into a chart that looked like a crossword puzzle, but indicated all our secret agents' activities," and attending Twenty Committee meetings from the autumn of 1943. He participated in maintaining bogus order of battle when that became important.

The month after Petavel's arrival, in September 1942, Major Sir Ronald Wingate, Bart., joined the LCS as the Army member of the operations subsection. Wingate had inherited his baronetcy from his father, Lieutenant-General Sir Reginald Wingate, "Wingate of the Sudan," one of those legendary figures of Empire days, a fluent Arabist who had directed intelligence in Egypt, fought the Dervishes, succeeded Kitchener as Governor-General of the Sudan and Sirdar of the Egyptian Army, and served as High Commissioner for Egypt. Fifty-four years old when he joined the LCS, Sir Ronald had himself had a distinguished career in the Imperial service, beginning in First World War days when as a young officer he had negotiated the British protectorates over the oil sheikdoms of the Middle East, and ending as Governor of Baluchistan. At the outset of the war he had been with the Ministry of Economic Warfare, had been political agent with the unfortunate British expedition to Dakar in 1940, and then had been with SOE in West Africa before Lumby went out there.

Described by Bill Baumer as a "bland likeable" man with "a rod of steel under that smiling, handlebar mustache exterior," and by Wheatley as a "charming and clever old fox" with twinkling eyes, Wingate had been eager to be assigned to the LCS and had lobbied energetically with his old friend Ismay to be recommended for the job. Bevan he may already have known, for they were fellow members of Brooks's; certainly they became great friends. He had an "encyclopedic" knowledge of politics, an "amusing cynicism about his ex-fellow civil servants," and "a vast experience of 'the working of the protocol,' as he termed it; thus, by using exactly the right approach, he was often able to achieve results which would have been beyond the scope of anyone lacking such highly specialized knowledge coupled with his particular form of guile."

On top of which, he was excellent company. "He used at times to declare that his one aim in life was to 'give everyone pleasure,'" said Wheatley, "and his unruffled calm, polished wit and unfailing good humor were in times of stress a blessing to us all." From October 1942 he regularly attended Twenty Committee meetings for Bevan, until Petavel took that over. In December 1942, Bevan left Wingate in charge when he took his first prolonged absence from the office for his initial visit to Washington. In March 1943, Wingate was formally appointed Deputy Controlling Officer.

Never quite comfortable with any service but the Army, Bevan was slow to fill his one Navy slot. His excuse was that there would not really be enough for a Navy officer to do. But deception for the North African landings involved some naval activity, and he could put it off no longer. He consulted Ewen Montagu. Montagu assured him that he could find plenty of work to fill any spare time a Navy officer assigned to Bevan might have. The Admiralty hunted up an officer, and the upshot was that not long after Wingate had reported for duty, Commander James Arbuthnott, RN, joined the group.

Arbuthnott was a slim, tallish, gray-haired man with smile-wrinkles round the eyes and mouth, quiet and shy in bearing; neither imaginative nor inventive, but with solid good judgment, possessing "in a very high degree," in Wheatley's words, "an inbred sense of justice, toleration, and sympathy for the underdog," and, as it proved, a good man in a conference, always bringing the discussion round to the main point when it had wandered off to side issues. He had been a regular Royal Navy officer, serving through the First World War and after. In 1926 he had retired to take up the growing of tea in Ceylon, rising to become chief executive of the biggest tea-planting corporation in that colony. Recalled to active duty when war broke out, he had first been posted to the Royal Navy's Indian Ocean Command, and then had spent a year at Middle East Command in Cairo before transfer to London.

Arbuthnott's indoctrination into the LCS's work showed how effective Dudley Clarke's security was. New officers generally got an initial briefing from Bevan and then a more detailed breaking-in from Wheatley. When Bevan gave him his opening talk, Arbuthnott observed: "Of course, we had nothing of this kind in the Middle East." It transpired that he had heard of "A" Force but did not know what it did.

Arbuthnott and Wheatley hit it off and, as the only non-Army men in what was otherwise an Army shop, they tended to share their work and cover for each other when one was away; helped in this by the fact that neither was involved in intelligence or order of battle. What was perhaps even more important, Arbuthnott was able to work amicably with Ewen Montagu.

The coming of Arbuthnott meant that all five desks crammed into the little room were filled. Bevan sat in the middle, underneath the clock that Mr. Rance set each morning. Wingate and Arbuthnott sat on his right, Petavel and Wheatley on his left. Arbuthnott and Wheatley dubbed it "the schoolroom." Gone were the days of thumb-twiddling and crossword puzzles. Visitors came and went, the telephones rang constantly, several telephone conversations going on at once was the norm.

Under these sardine-can conditions was done all the work for the cover plans for the North African landings, security for the Casablanca Conference, and more besides. Then in early 1943 the staff of the Commander-in-Chief Home Forces was moved out of its basement quarters, opening up more space in the underground War Rooms. Bevan was away and had delegated Petavel to acquire more elbow room for the LCS. The officer in charge of allocating space brushed Petavel off, saying that they had managed with two rooms before and they should be able to manage with two rooms in the new section. Wheatley tried his hand. The officer, a lieutenant-colonel, told Squadron Leader Wheatley that he was being impertinent.

Wheatley was up to a challenge like that. He went upstairs and put the case to the Deputy Secretary of the War Cabinet, who had enjoyed more than one lunch at the Hungaria and dinner at Chatsworth Court. Leave it to me, said the Deputy Secretary. The LCS wound up with a comparative embarrassment of riches: nine rooms, which Wheatley reduced to five by knocking out some partitions. Under the new dispensation there was a single room each for Bevan and Wingate, a double for their ATS assistant and Petavel, another double for the typists and a filing clerk—clerical help had expanded since the days of Miss Eden—and a triple for Wheatley, Arbuthnott, and a conference table, with room for a couple of bodies more.

Wheatley was a believer in comfort. He also saw a way to protect some of his most valuable possessions from air raids. So he turned the triple into one of the notable spaces in the Cabinet War Rooms. He brought in his Chippendale chairs and his big dining table for the conference table, with a bronze of the Dancing Faun to adorn it, papered the walls with maps, and covered the linoleum floors with his rare silk Persian rugs.

The down side of the move was the loss of access to the little Joint Planning Staff mess. It had been a valuable way to keep up the old-boy net, but now it was too crowded. Wheatley managed at least to preserve for LCS members the right to buy drinks there.

The LCS's establishment was further enlarged in March 1943, and two more Army officers joined the section.

The first was Major Neil Gordon Clark. In civil life he and his brother were heads of the firm that distributed Martell brandies, so he and Wheatley hit it off at once. He had served in tanks in the last part of the First World War, had rejoined the Army in 1939, and had been stuck with a Home Guard unit until his LCS assignment. He worked with Petavel on bogus order of battle, and with Wheatley and Arbuthnott on planning, and eventually he took over from Wheatley attendance at ISSB meetings. He occupied a third desk in the triple room, with Wheatley and Arbuthnott.

The second was Major Derrick Morley. In civil life "a strange cross between a playboy and a very able financier," said Wheatley, he was a well-traveled member of one of the great City banking firms. The most unmilitary of men, he had had to pull strings to get into the Army because of flat feet. The feet had given out on him in the Middle East and he found himself with a desk job in the London Controlling Section. He and the American Carl Goldbranson, whom we will meet later, were assigned to be trained together in deception work after the invasion convoys sailed for North Africa in the autumn of 1942.

Morley was originally supposed to be Petavel's helper on the intelligence side, but Petavel was so secretive that Morley found himself with little to do. So Bevan put Morley's experience in foreign capitals to work by using him to convey deception instructions to British ambassadors abroad. In particular, Morley went to Stockholm, Lisbon, and Madrid under cover in early 1944 to brief the British missions in those capitals on BODYGUARD, the overall cover plan for the invasion of Europe. Bevan also made Morley the LCS liaison with General Freddie Morgan's planning for the Normandy invasion.

Less successful were two civilian specialists that were attached to the Section in due course, one as a representative of the Foreign Office and the other as a scientific advisor. Bevan had asked for experts in both these fields; but when he got them his distrust of anybody not in uniform got the better of him and he did not use either one properly.

The Foreign Office representative, added at the time of the March 1943 augmentation, was Sir Reginald Hoare, "an elderly, gray-haired man with a pronounced stoop," who moved in with Wheatley, Arbuthnott, and Gordon Clark. Sir Reginald had had a distinguished career in diplomacy and most recently had been minister to Rumania. He was perhaps the only person who worked with Johnny Bevan who came to dislike him. The feeling was mutual. Sir Reginald was not interested in the work—he admitted to Wheatley that he had only taken the job so he could keep up with the secrets of high policy—and came into the office only for a few hours a day just to catch up with the news. He spoke slowly and with great deliberation. This drove Bevan to distraction and sometimes he lost patience. "He treated me like a footman," complained Sir Reginald to his office-mates once. They in turn were fond of him and of his subtle wit and sense of humor. They called him "Ambassador," respectfully and affectionately. But he made no contribution to the work of the Section at all.

The scientist was no less a personage than the President of the Royal Society himself, Professor H. A. de C. Andrade; "Percy" to Wheatley and company, at his own request. Andrade was a part-timer who only came in for a few hours a few days a week, to be consulted on matters like chemical and bacteriological warfare and the prospects for an atomic bomb. He could have contributed much more, for it transpired that he had an immense knowledge of the Germans and their psychology and probable reactions, and he often came up with suggestions of ways to deceive them, some of which were preposterous (one was to suggest to the Germans that a floating roadway was being built across the Channel). Bevan never put him to proper use, and directed that he be given no information about future plans.

As the staff grew, so did the support staff. By the winter of 1943–44 there were three civil service clerk-typists and an ATS officer, Margaret

Donaldson. In February 1944, another ATS officer, Junior Commander Lady Jane Pleydell-Bouverie, the twenty-year-old daughter of the Earl of Radnor, joined the office, and when Margaret Donaldson moved on, she became Bevan's personal assistant. She, Bevan, Wingate, and Petavel apparently were the only members of the office cleared for ULTRA; she shared Petavel's room, maintained the closely held files kept in Bevan's office, and did whatever typing was too highly classified for the civil service people to handle.

The London Controlling Section was on the whole a happy shop, with only the occasional frictions normal to human relations. Not a week passed by, Wheatley recorded, that several of them did not lunch or dine together, and Lady Jane remembered how much business was done informally over the telephone or at Rules or the Hungaria, Brooks's or White's. Bevan was a good if sometimes difficult man to work for, modest and unassuming, always giving the credit to somebody else. He was gracious and kindly—not quite unfailingly so, for he was an intense worker and perhaps over-conscientious. "Whenever a crisis arose, Johnny became harassed and displayed ill-temper," recalled Wheatley. Bevan knew this and felt badly about it. It could not have been too serious, for Lady Jane remembered him as "very kind, very quiet with a nice, gentle sense of humor." And even at one of these bad times, if one of the staff offered a witticism or suggested he take a break and come to dinner, "that smile which lit up his whole face would immediately flash out." And in fact he loved the work.

Bevan made a point of telling each new recruit that he saw no reason to run the shop on a rigid military basis; they were all middle-aged men, officers only for the duration, who had been successful in civil life and could be counted on without clock-punching or constant supervision. Nobody was trying to advance a military career, everybody lived at home or in clubs, hotels, or flats; life in many ways was as it would have been in some civilian office in London. All this led to a congenial atmosphere as compared with most military organizations, notwithstanding the occasional tenseness of the boss.

Congenial but hardworking. Like all such compulsives, Bevan stayed in the office in the evenings, or came back after dinner (often at ten o'clock at night), even when there was really nothing current to do. Loyal

Petavel would stay as long as the boss did. The rest of them were perfectly willing to work all night if they had to, and put in whatever long hours were needed when there was a crunch on; but staying late for no reason made no sense, especially to Dennis Wheatley, who believed in enjoying life. For a while he would stay late unless he had a dinner party. Then he and Arbuthnott decided that this was pointless and that at six o'clock they would get up and leave. Wingate joined them in their "unannounced strike" after waiting a few days to see which way the wind blew. This was after the surge of activity in connection with North Africa was over. It remained the pattern thereafter.

Bevan, said Wheatley, was "obsessed by security." "Security," said Baumer after spending some time with LCS in the spring of 1943, "is a fetish with Colonel Bevan. He is practically unknown in London and only two of his officers know everything that is being done. He accomplishes most of his work by dealing only with persons in the highest echelon and . . . it is not unusual for him to meet with the Prime Minister two or three times a week." How tight security was is shown by the fact that Petavel, who was with Bevan from August 1942 to the end (and was cleared for ULTRA), and his wife, who worked in ULTRA at Bletchley Park, had been married for two years before they discovered that their work was related.

Wingate took charge when Bevan was away. His style was more relaxed than Bevan's. He knew how to get the best out of people by a touch of flattery; how to charm one of the officers by suggesting that he take a day off and think about a new problem. When he was not in charge he went home promptly at six; when he was, he hardly ever left the office. But he remained unperturbed no matter what happened. Once when Wingate was holding the fort a crisis came up when Wheatley had planned a few days' leave. Wheatley offered to postpone it, as Bevan would have expected him to do. Nonsense, said Wingate. "If any one of us is not competent to run this section on his own, he has no right to be here at all."

To ensure that deception plans and real plans were kept in harmony with each other, a good deal of the London Controlling Section members' time was spent in liaison with other offices. Liaison with the uniformed services was conducted mainly through Wheatley's, and later Gordon

Clark's, attendance at ISSB meetings. The crucial liaison with MI5 and MI6 over the double agents was carried out through Bevan's membership on the Twenty Committee. Liaison with SOE was chiefly important in connection with generating and spreading rumors, for SOE was in a position to launch them in neutral countries and through the Resistance in Axis-occupied territory.

Bevan constituted four ad hoc committees of people who from their experience in various lines of work might help to cook up "themes," as they were called, to be leaked to the Axis, or help to put them over. Called the Oliver Committee, the Twist Committee, the Tory Committee, and the Racket Committee, they met weekly or fortnightly with representatives of SOE, MI5, MI6, and other offices, to make sure the themes were consistent with—but not too obviously similar to—the circumstantial messages being passed by the double agents, and to allocate misinformation assignments among the available channels. The Twist Committee dealt with allocation of channel assignments by way of double agents. The Tory Committee dealt with assignments to other channels. The Racket Committee (chaired by Wheatley) devised new *ruses de guerre* and fresh methods of passing misinformation to the Axis in conjunction with MI6, MI9, and other agencies. (What separate function the Oliver Committee may have performed is not presently clear.) Early in 1944, as part of the reorganization of deception mounted from the United Kingdom to cover the Normandy invasion, the Twist Committee was abolished and its responsibility was passed to SHAEF; the Racket Committee was abolished soon after. Bevan retained authority over assignments to other channels, for which purpose the Tory Committee was retained.

Perhaps even more useful was a committee, apparently without a name, which Bevan formed in early 1943, perhaps inspired by his experience in 1918. Composed of representatives from each of the service ministries, from MI6, and from the Radio Security Service, all cleared for Ultra, it met once a month (fortnightly from two months before D-Day to a month after it) and produced a mock telegram from the German high command to their theater commanders, setting forth their best estimate of the German appreciation of the war situation.

Close liaison with the Political Warfare Executive was also necessary, to make sure that propaganda did not inadvertently lead the Axis to draw conclusions inconsistent with deception plans. That was Wheatley's assignment. The next week's propaganda line was drafted on Tuesday nights. Every Wednesday morning Wheatley would visit PWE's chief, an old friend, to get the new line (and to stop by Harrods, Jackson's, and Fortnum & Mason to shop at opening time their limited supply of unrationed luxury items).

Relations with the Foreign Office were particularly important. The deceivers had always to be concerned lest the British mission in a neutral country say or do something inconsistent with a deception plan. And on occasion the Foreign Office could be directly helpful by making an inquiry or presenting a note to a neutral government from which the Axis might draw an inference when they heard about it. Similar considerations would occasionally arise in connection with the Ministry of Economic Warfare.

Wingate and Wheatley sometimes played a "Mutt and Jeff" game with recalcitrant Foreign Office bureaucrats they wanted to wring something out of. Wheatley enacted the ignorant, no-nonsense civilian in temporary uniform and Wingate the smooth, deferential former civil servant who tried to restrain his colleague's outbursts about total war and what the public would think if they knew British soldiers were getting killed because neutrals were being treated in so lily-livered a way. They usually went away chortling over getting much of what they had wanted. Relations with the Foreign Office remained excellent nevertheless.

A different order of liaison was needed with the Americans in Washington and with the theater commands—"A" Force for the Mediterranean and the Middle East, Peter Fleming's organization in India-Burma, and eventually Eisenhower's staff on the Continent. There was of course direct and constant communication with the theater commands. Liaison with the LCS's opposite number in Washington, the Special Section of Joint Security Control, was conducted through an officer with the British military mission in the American capital: initially Major Michael Bratby, then Colonel H. M. O'Connor, and at the very end of the war Group Captain E. M. Jones. On occasion, Bevan corresponded directly with the United States Army Assistant Chief of Staff, G-2, in the latter's capacity as senior member of Joint Security Control, and twice he went to Washington in person. When the office was fully established with a regular routine, Bevan instituted a "Weekly Letter" summarizing current activity. This was sent to the theater commands and to the liaison officer in Washington, who passed appropriate items on to Joint Security Control.

There was a great deal of paper to read in order to keep up with the war and with current planning, and as the office became fully staffed Bevan divided up this responsibility. Wingate read Joint Planning Staff papers on future plans. Wheatley read minutes of meetings of the War Cabinet, the Defence Committee, and the Chiefs of Staff, together with Foreign Office cables inbound and outbound. Hoare took over the latter task when he joined the office. Petavel read intelligence summaries and Joint Intelligence Subcommittee appreciations of Axis intentions. Arbuthnott read directives to commanders and reports of operations. Gordon Clark read papers about the movement of forces and resources, and reported on ISSB meetings. Morley kept posted on activities in preparation for the invasion of Europe.

Every morning at 9:30 there would be a staff meeting around Wheat-

(Julian Bevan)

The London Controlling Section
Left to right: Derrick Morley, Neil Gordon Clark, Harold Petavel,
Lady Jane Pleydell-Bouverie, John Bevan, Dennis Wheatley,
Sir Ronald Wingate, James Arbuthnott, Alec Finter

ley's dining table, at which each would report the substance of his day's reading. The whole team was thus kept uniformly current with developments and prospects. For the LCS did function as a team. Bevan would put to the group the real operations to be carried out, and they would brainstorm ideas for corresponding deceptive activity. Slowly a plan would develop. It would go through draft after draft before finally going up through the Joint Planners to the Chiefs of Staff. The same kind of open, freewheeling brainstorming was applied to everything the Section did; to the work of the Twist, Tory, and Racket Committees, and to the occasional assignment to devise cover for the travels of Churchill and other high dignitaries to attend Allied conferences.

It was an exceptional group of people with an exceptional task in which they delivered an exceptional performance. Over time the major players moved up in grade, Bevan to full colonel, Wingate to lieutenant-colonel and then full colonel, Wheatley to wing commander, Petavel to lieutenant-colonel. After the war the government was niggardly with honors, but in partial compensation, the LCS twice received specific resolutions of commendation from the British Chiefs of Staff. No other organization received even one.

And no one questioned that the success was due to Johnny Bevan. Dennis Wheatley said it best in a letter to Bevan when he, Wheatley, left the Section, its great work done, at the end of 1944. He had thought of writing a history of the LCS, and this, he said, would have been its final paragraph:

> If in some future time our country is again faced with similar perils to those we have passed through and can still find, among her ordinary citizens, a leavening of men having the qualities of John H. Bevan to fill key posts in the direction of the war, Britain will be well served indeed.

There is a serpent in every garden. The one in this garden continued to be Ewen Montagu. At the end of 1942, he lost his mentor, Admiral Godfrey, whose tactlessness got him kicked upstairs to naval command in India. But Montagu himself remained.

The previous August it had been Montagu and Godfrey who suggested that Bevan be made chairman of the Twenty Committee. Now, however, whatever his motive—perhaps he wanted Bevan's job for himself—Montagu started trying to undercut Bevan behind his back. On March 1, 1943, he circulated within the naval staff a sneering and vicious dismissal of the London Controlling Section and a personal attack on Bevan, in effect calling him a liar. Colonel Bevan, he said:

> is almost completely ignorant of the German Intelligence Service, how they work and what they are likely to believe. He is almost completely inexperienced in any form of deception work. He has a pleasant and likeable personality and can "sell himself" well. He has not a first grade brain. He is extremely ambitious and is not above putting up to the Chiefs of Staff a report which he must have known presented an entirely false picture of his work in connection with TORCH. He can expound imposing platitudes such as "we want to contain the Germans in the West" with great impressiveness. I have dealt with his character in this way as I am sure he will not improve with experience.
>
> The remainder of the staff of the London Controlling Section are either unsuited to this sort of work (in which they are all wholly inexperienced) or are third rate brains.
>
> Col. Bevan, in spite of all efforts by those who have experience in deception to co-operate with him, persists in devising plans and papers, and submitting them to the [Directors of Plans] or the Chiefs of Staff either *before* mentioning them to "the experts" or without ever mentioning them at all. . . . He can do practically nothing without drawing up long documents . . . which delay action (but look well to superior officers).

There followed a long fifteen-point catalogue of Bevan's alleged shortcomings. It charged him with various alleged failures in connection with cover and deception for the North Africa landings; with submitting to the Chiefs of Staff with respect to that operation "a report which indicated that he had helped to deceive the enemy, which was partly untrue and partly tendentious"; with delay and lost opportunities in subsequent deception; with alleged absurd suggested deception projects; and with a wasted trip to the United States:

He went to America to try and fix up liaison. The grand opportunity of someone with his backing "selling" to the Americans the real advantages of deception work was missed. He has brought back a promise that they will not trespass in Europe with any deception of theirs (which promise may or may not be kept by authorities he has not seen) but with no real knowledge of what they are doing (at least none that he has mentioned to us).

Bevan, said Montagu, "often insisted on getting authority from an unnecessarily high level. . . . This brought him to the attention of high authority but did not get on with the job as people on that level are too busy on other matters."

Bevan probably never saw this paper—if it ever got outside the naval staff it is likely that Ismay or Hollis would have suppressed it—but he could not but have sensed Montagu's treachery. By Montagu's account, there was "considerable friction" between the Twenty Committee and the London Controlling Section after the North African landings and during 1943 "when the Twenty Committee group chafed at the fact that no strategic deception went over from the U.K. between then and OVERLORD."

Even Montagu had grudgingly to admit that over time matters improved. Bevan "lost much of his original touchiness," perhaps, Montagu surmised, because he, Montagu, was furnishing Bevan with ULTRA material. But even after the war, Montagu was complaining that Bevan did not always call on the help of naval intelligence in time.

The simple fact appears to be that Montagu's vision was limited to ULTRA and to double agent work; that he knew less than he thought he did about matters above his level and nothing about bureaucratic or operational realities at the top echelons; that he was not as important as he probably thought he was or should be; and that he resented Bevan. As for his own personality, his actions speak for themselves. In any event, his backstabbing had no perceptible result.

Tracing the growth and maturity of the London Controlling Section has taken us ahead of our tale.

It will be remembered that the months when Bevan and Wheatley held the fort alone saw the turning-point for the two key tools of deception, the double agents and ULTRA. It was in June 1942, a few weeks after Bevan came on board and the same month in which Bevan's directive was issued, that MI6 finally agreed to share ISOS decrypts relating to the double agents. June and July saw the debate over the role of the double agents that would end with Bevan a member of the Twenty Committee and the double agent system placed at the disposal of the deceivers. Perhaps it was when Bevan was beginning to be inducted into these mysteries, as Oliver Stanley had never been, that he opened a file that he kept himself, marked "Lt. Colonel J. H. Bevan—Personal," and said casually to Wheatley that it was just intelligence material that he need not concern himself with.

The unwitting cause of all this was a singular young man who arrived in England four weeks before Bevan was appointed Controlling Officer. His name was Juan Pujol García. But as one of the four great double agents of the Second World War, and the greatest of them all, he will always be known as GARBO.

While Dennis Wheatley was losing his voice bawling orders on the parade-ground at Uxbridge that December of 1941, Dillwyn Knox and his team broke the Abwehr's machine cipher and MI6 began to read traffic between Berlin headquarters and the Abwehr offices in neutral countries. While Wheatley and Lumby sat twiddling their thumbs and doing crossword puzzles the next few months, MI6 was finding that the traffic between the Madrid office of the Abwehr and Berlin contained some very odd stuff. Evidently Madrid had an agent, or thought they had an agent, in England, called by the Germans ARABEL, with three subagents: a Portuguese travelling salesman named CARVALHO who operated out of Newport, in South Wales, where he could observe shipping in the Bristol Channel; a German-Swiss businessman named GERBERS, living in Bootle near Liverpool and reporting on shipping in the Mersey; and a rich Venezuelan student in Glasgow.

Apparently by way of letters written from England and taken to Lisbon by an employee of KLM, the Dutch airline, ARABEL was flooding the Abwehr in Madrid with information which was duly radioed on to Berlin. It was plainly wholly imaginary. There were tales of naval maneuvers on Windermere. There were specifics of regiments quite unknown to the British Army. There were reports of the sailing of nonexistent convoys. Some of ARABEL's information was so preposterous, indeed, as to suggest that he was not only not in Britain now but that he never in his life had been. He advised that there were "drunken orgies and slack morals at amusement centers" in Liverpool. Glasgow longshoremen, he reported, would do anything for "a liter of wine." Foreign embassies were said to move to Brighton in the summer to escape the heat of London. He could not manage pounds, shillings, and pence.

Particularly troublesome was a report about a fictitious convoy from Liverpool sailing to the relief of Malta, which evoked a substantial Axis operation to intercept it. This did no harm, but another such report might well result in U-boat activity that would endanger real convoys. Late in February 1942, MI5 decided to take action through BALLOON, one of its controlled double agents, to discredit ARABEL.

Just at that time, however, the Lisbon representative of Section V of MI6, a peacetime merchant banker named Ralph Jarvis, was visiting London, and in a chance discussion with an officer of MI5's B1A he disclosed the existence in Lisbon of a Spaniard with a curious history, named Juan Pujol. Pujol, it transpired, had repeatedly offered his services to Britain and been rejected. He had now tried yet again, through the good offices of the American naval attaché's office, assuring them that he was accepted by the Abwehr as an agent and that he wanted to help the Allies. Plainly, Pujol might be ARABEL.

MI5 was understandably annoyed at not being told of Pujol earlier. Liddell wrote to MI6 asking with some asperity for a full report, underscoring that Pujol might well have received important questionnaires from the Germans and that his reports had to be coordinated with other double agent operations. He urged that Pujol be brought to Britain, where his story could be checked and it could be decided whether he should stay there under MI5 control or return to Lisbon under control of Section V of MI6. MI6 did not like this; it tried to get assurance that if brought to England Pujol would be returned to Lisbon under their control; but it finally agreed. Pujol arrived in England on April 24, 1942.

He proved to be a thirty-year-old man of short stature, with dark receding hair slicked back and "warm brown eyes with a slight mischievous glint." His story was a bizarre one. A Catalan by origin (Pujol is a common Catalan name; the "j" is pronounced "zh," as in French, rather than "kh" as in Castilian), he had grown up in the comfortable middle class in

Barcelona. The Spanish Civil War broke out in 1936. By his own account, he had spent much of that war in hiding so as to avoid military service; then in 1938 had joined the Republican army in order to desert; had deserted to the Nationalists in September of that year, spent time in a detention camp, had been put into Franco's army, and managed to survive the war without ever firing a shot. Then he got a job managing a hotel in Madrid.

The Second World War broke out soon after. By Pujol's account, he had become disillusioned with Fascism and was increasingly repelled by the Nazis and their doctrines; so he decided to do something to help their enemies. In January 1941, through his wife, Araceli, he offered his services as a secret agent to the British consulate at Madrid. The British diplomats were under strict orders to do nothing to endanger Spanish neutrality, and the ambassador was hostile to any kind of secret service activity, so the offer was fobbed off.

He decided that if he could get himself accepted by the Germans he would then have something to offer the British. He read up a bit on Nazi doctrine. Then he got in touch with the German Embassy and offered his services to the New Order. The Abwehr were doubtful at first, but, by various shenanigans, Pujol persuaded them to take him on.

Armed with a bottle of secret ink, a cipher, a list of cover addresses in Lisbon to send reports to, and three thousand dollars in Abwehr funds, Pujol left Spain in July 1941 with Araceli and their small son, Juan Fernando. The Germans thought he had gone to England; in fact he established himself in Portugal. There he tried to interest the British Embassy, without success. He decided to go ahead and establish his bona fides with the Abwehr.

Genuine reports mailed from Britain would have borne British postmarks and censorship stickers. There was no way Pujol could get these, so he devised a plausible substitute. He took out a safe deposit box in a Lisbon bank and mailed the key to the German Embassy in Lisbon to be forwarded to the Abwehr in Madrid. He explained that he had recruited a steward on the plane on which he had traveled to England to mail his letters to the cover addresses in Lisbon with which he had been supplied, and to pick up replies from the bank box. He tried again unsuccessfully with the British, offering to produce his secret ink and questionnaires. He bought himself a guidebook to England, a Bradshaw railway timetable, and a large map of Great Britain, and with the aid of these and magazines and reference books in Lisbon libraries he began composing spy reports. The first went to Madrid in October.

Time passed. Pujol continued grinding out messages to Madrid. (By the time he left Portugal in April 1942 he had sent nearly forty of them.) He tried again with the British and failed again. In November, through an ingenious subterfuge Araceli got in touch with the American naval attaché's office in Lisbon. That office called in Captain Arthur Benson of the British naval attaché's office. Benson met with her that month, sent a report back to London through his own channels, and met with Pujol himself in January.

The next month the British began reading the Abwehr traffic from Madrid and discovered ARABEL's existence. Section V sent word to Jarvis in Lisbon to try to find him. Jarvis turned the job over to one of his best men, Eugene Risso-Gill. Coincidentally, Benson's report reached Section V in England. Pujol was clearly the man. Risso-Gill met with Pujol and Pujol agreed to go to England. Jarvis and Risso-Gill smuggled him on board a British steamer headed for Gibraltar. There he was met by the local MI9 man, Donald Darling. Darling arranged for air transport to England, and assigned Pujol a codename: BOVRIL, after the popular English beverage concentrate, whose well-known advertising slogan was that it prevented "that sinking feeling coming over you"; Pujol, he hoped, would have the same effect.

BOVRIL arrived in Britain on April 24. He spent his first fortnight in England being debriefed in secure premises by two MI5 officers, Cyril Mills and Tomás Harris, and an MI6 officer, Desmond Bristow. Mills was initially BOVRIL's case officer; but he spoke no Spanish and quickly dropped out of the picture. His main contribution was to suggest, after the truly extraordinary dimensions of Pujol's imagination and accomplishments had become apparent, that his codename should be changed as befitted "the best actor in the world"; and BOVRIL became GARBO. By contrast, both Harris and Bristow spoke Spanish like natives. "Tommy" Harris, who took over from Mills as case officer, was half Spanish and educated in Spain. The thirty-five-year-old son and business associate of a prominent and wealthy Mayfair art dealer, he was a talented artist in his own right. Bristow, a twenty-five-year-old Cambridge graduate, had grown up in Spain, the son of an English engineer who managed copper mines in Andalusia.

After two weeks of intense debriefing, Harris and Bristow were absolutely satisfied that GARBO was ARABEL; that he was exactly what he claimed to be; and moreover that he was a man of inexhaustibly fertile imagination who could be turned into a nonpareil among double agents. Bristow met twice with the Twenty Committee, and he and Harris twice more; and after some debate—GARBO's story was fantastic, but the evidence of ISOS, and particularly of German operational moves against the imaginary convoy from Liverpool, was incontrovertible—GARBO was accepted as a full double agent.* MI5 took him over; MI6 dropped out of the picture except to manage pickup and delivery of material in Lisbon. Bristow was soon transferred to the Mediterranean, where we will meet him again. Araceli and Juan Fernando were smuggled out of Portugal to join GARBO, and the Pujol family were set up in a safe house in Hendon with a trusty housekeeper and a nanny, both on MI5's payroll. GARBO was provided with documents identifying him as Juan García, a BBC translator.

GARBO and Tommy Harris, double agent and case officer, formed one of the great teams of intelligence history. Between them, over the next three years they saturated the Abwehr with an enormous and unceasing flood of material as extraordinary as any ever turned out by a spy real or imaginary. GARBO's messages to the Germans had already developed a unique style, a sort of churrigueresque language, or perhaps a verbal equivalent of the extravagant confections of Antonio Gaudí in GARBO's native Barcelona.† And he was industrious beyond belief. During his three years' service he churned out no fewer than 315 messages in invisible ink, averaging 2,000 words each, not to mention the equally long-winded

---

* As already noted, one of the benefits of the GARBO debate was that MI6, finally and reluctantly, consented to share ISOS with all members of the Twenty Committee and with MI5.

† No brief samples can do GARBO's style justice, but here are two anyway: "I do not wish to end this letter without sending a Viva Victorioso for our brave troops who fight in Russia, annihilating the Bolshevik beast." Again: "This unfolding of confidence can only be made between comrades. One could not say to an English Lord what one may say to a National Socialist Comrade. The former would consider himself ridiculous if he had to accept an observation from a subordinate. We accept, within the discipline of hierarchy, the advice of subordinates. Thus, the great Germany has become what it is. Thus, it has been able to deposit such great confidence in the man who governs it, knowing that he is not a democratic despot but a man of low birth who has only followed an ideal. The Fatherland! Humanity! Justice and Comradeship!"

letters in ordinary ink that covered them; supplemented, after March 1943 when he acquired a wireless operator, by 1,200 more radio messages. Over and above all this composition was the tedious labor of encipherment. To this industriousness was added imagination and inventiveness as flamboyant as his prose style. Tommy Harris, himself almost as creative, was nevertheless a steadying influence, guiding, inspiring, and shaping the total output.

MI5 rented for Harris and GARBO a small office near the Jermyn Street shopping arcade, convenient to MI5's main premises in St. James's and handy to Section V's Ryder Street office. There Harris and GARBO would meet each day, devising false information, composing messages, inventing new imaginary subagents, breaking occasionally for lunch or dinner at the Martinez Restaurant in Swallow Street or the Garibaldi in Jermyn Street, and beyond question having the time of their lives. Outbound letters, notionally handled by GARBO's KLM courier, went by British diplomatic bag to MI6 Lisbon; the latter took care of mailing them and picking up return messages at the bank.

The first few months with GARBO on board, it will be recalled, were the months in which the final policy of using the double agents for deception was decided on, and in which Bevan was gradually taking hold as Controlling Officer and being read into the mysteries of ISOS and the double agent system; and no systematic deception policy had yet evolved. So to keep the pipeline filled, as Tommy Harris put it later on, "we tried to report in as much confusing bulk as possible and, in the absence of another objective, to increase our network of notional agents." The Germans were told that GARBO was being allowed to stay in Britain as a political refugee, and was working freelance for the BBC and the Ministry of Information. This enabled him to gain information from unwitting sources and to recruit conscious agents (all imaginary, of course). The conscious agents were designated by MI5 as ONE, TWO, etc. Unwitting sources cultivated directly by GARBO were labeled J1, J2, etc. (standing for "Juan's 1," etc.), while those milked by numbered agents were designated "3(1)," "3(2)," etc.

ONE, TWO, and THREE were CARVALHO, GERBERS, and the unnamed Venezuelan in Scotland, who had already been well established before GARBO came to England. J1 was the KLM courier; the original J2 was a KLM pilot who was supposedly a backup courier; he was soon dropped and a new J2, a talkative RAF officer, was substituted. The first important product of GARBO-Harris teamwork, one that over time became extremely useful, was J3, a highly-placed official in the Spanish section of the Ministry of Information, who believed GARBO to be an anti-Axis Spanish Loyalist refugee, spoke freely with him, and often shared sensitive information; he was first produced in mid-May. At the end of that month came a new consciously collaborating agent, FOUR, a Gibraltarian waiter in Soho, violently pro-Axis because of resentment against the forcible evacuation of civilians from Gibraltar, who soon was assigned to work in (imaginary) secret underground munitions depots in the Chislehurst caves. (GARBO's eloquence soared in describing these.) He was soon followed by FIVE, THREE's brother (the one who, as MOONBEAM, moved to Canada the next year). Then came SIX, a fiercely anti-Russian South African with good contacts in various government departments, and at the end of the year came SEVEN, a Welsh seaman. Many of these had subagents and regular sources of information witting and unwitting.

The network kept growing through 1943. There was a guard at a munitions facility. There were disgruntled American soldiers and sundry other servicemen and women. There were Welsh fascists, a Communist sailor, a leader in something called the Brothers in the Aryan World Order, an Indian poet and his mistress. Before it was all over there were no less than twenty-seven agents and sources in GARBO's network, embroidered with a richness of detail that flowed without letup from his and Harris's imaginations.

Secret ink communication was slow and irregular and had to be correlated with genuine KLM flights to Lisbon. So in early 1943 a radio link was opened. The Germans had asked for a sample of the chemical crystals from British gas masks; these GARBO sent in the spring of 1943 in a tin of Andrews Liver Salts, together with a long letter reporting that FOUR had a friend with access to a ham radio transmitter, which could be acquired to send messages purportedly on behalf of Spanish Republicans to their underground comrades in Spain. This link started operation in March of 1943. Supposedly operated by a Spanish Republican who did not know what his enciphered messages meant but thought they were being directed to an anti-Franco underground, and actually operated by a peacetime bank clerk named Haines, this link became GARBO's main channel of communication with the Abwehr in Madrid.

The Abwehr in Madrid were swamped with GARBO's information; so much so that they became bemused by him. Although they sent agents to South America and the United States, they never again bothered sending any new ones to the United Kingdom. They never doubted their ARABEL. In the felicitous words of the official historian, the Madrid office "became entirely dependent on GARBO, regarding him as a sensitive quarrelsome genius of priceless value who had at all costs to be humored and satisfied. By the end GARBO and his British case-officer were able to treat his German case-officer as a temperamental mistress might treat an elderly and besotted lover. No assertion that GARBO made might be questioned; no demand unmet."

No double agent operated out of England would ever approach GARBO in importance, but late 1942—the crucial formative months of the London Controlling Section—saw the arrival of another who would prove to be second only to GARBO in his utility to Bevan's operations. This was Roman Garby-Czerniawski, codenamed BRUTUS.

Garby-Czerniawski was a Polish fighter pilot who had escaped to France after the German conquest of his country in 1939 and became an officer in a Polish division in exile. After the French surrender in 1940 he stayed behind and built up an intelligence network, and in October of 1941 he was flown to Britain and decorated for his achievements. Returning to France, he turned up nearly a year later in Madrid saying that his network had been broken up by the Gestapo and that he had been arrested but had escaped. He was brought to England, where after interrogation it transpired that when captured by the Germans he had offered to come to England as a German agent in exchange for the safety of his mother, his brother, and his friends in his network in France; and that in fact he had done this in order to act as an Allied double agent. After some consideration—it was one of the rare cases that the Twenty Committee bucked up to the W Board—it was decided in early 1943 to take BRUTUS on. Christopher Harmer, a young Birmingham solicitor in civilian life, was assigned as his case officer.

So in the latter months of 1942, London control was in place and its principal tools were being laid to hand. It is time to return to Dudley Clarke and the Middle East.

CHAPTER 6

The Turning of the Tide

We left Dudley Clarke at the beginning of March 1942 with his authority confirmed and a happy issue out of the afflictions of the Bagnold affair.

His first order of business was to reorganize "A" Force. The escape and evasion side under his friend Tony Simonds had already been reorganized in January. The deception side was enlarged and divided into a small headquarters or "Control Section" under Clarke—now a full colonel—with a Captain Flynn of the Royal Army Service Corps in charge of office management and paperwork, and Subaltern Margery Hopkins serving initially as the intelligence officer. Under "Control" were an Operations Section responsible for physical implementation, including supervision of the tactical deception officers in the field, and an Intelligence Section responsible for intelligence implementation of tactical and strategic deception plans, including maintenance of existing channels and building up of new ones.

Filling an elite specialized unit with fully qualified officers was not easy that spring of 1942, for most of the best men were with the fighting commanders out in the desert. Clarke did remarkably well under the circumstances. He was attracted to people, often cavalrymen, whom he had known socially before the war. This lent a distinctly Establishment touch to "A" Force.

Four officers made up the Operations Section (plus a space held for Major Wintle). Its original chief, Lieutenant-Colonel H. S. K. Mainwaring, served for only a few weeks. He was replaced by Lieutenant-Colonel Noel Wild. Under Mainwaring and then Wild were Major Oliver Thynne, Captain—as he now was—Bromley-Davenport, and Captain, as he too now was, Philip Druiff, still on Cyprus. Later, in the summer of 1942, Druiff moved on to Beirut. The Intelligence Section was set up for three officers: a lieutenant-colonel's slot that was not initially filled, and two majors, Major Count P. de Salis and Major V. E. Simcock; plus an ATS subaltern's slot, filled initially by a civilian, Alice Hopkinson, and in June by Subaltern M. F. Lassetter of the ATS. In addition to these there were "A" Force officers responsible for tactical deception at the headquarters of Eighth, Ninth, and Tenth Armies and XIII and XXX Corps. As has been seen, these soon faded away. Adding to this the six officers of Major Titterington's Technical Unit, and fifteen more on the escape and evasion side, plus Victor Jones's dummy units and support depot, gave Clarke a substantial "private army" for the crucial campaigns of 1942.

Clarke's new operations chief, Noel Wild, was an old Etonian, small and dapper and somewhat prim-looking—"a slim elegant little man," recalled Went Eldredge—whom Clarke had originally met through a colleague from his days as a war correspondent in Morocco, and encountered again in Palestine in the 1930s. Too young for the Kaiser's war, after finishing Eton Wild badly wanted to get into the Army but had his doubts about passing the entrance examination. An uncle who was a retired general pulled strings to get him a Territorial commission, and subsequently, in 1924, a regular commission in the fashionable 11th Hussars. He was a tank instructor in the 1930s, commanded a squadron of yeomanry in England after Dunkirk, and in 1941 he missed the opportunity to command a squadron in his own regiment, now distinguishing itself in the Western Desert, by being selected for the Staff College at Camberley. When he finally reached Egypt in late 1941, he found himself miserably pushing paper in Grey Pillars.

The only light in the gloom was the discovery that his old friend Dudley Clarke was in Cairo too; doing something mysterious, for his name was not even in the headquarters telephone book. Clarke and Wild renewed their acquaintance; they would meet after work for a late supper and then watch a film at one of the two open-air cinemas in Cairo until the wee hours. Then one evening in April of 1942, when Major Wild was in Shepheard's cashing a check, he was told by a porter that Colonel Clarke was looking for him. He joined Clarke in the bar.

"To your promotion," said Clarke, raising his glass.

"What promotion?" asked Wild.

"To lieutenant-colonel, as my deputy," said Clarke. He would not tell Wild anything about the work, but Wild decided to take the plunge and told Clarke he would do it—whatever it was.

When Wild's boss, General McCreery, heard about this he chewed Wild out. Clarke's operation was just a racket. There were better ways to get promotion. Or was Wild just looking for an easy war? But the die was cast; and at 8 A.M. the morning of April 20, 1942, Wild climbed the stairs to the first upper story of No. 6 Sharia Kasr-el-Nil and learned from Dudley Clarke what he had got himself into.

Noel Wild remained Clarke's right-hand man for deception until he left at the end of 1943 to take charge of deception for OVERLORD. Right-hand man, friend, and confidant; and thirty-two years later it would be Wild who gave the eulogy at the memorial service for Clarke at Chelsea Old Church. But Clarke was almost alone in liking Wild. Nearly everybody else detested him; thought him humorless and arrogant, with a "fancy for his own (and to my mind *very* inferior) intelligence," as one put it.

Oliver Thynne had joined Clarke three weeks before Wild reported. The son of Lieutenant-Colonel Ulric Thynne, who commanded the Wiltshire Yeomanry throughout the 1914 war, he was another Old Etonian and an Oxford graduate as well. Thynne had known Clarke ten years before in the Transjordan Frontier Force days. He had served with distinction in the fighting in Syria and again in Iraq, and then found himself looking for a posting. "Having appreciated [Clarke's] brain ten years earlier," wrote Thynne after the war, "no wonder I applied to him for a job when I wanted an interesting one in 1942!" Though nominally assigned to the escape and evasion side, he found himself active on the deception side as well.

Bromley-Davenport, it will be recalled, had been an early recruit. Physical deception was to remain his particular field. He was less colorful than many of the "A" Force characters, but everybody liked him; the Americans in particular remembered him as "a nice guy."

Over on Cyprus, Druiff—actually his full name was Jessel-Druiff—was cut from yet another cloth. One of his fellow officers, plainly fond of him, recalled him as "a very overdressed man who appears about to burst"; another described him as "of high color, yellow and red, with eyes like mussels about to fall into his lap. He wasn't exactly fat but being shaped like a pouter pigeon with a high bow window in front, no part of his expensive uniform seemed exactly to fit, with the result that every button was manifestly doing its duty." Druiff was Jewish and a strong

Zionist, an intellectual who before the war had been an assistant to Sir William Beveridge, the architect of the British welfare state. He was perceived by his fellows as "a decent enough lad but terrified of his seniors," with "a pathological horror of having to take the tiniest share of responsibility for anything," going through agonies lest he make a mistake; but he was also "a charming, if eccentric, companion," and they grew very fond of him. He proved unexpectedly useful, for meticulous research and detail-checking brought him to peaks of happiness, and he would dictate memoranda by the hour as long as his name did not have to go on them.

The Intelligence Section got off to a slower start. The top slot was not filled for some time; Major Thynne carried out some of its duties. Count de Salis was with "A" Force for only three months and left little mark. Simcock remained till May of 1943 but seems likewise to have left little mark.

More significant in the long run was Alice Hopkinson. She was the wife of Henry Hopkinson, a career diplomat and the Foreign Office advisor to Oliver Lyttelton, the Minister of State whom Churchill had sent out to be the "highest authority on the spot" in the Middle East. His current assignment was not unrelated to "A" Force, for among his functions for Lyttelton was chairmanship of the Middle East Intelligence Committee. Hopkinson wanted to bring his wife out with him, but wives could not accompany their husbands to the Middle East unless they had military or nursing duties. The Hopkinsons had known Dudley Clarke in Palestine in the 1930s, and he came to the rescue: He needed a personal assistant to run his private office and indeed had already put in a request for one; Mrs. Hopkinson would fill the bill nicely. She came out in the autumn of 1941 and went officially on the books of "A" Force in January 1942, working initially both as Clarke's personal assistant and with some intelligence duties. She remained until September 1943. Then she left for Lisbon to join her husband, who had been transferred there in May to be minister of the British embassy.

Hopkinson was another Old Etonian, a contemporary of Noel Wild's. Alice Hopkinson was an American, née Alice Labouisse Eno, of a Louisiana family on her mother's side, and on her father's side an heiress to a substantial interest in Procter & Gamble. In her husband's words,

"Alice had a direct manner and never hesitated to say exactly what she thought or felt. She was golden-haired with large blue eyes, a high rounded forehead and a small but firm chin." But the Hopkinsons were not a happy couple. Alice and Noel Wild became sufficiently fast friends to cause gossip.

In June 1942, Subaltern (later Junior Commander) M. F. Lassetter took over Mrs. Hopkinson's intelligence duties, while the latter continued as Clarke's personal assistant. Lassetter was supposed to be the office expert in Italian, though her knowledge of that tongue proved to be less deep than advertised. She was to remain with "A" Force in Cairo till January of 1945.

Margery Hopkins was to continue with "A" Force till the summer of 1944. She was courted by, and eventually married, a Yugoslav by the name of Radulovic. He was a member of the minor nobility and in due course Clarke dubbed her the "Duchess of Split." Another of the female contingent, formally assigned to the escape and evasion side but sometimes active on the deception side, was a young Englishwoman brought up in Cuba named Helen Mason, known as Peggy or more frequently as "Cuba." Dark and mildly exotic, she was a great horsewoman who sometimes brightened up the office by reporting for duty in jodhpurs.

But the officer who would prove in the long run to be the most important recruit to deception that early 1942 was initially brought on board in late February not on the deception side but on the escape and evasion side. His name was Michael Crichton.

Crichton was yet another Old Etonian, in civil life engaged in business in the City of London, and in military life a major in the Horse Guards. He had solid roots in the aristocracy; the Earl of Erne was a close relation, and his father had been equerry to King George. He had distinguished himself in the desert with the 7th Armored Division, the legendary Desert Rats, and had been closely connected with the Long Range Desert Group. Clarke had met Crichton years before, at about the same time he met Wild. He sounded Crichton out about joining him; but Crichton was not interested. His most recent assignment had been as a staff officer with the Desert Rats, but early in February his boss, the much-loved General "Strafer" Gott, had moved up to corps command, and he was anxious to get back into the field.

A favorite Cairo nightspot was the roof-garden restaurant of the Continental Hotel. It had a dance floor and a simple cabaret. The mistress of ceremonies was a pretty American blonde who introduced each act and then finished the show with a solo dance, prefaced always by the same little speech: "And now, introducing myself, Betty to you, in a rumba"—or whatever dance she was doing that night. So her fans always called her Betty-to-You.

Betty-to-You, officially named Alice Sims though always called Betty, was the daughter of schoolteachers from Wabash, Indiana. She was very much part of the Cairo scene. She had a convenient flat, with bath, which many of the young women in uniform would borrow, for it was not always easy to balance social life with military restrictions. After the show at the Continental, the regulars would gather with Betty in the lobby and sit talking into the small hours. Officers who had come in to Cairo from the desert made it a point to join Betty's soirées. Clarke always declared that the best way to find out what was happening in the desert was to buy Betty a drink and tell her what you wanted to know.

One evening Clarke and Betty were joined by Crichton for a drink. It appeared that Crichton and Betty were good friends. When Crichton had left, Clarke asked Betty if she knew him well. Learning that she did, he said:

"I want him in my office. He just asked you to lunch tomorrow. Do it and tell him I want him."

Betty demurred but finally agreed, and at lunch next day at Groppi's she told Crichton: "That man Clarke wants you."

Crichton said he did not want an office job.

"I promised you would be there at three," said Betty.

So, reluctantly, he went. And got a full dose of the Clarke charm; and before he knew it he found himself with an office job, in "A" Force.

Nobody ever met Michael Crichton who did not like him; and in addition he was a most efficient officer. "An individual of great charm and kindness," who nevertheless "suffered fools neither gladly nor at all," recalled one colleague. Tall and handsome, he was aristocratic with no trace of snobbery, courteous and gracious with no trace of stuffiness. He got along as well with Americans as with Englishmen. He never returned to the front as he had wanted to do; he became Clarke's No. 2 after Wild left for SHAEF, and ended the war as Bevan's successor as Controlling Officer in London.*

This was the organization that took "A" Force through the great year of 1942, with one addition. In the summer of that year the Germans resumed their offensive in southern Russia, and by early August the panzers were in the Caucasian foothills. With this potential threat to the "northern flank," Iran and Iraq were broken off from the Middle East theater in late August and a new Persia and Iraq Command was formed. To go with it, a three-officer branch of "A" Force christened "Advanced Headquarters 'X' Force" was created. Initially only two were assigned to it, both transferred from Cairo: Major Thynne to handle deception, and an officer for escape and evasion. The senior slot was never filled. A Lieutenant-Colonel L. W. G. Hamilton was appointed to it in December 1942 but was diverted to other duties. Major David Strangeways, whom we will meet later, was then destined for the post, and left London in January 1943 to take it up; but when he reached Cairo he was diverted to become "A" Force representative with Alexander's new 18th Army Group in Tunisia.

As the organization expanded, so did the flow of paperwork to control it. It was in March 1942 that Clarke inaugurated the system of " 'A' Force Instructions" and "Strategic Addendum" already described, through which he managed "A" Force's deception activities for the rest of the war.

Physical deception saw a reorganization and expansion in the first part of 1942 as well. Victor Jones acquired a new third dummy tank regiment, and organized his force into the "33d Armored Brigade." It saw a great deal of action in May and June; the frequency with which it came under enemy attack testified to Jones's skill. By the beginning of July and the last stand at El Alamein, Jones had no serviceable dummies left, his vehicles were in bad shape, and the crews of one regiment were withdrawn to fight as infantry; nevertheless he was ordered to put a new dummy

---

* Another personnel change with an indirect impact on Clarke and on deception in general was the dismissal of Brigadier Shearer in late February as theater Director of Military Intelligence, having been made a scapegoat for the unexpected disaster of Rommel's attack. His immediate replacement, Lieutenant-Colonel "Freddie" de Guingand, would eventually be Montgomery's chief of staff and undoubtedly learned useful lessons in deception from his tenure as DMI Middle East.

brigade back in the field as soon as possible. By drawing heavily on the dummy units in Cyprus and Iraq and on the dummy glider detachment, he was able by the second half of July to field a renewed and renumbered "74th Armored Brigade." It saw valuable service in the fighting of August and early September, and had a vital role at El Alamein in October.

A dummy tank brigade, the "303d Indian Armored Brigade," was also formed in the Persia and Iraq Command, but with the fading of the threat from the Caucasus its sole remaining regiment was transferred to India in April of 1943, having never carried out any operations.

There was other activity in the dummy business that eventful spring and summer of 1942. The prototype DRYBOB and WETBOB dummy landing craft were developed. SAINT dummy parachutists began to be produced, and were first tested in September in an unsuccessful effort to divert the Germans away from a commando raid on Tobruk.

Building on Bromley-Davenport's earlier efforts at Halfaya, "A" Force was authorized in May to equip an armored car with sonic deception equipment. Christened SONIA, it was ready at the end of June but encountered a decree that no sonic deception equipment was to be used without express authority from London. In the July crisis Auchinleck himself interceded with the War Office to obtain grudging permission to try SONIA out in the defense of the Delta. Late that month Bromley-Davenport ran an encouraging experiment with simulated tank noises in company with some of Jones's "tanks." SONIA was taken back for further improvements, while work was started on a sister car, to be named ORA. SONIA did good service in the opening phase of Montgomery's attack at El Alamein, when on several occasions German units were persuaded by SONIA that an infantry attack had tank support, and once her loudspeakers were used to persuade a German unit to surrender. But by the time ORA was fully ready, in early November, Rommel was in retreat and further sonic deception by "A" Force was suspended, to be resumed only under unsatisfactory circumstances in Italy in 1944.

March and April 1942 saw also one small deception plan, one small "dirty trick," one large deception plan, and, most important by far, the beginning of large-scale order of battle deception.

The small deception plan was called CRYSTAL. It related to the arrival, beginning the previous November, of the new American M-3 "General

Grant" medium tank. CRYSTAL was designed to lull the Germans into failing to prepare to meet these new tanks, by spreading rumors that most of them would be held back for training purposes until problems with their ammunition were solved. CRYSTAL came to a sudden end with Rommel's resumption of the offensive on May 27, and no results were ever known.

The dirty trick was referred to as "H.D.F.," standing for HEADACHE FOR DER FÜHRER. Its object was to impair the efficiency of one particularly troublesome regiment (the 8th Panzer Regiment) whose most recent commander, a known anti-Nazi, had been transferred to Germany, by sowing suspicion that his successor was in on some kind of anti-Hitler plot. Titterington's Polish forger prepared a letter purportedly emanating from the colonel in Germany to his successor, saying that he had seen two (named) senior generals, who had told him that "the move they had often discussed" was coming along satisfactorily, and that they had asked him to dissuade his officers from any premature step prejudicial to "this great affair." It even hinted that Rommel was showing signs of coming over to their side. With a cover letter purportedly smuggled out by a German prisoner, this was mailed in Istanbul to a known German agent. There was no way to confirm the result, if any, of what Clarke after the war called "this admittedly shabby trick on some good soldiers"—the only one of its kind that "A" Force ever attempted—though it was known that the commander and more than a dozen other officers of the regiment were replaced.

The large deception plan was designed to support a counteroffensive planned by Auchinleck for mid-May or early June once the front stabilized at the Gazala line. On March 21, he directed Clarke to prepare a deception plan—which received the name FABRIC—the object of which would be, on the tactical or operational level, to induce Rommel to keep his reserves well forward in the northern sector of the front, so as to open his right and rear to an armored thrust from the south; and, on the strategic level, to cause him to relax his guard in the belief that there would be no British offensive before the beginning of August.

Rommel could hardly be expected to believe that Auchinleck would sit on his hands for four months. The best Clarke could come up with was the well-worn stratagem of stressing British worries over the Iraq-Iran "northern flank." So the strategic "story" adopted was that British policy was to remain on the defensive till the beginning of September, when reinforcements would have brought their strength back up to the level as of the end of 1941. Meanwhile, as many troops as possible would be withdrawn from the Gazala line to guard Cyprus and the northern flank, while the "Desert Rats" would be held as a GHQ mobile reserve.

This "story" was spread as usual through ENTWHISTLE, SIME's rumormongering system. Commander Wolfson opened up for FABRIC the channel to a German-language newspaper in Central Europe (presumably the *Pester Lloyd*) which has already been mentioned. Order of battle deception from Operation CASCADE, presently to be described, encouraged the idea that formations were being withdrawn from Egypt for use on the northern flank. All shifts of units eastward were made in daylight with no effort at concealment, while movements toward the front were conducted in strict security. Arrangements were made (Plan MAIDEN) for a seeming visit to London by Auchinleck, timed for the eve of the offensive, complete with letters directed to him at suitable addresses marked "To await arrival."

In addition, in consultation with London—for the lessons of ADVOCATE were fresh in mind—the correspondent of the *Daily Mail* was enlisted to write a story on how difficult it would be for Rommel (hence by implication the British) to mount an offensive in the intense heat of May through July. Unfortunately, his competitor of *The Times* ran a story a few days later saying that the idea of a summer lull "in no way represents the views of the senior military commanders." The experience reconfirmed the general unwisdom of trying to use the Allied press for deception.

The tactical situation kept shifting, and as a result the tactical and operational side of FABRIC ran through a bewildering set of changes. FABRIC generally became more and more frayed as April and May wore on; for signs of a renewed German offensive increased, and Eighth Army engaged in increasingly energetic activity that made it ever harder to maintain the pretense that the desert force was being weakened. On May 28, Rommel burst forth in a new offensive that was to take him to the gates of Cairo, and FABRIC became irrelevant.

Nevertheless, FABRIC taught some useful lessons about tactical deception: That it is pointless to try to make long-range plans on that level, for the situation keeps changing; and that tactical deception should be planned and executed on the operating level, with the theater exercising only general supervision.

Infinitely more important was CASCADE. It initiated the large-scale long-term order of battle deception that was Clarke's and the Allies' crowning achievement.

Clarke had learned that it was a tedious process to build up a notional division to the point where Axis intelligence would regard it as accepted, and then to continue to keep its existence alive in enemy minds by passing on occasional false pieces of news. He decided that the time had come to consolidate his ad hoc planting of fictitious units on Axis intelligence into a single long-range plan, creating in the enemy's mind a picture of the Allied order of battle in the Middle East that would give "A" Force a comfortable margin to play with in any future deception plans. It was dubbed CASCADE.

CASCADE was first sketched out in early March of 1942. Its original main object was to deter the Germans from threatening the Middle East in the event of a spring offensive against the USSR, by exaggerating in German minds the total strength available to the British in the Middle East theater. In its initial version, CASCADE was designed to add—in the enemy's mind—eight new notional divisions, two imaginary armored brigades, and the 1st SAS Brigade to the fifteen actual divisions (plus a miscellaneous collection of nondivisional formations) already at Auchinleck's disposal.

After a struggle to get such an audacious idea accepted by the military bureaucrats in London, Cairo, and Delhi, Clarke found himself with an agreed overall order of battle and carte blanche to put it over to the Axis by any and all means. Chiefly these included planting through intelligence channels, physical implementation, and administrative measures.

Administrative measures involved getting the names of the fake divisions on as many official documents as possible, sowing general belief in the existence of these formations, and recognizing that sooner or later the Axis would capture some of these documents. Physical implementation involved setting out dummies to some extent, but a more important physical measure was to display the insignia of the fake units where Axis agents and witting or unwitting Axis informants would see them. A distinguishing insignia, or in British parlance a "divisional sign," was allocated to each notional division, and a genuine smaller unit "adopted" each notional division's sign and displayed it on its vehicles and in other ways. Thus, the vehicles of the Royal Armored Corps Base Depot in Cairo carried a white unicorn denoting the fake "15th Armored Division"; the role of the bogus "2d Indian Division" in Iraq was played by the Tenth Army Lines of Communications troops, bearing the divisional sign of a yellow hornet on a black rectangle; and so on.

Each of the notional divisions and independent brigades was equipped with a detailed history setting forth its origins and accounting for its current role and location. Appropriate notional subsidiary regiments and battalions were designated for each. A complete "Book of Reference" of the notional order of battle was provided to all persons responsible for feeding false information to the enemy, so that they could operate freely without concern that information passed through one source would be contradicted by information passed through another. Headquarters maintained looseleaf notebooks on all units both real and bogus, showing for each its current real situation; what views as to its role and location it was desired for the enemy to hold; and what was known about the enemy's present state of knowledge and belief about that unit.

Though CASCADE began in the spring of 1942, it did not go fully into effect until July. Its "July 1942 edition" continued until March 1943, when six more imaginary British divisions and a bogus American armored regiment were added to the roster. In May 1943 two more were added; in July 1943 a fictitious army, the Twelfth, was introduced, together with a new bogus division (while two others were removed). Then in December 1943 a bogus British corps composed of three phony British divisions was added. By the time CASCADE was superseded by a similar new plan called WANTAGE in February 1944, it was supporting one bogus army headquarters, three bogus corps headquarters, seventeen bogus divisions, and two bogus independent brigades.

CASCADE succeeded beyond all expectation. As early as May 1942, Axis intelligence was overestimating British strength by 30 percent, and three CASCADE divisions were already carried in their order of battle. By November 1942, Liss's Fremde Heere West had accepted seven out of the original eight bogus CASCADE divisions, plus the 12th Indian Division and an American tank regiment. By January 1944, Fremde Heere West, now under Roenne, had accepted almost every notional formation that CASCADE had offered, plus others.

The first full flowering of Dudley Clarke's genius, CASCADE made possible all the successful deceptions in the Mediterranean from 1943 to the end of the war. And it taught some important lessons: notably, that if you are going to enter upon strategic order of battle deception, both your administrative and deception staffs have to be prepared to go through with it thoroughly and exhaustively. "The amount of inconvenience and extra work it involves should deter anyone from embarking upon it lightheartedly," said Clarke looking back after the war; "for, once started, it remains a standing dish which can never be neglected nor abandoned, and which may imperil both military plans and delicate Deception machinery if not tended with sufficient and regular care."

The German offensive that put an abrupt end to FABRIC went from victory to victory. When at the end of June Rommel broke through and began racing eastward it looked as if the Germans would be in Cairo and Alexandria any day. The Royal Navy weighed anchor and left Alexandria precipitately, and this set off what was always known as The Flap. As thousands of truckloads of weary and defeated soldiers streamed into the city from the desert, shopkeepers began readying black-white-red bunting and pictures of Hitler to adorn their windows. Nationalist groups headed by Anwar Sadat and others prepared to welcome the liberators. Taxi drivers jovially told their British passengers: "Today, I drive you to Groppi's, tomorrow you drive me." British women and children jammed the railway station trying to get on the few trains for Palestine or upriver. There were runs on the banks. On July 1, known thereafter as "Ash Wednesday," a blizzard of paper ash filled the air and columns of smoke rose over Grey Pillars and the British Embassy, as file after file went into the flames.

"A" Force was caught up in The Flap like everyone else—especially its escape and evasion side. If Rommel broke through, many British soldiers would be cut off and left to fend for themselves. It would be "A" Force's task to round them up and help them rejoin the main body or, if necessary, form them into a guerilla force based in the bewildering maze of the Nile Delta.

Plans were swiftly laid. If Cairo should fall, a new "Advanced Headquarters 'A' Force," including a cadre of Titterington's unit, would be set up in Jerusalem. Clarke, Wild, and Simonds would stay behind, with a small technical unit run by a Polish cadet officer named Redo, a signal party, a demolition party, and a body of dummy tanks. Stocks of ammunition, food, and fuel were buried at three "hideouts": an initial one at the shop at Rod-el-Faraq on the northern edge of Cairo where the dummy tanks were made; an intermediate one in a school building five miles further north near the village of Shubra; and a final headquarters for guerrilla operations at a site near the only possible landing-place for a light airplane inside the Delta. In Cairo itself, a perfume factory and a bookstore were taken over to serve as communication centers and safe houses, a laundry van was purchased for covert transportation, and a yacht and some motor launches were acquired for escape southwards up the Nile.

An astonishing number of civilians undertook to help find stray soldiers if the city should fall. Outside Cairo and Alexandria, the Freemasons offered to set up and operate secret escape centers in every village. A joint team of smugglers and police volunteered to run an escape route across the Red Sea hills to the Gulf of Suez via the established smugglers' trails.

For two days, files of "A" Force went up in smoke at No. 6 Kasr-el-Nil. On "Ash Wednesday" the most important permanent records went off on the last southbound evacuation train, bound for Aswan and ultimately for London. That same afternoon, "Cuba" Mason and Margery Hopkins left by train for Jerusalem with the remaining operational files. By the end of the week Clarke and crew were poised to move out on twelve hours' notice. Their transport was loaded, equipped with paint and license plates to convert them into civilian vehicles at the first hideout. (The license plates had not been stocked up in advance; Major Simcock rectified the omission by stealing them off of cars parked outside the Muhammad Ali Club while their owners were at lunch.) They even had a captured German car, laden with a supply of Axis uniforms, weapons, and traffic signs.

And then The Flap began to wind down. Auchinleck had held Rommel at the First Battle of El Alamein. Clarke called back the Jerusalem party on July 6. Within another fortnight it was clear that the Auk had stopped Rommel for good. As the scattered elements of "A" Force trickled back to No. 6 Kasr-el-Nil, Clarke had to confess a certain regret that there had been no chance to put his emergency scheme to a practical test.

The Flap made one happy contribution to life at "A" Force: Michael Crichton and Betty Sims got married. A romance had been blooming, and with Rommel at the gates and no knowing what the next days or weeks would bring, they decided to tie the knot. Betty went to the American consulate for some necessary documents. It swarmed with pushing, shouting Americans desperate to get out. She managed to make her way to the harried consul and stated her business. The consul could hardly believe his ears. He leaped to the top of the table.

"Look at this lady!" he called out to the crowd. "You're all frantic to get out, and she's not trying to get out, she's staying and getting married!" But it did no good. Betty got her papers and made her way through the crowd, still pushing and shouting, back to Grey Pillars. There she and Crichton were married by the Judge Advocate General; and then Crichton had to go back to work. "I may not see you for a while," he told her. She in turn was supposed to go up the Nile in one of the motor launches that Clarke had acquired. But The Flap subsided, and a few days later they had a proper wedding in a little church in Cairo, with just a few witnesses, including a Free French pilot. Dudley Clarke gave the bride away, and the English aristocrat and the girl from Indiana lived happily ever after.

The Flap brought professional as well as personal happiness to "A" Force; for Cheese came back to life.

It will be remembered that after the Crusader offensive in November 1941, Cheese had seemed to be irretrievably blown. But Captain Simpson was convinced that with imagination and perseverance the channel could be rehabilitated, and he set out to do so.

First the fictitious Nicossof sent a series of messages through which he little by little dissociated himself from the friends who had given him such bad information. Then he began to take the line that he was continuing to operate solo in the hope that the Axis could pay him. He told his control that he was alone, had no funds, and could not hire agents. On this basis Simpson kept him going with only the lowest-grade chicken-feed. He complained endlessly about his poverty. He had supposedly been left with £230 by Levi when the latter went back to Italy. From this he had supposedly paid £75 to Piet in November 1941; and by early 1942 his money had run out. In June he told the Germans he needed £1,000 to pay his debts and line up some new agents.

His control in turn kept promising to get money to him. But no money ever came. NICOSSOF grew sarcastic and insubordinate. He threatened to sell his radio set to pay the debts he had piled up in the Axis cause. This went on for months, with NICOSSOF sending two transmissions each week. Through it all, Simpson applied his novelist's skill to developing a coherent picture of NICOSSOF's personality, "gradually building up," in Clarke's words, "the picture of a moody, mercurial Slav-with-Levantine-blood, who sank easily to the depths of despair when no money came his way but revived without fail on every specious promise. So well did he handle the fiction . . . that he established a situation in which it began to look as though only pressure and encouragement from the enemy was preventing an unwilling agent from chucking up the whole business. This, he argued, could never come about with a channel which was working under British control!" The main concern was that the Axis might write NICOSSOF off, or that it might seem too implausible that he would keep working without getting paid.

This last concern was dealt with by perking up NICOSSOF's interest when Rommel began his offensive. Enthusiastic over the imminent arrival of the Desert Fox in Cairo, NICOSSOF began sending more and more good information. By late June it appeared the CHEESE channel was sufficiently rehabilitated to be used once more for strategic deception material; and this was confirmed, to the joy of SIME and "A" Force, when on July 2, the day after "Ash Wednesday," CHEESE's control radioed to him: "Be very active these days. Good information will be well rewarded. From now onwards we are going to listen in every day for your signals."* On July 4, ISOS revealed that the Abwehr regarded NICOSSOF as "credible," and by July 12 that he had been upgraded to "trustworthy"

---

* This appears to have been the point at which control of CHEESE was taken up by the Abwehr in Athens directly rather than through Bari. At the end of June, the Americans discovered that the Axis were reading the code used by Colonel Bonner Fellers, their military attaché in Cairo, and his comprehensive and informative messages about British plans and the situation in Egypt abruptly ceased. Rommel had relied heavily on this priceless source of intelligence since it first became available in January 1942; the sudden new German interest in CHEESE was presumably part of an urgent effort to come up with something to replace it. To add to CHEESE's importance, Rommel lost his other most important source of intelligence a few days later, when Australian troops killed the highly efficient chief of his "Y" service, Captain Alfred Seebohm, and captured his files, leading the British to tighten their hitherto lax signals security.

and that CHEESE reports were henceforth to be passed directly to Rommel's headquarters.

It proved to be a permanent rehabilitation. From that point until it was finally shut down in October 1944, CHEESE continued as one of the four top special means channels for Allied deception, along with GARBO in England, GILBERT in the western Mediterranean, and SILVER in India. A committee, chaired by Clarke, met each day at No. 6 Sharia Kasr-el-Nil to prepare the day's message; to Major Thynne, until he left for Baghdad, fell the responsibility of supervising the transmissions (almost always, alas, at dinnertime). By contrast with the first half of 1942, when CHEESE sent no valid military information other than three worthless items, in the last half of the year he sent no less than ninety substantive messages and dealt with six major subjects of strategic deception. With the channel rehabilitated* it was time for NICOSSOF to spread himself a little. Sergeant Shears was being worked to death and at least once had had to be carried from the hospital on a stretcher to send a NICOSSOF message in the authentic fist. So, beginning in July, NICOSSOF was furnished with a notional backup wireless operator and general factotum in the person of an imaginary girlfriend codenamed MISANTHROPE by the British. Identified to the Germans as "Marie," she was notionally a clever young Greek of good education who hated the British and had a wide circle of acquaintances in Cairo military circles; she not only picked up information for NICOSSOF but over time learned to encipher his messages and operate his transmitter.

If Rommel was not to reach Cairo, NICOSSOF would have to continue importuning the Abwehr for money. In July he told his control he now needed fourteen hundred pounds. On August 4, the Royal Navy dis-

---

* Clarke was given to producing light verse. At some point he celebrated NICOSSOF in rhyme. (Note to American readers: The pun is on "knickers off," "knickers" being the British term of the era for what Americans call "panties" or "drawers.")

> Nicossof's a Russian name
> And not what you might think,
> A form of Oriental vice,
> Or buggery, or drink.
> A scion of this noble house,
> An unattractive sod,
> Was Stanislas P. Nicossof
> Of Nizhni Novgorod.

abled a U-boat off Haifa and rescued the survivors. One turned out to be a Lebanese student named Hamada, a known German agent, who had in his possession roughly fourteen hundred pounds in a variety of currencies, together with instructions to hand it over to someone named "Paul" in Cairo. While reassuring, this incident would not help to pay NICOSSOF's bills. He kept begging for money, and after Rommel's last effort to break through failed in the first part of September, "the cries were raised to a scream," in Clarke's phrase.

The Germans kept trying. On October 5 they alerted NICOSSOF that they were planning a new means of getting money to him and needed an address for a rendezvous; and on October 12 they told him delivery had been arranged. "A" Force and SIME did not want an Axis courier, and maybe confederates of his, poking into CHEESE's affairs. A rendezvous in a public place, but not too public, seemed safest. The Café Bel Air on the Rue des Pyramides a few miles outside Cairo was finally selected. NICOSSOF radioed back to the Abwehr that his *amie* Marie would meet the courier at a rendezvous that would be fixed when the date was set.

For this adventure Clarke and SIME produced a real-life "Marie." She was, in Clarke's words, "an attractive Cretan girl with some past experience with espionage whose personal experiences under German occupation had imbued her with a determination to make her own contribution to the defeat of the Axis," and who "had given a good account of herself already before fleeing from Crete to Egypt and proved well capable of meeting any emergency." It transpired that she carried a small revolver in her handbag and knew how to use it, so at "A" Force she was known always as the "B.G.M.," standing for "BLONDE GUN MOLL." (She was said to have disposed of at least one man with her revolver, though there were those who claimed that she had actually only pushed him off of a roof.)

For a full month, beginning on October 9, the B.G.M. lived in a SIME safe house, absenting herself from her customary haunts, changing her appearance, and undergoing intensive coaching on her role and on comprehensive details of her imaginary history and her relationship with "Paul Nicossof." She displayed marked talents for the game, and offered many useful suggestions of her own. On October 25, Athens set a date for between November 10 and 15 and provided a password. NICOSSOF responded that Marie—who he said would be "dressed in white, with a red belt and red bag"—would be waiting each day from 8:00 to 8:30 in the Café Bel Air.

A meticulous scenario was worked out. The B.G.M. was to learn what she could from the courier about himself and his confederates, and then invite him to come back to Cairo with her in her taxi to meet NICOSSOF at his flat, or make an appointment to meet the courier there later in the evening. Depending on the degree of confidence that any Axis agents tailing her had been shaken off, the courier would be arrested either on the road to the flat or in the flat itself (a SIME safe house, of course). The B.G.M. rehearsed the scenario until she was letter-perfect.

Early in the evening of November 10, the B.G.M., properly dressed in white with red belt and handbag, drove out to the Café Bel Air in a taxi. The cabdriver was a SIME agent, and two more were hidden nearby to watch for signals from the B.G.M. But nothing happened. She waited till 8:45 and then went home. It was the same story next evening, November 11. NICOSSOF radioed a complaint to Athens. Athens responded plaintively: "Do you think this is all so simple?" They promised a new date, but did not deliver. On November 24, they radioed that something had gone wrong.

After a couple of uncomfortable weeks for SIME and "A" Force, another message came in from the Abwehr: "The money is already in Cairo. Give us the address of your *amie* and it will be delivered there inside a bottle of milk on an agreed date." This time there seemed no way out of giving the Abwehr a real address. So NICOSSOF radioed to them the address of a small flat kept by SIME at 20 Rue Galal; the residence, he said, not of Marie, who for lack of money had had to give up her own place, but of a reliable Italian friend.

A genuine Italian was installed in the Rue Galal flat, a questionable character over whom SIME exercised control. But before the delivery of the milk bottle—which the Abwehr had set for December 27—the Egyptian police suddenly raided the flat and searched it, apparently looking for something on the Italian. This rather rattled SIME and "A" Force, but they decided it was safest to ask no questions. NICOSSOF radioed word of the raid to the Abwehr and asked that delivery be stopped. Athens responded that it was too late. But December 27 came and went, with no milk bottle and no money. Athens then said that they would try again on January 25. SIME opened up a new flat and installed the

B.G.M. in it to take delivery. But on January 25 delivery was postponed once more.

Meanwhile, NICOSSOF was now desperate for money. Angry at his German employers—but not angry enough, of course, to quit sending them information—he began to look for a job.

German efforts to develop new intelligence sources after losing Colonel Fellers's messages produced SLAVE, a young Egyptian journalist recruited by the Germans, through whose transmitter some "A" force items were sent in May–July 1942; an Egyptian gang dubbed THE PYRAMIDS, who never got in satisfactory touch with the enemy; and two more famous agents named Eppler and Sandstetter (or Sanstede), codenamed FLESHPOTS by the British, who were unusable as double agents because of publicity attending their capture. Then in October two channels, known as QUICKSILVER and THE PESSIMISTS, opened up that were to provide invaluable service for the next two years.

QUICKSILVER was George Liossis, a Greek air force officer and a nephew of General Liossis, a top-ranking Greek officer in Cairo. He may have been in touch with British intelligence as early as the collapse of Greece in April 1941; in any event, he offered himself as an agent to the Abwehr's Athens station as a means of getting out of occupied Greece. The Abwehr gave him some rudimentary training, and on August 5 dispatched him with a transmitter and two companions by caïque from Athens bound for Syria with orders to transmit from somewhere near Beirut.

The companions were a young Greek seaman codenamed RIO by the British, and a young Greek woman codenamed GALA. Both were unsavory. RIO, who real name was Bonzos, was a thug who had worked for the Gestapo in Greece but whose methods and extracurricular activities—extortion, blackmail, murder—had dismayed even that organization; they were happy to fob him off on the Abwehr and get him out of the country. GALA, whose real name was Anna Agiraki, was the mistress of the local chief of the Italian SIM. She had grown difficult, to the extent of attracting the OVRA's attention; SIM was glad to concur with the QUICKSILVER project provided GALA was sent along too.

Forewarned by ULTRA, the Royal Navy intercepted Liossis's party off the coast of Latakia on August 20. Liossis readily undertook to work for

SIME. As QUICKSILVER he established radio contact with Athens in September and was transmitting on behalf of "A" Force from October 16. He was to be a mainstay of the double agent system for the next two years, transmitting several times a week and receiving frequent questionnaires from his control. "Military and naval information was sent across," as Clarke recorded after the war, "which originated, notionally, either from QUICKSILVER's own observation, or from indiscretions which the attractions (also, alas, largely notional) of GALA might draw from an ever-widening circle of Allied officers around the cafés and cabarets of the Port of Beirut." Notionally, GALA operated as a high-level prostitute. The lady herself was under lock and key in Palestine.

So, naturally, was RIO. For the Germans' benefit, however, he was notionally called up for service in the Greek Army, and then transferred to the Greek Navy for service aboard a (genuine) Greek destroyer. He would notionally send coded letters from Mediterranean ports at which his ship called to QUICKSILVER, who would pass on the information to Athens. According to Clarke, RIO "acquired quite a reputation as an expert on the signs which heralded amphibious operations," and later "became also an astute observer of new anti-submarine weapons."

QUICKSILVER himself, Liossis, though unquestionably loyal, was not allowed to run loose; he was kept effectively under house arrest in a small villa on a steep stretch of the road from Beirut to Aley (the town in which Ninth Army headquarters was situated), guarded by a company of Africans who spoke no English or indeed any other tongue known north of the Sahara. He was a rotund little man who reminded one observer of the Tenniel illustrations of Tweedledum and Tweedledee, complete with an expression of "glowering concentration" which would break briefly into a cheerful smile by a show of friendliness, a Greek treat for dinner, a victorious card game, or a joke, "preferably at the expense of the Italians." His semi-imprisonment made him fretful. Not only was it boring, but lack of feminine solace caused him much misery. He would occasionally attempt to escape. He succeeded at least twice, finding on both occasions the comfort he sought.

Almost from the beginning, QUICKSILVER had in his villa the companionship of a second valuable double agent, one of another three-man party of German agents known to the British as THE PESSIMISTS. Officially codenamed PESSIMIST Y and often referred to by the British as JACK (to the Germans he was MIMI, short for Demetrios), he was an Alexandrian Greek, a professional singer by trade, trained as the wireless operator of the party. Of his two colleagues, PESSIMIST X, the leader, was an Italian, half Swiss in ancestry, while PESSIMIST Z was another Alexandrian Greek, yet another thug, with a drug-smuggling conviction on his record. PESSIMIST X was young, the other two middle-aged. All spoke three or four languages, they had received better training than had Liossis's team, and in Clarke's words they "were altogether a superior gang." Perhaps because the Germans recognized this superiority, they financed THE PESSIMISTS with a generous fund of six thousand dollars U.S.; thus eliminating most of the payment problems that troubled the CHEESE and QUICKSILVER channels.

Assigned to establish themselves in Damascus, THE PESSIMISTS landed near Syrian Tripoli in a small boat from a refugee schooner in the small hours of the morning of October 20. The British, alerted by ULTRA, arranged for the French Sureté Générale to pick them up and hand them over to SIME. PESSIMISTS X and Z were imprisoned. PESSIMIST Y had been in touch with MI6 in Athens before signing up with the Abwehr, and was soon established as a double agent working back to Athens; contact was established by November 14. He was moved in with Liossis in the house on the Beirut-Aley road, which came to be called "Q and P Hall."

PESSIMIST Y resembled QUICKSILVER physically; indeed, viewed together the Tweedledum and Tweedledee effect was remarkable. Morally there was less resemblance. Liossis was a genuine patriot; PESSIMIST Y, said a British officer who worked with him, "was a fat, oily, plausible slob, the blueprint and prototype of the universal spy. . . . I think that was why we called him Jack—on the basis of F— you, Jack." He wore a permanent smile and tried unsuccessfully to ingratiate himself with all and sundry while privately goading QUICKSILVER to periodic revolt. His original case officer was satisfied that he would sell out the Allies as he had the Axis, unless regularly reminded of the possibility of a firing squad.

Q and P Hall was so located that German direction-finders might plausibly confirm both that QUICKSILVER radio transmissions emanated from Beirut and that PESSIMIST transmissions emanated from Damascus. It bristled with radio masts, most of them dummies. The intended effect was of a radio intercept station, accounting to the curious for the tight security.

The case officer for the QUICKSILVER and PESSIMIST rings was Captain John Wills, a competent linguist who before the war had been the Paris representative of the Federation of British Industries. Under his supervision and the guidance of the local 31 Committee, THE PESSIMISTS soon ranked with QUICKSILVER in importance. PESSIMIST X notionally traveled round Syria gathering information for PESSIMIST Y, and from time to time, at the request of Athens, wrote letters in secret ink to Axis contacts. PESSIMIST Z notionally established a link with a smuggler operating into Palestine, and notionally got a job with a cross-desert transportation company which enabled him to visit Iraq frequently. THE PESSIMISTS were restricted to relatively low-grade information—it would not have been plausible for them to gain access to higher echelons—but their material was no less useful for that reason, particularly in connection with order of battle information.

In November 1942, an additional channel in this region was made available to the British through the good offices of the Turkish secret services: DOLEFUL (also called DOMINO), a Wagon-Lits attendant on the Taurus Express between Istanbul and Baghdad. A Turk who had served the German Army in 1918 and now passed information from Baghdad to a former German officer in Turkey, he was useful but never fully trusted, and utilized only for low-level information.

In June and July, notwithstanding The Flap and the crucial fighting, "A" Force mounted two strategic operations that set the pattern for maintaining a threat to the Axis in the eastern Mediterranean.

The first, called PATENT, followed on the threat to the Dodecanese that had been mounted in April in connection with the invasion of Madagascar. ISOS revealed that Axis apprehensions over the Dodecanese continued, and in early June "A" Force got orders to heat up the threat in hopes of diverting Axis attention from an important convoy to Malta. PATENT was implemented mainly through SIME's familiar intelligence channels; in addition, when assault ships were notionally setting out for the Dodecanese, all telecommunications and travel from Egypt, Palestine, Syria, and Iraq were suspended for forty-eight hours—the first use of this trick. After the Malta convoy got through—with severe losses, though PATENT did succeed in creating alarm throughout the Aegean—PATENT was kept going under the "story" that Rommel's offensive had forced a last-minute postponement of the Dodecanese operation.

In early July, the focus of PATENT was quickly shifted to Crete, in hopes that German forces would be kept there rather than reinforcing Rommel. While that proved a false hope, the idea itself seemed sound for the longer term. German sensitivity to threats to Crete had been made apparent in the first months of 1942. After Pearl Harbor, "A" Force had mounted a modest operation, called ANAGRAM, to cover the recall home of Australian forces in the Middle East by advising Axis intelligence that the conspicuous preparations for that move were in fact preparations for an invasion of Crete. Apparently as a result, the Germans did in fact reinforce the island. It seemed worthwhile to play on their concerns again.

So late in July, "A" Force launched RAYON, a more elaborate threat to Crete the goals of which were once again to distract attention from a Malta convoy, and in the longer run to discourage reinforcement of Rommel. Its "story" was that a combined seaborne and airborne operation would be mounted against Crete in mid-August, employing Greek and British troops, with Cyprus-based American planes contributing to air support. Reliance on Greek troops and American air in the "story" was designed to avoid any suggestion that forces might be withdrawn from Ninth Army on the "northern flank" or from Eighth Army before Cairo.

Physical implementation was important to RAYON, for Axis reconnaissance planes easily reached as far as Suez. The new DRYBOB and WETBOB dummy landing craft were used for the first time, while dummy gliders simulated airborne preparations. Greek troops accompanied by a Greek destroyer were put through actual amphibious training, and American air forces in Cyprus mounted a display of activity. Intelligence implementation focused on rumor-spreading in the extensive Middle Eastern Greek community, while the newly rehabilitated CHEESE played a leading role.

The Malta convoy got through with less disastrous loss than the preceding one (though it was still very bad), for which the admiral commanding expressed to "A" Force his satisfaction with RAYON.

Contemporaneously with RAYON, "A" Force conducted an operation of a more tactical or operational nature called SENTINEL, which was designed to exaggerate the strength of the Nile Delta defenses. Jones's dummy tanks were initially used for the purpose in late July. When they were preempted for more direct use, Clarke fell back on intelligence implementation, planting word that Eighth Army had been reinforced by three of the CASCADE notional divisions, plus a real division that had in fact not yet arrived from England; and that the British forward defenses were thin, relying heavily on minefields, while a strong defensive line was being built in the rear. This latter was swiftly created from dummy tanks, guns, tents, and trucks; signs warning of (imaginary) minefields were erected, similar warnings were issued to the press, and the Cairo-Alexandria road was closed to civilian traffic.

These activities coincided with major changes at the top. Just as Churchill had lost confidence unjustly in Wavell, so now did he lose confidence unjustly in the Auk. At the end of July he came out to Egypt on his way to visit Stalin in Moscow. He fired Auchinleck and brought General Sir Harold Alexander from Burma to be Commander-in-Chief, Middle East, while Michael Crichton's old boss, "Strafer" Gott, became Eighth Army's new commander under Alexander. Gott was killed in a plane crash before he could assume command. In his place General Bernard Law Montgomery, a protégé of Field-Marshal Brooke, got the job.

"A" Force mounted a minor operation called GRANDIOSE to cover Churchill's travels. A false itinerary was circulated through the usual rumor mill, while reasonably accurate—but belated—reports as to the Prime Minister's whereabouts were passed to the Germans by CHEESE and the Turkish channels. Clarke contemplated continuing with GRANDIOSE II, suggesting that the Cairo meetings had resulted in a plan to invade Greece after Crete, but this was dropped after learning of the planned North African landings.

Montgomery took command on August 13. Rude, tactless, ungracious, ungenerous, and self-righteous, much given to bluster and to insulting denigration of others, he was a singularly unattractive human being, but he was an effective leader and a highly qualified technical soldier; and he quickly appreciated the value of Clarke's work. Six days after taking command he summoned Clarke to his headquarters. ULTRA showed Rommel building up for one last effort to break through at El Alamein; "A" Force was to do what it could to delay that for a fortnight. Clarke modified

Sentinel by adding a "story," passed by Cheese to Athens, that the British planned to lure Rommel on to their minefields and concealed antitank guns. Rayon was continued in the hope that it too would help induce Rommel to delay. More dummies were set out; additional Greek troops were dispatched to the training site; amphibious exercises were conducted in Cyprus as well.

But Rommel struck when he was ready, on the evening of August 30. For eight days the battle raged around the ridge of Alam Halfa, but by the evening of September 7 it was over, the Desert Fox had shot his last bolt, and he could only wait at El Alamein and prepare for the British counteroffensive.* Three days after that, Clarke was told for the first time that an Anglo-American operation to seize French North Africa, codenamed Torch, would be carried out in early November. And on September 14 Montgomery briefed Clarke over lunch on his plan for an offensive at El Alamein. Both Torch and the attack on Rommel would call for deception plans from "A" Force. So it would be a busy autumn.

The deception operation for Montgomery's offensive was one of the great success stories of the war. It fell into two parts: a strategic operation called Treatment, and a tactical operation called Bertram. Clarke conceived them, but he did not participate in their execution, for he was called to Washington and London to work on the Torch cover plans. So they were the test of the organization he had put together, and their success confirmed its soundness. Clarke had sketched out Treatment by September 16. Then he had to leave, and turned it over to Noel Wild. Wild went out to Eighth Army headquarters for further discussions on September 21, and by September 26 it had been approved. Bertram was

---

* A haversack ruse devised by Major E. T. Williams of the Eighth Army staff (subsequently a brigadier and Montgomery's chief of intelligence) may have contributed to the German failure. To encourage the Germans to attack where Eighth Army was best prepared to receive them, at Williams's request the topographical staff prepared a falsified version of the official map of the area which relabeled an area marked "firm and fast" as "generally impassable," while another area described on the real map as "reconnaissance essential before movement, continuous low gear" was relabeled "fair going." Copies were allowed to fall into German hands, and General von Thoma, later captured at El Alamein, stated that the Germans had been misled into sending tanks into sandy terrain, trebling their consumption of precious fuel.

planned and implemented by Lieutenant-Colonel Charles Richardson of Eighth Army operations.

Montgomery planned to launch his attack by the light of the full moon the night of October 23. On the strategic level, the object of TREATMENT was to lower the Germans' guard in the belief that no attack would come till November. On the operational level, the object of BERTRAM was to induce Rommel to hold his main strength in the south in the belief that the British *Schwerpunkt* would fall there; whereas in reality the weight of Montgomery's assault would fall on the northern sector of the front, with only a secondary attack in the south designed to keep Rommel from reinforcing his northern wing.

The "story" for TREATMENT built on the existing threat to Crete through RAYON, and on British concern for the "northern flank" in light of the German invasion of the Caucasus. The British, the "story" ran, would not mount any major operation in the desert until winter should close down German activity in the southern USSR, at which time they would be in a better position to assess long-range prospects on the northern flank. Accordingly, they would continue preparations for an attack on Crete, while in the desert there would be only a minor offensive, beginning with a night attack during the moonless period around November 6. The "story" for BERTRAM added that the delay until November was attributable at least in part to problems with the new American Sherman tanks; and that once the offensive came, it would begin with a feint in the north, after which Montgomery would hit the German southern wing with his main armor.

Implementation of both plans was divided roughly between "A" Force under Wild for intelligence implementation, and the operations staff of Eighth Army under Richardson for the physical side.

Intelligence implementation relied heavily on CHEESE, in addition to the customary ENTWHISTLE activity. (It was an active time for NICOSSOF; he was concurrently trying to get money through the B.G.M., and busy with the cover plan for the North African landings.) Wolfson's channels in Istanbul and Arnold's in Ankara pitched in. On October 6, Wolfson's special journalistic channel produced a story in a German newspaper revealing that troops were being transferred from the desert to Iran.

Clarke, who was in the United States by then, arranged for an American traveler to drop appropriate indiscretions in Lisbon.

In Cairo, military, diplomatic, and other civilian officers were asked to drop hints that nothing would happen till early November, to schedule trips and social events for late October, and to speak slightingly of the new American tanks. Reservations for several prominent desert generals were made at Cairo hotels for late October, and the course schedule at the staff college at Haifa was extended to the middle of November. For the benefit of the extensive network of German informers in Iran, elaborate arrangements were made for a conference in Tehran supposedly to be attended by Alexander, Wilson, and Wavell.

All this culminated on October 18 with a message from CHEESE to Athens predicting an attack in the desert in November. Athens sent in response a gratifying request for information as to British plans for attacking in the south.

Montgomery had three army corps. His right or northern wing was held by XXX Corps and his left by XIII Corps. Behind them was the newly-formed X Corps. His plan was for XXX Corps to punch two holes in the German left wing through which X Corps's 10th Armored Division (there was now a genuine one) and 1st Armored Division would pass, while in the south XIII Corps would punch a single hole, clearing the way for the "Desert Rats" of 7th Armored Division to pass through and do all they could to divert Rommel's attention from his left. The goal of BERTRAM was to ensure that the armor on Rommel's southern wing— half of his total—stayed there. The key was to keep the Germans from learning that X Corps was being positioned in the north behind XXX Corps.

Dumps behind the XXX Corps front to support the main attack were laid down at night and camouflaged as vehicle parks and camps during the day. Then, beginning on October 6, a massive concentration of vehicles—four thousand real ones and twelve hundred fakes—was accumulated behind the lines in the north, and some sixteen hundred more vehicles were gathered at a staging area behind the center of the line. The purpose was to get the Axis accustomed to the sight of masses of transport vehicles in those locations.

Meanwhile, a fake buildup was started behind the XIII Corps front in the south. A dummy water pipeline was laid across twenty miles of desert, complete with dummy pump houses and reservoirs (Operation DIA-MOND); the work moved at a rate designed to lead German observers to calculate that it could not be finished before early November. Dummy supply dumps were built; together with dummy camps and associated bogus transport, field kitchens, and such (Operation BRIAN). In a rare double bluff, obvious dummy guns were emplaced, subsequently osten-tatiously replaced by real ones (Operation MUNASSIB). An apparent stag-ing area for 10th Armored Division was laid out behind the southern wing.

The main concentration began on October 18. One at a time the ar-mored brigades of X Corps were surreptitiously moved to the northern assembly area, where the tanks and guns were covered with SUNSHIELDS and CANNIBALS, replacing real vehicles and dummies in the bogus vehicle pool, so that Axis reconnaissance would notice no difference. This began with the movement of an armored brigade of 1st Armored Division to a staging area behind the center of the line, remaining there in full view for a day. That night it moved on to the north and was replaced by another armored brigade of that division, which in turn remained in full view for a day; seeming to Axis eyes to be the same brigade as before. A third brigade, with division headquarters, moved out and exchanged places with the dummy motor pool that had been planted behind the center a few weeks before. That same day, October 19, 10th Armored Division moved to its staging area in the south. By that evening, therefore, Axis re-connaissance would see that a new armored division had taken up a posi-tion in the south and a new armored brigade in the center.

Over the next two nights the two remaining brigades of 1st Armored Division, and the whole 10th Armored Division, moved to the north, where SUNSHIELDS and CANNIBALS converted them seemingly into ordi-nary transport (Operation MARTELLO). The real and dummy vehicles re-placed by the tanks and guns of 1st Armored Division in turn replaced them behind the center (Operation MURRAYFIELD); while 10th Armored Division was replaced in the south by a huge display of dummies, in-cluding Colonel Jones's entire dummy armored brigade (Operation MELTING POT). Meanwhile, for four weeks an elaborate radio deception plan had simulated the communications net of 10th Armored Division (Operation CANWELL), and this opened in full voice on the 20th with the MELTING POT dummies, using twenty-five transmitters to mimic

Eighth Army Tactical Headquarters and the headquarters of a corps, two divisions, and five brigades.

So on October 23, as far as the Axis could tell, Montgomery had one armored division in a holding area twenty-five miles from the southern end of the front and another forty miles behind the center, and no armor in the north.

That night, Montgomery attacked.

Both TREATMENT and BERTRAM were as successful as anyone could have wished. The attack was a total surprise. Rommel was away on sick leave. Not until midnight the next night did Hitler order him to return to the front; only then did he regard the situation as "serious." Numbers of German prisoners were captured in their pajamas. Rommel's deputy, General Stumme, had told his corps and division commanders on October 7, and repeated on October 20, that the focus of the British offensive would be the northern part of the southern sector, with a secondary advance in the north. General von Thoma, taken prisoner in the battle, told his captors that German reconnaissance had picked up no increase of vehicles in the south, only in the north, and that he had been certain that the main attack would come in the south. Not until the battle was three days old, with Eighth Army deep into the Axis position, did Rommel finally decide that the British offensive was focused in the north, and shift thither one of his two armored divisions in the south.*

By November 4, Eighth Army had broken the German position. Rommel pulled up stakes and retreated westwards.

Four days later the Allies landed in French North Africa. The tide of war in the Mediterranean had turned for good and all.

Victory at Alamein put the final seal of acceptance on deception. Already during the run-up to the offensive, on October 8, "A" Force had put on

---

* At the Royal Navy's request, Bromley-Davenport and his sonic deception team faked an amphibious assault on German lines of communication at a place called Baggush, where the railroad and road ran close to the sea. During the night of the opening attack, October 23–24, soundtracks of winches running and anchors being dropped and weighed were played off Baggush through sound equipment from the new armored car ORA loaded on a motor torpedo boat. No results were detected, but the exercise was good practice for the more elaborate waterborne sonic deceptions of such units as Douglas Fairbanks's "Beach Jumpers" over the next two years.

in Cairo a highly successful demonstration of dummy tanks, SUN-SHIELDS, SONIA, SAINTS, and similar contraptions. After Alamein, Wild's old boss General McCreery sought him out and retracted everything he had said. "A" Force was very clearly neither a "racket" nor a place to spend an easy war.

McCreery was not alone. Montgomery had seen deception work. For the rest of the war he was its enthusiastic proponent, giving its practitioners his full backing and all the authority they needed. The same was true of "Freddie" de Guingand, who had moved from his brief tour as DMI Middle East to be Monty's chief of staff for the rest of the war, and who supervised Richardson's planning and execution of BERTRAM.

TREATMENT and BERTRAM justified everything Clarke had gone through for nearly two years. If there was any shortcoming, perhaps it was that with the abrupt and clearly permanent turning of the tide, the organization he had put together was now too large for the Middle Eastern theater. No matter; in the broadened Mediterranean theater there would be plenty to do: new deception operations, and initiating into the mysteries of deception the Americans who had now come on to the scene.

CHAPTER 7

Enter the Yanks

Stonewall Jackson, great-great-grandfather of modern British deception, had little progeny in his own land.

Not that Americans had no experience with military ruses of the traditional kind. In 1779, George Rogers Clark approached Vincennes carrying extra flags so as to multiply his apparent strength. When the British Admiral Warren harassed St. Michaels, Maryland, in the War of 1812, the townsfolk are said to have hung lanterns in distant trees to make his gunners overshoot their mark. In the Civil War, besides Jackson's efforts one thinks of the wooden "Quaker guns" and scarecrow sentinels at Manassas and Harrison's Landing in 1862, of Magruder's theatrics at Yorktown and Gaines's Mill, of the empty trains bearing "reinforcements" that covered Beauregard's retreat from Corinth, of Forrest marching the same two guns round and round till his opponent thought himself outnumbered; of Colonel Albert Myer, the Union Army's chief signal officer, fooling Lee before Chancellorsville into opening a dangerous gap in his front by sending deceptive messages by signal flags known to be read by the Confederates.*

In 1918, the Americans conducted the "Belfort Ruse," an elaborate effort to mislead the Germans into expecting an American attack through the Belfort Gap, far from Pershing's real objective at St. Mihiel. An officer whom Pershing had relieved of command was duped into believing that he was getting an opportunity to redeem himself by leading a drive to Mulhouse and the Rhine through the Belfort Gap. A swarm of

---

* At least one Civil War deception backfired as CAMILLA did with Wavell. When in June 1863 Lee was advancing towards Pennsylvania, to confuse Hooker's pursuit Confederate troops spread the rumor that Longstreet—actually far in the rear—was already across the Potomac. Hooker, persuaded by this that Lee's whole army was already over the river, crossed it two days sooner than he would otherwise have done—so that Lee was surprised to find the Yankees at Gettysburg.

staff officers descended on Belfort. Reconnaissance flights were flown over the German lines. The airwaves were filled with American radio. Tanks drove around leaving tracks for German aviators to see. An American officer left carbon paper carrying details of the phony plan in a wastebasket at a hotel frequented by German agents, and observed that it disappeared. And in fact the Germans reinforced the Belfort front and prepared to defend Mulhouse.

But this inspired no systematic further development, as Allenby's ruses in Palestine had done.* A 1934 *Infantry Journal* article entitled "Deception in War" seems to have had no effect. In peacetime maneuvers the Navy developed skill in deceptive radio signaling. At the decisive battle of Midway in June 1942, the Japanese were ambushed by three American carriers when they expected to encounter only one, having been fooled by radio signals imitating an American carrier group in the faraway South Pacific. But none of this was high-level strategic deception. For that, the Americans had no experience and no organization.

Indeed, by British standards the Americans had little organization for anything above the level of the two independent services, Army and Navy. The War Department for the Army† (including the Army Air Forces, which was to become de jure an independent United States Air Force only after the war, but was already in practice a separate, and cantankerous, service), and the Navy Department for the Navy, were independent and fiercely turf-conscious; there was no unifying Department of Defense. When Churchill and his Chiefs of Staff came to Washington after Pearl Harbor for what came to be known as the ARCADIA Conference, it was agreed that representatives of the British Chiefs would be stationed in Washington to meet at least weekly with the American Chiefs for common management of the war effort. This arrangement was known as the Combined (*i.e.,* inter-Allied) Chiefs of Staff ("CCS"); under it were various committees including a committee of Combined Staff Planners. The British membership was presided over by the former

---

* With one possible indirect exception. Brigadier General Fox Conner, who supervised the Belfort Ruse, became the mentor of the young Dwight Eisenhower, and perhaps Ike's enthusiasm for Dudley Clarke's work reflected memories of Conner's accounts of the operation.

† For simplicity, however, it is referred to anachronistically as the Air Force in this book.

Chief of the Imperial General Staff, Dudley Clarke's old mentor Field-Marshal Sir John Dill.

This in turn forced the Americans to develop the first joint (*i.e.,* interservice) organization of the heads of the American armed services; which came to be called the Joint Chiefs of Staff ("JCS"). It was composed of General George C. Marshall, the Chief of Staff of the Army; Admiral Ernest J. King, the Chief of Naval Operations ("CNO") and Commander in Chief, United States Fleet ("COMINCH"); and Lieutenant General Henry H. "Hap" Arnold, the Commanding General, Army Air Forces, as a de facto third chief of staff; plus Admiral William D. Leahy, a former CNO, who was designated "Chief of Staff to the Commander in Chief" (*i.e.,* Roosevelt) and was in fact though not in title the first Chairman of the Joint Chiefs. Under its committee structure, as reorganized in the spring of 1943, outline plans for future operations were drafted by planning teams of a Joint War Plans Committee ("JWPC"); reviewed by the Joint Staff Planners, composed of the four chief planning officers for the Army and Navy and their respective air arms; and passed on to the Joint Chiefs and, where appropriate, by them to the Combined Chiefs. Detailed planning of operations fell to the staffs of the overseas theaters.

The American joint planners did not develop a separate deception planning entity like the London Controlling Section. Deception plans, like other plans, emanated from the regular planners. Until the final phase of the war, strategic and operational deception planning was of more interest to the Army than to the Navy, so that the officers most directly involved were Army officers from the Operations Division (until March 1942 called the War Plans Division) of the War Department General Staff. Known always as "OPD," this was General Marshall's command headquarters, headed by Brigadier General Dwight D. Eisenhower until June 1942, when he was sent to Europe as commanding general of the European Theater of Operations, U.S. Army ("ETOUSA"); then by Lieutenant General Thomas T. Handy till late October 1944; then by Lieutenant General J. E. Hull to the end of the war.

The most important subdivision of OPD was the Strategy and Policy Group ("S&P"), described by Hull as "the brain trust" of the Army. Handy headed S&P under Eisenhower, until he moved up to succeed Ike in June 1942. His successor was Brigadier General Albert C. Wedemeyer. When Wedemeyer left for Southeast Asia in September 1943,

Brigadier General F. N. Roberts took over until the end of November 1944; then Brigadier General G. A. Lincoln (known, inevitably, as "Abe") headed S&P till the end of the war. Until moving into the new Pentagon Building in September 1942, OPD was housed in a flimsy 1917-era "tempo" on the Mall side of Constitution Avenue, not far from the present-day Vietnam Memorial, called the Munitions Building.

Navy planning, less elaborately organized, was the responsibility of the Assistant Chief of Staff (Plans), known under the Navy's system as "F-1," in the headquarters of Admiral King in his capacity as COMINCH. In the first half of 1942, F-1 was Rear Admiral Richmond Kelly Turner, succeeded in mid-1942 by Rear Admiral C. M. Cooke, Jr. In October 1943, "Savvy" Cooke moved up to be Deputy Chief of Staff, and was succeeded by Rear Admiral B. H. Bieri, who had been his Assistant Plans Officer; Bieri was succeeded in May 1944 by Rear Admiral D. B. Duncan, who in turn was succeeded in July 1945 by Rear Admiral M. B. Gardner. Under F-1, the Assistant Plans Officer (called F-12 till April 1943, thereafter F-10) did much of the Navy work for the Joint Planners.

The topmost Army brass—the Secretary and Assistant Secretaries of War, and General Marshall—were in the magnificent Second Empire wedding-cake of a structure adjacent to the White House, known then as "State-War-Navy" and nowadays as the Old Executive Office Building. The corresponding Navy brass were in another "tempo" (a rather more durable one than the Munitions Building) on the Mall side at 18th Street and Constitution Avenue known as "Main Navy." Both the Joint Chiefs and the Combined Chiefs met on the neutral ground of the Public Health Building, on the north side of Constitution Avenue between 19th and 20th Streets, where the Department of the Interior now stands.

It was TORCH, the invasion of North Africa in early November 1942, that introduced the Americans to strategic deception.

Viewed with skepticism by the Americans, who thought priority should go to defeating Germany in Northwestern Europe, this operation was nevertheless decided upon in July as the best available option for early offensive action. Eisenhower was chosen to command it. Total surprise was crucial to its success; and, as we have seen, under date of July 27 Bevan was directed to prepare a cover plan for it and a plan for inducing the Germans to keep as large a force as possible in Northwest Europe in the autumn of 1942.

Soon thereafter, the Americans took their first tentative steps towards setting up an approximate equivalent to the London Controlling Section. The initial impetus reflected concern not for deception but over security leaks. In the spring, the Joint Planners had shot down a proposal to form a "security committee," expressly modeled on the British ISSB. But after TORCH planning was under way, a similar move came as a sort of preemptive strike from one of the most aggressive players in the military bureaucracy, Major General George V. Strong, since June the chief of Army intelligence (officially, the Assistant Chief of Staff, G-2, and head of the Military Intelligence Division of the War Department General Staff).

Strong was a tough sixty-two-year-old cavalryman, a Chicago native and 1904 West Point graduate. Broader-gauged than many Army officers, he was also a graduate of Northwestern University law school, and besides troop duty (and staff duty overseas in 1918–19) he had held various judge advocate posts and had taught law at West Point. His intelligence experience dated back to an early tour as military attaché in Tokyo, and included service as military advisor to the American delegations to various disarmament conferences and a tour as chief of the Intelligence Branch of G-2 in 1937–38. In 1938–40, he headed the War Plans Division, and then had commanded a corps till Marshall summoned him back to take over G-2 in June 1942.

Strong was "no fool but filled with a tremendous sense of his own importance and very slow of speech," said Guy Liddell of MI5 after meeting Strong in 1943. "I should say that if his stenographer could do 3 words a minute she would suit him admirably." Secretary of War Stimson thought him "a very good man but the wrong man temperamentally for G-2," "able but pretty impossible." Adolf A. Berle of the State Department, who because of Strong's middle initial referred to him as "George the Fifth," considered him "a sound, solid citizen." Strong and his Navy opposite number had a running battle with "Wild Bill" Donovan over the control and role of the OSS, and to one of Donovan's earnest partisans, Strong was "the only affirmatively vicious man I have met in Washington, for all his ability."

Strong was in touch with the deception staff in London even before the final organization of the London Controlling Section, and he visited London in early August. There he conferred with Eisenhower while his deputy worked the bureaucracy in Washington; the upshot of which was a decree assigning security for TORCH to the Joint Intelligence Committee, of which Strong was chairman. But this decision was promptly reconsidered. Admiral King was appalled to overhear officers discussing the forthcoming operation at a cocktail party. He summoned his intelligence officer for plans and operations, Captain George C. Dyer, a forty-four-year-old Oregonian of the Annapolis class of 1918, and demanded to know how many headquarters personnel knew something about TORCH. Sixty-one, reported Dyer after investigation. King was as tough and salty as they come—it was not just that he did not suffer fools gladly, he did not suffer *anybody* gladly—and he discharged his wrath in Dyer's direction; after which he informed Dyer that he, Dyer, would have Navy responsibility for correcting the situation.

Things were even worse with the Army. Marshall's "War Department Security Officer" for overall security control with respect to TORCH, Colonel Norman E. Fiske, a former military attaché in Rome, reported "an appalling laxity in almost all commands and divisions of the War Department as regards security for such a highly important and delicate undertaking"; between four thousand and five thousand people probably now had some "knowledge of a projected immediate operation in the particular area." "The Chief is disturbed over this matter, has instituted an investigation and is contemplating drastic measures," Strong cabled to Eisenhower on August 19.

These concerns fed the feeling that security should not be the province of a body that included civilian membership, as did the Joint Intelligence Committee. King's intelligence and planning chiefs suggested that the Joint Chiefs create a "Security Command," "clothed with the authority to guide outside agencies (outside of the Army and Navy) in a way that their activities would not operate to compromise military operations, without disclosing to these outside agencies that any 'Special Operation' is contemplated or projected." The Chiefs rescinded their earlier decision and on August 25 decreed such an entity, originally officially called "Security Control" but almost always known as "Joint Security Control," to be composed of one senior officer each from the Army and the Navy.

Joint Security Control was not an American London Controlling Section. Its original emphasis was on security, deception was little more than an afterthought; but it was understood from the outset to be part of the new entity's mission nevertheless. Its function was not planning but "regulating and coordinating" and assigning missions, "under the Joint U.S. Chiefs of Staff, and under the guidance of the Joint U.S. Staff Planners"; moreover, its directive emphasized the primacy of the theater commander: Its activities were to be "coordinated with, and when appropriate will be under the control of, the supreme commander concerned." Not a London Controlling Section, indeed more an ISSB than an LCS; but a start.

Marshall designated General Strong as the Army member, while King designated Captain Dyer; which meant that Strong as the senior would dominate the new entity. Admiral Leahy wrote to the heads of the civilian agencies whose activities might fall within Security Control's field apprising them of the new agency and its responsibilities, while Strong notified Lieutenant-Colonel E. F. B. Cook (known to his friends as "Peter"), the Washington representative of British military intelligence. (Apparently the FBI was not notified; obviously, nobody was yet thinking of the uses to which their double agents could be put.)

Strong designated Colonel James R. Pierce to act for him as Executive Officer of Joint Security Control, and to be Fiske's alternate as "War Department Security Officer" for TORCH. Fiske soon went to London, probably as liaison to ISSB for the TORCH cover operations. "A most intelligent man," recorded Dennis Wheatley, "who did all he could to help us sell our new-fangled toy by bringing several American Generals down to see Johnny, so that he could explain it to them personally."

In early October, General Handy of OPD informed Strong that deception should be planned by the Joint Planners along with the real plan; that Joint Security Control should "attend the portions of the meetings of the Joint Staff Planners during which security control planning is discussed"; and that Joint Security Control should be responsible for liaison with the British.

When Eisenhower formed an Anglo-American staff for his new job as supreme commander of the Allied expeditionary force for TORCH, the post of chief intelligence officer fell to an experienced British officer, Brigadier E. E. Mockler-Ferryman. Mockler-Ferryman designated an

American officer who had been assigned to his staff, Lieutenant Colonel Carl E. Goldbranson of the 34th "Red Bull" Division, Iowa National Guard, as his liaison with the London Controlling Section with the designation "Cover Officer." Along with Newman Smith, Bill Baumer, and "Baron" Kehm, Goldbranson was destined to be one of the four key American officers in strategic deception.

Carl Goldbranson, forty-two years old in 1942, had been an officer in the Red Bull since 1927 and had been on active duty since December 1940. In civil life an official with the Union Pacific Railroad at its headquarters in Omaha, Nebraska—his title was General Clerk, which was higher up in middle management than it sounds, and he had worked for the company since he was eighteen—, he was tall and blond, befitting his Norwegian ancestry, with a ready smile and the uncomplicated, open nature of a son of the prairie states. Happily married (also since he was eighteen) and still much in love with his "Leafy," father of three children, a Boy Scout leader, he was in every respect one of those plain, solid products of Middle America that most public-school Englishmen had never met and that many could not relate to.

Goldbranson lived across the river from Omaha in Council Bluffs,

(U.S. Army Signal Corps)

Carl Goldbranson

Iowa, and accordingly was a member of the Iowa National Guard. His son Carl Jr. was himself a non-com with the 34th Division and indeed crossed the ocean in the same convoy as his father in February 1942 as part of the second contingent of United States troops to go to Europe. After participating in amphibious maneuvers in Scotland, Goldbranson found himself detached for duty with the new headquarters in London, and in September was initiated into the mysteries of the London Controlling Section. Wheatley found him "both competent and cordial," and others who worked with him later were equally complimentary. "A pretty solid man and a very decent man," recalled Arne Ekstrom. "He didn't take long to pick up the scent, as it were," recalled Desmond Bristow; "very congenial . . . always enthusiastic." Dudley Clarke, as will be seen, could not overcome his prejudices against this unsophisticated son of the western plains, and treated him unjustly. Nevertheless, he became one of the most skilled deception planners in the Allied service.

Eisenhower was determined to create fully integrated binational staffs in the two Anglo-American headquarters organizations that he set up—the one that began in London that summer of 1942 and went on to Algiers, called AFHQ (pronounced "Affkew") for Allied Force Headquarters, and SHAEF, the Supreme Headquarters, Allied Expeditionary Force, that conducted the OVERLORD campaign in 1944–45. His fellow countrymen, at least, were left in no doubt that he intended to make it work. You could call your British opposite number an S.O.B., the saying ran, but call him a Limey S.O.B. and you would be on the next plane home. Every staff element, every committee, was to have both British and American representation, and there were to be no all-American or all-British meetings. It was a measure of his success that not a few Americans, from General George Patton down, became and remained convinced that Ike had been hoodwinked by the wily British.

One problem that contributed to this feeling arose from the different approach of the two nationalities to staff committee work. A British interservice committee was a true committee; the Army, Navy, and RAF did not come as delegates for their parochial service viewpoints. With the comparatively inexperienced Americans, by contrast, an officer assigned to such a committee tended to behave as an ambassador of the Army or the Navy, charged to negotiate with the emissaries of the other interested parties. He was often unable to yield on a point and not infrequently found that his principal would not ratify what had been done.

The British found this frustrating. To counter it, they would caucus among themselves prior to inter-Allied committee meetings, whether Ike liked it or not, and work out an agreed view, while the American members, each representing only their service viewpoints, had none. The Americans referred to "the whistle"—an inaudible signal, as it were, like a silent dog whistle, upon which the British at some point in a meeting would adopt a united position on the question under discussion. And in the nature of things each American delegate would find that the united British position included points in his own presentation, so the British position was likely to be accepted in the end. This irritated and frustrated the Americans.

American resentments were not wholly unwarranted. Much British behavior was geared to perceptions of social class, and Americans did not fit into the structure. If they had been Norwegians or Dutchmen this would not have mattered, but speaking the same language made it hard to forget the class rules, because class perceptions were inextricably bound up with accent. This combined in many cases with a feeling that the Americans were ignorant newcomers to the war, inexperienced and unprofessional. "They treated us as colonials in 1942–43," said Bill Baumer, "but by the end of 1944 they were beginning to realize that we had so much raw power they couldn't cope."

British officers who could free themselves of these prejudices— Alexander, for example, or Montgomery's chief of staff, "Freddie" de Guingand, or Sir John Dill, or indeed many another—got on famously with the Americans; unlike Montgomery, for example, or his mentor Brooke, who could not. It has to be said that Dudley Clarke never quite broke out of the mold. He and his intimates laughed among themselves at most of the plain Midwesterners assigned to "A" Force (making little jokes such as why did they all seem to come from Council Bluffs?); though not at cosmopolitan Americans like Arne Ekstrom and Douglas Fairbanks, who were welcomed to Clarke's club.

Consistently with Bevan's fundamental directive, the British staff mission in Washington had advised the Joint Chiefs in July that British deception had been reorganized and strengthened, and proposed inter-

Allied cooperation with an exchange of liaison officers. When the Joint Chiefs looked at this suggestion, Admiral King growled that he had "decided objections towards the interchange of liaison officers for this purpose, particularly since the liaison officers appointed by the British usually lack authority to make any decisions." The matter was referred back for a collaborative arrangement to be worked out similar to other Combined Chiefs supporting agencies. There it sat for a while.

Meanwhile, Bevan was not deterred from sending a liaison officer to Washington anyway. The Director of Military Intelligence was reluctant to assign a fully-trained intelligence officer to this unproven business, but he was willing to allow an officer who was already going out to Washington to be a part-time representative of the London Controlling Section, working under Peter Cook. A major in his early thirties named Michael Bratby—in peacetime an announcer for the BBC and an amateur ornithologist and author of books on bird-watching—was selected, seconded to Bevan for ten days to learn something about the business, and sent out in mid-August. Bratby dismayed Bevan in some ways— Bevan thought a sloppy appearance betokened a sloppy mind, and Bratby took a casual, not to say slovenly, attitude to the uniform—but Bratby proved to be useful in Washington. "An interesting individual," recorded Colonel Bill Baumer; "not too smart and effectual but always pleasant and helpful."

"Pleasant and helpful" was not enough to make Bratby the right man to educate George the Fifth Strong. That would call for the master himself. In the same message delivered to Dudley Clarke on September 10 in which he was first apprised of TORCH, Clarke was directed to proceed to Washington and thence to London to work on cover and deception for the operation and read the Americans in to his work.

Flying in an American plane via the Sahara, West Africa, Ascension Island, Brazil, the West Indies, and Miami, he reached Washington at 6:30 the morning of September 28 (having fitted in a good bit of social life en route.) He spent the morning and lunch at the Army-Navy Club with Peter Cook, and that afternoon held his first meeting with General Strong. The next day he lunched with Field-Marshal Dill and his wife. On September 30, he met with the Joint Staff Planners, and prepared a memorandum requesting allocation of a notional American armored brigade and infantry division (he suggested it be called the 22d Infantry

Division) for the Middle East. On October 1, he spent two more hours with General Strong, and lunched at the British Embassy with the Ambassador and his wife, Lord and Lady Halifax. He managed to fit some night life into this busy schedule, and the next afternoon he took the train to New York for a long weekend packed with sightseeing and still more night life, returning on Monday afternoon, October 5. That day he met with Colonel Fiske and others, and in the evening he dined with Donovan of the OSS at the latter's house, caught a late train to Baltimore, and took off for London shortly after midnight. His flight to London took three days, with bad weather in Newfoundland and a damaged airplane in Ireland, arriving at 11 P.M. October 8.

Busy with TORCH cover plans, Bevan had to worry also about the longer run: How to bring the Americans in Washington in as full players, how to coordinate deception worldwide. So he had called a gathering in London for mid-October, to be attended by Clarke, by Peter Fleming from India, by Cook from Washington, and by appropriate people from Eisenhower's staff; and he invited Strong and Dyer from Washington as well.

"For days," recorded Wheatley, "all of us sweated blood getting out a vast agenda, writing long papers on the innumerable ways we might hope to deceive the enemy, upon grades of operators and spheres of influence, and arranging accommodation for our visitors." The gathering convened in the Chiefs of Staff's conference room on October 15— Wheatley enlivened the proceedings by wildly oversleeping and arriving when the first day's meeting was about to break up for lunch—and went on for three days. Useful in many respects to clarify relationships among deception organizations, the conference had as its most important long-run result an understanding with Strong that in general the British would have responsibility for deception in the European theaters and the Americans for deception in the Pacific.

Clarke, it would appear, did not attend all these meetings. He had many fish to fry on a visit to London. Over and above his usual busy social round, there were calls on the Chief of the Imperial General Staff and on General Ismay; a meeting with Mockler-Ferryman; meetings with MI9 and its chief; meetings with SOE; a meeting with the officer in charge of the special intelligence units that handled ULTRA; meetings with ISSB; a meeting with the Admiralty. On the evening of the 14th he was received at Downing Street by Churchill himself.

Particularly important for the future was a meeting with Goldbranson on October 20 at which Clarke enlightened the American, together with Mockler-Ferryman's ULTRA officer, as to many deception techniques. He described for them the structure of "A" Force. He reviewed the use and layout of the "Strategic Addendum." He imparted tips on such practical matters as W/T activity to simulate dummy units; mismarking of aerial photographs and maps used in training, so your own people will not be sure what actual place they are studying; use of dummy railheads, dumps, and air bases; calling for potential interpreters for a country no-tionally to be invaded, and tipping foreign exchange speculators to buy currency of such a country; sending instructions to resistance groups; various haversack ruses; and methods of inflating your perceived strength, including assignment of false unit designations to rear-area units and inclusion of phony units in casualty lists. False order of battle construction, he warned, should not attempt to inflate true strength by more than 30 percent (Clarke would later learn that with the Germans this figure could be exceeded).

On November 1, Clarke took off from England to return to Cairo by the Mediterranean route, accompanied by Peter Fleming on his way back to India. They spent the night and the next day at Gibraltar, where the TORCH invasion fleet was already gathering. Clarke and Fleming made a couple of business calls, drove to the top of the Rock, and flew on out that evening. Their Liberator touched down in Egypt at sunrise next day, November 3, and at 8:30 Clarke was back home at Shepheard's. Five days later, on November 8, American and British forces landed in French North Africa.

TORCH was a complex operation involving three different task forces landing simultaneously hundreds of miles apart. The Western Task Force, composed of American forces from the United States commanded by Major General George Patton, was to sail from Norfolk, Virginia, on October 23, directly to its objectives, Casablanca and other points on the coast of French Morocco. The Central Task Force, composed of Ameri-can forces from the United Kingdom commanded by Major General Lloyd Fredendall, sailed from Britain to its objective, Oran in Algeria.

The Eastern Task Force, initially under the Red Bull Division's commander, Major General Charles Ryder, and composed of British and American troops from the United Kingdom, sailed from Britain to its objective, Algiers in Algeria. The problem was to keep the Axis from learning of these armadas for as long as possible, and to make the Axis believe, once they were discovered, that they had some target other than French North Africa.

Cover planning for Patton's task force began in a pessimistic atmosphere. By mid-August, when it was learned that a French official in London had heard from a French officer in Washington that the Americans were working on plans for operations in French Morocco, Strong already felt that "it must be assumed that the German General Staff has knowledge of the operation excepting possibly date of action and composition of forces," and that the best to be hoped for might be that the Germans would think a deliberate plan to deceive them was afoot. Admiral Cooke, the Navy planner, was almost as pessimistic. By mid-September the very codename "TORCH" was so badly compromised that consideration was given to changing it, first to BEACH HEAD and then to WEDLOCK.

Things got worse. In late September a British flying boat crashed off Cadiz. In the pocket of one of the victims, whose body was turned over to the British by the Spanish authorities, was a letter from the American General Mark Clark to the Governor of Gibraltar saying that he and Eisenhower would arrive shortly before D-Day, and that "the target date has now been set as 4th November." No one could be sure whether the Spaniards had let the Germans read it. Soon afterwards, a memorandum by Churchill dealing with TORCH was picked up by a charwoman on the London streets. Across the Atlantic, on October 2 the Hollywood gossip columnist Hedda Hopper published an item that the film mogul Darryl Zanuck was headed overseas where he would probably join up with General Jimmy Doolittle; Zanuck and Doolittle were in fact attached to Patton's task force. That same day an Air Force officer was arrested drunk in Washington just after being briefed on the operation.

Two weeks later it transpired that one of Patton's aides had blabbed to friends in the Washington horsey set that Patton's command was "all equipped for overseas duty, and would 'leave Norfolk next week' "; this got back to Mrs. Patton, and the aide got an official reprimand and Patton fired him. (Yet Patton himself unwisely held a reception for the officers of his headquarters staff at his house in Washington.) A plain-language radio message from Camp Pickett in Virginia was picked up saying "The American staff can leave for Algeria October 24." Some Air Force officers wrote to their wives that their squadrons had been loaded on a carrier in Norfolk. Another officer tried to write to his wife in code, naming the target cities.

Nevertheless, work went forward on three cover plans. SWEATER (originally STAB) was designed to give the impression that the amphibious forces for Patton's task force were destined for amphibious training in Haiti; QUICKFIRE (originally PADLOCK) was designed to give the impression, once it became obvious that Patton's force was not bound for Haiti, that it was destined via the Cape of Good Hope and the Red Sea for the Middle East (Eisenhower's headquarters had suggested Australia, but this was vetoed in Washington); and HOTSTUFF, a naval radio deception plan, was designed to imitate the radio traffic of a convoy proceeding along the QUICKFIRE route from Norfolk around Africa to Syria.

SWEATER got elaborate implementation. The Hydrographic Office of the Navy distributed charts of Haitian waters. Permission of the Haitian government to conduct amphibious training in their territory was secured, and a party of officers went out to Haiti at the beginning of October to survey the supposed site. A senior general, Major General Daniel I. Sultan, was appointed commander of the SWEATER task force, was not told that it was only a cover, and was directed to plan it as an alternative to TORCH. Four days after Patton's task force sailed, Sultan was told that SWEATER would not be used but that he should finish the plan for possible future use; the Haitians were not told that it had been canceled until November 15, a week after the TORCH landings.

Once the task force set sail, QUICKFIRE took over. Its "story"—based on a possible American reinforcement of the Middle East that had actually been considered—was that an American force commanded by General Sultan, drawn chiefly from the 3d and 9th Infantry Divisions and divided into three subsidiary forces under Generals Ernest Harmon, Jonathan W. Anderson, and Lucien D. Truscott (this more or less matched the actual composition of Patton's force) would be sent to Haifa, Cyprus, and Beirut, relieving British forces preparatory to a buildup for offensive operations. (Originally, the QUICKFIRE force was notionally to occupy the Dodecanese as well, but that was eliminated at British request.) An American escort would protect the convoy to a rendezvous off Sierra Leone; the Royal Navy would take over from there around Africa to the eastern Mediterranean.

QUICKFIRE, like SWEATER, was the subject of actual planning by Sultan and his staff. Administrative implementation included arrangements at Recife, Brazil, to stock up enough fuel to accommodate the notional task force, and a message to the American commander in the Middle East on October 29 asking for information on the port of Aqaba and on rail and road connections to Syria. British double agent channels fed appropriate false information to the Germans.

The most elaborate implementation of QUICKFIRE was by HOTSTUFF, the associated naval radio deception operation. An imaginary Task Force 23 was set up (the real one was Task Force 34). Two days before the real sailing date, dummy traffic began to be transmitted to simulate its designation, organization, assembly, routing, and actual sailing, with timing arranged to permit notional refueling of destroyers at sea, and fictitious air and submarine coverage suggested by phony messages to appropriate bases. Traffic peaked at Guantánamo and San Juan on October 25, at Trinidad on October 31, and at Recife beginning November 1; at about which time the British Admiralty joined in with traffic from Freetown and Cape Town, suggesting the forthcoming transfer of escort responsibilities to the Royal Navy. The imaginary task force should have reached Recife just at the time the real one arrived at Casablanca on November 8. HOTSTUFF was abruptly terminated at that point; an amateurish move, marking it as a deception to any Axis intelligence service that had followed it.

HOTSTUFF would have been implausible, of course, without corresponding radio security for the real task force. Once it left Western Atlantic waters, its outgoing messages were routed via Gibraltar or another British station, using British procedures and ciphers and giving in all respects the impression of emanating from a British ship. Messages to it from Eisenhower's headquarters in London were first cabled to Washington and sent from there. As it approached the African coast, messages to it from Washington were first cabled to London and sent as British traffic, while messages from London were passed as British traffic.

There were blunders in the American deception program. Donovan's new OSS got into the act with an effort to fuel German and Vichy belief in an Allied operations against Dakar—a "story" that had been early mooted and soon abandoned,* but evidently OSS did not get the word. Worse, in mid-August, without checking with anyone else, a psychological warfare officer and someone at the Office of War Information decided to mount a deception of their own. Major George Fielding Eliot, a widely syndicated military affairs columnist, was induced to publish an article in the New York *Herald Tribune* emphasizing Allied interest in Norway. American censorship would be instructed to prevent the article from leaving the country, and news of this censorship action would then be leaked to the enemy. The Washington representative of the British Psychological Warfare Executive learned of this at the last minute by accident, and was dismayed; suppose these clowns had chanced upon North Africa instead of Norway? Fortunately, Elmer Davis, the Director of OWI, grew concerned when he heard of the matter and consulted John J. McCloy, the Assistant Secretary of War. Davis dispatched an apology to Eisenhower, but the incident did nothing to help the Americans' image in British eyes.

Much more elaborate—and more professionally executed—were the British cover and deception operations for TORCH. Eight in number, called GIBRALTAR COVER PLAN, SOLO I, SOLO II, OVERTHROW, KENNECOTT, TOWNSMAN, PENDER I, and PENDER II, they had three objects: to induce the Germans to keep as many troops as possible in western Europe and Norway, far from the scene; to prevent the Axis from learning about the TORCH convoys for as long as possible; and thereafter to mislead the Axis as to their objective.

The GIBRALTAR COVER PLAN was set in motion early. Its goal was to account for the increased activity at Gibraltar by spreading the word that a massive expedition for the relief of Malta would soon be mounted; and

---

* Dakar was suggested as the cover destination in the early planning stages. But this was risky, for it could lead to reactions by both Vichy and the Germans that would jeopardize TORCH; moreover, as soon as the convoys from Britain turned east it would be clear that Dakar was not the objective, while a cover target within the Mediterranean could be maintained to the last minute. Furthermore, German propaganda was floating the idea that the Allies might attack Dakar, and there was much diplomatic speculation about it. So after the opening phase, Dakar was downplayed. At the very end of the cover and deception period, when it was too late for any major Axis or Vichy reaction, Patton's convoy maintained a track headed towards Dakar as long as possible before turning northeast for the final rendezvous.

that the landing craft being gathered at Gibraltar were destined for West African ports and eventually round Africa to the Middle East. Major David Strangeways of the Duke of Wellington's Regiment was assigned to Bevan and sent out to Gibraltar, Malta, and the Middle East to brief the appropriate authorities. He left London on August 19, carrying with him an autographed copy of Wheatley's latest book as a present for Henry Hopkinson, with a note from Wheatley to Hopkinson containing "indiscreet" false information—just in case his luggage should be checked by Axis agents in Gibraltar or Cairo. He briefed the Governor of Gibraltar in detail, and remained until early September to assist him in putting the story over.

The Governor joined in the game with a will. He broadcast to all troops on the Rock that Gibraltar would become a transit base for the relief of Malta. He flew to Malta himself to coordinate arrangements. Maps of Malta were sent in quantity to Gibraltar. The story was generally believed by the troops and the civilian population of Gibraltar until the end of October, by which time they were coming to believe that this was a cover story and the real goal of the preparations was Dakar.

Strangeways flew on to Egypt. ("A somewhat dicey journey," he recalled, "as our pilot made an approach for landing on an enemy airfield.") In Cairo, on September 10, it was he who told Dudley Clarke about TORCH and London's cover plans for it; and conveyed to Clarke instructions to go to Washington, together with an invitation from Bevan to contribute the "story" to be followed once the Germans learned that the convoys were in the Mediterranean and were invasion convoys.

For this last there were two main possible "stories." One would be that a landing was intended somewhere in Rommel's rear, perhaps at Benghazi. Clarke ruled that out on the ground that the Germans would not believe that the Allies meant to run the gauntlet of the Sicilian Channel. So he recommended, and Alexander concurred, that the "story" be an intended landing on Sicily and the "toe" of Italy. The overall plan for putting this across received the codename KENNECOTT; "A" Force's implementing plan was dubbed TOWNSMAN. The latter was approved by Alexander on September 13, and Strangeways left with it the same day for London.

Meanwhile, London had under way two broad strategic plans, SOLO I and OVERTHROW.

SOLO I, with the object of inducing the Germans to retain as many troops in Norway as possible, was essentially a recycling of the old HARD-BOILED threat to Norway. Its "story" was that a landing, this time in the Trondheim-Narvik area, would be mounted from Scotland in early November. Double agents reported public speculation that Norway would be invaded and retailed tales of energetic military activity in Ayrshire, of special Commando training going on, and of stockpiling of tire chains and antifreeze, plus hints that Swedish collaboration was being sought. Rumors were spread in Stockholm that the British were training special pioneer companies of Canadian lumberjacks while the United States Army was on the lookout for Scandinavian speakers among its personnel.

An elaborate program of W/T simulation of substantial movements of warplanes to Lossiemouth and Kinloss in Scotland (natural air bases for a Norway campaign) began in late October, together with intense antisubmarine and reconnaissance activity over the North Sea. Specimen billeting and requisition forms were printed, twenty thousand shoulder flashes reading *Norge* were ordered, arctic kits and maps of Norway were issued, the troops were lectured on the dangers of frostbite, and rumors were spread about woolen underwear and special windproof jackets. The Americans were asked to circulate a report that Lascar merchant seamen, whose contracts specified that they not be required to serve in the cold climate above 60 degrees north latitude, were being offered bonuses to do so. Practice exercises in landing against a rocky coastline were conducted. When the imaginary expedition was scheduled to depart, fourteen American merchant ships being assembled for a convoy to Murmansk made a dummy run as far north as the Orkneys.

Further south, OVERTHROW* was designed to induce the Germans to retain as many troops as possible in France. Its "story" was that the Allies would launch an operation to seize a bridgehead in the Pas-de-Calais

---

* The original codename for OVERTHROW was PASSOVER. This, it was discovered, was on the list of codewords reserved for American use, so it was changed to STEPPINGSTONE, and then eventually to OVERTHROW. Bevan chose SOLO in the hope that if the Germans learned of it they might perceive that it was an anagram of "Oslo," and PASSOVER and STEPPINGSTONE for their obvious connotations; not realizing—being still relatively new to the game—that if the Germans noticed these implications at all, they might well assume that the principle that a codename must never suggest the actual operation would in fact rule out Norway and a Channel crossing as their objectives.

at about the time of the actual TORCH operation. It was supported by extensive activity on the south coast of England, including construction of additional "hards" (paved stretches of beach from which landing craft could be loaded) for invasion craft, the use of coastal shipping to simulate the marshalling of invasion craft during the first third of October, an exercise by the British 3d Division, the strengthening of anti-aircraft defenses, and increased aerial reconnaissance over the French coast.

On the nonphysical side, British Airways was directed to shift the terminal for its Lisbon flying boat operations from Poole on the south coast of England to Shannon in western Ireland. French dictionaries were purchased in quantity. Enquiries were made in Lisbon for engineers familiar with the coal mines in northern France. The double agents reported the formation of a new separate command in Kent and general unusual activity in the south, suggestions that concentrations for SOLO I involved simultaneous attacks on Norway and France, and references to the great August 19 Commando raid on Dieppe as a "dress rehearsal."

The climax was supposed to be an actual combined exercise in the Channel just before the TORCH landings, called CAVENDISH. But the ships and men for this could not be assembled before November, when bad weather was expected to render such an operation implausible; so CAVENDISH was canceled.*

SOLO II was the equivalent of SWEATER for the troops coming from the United Kingdom; they were given the impression that they were destined for tropical climes and allowed to believe that the ultimate objective of the operation was the Middle East, seizing Dakar on the way. They were told that tropical kits and mosquito netting would be issued once they were on board ship; that cholera shots would be given en route; that they must not buy unwashed fruit at ports of call. They got lectures on the dangers of lice, malaria, and dysentery. Their cooks were given special instruction in preparation of Middle Eastern foodstuffs.

KENNECOTT, the last-phase plan suggesting Italy and Sicily as the fake objectives, was implemented with carefully controlled timing, for premature identification of Italy and Sicily would enable the Germans to

---

* Originally planned also as an aid to OVERTHROW was an actual operation AFLAME (later renamed COLEMAN), designed to bring the Luftwaffe to battle. But it was canceled.

concentrate U-boats off Gibraltar. The first information, therefore, was designed to reach the Axis on D minus 13, with no direct indication of the convoys' target before D minus 7. The British embassies to Egypt and Turkey were asked urgently to report on the attitude of the Italian communities in those countries if Italy and Sicily were to be attacked. Comments meant for Vichy ears were made to the effect that the Allies were confident that their offensive against Italy and the Central Mediterranean would meet with at least benevolent neutrality from the French in North Africa. British consulates in Spain made enquiries for Spanish merchant sailors acquainted with the north coast of Sicily and the southwest coast of Italy. The British consul-general at Barcelona was asked to report on affairs in Sicily; the former British consul at Palermo was told to be ready to proceed to Gibraltar; inquiries were made at the Vatican about the presence of church dignitaries in Sicily.

On October 25, Noel Wild in Cairo began implementing TOWNS-MAN. On that day, CHEESE reported: "There are rumors of a Second Front in Italy next month, though I believe Crete is a more probable target." On October 31 he sent a strong warning of a threatened attack on Italy, in which Crete now appeared to be a secondary target for a diversionary raid.* On November 2, D minus 6, he reported: "The Americans are preparing to launch huge air raids on Italy next week. These are to be the prelude to invasion." On November 3 he reported excitement in Egypt at the imminent invasion of Italy, and that the Vichy French fleet in Alexandria would come over to de Gaulle. On November 4: "The American raids are to be directed on Naples in consequence of an unwillingness to bomb Rome."

The convoys from the United Kingdom set out on October 22 and 26. As they plodded towards Gibraltar, PENDER was brought into play. Its aim was to suggest that the overall commanders of TORCH, Eisenhower and the British Admiral Cunningham, who were actually due to arrive at Gibraltar about November 1, were far from the scene. PENDER I, implemented entirely on the American side, involved spreading word that Eisenhower had been called to Washington for a conference early in November. PENDER II, implemented by the British mainly through the

---

* Crete was employed in TOWNSMAN for smooth integration with TREATMENT, the concurrent strategic deception plan for the El Alamein offensive, which, as will be recalled, featured a notional attack on that island.

Governor of Gibraltar, put out the "story" that Cunningham was en route to take up command of the British Eastern Fleet.

By November 2, when Clarke and Fleming passed through Gibraltar and viewed the scene from the top of the Rock, they found spread before them an extraordinary sight. As Clarke described it:

> On the aerodrome aircraft were stacked in closely packed ranks as for the flying deck of some gigantic aircraft carrier, leaving only a single narrow runway down the center; while the solid walls of stores and supplies flanked the streets leading into the town. Inside it the startled population of the Rock rubbed shoulders with the crowds of staff officers, airmen and supply personnel of both Britain and America, straining the packed quarters of the fortress as never before. It seemed amazing that German agents across the narrow frontier-line to Spain could not have given Berlin a proper account of what was in store.

Yet no German agent did so.

Clarke had planned that KENNECOTT and TOWNSMAN peak on the night of November 6 with a dummy paratroop drop on Sicily. This finale was canceled for lack of airplanes, but a second last-minute move, designed to suggest Sardinia as the objective, was implemented on November 7: Two radio messages, purportedly from an aircraft and a Vichy French merchant vessel, reporting the sighting of a convoy.

TORCH cover and deception was almost too successful. Axis intelligence was swamped with contradictory reports and rumors—much of which could be tracked through ULTRA, and this raised concern that the desired message would be lost in the noise. Attacks were predicted from Dakar to Narvik; one report from the Vatican, that between mid-October and mid-November the British would land at Algiers while the Americans landed at Dakar, was fortunately not followed up.

SOLO I and OVERTHROW were, in the happy phrase of the official historian, "pushing at an open door." As for SOLO I, Hitler himself was endlessly concerned over Norway—it is "the zone of destiny of this war," he had told his Navy chief, Grand Admiral Erich Raeder, in March 1942. On August 26, Raeder referred to "the constant threat of an enemy invasion" as a reason for keeping strong naval forces in northern Norway. In late September, there were reports that the Germans anticipated an attack by ten divisions said to be concentrating in the Orkneys. On October 19, the Führer ordered Narvik reinforced and its approaches specially protected. On November 2, all central and northern Norway went on full alert. As the winter nights lengthened, Hitler worried lest the Allies take advantage of them for a surprise descent upon northern Norway; as late as December 22, he fretted that January would be the most dangerous month for Norway. The bottom line was that in 1942 no German ground forces were withdrawn from Norway for service elsewhere; and that was the object of SOLO I.

Things went equally well on the OVERTHROW front. As early as the beginning of August, Rundstedt's weekly situation report opined that the U-boat campaign would force the Allies to move against northern France. A week later, small craft concentrations on the English south coast had confirmed his expectations. After the Dieppe raid, he thought that that operation had been only "feeling the way" and that Brittany would soon be struck; and a week after that he was referring to Dieppe as a "dress rehearsal"—the double agents' very language—with "every possibility of renewed large-scale landing attempts based on the experience gained in that operation." On September 14, he reported that enemy activities confirmed "his opinion that further enemy operations are imminent."

On October 5 Hitler weighed in, perceiving a threat to the Channel coast "on the basis of numerous reports from agents"; he ordered the coast put on alert and its defenses strengthened, and on October 9 he decided Cherbourg was the likeliest target. "Enemy attack possible at any time and in different places," Rundstedt reported on October 12; he had picked up the exercise by the British 3d Division as part of OVERTHROW, and recorded that the British "are more occupied than hitherto in carrying out landing practice in divisional strength." On October 19 he was "still firmly of the opinion that British are trained and in complete readiness for landing operations on a fairly large scale." On October 31 and November 1, the Luftwaffe carried out heavy raids on Canterbury— probably to disrupt the new command in Kent reported by the double agents. On November 2, Rundstedt considered Dieppe-style operations likely in November, telling his army commanders that "there will be no slowing down for the winter months for the Western Front." On November 16, eight days after the Torch landings, he remained certain of an assault on France; and he continued of that opinion until he had to recognize in mid-December that the weather now made such an attack impossible.

Once again, the bottom line came out right. Rundstedt stayed on alert, and no appreciable force, if any, was withdrawn from him to reinforce any other front—the object of Overthrow.

Operations directly covering the landings went even better. The Axis seems never to have suspected that Eisenhower and Cunningham were in Gibraltar, so Pender must be counted a success; indeed, the story that Ike was in Washington for conferences reached the German press, via a dispatch from Stockholm to a Berlin newspaper. The Kennecott/ Townsman "story" that Italy and Sicily were targeted struck home—although, contrary to Clarke's expectations, the Axis considered a landing in Rommel's rear as the likelier goal. On November 5, Rome radio called on the populations of Italy and Sicily to be ready for an immediate invasion. Incredibly, the Axis never spotted the Torch convoys till agents saw the leading formations pass Gibraltar on November 5, three days before D-Day; and Axis reconnaissance did not locate them till next afternoon, well inside the Mediterranean. The German military attaché in Madrid radioed to Berlin—read by Ultra—that Spanish opinion was that they would reinforce Malta and then land behind Rommel, expressly rejecting any suggestion of Italy or French North Africa. (Meanwhile, the German embassy was reporting that the Spaniards thought Italy was the objective.) The German Navy advised Hitler that Rommel's rear was the likeliest objective, followed in descending order of probability by "Sicily, Sardinia, the Italian coast, in last place, French North Africa."

Fremde Heere West's daily situation report, early on November 7, might well have been written by the Allied deceivers:

> Agents' reports have been received giving the composition of the first convoy to go into the Mediterranean on the 5th and 6th November. . . . Given a maximum speed of 14 knots, the enemy could be at Cape Bon by about 1600 hours on the 8th. If it is intended to break through the Sicilian Channel, having regard to the strength and makeup of the formation, we must take into account the possibility of a landing being made in the Tripoli-Benghazi area, in Sicily or Sardinia, apart from supplying Malta.

The Italian Navy on November 6 ordered Axis submarines in the Mediterranean to concentrate at the Sicilian Channel, and next day Field Marshal Kesselring, the German commander in the Mediterranean theater, ordered all available bombers of the Greece-Crete command transferred to Sicily. There the Axis lay uselessly in wait, making no effort to attack the Allies (apart from one torpedoing by a submarine in a chance encounter 120 miles from Algiers) before they stormed ashore in French North Africa on November 8.

"When the African landings were made, they were a complete surprise," Jodl told his Allied interrogators after the war. "The first news was received by air reconnaissance from the Gibraltar area. Even then it was assumed that the operation was directed towards a landing in Rommel's rear." Literally caught napping, the Italian Armistice Commission in Algiers were seized in their hotel, and the German commission escaped by the skin of their teeth. Far away in East Prussia, Hitler was not even at his command post; he had left the afternoon before to attend the annual Nazi Party reunion in Munich. Algiers fell the first day, Oran two days later, all Vichy resistance ceased the day after that. The first token German units did not appear in French North Africa until the 10th.

And, as has been seen, the failure to predict both the El Alamein offensive and the North Africa landings marked the beginning of the decline of Canaris and the Abwehr.

It will be recalled that that late summer and early autumn of 1942 had seen MI5's double agents finally integrated with the deceivers' work. The TORCH deceptions marked their first fully orchestrated employment for deception. The agents used were FATHER, MUTT, JEFF, TATE, GARBO, DRAGONFLY, CARELESS, BALLOON, and GELATINE. TATE, for example, could report on troop movements in southeastern England; MUTT and JEFF reported from Scotland; GELATINE picked up gossip from her military friends; GARBO milked his budding network for all it was worth.

Bevan and his colleagues worked out a detailed timetable for the passage of information and rumormongering. August emphasized the relief of Malta. September saw a shift of emphasis to Scotland and SOLO I. Mid-September saw the beginnings of troop indoctrination under SOLO II, and the beginning of OVERTHROW with reports of troop movements and other activity and the characterization of Dieppe as a "dress rehearsal." The end of September would see a wave of diplomatic indiscretions about Allied interest in Scandinavia, relations with de Gaulle, and intentions to relieve Malta. October would bring a crescendo: reports of civilian evacuations from the south coast, army drivers practicing driving on the right-hand side of the road, issuance of mountain clothing in Scotland, demand for Italian interpreters and guidebooks, and imminent air raids on Italian cities.

GARBO starred in all this. His network had not yet attained its full splendor, but he had enough to turn out quantities of useful information. He reported on Norwegian, Scottish, and Canadian troops training in the Highlands for mountain warfare, and his friends at the Ministry of Information enabled him to report that rumors of an attack on Dakar were nothing but cover for an attack elsewhere, perhaps France or Norway. Unfortunately, one of his long-standing subagents, GERBERS (AGENT TWO), the German-Swiss businessman living near Liverpool, had to be sacrificed. There was no way AGENT TWO could not have noticed the great buildup of shipping in the Mersey. So he was notionally sent to the hospital, where he eventually died. MI5 placed an obituary notice in the *Liverpool Post:*

> GERBERS. Nov. 19 at Bootle, after a long illness, aged 52 years, WILLIAM MAXIMILIAN. Private funeral. No flowers, please.

GARBO sent a clipping of this item to Lisbon.

Under date of October 29, GARBO reported that a convoy had sailed from the Clyde (this was true), and that "an operation of great importance is imminent and I think I fulfill my duty by advising you of this danger." Under date of November 1, he wrote that he had learned at the Ministry of Information that the Allies were about to land in French North Africa. Alas, through some unaccountable mishap in the post office these vital messages were delayed and only reached the Germans on November 7, just before the landings. Though the information was now useless, the Abwehr was overwhelmed by the skill of their star agent. "Your last reports are all magnificent," GARBO was told on November 26, "but we are sorry they arrived late."

*   *   *

Axis self-deception could not detract from TORCH as a cover and deception triumph, for Bevan and his section in particular. The British Chiefs of Staff noted their appreciation by formal action. Bevan was promoted to full colonel the next day. "So the next time you see me," he wrote to his mother, "I shall be all dressed up in red tabs as the typical 'Whitehall Warrior.' It is rather awful! . . . I thought you would like to know that your youngest son is at the moment very much 'above himself.' "

Remarkably, the whole elaborate project had been planned, and its execution supervised, by only five men, Bevan, Petavel, Wheatley, Wingate, and Arbuthnott; and the last two had only joined in September. It had been "literally a day and night affair," said Bevan. It was shortly after TORCH that Bevan sought authority to enlarge his staff somewhat, with "operations" and "intelligence" branches patterned (on a smaller scale) after Clarke's structure in Cairo.

Other "Lessons Learnt," according to a paper by Bevan on that subject, included the need for important commands to have their own deception staffs; the need to have a clear "object" in terms of desired enemy action; the importance of close collaboration with the intelligence services and of careful preparation and timing of messages to be fed to the enemy (which, said Bevan, "must be undertaken by individuals endowed with an imaginative and tortuous mind"); and the need for improved methods of spreading rumors at home.

A particular "Lesson Learnt" dealt with American deception efforts. "Duplication and crossing of lines occurred between the two sections in the implementation of 'TORCH' cover and deception plans," recorded Bevan. "There are a number of problems affecting the two sections still undecided." He was blunter off the record. "Apparently," said Hollis in a note to Ismay, "one of the major shortcomings in our cover and deception is the very poor showing which the Americans make and a lack of co-ordination of effort between them and our organization." "Colonel Bevan considers that the only way to improve matters is for him to go to Washington, and I am inclined to agree." The British Chiefs approved such a trip on December 8. So Bevan made arrangements to go to the States with Goldbranson over the Christmas and New Year period to meet with Strong and other interested parties.

First, however, he drafted a proposed overall deception policy for the winter of 1942–43. In tentatively final form under date of December 6, approved by the British Joint Staff Planners, it assumed that Sardinia would be the next target, in March. It proposed that to contain enemy forces in northern Europe (and contain their U-boats in the North Atlantic) a new threat to Norway be built up, and staff talks might be initiated with Sweden; strength in the United Kingdom should be exaggerated, indicating vast preparations for spring 1943; and a large increase in movement of American troops to Britain should be indicated. To induce the Axis to withdraw forces from the Russian front, Russian losses should be exaggerated. In the Mediterranean, the enemy should be encouraged to reinforce the Balkans and the French coast, rather than Sardinia and Italy, by maintaining the threat to Crete, hinting at landings in southern France, and suggesting that TORCH aimed merely to free the Mediterranean for convoys to the Middle East and India. In the Far East, the Japanese should be discouraged from operating in the Indian Ocean by building up their impression of the strength of the British Eastern Fleet.

Goldbranson and Bevan arrived in Washington the afternoon of December 26. By then there had been some changes at Security Control. It had received authority to veto publicity that would endanger military operations. Probably feeling that the Navy should be represented by an officer of equal rank to Strong, Admiral King had designated Rear Admiral Harold C. Train, the Director of Naval Intelligence and General Strong's opposite number, as Navy member, though Captain Dyer continued to do most of the actual Navy's work. Strong had added two more Army officers, a Colonel R. C. Jacobs, Jr., and Colonel Clarence E. Jackson. Only Jackson was to play any significant role in deception. A "plodding, good-natured, Clydesdale type of old infantryman," in Bill Baumer's words, he had been the American military attaché in Bangkok from February 1941 to Pearl Harbor, and spent seven months in Japanese internment before being repatriated.

A preliminary meeting on December 30 reviewed Bevan's draft winter 1943 policy; minor changes in phraseology were made, and an additional object of deception was added for the Far East: "To contain the maximum of Japanese naval strength in home waters by threatening raids in force and by increased submarine attacks." The chief meeting was held in the new Pentagon building at 10 A.M. on December 31, with Captain Dyer presiding. Bevan was accompanied only by Bratby for the British side. Present for the Americans were Pierce, Jacobs, and Jackson for Security Control; Goldbranson; a Navy planner and an Army planner; and Captain L. J. Abbott, who would replace Goldbranson as ETOUSA's liaison with the London Controlling Section when Goldbranson moved on to North Africa as was planned, and would later be involved in the non-deception, straight security work of Joint Security Control.

After an introductory discussion of PURPLE WHALES and its subsidiary FLOUNDERS (British deception operations through the Chinese, to be described in the next chapter), Bevan outlined his own organization. He stressed the importance of having both an operations element with the responsibility of making plans in close consultation with the service planners, and an intelligence branch responsible for knowing the enemy's thinking and for collecting and disseminating information to the enemy by "special means."

He then proposed a formal division of control over deception worldwide: London controlling deception in Europe, North Africa, the Middle East, and India, and Washington controlling the Americas, Australia, the Pacific, and China less the PURPLE WHALES channel. The controlling section or theater commander who initiated a deception plan would be responsible for its implementation, and no other controlling section or theater could implement a plan unless requested to do so. Coordination of plans was an absolute necessity.

To this, the meeting agreed. The one matter left open was whether a controlling section should be allowed to implement a plan outside its own sphere. "For instance," asked Bevan, "would it have been in order for the London Controlling Section to implement TORCH cover plans in South America?" The group could reach no decision on this, because it depended mainly on the extent and efficiency of American special means channels and their cooperation with British channels. Bevan said that he knew nothing of American channels to the Axis and would much appreciate information on that subject. He underscored the importance of diplomatic channels and their coordination. He put a British double agent in Canada at American disposal, through the Mounties; and told them that it was his understanding that the same had been done with certain British channels in South America (this seemed to be news to Pierce, who undertook to confirm it).

The group discussed Eisenhower's immediate need for deception plans, and Goldbranson's efforts to get notional American units for order of battle deception. Pierce and Dyer "emphasized the fact that the U.S. Security Control was very definitely in an embryo state and they wished to make certain that Colonel Bevan appreciated this." Pierce undertook to arrange for Bevan to meet the key American planners, Generals Handy and Wedemeyer and Admiral Cooke. The meeting agreed that the modified deception policy for winter 1943, and the FLOUNDERS plan, should go forward to the Combined Chiefs and thence to theater commanders.

No doubt Bevan knew nothing about American special means channels because outside of diplomatic contacts there were almost none. The FBI had put RUDLOFF on the air earlier that year, and may have been running ASPIRIN at this time; PAT J was in their hands, by courtesy of the British, but he did not go on the air till February 1943. Colonel Pierce recorded that there were "established American channels," information released through which would reach "the highest Axis echelons" in four to five days; but this probably referred to contacts of military attachés abroad.

In any case, Bevan's veiled hints that the British might implement deceptions in South America made Pierce, and through him Strong, sit up and take notice. Pierce met with Strong that afternoon. Strong said it was unwise to use any channel on a shared basis, and that only American channels should be used in Latin America. "In other words, if the British wish to release anything in Latin America they should turn the material over to the United States for release through channels established by the United States and in no case should they release anything through established channels of their own."

Bevan and Strong met on January 5 and settled the question. The two controlling sections would not undertake implementation of deception outside their respective spheres of responsibility. "For instance," in the words of Bratby's memorandum for record, "L.C.S., when desiring to implement deception in South America, would request Security Control to undertake such implementation. Similarly, Security Control would not implement in Europe or within the British sphere."

Bevan's proposed winter deception policy had gone forward from Joint Security Control to the Joint Chiefs through the Joint Staff Planners immediately after the December 31 meeting (with a few touches later added at Navy request to downplay its British origin). The day after his meeting with Strong, Bevan discussed it with the American Joint Staff Planners. When it was made clear, in the words of the minutes of the meeting, "that the policy had not yet been approved by the British Chiefs of Staff nor was it definitely based on any agreed strategic concept," consideration of it was "deferred pending agreement on a basic strategic concept."

It is hard to resist the conclusion that the deal with Strong about implementation within respective spheres of responsibility was the main thing Bevan had in mind in coming to Washington, and that his half-baked winter deception policy paper—which, after all, was certain to be overtaken by whatever decisions should emanate from the Casablanca Conference—was little more than an excuse. The deal with Strong meant that there would be no more George Fielding Eliot capers, Dakar-focused OSS leaks, and other blundering around in the British sphere by the inexperienced Americans. Bevan was giving up little by turning over British Latin American channels; Latin America meant little to the British and a great deal to Strong and the FBI, and Bevan's references to that region were probably designed to encourage Strong to hasten to protect his turf. Bevan's repeated emphasis on the need for American help with FLOUNDERS served, of course, to emphasize British generosity in sharing their assets. Johnny Bevan was nothing if not smooth.

While this bureaucratic minuet was being danced, Goldbranson was working on American fake order of battle activity. At least two bogus American units had already been introduced into double agent reports from the Middle East: a phony "9th Armored Regiment," which had supposedly been moved into Egypt in increments from June to November, and a phony "22nd Infantry Division," which had supposedly arrived in the Middle East regiment by regiment between September and November. It is not presently clear how many fake units had been set up as of the end of 1942, but the extensive activity of this kind thereafter had its origin in Goldbranson's work with OPD on the New Year's visit of 1942–43; and a table of notional American units available for deception purposes was sent to the London Controlling Section on January 16.

During all this, the American planning staffs had been working night and day getting ready for the forthcoming summit conference between Roosevelt and Churchill and their staffs at Casablanca. Once that was completed—the final preparatory meeting in the White House took place just two days after the Strong-Bevan meeting of January 5—there was an opportunity to turn to the future and designate officers in OPD to specialize in deception. The choice fell on two men: Major William H. Baumer, Jr., and Colonel Harold D. Kehm.

Born in 1909 in Omaha, son of an insurance salesman, Bill Baumer had had a solid old-fashioned Midwestern Catholic upbringing, short on luxuries but long on hard work, and on the enjoyment of simple good things. There were summers on an uncle's farm in South Dakota, mud fights at Boy Scout camp, helping bring up six brothers and sisters after his mother died when he was twelve, working every night as a teenager selling popcorn at the local movie theater and driving an ice-cream truck

(Catherine R. Baumer)

William Baumer
(a postwar photograph
as a brigadier general)

on weekends; and there was a no-nonsense education at Creighton Academy, the local Jesuit school. When his father's health broke at about the time he finished high school, a West Point appointment from the local congressman, a friend of his father's, was too good to resist. He took classes at Creighton University in Omaha till the place at the Academy opened up, and arrived at West Point in July of 1929, a bit older than many of his classmates of the class of 1933.

The old all-male West Point was "Hell on the Hudson," with a tough and merciless regime. But Bill Baumer flourished there. The Academy's director of public relations soon found he had a knack for writing. He got assignments for stories on Academy athletics for papers all over the country, and worked in the radio press box at Army football games with such legendary figures as Floyd Gibbons, Ted Husing, and Bill Stern. His first three years after graduation were spent at posts near New York City, and he became a stringer for the *New York Times*. Then, newly married to Alice Brough, a girl from the New Jersey suburbs whom he had first met on a blind date at West Point, he spent tours first at the Infantry School at Fort Benning and then at Fort Snelling in Minnesota; continuing at intervals to file *New York Times* stories on Army life.

In 1938 he was picked for a tour as an instructor in history at West Point. Much of the teaching at the Academy was traditionally done by young serving officers who as cadets had done well in the subjects they taught. The redoubtable head of the social sciences department, Colonel Herman Beukema, took the long view. "Gentlemen," he would tell his instructors, "there is a war coming. Every one of you will be on the General Staff, and I want every one of you to go to graduate school and get a degree." So while instructing at the Academy Baumer commuted to New York, where he took a master's degree in political science at Columbia.

I was at West Point almost four years from 1938 to 1942 [recalled Baumer in later days] in what must have been the busiest four years of my life. I taught history six days a week, till noon on Saturdays, was active in handling sports publicity in the afternoons on a voluntary basis, went to Columbia two nights a week for three years. Also I wrote six books and probably at least a hundred articles during that period and last but not least, my wife and I had two wonderful girls, one born in 1939 and the other in 1941.

The books were about West Point, West Pointers, and Army life. Several went through a number of editions.

Then, early in 1942, Colonel Beukema's prophecy was fulfilled: Major Baumer was assigned to OPD. In late May he reported for duty at the Munitions Building.

He worked initially on plans for operations in Africa: BARRISTER, a potential operation to seize Dakar, and GYMNAST, the original designation for what became TORCH. He was deeply involved in that operation. It fell to Baumer to break the news to Patton that he was going to put an invasion force ashore in French Morocco. "His face fell a foot" with astonishment, recorded Baumer, for he had expected to be sent to Cairo to help fight Rommel, but he recovered swiftly. "Get me every son of a bitch who knows anything about North Africa," he ordered Baumer.

Baumer worked closely with Patton until the latter left Washington to ship out on October 23. He grew to know Patton well, was often invited home by the General for a drink, and had regretfully to turn down an offer to join his staff.

Early in September, OPD moved across the Potomac into the new Pentagon Building. It was still under construction, so Baumer and his colleagues "were walking over piles of wood and all sorts of plaster materials and so on; it was a mess week after week and month after month." Still, they were more comfortable and functional quarters than the crowded and primitive conditions in the Munitions Building. Comfort and functionality were needed, for OPD was a pressure cooker of what Baumer called "seven-day-a-week, twelve-hour a day milling around." He had no real time off from August of 1942, when he went to New Jersey to bring Alice and the children to a house he had finally found in Chevy Chase, until a three-day weekend the following March.

Baumer liked his bosses. With Wedemeyer he shared the common background of Omaha and Creighton Academy. Handy he found impressive and Hull he particularly liked: "Always kind, yet he was certain of what he wanted." His immediate boss, Colonel J. C. Blizzard, he regarded as "a nice gentleman," but ultimately Blizzard faded from OPD, leaving the impression that he had not had "what it took in the planning game when the chips were down." The work itself, of course, was endlessly interesting: preparation of weekly strategic summaries, analyses of possible future operations all over the world, drafting of Roosevelt's responses to Churchill's "Former Naval Person" messages.

Then at Christmastime came what Baumer described as "a real blitz" lasting for "two solid weeks," drafting papers for the Casablanca Conference. If Baumer thought that after that he could catch his breath he was wrong. For Johnny Bevan's visit had ended only a few days before, and Baumer found himself tapped for the deception business. As he wrote in his diary on January 28:

> Immediately after the bigwigs left for their rendezvous, in fact the following Monday morning, Jan. 11, I was turned over to a special project with Colonel Kehm. It is fascinating work and something new to the War Department and to our Army; the British have a start on us of perhaps a year. We have spent the intervening time studying all about the subject—and there is a meager documentation—and doing some thinking on our own. . . . No one knows where it will lead, and there is no path that is clearcut for us to follow. All of which pleases me greatly. We can therefore strike out for ourselves and see what happens.

The "Colonel Kehm" (pronounced "Kem") mentioned by Baumer was a new recruit to OPD, a 44-year-old graduate of the West Point class of 1923, named Harold D. Kehm. A Pennsylvania Dutchman from a poultry farm near Allentown, Kehm had a genuine German mother and a grandfather who had fought in the Prussian army in 1870. This equipped him to do an imitation of an arrogant Hun, "Baron von Kehm," strutting and exploding in exasperated German, which his classmates found sidesplitting; the nickname "Baron" stayed with him all his Army career. An artilleryman, Kehm had served with troops in the States and Panama until tapped in 1932 for Colonel Beukema's department at West Point. During his six years on the Academy faculty, conformably to Colonel Beukema's rule Kehm acquired an M.A. in political science from Penn State (which he had attended before going to West Point). From West Point he went to the Command and General Staff School at

Fort Leavenworth; from there to be an instructor at the Field Artillery School; and thence, in January 1943, to OPD.

Baumer found him a "colloquial-speaking, thought-skipping, likeable gent, whose first love was the Field Artillery and second, the Economics, Government and History Department at the U.S. Military Academy." Kehm was an important figure in early American deception planning— a favorite term of his was "overplay"; whenever a paper in the planning files warns that a scheme might be "overplayed," you suspect the Baron's hand—but he was to leave less of a mark than did Baumer.

At about this time, presumably, the Navy planners were likewise read into deception work. But the Navy was not to play a major role in strategic deception planning for some time.

So now for the first time there were American officers formally assigned to deception planning. George V. Strong, however, would not have been content with that. He would unquestionably have wanted as a close link to OPD a military intelligence officer ultimately answerable to him. Colonel Jackson was in bad health, made worse by his treatment in Japanese hands; moreover, a "plodding, goodnatured, Clydesdale type of old infantryman" was not what was called for. The ideal would be an American Bevan: a mature combat veteran who would command the respect of fighting officers, and with high-level international experience as well.

As luck would have it, just the right man had newly been commissioned as a lieutenant colonel in military intelligence. Destined to become as near to an American opposite number as Johnny Bevan would have, he was a fifty-three-year-old bantam rooster of a man, five feet six with gray-blue eyes, light brown hair, and a General Pershing mustache—indeed, a "miniature Pershing" in appearance, consciously modeling himself on the great Black Jack, according to one who knew him—an Alabama native with an outstanding combat record and years of experience in Europe on a sophisticated level. His name was Newman Smith.

Born in 1889 to a merchant and his wife in the lumber boomtown of Dothan in southeastern Alabama, Newman Smith got his early education from private tutors and at a nearby academy. Always attracted by the military life, after clerking briefly in his father's store he joined the Army at eighteen (lying about his age, for the enlistment age was twenty-one in those days), and served two years in the 6th Cavalry, rising to corporal and seeing combat against the Moros in the Philippines. Enlisted men led a hard and brutal life in the old Army, but Newman appears to have survived it well. His officers rated him as of "excellent" character, "honest and faithful" in service.

After his discharge in 1909, he attended Howard College (now Samford University) in Birmingham. But after two years he dropped out and read law in a lawyer's office in Montgomery. At some point he married a girl named Mildred Beasley. Perhaps therefore needing a steady job, he took some elementary business administration courses at a local business college and went to work for the First National Bank.

The marriage to Mildred ended in divorce. It was probably at the bank that he met his true love. She was the cashier for the women's department of the bank, a remarkable young woman of exactly his own age

Newman Smith

(Eleanor Lanahan)

named Rosalind Sayre.* The second-oldest of the four daughters of a justice of the state Supreme Court, Rosalind was an independent woman
who insisted on earning her own keep in an era when well-born young
ladies just didn't do that. She had done newspaper work in Birmingham
and Montgomery; she and her next-younger sister Clothilde, known as
"Tilde," wrote the first film scenario written in Alabama, and it was actually produced and released.

While waiting for his divorce and courting Rosalind, Smith still felt
the lure of the Army. He joined the Alabama National Guard and in
April 1916 was commissioned a captain in the 4th Alabama Regiment,
the "Old Fourth," once one of the great regiments of the Army of Northern Virginia, serving from First Manassas to Appomattox with a side trip
to Chickamauga. The war in Europe had shown that the machine gun
was now queen of the battlefield, and he was one of the organizers of the
new machine-gun company of the Old Fourth, with himself as its captain. Shortly thereafter, he and his company were called to active duty
with Pershing's expedition to Mexico. That adventure lasted till the
United States entered the First World War in April 1917. As the 167th
Infantry, the Old Fourth became one of the regiments of the soon-to-be-
legendary 42d "Rainbow" Division.

Newman's and Mildred's divorce became final on August 1, 1917.
Later that month the Old Fourth left Montgomery to join the rest of the
Rainbow Division at Mineola, Long Island. Within a few weeks came orders to go overseas. Rosalind journeyed to New York, and she and Newman were married by a justice of the peace in Hempstead, Long Island,
on October 8. Captain Smith shipped out for France on November 3.
The couple spent a last night together at the old Hotel McAlpin at
Broadway and 34th Street, and to the end of her life Rosalind saved the
hotel bill in a scrapbook, with "Goodbye for France!" written under it.
After three weeks of marriage they would not see each other again for almost two years.

It was to be the happiest and tenderest of marriages, though they
never had children. She referred to him as "Capitán" and he called her
"my sweetheart" all his life. Rosalind's niece, who would visit the Smiths
as a child when they all lived in Europe, remembered their home as "a

---

* Pronounced to rhyme with "fair."

protected little nest" where she "always felt very safe and secure," and when they were an elderly retired couple in Montgomery after more than forty years of marriage she recalled "marveling once again at the way they had turned their lives into a little fairy tale."

Smith reached France on November 17. Three months later the Rainbow Division was in the line in the Lunéville-Baccarat sectors. No division of the AEF covered itself with more glory. It fought in the Second Battle of the Marne, helping turn back Ludendorff's last desperate effort to win the war, and in the Aisne counterattack, the St. Mihiel offensive, and the big push into the Meuse-Argonne.

Smith served with distinction in these engagements, and in May he was put in command of one of the division's three machine-gun battalions. "You deserve everything you have been recommended for and more," his colonel told him. "You won your promotion on absolute merit and hard work, and your old regiment is proud of you." He became a major on August 1. "I have lived centuries in a few days," he wrote home during the hard fighting that summer, "and I wonder if I ever shall be older than I am tonight." In the autumn he was gassed and spent some time in base hospital. When the Armistice came he could boast five battle stars on his campaign ribbon, for the Champagne-Marne defensive, the Aisne-Marne, St. Mihiel, and Meuse-Argonne offensives, and the Lorraine defensive sector.

The Armistice did not release him to rejoin his bride. From November 1918 to April 1919 he was with the operations staff, G-3, first of the American Second Army and then of IX Corps; he then served as an aide to Herbert Hoover in the latter's postwar relief work in Europe. This was a high honor, for Hoover drew these assistants from young officers suggested by Pershing as outstanding. But it meant that when the Old Fourth marched up Dexter Avenue in Montgomery for its grand homecoming in May, Major Smith was doing relief work in Rumania. He finally landed in New York and was mustered out at Camp Dix in New Jersey at the end of August 1919. Rosalind, who had lived at home in Montgomery while her husband was overseas, was there to meet him; for they had decided to start a new life in New York and Europe.

Smith went to work for the Guaranty Trust Company, one of the major banks of New York (predecessor of the present-day Morgan Guaranty), and took courses in economics, accounting, and finance at New York University, while Rosalind edited advertising copy for a printing company at which sister Tilde's husband worked, and sold little stories to the New York *Sun*. In the summer of 1920 they moved to Constantinople, where Smith was part of a team opening a new branch of the bank, and after a year he was transferred to Brussels; the Smiths would live there until late 1933. Assigned at first to reorganize the Belgian branches of Guaranty Trust, he was soon in charge of all the bank's European operations.

In May of 1928 Smith left Guaranty Trust to become head of a new entity called Motor Dealers Credit Corporation, formed to finance European sales of Studebaker and Erskine cars. The next year it was taken over as a subsidiary of Commercial Investment Trust, with Smith in charge of all CIT operations in Europe, introducing the Europeans to American consumer credit.

Though bankers are notoriously ill-paid, and from time to time Smith had to borrow on his government life insurance or even from his brother-in-law, he and Rosalind lived a good life in Brussels. He took up fencing, polo, and golf; he became a notable fencer, a member of the Royal Fencing Club of Belgium, an honorary member of the Académie des Armes of Brussels and a member of the Fédération Internationale d'Escrime of Geneva; he is said to have once fought a duel for some reason with Barclay de Tolly, descendant of the great Russian field marshal of the Napoleonic era.

Their only problems were generated by Rosalind's sister Zelda, eleven years younger than Rosalind, and her husband. While Smith and his machine gunners were helping hold back Ludendorff's last onslaught that July of 1918, eighteen-year-old Zelda—who was somewhat "wild" by the standards of the time—met a lieutenant from Minnesota at a country club dance, a Princeton dropout and hopeful author named F. Scott Fitzgerald, and fell in love. Rosalind disapproved of Fitzgerald, deeming him unstable and too apt to play up to Zelda's "wild" side. But his first novel was accepted for publication and he began making money from selling stories to the mass magazines, and they married not long before the Smiths left for Europe.

The Fitzgeralds came to Europe in April 1924 and lived there till December 1926, and again from March 1929 to September 1931. Through them, the legendary embodiments of the Jazz Age, the Smiths witnessed the Lost Generation in full cry. There was much too much drinking; Zelda's behavior became erratic; in the spring of 1930 she broke down. She was briefly hospitalized near Paris. The Smiths came down from Brussels to help. Eventually they got her to let Scott and Newman drive her to a clinic in Switzerland, where she was diagnosed as schizophrenic.

Rosalind blamed Scott and "the mad world you and she have created for yourselves" for Zelda's condition, as did the whole Sayre family. Rosalind wanted to adopt Scott's and Zelda's eight-year-old daughter "Scottie," deeming Scott too unstable to be in charge of her. Scott took revenge on the Smiths in perhaps the finest of all his short stories, *Babylon Revisited,* which is based on this controversy. In it Rosalind and Newman Smith appear as "Marion and Lincoln Peters," the sister and brother-in-law of the protagonist's wife. "Marion" is depicted as unreasonable and neurotic, motivated by resentment that while the protagonist and his wife had been "tearing around Europe throwing money away," her husband "didn't touch any of the prosperity because [he] never got ahead enough to carry anything but [his] insurance." (This last was a particularly pointed dig, for Smith had once had to borrow from Fitzgerald to reinstate a lapsed insurance policy.)

In 1933–34, in view of the Depression and the uncertain state of Europe, CIT phased out its European operation. The Smiths returned to the States, where Newman was initially in charge of sales promotion work in New York, and then from the latter part of 1938 in Atlanta, where he was vice president in charge of the company's southern division. By now it was clear that Zelda would be in and out of institutions for the rest of her life, while the grim 1930s had turned Scott Fitzgerald into a back number. His last great completed novel, *Tender Is the Night,* was a commercial failure. He struggled to publish short stories and to write Hollywood scripts. Drinking continued to be a very big problem. In 1940 he died, on the fringes of Hollywood, holding the bottle at bay, depressed that Rosalind felt that he was keeping Zelda in a sanitarium for his own reasons, forgotten by the general public. Zelda did not feel well enough to attend Scott's funeral in Rockville, Maryland, and asked Newman to go in her stead; he traveled up from Atlanta to do so. There was more than a little irony in that, given the way Fitzgerald had felt about the Smiths. Scottie wrote years later:

I can easily understand why my father always had such a strong urge to administer a swift kick to Uncle Newman's smug and self-satisfied persona. He was one of those people who are always proper and always in the right, stuffy, unimaginative, and yet thoroughly commendable, handsome in a stiff-backed sort of way, a good citizen and a devoted husband always to Aunt Rosalind. He naturally backed her up in the quarrel, though I don't believe he felt quite the hostility she did toward my father.

In his years as a banker, Newman Smith did not forget his first love, the Army. He held a reserve commission as a major from 1920 till he resigned it when approaching age fifty in 1939. In the Guaranty Trust years he acted as an informal eyes and ears in Europe for U.S. Army intelligence. When Pearl Harbor came, nothing would do but that he get back into uniform despite his age. "Available immediately," he advised the Army, "and eager to serve in any capacity and in any arm or branch of the service. Believe, however, experience and background qualify me to be of most useful service in the Military Intelligence Division." It took a while, and probably a little wire-pulling. But eventually a commission as lieutenant colonel in military intelligence came through, and orders to report to General Strong himself.

Smith left Atlanta for Washington on November 19, 1942, and met with General Strong the morning of December 1. Strong must have seen as providential the arrival of a mature, combat-tested officer, speaking French, German, and Spanish, with years of practical experience in international business. On December 23, Smith was formally sworn in as a lieutenant colonel, and at the beginning of 1943, at the same time as Kehm and Baumer were being brought on board as planners, he took up duties as an operations officer of the General Staff with responsibilities in strategic deception.

With Joint Security Control established; with operations and planning officers assigned to the work; with a man of Smith's caliber and background in place in Washington; with Goldbranson destined for the Mediterranean, and another officer soon to be assigned to deception work in the China-Burma-India theater; and with an agreement with the British as to division of labor, the Americans could fairly be said to have arrived on the strategic deception scene. One task remained: To establish a clear understanding with the British as to procedures and responsibilities in the war with Japan. For British deception was already well under way in the Japanese war, supervised by a remarkable man named Peter Fleming.

CHAPTER 8

Hustling the East (I)

*At the end of the fight is a tombstone white with the name of the
late deceased,
And the epitaph drear: "A Fool lies here who tried to hustle the
East."*

So wrote Kipling. But you did not have to be a fool to fail at hustling the
Japanese. Part of the problem was that the Japanese intelligence services
were too dull-witted to be hustled. Another part was that until late in the
war the Allies could mount no significant operation in Southeast Asia for
a deception plan to support, while strategy in the Pacific was long a subject of debate. "It is impossible, or at least highly dangerous," lamented
Peter Fleming afterwards in a comment already quoted, "to attempt to
tell a lie until you know what the truth is."

There were broader problems, too. Strategic deception of the Japanese
was primarily an American responsibility, but the Americans found
themselves unable to set up the kind of smooth-running machinery that
the British had developed. MacArthur's command in the Southwest Pacific was a fiefdom of its own. And the United States Navy was reluctant
to share the Pacific war with anybody.

The clouds of war were already gathering over eastern Asia when Wavell,
sacked by Churchill from the Middle East, arrived in Delhi on July 11,
1941, as Commander-in-Chief India—which command covered Burma
as well, so that if war came, an obvious key Japanese objective would lie
within Wavell's theater: the Burma Road, the last useful link of Chiang
Kai-shek's China with the outside world.

For the time being, Wavell let "A" Force act as in effect the deception
organization for India—and for the Persia and Iraq Command, which at

that time came under the India Command. Liaison arrangements between Clarke and those theaters were regularized at Clarke's meetings in London in early October; but provision was also made for the possibility that a new war zone might soon open up in the Far East, with India then to become a separate theater for deception purposes.

On December 7—December 8, in the Far East—Japan struck the United States Pacific Fleet at Pearl Harbor. A British general passed through Cairo on December 9, on his way to Singapore, and Clarke took the opportunity to send with him necessary lists and documents to enable coordinated strategic deception to be opened in the Far East when required, and to be supported by the implementing resources of the Middle East, Persia-Iraq, and India. Effective as of January 1, 1942, India became officially a separate entity for deception purposes, independent of "A" Force.

The Japanese moved against American, British, Dutch, and Australian possessions. Hong Kong fell on Christmas Day, Manila on January 2. Wavell was appointed supreme commander of a combined Allied command, and set up headquarters in Batavia, the capital of Java. He decided that he needed another Dudley Clarke. On January 8 he again asked London for a specific individual for deception; not a regular soldier this time, but a wartime officer of many talents, who had been included on a list of officers with Far Eastern qualifications sent by the War Office. He was Major R. P. Fleming, Grenadier Guards, better known as Peter Fleming, best-selling author, traveler, and all-round adventurer.

Peter Fleming had led a life most people would hardly dare dream of, building an unbroken series of accomplishments upon a very advantageous original foundation. Born in 1907 to considerable (if relatively recent) wealth, he grew up in the Establishment, and was a great success at Eton and at Christ Church, Oxford. He was a world traveler; the Continent, North America, Central America; a journey to China via the Trans-Siberian Railway and back via Malaya, Burma, India, Iraq, and Turkey; an expedition across the uncharted heart of Brazil; again via Russia and Japanese-occupied Manchuria to China, where he obtained an interview with Chiang Kai-shek and visited the Communist-held areas, thence home via Japan and the Pacific and across the United States; once more across the USSR to Manchuria and China, this time by way of the Caucasus and Central Asia, thence with an equally intrepid Swiss woman on sundry journeys to Mongolia, crowned by an epic seven-month, thirty-five hundred-mile trek with her from Peking across Sinkiang to India; to China again via India and Burma and up the Burma Road, covering the war, and with two more interviews with Chiang, thence to Japan.

He had published best-selling books about his travels. He had been assistant literary editor at *The Spectator,* and a regular contributor to that magazine thereafter; roving special correspondent, leader-writer, and China war correspondent, for *The Times;* a staffer at the BBC; contributor to a short-lived glossy magazine patterned on the *New Yorker.* The

(Kate Grimond)

Peter Fleming

Chinese knew him as Fu Lei-ming, which means literally Learned Engraver on Stone.

Fleming was of middling height, lithe and sinewy and dark-haired, with a square, bronzed, tight-skinned face, reminding Dennis Wheatley of a jaguar until he broke into his wide smile. A perpetual pipe emitted clouds of smoke which a friend said smelled like "motor-tires burning in syrup"—but then Fleming had lost his own sense of smell to a childhood illness. Regarded by ladies as devastatingly handsome, by gentlemen as a good fellow, and by reviewers as "A Modern Elizabethan," he was happily married to the exquisitely pretty and highly intelligent Celia Johnson, one of the most talented actresses of her generation, lived in great comfort on a country estate, and passed time shooting pheasants, grouse, and Scottish stags when not roaming the world or penning books and essays.

Perhaps the self-deprecation was just a bit thick; perhaps the assumption that his elders and the mighty would accept him as one of them was just a bit confident; perhaps the pose as a figure out of John Buchan or Dornford Yates was just a bit self-conscious; perhaps the prose style was just a bit arch: nevertheless, he was so transparently decent that all this never seems to have mattered to most people. Joan Bright, who ran the Commander-in-Chief's Special Information Centre in the War Office and knew absolutely everybody who was anybody in the British war effort, recalled him as "a four-square, basic, solitary sort of person, immune to luxury, to heat or to cold, with a rock-like quality that made him the most staunch of friends and a kindness which made him the least vindictive of enemies." Even people who wanted to loathe him on principle usually found that they could not.

Fleming's father had fought and died on the Western Front in the Kaiser's war. He himself joined the Grenadier Guards (it *would* of course be the Guards) as a reserve officer. Called to active duty when war broke out,* he spent a few frustrating months in Whitehall, and then when the Germans invaded Norway he was the first British soldier to set foot on Norwegian soil. There he saw adventurous service at the right hand of

---

* Peter Fleming was the eldest of four brothers, all of whom saw service. The second, Ian, was, as has been seen, personal assistant to Admiral Godfrey, the Director of Naval Intelligence. The third, Richard, served in the Lovat Scouts and the Seaforth Highlanders, was wounded and awarded the Military Cross. The youngest, Michael, died of wounds as a German prisoner in 1940.

General Carton de Wiart; there he and Johnny Bevan carried out their little experiment in deception. After Dunkirk, he trained Home Guards to act as staybehind guerrillas if the Germans should invade. In early 1941 he took a party of Commandos to Egypt on an abortive scheme to recruit Italian prisoners of war, and while in Cairo he was introduced to Clarke's work. When the Balkan campaign broke out he took his little force to Greece in time to be of some help blowing bridges and wrecking trains in the retreat; he was wounded in a Luftwaffe attack on the ship that carried them to Crete. He spent the rest of 1941 in London, running a training school for street warfare, and fretting to be overseas. Finally came Wavell's signal on January 8, 1942: "Should be glad of Peter Fleming as early as possible for appointment my staff."

After a month-long journey from Britain by sea and air, including a week's stopover in Cairo where he immersed himself in the files of "A" Force and educated himself as much as possible on Clarke's developments, Fleming reached Delhi on March 17. Delhi was now again Wavell's headquarters, for Java had fallen, the Japanese had invaded Burma, and Wavell had returned to India, while General Alexander had flown out to take over the Burma front under him. Waiting for Fleming was an invitation to be Wavell's guest at his own house till he should find quarters; the general himself was in Calcutta for a few days.

When Wavell returned it was at once evident that he and Fleming would hit it off well. As a mutual friend observed, they both "found the long silences which punctuated their conversations intensely satisfying." As a practical matter, Fleming was to have with Wavell much the same independence and direct access that Clarke had had back in Egypt, though his nominal superior was Brigadier Walter (Bill) Cawthorn, Wavell's Director of Military Intelligence, an Australian who proved to be thoroughly sympathetic with the work of deception. Fleming's operation was established in the staff of India Command as a section of the Military Intelligence Directorate called "GSI(d)." It was "a one-horse show, and I am the horse," as Fleming wrote to Dennis Wheatley, with his office in a cubicle on the ground floor of one of the huge red sandstone buildings that housed the Raj.

By that time, Rangoon had fallen and the Japanese were advancing up the great central valleys of the Irrawaddy and the Sittang. Alexander was conducting a fighting retreat up the east bank of the Irrawaddy towards Mandalay with British Empire troops while Chinese divisions under overall command of the American General "Vinegar Joe" Stilwell did the same up the Sittang. At the end of the month Fleming accompanied Wavell and Alexander on a visit to Burma, and on April 4 he flew on to Chungking to "have a look round," as Wavell put it. There Fu Lei-ming evidently renewed old acquaintances that would be useful in the ensuing years. Soon after his return, he was called upon to prepare the way for the version of the Madagascar cover plan already described, GOLDLEAF-HERITAGE, in which the cover destination of the Madagascar expedition was to have been Trincomalee; he flew down to Ceylon and requisitioned dozens of buildings to accommodate the force during the time it was supposed to be there.

Then Wavell launched him upon the first deception project of his own, Operation ERROR.

One evening at dinner, Wavell told Fleming and another officer the story of Allenby's deception operations and Meinertzhagen's Haversack Ruse. The idea presented itself: Why not do something similar to try to make the Japanese advance more cautiously in the belief that India was defended in great strength? Wavell and Fleming sat up until one o'clock in the morning planning a haversack ruse of their own; which came to be called ERROR. The idea was to persuade the Japanese that Wavell himself had been directly involved, and injured, in the retreat in Burma, leaving behind secret documents indicating that the Japanese faced more formidable Allied forces than was really the case.

Concocting the bait was a congenial task; for one of Fleming's temperament, pastiche is endlessly delightful. Fleming persuaded Wavell to give him a few private letters and even to part with a much-loved photograph of his daughter. He confected a seeming personal letter to Wavell from Joan Bright, hoping that the Japanese would deduce from "veiled and slangy" references in it that she had an uncle who was an officer in an armored division that had sailed for India. He constructed other documents indicating that Wavell had been wounded in the retreat, and implying that India was heavily garrisoned and that further reinforcements were on the way; including some "Notes for Alexander" advising that two armies would be available for Burma and hinting at some sort of secret weapon. All this material was placed in a briefcase initialed "APW."

Early in the morning of April 29, Fleming took off from Delhi in company with Wavell's aide-de-camp Captain Sandy Reid-Scott of the 11th Hussars, armed with the briefcase, a tunic of Wavell's with his medal ribbons, and a valise containing a blanket marked with Reid-Scott's name and regiment. The plan was to leave the evidence at the landing ground at Mandalay, where the Japanese would be sure to find it. After losing a day waiting at Dum Dum airport outside Calcutta, they finally reached the deserted airfield at Shwebo in Burma, Alexander's command headquarters. Smoke was still rising in the distance from the bazaar of the town, obliterated by Japanese bombers the day before. "Everyone seemed a little dry in the mouth and absent-minded," noted Fleming.

They found Alexander and explained ERROR to him, his chief of staff, General Winterton, and his intelligence officer. This was the first Alexander had heard of it, for a message from Wavell had never arrived. He approved (though, as Fleming noted, the generals "had too much urgent business of their own to be really interested"), but told them they were too late to carry out the plan at the Mandalay landing ground. Winterton suggested leaving the bait in an abandoned car at the great Ava Bridge across the Irrawaddy at Sagaing, not far downstream from Mandalay, which was to be blown at seven o'clock that night.

Sagaing was seventy miles away and there was not much time. They rounded up a jeep and a large green Ford sedan. Joined by an engineer major named Michael Calvert, an old comrade of Fleming's from the days of training Home Guardsmen as guerrilla fighters, and by one of Calvert's toughest men, a private named Williams who had recently escaped from Japanese captivity just before he was due to be bayoneted to death (a favorite sport of the Japanese Imperial Army), they made their way down the road through "a few wretched Indian refugees and a lot of Chinese transport . . . drifting back in fairly good order. Officers young and relatively smart, some with field glasses and map cases." The jeep was loaded with high explosives and booby-trap devices; Williams, who drove it, did not know the real purpose of the mission.

They reached Sagaing just before seven o'clock and presented themselves to the division commander in charge. "Most sympathetic and sensible," he told them that withdrawal across the bridge was to be completed by ten o'clock and the bridge was to be blown not later than midnight. He suggested the far bank of the river as a good site, and undertook to let it be known that Fleming's party was planting a booby trap; this would keep Allied troops from interfering with the car once it was planted. They nipped quickly across the bridge so as to make the most of the last hour of daylight, moving against a flow of "Gurkhas doing a timed withdrawal like clockwork, silent and alert." Something was burning in the middle distance; the Japanese were drawing near.

A suitable place was found at a curve in the road about four hundred yards from the bridge. Calvert made some skid marks and they sent the Ford over the embankment. "It flounced down, crossed a bullock cart track and then down again into a crater-like depression, where it remained right way up with the engine still self-righteously running. Highly unsensational. We laughed and followed it up." They arranged the contents of the trunk (Wavell's briefcase, his jacket with ribbons, Reid-Scott's valise, three novels snitched from the Shwebo Club library, and a loose blanket), punctured one front tire, let the air out of the other, removed the ignition key, and inflicted minor damage on the car body. They admired their handiwork; the car was right side up, in good condition, and readily visible from the road; it looked as if it had been abandoned in a hurry, and it was more likely to attract attention than most of the wrecks that marked the line of retreat.

An armored car appeared down the road. They did not wait to find out whether it was Japanese. They recrossed the bridge in their jeep, "bumping over planks where the charge was laid." It was eight o'clock and getting dark. They got something to eat at division headquarters and went back down to the riverside. From across the river in the still hot night came the sound of dogs barking and men and women talking excitedly. A Chinese patrol told them that the Japanese had arrived. At 11:20 the bridge was blown up.

They slept that night, much tormented by mosquitoes, in the bungalow that had been division headquarters. (Next morning Fleming found there an old copy of *Punch* with a piece by him in it.) They went down to look once more at the bridge. Two spans had been cut by the blast. There was some shellfire on the other side of the river but Chinese guarding the bridge said all was quiet.

They drove back to Shwebo. Army headquarters had already pulled out. "Typewriters, stationery, scattered clothes, foundered cars, the bazaar still smoking from its ashes, wires down across the roads; a nasty, too familiar smell." He left among the litter in the abandoned operations building a note to "Dear G" from "Henry" urgently asking help in finding a dispatch case initialed APW and a valise marked for Reid-Scott, believed left on the far side of the Ava Bridge. He spent the night with Calvert and made his way back to Delhi.

Wavell was pleased. The only thing that might give the mission away, he reckoned, was that someone on the Japanese side, some German liaison officer perhaps if there were any (there were none), might remember Meinertzhagen's haversack ruse. He fretted a bit that one of the items planted in the briefcase made mention of his biography of Allenby. But, he noted on Fleming's report, "Any harm done, if the Jap decides it was a plant? I don't think so." "Might help if impression of anxiety over seriousness of loss of papers in Burma were fostered," he cabled to London.

Alas, Operation ERROR was casting pearls before swine for all anyone ever knew. There was never any evidence that the Japanese had paid any attention to the car, much less that they drew any conclusions from its contents. Fleming was told in May 1944 by Cheng K'ai-min that during the 1942 Burma campaign the Japanese had captured important documents indicating that India's defensive potential was stronger than they had believed; and it was gratifying to suppose that this referred to ERROR.

On May 19, the last of Alexander's force crossed the border into India, as the monsoon rang down the curtain on operations by either side.

In time, Fleming would realize that even so mildly subtle an item as Joan Bright's forged letter was "far above the heads of the Japanese, who could in no circumstances be relied upon to make the required deductions from even the most obvious hints." But at that stage, prudence suggested a more conservative assumption. As Fleming wrote to Dudley Clarke on May 11:

> The situation in this theatre appears to me to be rather similar to the situation in Western Europe after the collapse of France. The Japanese, like the Germans in 1940, have to a very large extent outrun their own Intelligence organisation and are (in my guess) up against something like a blank wall.

Perhaps the way through the blank wall was to rub their noses in something that did not require them to draw conclusions. "What we want," Fleming cabled to the ISSB back in London, "is not red herrings, but purple whales." Perhaps direct evidence that India was more heavily defended than it really was could be put into their hands, seemingly from such a high level that it would unmistakably reflect Allied policy. (High-level forgery appealed to the perpetual schoolboy in Fleming. As far back as 1939, he had proposed a forged update of the "Tanaka Memorial," the supposed blueprint for Japanese world power.) There was reason to believe that the chief of Chiang Kai-shek's secret police, General Tai Li—unkindly referred to by Fleming as "the local equivalent of Himmler"—would have ways and means of conveying the right item to the Japanese.

Fleming was musing upon this problem when he chanced to read the verbatim transcript of an inter-Allied conference in Chungking. He was struck by the advantages that such a record offered to a deceiver. "The verbatim medium has an allusive quality," he reflected, "which enables the forger to avoid over-definitely committing himself to facts or figures which may later be revealed as false." A forged order for an operation would irrevocably commit the deceiver to propositions that would be disproved; whereas a forged *discussion* of an operation could convey a vivid impression without ever actually pinning the forger down to a lie; and after events demonstrated the error of the initial impression, review of the document would produce no concrete evidence that it was inauthentic. "This method of approach, incidentally," Fleming noted, "is one well calculated to disturb the balance of the Japanese mind, on which any suspicion that there is a wrong end to the stick preys with disproportionate effect."

Fleming's initial thought was to concoct the imaginary transcript of a summit conference on Allied grand strategy, bearing the signatures of Roosevelt and Churchill themselves. London regarded this as rather too high-flying, and Wavell decided that the bait should be the record of a meeting of top brass in Delhi. The propositions to be suggested in the text were the subject of some back-and-forth between Fleming and the Joint Planning Staff in London between March and June. Fleming continued to hope to get at least Churchill's signature certifying to Chiang Kai-shek a fake estimate of the situation, and Oliver Stanley continued to reject this as "rather grandiose." ("Better a tamer whale than an unmanageable Leviathan!!" noted Ismay.) As late as mid-May Fleming was still hoping, as he put it, "that London, given a little more time, might produce something—in the words of Wodehouse—both louder and funnier," and they did come up with something a bit better from his perspective by early June.

At the same time, another uncertainty had to be dealt with, in that it might not be possible to plant the paper through China after all. British prestige with Chiang Kai-shek was at a low ebb, and with the steady Japanese advance, air communication between India and China might virtually cease. Fleming visited Kabul to investigate the prospects for passing deceptive material through neutral Afghanistan (chartering a taxi in Peshawar to take him the whole five-hundred-mile journey through the Khyber Pass and back). He made a useful contact with Cornelius Van H. Engert, the United States minister to Afghanistan, who already had a channel for passing information to the Japanese; but he determined that Kabul channels were not suitable for this operation. By early June, General J. C. Bruce, head of the British mission at Chungking, could report that things were looking up and that the Generalissimo had approved the project in principle, subject to minor conditions.

Meanwhile, Fleming turned his hand, "at odd moments and under the pressure of other duties," to construction of the purported verbatim transcript of a conference held on May 31 among top Allied commanders—British, American, Chinese, Dutch—from all over the Far East. Propositions implied in the final document included that the Americans were strengthening their South Pacific defenses and planning an offensive in the Central Pacific; that Britain would be ready to invade the Continent before the summer was out; that the British were running an extensive network of spies behind Japanese lines; and, most important to Wavell, that India was very powerfully garrisoned. (There was also a passing reference to mysterious "V-bombs.") The first page was endorsed with appropriately portentous nonsense: *This document is Category A 1. (SEC) and if despatched by air must be conveyed in a jacket, self-destroying, Mk. IV by an officer of the rank of Major or above.* Attached to the document was a cover letter, purportedly a chatty personal note from Wavell's deputy to Bruce at Chungking, enclosing a "spare copy" of the transcript.

Like any such transcript, in Fleming's words, it "approximate[d] in *genre* to an unusually tedious and ill-constructed one-act play." It was, perhaps, not a perfect forgery. An American would have noticed something distinctly odd in the way Vinegar Joe was made to talk.* One potentially disastrous error crept in, by way of reference to a German air raid on England of which Delhi could not have known on May 31; this was spotted and corrected in time.

On June 22 Fleming flew to Chungking with his forgery and a personal letter from Wavell to Chiang asking the Generalissimo's help. The next day Bruce took him to see General Ho Ying-chin, chief of Chiang's general staff. Ho was courteous but noncommittal; the forgery and Wavell's letter were left with Ho's translator.

There then ensued more than a week of the unexplained waiting that often attended efforts to do business with the Chinese, punctuated only by an inquiry from the translator as to the meaning of the phrase "Japan is your pigeon," ascribed in the forgery to a British general. The delay proved fortunate, for an intercepted message from the Japanese legation at Kabul to Tokyo referred to a published item that required the date of the fictitious conference to be moved back to May 30; two pages were retyped and flown in from Delhi.

Eventually (with the help of a close associate of Chiang's whom Fleming knew), word came back that the Generalissimo had given his approval. On the evening of July 6, Fleming gave the amended forgery to a Colonel Chen of Ho's staff, who showed a lively appreciation of the project and asked to be used as intermediary and interpreter in future projects. Tai Li never appeared, but it was evident that he would handle the passing of the document to the Japanese. On July 7, one of his men left Chungking for South China; at dawn that same day, Fleming departed for Delhi.

On July 27, a high Tokumu Kikan officer in Canton paid the very large sum of ten thousand Chinese dollars for Fleming's forgery.

---

* GENERAL STILWELL: Can you tell us, General, whether the outcome of this major engagement in Libya is likely to, or perhaps I should say is capable of modifying the decision recorded in Paragraph B.

GENERAL STILWELL: I asked because, although I am not fully in the picture yet, I gather from our people here that MacArthur sets a good deal of store by air-borne troops. . . .

GENERAL STILWELL: I certainly had no idea that they had got on as far as this. Those people are darned cagey.

The channel thus established—Fleming christened it PURPLE WHALES, in line with his earlier message to the ISSB—was used a total of six times during the war, only for false documents of an apparent extremely high grade. Directly in charge of the operation was Tai Li's right-hand man for such matters, Lieutenant General Cheng K'ai-min—who proved throughout the war to be a good friend to Fleming's efforts. But for the time being at least, there was nothing to show that PURPLE WHALES was not another wasted effort. It sank into the Japanese intelligence system without a ripple.*

Some modest but fruitless attention was paid to strategic deception of the Japanese back in London during these early months of the Far Eastern war. Stanley mooted in January the idea of trying to induce the Japanese to divert forces from their march of conquest in the belief that the Soviets were likely to attack them. But suppose they attacked Vladivostok in response? So nothing was done about that. The British planners suggested in April some rather obvious themes that might be dropped into the rumor mill, and Stanley added some of his own. He proposed soon thereafter a sketchy overall deception plan that included some items for the Japanese war. But nothing came of these and without American participation little or nothing could be done outside India.

GSI(d) clearly could not effectively operate as a one-man show. By orders dated June 2, Fleming was authorized to expand his shop to nine officers (including some naval and air officers) and a clerk. Moreover, slightly better accommodations were found, on the first upper story of the eastern block of the old India Army headquarters building, in rooms that had been condemned as unfit for Indian clerks.

Fleming did not take advantage all at once of his authority to expand; at the end of June, GSI(d) had grown only from one officer to two. But over time, he put together a varied and talented staff. In March 1943 he acquired Major Peter Thorne of the Grenadier Guards as second in com-

---

* It seems possible that the PURPLE WHALES channel, at least from 1943, was an agent of the Tonegi Kikan—but in truth an Allied double agent—referred to by the cover designation "GH" in Japanese intelligence reports intercepted by the Americans. GH, who may have been a former employee of the Japanese consulate-general in Hong Kong named Lin Ch'ing-shan, specialized in acquiring official documents.

mand. Thorne was another Old Etonian and Oxford graduate, the youngest son of Major-General Sir Andrew Thorne, for whom Fleming had worked during the invasion-scare days of 1940; he had known the younger Fleming brothers, but not Peter. He possessed a penetrating mind and a quick sense of humor, plus the most useful of all a soldier's skills: the ability to go to sleep whenever the opportunity offered. He had finished the short wartime course at the staff college at Quetta and was about to be posted to a dreary job at Army headquarters when Fleming invited him to join GSI(d). Fleming told Thorne that Thorne was to run the office in Delhi; that he, Fleming, would often be away and this meant that Thorne must stay at home and not take part in fun and games in the field. Thorne agreed, hoping that this would change later on (but it never did). Thorne received the particular assignment of developing and putting over a false order of battle.

As time went on, Fleming acquired a cast of characters at least as colorful as those at other deception headquarters. There was Major André Bicat, a peacetime artist, who, frustrated by his original assignment to develop camouflage, persuaded Fleming to put him in charge of tactical deception devices; starting out in the former quarters of the Viceroy's band, Bicat's activities grew into a small network of factories turning out a full range of contraptions, including the PINTAILS that were Fleming's own invention. There was Major Lucas Ralli, a wealthy Old Etonian W/T expert; he was able to increase the range of the hopelessly inadequate transmitters that captured enemy agents brought with them, without the Japanese realizing that they had been altered. In 1943, Major Frank Wilson of the 19th Lancers joined Fleming to run a tactical headquarters in Calcutta. Wilson's accomplishments included fluency in Urdu—a useful skill in connection with deceptive messages—plus talent as a caricaturist, which Fleming would also put to use. At the end of 1944, Major A. G. "Johnnie" Johnson came to Delhi from "A" Force in East Africa, bringing with him the know-how to operate special means channels out of that region that Clarke no longer needed.

For the Navy, Commander Robertson MacDonald came out in the spring of 1943, supposedly fully familiar with the workings of the London Controlling Section; but, said Thorne long after, he "never really understood deception." He was replaced in March 1945 by Lieutenant-Commander the Earl of Antrim, formerly of "A" Force; "Ran" Antrim was an old friend of Fleming's, and proved to be more useful. For the RAF there was Wing Commander Mervyn Horder, son of Lord Horder, physician to the royal family. Horder was in civil life a director of the mildly eccentric publishing house of Duckworth; a preparatory-school schoolmate of Thorne, an Old Wykehamist, a Cambridge graduate (Trinity College), and an aviator of sorts, for he was the only man licensed to fly an Autogiro without holding a regular pilot's license. Horder was, in Thorne's words, "the most literate member of GSI(d) after Peter himself," and he eventually got the job of writing GSI(d)'s, and later D Division's, weekly newsletters.

Clerical and secretarial work was handled by young women drawn mainly from the families of the headquarters staff, including two of Wavell's daughters.

Of essential importance, Fleming established liaison with the office of the Director, Intelligence Bureau, Home Department, Government of India—known always as DIB (for "Delhi Intelligence Bureau"). As the war wore on, DIB was to furnish the chief double agent channels to the enemy in that part of the world, much as did MI5 in Britain and SIME in Egypt.

The Director of the Intelligence Bureau himself was a certain Sir Denys Pilditch. His staff was a small and interesting group and a skillful one, though woefully short of both personnel and equipment. Pilditch's deputy, John Jenkin, was an interrogator of redoubtable skill; Malcolm Johnston was a former officer of the Indian Police; William Magan was a cavalry officer from the Indian Army; Philip Finney, formerly of MI5, claimed the distinction of having been bitten by more mad dogs than anyone else still above ground. After visiting the theater, Hugh Trevor-Roper of MI6 remarked to Thorne, who had known him at Oxford before the war, that Fleming and Thorne were lucky to deal with "such a very professional staff as DIB instead of some of the extraordinary people in MI5 at home."

Fleming's relations with DIB were never ideal. John Marriott of MI5, accompanied at least part of the time by Trevor-Roper, spent January to August of 1943 visiting DIB and the Indian Police. He found, in Guy Liddell's words, "quarrels between DIB and the Army and nobody had a very clear picture of precisely what was wanted or how the job should be carried out," with Fleming and Cawthorn "thinking entirely in terms of deception, not realizing that it is necessary to build up on firm foundations a CE network in which the enemy had full confidence." When Cawthorn and Fleming passed through London in June of 1943, they got a lecture from Tar Robertson to this effect. And by the time Marriott returned to England in August, he felt, according to Liddell, that he had "finally succeeded in knocking the heads of the military and police together."

He could not so easily smooth over personal animosities; after the war, Magan—who was eventually to be case officer, along with Johnston, for the great double agent SILVER—stated his view of Fleming as "an irresponsible, ambitious and irrational man who was always trying to persuade us to pass messages which we believed would 'blow' the channel." Right up to the last months of the war, DIB was still fretting with Fleming over the contents of items in the PURPLE WHALES and INK channels.

Peter Thorne (who got on well with the DIB people) carried out much of the day-to-day working contact with DIB after he came on board in March 1943, dealing primarily with Johnston, sometimes with Magan and Finney, and with Jenkin on one or two special occasions.

Fleming's operation dealt only with DIB and had no contact with the Indian Police; DIB dealt with the local police in such matters as choosing suitable locations for the transmitters used by double agents.

In addition to DIB, the British secret services, SOE and MI6, had a presence in India, and Fleming and Thorne maintained liaison with them as well. SOE was in the charge of Colin Mackenzie, who as a Scots Guards officer had been badly wounded in the First World War. MI6, which in India as in the Middle East went locally by the name Inter-Services Liaison Department or ISLD, was headed by a Colonel Steveni, "a much experienced spymaster," in Peter Thorne's words; his deputy was Harry, Lord Tennyson, the poet's great-grandson and a friend of Thorne's.

During his three and a half years in India Fleming lived the life of a sahib under the Raj. After a few weeks as the Wavells' houseguest, he settled in with the Adjutant of the Viceregal Bodyguard in a bungalow "extremely comfortable and bursting with servants" and a garden full of English flowers. Peter Thorne moved in with them when he arrived in the spring of 1943; and for a few months in the winter of 1943–44 the "three Peters" were joined by a fourth officer, Ralph Arnold, in civil life a publisher and currently on Mountbatten's public affairs staff. When their host was posted elsewhere, Fleming and Thorne found lodgings for the rest of the war in the adjoining bungalow of the commander of the Viceregal Bodyguard and his wife. It was a nabob's life: a ride every morning accompanied by an Indian bearer, lunchtime swims at the pool on the grounds of the Viceroy's residence, squash games and duck-shooting.

While getting PURPLE WHALES under way in mid-1942, Fleming was also preparing a long-run operation of a different nature, called HICCOUGHS.

In June, Wavell told Fleming to consider using radio messages ostensibly directed to agents operating behind the Japanese lines. "Mystify and mislead the enemy": Wavell was fond of quoting these words of Stonewall Jackson's; at a minimum such messages would mystify the Japanese; at best, "by trading on the spy fever endemic in the Japanese nation," to quote Fleming, they might lead the Japanese commander in Burma to "doubt the security of his base and thus introduce a deterrent factor into the various factors on which he might base a decision to advance on India." So Fleming concocted HICCOUGHS, a notional staybehind network of Burmese agents—their supposed number fluctuated, but averaged around eight; one was supposedly a woman in Rangoon—engaged in espionage, subversion, political warfare, and sabotage. They supposedly reported to Delhi by radio and occasionally by courier, receiving their instructions in cipher messages broadcast twice daily over All India Radio, the Indian equivalent of the BBC.

At the end of the All India Radio morning news on June 25, the announcer read out an excerpt from *King Lear*, giving exaggerated emphasis to certain words. After the evening news he similarly read out an excerpt from *Macbeth*. This was repeated for five more days. Then on July 1 the announcer intoned: "RRUTE UIFII UERIE CTTRN. . . ." This was the first HICCOUGHS cipher message; the emphasized words of Shakespeare had, of course, been the keys to the cipher; it was designedly one that any mildly competent Japanese cryptographer should be able readily to break. (Similarly, the regions in which HICCOUGHS agents were supposedly operating were referred to by transparent codenames, such as

"Mytchett" for "Myitkyina.") The keywords were changed once a month, always from Shakespeare; the cipher remained readily breakable.

Twice each day for the rest of the war, HICCOUGHS messages were read out on All India Radio. Not for two years would there be any feedback indicating Japanese reactions, and there was little hard intelligence from behind Japanese lines from which to derive touches of verisimilitude. Concocting them soon became at best a bore and sometimes a real burden, and little serious thought could be given them. "We were often reduced to puerile shifts," said Fleming later, "such as telling an agent in Rangoon to 'put it in the swimming bath' and announcing to our man at Mandalay the birth to his wife in India of twins, one of which subsequently expired." This lack of substance was compounded by such procedural implausibilities—the use of a weak cipher, the fact that the Japanese could search the airwaves till Judgment Day without finding any trace of the agents' supposed lengthy messages back to Delhi, the lack of any real incidents of subversion and sabotage that could be correlated with the HICCOUGHS traffic—that an effort of faith was required in order to believe that the Japanese could take HICCOUGHS seriously.

Fleming had that faith, for he became convinced that the Japanese intelligence service was an easy mark. "Only the Japanese—with their superstitious dread of espionage, their strange faith in the cunning of older Anglo-Saxon races and their lack of intellectual self-confidence—would have fallen for this ruse," he wrote of HICCOUGHS after the war.* Moreover, he sensed that with this mind-set the Japanese would ascribe ordinary accidents to the nefarious workings of the mysterious network in their midst. So he plodded ahead with HICCOUGHS; and in the end his faith was justified in full measure.

---

* This was casting the first stone, for in 1940 the British had fallen for the same ruse at the hands of that experienced deceiver, Joseph Goebbels. At the opening of the offensive against the West in May 1940, Goebbels had arranged for meaningless code phrases to be broadcast to a nonexistent underground organization in France; and again after the fall of France, strings of meaningless ciphers were broadcast to imaginary Fifth Columnists in Britain. These stunts were rousing successes, contributing strongly to the Fifth Column panic that swept Britain that summer. A special committee of the Security Intelligence Center set up to combat Fifth Column activities swallowed Goebbels's bait whole, concluding that in addition to cipher communications the Germans were almost certainly using plain language transmissions in this way for Fifth Column purposes.

*   *   *

On September 12, Fleming left India to attend the London gathering organized by Bevan at the height of the Torch preparations. Besides seeing Celia, family, and friends, the visit home gave him an opportunity to get in some pheasant shooting; and on October 19, he was invited down to Chequers for lunch with Churchill, who had been a friend of his father. On November 1, with Bevan's meetings over, he took off with Clarke for Gibraltar and thence for Cairo. In Cairo he and Clarke met with Malcolm Muggeridge of MI6's station in Portuguese East Africa to discuss certain special means channels at Muggeridge's disposal. Muggeridge (who "had vaguely known [Fleming] for a number of years") always remembered how powerfully impressed Fleming was by Clarke's ability to secure transport seemingly at will, "having as he put it, God-high priority."

Fleming then proceeded on back to Delhi. Waiting for him was news that DIB was putting at his disposal for deception a very special double agent, codenamed Silver.

To put Silver into context it is necessary to review the story of an extraordinary figure, Subhas Chandra Bose.

Bose was born in 1897, son of a respected Bengali lawyer. Educated at Cambridge, he placed highly in the examination for the elite Indian Civil Service. But in 1921, in the upsurge of nationalist feeling after the Great War, he resigned; the first of his countrymen to do so in the modern period. Repelled by Gandhi's doctrine of nonviolence, he became a prominent figure in the Congress Party in opposition to the Mahatma.

He was in and out of prison, and traveled to Europe, where he met a variety of statesmen—Mussolini, Hitler, Benes, Attlee, Sir Stafford Cripps—and impressed them all. In 1938 he became president of the Congress Party. Jailed by the British in July 1940, he went on a hunger strike. Not wanting him to die in their custody, the Government of India released him under house arrest in his Calcutta residence. On the night of January 16–17, 1941, he was spirited out of Calcutta by his nephews and made his way to Peshawar disguised as a Pathan. At Peshawar he came under the care of a man named Bhagat Ram Talwar. Before dawn on January 26, Bose, Talwar, and three companions drove out of Peshawar and headed for the Afghan border. Reaching it safely, they slipped across the frontier on foot. When that night Bose's disappearance was discovered he was already out of India—his ultimate destination Berlin.

On foot, by mule, and by truck, Bose and Talwar and their companions made their way through the tribal hill country to Kabul. Initially, it would appear, Bose wanted to go to Moscow and see Stalin. For a week, he and Talwar stood in the snow outside the Soviet Embassy, but could not gain admission. The German legation showed interest, and Bose decided to go to Berlin instead. But to reach Germany, he would need permission to pass through the then-neutral USSR. More weeks passed. Then word came to get in touch with the Italian legation. The Italians produced a diplomatic passport in the name of Orlando Mazzotta, wireless operator. As Mazzotta, Bose entered the USSR and entrained for Moscow and thence for Berlin, where he arrived on a cold wet April evening.

After cooling his heels while the Germans decided what line to take with him, he was received by officaldom, established in comfort, and encouraged to take an interest in an "Indian Legion" organized from disaffected prisoners of war. When the Japanese entered the war in December, he began to broadcast anti-British propaganda as the voice of "Azad Hind," *i.e.,* "Free India."

Eventually Bose would make his way to Rangoon to set up a pro-Axis "Provisional Government of Free India," thereby multiplying Fleming's deception channels. For the present, however, we will turn out attention to Bhagat Ram Talwar, the man who spirited Bose out of India. For Talwar was not what he seemed; he was in fact a Soviet, and later a British, agent, who would be known to Fleming and DIB by the cover name Kischen Chand, and by the codename SILVER.

SILVER was a Hindu from the up-country, thirty-two years old at the time of the Bose escape; short and lean, but handsome and physically fit; intelligent, able, cool-headed, quick-witted, and self-assured. (It is said that once when his double agent operation was put at risk because an Afghan who knew him encountered him in Delhi with British officers, he disposed of the problem by calmly meeting the man for dinner and slipping a lethal dose of chopped tiger whiskers into his curry.) He knew his way about the hills of the North West Frontier and Afghanistan, and spoke Pushtu so fluently as to be able to pass as a Moslem of the frontier tribes. With this skill he assumed at will the false identity of a tribesman named Rahmat Khan, and in that guise he was able to make his way on camelback, by foot, and by truck between Peshawar and Kabul.

The major powers maintained legations in Kabul and it was a center of intelligence activity in a rather Kiplingesque style; indeed, the Allies resisted the temptation to induce the Afghan government to close the Axis missions because they were so useful as special means channels, for almost all the agents who supplied information to the Japanese and Germans were under Allied control. Nonofficial Germans had been expelled from Afghanistan in the autumn of 1941. As of mid-1942 only the ten members of the legation staff were left in the country. Six of these were conventional Foreign Office personnel: the minister and his No. 2 and the latter's wife, the legation physician, a stenographer-typist, and a wireless operator. The other four were secret service men.

One of these, Carl Rasmuss, had the title "commercial attaché." He was an old India hand, having served in the same capacity in the German consulate-general in Calcutta from 1936 to 1939; then for two years he had been with the German legation in Bucharest before coming out to Kabul in May of 1941. The remaining three were not diplomats but army men, who formed an Abwehr station: Lieutenant Dietrich Witzel, and two enlisted radio operators named Doh and Zugenbühler. They too had come out to Kabul in 1941 (Doh as a Foreign Ministry courier, lugging a mass of wireless equipment), to set up operations targeted on India codenamed FEUERFRESSER ("FIRE-EATER") and TIGER. These were never carried out, though Witzel may have had some other obscure operations going. Meanwhile, Rasmuss and Witzel recruited Talwar for covert activity in India on behalf of Germany, and paid him and companions a substantial sum.

Unknown to Rasmuss and Witzel, Talwar was a Communist. (His brother had been hanged for the assassination of the British governor of the Punjab in 1930; perhaps he felt thereby some affinity to Lenin.) After Hitler attacked the USSR, he reported this recruitment to the Soviets. They encouraged him to continue (and took the money the Germans had given him). He presented himself to the Germans, in Sir Michael Howard's words, "as a kind of Lawrence of Arabia, a master of disguise, held in numinous respect by the hill tribes of the North-West Frontier and deeply knowledgeable about the various revolutionary movements in India itself." He kept the Soviets posted on his work as the link between the Axis legations in Kabul and Bose sympathizers in India.

In 1942 the British persuaded the Afghans to expel Talwar, and arrested him crossing the border. The Soviets were now cobelligerents against Germany. Soviet intelligence in Moscow approached the British there, disclosed Talwar's status as one of their agents, and offered to make him available to be run by them and the British jointly. (Their motivation appears to have been to obtain from the British, in exchange, information about subversive movements on their frontier with China. "They seem to have a pathetic belief in the British Intelligence Service which in fact has no such information," noted Guy Liddell to his diary. "This the Russians find it hard to believe.") DIB officers visited Moscow to work out the details, and from October 1942 Talwar, as SILVER, became a DIB double agent.

DIB in turn made SILVER available to Fleming as a special means channel. Fleming contributed military deception material and some technical and administrative facilities, but throughout SILVER's career it was DIB's Special Section who nursed the case along, extricated it from various contretemps, and introduced SILVER into the Axis intelligence system as a high-grade source. (But SILVER remained a Communist first and foremost, and whenever he entered Afghanistan, practical control passed to the Soviets.) He made at least five such journeys after his work for DIB began, departing from India in January 1943, September 1943, April 1944, August 1944, and March 1945, and was setting out on a sixth when the war ended in August 1945.

To feed German interest in political affairs, DIB concocted a notional "All India Revolutionary Committee," a super-secret body of extreme nationalists with headquarters in Delhi and branches blanketing the country and controlling all revolutionary activity in India, in which SILVER would be an important cog. If swallowed by the Germans the Committee would not only afford imaginary sources comparable in scope to the GARBO network in England, but might help convince the Axis that it was unnecessary to try to open further networks. When he returned to Kabul in January 1943—and on every visit thereafter—SILVER took with him a report of the Committee's activities. The January 1943 report was a two-part document, one reporting on the political situation and the other a carefully constructed mixture of truth and imagination, setting forth both strategic speculation and detailed order of battle data for all three services. It was the opening of what was to prove Fleming's and Thorne's major overall deception success, the planting on the Japanese of a hugely inflated order of battle.

Rasmuss and Witzel took the bait, though to Fleming's disappointment they were more interested in the potential for insurgency and sabotage than in opportunities for military intelligence, and made no move to pass the military data over to the Japanese. They wanted to make arrangements to fly German advisors into the North West Frontier tribal area to support an insurgent movement. They wanted help in sabotaging railway tunnels in Iran through which moved Allied supplies for the Soviets; SILVER demurred to this, but it was tentatively agreed that the All India Revolutionary Committee would try to put an agent into Tehran. It was agreed that a wireless link would be established between the Committee in Delhi and the German legation in Kabul. SILVER left Kabul on April 2, 1943, with four thousand pounds, the first of several generous payments of precious hard currency he would receive.

ULTRA and MAGIC were intercepting and reading both the German and the Japanese traffic to and from Kabul; and Fleming and DIB were able not only to keep tabs on the Axis legations there but to follow events daily during the course of SILVER's visits as they were reported to Berlin and Tokyo.

A fourth channel to the Japanese opened in the last weeks of 1942. In November, a three-man party of agents—a team leader, a radio operator, and a courier—was parachuted by MI6 into Japanese-held territory near Rangoon. All were Karens, a minority nationality of eastern Burma. From the first radio message back to India it was plain that they had been caught and were working under Japanese control. The channel was turned over to Fleming to use for deception. Christened BRASS, it proved to be useful—and amusing—but its full development did not come till 1943.

Fleming himself finished the year 1942 on a pleasant personal note: a tiger hunt, on which he bagged a handsome specimen whose pelt graced his study floor back in England to the end of his days. For Wavell, too, the year ended agreeably, for he became a field-marshal in the New Year's honors. But as the first anniversary of his arrival in the theater approached, Fleming—a lieutenant-colonel now—had little to show for a year's work. ERROR and PURPLE WHALES had vanished without a trace. A beginning had been made at opening up special means channels, and at feeding order of battle material through them. But since the end of the retreat from Burma in May of 1942, the front had been quiet, and there had been no genuine operations for deception to support. Moreover, he fretted "doing this ungentlemanly stuff far from the heat and dust," while friends were being killed and wounded; "high time I shot one of our opponents or at least put myself in a position to do so."

In December 1942 Wavell opened a modest offensive down the Arakan coast of Burma. Fleming went down to the front, and got shot at by the "opponents"; and correctly predicted that the offensive would not get very far.

Then in January 1943 the Casablanca Conference tentatively authorized Wavell to mount in the autumn an operation to recapture Burma, called ANAKIM. For it, Fleming constructed a deception plan that showed that he still had a great deal to learn. Its "story" was that ANAKIM itself—a land offensive against northern Burma, followed by an amphibious offensive against southern Burma—would be a deception operation to cover the supposed real offensive, which would be an attack mounted from India and Australia against Sumatra and Java respectively, aimed at recapturing Singapore. This would be "put across to Japanese as real plan by secret channels while we use open channels such as manipulation of press, radio and utterances by prominent individuals to put across real plan as cover plan." The open aspect was called KINKAJOU and the secret aspect WALLABY.

Wavell should have known better than this. (So should General Wedemeyer of OPD, who gave the plan his general approval when visiting India on his way home from Casablanca.) Aside from the danger inherent in any double-bluff operation, KINKAJOU-WALLABY was not only too cumbersome and insecure, but there was not the slightest chance that the Chiefs of Staff, American or British, or their civilian masters, would approve such manipulation of a free press. The official historian has correctly characterized "this ambitious proposal for global thought-control as unrealistic and absurd, the product of inexperience and despair." Bevan certainly so reacted when he read it; but the response he, or in his absence Wingate, drafted for the Chiefs of Staff, was diplomatically phrased. By that time, the modest Arakan offensive was fizzling out and ANAKIM itself was in doubt. London's response was therefore explicit that no action was to be taken to implement any ANAKIM deception plan until the Combined Chiefs had confirmed that that operation would in fact take place. (It never did; it was abandoned for practical purposes at the TRIDENT Conference in May.)*

With one aspect of Wavell's message transmitting KINKAJOU-WALLABY, however, there was agreement. Recognizing that the plan would require world-wide coordination, Wavell had recommended that this be done from Washington, and that a preliminary conference to arrange inter-Allied deception machinery for the Japanese war be held as soon as possible. In that, London concurred. In Washington, Peter Cook, Baron Kehm, and Jackson of Joint Security Control arranged for such a gathering to begin on May 20, to coincide with a planned summit conference in Washington called TRIDENT. Though in light of uncertainty about ANAKIM there were last-minute suggestions to cancel the meeting, it was "agreed," as Newman Smith put it, "that considerable value should be obtained from a free interchange of ideas and thoughts on this subject which was relatively new to American representatives."

In addition to officers already in Washington, Wingate would come over to represent the London Controlling Section. MacArthur would have a representative. From India would come Cawthorn and Fleming. So would a new figure on the deception scene: the recently appointed deception officer for "CBI," the American China-Burma-India Theater.

By late 1942, the CBI theater organization had been fully established. Stilwell in Chungking commanded it, while Brigadier General Benjamin G. Ferris commanded a rear echelon headquarters in Delhi. Most personnel were Air Force. Brigadier (soon Major) General Clayton L. Bissell was their overall commander, with the title Commanding Gen-

---

* In March Fleming proposed another abortive plan, RASPUTIN, to convey to the Japanese word that the War Cabinet had decided to employ gas warfare against Japan when expedient after the elimination of Germany, thereby creating uneasiness at high levels and diverting Japanese shipping space and trained personnel. Bevan rejected it because it would cut across Plan LANGTOFT, described in Chapter 12, and might induce the Germans and Japanese to initiate gas warfare themselves.

eral, Tenth Air Force. Under him, Brigadier General Claire L. Chennault's China Air Task Force—soon to become a separate Fourteenth Air Force—was in fighting contact with the Japanese in China. The entry of the Americans into the deception business in late 1942 had two effects in the CBI: a half-hearted effort to engage the Americans in PURPLE WHALES, and appointment of a deception officer for the theater.

The American PURPLE WHALES was an abortive project codenamed FLOUNDERS. As early as the end of August 1942, Peter Cook in Washington approached Brigadier General Hayes Kroner, head of the Military Intelligence Service under Strong, with an explanation of PURPLE WHALES and a proposal from "our representative in India" that a sequel be marketed, emanating from Washington and ostensibly taken to Chungking "by a prominent personality not noted for guile or associated with propaganda." This should be a document with "a major strategical aim," and meanwhile the market could be kept alive by interim sales of material having a verifiable true basis. Kroner acknowledged and said the matter was being referred to the Joint Planners; where it apparently disappeared.

In late October the British mission in Washington approached OPD with an explanation of PURPLE WHALES, advising that London would be in touch with Wavell about similar activity by the Americans, and Wavell in turn would be in touch with General Ferris. Strong had told them Stilwell must approve anything that might be done, and they had been told by their people in Chungking that Stillwell had appointed Ferris head of his "countering organization" there—news that might have surprised Stilwell—and that Stilwell would not allow the use of any American units for deception. In November and again in December, Ferris spoke with Wavell, who told him that he had heard nothing on the subject from London. Cook raised the matter with General Strong again in early December.

When Bevan came to Washington at the end of 1942, one of his chief topics, as has been seen, was PURPLE WHALES and FLOUNDERS. At the Pentagon meeting on December 31, he emphasized that the principal object at the moment was to keep the PURPLE WHALES channel open, and it was agreed that American material would be furnished to help do so and that a courier should be sent to MacArthur in the Southwest Pacific, and Halsey in the South Pacific, to sell them on the scheme. But still nothing seems to have been done. In April of 1943, Bill Baumer's boss, Colonel Blizzard, pointed out to Wedemeyer that the channel could be critically important in connection with ANAKIM and plans in the Pacific. But there is no record that FLOUNDERS ever took on any life.

Meanwhile, out in CBI a deception officer was appointed in March 1943: Lieutenant Colonel E. O. Hunter of the Air Force.

Ormonde Hunter (sometimes called jocularly by his friends "Hormones" Hunter) was a fifty-year-old bachelor, something of a bon vivant, trim and handsome, with a dashing mustache. In civil life he was a prominent lawyer in Savannah. A graduate of Yale and the University of Georgia law school, he had served overseas as an artillery captain in 1917–18. He had been a state legislator and vice chairman of the Board of Regents of the state university system, where he had clashed with Eugene Talmadge, the powerful Governor of Georgia. Hunter could be overbearing and was a man of principle and of direct action: He had once

(Nell Pillard)

Ormonde Hunter

been prosecuted (unsuccessfully) for assault and battery by a loan shark who had insulted him for taking the case of one of the loan shark's victims—a poor black man—and had found himself at the wrong end of Hunter's fists. His brother, a World War I ace, was an Air Force general, and perhaps for this reason when Ormonde returned to active duty in 1941 it was as an Air Force officer.

Hunter was to find hustling the East as frustrating as did everyone else who tried it. In his case, part of the problem was that he was a bull in a china shop, bringing to his assignment a great deal of energy but not much subtle deftness.

Hunter's assignment started out bumpily. On March 22, he met with Bissell (just promoted to major general), who outlined his duties and told him he wanted aggressive action and to be kept advised. Two hours later, he met with Ferris's G-2 and G-3, from whom he gained the impression that his duties were to "differ considerably" from those described by Bissell. They gave him various secret papers to peruse, which he returned ten days later; no doubt puzzled as to what his job really was.

He appears to have heard nothing more until summoned to see Ferris on April 16. Ferris told him that he would work as originally planned; that he should see Cawthorn at British headquarters; and that he should discuss the work with no one—including Tenth Air Force—but Ferris, Ferris's G-3, Stilwell himself, and such British officers as duty might require. So Hunter repaired to British headquarters and met Cawthorn and Fleming. They filled him in generally and said they wanted him to keep his desk there rather than carrying documents to and fro. Back to Ferris, who approved.

Hunter's next weeks were filled with preparations for the Washington conference; he kept Fleming apprised of Washington's strong desire to see it through, and of the progress of thinking towards building up the American order of battle in the CBI. For security, his title was "Air Officer Rear Echelon CBI."

Early in May they all departed for Washington. GSI(d) was left in Peter Thorne's capable hands. The British went by way of England; Fleming stopped off in Cairo to look in on Dudley Clarke before going on to London. In London, he and Cawthorn met with key people at MI5 to review the handling of double agents in light of suggestions made by

Marriott. The trip gave Fleming a glimpse of his country estate in spring-time, and a fleeting visit with Celia and their little boy. Then Fleming and Cawthorn, joined by Sir Ronald Wingate, set forth for North America with the British delegation to the TRIDENT conference, headed by Churchill, aboard the *Queen Mary.* (Fleming had covered her maiden voyage for *The Times* back in 1936.)

For Wavell it was an ominous journey. He was still, or again, in Churchill's black books; the Arakan campaign had sealed his fate. For Fleming, Cawthorn, and Wingate, the decisions to be made concerning management of deception in the Japanese war would be fundamental. They meant to try to get control over that management; but the Americans were determined not to be outplanned, outstaffed, and outargued at this conference, as they felt they had been at Casablanca. This was what led to the reorganization of the Joint Chiefs committee system already mentioned. It led also to American determination not to cede control over strategic deception in the war against Japan; and ultimately to the first general policy for deception in the Japanese war.

Early 1943 had seen a good deal of activity in Washington on the subject of deception. But it dealt with dull matters of organization, authority, and policy rather than with actual operations.

Kehm and Baumer dominated the planning scene. An Air Force lieutenant colonel named George C. Carey, and Commander Henri H. Smith-Hutton, the former naval attaché in Tokyo, had Air Force and Navy responsibilities, but appear to have been largely inactive. General Strong continued to dominate Joint Security Control, which he was to describe to the Joint Chiefs themselves as "a small group headed by the Director of G-2," *i.e.,* himself. Now ensconced in Room 2E816 of the new Pentagon building, it acquired two "executive officers" from the Army and the Navy; there is no indication that they spent any time on deception, as opposed to security policy—handling of classified material, codenames, and the like. Colonel Jackson continued as the Army's functioning member for deception, though his health was declining rapidly. Newman Smith, who shared Jackson's office, evidently acted in these first months as his alter ego and *éminence grise;* he would not take over fully from Jackson till June.

On the Navy side, Captain Dyer moved on to Mediterranean combat duty in January 1943. Smith-Hutton succeeded him in his strictly intelligence functions; his place as "Assistant to the Naval Member of Joint Security Control and Security Control Officer, Headquarters of COMINCH" (F-16)—which had become a full-time job—was taken by the capable and highly regarded Captain Homer L. "Pop" Grosskopf, a forty-nine-year-old Minnesotan of the Annapolis class of 1915. Grosskopf held that position (later redesignated F-24 in a mid-1943 reorganization of COMINCH intelligence) until August 1944, when he went to the Pacific as captain of the battleship *Nevada*.

Joint Security Control was caught up in the reorganization debates of that spring; especially in the inevitable bureaucratic fights over turf. It will be recalled that in its original charter of August 1942 it had been kept strictly out of planning, though it could make recommendations; and for implementation it had only coordinating authority. (Joint Security Control had nevertheless in fact planned QUICKFIRE and SWEATER, which were then approved by the Joint Staff Planners.) Strong wanted it to have formal authority both to plan and to implement deception.

He was rebuffed as to planning, when in February the Chiefs rejected a proposed cover plan for an operation to evict the Japanese from the Aleutians that was sent to them by Joint Security Control directly, and only approved it when submitted though the Planners.*

At about that time, Peter Cook was reporting to Bevan that American deception was "in a complete muddle" and had "come to what I hope is only a temporary halt," with Jackson "fed up to the teeth as he does not know where he stands," getting little or no cooperation from the Navy or the Air Force, "whilst Kehm and Baumer situated on the floor above, are also wallowing about from the O.P.D. point of view but without any charter." Then in early June came a productive exchange among Kehm, Wedemeyer, Strong, and Admiral Train. Strong urged centralized control of deception through what would in effect be an American Bevan. (He no doubt had Newman Smith in mind for the job.) Bill Baumer had

---

* Called BUXOM, it simply entailed leading the troops to believe they were being trained for the South Pacific. It was implemented with gusto in the field. In the words of the official Navy historian: "Sundry 'cloak and dagger' measures were taken to keep the destination secret. A false training order was given wide circulation; medical officers lectured on tropical diseases; stacks of winter clothing were hidden; commanding officers allowed themselves to be seen studying sailing directions for the North Atlantic and charts of the Argentine Republic."

returned from a visit to London and Algiers, powerfully impressed by the authority and functioning of Bevan and his London Controlling Section. The result was general acceptance of the idea—mooted already in OPD—of a permanent deception planning subcommittee of the Joint Planners.

Though no such permanent committee seems ever to have been formally established, this exchange reemphasized the importance of Joint Security Control participation in deception planning. Crucially, on May 25 a "Third Section" of Joint Security Control, consisting of Jackson and Smith, had been designated specifically for "Cover, Deception, and Task Force Security."

Meanwhile, Joint Security Control's authority over deception implementation had been markedly enhanced* by an April 15 decision of the Joint Chiefs, proposed by General Marshall (no doubt at Strong's urging). Under it, deception *policy* was to be prepared by the Planners and approved by the Chiefs; cover and deception *plans* were to be prepared by the Planners or by task force or theater commanders, as directed by the Chiefs; and *implementation*—not merely the timing of implementation measures, and not just in the United States—was assigned to Joint Security Control (except that theater commanders would have clear authority to implement within their own theaters.) On the inter-Allied level, *planning* coordination would be effected through the Combined Staff Planners; Joint Security Control would coordinate *implementation* directly with the London Controlling Section.

Inextricably intermixed with this bureaucratic jockeying about machinery were some efforts at actual deception policy planning. But it was entangled in more bureaucratic jockeying.

Bevan's draft 1943 deception policy was withdrawn early in February in light of the Casablanca decisions, and replaced by a draft 1943 deception plan for Germany and Italy only. At the Joint Chiefs' direction, the Planners on March 5 designated an interservice subcommittee presided over by Pop Grosskopf to study the proposal and make recommendations. Kehm, Baumer, Jackson, and the Joint Security Control executive officers were members. The subcommittee recommended a substantially

---

* After an apparently inadvertent reduction in its authority, soon corrected.

revised version that, among other things, added a discussion of Japanese conditions and a suggested deception policy for the Pacific. The Navy moved quickly to fend off any interference with its control over the Pacific war. The Joint Planners deleted the discussion of the Japanese war and sent to the Chiefs only the British proposal with minor modifications. It was approved by the Joint Chiefs on March 26, and by the Combined Chiefs on April 2.

The Army tried again. At Wedemeyer's suggestion, on April 7 the Planners directed the same subcommittee to prepare a separate deception policy for the Pacific theater for 1943. With Carey of the Air Force sitting in for Baumer, who was off on a visit to London and Algiers, the subcommittee produced a fortnight later a proposed policy that amounted to a fine-tuning of the sections that had been removed from the paper on Germany and Italy. The Navy still resisted stubbornly. The matter bounced back and forth for a month, with Wedemeyer and Kehm for the Army, and Admirals Cooke and Bieri, the chief Navy planners, all joining in. Army participants, especially Kehm, played on Navy mistrust of the British and concern that the Pacific war would be made a stepchild of the European one. At a May 19 Planners meeting the Navy eventually agreed that a broad deception policy for the Pacific theater, without suggesting specific measures, should be directed toward two "objects": containing the Japanese in their homeland, and keeping them away from areas where the Americans expected to launch operations.

This at least gave the Americans something to present at the conference with the British; which was already getting under way.

Baumer was back, just in time. His account of what he had seen, particularly at the Anglo-American headquarters in Algiers, could only heighten American resistance to a combined Anglo-American deception staff. The British had more experience than the Americans, he reported; they recognized better the importance of planning, selected planning officers with care, imposed their own staff system, took full advantage of their system of "local rank," and had "a certain cohesiveness and singleness of purpose which we have as yet failed to obtain." Americans, he said, were "particularly at a loss in the committee system. Invariably it is reported that the British overload the committees and, speaking as many of them do in their very low voices, oftentimes with a stutter, things are accomplished by two officers at the end of the table. No one knows the decisions until they later appear in the minutes."

Ormonde Hunter reported in on May 16 and was directed to report to Kehm. Captain Herbert Ray of the Navy showed up to represent MacArthur's headquarters; "a quite capable man," noted Baumer, who had served at Bataan and Corregidor, and had been one of the PT boat officers who brought MacArthur out. He had just been detached from MacArthur's staff for service in Admiral King's headquarters, and was probably there simply to warn MacArthur's headquarters of any undesirable decisions. By the next day, Fleming was in town, and there was a two-hour informal gathering of Fleming, Jackson, Kehm, Baumer, Bratby, and Hunter in Jackson's office. ("I liked [Fleming] very much at first," noted Baumer cryptically. "He's very sure of himself.") "Complete accord seemed to exist," said Hunter innocently to his diary.

Serious meetings began May 19 on an informal basis—the same day that Kehm wrung a grudging accord out of the Navy. Formal sessions ran from May 25 to May 28 and informal ones continued to May 30. The active American participants were Grosskopf, Jackson, Kehm, Baumer, Hunter, and Ray; Newman Smith sat in sometimes; Carey turned most Air Force representation over to Hunter. For the British there were Cawthorn, Fleming, Wingate, and Bratby.

Covert smiles must have been directed at Baumer when at the first formal meeting the British produced with a flourish a proposed "Machinery for Deception of Japan." Under it, "a British-American Deception Planning Staff" would be established in Washington to formulate deception policy, draw up overall plans, and approve theater plans; while a British-American "Implementation Staff" would arrange and coordinate "the transmission of information and misinformation for deception purposes"; this could be located in Washington but would preferably be located in India.

Though presented with all Fleming's charm, this was not going to fly. Emphasizing the difficulty of selling deception, or any new committees, to the Joint Chiefs, but assuring the British blandly that American deception machinery existed and was just waiting for a suitable opportunity, Jackson brought out an American counterproposal that amounted to restating the existing machinery of the Joint Staff Planners and Joint Security Control, plus appropriate liaison among the theaters involved.

A subcommittee, composed of Jackson, Hunter, Fleming, and Wingate, met next morning and produced a "reconciliation" whose main concession to the British was a vague undertaking that the question of coordinated control for implementation would be considered after a visit by an officer from Washington to the theaters involved. This was adopted with minor changes that afternoon, and formally adopted with further minor changes (and some elaborate but essentially meaningless maps and charts) on May 27.

Having lost their power play, the British delegation had relatively little interest in the rest of the conference and it ran smoothly. Two more papers were adopted. One was another recycling of the paper on recommended deception policy for the Pacific area for 1943, adding to its two "objects" a third: "To exploit all elements of disunity between Germany and Japan." This of course was not a deception object at all, but a psychological warfare one; it was included at Hunter's suggestion and reflects a confusion in his mind that may have never entirely been straightened out, but it is surprising that the more sophisticated participants let it pass. Hunter did at least correctly recognize that the paper was nothing but "glittering generalities." It was formally passed to the Joint Staff Planners as well as being adopted by the conference.

The final order of business was adoption of a recommended "General Deception Plan for the War Against Japan," which included a long list of "suggested methods of accomplishment."

By May 30 the conference was over, with a clear victory for the Americans (as on the whole the full-scale Trident conference had been). Fleming's disappointment was alleviated by the discovery that his brother Ian had come to Trident, as had Joan Bright. The three of them decided to slip off to New York. There Ian disappeared on business of his own, while Peter and Joan went to see Ethel Merman in *Something for the Boys*. "An attractive pair, amusing, good-looking, sure of themselves and devoted to each other," said Joan Bright of the Fleming brothers.

The American participants suspected that the British might try to slip through a decision for a combined deception organization via some other channel. On June 2, Kehm and Baumer warned the Joint Staff Planners that a seemingly innocuous resolution which the Combined Chiefs had adopted, and a British proposal for a combined plan for intelligence and psychological warfare operations in the China-Burma-India theater, appeared to be efforts to ease the British plan in through the backdoor. Jackson alerted Strong on June 4 that "The British lost their point but it is believed they will make further efforts to attain their end through other channels." (This was one of Jackson's last acts before retiring for reasons of health; Newman Smith took his place a week later.)

The British in fact gave the American planners no further trouble along these lines. But the United States Navy stalled deception planning for the Japanese war throughout the summer and into the autumn. A wrestling match was going on between the Army and the Navy over the proposed Central Pacific offensive and its relationship to MacArthur's plans in the Southwest Pacific, and the Navy was not inclined to give an inch anywhere till that matter was settled to its liking. "In June, there was work to be done on deception with Baron Kehm and with Lt. Col. Newman Smith, who had taken Colonel Jackson's spot in Joint Security Control," Baumer wrote in his diary. "We kept trying to put in one paper after another on Pacific deception measures but to no avail—the Navy seemed bent on stopping us. Also, we lacked the Army backing which would have helped." "The Navy wouldn't have anything to do with us, told us to go to hell," added Baumer long after the war. "MacArthur's headquarters practically the same thing." There was much back and forth between the working subcommittee, notably Baumer, Kehm, and Smith, supporting the proposals that had emerged from the conference—sometimes with augmentations—and Admiral Bieri, the Navy deputy chief planner, who wanted the simplest possible document.

The logjam was eventually broken through some smooth bureaucratic management by General John Deane, Secretary of the Joint Chiefs and American Co-Secretary of the Combined Chiefs, impelled by Newman Smith, both directly and through the British mission. Deane evidently worked out a deal under which only a modest three-page paper, limited to procedures, would be adopted at this time, in exchange for an undertaking by the Navy to address deception policy and measures once the Central Pacific offensive should be finally settled. The British went along, though disappointed "that the work of the Washington Conference should have to be set aside so completely, though this was possibly unavoidable." After review and minor changes in London, the paper was finally adopted as a directive by the Combined Chiefs on August 6.

"I have the impression that the Americans do not attach the same importance to deception measures as we do ourselves," remarked Field-Marshal Brooke drily when the matter came before the British Chiefs. At his suggestion, Bevan was directed to prepare for possible American consumption "a short account of deception schemes which had been employed with success."

Procedurally, this otherwise pointless exercise benefited Joint Security Control, by setting forth its authority unambiguously. It now had explicit responsibility for coordinating the implementation of "general overall deception plans approved by the Combined Chiefs of Staff." It also had explicit responsibility to "provide for the implementation of such features of approved deception measures as require implementation outside the theater of origin," and for coordinating such implementation; and theater deception plans prepared in the theaters were required to indicate the parts of the plan that fell within this category. Direct communication between theaters with respect to implementation was authorized, but Joint Security Control and the London Controlling Section were required to be kept advised. And one more possible loophole was plugged by authorizing Joint Security Control to "provide for continuity of deception during those periods when no deception plans relating to specific operations are in effect," with the understanding that this did not prevent theater commanders from making plans during such periods.

The two actions by the Joint Chiefs of March 17, 1943, and April 15, 1943, and that by the Combined Chiefs on August 6—their official designations were JCS 234/2/D, JCS 256/1/D, and CCS 284/3/D—were to remain for two years the official charters under which Joint Security Control would function.* Joint Security Control was still not a London Controlling Section and never would be. But it now had express implementing authority, an implicit direction to the Planners that they plan cover and deception in conjunction with Joint Security Control, express authority to deal directly with Bevan, authority to provide for the continuity of deception, and the recognition that deception was a special function separate from security in the strict sense.

* JCS 234/2/D became JCS 234/3/D in April 1944 as the result of an amendment not relevant to deception; and JCS 256/1/D was augmented in 1944 and became JCS 256/2/D, as described in Chapter 11.

*   *   *

Deane now turned to the next step. On August 11—the opening day of the next conference with the British, QUADRANT, held in Quebec—in the Chiefs' name he directed the Planners to "prepare an overall deception plan for the war against Japan" for consideration by the Chiefs no later than September 15. The Navy tried unsuccessfully to sidestep this; then QUADRANT settled the Central Pacific offensive satisfactorily to the Navy, and this broke the logjam. The project was turned over to the Joint War Plans Committee, where the right people were working on it. Baumer had been delegated to that committee for two months in July. Moreover, as will be seen, by now Newman Smith had formidable reinforcement in the experienced person of Carl Goldbranson. "Deception came up again—the old bugaboo of the Pacific," Baumer told his diary. "We are now directed to write an overall plan. So here I go."

Baumer and Smith were probably also working with the British—presumably with Lieutenant-Colonel H. M. O'Connor, who had replaced Bratby in early July. Unlike Bratby, "Tim" O'Connor—who shared with Bevan a background in the financial world—reported to Bevan immediately rather than through Cook. Back in London, in September Bevan added Bratby to his staff as Petavel's assistant.

In short order a tight, simple document was turned out. Dudley Clarke might have laughed at its lack of sophistication, but it was a notable advance over anything that had preceded it. Gone were "objects" of a psychological warfare nature; there was nothing (with one lapse, and that due more to Fleming than to the Americans) about what the Japanese should be made to *think,* only what they should be made to *do:*

6. *General* . . . The basic objective of over-all deception in the war against Japan is *to cause her to dispose her forces in a manner favorable to our plans.* In order to carry this over-all deception plan into effect, we should exaggerate our strength in areas where no operations are impending, while deflating our capabilities in areas where operations are scheduled. In summation, our plan is to draw enemy forces into an area when no operations are scheduled therein, but to draw Japanese strength away from that area when operations are about to be launched. Our objectives during 1943–44 should be:

a. To cause Japan to maintain a considerable proportion of her forces at home.

The following threats to the Japanese homeland may be exploited in order to obtain the effects desired:

(1) Air attack from the Aleutians.

(2) Raids by carrier-borne aircraft.

(3) Air raids from Eastern China.

b. To cause Japan to dispose, in a manner favorable to our plans, such of her forces as are not in Japan proper.

Due to the fact that numerous operations are projected throughout the Pacific-Asiatic area, it is considered that this objective should be attained, so far as possible, by deception as to the sequence and timing of projected operations rather than by relying on cover operations.

A supplementary aim, to cause wastage of Japanese air and shipping resources, is of sufficient importance to warrant special attention. The following means can be employed to attain this aim:

(1) Causing the enemy to build up defenses by threatening areas which we do not intend to attack.

(2) Causing the enemy to withdraw defense forces, particularly air, from areas which we intend to attack.

(3) Sporadic intensification of our air activity at widely separated points to cause dispersal of effort.

(4) Enticing Japanese air and sea forces into areas where we can bring them to battle under conditions favorable to ourselves. . . .

7. . . .

b. Objectives for each theater should be as follows:

(1) *Central and North Pacific.*

(a) To contain maximum Japanese forces in the homeland.

(b) To draw Japanese air and naval strength into the Kuriles, except at the time of an operation in that area.

(2) *Southwest Pacific.*

(a) To induce the Japanese to withdraw mobile strength at the time of initiation of certain operations.

(b) To draw Japanese strength into various parts of

the Area when operations are being launched in
Burma, the Gilberts, the Marshalls, and the Carolines.

(3) *Southeastern Asia Command, India and China.* To
cause the Japanese to believe that after 1 January 1944 no
large scale operations will occur in Burma during the
1943–44 dry season.

Probably more important in Baumer's and Smith's minds than the actual text—which, after all, contained little that was very startling—were two appendices, essentially first-grade primers for the theater staffs: one setting forth some basic desiderata of deception plans and another setting forth a general checklist of form, design, and implementation for such plans.

This paper sailed through the bureaucratic process. Admiral King himself presented it to the Joint Chiefs on September 15, noting that Admiral Leahy recommended that the appendices be omitted when it was presented to the British, "in view of the fact that they had gone farther than we in the matter of deception planning." So the Chiefs approved the paper, without the appendices. When it came before the Combined Chiefs on September 17, Leahy, King, and Sir John Dill all agreed that while, in Leahy's words, it "would not carry the subject very far," it would at least draw the attention of the commanders in the areas concerned to the importance of formulating a deception policy. So it was approved, under the title "General Directive for Deception Measures Against Japan, 1943–1944."

When news of this reached London, Bevan was annoyed that the paper had not been referred to the British Chiefs before final action, "an unusual and probably undesirable practice." However, "On this occasion, no harm has been done as the directive is of a nebulous character and devoid of concrete proposals as far as [India/Burma] is concerned."

"Nebulous" it might be, but there was at last some kind of charter for deception in the Pacific. Newman Smith would soon show that that would be all he would need to run with the ball.

It is time now to return to the real world where people were actually doing deception and passing from success to success. It is time to get back to "A" Force and the Mediterranean.

CHAPTER 9

The Soft Underbelly

The year 1943, said Dudley Clarke, was "not only the most active and perhaps the most fruitful year for 'A' Force, but also without doubt the most instructive." It saw the full flowering of Clarke's order of battle deception, with the Germans constantly on edge over an invasion of Greece and the Balkans by an imaginary host. It saw the development of regular tactical and operational deception, and successful deceptions for the landings in Sicily and Italy. It saw the training of Americans by "A" Force, and the coming of the United States Navy's tactical deception teams, the "Beach Jumpers."

On November 6, 1942, with Rommel's army beginning a westward retreat that would take it to Tunisia, and the TORCH convoys poised to strike, Clarke met with General Alexander in Cairo. Alexander wanted a continued threat to the Balkans, with Crete as a first step, so as to induce the Germans to keep reinforcements away from Africa and from threats to Iraq and Iran. So Clarke produced Plan WAREHOUSE. Implemented largely by special means, its "Stage A" threatened a landing on Crete, originally for January 7, subsequently "postponed" several times. On the assumption that all North Africa would be quickly occupied and the next Allied target, around mid-March, would be Sardinia (codenamed BRIMSTONE), WAREHOUSE contemplated a "Stage B," the "story" of which would be that after clearing North Africa, Eisenhower would invade Sicily and thence hop across to Greece, landing on the west coast of the Peloponnese while Alexander, staging from Crete, landed on its east coast.

This proposal went out to London in mid-November. There it was overtaken by events. The Germans swiftly occupied Tunisia and were not to be driven out till May. And Bevan was instructed to prepare a cover plan for BRIMSTONE based on the assumption that Sardinia would be assaulted by forces direct from Britain and the United States. He produced a plan in two parts, CANUTE and GARGOYLE, aimed at inducing the Germans to reinforce areas distant from Sardinia. CANUTE, managed from London, would be designed to make the Germans deploy to meet a threat either to Norway or to northern France. GARGOYLE, managed by "A" Force, would essentially continue WAREHOUSE "Stage A." (A third, minor, element, JIGSAW, would lead the Allied troops to believe they were bound for North Africa.) At about the same time, Bevan drafted his proposed overall deception policy for 1943. Then the Casablanca Conference decided that the next operation, codenamed HUSKY, would be Sicily rather than Sardinia, not to be mounted until summer. Bevan was directed by his Chiefs of Staff to draft a new 1943 deception policy for the war with Germany and Italy on this basis.

As approved on February 4 by the British Chiefs, this new policy would be to mount as many threats as possible to the enemy to induce them not to transfer forces to the Russian front,* and to waste effort and materiel. The Axis should be led to believe that the main Allied operations would be three invasions of the Continent: cross-Channel, in the south of France, and in the Balkans. To contain German forces in Norway there should be a threat of attack in the spring, which might be renewed in the autumn. To contain them in France a threat of invasion in July should be mounted, later postponed to September. Two more objects were, first, to bring the Luftwaffe to battle by deception operations and feints, featuring an apparent approach to the French coast by swarms of small craft; and, second, luring the U-boats into the North Atlantic and away from the Mediterranean, by means of the exaggeration of BOLERO, the American buildup in the United Kingdom.

The prime object in the Mediterranean would be to induce the Axis to retard reinforcement of Sicily. Allied forces in North Africa should be seen as preparing to invade the south of France after Tunisia was cleared;

---

* This object was not universally favored. With the tide now turned on the Eastern Front after Stalingrad, the Foreign Office representative who examined the proposal expressed the view that "it is now in our interest to allow the Germans to believe" that they could reinforce the Eastern Front freely. "Thus the Russians and the Germans will be killing each other, and Russia, who, however sentimental we may be about her at present, is likely to be a troublesome customer at the end of the war, will be thoroughly weakened." (The Navy planner noted in response: "And if the Russians take a knock and sign a separate peace—what then!")

the chief goal of the North African campaign should be seen as being to clear the Mediterranean for Allied shipping, and the heavy air attacks on Sicily in preparation for HUSKY should be depicted as designed to neutralize the Axis airfields there. A second object (not of course a proper deception object) should be to lower Italian morale and will to resist. In the eastern Mediterranean the object should be to contain enemy forces in the Balkans, emphasizing that the clearing of North Africa and the elimination of the threat to the Caucasus after Stalingrad would release considerable forces for eastern Mediterranean operations.

As has been recounted in Chapter 8, when the Americans studied this proposal in March there ensued a debate over whether to add a deception policy for the war with Japan. Apart from that issue, Pop Grosskopf's committee proposed minor revisions based on perceptions that the threat to southern France should not be stressed, nor the buildup in the United Kingdom be "overplayed." The Joint Chiefs sent revisions, based largely on the committee's report, to the British Chiefs on March 22, adding as an "object" the deterrence of any German move through Spain. The British Chiefs accepted these changes with minor procedural modifications, and the policy was formally adopted by the Combined Chiefs on April 2.

In northern Europe this policy led directly to the fiasco of Operation COCKADE, to be described in Chapter 12. In the Mediterranean, consistent operations were already well under way.

In mid-January, Clarke had devised Plan WITHSTAND, designed to exaggerate, in both Turkish and German eyes, the Allied perception of the Axis menace to Turkey, so as to help pin down German forces in the Balkans and, hopefully, encourage Turkey to consider joining the Allied side. Its "story" was that the Allies anticipated a German invasion of Turkey through Bulgaria and intended to divert German forces from such an operation by invading Crete and the Peloponnese and strengthening forces in Iraq and Iran. But the Middle East Defense Committee disapproved any efforts to deceive the Turks themselves, so the project simply merged into the existing WAREHOUSE.

WAREHOUSE/WITHSTAND was implemented by physical methods along the Turkish-Syria frontier and by intelligence means. The dummy tank regiment was displayed, along with real troops, from January to

May. The operation fooled the Germans remarkably successfully, peaking first on March 2 when the German Navy canceled all leave in the Aegean in the expectation of imminent action, and again on March 29 with an order from Keitel that secret arrangements with the Bulgarians to meet an "Anglo-American-Turkish invasion" be continued till June 1. Confirmed too was the vital importance of CASCADE order of battle deception in making such threats plausible. "At the present time," said Fremde Heere West on February 4, "the available British forces in the Near East are certainly sufficient for a landing in Greece and . . . for partial occupation of the country"; their estimate of forces for a Crete landing was four and a half divisions, with another four and a half poised to cross from Syria into Turkey.

In March, with the new general policy approved, WITHSTAND was incorporated into an updated "WAREHOUSE (1943)." Its "story" was that the Balkans would be invaded from the Middle East in early summer, with synchronized landings in western Crete and the Peloponnese. This was supposedly expected to induce Turkey to join the Allies, who could then bypass the Dodecanese and move air forces and two armored divisions through Turkey against Thrace and to support Turkish operations against Bulgaria. But this plan was never implemented. Discussions in March at AFHQ, Eisenhower's headquarters in Algiers, made it clear that HUSKY would require something more elaborate. So WAREHOUSE (1943) merged into a theater-wide plan called BARCLAY, presently to be described.

Deception in the Western Mediterranean after TORCH was originally to be handled by Goldbranson and a small team at AFHQ, with Bevan coordinating Goldbranson's work with "A" Force. Goldbranson had worked closely and well with Bevan and his people in implementing the TORCH deceptions; as Goldbranson said, "we all did much 'groping' together." Once the TORCH convoys had sailed, Derrick Morley had joined Goldbranson so that an American and a British officer could train together. But then Goldbranson got sidetracked on the Washington trip with Bevan, and did not start in Algiers till January 23.

Clarke evidently took advantage of this delay to secure theater-wide deception in the Mediterranean for himself. London in December divided deception units into "Operators," responsible for a whole theater or more; "Sub-operators," for specific areas within the theaters; and "Special Correspondents," planting deceptive information when directed by an Operator or Sub-operator. The Operators were "A" Force in the Mediterranean and Middle East; Fleming's GSI(d) in India and Burma; and the two governing bodies, the London Controlling Section and Joint Security Control. Goldbranson at Algiers was to be merely one of "A" Force's Sub-operators, along with Advanced Headquarters "X" Force in Baghdad, under Oliver Thynne to February 1943 and then under Captain David Mure; Rex Hamer in Syria; and officers at Eighth Army, XXV Corps on Cyprus, East African headquarters at Nairobi, naval headquarters at Cape Town, the fleet base at Kilindini in Kenya, and Sudan headquarters at Khartoum. Its Special Correspondents were Arnold at Ankara and Wolfson at Istanbul; officers at Malta and at Asmara in Eritrea, and Malcolm Muggeridge at Lourenço Marques. (Responsibilities in South and East Africa reflected the Royal Navy's shift of its Indian Ocean headquarters and chief base from Colombo in Ceylon to Kilindini, owing to Japanese intrusion into the Indian Ocean after the fall of Singapore.)

Hamer at Ninth Army, in civil life a housemaster at Stowe, was perhaps the only officer to go through the war still wearing the badges of a school officer training corps, never permanently assigned to any regiment. He first encountered Clarke's work, unwittingly, on Cyprus when "Headquarters Troops Cyprus" suddenly became "Headquarters 7th Division" as part of the CYPRUS DEFENSE PLAN. In early 1942, he was summoned to Cairo and met Clarke for the first time. He later realized that he was being "vetted" for "A" Force. In September, Clarke arranged for him to be sent to the staff of Ninth Army. There he worked on deception under the supervision of "A" Force in Cairo, though not yet officially on their rolls. He was to be one of Clarke's stalwarts for the rest of the war.

With its new responsibilities "A" Force needed more space in Cairo. By the end of 1942, presumably to the inconvenience of the ladies of No. 6 Sharia Kasr-el-Nil, it had expanded to occupy six flats from its original two.

The stated justification for subordinating Goldbranson was the inexperience of the Americans in deception. Clarke did not, however, get away from Cairo to establish a new organization in Algiers until March. In the interim he was busy with his usual variety of things. One of them was something he had longed to do: Create a force for tactical and operational deception (plus escape and evasion duties) at army group level, reporting directly to the commander and his chief of staff, called "Tactical HQ 'A' Force," or usually "Tac HQ."

In February 1943, Alexander was appointed to command the newly-formed 18th Army Group in French North Africa, consisting of the British Eighth and First Armies and the American II Corps. General Sir Henry Maitland Wilson, known always as "Jumbo," succeeded to the Middle East command. London suggested that Alexander move all of "A" Force with him, noting that the Americans had "virtually no experience in deception." But the established Middle Eastern channels could not readily be managed from such a distance, and no doubt Clarke would not willingly have subordinated the strategic element of "A" Force's work to a mere army group headquarters. So it was agreed that Alexander would take only a "small team" and gradually build it up. Clarke designed a tight structure with a commanding officer, two officers each on the deception and escape sides, ten enlisted men, and suitable vehicles.

Lieutenant-Colonel David Strangeways, the officer whom Bevan had sent out to work on TORCH cover in Gibraltar and Malta, and who then brought word of TORCH to Clarke and carried TOWNSMAN back to London, had struck Clarke favorably. It will be recalled that Strangeways had been designated for the senior slot at "Advanced Headquarters 'X' Force" in Baghdad. Strangeways was in Cairo at the end of January on his way to take up this duty when Alexander's "small team" was decided upon. "X" Force seemed no longer important, and Clarke decided that Strangeways was the right man for Alexander.*

David Strangeways was a son of Dr. T. S. P. Strangeways, founder of the Strangeways Research Hospital in Cambridge. A Cambridge graduate of 1932 and a regular officer in the Duke of Wellington's Regiment since 1933, he was no deskbound planner but a front-line soldier. Even after he got into the deception business, if his troops were up front he too

---

* Thynne left "Advanced HQ 'X' Force" in February 1943. It continued until the end of August 1943 as a one-man operation under David Mure. When it was shut down at the end of August 1943, Mure, now a major, took over the "A" Force outstation at Beirut from Rex Hamer, who went to Cairo.

went up front at least once a day; Clarke chided him for this, but his view was that "if you've got troops under you, you must be seen with them. . . . You can't ask men to do what you feel you're too precious to do yourself." At Dunkirk he had saved his men by wading and swimming through the surf to an abandoned barge, loading them on it, and, clad only in a doormat he found on board, sailing it to England with a compass and a school atlas.

"He was a small, good-looking man, with a brisk, efficient manner and a very quick mind," recalled Wheatley; "so beautifully turned out that, even in battledress, he looked as if he had stepped straight out of a bandbox." He had an impish sense of humor, he was a master technician at whatever he turned his hand to, a perfectionist who did not hesitate to speak his mind to superiors and did not suffer fools gladly; and he liked to do his job, "get on with it," without worrying about what others did or said. "A nice man, but a bit nature's head prefect sort of chap," recalled Montgomery's chief of intelligence. Noel Wild loathed him; but Strangeways—who according to his intelligence officer "thought Noel Wild was a nincompoop"—liked to torment Wild, whom he regarded as a "pain in the neck" who contributed nothing. "Both these British offi-

David Strangeways

(David Strangeways)

cers are most capable and efficient," Douglas Fairbanks reported back in Washington in early 1945, "but they are not too compatible personally."

Strangeways's initial deception officers were Major the Earl Temple of Stowe (known as Chandos Temple), who would be his No. 2 and alter ego for the rest of the war, and a captain named Fillingham. By train, ship, and air, they made their way with Tac HQ to Tunisia, and set up shop on February 17 with Alexander's headquarters. Strangeways surveyed the front, and prepared his first deception plan, approved March 9. Called LOCHIEL, it supported Eighth Army's attack on the Mareth Line, the old French fortified line on the Tunisia-Libya border, behind which Rommel had made a stand.

LOCHIEL picked up on a plan called WINDSCREEN that Clarke had operated in connection with Montgomery's advance towards Tripoli. In it, double agents had reported that there would be a landing at Tripoli in Rommel's rear because administrative difficulties had slowed Montgomery down. The "story" of LOCHIEL was that while Montgomery paused to bring up supplies, the Americans and French would attack further north towards Cairouan, with a diversionary landing in the Gulf of Gabès. Strangeways's next operation was HOSTAGE, designed, by radio deception and a display of dummies, to aid an American attack further north. Early in April came HENGEIST, in which the approach of the British 6th Armored Division to Fondouk was simulated by dummy equipment, a few real tanks, and radio deception, while the real division approached by another route.

Next was COWPER, supporting Alexander's final assault against Tunis, made by the British First Army and American II Corps from the west. The "story" that it would be made by the British Eighth Army in the south was passed by a notional French airman in Algiers and by CHEESE, and supported by covering the shift of a British armored division from Eighth Army to First Army, replacing it with dummy equipment in the south while disguising it by radio deception as the American 1st Armored Division in the north. A variant, COWPER II, covered a revival of the offensive after a German counterattack. This was the last tactical deception in Africa; for Tunis fell on May 7.

It was not, however, David Strangeways's last action in Africa. Commanding a special ad hoc unit denominated "S" Force, he dashed into

Tunis and Bizerte at the moment resistance ceased, to seize documents, cipher machines, and other valuable intelligence material before the Germans could destroy them. He carried this out with his usual dash and flair, and received the Distinguished Service Order for it.

General McCreery—he who had once thought "A" Force a "racket"—was now Alexander's chief of staff. When the Tunisian campaign was over, he wrote to Dudley Clarke: "This is a line to tell you how splendidly Strangeways has done from my point of view. Really our tactical deception, considering the short time that was available, has been amazingly successful. . . ." Strangeways moved his headquarters to Bouzarea near Algiers, and began planning cover and deception for the landing in Sicily.

Carl Goldbranson finally reached Algiers on January 23, bringing with him Derrick Morley and two fellow Iowans. One was Captain Darwin A. Dunn, known as "Chunky," a short, stocky reserve officer from Villisca, Iowa, who was a great hand with the ladies (he soon became good friends with a handsome Frenchwoman in Algiers). The other was Captain Robert Rushton, known as "Bobby," whose main thought in life was to get back to Red Oak, Iowa, and marry a girl named Mary Rose. His fellow deceivers liked him but always regarded him as a lightweight.

When Goldbranson reported at AFHQ to Mockler-Ferryman, the latter told him that Clarke would be taking over for the whole Mediterranean and that he should visit Cairo. So Goldbranson spent the first two weeks of February in the Egyptian capital learning the ropes. There he first met Noel Wild, with whom he got on surprisingly well. During this time, Clarke worked out a new organization. Cairo would be "Main HQ 'A' Force," with Noel Wild in charge. "Advanced HQ 'A' Force" would be opened in Algiers, with Michael Crichton and Goldbranson in charge. Tac HQ would continue at 18th Army Group under Strangeways. "Rear HQ 'A' Force" at Nairobi would handle Indian Ocean matters. Each headquarters would have control, deception, and escape sections. The deception section would be divided into operations (physical) and intelligence (special means). Main HQ would control outstations at Beirut, Cyprus, Baghdad, Tehran, and Smyrna, and Special Correspondents at Istanbul, Ankara, Khartoum, and Asmara. Advanced HQ would control outstations at Malta, Oran-Casablanca, and (eventually) Tunis. Rear

HQ would control Special Correspondents at Lourenço Marques, Madagascar, Durban, and Cape Town. Jones's dummy armored brigade and Titterington's Technical Unit would stand outside the system.

The control section at Algiers would consist of Goldbranson and Crichton; a naval officer and an officer for the Oran-Casablanca outstation, to be selected; Max Niven, David Niven's brother, who had recently joined "A" Force, at the Malta outstation; and Chunky Dunn as the adjutant. The deception section would consist of a new recruit from Cairo, Major Freeman Thomas, and the reliable Margery Hopkins for intelligence; and Derrick Morley and Bobby Rushton for operations. The escape section need not concern us, except to note that Cuba Mason came over from Cairo with it.

Thomas did not fit Dudley Clarke's usual Establishmentarian pattern. He was a strapping young Welshman who spent a great deal of time taking care of his hands and in civilian life had been a men's clothing department manager at one of the big London department stores. Some of his fellow officers disliked him intensely. This was probably snobbery.

When Goldbranson returned to Algiers he, Derrick Morley, Dunn, and Rushton held the fort for a full month, for Clarke did not arrive to set up the new "Advanced Headquarters" until March 15. Meanwhile, they acquired a new boss, for on February 20 Eisenhower replaced Mockler-Ferryman with Brigadier Kenneth Strong, who had served as the British assistant military attaché in Berlin just before the war.

While he waited for Clarke, Goldbranson managed to collect a few low-grade double agents from the French and began building them up. He unsuccessfully sought direction from Mockler-Ferryman as to what might be desired in the way of order of battle exaggeration. During that dull period he did have one of the happiest days of his life. His son Carl, Jr., was commissioned a second lieutenant, and on March 7 Goldbranson pinned his bars on him. For the rest of Goldbranson's time in Algiers, father and son shared quarters as fellow officers.

By that mid-March of 1943, Allied headquarters had been in Algiers for four months, and the French-Arab city had been transformed. AFHQ itself was centered in an old French hotel, the St. Georges (now the El-Djazaïr), a grandiose pile on the crest of the hill overlooking the port area, designed in an eclectic Mediterranean architecture and set amidst palms and other semitropical trees. With more than two thousand officers and enlisted men by the end of March and still growing, it spilled over to other hotels and into commandeered office buildings; a Grey Pillars of the west.

And Algiers had become something of a Cairo of the west. Soldiers milled about, and military traffic filled the streets, honking their horns endlessly; a "tremendous chaotic energy." remarked Malcolm Muggeridge, who arrived there from Lourenço Marques in the summer. It was a foretaste of liberated Paris, too. "What particularly excited us all," remembered Muggeridge, "was that Algiers seemed so unmistakably French, with policemen in blue carrying white truncheons, blowing whistles and directing traffic, just as if they were in the Place de la Concorde; with cafés sprawling onto the pavements . . . and, above all, that composite smell of *gauloise* tobacco smoke, cheap scent and *pissoirs*. . . . Algiers was a half-way house to France."

Social life, however, was thin by comparison with Cairo. There were restaurants and bars, but no wealthy cosmopolitan society, no Groppi's or Shepheard's or Mohammed Ali Club. Until after the fall of Tunis, the blackout was as total as in London; there were no cabs and nothing much to do. Night life picked up after the war moved further away. The center of social activity was the Cercle Interallié, a club set in an old Roman amphitheater and serving a sort of wine punch like sangría. Hundreds of people went there every day after work, admired the girls, and, recalled Arne Ekstrom, "everybody in the end knew everybody else."

Clarke finally showed up in Algiers on March 15 with a planeload of people from Cairo; exercising no doubt his God-high priority, he had laid on a special C-47 for the trip. The same day, Bevan arrived from London for consultations on deception planning. Goldbranson and Morley were working from one room in the St. Georges Hotel, with no space for visitors, so Clarke and Bevan set up shop in the garage of a nearby house. There they blocked out the first draft of what would become Plan BARCLAY.

AFHQ was organized on the inflexible American staff pattern, which had no room for Clarke's preferred arrangement, under which deception should report directly to the chief of staff and the commander; it must be under either G-2, Intelligence, or G-3, Operations. So Advanced HQ did not get an independent position, to Clarke's lasting dissatisfaction. "He still doesn't like it much," commented the G-3, the American Brigadier General Lowell W. Rooks, nearly two years later.

To compound his unhappiness, pursuant to current American thinking, Advanced HQ was originally placed in the intelligence staff. Clarke had stayed with Strong, the new G-2, in his Berlin flat on his last visit to Germany in 1939 and described him as an "old friend," but for some reason he could not get along with him now. In July, Clarke pressed for a change—"A" Force was in Operations everywhere else, he pointed out, and it did not gather information—and finally Advanced HQ was switched to General Rooks's G-3, with appropriate assurances that G-2's prerogatives would be respected.

Suitable quarters were soon found: a pair of newly-built villas at 20 rue Mangin, a few yards off the Blida road on a little plateau not far behind the St. Georges on its hilltop. One was occupied by the control and deception sections and the other by the escape section. They were little more than small one-story bungalows of three or four rooms, with no garden and an unfinished driveway; Clarke's office was in the sitting room of one of them, and the files were kept in the kitchen. For the next sixteen months this was to be increasingly the base from which the chief activities of "A" Force were to be conducted. Clarke himself would spend more and more time there. Advanced HQ moved in officially on March 28.

That same day, Clarke, Bevan, and Rushton set out on a three days' visit to Alexander's headquarters, and to meet with Alexander, Strangeways, Temple, and others. After returning, Clarke remained in Algiers for over a fortnight longer. He worked on BARCLAY. He established good relations with the French and the special means channels under their control, and to handle these relationships he recruited a remarkable American officer whom we will meet shortly. On April 11 he flew out for Cairo with stops to see Strangeways and to visit Malta (sharing a plane for part of the way with the American Archbishop Spellman, who must have had God-high priority if anyone did), and reached Cairo on April 16.

The next three months were frustrating for Carl Goldbranson. Clarke had made it clear that Crichton was in charge and that Goldbranson was there to learn. But he found himself with little to do and not learning much. He lost control over his handful of French double agents; he had not deemed them worthy of use for deception purposes, and was proved correct when the only two W/T agents got "blown" when they were so employed by Crichton and Thomas, on one occasion over Goldbranson's protest. He had no dealings with Strangeways, although Strangeways had told Goldbranson in Cairo that he would keep him informed and ask for assistance when needed. (In this Goldbranson was not alone, for Strangeways operated as a law unto himself, dealt directly with Cairo, and Crichton and Thomas had no more contact with him than Goldbranson did; Clarke had told Goldbranson in Cairo that he should do nothing about Strangeways unless called upon.) Goldbranson got to write no plans himself, though Crichton did bring him in on the planning. He was not yet cleared for ULTRA and therefore could not understand why he was not permitted to sit in for Crichton or Thomas at meetings when they were absent. He did, however, do useful work in connection with psychological warfare, in civil affairs, in getting the British and American camouflage people together, and especially in connection with the setting up of the interservice wireless deception committee.

Goldbranson was a good soldier and he stuck to it despite the frustrations. Imbued with Ike's determination to make a truly Allied headquarters work, he voiced no complaints to his American superiors. He had no way of knowing, of course, that behind his back Clarke constantly made fun of him, mispronouncing his name in Clarke's notion of a Midwestern accent and then going into convulsions of laughter.

More pleasant were the experiences of the American officer whom Clarke had recruited at the very end of his sojourn in Algiers.

It will be recalled from Chapter 4 that after the Germans occupied the whole of France following the TORCH landings, Paul Paillole made his way to North Africa, arriving in January 1943, joined the Free French, and put the double agents of his clandestine Direction des Services de Renseignements et des Services de Securité Militaire ("DSM") and its subsidiary "Travaux Ruraux" ("TR") at the service of "A" Force. French security was never wholly trusted by the "Anglo-Saxons"—not until late in the war were the French allowed officially to know of ULTRA—but

Paillole himself was highly regarded and his double agent network was a godsend.

To supervise these new channels, Clarke urgently needed an officer with the right knack who could work well with the French and was fluent in the language. Early in April, he met with Eisenhower's chief of staff, Bedell Smith (characteristically bypassing the G-2), and asked him to ask Donovan of OSS to recruit some suitable American officers for Advanced HQ. He thereby put his foot in it, for Donovan was General George V. Strong's particular bête noire. When Smith cabled the request to Donovan, Strong saw the message and exploded. Smith, evidently unperturbed and mildly amused, warned Donovan off. Strong promised to send some qualified officers, but they seem never to have materialized.

(Clarke soon distanced "A" Force from Donovan. OSS tried to run its own counterespionage operation in North Africa in competition with Paillole, and from Clarke's perspective did it very clumsily. "These people that the OSS had running these separate agents . . . were very spoiled characters, set up in expensive villas," recalled Arne Ekstrom long afterwards. "And I wasn't allowed to speak to them because it was thought by the British that they were totally insecure . . . just couldn't have anything to do with them.")

Meanwhile, Desmond Bristow of MI6 Section V, whom we last met when GARBO was being broken in in London, found an ideal American officer for Clarke.

Bristow was transferred to Gibraltar in June 1942, after GARBO was established. In November he was transferred again to Algiers. (There he would remain until November 1944, when he was transferred yet again, to Lisbon.) He worked with Paillole's people and with Goldbranson's skeleton operation before Advanced HQ opened up. Clarke met with him and Trevor Wilson, the MI6 station chief, on March 20, and perhaps then told them of his needs. Bristow may even then have thought about an American officer he had met by chance not long before, when he was a passenger in an American military plane for an exceptionally bumpy trip from Casablanca to Algiers.

Most of his fellow passengers on that flight were throwing up into their helmets. One second lieutenant, however, sat calmly reading Proust in the original French. This is not an American such as one sees every day, said Bristow to himself. He struck up an acquaintance. The lieutenant proved to be a native of Sweden, now an American citizen, rather old (thirty-four) to have joined the Army at the lowest officer level, a slightly-built, dapper man with a twinkling and often wicked sense of humor, a highly cultivated and civilized man who had lived in France and spoke perfect French, the aforementioned Arne Ekstrom.

Ekstrom was born in comfortable circumstances in Stockholm in 1908. Growing up he found Sweden uncongenial, and from the age of eighteen he lived in France, becoming fluent in the language and thoroughly at home. Eventually he went to work for an American paper company headquartered in the United States. On board the S.S. *Berengaria* from France to New York in 1932 he met, and fell in love with, a New York girl named Parmenia Migel, known as Pam, who had been studying art in Paris. Her father, M. C. Migel, made his fortune in the silk trade, retired from business at an early age, and devoted much of the rest of his life to helping the blind. Pam and Arne were married on the last day of 1933.

Arne Ekstrom
(in the uniform of the
Army Specialist Corps,
prior to commissioning
in the Army proper)

(Nicolas Ekstrom)

Ekstrom left the paper company and started a small publishing business in New York called Greystone Press, while Pam's great interest was the ballet; she founded the Ballet Guild and served on the board of the Ballets Russes de Monte-Carlo. He became an American citizen. It was an idyllic marriage ("We had a passionate love affair for 56 years," he said when he was eighty-five years old), with one son born in 1938. When war came, nothing compelled a middle-aged married man with a child to leave Pam and their little boy and their loving marriage and their civilized, privileged life. But he had become an American citizen, and he disliked the Germans and loathed the Nazis and all they stood for; so he volunteered for the Army.

Though the most unmilitary of men, Ekstrom found to his surprise that he did everything right in his Army training. He could hit the bull's-eye like a marksman even when shivering with winter cold, could strip his gun blindfolded in a few seconds, and do all the other things a proficient soldier is supposed to do. Except one: He never caught on properly to map-reading. An officer was supposed to carry a map case. He filled his with books.

On his way to North Africa in one of the early convoys, his ship fell behind with engine trouble and limped into Casablanca in late January. Ekstrom was directed to fly to Algiers. He got a flight after a long wait. Immune to motion sickness, he fished Proust out of his map case while the aircraft bumped across the Atlas Mountains and was calmly reading it when Bristow spotted him.

In Algiers he waited for an assignment. Eventually—probably on April 9—he encountered Bristow again in an Allied officer's mess in Algiers. They chatted further. Finally, Bristow suggested that he had a jeep and would be glad to show Ekstrom around. Ekstrom thought that sounded a little fishy, but he went along to see what was what. They pulled up in front of a building on a hill overlooking Algiers. A small British officer came out and Ekstrom stood up in the jeep to salute him. It was Dudley Clarke.

Clarke and Ekstrom hit it off famously. Clarke called him in again the next day and Ekstrom was recruited to "A" Force. (This was the occasion when Clarke, accustomed to the British system of local rank, told Ekstrom he would be a major, of course; only to be informed at AFHQ that when the United States army needed a major, they sent a major. Ekstrom did get promoted to first lieutenant, though it took a while before he had time to obtain a pair of silver bars.) He was given a desk in the "A" Force villa facing Freeman Thomas, and was expected to pick up the business of running double agents in a couple of weeks by a sort of osmosis; complicated by the fact that he could not yet be told about feedback from ULTRA, though he could tell that something mysterious was being held back.

On April 25, Crichton and Captain Germain of Paillole's organization took Ekstrom to Oran to open up an "A" Force outstation and a "41 Committee" (consisting of Ekstrom; Captain Bobby Barclay, of the great British banking family, MI6 Section V's man in Oran; and one Édouard Douare for the French) to manage the double agents the French had made available there and in Casablanca. Crichton and Germain remained just long enough to get things set up; then Ekstrom was on his own, thrown into this highly sensitive business to sink or swim.

He swam. The Oran-Casablanca double agents were soon known as the "Ekstrom Team." " 'A' Force was to reap great benefit from Ekstrom's arrival," said Clarke after the war, "for he was to prove one of the ablest officers in the theater in the management of secret intelligence channels, while his intimate knowledge of the French people and their language did much to further the friendly liaison with the officers of the Deuxième Bureau."

Living conditions in Oran were dreary. There was no fresh water; one had to bathe in brackish water. When Bill Baumer passed through the city in May he found that it "lives up to its reputation as being one of the four dirtiest seaports in the world." At first Ekstrom had to share a rather unattractive bed with Bobby Barclay. Through Crichton's good offices he was soon billeted in a spare room in the apartment of two wealthy Jewish sisters (one of whom wore a neck-piece of silver foxes around the apartment at all hours, on the theory that good furs need constant airing).

He moved in on the day he and Crichton returned from an exhausting thousand-mile trip to Casablanca and back, bumping along execrable Moroccan roads in an open weapons carrier through cold, pelting rain. The *bonne à tout faire* from the adjoining apartment, a barefooted young French-Arab named Marie, tapped on the door and asked shyly whether she could do anything, perhaps something hot to eat? She produced a simple dish of vegetables fit for the gods, and Ekstrom realized that he had stumbled upon a great natural chef. He quickly arranged with Marie for her to take care of his needs while he was in Oran; rather to the annoyance of her regular employers. With a batman named Jenkins shining his shoes, access to the American commissary, and the culinary ministrations of Marie, Ekstrom lived the good life in that disagreeable city. (And outside it too; sharing in Clarke's God-high priority. "If I had to get somewhere I could kick a four-star general off the plane," he recalled long after. "This happened once or twice.")

Then after several months Clarke brought Ekstrom back to Algiers and put him in general charge of double agent work under Crichton. At this point, too, Ekstrom was initiated into ULTRA; "vaccinated," in the local slang. Cuba Mason became his secretary. In Algiers, Ekstrom was billeted at first in the house of a French family named Delacroix whom Crichton knew, where the family meat locker, with a few sausages hanging from hooks in the ceiling, was converted into a rather cozy little bedroom for him. After a few months of this he decided that he deserved more comfort. So on a business visit back to Oran he got in touch with Marie and persuaded her to move to Algiers and set up a mess for him and selected colleagues.

Desmond Bristow found some MI6 funds that could be expended for such a purpose. Ekstrom turned up a charming cottage surrounded by orange trees and mimosas in the garden of a large villa belonging to a Frenchwoman. He lived there; Marie was housekeeper and cook. Ekstrom, Clarke when he was in town, Crichton, and Freeman Thomas lunched there every day. Guests were regularly entertained at their mess. One in particular, who came fairly often when he was in town, was the second exceptional American to join the orbit of "A" Force that summer: Lieutenant, later Lieutenant Commander, Douglas Fairbanks, Jr., who came to the Mediterranean with the first American tactical deception organization, the United States Navy's "Beach Jumpers."

Douglas Fairbanks, Jr., was an authentic matinée idol like his father, the great silent film star, before him. And like his father, he moved in rather exalted circles. In England, he counted among his friends not merely the likes of Noel Coward and Laurence Olivier, but the Duke and Duchess of Kent and Lord Louis and Edwina Mountbatten; the Mountbattens had stayed with Douglas Senior in Hollywood on their honeymoon voyage to the United States in 1922. At home, he had been conspicuous in pro-Allied causes, and had been a vigorous supporter of Roosevelt's third-term campaign in 1940; he and his wife Mary Lee had more than once been guests at the White House and at Hyde Park, and he had been master of ceremonies at the White House dinner celebrating the third inauguration.

In the spring of 1941 he wangled his way into a reserve commission in the Navy notwithstanding his lack of a college education, and in October he was called to active duty. He had sea duty in a supply ship out of Boston, a destroyer to Iceland, a battleship in the North Atlantic. After brief staff duty in Washington, he was off to London, where he caught up with old friends including Mountbatten, now head of Combined Operations, which is to say the Commandos; thence to another battleship as an admiral's flag lieutenant, a carrier running planes to embattled Malta, and a cruiser covering a disastrous convoy to Murmansk.

After that, Mountbatten asked for Fairbanks's assignment to his Combined Operations headquarters, then to an installation in Scotland where tactical deception devices were developed, then to training on the

Douglas Fairbanks

(U.S. Navy)

south coast of England in light boat raiding and diversionary tactics. At one point he was briefed on Bromley-Davenport's efforts at seaborne sonic deception at the time of El Alamein. Mountbatten encouraged him to press for tactical deception in the United States Navy.

In the autumn of 1942 Fairbanks was ordered home, and at the turn of the year reported to Vice Admiral H. Kent Hewitt at Norfolk Navy Yard. Hewitt, "a shy and soft-spoken brass hat" in Fairbanks's words, a big, heavy-set man with a flair for mathematics, had commanded United States Navy participation in TORCH and was refitting to return to the Mediterranean as Commander, United States Naval Forces, North African Waters ("ComNavNAW," in Navy jargon) and Commander, Eighth Fleet.

Hewitt was receptive to the idea of tactical deception units and cleared the way with Admiral King's office in Washington. Particularly receptive was Commander, later Captain, Jeffrey Metzel, chief of the Readiness Section of the Readiness Division of King's COMINCH staff, one of whose functions was handling special projects and developments. Of especial relevance to the Fairbanks project, the Readiness Division and the War Department Planning Board were sponsoring a study of the possibilities of sonic warfare.

In charge of physical research for the sonic aspects of the project was Professor Harold Burris-Meyer of Stevens Institute of Technology. Burris-Meyer was a humorous, voluble, ebullient man, six feet four, with a pencil mustache, a lively imagination, and a wholly unacademic outlook on life. ("An amusing fey guy," recalled Captain Went Eldredge of the Army.) The son of a Methodist minister, forty years old in 1942, he had grown up in New York, graduated high up in his class from CCNY, done graduate work in English and theater at Columbia, and taught English at a college in Pennsylvania. At Stevens he had intermingled engineering and the arts by developing a creative department of theater, focused on techniques of theatrical illusion. He was an early pioneer in stereophonic sound reproduction; his work contributed to the first theatrical stereophonic system, used in Walt Disney's *Fantasia;* he had been an acoustical consultant to the Metropolitan Opera; the disquieting subliminal bass notes that accompanied Paul Robeson's performance in *The Emperor Jones* were his doing; he had developed something called the "acoustical envelope" that enabled a singer to control his sound production. Still classics are his textbooks on *Sound in the Theater, Theaters and Auditoriums,* and *Scenery for the Theater* (with Edward Cole); to this day the United States Institute for Theatre Technology annually grants the "Harold Burris-Meyer Distinguished Career in Sound Design Award."

In the first stage of the research project Burris-Meyer's team explored possible ways of generating sounds so terrifying, or ear-splitting, as to be weapons in themselves—at one point he was dropping beer bottles out of an airplane to see whether they would whistle like falling bombs (they did not)—but got no results worth following up. He then explored, without useful results, the possibility of a "sonic bomb," generating noise so loud as to injure enemy soldiers. Soon the focus of investigation was sonic deception. Specifically, the Navy was interested in sonic deception in connection with amphibious operations.

The crucial feasibility test was the "Battle of Sandy Hook," fought on the night of October 27, 1942. A three-mile strip of beach at Sandy Hook, New Jersey, was designated an "island" and protected by three hundred infantry. Another three hundred men in six landing craft were

Harold Burris-Meyer

(Anita Burris-Meyer)

to establish a beachhead. Equipment was improvised to project the sound of small boats. While Burris-Meyer and his technicians and military men watched from a commandeered yacht, the attackers feinted towards the south end of the "island" with recorded sounds and one landing craft. The defenders rushed to oppose them, and the main force landed unopposed at the north end. It looked as if there might clearly be a place for sonic deception in amphibious operations.*

The Navy took a few months to mull it over, but by the time Fairbanks reported to Hewitt in January 1943, it had decided to train special small-boat deception teams. They came to be called "Beach Jumpers." This odd cover name had its roots in the sonic bomb experiments. When someone asked the purpose of these experiments, Burris-Meyer said it was to scare the BeJesus out of the enemy. From that time on, the effectiveness of a sonic device was referred to as its "BJ factor." The new outfit was accordingly referred to as the "BJ Team"; "Beach Jumper" was then devised to fit the ready-made acronym.

Fairbanks was put in charge of Beach Jumper training at Camp Bradford, near Virginia Beach. He made the rounds of colleges and training camps recruiting officers and men with a technical background and small-boat experience for undisclosed hazardous duty. Meanwhile, Bell Telephone Laboratories was asked to produce a serviceable sound projector to replace the jury-rigged equipment used at Sandy Hook. This was HEATER, the first of which was delivered to Camp Bradford in April. Burris-Meyer and one of his colleagues were commissioned lieutenant commanders and moved to Virginia Beach to aid in training.

Sonic deception, however, was not the only subject of experimentation in those formative months. Fairbanks himself developed his self-destructing dummy paratrooper during this period, and experimented with such other projects as dummy landing craft, GOONEY BIRD dummy infantrymen, an expendable transmitter for wireless deception, and drone boats.

On May 15, an Annapolis graduate was appointed to command the Beach Jumpers. Fairbanks reverted from Officer in Charge to Training Officer. The first units shipped out for the Mediterranean soon after and

---

* By chance, the Battle of Sandy Hook took place only four days after Bromley-Davenport's experiment with amphibious sonic deception at the opening of El Alamein.

established a base at an old French seaplane base at Ferryville, on the southern shore of Lake Bizerte. Burris-Meyer went with them to observe and supervise the first employment of HEATER, and conducted trials and experiments on Lake Bizerte during the preparations for HUSKY. Fairbanks followed by air in June and reported to Hewitt at AFHQ. Hewitt assigned him to his planning staff.

The original Beach Jumpers commander proved to have mental problems and was removed from command after HUSKY. He was replaced by Captain Charley Andrews, "an experienced old crony of Admiral Hewitt's from Annapolis days," said Fairbanks, "a wonderful, swaggering hearty." Hewitt enlarged the Beach Jumpers' role to include raiding, beach reconnaissance, and infiltrating agents. He made Fairbanks Andrews's Assistant Chief of Staff for Operations, and constituted him also as the Special Operations Subsection of the War Plans Section of his own staff, designated "N-31(S)," supervising training, supply, and planning for raids, special assault landings, diversionary activities, and other special operations.

Put in touch in this capacity with Advanced HQ "A" Force, Fairbanks was welcomed with wide-open arms by Dudley Clarke, film buff and social lion-hunter. Eventually, in February 1944, he was designated to fill the naval slot at Advanced HQ, with the title "Allied Naval Liaison Officer With 'A' Force."

Burris-Meyer remained in the theater through the summer. Among other things, he consulted with Michael Crichton about the possibilities for simulating unit headquarters by prerecorded voice radio transmissions, and late in August Crichton proposed further research along those lines.

On May 2, Bill Baumer showed up in Algiers for a brief visit.

At the beginning of April, after once canceling such a trip, Wedemeyer had approved a visit to England by Baumer to meet the deception agencies there; and Kehm and his boss, Colonel Roberts, insisted that he go to North Africa too. With another OPD officer, he was to travel to Britain by Pan American flying boat out of La Guardia Field in New York, via Bermuda, the Azores, Portugal, and Foynes in Eire, wearing civilian clothes so as not to ruffle the sensitivities of neutrals. After four days' wait in New York for good weather, they took off on Friday morning, April 10. There were unexpected breaks in the journey: a deep-sea fishing trip when grounded in the Azores by rough water, a hotel with bedbugs in Estoril, Portugal, a stopover in Ireland. He reached London on April 14. The next day, after courtesy calls on the generals commanding the American headquarters, he was taken to Storey's Gate to meet the London Controlling Section. He spent the next week there, steeping himself in their work.

Wheatley and Baumer took to one another at once. "A delightful fellow and a very able one," recorded Wheatley. "Of them all, Dennis Wheatley, England's leading thriller writer, was most kind," remembered Baumer. Wheatley took Baumer home to dinner, and "arranged several luncheons at which he always trotted out a variety of cabinet secretaries and air vice marshals." At one of these, when Baumer expressed regret at not being able to see anything of England apart from London, the Vice-Chief of the Air Staff told Wheatley to give Baumer a tour, and made his car available—something indeed special in view of tight fuel rationing. Wheatley took Baumer to visit not only several RAF bases but Hampton Court, Sandhurst, Aldershot, Windsor Castle, and Eton. Wheatley sadly remarked at one point, as one author to another, that he regretted that he would never be able to write about deception, even though it was the type of thing that he did best.

"From such an officer we had 'nothing to hide,' " as Wheatley said (referring to a story about Roosevelt encountering Churchill in the bathtub in the White House), "so he was duly indoctrinated into our mysteries." One of these was the planning and mounting of MINCEMEAT, presently to be described; and Baumer took a modest part in BARCLAY, communicating with Washington over the question whether men of the 45th Division, sailing from the United States to the HUSKY landings, should be told that their destination was French Morocco.

On Easter Sunday, April 25, after ten days in London with the London Controlling Section and with the staff planning the invasion of France, Baumer left London for Algiers. It was another typical wartime journey: a rough flight to Prestwick in Scotland, a two-and-a-half-day wait there for a plane to North Africa, another wait at Marrakesh. Eventually, on May 1, he reached Oran, which he found lived up to its reputation for dirtiness; the hotel had "as usual, no sheets, no nothing—and the bedbugs!" The next day he finally made it to Algiers.

After reporting in to AFHQ, where Eisenhower greeted him and his companion genially "and treated us as old OPD members," Baumer spent five days with Goldbranson. Goldbranson filled Baumer in on such matters as the use of double agents; the need for communications security officers and for American dummy armored units; the importance of coordinating propaganda and deception; shoulder patches for notional American divisions; and developments in dummy devices. He gave Baumer a set of notes on Dudley Clarke's theories and methods. He suggested to Baumer a conference in Washington once HUSKY preparations were completed; Rushton, he said, should by then be able to hold the fort.

With a very black blackout and no taxis at night, Baumer found like others that there was nothing much to do in the evenings but "drink *vin rouge* and turn in early." But his last night in Algiers, Friday, May 7, was different. Tunis had fallen that day, the war in North Africa was substantially over, and a friend invited Baumer "to a dinner at a black market place called Celeste's. The Americans under General Nevins had taken over the place," wrote Baumer, "and the proprietress . . . said nothing was too good for the Americans and went to the bottom of the cellar to bring out a gallon bottle of Hennessy and numerous liqueurs."

The next morning, May 8, he took off on a dry and hungry flight for home; via Marrakesh, where the hotel had run out of champagne and brandy, thence across the Sahara, thence across the Atlantic to Brazil on a flight on which the pilot had forgotten to pack any food, thence via British Guiana to Miami, finally arriving in Washington at 9 A.M. May 15—just in time to join the conference on Pacific deception.

Dudley Clarke came back to Algiers on June 18 and stayed till July 3. While there he in effect fired Goldbranson on specious grounds.

It came as a bolt from the blue. On June 28, saying that he wanted to be fair and wanted Goldbranson's comments before submitting it officially, Clarke handed Goldbranson a purported efficiency report. It was highly negative. It charged Goldbranson with insufficient staff training as compared with British officers, inability to speak French, and lack of aggressiveness for not having written any plans in London, done anything in Algiers before Advanced HQ opened, or given direction to Strangeways.

Goldbranson, dumbfounded, pointed out that Clarke knew nothing about him or about what had gone on in London. Goldbranson had done what little could be done before Advanced HQ opened and had not been allowed to do much since. *Nobody* could give direction to Strangeways; nobody even tried. Neither staff training nor speaking French was important to his position.

Clarke was visibly displeased to learn that not he but the American deputy G-2 was to make out Goldbranson's efficiency report. But Clarke could not be forced to keep Goldbranson on his team, nor would Goldbranson have wanted to stay under such circumstances; and Goldbranson was ordered transferred to G-2 in Washington. From there, General Strong cabled his puzzlement: "Colonel Goldbranson well known to this HQ, but action your HQ not understood." The Algiers deputy G-2 responded with the diplomatic excuse that Bedell Smith had decreed that some promising American intelligence officers with good experience in Tunisia should be exchanged with War Department G-2. "Dudley Clarke wishes to place at disposal of Security Control experiences gained in this very specialized work during Tunisian campaign. In addition . . . Clarke considers French language essential qualification senior officers this particular section."

Goldbranson had learned the work thoroughly, was highly competent, and was of inestimable value to American deception for the rest of the war (and in a postwar career in the United States Air Force as well). Perhaps Clarke's motivation was no more than prejudice. Perhaps something more devious was involved; Goldbranson noticed that June 30, the due date for American efficiency reports, was also the date for resubmission of the British "war establishment," the authorized table of organization. Perhaps it had something to do with the falling-out between Clarke and General Kenneth Strong that led to Advanced HQ being put under General Rooks.

Goldbranson was not transferred until shortly after the HUSKY landing in early July. (Nor was he the only deceiver to leave Algiers; Derrick Morley had been transferred to Bevan's office in late May.) Goldbranson was back in the States by July 25 and in Washington by July 27. After a home leave in Council Bluffs he joined the deceivers in the Pentagon, where for the next two years he would be Newman Smith's right-hand man. Waiting for him was a cordial and thoughtful letter from Johnny Bevan.

*   *   *

Goldbranson's eventual replacement was another American lieutenant colonel, named George Francis Train III. He did speak French; and, what was perhaps in truth more important to Clarke, he was an Ivy League Easterner. A 1915 Yale graduate, he had worked for various Wall Street firms, traveling extensively in Europe, Latin America, and South Africa. In the First World War he had seen action in France in the Tank Corps. He had joined the Reserve in 1941, was called to active duty in June 1942, and landed at Oran on TORCH D-Day. He had known Crichton through business dealings before the war. On September 20, 1943, now a lieutenant colonel, he was assigned to Advanced HQ "A" Force; in November, he was to succeed Crichton as its commander.

Train was forty-eight years old, over six feet tall, and lean, with blue eyes and medium brown hair. Though more suited to Clarke's social prejudices than Goldbranson had been, behind his back he got less respect from some of his colleagues than Goldbranson had; they regarded him as a sort of fifth wheel. "George Train, the American in nominal charge of the Algiers office, was once asked (not surprisingly) how he became a Lieutenant Colonel," said Rex Hamer years later. "He replied 'Gee, I ranked Lieutenant Colonel from my civilian job.' " "I don't think he ever made any decision, that he ever even made any effort to learn what it was all about," was Ekstrom's recollection. "He was a man of some authority and lots of impatience. And he very quickly acquired a mistress, who was the daughter of the Swedish king's dentist."

A variety of new special means channels opened up in the Mediterranean in 1943–44.

The "Ekstrom Team" in Oran and Casablanca brought in by Paillole were a typical motley lot. In Oran were ARTHUR, a rich Spanish Jew; EL GITANO, a Catalan hairdresser, smuggler, and pimp; LE PETIT, a Czech interpreter on the Oran docks; and CHER BÉBÉ, a Spanish mechanic. In Casablanca were CUPID, "an attractive and intelligent young German Jewess who ran a bar," to quote Dudley Clarke, and DAVIL, an employee at the air base.

ARTHUR moved in high circles and could plausibly pass high-grade information. He had some shady financial dealings with the Spanish consul at Oudjda in French Morocco, headquarters of the American Fifth

Army. His four subsources included HARRY, a sometime warrant officer in the French Army. Employed for deception from March to October 1943, he was particularly useful for BARCLAY.

EL GITANO had been recruited by the Germans in December 1942. His routine contacts were the Spanish consuls in Oran and, sometimes, at Melilla in Spanish Morocco. This activity was what gave the DSM their hold over him; he was a double agent from the opening of the Oran station. While lying low in Melilla after a brush with the law involving his pimping activities, he was recruited by the Italian SIM and given a transmitter. From June to October he made almost daily wireless reports from Oran to Melilla, mainly low-level information about shipping and the French Army. After the Italian surrender he continued to be used on a much reduced scale, with no transmitter, until February 1944.

LE PETIT, a stocky, self-opinionated little man who nominally worked for the Spanish secret service, was a valuable channel from April 1943 to August 1944, passing high-grade material through the Spanish vice-consul at Oudjda.

(Susan Train)

George Train

CHER BÉBÉ—he owed his codename to a song currently popular in Algerian music halls—had cooled his heels in jail as a German agent for over a year before agreeing to work for the DSM; then served as a deception channel for some months from May 1943, reporting through the Spanish consul at Oran.

CUPID wrote a weekly letter in secret ink to the Germans in Barcelona, but her information was low-grade, and efforts to use her for deception ran only from March to June. A far more useful Casablanca channel was DAVIL, a Frenchman working for Paillole's organization who had gotten himself recruited by the Abwehr in Madrid and sent to Casablanca with a W/T set, working back to Hamburg. Employed at the air base, from January to July 1944 he sent misleading aviation information and order of battle and shipping data.

Besides the "Ekstrom Team," Paillole and his colleagues brought to "A" Force a group run out of Algiers whose traffic was managed by the 40 Committee (originally called the 37 Committee), chaired by Crichton, with Freeman Thomas sitting in, Desmond Bristow representing MI6, and Captain Doudot of Paillole's staff representing the French.

The first of these was RAM, a French sergeant, working in fact for Paillole's "TR," whom the Germans recruited in Paris and sent to Algiers in early 1942. Notionally employed in communications at French Army headquarters in Algiers, from March to September he radioed to Paris low-grade material on Algiers port activities. He was the W/T operator also for NORBERT, an extremely intelligent flight sergeant in the French Air Force with years of experience as a double agent, and ÉDOUARD, a sergeant in the intelligence office of the French XIX Corps.

The RAM trio were soon followed by WHISKERS, an exiled Spanish officer who passed material to the Abwehr in Spanish Morocco from April to August 1943 via the Spanish vice-consul at Algiers. Every Monday, WHISKERS would hand him a report that had in fact been written by his French case officer, based on material furnished by the 40 Committee the Thursday before. The 40 Committee built up a network of high-grade notional sources for WHISKERS all over Algeria, and in July 1943 he found himself appointed head of the Spanish secret service in Algeria. But he was suddenly cut off when the vice-consul was removed from his post for black marketeering.

An agent codenamed COCAINE radioed daily from Algiers to the Germans in Melilla from late July 1943 to September 1944. In August 1943, a former wireless operator for the French Air Force, whom the Gestapo had recruited in Paris, parachuted into Algeria and turned himself in. Under the codename GAOL he transmitted two or three times a week to the Germans in Dijon. Notionally employed at the Algiers airport, GAOL was a good channel for reporting fictitious travel by VIPs. He was shut down at the end of August 1944 by notionally transferring him to France. A low-grade W/T channel called RUBY transmitted to Dijon from December 1943 to February 1944. Last of all in Algiers came BYZANCE, who worked by W/T to Berlin but did not arrive till the landings in Southern France were under way in August 1944; he was only run for two or three weeks.

From its formation, the 40 Committee provided the material for a channel that Goldbranson had brought from London and run since January. This was OLIVER, a fictitious South African field security policeman who had notionally come with the TORCH landings in November. He sent information in secret ink semimonthly to a supposed German agent in London. His letters, sent by courier to the Abwehr in Lisbon, were prepared in London from text provided by Goldbranson and subsequently by the 40 Committee. Useful mainly for order of battle because of the time delays involved, OLIVER came to an untimely end in July 1943 when the actual writer of the letters in London was killed in a plane crash.

The 40 Committee also provided the material for MOSELLE, a party of Spanish and Sardinian saboteurs that the British had infiltrated into Sardinia, communicating by wireless with Algiers, whom the Italians had evidently captured and were working under control. It was useful for keeping alive Axis expectations of an attack on Sardinia; so useful that in May a special flight was laid on to drop money, weapons, and supplies for which the "saboteurs"—actually a certain Sergeant-Major Silvestri, who was running the operation on the Italian side—had pleaded. MOSELLE was run until the Axis evacuated Sardinia early in September 1943.

One special means channel had been picked up in Libya: LLAMA, a Libyan wireless operator in the Italian Army who had been left behind to operate a transmitter from the hills behind Tripoli. Tracked down and turned by SIME, from March 25 he was used to send order of battle information and fake troop movement reports to the Italians.

But none of these could hold a candle to Paiolle's prize catch. The day after Tunis fell, there presented himself to the DSM an athletic-looking, fiftyish Frenchman, "of medium height, sturdily built, with gray hair and military mustache." He identified himself as André Latham, a French Army officer, a St. Cyr graduate, a colonel in the Legion of French Volunteers Against Bolshevism, and currently the chief of a team of staybehind German agents, who was ready to turn himself in and work for the Allies. Under the codename GILBERT,* Latham was to join CHEESE, GARBO, and SILVER as the fourth of the great quartet.

The Lathams were a well-known Protestant family of the minor French aristocracy. Young André attended St. Cyr and served with distinction in 1914–18, rising to the rank of major. The family lost its money during the war, and in 1919 he left the Army hoping to recoup the losses in business. He promoted various ventures, successful and not; moved in cosmopolitan circles; and managed to support a succession of mistresses, one of whom presented him with a son.

When the Second World War began he returned to the Army and received command of a battalion. It fought well in 1940 before the collapse. After the armistice, with nothing to do, he was offered a position as Intendant of Police in Lyons and subsequently in Marseilles. He left this service under an unexplained cloud. He was then commissioned a colonel in the "Legion of French Volunteers Against Bolshevism," an organization of Frenchmen fighting with the German Army on the Eastern Front. Before he could take up his command—on November 9, 1942, the day after the TORCH landings—a fellow officer got in touch with him and persuaded him to go to Paris to meet a man named Albert Beugras.

Beugras was head of the special intelligence branch of the PPF, the Parti Populaire Français, a French fascist party collaborating with the Germans. He passed Latham on to Colonel Reile of the Abwehr station in Paris. After several interviews, Reile proposed that Latham go to French North Africa as chief of a party of secret agents gathering military and naval information for the Abwehr and political information for the PPF. After setting up a headquarters in Tunis under his second in com-

---

* Not to be confused with the SOE operative GILBERT, Henri Déricourt, whose conduct in the French underground has been the subject of controversy.

mand with two W/T transmitters, a radio operator, and an expert in sabotage, he would be parachuted behind Allied lines with another transmitter and radio operator. Thence he would radio information to Tunis, to be passed on to the Abwehr in France.

Not long afterwards, Latham was introduced to his team at a rendezvous above the Elizabeth Arden beauty parlor at 52, rue Faubourg St. Honoré. His second in command was to be a Captain Dutey-Marisse, codenamed LE DUC, "The Duke." LE DUC was a rich playboy some forty years of age and not especially bright, whom Latham already knew slightly. The wireless operators were a nonentity named Falcon, and a dull-witted and gullible former petty officer in the French Navy named Blondeau, codenamed ALBERT. The saboteur was an unsavory character, Lieutenant Duteil by name, alias Joseph Delpière. Duteil had been a pimp in civil life; recruited to the PPF by Jacques Doriot, its chief, he had served with Doriot on the Soviet front and had been secretary to Beugras. Duteil had two assignments besides sabotage. One was to gather information and conduct PPF propaganda. The other—not disclosed—was to kill Latham if he thought he was double-crossing the operation.

They left Paris late in March 1943 and made their way to Tunisia via Germany and Italy, escorted by German officers and accompanied by Beugras. There they waited for some weeks while nothing happened. Early in May, on the eve of the German collapse, Latham and Beugras arranged for the team to remain as staybehinds when the Germans pulled out. The plan was that three transmitters would eventually be set up: a main one in Tunis, codenamed ATLAS I, and outlying stations in Algiers (ATLAS II) and Oran (ATLAS III). Beugras then got out and went back to Paris.

Latham told LE DUC—but none of the others—that he planned to turn himself in when the Allies came. On May 8, the day after Tunis fell, Latham sought out the DSM and told them his story. Paillole, who had known Latham in France at the end of 1941, reflected for twenty-four hours and decided to break his usual rule against accepting walk-ins. Latham was set up as a double agent, codenamed GILBERT. He was installed in a flat in the center of Tunis along with LE DUC, and assigned to the French staff with the rank of major. ALBERT was taken into French naval headquarters; Falcon was taken into the French Army and a replacement, known as JEAN, was found for him. Duteil joined the French

Army, planning to desert and make his way to France through Spain. He was eventually arrested, his information was shaken out of him, and he was executed.

With ALBERT as his radio operator—ALBERT had been told that he was working now for a group of right-wing French officers—GILBERT was on the air from June 10 sending traffic to the Germans in France. It was soon plain that the Germans considered him an agent of very high class. "A" Force had been brought in early on, when Clarke sent Freeman Thomas to explore double agent opportunities in Tunis not long after its fall. On June 27, a new 43 Committee was opened in Tunis largely for GILBERT. Its "A" Force representative was Major P. A. T. Grandguillot, a Frenchman resident in Alexandria—and a tennis player of international stature—who had joined the British Army and had earned the Military Cross on the escape and evasion side of "A" Force.

GILBERT was a *bon viveur,* athletic, a boon companion, fond of horses and shooting and of good food, wine, and women. Transmitting almost daily for fifteen months from North Africa, and almost as often from France after the invasions, he built up a personality as distinctive as that of GARBO, and one that was consistent with the real André Latham. There was an occasional touch of bombast; and his messages seldom explained or hedged but simply stated facts in a confident downright tone befitting his supposed access to high Allied military circles. Central to GILBERT's style, especially after he moved to France in the latter part of 1944, was a sardonic tone in his allusions to French political and military efforts. This served as psychological justification for his own treason; more important, it enhanced his credibility in German eyes. As the official report put it, "The absence of effusiveness, the cool cynical realism of his attitude dispensed with any need for lip-service to 'the cause' or verbal manifestation of loyalty, making for a different and no doubt much more effective relationship between this agent and the enemy than existed in the majority of other cases," making it "possible to have him put insistence and emphasis into his messages without giving the impression of an 'axe to grind,' or anxiety to be believed, or even a predominant desire to serve the German cause."

Since GILBERT's messages thus might often trespass on French sensitivities, a combination of humor and tact on both sides was called for, even though he was to the French "the greatest prima donna of their counter-espionage system," and French counterintelligence officers "stood in awe of his importance and his powers."

GILBERT's immediate case officers were French, headed by Captain Germain of Paillole's staff, in civil life a professor at the University of Metz. They handled him with such skill that he never realized that his traffic with the Germans was contrived by a Franco-Allied deception organization. He knew only that the information he sent came from what he smilingly called a *"consortium d'intelligence."*

In that spring and summer of 1943 two additional useful channels, THE LEMONS and THE SAVAGES, were added in the Eastern Mediterranean.

THE LEMONS—BIG LEMON and LITTLE LEMON—rowed ashore in Cyprus from a Greek caïque in mid-May. BIG LEMON, the leader, was a shifty-looking middle-aged Istanbul Greek who had volunteered to work for the Germans and after minimum training was shipped out of the Piraeus with nearly four thousand dollars and directions to land in Syria and join a supposed spy ring in Beirut. LITTLE LEMON, his wireless operator, was a young man, tall and blond, who had worked in his father's Athens restaurant and had been lured into working for the Abwehr by "a boastful Quisling and black-market operator."

After a voyage marked by squabbling and outbreaks of cold feet, THE LEMONS were put off at Cyprus rather than Syria. They told the villagers who met them that they were refugees. But after three days LITTLE LEMON broke down and confessed. BIG LEMON was relieved of the Abwehr's money and sent to prison. LITTLE LEMON was set up as a double agent and installed in a Nissen hut in the Athlassa Forest, where, "but for an endless repertoire of sentimental ballads with which he enlivened the intervals between transmissions, he gave no trouble at all."

THE LEMONS made contact with the Abwehr in late June. Managed by a 36 Committee subordinate to the 31 Committee in Beirut, and handled by a SIME case officer, a small, dapper, and talented Greek-English officer, Lieutenant Klingopoulos, the channel developed into one of the most effective in the Middle East. BIG LEMON was depicted to the Abwehr as being at large and an agitator for union of Cyprus with Greece. LITTLE LEMON was depicted to them as rather a ladies' man, suffering on occasion from venereal disease. His key notional sources were a cosmopolitan bevy of genuine chorus girls working the cabarets of

Nicosia, known to "A" Force as GABBIE, SWING-TIT, MARKI, TRUDI, and HELGA.

By September 1944 it was decided that little more useful information could be seen as emanating from Cyprus. So THE LEMONS were shut down—because of LITTLE LEMON's health, the Germans were told. The story had an unusually happy ending; Klingopoulos attended the homecoming celebration when LITTLE LEMON returned to the family restaurant in Athens in July 1945. And the Germans, it transpired, had faithfully paid his parents the maintenance and gifts they had undertaken to provide while their son was away in their service.

A more surprising group was THE SAVAGES, a pair of newlyweds and their best man, who arrived in Cyprus, again in a caïque from the Piraeus, two months after THE LEMONS, bringing with them a wireless transmitter, three thousand dollars, and a small sum in British notes and gold. The bride, SAVAGE III, and the groom, SAVAGE II, were both physicians. SAVAGE I, the best man, was a young former law student in Athens, and former prisoner of war, who had been recruited into the German service by SAVAGE II in January 1943. They reported themselves as refugees after burying their funds and equipment. Under interrogation, SAVAGE III finally told the truth.

The British concluded that their primary motivation was to get out of Greece, but their initial lack of candor told against them nevertheless. The bride and groom were sent to a detention camp in Egypt for the rest of the war. SAVAGE I was chosen as the wireless operator for a new channel. THE LEMONS were all that was needed out of Cyprus, so SIME "produced a fiction of daring ingenuity" to justify a move of THE SAVAGES to Egypt. The Abwehr swallowed it, and SAVAGE I and his transmitter were installed in a villa outside Cairo, where he was supposedly employed in the Allied Liaison Branch at GHQ Middle East. He sent a message nearly every day from late August 1943 until the very eve of the German surrender.

THE SAVAGES became a valuable backup channel for CHEESE, specializing in order of battle material. In the words of a SIME memorandum, SAVAGE I's "biggest 'scoop' was in May 1944 when he saw an instruction on military security in his office and copied the distribution list. This included two armies, four Corps, and twelve divisions. He passed this information to the Germans, who warmly congratulated him." From the autumn of 1944 his main use was to aid Peter Fleming's order of battle deception (there was an abortive plan called WHISKY, for putting him in direct W/T touch with the Japanese). THE SAVAGES ended on as happy a note as did LITTLE LEMON. The newlyweds were released in the summer of 1945 to settle down in Cyprus, while SAVAGE I returned to his interrupted law studies in Athens.

Meanwhile, the oldest channel of all, CHEESE, kept up a steady flow of information—and money complaints—to the Abwehr.

We left NICOSSOF in January of 1943, still desperate for money, with the B.G.M., his *amie* Marie, waiting in her flat for a delivery of cash. Weeks dragged by, with Athens begging him to have patience. NICOSSOF told them he was struggling to get along by giving occasional private lessons in Arabic and English.

There was a disturbing period in March when the Germans unaccountably warned NICOSSOF to "exercise the very greatest prudence," while contemporaneously strange things happened at the B.G.M.'s flat. She had fallen ill and SIME installed a new tenant in the flat, a man this time. The day after he moved in, March 23, an old lady came to the door looking for "Marie" and claiming to be her aunt. The next day a young woman appeared, claiming to be Marie's cousin. On the third day came another young woman, claiming to be her friend. All were told that Marie was in hospital and could not be troubled. Still no money. Then for a time almost no messages came from Athens, and when traffic resumed, the Germans appeared to be less interested in NICOSSOF. It was May before things seemed to be back to normal.

This business may have had a perfectly simple explanation; but what it was was never found out. It came at a bad time for BARCLAY, for the CHEESE channel was less useful in putting that over than it should have been. SIME was finally compelled to find NICOSSOF a job. He told Athens that he had found work as an interpreter for the Cairo office of the "Occupied Enemy Territory Administration."

In May, an Armenian businessman of doubtful character who had been working for the Abwehr in Istanbul offered his services to MI6, receiving the codename INFAMOUS. (The name was merited on several grounds, one of which was that it later transpired that he was passing unauthorized phony information to the Germans.) INFAMOUS told MI6 that he had smuggled several thousand pounds to Egyptian bank accounts on behalf of an unidentified "Swiss" (MI6 suspected that this "Swiss" was in reality another Armenian employed by the Germans), and that he had been instructed to send £415 by courier to someone named Marie in the Rue Galal in Cairo. The money was to be picked up from INFAMOUS in Istanbul by an Italian priest from Cairo bearing a message written inside the lining of his coat. The priest—if priest he was—was to pass the money to another courier in Aleppo; that courier was to arrange for someone named Abbas to deliver the money by milk bottle to the Rue Galal.

On May 24 the "Swiss" gave INFAMOUS $1,700 U.S., and told him that the courier would not now be the priest but another individual, who would show up in early June, likewise bearing a message in the lining of his coat and identifying himself as well by presenting a Turkish pound note with a specified serial number. But on June 4, before this new courier appeared, the "Swiss" showed up again, told INFAMOUS there was danger, took back the packet of money, and departed, allegedly for Zurich, leaving with INFAMOUS only a letter reading: "Shortly you will receive £415 for—you know who." On June 7, a Turk named Aslan bearing the correct Turkish pound note presented himself, took the letter, and vanished. Neither he nor the letter was ever seen in Cairo.

On July 11, the Abwehr reported to NICOSSOF that on March 26, the day after the third of "Marie's" visitors had turned up, £450 in an envelope had been handed to an old woman at the address he had given. NICOSSOF angrily denied this, of course. The Abwehr stood by their story. Evidently, however, they maintained their faith in NICOSSOF and assumed—no doubt correctly—that the old lady and her confederates had swindled them.

SIME and "A" Force were never able to get to the bottom of all this, but they attributed it to nothing more worrisome than German incompetence. They concluded, however, that if NICOSSOF was ever going to get paid—as he must, to maintain his credibility—they would have to make the arrangements at their end. A complex plan, codenamed HATRY, was devised, pursuant to which NICOSSOF suggested to Athens late in August that he could get money from them through a financier in Cairo named Cohen, who could buy Swiss francs in Istanbul for a daughter in Switzerland. The Germans agreed at once. SIME lined up a group of unsavory characters to enact the necessary roles in what proved to be a complex drama; one of these in particular, codenamed GODSEND, played his role splendidly, and eventually NICOSSOF was able to radio to Athens on January 6, 1944, that after waiting for two years to be paid he had at last received £1,400.

(Two months later, Renato Levi, the genuine original CHEESE, arrived back in Cairo, having been liberated in October 1943 by the Allied advance in Italy. But he took no further part in his own supposed espionage.)

Two less gratifying channels, known as CAPRICORN and LILOU, showed why Clarke respected the Italians as opponents far more than he did the Germans.

CAPRICORN was an MI6 W/T agent in the Peloponnese, working back to Cairo. In late 1942, MI6 concluded that he was operating under Italian control, and made the channel available to "A" Force. From mid-February 1943 on, he was sent questionnaires about Axis defenses in the Peloponnese to help create a threat in that area. Soon, however, he began pleading for funds. These had to be supplied to keep the myth alive. The intermediary was PIG, a German agent in Istanbul who had been ordered to penetrate the British secret service (but did not know that the British knew this). A package of cash, including one-pound notes surcharged "Greece," with a cover letter telling CAPRICORN that the surcharged notes were for petty cash in the forthcoming British invasion, was confided to PIG, in the knowledge that he would deliver it to the Abwehr in Sofia. PIG returned to Istanbul at the beginning of September, reporting that his mission was accomplished. Then Italy surrendered, and no more was heard of CAPRICORN—except that after Italy changed sides, the Italians who had run the case said they had suspected that the Allies had tumbled to the fact that they were running CAPRICORN, but that his traffic was sent to the Italian general staff nevertheless.

LILOU was another British agent, working from Sicily by W/T to Malta, known originally as MISCHIEF (also as QUACK). In February 1943, MISCHIEF sent a message saying that he had been under Italian control from the beginning, that he had evaded surveillance long enough to make this transmission, and that any message he might send signed LILOU was genuine. This opened the door to a maze. Was LILOU the real thing? Even if he was, did the Italians know about him? If so, did the Italians know that the British knew that the Italians knew? If so, efforts to direct attention towards Sardinia or Corsica would in fact point to Sicily. There followed an intricate back-and-forth quadrille of cautiously worded questions and equally cagey responses. In April, Max Niven was installed in Malta as a new "A" Force outstation to handle the case.

Gradually "A" Force grew convinced that the Italians were running both MISCHIEF and LILOU. After the landings in Sicily, both went off the air. Early in 1944 the true story was learned. MISCHIEF had been caught when he landed and had been run throughout by the Italians. The first LILOU message had been real, but had been monitored by the Italians. MISCHIEF was "shot while trying to escape," and both channels had been taken over by an Italian operative who imitated MISCHIEF/LILOU's fist. But they could not be sure that the Allies had not noticed the change in operators, and reposed no confidence in the channel thereafter.

BARCLAY, the theater-wide plan to cover HUSKY, was to prove, in Dudley Clarke's words, "the peak of the Deception effort in the Mediterranean theater. . . . 'BARCLAY' was neither the biggest nor the longest of the major Strategic Deception Plans of the War, but it was one of the most straightforward in the almost 'classic' style and it illustrated more clearly than most the principal lessons both of planning and implementation."

The odds were against BARCLAY, for it was almost self-evident that the next Allied move after North Africa must be Sicily or Sardinia, probably the former. Sicily commanded the Sicilian Channel linking the two halves of the Mediterranean, flanked any approach from French North Africa towards Italy or Greece, and was the base for the aerial siege of Malta. Preparations for HUSKY—a buildup in Malta, reconnaissance, the seizure of Pantelleria—would confirm Sicily as the goal. Even so, BARCLAY was to achieve a remarkable degree of success.

BARCLAY had three stated objects: To induce the Germans to keep as many forces as possible away from the central Mediterranean by threatening the south of France and the Balkans; to deter reinforcement of Sicily and weaken its garrison; and to keep down naval and air attacks on the shipping being assembled for HUSKY.

The "story" was complex, spelled out in extraordinary detail. The bogus British Twelfth Army would invade the Balkans from the Middle

East in early summer, landing initially in Crete and the Peloponnese, thereby bringing Turkey into the war; operating thence with the Turks against Bulgaria, and in support of the Yugoslav resistance; and ultimately linking up with the Soviets. When the Germans were well committed in Greece, diversionary landings would be made in northern and southern France. Those in southern France would be made by Alexander, covered by the Americans under Patton landing in Corsica and Sardinia; the British Eighth Army and a French army from North Africa would move up the Rhone Valley. Italy and Sicily would be bypassed; the big air buildup in North Africa and Malta was aimed at an extended bombing campaign, neutralizing Sicilian airfields and attacking Italy generally.

The familiar device of "postponement" of the notional D-Days was used twice, from late May and early June to late June and early July, and again to late July and early August. The true Husky D-Day was July 10. All three notional dates were in the dark of the moon, in hopes that the Axis would conclude that this reflected an element of Allied landing technique and relax their vigilance at other times.

The aim was to induce the Axis to reinforce the Balkans and the south of France during the first two phases; then, after the second "postponement"—when evidence would begin to mount that Sicily was the target, too late for them to shift troops to Sicily from Greece and France—to reduce alertness in the island on the basis that the attack would not come until a moonless night later in July. Threats to Crete and Greece played up to known Axis sensitivities: Ultra had revealed concern by the local commanders throughout the winter, and had disclosed that staff conferences in Rome in February had concluded that Greece would be the most vulnerable target after Tunisia was lost.

Barclay was approved by the theater commanders in mid-April, and by Churchill himself soon after. ("Be careful not to alarm Turkey," he wrote in the margin.) All the customary systems were engaged to put over the "story." The usual double agent channels were busy. Wolfson in Istanbul planted another story in the German-language newspaper to which he had access. The rumor mill was set in motion. A psychological warfare plan tailored to the "story" was prepared and implemented, with due attention to the need to avoid a premature rising by resistance groups. (Goldbranson assisted by bootlegging the Barclay plan to the

psychological warfare people, contrary to standing procedure.) Bevan provided the Foreign Office with appropriate tidbits to be dropped by diplomats at dinner parties in Bern and Stockholm. Men were landed from submarines on beaches in Sardinia and the Greek island of Zante to leave behind evidence of their presence. Greek troops in Egypt and French troops in Algeria were conspicuously given amphibious training.

A massive display was mounted in Cyrenaica—the only Eastern Mediterranean area within range of German reconnaissance aircraft—to simulate the gathering of a host to invade the Balkans. Called WATER-FALL, this project included over one hundred dummy landing craft in the harbors, enough real and dummy gliders for an airborne division, an entire dummy "8th Armored Division" of Colonel Jones's contraptions, dummy camps and training areas, and eleven "squadrons" of fighters on seven separate airfields, protected by real antiaircraft batteries that opened fire, and a handful of real fighters that were scrambled when German planes approached. Dummy forces moved forward and back in keeping with the notional postponements. An elaborate wireless deception plan simulated the radio traffic to be expected from such a host.

The headquarters in Egypt planning Montgomery's operations for HUSKY was formally designated as "Twelfth Army," with gratifying results in terms of Cairo gossip and rumors. Calls were issued in the Middle East (and in India, through Fleming) for Greek-speaking British officers and men. The theater paymaster's office started heavy buying of Greek drachmas on the Cairo exchange. Bevan sent out fifty strongboxes labeled as Greek bullion; they were unloaded and shipped under armed guard to a Cairo bank in the name of the civil affairs branch of the Army. Pound notes overprinted "France," "Greece," or "Bulgaria" were "mislaid" in various countries. Pamphlets on Greece, leaflets on hygiene in the Balkans, Polish-Bulgarian phrase books, and maps of the notional target areas were distributed. In Algeria, inquiries were made for fishermen acquainted with the waters around Sardinia, Corsica, and southern France. Goldbranson and Freeman Thomas ostentatiously sought guidebooks to those areas in Tunis bookshops. Signboards in Greek were set up at Tripoli in Libya, and some Greek officers were attached to the British HUSKY forces in that port and in Malta. (One patriotic Greek managed to remain with a British unit and was no doubt amazed to find himself landing in Sicily instead of his homeland.) A genuine sabotage operation by the Greek resistance under SOE sponsorship, codenamed ANIMALS, helped keep German attention directed eastward in late June.

The Turkish correspondent of a London newspaper helped lend credibility to the second "postponement" by gossiping freely about his editor's putting off for a month an alert to be ready to cover an event in the direction of the Balkans in late June. Notice was given that on June 16, Middle Eastern frontiers would be closed and the broadcasting station in Syria taken off the air; this was rescinded on June 15, but not before gratifying publicity had been given to the resulting postponement of a major medical conference in Beirut. In North Africa, Eisenhower canceled all leave after June 20, and rescinded the order on June 15. When Montgomery came to Cairo in early July to review final plans, a notional leave in Palestine was arranged for him and canceled at the last minute.

The radio systems of two allies, each with three services, badly needed security coordination. In May, a British wireless expert, Major S. B. D. Hood, known always as "Sam," joined "A" Force, working primarily in Cairo, while the Americans sent two communications security experts to Algiers, Commander J. Q. Holsopple of the Navy, and a Lieutenant Colonel Handy of the Air Force. Holsopple, Handy, and Hood organized a wireless security and deception system for the whole Mediterranean, and at Algiers a special interservice committee was formed to prepare for BARCLAY the first comprehensive signal deception plan. "Their contribution towards 'HUSKY' had the most far-reaching results," said Clarke after the war.

Less satisfactory were "A" Force's efforts to induce the Allied air forces to distribute their bombing and leaflet-dropping in a pattern supporting the BARCLAY "story." Especially in June, the pattern of air attacks clearly pointed towards Sicily. Though the BARCLAY "story" tried to deal with this by its claim that Sicily would be neutralized by air bombardment, substantial attacks on Corsica and Sardinia would nevertheless have been highly desirable. But Clarke's pleas for these were in vain. He decided that for the future any large-scale deception plan should have from the outset an appendix detailing supporting air operations, to which the air staffs and commanders would be committed.

One element of BARCLAY implementation was a haversack ruse codenamed MINCEMEAT. Writing when it seemed that BARCLAY would never be declassified, Ewen Montagu managed to give the impression, in a book (subsequently filmed) called *The Man Who Never Was,* that he was single-handedly responsible for the entire deception scheme in support of HUSKY, and that his MINCEMEAT by itself fooled the Germans into being totally surprised by the landings in Sicily. In truth, it was Flight Lieutenant Charles Cholmondeley* "who really triggered off the whole concept of MINCEMEAT," wrote Tar Robertson after the war; and Bevan recognized that Cholmondeley and Montagu were due at least equal credit and recommended both for decorations. And while highly effective, MINCEMEAT was but a single cog in the elaborate machinery of BARCLAY.

The germ of MINCEMEAT was the incident, mentioned in Chapter 7, when a British flying boat crashed into the sea off Cadiz a few weeks before TORCH, killing a passenger who was carrying a letter referring to D-Day for the North African landings, and nobody could be sure whether the Spanish authorities who recovered the body had shared this with the Germans.

Attached to Section B1A of MI5 in London was a twenty-five-year-old RAF officer, Flight Lieutenant Charles Cholmondeley, "a most extraordinary and delightful man," wrote Tar Robertson, "who worked in my section largely as an ideas man." It occurred to Cholmondeley that a similar "accident" might plant deceptive information on the Germans. He proposed to the Twenty Committee on November 5 a scheme originally labeled TROJAN HORSE, under which a dead body would be procured from one of the London hospitals ("normal peacetime price £10"), dressed in uniform with secret documents in an inside pocket, and dropped from an airplane into the sea where currents would carry it to enemy territory. "Whilst this courier cannot be guaranteed to get through," he pointed out, "if he does succeed, information in the form of the documents can be of a far more secret nature than it would be possible to introduce through any other normal B.1.A channel." Further investigations, he noted, would have to include medical advice, whether it would be apparent that the body had been dropped from an airplane, the location of a suitable drop site, the nature of the documents, the rank

---

* Pronounced, of course, "Chumley."

and service of the "courier," whether he might double for an actual officer, and the possibility that the Germans might pick up the plane that
made the drop. Cholmondeley was encouraged to proceed.

Cholmondeley brought Montagu into the project. Montagu entered
into the scheme with a will. Over a glass of sherry at the Junior Carlton
Club he consulted Sir Bernard Spilsbury, the most celebrated forensic
pathologist of the time. Spilsbury assured him that because people die in
air crashes for many reasons, it was not necessary to find a corpse that
had died from drowning. Further inquiries of Spilsbury's friend and colleague W. Bentley Purchase, the coroner for the St. Pancras district of
London, confirmed these views. Purchase undertook to look out for a
suitable corpse.

On January 26, 1943, a homeless and perhaps mentally ill derelict,
Glyndwr Michael by name, born out of wedlock thirty-four years before
to an illiterate woman in a Welsh mining community and a colliery
hauler, crawled into a warehouse in London and put an end to his sad life
with rat poison, of which he died in hospital two days later. Purchase
alerted Montagu that this body would do; and there were evidently no
relatives to raise questions. Purchase held a brief inquest without postmortem on February 4, with a verdict of death by suicide. He gave official notice that the body was being removed out of England for burial; in
fact he kept it under refrigeration at Hackney Mortuary for Cholmondeley and Montagu. It would keep for about three months, he told them.

Cholmondeley and Montagu presented the project, now codenamed
MINCEMEAT, for approval to the Twenty Committee at a meeting that
same February 4. Specifically, they proposed that the body, dressed in an
officer's uniform, together with fragments of aircraft wreckage, be
dropped from an airplane at some place where it would wash up on
Spanish territory. It would be carrying a seeming draft of the real plan for
the next operation after North Africa, and a seeming draft of the cover
plan for that operation. The supposed real plan would go into extensive
but ambiguous detail and specify a date one month sooner than the true
D-Day. "If the real target is to be, say, Sicily, the target in this document
should be, say, Sardinia." The supposed cover plan would give a date for
the false operation a week earlier or later than that in the supposed real
plan. "A different but credible target should be given say the Balearics

and Marseilles, and give Sicily as an alternative for consideration. . . . If Sicily is the real target and omitted from both the 'operation plan' and the 'cover plan' the Germans will almost certainly suspect. . . ."

It was pointed out that expert opinion was that the body would pass inspection. ("This is made the more probable as the Spaniards will not be likely to hand the body over to the enemy (but to a British Consul) and do not approve of post mortems.") Difficulties were that the body must be dropped within twenty-four hours of being removed from cold storage, so that the operation once begun could not be canceled or postponed; and that the body would be buried in Spain, and the military department whose uniform it wore would have to deal with inquiries about it.

The Twenty Committee approved the project and parceled out assignments: The Air Ministry representatives to lay on a plane, the Navy representatives to determine a good spot for the drop, Bevan to get the Planners' approval and have the fake documents drawn up, the Army representative to provide a name and identity papers; and so on.

The project was fine-tuned six days later at a gathering in Tar Robertson's office with John Marriott (soon to leave on his visit to India), Bevan, Cholmondeley, and Montagu. It was agreed that the papers should be carried in a separate container such as a briefcase, since "papers actually on the body would run a grave risk of never being found at all due to the Roman Catholic prejudice against tampering with corpses." But how to ensure that a briefcase remained with the body? Attaching it by a chain, as was done by bank messengers and the like, might endanger the whole operation, because this was not done by British officer couriers. Robertson undertook to experiment with various sorts of bags. (Eventually, however, a chain was used, hoping the Spanish and Germans would not realize that this was irregular.)

The Balearics, being neutral Spanish territory, were eliminated as notional objectives; the fake operational plan would be directed at Sardinia and the fake cover plan at Marseilles and Sicily. The meeting agreed, finally, that the body would be that of an army officer, in battle dress. Army officers, unlike those of the other services, did not carry their identity cards when traveling abroad; identity cards bore photographs, and it was proving impossible to take a photograph of the corpse of Glyndwr Michael that looked like anything but a corpse. Moreover, a regular uniform would be too difficult to fit upon the corpse; but Army officers traveled in battle dress. And, of course, the fake documents contemplated were appropriate for an Army officer to carry.

Back at Storey's Gate, Bevan discussed the scheme with the Planners. They felt that since strategy might change, the project should not be carried out more than two months before the real HUSKY. This ruled out the idea of fake plans. They proposed a letter instead, from an officer in London to his opposite number in Algiers; and, Bevan told Robertson, they "thought that the contents of such a letter should be of the nuts and bolts variety and not on a high level."

When Bevan discussed it with the Twenty Committee it was generally agreed that a sufficiently plausible letter from a very high officer would carry more weight with the Germans than would a letter on, say, a deputy director level. Montagu suggested a personal, off-the-record letter from General "Archie" Nye, the Vice Chief of the Imperial General Staff, to Alexander, referring to operations in Sardinia, implying a debate over whether Sicily or Marseilles should be the cover target, and including some personal chitchat. There had been some debate as to where to make the drop, but it was generally agreed to stick to the original idea of doing it off the coast of Spain. There was too much risk of a careful postmortem if it were to wash up in German-controlled territory.

There the project rested when on March 11 Bevan left for his consultations in Algiers.* In the course of his meetings there with Dudley Clarke the documents to be carried by the corpse were discussed. Clarke favored a low-level item, perhaps no more than a letter including a definite false indication of HUSKY D-Day. He sketched a draft of what he thought would be suitable, but undertook to keep an open mind.

Bevan got back to London on March 28 to find that Cholmondeley and Montagu had been busy. The assistant naval attaché in Madrid had come to England and been briefed; and he had recommended that the drop be made just off Huelva, where the Spanish police were known to be most intimate with the Germans. The Navy had decided that the drop should be made by submarine and had assigned the job to HMS *Seraph,* commanded by Lieutenant N. L. A. Jewell—an experienced hand in ir-

---

* It will be remembered that during this same period Montagu was circulating a paper that charged Bevan with incompetence, self-aggrandizement, and mendacity.

regular work, for it was he who had spirited General Giraud out of France and smuggled General Mark Clark into French North Africa before TORCH. *Seraph* was due to depart for the Mediterranean soon, so there was no time to be lost; and since HUSKY was now firm, the Planners were willing to let the operation go ahead now.

Sir Bernard Spilsbury had confirmed that the body should keep sufficiently fresh in an airtight container if as much oxygen as possible was removed. Cholmondeley had had a container made that would fit through the torpedo hatch of the submarine. Montagu had made some revisions in the proposed letter from Nye to Alexander, and had drafted a letter affording a reason why "Major Martin" should be carrying the Nye letter. For Glyndwr Michael, pathetic homeless derelict, had become "Major William Martin," Royal Marines. (The Army had been ruled out after all, because the finding of the body would be too widely reported under Army procedures. Of the other services, only the Marines traveled in battle dress. The Germans would find at least one "William Martin" among the Marine officers in the Navy List.)

Major Martin then received a full personality, in the form of items to be found on the body. There were cigarettes, keys, a watch, bus tickets, and the stubs of London theater tickets dated April 22, showing that he had left the capital after that date. The Major, it appeared, was engaged to a girl named Pam. He was carrying a snapshot of her in a bathing suit (it was actually Jean Gerard Leigh, an MI5 clerk). There were love letters from her ("Bill darling, do let me know as soon as you get time & can make some more plans, & dont *please* let them send you off into the blue the horrible way they do nowadays—now that weve found each other out of the whole world I dont think I could bear it"); they were composed by another female employee and written out by a third, one of the participants being Ian Fleming's secretary, the future Lady Ridsdale and by some accounts the real-life model for James Bond's Miss Moneypenny. There was correspondence with his solicitors about his will; a letter from his father to the solicitors about a marriage settlement; a properly Edwardian letter to Bill himself from his father ("Your cousin Priscilla has asked to be remembered to you. She has grown into a sensible girl though I cannot say that her work for the Land Army has done much to improve her looks"). There were a receipt from a Bond Street jeweler for an engagement ring, a dun from his banker about an overdraft, a receipt for some articles of clothing, a receipt from the Naval and Military Club for five nights' lodging, an admission card to a nightclub.

Most important, there was an identity card—Marine officers carried them—for Captain (Acting Major) William Martin, bearing a photograph that looked remarkably like the corpse (it was in reality a certain Major Ronnie Reed, whom Montagu had encountered by accident; he had a career in MI5 after the war), together with a recently expired pass to Commando headquarters. In fact, three identity cards with different photographs were prepared, so that the one that most resembled the body in its final condition could be put in place at the last minute; that was Reed's. There were a silver crucifix on a chain for his neck, a St. Christopher medal, and identity tags showing his religion as Roman Catholic; it was hoped that this evidence of his faith would help discourage the Spaniards from performing an autopsy.

In Major Martin's briefcase were three items. There were proofs for a new book on the Commandos, supposedly being sent by Mountbatten to Eisenhower with the request that he contribute an introduction; this bulky item was the excuse for his carrying a briefcase. There was an informal note from Mountbatten introducing Major Martin to Admiral Cunningham, the Allied naval commander at AFHQ. In a rather broad hint, it concluded: "Let me have him back, please, as soon as the assault is over. He might bring some sardines with him—they are 'on points' [i.e., rationed] here." And there was the letter from Nye to Alexander, delivery of which was the object of the whole exercise.

The Nye letter went through several drafts, coordinated with Dudley Clarke, and was rewritten in its final form by Nye himself. The original idea was to refer to an operation in Sardinia and suggest a debate over whether Sicily or Marseilles should be the cover target for it. A second draft referred instead to operations in the Peloponnese with a debate over whether Sardinia or Sicily should be the cover target. When the Chiefs of Staff looked at this draft in early April, they felt that it needed to be more personal in nature, that the codename Husky itself should appear in the letter, and that perhaps Nye might be asked to draft the letter himself. Bevan tried a new draft along these lines. Evidently after a conference with Nye, he tried again, implying that Husky was the codename for the

Peloponnese operation, that there was another unidentified operation codenamed Brimstone, and that Sicily would be the cover objective for Brimstone.

Nye then tried his own hand, and produced a masterly draft along the lines of Bevan's last effort; including personal chitchat about such matters as an appointment to a command in which Alexander had been interested, and the problems raised by an American suggestion that Purple Hearts and other American decorations be awarded to British troops serving in allied commands. Bevan forwarded this to the Chiefs of Staff on April 12. They approved it next day, with one change. As a personal item to lend verisimilitude, Montagu's original suggestion had included a reference to Montgomery's self-importance ("Is Alexander taking as big a size in hats as Montgomery yet?"). While this pleasantry had not survived all the drafts, a dig at Montgomery evidently appealed to Nye, who concluded his own letter: "But what is wrong with Monty? He hasn't issued an 'Order of the Day' for at least forty-eight hours!" The Chiefs humorlessly ordered this taken out. Nye signed the final version, dated April 23, on April 14. "Now I hope your friends will ensure delivery," he told Bevan.

The Chiefs also directed Pug Ismay to clear the plan with Churchill. Bevan's own account of how that was done cannot be improved upon:

> I was instructed by Lt Gen Ismay to see the Prime Minister at 10 a.m. on [15] April and explain Operation Mincemeat to him. To my surprise I was ushered into his bedroom in the Annex, where I found him in bed smoking a cigar. He was surrounded with papers and black and red Cabinet boxes.
>
> After explaining the scheme, in which he took much interest, I pointed out that there was of course a chance that the plan might miscarry and that we would be found out. Furthermore that the body might never get washed up or that if it did, the Spaniards might hand it over to the local British Authority without having taken the crucial papers. "In that case," the P.M. said, "we shall have to get the body back and give it another swim."
>
> He agreed [to] the plan but directed that permission for its execution must be obtained from Gen. Eisenhower. This was duly obtained.

In Algiers, Bedell Smith gave approval on behalf of Eisenhower the next day.

The day after that, April 17, Montagu, Cholmondeley, and "Jock" Horsfall, a former racing driver now with MI5, met Bentley Purchase at the mortuary, dressed and prepared the body, inserted it and the briefcase into its container packed in dry ice, and drove it to Greenock in Scotland. There the canister was transferred to *Seraph*. Only Lieutenant Jewell knew the "story"; the container was labeled "Handle With Care—Optical Instruments—For Special F.O.S. Shipment."

Seraph left the Holy Loch submarine base the evening of April 19. At 4:15 in the morning of April 30, she was close inshore off Huelva. The crew were sent below and only officers participated in the launching of Major Martin. Jewell told his officers that the operation was a test to check on reports that the Germans were getting at papers on bodies washed ashore. The container was hoisted through the torpedo hatch and opened, and the Major was lifted out. The lower part of his face was covered with mold and the stench was powerful ("The body was very high," noted Jewell drily in his report). They inflated his Mae West life jacket, and launched him towards the shore together with a capsized rubber dinghy such as were carried on Catalina flying boats. When *Seraph* was well out to sea Jewell had the container dumped overboard and sunk by machine-gun and revolver fire.

On May 1, the British naval attaché in Madrid was notified of the finding of the body. A funeral was held at Huelva on May 2. The Spanish Chief of Naval Staff returned Major Martin's effects, including the documents in their seemingly unopened envelopes, to the attaché on May 11. Major Martin's name was included in a casualty list published in *The Times* on June 4; by good luck, some prominent officers who had been publicly identified as having died in air crashes were included in the same list. Later that month the attaché was directed to place on the grave a wreath with a card "From Father and Pam," and to have erected a tombstone bearing the legend: "William Martin, born 29th March 1907, died 24th April 1943, beloved son of John Glyndwr Martin and the late Antonia Martin, of Cardiff, Wales. Dulce et decorum est pro patria mori. R.I.P." It is still there.

The attaché had an agent in the Spanish Navy or Ministry of Marine codenamed ANDROS, whom he asked to get the story of Major Martin's briefcase. On June 8, ANDROS provided a circumstantial report of the finding of the body, the performance of at least a modest postmortem examination which concluded that Major Martin had died by drowning and had been in the sea for 15 days, and the fact that the documents had been opened, copied, and resealed before being returned; with an account of the ultimately successful efforts by the Germans locally and in Madrid to get hold of them.

This last was old news. As early as May 12, ULTRA intercepts pointed unmistakably to the conclusion that the Germans had read the Nye letter and had bought the "story" hook, line, and sinker.

BARCLAY was broadly successful, even though the Axis recognized that limitations on the available shipping would not permit operations on the extravagant scale that the full BARCLAY "story" proposed. The Germans expected rather a limited attack on Crete or the Peloponnese or the Dodecanese, plus an attack in the west on either Sicily or Sardinia. The MINCEMEAT documents were swallowed whole when they arrived in May; first by the Abwehr, who sent them to OKW with detailed reports, and then by Fremde Heere West itself. Even the reference to "sardines" was pounced upon as showing Sardinia to be the target. "The circumstances of the discovery, together with the form and contents of the dispatches, are absolutely convincing proof of the reliability of the letters," Roenne told Jodl on May 11. The next day, Jodl directed all commands and headquarters concerned with the Mediterranean to strengthen their defenses as quickly as possible, with measures for Sardinia and the Peloponnese "to have priority over everything else," and warned the Commander-in-Chief, Southeast, that "an absolutely reliable source" indicated that the Allies in the east would land at the very points in the Peloponnese mentioned in the MINCEMEAT documents. At his regular conference on May 14, Hitler said that he disagreed with Mussolini's view that Sicily was the most likely invasion point, and that "the discovered Anglo-Saxon order"—which he identified as "the papers found on the body of a British courier washed up on the southern coast of Spain"—"confirms the assumption that the planned attacks will be directed mainly against Sardinia and the Peloponnese." Within a week Mussolini had received the MINCEMEAT documents and had come round to the same view. In late May, Jodl told the German military attaché in Rome: "You can forget about Sicily, we know it is Greece." *

Everything fed to the Axis under BARCLAY for the next two months reinforced the Germans' expectations. The very night before the HUSKY landings, Keitel dispatched a "Most Immediate" assessment of Allied intentions that forecast, in addition to a move against Sardinia, Sicily, and Corsica, a landing in Greece by the powerful forces that the Allies had supposedly transferred from North Africa to the eastern Mediterranean. He reckoned Allied forces in the Mediterranean at some thirty-eight divisions and seven armored brigades—double the actual number and a striking tribute to the effectiveness of CASCADE.

Even after HUSKY, the Germans, and Hitler in particular, remained focused on the Balkans. ULTRA revealed the Abwehr making elaborate plans for possible evacuation of Greece, including staybehind agents and saboteurs. Late in July, Hitler sent Rommel to Salonika to take personal charge if the Allies should strike in Greece or Crete. Rommel arrived at his new headquarters the morning of July 25; twelve hours later, with the fall of Mussolini, he was recalled to Hitler's Wolf's Lair headquarters to plan the takeover of Italy. Next day, Roenne issued a report suggesting that the Peloponnese operation had been dropped at least for the time being.

At HUSKY D-Day there were only two German divisions in Sicily in addition to the Italian forces there. This did not, however, necessarily reflect BARCLAY, but rather Hitler's suspicion that Italy might quit the war, and his unwillingness therefore to put German troops out on a limb. By then there was general recognition, especially by the Italian intelligence services, that Sicily was certain to be attacked. Indeed, throughout the run-up to HUSKY the Italian intelligence services performed with notable efficiency and accuracy, including spotting the carefully guarded movements of airborne forces to their assembly areas.

As the true HUSKY D-Day approached, DERRICK, the operational deception plan for the invasion itself, took priority over BARCLAY. The landings

---

* But Goebbels suspected a British plant when Canaris boasted to him about the recovery of the Nye letter revealing Sardinia as the Allied goal. It takes a liar to catch a liar.

were to be made on the southeast coast of Sicily, so the "object" of DER-RICK was "To contain enemy forces in the western portion of Sicily as long after D-Day as possible, and in any case until first light on D plus 3." Accordingly, a landing would be threatened in the Marsala area on the western tip of Sicily, seemingly to take place on July 12, two days after the real landings; it was hoped that the Axis would deem the real landings a feint and hold forces in the west to meet the notional main threat.

Strangeways planned DERRICK in detail. Beginning on July 9, there would be at Bizerte a display of dummy landing craft together with genuine activity by troops apparently preparing to embark. The followup convoys for the main landings would set out July 11 on a course for Marsala and would only change course after darkness fell; and that night, ships and aircraft would bombard the Marsala area (Operation FRAC-TURE), while the Beach Jumpers simulated landing preparations and dummy paratroopers and PINTAILS were dropped inland.

Part of the deception in advance of D-Day was a fake move to Oran of the headquarters planning HUSKY. It was less than successful, because the advance party was in on the secret and did not want to bother the people in Oran. Bromley-Davenport set up a wireless detachment that from June 27 to July 10 regularly communicated dummy traffic—identified by the codeword UNBRIDLE—with England, Algiers, and Cairo, as a major force headquarters would do. A related deception was LOBSTER, in which a decoy airfield was set up near the real air base at Djidjelli; it drew off a portion of a German air raid on June 18.

DERRICK and its associated operations were dogged by bad luck. The anchorage at Bizerte was not vacated in time by naval vessels; no more than nineteen of the planned one hundred dummy landing craft ever got into place, and these were blown about by the wind. The Beach Jumpers' first trial was unimpressive. The night of July 11, twelve self-destroying dummy paratroops and twenty PINTAILS were to be released from each of four C-47s, one each in charge of Strangeways, Goldbranson, Rushton, and Alexander's chief camouflage officer, Major F. G. Baxter (who had just returned from a trip to the United States to procure American dummy vehicles and weaponry). Just after takeoff a PINTAIL went off in Baxter's plane and it caught fire and crashed, killing all on board.

Less tragic, but perhaps more disappointing, was the last-minute cancellation of the FALSE ARMISTICE PROJECT. Suggested to Clarke by Bevan, who remembered how the Italian collapse at Caporetto in 1917 had been fueled by the rumor that an armistice had been signed, this project would have entailed dropping of quantities of a leaflet in Italian, bearing the royal arms and a facsimile of King Victor Emmanuel's signature, purporting to be a proclamation by the King ordering his brave soldiers to lay down their arms because Germany had abandoned Italy. Major Titterington's forgers produced a document pronounced by an Italian exile as "indistinguishable from the real thing," (To obtain an example of the royal signature, they ransacked the secondhand bookshops of Cairo and dug up a six-volume history of Italy in the First World War containing a facsimile of an autographed portrait of the King). A supply of leaflets was flown to Algiers six days before D-Day; but at the last minute London canceled the project, presumably because of its political sensitivity.

A more unaccountable cancellation befell an effort to bolster the key double agents in North Africa. When the invasion force put out they were certain to know about it, and if they did not report it they would surely be blown. Crichton and Thomas worked out a plan to enhance their status by passing word of the invasion just too late for it to make any difference. ARTHUR and LE PETIT would separately go to Oudjda and say that a large convoy had left Oran. NORBERT had already sent via RAM on June 30 a message that he thought the large concentration in Oran and Algiers was destined for the south of France. RAM would now send word that NORBERT was coming with an important new message and to stand by; the message would be that NORBERT had heard a French officer expressing anger that the invasion would be not of the south of France but of Sicily. And at 11:00 P.M. the night before D-Day, GILBERT, in his downright fashion, would send word flatly: "Most important. Have learned from reliable source that large force now on its way to Sicily. Invasion may be expected hourly."

This was all orchestrated with an elaborate timetable and submitted to AFHQ for approval. After unaccountable delay, Crichton received word two days before HUSKY D-Day that Eisenhower had turned it down. "I feel I must warn you," wrote Crichton urgently to Clarke, "that the whole framework of our Special Agents—never very strong—is now in considerable danger." The eventual outcome of this flap is not presently clear. (But the agents did survive.)

Despite these misfortunes, DERRICK and its associated operations were successful. Even three days after D-Day, Italian commanders in the western part of the island believed the main landing would be in their area, and the 15th Panzer Division, one of the only two German divisions in Sicily, remained in the west for the first two days of the invasion. And a captured situation report of the Italian Sixth Army, plus interrogation of the Italian commander in charge of the southeastern corner of Sicily, showed that BARCLAY had successfully persuaded the defenders that the Allies would attack only on a moonless night.

A fortnight after HUSKY D-Day, Mussolini was toppled, and the new Badoglio government began groping towards getting Italy out of the war. The London Controlling Section and "A" Force played modest roles in the Italian surrender.

The LCS's role was merely an abortive suggestion proposed in February by Bevan that appropriate circles in Italy be advised through secret channels that an honorable peace for Italy was possible notwithstanding the "unconditional surrender" announcement at Casablanca. He pointed out that even if the matter leaked to the Germans it would be admirable cover for HUSKY. But this was vetoed by the Foreign Office.

"A" Force's role was more substantial. After overthrowing Mussolini, the Badoglio government was making clumsy and disjointed efforts to conclude an armistice. Since the LLAMA channel in Libya was a direct line to the Italian military, using a cipher presumably unknown to the Germans, and had lost most of its value with HUSKY under way, Clarke offered to "Jumbo" Wilson to make it available to communicate with the enemy. On the morning of August 5 LLAMA dispatched a message to the Italians that their agent had surrendered, that this was the British General Staff calling, and that the opportunity was being offered to use this channel for secure communications. The Italians responded that they did not trust the cipher being used and asked how they could get a secure one to the Allies. The British replied that an aircraft to Benghazi would be given safe conduct, and Wilson alerted Eisenhower. But London vetoed use of the channel. It was kept open for possible emergency use—a few days later it seemed briefly that it might be needed after all—but was finally shut down on August 23. LLAMA himself was released with a seventy-pound bonus and a job in the civil affairs department of Tripolitania.

Even before the Sicilian campaign was over, Eisenhower had begun planning Operation AVALANCHE, a landing by the American Fifth Army (including the British X Corps) at Salerno to take Naples, and Operation BAYTOWN, a crossing to the toe of Italy by Montgomery's Eighth Army. Clarke prepared a continuation of BARCLAY to cover these moves, codenamed BOARDMAN. Its object was to weaken enemy forces in southern and central Italy, and contain maximum enemy forces in Greece. Its original "story" was that Sardinia and Corsica would be invaded on September 5, the heel of Italy on September 10, either southern France or the Genoa-Leghorn area of Italy from Corsica thereafter, and the Peloponnese at the end of September. Implementation was largely a continuation of BARCLAY, with some refurbishment of the WATERFALL dummy displays in Cyrenaica. The MOSELLE party in Sardinia was useful. And a remarkable new channel opened up, codenamed GUINEA.

GUINEA was James Ponsonby, a forty-two-year-old Englishman for some years resident in Tangier; commercial attaché to the British Consul-General, he was well known in that tightly-knit international community. He had joined SOE in July 1941 and had done good work for it. It was common knowledge in Tangier that Jim Ponsonby was a generous soul, no good at keeping his finances straight, and deeply in debt. In July 1943 the MI6 man in Tangier came up with a plan for putting this fact to use to establish Ponsonby as a channel for "A" Force.

Ponsonby and his friend the British military attaché began conspicuously frequenting the bars and restaurants of Tangier, seemingly drinking much too much. This went on for some weeks, until it was reasonable to assume that the Germans in Tangier had taken notice. Then Ponsonby called upon one Goeritz, an official in the German consulate. He needed money badly, he said, and to get it he was prepared to sell information to the Germans. Goeritz, knowing of Ponsonby's financial situation and his intimacy with the military attaché, and believing that he had access to high-level information, undertook to pay generously for really high-grade material.

On September 4, Goeritz told Ponsonby that he would be well paid for information as to what the Allies would do next. AVALANCHE was scheduled to take place five days later, but of course Ponsonby did not know that. He said he would meet Goeritz again in a day or two. "A" Force puzzled over what information to convey to the Germans on this last-minute basis, and finally directed Ponsonby to say that the Allies had two plans, depending on whether Italy should surrender soon enough. One was to land in Sardinia and Corsica on September 12 and thence descend upon the Leghorn coast; the other he had not yet learned. Moreover, the Peloponnese operation had just been called off at the QUADRANT conference in Quebec.

The night of September 7–8, Ponsonby drove into the hinterland southeast of Tangier, met Goeritz in the small hours after midnight, told him his news, and received 2.5 million francs. (Neither left his car; they spoke and passed the money through their respective open windows; each had an armed guard hidden in his trunk, unknown to the other.) Next morning the Abwehr radioed this information to Berlin, adding that the possibility that this was a deception was "very slight." That afternoon the Italian surrender was announced and next morning the Allies landed at Salerno. This Ponsonby explained at his next meeting with Goeritz on the ground that the Sardinia force had been diverted at the last minute.

Ponsonby met a number of times with Goeritz thereafter, and with his successor, one Petersen. Incredibly, they continued to regard him as a valuable source. They even sent him to Lisbon to meet with the chief of German counterespionage there. The British appointed him (notionally) "Coordinator of the War Effort" at Tangier, which enabled him to talk (notionally) with such personalities as the King of Greece and the Governor of Gibraltar, from whom he of course (notionally) obtained important information. Notwithstanding ill-health, he continued in this dangerous role till the spring of 1944, at which time, with German sabotage agents appearing in Tangier, it seemed prudent to withdraw him to England. He was awarded the MBE for his service.

After DERRICK, Strangeways reorganized Tac HQ somewhat; Fillingham went to the escape and evasion side, replaced by a Captain P. W. Laycock; Temple went to England, and was temporarily replaced by Major Harry Gummer. Gummer was an old acquaintance of Clarke's, a fellow artilleryman with but one eye, rather hostile to the Americans at "A" Force.

They crossed to Sicily at the beginning of August and set up shop in the olive groves south of Syracuse. Strangeways's first assignment was the trivial Plan FLATON, involving little more than belittling the effectiveness of Allied intelligence through special means channels in order to protect certain intelligence sources that had been useful in Sicily. Deceptions for the landings on the Italian mainland had more substance. For Salerno and Montgomery's crossing there was BOOTHBY, whose "story" was that Montgomery would land on the "ball of the foot" near Crotone, supplementing the BOARDMAN threat to the "heel." Implementation included W/T deception; naval demonstrations; dropping of PINTAILS, dummy parachutists, and carrier pigeons bearing messages purportedly from Allied agents in the "heel," and an all-out effort by the special means channels. GILBERT reported that Montgomery's staff was particularly interested in the Crotone area on the "instep," and that a tipsy British officer in a Syracuse bar had told LE DUC that his unit would be operating in the Gulf of Taranto. Montgomery's crossing was in fact essentially unopposed; the Germans had no intention of offering resistance that far south. Soon afterwards, on September 9, there really was a hastily-mounted landing at Taranto (Operation SLAPSTICK).

Closer to the Salerno beachhead, a special task group under Captain Charley Andrews, accompanied by Douglas Fairbanks, and including a Beach Jumpers unit, simulated a diversionary landing on September 9 with little result; more successfully, on September 17 they accepted the surrender of Capri, where Andrews and Fairbanks ensconced themselves in the villa of Countess Ciano, Mussolini's daughter.

Strangeways, with Tac HQ and an improvised dummy tank company, reached the Salerno beachhead seven days after the landing. Bromley-Davenport moved over from Algeria to join the forces at Taranto. Gummer stayed behind at base headquarters in Sicily. Strangeways first implemented Plan CARNEGIE, exaggerating Allied strength in the beachheads by wireless deception and display of insignia of absent and imaginary units. Almost simultaneously, at Alexander's request a notional threat (Plan COLFAX) was mounted against the Gulf of Gaeta, north of Naples, to induce two German divisions to remain in that neighborhood. Conducted almost entirely through GILBERT and the other North

African double agents, COLFAX came too late, for the divisions were already on the move for Salerno. CARNEGIE too was overtaken by events, and for the moment Strangeways had to be content with simulating tank activity on the right flank of the Salerno beachhead.

An extraordinary self-deception—in which the Allied deceivers had no role—seized the Germans at Salerno the next few days, when they became convinced that the Allies were about to give up and evacuate the beachhead. Yet they did not attack, fearing among other things that that would give the Allies a chance to land further north—"a prospect that would remain an obsession of the German commanders," in the words of a leading history, and one upon which Allied deception plans would play again and again over the rest of the Italian campaign.

Notwithstanding mismanagement by the overall commander, the American General Mark Clark, the Salerno beachhead was held, and eventually Clark and Montgomery linked up while the Germans made a fighting withdrawal to the Gustav Line, an extremely strong defensive position running across the peninsula north of Naples. Alexander moved his army group headquarters to Bari on the Adriatic at the "ankle," and Strangeways set up shop at San Spirito, a suburb of Bari, on October 2. Gummer returned to Algiers and Bromley-Davenport took his place with Tac HQ.

There followed a series of short-term deception plans in aid of the indecisive fighting of the next two months until winter shut down operations. GARFIELD was a W/T deception to help Fifth Army's crossing of the Volturno in mid-October; its "story" was that the British 7th Armored "Desert Rats" Division was being withdrawn for an amphibious assault behind the German right flank. HARDMAN supported Montgomery's crossing of the Sangro on the Adriatic side in November. Its "story" was that there would be amphibious and airborne landings by the British 1st Airborne Division at Pescara behind the German left; it was implemented by conspicuous preparations for such a move, including W/T deception, together with messages from the North African double agents. BROADSTONE was a brief repeat of HARDMAN on the western coast, threatening a landing by an American regimental combat team in the German rear to support preparations for the first unsuccessful approach towards the Gustav Line. It was followed by CHESTERFIELD, to support the attack that finally resulted in German withdrawal to the main Gustav Line. Threatening a landing at Gaeta by a brigade of the

Desert Rats, CHESTERFIELD was implemented by a comprehensive report to the Abwehr by GILBERT, plus W/T activity and visible preparations for embarking a force from the Bay of Naples. None of these could point to any immediate German reactions, but CHESTERFIELD aided a long-lived German concern over possible Allied seaborne attacks behind the western anchor of the Gustav Line around Gaeta.

On the strategic level, BOARDMAN was followed in late September by FAIRLANDS, with the objects of discouraging German reinforcement of the Gustav Line, inducing them to evacuate Crete and Rhodes, and containing as many Germans as possible in the Balkans. It continued the threat of landing behind the German lines in the Leghorn-Pisa area, this time by the American Seventh Army under Patton—now little more than a headquarters, but the Germans did not know that—staging out of Corsica (recently evacuated by the Germans). Implementation included a conspicuous visit by Patton and his staff to Corsica, and wide distribution of a pamphlet about the ancient buildings and works of art in the Leghorn-Pisa area.

An ill-fated British expedition seriously affected the second object. In mid-September, the British seized Cos, Samos, and Leros, intending to proceed to Rhodes. FAIRLANDS was accordingly modified to replace threats to Crete and Rhodes with threats to Corfu, Cephalonia, and Zante. But the Germans counterattacked and by the end of November had retaken the islands the British had seized, and had occupied the Cyclades to boot; controlling thereby the whole Aegean, and sharply reducing any chance that Turkey might join the Allies.* This undercut the whole logic of threats in the eastern Mediterranean. Where, the Germans might well ask, was the Middle Eastern host that was supposedly poised to invade Greece and march into the Balkans? "A" Force prudently pulled in its horns in that region for the time being.

FAIRLANDS merged in November into OAKFIELD, whose "story" continued the Leghorn-Pisa threat and enlarged it to involve as well a land-

* There had been some hope of this, and British antiaircraft units and RAF ground personnel had been concentrated in Syria on the Turkish border ready to move into Turkey on short notice. A short-lived Plan WHITWOOD, implemented by W/T and special means, explained the concentration to the Germans as being for possible use in the event of unrest in Aleppo—which was in fact a genuine problem.

ing around the head of the Adriatic, the two invasions forming a pincer to cut the Germans completely off in the rear; the Adriatic landing was notionally to be conducted by Middle East forces, thus justifying their failure to respond to the German counterattack in the Aegean. OAK-FIELD came to be particularly closely related to the Anzio landing in January 1944, and we will defer its tale until a later chapter.

There were developments in the GILBERT case after the Sicilian campaign.

In late August, LE DUC left Tunis for a sojourn in Sicily as an employee of the commission for repatriating Italians (first notionally, later in fact). From Sicily he notionally sent information by letters in secret ink to GILBERT in Tunis. Soon after LE DUC's return, in late October, the Germans informed GILBERT that an addition to his team would be parachuted in. This was one Charmain, alias Caron, codenamed LE MULET ("The Mule"). He was a member of the PPF and might know ALBERT; so ALBERT was disposed of by having the French navy order him to Casablanca and eventually to Dakar, where he ended in the brig and played no further role in the GILBERT saga.

The German plane bringing LE MULET missed the signal light and dropped him a long way from Tunis; but two days later, on October 25, he showed up at GILBERT's flat; he had hidden in a ditch for twenty-four hours and hitchhiked to Tunis. LE MULET was a mechanic in ordinary life, a true believer in the PPF, convinced that collaboration with Germany was best for France, who had left a wife and children behind in Paris. He brought GILBERT two hundred thousand francs,* a new cipher, confirmation of how highly the Abwehr regarded him—according to LE

---

* Malcolm Muggeridge noted that it gave Paillole and his people particular joy to milk the Abwehr for funds—which they tried always to have dispatched in gold rather than paper, GILBERT telling the Germans that banknotes were dangerously traceable. Once, according to Muggeridge, they were short of automobile tires, and GILBERT induced the Abwehr to parachute in a set of Dunlops. This led Muggeridge to suggest that GILBERT tell his control that he was so gravely troubled by lack of feminine consolation that he could not concentrate on his work, yet "hesitated to avail himself of local facilities for fear of giving himself away." Muggeridge envisioned "Rhine maidens floating down from the sky, but our French colleagues were more skeptical. The most that could be expected by way of response, they insisted, would be a supply of bromides"—bromides having, presumably, the qualities that Americans more often associate with saltpeter.

MULET, the Germans refrained from bombing Tunis lest the power plant be destroyed and GILBERT's transmissions thereby silenced—and a message from Beugras back in Paris that he was unhappy that GILBERT was not sending more political information. LE MULET himself was supposed to set up a PPF network. He was to contact various unsavory characters, all of whom were gently eased out of the way.

He was allowed to operate under GILBERT's eye, without knowing what the team was really up to. He had obviously kept some money for himself; much of his time was spent relaxing with ladies of the night. Paillole was concerned lest "between two sessions of debauchery he might perceive what our game was and compromise its success." He resolved to dispose of LE MULET at an appropriate opportunity; as will be seen in Chapter 14.

An odd special aspect of this great year in "A" Force's history was a group of naval deceptions.

In December of 1942, the British Admiral Somerville, commander of the Royal Navy's Eastern Fleet, operating in the Indian Ocean, had visited Cairo and had proposed three deceptions to Clarke, codenamed WORKHOUSE, BIJOU, and WYANDOTTE.

The object of WORKHOUSE was to protect the Persian Gulf against submarines by leading the enemy to believe that the Straits of Hormuz were mined. The original idea was to sow dummy mines, and lend a touch of verisimilitude by actually blowing up some expendable craft. But it transpired that the shore was so thinly populated that nobody would be likely to notice, and the local fishermen would probably freely sail through the notional minefield. The submarine threat faded, and nothing much ever came of WORKHOUSE. Later in 1943, responsibility for the area was switched to Peter Fleming. On one isolated occasion, word that the Straits were mined was passed through one of the channels in Turkey at Fleming's request.

BIJOU was a naval order of battle deception emanating from the London Controlling Section. Its object was to make the Japanese believe that the aircraft carrier HMS *Indefatigable* had joined the Eastern Fleet. *Indefatigable* was in reality still under construction and it would be a year before she was completed, but ULTRA indicated that the Japanese thought she was in commission. Her leaving the Clyde was simulated by Admiralty W/T a few days before Christmas 1942, while one of B1A's channels told the Germans that she was off for the Indian Ocean, stopping at Cape Town on the way. Wireless deception tracked her to Cape Town; there "A" Force's local contacts put out stories that she was refueling at the Simonstown naval base. Noel Wild on a visit to East Africa in late December arranged for Malcolm Muggeridge at Lourenço Marques to spread appropriate misinformation. A few weeks later, QUICKSILVER reported to the Germans in Athens that one of his sources had overheard "Sub-Lieutenant Ravenhead" in Beirut talk about having gone out to Cape Town in *Indefatigable.*

East of Cape Town, Fleming took over, and for months thereafter the imaginary *Indefatigable* sailed around the Indian Ocean like the Flying Dutchman. In July 1943, "A" Force rejoined the charade at Fleming's request by sending to an imaginary naval officer in a German POW camp a letter from a brother officer, written on the crested stationery of *Indefatigable,* referring to his arrival in the Indian Ocean early in the year; and again contributed some leakage at Lourenço Marques in August. At the end of the year the ghost ship returned to the Clyde for a refit; and when she put out again she was the genuine *Indefatigable.* ULTRA confirmed that both the Japanese and the Germans believed throughout 1943 that there were two British carriers in the Indian Ocean, when in reality there was never more than one and more often none.

WYANDOTTE was a project to put out the story that the British were operating Q-ships against Japanese submarines in the Indian Ocean, which it was realized on reflection would tend simply to make the submarines more ruthless.

On New Year's Day of 1943, "A" Force welcomed its first naval recruit, Lieutenant-Commander Alec Finter, formerly of the intelligence staff of the Mediterranean Fleet. Lady Jane Pleydell-Bouverie, who knew Finter after he was transferred to London, remembered him as "charming, but rather sad in a way. . . . He was a small part character actor, and I imagine that is a hard life. He was a bachelor, friendly and nice." Rex Hamer remembered him as temperamental. In his time, "A" Force ran six naval deceptions, called BROADARROW, PLAYBOY (renamed COMPASS), FERNBANK, AXTELL, ERASMUS, and BARDSTOWN.

BROADARROW was another naval order of battle deception, with the object of adding five notional ships to the antisubmarine force in the eastern Mediterranean. It was implemented entirely through the reports of QUICKSILVER, passing on information from GALA and RIO. It ran for a short time in early 1943, after which the antisubmarine force received enough genuine reinforcements to make it unnecessary.

FERNBANK ran for some months from October 1943. It was designed to induce U-boats to stay above three hundred feet depth by persuading the Germans that the Allies had a secret rocket-propelled depth charge called the "M. F. R." that could not be used at depths of less than three hundred feet because it would endanger the attacking ship. The "story" was fed piecemeal through the autumn by items from RIO sent by QUICKSILVER. It was handed off at the end of the year to London, where TRICYCLE reported that METEOR, one of his sources from the Yugoslav Navy, had heard about it in an antisubmarine warfare course; and again in February Commander Wolfson in Istanbul arranged for a British instructor in an antisubmarine course for Turkish officers to drop a hint about "M. F. R." to a pro-German students. Whether FERNBANK had any effect on U-boat tactics was never learned.

Finter drew up in January a plan, BRUNETTE, that was turned down in Algiers. Its object was to induce the German Navy to persuade the German Army not to reinforce Tunisia because loss of Axis shipping would soon reach unreasonable levels. Four methods were suggested. One was a BIJOU-style notional running of eight submarines to Malta, where they would be simulated by dummies. Another was to suggest by special means, and by displaying dummy torpedoes at RAF bases, that several genuine RAF squadrons were to be converted to torpedo planes. A third was to use BIJOU or BROADARROW methods to add a notional fast minelayer to the British fleet off North Africa. A fourth was to bring out a squadron of notional destroyers from England.

All these ideas but the first were shot down by the naval staff in Algiers: There really would be RAF torpedo planes; the Germans might react by trying to sink the only genuine fast minelayer; there were no dummy destroyers available. The first idea led to a comical incident in which a collapsible dummy submarine (christened SLY-BOB) was constructed at Colonel Jones's workshop. During trials on the Gulf of Suez she was sighted by a passing destroyer, who called in airplanes and a destroyer with depth charges; SLY-BOB came within an ace of being attacked and sunk. This was not a bad verification of her plausibility; but no more were constructed and BRUNETTE never got off the ground.

PLAYBOY, renamed COMPASS, was not really a deception plan but merely a scheme to lure a U-boat, by messages from QUICKSILVER or THE PESSIMISTS, to the Lebanese coast to meet a fishing boat supposedly carrying a German agent, and sink her by means of limpet mines covertly attached to her hull from the fishing boat. After a good deal of experimentation presided over by Rex Hamer, the idea was given up.

The "A" Force station in Nairobi, consisting of Major A. G. Johnson, presided over three deception plans in 1943, all naval in origin. The first, ERASMUS, in the first part of the year, began as a naval deception designed to divert the attention of enemy submarines from two convoys carrying brigades of the 11th East African Division from Mombasa to Ceylon, by feeding the "story" that they were destined for the Middle East. The Navy decided that the submarine threat was not serious enough to warrant the effort, and ERASMUS ended as a conventional order of battle deception for Peter Fleming, designed to add a notional East African division to India Command. The second, AXTELL, was an order of battle plan prepared in the late summer, designed to make the antisubmarine forces available to protect shipping in the Mozambique Channel seem to be greater than they really were, and differently disposed; it was overtaken by events, since with the opening of the Mediterranean, the U-boats moved on to more profitable hunting grounds off Aden. Finally, the original ERASMUS "story" was implemented (Operation BARDSTOWN) in late 1943 and early 1944 for another move of East African troops to Ceylon, for by then the Japanese had made available to the Germans a base at Penang in Malaya from which the U-boat threat in the Indian Ocean was serious.

After BARDSTOWN, "Rear HQ" in Nairobi had served its purpose for "A" Force; it was technically moved to Cairo, while Johnson joined Peter Fleming.

In November 1943, Finter was transferred to the London Controlling Section—leaving in something of a huff, according to Rex Hamer—and in exchange James Arbuthnott came out to "A" Force. Arbuthnott—who impressed the old hands at "A" Force as charming, experienced, and wise—remained until April 1944, when he returned to London. Finter did not return; he joined the planning staff for OVERLORD.

Meanwhile, as early as September, Clarke was giving thought to the coming year. OVERLORD, the great invasion of Europe, would clearly be mounted in 1944, and for operations on the Continent he recommended an organization structured along the lines of his own, coordinating its activities with "A" Force; moreover, operations in Italy would need greater special means support through channels controlled either from London or by the new entity. Then at the Tehran summit conference with Stalin in November, the Western Allies committed to mount OVERLORD in the spring. At Anglo-American talks in Cairo immediately thereafter—which both Bevan and Dudley Clarke attended—Roosevelt named Eisenhower as supreme Allied commander for OVERLORD. Montgomery in turn was designated as the ground commander for NEPTUNE, the cross-channel assault and initial operations in Normandy.

Himself a great believer in deception, Montgomery had seen Strangeways at work from the moment Tac HQ was activated, and wanted him as the deception officer for his new army group in Northwest Europe. Eisenhower too was a believer, and in line with Clarke's September recommendations he confirmed that there would be a deception section in his new SHAEF (Supreme Headquarters, Allied Expeditionary Force). Clarke offered Noel Wild to head it up.

At the top level, "Jumbo" Wilson was moved from the Middle East command in Cairo to the AFHQ command in Algiers, and the American General Jacob Devers, who had commanded ETOUSA in London, went to Algiers as Wilson's deputy and commander of the North African Theater, United States Army (NATOUSA). "A" Force was reorganized as of December 19. "Main HQ" was split, between a modest and highly mobile "Main HQ" consisting in effect of Clarke himself, and an "Advanced HQ (East)" in Cairo for Middle Eastern deception; the Algiers operation became "Advanced HQ (West)." Tac HQ officially became "Tac HQ (West)"; no "Tac HQ (East)" was ever activated. At the end of October, responsibility for deception activity in Iraq, Iran, and Africa south of Egypt had been reallocated to Fleming's command, but Nairobi continued until April 1944 as "Rear HQ."

Michael Crichton went back to Cairo to take over Advanced HQ

(East) from Wild. George Train succeeded him as head of Advanced HQ (West). Sam Hood went from Cairo to Bari to take over Tac HQ (West) from Strangeways. Max Niven's Malta outstation had been closed down after the Sicilian campaign was over, and after a brief tour in Cairo Niven replaced Bromley-Davenport under Hood, moving shortly thereafter to Naples to open an outstation there. Harry Gummer opened a new outstation and 42 Committee at Gibraltar, looking towards more extensive use of Spanish and Portuguese channels in 1944. In Lisbon, Henry and Alice Hopkinson had at Bevan's request built up a small deception network in the Portuguese capital; in the spring of 1944 this was integrated with "A" Force by the establishment of a 60 Committee working with Gummer at Gibraltar.

The year closed on an especially high note for Dudley Clarke: In December he was promoted to brigadier. Unlike American brigadier generals, a brigadier in the British Army was not a general officer. Clarke was no more troubled by this than by the regulations governing wearing of his Royal Flying Corps wings. When he was in Algiers he managed to wangle himself a chauffeured staff car flying a one-star flag, like an American brigadier general.

So ended for "A" Force the great year 1943. It cannot be too often repeated that the foundation for almost everything was Clarke's great accomplishment, the systematic order of battle deception that through 1943 continued as CASCADE. CASCADE would continue, under the name WANTAGE; and more great things for "A" Force were to come in the following year. But with the final commitment to OVERLORD the main focus of the war would shift at last from what Churchill had called the "soft underbelly" of the Mediterranean to the hard carapace of northwestern Europe. And on that decisive front would be played out the most significant of all deception efforts in the Second World War, FORTITUDE SOUTH, the masterpiece of David Strangeways.

Before taking up that story, however, we will return to the travails of Peter Fleming and Ormonde Hunter in the Indian subcontinent.

CHAPTER 10

Hustling the East (II)

After the Washington conference, the frustrations of Peter Fleming and his British colleagues, and of Ormonde Hunter for the Americans, continued through 1943 and into 1944. In India-Burma and China there was no clearly defined Allied strategy, and hence there were few real plans for deception planning to support. Additionally, Hunter misapprehended his own position and the command structure of which he was a part. His odd initial reception by General Ferris should have tipped him off at the outset that his position was unsure. But Hunter did not take hints readily.

A summary of the strategic backing and filling in the theater in 1943–44 is essential to put the frustrations of Fleming and Hunter into context.

The British, the Americans, and Generalissimo Chiang Kai-shek each had their own agenda. Until early 1944, American grand strategy was to aid Chiang to raise and train a modern army and use it to drive to the sea in south China and advance thence northwards. Air bases would then be established in northern China; from them a fleet of huge long-range B-29 bombers would pound Japan into submission, or at least soften her up for invasion. The role of the American China-Burma-India theater, and of Stilwell as its commander and chief of staff to Chiang, was to build the Chinese Army that would accomplish this task.

American concern with Burma and India was related solely to this purpose. As long as the Japanese occupied Burma, only a trickle of supplies could be delivered to China, by air over the "Hump," the range of high mountains, extensions of the Himalayas, that separate Burma from China. The Americans were interested in liberating Burma only so as to open an overland supply line to China. They were distinctly *not* in the region to aid the British to recover their imperial possessions.

It took Washington—and specifically Roosevelt, who clung to unrealistic views about China—a long time to realize that Chiang Kai-shek had no enthusiasm for creating a modern force that might provide a base for some rival; nor for seeing his army expended against the Japanese and diverted from keeping watch on the Communists and rival warlords. He would be delighted to have the Americans take care of the Japanese for him. General Chennault of the American Fourteenth Air Force in effect undertook to do that. Like most early airmen, he was sure that airpower could do anything. He maintained that, given a relative handful of fighters and bombers plus the lion's share of tonnage over the Hump, he could first destroy the Japanese air force in China and then bomb the home islands into submission. Since this would call for little effort on Chiang's part, the Generalissimo liked it, and therefore liked Chennault, very much. By the same token, he disliked Stilwell, who not only despised him personally but insisted on trying to tell him things he did not want to hear. He disregarded Stilwell's warning that the Japanese would react by simply overrunning Chennault's bases, and that he needed a modern army to keep them from doing so.

As for the British, Wavell and others in India wanted to retake Burma. Their chiefs in London, however, preferred to soft-pedal Burma and direct British strength towards retaking Singapore, and thence returning British power to the Pacific. They did not share Roosevelt's enthusiasm for Chiang or for China; for them, there was little appeal in expending blood and treasure to reconquer Burmese territory in order to support China.

At the TRIDENT conference in Washington in May 1943, Wavell's proposed full-dress Burma campaign, ANAKIM, was dropped, as noted earlier; it was replaced by a less ambitious effort initially codenamed SAUCY. This was essentially reaffirmed at the QUADRANT conference in Quebec in August. At TRIDENT it was also agreed that there would be established in Southeast Asia under a British supreme commander a new combined Allied theater, Southeast Asia Command ("SEAC"), covering Burma (but not India), Ceylon, Sumatra, and Malaya (plus, later and informally, Thailand* and French Indochina). At QUADRANT, Admiral Lord Louis Mountbatten, Douglas Fairbanks's friend, heretofore chief of the

---

* It is often forgotten that Thailand was occupied by the Japanese and indeed its puppet government was technically at war with the Allies.

British Commandos, was approved as Supreme Allied Commander Southeast Asia ("SACSEA"). He formally opened his headquarters in New Delhi on November 1. Wedemeyer moved from OPD in Washington to become his deputy chief of staff. The American CBI Theater continued independent of both Mountbatten and Chiang Kai-shek; to Stilwell's role as Commanding General of CBI and chief of staff to Chiang was added that of Deputy SACSEA.

There had been changes in both the American and the British high commands even before SEAC was opened. In June, Churchill kicked Wavell upstairs to become Viceroy of India. (Fleming's close personal relationship with him continued.) Auchinleck succeeded Wavell as Commander-in-Chief India, but this did not entail operational responsibility for the war against Japan. On the American side, General Bissell, with whom Chennault had feuded and who had thereby incurred Chiang's displeasure, was recalled to Washington in July; serving first as "Hap" Arnold's chief of air intelligence, he would succeed George V. Strong as G-2 of the Army in January 1944. Arnold's chief of staff, Major General George E. Stratemeyer, went out to India in August in the new position of Commanding General, Army Air Forces, India-Burma Sector, with Bissell's old Tenth Air Force and training and service commands under him. In December, he became additionally commander of SEAC's "Eastern Air Force," which included Tenth Air Force and the RAF Bengal Command.

The opening of SEAC led to a new round of operational planning. The old SAUCY was dropped and succeeded by a confusing variety of plans (with a multiplicity of codenames), none of which was approved. But by the next summit conference—SEXTANT, held in Cairo in November and December—American strategists were beginning to see the possibility of basing the B-29s in the Marianas, rendering operations in Southeast Asia and China essentially diversionary.

The net result with respect to strategic deception was that there was never any overall theater deception plan because the deceivers could not catch up with the planners. The only operations in Burma for the 1944 campaigning season, beginning in November 1943, were an overland advance on the Arakan coast, an advance in the north by Chinese divisions under Stilwell supported by penetrations behind the Japanese lines by two long-range special forces, the British "Chindits" under the eccentric Orde Wingate (a cousin of Sir Ronald of the London Controlling Section) and the American "Merrill's Marauders"; and an advance to the Chindwin River on the main front in Assam.

The North Burma and Arakan campaigns began in November 1943. In February 1944 the Japanese counterattacked on the Arakan front and were turned back. In March they opened a major offensive against the Imphal-Kohima area in Assam on the central front, which the British under General William Slim fought to a standstill by June. In the north, after a few weeks of stalemate, Stilwell took personal command and the Japanese were outfought and pushed to the south; by August 1944 Stilwell had taken the crucial North Burmese town of Myitkyina.

For a time in late 1943 and the first part of 1944, B-29s, formed into XX Bomber Command, were to be based in India, staged through China, and used in a bombing campaign against the Japanese steel industry. Construction in anticipation of the big bombers' arrival in India began in November 1943; the first raids on Japan from China were carried out in June 1944. But the role of China would steadily fade after the conquest of the Marianas in mid-1944. The B-29s of XX Bomber Command eventually joined their fellows there.

Finally, all this took place against a backdrop of insecurity in India. The summer and autumn of 1943 saw one of the most hideous famines of modern times in Bengal; more than a million people died. Taking advantage of the hard-pressed position of the British, Mahatma Gandhi's "Quit India" campaign produced widespread insurrection. It seemed a time of opportunity for the likes of Subhas Chandra Bose, and the Axis was not slow to take advantage of it.

At the deception gathering in Washington in May 1943, with ANAKIM reduced to SAUCY, the deceivers went back to the drawing board. On May 28, Cawthorn, Wingate, Fleming, Bratby, and Hunter caucused and sketched an "Outline Cover Plan for Operation SAUCY." Its objects were to cause the Japanese to underestimate the tonnage over the Hump and the American air buildup in China until September 1, and to overestimate it thereafter; and generally to induce them to focus on defending Sumatra and Singapore rather than Burma. The method suggested was a generalized repetition of Fleming's KINKAJOU-WALLABY: Encourage publicity stressing the need to reopen the Burma Road, while informing the Japanese by special means "that this public clamor is an elaborate bluff" to cover a decision to direct the major thrust against Northern Sumatra. The plan would be "materially assisted" if MacArthur could threaten the Netherlands East Indies; "India can, if necessary, transfer two dummy divisions to S.W.P.A. to assist in this feint."

Cawthorn got back to New Delhi on June 28 and Fleming on the 30th. By July 12, Fleming had elaborated the sketch prepared in Washington into a full-dress plan called RAMSHORN. Listing the same objects as had been agreed upon in Washington, RAMSHORN did have the virtue of cutting back the two-pronged approach of KINKAJOU-WALLABY and the Washington outline. The "story" would be that the main object of the British was to bring the Japanese air force to battle and divert it from the more decisive operations of Chennault and MacArthur; land operations in Burma would essentially be feints and harassing operations. Soon thereafter, a seaborne assault would be mounted against the Andamans, and Sumatra would be invaded in early 1944 if reinforcements from the Mediterranean and support from MacArthur were forthcoming. Fleming recognized the risk that in the long run the plan might cause the enemy to reinforce the Andamans and Sumatra, objectives of future operations.

When RAMSHORN reached the London Controlling Section, Bevan, with characteristic courtesy, treated it with respect but in effect shot it down in flames. "The main difficulty at the moment," he said, "is that Allied strategy for the war with Japan is not yet settled. Consequently there is no broad deception policy and no deceptive threats are being undertaken in other theaters of the Far East." He recommended that the Chiefs dispatch messages along these lines. But when he met with the Chiefs on August 6, they directed that Fleming simply be told that until operational plans were firm it was unwise to decide on a cover plan. Fleming appears to have taken this as allowing him to implement RAMSHORN for the rest of the year with such means as he had available, and felt that it at least imposed upon the Japanese "the illusion that we were in a position to mount at any time large-scale amphibious operations against a number of objectives."

At the end of October the London Controlling Section produced a plan called DUNDEE. At that point, an amphibious assault in the Bay of

Bengal was contemplated for the spring of 1944. On December 1, a substantial force—battleships, carriers, destroyers and escort vessels—would set out from Britain via the Mediterranean and the Suez Canal to reinforce the Royal Navy in the Indian Ocean. The movement of the flotilla through the Mediterranean would inevitably be observed, and DUNDEE proposed to put this fact to use in aid of the Bay of Bengal operation. Its object was to induce the Japanese to send reinforcements to the southern Netherlands East Indies rather than to the Bay of Bengal, by feeding the "story" that the flotilla was for use in an American-Australian operation against Timor.

Soon thereafter, Mountbatten opened SEAC, and plans moved forward for campaigns in northern Burma, and in the Bay of Bengal against the Andaman Islands. To cover these, late in November Fleming outlined a plan that included and built upon DUNDEE. With no clear object, it involved an elaborate "story" culminating in a notional British invasion of Timor in the spring of 1944 in conjunction with Australian forces. The British, the "story" would run, were gravely concerned over growing American influence in Australia; the stroke against Timor would be valuable to the Dutch, who were likewise concerned over American influence in the Pacific; a secret clause that Portugal, which ruled half of Timor, had included in an agreement about Allied use of the Azores was an important consideration. At this time, GSI(d) was still part of India Command and not under Mountbatten; but Mountbatten approved the plan and approved GSI(d)'s conducting it on his behalf.*

"Tim" O'Connor, who it will be recalled had arrived in the summer as Bevan's man in Washington, passed DUNDEE to Newman Smith for American consideration, suggesting that its timing and implementation be controlled by GSI(d). Smith recommended informal concurrence to General Strong. Nothing came of Fleming's project, but DUNDEE, as will be seen, played a major part in cover and deception planning for the Anzio landing.

With the seemingly endless succession of genuine plans that got nowhere, Fleming's attention henceforth was focused not on strategic

---

* Presumably this was the otherwise unidentified plan called GLOSSOP, which SEAC approved and sent to London and Washington in December 1943. Wingate put it on ice till the SEXTANT decisions should become clear; and, like the real operations it was designed to cover, it was eventually shelved for good.

deception planning but on short-term tactical and operational deception, and on generalized order of battle activity.

Concrete plans or no concrete plans, the special means channels had to be kept operating. As Fleming put it, "just as infantry in a quiet sector can and should dominate No Man's Land by offensive patrolling—a deception staff, even when they have no policy to go on, can and should seek out and impose their will on Japanese intelligence." For such purposes the established channels continued, and new ones opened up. Four of these—ANGEL ONE, BULL'S EYE, OWL, and MARMALADE—had opened in the spring shortly before the departure for Washington.

ANGEL ONE was first cousin to HICCOUGHS, the stream of messages to nonexistent agents in Burma; the imaginary ANGEL ONE agents were in Thailand. They were a notional group of spies and saboteurs, supposedly equipped with transmitters for sending messages back to India. The leader, ANGEL ONE himself, was notionally located in Bangkok. Every day a message in an elementary cipher went out to an imaginary team member through the Siamese-language broadcasts of All India Radio, much in the style of the HICCOUGHS messages. For the time being there was nothing much to do with ANGEL ONE except build it up against the day when it might be useful.

April 1943 saw the beginning of BULL'S EYE, a multiple channel to the Japanese via "A" Force and thence through Brigadier Arnold in Ankara. The first BULL'S EYE channel was a Japanese journalist with whom Arnold had made connection, but the Turks arrested him before he could be put to use. Two months later, Arnold opened a channel via the Polish military attaché, to the Germans and thence to the Japanese. Fleming had already passed some material by this channel when in August the Pole declared his inability to get further material to the Germans. But a few weeks later, Arnold reopened his connection, this time to the German naval attaché; and in October, a month after the Italian surrender, the Italian military attaché in Ankara told Arnold he could get information to both the Germans and the Japanese.

A fair amount of BULL'S EYE traffic was passed until Turkey broke relations with Germany in the summer of 1944. It was well received by the Axis attachés—one item (information about a new American bomber; perhaps one of Hunter's reports about the poor performance of the B-29, or perhaps chickenfeed that Smith provided Hunter about the new B-32) caused the Japanese naval attaché to "jump four feet into the air"—but no evidence was ever found that BULL's EYE material actually got to Berlin or Tokyo.

On April 21 came the first Japanese effort to insert radio-equipped spies into India, when seven different three-man parties (each with one leader, one radio operator, and one courier), collectively codenamed BATS, were parachuted into Assam with orders to set themselves up in sundry areas of Assam and Bengal. Most were caught or gave themselves up soon after arrival. Interrogation showed that they were all low-grade, with limited education and only rudimentary training in spycraft; indeed, most had clearly signed up to get back to India, with little or no intention of actually doing anything for the Japanese.

Two of these parties were chosen for double agent use. One was led by an Indian peasant in his late twenties named Adyuda Das, who had been captured at Singapore and given some superficial training; once in India, he promptly turned himself in at the nearest police station and asked what he could do. Christened OWL, he was turned into a one-man operation and went on the air from Calcutta on May 5. The other, called MARMALADE, went on the air from Dibrugarh in Assam on May 16.

The first months were devoted to establishing their credibility. It was not easy working with them, for they turned out to be even more poorly educated than had been realized, the MARMALADE party in particular; its radioman was surely the only W/T operator in the world who could not write; and they knew very little of the language—Urdu, but written in the roman alphabet—in which they were supposed to communicate. (This was compounded by the fact that the Japanese control for both parties had only a shaky command of the language himself, and his messages were filled with transmission errors as well.) Most of MARMALADE's messages were devoted merely to keeping them active on the Japanese books, with complaints about problems with their radio set and pleas for money. OWL did better, and began to build up a notional network of sources in Calcutta—an Indian Navy sailor, a Bengali with a radio shop and a brother in Ranchi, a Sikh cabdriver named Hari Singh, who became OWL's "guide, philosopher, and friend," as Fleming put it.

*   *   *

The channels previously established continued to be judiciously used.

Two forged copies of telegrams from Anthony Eden, the Foreign Secretary, were sold to the Japanese through PURPLE WHALES. One, called BARONESS, gave advance information (too late, of course) as to the timing of TORCH; another, called MARCHIONESS, did a similar service for HUSKY. There appears to have been a project to sell the Japanese a fake cipher in early 1943, presumably through PURPLE WHALES; it is not clear whether that was carried through. A forged report of the QUADRANT conference in Quebec was considered in late 1943; this idea seems to have originated with General Strong, who asked O'Connor to ask Bevan to provide a draft, which he reviewed and approved with a minor amendment. Bevan appears to have blocked it from going further, since the American planners had not approved it; it was "an operational matter purely for the planners to determine and has nothing whatever to do with General Strong," he wrote to O'Connor. In the spring of 1944, Fleming floated the idea of generating a PURPLE WHALES item in connection with a forthcoming visit of Vice President Wallace to China, but nothing seems to have come of it. The Japanese paid steadily higher prices for PURPLE WHALES material, to the profit of the agent involved (with, no doubt, substantial cuts for Cheng Kai-min and for Tai Li).

During Fleming's visit to London in October 1942, he and Ewen Montagu had concocted a scheme to pass to the Japanese the blueprints of a hybrid battleship and aircraft carrier, several of which the British were allegedly building. The Royal Navy had considered converting some battleships under construction to such "battle carriers," but had concluded that they would be worthless; it was known that the Japanese were building them, and it seemed desirable to encourage them to continue in the belief that their enemies were doing so. A great deal of information to this effect was passed through special means, and PURPLE WHALES seemed to offer a particular opportunity. So Montagu provided Fleming with blueprints of one of the actual proposed conversions, doctored with such touches as penciled notes saying "Who's that blighter with the beard next to Joe?" and "I'm not sure but he's representing FOCRIN." Whether this was ever actually passed to the Japanese is not clear, though Fleming did record that PURPLE WHALES passed "a personal letter from C-in-C, Eastern Fleet dealing with aircraft carriers."

The overall "battle carrier" deception was terminated in July 1943, by which time it was clear that the Japanese were well committed to building such ships.*

SILVER got back to Delhi from Kabul at the end of April 1943. As had been agreed with the Germans, the "All India Revolutionary Committee" set up a wireless station at Delhi, dubbed MARY; the station at the German legation in Kabul to which it worked was dubbed OLIVER. MARY-OLIVER contact was established in July. The project to put a "Committee" agent into Tehran was postponed and eventually shelved.†

At the same time as MARY-OLIVER opened up, the Afghan government, under pressure from the British and the Soviets, demanded the recall of two of the three Abwehr men at the German legation, Lieutenant Witzel and the radio operator Doh; presumably there was a connection between the two events, though the details presently remain obscure. They left Kabul on September 25, along with the legation's stenographer-typist, whom Witzel had married; passed through India on safe-conduct, and reached Berlin by way of Iraq and Turkey in late November. Meanwhile, SILVER went again to Kabul in September 1943, bringing with him a report of "Committee" activities. Portions of this the Germans shared with the Japanese in Tokyo, the first SILVER material they passed to their allies. That visit resulted in an agreement to open an additional wireless contact from the "Committee" direct to Berlin (or, to be precise, to a control station at Burg near Magdeburg). Christened TOM, this link opened when SILVER returned from Kabul. It was destined to carry much traffic and to be valuable to the Allies right through to April of 1945.

For no presently clear reason, in October the Soviets blew the SILVER

* No doubt unaware of this earlier British deception, the Americans appear to have launched a similar deception on their own in 1944–45. There appears to have been a suggestion in December 1944 that the American-controlled double agent PAT J in New York should send a message that a seaman informant had told him that "CVB" meant "carrier battleship," but this apparently was not sent. The American special means channel GUS in Buenos Aires told his Japanese contact in February 1945 that work on the "carrier battleship" *Brooklyn* was proceeding rapidly and the ship was 50 percent completed.

† Putting an agent into Iran would have fit with Fleming's charter as of October 1943, when the former "A" Force activities in Iran, Iraq, East Africa, Portuguese East Africa, the Rhodesias, and the Union of South Africa were reassigned to GSI(d).

case by telling Rasmuss, who it will be recalled was overtly the German commercial attaché but was in fact the legation's senior intelligence officer, that they had detected the MARY-TOM and MARY-OLIVER links and had read and deciphered all its traffic. The Soviets ostensibly dropped SILVER (Fleming and DIB suspected that in fact they had not), while the Afghans demanded Rasmuss's recall. SILVER managed to weather this crisis; indeed, he professedly was unaware of it till he returned to Kabul in April 1944, when he was probably told of it by the Soviets themselves. By that time Rasmuss too had returned to Germany, having left Kabul on February 11 and reaching Germany at the end of March. This left only the other radio operator, Zugenbühler, "by nature a dull and gloomy character," in Fleming's words, slow of wit and of mediocre abilities, to act as SILVER's German control. Zugenbühler hinted to SILVER that he had been betrayed by a member of the "Committee"; SILVER himself was evidently back in the Germans' good books.

HICCOUGHS, the notional team of spies and saboteurs receiving their instructions over All India Radio, plodded on. In the summer of 1943, Fleming decided that it would help plausibility to airdrop to HIC-COUGHS a container of supplies. In addition to such predictable items as explosives and related paraphernalia, a revolver with cartridges, and medical supplies, the contents included orders in a new cipher, a Japanese uniform, a letter in Burmese to one of the agents from his wife in India, and the ribbon of the MBE (Civil Division) for the team leader, together with a personal note in Auchinleck's own handwriting. Cipher messages advising the team to expect the parcel were broadcast, and on the night of August 16, 1943, the container was dropped at Henzada, a town on the Irrawaddy, halfway between Rangoon and Prome, where the Japanese maintained a fort. It landed smack in the main street and had the appropriate effect on the Japanese.

Indeed, HICCOUGHS was having an abundant effect on the Japanese overall. By the spring of 1944 it was clear that they took HICCOUGHS messages very seriously; an intercepted message showed that HIC-COUGHS was one of the two factors forming the basis for the Japanese Burma Area Army's appreciation of Allied strategic intentions. In fact, HICCOUGHS messages had been studied from early on by the Kempei Tai and the Japanese army and navy at Rangoon headquarters and elsewhere.

Decipherments were distributed within twenty-four hours of the transmission; but with no centralized analysis, so that different headquarters made their own interpretations and took action accordingly. A gratifying diffusion of effort resulted. All sorts of ordinary accidents, mishaps, and breakdowns were attributed to HICCOUGHS sabotage, and the Kempei Tai wasted much time and trouble chasing imaginary spies. There was what Fleming called a "Jabberwockian atmosphere which distinguished the Japanese attempts to wrestle with HICCOUGHS. Some of the comment and deduction based on them was sound," he added "but as often as not the sense was wildly garbled. Sometimes additions were unaccountably made to the text and in many cases the deductions drawn were so arbitrary and far-fetched as to give an impression of hysteria."

BRASS, the three-man team that had been caught by the Japanese near Rangoon in November 1942, also developed satisfactorily. The Japanese seemed to have no idea how to make a notional agent seem real. BRASS never complained about money or logistical difficulties, as real-life agents do. His field trips outside Rangoon were made with improbable celerity. He stayed on the air much longer than would a real agent who had to worry about hostile direction-finding equipment. The style of his messages was inconsistent. Only the Japanese could have believed that their opponents were being fooled. By the same token, they never caught on to the fact that their opponents were fooling *them*. BRASS ran right up to the recapture of Rangoon in May of 1945, and over that time the Japanese transmitted no less than 177 messages purporting to come from BRASS and received 127 messages in return. Every seven or eight months a consignment of supplies—pistols, ammunition, money, clothing, watches, radio spare parts—would be air-dropped to the BRASS party. The airplane was never intercepted or fired on; but a party of Japanese would lie in wait at the drop zone in case further agents should be parachuted in.

To support deception, Fleming would direct BRASS to take steps or ascertain information appropriate to a notional operation, and to spread rumors whose seeming falsehood was consistent with purported cover for such an operation. In dealing with the Japanese, subtlety was out; such instructions and rumors had to point unambiguously to the desired goal. Thus, if a deception included a notional airborne landing in a particular area, BRASS would be directed to reconnoiter the area for appropriate drop zones, recruit locals to act as guides, and spread the rumor that the Allies had no airborne troops in the theater. Rumor-spreading could be fun. "Once, in a lean period," recorded Fleming, "the rumor ran 'The Emperor of Japan is NOT a monkey but he and all his family have short furry tails of which they are very proud'; this is known to have reduced the Kempei Tai to a state of near-apoplexy which lasted, with the case officer, for two days."

That same case officer, one Captain Seruta, ripened into an old friend. (Once he asked that a quantity of medication for venereal disease be included in one of the supply drops for BRASS; postwar inquiry confirmed that this was for his own use.) That he, and through him the Kempei Tai and the Japanese command, was wholly taken in, was abundantly confirmed. Headquarters of the Japanese Burma Area Army called BRASS "our spy who is operating among the enemy," "our special source of information." The Japanese frequently took action based on messages to BRASS. In June of 1943, for example, it was desired to draw Japanese attention to the Prome area. BRASS was encouraged to open a subagent there, and dutifully did so, notionally of course; the Japanese actually recruited a man to visit Prome from time to time to garner appropriate genuine chickenfeed. In early 1944, to implement deception in aid of one of Orde Wingate's expeditions, they were induced to establish a second notional subagent at Mandalay.

The eternal schoolboy in Fleming never ceased to rejoice in confecting haversack ruses. He asked RAF pilots to drop an endless variety of things where the Japanese would find them. There was a scratched and battered map case containing a worn and shabby brigadier's cap, a pair of pajamas and slippers, a half-squeezed-out tube of toothpaste and a well-worn toothbrush, and of course some fake documents. (The map case and cap were genuine; Fleming had swapped the brigadier six bottles of Scotch for them.) Carrier pigeons bearing phony messages, a few wing feathers broken as if by accident, were dropped behind enemy lines to be caught and hopefully turned in to Japanese intelligence. There were sacks of mail; quite a few sacks of mail, generally dropped near where an airplane had crashed, as if they had been in the plane's cargo. The ladies in the office, and a number of them back in England too, were pressed into service writing "letters from home" to fill these sacks. It did not matter what the letters said; the important thing was their addresses, for these ruses were intended to build up the false order of battle.

Dropping such jetsam came to be a normal feature of life for the pilots who supported the clandestine services in Burma. "If any pilot from [the air base at] Jessore were to tell me now," wrote Squadron Leader Terence O'Brien, who flew many of these missions, "that he had . . . a visit one day from a charming colonel who had asked him to drop a dead goat wearing a collar and a regimental badge, I could well believe him. Contact with D Division could permanently unsettle a skeptical nature."

It was but a short step to a more ambitious project. Perhaps, reflected Fleming, a channel similar to BRASS could be planted on the Japanese commander on the Arakan front, in direct aid of the offensive planned for November 1943. MINCEMEAT in the Mediterranean was suggestive. So was an abortive SOE project to lure the Japanese away from the trail of a British officer working deep behind their lines by parachuting in a corpse that the local natives would identify as his. Suppose the Japanese were to find behind their lines the body of an Indian spy whose parachute had not opened, together with his working radio transmitter, a suitable cipher, and a questionnaire, plus evidence to suggest that he had had a Burmese companion? Surely they would see this as an opportunity to open a deception channel back to the British, which could then be run like BRASS. This project, called FATHEAD, began to germinate in July of 1943, after Fleming got back from Washington.

Finding an appropriate corpse should have been child's play. The great famine was at its height. People were dying like flies in a land of nearly half a billion souls. Yet no suitable body could be found. (One problem was that an emaciated famine victim would not make a plausible secret agent.) Weeks passed, and autumn came on. The launching of the offensive drew near. But no body.

Everything else was ready. After extensive testing it had been concluded that the least suspicious way to show that the parachute had failed to open would be to have a broken clip attached to the static line. The wireless set, the cipher, the questionnaire, the remaining paraphernalia, were ready and waiting; details of what should be found on the body and how it should be prepared had been worked out to the last detail. But no body.

November arrived, with D-Day for the offensive imminent. The full moon, the only plausible time for such a drop, was coming on. The airplane was waiting. But no body.

On November 11, Fleming detailed a South African officer named Gordon Rennie and told him: Go to Calcutta, find a body, carry out the operation, and don't come back till you have done so. So for five gruesome days Rennie searched the inferno of famine-racked Calcutta. He visited the morgues and hospitals. Nothing suitable. The police promised to produce a body. They did not deliver. He visited the mortuaries. One of these came up with a suitable corpse, that of a rickshaw-puller, but at the last minute the family showed up to claim it. He rode round the streets and through the slums and back alleys with the trucks that went out early every morning to collect the cadavers of famine victims from the streets and sidewalks. Plenty of bodies, but all too gaunt and wasted or too rotten. At long last, the army hospital at Fort William came through with the corpse of a Bengali Hindu; and "two hours later in a thicket at the back of an Indian General Hospital a full Colonel in the Indian Army Medical Corps might have been seen helping to man-handle the naked body of an Indian out of an ambulance into the back of a 15 cwt. truck."

The body was taken to a garage at Alipore. There it was washed and dressed in Indian clothes. In the pockets were planted cigarettes, matches, train tickets, a wallet, small change, and similar odds and ends; everything was wiped clean and handled with rubber gloves, and the cadaver's fingerprints then placed on shiny surfaces. Overalls and helmet were added, and the body strapped into a suitably misfolded parachute. At 4 A.M. on November 17, the plane, carrying the corpse, two containers packed with the radio and other appropriate agent's equipment, and Rennie, took off from Dum Dum airfield and headed for the drop zone thirty miles south of Akyab. The pilot recalled that when he got into his plane he looked back at his cargo covered with a white sheet, and saw an arm flop out and dangle down. He strapped himself in the pilot's seat, "and, like the Ancient Mariner, 'turned no more his head that night.' "

The night was clear and fine; the flight was smooth; no Japanese were encountered. But a smell began to be noticeable. It got worse. It filled the plane and penetrated to the cockpit. Fortunately, this possibility had occurred to Rennie at the last minute, and he had brought along a bottle of cologne and some cotton masks for himself and the aircrew. Masked and fragrant, they located the target area with some difficulty, dropped down to eight hundred feet, and unloaded their cargo. In the moonlight they could see the container parachutes open while the corpse plunged to its notional doom. They turned and headed for home, and no doubt for a long hot shower.

After all this effort, FATHEAD proved to be one more anticlimax. The Japanese found the corpse and searched earnestly for his imaginary Burmese comrade. But they never put the radio on the air. A backup transmitter with a duplicate set of instructions was parachuted in, in case the first had been damaged beyond repair in the original drop. Nothing. For six months, FATHEAD's control at Calcutta called and listened on the appropriate frequencies. Still nothing. Nor was there ever anything. "A project too boldly conceived for the timid and bungling Japanese," sighed Fleming. "The European 'man who never was' did apparently achieve an important deceit of the enemy," wrote Terence O'Brien long after. "In the hierarchy of nonentity our man was far superior—he not only never was, he also never did."

In July 1943, Fleming inherited from MI5 in London a channel known as FATHER. FATHER was Henri Arents, a Belgian test pilot attached to the RAF, who was recruited by the Abwehr after the fall of France and sent into Britain to collect operational and technical RAF intelligence. He turned himself in on arrival and was used as a special means channel by MI5, communicating with the Germans by the familiar method of messages in invisible ink to a cover address in Madrid, and receiving his instructions by wireless code messages that could be heard with an ordinary household radio receiver.

By early 1943 the Germans were asking FATHER technical questions which he could not plausibly ignore or give false answers to. So he was transferred out of Britain to the RAF base at Chittagong, and from there he dispatched his first secret-ink messages on August 31 and September 1. That same month he was physically transferred to an RAF squadron near Calcutta. Notionally he remained at Chittagong till November, and was then notionally transferred to a Spitfire squadron at Alipore. There he acquired a notional source in the person of an indiscreetly talkative female friend, supposedly the secretary to the Military Secretary of the Eastern Army.

FATHER's utility was impaired by the two months or more that it took for his letters to reach Madrid. By good fortune, the Germans came to the same conclusion. In January 1944, ULTRA revealed that the Germans were trying to get a radio transmitter from Istanbul into Bombay, and in the middle of February, when they inquired of FATHER whether he could take delivery of "an object" at a place eighteen hundred kilometers west of Chittagong, it became evident that he was the intended recipient. After a long, complicated, careful process given the codename DUCK, the set was collected by FATHER later in 1944 and went on the air in August.

On April 20, 1943, as Fleming was making ready for the journey to Washington, the Japanese submarine *I-29* quietly left Penang and headed out into the Indian Ocean. Six days later she reached a rendezvous point four hundred miles south-southwest of Madagascar. The next day, a German U-boat made contact. She had slid out of Kiel some weeks before, eluded Allied destroyers, and broken free out into the Atlantic and on south around the Cape of Good Hope. On board were Subhas Chandra Bose and one follower.

The weather was raw and the seas were high, for late April is autumn in the southern hemisphere. The two submarines adjourned to the northeast in search of smoother water. In the afternoon a German officer swam over to *I-29* to confer with the Japanese captain. They agreed to wait overnight. The next morning was better. The Germans inflated a rubber dinghy and ran a line across to the Japanese boat. Bose and his aide got into the dinghy and pulled themselves across. A week later *I-29* put back into Penang, and on May 16 her human cargo was in Tokyo.

Bose was received by Tojo and formally introduced to the Diet. He gave a triumphant press conference and broadcast a fiery speech. He proceeded to Singapore, where at the beginning of July he took over leadership of the "Indian Independence League" and of an "Indian National Army" that had been recruited from Indian prisoners of war. On October 23, he proclaimed a "Provisional Government of Azad Hind." It promptly declared war on Britain and the United States. In January 1944, its seat was moved to Rangoon. Bose got control of the Japanese-run "spy schools" at Rangoon and Penang and began the training of his own agents.

Fleming and DIB received this news with pleasure. "His friends are always pleased when a man who is rather deaf acquires a hearing-aid," observed Fleming after the war,

> and the arrival of Bose in the Japanese camp was for analogous reasons welcomed by the small Deception Staff in Delhi. . . . [T]hey had good reason to hope that his quick wits, his dynamic personality and his long experience of under-cover activities would before long widen the front, hitherto disappointingly narrow, on which they were in contact with the enemy's Intelligence; they looked forward also to dealing with a sophisticated adversary who could be relied on—as the Japanese could not—to see the point of the information they gave him. These hopes were only partly fulfilled.

In December came Bose's first substantial effort at espionage, when a Japanese submarine put eight men equipped with four radio transmitters ashore on the Kathiawar peninsula north of Bombay. Trained in Germany, they reflected a fair sample of the ethnic variety of the subcontinent—Sikh, Punjabi, Bengali, Madrassi, Gujarati—and they had orders to split into four two-man groups and set up stations, each with a transmitter, at Calcutta, at Bombay, at Benares, and on the North West Frontier. From these bases they were to establish contact with subversives and revolutionaries throughout India and pass political and military information back to Bose and the Japanese. The Calcutta station was to be the headquarters and the contact with Rangoon.

All eight were promptly caught. In January 1944, the case—dubbed the PAWNBROKER party—was turned over to Fleming and DIB for deception. It was decided to set up transmitters at Calcutta (PAWNBROKER) and Bombay (THE TWERP) and notionally eliminate or disperse the rest of the group. Calcutta went on the air in late March and transmitted regularly. No reply came from Rangoon, but PAWNBROKER kept trying; and, as will be seen, it eventually became a valuable channel.

Bose's next effort was a four-man group, recorded as the PURI party, which landed from a submarine at Puri in March 1944. It called its control for several months. No contact was ever established, and the group was rounded up from September 1944 to January 1945.

In February of 1944 a triple-cross channel to Bose was attempted, codenamed BACKHAND. BACKHAND was Captain Mohammed Zahiruddin of the Indian Army, who had been cashiered in August of 1940 for anti-British activities in Singapore. He had changed his views, and had sought to rejoin the Army. The Army offered him to Fleming, and Fleming had him dropped by parachute near Taikkyi, seventy miles northwest of Rangoon, on February 14, 1944, with orders to give himself up without disclosing that he had been ordered to do so, and function as a triple-cross agent. The Japanese believed him, but did not use him as a W/T agent as Fleming had hoped. Instead, they sent him to join Bose's entourage. There he would rise high and then fall far, as will be seen.

One useful way of "patrolling No Man's Land" with this substantial array of special means channels was to build up a notional order of battle in enemy minds. All during this seemingly quiet period, GSI(d) worked on that: an item here, an item there, all in good Dudley Clarke style, with Thorne in direct charge. Through PURPLE WHALES Fleming added a new twist suited to the literal-mindedness of the Japanese, selling to them what purported to be the sketchbook of an artist employed by army public relations—it was in fact prepared by Frank Wilson, who was a talented caricaturist—filled with sketches of soldiers of various nationalities in the uniforms of fictitious units, with their insignia prominently displayed and notes in pencil that identified and located their units. ("This was much appreciated by the Japanese," noted Fleming, but nevertheless "it is thought to have been a little too difficult for them.") In aid of order of battle deception were the sacks of phony mail and other haversack ruses; and there were additional stunts such as the occasion when Fleming had printed up a program and a number of posters announcing a big show featuring Bing Crosby, Bob Hope, and Dorothy Lamour, which would be held on several nights for the men of various fictitious divisions supposedly on the Arakan coast.

Documents captured in June 1944 demonstrated that in November 1943, Imperial Headquarters believed that the Allies had 51¾ divisions in Southeast Asia and India, when the true total was more like thirteen plus half a dozen independent brigades. Some of this wild exaggeration was a product of what Fleming called "spontaneous self-deception," but an appreciable proportion of the bogus strength (three-fourths, Fleming thought, but considerably less, in the skeptical view of Joint Security Control) was directly attributable to Fleming's work.

The year 1943 saw also the establishment of the first substantial tactical deception force in the theater. It will be recalled that the 303d Indian Armored Brigade, a dummy tank unit, was transferred from the Persia and Iraq Command to India in April 1943. Fleming set it up in six small thirty-man field sections, called for cover purposes "Observation Squadrons," to conduct tactical deception. Under the command of Lieutenant-Colonel "Pat" Turnbull, with Frank Wilson as second in command, it participated in most of the 1943–44 operations, largely right in the front lines. The three squadrons on the Arakan front worked smoothly. The single squadron on the Imphal front, after an initial period when its value was not appreciated, covered the retreat of 20th Indian Division in the face of the Japanese offensive and earned high praise from the division commander. Two squadrons that supported Wingate's Chindits had less success. But on the whole it was a successful trial that laid the foundation for the highly effective tactical unit of the last year of the war in Burma called "D Force." In December of 1943, the tactical deception force was augmented when two "Light Scout Car Companies," outfits trained and equipped in England for sonic warfare, came out to India from England, together with a "Light Scout Car Field Park," which produced sound tracks for their POPLIN equipment.

Fleming's frustrations over having no major strategic plan on which to use his extensive array of deception tools were as nothing beside the frustrations of Ormonde Hunter.

Hunter got back to New Delhi on June 22, 1943, from the conference in Washington. Next day he reported on the conference to General Ferris, and on the 28th to Colonel Frank Merrill, Stilwell's right-hand man since the early days in Burma, who was just finishing his report on a comprehensive tour of the unsatisfactory situation in India-Burma for Vinegar Joe. Merrill "exploded," in Hunter's words. "There is not————use in doing anything on that until we find out what the British intend to do in [Burma]," quoted Hunter to his diary. "They are unwilling to commit themselves in any serious operation but are perfectly willing and anxious to go ahead with this funny business." "Under the circumstances," a dismayed Hunter reported to Washington, "it can be understood readily that the U.S. Staff is not in any hurry to give consideration to my type of work with the real plan in such a chaotic state."

This was the beginning of a long-drawn-out time of misery for Ormonde Hunter. His was the most uncomfortable situation that an able man can have in an organization: He thought he had an assignment, he believed it to be important and wanted to do well in it, but nobody else knew anything about it or gave it much thought and he had nothing to do. His trip to Washington had made things worse. He had gone round calling on people up to and including Chennault—he tried to see Stilwell, but failed—all of whom politely expressed interest in his work; he deluded himself that this meant that his job was considered important. Perhaps worst of all for his own future misery, he had gained the impression from Colonel Jackson of Joint Security Control that "the General Staff" had big plans for him to set up a deception operation in China. This was surely a misunderstanding; and in any case Jackson was gone for good soon after.

Hunter compounded his problems by persistent misapprehensions as to elementary aspects of the command and staff structure. He operated for months under the impression that Joint Security Control was part of OPD. He was convinced that Lieutenant Colonel George Carey, the air force officer who had worked with him in Washington in May, was "Air Corps Member, Deception Committee," or "Air Member, Deception Committee," and kept sending material to him; as late as March 1944, Newman Smith had to remind Hunter tactfully that Carey was not a member of Joint Security Control. He could not get clear the distinction between the General Staff of the Army and the Joint Chiefs of Staff. He called the London Controlling Section the London Controlling Office, and thought it was the "London Office of GSI(d)." He got into a dispute with Mountbatten's staff, contending that all deception plans against Japan should be submitted to the American Joint Chiefs; Smith had to explain that plans of a British theater commander such as Mountbatten went first to the British Chiefs. He communicated sometimes with Baumer at OPD and sometimes with Joint Security Control, and continued addressing communications to Jackson long after being told that Jackson was gone.

Hunter worked hard; as bulls in the china shop often do. In January 1944, nearly a year after he arrived in the theater, he wrote to George Carey that he had had only five afternoons off since coming to India. He communicated outside proper channels in a fashion that regular officers would have found irregular, to say the least. He did not hesitate to write directly to "Hap" Arnold himself, urging that a full-time air officer be assigned to Joint Security Control. Nor did he always apply the tact to be expected of a subordinate officer. "I was delighted to receive the radio intelligence report with your name at the top in place of General Strong," he wrote to Bissell when the latter became Army G-2. "As far as the Armed Forces are concerned it is a distinct step forward."

Unaware, apparently, of General George V. Strong's determination that "Wild Bill" Donovan's hated OSS be kept at arm's length from deception, Hunter went out of his way to involve it in his work. During the Washington conference he had called on Donovan's deputy and on Donovan himself, and had been welcomed effusively; Donovan, recorded Hunter with satisfaction, "offered his complete facilities 'including any funds needed' "; he called in his people involved in gadgets and forged documents and "told them to show me the works." Over a three-day period the OSS briefed Hunter and introduced him around. Obviously Donovan rejoiced at this opportunity to end-run Strong, not only on general principle but because just at that moment a struggle was going on over control of intelligence operations in CBI; but Hunter seems never to have caught on.

From India he urged Washington to have the OSS print phony Japanese money to be used in connection with WALLABY. He held long meetings with Lieutenant Colonel Carl F. Eifler, the 250-pound, stentorian-voiced former border patrol agent and Stilwell protégé who was in charge of OSS sabotage and paramilitary activity in north Burma, and in early October he paid a visit to Eifler's OSS base camp in northeastern India; from which, he concluded, there were possible channels for deception in the form of local rumors, haversack ruses, and a compromised or low-level cipher. In August of 1943 he had a long meeting with Eifler's rival, Donovan's darkly handsome former law partner Lieutenant Colonel Richard Heppner, who had recently been put in charge of OSS activities in the CBI, to discuss possible channels for passing misinformation. (When Heppner showed up in Washington the following March bringing rumors about deception in India, and trying to acquire HEATER equipment in an attempt at sonic warfare, Newman Smith had to warn Hunter that "deception of a strategic or tactical nature should not be performed by other than military units directly under control of recognized deception agencies.")

Hunter busied himself in trying to line up special means channels. He got in touch with the Assistant Consul General at Lourenço Marques (which he appears to have thought was in Madagascar). He made contact with Engert, the United States Minister to Afghanistan, to whom Fleming was passing suggestions for rumors. He bombarded Joint Security Control with small-scale bright ideas for planting phony information. By and large these were reinventing the wheel; many of them were borrowed from Fleming.

The tone of his correspondence on such matters is plaintive, as if he were pleading for someone to remember that he existed and perhaps to tell him that he was doing well. He told Washington about a night club in Kweilin that was known to be a clearinghouse for Japanese agents. He told them that one Chang, a servant of a United States embassy official in Kabul, had been found to be a Japanese agent, and suggested setting up a chatty correspondence with Chang's employer containing appropriate indiscretions for Chang to find. (It turned out, however, that Chang's name was Yang, his employer had been transferred to London taking Yang with him, and Yang was not "a professional agent with channels to the enemy," as a slightly crestfallen Hunter reported to Newman Smith, "although the Chinaman has undoubtedly committed indiscretions nightly in the company of a comely Japanese wench who lived next door.")

He reported on Fleming's experiments with dropping crippled homing pigeons (Scheme POT-PIE, Newman Smith dubbed them in reply). He compiled a "pamphlet" on deception ideas for air units and sent it to Carey, and followed it up with some additional ideas which he suggested be passed to General Bissell when the latter succeeded Strong as G-2. When Mountbatten was announced as SACSEA Hunter prepared and sent to Washington—evidently to Baumer rather than to Joint Security Control—an "Estimate of the Situation" that recommended using the old full-dress ANAKIM plan as "the cover plan for *all* Burma operations."

Smith acknowledged Hunter's suggestions politely. From time to time he had to warn Hunter to be careful, or to explain why an idea would not be feasible. Particularly unsettling was a casual-seeming approach to order of battle deception in Hunter's "pamphlet" on air deception. Smith warned him that unless based on an approved list of notional units this could be "extremely dangerous." Again, Hunter worked out a scheme with the theater censor for adding deceptive material to letters that might appear to be deliberate attempts to communicate information to the enemy; Smith warned him that this "would seem to require extremely delicate handling."

Hunter suggested that a double agent channel be opened to the Japanese through Argentina for his use; Smith had to explain that while such channels already existed, any such material would have to be pursuant to an approved theater plan and passed through Joint Security Control. Hunter suggested planting information through Japanese civilians being returned home via the neutral Swedish ship *Gripsholm;* Smith had to explain that repatriation was a civilian matter handled entirely by the State Department and that, once again, deception could only be pursuant to an approved theater plan. Perhaps by way of a gentle hint, Smith sent Hunter a set of Joint Security Control's "Definitions and Principles of Deception," "General Checklist for Deception and Cover Plans," and "Suggested Channels and Methods for Effecting Deception," but this did not slow the flow.

Fleming's Plan RAMSHORN brought to light some confusion in Hunter's mind as to his working relationship with the British. He sent to Washington a packet of ideas he had brainstormed for implementing it (he still called it WALLABY), without advising Cawthorn and Fleming. When they learned of it they asked for a meeting with Hunter and Ferris. "Are we going to form a combined committee to get ahead with the matter," asked Cawthorn, "or is Colonel Hunter building up his own organization and if so what has been organized?" After reflection, Ferris advised Cawthorn that Hunter would "act with GSI(d) as an Inter-Allied Section" in planning for the winter offensive, and concurrently would "organize a U.S. Deception and Cover Planning Section in this headquarters to function in other and future operations as an independent section, in very close liaison with GSI(d)." Hunter's new one-man

Special Planning Section was placed under Ferris's G-3 for plans, Lieutenant Colonel Dean Rusk.

Just at this time, July 1943, General Stilwell descended on India for a tour of inspection. Annoyed with Chiang Kai-shek, Stilwell was in no receptive mood when Ferris raised with him both RAMSHORN and Hunter's desire that a deception organization be started in China, with part-time officers in Chungking and at Chennault's headquarters in Kunming. Vinegar Joe responded, Ferris told Hunter, that he "thought it all foolishness and wanted no part of it. That he would not have any of his staff in Chungking or Kunming bothered with it." After sleeping on this, Hunter visited Ferris again and asked if he could get an appointment to present to Stilwell directly what he characterized as the views of the "Washington General Staff." Ferris said he would try.

Hunter then went to see General Bissell and asked point-blank if Bissell could tell him the circumstances under which he had been ordered out to India. Yes, he could in a few words, said Bissell. At the meeting in February when Wedemeyer, "Hap" Arnold, and Sir John Dill had reviewed ANAKIM and KINKAJOU-WALLABY, Bissell had told Arnold that he did not have anyone he could put on the cover plan. Arnold had told him that he had just the man, Hunter, and would send him out. Stilwell would have known nothing about this. Hunter had been meant for Tenth Air Force "and when you got here Ferris did not know what you were here for. His nose got out of joint and he would not turn you loose."

After two more days, on July 31, he got his interview with Stilwell. To his pleased surprise, the General talked at some length, told Hunter (or so Hunter thought) that he would take him to China with him, and intimated agreement with the idea of part-time officers in Chungking and Kunming. He said he had approved RAMSHORN because nothing better suggested itself; the situation pointed so clearly to the intended lines of action that there was no room for an overall cover "story."

Hunter's hopes rose. "Very much impressed, I was," he told his diary. "Given a good plan, I am sure the General would be a strong supporter." It was more self-delusion. His invitation to China never came. Stilwell did not return to Chungking till the end of August, and did not offer to take Hunter with him when he did.

Cawthorn and Fleming themselves got tired of waiting for something to happen in China, and in the second week of September, to Hunter's intense embarrassment, Fleming—disregarding the fact that the China theater was American territory, with the agreed exception of PURPLE WHALES—left on a three weeks' visit to Kunming and Chungking to confer on deception channels. As a followup to his visit, Chennault's command suggested to Fleming deceptive information that they would like to see implemented, and in October he stationed one Major S. C. F. Pierson as a permanent GSI(d) liaison officer in the Chinese capital.

This incident deepened Hunter's dissatisfaction. General Stratemeyer, who had now arrived in the theater, and Brigadier General Howard C. Davidson, Bissell's successor at Tenth Air Force, felt Hunter out to see if he would join one of their staffs. Hunter went to see Ferris and "told him of Fleming's trip to Chungking, my embarrassment at his entering our field because we had nothing . . . and although I liked to stick to a job, if I couldn't get requisite support to do it, I would go back to Air Corps. General F. seemed disturbed at Fleming going to China and said 'I guess you better stick around with us awhile.' " Hunter consoled himself with his visit to Eifler's OSS base camp in the foothills of upper northeastern India.

Then his hopes were raised again. Late in September he was given an assistant, in the person of First Lieutenant Cornelius Warren Grafton. Grafton—his friends knew him as "Chip," and as "C. W. Grafton" he had published prizewinning murder mysteries—was a thirty-four-year-old lawyer from Louisville, Kentucky; born and raised in China, he spoke the language fluently. Hunter hoped to send him to Chungking or Kunming as his representative; later, however, he designated Grafton as his deputy. By November, a proposed table of organization for Hunter's section, which currently consisted of himself and Grafton plus a clerk-typist, T/3 Julian Levinson, had been prepared that would enlarge it to a full colonel, two majors, and a captain. He finally got a trip to China that same month, in the shape of a brief visit to Fourteenth Air Force at Kunming. By January 1944, expecting that his new table of organization would be approved, he was cabling Joint Security Control to ask for a major, preferably with some knowledge of Chinese, to be assigned as his man in China.

*   *   *

When Hunter had met Chennault in Washington at the time of TRI-DENT and the May gathering on deception, he had found the general "more deceptive minded than any of us." Presumably with Chennault's encouragement, in the last half of 1943, as Hunter described it, there was "a tendency in our general deception to stress the inefficiency of [United Nations] air strength, our failure on Hump lift, poor morale and lack of cooperation between British and U.S.," the purpose being to make the Japanese "so overconfident that they would fall prey to a really efficient force." By January, Chennault was telling Hunter that "what was done along this line before was quite successful in luring the Japs to combat with . . . quite excellent results," though Hunter was wondering whether this activity should be continued, since "a lot of what the Japs have been told as deception has turned out to be true."

After Hunter lamented to Newman Smith the lack of usable chicken-feed from Washington, Smith sent him material from time to time there-after with respect to such matters as B-29 production and characteristics, command developments in the North Pacific, new fighters, and a new B-32 bomber.

Hunter, not understanding the difference between military deception and psychological warfare, had contributed to the May meeting in Washington the idea of "exploiting disunity between the Western Axis and Japan" as a deception goal. Pleased as Punch with this—"one of the duties of my section as laid out by the General Staff is to foment discord between the Japs, Germans, and Italians," he told Chennault proudly—when he got back to India he set to work doing something about it. No doubt inspired by Fleming's fun and games, he doctored up a Tenth Air Force situation report so as to include a fictitious account of a statement critical of the Japanese purportedly made by a (genuine) high-ranking captured German air force officer. He had grandiose hopes for the effect this would have when allowed to fall into Japanese hands. "Enough fuel added to the flame from time to time might result in the Japs expelling German technical air experts and observers from Japan," he said. In mid-October he sent this to Chennault to be planted on the Japanese and, as he put it, "well launched on its voyage to the Chief of Intelligence, Japanese Imperial Headquarters."

A few weeks later, on his first trip to China, he learned that Chennault's intelligence officer had passed the document to the British for transmission to the Japanese. The British had in turn sent it back to Colonel Steveni, head of MI6 in Delhi. Steveni held it up, and gave it back to Hunter when the latter came to see him after getting wind of what had happened. Hunter's chagrin was compounded by what he saw as failure by Chennault's intelligence officer to understand its delicate nature. The whole business showed once more the need for an independent full-time American deception section for the theater, he told Ferris; without result.

He sent his document back to Fourteenth Air Force with instructions that it be passed to the Japanese through an American channel. It seems never to have been heard of again.

Next to his frustration over China, perhaps the worst of Hunter's travails was a classic snafu over the B-29s of XX Bomber Command that ran well into the spring of 1944.

At the end of 1943, not long after construction of airfields in India for the big bombers had started, Hunter suggested a deception campaign either to exaggerate or to understate their performance; and urged that a full-time air officer be designated to plan deception for their offensive. These thoughts got no response.

Meanwhile, without coordinating with anybody, beginning the previous October "Hap" Arnold and the War Department's Bureau of Public Relations had launched a deception scheme of their own. Through public statements and press contacts they set out to link the first combat use of the B-29s with England rather than with the Far East, to sow confusion as to the date of this first use, and to put out the story that their employment in Asia would be not as bombers but as armed transports to increase the tonnage being flown over the Hump.

The public relations people preened themselves on their success with this project. They got press comment on the building of superbomber fields in England. They set up a flight of a single B-29 to England with appropriate leaks to the press. They arranged for a prominent B-17 pilot making a public relations tour to tell reporters that he and his crew were going to ask for B-29 assignment. They sought to encourage press think-pieces on the use of superbombers in shuttle service between the USSR and bases in England and Italy and on German dismay over the prospect of an air onslaught reaching to the furthest corners of the Reich. They arranged for Stilwell to tell the press in Chungking that the largest and newest cargo carriers would be used to fly more supplies into China, for Arnold to mention in a speech that B-29s could be used as cargo carriers, and for Stratemeyer in a press conference to describe "armed air freighters twice as large as the B-24" and say that airfields were being prepared for these giants. Prominent military analysts discussed Stilwell's and Stratemeyer's statements in their columns. "Treatment of both statements could be described as almost perfect," gloated the Bureau of Public Relations.

Stratemeyer's press conference alerted Hunter. He fired off a testy complaint, not to Smith but to Baumer. His feelings were hurt. Here was the first deception plan in his theater since he arrived, and nobody had told him about it. He sent to Newman Smith a cable suggesting that he pass a supporting "story" through special means channels.

That was the first that Smith himself had heard about this project. He was as annoyed as Hunter was. BODYGUARD, the great deception plan for the invasion of the Continent, was just getting under way, and the B-29s played a significant role in it; moreover, Smith already had to deal with an uncoordinated release of B-29 information by "A" Force. He made inquiries, and was dismayed at what he found. Pleased with themselves, the public relations people were on the point of broadening their activities by releasing information that big air bases were being constructed in the Aleutians; "this could cause division of Japanese defenses," they gloated. It also would impinge directly on the first major wholly American strategic deception, just getting under way: Plan WEDLOCK, a notional threat to the Japanese northern flank from the Aleutians.

Smith alerted General Bissell and Admiral Schuirmann, Admiral Train's successor as Director of Naval Intelligence. In their capacity as the full members of Joint Security Control they dispatched to Arnold and the Director of the Bureau of Public Relations what amounted to a strong rebuke, drafted by Smith, though couched in diplomatic language, suggesting "that in the future, if for any reason special deception plans are desired, they should be initiated and implemented in accordance with" the relevant Combined Chiefs and Joint Chiefs directives.

This was only the beginning of Ormonde Hunter's chagrin over the B-29s. His proposal to Washington after the Stratemeyer statement was that he be authorized to send through special means channels the "story" that the B-29s had proved to have insufficient range for use as bombers in Asia and were accordingly being sent to England, while modified models were going to China as armed transports. Smith had approved, provided the "story" was amended so as to be "more appropriate for over-all policy" by deleting any reference to inadequate range and saying instead simply that very-long-range bombers were needed in Europe. Hunter protested, but was overruled. (He also requested cover designations to use for XX Bomber Command and the B-29s in their notional cargo-carrying role, and was authorized to use "20th Troop Carrier Group, India-China Wing, Air Transport Command" for XX Bomber Command, while the cargo B-29 could be called the "C-75-A.")

Irritated by what he saw as second-guessing, Hunter was even more irritated on finding that extensive information about the airplane had appeared in open publications, including *Time* and *Newsweek*. "I am of the opinion that someone is trying to fool the Japs as to the plane's performance and armament as I cannot believe that true information of this kind would be put out about a plane that has never been in action," he groused to Smith. "It may be thought that the best way to fool the Japs is to mislead the Deception Officer in this theater. Naturally, I cannot subscribe to such a doctrine." Smith sent Hunter a mollifying letter. "Please be assured, old man, that I would have brought you into this picture earlier had I known anything about it myself."

Hunter continued to nurse his grievances. His former contact Colonel Carey passed through India. From him he discovered that a certain Colonel Bentley was now the Air Force contact man for deception matters. Hunter blamed Bentley for the rejection of his "story" about inadequate range. "I would be very much interested in knowing why he thought that it would be good deception to convince the Japanese that the B-29 had a very long range," he wrote sarcastically to Smith. That was not the purpose of the change, Smith replied, in a letter drafted by Goldbranson. "An overall deception plan was in effect at that time which we hoped would lead the Japanese to believe that the B-29 had a restricted range despite all the ballyhoo of previous radio and newspaper publicity. It was felt that passing on a silver platter by secret channels of the word that these craft were shorter range than advertised would be too obvious and probably do more harm than good."

This drove Hunter to near-apoplexy. Silver platter, indeed! He fired back a long handwritten letter. "If my method after a year out here would still be 'too obvious' then I have learned nothing." He had planned, he said, to put his "story" over by a variety of special means that he listed. These might well have been thought by a more experienced deceiver to qualify as "silver platter"; but Smith wisely let the matter rest.

Mountbatten arrived in Delhi in October 1943. He soon became known derisively as "Supremo." He got off on the wrong foot with an opening address to the assembled officers in which to their horror he denigrated the much-loved Wavell and, in the words of one witness, "more or less promised to win the war with his superior knowledge of Combined Operations and his determination to brook no delays." Fleming knew this was nonsense. He had told Peter Thorne in early 1943 when Thorne first came on board that only after victory in Europe would there be landing craft enough for any but land-based operations in Southeast Asia. But he had no choice but to go along with Supremo's fantasies.

That same October, just before SEAC opened, the authorized strength of GSI(d) was increased, with five subsections: headquarters and planning, liaison with DIB, and technical matters, plus sections with the army in the field and for liaison in Chungking. Goldbranson viewed this with a skeptical eye when Hunter sent it to Washington. There was no assurance that GSI(d) would be moved from India Command to SEAC, so this looked like "a definite attempt to 'collar' deception activities in CBI and freeze out Hunter and the Americans," he told Smith. "Wedemeyer should be urged to protect our interests." Clarification came a few weeks later. Cawthorn, and through him GSI(d), would be responsible jointly to SACSEA and India Command. A combined committee of GSI(d) and Hunter's Special Planning Section would initiate, coordinate, and implement cover and deception plans for specific operations. Moreover, the proposed new table of organization would enlarge Hunter's section appreciably. Hunter reported that he was generally satisfied. Meanwhile, he was pleased to be included in the daily meetings of SEAC theater planners.

But no final decision on the transfer, the new table of organization, or indeed on Hunter's position in his own command ever came. Through the autumn and winter Hunter remained more or less in limbo. Stratemeyer offered to make him his chief of intelligence; Arnold had suggested him as an air intelligence officer for SEAC; Ferris had nothing for him but would not let him go. When an important visitor from Washington G-2 who was in touch with deception matters expressed surprise at finding him in Delhi, he wondered plaintively to Newman Smith "whether General Strong knows that he has a deception officer in the CBI theater."

In mid-December, Hunter had a visit from Baron Kehm. Kehm had left Washington in September on a four-month overseas journey that took him to Britain, the Mediterranean, the Middle East, the India-Burma theater, and return via Turkey and Cairo. His tour was only incidentally concerned with deception; his primary purpose was to observe British joint planning and overseas operations of the Office of War Information. From September 21 to October 21 he was in London, with side visits to the field. He visited "A" Force in Algiers and Cairo, in Italy discussed deception implementation with Strangeways and Fairbanks, and met on December 10 with Hunter and GSI(d). He returned to Washington in January 1944, impressed by the high development of British deception planning and implementation. American operating agencies, he concluded, needed more interservice coordination and a "central authoritative source of help"; Joint Security Control should be charged with developing devices and equipment and organizing and training personnel for deceptive operations.

Kehm passed through Turkey on his way home and shot down one of Hunter's ideas on the way. Hunter had proposed to write letters containing indiscretions to American officers interned in Turkey. They would leave the letters lying around for prying eyes to read, and perhaps spread a rumor or two. Hunter had asked Smith for the names and nicknames of two or three such officers for this purpose. Smith held off until Kehm's return, when Kehm told him that after canvassing the idea in Turkey and Cairo he had concluded that the project should be dropped and had so advised Hunter.

In January 1944, Fleming proposed a more far-reaching reorganization than his October one. Under it, GSI(d) and Hunter's Special Planning

Section would be transferred to SEAC and consolidated into one entity to be called "D Division," situated in the administration section of the staff, independent of operations, planning, and intelligence, but with easy liaison to them; much in line with Dudley Clarke's thinking. A small element would be left with India Command. D Division would be divided into substantially the same sections as had been envisaged for GSI(d) in the October proposal, plus a section for liaison with MacArthur. Hunter discussed this proposal with Wedemeyer and reported to Smith that Wedemeyer "expressed the definite desire to have me accompany him to Ceylon as a part of the Supreme Allied Commander's Staff." By the end of the month, however, he was reporting that it did not look as if SEAC would be organized on the scale envisioned, nor that final action would take place on the plan at any early date.

Perhaps this uncertainty was the last straw for Hunter. At the beginning of February he applied formally for transfer of himself and his section to Stratemeyer's staff. He told Newman Smith that he had "definitely decided that there is no chance of getting ahead with the work as a CBI Theater Section or with the ground forces." Goldbranson and Smith called off their search for the major that Hunter had requested.

Hunter's application was granted on March 8, and he was designated as deception officer for the USAAF, India-Burma Sector; at the same time the major Air Force commands in India-Burma were directed to designate an officer responsible for deception, with full-time officers requested for the future; and Chennault's Fourteenth Air Force was invited to establish formal liaison. "Already I feel as if I am part of an organization," wrote Hunter happily to Newman Smith, "which feeling I never had with Rear Echelon."

Soon thereafter, Hunter acquired a third officer, Captain Laurence Sickman, to join himself and Grafton. Sickman was a thirty-seven-year-old air intelligence officer who in civil life had been a specialist in Chinese art and archaeology, had lived for some years in China, and could both read and speak the language. Hunter sent him to be the deception officer at Chennault's headquarters.

In March 1944, Fleming's craving for action got the better of his judgment. Over the vigorous protests of Peter Thorne, he went personally along on Orde Wingate's expedition in support of the offensive in northern Burma, in which his friend Mike Calvert played a substantial role. The glider in which he was riding crashed in the jungle and he and his party had an adventurous few days making their way back to the British lines. He should have been court-martialed for this, for it was a sacred rule that no one cleared for ULTRA must ever get into a position where he might fall into enemy hands. But all he got from Mountbatten was a mild dressing-down.

At the end of the month he was off again in connection with a deception called BLANDINGS. This was a tactical continuation of an operation called UKRIDGE, which had run since January. UKRIDGE had been aimed at deterring the Japanese from reinforcing Burma from elsewhere, and focusing their attention on central Burma, by threatening landings in the Andamans and on Ramree and Sumatra; together with an airborne threat (MALINGER) to Mandalay. The "story" of BLANDINGS was that Allied airborne forces in conjunction with seaborne forces would attack the Taungup-Sandoway-Prome area between Rangoon and Akyab, to cut Japanese communications to the Arakan area and to establish airfields to provide fighter cover for bomber operations to the east and south. Grafton had been working closely with GSI(d)—he was spending at least half of every day with them, and Hunter felt that he had "their complete confidence"—and Fleming offered to take him along; Hunter approved, as did Lieutenant Colonel Rusk.

Grafton's specific assignment was to implement a haversack ruse code-named ENVY, planting a glider containing documents and equipment on a sandbank in the Irrawaddy somewhere near Prome. He and Fleming flew to the headquarters of the legendary American aviator Colonel Philip Cochrane, commander of the aggregation of aircraft that supported Wingate's and Merrill's long-range penetration groups. Fleming moved on and Grafton went over the project with Cochrane. Cochrane concluded that the job could be tried by sending in two pilots and two gliders, one of them the "plant," and snatching off both pilots in the other glider. But the operation was extremely risky, for gliders landing on soft sandbanks were usually wrecked; and the target area was some three hundred miles from the nearest Allied landing fields, far beyond light plane range.

"Tell Colonel Fleming," said Cochrane,

that if he says so, after knowing exactly what is involved, I will ask for volunteers, will get them, and will try it. The men who volunteer will have to be told in advance that in all probability they will have to take their chances walking out of the Prome area and frankly, I would feel like recommending them for decorations regardless of the results. We would, of course, take every precaution and might be lucky enough to make the landings and the snatch without incident. I would also take considerable risks to pick them up in a light plane as soon as they get back within range. There's the story.

Fleming decided that ENVY was not worth these risks, and canceled it. He suggested that Grafton join the tactical deception unit assisting the Chindits' operation in north central Burma. Grafton spent something over two weeks with them, participating in a variety of tactical missions. "I'm afraid I didn't contribute much to anything, but I feel that I learned a great deal," he reported to Hunter. "I also had a good time, and made a lot of very good friends."

To underscore his devotion to amphibious warfare (and hence the threat to the rear of the Japanese in Burma), Mountbatten had intended from the outset to locate SEAC headquarters at Kandy in Ceylon. The move from Delhi to Kandy finally took place in April 1944. This created a problem for Fleming. Much the most important machinery for strategic deception was in Delhi. It, and Fleming himself, must stay in Delhi to keep in close touch with DIB. Yet Kandy was where strategic planning would take place. Initially (though recognizing that Mountbatten would resent it), Fleming simply sent a token staff to Kandy, composed of Robertson MacDonald for naval matters and Mervyn Horder for air; Thorne for the Army went down to Kandy from Delhi every six weeks or so. At Kandy, D Division had offices in one of the "lavish array" of huts that sprang up in the Botanical Gardens, hard by the King's Pavilion in which Mountbatten set up his headquarters. On the positive side, the move opened up much more adequate office space in the western block of the Army headquarters building in Delhi, and brought about the enlargement of Fleming's command, with promotion for himself and others. He became a full colonel as of March 10—over his own protest, for he knew this would reduce his chances of getting a line assignment—

while Wilson, and in due course Thorne, were upgraded to lieutenant-colonel.

These changes were not authorized without some difficulty. More than one senior officer choked on the lavish scale of Mountbatten's headquarters, which quickly reached nearly ten thousand men and women, with what many thought an unseemly level of opulence for a wartime headquarters; Fleming found it "a perpetual *fête champêtre,*" while his old boss General Carton de Wiart, now British liaison in Chungking, compared it to the court of the Archduke Charles of Austria, Napoleon's foe. With this attitude went a lax approach to security that horrified American visitors. A very senior Treasury official was sent out from London to look into aspects of the matter, including Fleming's proposal to enlarge his domain. Thorne, with his staff college training, found himself delegated to defend the proposal. The senior official proved to be well-informed and skeptical, but to Thorne's surprise and delight the proposal was accepted in full.

At the same time as SEAC moved to Kandy, Stratemeyer moved his headquarters, including Hunter, to Calcutta. Hunter set up shop there in mid-April in the old Hastings Jute Mill.

New organizations, new headquarters, that spring of 1944; perhaps their luck would change. Perhaps at last the deceivers in the Far East could accomplish something comparable to what their colleagues were achieving in the West, where what Dudley Clarke had begun in the winter of 1940 was moving towards its climax.

CHAPTER 11

American Deception Grows Up

Soon after the designation of Joint Security Control's "Third Section," about the middle of June 1943, Colonel Jackson's health forced him to step down and Newman Smith took over deception for Joint Security Control in name as well as in fact. He was joined in August by Carl Goldbranson. In September, Lieutenant Eldon Peter Schrup, known always as "Pete," a Naval Reserve officer, reported for duty. Schrup was another Iowan, from Dubuque; a big man, six feet three, in civil life a lawyer with the Federal Trade Commission. He had attended the Naval Intelligence School at Norfolk, Virginia, before being assigned to Smith's section. Hal Burris-Meyer joined the section at about the same time or soon thereafter and thereby balanced out the Navy's representation.

Matters were bureaucratically regularized in November 1943, when the component officers of the staff of Joint Security Control were officially detached from their services and assigned to the Office of the Joint Chiefs. The Special Section—no longer the "Third Section"—of Joint Security Control was formally designated for "Cover, Deception, and Task Force Security." By March 1944 Joint Security Control was housed in Room 2B656 of the Pentagon building—that is, in the second ring of offices from the center, on the ground floor.

---

* In November 1945, Newman Smith prepared an "Informal Memorandum of the Origin, Development, and Activities of the Special (Deception) Section, Joint Security Control, 1942-1945." Long believed lost, it was finally found by Lt. Col. Robert R. Mackey USA and declassified in 2007. It settles a few earlier uncertainties about the timing and progress of the early development of Smith's Special Section. Otherwise it succinctly summarizes the history of American deception as told in this book; and in particular it reconfirms that the British gravely erred in thinking that Smith was not alive to all the nuances of deception and to American shortcomings in that field. See further the Addendum at p. 807 below.

Smith—a full colonel from January 22, 1944—was in general charge. He handled contacts with high-level military, civilian, and diplomatic agencies, and represented the section on departmental and combined-level deception matters. Goldbranson was his No. 2 and was in charge of analyzing overall and theater plans; developing and recommending plans for continuity of deception; coordinating implementation, research and development, and theater personnel and materiel requirements, selection, and allocation; and compiling information for dissemination to theaters. Burris-Meyer was Goldbranson's Navy counterpart. Schrup monitored the development of double agents, coordinated information passed through these channels, developed sources of information analyzed publications to pick up suitable chickenfeed, maintained separate records for each double agent, kept a journal of activities with respect to implementation of each approved deception plan, analyzed propaganda directives, and studied reports on the monitoring of enemy communications.

Smith personally conducted liaison with Robert Cunningham, head of the FBI's Espionage Section, and personally signed off on proposed messages to be sent through FBI channels. Dorothy Stewart, a WAAC lieutenant who had been an enlisted woman with Goldbranson at "A" Force in Algiers, assisted in this and took care of other administrative duties. Colonel William C. Bentley, Jr., of the air staff—he whom Ormonde Hunter blamed for undercutting his ideas in connection with the B-29—though never formally a member of the Joint Security Control staff, served in effect as its air member, and had two lieutenant colonels to assist him as needed; Bentley had personal experience of tactical deception in North Africa. Lieutenant Commander J. Q. Holsopple of the Navy and Lieutenant Colonel Handy of the Air Force, who had participated in setting up the wireless security and deception system in the Mediterranean the previous May, and Major Lawrence C. Sheetz of the Army, were technical advisors in signal matters. Captain Pop Grosskopf, as Security Control Officer under the Director of Naval Intelligence, remained the key high-level contact on the COMINCH staff. In May 1944, nearly a year after taking over, Smith could record that although there had been a "lack of adequate assistance in the Deception Section in the early days of Joint Security Control, when this activity was new in this headquarters as well as in the theaters, and we were pioneers in the entire subject, . . . the various deficiencies have been, to a large extent, overcome, and . . . the deception staff of J.S.C., both Army and Navy, is quite adequate for current work."

As the unfinished official history observed, "with Joint Security Control, as with almost every other deception organization in the United States armed forces, function and activity outran formal authority. In almost all cases the obvious organization did the job and recognized the responsibility later. This is particularly notable in the case of the development of an informal planning section." Outside Joint Security Control proper, Kehm and Baumer continued to be closely engaged in deception planning, in constant touch with Smith and Goldbranson, until both were transferred out of OPD to the SHAEF staff in Europe—Kehm in April 1944 and Baumer the following September—where both were in-

(Harriet K. Schrup)

E. P. Schrup

volved to the end of the war not with deception but with the OSS psychological warfare, and other such oddments.

Smith was very clearly in charge, and took his job seriously; maybe too seriously. He was older than the others; and unlike them, he had seen real combat, at its worst, on the Western Front of 1918. They regarded him as solemn and humorless. (Some of them would put him on and he would not realize it; Burris-Meyer told them to stop that, it was not gentlemanly.) We know how playful he and Rosalind were together, but at the office only the "smug and self-satisfied persona" that so irritated Scott Fitzgerald seems to have been visible. "Everyone regarded him as a pain in the ass," said Went Eldredge. "Absolutely no sense of humor at all," recalled Mrs. Burris-Meyer. "Very dolorous. He lived by the book, and that was it." She remembered that he breakfasted every day at the Army-Navy Club, looking gloomy, and everybody would wonder what his wife had done to vex him. (It was probably simple convenience; the club was only two blocks from the Smiths' apartment on 16th Street between K and L Streets.)

Burris-Meyer, by contrast, was a fellow of infinite jest who did much to lighten the atmosphere. Goldbranson brought his good nature, hard work, and "A" Force experience to the team. He and Smith, though different in background and personality, made an effective combination; Goldbranson drafted many of the official communications that went out over Smith's name. Schrup was something of a foil to the rest. To their alternating amusement and exasperation, he brought nine-to-five habits to an office of workaholics. They swore that every day at 4:55 P.M. he closed down all his files, eyed the clock with folded hands till the stroke of the hour, and flew out the door. Johnny-on-the-spot promptness became known as "Schrupping."

George V. Strong laid down the job of G-2 of the Army and ex officio Army member of Joint Security Control to become an advisor to the Dumbarton Oaks Conference, and was succeeded, as of February 7, 1944, by that same Major General Clayton Bissell who had commanded Tenth Air Force in India. A quiet man, forty-seven years old when he became G-2, Bissell was a Pennsylvania native who had officially qualified as an ace in the First World War for shooting down five German planes in single combat. Emerging from the war with many decorations (including the British Distinguished Flying Cross, the medal that meant the most to him to the end of his life), he worked closely with the legendary General "Billy" Mitchell in the 1920s, held successively higher positions, went out to the Far East with Stilwell in January 1942 as the latter's principal air officer, and went thence to the command in India. He became G-2 at an important time, just when BODYGUARD, the great deception plan for the invasion of Europe, was getting under way. Smith briefed him on BODYGUARD on February 9, just two days after he took over his new position, and he directed Smith to report to him every Monday on its progress.

The coming of Bissell was one of the turning points in American deception. Strong, it would appear, had been interested only off and on in actual deception as opposed to bureaucratic authority, but Bissell played an active role in the Special Section's work throughout the rest of the war. "[T]here has been a general increase in interest in [the] high command here in Washington in deception matters and we are much busier now than we ever were before," Smith wrote happily to Ormonde Hunter the day after he briefed Bissell on BODYGUARD. The negative side of Bissell's active interest was that Bevan at a distance, and Bevan's man O'Connor locally, tried on occasion to deal directly with him, bypassing Smith. But Bissell appears to have kept this under control.

One important consequence of Bissell's arrival may have been to clear Smith and at least some members of his team for ULTRA intelligence. O'Connor told Bissell on May 1, 1944, that Strong had been unwilling to keep Joint Security Control fully in the picture with respect to PURPLE WHALES "on account of the possible connection of this channel with Special Intelligence." But Smith and Goldbranson were plainly cleared for ULTRA by then or very soon thereafter.

Until the latter part of 1944 and the looming shift of emphasis to the Pacific, the Navy played a relatively minor role in the Special Section's work. In July 1943, Admiral Train was succeeded as chief of Navy intelligence and ex officio Navy member of Joint Security Control by Rear Admiral Roscoe E. Schuirmann, an Illinoisian and 1912 Annapolis graduate whose most recent assignment had been as captain of the battleship *Idaho*. Upon the whole, Schuirmann appears to have been content to leave such matters to Bissell, though his attention was aroused when Newman Smith and OPD embarked upon Operation WEDLOCK.

Schuirmann's successor in the autumn of 1944 was Rear Admiral L. H. Thebaud. He was to take a more active interest in deception.

Liaison with the London Controlling Section through O'Connor—he became a full colonel in February 1944—was generally efficient. As will be described later, this was misleading; in his private communications to Bevan, O'Connor was contemptuous of American deception generally and of Smith in particular. Even on this smooth surface there were occasional bumps. Smith and Goldbranson suspected that O'Connor was not clearing with them information about American forces for special means transmission. Eventually, especially in light of some items Dudley Clarke had passed without proper clearance, Smith got Bissell to nail down with Bevan in writing the understanding that neither country's deception agencies would make any mention of forces, materiel, or activities of the other unless they were specifically incorporated in an approved deception plan or allotted to the theater commander charged with implementing the plan. But Smith remained suspicious that Wild, Clarke, or Fleming might occasionally introduce fictitious American units into order of battle without clearance with Joint Security Control. His own ability to supply American foodstuff for British channels was hampered by the sensitivity of the theater staffs and the Joint Planners to anything that might look as if Joint Security Control were trespassing on their territory or exercising command authority. He could initiate nothing, not even minor messages for double agents, for use in a theater except pursuant to an approved theater deception plan; and in Southeast Asia there were none.

The Strong-Bevan agreement of 1943, that without permission neither ally would run double agents in the territory assigned to the other, was generally observed. The British cleared with Joint Security Control MI5's transmission of deception material through controlled sources in Canada, and Felix Cowgill, the head of Section V of MI6, personally visited the States in May 1944 to clear opening up a British double agent in Montevideo (presumably LODGE, to be described in a later chapter); in both cases subject to the understanding that all material passed would be cleared through Joint Security Control. Except for a period of misunderstanding when Section V thought it was free to distinguish "penetration" from "deception" material and pass the former without clearing with

Smith, this arrangement worked satisfactorily. But Smith would still decline to clear material for LODGE unless it was required to implement an
approved plan.

Promptly after taking charge, Smith moved to put American notional
order of battle on an organized basis after it transpired that Army Ground
Forces had activated a real division with the same number as one that had
been provided to Dudley Clarke as a bogus one for the Middle East. It
was clear that this aspect of deception could not be held as closely as it
had been, so Smith let two Army Ground Forces officers in on the secret,
and they supervised preparation of a new master list of notional divisions. Each such division was equipped with a full roster of bogus component units—infantry and artillery regiments, engineer and medical
battalions, signal and quartermaster companies, and the like, all bearing
appropriate numerical designations—for use as needed (although it was
recognized that it was generally best to identify only the division to the
enemy). In addition, a detailed record of the imaginary previous activities
of each notional division was compiled, for use in adding bits of
verisimilitude to controlled agents' reports. As and when required, commanders were designated for such units, and fictitious unit histories were
provided. Later, notional corps and armies, and even an army group,
were added. The master list was in Smith's keeping, and he would release
units for use by the theaters as needed.

It soon became desirable to provide shoulder patch insignia for these
units. Bill Baumer, accompanied by Goldbranson and by an old West
Point friend, Colonel John Stanley, the chief of the Propaganda Section
of the Military Intelligence Service, went down personally to the
Heraldic Section of the Office of the Quartermaster General—located in
that era in a temporary building at Fort Humphreys (now Fort Mc-
Nair)—and discussed with "the fine old gentleman whose job was creating insignias" (his name was Arthur E. DuBois, chief of the section) the
designing of insignia for notional units on request of Joint Security Control or OPD. DuBois must have suspected the purpose of this request;
and eventually he, and no one else in his office, was let in on the secret.
Over time, shoulder patches for nearly every notional unit from division
up were designed, on exactly the same basis as for real units, including official explanations of the symbolism involved in each patch. Patterns

were released to manufacturers and patches were duly made up. (They can still be bought by collectors.)

In June 1943, by which time a number of notional outfits had been designated, *National Geographic* magazine published a comprehensive full-color booklet of American insignia; and the deceivers were appalled to see that none of their fictitious units was included. Bevan had just designated the bogus 46th Division as notionally arriving in England pursuant to Operation LARKHALL presently to be described, and sought permission to use a real division instead; but Newman Smith would not permit this. He ensured that in the second edition, published early in 1945, this oversight was corrected; and copies of this were set in motion towards Axis intelligence.

On one occasion, a divisional patch was designed in the field. The American deceivers in England devised a patch for the bogus 59th Division, obviously inspired by the "Don't Tread on Me" flags of the Revolutionary War, and GARBO informed his masters that one of his (imaginary) subagents had discovered a division "which has the insignia of the serpent which the Americans call rattlesnake." Their design turned out to be in reverse by customary standards. The Heraldic Section redesigned a proper patch amid some concern lest samples might have been actually worn in the field; apparently they had not, but this contretemps held up production of the second edition of the *National Geographic* booklet for two months.

Smith did his best to impose some centralized direction on the burgeoning but helter-skelter development of American tactical deception skills and devices during 1943.

After the first Beach Jumper unit had been sent to the Mediterranean, a Marine unit was sent to the South Pacific, and a second naval unit to Nimitz's command. In January the Beach Jumper training base was moved to Ocracoke, North Carolina, and Navy tactical deception training continued there till the end of the war. As the Navy took the lead in developing sonic deception, the Army Engineers took the lead in developing inflatable dummies (called "targets" as a cover term)—tanks, guns, vehicles, and landing craft. The Air Force Camouflage School at March Field, California, developed dummy aircraft and equipment, and its Third Air Force Deception Unit at Esler Field, Louisiana, developed a schedule of dummy equipment and produced an extensive two-volume manual on its use. But none of the services was fully aware of what the others were doing.

Looking ahead to the invasion of the Continent, in October 1943 American headquarters in Britain ("ETOUSA") sent Major Ralph Ingersoll to the States to explore obtaining trained tactical deception units for the theater. Ingersoll found his way to Joint Security Control. Smith convened an informal conference of various agencies, one upshot of which was an agreement by the Army Engineer Board to produce pilot models of various dummy devices.

After Ingersoll reported back in London, his theater requested a tactical deception unit capable of representing an entire corps of one armored and two infantry divisions. Under Goldbranson's prodding, the Army developed "23d Headquarters Special Troops" to fulfill this requirement. Composed of a three-company camouflage battalion (its personnel largely young commercial artists from New York) capable of representing two infantry and one armored division respectively, a company to represent antiaircraft and medical installations and corps artillery, an engineering company to do the spadework, and its own transportation, its complement totaled 75 officers and 908 enlisted men. Formally activated on January 20, 1944, the 23d trained at Camp Forrest, Tennessee, shipped out for England on May 2, and first went into action in Normandy in July. An associated sonic unit, the 3132d Signal Service Company (Special) was equipped to make the noises of an armored division, and troops for antitank and antiaircraft protection were attached. The 3132d was one of three sonic deception companies (the others were numbered 3133 and 3134) trained in 1943 by the Army Experimental Station, an organization for research and development in sonic and allied deception at Pine Camp, New York—an outgrowth of the early experiments at Sandy Hook.

Closely related was the 3103d Signal Service Battalion, a radio communications deception unit designed to be able to simulate a corps, equipped to mimic the traffic of an infantry division, an armored division, and a corps network simultaneously. Organized under the Chief Signal Officer at the theater's request, it trained at Fort Monmouth, New Jersey in January 1944, and was to play a leading role in the deception for the Normandy invasion.

These were tangible accomplishments, but there was lack of coordination and wasteful duplication, and not enough information was disseminated to the field. Army, Navy, and Air Force radio traffic was not coordinated, which itself could compromise deception efforts. Perceiving Joint Security Control as the logical central entity for bringing some order to the situation, Smith, Goldbranson, and Burris-Meyer did their best to coordinate all this activity. As but one example, when in November AFHQ wanted one hundred dummy LCTs for the Mediterranean for the spring of 1944, Joint Security Control brought the Navy's Readiness Division and the Army Engineers together; it transpired that the Navy had been working for over a year on dummy landing craft, unaware of the British BIGBOBS developed in the Middle East. But without clear authority from the Joint Chiefs, Smith, Goldbranson, and Burris-Meyer could only cajole.

Then, indirectly, Johnny Bevan came to the rescue.

At this point Bevan was under some pressure and he may well have felt that a summary of deception successes would be helpful. On November 13, he pointed out to the British Chiefs of Staff that "So far this war planning and implementation of strategical and tactical deception has, by and large, been limited to the Mediterranean Theater," and that "a report on the experiences gained during the last year would prove most instructive to Theater Commanders and senior British officers in the United Kingdom and India." As a result, the Combined Chiefs asked Eisenhower at AFHQ to report on deception and deception organization in the Mediterranean.

Under date of December 21, AFHQ responded with a succinct account of deception in the Mediterranean theater over the past year— probably drafted initially by Dudley Clarke—with an appendix consisting of a report dated November 11 from Admiral Hewitt to COMINCH, drafted in large measure by Fairbanks. It described "A" Force under its current organization and procedure. It described how, once "A" Force prepared a plan, it was referred to G-3 to initiate the necessary action and keep a check on progress made. It described the strategic, tactical, and order of battle deception plans carried out and the extent of their success, and pointed out the fundamental importance of order of battle deception in leading the enemy to believe that strategic deception plans were within the Allies' capacity.

The report recommended, first, that "while 'A' Force has functioned very satisfactorily, any organization of a similar character hereafter established be constituted along 'Inter-Service' lines with personnel drawn from Army, Navy, and Air Force"; and, second, that "due to the extent to which strategic and tactical cover plans have assisted the attainment of real objectives, no major operations be undertaken without planning and executing appropriate deception measures."

Admiral Hewitt's report went even further, recommending not only an interservice organization, but that one be established by the Joint Chiefs to develop and present overall strategic deception plans; supply theater commanders with trained personnel; develop and furnish special equipment; and be prepared to help with other special service such as prisoner escape, agent infiltration, and other clandestine work.

This gave Newman Smith and his colleagues their opening. They got the Eisenhower report referred to them by the Joint Chiefs for comment and recommendation. They then prepared for the Chiefs an account of the problems created by lack of centralization and coordination. They pointed out that no joint level organization had been specifically charged with responsibility for interservice coordination, selection, and training of personnel; priorities and allocation of deception personnel and equipment to theaters; preparation and promulgation of deception information; and inter-Allied deception affairs. They proposed accordingly that their charter be amended to confer these responsibilities upon Joint Security Control. With some watering down at the Navy's insistence—mainly to ensure that Joint Security Control would have no executive authority—an amended version of their recommendations was adopted by the Joint Chiefs on February 1, and distributed to the theaters and major commands and to the British Joint Staff Mission.*

While this augmented charter for Joint Security Control—it was officially "JCS 256/2/D," replacing the old "JCS 256/1/D"—was less than Smith and his colleagues had hoped—and less than Admiral Hewitt had

---

* Douglas Fairbanks, a lieutenant commander since October, was in the States on home leave around the turn of the year and did some lobbying in Washington for the ideas in Hewitt's memorandum. (En route through London, he had come down with jaundice, and had shared a hospital room with Ewen Montagu.) Some of the Washington brass were suspicious of a movie star's credentials as a genuine fighting man, and it is at least possible that some of the Navy's opposition stemmed from this.

recommended—it was a major improvement. One immediate favorable development was a formal offer from another joint body, the Joint Communications Board, to make experienced communications officers available as advisors to Joint Security Control.*

Smith moved promptly to establish liaison with the FBI concerning the use of double agents for deception purposes, and reserved this function to himself, with Goldbranson's backup.

For reasons of simple geography, the FBI's double agents were far less numerous than those caught and run by the British in England or India. They were all German agents; the Japanese were simply not in a position to attempt any appreciable penetration of the United States mainland. Unless they failed to recognize the significance of an item, the Germans were generally conscientious in passing to the Japanese intelligence material of use to them. Accordingly, for much of the war, most information destined for Japan was in fact passed through Germany. Not until the winding-down of European deception in the autumn of 1944 was a major effort begun to develop a substantial network working directly to Japan.

Four long-term notional networks were run by the Bureau during the war. Their heads were double agents codenamed RUDLOFF, PAT J, BROMO, and PEASANT. The Bureau also ran several individual agents with no notional network under them, the most important of whom was codenamed TOM X but was generally referred to by his file name, COCASE.

The head of the earliest and perhaps most successful of the notional networks, RUDLOFF, known to the Abwehr as "A.3778," was a native of Argentina, "a nice-looking tall pleasant-looking businessman with some glamor effect," according to an FBI description, forty-six years old when he arrived in the States in late 1941. His real name was Jorge José Mosquera. The son of a prosperous merchant in the leather export trade, he took over the family business in 1918 at the age of twenty-three and moved its headquarters to Montevideo. He developed trade with Germany and lived in Hamburg from 1929 to 1941. With war looming in

* The general cause of deception education was additionally furthered by a summary of deception to date, which the British Chiefs of Staff directed Bevan to prepare after receiving the AFHQ report. Bevan's summary was submitted to the Chiefs on February 10 and directed by them to be sent to major commanders and to Washington.

the late 1930s, he decided to liquidate his German business and return to Montevideo. But he found that his funds were blocked. To get them released he agreed to become an Abwehr agent. After three trial missions as a clandestine courier to Holland, Portugal, and the Canaries, he received training in secret ink and wireless telegraphy. He showed no aptitude for the latter, and was told to hire an operator when he needed one.

The plan was for him to set up a transmitter in Montevideo, leave it in charge of an operator, and then proceed to the United States and set up a second station there; he would send information from the United States to Montevideo and Montevideo would pass it on to Germany (thus taking advantage of the greater reliability of north-south communication in the current phase of the sunspot cycle, as was noted in Chapter 3). He did not get out of Germany until early 1941, and reached Montevideo in June. There he waited for some time and did nothing. Finally he decided to talk with a business friend and seek his advice. The friend called in a lawyer. The lawyer suggested that they go together to the United States consulate and tell the whole story. The consulate got in touch with the FBI. After some discussion that the British might run him, it was decided to bring him to the United States and run him directly from there without the Montevideo link. Mosquera sent a cable to Germany advising that it was impossible to operate from Montevideo but that he could go on to the United States. His control authorized the change.

Mosquera arrived in the United States in November 1941. The FBI took him off the ship at quarantine to ensure that no one on board would tail him. He received the designation ND98 and the codename RUDLOFF (the reason for this unusual codename is not presently clear; the British had codenamed him MINARET; he may indeed have been given a full name, "Max Fritz Ernst Rudloff"), and was set up in New York as "La Plata Trading Co.," dealing in leather and musical instruments. La Plata Trading Co. had been designed as a cover for drops and visits from other German agents, in line with the successful sting operation against the Fritz Duquesne ring described in Chapter 4; but the publicity given the Duquesne case spoiled any chance at a repeat performance. E. E. Conroy, Special Agent in Charge of the New York Field Office of the Bureau, supervised the case.

Just as the United States entered the war, through a channel in Montevideo RUDLOFF received money to set up a transmitter. A station was established on Long Island—very possibly in the same location that was contemporaneously used for TRICYCLE's communications—and RUD-LOFF (really, of course, the FBI) went on the air on February 20, 1942. His first message advised that he had established an important contact with an engineer at the Republic Aviation aircraft factory on Long Island and at the Brooklyn Navy Yard, and could obtain detailed information from him and his friends for fifty to one hundred dollars per contact. By the next month he was reporting that he had important material and blueprints and needed a courier on a Spanish or Portuguese ship; and soon afterwards he was making arrangements for money. In April he advised that he had added another source, the brother of the Republic Aviation engineer, a civilian with the War Department in Washington. By May he was advising of a third source, a civilian in the Navy Department in Washington.

The FBI transmitter was in a secluded farmhouse on a cliff overlooking a beach on Long Island, more than two miles from the nearest road, visited only by electric meter readers and the like, who showed no suspicion of what went on inside. The antenna was strung in an adjacent wood, invisible to the casual viewer. Power for the transmitter (it was operated at much lower than its rated capacity, not over eight hundred watts) came not from the commercial electricity to the house—that might have made the meter reader curious—but from a gasoline-driven generator in a blind part of the cellar, vented silently to the open air through an oversized Buick muffler. The establishment was manned by three full-time FBI men. One, the purported "owner" of the farm, gave the appearance of being retired for poor health; he engaged in local town affairs, while the other two posed as "workmen" on the place, holding regular jobs in war work somewhere but working different shifts. There were guard dogs and other inconspicuous security.

As an independent businessman RUDLOFF was plausibly free to radio messages to Hamburg daily; he in fact averaged three or four transmissions a week. The FBI changed radio operators occasionally; RUDLOFF would report to Germany that his old operator had been drafted. He himself was kept under constant surveillance and supervision, just in case. In the early days RUDLOFF himself deciphered the inbound traffic from Hamburg. This ended when he discovered from deciphering a message that the Germans were sending him twenty thousand dollars. He must have half of it, he said, or he was going back to Argentina. Whether he was given the half does not appear from the available records, but that was the end of his deciphering.

RUDLOFF was a big spender. The Germans encouraged this, and paid him accordingly. This supported the FBI's conclusion that the Germans believed him implicitly. He took in more than fifty thousand dollars from early 1942 to March 1944. The money came by various channels, usually through a cutout in Montevideo. Once, as has been related in Chapter 4, sixteen thousand dollars meant for him was dropped from an airplane out of Buenos Aires at an agent's place in Montevideo—an agent friendly to the United States.

One reason for the big spending was that RUDLOFF was a ladies' man. He had a wife in Montevideo and had had a mistress in Hamburg. In New York he acquired a replacement for the latter. The lady was thirty-nine years old, of Italian extraction, had ambitions to sing opera, and did not know that RUDLOFF was married. RUDLOFF set his lawyer in Montevideo to work on a divorce, and pressed the FBI for money to help his new friend's ambitions. The Bureau actually arranged an audition at the Metropolitan Opera for her. By 1944, RUDLOFF was again threatening to go back to Argentina if the Bureau did not give him fifteen thousand dollars to further her career. Whether they came through with the money does not appear from the available records. He would have found it hard to leave the country without their consent.

The original notional trio established in 1942—they were codenamed REP (the Republic Aviation engineer), WASCH (his brother the War Department employee), and NEVI (the Navy Department employee)— were to be RUDLOFF's primary notional sources till the end of the war. A few notional subsources were added to these over time: WASCH had an officer friend in the Army Service Forces in San Francisco, and another friend who wanted to set up a radio operation to the Japanese from the West Coast (this was offered in response to Newman Smith's expressed desire to try getting RUDLOFF in touch with the Japanese), while NEVI had a notional subsource called OSTEN. RUDLOFF sent a steady stream of messages right up to the fall of Hamburg at the beginning of May 1945. The bulk of his traffic was technical chickenfeed dealing with airplane characteristics and such matters, seeded, after Newman Smith came into the picture, with deceptive material; plus money concerns, and complaints that he needed reliable help. As was noted in Chapter 4, his case was supervised by Special Agent Gil Levy on the Major Case Desk in Washington.

There were contretemps between the FBI and the British with respect to RUDLOFF. At the outset, the British undertook to supply chickenfeed items for him, all of which would be true. In April 1942, they furnished an item for RUDLOFF to use suggesting that Tokyo would be bombed. Three days later Tokyo was indeed bombed, by Jimmy Doolittle and his raiders. Mickey Ladd was enraged. How had they known about the Doolittle raid, and why would they include it in foodstuff? The British had to admit somewhat shamefacedly that the item was of course false (or had been thought to be), and had been provided in hopes that it would help draw off Japanese airplanes and carriers from Burma and the Indian Ocean. Ladd came away from this convinced that the British, in Guy Liddell's words, "were trying to pass off a piece of dud information on Mosquera which would have jeopardized his position."

The British got their own back a few months later, or thought they did, when their monitors detected a RUDLOFF message informing the Germans of a change in British command that had not been publicy announced. They protested, in view of an agreement Liddell had made with Tamm and Ladd to cooperate in the running of double agents of mutual interest and share copies of their traffic. The Bureau undertook to furnish copies of all RUDLOFF traffic to Stephenson's British Security Coordination in New York—or Stephenson thought they did. By this time, Hoover loathed Stephenson and all his works, and it is to be doubted that this was in fact done. Certainly, as late as 1945 the British were intercepting and deciphering RUDLOFF messages.

The second major double agent run by the FBI was a Dutchman named Alfred Meiler, referred to as Walter Köhler by the Abwehr and code-named PAT J by the Allies. His control was Lieutenant Commander Thomas Hübner of the naval branch of the Abwehr's Paris main post.

Meiler, as was mentioned in Chapter 4, was identified through ULTRA and turned over to the Americans by the British. A diamond cutter by trade, he was a heavy-set man of middle height with horn-rimmed glasses, "expressionless to hint of pleasantness," noted the FBI. "Looks very much like normal Hollander in earlier [1914–18] photograph. Looks now more like mixed Irish." His English was poor; and, as was noted in Chapter 4, his German was filled with Dutch locutions that made it difficult to compose authentic-sounding messages for him.

Meiler's own story was that he had been recruited in 1941 by an Abwehr officer named von Bonin. He had been a German operative in the 1914–18 war, and the FBI was inclined to believe that he might have been one continuously since then; or he might have been contacted and reemployed at any time from 1923 on. Whatever the truth, the Abwehr trained him as a radio operator at a hotel in Paris used by them as a training center, and gave him the assignment to go to the United States and send back information. He apparently had some engineering or scientific training and his questionnaire included a list of desired information about American atomic research. In Paris in May 1942 he and Hübner received a lecture on basic principles of atomic physics from one Henry Albers, a scientist with the Reichsforschungsrat, or Reich Research Council.

Furnished with money—five thousand dollars, he claimed, but it might have been as much as fifteen thousand dollars, in view of what transpired later—he and his wife were smuggled out to Spain. When he got to Madrid in April 1942 he turned himself in to the Dutch consulate. Through the good offices of MI6 he was directed to the American consulate. When that office queried Washington the FBI told the State Department to send him on. With his wife he took a Portuguese ship for the United States. Contracting pneumonia on the voyage, he was at death's door off the coast of Florida when the United States Navy took him and his wife off the ship and rushed him to a hospital. He recovered, but lost some twenty-five pounds and was in poor condition thereafter; at one point he became so despondent that he tried to commit suicide.

In mid-July, the FBI took the couple to New York and installed them in a hotel. PAT J claimed to have been given only five thousand dollars and to have brought only half of that with him. His wife had smuggled the other ten thousand dollars sewn into her girdle and had buried it before they left Florida. The FBI found out about this; PAT J tried to deny it but eventually admitted it and claimed that the money was his own from salvage of his business. The FBI kept the money and held it over PAT J's head as earnest of good behavior. After writing a "sign of life" letter to show that he had arrived in the United States, directed to "Uncle" at an address in Spain, he was set at liberty—under constant surveillance, of course—and a job was found for him in New York as a diamond cutter at twenty dollars a week, later raised to twenty-five dollars. His wife worked off and on, sometimes as a waitress. He had wanted to be set up in a radio parts shop, but the FBI had no intention of giving him access to materials he might have used to build his own transmitter.

PAT J had brought with him on a microphotograph the plans for a transmitter. The FBI built a set according to these specifications, so that the characteristics of the signal would match German expectations. He was required to make a number of phonograph recordings of his sending style so that his fist could be imitated. His radio went on the air from the Long Island house on February 7, 1943, and it continued in communication with the Abwehr until the collapse of Germany, originally only one afternoon a week, and later on twice a week; these were supposedly the only times he could get off work. His transmissions went to Hamburg, but until the liberation of Paris he continued to be controlled by Hübner out of the Paris Abwehr main post. The real PAT J took no part in this.

Over time the notional PAT J acquired notional subagents, all evidently billed as persons of German extraction sympathetic to the Fatherland. There was HOLTZ, an imaginary seaman second class in the Navy. There was HERMAN, an imaginary seaman first class in the Philadelphia Navy Yard. There was OTTO, a supposed laborer in the Brooklyn Navy Yard. Given PAT J's station in life, the information he could pass was low-grade: barroom talk, weather reports, tidbits picked up by low-level informants, and items seen on the streets. In October 1943, for example, the FBI advised Smith that PAT J had been asked to find out what divisions or regiments were being shipped to England or North Africa, and to report letter and number groups seen on the bumpers of Army vehicles in New York, with the place and time of observation. In regard to the first, said the Bureau, "the informant can explain that he does not have access to such information, or we can furnish some information if the Army so desires, and will make it available to us." As to the second, "The informant cannot give any logical excuse for not furnishing such information, as anyone in New York City may observe these vehicles without arousing suspicion."

As with all agents, a good portion of PAT J's traffic dealt with money. Diamond cutters were often paid on a piecework basis, and need for money was the reason he gave for getting off work only once a week to operate his radio. Uncle proposed to send him $2,000 through a Swiss bank; this was turned down as too risky. Eventually that sum did reach him, apparently through an innocent third party who thought he was helping a fellow Jewish refugee. After this he was able to send twice a week, on Tuesdays and Fridays at 1 P.M. Over time the Abwehr sent him two batches of rare stamps by way of South America. Each was allegedly worth $2,000 but in New York they fetched only about $150.

He continued to howl for money. Eventually the Germans sent over a second agent (briefed by Hübner's assistant, a certain Lieutenant Colonel Walter Stockmann), who carried six thousand dollars in jewelry to pay PAT J. This was another Dutchman, a motion-picture executive named Letsch. (He may have been accompanied by a man named Kusters.) Letsch likewise turned himself in to the Americans in Madrid and volunteered to serve the Allies. Codenamed MIKE by the FBI (presumably as in "Pat and Mike"), he arrived in New York in November 1944. The Bureau concluded that he was "forthright, apparently intelligent, and courageous," in Hoover's words. He told them about his assignment to deliver jewels to PAT J; they pretended never to have heard of PAT J but told him to keep the rendezvous. It was held in the lobby of the Astor Hotel; an FBI man whispered the password, grabbed the package of jewels from MIKE, and vanished. A mortified MIKE apologized profusely to the FBI for letting them down. They reassured him that this was the breaks of the game. "Had trustworthy person get jewelry from L," PAT J radioed to the Abwehr on December 1, "and have it in my possession now."

In late March 1945, when it appeared that MIKE might soon be returning—he in fact stayed in the States till the end of the war—he proposed to give Hübner an oral report tailored to Nazi preconceptions and designed to discourage any thought that the Allies would quit short of unconditional surrender or that the Western Allies and the USSR might fall out. When Special Agent Robert Newby of the FBI consulted with Joint Security Control, Smith and Goldbranson disapproved most elements of this proposal. MIKE did not in fact leave for home until after the German surrender. He wanted then to talk with his German connections about the sudden growth of anti-Soviet feeling in the United States, and the Bureau sent this too across to Joint Security Control for comment.

Pat J played a minor role in the run-up to Overlord, as will be seen. He frequently contributed to order of battle deception through reports of observations of unit insignia. Towards the end of the war he furnished some data that it was hoped would reach the Japanese. He last heard from Germany on April 27, 1945. "Uncle will protect your interest in the future, as before," said the Abwehr, "and continue to care for you." But Uncle never did.*

Pat J's assignment to get information on American atomic research connected him—notionally, at least—with one of the leading scientific figures of the age; and may have done that figure a substantial injustice. A manuscript chart of double agents and their networks in the United States as of the spring of 1945 in Pete Schrup's handwriting in the declassified files shows, along with the notional Holtz, Otto, and Herman, two sources for Pat J marked as "Real," one of them "Dr. De Bye". The implication is that Pat J was sending information—presumably obtained from an intermediary, since Pat J himself operated on a far less exalted level—obtained from the great physical chemist and Nobel laureate Petrus Josephus Wilhelmus Debye of Cornell. A Dutch subject, Debye had worked mainly in Germany till 1939, had a German wife and a daughter in Germany, and was denounced as pro-Nazi to the FBI by several refugee scientists; so throughout the war, although he would logically have been a member of the world-class team that was building the atomic bomb, he was denied clearance to do any classified work.

Debye's was one of three names listed on Pat J's questionnaire as scientists with whom he might get in contact about American atomic research, and Hübner told Mike in 1944 that Debye was pro-Nazi. Pat J in fact sent worthless atomic information, and presumably Schrup's indication of Debye as a real source for Pat J meant that the Abwehr was being led to believe that this information originated with him. If this was why Hübner in 1944 told Mike that Debye was pro-Nazi and that information could easily be obtained from him, retention of that fact as

* Declassified documents currently available offer no support for the claims made by the late Ladislas Farago in his book *The Game of the Foxes* that Pat J was in fact a successful triple agent, "The Spy Who Fooled J. Edgar Hoover."

derogatory information in Debye's file worked a substantial injustice. (He in fact appears never to have shown any sign of disloyalty, and was cleared for secret work by 1952.) In any event, the net result was that the Germans got no real information about the Allied development of the atomic bomb from the Abwehr (or indeed from any other source), and none from PAT J in particular.*

The third—and considerably less important—of the main double agent channels to be opened up from the United States was a Spanish Basque named José Laradogoitia, codenamed BROMO and sometimes referred to as LITTLE JOE by his FBI handlers.

Raised on a farm in Urduliz in the Basque country, Laradogoitia had first come to the United States in 1930 at the age of 18, when he jumped the ship on which he had been working as mess boy and joined a brother in Idaho. During the 1930s he worked on ranches and as a small-scale trucker in Boise. At the time of the Spanish Civil War he distributed literature and collected money for a Basque Republican group. In 1940 he was arrested and convicted for passing bad checks, and was subsequently deported back to Spain.

In Bilbao he was approached by a German named George Lang and recruited for espionage against the United States and Canada. He was trained in the preparation of secret ink and was given three ciphers to use. His first assignment, from May to October 1942, was to send shipping information from Rio de Janeiro. His next was to report from the United States. Shipping out as a fireman on the Spanish ship *Manuel Calvo,* which as noted in Chapter 4 was a main vehicle for German couriers, Laradogoitia reached Philadelphia in May 1943. The British had alerted the Americans to his coming. Interviewed first by Navy intelligence and the OSS, he was turned over to the FBI late in the month. He was easily persuaded to become a double agent. Investigation of his years in the United States in the 1930s depicted him as of above average intelligence and a smooth talker despite his lack of education, but a heavy

---

* The Schrup chart also shows one "Dr. Van de Grint" as another real source for PAT J. He has not been identified. In a message of December 22, 1944, PAT J told Hamburg that MIKE had told him that he, MIKE, was supposed to call for money from "V. G." and that "I told him not to do this for the time being as you warned me against him." Conceivably "V. G." and Van de Grint might have been one and the same.

drinker and unreliable. Nevertheless, it was decided to let him participate in his own running—unlike, say, PAT J.

Handled by Special Agent Tom Spencer in the New York Field Office
of the Bureau, BROMO was established in New York and reported to Lang
that he had a job there as a mechanic and that his sources included a
Basque friend working in the ship chandler business, together with what
he could pick up on the waterfront and from newspapers and magazines.
Over the rest of 1943, he sent back to Lang three letters in secret ink addressed to "Serafina Elorriaga, Heros 26, Bilbao," one of his mail drops,
plus a letter by seaman courier. At the end of the year he received a secret
ink letter from Lang implying that he should arrange for radio communication and asking for reports on rumors and data concerning the invasion
of Europe, troop concentrations, shipping information, and the like. At
about the same time he made contact in Philadelphia with a Spanish seaman from the *Manuel Calvo* named Hernandez, who brought money
and a further questionnaire.

Arrangements were therefore made to set him up with a transmitter
and an operator. From March 1944 on, he worked back to Madrid on a
regular schedule, sending chiefly information about equipment and
shipping. In early 1945 he acquired a modest network with at least two
notional subsources. One, ALBERTO, was supposedly a military policeman of Spanish descent in New York. Another was Luís, supposedly an
employee of the War Production Board. BROMO's information was never
very high-grade.

In August 1944 the FBI put on the air from Washington a channel codenamed PEASANT. Though less important than RUDLOFF or PAT J—and
apparently not rated very highly by the Germans—he was perhaps the
most interesting of the four main long-term double agents run from the
United States; not the least interesting thing about him being that
though he was real, he was not in the United States at all. Another was
that this agent of Nazi Germany was a Jew by ancestry and early upbringing. He was an adventurous, irresponsible rolling stone, a great
hand with the ladies, named Helmut Siegfried Goldschmidt.

Goldschmidt was born in Holland in 1895 to Orthodox parents of
German origin, though they had lived all their lives in the Netherlands.
His father, a well-off property owner, died when he was fourteen. His

mother remarried another Orthodox Jew, from Munich, and moved to Germany, and young Helmut attended the Gymnasium in Wiesbaden. Under the influence of a teacher there he renounced Judaism, to his mother's and stepfather's dismay; they fell out, and when he finished school he went back to Holland for his military service. He spent 1914–19 in the Dutch Army, rising to the rank of captain in the reserves.

He was demobilized in 1919 and that same year his mother died, substantially disinheriting him. From this point on he lived a rather aimless and happy-go-lucky life, moving from one job or get-rich-quick scheme to another and continually nicking friends and acquaintances for loans large and small. He worked as a minor government clerk in Java; for Shell Oil in Oklahoma, beginning as a general roustabout and advancing to a desk job involving routine accounting matters; in a gold mining venture; with Fokker Aircraft, in Holland and in South America; and again in Java. This nomadic career was enlivened by problems with women and sometimes with his expense account. Then when war loomed he was called back to active duty with the Dutch Army.

When the Germans invaded Holland in 1940, Goldschmidt got separated from his unit and was charged with desertion. He was convicted but never served time, for in March 1941 he was recruited by the Abwehr. After a full course of agent's training in Hamburg, he was smuggled out of Holland in the spring of 1943, bound for Lisbon, with six thousand dollars and instructions to make his way to North or South America and send back primarily aircraft, naval, and shipping information.

Goldschmidt reached Lisbon on May 27, 1943, and went straight to the Dutch Embassy, asking for help as a Jewish refugee and offering his services as a double agent. Turned over to the British by the Dutch, he was taken to England and came under MI5 auspices. MI5 quickly concluded that he was "an extremely selfish individual, arrogant, extremely difficult to control, and of a very low grade moral character," to quote an FBI account. Within one two-week period he was caught *in flagrante delicto* with "a woman of definite class," with a hotel chambermaid, and on the grass in a park with a "pickup." No doubt this behavior underlay his codename, Peasant. MI5 happily offered him to the FBI. The case fell to the lot of Special Agent Mark Felt of the Major Case Desk. Felt recommended to Hoover that under the circumstances Peasant should be kept in England, telling the Germans that he was going to the United

States, and be run as a notional from the States. Hoover concurred, though not without some concern over the fact that the real PEASANT would be at large in England.

Samples of PEASANT's handwriting and printing and his secret ink materials were obtained for the FBI, together with recordings of his fist from which a Bureau operator could learn to imitate his transmissions. It was decided to put to use his prior employment with Shell Oil by notionally giving him a job in Washington with that company. Shell had elaborate and well-staffed offices on the eleventh floor of the Shoreham Building in the capital, which maintained contacts with such major agencies as the War and Navy Departments and the War Production Board. With such a job and an active social life, PEASANT could plausibly get information of all kinds, of a sort the existing channels based in New York could not provide.

So he wrote to the Abwehr in November that the Dutch minister in London advised against his going to Argentina "because attitude of Argentine Government not very liberal. . . . Unhappily my name not Dutch and not Aryan." On December 8, 1943, he—*i.e.,* the FBI—sent a secret ink letter from New York announcing that he had reached that city. After some months pleading for money and arranging for communication, PEASANT—*i.e.,* the FBI—went on the air from a transmitter near Washington on August 26, 1944, advising that he had a job with Shell in the capital.

Felt had already alerted Joint Security Control that this new channel might soon open, and now he asked for some good foodstuff. "Inasmuch as time is of the essence," he wrote—Paris had been liberated a week before, and the Allies were sweeping across France; many people were sure the war in Europe would be over by Christmas—"it might be desirable, therefore, to give the Germans something apparently of considerable value at an early stage in the traffic."

The most valuable thing PEASANT could give the Abwehr at an early stage was a notional network of good sources. "Have good contact who would furnish details Chrysler production if adequately paid," he radioed on September 16. A week later he sent information on the Pratt & Whitney R-4360 airplane engine, obtained from an aviator friend of Shell's Washington manager. Before the month was out—playing up to the Abwehr's undoubted knowledge of Goldschmidt's womanizing—he reported that at a cocktail party he had met a Wave who was "secretary to a high Navy official," and had "reason to believe that she may develop into a good source of information." By early October he was passing information about airplane production at Ford's Willow Run plant allegedly obtained from a friend at the War Production Board.

From then till the end of the European war, under Felt's management, with the assistance of Special Agent Taylor of the Bureau's Washington Field Office, the imaginary PEASANT passed a steady stream of information—mainly technical and statistical—notionally derived from these subsources. The "Wave friend" never acquired a name other than simply WAVE; the Chrysler man was KLEIN (originally GANTZ); the War Production Board source was ROBERTS; the manager's aviator friend was BATES; the manager himself, never named, was often an unwitting source; an Air Force captain named SAUNDERS was briefly mentioned. Though PEASANT's messages went to Germany, from the beginning the real intended customer for most of his information was the Japanese.

Like all agents, PEASANT nagged constantly for money. Evidently he had never been paid by the time the Abwehr went off the air with him for good on April 28, 1945. He had other troubles with Hamburg. Sometimes they chided him for loose tradecraft. They failed sometimes to recognize the significance of information that was really intended to be passed to the Japanese. Thus, when he advised that the WAVE was unavailable many evenings because of extra work caused by strong Japanese resistance in the Philippines (this was part of a deception effort to make the Japanese expect that MacArthur would not move against Luzon as soon as he in fact planned to do), Hamburg chided him for sending such trivia "since they are of no use to us." And for some reason his messages were often in Spanish, which could get garbled in transmission.

In one amusing flap, Newman Smith decided to send to the Axis via PEASANT the true figures for American production of multiengined aircraft so as to demonstrate the power of the American industrial leviathan. PEASANT accordingly advised Hamburg that he had learned from ROBERTS that in the month of September 1944 Douglas had built 49 four-engined and 323 two-engined heavy combat and transport planes, while Curtiss had built 181 two-engined transports and Beech had built over 100 two-engined light transports. Hamburg retorted a few weeks later (presumably reflecting the reaction of Albert Speer's production experts) that this perfectly accurate portrait of the might of American industry was "*unglaubwürdig*"—incredible, or unworthy of belief. "Your suggestion as to an appropriate response will be appreciated," wrote Felt to Newman Smith. "It would appear logical for PEASANT to be somewhat indignant." "My friend was greatly incensed when I questioned accuracy of his report," PEASANT radioed back. "At great risk to himself he removed record sheets from WPB files to show me and I was able to verify the production totals which I sent you."

These were all minor bumps in the smooth course of dealing with the Abwehr. "Hearty Christmas greetings and much luck in the new year," radioed Hamburg on December 21. "We hope that transmittal [*i.e.,* money] arrives on time." When early in the new year Hoover announced in a blaze of publicity the arrest of two German agents named Colepaugh and Gimpel, who had been landed from a U-boat on the coast of Maine, PEASANT was appropriately troubled. "Did they have money for me or did they know about me?" he asked Hamburg. "I am greatly concerned." "No reason to worry," responded Hamburg, "since both those named were without any connection to you."

The most significant FBI-run double agent in this period with no imaginary network of informants was a distinguished French aviator named Dieudonné Costes, codenamed TOM X but more commonly referred to by his file name, COCASE.

Costes was a First World War ace who had made a famous round-the-world flight in 1927 and a notable east-to-west transatlantic crossing in 1930. Retiring from the French air force in 1939 as a lieutenant colonel, he became a vice president of the Hispano-Suiza aircraft company and manager of its Paris factory. In 1942, under the German occupation, he was pressured into undertaking an espionage mission to the United States, and after the usual training, was sent off that autumn to the United States with his Russian-born wife. In Spain he turned himself in to the Americans. He was sent on to the States, though apparently under some suspicion of intending to be a triple agent.

Costes was met at Miami by two legendary FBI special agents, Earl J. Connelly of the New York Field Office, and Robert Maheu, one of the Bureau's rising stars (in later years he would achieve prominence as one of Howard Hughes's right-hand men). They grilled him for two or three days—Maheu posed as "Robert A. Marchand," a New York black marketeer of French-Canadian origin—and finally accepted him as a double agent. He went on to New York, set himself up in some comfort in the Park Lane Hotel, and began sending secret ink messages back to mail drops in Portugal.

These were mostly chickenfeed; the FBI's main interest in him was for penetration purposes. Not till early 1944, apparently, did they get in touch with Joint Security Control about him. In February of that year, the FBI and Costes concocted a proposed communication to the Abwehr with respect to the forthcoming invasion of France. "If you will give us further detail as to objects to be accomplished," Newman Smith responded diplomatically when this draft was sent to him, "we could perhaps work out a different type of message which would fit in better with present plans and establish a higher degree of credibility." The Bureau replied that the chief goal was to provoke a reply from the Germans, since Costes had not heard from them for some time. This opened a series of COCASE messages through the late spring and early summer of 1944, some of which formed part of Joint Security Control's contribution to the BODYGUARD deception leading up to the Normandy invasion.

The FBI was not greatly enamored of Costes. "Although he might not deliberately try to assist the Germans," they advised Joint Security Control when passing along a copy of a report Costes had prepared, "he would not hesitate to lie or include false information in preparing such a report if he believed the report would enhance his own importance and indicate him to be particularly learned and intelligent. He does not hesitate to resort to falsehoods in any situation and he is especially prone to elaborate upon his own superior qualities and position." In the long run his most useful contribution was to entrap a major German agent, a Frenchman named Paul Cavaillez.

The running of these double agents was carefully tracked and coordinated between Cunningham and Smith. Cunningham would send to Smith a memorandum about each agent that might make a profitable double, with an indication of the kind of data that might be useful for him to send. When Joint Security Control wanted a particular message transmitted, it would be passed to the Bureau on a separate sheet. Similarly, Cunningham sent to Smith each proposed message on its own sheet; Smith would clear it by initialing a copy, or on occasion criticize it, offer suggested changes, or ask that it not be sent. Each message or proposed message was assigned a number—"A-1," "A-2," and so on for messages proposed by Joint Security Control to the FBI, "B-1" and so on for Bureau proposals—and tabulated with a record of when they were sent and through which agent. (There were also "C" and "D" series for proposals to and from agencies other than the Bureau.)

Smith nursed the process closely and from time to time issued memoranda for guidance; noting, for example, that care should be taken not to send too much information which appeared shortly thereafter in the press, and that rather than advise Hamburg that an item of information could not be obtained, "it is preferable to let the matter drag along, on the basis that an attempt is being made to obtain it."

Joint Security Control also maintained close liaison with the communications intelligence people to monitor any appearance in ULTRA and MAGIC intercepts and decrypts of items sent to the enemy. This was not only to track the progress of misinformation through the enemy system and evaluate the extent to which it was accepted, but also to enable Allied intelligence analysts to discount such material when they picked it up. Smith systematically sent a list of special means items to Colonel Carter Clarke, chief of the Army's signals intelligence. In addition, for the same purpose the British furnished to Joint Security Control deception material which they passed to the Japanese, including material through the channel known as SUNRISE, together with a series known as MARBARS which contained other deception material deemed likely to appear in ULTRA traffic. To these was added in the spring of 1945 a service from Bletchley Park called AGONY, through which a specialist reported all deception items spotted by the British in Japanese traffic. In turn, O'Connor was regularly provided with a list—not necessarily always complete, it would appear—of American items released through special means.

It will be recalled from Chapter 8 that the Combined Chiefs adopted on September 17, 1943, a "General Directive for Deception Measures Against Japan in 1943–1944." The ink was barely dry on this document—its official designation was CCS 284/5/D—when a bureaucratic question arose. Joint Security Control had responsibility for "continuity of deception measures during those periods when no deception plans relating to specific operations are in effect." But who should initiate plans for such "continuity"? Bill Baumer, almost certainly in conjunction with Smith and Goldbranson, furnished an answer, adopted without amendment by the Chiefs on October 9, which in effect authorized Joint Security Control to write its own ticket by recommending measures for "continuity of deception" to the Chiefs through the Joint Planners, accompanied by draft directives to theater commanders or other agencies.

Another matter left over from the long gestation of CCS 284/5/D was the material on practical suggestions for deceiving the Japanese, definitions and principles, checklist for plans, and channels and methods of implementation, which had been removed from the paper at the last minute. Smith and friends had worked hard on these and he was not inclined to give them up. He and Goldbranson proposed to send a modified version of them to the theater commanders. After Goldbranson consulted with Baumer as to how to proceed without trespassing on the prerogatives of the theater staffs, the paper went out on October 4.

But no response came in from the theaters, even though crucial offensives were planned for the next few months by both Nimitz and MacArthur.

The clarification of Joint Security Control's charter with respect to "continuity of deception" was followed by a revision of CCS 254/5/D, the general directive for deception in the war with Japan.

On January 6, the day before departing for a second tour of duty with the British planners, Baumer sketched out for Kehm's benefit on the latter's return some notes proposing revision of CCS 284/5/D so as to take into account the decisions at Tehran and Cairo and the changing strategic situation in the Pacific. ("An additional reason for bringing this matter to the attention of the JCS," Baumer noted, "is to forestall proposals by the Deception Staff of the Southeast Asia Command.")

Kehm got back on January 11 and worked on this project for some weeks, consulting with the Joint Intelligence Staff, OPD, Grosskopf, Carey, and Newman Smith. His draft revision—which was not greatly different from the existing directive—was approved by the Joint Chiefs on February 20 and forwarded to the British. There it languished for weeks, somewhat to O'Connor's embarrassment, because of the British Chiefs' uncertainty as to future strategy in Southeast Asia. Finally, in mid-April, Wingate on behalf of Bevan recommended acceptance of the American proposal with only minor suggested changes, mainly that any objective of drawing Japanese troops to Malaya and Sumatra should be subject to decisions as to future operations in Southeast Asia. By informal action the new directive was adopted by the Combined Chiefs on May 26, 1944.

So from the time Baumer first broached the subject it had taken nearly five months for a revised directive—it was officially "CCS 284/10/D," and replaced the old "CCS 284/5/D" of the previous September—to be adopted; and it cannot be said that it worked any noteworthy changes.

Besides organizational and procedural matters, plenty of substantive work got done that first year under Smith's leadership. He and his team grew adept at keeping many balls in the air at once. In the words of the unfinished official history, "on a typical day the same officer might well arrange for the distribution of shoulder patches for notional units in Seattle, clear the blueprints and operating instructions for a useless but impressive piece of equipment for sale to an enemy agent in South America, and originate the text of disclosures to German intelligence representing the characteristics, armament and purpose of the B-29"; and, it might have been added, write a mollifying letter to Ormonde Hunter, try to read between the lines of an O'Connor memorandum to figure what the British were up to, and have a quick lunch with Bill Baumer to get the latest scuttlebutt from OPD and the Planners.

Actual deception plans and operations on which they worked in that period included the American ends of LARKHALL, an effort to inflate the American order of battle in England; of WADHAM, a component of COCKADE, directed against the Germans in France in 1943; and of BODYGUARD and its subsidiary operations, the overall deception plan for the invasion of Europe. On them, Joint Security Control played only second fiddle to the London Controlling Section. But closer to home, Smith and his colleagues were making history in 1944 with the first all-American strategic deception of the war. Codenamed WEDLOCK, its object was to cause the Japanese to reinforce the Kuriles, the island chain that stretches from Hokkaido, the northernmost of the four main Japanese home islands, up to the tip of Kamchatka, drawing off forces that might otherwise oppose Nimitz's Central Pacific offensive and Mac-Arthur's drive up the New Guinea coast towards the Philippines.

In the spring and summer of 1943, while Attu and Kiska, the Aleutian islands which the Japanese had occupied in mid-1942, were being cleared, American planners looked into the possibility of moving on to the Kuriles. Though such an advance was strongly urged by the American commanders in the region, staff studies in Washington were less sanguine; and by the end of September it was clear that there would be no Kuriles operation before 1945 at the earliest.

This was Smith's cue. Stated objectives in the first general directive for Japanese deception, CCS 284/5/D—which at that time had just been adopted—included "To contain maximum Japanese forces in the homeland" and "To draw Japanese air and naval strength into the Kuriles, except at the time of an operation in that area"; and now there would be no operation in that area. On September 30, Smith sketched out an "Outline Plan for Attaining First Objective in Deception Against Japan." It proposed "a gradual buildup of forces, supplies and equipment in Northwest United States, Alaska and the Aleutian Islands to indicate forthcoming operations in the direction of the Kuriles and the Japanese homeland," involving such means as exaggeration of forces, emphasis on the participation of Canadian troops, and use of dummies as appropriate. Though no formal "story" was drafted then or later, the basic idea was a notional invasion of the Kuriles; at this early stage the notional target date was August 1, 1944. Such an operation would meet the basic criterion of playing up to the enemy's own fears and expectations; for the Japanese had built such substantial defensive bases in the Kuriles that it was evident that they were worried over an attack from that direction.

A deception plan for the Pacific should appropriately have emanated from the appropriate theater staff, at Nimitz's headquarters in Hawaii. None was forthcoming. Smith—very possibly in collusion with Baumer—appears to have decided to get the ball rolling through Army channels. The crucial deception directives—CCS 284/3/D, the basic charter, and CCS 284/5/D, the directive for Japan—were sent to the Army's Alaska Defense Command (soon to be renamed the Alaskan Department) and its parent organization, Western Defense Command, and OPD sent to Alaskan Defense Command on November 3 a directive, inspired by Smith and Goldbranson, to prepare a detailed deception plan to deceive the enemy as to plans for the disposition of forces in Alaska and the Aleutians, and support the objectives assigned to the Central and North Pacific areas in CCS 284/5/D. The OPD directive perked up Navy interest in the project as well.

Meanwhile, Smith jumped the gun a bit, approving on that same day an item to be transmitted by RUDLOFF to the Germans—and hopefully passed by them to the Japanese—to the effect that a high-ranking naval officer had made an inadvertent remark at a social gathering implying that there would relatively soon be a large-scale attack on the northern Kuriles, perhaps in conjunction with a feint, also on a large scale, in late November. (This "feint" would of course be the opening of the Central Pacific offensive at Tarawa and elsewhere beginning November 22). Almost simultaneously came an unwitting contribution to the project by Lieutenant General Simon B. Buckner, Jr., commander of the Alaskan Department. In an Associated Press interview—repeated in Japanese news broadcasts—Buckner said that the Aleutians formed the shortest route to Asia and "dominate the Pacific."

In response to the OPD directive, Buckner's staff set to work and forwarded to Washington on January 9, 1944, an outline plan denominated "AD-JAPAN-44," which eventually received the codename WEDLOCK. It was a reasonably imaginative outline, given their lack of experience in such matters; perhaps reflecting the paper Smith had distributed in October. It proposed such devices as an extensive program of news releases and propaganda focusing attention on the northern approach to Japan, heavy increase in Aleutian radio traffic, establishment of a notional "I Alaskan Corps" in the Aleutians, shifting around of real and notional troops, issuance of Arctic clothing to soldiers passing through the Seattle port of embarkation, extensive construction of dummy facilities, increased scouting and raiding against the Kuriles, and dissemination of suitable information to the masters of Soviet vessels calling at Aleutian ports, all with a rough suggested timetable.

After the usual review and comment by the Joint War Plans Committee—meaning, in effect, Baron Kehm—plus some last-minute input from Nimitz on which Newman Smith commented, the Chiefs approved the plan on February 17. OPD formally transmitted it to the Alaskan Department, noting that methods of implementation outside of the Department were to be left to Joint Security Control, and warning that propaganda and publicity were weak means of implementation that must be handled most carefully. Newman Smith discussed it with Bissell on February 22.

The project would depend heavily on deceptive radio traffic, for the vile climate and high seas of the region impaired Japanese air and submarine reconnaissance. Smith and Goldbranson met with Army and Navy signal officers on February 22 to take a first look at the possibilities. On March 15, a message from Smith to Alaska discreetly encouraged further efforts in that direction.

Alaskan Command entered into the game with a will. In February, Captain (shortly to be Major) Josiah Collins, Jr., of Buckner's staff visited Washington and on February 12 got a briefing on cover and deception implementation. On March 18, Buckner's G-2 submitted a meticulous scenario, prepared by Collins, for the notional WEDLOCK campaign. The assault force would total 109,000 men, consisting of five American divisions and one Canadian division, plus headquarters and corps troops. One American division each would stage out of Attu, Amchitka, and Fort Mears (Dutch Harbor), and two more out of Fort Greeley on Kodiak; the Canadian division and the corps troops would stage out of Adak. The objectives would be the key islands of Paramushiro and its close neighbor Shumushu, at the northern end of the Kuriles chain. On D-Day an American division would land on the northeastern shore of Paramushiro (to be followed by another the next day), while a third American division landed at two beaches on the southern end of the island and pinched off the Paramushiro air base. Meanwhile, a fourth American division would land on the southeast shore of Shumushu, on D plus 2 the Canadian division would land on its northwestern shore, and the two would converge on the center of the island. The fifth American division would be held in floating reserve.

This scenario included no naval activity. That was soon remedied. Nimitz visited Washington in early March to discuss Pacific strategy. The upshot was a Joint Chiefs decision on March 12 decreeing, among other things, seizure of the Marianas with a D-Day of June 15 (Operation FORAGER). Two days later, on his way back to Pearl Harbor, Nimitz conferred personally with Buckner at San Francisco on the latter's initiative. He appears to have recognized from this meeting the opportunity that

WEDLOCK might offer as a diversion for FORAGER. On March 22, he sent to Washington a proposed "propaganda"—later corrected to "deception"—plan for the latter operation, the thrust of which was to emphasize threats in the Southwest Pacific and the Kuriles until the Marianas invasion was launched. The Joint Chiefs approved with minor changes—including a direction that the plan be coordinated with WEDLOCK—on April 18. (On May 2 they approved a further amendment, to mollify MacArthur by making it clear that the project must not give away his actual planned axis of advance.)

Meanwhile, the meeting between Nimitz and Buckner had led to a conference in San Francisco on March 23, presided over by Buckner himself, attended by his chief of staff and chief signal officer, signals officers from Nimitz's staff and from the Navy's North Pacific command, plus Lieutenant Commander Holsopple from Washington. It was agreed that, in general, WEDLOCK would be executed under Buckner's tactical direction and Nimitz's overall strategic direction. A joint Army-Navy communications center for deceptive radio traffic would be set up on Adak; a notional Ninth Fleet would be established for the North Pacific, and an associated notional IX Amphibious Force ("IX 'Phib") would be activated, comprising the five notional Army divisions with supporting services. The notional D-Day was changed to June 15 to match the D-Day for FORAGER. The preparatory and initial phases of WEDLOCK would run till May 5; then radio traffic would indicate fleet movements and logistical preparations till May 20, fleet concentration in the North Pacific from May 21 to June 1, amphibious rehearsals from June 3 to June 5, embarkation June 6 to 8, and overseas movement towards the objectives June 10 to 14. Nimitz and Joint Security Control promptly concurred with this outline plan.

Collins visited Washington again in April. He sat in on several interservice conferences, got a thorough education from Joint Security Control, and visited a Navy dummy shore installation. He took back to Alaska a letter from Newman Smith with suggestions and questions, including a query how the breakoff would be accomplished, and a warning that no action should be taken to involve the Canadians without first handling with Joint Security Control.

Joint Security Control furnished five bogus infantry divisions for use in WEDLOCK: the 108th, 119th, 130th, 141st, and, on the first of June, the 157th. Radio traffic depicted them arriving during May at Attu, Amchitka, Dutch Harbor, Kodiak, and Adak respectively. Use of a Canadian division proved too sensitive to attempt, but by good fortune the *Vancouver Sun* published a story, with photographs, at the beginning of May concerning twenty-five hundred Canadians who had volunteered for overseas duty being reviewed by a Canadian general accompanied by an American colonel, and these were put to use and notionally assigned to Adak.

The actual sea and air raids on the Kuriles turned out to be the least satisfactory element of the operation. Cruisers and destroyers shelled Paramushiro and Shumushu several times during the spring, and air strikes were launched against the Kuriles as often as weather permitted—which was not very often.

More significant was the series of special means items launched by Joint Security Control through the double agents and other channels. This began with a message from RUDLOFF to Hamburg on March 22 that NEVI, his Navy Department source, had learned that an operation of considerable magnitude was being prepared in the Alaskan area and that this information was supported by the fact that Alaska was being removed from the Thirteenth Naval District to form a new Seventeenth Naval District with headquarters at Kodiak and Adak. (Establishment of the new naval district was confirmed through one of Fleming's channels early in April.) Eight days later, RUDLOFF reported that NEVI advised that the Navy Department was greatly concerned over the priority being given to the Mediterranean in allocating newly designed invasion craft that had been tested for use in rough Pacific waters. On May 2 he told the Germans that WASCH had learned from a confidential War Department source that Eisenhower had complained over diversion of landing craft to the Pacific.

The channels through which other special means items passed cannot now be clearly determined; but certain it is that the Axis was told that Army tonnage through Seattle in March 1944 was 175 percent greater than in December 1943; that the United States had fifty aircraft carriers in the Pacific; that four big floating docks had been built at Vancouver, Washington; and that special devices to aid aircraft takeoff and landing on muddy terrain had been developed, including half-track landing gear and an anti-mud compound called Stabinol. In the first part of May, items were released to the effect that Army officials were pleased over the report that twenty-five hundred Canadians on the Pacific coast, excited by a visit to Kiska, had volunteered for overseas duty, and that Canadian officers had been heard discussing the matter at a social function. Later in May, in response to a show of interest by the Germans, an item confirmed that the Navy had established headquarters at Adak for Ninth Fleet and IX Amphibious Force, and another reported that soldiers in West Coast bars had been heard to say that they were on embarkation leave and were to report back to camp by June 1.

Additional stratagems included issuing Arctic clothing to all soldiers leaving the Seattle Port of Embarkation, requiring that they sign a statement not to divulge this fact. The Canadians cooperated to the extent of making inquiries for a supply of Arctic clothing sufficient to equip a division. Press speculation concerning the Aleutians as the great-circle route to the back door of Japan was encouraged throughout the winter and spring. In May and June a supply of shoulder patches for the fictitious divisions was ordered and sent on to Alaskan Command. (It is not clear what use was in fact ever made of them.) Even the USSR was utilized as a channel. Joint Security Control sent to the American military mission in Moscow a list of items to be passed on to the Soviets. The USSR being neutral in the Japanese war, Soviet fishing boats plied the North Pacific and were often intercepted and interrogated by Japanese naval patrols. On April 29, a Soviet ship visited Adak, and care was taken to ensure that a new large "Adak Reserve Depot" was visible to the crew, while the captain was fed gossip about feverish activity, fast turnover, and return Stateside of four thousand civilians employed on the island.

In May there was speculation in the press as to the future employment of Admiral Halsey after the formal closing down of his South Pacific theater, in light of a publicly announced conference between King, Nimitz, and Halsey. Halsey was in fact to command Third Fleet in the Central Pacific, but at Nimitz's staff's suggestion a double agent report suggested that he was to undertake a new assignment in the Northern Pacific; and mention of him was suppressed from all publicity for the time being (though there was an uncontrolled, but helpful, United Press story speculating that Halsey might be "going to a new command where his aggressiveness, ability and hatred of the Japanese would find new outlets.")

For the all-important communications deception program, the Army formed up a special "Task Group Nan" of eighteen specially trained officers and forty enlisted men chosen for their cryptographic abilities. It was physically consolidated at Adak with the Navy security unit for the North Pacific to constitute a Joint Army-Navy Communications Center for the project, under the overall control of Nimitz's Navy security unit at Honolulu; outposts were set up on the islands where the fictitious divisions were notionally to be stationed, and at Seattle and other key logistical points. A special cryptographic system was introduced so the Japanese would notice that something novel was afoot, and special call signs for the imaginary Army divisions and for Ninth Fleet task forces were established for the benefit of Japanese monitors. Dummy traffic was carefully planned—analysis of traffic patterns leading up to the invasion of Sicily was used as a base—and the program was well under way by mid-April.

As May progressed, the Japanese could follow the establishment of new divisions; could see Air Force traffic build up between the States, Alaska, and China; could see the buildup of traffic between Alaska and Hawaii; and could observe that the radiotelephone circuit between Anchorage and Seattle was discontinued at the beginning of May, as if for security reasons (this last was an unplanned happy accident). A call sign for the commander of IX 'Phib and a personal one for General Buckner appeared at Adak the latter part of the month. Port traffic at Seattle, Prince Rupert, and San Francisco was augmented to indicate a general logistics buildup and the specific movement of convoys to the staging areas.

Radio traffic in the Aleutians rose to its peak in late May and then fell off as June 15 approached, suggesting that the radio silence customary when an operation is launched had begun. Division call signs began disappearing from their established locations, in the order of their distance from Paramushiro and at times consistent with the sailing time required by each division to reach an appointed rendezvous at the same time, and the traffic of notional transport ships began to appear. With assault forces notionally now afloat, garrison troops and following echelons were indicated as progressively embarking up through D-Day.

The breakoff story finally decided upon was that at the last minute the operation had been canceled for the time being in order to send shipping to the Central Pacific, but the buildup would be maintained looking to-

wards a Kuriles operation in the fall. The radio traffic pattern was therefore designed so as to give the Japanese the impression that a Kuriles assault was contemplated correlative to, but not necessarily simultaneous
with, D-Day in the Marianas; that the plan had been that upon Japanese
discovery of, and reaction to, the force approaching the Marianas, the
WEDLOCK commander would be ordered either to make the assault or
postpone it; and that postponement had been decided upon. On June 13
a few urgent messages went out from the Joint Army-Navy Communications Center to IX 'Phib (implying a change in plans); whereupon one at
a time, based on an average convoy speed of ten knots, the division call
signs began to reappear at their respective bases and the volume of traffic
went back to normal.

The breakoff of WEDLOCK flowed smoothly into a new deception operation designed to cover Nimitz's next move. His objective after the Marianas was Palau (Operation STALEMATE), with a D-Day of September 15.
On June 9 he sent to Washington a proposal for STALEMATE deception. It
should focus, he said, on the timing of his next move rather than on its
direction, for the latter would be obvious to the Japanese. He therefore
proposed a deception "story" to the effect that since shipping shortages
and logistical strain on Central Pacific bases would require six months'
delay before the advance on that front could be resumed, the next American move would be against the Kuriles during the favorable weather to
be expected in that region in September. The Chiefs approved the proposal on June 25.

Joint Security Control organized a planning conference of signal officers, held at Pearl Harbor on July 10 with representatives from the staffs
of General Delos Emmons (who had succeeded Buckner in June),
Nimitz, and MacArthur, with Holsopple and Sheetz representing Joint
Security Control. They worked out a plan to indicate a new threat
against the Kuriles with a probable target date between September 5 and
10; this time, the traffic pattern would depict a generally high level of activity to imply a threat, but with no effort to depict actual embarkation
of troops. Nimitz would again be in overall control. MacArthur's command would be kept informed and would "continue to maintain the appearance of normal operational and administrative communications to
indicate a period of relative quiescence." After the usual staff reviews,

their plan with minor revisions was approved by the Joint Chiefs on August 28. The operation received the codename HUSBAND.

Like WEDLOCK, HUSBAND was implemented chiefly by deceptive radio traffic. The departure of the WEDLOCK supporting task force from the North Pacific to return to the Central Pacific was simulated on July 17, and a move of the 141st Division from Kodiak to Umnak was suggested in early August. Army traffic showed a gradual rise in volume between Washington and Alaska, while Navy traffic indicated continuing activity of Ninth Fleet and IX 'Phib, with an increase in the movement of surface units from Hawaii to the North Pacific. The tempo picked up as the September target date approached, and on September 11 radio traffic began simulating a fast carrier striking force leaving Pearl Harbor for Alaskan waters, fueling and rendezvousing with North Pacific forces, and then remaining in the region as a threat to the area between the Kuriles and the Bonins. Thereafter the tempo fell off, though a high enough level of traffic was maintained to continue an impression of strength.

Dummy construction continued at Attu throughout June and July. ACORN and CUB FILBERT installations were put up, including an airstrip, antiaircraft positions, dumps, a tank farm, and supply depots; and fifty BIGBOB dummy LCTs diverted by Joint Security Control from the European theater began to be in evidence in Attu Harbor. Additional efforts at visual deception ceased thereafter, on Emmons's orders, no doubt in recognition of the total lack of Japanese air reconnaissance. The Attu installation was the only complete FILBERT erected in the course of the war. The BIGBOBS created some embarrassment when not only were eight lost in storms during July but one was blown some three hundred feet onto dry land.

Only a handful of special means items were released in support of HUSBAND. Schrup suggested a series of such items involving an agent meeting a soldier on emergency leave from the 119th Division on Amchitka who would mention a big operation coming up, and meeting the father of an officer from the 157th Division who was taking Japanese language training, but no more seems to have come of these than that RUDLOFF sent in September a report from WASCH concerning a newspaper story on increased language instruction for civil affairs officers; he sent also a report that a submarine was overdue in the North Pacific. In early November, as part of the termination of HUSBAND on October 31, an item was released through South American channels to the effect that swiftly-moving events in the Central and Southwest Pacific had delayed plans for action in the North Pacific.

Secondary deceptions in the Aleutians continued, called BAMBINO, followed by VALENTINE; these will be described in a later chapter.

Postmortems after the war criticized some aspects of WEDLOCK. A study of the wireless deception program fingered a number of technical weaknesses. Another analysis faulted the special means releases for lack of apparent pattern, absence of operational intelligence, and failure to counter contrary intelligence known to be influencing the enemy; but, as the analysis recognized, such shortcomings were inescapable in the absence of a controlling "story" and in view of the newness—and, it might be added, the indirectness—of the channels available.

But the proof of the pudding is in the eating. Japanese intelligence absorbed much of the deceptive information fed to them in WEDLOCK. Their communications analysts drew many of the desired conclusions from intercepted traffic, locating five American divisions in the theater—plus up to two Canadian divisions and an uncertain number of Marines—and reaching the desired appreciation with respect to the volume of shipping in the North Pacific. By June, Japanese intelligence credited the Americans with four hundred thousand men in the theater—the real number had in fact fallen from one hundred thousand to sixty-four thousand during that period—and with from 400 to 700 planes when the real number was 343; believing also that there was a powerful naval force in the region, including battleships and carriers. Until the end of April and early May, local intelligence estimates as well as those distributed by the Imperial General Staff showed great concern over the offensive threat from the Aleutians. The Naval General Staff in Tokyo estimated on June 4 that the Americans were planning operations with one attack group in the Kuriles area coordinated with large-scale operations in the Central Pacific. On June 7, the Vice Chief of the Imperial General Staff alerted all Japanese attachés in Europe to be particularly alert for information about the Alaskan and Aleutian area in view of steady progress in enemy preparations for operations there. Even Radio Tokyo's news broadcasts speculated that the Americans were preparing for an offensive from the Aleutians and predicted that the Kuriles "will one of these days become a main war theater."

Then genuine Central Pacific traffic rose sharply in connection with FORAGER. As a postwar American analysis concluded, "the communication traffic peaks rising in two areas posed simultaneous threats to the Japanese. Under these circumstances, it can be stated that the communications deception practiced under WEDLOCK made the maximum possible contribution. In order to evaluate the relative importance of the two threats the Japanese would have to, and did, rely upon other sources of intelligence."

Their most significant such "other source" was the data fed through special means. During the first half of 1944, the Japanese received from their German allies and from other European sources such bits of information, matching Joint Security Control's special means program, as that there was increased shipping out of Seattle; that Canadian troops would be moved out of British Columbia to the Aleutians; and that the 17th Naval District had been established. In addition, in their 1944 appreciations Japanese intelligence often cited the press releases and general publicity put out by the Alaskan Department as evidence of the growing importance of the North Pacific.

But what mattered, as always, was not what the Japanese *thought* but what they *did*. And here WEDLOCK was most gratifying. Although WEDLOCK does not seem to have convinced Tokyo that an actual operation had been launched and recalled—the Japanese consensus seems rather to have been that the main American thrust would come first in the south on MacArthur's front—there was an all-out effort to reinforce the Kuriles in the first half of 1944. Two additional divisions of Army troops, plus independent brigades and an independent expeditionary force, were moved into the Kuriles in the February-July 1944 period, and a new army headquarters, Twenty-Seventh Army, was formed in April to take over control of Japanese forces in the island chain. The total number of ground troops in the Kuriles rose from some fourteen thousand in 1943 to twenty-five thousand in January 1944 and at least sixty thousand by mid-June.

Out of concern that the Americans would occupy one of the Kuriles from which Tokyo could be bombed, four air regiments had been moved to Hokkaido from Osaka and Manchuria in February 1944, replacing naval aviation units. Then, from late April to June, two air flotillas, composing the Twelfth Air Fleet with headquarters on Paramushiro, were moved into the Kuriles. The total of Japanese aircraft in the region rose from 38 in January 1944 to 589 at the end of June (although thereafter Japanese sea and air strength in the region was drastically reduced, in order to make up for the disastrous losses in the Battle of the Philippine Sea). All these forces were urgently needed to oppose Nimitz and MacArthur. Moreover, a good deal of shipping was lost in attempting to supply these enlarged garrisons; indeed, it was mainly these shipping losses, rather than any lessening of the perceived threat, that led the Japanese to cut back their forces in the Kuriles in the last months of the war.

And finally, when in 1944 the Japanese prepared their SHO or Victory plans for counterattack in the event of four different possible major American thrusts, one of the four anticipated was an attack on the Kuriles and northern Japan.

That the Japanese reinforcement of the Kuriles in 1944 reflected a decision already reached to treat the islands as the northern anchor of their grand defense line does not detract from the success of WEDLOCK; rather, it underscores the extent to which the operation met a prime requisite of deception by playing up to the enemy's concerns. The bottom line was that WEDLOCK encouraged the Japanese to expend a substantial portion of their rapidly declining resources, badly needed elsewhere, to defend an area that their enemies had no intention of attacking. Not much more can be asked of any strategic deception operation than that.

Smith and his colleagues at Joint Security Control had of course nursed and nudged WEDLOCK along from its very beginning, and tracked its course day by day, with meticulous charts and tables maintained by Pete Schrup and Dorothy Stewart.

By the autumn of 1944, with WEDLOCK under their belts and HUSBAND running smoothly, having given vigor and direction to American tactical deception, having established smooth cooperation with the FBI in the double agent program, and having played a small but significant role in the European deceptions of the past year (as will be seen), Newman Smith, Carl Goldbranson, and Hal Burris-Meyer could congratulate themselves on a job well done. With the backing of Clayton Bissell, and the steady support of Bill Baumer and Baron Kehm, they had nursed

American deception to maturity and brought off a strategic deception operation that could be compared without embarrassment to British accomplishments. American deception had grown up.

But not, it must be said, in the British view; then or later.

Unfortunately for the cause of Allied solidarity, O'Connor never fully appreciated this progress, or any of the further progress that Newman Smith would make, right through to the end of the war. He seems to have decided early on that Joint Security Control in general, and Smith in particular, were hopeless, would do nothing, and indeed could do nothing. He communicated this feeling to Bevan, who in turn communicated it to Fleming in India.

Smith was the particular butt of O'Connor's sarcasm. "He knows very well he is not capable of drafting any such document" as a proposed scenario, O'Connor told Bevan in May 1944, "and you will in effect have to do it for him. This, I think will always be the case as long as he is in the chair." Again, he told Bevan that Smith would find much difficulty working out anything concerning a particular proposal that made any sense, and that if Smith put it up to Bissell, that might show Bissell "the nakedness of the land." Again, even if Bissell were to consent to a proposed Fleming operation, said O'Connor, Joint Security Control would never supply suitable material. (All this, when WEDLOCK was running at full throttle.)

When Smith told him that Joint Security Control would consider using British channels against Japan once an overall deception policy for the Pacific War had been produced, O'Connor deemed this a "feeble bromide." When arrangements were being made that American deception items likely to appear in ULTRA would be furnished to the signals intelligence people through G-2 so they could keep an eye out for them in Axis communications, O'Connor reported with apparent relish that "G-2 have no expectation that JSC will in fact produce anything or that they have anything to produce." (But, of course, apart from the G-2 himself and a few other officers, G-2 was not privy to Joint Security Control's activities). "They speak of [their] channels as good," O'Connor told Bevan, "but do not tell me enough to convince me that they are in fact worth much." When ULTRA showed that the Japanese were reading Chinese military attaché messages, O'Connor fretted that "our friends here" might try to use this as a deception channel "in an amateurish way." In the spring of 1945 O'Connor told Bevan that any British recommendation of officers from Europe for work with Joint Security Control would do them more harm than good. O'Connor appears to have derived some pleasure from a report that Smith had opposed Bevan's visit to Washington in 1945 and had been "overruled on all counts."

Bevan naturally concluded from all this that Smith and his colleagues were incompetent. When the possibility arose that officers who had worked in deception at SHAEF in the summer of 1944 might be sent to Washington, Bevan's reaction was that "certain personnel at J.S.C. would be quite upset at the possibility of such officers going and telling them that they know nothing about the job." By extension, he took a dim view of American deception generally. He remarked in September 1943 on "how hopeless the present set up is in Washington." "Washington simply does not understand counter-action," he told Fleming a year later. He prefaced a message to O'Connor in April 1945 with the words "If, therefore, J.S.C. ever do wish to take on strategical deception," and again commented that there were no "signs of the Americans playing on this type of business," and yet again that "the Americans seem to pay little or no attention to strategic deception."

Peter Fleming understandably followed Bevan's lead. In the summer of 1943, concerning a proposed activity he cabled to Bevan: "Judging by local talent and machinery available and by performance during last 3 months Washington are incapable producing a comprehensive adequate scenario." "Joint Security Control are not normally either influential or well-informed," he told Mervyn Horder a year later. In 1945 he described Joint Security Control as "a body in whose capacity to either (a) appreciate Japanese mentality or (b) take a sensible decision in any context I have never heard confidence expressed."

All this was quite unfair. As will be seen at length, Smith did a remarkable job under the constraints imposed on him. These constraints included, not necessarily in order of importance, the placement of deception in intelligence rather than in operations; the American staff structure, interservice jealousies, and bureaucratic turf protection that prevented an American London Controlling Section from ever being created; uncertainties as to strategy in the Japanese war; General Strong's

sense of his own importance, which led him to keep matters to himself; *
General Bissell's suspicions that the British would try to take control of
deception given half a chance, leading him to play matters close to his
chest with them; and—what was perhaps the most frustrating single factor—the rigid requirement that nothing be done absent an approved
theater plan. (Sir Michael Howard has justly called this last "a recipe for
inaction.")

Though O'Connor was daily in the Pentagon, he (and through him
Bevan) seems to have been insufficiently aware of these limitations on
Joint Security Control's freedom of action. That is understandable, for
many of them were internal, and American officers would have been reluctant to discuss them with a foreigner. And surely Newman Smith's
rigid, by-the-book persona could not have helped at all.

It is curious that so far as the declassified files show, Bevan made little
or no effort to communicate directly with Smith. Instead, misappre-
hending the problem, he, and on occasion O'Connor, seem to have
thought they could improve matters by end-running Joint Security Control. At one time or another Bevan approached, to no avail, Strong,
Kehm (and surely Baumer as well), Bissell, Douglas Fairbanks, and Captain Thurber after the latter joined Joint Security Control; sometimes
about specific matters, sometimes to urge the creation of an American
London Controlling Section. O'Connor for a time thought that Bissell
would overrule Smith if only he were made aware of the situation, or that
the Navy would cooperate if only Bissell would ask them. Toward the
end, they seem to have given credence to speculations that the root of the
problem was that an ambitious Bissell wanted to make a name for himself and probably thought Bevan was similarly "far more concerned to
maintain [his] own position than to co-operate with the Americans."

Perhaps American deception was indeed lacking in the imaginative
flair and the insouciance, one might say, of the British. And there were
still things that wanted doing as of the autumn of 1944. The Navy

---

* Guy Liddell recorded that when he explained the work of the Twenty Committee to
Strong, the General "told me that deception needed very special qualifications—dramatic instinct and imaginative intuition—it was very rare that such qualifications could be found in
one man. I knew of course what was coming and was not in the least surprised when it was
made clear to me that the combination of these qualifications was to be found in the person
of Gen. Strong alone, although he did say that he also consulted the naval authorities."

needed to get more closely engaged in the work of Joint Security Control. Most important was the need to place officers with an understanding of the value of strategic deception, and trained in its planning and execution, on the staffs of the major American theater commanders in the war with Japan. But before turning to the story of how Newman Smith and his colleagues addressed these needs, we will return to the European theater and the climactic deceptions of the war with Germany.

CHAPTER 12

BODYGUARD

When the Casablanca Conference of January 1943 postponed OVER-
LORD, the Allied cross-Channel invasion, until 1944, it was agreed that
an inter-Allied staff would be formed to begin planning the operation.
The head of the new planning staff, and the staff itself, were called
"COSSAC," for Chief of Staff to the Supreme Allied Commander (Des-
ignate), no Supreme Commander having yet been named. In March, the
British Major-General Frederick Morgan was appointed COSSAC.
Known to all as "Freddie," he was a capable officer admired and liked by
the Americans as well as by his countrymen.

Morgan's first weeks were spent organizing the COSSAC staff in London at Norfolk House, St. James's Square, and establishing smooth relations with ETOUSA, the United States Army establishment in Britain
headed by Lieutenant General Jacob L. Devers. Not until April 23 did
the Combined Chiefs of Staff issue him his formal directive. It included
not only planning for Allied reentry to the Continent as early in 1944 as
possible, but also preparation of "an elaborate camouflage and deception
scheme extending over the whole summer with a view to pinning the
enemy in the West [during 1943, when the Allied focus would still be in
the Mediterranean] and keeping alive the expectation of large scale cross-
channel operations in 1943. This would include at least one amphibious
feint with the object of bringing on an air battle employing the Metropolitan Royal Air Force and the U.S. 8th Air Force." This was consistent
with the action of the Combined Chiefs some three weeks before, as will
be recalled from Chapter 9, when they had signed off on an overall deception policy for 1943 in the war against Germany and Italy that included threats to Norway and France with the objects of holding
German forces in the West and bringing the Luftwaffe to battle over the
Channel. This last object was a pet of Churchill's—as he confirmed in a
memorandum to Morgan even before the Combined Chiefs' directive

was issued, exhorting "Camouflage and pretense on a most elaborate scale."

Johnny Bevan, fresh from the gratifying experience of the TORCH cover plans and his own promotion to full colonel, played a role in this setting up of COSSAC. Morgan organized his staff along American lines. Bevan urged a deception section, reporting directly to the chief of staff on the "A" Force model; but, as always, the rigid American structure would not permit this. Provision was made instead for a deception section of the G-3 or operations division, largely designed by Bevan, dubbed "Ops (B)" ("Ops (A)" was the section for regular operational planning), with Lieutenant-Colonel J. A. Jervis-Read, known to his friends as "Tony Read," in charge. At that initial stage Bevan expected that Ops (B) might need a substantial physical deception staff, and it was on this that Jervis-Read was to concentrate. Expecting that special means implementation would be handled by his own TWIST Committee, Bevan did not foresee any need for a substantial intelligence component of Ops (B). Its only intelligence officer was Major Roger Fleetwood-Hesketh—in civil life a barrister, "a charming and witty man," according to Wheatley, "with one of the best cellars of claret in England"—who was seconded from COSSAC G-2 to serve as a link between Ops (B) and the TWIST Committee.

COSSAC had only planning authority. Implementation of physical deception and deceptive troop movements would have to come from the service departments. Joe Hollis suggested a special conference, presided over by the Prime Minister, to impress the service commanders with the importance of deception, but the Chiefs of Staff rejected the idea.

Morgan had an outline scheme for the deception operations within a week of his formal appointment, and for the next several weeks COSSAC, with help from Bevan and his London Controlling Section—currently also deep into BARCLAY and MINCEMEAT—worked on a plan. There emerged a family of three operations, called STARKEY (originally called DOMESDAY and then BROADSWORD), WADHAM (originally called BLAST), and TINDALL (originally called UPSHOT), which received the overall name COCKADE. They were submitted to the Chiefs of Staff on June 3 but were not approved by them till June 23. "There has been an appalling delay in getting these plans approved by COSSAC," wrote Bevan to Dudley Clarke, "which was largely due to the fact that General Morgan was not appointed to the job until a comparatively short time ago."

TINDALL was a threat to invade Norway from Scotland, landing at Stavanger to seize the airfield there and advancing on Oslo. It would continue till late August and then be notionally called off to release forces for STARKEY and WADHAM. STARKEY and WADHAM were coordinated as a notional one-two punch across the Channel in September, to be called off at the last minute. TINDALL would then be revived, to last as a standing threat until winter weather should render it implausible.

STARKEY was the feint designed to bring the Luftwaffe to battle: A threatened assault on the Pas-de-Calais on September 9 by fourteen British and Canadian divisions. According to COSSAC's plan, beginning June 15, the main Allied air effort would be directed at the Pas-de-Calais and Brittany, with the heavy bombers joining in for the last two weeks. As many landing craft as possible were to be gathered in the Channel by the end of August, plus heavy shipping concentration at Southampton and in the Thames. Naval support would be provided for the "invasion," including two battleships for shore bombardment. Security measures would be put into effect along the southeast coast from early August. As D-Day approached there would be a substantial buildup of troops in southeastern England, some of whom would actually embark—and turn back—on the morning of September 9.

WADHAM was a threat that in late September, some three weeks after the STARKEY landing in the Pas-de-Calais, when German reserves would have been drawn to that front, the Americans would invade Brittany, seize Brest, and move on eastwards. When STARKEY was notionally called off, WADHAM would be "called off" too.

The threat to Norway, TINDALL, would then be revived, threatening a landing in the Stavanger area in September—changed to November when Bevan urged that this was a more plausible timing. He urged further that a threat to Narvik and points north would be more effective than one against Stavanger; but the Chiefs of Staff disagreed, holding (erroneously, in fact) that the Germans would recognize that Allied inability to provide fighter cover over northern Norway would render such a threat implausible.

It will be recalled that this was a period in which Bevan and his London Controlling Section had to endure sniping and backstabbing from

Ewen Montagu of the naval intelligence staff. One of the elements of TINDALL was the concentration of shipping on the east coast of England; the fact that the Navy preferred that the double agents not draw attention to this gave Montagu a new excuse to snap at Bevan's heels.

It was hardly fair to load upon the new and inexperienced COSSAC staff, in addition to its main job, the chore of planning this hefty package and cajoling the operational authorities into carrying it out; plus taking care that resistance forces on the Continent were not misled into a premature uprising, and forestalling reporters' misinterpretation of the notional operation as a genuine failure by telling them in confidence that what was about to take place was a rehearsal.

The necessity for convincing the Germans that Allied forces were strong enough to carry out operations of such magnitude brought home to the United Kingdom for the first time the need for order of battle deception in Dudley Clarke's style. Separate projects were conducted for the American and British forces, carried out entirely through the double agents and known as LARKHALL and DUNDAS respectively.

LARKHALL, designed to inflate the progress of BOLERO, the American buildup of forces in Britain, began in May. A memorandum of April 23 from the London Controlling Section to Joint Security Control set forth in considerable detail the kinds of help the Americans might provide in terms of "special means" messages and dissemination of rumors in Latin America, with some suggestions as to detailed methods of implementation.* The project was concurred in by OPD on May 7, and approved by General Strong on May 12.

At that time there were 107,000 American troops in the United Kingdom, mostly Air Force supply personnel. The only combat force was the 29th Division; it was due to be joined in July by the 5th Division from Iceland, and in September by the 101st Airborne and the 3d Armored,

---

* Included as models were some examples of gossip concocted for dissemination by British diplomats in neutral countries. A fair sample is one for Switzerland in aid of BARCLAY that was obviously meant to appeal to Nazi ears: "I'm told that Israel Sieff has persuaded Ralph Grimthorpe into selling that lovely villa of his at Amalfi for £75,000. These Jews are amazingly shrewd. No doubt Sieff realizes that the end of the war is in sight already and he realizes that however hard we blitz the Italian cities we should never bomb a little place like Amalfi; whereas a villa anywhere on the South Coast of France is liable to suffer from serious war damage in the next few months."

from the United States. The double agents told the Germans in May and June that the 3d Armored was already arriving, together with the (bogus) 46th Infantry Division. When the 5th Division moved to England it was replaced in Iceland by the (bogus) 55th Division, supposedly specially trained in mountain warfare, while the coming of the 101st Airborne to England was reported at the same time. The Germans were further given an overstated number for Air Force personnel and told that they were mostly bomber crews.

One minor way of exaggerating American strength, at least in the eyes of stray observers, was by confusing British with American troops. British vehicles were marked with the white stars customarily used for American vehicles. (All Allied vehicles were eventually to be so marked, but this fact was not yet announced.) United States troops in the rear areas were dressed in British battle dress with American insignia, passing the "story" that this was being tested against American battle dress.

LARKHALL moved along smoothly, although Newman Smith had to remind the British that such activity should be coordinated by having requests for American action passed to Joint Security Control through ETOUSA. By August, when LARKHALL was shut down, the Germans had been told that 570,000 Americans were in Britain when the true number was 330,000.

DUNDAS, designed to inflate British strength, was less satisfactory. Overestimating the German "Y" service, MI5 feared that the double agent system would be blown unless reports of bogus British divisions were confirmed by corresponding bogus radio traffic. Neither personnel nor equipment sufficient for this was available in 1943, so COSSAC and MI5 would not allow notional British divisions to be activated for DUNDAS. The deceivers had to be content with the "story" that British forces in the United Kingdom had been reorganized for an offensive role, with the former Expeditionary Force renamed (prophetically) 21 Army Group, composed of two British armies and one Canadian army; Scottish Command and Eastern Command became the British Fourth and Sixth Armies respectively; plus a "Home Forces Static," underequipped and largely for home defense.

The Germans greatly exaggerated Allied strength in Britain in 1943; though as usual it is hard to tell how much of that exaggeration resulted directly from LARKHALL and DUNDAS. Fremde Heere West estimated in

October that forty-three Allied divisions were available for a landing in northwest Europe. At the beginning of November, Jodl gave a comparable estimate to an audience of gauleiters in Munich, adding that there was sealift for an assault force of eleven to twelve infantry and one to two armored divisions, plus airlift for twenty thousand airborne troops.

STARKEY and TINDALL were for COSSAC. WADHAM was an all-American show, handed off to ETOUSA.

Lieutenant General Frank Andrews, Eisenhower's replacement as commander of ETOUSA, was killed in an airplane crash on May 3. Lieutenant General Jacob L. Devers succeeded him. Devers assigned WADHAM to his G-3, Major General Daniel Noce. Logically, Noce should have given the action to Major L. J. Abbott, who had been ETOUSA's liaison with the London Controlling Section since the beginning of the year. Instead, he gave it to an able civilian-in-uniform who handled his odd jobs, a bald, six-foot-two, forty-two-year-old captain,

(Harriet Harris Stroup)

Ralph Ingersoll

soon to be a major, who in civil life was one of the most prominent journalists in the United States, Ralph Ingersoll.

Ralph McAllister Ingersoll, known to his friends before the war as "Mac," was born in New Haven in 1900 and brought up in New York City in the best circles (Ward McAllister, the social arbiter of Gilded Age New York, was his great-uncle); though notwithstanding these advantages he was an unpopular boy at Hotchkiss and a social outcast at Yale, from whose engineering school he graduated in 1921. He worked only briefly as an engineer; he really wanted to be a writer. After a few years in New York as a reporter, he was taken on by Harold Ross, editor of a fledgling magazine with a doubtful future called the *New Yorker*. Ingersoll quickly became the managing editor of the magazine and played a great role in its success. But he fell out with Ross after a few years, and in 1930 left the *New Yorker* to work for another successful magazine entrepreneur, Henry R. Luce of Time Inc. He was the prime mover in bringing Luce's newest magazine, *Fortune*, to a commanding position of influence, respect, and financial success. By 1936, Ingersoll was general manager of Time Inc. and the publisher of *Time*. In the latter role he could legitimately say that he rejuvenated that magazine. He also steered its editorial policy towards the leftish views of many of its staff.

Ingersoll and Luce drifted away from each other. Ingersoll evolved from socialite journalist to a full-blown specimen of what Luce would have called a parlor pink. During much of 1938 he attended meetings of a Communist study group before deciding that slavish following of the Party line was not for him. He had love affairs with the Stalinist fellow-traveling Lillian Hellman, and with the more conventionally liberal Laura Z. Hobson. He dreamed of starting a new newspaper, dedicated to fighting "people who push other people around," leftish in tone, carrying no advertising, structured in a novel way, reading like a magazine rather than a newspaper, and telling the news exactly as it, or perhaps more accurately Ingersoll, saw it. In April 1939 he and Luce finally fell out, and he left Time Inc. After an agonizing search for financial backing, his newspaper, called *PM*, went on the streets in June 1940. It had many problems, not least of which were Communist infiltration of the staff and the public's awareness thereof—the *Uptown Daily Worker*, wags called the paper—and what a recent historian of the paper has aptly called Ingersoll's "egocentric hysteria." But with all its faults, it was very much a presence on the New York scene, and it made Ingersoll, in his biographer's words, "perhaps the most celebrated—and satirized—reporter in America."

From the spring of 1941 on, *PM* was beating the drums for the United States to get into the war against fascism. When the United States did join the war, some people began asking why Ralph Ingersoll, who had crusaded so vigorously to get American boys shot at, was not getting shot at himself. Among them were the members of his draft board, which classified him 1-A and sent him its greetings in June of 1942. Since he was forty-one and hardly normal draft material, Ingersoll thought, possibly correctly, that this reflected some sort of right-wing plot to get him. But he announced that he was ready to serve, and enlisted before they could draft him.

He asked for assignment to the Engineer Amphibian Command, where his long-rusted engineering training would enhance his prospects for advancement. At basic training at that unit's camp on Cape Cod, he fell under the eye of its chief, Brigadier General (later Major General) Daniel Noce. Noce, an engineer officer of the West Point class of 1917 who had served in France in 1918 and had had a distinguished career between the wars, decided that Ingersoll was all right after watching his dignified handling of the media blitz that descended on his camp when Ingersoll joined it, and the sportsmanlike way he took basic training as if he were just another young recruit; and set in motion a lieutenant's commission for him. He set Ingersoll to producing training manuals, and then made him his public affairs officer; an assignment Ingersoll had feared and which he tried to get out of as quickly as possible.

In early 1943, Noce went to North Africa to review lessons learned from the TORCH amphibious operations and took Lieutenant Ingersoll with him. While Noce conferred in Algiers, Ingersoll managed to get himself attached to a Ranger company for the fighting around El Guettar. When he returned to Algiers the general told him that his old friend Jakie Devers wanted Noce for his G-3 and that he, Noce, wanted to take Ingersoll to ETOUSA with him. At the end of March and the beginning of April Noce, Noce's aide, and Ingersoll spent a week in London—staying at the Dorchester, at Ingersoll's suggestion, rather than in regular Army billets—while the general got himself settled in. Then he told Ingersoll to take a home leave till sent for. Back in New York, Ingersoll dashed off a book about the El Guettar fighting called *The Battle Is the Payoff.* A best-seller, it remains one of the outstanding books about combat in the Second World War.

He had just finished the manuscript when he got his orders for ETOUSA headquarters. He had barely reached London again when he was put out of action for several weeks by a traffic accident (forgetting that the British drive on the left, he stepped off the curb to approach a pretty girl and was knocked down by a taxi). Planning for COCKADE was getting under way when he got back on his feet. Ordered to start on WADHAM, Noce sent for Ingersoll—now a captain—and told him to report to a hush-hush conference at a country house somewhere. When Ingersoll arrived, the British sentry refused him admittance: Nobody lower than colonel allowed. He finally gained entry, though he was treated rather loftily, he thought. This was the beginning of his full-time occupation for the rest of the war.

The Ralph Ingersoll that entered upon the deception business that summer of 1943 was a different man from the one that had set New York on its ear. Accounts of his tenure both at the *New Yorker* and at Time Inc., even a sympathetic biography, are larded with phrases like "hard-driving, ambitious, and abrasive"; "insulting, abrasive, mean, and rude"; "manipulative"; "capricious"; "erratic, hard to work for, get[s] involved in trivia, too literary and academic to edit a newspaper, ruthless, self-centered and selfish, subject to the last man seen"; "conceited egoist . . . snake-in-the-grass." (He was a terminal hypochondriac to boot.) But he seems to have shed most of these aspects of his nature for his military service. He became a team player and lived reasonably well with authority. He learned to value the career military officers upon whom he would not have wasted a glance a few years before; indeed, two of his professional colleagues, General Noce and Lieutenant Colonel Clarence Beck, became fast lifelong friends. Not that he wholly shed the old skin. "Ingersoll was the trickiest, most elusive person I've ever dealt with," said his intelligence officer with admiration. "I've never met anyone who was such a bright guy who was such a goddamned liar. He'd say anything to get what he wanted."

Two other characteristics from civilian life carried over. He was a marathon, and verbose, memorandum-writer. And, in the words of another of Luce's right-hand men, "He blew his own horn in the most outrageous way," while yet another former Time Inc. executive spoke of his "annoying habit . . . of trying to take credit for things he did not do." This characteristic shows in his accounts of his own role in strategic deception in the Second World War. He did not in fact accomplish much, and was involved more in tactical than in high-level work; but he raced back to Washington at the end of the war in Europe to be first to file a report, and for the rest of his life, to hear him talk you might have thought that he had pulled off the great OVERLORD deception single-handed.

TINDALL, the threat to Norway, got under way at the end of June with a conference, codenamed RATTLE, of senior commanders at Largs, the Scottish headquarters of the Commandos. It was notionally to be conducted by the British Fourth Army—the disguise adopted by Scottish Command for the purpose—and was to be mounted by four seaborne divisions (including an American one from Iceland) and an airborne division. The British units were real, already stationed in Scotland for training. The American one was the bogus 55th Division, supposedly in Iceland; it replaced the genuine 5th Division, which moved from Iceland to England in August. Elaborate notional orders were prepared, and the deception itself was mounted through double agent reports, displays at air bases, and radio traffic. At the end of August, the operation stood down to make room for STARKEY and WADHAM.

TINDALL was relatively easy to put on. By contrast, STARKEY was almost too much for COSSAC. Not enough landing craft, even fake ones, could be scraped together. In July, COSSAC suggested to the Chiefs of Staff that maybe the amphibious feint should be called off; without success. The civilian economy of southeastern England was disrupted by pulling together the genuine light shipping required, making ready the actual movement of troops, and the various security precautions that were applied in the area from early August. Air defenses in the southeast were strengthened and dummy installations sprang up. The Army began putting together a major exercise, called HARLEQUIN, involving the equivalent of two and a half British divisions (British XII Corps) in the Newhaven-Dover region, and a similar force (I Canadian Corps) further west around Southampton.

The Commandos scheduled for July and August a series of fourteen raids on the French Channel coast—codenamed FORFAR, with individ-ual raids called FORFAR BEER, FORFAR EASY, etc.—to suggest that infor-mation was being sought in preparation for an invasion. Deceptive letters were left behind on these raids for the Germans to find, and, for the benefit of German interrogators in the event of capture, the troops were told that all leave in the Commandos would be stopped after Au-gust 1. Only eight of these raids were actually carried out. Still, this was more success than COSSAC had inducing the other services to play the game. A proposal that the Navy supply two old battleships for shore bombardment in aid of the "landing" was greeted, recalled Morgan, with an "explosion that shattered the cloistral calm of the Chiefs of Staff Committee when I put forward the suggestion." Nor were the air com-manders willing to divert their big bombers from the offensive against the Reich, except for limited activity around the notional D-Day. They were willing to put training flights to work over the notional objective area, however, and thus nearly doubled the number of daily sorties over northern France. Meanwhile, no less than eighty-one squadrons made ready for the great air battle that the feint was supposed to bring on.

Bevan and the Twenty Committee presided over a steady stream of messages from the double agents to alert German intelligence to the forthcoming assault. GARBO in particular was busy. He recruited no-tional new subagents (an airman, a censor in the Ministry of Informa-tion, and a soldier on guard duty reporting to the Gibraltarian waiter at Chislehurst). He himself visited Scotland in August; receiving there re-ports from his agents in the south about big doings in their territory, he betook himself to the Brighton region, where he found seven divisions marshaling, with supporting aviation and amphibious vehicles. To ease the way for the forthcoming breakoff, he opined for the Germans' bene-fit that it might be just an exercise; maybe the fall of Sicily would lead to a change in plan. But on September 5, shortly before D-Day, he radioed that his AGENT THREE in Glasgow reported that a "very certain source" advised that the operation was scheduled for September 8, and on the 8th he reported that troops were confined to barracks and landing craft were concentrating.

COCKADE enabled the rebuilding of TRICYCLE's credibility. During the winter he had not sent much information and the Germans were begin-ning to lose patience with him. But in May he began sending a flood of reports about such matters as troop concentrations, assembling of landing craft, and preparations to receive casualties; in July he met again in Lisbon with his handlers; he bombarded them with information after his return to London.

Less significant double agents did their part too. MUTT and JEFF, the Norwegians, reported on threats to Norway. TRICYCLE's subagents BALLOON, supposedly a cashiered British officer in the arms business, and GELATINE, supposedly a lively lady with many friends in the military, contributed their share. BRUTUS, the Polish Air Force officer, pitched in, as did PUPPET and MULLET, two businessmen with connections to an Austrian Jewish businessman employed by the Abwehr in Lisbon dubbed HAMLET, and the elegant BRONX with her wide connections in the diplomatic corps.

There was an established practice of dropping carrier pigeons in France, bearing questionnaires to be filled out and returned via the pigeon by any Frenchman who found them. Since many of these were known to fall into German hands, questionnaires focusing on COCKADE target areas were included throughout the summer.

From August 25 the troops concentrated on the south coast. Three flotillas of minesweepers set to work clearing a passage across the Channel. The troops were brought forward and their motor transport was actually embarked on September 8. On D-Day the weather was beautiful, and Morgan and a bevy of fellow generals, British and American, who had come down by special train, watched from the Kentish beaches while a thirty-ship convoy assembled off Dungeness and put out towards France, while another twenty ships carrying the motor transport set off from the Solent. The earth shook with the thunder of Allied fighters racing to the great air battle that STARKEY was meant to bring on. "It was an inspiring sight to see everybody doing his stuff to perfection," wrote Freddie Morgan after the war, "except, unfortunately, the German."

For the Germans barely noticed. They alerted their defenses, but did not report anything noteworthy; Morgan recalled that "a German coast-artillery subaltern on the far shore had been overheard calling his captain on the radio to ask if anybody knew what all this fuss was about." In particular, the great air battle never happened. By 9 A.M. the "invasion fleet" was within ten miles of the French coast. Then, according to plan, it turned about under cover of a smoke screen and headed back home.

So STARKEY, on which so much effort had been expended, was a party to which nobody came.

"It appears that the operation has been suspended," GARBO radioed to Madrid on September 8. "Troops surprised and disappointed." (This last was quite true, as Morgan and his colleagues noted with great satisfaction.)

"Suspended" at the same time was WADHAM, the American invasion of Brittany, which had been notionally due to take place some two weeks later. The "story" of WADHAM was that once German reserves had been drawn to the Pas-de-Calais to oppose STARKEY, an American force under Devers would invade Brittany. (In the earlier planning stages it had been anticipated that the opening of an American field army headquarters in England would be announced—General Andrews, and subsequently Devers, had been urging this, initially as an aid to deception, then phasing into actual OVERLORD planning—but that idea was dropped.) The American V Corps of five divisions from England—building upon the genuine V Corps headquarters already actually in the country—commanded by Major General Leonard T. Gerow, would form "Task Force A," landing near Morlaix on the beaches west of St. Brion to attack Brest from the rear and seize it and the beaches to the mouth of the Odet River. Some days later, V Corps would be followed by "Task Force B," the American VII Corps of seven divisions direct from the United States, commanded by Major General Roscoe B. Woodruff, and escorted by a Navy task force. VII Corps would relieve the V Corps troops on the southern portion of the front and follow up with an attack eastwards south of the Brest-Rennes highway, seizing the portion of the peninsula in its sector and driving the Germans east of the line Nantes-Rennes.

Never very plausible, WADHAM was amateurishly handled in both England and Washington. It came at a time when Newman Smith was just beginning to take hold at Joint Security Control: Goldbranson did not join him till August. But the blame belongs not to Smith but rather to COSSAC, to Ingersoll at ETOUSA, and in some measure to Kehm at OPD for not ensuring that Smith got the needed coordinating authority. (Kehm did warn his bosses at OPD that "Should the U.S. fail to partake of its share in deceptive operations it is very likely that the British will handle them for us.")

As an operation under a Combined Chiefs directive, WADHAM should have been conducted under combined and joint procedures. It should have been assigned to Joint Security Control for advance study and integration with the rest of COCKADE. There should have been directives from the Joint Chiefs covering the participation of naval and ground task forces; but there were none. Instead, ETOUSA—meaning Ingersoll, assisted by a first lieutenant who had no greater background in the field—seems to have been wholly ignorant of joint procedures and evidently was not aware that Joint Security Control even existed; they dealt with VII Corps and the stateside aspect of the operation as a matter to be controlled from London by liaison with OPD. OPD in turn—meaning, of course, Kehm—undertook implementation through informal memoranda and correspondence, Joint Security Control being merely kept informed. Much of the operation's desired realism was thus lost, as was the staff training, including civil affairs planning, that should have resulted from going through a proper exercise.

Much of WADHAM consisted of feverish planning by corps and division staffs, few of whom were aware that it was all just an exercise. That was true not only of V Corps in England but of VII Corps at Jacksonville, Florida. Its staff was enlarged, and arrangements were made for divisions to be assigned to it. Its commander, General Woodruff, made a publicized visit to England and on return issued a press release declaring his opinion that "the European fortress could be cracked wide open."

Implementation was presided over by Ingersoll. He prepared in mid-July, and regularly updated, a detailed "scenario" for WADHAM laying out a spreadsheet of events to take place, their dates, the "story" of each for enemy consumption, the date by which the enemy should know the story, and remarks. In July, BIGBOBS started appearing along the southwest coast and dummy Mustangs, Thunderbolts, and gliders began to show up by the hundreds in airfields that had previously been quiet, and a genuine training exercise in South Wales (Operation JANTZEN) was passed off as part of training for the Brittany attack.

By August 12, General Strong in Washington (obviously prompted by Smith) could report to Devers that special means steps on the American side of the ocean were substantially complete. Double agents had passed to the Germans information that Woodruff had visited England and had been called home hastily; that training activities on the east coast of the United States were being speeded up; that it was beginning to be clear that the principal task force for the coming invasion would sail directly from American East Coast ports; that elements of the 101st Airborne were in the theater; and that Woodruff had been designated to command an important task force on the East Coast. RUDLOFF was a channel for some of this information.

It was well into August before VII Corps noticed that nobody had brought the Navy in on an operation that was supposed to involve a major convoy from the East Coast. They brought this problem to OPD rather than to Joint Security Control where it belonged. Kehm dealt with Admiral Bieri, and subsequently Captain Wetzel, to arrange issuance of a directive for formation of a naval task force to convoy VII Corps; a "Task Force 69" was designated, under Rear Admiral Alan G. Kirk, with a notional sailing date of October 24, and a naval officer was then attached to VII Corps headquarters. But all this was, as Smith lamented to Strong, "done too late for full value to be derived from combined planning. No complete naval planning staff was ever established."

So the operation limped along. As D-Day approached and the men and supplies for the operation did not make their appearance, "a mounting wave of desperation rose," as Wingate recorded, in view of the fact that hardly any of the people concerned with the operation knew it was not real. Abruptly, on September 8, it was announced that the operation was canceled. "The explanation," in Wingate's words, "was simply that the fickle high command had decided on something else." The "story" for German consumption had always been that WADHAM was contingent on the success of STARKEY.

And that was that for WADHAM, except for some postmortems back in Washington. One of Goldbranson's first assignments from Smith, even before WADHAM was over, was to critique it, with recommendations for the future. The WADHAM experience assuredly contributed to Joint Security Control's efforts in 1943–44 to enlarge and clarify its charter. As for VII Corps, both General Woodruff and his G-2 reported to OPD a long list of problems encountered, most of which could have been avoided had Joint Security Control been properly employed.

With STARKEY and WADHAM terminated, TINDALL, the threat to Norway, was brought back to life; this phase being sometimes called TINDALL II. It was maintained as a threat in being until November, when the onset of winter rendered it no longer plausible.

Neither COCKADE as a whole, nor its components STARKEY, WADHAM, and TINDALL, produced anything like the results hoped for. "The operational reactions of the enemy to this group of operations can only be described as disappointing in the extreme," drily noted MI14, the branch of British military intelligence that kept up with the German Army. The WADHAM component seems not to have been noticed by the Germans at all. They were aware of the STARKEY component, but they did not even come close to scrambling the Luftwaffe for the air battle over the Channel that was a prime aim of that operation. Nor did STARKEY succeed in holding German forces in France. Rundstedt, the German commander-in-chief in the West, was appropriately concerned by the preparations he saw going forward in England. But Jodl's staff back in Germany saw the situation more clearly. "The *Schwerpunkt* of the enemy attack on the mainland of Europe lies in the Mediterranean and in all probability will remain there," it told Rundstedt in July; and it pulled out ten of Rundstedt's forty-five divisions between May and October.

Rundstedt protested, of course. He warned OKW at the end of August that supply drops to resistance forces and heightened activity in southern England kept alive "the constant possibility of far-reaching surprise operations." By early September, when STARKEY was at its height, his staff was beginning to be a bit suspicious that the flood of agents' reports and preparations for a Channel attack were perhaps too obvious; but this, they worried, might just mean a feint, with the real attack coming anywhere from Brest to the Heligoland Bight.

Even after STARKEY was called off, Rundstedt's staff considered that it had been an elaborate rehearsal. "Transition to a real invasion attack is possible at any time," they fretted. OKW, however, kept its eye fixed on the Mediterranean and the Eastern Front. Only in October, when the situation in Italy was stabilizing, did they begin sending forces back to Rundstedt. He in turn continued to fret over the possibility of an attack until the end of November; by which time Hitler and OKW had come to agree with him.

TINDALL had a bit more direct success. "It never took much to make

Hitler believe in a threat to Norway," as the official historian has observed. This time Baron von Roenne, who it will be recalled had taken over Fremde Heere West in March, agreed with the Führer. At the end of August he judged that "all the evidence indicat[es] an early beginning" of an Allied attack on Norway, estimating the enemy force quite correctly at from four to six divisions. Throughout 1943, the Germans kept twelve divisions idle in Norway that would have been far more useful in Italy or the Ukraine.

Notwithstanding this modest success, there was no blinking the fact that all the immense effort expended on COCKADE had produced few tangible results and had failed in its primary goals. Not that the effort was wholly wasted. HARLEQUIN, the Army component of STARKEY, taught a number of lessons that were valuable when the real Normandy invasion came. And the heavy reliance on GARBO produced a procedure set up by Hesketh, modeled on the system developed by Dudley Clarke, that became the standard method of handling major double agent traffic in London. The basic "story" was broken down into a string of "serials." For each serial there was a sub-"story" for the Germans, the date by which it was to reach them, and the evidence (real and notional) to support it. This evidence was then distributed among the members of GARBO's network according to their skills and their individual situations. Suitable messages were drafted and checked with the appropriate authorities in the various services—initially through the Twenty Committee, but more and more directly as time went on—and duly passed along by GARBO.

STARKEY not only rehabilitated TRICYCLE, but elevated GARBO to even higher rank among German sources. "Your activity and that of your informants gave us a perfect idea of what is taking place over there," his case officer wrote to him ten days after the climax of STARKEY; "these reports, as you can imagine, have an incalculable value and for this reason I beg of you to proceed with the greatest care so as not to endanger in these momentous times either yourself or your organization."

Most important, FORTITUDE in 1944 could not have run as smoothly as it did if the London Controlling Section and its fellows had not gone through the exercise of COCKADE the year before. But such a silver lining was not visible in the autumn of 1943. With the failure of COCKADE following upon Montagu's vicious attack, the autumn and early winter of 1943 was a dark time for Johnny Bevan.

*　　　*　　　*

One of the early products of COSSAC planning was the decision, reached in June after agonized wrestling with the problem, that the OVERLORD landing in France should take place in Normandy rather than the obvious alternative, a direct assault across the Straits of Dover against the Pas-de-Calais. The planners recognized that the latter region offered many advantages. It was close to England, and it afforded a short route for an advance athwart the German lines of communication to occupied France. But for these very reasons it was "the most heavily defended area of the French coast and the pivot of the German coastal defense system." There were only four suitable beaches, all strongly defended, and with little supply potential. The excellent French road and rail net in the area would facilitate bringing up German reinforcements and make it hard to isolate an Allied bridgehead through airborne drops and Resistance activity. The ports in the area—Boulogne and Calais—had but limited capacity. The topography of the area offered no natural defenses for holding the bridgehead during the buildup. So, the COSSAC planners concluded, "In spite of the obvious advantages of the Pas-de-Calais provided by its proximity to our coasts, it is clear that it is an unsuitable area in which to attempt our initial lodgement on the Continent." The necessary choice, therefore, was Normandy.

Soon after this decision was reached, and while COCKADE was getting under way, Bevan and his staff began to address the question of a deception plan for OVERLORD. "Our first survey of the position," recorded Wingate, "showed a picture of almost unrelieved gloom." It was obvious that Northwestern Europe would be the Allied target for early 1944. The Germans could not fail to observe the massive buildup that would take place in England during the winter and spring, or the shift of seasoned divisions from the Italian front to England. "The sole favorable factor," as Wingate put it, "seemed to be the necessity for the enemy to safeguard himself at every vulnerable point on the perimeter of the European fortress, and to this extent the continuance of our deception policy of keeping him stretched by threatening the Balkans, Southern France and Norway and Denmark seemed plausible."

London Controlling Section's initial product, a paper circulated on July 14, was appropriately entitled "First Thoughts." But even at this early date, Bevan perceived what was to be the pivotal theme of deception for OVERLORD: The fact that after a certain point in time—X weeks before D-Day, "D minus X weeks" in Bevan's terminology—it would be impossible to conceal from the Germans the fact that a cross-Channel assault was coming soon; so that the "object" of the deception would fall into two stages. Stage 1 should be "to induce the enemy to weaken his forces in Northern France and in areas (excluding Southern France) where they would be available as immediate reserves for Northern France," and "to contain the maximum enemy forces in those theaters from which their transfer to Northern France would occupy more than X weeks." Stage 2 would be "to induce the enemy to dispose his forces in Northern France and Belgium in such a manner as will cause the least interference with OVERLORD." This would remain the basic structure of OVERLORD deception for more than a year.

"First Thoughts" assumed that Italy would collapse after HUSKY and that the Germans would accordingly feel threatened in the Balkans and Southern France, and that an operation against the Azores which was currently contemplated* would make them nervous about Allied intentions towards the Iberian peninsula. The suggested "story" for dissemination after STARKEY and WADHAM was that those operations had been canceled because of Allied realization of how strong the German defenses were, together with a shortage of landing craft. The Allies had decided instead to concentrate on crushing Germany by aerial bombardment (the combined bomber offensive, known as POINTBLANK). There would be subsidiary attacks on the Balkans, with a diversionary attack on the south of France, from Italy and the eastern Mediterranean; and from England on Denmark, Norway, and the Bilbao-Bayonne area of Spain and France. Once the Germans fully appreciated the magnitude of the preparations in England, the "story" would be that no cross-Channel landing would be made if it were not certain that there would be only limited opposition, and that even then the purpose would merely be to try to open a bridgehead in the Pas-de-Calais looking towards further operations in the autumn of 1944.

---

* In the spring of 1943, consideration was given to seizing the Azores by a sudden descent (Operation LIFEBELT, formerly BRISK). For the operation Bevan and staff prepared a modest cover plan called DUMMER, consisting mainly of telling the troops involved that they were bound for the Mediterranean to attack Corsica. However, in August the Portuguese agreed to allow the Allies to open bases in the islands, and they were peacefully established in October (Operation ALACRITY).

"First Thoughts" produced in COSSAC a reaction "more despairing than unfavorable," in Wingate's words. This was the occasion, mentioned in Chapter 2, when Morgan suggested that perhaps the thing to do would be to exaggerate Allied strength to such an extent as to frighten the Germans into withdrawing from France, and Bevan had gently to remind Morgan of the CAMILLA principle: Suppose what the Germans did in response to such an exaggeration should be to strengthen their defenses in France to such an extent that OVERLORD would be hopeless?

"First Thoughts" was at least a start, and "resulted in a good deal of hard thinking by all concerned," to quote again from Wingate. Among those thinking hard were Goldbranson, Smith, and Kehm, all of whom made comments. While the hard thinking went on, COCKADE was mounted and failed, and it was in an atmosphere of some discouragement that Bevan and his colleagues worked up a full-dress proposal for a deception policy for OVERLORD, dubbed JAEL.

JAEL was prepared in light of the decisions reached at the QUADRANT summit conference in Quebec in August, the most important of which was that OVERLORD in the spring of 1944 would be the main object of Allied efforts for that year; and it reflected several conclusions that the planners had reached during the late summer. A significant one was that the Germans were likely to be ever more concerned lest the remaining neutrals apart from Switzerland—Turkey, Spain, Portugal, and Sweden—might turn against them. Another was that the London Controlling Section should address in detail only the first or strategic phase, leaving to COSSAC the specifics of the second or operational phase.

JAEL, after gently dismissing the idea of exaggerating Allied strength, stated its object as being to cause the enemy to dispose his forces so as to keep German strength in France below the level beyond which COSSAC currently believed that OVERLORD would not be feasible. The "story" would be that while Allied forces in Britain were substantial, there were few divisions trained in amphibious operations and several of the divisions were training in mountain rather than amphibious warfare; some experienced divisions would be transferred from the Mediterranean but they would be replaced by new divisions; landing craft transferred from the Mediterranean would be largely used for training; though Americans would be arriving in England during the winter in great numbers, a large proportion of them would be bomber crews and ground personnel for POINTBLANK, plus administrative and garrison troops; and the British were so short of manpower they were cannibalizing their divisions.

Allied strategy, the "story" would run, was dictated by caution; the Allies placed great faith in POINTBLANK, and hoped that POINTBLANK, a Soviet summer offensive, and their own Mediterranean operations would so weaken German strength in the West as to ease their entry into France; they would therefore not be likely to try a cross-Channel operation till the summer of 1944. They intended to invade southern Norway and Denmark, knowing them to be less well defended than France. Allied forces in the Mediterranean were greater than was in fact the case, and would be used to secure northern Italy and establish bomber bases there, to invade the Balkans through Istria and across the Adriatic, to mount a simultaneous amphibious attack in the Eastern Mediterranean (this would be assisted by American forces direct from the United States), and to attack the Balkans through Turkey; while southern France would only be attacked if the Allies were satisfied that little opposition would be met.

It was to be assumed that at about D minus 30 the Germans would recognize that a cross-Channel assault would come soon. The "story" at that point should in general be that the attack would fall mainly on the Pas-de-Calais and Belgium; that Allied forces were still concentrated for an attack on Norway and Denmark, and the Allies were trying to enlist Sweden in this operation; that the Balkan, Eastern Mediterranean, and Turkish operations were imminent; and that the Allies had approached Spain for permission to land in the Biscay ports and Barcelona and to transit troops across Spain to attack southwestern France.

Bevan submitted JAEL to the British Planners on September 22, suggesting that representatives of COSSAC and AFHQ sit in with them when they considered it, and noting that Dudley Clarke was expected in London in the next few days. When COSSAC studied JAEL his chief concern, aside from nuances of emphasis and timing, was doubt as to the wisdom of diplomatic approaches to the neutrals. Clarke did indeed pay a long visit to London from September 26 to October 28, and JAEL was necessarily one of the matters he discussed; the exact course of staff consideration of JAEL is not presently clear, but a final draft was produced on October 8, with which COSSAC concurred on October 23.

On the other side of the Atlantic, Goldbranson prepared on October 8 a cogent critique of the first draft for Newman Smith. "Generally sound," he thought, "with one exception, namely, it has minimized the threat from the Mediterranean and has not fully exploited its potentialities from a deception standpoint." He felt particularly that a threat to southern France by troops direct from the United States would be more plausible than one against the Balkans. Baumer agreed with this analysis. Other American reactions were even less favorable, and further consideration was deferred pending final decision by the British Chiefs. Bevan and his colleagues were themselves not greatly impressed with their own handiwork. All this proved to be relatively abstract, for on October 26 the British Chiefs agreed to defer consideration of JAEL for approximately one month, *i.e.,* until after the forthcoming Tehran and Cairo conferences.

Bevan and his colleagues worked up a stopgap plan while things hung thus in limbo, whose objects were to prevent any further reinforcement of the Germans in Italy, and to induce the Germans to disperse their forces as widely as possible by creating short-term threats to any weak spots that might conceivably be attacked during the winter. LARKHALL, dealing with American forces in Britain, was also revised. Since August the policy had been changed to play down the arrival of American troops. Now it seemed prudent to bring German information as to American strength back to accord with reality, so as to ensure German belief in Allied ability to take advantage of any German weakness that might appear.

The most significant matter that Bevan worked on in this interim period was Operation FOYNES. This was a plan to conceal from the Germans the weakening of the Allied position in the Mediterranean by the transfer of eight veteran divisions (the American 1st, 9th, 2d Armored, and 82d Airborne; the British 50th, 51st, 7th Armored, and 1st Airborne) from that theater to the United Kingdom to serve as the experienced backbone of the host gathering for OVERLORD. Under FOYNES, Allied strength in the Mediterranean was maintained in German eyes by sending out to that theater three genuine divisions (the Canadian 5th Armored, and the American 85th and 88th) and adding to the Allied order of battle in the theater four notional divisions (the British 40th,

42d, and 57th Infantry and 5th Airborne); two more notional divisions (British 61st, later 68th, and American 46th) were held in readiness to be "sent" but were never used. The notional movement to the Mediterranean was meticulously worked out to coincide with the passage of real convoys, and appropriate administrative measures were taken as if real movements were involved.

Associated with these notional troop movements was a plan that fizzled rather comically. The Germans maintained twelve observation posts on both sides of the Straits of Gibraltar, which were brightly lighted at night to show the silhouettes of all passing ships. To suggest to them that the highly visible westward movement of landing craft was being counterbalanced by an eastward movement of replacements from the States, Douglas Fairbanks and Harry Gummer (who was newly posted to Gibraltar to form an "A" Force outstation and 42 Committee at the Rock), came up with the idea of laying down a smoke screen and projecting false silhouettes on the smoke by a sort of magic lantern device. This proved unfeasible. Eventually the London Controlling Section drew up Plan GOTHAM, under which a number of dummy inflatable landing craft would be visibly carried on the decks of merchant ships passing eastwards through the Straits. Alec Finter tried this on a convoy that left Liverpool at the beginning of January. But when the ship encountered heavy weather the dummies bounced around like the large balloons that they really were, so that was the end of GOTHAM.

Bevan proposed a "double bluff" for concealing the move of the veteran divisions to England, by giving publicity to the homecoming of famous British units and informing the Germans by double agent channels that this was in fact deceptive cover for a buildup in the Mediterranean for an invasion of the Balkans; but the British Chiefs rejected it, and all that could be done was to keep the move as secret as possible. In April 1944, thinking that the Germans must surely know by now that the eight divisions were in England, Bevan so advised them through the double agents and authorized public announcement of their presence. The Germans had in fact not suspected the movement.

This reflected the tenacity of Roenne at Fremde Heere West. Once a unit was entered into his order of battle it stayed there until it was proven to be elsewhere. Kesselring's intelligence people at the front reported the disappearance of FOYNES units; Roenne would not budge. His situation report for December 31 (promptly read by ULTRA) advised that there was "no indication of any kind of considerable forces being transported away from the Mediterranean area," and in February he actually asked the Abwehr to locate the supposedly missing divisions.*

As for landing craft, he in effect threw up his hands entirely, admitting in late October 1943 that in view of the decline of German aerial reconnaissance in the Mediterranean he "had no information whatever about their present dispositions." And even six months later, with the Allied hosts gathering in England, he simply reported that he could not estimate the number of their landing craft; they had "a masterly grasp of the art of camouflaging their landing craft over a wide area," and "we must therefore expect these craft to be still in the Western Mediterranean."

There were odds and ends of other deception projects that Bevan had to attend to while the major decisions with respect to OVERLORD hung fire. Some were attempted and most were not. One called LANGTOFT had started in February 1943 and ran till 1945. Conducted almost entirely through the double agents, its object was to conceal the fact that the British had cut back on production of poison gas, discourage the Germans from initiating chemical warfare, and still give them no excuse for cutting back their own expensive preparations for defense against such warfare, by telling them that the British were well stocked with new-type gases, improved gas masks, and special clothing. Another, called LEYBURN, intended to show general Allied interest in returning to the Continent in support of BARCLAY and STARKEY, involved a pattern of inquiries in neutral countries with respect to protecting works of art and ancient monuments in France and the Low Countries.

Various other schemes, mostly variations on the haversack ruse, never came to pass. SHOTGUN was an abortive project to pass to the Germans a forged technical document, exaggerating the manpower, war potential,

* An example of Roenne's stubborness: The American 82d Airborne disappeared from Italy. Some three weeks later the German radio intercept service picked up on an unidentified net in England a reference to a soldier wanted in connection with paternity proceedings in the United States. His identification tallied with the code designation used by the 82d, and German communications intelligence reported to OKW that the 82d might have been transferred to England. The sarcastic response was that they must have sneaked there by submarine.

and equipment of the British Army. GILMERTON was a project to brief a returning German prisoner of war, General von Cramer, to the effect that the Allies intended to destroy Germany utterly by bombing. (It was dropped because of opposition from the Deputy Prime Minister, Clement Attlee.) BLARNEY STONE proposed to let a POW escape by way of Eire with important-seeming phony papers. CHRISTMAS CAKE and IDES likewise proposed to let a POW escape by stealing a plane aboard which fake documents would be found. CATSPAW proposed to let the Swedish air attaché find such documents in a taxi. RISSOLE envisaged a double-bluff version of MINCEMEAT, planting an obviously faked body carrying papers pointing to the real Allied objective. There were unimplemented projects to fake conversations over the transatlantic telephone (TRANSATLANTIC TELEPHONE PLAN); to smuggle into Lisbon a specially altered copy of some inconspicuous newspaper after the export of such newspapers from the United Kingdom was forbidden (NEWSPAPER PLAN); to lose phony documents in Madrid and Lisbon; and one to focus attention on specific areas by making inquiries through the international fire insurance market. (This last idea would bear fruit in the spring of 1944 in a project called PREMIUM).

There were personnel changes too. Michael Bratby moved on from the London Controlling Section to become an instructor at an intelligence school. Alec Finter arrived in November for his relatively brief stay. Captain Gordon Waterfield, in civil life a *Times* correspondent, came, stayed a few months, and went back to regular intelligence work at his own request.

One of the unfortunate results of COCKADE was that it gave strategic and operational deception a bad name with a number of important people. One was General Devers. William J. Casey, a young OSS officer in London, recalled that "Devers watched and shook his head" at STARKEY, and that thereafter Devers's attitude to large-scale deception was that he "had seen it, he didn't like it and it had failed." In 1944–45, as will be seen, Gene Sweeney and Arne Ekstrom would turn Devers totally around. But in the latter part of 1943 his attitude was profoundly negative, and so was that of General Noce.

This prejudice did not extend to tactical deception. It was after COCK-ADE that Noce sent Ingersoll to the States to look into tactical deception equipment and personnel. (After the war Ingersoll characteristically gave himself credit for devising the inflatable dummies used in American tactical deception.) It extended in full measure, however, to COSSAC's efforts to devise an operational cover and deception plan for the Normandy landing itself. And with good reason.

The COSSAC planners' study ruling out the Pas-de-Calais had closed with the admonition: "These conclusions are, however, without prejudice to the importance of the Pas-de-Calais area as an objective for feints and diversions." In line with this, COSSAC worked that summer and autumn of 1943 on its own operational deception plan for the Normandy landing, based on a threat to the Pas-de-Calais. Initially merely called "Tactical Cover for Operation OVERLORD," subsequently codenamed TORRENT, and eventually referred to simply as APPENDIX Y to the OVERLORD plan, it ran through some seven revisions between September 4 and its presentation to the British Chiefs on November 20. The very first concept, prepared before STARKEY, envisaged a STARKEY-style feint against the Pas-de-Calais about D minus 14. That was quickly dropped after the STARKEY fiasco, and successive drafts played variations on a simple basic theme: To conceal as much as possible, by camouflage and whatever other means were available, the fact that the main preparations for OVERLORD were taking place in the southwest of England, while exaggerating, by display of dummies and actual troop movements, the extent of preparations in the southeast; with the modest goal of making the Germans think a six-division force was poised in the southeast for a descent on the Pas-de-Calais.

TORRENT, or more properly APPENDIX Y, was reviewed for the Americans by Ralph Ingersoll and a regular officer whom Noce had brought over from the States in the late summer who was destined to become Ingersoll's boss and a major player of the deception game. This was a short, cheerful thirty-two-year-old West Point classmate of Bill Baumer's, Lieutenant Colonel William A. Harris, known to all as "Billy."

Harris came of an accomplished Georgia family. There were Confederate officers in his ancestry; one of his uncles was Dr. Seale Harris, the great pioneer in the study of blood sugar and discoverer of hypoglycemia; another was long a distinguished United States senator from Georgia; another was superintendent of schools for the deaf in that state;

another had been the Adjutant General of the United States Army. The Army was in his blood and that of his elder brother, Hunter Harris; their father was an Army officer, retired early for reasons of health, their mother had introduced Lieutenant "Ike" Eisenhower to Mamie Doud in 1917, and Billy had himself been born in the Philippines when his father was stationed there. Both Hunter and Billy went to West Point, in the classes of 1932 and 1933 respectively. Hunter joined the Air Corps and by late 1943 was commanding a bomber wing in England (he would rise to four stars and command of all air forces in the Pacific in the 1960s). Billy joined the artillery, and had risen to the rank of lieutenant colonel commanding an artillery battalion in a training division when, in August 1943, he was one of ten officers selected by Noce for his staff, to assist in American planning and in particular to study and report on COSSAC plans from the American perspective. In September, when Omar Bradley was selected for army group command in OVERLORD, it was Harris whom Noce detailed to brief Bradley on the plan and accompany him to Washington. On return he continued with the study of COSSAC

(Harriet Harris Stroup)

William Harris

503

plans, and worked with Ingersoll specifically with respect to deception plans.

Harris and Ingersoll reported so unfavorably on APPENDIX Y that Devers refused to go along with it. It was like "putting a hooped skirt and ruffled pants on an elephant to make it look like a crinoline girl," said Ingersoll.* Dudley Clarke likewise took a look at it and pointed out that it violated the basic CAMILLA principle: It did not suggest what the planners wanted the Germans to *do*. He suggested an object such as "to weaken the German forces in the assault area"; this he deemed "the most important thing of all," and suggested a separate plan for the post-D-Day period, the object of which "might well be to prevent German reserves moving from a specific area."

Bevan's memorandum on APPENDIX Y for the Chiefs of Staff pointed out as well that nothing should be decided until the overall deception policy for OVERLORD was settled (though there was nothing against COSSAC's making "long-term preparations for visual misdirection, W/T deception, etc., in anticipation of events"). So, except for three features of this type that needed an early start (the introduction of wireless silence periods, so that the wireless silence before D-Day would not seem out of the ordinary; exaggeration of the capacity of southeastern England to accommodate troops; and a general camouflage policy of concealment in the west and display in the east), APPENDIX Y was held up by the British Chiefs of Staff pending the decisions to be made at Cairo and Tehran.

Towards the end of November—officially, on November 21—Harris and Ingersoll were reassigned to the G-3 section of the recently formed headquarters for the American army group that was expected to be activated on the Continent once the buildup justified it—always called "FUSAG," for "First United States Army Group." Commanded by Omar Bradley, and for the time being only a skeleton headquarters, FUSAG began taking over American OVERLORD planning from ETOUSA.

There were two major summit conferences at the end of November and the beginning of December 1943. The first, SEXTANT, was held in Cairo

---

* Not that Ingersoll himself did any better. He came up with a truly bizarre idea to focus German attention on the Straits of Dover, involving a "reverse Dunkirk": A swarm of small boats gathering in southeastern England ostensibly to ferry two million men across to Calais.

between Roosevelt, Churchill, and Chiang Kai-shek from November 23 to 27, and resumed at Cairo without Chiang from December 2 to 7; it was there that Roosevelt announced his selection of Eisenhower to command OVERLORD. The other, EUREKA, was the great conference held at Tehran from November 28 to December 1 between Roosevelt, Churchill, and Stalin. At EUREKA it was decided, among many other things, that OVERLORD would definitely take place in May, with an associated landing (ANVIL) in southern France; while the Soviet summer offensive would open in coordination with D-Day in the West. When Churchill asked Stalin whether any arrangements had been made among the three great powers to provide a combined cover plan for these operations, Stalin responded with a eulogy of Soviet tactical deception. "Truth deserves a bodyguard of lies," said Churchill. "This is what we call military cunning," said Stalin. Churchill replied that he considered it rather military diplomacy.

Bevan flew out to Cairo for SEXTANT, arriving on the last day of the first portion, November 27, and remaining at least till December 8; taking the opportunity to confer with Dudley Clarke about the Mediterranean. On December 6, he received his formal directive to prepare the deception plan for OVERLORD. He returned to London and set at once to work on a plan. From the outset it was codenamed BODYGUARD, in light of the Prime Minister's aphorism.

It had been a difficult year, what with the trouble fomented by Montagu and the failure of COCKADE; and the forthcoming operation would be the decisive event of the war. So Bevan was under considerable pressure. By Wheatley's account, day after day, he and the rest of the staff struggled with the problem, only to have Bevan, "haggard from sleeplessness," bounce their drafts back with revisions and additions. The paper kept growing; and gradually, said Wheatley, he realized that "it had become a hopelessly depressing document, virtually informing the Chiefs that the chances were ten to one against the cover plan we had produced succeeding."

So one night, as he told the story, Wheatley stayed late at the office; and as he expected, Bevan returned to work far into the evening and wondered in surprise why Wheatley was still there. Wheatley told him he was "trying to rewrite his bloody BODYGUARD paper, because, if it goes in as it is, you are going to get the sack." To Bevan's puzzled question, he replied that "it has become so long and complicated that no one will be able to understand it." And "four-fifths of it is devoted to pointing out every sort of unexpected happening that may cause it to fail." The Chiefs will think that "we have let them down and are incompetent to produce a sound cover plan. Then they'll sack the lot of us and get in a new team in the hope that it will produce something acceptable." Instead, he urged Bevan, give them a brief paper that sets forth your conclusions, and put the reasoning in an annex. "It is a good plan and, believe me, they will swallow it like lambs. No one could possibly guarantee complete success and, should we have the ill-luck that something goes wrong, you will still be covered by the annex." So next day the plan was completely redrafted; and Wheatley congratulated himself that although he might be "lazy and indifferent about all minor problems to do with deception, . . . at least I saved dear, brilliantly capable but over-conscientious Johnny from being given a bowler hat by pushing him into redrafting plan BODYGUARD."

Well, it is a good story, like all Wheatley stories. The fact is that BODYGUARD in its final form has no "annex" and is plainly a revision of JAEL. But as submitted to the British Chiefs it was accompanied by a "Report of the Controlling Officer," which may have been what Wheatley had in mind; and there is no reason to think his story is not true in principle. Certainly, the drafting process was difficult—"drafted, redrafted, amended, altered," wrote Gordon Clark many years later—and skepticism had to be overcome. "It is difficult at this late date to recapture the atmosphere of those days, the sense of urgency, the sudden problems and much sheer hard work," wrote Gordon Clark. "I do recall that when I first presented the outline BODYGUARD to ISSB they flatly refused to believe that it would be possible to deceive the enemy over the Normandy Landings which in their view, at that time, must become obvious."

The British Chiefs of Staff approved BODYGUARD with minor revisions on December 27. Masterman and the Twenty Committee doubted "whether BODYGUARD would do more than deceive the enemy in vacuo, but it was agreed that the Plan represents the best that could be done in extremely difficult circumstances." On the American side, the Joint War Plans Committee did not have great hopes for the plan; but admitted that they were "unable to propose a more acceptable alternative concept. . . . BODYGUARD can do no harm and may do considerable good by misleading the enemy for several months as to our intentions." They submitted a draft memorandum from the Joint Chiefs to the Combined Chiefs very much in the style of Baron Kehm (who himself was still abroad):

> We feel that the over-all deception policy in BODYGUARD is so ambitious as to be subject to some question as to its general plausibility. It is our view, however, that a maximum of success may be forthcoming if a considerable degree of reserve characterizes its execution. We feel, in brief, that it could easily be overplayed.

The American Chiefs approved their recommendation with minor changes on January 18. By informal action the Combined Chiefs approved it on January 20, and it was officially transmitted to Eisenhower as Supreme Commander on January 22.

It was the most important single strategic deception policy document of the war.

BODYGUARD followed JAEL in most respects. Its object was "To induce the enemy to make faulty strategic dispositions in relation to operations by the United Nations against Germany agreed upon at EUREKA," specifically, "to contain enemy forces in areas where they will interfere as little as possible with operations on the Russian Front and with OVERLORD and ANVIL," namely, northern Italy, southeastern Europe, and Scandinavia; then, "As soon as our preparations for OVERLORD and ANVIL clearly indicate to the enemy our intention to undertake a cross-Channel operation and an amphibious operation in the Western Mediterranean, Theater Commanders concerned must implement their tactical cover plans to deceive the enemy as to the strength, objective and timing of OVERLORD and ANVIL."

As for the Normandy landings, since it would be impossible to conceal the Allied buildup in England from the Germans, the best policy would be to indicate that an adequate force could not be built up until late summer. It would help to suggest that OVERLORD would not begin until after the Soviet summer offensive had opened, and that it would not start before the end of June. As for ANVIL in the south of France, the buildup should be less evident than the OVERLORD buildup in England, so that for it no strategic plan prior to the theater operational plan for the assault itself should be necessary.

The "story," accordingly, would be that the following was the Allied plan for 1944:

a. POINTBLANK operations were seriously affecting the enemy's war potential and, if continued and increased, might well bring about his total collapse. Consequently, reinforcement of the United Kingdom and the Mediterranean by long-range American bombers has been given such a high priority that ground forces buildup in the United Kingdom has been delayed.

b. The Allies must be prepared to take advantage of any serious German weakening or withdrawal in Western Europe and preparations to this end must be put in hand forthwith.

c. To concert in Spring an attack on Northern Norway with Russia with the immediate object of opening up a supply route to Sweden. Thereafter to enlist the active cooperation of Sweden for the establishment of air bases in Southern Sweden to supplement POINTBLANK with fighter-bomber operations and to cover an amphibious assault on Denmark from the United Kingdom in the summer.

d. Since no large-scale cross-Channel operation would be possible till late summer, the main Allied effort in the spring of 1944 should be against the Balkans, by means of—

i. An Anglo-American assault against the Dalmatian coast.

ii. A British assault against Greece.

iii. A Russian amphibious operation against the Bulgarian-Rumanian coast.

iv. In addition Turkey will be invited to join the Allies to provide operational facilities including aerodromes to cover operations against the Aegean islands as a prerequisite to the invasion of Greece. Her refusal would not materially modify the Allied intentions.

v. Pressure against the satellites to induce them to abandon Germany.

e. Anglo-American operations in Italy would be continued, and in order to hasten their progress, amphibious operations against the northwest and north-east coast of Italy would be carried out. Provided these were successful, 15 Army Group would later advance eastwards through Istria in support of the operations mentioned in d. above.

   Note. The operations in c. d. and e. above would enable us to employ our amphibious forces and retain the initiative until preparations for the final assault in the late summer were completed.

   f. Though Russian operations would presumably be continued this winter it would not be possible for them to launch their summer offensive before the end of June.

   g. In view of the formidable character of German coastal defenses and the present enemy strength in France and the Low Countries, possibly as many as twelve Anglo-American divisions afloat in the initial assault and a total force of about fifty divisions would be required for a cross-Channel assault. This operation would not be launched until the late summer (*i.e.* after the opening of the Russian summer offensive).

(The figure of fifty divisions may well have been inspired by the MAGIC decrypt, in early November, of a message from Japanese Ambassador Oshima in Berlin reporting to Tokyo that the Germans thought that a force of that size would be required in order to conduct so hazardous an enterprise.)

   As to Allied strength and dispositions, the "story" should be:

   a. *United Kingdom*
   i. Shortage of manpower has obliged the British Army in the United Kingdom to resort to cannibalization, while several of their formations are on a lower establishment, or still lack their administrative and supply units. The number of Anglo-American divisions in the United Kingdom available for offensive operations is less than is, in fact, the case. Some United States divisions arriving in the United Kingdom have not yet completed their training.
   ii. Personnel of certain Anglo-American divisions in the Mediterranean with long service overseas are being relieved by fresh divisions from the USA. British troops will, on relief, return to the United Kingdom where they will re-form and be utilized for training inexperienced formations.

iii. Invasion craft remains the principal bottleneck due to operations in the Pacific and the full number required for the initial assault cannot be made available from home production and the USA before summer.

b. *Mediterranean.*

i. Anglo-American forces in the Mediterranean, especially in the Eastern Mediterranean, are greater than is in fact the case.

ii. French forces are taking over responsibility for the defense of North Africa, thus leaving Anglo-American forces free for offensive operations elsewhere in the spring of 1944.

iii. Certain British divisions and landing craft are being transferred from India to the Middle East.

iv. Fresh divisions from the United States of America are expected to arrive in the Mediterranean.

From this master blueprint some eight subsidiary plans were eventually to be drawn up. ROYAL FLUSH would implement the diplomatic approaches to the neutrals, while GRAFFHAM would implement the specific opening to Sweden. ZEPPELIN would be the overall plan designed to keep the Germans from moving forces from the Mediterranean; within it, TURPITUDE would threaten the Balkans while VENDETTA threatened the south of France; IRONSIDE, threatening the Biscay coast, would supplement them. FORTITUDE NORTH would implement the threat to Norway, while FORTITUDE SOUTH would be the crucial operational deception aimed at inducing the Germans to hold forces away from Normandy to meet a notional threat to the Pas-de-Calais. Underlying them all, and making them plausible, would be the order of battle deceptions, CASCADE and its successor WANTAGE in the Mediterranean and the order of battle components of FORTITUDE in the United Kingdom. And from Washington, Newman Smith and Joint Security Control would support the "stories" that the Allies were short of landing craft and were relying heavily on the success of POINTBLANK.

Before these plans could be laid, in light of the discussions at Tehran it was deemed wise for the Controlling Officer to present BODYGUARD to the Soviets in person. Bill Baumer, roped into accompanying Bevan to Moscow much against his will, was to find the journey much the most memorable of his wartime travels.

*　　*　　*

Baumer, it will be recalled, was sent to London in January 1944 for a second tour of duty with the British planners. Flying this time by way of Newfoundland and Greenland to Prestwick in Scotland, he was billeted at the Cumberland Hotel and reported to ETOUSA at 20 Grosvenor Square. He was introduced to General Noce, who arranged temporary quarters sharing an office with Ralph Ingersoll and Billy Harris. This was pleasant, Harris being his classmate, while he had met Ingersoll on the latter's trip to Washington and regarded him as "a good man." On January 11, he reported to the Joint Planning Staff at the War Cabinet Offices; looked in on Bevan and the rest, and was promptly swept off to lunch at the Hungaria by Dennis Wheatley. "He must have bought at least the decorations of the place by now," Baumer noted, "and gets service in line with his patronage."

The socializing typical of London duty went on thereafter. Two days later it was back to the Hungaria with Ronald Wingate and wife, and Noel Wild, whom Baumer had met in Algiers the previous year, and his wife. The next day he took the Wheatleys to lunch at the American mess in Park Lane, where good things were obtainable that could not be had on the British ration; and did the same for Michael Bratby and his wife the next day.

For anyone who was in London that winter of 1943–44, with the Americans streaming in by the scores of thousands, the desperately hard work during the day and the frenetic partying behind the blackout curtains during the long dark nights, with the pea-soup fogs and the rain glistening on the black streets, and always the tension of knowing that one of the great events of the twentieth century, in which thousands would die, was imminent, it was a time that would never be forgotten. "London is the Paris of this war," remarked Baumer one night after a round of club-hopping with friends.

But after two weeks of this, on Saturday, January 22, Baumer received a summons to proceed to Moscow with Bevan, on behalf of their respective Chiefs of Staff, to brief the Soviets on BODYGUARD and secure their concurrence and cooperation.

He had feared that he might be stuck with that job, believing that it "would only mire me deeper in the deception game." Before leaving Washington he asked his boss General Roberts not to send him on such a mission if it should come up. Roberts had reassured him. Major General John Deane—the former Secretary of the Joint Chiefs and American Secretary of the Combined Chiefs, who since October had been head of the American military mission to Moscow—could handle the American side, he said. The day before the cable came in, Bevan had told Baumer that he, Bevan, had suggested to Washington that Baumer accompany him. "He laughed quietly," wrote Baumer, "and said that he hadn't mentioned it to me because he knew I would refuse. He then added: 'But Washington has turned down the request.'" So Baumer was "flabbergasted"—his word—when his orders came. (He was told later that the British Chiefs had taken up the matter and that Pug Ismay had repeated the request in a cable to Roberts.)

So he got a passport from the American Embassy and sent it to the Soviet Embassy for a visa. At noon, Dennis Wheatley feasted him and Wingate at the Hungaria, presumably as a bon voyage party (they "tore a huge hole" in a bottle of 1908 brandy). Back at the office he did not feel much like working, and the regular 4 P.M. cup of tea seemed remarkably good. But nearly a week passed, and no Soviet visa. At the end of the day on Friday, January 28, he had a drink in the little mess in the Cabinet War Rooms and General Ismay struck up a conversation. I won't give you any advice, said Pug, but if I were in your shoes I would go anyway, visa or no visa. That just about made up Baumer's mind for him, for he had been thinking along those lines himself. He spent a sleepless night, knowing that Bevan was already en route for Prestwick, from which a special plane was due to leave for Moscow next afternoon.

The next morning he told the American Embassy that he was going without a visa; if the Embassy had any objection, wire it to him at Prestwick, for he was off. He sent a messenger to retrieve his passport from the Soviet Embassy while he grabbed his bags from the hotel and told Bevan's office to call Prestwick and tell them he would be there by five o'clock, which he understood to be the departure time for the plane.

A bumpy flight in a bucket seat on a DC-3, and he was at Prestwick at 5:15. There was no message from the Embassy saying not to go. Bevan was waiting, as were the British ambassador to the USSR, Sir Archibald Clark Kerr, a lively and pleasant man (though he did seem to have memorized every threadbare joke about Americans and British being separated by the same language).

After some delay, the passengers—including several RAF men and a Soviet diplomat and his wife—all swaddled in their bulky flying clothes, clambered into the bottom hatch of the unconverted B-24 Liberator bomber that was to take them to Moscow, and disposed themselves around posts in the bomb bay on pillows and blankets. It proved to be a memorable journey through the sub-Arctic winter darkness. The oxygen system failed to deliver full pressure—perhaps because the crew had neglected to open a crucial valve. All the passengers suffered severely, Bevan worst of all. Baumer vaguely remembered feeling sharp stabs of air coming up at the plane at one point of the journey and realizing that it was antiaircraft fire as they passed over German-occupied Norway.

Eventually they felt the plane bump to a landing. Dazed, they lowered themselves out of the bomb bay and found themselves on a huge, flat, snow-covered field, with Russians in uniform all around them, smiling in a friendly way but plainly curious. They had not been expected at this field; it turned out that the radio operator and the flight engineer had passed out for lack of oxygen, and the remaining crew, unable to communicate with Soviet air traffic controllers, had set the plane down at the first airfield they saw near Moscow. It was ten o'clock in the morning local time on Sunday, January 30. They waddled in their clumsy flying suits half a mile to the administration building, half carrying the still-groggy Bevan.

There they waited amidst much chattering and telephoning in Russian. Eventually several British Embassy staff appeared, including the British minister, a friend of Bevan's. They were all taken to a mess hall, where the Russians fed them generously and plied them with vodka. Kerr warned Baumer that getting foreigners drunk was standard Soviet procedure. Then Bevan, still in poor shape, left with his friend the minister for the British Embassy, where he was to stay, while Baumer got a lift to the United States Embassy from Kerr in the latter's Rolls-Royce. In due course Baumer reached his own quarters in the former residence of the German military attaché, now occupied by several American officers, some of whom he knew. After a hot bath he slept all afternoon.

That evening Baumer attended a reception at the residence of the American ambassador, Averell Harriman. Harriman told Baumer with evident amusement that he wished Baumer—the first American to have come into the Soviet Union alone without a visa—had been jailed for it so he could have a report on their prison system; but he was pleased that Baumer might have created a small incident that he could put to good use.

The next morning, Baumer reported to General Deane's apartment—the general was laid up with a broken leg—and explained BODYGUARD to Deane and Harriman. "It took several hours to explain the plan to me," recalled Deane, "despite the fact that it was fairly simple once I saw the light. I could foresee the difficulty we were going to have in making the Soviet representatives understand it through the medium of interpreters."

Then followed a week of inaction. Bevan, restored to health, had the basic BODYGUARD directive translated into Russian at the British Embassy. There was nothing else to do until the Soviets should call a meeting. Baumer signed up to take Russian lessons every morning. He took long walks around the Kremlin in the snow with other American officers and imparted to them such news from home as the plans for the Central Pacific offensive. From them in turn he learned about daily life in wartime Moscow, with its astronomical black market prices and the buying power of a few pounds of sugar or cartons of American cigarettes.

Days passed and still nothing happened. Bevan cabled to London that he was "living in hope." "The days at the office," recorded Baumer, "were sort of sorry." Evenings were spent at the opera or ballet at the Bolshoi Theater, which impressed Baumer with its beauty; and he noted that foreigners were segregated from the Soviet citizenry, but since the foreign ghetto was "in the first tier of the loges, [they] really had about the best seats in the house." There were late-night parties with the American colony—officers, diplomats, correspondents, secretaries from the American mission—with caviar, smoked herring, much vodka and grapefruit juice, and dancing to a piano if someone could play it.

Deane had warned Baumer that it would take the Russians a long time to decide to hold the first meeting. So it was a pleasant surprise when Baumer and Bevan were summoned to a General Staff building in Karl Marx Place on Monday, February 7, only eight days after their arrival. Present for the Americans were Baumer, Deane, the American minister, and one of Baumer's housemates as interpreter. For the British were Bevan, his friend the British minister, and a colonel from the British military mission. On the Soviet side were a Foreign Office representative named Dakanosov, Lieutenant General Fedor F. Kuznetsov of the General Staff, Major General Slavin (of the General Staff's external relations branch, without whom no contact with foreigners was allowed), and their interpreter, an unidentified captain, "a red-headed kid who looked as if he had just come from an Indiana farm," recorded Baumer. "He was friendly, knew his English grammar from the book, but had never translated before. Naturally the special language used to describe our business completely floored him." "He started every sentence," recalled Deane, "with 'General, he says,' and then would proceed with a jumbled mass of English construction which revealed nothing that anyone could possibly have said. To his great displeasure, our interpreter would then whisper to us the gist of Kuznetsov's remarks." But the American interpreter too was "out of his depth in the special military language," said Baumer. Under the circumstances, Deane considered Kuznetsov "a mental giant" when he nevertheless seemed in a very short time to have mastered the intricacies of BODYGUARD.

The meeting started badly. Bevan got off into "a long-winded, polite British conversation about why we had come," which got garbled in translation. Finally, "General Deane stepped in as he did from then on, with a short summation, a quick word or two, a joke and what have you. You could see that the Russians liked and admired him. We in turn came to like the Russians very much before the meeting was concluded and it ended amicably enough." The Soviets seemed at first to be somewhat put out that the Western Allies had already approved BODYGUARD without consulting them, but were reassured that the plan could be amended to fit Soviet needs. Mollified, they warned the westerners that no immediate decisions could be made, as they would have to consult their three fighting services, but gave assurances that the matter was considered important. They agreed to meet again on Thursday, February 10.

The days dragged on. There was more ballet. There was the Moscow Circus, with a "poor emaciated elephant." There were the usual cocktail parties. The February 10 conference with the Soviets was held as scheduled. At this one, Baumer thought, "things seemed to go swimmingly." A number of substantial matters were discussed. The Soviets urged a handful of changes. Bevan and Baumer undertook to cable their Chiefs of Staff asking them to go along with these requests. This they did, and both sets of Chiefs approved.

More days passed; more parties; more ballet. On February 14 came another meeting with the Soviets, attended at Soviet request only by Bevan and Baumer and their translator. The Soviets suggested several amendments to the plan. In particular, they seemed unwilling to mount a threat to Bulgaria, contending that a threat to Rumania was more plausible and politically acceptable (the USSR was not at war with Bulgaria). Again cables went back and forth to London and Washington. Back in London Dudley Clarke, who was in town, objected strenuously to Wingate that the Bulgarian threat was essential to his share of BODY-GUARD; moreover, a plausible threat to Rumania implied an opposed landing, for which Soviet landing craft simply were not available.

A near-final meeting was supposed to be held on the 16th. The Soviets postponed it. Then Bevan and Baumer heard nothing more, other than a meaningless casual assurance by Molotov to Harriman, and another to Kerr, that Bulgaria was not really a problem and everything would be taken care of "when Stalin got back from the front." (They heard later that Stalin had been very sick, rather than at the front; though no one ever verified it.) Two weeks passed in the usual time-killing of the foreign colony: parties, weekly movies at Spasso House, a concert (excellent) by the Red Army Ensemble, a concert (less good) by the Red Navy Ensemble, back to the Moscow Circus, always trailed by the NKVD.

The high point was attendance as guests of the Molotovs at the annual Red Army Day reception at the State Guest House on February 23. It provided an eye-opening glimpse of the way the Communist rulers lived in the depths of the third winter of the war, where, as Baumer recalled long afterwards, "every morning when we walked to work, we would see several dead bodies being pulled on sleds to some sort of burial place. Also we would see people pulling sleds from out of the city with one little piece of wood." None of this tragedy and squalor was visible in the State Guest House. There was champagne, and music by Honored Artists of the Soviet Union, and much milling about of the guests while tuxedo-clad NKVD men took care to keep American and Japanese guests apart; and, as Baumer recorded in his diary, "about as sumptuous a spread as has ever been shown even in the moving pictures. There was game, roast beef, caviar, smoked herring and wines and vodka in profusion."

Notwithstanding awareness of the Soviet predilection for getting their foreign guests drunk, Baumer and Bevan had several shots of vodka with General Kuznetsov, and "when at midnight they brought in the hot food, turkey, grouse etc. I wasn't interested. I just had some more drinks and ran into Marshal Budenny," the mustachioed hero of the Russian civil war. "They tell me I had my arm draped around him drinking with him and insisting that he sign my short snorter bill." Off now to a good start, Baumer spent some time talking pidgin Russian to a low-level official; then:

> When I saw Molotov making the rounds drinking to the Red Army, I felt that I was only going to be there once so boldly walked up to him, clinked glasses and drank to the Red Army. Much later blini and ice cream were served and I remember getting out the door, after talking to Kathy Harriman [the Ambassador's daughter and official hostess]. Her father came out while I was looking for the car and he was one mighty sick man, throwing up everywhere. Some business for a man who doesn't drink but lightly in the U.S.

Six more days went by. On Tuesday evening, February 29, the Soviets called Bevan, saying they wanted to meet. Bevan insisted that Baumer be invited as well. They rushed over to Karl Marx Place to find that all the Russians wanted was another copy of the plan.

Another four days passed. Then, at 10:30 Saturday night, March 4, a call came from the Soviets to say there might be a meeting later that night. (Baumer had learned by now that the Kremlin workday started at 5 P.M. and ran till 5 A.M.) At 12:40 A.M. they were asked to come over to a 1:30 A.M. meeting. Upon arrival General Kuznetsov dumbfounded the westerners with the announcement that BODYGUARD was accepted lock, stock, and barrel in its original form, and that the Soviets would cooperate in its execution. "We didn't know whether to shout or be suspicious," said Baumer later. "General Deane reacted immediately with 'ochen khorosho,' and John Bevan took the cautious line." After some friendly chitchat with Kuznetsov, the westerners went home and fell into bed.

That afternoon they returned to Karl Marx Place to complete the protocol agreement. The interpreters bungled some diplomatic niceties in the English translations and Baumer undertook to retype them himself on an old American typewriter, while Bevan and Deane chatted with

Kuznetsov about using the press for deception, making it clear that the Western Allies did not do so. Baumer always remembered that "when we said that in a democracy you couldn't use the press to fool your own people, the Soviets said 'Oh, well, we do it all the time.' "

That was Sunday, March 5. Bevan and Baumer left for home on Wednesday, March 8. The night before their departure, Baumer's messmates taught him the rules of the Moscow delegation: He could not take anything out, for such items as gloves, socks, and underwear were hard to come by in Moscow, whereas he could replace them back home. So he sold them his belongings.

This time Baumer and Bevan took the southern route, by way of Baku, Tehran, Abadan on the Persian Gulf, across the desert, Palestine, and the Sinai, landing at Cairo at 5:30 P.M. on Friday, March 10. Baumer checked in at Shepheard's; Bevan stayed with Michael and Betty Crichton.

Baumer had a pleasant visit in Cairo, getting at last a firsthand look at the original home of "A" Force, and meeting Colonel Jones, the master of the dummy tank business; "one of those who spends his every word trying to impress with his acquaintances and lists of names," said Baumer. Next day, Saturday, March 11, they went shopping for a few exotic gifts to take home and some more practical items to be sent as thankyou presents to the delegation back in Moscow. Lunch with Rex Hamer and the Crichtons at the apartment of Terence Kenyon, an old friend of Clarke's who had joined "A" Force the previous October; an afternoon at the races; tea and dinner at the Crichtons', where Baumer met the British Ambassador and others, and brought out some of his Moscow caviar for hors d'oeuvres. Betty Crichton told them that they could not come to Cairo without seeing the Pyramids, so it was decided to drive out to them after a scheduled visit to the workshop where dummy tanks and such were made.

Next day, Sunday, March 12, a sand-laden wind was blowing that "got into your eyes and clothes and made even the city gasp for breath." After visiting the workshop—Baumer thought their dummies were "excellent improvisation and better than our hastily built light metal pieces"—they drove through the sandstorm for some fifteen miles in an open car, dined reasonably well at the hotel where the SEXTANT conference had been held, and then drove as close to the Pyramids as possible, where Baumer and Betty trudged through the swirling dust for a closer look at the Great Pyramid and the Sphinx while Bevan and Crichton, less intrepid, waited in the car.

After a day's shopping, ice cream at Groppi's, and dinner with Rex Hamer, Bevan and Baumer flew on via Tripoli to Algiers. There they received a "royal welcome" from Dudley Clarke (just back from London) and George Train, and discussed Russian cooperation on BODYGUARD with Clarke and General Burrows, the newly designated head of the British military mission to Moscow. Next day Baumer called on General Noce, who had just joined Devers in Algiers, and that evening they visited the Cercle Interallié and repaired to Ekstrom's for supper. From Algiers they flew to Casablanca and Prestwick and finally arrived back in London on Sunday, March 19. There Baumer, billeted this time in the Mayfair Hotel, reported once again to the British Joint Planners.

Everything had changed in his absence. ETOUSA was out of the invasion planning business, FUSAG had taken over but "was working in a nebulous sphere," the United States First Army headquarters had gone to Bristol, and Eisenhower's SHAEF headquarters was moving to Bushey Park. He remained for three weeks before getting orders home. He had tea at the Connaught with Ingersoll; Ingersoll "was interested in Russia and pumped me dry while I also got from him the happenings of the previous two months." Billy Harris had been made chief of Ingersoll's section. (Ingersoll always maintained airily that Harris was only "titular boss," that he, Ingersoll, was the head man, having "obtained" Harris for "my" Special Plans section "because I needed a regular officer of rank to carry the directions I got from the British conceivers of the cover plan to the American HQs who would be required to implement them"; a view of their relationship that would no doubt have astonished Harris and dumbfounded Noce.) That pleased Baumer, though he reflected wistfully that it meant a colonel's eagles for Billy, which was not in the cards for Baumer as long as he was stuck in the planning business.

There was socializing with Robert E. Sherwood, whom Baumer had known in connection with propaganda work in Washington, and through Sherwood with Edward R. Murrow and Pamela Churchill; with Joan Bright; with the Dennis Wheatleys. One evening he and Bevan broke out their Russian caviar and hosted a cocktail party in the Cabinet War Rooms. His old boss Wedemeyer was passing through on his way back to SEAC from a journey to Washington, and Baumer visited with him and with old OPD friends on his staff. He visited Hunter Harris at his air base in Suffolk. He spent a pleasant weekend with the Bevans at their country house, milking a goat and chopping wood. And he interviewed and approved Captain H. Wentworth Eldredge, a candidate for the intelligence officer's slot in Billy Harris's section.

He wanted to stay in London—"everything there was full of life and there was so much activity, I didn't want to miss the biggest show in history." But that was not to be. General Bull, the SHAEF G-3, took him to lunch with Bedell Smith and another general and told Baumer he had asked for Baumer to be assigned to him, but General Handy, the head of OPD, would not let him go.

Baumer arranged to go home on the *Queen Mary* or the *Queen Elizabeth,* but had to give that up too when he was ordered home by air. Back in the Pentagon he found himself assigned again to the Joint War Plans Committee. He was losing his longtime colleague, Baron Kehm, who was off to the propaganda section of SHAEF. He was getting anxious to get back to the real Army; he noticed that others of his contemporaries besides Billy Harris were making full colonel.

As for Bevan, Wingate had held the fort during his protracted absence; and while he was fretting away those vital weeks in Moscow, the centerpiece of the OVERLORD deception, Operation FORTITUDE, had been approved and was under way.

CHAPTER 13

QUICKSILVER

Nobody, as Carl Goldbranson had protested to Dudley Clarke, could give direction to David Strangeways; nobody even tried. Strangeways was a law unto himself, brashly sure of his own judgment, cheerfully interpreting his orders to suit that judgment, and able to get away with it as long as he enjoyed the confidence of his superiors. And throughout the war he did enjoy the total confidence of his two immediate superiors, Montgomery's chief of staff "Freddie" de Guingand, and Monty himself. With this self-confidence and support, Strangeways pushed through and guided the most important deception plan of the war.

Morgan had been directed to assume limited resources that enabled COSSAC to plan an invasion entailing an initial landing in Normandy of only three divisions on a relatively narrow front, with a target date of May 1, 1944 (Operation NEPTUNE). Simultaneously with this landing, the Allies would mount a landing in the south of France, Operation ANVIL. When at the end of 1943 and the beginning of 1944 Eisenhower became the actual Supreme Allied Commander and he and Montgomery took over from Morgan, they and their staffs quickly concluded that NEPTUNE must be enlarged to a five- or six-division assault. Since sufficient landing craft were not available to mount both an enlarged NEPTUNE and ANVIL simultaneously, D-Day for NEPTUNE was postponed for a month and that for ANVIL was postponed until mid-August.

At the head of the Allied force for OVERLORD would be a Supreme Headquarters, Allied Expeditionary Force ("SHAEF") headed by Eisenhower, with a staff organized on American lines and built on Morgan's COSSAC staff as its original nucleus. Under SHAEF would be separate air, sea, and land commands, each headed by a British officer: the Allied Expeditionary Air Force under Air Chief Marshal Sir Trafford Leigh-Mallory; naval forces under Admiral Sir Bertram Ramsay as Allied Naval

Commander, Expeditionary Force; and on land, initially 21 Army Group under Montgomery. For the assault and the opening phase of operations in Normandy, 21 Army Group would initially be composed of the British Second Army (including the Canadians to begin with) under Miles Dempsey, and the American First Army under Omar Bradley. The American First Army would steadily be built up, followed by divisions to compose the American Third Army under Patton. When a critical mass was reached, Third Army (and Canadian First Army) would be formally activated, Bradley would turn over his First Army to Lieutenant General Courtney H. Hodges, and First and Third Armies would be incorporated into the American 1st Army Group, FUSAG, under Bradley. Montgomery would continue as overall ground commander of the two army groups until such time as Eisenhower should elect to assume direct command. In the much longer run, the American Ninth Army under Lieutenant General William H. Simpson, which was slowly gathering in England, would cross into France and join Bradley's army group.

The crucial period for NEPTUNE would be the first two weeks. The assault divisions must make a successful lodgment and hold on while the follow-up divisions built up to a force that could withstand German counterattacks.

The goal of the deceivers, therefore, was twofold. First, to mislead the Germans as to the time and place of the invasion. Second, to keep them from concentrating at the bridgehead long enough for the initial buildup; specifically, to deter them from reinforcing their Seventh Army, which held the area of the D-Day bridgehead, with forces from their Fifteenth Army further north in the Pas-de-Calais* and Belgium.

Strangeways left Italy on December 11 and arrived in London before Christmas, set up shop with Montgomery's London headquarters in St. Paul's School in Hammersmith, West London, and busied himself putting into place his deception organization for Montgomery's 21 Army Group.

---

* For simplicity, throughout we use "Pas-de-Calais" not in its strictly correct sense but to denote the whole stretch of coastline from the Seine northerly to Belgium (in other words, the departments of Nord, Somme, and Seine-Maritime in addition to the department of Pas-de-Calais); and "Normandy" to refer to the stretch from the Seine westerly through the Cotentin peninsula (departments of Calvados and Manche).

Called originally Ops (D) but renamed G(R) as of February 22 (the "R" was intended to suggest "reconnaissance" to those not in the know), it was established on the "A" Force rather than the American staff pattern, reporting directly to de Guingand and to Montgomery himself, and working closely with Monty's intelligence officer, Brigadier E. T. Williams, known always as "Bill." And again on the flexible British pattern, Strangeways was not only an army group staff officer but directly commanded a tactical deception unit, known as "R" Force, with the usual contingents of dummy equipment, sonic devices, and radio apparatus.

Even before leaving Italy, Strangeways had put together his key staff—a lean group of himself and three other officers. Chandos Temple of his old Tac HQ was his second in command and general alter ego. A quiet-voiced, lanky, somewhat stoop-shouldered cavalry major ("a drippy, lanky, sloppy, drawling kind of chap," reminisced Strangeways fondly), in civil life the head of a road-building firm—"probably the biggest muck shifter in England," he said of himself—he was remembered by his colleagues as "a wonderful chap." As his intelligence officer, whose job would chiefly be to maintain touch with Bill Williams's appraisals and keep up with ULTRA, Strangeways picked an old neighbor from Cambridge, Captain Philip Curtis, in civil life a stockbroker. Curtis's elder brother had been an officer in Strangeways's own regiment and had been wounded; visiting him in the hospital, Strangeways asked what Philip was doing, learned that he was in intelligence with an armored unit, and brought him aboard. With the addition of a quartermaster officer for administrative work, that was the totality of Strangeways's staff.

Ike brought with him from the Mediterranean his enthusiasm for cover and deception, and in January Bedell Smith began a series of conferences that led eventually to formalizing an American deception staff. Billy Harris had been detached to work with the 21 Army Group staff on overall planning. When that was completed, at the beginning of March, Harris and Ingersoll were informally assigned to coordinate the American part in OVERLORD deception. This was formalized by the official establishment of a "Special Plans Section" of FUSAG G-3 on March 26. Harris was promoted to full colonel and Ingersoll to major, and under date of April 7 orders were issued designating Harris as "Chief of the Special Plans Branch, G-3 Section, FUSAG," and assigning Ingersoll to the Branch. For the Normandy deception, Special Plans took directives from Strangeways and was in effect a part of Strangeways's G(R).

The deceivers had the usual problem being taken seriously in the American staff hierarchy. When Harris briefed Brigadier General Kibler, the G-3, on the plan to focus German attention on the Pas-de-Calais, the general's first response was: "But we are not *going* to land in the Pas-de-Calais."

Harris's senior officer and deputy during the early part of 1944 was a graduate of the West Point class of 1938, Alabama-born Lieutenant Colonel Clarence Beck of the 1st Infantry Division. He had apparently assisted Ingersoll in connection with the organization of tactical deception in the fall of 1943; he went back to the States in January 1944 to round up signal deception officers and to prod along the formation and training of the 23d Headquarters Special Troops, and followed up from London on his return. Subsequently, in the weeks leading up to D-Day, deception operations involved a good deal of detailed, if notional, manipulating of American units, and Beck's familiarity with the nuts and bolts of the Army was essential. It helped too that he had combat experience in the Mediterranean, for it sometimes seemed that nobody without such a record would be taken seriously in Bradley's army.

"Clare" Beck was a fighting officer who had already won the Silver Star for gallantry in North Africa, and he was not interested in an extended tour in this funny business; so after the invasion he returned to the Big Red One. But he and Ingersoll became fast friends during their time together; after the war he would marry in Ingersoll's living room an English girl whom he met while she was working at MI5 and he was working with Special Plans; they were godfathers to each other's children, and two decades later Ingersoll's son would marry Beck's daughter. After Beck left, Lieutenant Colonel Olen Seaman of the 23d Headquarters Special Troops briefly acted as Harris's deputy for FORTITUDE SOUTH II.

Ingersoll's private life was somewhat complicated that winter and spring. He had shared a flat with General Noce. He got his own quarters after Noce left to join Devers in the Mediterranean. He met an attractive German Jewish woman who sold theater tickets in a broker's stall at Claridge's. One thing led to another, and she moved in with him. It was a difficult relationship, for she proved to be a kleptomaniac. (Moreover, after he went to France she announced that she was pregnant by him; but it turned out she had been seeing several other officers too.) It may well have been a relief to leave on April 2 for a brief trip to the States to visit the sonic deception establishment at Pine Camp, New York, and the 23d at Camp Forrest, Tennessee, to expedite its movement to Europe, and incidentally to talk about additional officers for Special Plans.

Meanwhile, an intelligence officer to look after American interests in special means had been acquired in January. Dennis Wheatley had found him. Knowing of the need for such an officer, he asked his stepdaughter Diana Trench, who was separated from her husband and effectively single and had seen young American officers a good bit, whether she knew "some bright American who did not have a big mouth." She suggested an Air Force captain whom she had come to know extremely well, named H. Wentworth Eldredge, known to his friends as "Went."

Eldredge was a thirty-four-year-old bachelor, dark-haired and dapper, in civil life Assistant Professor of Sociology at Dartmouth. The descendant of a long line of Cape Cod and Sag Harbor whalers and grandson of

(James Eldredge)

H. Wentworth Eldredge

a Scot who immigrated to Brooklyn before the Civil War and made a fortune in mattress ticking, he grew up in the old-fashioned Brooklyn of the early part of the century, graduated from Dartmouth in 1931, took his Ph.D. in sociology from Yale in 1935, and returned to his alma mater to teach. Travel in Europe in the early 1930s had taught him to despise the Nazis, and a youthful anglophobia was overcome by admiration for the British stand of 1940–41. So when the United States entered the war he offered his services. In 1942, he served briefly with the War Division of the Department of Justice, and visited South America for the intelligence section of the American Republics Division of the State Department before being commissioned into the Air Force. Trained as an intelligence officer, he reached England in February 1943, assigned to headquarters of Eighth Air Force at Teddington, near London. After a boring few months working on a little classified magazine put out for the air crews, and a bout with jaundice that sent him to Oxford for a hospital stay, he was assigned to a more congenial job in the Eighth Air Force historical section. But, as he later said, he "wanted to be doing something, not recording what other people had done or might do."

On his first night in London, Eldredge had been taken to a party at which he met a number of charming English girls. One of them was Diana Trench, and he had seen a good deal of her thereafter. Eldredge had known that her stepfather did something hush-hush and interesting; and in January Wheatley suggested to Eldredge that he take to lunch a Colonel Baumer who was visiting from Washington, and something worthwhile might turn up.

Eldredge got in touch with Baumer and took him to lunch at a fancy restaurant that with a bottle of claret set him back a couple of pounds—a substantial sum in those days. They hit it off well. Baumer asked if he knew Ingersoll; no, said Eldredge, but he knew the story of *PM* and thought Ingersoll a very bright and tough guy. Good, said Baumer; he would make an appointment with Ingersoll for Eldredge.

Two days later, Eldredge went to Ingersoll's office at 20 Grosvenor Square. "God he was ugly!" remembered Eldredge. "I buttered him up, of course, and he glowed." Eldredge's politics at that stage of his life were similar to Ingersoll's and they got along famously. Ingersoll told him he would find out what the job was once he was transferred from Eighth Air Force. That did not take long; the next week, Eldredge moved from Teddington to a new billet in a hotel in Kensington High Street, and became with Harris and Ingersoll the third member of the three-man team that would carry deception for Omar Bradley's army group for the rest of the war.

Eldredge was introduced to Ops (B), the SHAEF deception section, and he and Hesketh became good friends. He met Bevan and the rest of the London Controlling Section and frequently visited their underground rooms. His initial responsibility was for the American element of the material being put over by special means, and he got to know Tommy Harris, Hugh Astor, and the other case officers. After a few weeks he began to think it was all a waste of time. He went to see Bevan and said he could not go on without some concrete evidence that this chickenfeed was believed by the German staff. "Johnnie tried to feed me a snow job," he recalled long after the war, "stating they (the Krauts) were running aerial recce and they had a pretty good idea from that. I replied that aerial recce was practically impossible and this was not enough. I said I would quit unless I was shown some regular evidence. Johnny looked at me and said to come back in a couple of days."

This led to an appointment with Colonel Telford Taylor, in charge of ULTRA matters for the U.S. Army in London. As Eldredge told the story, Taylor drew the curtains, told Eldredge he would be shot if he revealed what he was going to learn, and indoctrinated him into the ULTRA secret. "I thought, no drink, and I hoped I did not talk in my sleep," recalled Eldredge. After that he read some of the traffic relevant to the double agents and became a believer. For a long time he was the only member of Special Plans who was privy to ULTRA; neither Harris nor Ingersoll was let in on the secret until after the invasion.

For the rest of the spring, most of Eldredge's work consisted of reading and sharpening up double agent traffic from an American perspective. At first he said to himself that there was little in it that a bright sixteen-year-old could not do; but over time he realized that he was part of a complex and subtle system in which no slips could be tolerated. He got along well with it, and with his English colleagues. "A gentleman," recalled David Strangeways. "A breath of fresh air, in point of fact." Eldredge reciprocated this admiration. Strangeways's force "were a ragamuffin lot of soldiers," he recalled after the war. "Desert boots (soft shoes), old sweaters, stained khaki pants." But he was greatly impressed by their professionalism when he went down to visit their field headquarters near Portsmouth.

Above G(R) and the Special Plans Section—at least in theory—was the theater deception staff, SHAEF Ops (B). It was clear from the beginning that it would need someone in charge with broader experience than Jervis-Read had. Clarke offered Noel Wild. In late November, Bevan discussed the matter with the senior generals in COSSAC G-3, and it was agreed that Wild would be brought to England in mid-December, be sized up, and be given the job if they were satisfied with him. On December 17, Clarke sent Wild to England without mentioning SHAEF; ostensibly it was simply Wild's first home leave in two and a half years. He passed the tests, and just after Christmas, Bedell Smith, who came to SHAEF with Eisenhower as his chief of staff, summoned Wild to Norfolk House and told him of his new assignment; with promotion to full colonel into the bargain.

Wild took over at the beginning of January. Under him, Ops (B) continued to be divided, on the "A" Force pattern, into a physical deception subsection under Jervis-Read, and an intelligence subsection, which he assigned to Roger Fleetwood-Hesketh on the recommendation of the head of COSSAC's intelligence section. Two Americans who had been temporarily assigned to COSSAC Ops (B), Lieutenant Colonel Percy Lash and Major Melvin Brown, returned to Ops (A), the regular planning section. As time passed, the staff was enlarged. Several more Americans joined in early May. Lieutenant Colonel Frederic W. Barnes and Major Al Moody were West Pointers, and both cavalrymen; Barnes of the class of 1934, while Moody had been No. 1 in the class of 1941 and by choosing the horse arm had caused some distress in the Corps of Engineers, who considered that they had a prescriptive right to the top men of every class. A third American was Captain John B. Corbett, known as "Jack." None contributed greatly to the OVERLORD deception, and they remained for only a few months, but on the strength of their experience they would later be picked for further duty with American deception in the Pacific and Far East. At the end of the month Sam Hood left "A" Force Tac HQ in Italy to join Wild. At the end of April, Bobby Rushton was urgently sent for from Algiers; originally destined for Ops (B), he was soon taken by Harris to be the administrative officer for Special Plans.

On the intelligence side, Hesketh was joined by an MI5 officer seconded to act as liaison with that agency, Major Christopher Harmer, the young solicitor from Birmingham who had been case officer for the double agent BRUTUS. He remained until the end of March, after which he spent much of his time training the special counterintelligence unit that he would take to the Continent, moving on or about April 20 to Montgomery's headquarters near Portsmouth, though he traveled back and forth to London and kept in close touch with Hesketh. An MI5 civilian, Phyllis White, joined Hesketh and Harmer to provide secretarial and filing services with respect to highly classified ULTRA and double agent material; and Wild arranged for Hesketh's brother Captain Cuthbert Fleetwood-Hesketh, who spoke German (not that he spoke it or anything else very much; he was notably taciturn), to be brought in from Military Intelligence liaison in the War Office. Cuthbert's primary function, as it transpired, would be to act as a high-level courier between Ops (B) at Norfolk House in London and Montgomery's headquarters during the busy days from April onwards.

It was evident that the double agents would play a great role in OVERLORD deception, so Bevan made Wild a major player in that game from the outset. The day after New Year's he took Wild to lunch at Brooks's after the two had paid a call on Freddie de Guingand to discuss physical deception and the APPENDIX Y deception plan. Wild asked Bevan innocently whether MI5 had any tame agents feeding false data to the Germans comparable to those that SIME and "A" Force had run from the Middle East. Bevan assured him with some amusement that they did. The next day he took Wild to a meeting of the Twenty Committee and opened his eyes to the menagerie of double agents being run out of England.*

Bevan then suggested that his TWIST Committee be abolished and replaced by biweekly meetings of Ops (B), the London Controlling Sec-

---

* Wild—no doubt reluctantly—advised the Twenty Committee, at the first meeting he attended, that Brigadier Bill Williams, Montgomery's intelligence chief, "had suggested, in addition, that the assistance and co-operation of Lt. Col. Strangeways should also be sought." "Williams does not think a great deal of Wild and has a high opinion of Strangeways," said Guy Liddell to his diary after dining with Williams and Masterman. "He says that Wild works rather day to day whereas Strangeways looks much farther ahead. He regards Strangeways as being extremely intelligent as well as very brave and practical as a regular soldier."

tion, and B1A. This was formally agreed to on January 12, and Bevan and Tar Robertson agreed that Ops (B) would coordinate all deception output through the double agents. Six weeks later, the RACKET Committee was abolished as well. Wild became a full member of the Twenty Committee so that he could keep current with its work; and in addition to his regular office in Norfolk House, Bevan found a desk for him in the London Controlling Section's cramped quarters. Ops (B) would deal directly with the Diplomatic Section of MI5 for planting leaks through diplomatic channels in London, and through the London Controlling Section with the Foreign Office for diplomatic leaks abroad. The London Controlling Section remained the central clearinghouse among the theaters: On January 27, Bevan decreed that Ops (B), "A" Force, and Joint Security Control would communicate with one another through the London Controlling Section and not directly.

Wild thus had abundant authority. But, as it turned out, he could not do much with it. In part this was because of his own limitations. "Noel Wild was useless," a participant recalled many years later. "I never knew him to come up with a decent idea." In part it was because *nobody* could do anything with David Strangeways.*

"Anyone looking back on those years has to relive the atmosphere that existed when the Cairo/Eighth Army gang came back to England at Christmas 1943," said Christopher Harmer fifty years later. "They were, quite literally, on top of the world in their confidence, self-assurance and self-estimation that, so far as the U.K. was concerned, they were the fighters while we at home were the lay-abouts."

Strangeways brought to London a full share of this bumptiously self-confident attitude. He had looked at APPENDIX Y, and he did not like it. Five days after Christmas, Jervis-Read was already noting drily that "APPENDIX 'Y' had a certain amount of comment at 21 Army Group and they have some new ideas about timing and control." Nobody else had liked it either, of course, but Strangeways had special reason for his dislike, for implementing it would be his job.

* When Niall Macdermot, a "tough-minded and brilliant" peacetime barrister, was selected as MI5's representative to head counterintelligence at 21 Army Group, John Marriott commented to Harmer and Tar Robertson: "The great thing about Niall is that he's not afraid of David Strangeways and stands up to him."

The basic policy laid down by BODYGUARD, it will be recalled, was what came to be called the "postponement" or "delay" theme, that no cross-Channel attack could occur until late summer, if then. Then, when the Germans finally realized that cross-Channel operations were imminent, "the tactical cover plan prepared by Supreme Commander, Allied Expeditionary Force will come into force with a view to deceiving the enemy as to the timing, direction and weight of 'OVERLORD.' " The notional D-Day for this "tactical cover plan" was to be chosen by the Supreme Commander, but BODYGUARD recommended that the date be later than the Overlord D-Day, "with a view to delaying the dispatch of enemy reinforcements for as long as possible." This "postponement" theme pestered the deceivers for months until it was simply scrapped.

APPENDIX Y had come up with nothing better for attracting German attention to the Pas-de-Calais than making the southeast of England seem somewhat more strongly held than it really was by means of "discreet display"—*i.e.,* sloppy camouflage—of dummies and the like, and then, on the assumption that the Germans would see that an invasion was imminent when the troops moved to preparatory positions and were loaded on the invasion craft, representing a six-division force in southeast England for a fortnight after D-Day. There was a general notion that GHQ Home Forces would play the role of this force.

Through January there was a series of meetings to concoct a new plan for the Pas-de-Calais threat and also for a threat to Scandinavia, to both of which collectively the codename MESPOT was assigned. A first sketch went out on January 3; a preliminary meeting was held January 7; a draft was circulated January 11, with a head planners' meeting January 14. On January 17 a draft was circulated under which, by phony activity in the southeast and otherwise, the Germans were to be led to believe that the preparations they would observe during the spring were for a massive invasion of the Pas-de-Calais planned for mid-July (D plus 45); they would thus hopefully be taken by surprise when the real landing came six weeks sooner and in Normandy. The proposal paid little attention to the situation after the real D-Day in Normandy, suggesting only a "story" that a six-division force was assembled in the southeast to carry out a subsidiary operation in the Pas-de-Calais area "with the object of drawing German forces away from the main target area."

Little better than a warmed-over APPENDIX Y, this new proposal received no better reception than APPENDIX Y had. Two British generals who were not even directly involved in deception noticed one of the main problems with the "delay" theme at once. If the Germans were able to observe Allied preparations sufficiently to judge their state of readiness, the real activity in the south and southwest of England would lead them to expect an invasion about June 1, while pointing fake activity in the southeast toward July 15 would mean that they might not notice anything at all in that region till after D-Day itself. Even if they somehow did evaluate what they saw as suggesting June 1 in one place and July 15 in the other, they might well be inclined to split the difference; or they might decide that Normandy would be invaded on June 1 and that something would happen in the Pas-de-Calais six weeks later—six long weeks, in which they would have abundant time to deal with Normandy before turning to the Pas-de-Calais.

Strangeways most certainly saw these problems. By now he was beginning to make a nuisance of himself. Even before the January 14 meeting, he was so unhappy with this approach that he threatened to wash his hands of the whole matter. "Unofficially," wrote the deputy chief of the plans and operations section of SHAEF G-3 on January 13, "I believe 21 Army Group wish to free themselves from APPENDIX 'Y' to 'OVERLORD' and limit themselves to local tactical deception for 'NEPTUNE' area. I feel we should resist this as it is essential that APPENDIX 'Y' planning be tied in as an integral part of the 'OVERLORD' mounting plan." "I was not a much loved person," Strangeways himself said cheerfully many years later. He was "much disliked," said Christopher Harmer, who in fact admired him greatly. "Tried to act like Montgomery." "He was a 'snotty' character," recalled Went Eldredge, who nevertheless likewise admired him.

Unconcerned over stepping on toes as long as the job was done right (in fact, enjoying stepping on the toes of Noel Wild), Strangeways set out to push the plan in what he saw as the right direction. The emphasis was wrong (as Dudley Clarke had perceived months before). Some vague threat from southeastern England would not induce the Germans to do what the Allies wanted them to do: Hold strong forces away from Normandy. The focus should always be on the Pas-de-Calais; even NEPTUNE itself should be painted as only a preliminary bout before the main event.

Obviously at Strangeways's instigation, de Guingand sent to SHAEF G-3 a letter on January 25 urging in essence that the "delay" theme be played down; that the focus be on the Pas-de-Calais from the outset, and that that focus should suggest that the Pas-de-Calais assault would be the *main* attack:

> I consider that the enemy should be led to believe that from now on our target in Northern Europe is the Pas-de-Calais area. Once our efforts have failed to make the enemy believe that our attack will not take place before the late summer, we should concentrate on telling him that the Pas-de-Calais area is our *early* objective. This may mean that we shall have to tell him as early as D minus 7—in view of the fact that he will see the obvious preparations for 'NEPTUNE.'
>
> . . . I do not agree with the object which had been given for the attack on the Pas-de-Calais area. If we induce the enemy to believe the story he will not react in the way we want. I feel we must, from D-Day onwards, endeavor to persuade him that our *main* attack is going to develop later in the Pas-de-Calais area, and it is hoped that 'NEPTUNE' will draw away reserves from that area.

This produced results. The final draft, forwarded to the British Chiefs of Staff on January 30, embodied much of Strangeways's view. As approved on February 18 with minor revisions by the British Chiefs, the "story" was divided in two. Until NEPTUNE D-Day, "Story A" would be that a cross-Channel operation to seize Antwerp and Brussels and drive upon the Ruhr would be carried out:—

> by a total force of fifty divisions with craft and shipping for twelve divisions. The assault will be made in the Pas-de-Calais area by six divisions, two east and four south of Cape Gris Nez. The followup and immediate buildup will be a further six divisions. The force will be built up to the total of fifty divisions at the rate of about three divisions per day.

After NEPTUNE D-Day, the version of "Story B" for which Strangeways had pressed would take over. Under it, the Pas-de-Calais would not be a subsidiary operation but the main one, for which NEPTUNE was a preliminary. "For as long as possible, the enemy should be induced to believe" that:—

> the operation in the NEPTUNE area is designed to draw German reserves away from the Pas-de-Calais and Belgium. Craft and shipping for at least two assault divisions are assembled in the Thames estuary and southeast coast ports; four more assault divisions are held in readiness in the Portsmouth area and will be mounted in craft and shipping from NEPTUNE. When the German reserves have been committed to the NEPTUNE area, the main allied attack will be made between the Somme and Ostend with these six divisions in the assault.

The plan paid lip service to the "delay" theme by providing that the notional target date for the invasion would be NEPTUNE D plus 45 (July 15). It maintained the BODYGUARD "story" that fifty divisions would be required for the cross-Channel invasion plus eight more to threaten Scandinavia. Since there would only be fifty-three divisions in hand by mid-July, "we should, therefore, induce the enemy to believe that the deficiency of about five divisions will be made up from the United States during the operation. At the same time, in order to emphasize the later target date, we should minimize the state of preparedness of the NEPTUNE forces by misleading the enemy about their state of training, organization, equipment and their location."

But this was only lip service, for the plan also provided that if before NEPTUNE D-Day it should become apparent that the enemy did not believe in this later target date (which would almost surely be the case), "the threat to the Pas-de-Calais will be fully developed." And this still did not fully cure the "delay" problem, for if the threat to the Pas-de-Calais was "fully developed" and yet the assault came in Normandy, this would tend to discredit the very same channels of information that would be needed to put over the Pas-de-Calais theme after D-Day.

Effective as of its approval date by the British Chiefs, the plan had a new codename. Churchill disliked MESPOT. SHAEF was offered a selection of possible new names—BULLDOG, AXEHEAD, TEMPEST, SWORDHILT, FORTITUDE, LIGNITE—and the G-3 chose FORTITUDE. Henceforth the cross-Channel deception was called FORTITUDE SOUTH. That aimed at Scandinavia was called FORTITUDE NORTH.

* * *

Six divisions for the notional Pas-de-Calais operation after NEPTUNE D-Day, as APPENDIX Y had envisioned, would clearly now be inadequate for the new emphasis on that operation. On February 3, Wild promulgated a ten-division plan for implementing FORTITUDE SOUTH. This carried over from APPENDIX Y the "story" of a six-division British army under GHQ Home Forces; it would land east of Cape Gris Nez on the French side of the Straits of Dover, while the American V Corps and British I Corps—which were genuine components of 21 Army Group— would land west of the straits. These two corps, however, were part of the real NEPTUNE assault. Wild's plan accordingly contemplated that after NEPTUNE D-Day, the GHQ Home Forces six divisions would continue the threat "with four additional assault divisions which will be mounted in the Portsmouth area to assault South of Cape Gris Nez when the German reserves have been committed."

The trouble with this February 3 plan—a flaw built into FORTITUDE itself—was a failure once again to think through the post-D-Day story. As Hesketh asked after the war, "where were these four extra assault divisions to be found after V Corps and I British Corps had landed in Normandy? One could not start building them up before the invasion because until then the story would be that the two real assault corps would attack South of Cape Gris Nez. Yet one cannot produce four imaginary assault formations, ready to embark at a moment's notice, by the wave of a hand." Even more important, the "story" included no separate command for the Pas-de-Calais assault; the Germans were not likely to believe that Montgomery could control both a substantial offensive in Normandy and the definitive attack upon the Pas-de-Calais.

The February 3 plan contemplated extensive physical activity such as display of dummies and real troop movements, and as fleshed out in a SHAEF guidance letter to the War Office on February 9, it would even have included moving two real divisions into the southeast shortly after NEPTUNE D-Day to "spread themselves about." This did not go down well with Strangeways either. From the very outset, Strangeways had been opposed to any heavy dependence on such activity. It would eat into the limited resources available, and German reconnaissance was so scanty that he saw no need for anything other than dummy shipping in the invasion ports. He had cut the camouflage staff sharply.

Strangeways had not enhanced his popularity by pushing his ideas. When on February 1 the deputy G-3 sent de Guingand a draft directive for implementation of FORTITUDE, he noted that "we have worked on the principle of relieving 21 Army Group of as much responsibility as possible consistent with the efficient execution of the plan." (In other words, keep David Strangeways out of our hair.) In form a directive to the Joint Commanders (*i.e.,* Montgomery, Ramsay, and Leigh-Mallory; in practice, Strangeways and his navy and air force colleagues*), this draft would have assigned to Wild's Ops (B) all special means implementation; the concealment of the movement of naval and military forces from Scotland to the south coast; combined and joint wireless activity supporting the false order of battle and its concentration; installation of deceptive and decoy lighting; movement and administrative preparation; and real and dummy antiaircraft defense. Strangeways and his colleagues would be left responsible for nothing but "directing the threat created by the forces under their command towards the Pas-de-Calais area and for concealing the state of readiness of their forces so as to indicate 'NEPTUNE' D plus 45 as the real target date," plus "making preparations to continue the threat against the Pas-de-Calais after 'NEPTUNE' D-Day, until such time as the Supreme Commander allots the responsibility for control of this threat to Commander in Chief, Home Forces." In other words, nothing but dummy displays and troop movements.

Strangeways would not cheerfully have thus turned over to Noel Wild most of the responsibility, leaving him with nothing but the physical deception in which he did not believe. No doubt he called on de Guingand and perhaps on Monty himself; and this proposal did not last long. After successive drafts, the final directive, issued February 26, left Ops (B) with nothing (aside from meaningless overall coordination and control) but special means, political warfare and propaganda, and "occupation and Scandinavian operations," and eliminated any reference to Home Forces. It listed steps that had already been taken under Ops (B)'s aegis— mostly physical—but provided that they were "subject to adjustment as

---

* These were Alec Finter for Ramsay and Wing Commander F. E. W. Birchfield, Royal Canadian Air Force, for Leigh-Mallory. Finter had left "A" Force on November 7 to change places with James Arbuthnott and he was then seconded by Bevan to SHAEF; from which he did not return to the London Controlling Section.

a result of detailed planning." It authorized 21 Army Group to deal directly with the War Office and GHQ Home Forces.

As soon as it became clear that the Joint Commanders would be running the show, Strangeways took steps to scrap the idea of extensive physical deception. At a February 23 meeting, de Guingand (speaking, of course, for Strangeways) decreed that no more than routine camouflage would be practiced, there would be no special troop movements or special construction, and deceptive display would be restricted to dummy landing craft and the clearing of such "hards" as would not be actually used for OVERLORD. Wild protested, and on March 4 SHAEF overruled de Guingand and decreed that the former policy would be continued under the aegis of the War Office and GHQ Home Forces. Strangeways responded by procuring a reconfirmation of 21 Army Group's authority to "arrange direct with the War Office and GHQ Home Forces, for the continuation of the measures already initiated, adjusted as may be desirable and feasible. . . ." Then he coolly "adjusted" almost all the physical "measures already initiated" by simply canceling them, and this was directed on March 26.*

Though memories as to time were uncertain years later, it must have been about the end of February that a famous meeting was held, the decisive meeting for the OVERLORD deception plan. Strangeways took charge. Verbally if not literally, he tore up the prior FORTITUDE SOUTH planning, saying it was useless and he would rewrite it. It was the talk of the entire intelligence service, recalled Christopher Harmer vividly a half-century later. Everybody was furious. This bumptious so-and-so, who does he think he is?

"It'll be just the same with a few new ideas," said Hesketh to Harmer.

Some days thereafter, Harmer and Hesketh were sitting together in Hesketh's office.

"We are supposed to get Strangeways's new plan today," said Hesketh.

It arrived. Hesketh sat and read it in silence.

"What do you think about it?" he said, handing it to Harmer.

Harmer read it through in growing admiration.

---

* So if "The Needle," the German agent in the Ken Follett thriller *Eye of the Needle,* had existed, he would have found in East Anglia not canvas buildings and plywood airplanes, but rather—nothing at all, except the innumerable airfields already known to be there.

"I can't believe we will ever get away with this," he said to Hesketh.

"It was a revelation," he said years later. "Nobody in SHAEF had thought of it."

What Strangeways saw that nobody else had seen was that the Pas-de-Calais story would be truly plausible only if a major force appeared to be dedicated to that specific operation, and that this could be done with the *real* Allied units in England. The emphasis on fifty divisions could be scrapped. Monkeying about with six Home Forces divisions and four new ones out of thin air could be scrapped. And the actual training exercises and practice landings scheduled during the run-up to D-Day could be turned to deceptive account.

His FORTITUDE SOUTH "Joint Commanders' plan" was not recorded in its official written form till May 18. But that was simply bureaucratic delay; by that time it had been in full operation for weeks, having been launched in early April. That official version first recited the basic "story" of FORTITUDE:

> The story on which Plan FORTITUDE is based falls naturally into two phases:
> Phase (i)—Pre D-Day.
> Phase (ii)—Post D-Day.
> *(a) Phase* (i)
> The main Allied assault is to be made against the Pas-de-Calais area.
> In the first place, the notional date for the operation will be D-Day plus 45. There will come a time, however, when as D-Day approaches our preparations will indicate the imminence of the assault, and when the enemy will realize the approximate date of our attack. When it is estimated that this period has been reached, the imminence of an attack will be confirmed by special means, but the area of the main attack will remain the Pas-de-Calais.
> *Phase* (ii)
> NEPTUNE is a preliminary and diversionary operation, designed to draw German reserves away from the Pas-de-Calais and Belgium. Once the main German reserves have been committed to the NEPTUNE battle, the main Allied attack against the Pas-de-Calais will take place.

The enemy will be induced to believe for as long as possible after NEPTUNE D-Day that the main threat to the Pas-de-Calais is still to be carried out.

It then set forth Strangeways's implementing plan. Codenamed QUICKSILVER, it did not fit FORTITUDE in every particular. It was divided into six subplans codenamed QUICKSILVER I through VI.

QUICKSILVER I was the centerpiece: an order of battle making the basic "story" plausible. An ULTRA decrypt of January 10 had shown that from interceptions of plain language radio communications the Germans had discovered the existence in England of both FUSAG and the American First Army. Very well: Build on this knowledge by leading them to believe that FUSAG was not a mere skeleton headquarters but a real army group composed of genuine corps and armies. Then it would be entirely credible for the Allied plan to allocate separate roles to the two army groups: 21 Army Group would operate in Normandy while FUSAG waited to strike the main blow in the Pas-de-Calais when the time was ripe. But, significantly, there was to be no explicit mention of the Pas-de-Calais prior to NEPTUNE D-Day; the Germans were to be left to draw the inference themselves. During Phase I, before D-Day, they should be led to believe simply that Eisenhower had two army groups: Montgomery's 21 Army Group, and FUSAG, located in the east and southeast of England and composed of the Canadian First Army (Canadian II Corps and U.S. VIII Corps), and the U.S. Third Army (XX Corps and XII Corps), with the U.S. Ninth Air Force associated with it. Then in Phase II, after D-Day, they should be led to believe that FUSAG and a proportion of Ninth Air Force was ready to attack the Pas-de-Calais once 21 Army Group had enticed the German reserves in that area towards the NEPTUNE bridgehead, and that before the final assault a series of large-scale exercises would take place (of which the mounting of NEPTUNE would be one).

All the units mentioned—army groups, armies, corps—were perfectly genuine formations. The only differences were the force structure into which they would notionally be organized, and the fact that while the Canadians were in truth conveniently situated in the south and east, the American units were not where QUICKSILVER I said they were. FUSAG headquarters, for example, was not near Ascot but in Bryanston

Square not far from Marble Arch in London; and Third Army headquarters was not in the east but in Cheshire in the west.

This matter Strangeways addressed in QUICKSILVER II, his radio deception plan. The American forces in FUSAG would go on radio silence in their true locations, and be represented by dummy W/T and voice traffic as moving to their new concentration points in the south and east, from which they would seem to be poised to strike the Pas-de-Calais. FUSAG headquarters itself would be represented as being at Wentworth near Ascot. The American VIII Corps, notionally assigned to First Canadian Army, would be represented as moving from Marbury in Cheshire to Folkestone, convenient to the Canadian II Corps (represented as being headquartered in Dover), with its component divisions nearby. The American Third Army would be represented as moving from Mobberly in Cheshire to Chelmsford in Essex, with its XX Corps moving from Marlborough in Wiltshire to Bury St. Edmunds in Suffolk and its XII Corps from Bewdley west of Birmingham to Chelmsford; and so on. The dummy W/T traffic would be supplemented by double agent reports confirming the moves. By the end of May, hopefully, based on "Y" service direction-finding and traffic analysis confirmed by agent reports, German intelligence would recognize that FUSAG was concentrated in the southeast and east of England, in a position to descend upon the Pas-de-Calais, with the remainder of the Allied host poised further west— perhaps to support FUSAG, perhaps for some role of its own.

QUICKSILVER III, V, and VI dealt with the very limited physical display that Strangeways's plan allowed. QUICKSILVER III decreed a display of BIGBOBS and other dummy landing craft at southeastern and eastern ports from Great Yarmouth round to Folkestone, with associated simulated radio traffic and some limited posting of appropriate signs ashore. QUICKSILVER V directed special activity at Dover to give the appearance of extra tunnel construction and additional wireless stations, suitable for the Canadian II Corps headquarters. Under QUICKSILVER VI, from mid-May the tempo of seeming activity round the dummy craft on the east coast would pick up, by simulated beach lighting and vehicle lights suggesting round-the-clock busyness.

Finally, QUICKSILVER IV was the air plan. Prior to NEPTUNE D-Day there would be increased training flights and practice air-sea rescue missions in the southeast. Fighter cover for a substantial bombing mission would be flown from advanced bases in the southeast, with more frequent small-scale versions of the same activity. The basic OVERLORD air plan—stressing disruption of supporting facilities in the German rear, and interdiction of the Seine bridges—needed no adjustment, for it would appear to support a Pas-de-Calais assault equally with one further south.

All these, however, were mere sideshows. The centerpiece of Strangeways's plan was, and would remain, the QUICKSILVER I false order of battle, with its associated implementation from QUICKSILVER II and special means.

This plan had one shortcoming: It still preserved the notion that at some time shortly before D-Day the Germans might be told that an attack on the Pas-de-Calais was imminent, thus undercutting the plausibility of the post-D-Day threat to that area. But that was built into FORTITUDE.

Hesketh and Harmer could offer only one improvement to the plan, but that a major one. Omar Bradley was the commander of the real FUSAG, but Harmer pointed out that to carry him as commander of the notional one would compromise the plan from NEPTUNE D-Day on. The redoubtable General George Patton commanded the American Third Army, but the Germans did not know this. They were known to respect Patton's ability. (Indeed, they respected it more perhaps than the Allies realized. "Their best general," Hitler himself called Patton. "That's the best man they have.") And they knew that he was seen as Montgomery's rival. Harmer suggested that Patton be the commander of the notional FUSAG, Hesketh passed this idea on, and eventually it was adopted.

With physical deception largely scrapped, the "story" would have to be put over by radio deception and special means.

For British army radio deception a formation called "No. 5 Wireless Group" was organized by February as part of Strangeways's "R" Force. It was equipped to mimic the traffic of a corps of three divisions, from vans each of which could simulate the traffic of six transmitters representing a division and its component brigades. Its voice traffic was prerecorded on wire recorders; Strangeways would provide the specifications—to simulate, say, a battalion practicing an amphibious landing—and the "production" teams would study a genuine exercise, write a script, and round up some genuine troops to record it.

On the naval side, the Admiralty organized three units called "C.L.H." units, each of which was equipped to simulate the naval force associated with an assault division. They operated with transmitters on vans similar to those used by No. 5 Wireless Group, but broadcast live rather than through wire recorders. Two of these units were in action by the end of March, and a third at the beginning of June.

On the American side, the 3103d Signal Service Battalion was trained at Fort Monmouth, New Jersey, in December and January, shipped out of New York on January 30 on the *Mauretania,* and arrived in England February 12, setting up headquarters at Dudley in Warwickshire. Designed to be able to simulate a corps, producing the signals of an infantry division, an armored division, and a corps network simultaneously, during FORTITUDE SOUTH it managed at one time or another to mimic the traffic of nine divisions, three corps, one army, and one army group. Focused almost entirely on W/T rather than on radiotelephony, it did on rare occasions use voice radio to read scripts imitating air force operations. Its commanding officer was Lieutenant Colonel Arthur McCrary, a West Point classmate of Baumer and Harris.

The work of the 3103d was representative of the work of all the deceptive radio units in FORTITUDE. The Chief Signal Officer of ETOUSA directed a daily compilation of radio traffic of all U.S. units in the United Kingdom. These were analyzed for such characteristics as length, precedence, time, and security classification, while ETOUSA's Signal Intelligence Division monitored American radio nets for operators' fists and other telltale features. In March, the latter office organized a Radio Countermeasures Detachment that visited the signal and radio offices of American units to gather still more data. IBM machines were employed to generate quantities of text composed of random words, from which could be quarried messages of any desired length. By the time the 3103d was ready to begin implementing QUICKSILVER, there was abundant data on hand to design communications nets, schedules, and traffic to represent any American unit from division size up, in any state of training.

The Radio Countermeasures Detachment prepared the actual plans for showing the concentration of U.S. units in the southeast. Harris's Special Plans Branch would furnish a schedule of the notional movements and order of battle (designed, of course, by Strangeways). The Radio Countermeasures Detachment prepared schematic diagrams of the desired communications nets and set the traffic characteristics for each station for each day. It then prepared a message traffic schedule for each station, prepared the messages themselves by simply cutting segments of desired length from the supply of IBM-generated random text, and turned messages and schedules over to the 3103d. The 3103d's people would encipher the messages on their own machines and transmit them from the appropriate locations according to the schedule, with some help from operators from the real units whose fists were thought to be particularly distinctive. Before it was all over, 13,358 phony messages would have been sent by the Americans alone, averaging 230 a day.

For the signalmen themselves who sat at the Morse code keys tapping out the encrypted messages, it was all in the day's work. They were accustomed to sending unintelligible gibberish. They knew generally what they were working on, though they never got a detailed briefing. Rigid secrecy was impressed upon them; they were scared to talk not merely in the local pubs but in the latrines; they changed the identifying stencils on their vehicles so often the bumpers grew thick with paint.

By the last week in April all was ready, and on the 24th of the month the FORTITUDE SOUTH wireless net opened up. So from then on, any Germans listening would hear radio links bearing normal traffic patterns between and among the command nets of FUSAG—divisions, corps, armies, and the army group itself—and slowly concentrating in the south and east of England.

If any were listening. As we now know, few were; or if they were, they could not unravel what they heard. The elaborate wireless deceptions of FORTITUDE both SOUTH and NORTH were almost entirely wasted. For when German documents were analyzed after the war, not one instance was found of the Germans' deeming an agent's report to have been discredited by "Y" Service analysis (and only two in which such a report was noted as having "Y" Service confirmation). But nobody could know that till after the war.

There had been a lull for the double agents after the big—and wasted— effort in support of Operation COCKADE in the summer of 1943, and MI5 had been hard put to it to generate chickenfeed for them. Resort was had to a device called "Field Security Reports." British GHQ Home Forces directed all the field security sections in the United Kingdom to submit fortnightly reports of observations which they believed might help a spy make inferences about the invasion. The field security sections were told that this was simply a security measure; in fact, the reports were used to generate plausible grist for double agents' reports.

For FORTITUDE, special means meant as a practical matter mainly those double agents who were in touch with the Germans by radio; for mail and other communication with the outside world would be shut off during the run-up to D-Day. Of all MI5's stable, only five adequately filled this bill. They were TRICYCLE, BRUTUS, TATE, TREASURE, and of course GARBO.

Now fully rehabilitated in German eyes, TRICYCLE had recently acquired a W/T operator. During a visit to Lisbon from mid-July to mid-September of 1943, he had set up with the Abwehr an escape route to slip out of Switzerland Yugoslav officers marooned in that country who would undertake to work for the Germans. One of those exfiltrated was an officer, the Marquis de Bona, whom the Germans had trained in W/T and secret ink. Reaching England in December 1943, he became TRICYCLE's operator under the codename FREAK, using a transmitter that TRICYCLE brought back in January from a two months' visit to Lisbon. Radio contact with Lisbon was established in February.

It will be recalled that BRUTUS, the Polish air officer Roman Garby-Czerniawski, had been taken on cautiously as a double agent in early 1943, and had been useful in connection with COCKADE. He then got into trouble with the Polish authorities in Britain on a disciplinary charge stemming from political activities against the head of the Polish air force, from which he was not released till the end of the year. He resumed his radio transmissions, sending messages in a sort of "telegraphese French." For safety's sake, MI5 used their own radio operator; BRUTUS described him to the Germans as an elderly retired Polish air force officer who wanted to help the Germans for ideological reasons, having lost his family to the Soviets. (This fictitious individual was codenamed CHOPIN by MI5.) The transmitter was operated from Richmond in south London. BRUTUS's background made him an appropriate channel for high-level order of battle information, so in March he was added to the deception roster for FORTITUDE. As will be seen, he was employed first to describe notional preparations in Scotland seemingly targeted on Norway, and was then notionally posted to FUSAG itself. Harmer had been BRUTUS's case officer; when Harmer was seconded to SHAEF, BRUTUS was turned over to Hugh Astor of MI5.

TATE was a double agent of long standing. A Danish subject (born German) named Wulf Schmidt, who had parachuted into England in September 1940 and by the end of the war held the marathon record for radio contact with the Abwehr (from October 1940 to the fall of Hamburg in May 1945), he had since September 1941 been notionally working on a farm near Radlett in Hertfordshire, sometimes visited by a notional girlfriend, MARY, who was supposedly a cipher clerk in the Admiralty in London with contacts with the Americans. In 1943 he notionally moved for the summer to another farm near Wye in Kent, from which he was in a position to send a modest report on the preparations for STARKEY, and in late May 1944 he returned there, where he could observe the concentration in the southeast for QUICKSILVER. (The real TATE continued living in London; his radio messages were sent by land line to a transmitter site in Kent.) TATE appeared to be highly valued by the Germans—he had been awarded the Iron Cross, and his control was enthusiastic and encouraging—but since his control was in Hamburg reporting to headquarters by land line, there was no way to monitor the evaluation of his work through ULTRA.

The fourth usable W/T double agent as of the spring of 1944 was Nathalie, or "Lily," Sergueiev, codenamed TREASURE. She was a White Russian who had grown up in Paris and had relatives in England. She had volunteered her services to the Abwehr in 1940, and after training was dispatched to England by way of Madrid in June 1943. There she turned herself in to the American Embassy and MI6, and reached England in November. Originally a letter writer with a notional job at the Ministry of Information, in this capacity she went to Lisbon in February 1944, and when she returned in late March the Germans had given her a transmitter. She opened communication with Paris on May 10. Shortly thereafter she admitted that, in revenge for the fact that her pet dog had been kept in Gibraltar under the absurd British quarantine laws and had died there, she had withheld from MI5 the security signal to alert the

Germans to the fact that she was under control, and had intended to use it to warn them. MI5 dropped her promptly but kept the channel going with their own operator.

And then there was the old master, GARBO.

Travel to coastal areas would be banned for security reasons during the run-up to the invasion. Though the details were only settled on March 10, that there would be some such ban was sufficiently certain beforehand for it to be prudent for some of GARBO's informants to move into the forbidden areas before it would cease to be plausible for them to do so. On this project Pujol and Tommy Harris gave their imaginations free rein. One of GARBO's subagents, known officially as AGENT SEVEN and less officially as STANLEY (the Germans knew him as DAGOBERT), was a notional ex-seaman and ardent Welsh nationalist living in Swansea. Early in December, GARBO reported that SEVEN had rounded up some more recruits, active in an organization called "Brothers in the Aryan World Order." AGENT 7(2) was an ex-seaman, known to the Germans as DONNY; AGENT 7(4) was an Indian poet and Aryan brother, known as RAGS and referred to by the Germans as DICK; while AGENT 7(3) was his mistress, a Wren named THERESA JARDINE, known later on as GLEAM. AGENT 7(5) was a relative of AGENT 7(2), whom the Germans dubbed DRAKE; AGENT 7(6), who seems never to have received a name, was another Welsh fascist, as was AGENT 7(7), called by the Germans DORICK.

In February, some of these new recruits were fanned out around the south and southeast coast: 7(5) (DRAKE) to Exeter, 7(4) (DICK) to Brighton, 7(2) (DONNY) to Dover, 7(7) (DORICK) to Harwich, while SEVEN himself remained in Swansea and 7(6) also in South Wales. In addition, FOUR, the Gibraltarian waiter who had been working in the notional secret underground munitions depots in the Chislehurst caves, was transferred at the end of April to the encampment of the Canadian 3d Division, a genuine assault unit, in Hampshire. Up in Scotland, THREE, the Venezuelan (known to the Germans as BENEDICT), recruited, among others, a Communist Greek seaman deserter logged as 3(3), for use in FORTITUDE NORTH.

Over and above these reinforcements in the field, in late May GARBO himself was put in a better position to glean information at the center, by being notionally hired as a full-time employee of the Ministry of Information. Late in 1943 he had appointed THREE as his deputy; he now moved THREE to London to collect messages and arrange for their transmission, while MRS. GERBERS, the widow of the Liverpool agent who had had to be killed off at the time of TORCH, was brought in to help with encipherment.

It must always be kept in mind that these people were all figments of Juan Pujol's and Tommy Harris's imaginations.

GARBO was by far the busiest of the double agent channels during the FORTITUDE deception; between January 1944 and D-Day on June 6, more than 500 messages passed back and forth between him and his control in Madrid. But after the war, when the secrets of German intelligence were laid bare, it would transpire that BRUTUS's contribution to FORTITUDE had been perhaps greater even than GARBO's, while TATE's was disappointing. By Hesketh's count after the war, of 208 messages in the German intelligence summaries for 1944 denominated Lagebericht West or Überblick des Britischen Reichs that could be traced to double agent reports, 91 were due to BRUTUS, 86 to GARBO, and only 11 to TATE.

The letter-writing double agents were able to contribute little, owing to restrictions on mail. BRONX, GELATINE, METEOR, and SNIPER all played small roles. One small contribution was made through the channel codenamed PANDORA, which was a series of anonymous letters written to the German minister in Dublin, supposedly from an Irish hater of the English, into which deception items were occasionally inserted.

Strangeways worked out an elaborate and detailed timetable for the slow movement of units to the notional concentration points over the crucial weeks. Based upon this, the signals people simulated the movement of headquarters and their joining into new communications networks; while Hesketh followed it in working out a detailed plan for "special means," specifying what item of misinformation should be passed to the Germans and when. The best agent to put over each item of information would then be decided on by Tar Robertson and his colleagues. The agent's case officer—most often, naturally, Tommy Harris for GARBO or Hugh Astor for BRUTUS—would come to Norfolk House to meet with Hesketh and, as appropriate, Eldredge and perhaps Harmer and Macdermot; the general sense of the message would be decided on, and the case officer would then put it into appropriate form and arrange for its transmission.

Daily touch had to be maintained with Strangeways. When on April 26 Montgomery moved his main headquarters from St. Paul's School to Southwick Park near Portsmouth, Strangeways followed. Cuthbert Hesketh took on the role of courier, traveling almost daily between Southwick Park and Norfolk House (where Ops (B) intelligence remained, although SHAEF main headquarters had moved to Bushey Park, near Kingston-on-Thames some ten miles west of London, by April 15).

Ingersoll and Harris moved down to Southwick Park too. "General Montgomery has asked me to tell you," said Freddie de Guingand when he welcomed them, "that he is very glad you are here. He would like you to understand that this is not an *Anglo-American* headquarters. This is a *British* headquarters."

There is not space here to follow the slow, meticulous, detailed working out of Strangeways's choreography of QUICKSILVER. One sample must suffice.

An element of the plan was to show the American VIII Corps, composed of the 28th, 79th, and 83d Divisions, moving from its true location at Marbury in Cheshire to Folkestone on the south coast, where with the Canadian II Corps at Dover it would form First Canadian Army headquartered at Leatherhead, one of the two component armies of FUSAG. It was made sure first that the Germans knew that these divisions had been in the western part of Britain. GARBO reported the 28th in South Wales on April 25, and TRICYCLE's radioman FREAK reported seeing the 79th and 83d in Cheshire on April 26. On April 25, the 3103d opened up a seeming wireless link between VIII Corps near Folkestone in Kent and First Canadian Army. The next day, it opened up the notional radio of the 79th Division on the VIII Corps net, transmitting from Heathfield in Sussex; and similarly opened up the 28th Division on April 27 at Tenterden and the 83d on May 4 at Elham in Kent—all in southeastern England. On May 1, GARBO reported that his AGENT 7(2), the ex-seaman he had stationed in Dover, had seen troops bearing VIII Corps insignia in Folkestone and had seen officers of the 28th Division in Dover and Folkestone, adding that his AGENT 7(7) had seen the 28th in South Wales before he left for Harwich. On May 8,

TREASURE made one of her few contributions, advising the Germans that the American First Army was under Montgomery's overall command; this was to accustom them to the mixing of nationalities so that it would not seem strange that First Canadian Army was under FUSAG and the U.S. VIII Corps in turn under Canadian command.

In late May, SHAEF decided that VIII Corps would go to Normandy much sooner after D-Day than had been planned, as would its 79th Division. For QUICKSILVER, Strangeways decreed that steps be taken to replace the corps by XII Corps, and the division by the 35th Division, both notionally transferred from Third Army. The 3103d duly shut down the radio of the 79th on May 28, opened the radio of a forward command post for XII Corps at Folkestone on May 30, shut down the old XII Corps and 35th Division stations in East Anglia on the same day, and on the 31st shut down VIII Corps, opened up XII Corps fully at Folkestone, and brought the 28th and 83d into its command net. On May 27, GARBO reported that a car park with vehicles of the 83d Division had been spotted in the Deal area. To wrap up the new arrangement, on the night of June 6–7 BRUTUS advised that XII Corps, formerly of Third Army, had been reported at Folkestone, and that the 35th, now in Kent, and the 28th at Harwich were now part of XII Corps.

The unexpectedly early departure of VIII Corps and the 79th for Normandy raised another problem: Suppose the Germans took prisoners from those units who said they had never been anywhere near the south coast until they were shipped over to France? As a precaution, GARBO reported on June 5 that his informant 4(3), a talkative American sergeant working in an unspecified headquarters in London, had told him that most American troops arriving in England spent time initially in staging areas convenient to their arrival port of Liverpool before moving to their embarkation points. It was hoped that the Germans would conclude that a POW who said he had never been anywhere but Wales or Cheshire was a recent arrival who had gone directly from the staging area to the embarkation port.

Of scores of such bits and pieces was the great deception made.

Over and above this slow buildup of a mosaic, on three occasions the risk was taken of rubbing the Germans' noses in the big picture; twice before D-Day and once after it.

The first was a visit by TRICYCLE to Lisbon in March and April, when he gave the Germans a quantity of order of battle information which Fremde Heere West noted as "particularly valuable," and which set much of the background against which the QUICKSILVER deception was played out.

Then in late May, after the notional concentration of FUSAG in the east and south and shortly before NEPTUNE, BRUTUS gave the Germans a summary wrap-up of Allied dispositions. They were told that with effect from May 27 he was posted to FUSAG headquarters—not the main headquarters, notionally at Wentworth, where he might be expected to pick up too much information, but at an imaginary subsidiary headquarters that had supposedly been set up to plan the recruitment of Poles working in occupied territory on the Continent that was likely to be liberated by FUSAG. This was located at Staines, up the Thames from London and usefully inconvenient to CHOPIN at Richmond, so that messages could be plausibly delayed when necessary. From his new vantage point, BRUTUS sent messages on May 31 and June 6 describing the structure and locations of FUSAG and 21 Army Group and their constituents, confirming Patton and Montgomery as their commanders, and noting that "FUSAG gives the impression of being ready to take part in active operations in the near future."

The third time the risk was taken of rubbing the Germans' noses in the conclusion they should draw came in a message from GARBO three days after the landings, which will be described in due course.

The sites for the limited display of dummy landing craft under QUICKSILVER III were chosen at the end of March. A total of 255 BIGBOBS were set out from Great Yarmouth around to Folkestone during the three weeks beginning May 20. As it turned out, there was no German reconnaissance to be fooled, although by his major message of June 9 GARBO tried to tempt them into sending something over by telling them that a friend of AGENT 7(7) had seen landing craft on the rivers Orwell and Deben. It was just as well that the Germans resisted the temptation, for the dummies had a way of blowing out of place, and efforts to simulate realistic activity around the craft came late and were ineffective.

Closely related were QUICKSILVER V, the simulation of increased activity at Dover, and QUICKSILVER VI, dummy lighting to accompany the dummy landing craft. The first seems to have been relatively routine.

The second was hampered by the fact that the Admiralty was unconscionably slow in deciding what lighting would be used at the genuine embarkation points, so that the RAF's dummy lighting experts had little guidance as to what they were supposed to imitate. As things worked out, the lighting at the fake locations was brighter than at the real ones; this again did not matter since there was so little German reconnaissance.

QUICKSILVER IV, the bombing program, went smoothly. The bombing plan to interdict the Seine for the real NEPTUNE supported the idea of a Pas-de-Calais attack equally well. During the preinvasion period, more airfields were attacked in the Pas-de-Calais than in the real invasion area; railway junctions supplying the Pas-de-Calais were hit much harder than those in the direct target region; and Allied bombers made twice as many visits to coast defenses and radar stations in the cover area as in the target area. One concern was that Allied fighter strength was concentrated much more heavily in Hampshire, facing Normandy, than in the southeast. To show the Germans that these were able to operate against the Pas-de-Calais, at the end of May an exercise was carried out in which fighters took off against northeastern France from advanced bases in the southeast and returned to their regular bases in Hampshire.

One of the few elements of the old original APPENDIX Y plan to survive was the observance of several periods of total radio silence during the preinvasion period. It was expected that the Germans would interpret this as "crying wolf," designed to induce them not to read radio silence during the assault loading period as a tipoff that the invasion was on the way. This was something of a double bluff, for in fact radio operation would continue uninterrupted during the loading period. The American 3103d infiltrated the communications net of the American First Army and continued to simulate that force after the real one had gone on radio silence for the invasion (Operation WILLIAMS), and the British deceptive wireless troops did the same for the British (Operation QUAKER). Then from H-Hour of D-Day until July 6, radio silence was imposed on every nonoperational circuit in the United Kingdom (Operation ADORATION).

Of a different pattern were an exercise designed to simulate XIX Tactical Air Force linking up with Third Army in East Anglia (Exercise VAN DYKE), and the conducting of three notional amphibious training exercises, represented by fake radio traffic of the land and naval forces involved. The first of these, DRYSHOD (later called CENT), represented the Canadian 4th Infantry Brigade with naval Force "G" conducting a combined headquarters communications exercise on April 16, followed by the second, WETSHOD (later called DOLLAR), a bogus amphibious exercise involving the same units on April 25. The third, SEE SAW, represented the American 28th Division—supposedly, as will be recalled, part of the American corps of the Canadian First Army in FUSAG—conducting a landing exercise with Royal Navy support near Ipswich on the east coast on June 1 through June 3. Taken in conjunction with the real assault exercises that were necessarily conducted during the weeks before D-Day and which could not be concealed from the enemy, it was hoped that these notional exercises (in addition to confirming the notional location of the 28th Division) would suggest to him that the assault would be made at high tide, and would mislead him as to such matters as the order of the assault waves and the equipment in each.

The genuine assault exercises were trouble enough. Exercise TIGER, at Slapton Sands on the south coast on April 26, resulted in heavy losses when German motor torpedo boats struck it by surprise. Strangeways sent Ingersoll down posthaste to assess whether FORTITUDE SOUTH had been jeopardized by the German foray, and concluded it had not. The major invasion rehearsal, Exercise FABIUS, in early May, involved all the assault divisions and their naval support.

Such a massive exercise could not possibly escape the attention of GARBO's agents, so the opportunity was taken to fine-tune German perceptions of his network. Late in April, his AGENT FOUR reported that the Canadian 3d Division and others had been issued cold rations, vomit bags, and lifebelts, so that the invasion must be imminent; and followed this up with even more exciting news that the troops had left camp. GARBO reported to Madrid that either the invasion was on, or the Canadians had gone to reinforce the troops threatening Norway, and that he discounted information from his source J(5), the secretary in the Cabinet Offices, to the effect that what was going on was merely an exercise. When that was just what FABIUS proved to be, GARBO expressed to Madrid his annoyance at the "stupidity" of his "simpleton" FOUR. "You should give him encouragement," replied Madrid soothingly, "as, if not, it might happen that when the real invasion is about to take place he will not notify this owing to over-precaution."

The net effect was to raise in German eyes the credibility of GARBO's sources in the London government offices. And, of course, the stage was set for the Germans to interpret the launching of the real NEPTUNE force—if they learned of it—as another exercise.

FABIUS put an end to the "postponement" theme, inherited from BODY-GUARD, that the Germans might be persuaded that the invasion would not take place until much later in the summer or early autumn, if at all; for it would hardly be plausible that such a massive rehearsal would be held so long before the actual event.

"Postponement" had never been a useful concept and the themes relied on to support it—that maybe the POINTBLANK combined bomber offensive would make invasion unnecessary, that the invasion would require more divisions and equipment than could be amassed and trained before late summer, that the invasion would await the Soviet summer offensive—got only halfhearted attention even before FABIUS.

TATE reported on January 20 that MARY had told him that labor troubles in the United States had cut back landing craft production. The next day, GARBO reported that his contact in the Ministry of Information thought there would be no invasion for a long time and perhaps not at all because Germany might collapse under the air offensive. Two days after that, BRUTUS reported a widely-held view that Montgomery would probably take the time to train the troops all over again. Through MI6 and the Foreign Office, Bevan spread in diplomatic circles the POINT-BLANK theme and the idea that there would be no landings till the Soviet summer offensive had begun.* At the beginning of May, he floated with Joint Security Control the possibility of making preparations for a

---

* Particularly useful was "A" Force's old friend Henry Hopkinson, now in Lisbon. In February, he was summoned to London and briefed to suggest the Pas-de-Calais as the objective in social conversation. (It was on this visit that he worked out with Dudley Clarke arrangements for what became the 60 Committee to supervise Lisbon-based deception.) He was given a fairly clear idea of the real objective and timing of OVERLORD. "To say I was shocked to be given this vital secret is an understatement," he wrote later. "I went on the wagon for all those anxious four months!"

United Nations summit conference in Washington supposedly scheduled for a few days after D-Day; presumably the Germans would assume
that Churchill would not absent himself from Britain at the time of the
invasion. But nothing came of this.

After the beginning of March the "postponement" theme was substantially dropped from double agent traffic in the United Kingdom,
though as late as April 21—just before FORTITUDE SOUTH was
launched—GARBO was reporting "that only preparations but no indicative action of concentration is noted." It was just as well. German intelligence was clogged with predictions about the timing of the invasion. To
cover their own flanks, uncontrolled agents like JOSEPHINE and OSTRO
had been plugging away at the notion of an early assault. As early as February 23, Fremde Heere West noted that "enemy diplomatic sources
seem to be systematically spreading information about the postponement of the invasion," and on March 20 it advised that "Numerous reports of an alleged postponement of the invasion or of its complete
abandonment in favor of an intensification of the air war or of smaller
local landing operations are, in the opinion of this section, to be regarded
as systematic concealment of the actual plan."

"Postponement" was the theme of most of what Joint Security Control
was called on to do in aid of BODYGUARD.

In the early stages, Bevan had drawn up a list of items that Joint Security Control might pass through American special means. Aside from
suggesting a buildup of American forces in Iceland, most of them were
designed to promote "postponement." Included were such topics as the
forthcoming use of B-29s against Germany to augment the POINTBLANK
bombing program (entirely notional, for they were earmarked for the
Japanese war); shortage of landing craft because industrial unrest had
impeded the construction program; inadequate training of American
troops being sent to Britain; and similar items. The newly energized
Joint Security Control—this was just when Bissell replaced Strong—was
focusing much of its attention on WEDLOCK and the Kurile Islands, but
in close coordination with Bevan through O'Connor the suggested program was substantially carried out.

The FBI's double agent RUDLOFF made contributions on all the "postponement" topics. On March 2 he told Hamburg that his friend WASCH

at the War Production Board advised that B-29 production was increasing though still behind schedule, and that tests were being conducted preparatory to sending the first B-29s to England. On March 14 he passed along a story from WASCH that optimistic predictions that the war would end in 1944 had caused a lag in production, and that the administration hoped that the invasion would take place in time to influence the November elections. On March 23 he passed along some elementary data regarding the B-29's power and armament; reported on April 17 that the administration had called labor leaders in and threatened to draft strikers who had interfered with production "particularly of invasion craft"; reported on May 2 that WASCH had heard that Eisenhower had protested the diversion of landing craft to the Pacific; advised on May 4 that WASCH said that the British were criticizing the slow buildup of American forces in Britain and that the troops were inadequately trained; and on May 9 said that the landing craft program was still behind schedule. BROMO contributed to the B-29 story with a message on March 10—his second since going on the air—implying increased stress on heavy bombers.

The B-29 story received some physical confirmation as well. At the beginning of March a B-29 was flown to England. The crew were told that they were a pathfinder for a B-29 force to be based in the United Kingdom and that they were to treat this as secret information. Leaflets threatening the coming of the giant bombers were apparently dropped over Germany in the spring.

The B-29 story created a nuisance for Newman Smith. He had already had trouble with the overeager publicity people in Hap Arnold's office and the ensuing problems with Ormonde Hunter about the big bombers. Then in late April he learned that the previous December Dudley Clarke had leaked to the Germans that a new type of Flying Fortress was expected in large numbers in Africa in the near future. There followed an exchange between a somewhat testy Bissell and a properly contrite Bevan, leading in late May to the express agreement, already mentioned, that the deception agencies of neither country would refer to forces, materiel, or activities of the other without prior clearance or specific inclusion in an agreed plan.

At this period, Smith was developing special means channels in South America. He arranged for the POINTBLANK theme to be leaked by the

American military attaché in Argentina in March, and in late May the Chief of Staff and Commander of the Argentine Army made a speech characterizing the threatened invasion as "nothing more than a fairy tale," which the attaché thought might be credited to this leakage.

Almost as much effort as went into FORTITUDE SOUTH went into FORTITUDE NORTH, and it can fairly be judged a success. But it did not have the life-or-death implications of FORTITUDE SOUTH, and its tale can be quickly told.

The "story" of FORTITUDE NORTH was fundamentally a rerun of TINDALL the autumn before, augmented by a notional attack on Narvik. The notional British Fourth Army, under Lieutenant-General Sir Andrew Thorne,* commander since April 1941 of the genuine Scottish Command, operating in conjunction with the Soviets, would mount a two-pronged attack on Norway, one on Narvik, to secure the supply route to Sweden, and one against Stavanger, to march on Oslo and thence on Denmark. Narvik would be attacked by the notional British VII Corps headquartered at Dundee, composed of the genuine British 52d (Lowland) Division at Dundee, a Norwegian brigade at Callander, and the notional American 55th Division and the notional American 7th, 9th, and 10th Ranger Battalions, all in Iceland. The British II Corps (genuine, but being disbanded) at Stirling (British 3d Division on the Moray Firth and British 55th Division in Northern Ireland, plus British 113th Infantry Brigade in the Orkneys), would assault Stavanger, and together with the genuine American XV Corps (American 2d, 5th and 8th Divisions) from Northern Ireland as a followup force, would advance towards Oslo.† Full development for the Stavanger threat was targeted for May 1 and for the Narvik threat by mid-May.

Unlike the delegation of FORTITUDE SOUTH to the Joint Comman-

---

* "Bulgy" Thorne, father of Peter Fleming's colleague Peter Thorne, had served with distinction on the Western Front in the First World War. In the bloody fighting at First Ypres in October 1914, his unit had faced that of Corporal Adolf Hitler; and for this reason, when Thorne served as British military attaché in Berlin during the 1930s, Hitler picked him out for special attention. Hitler's respect for Thorne may well have contributed to the Führer's concern over Norway throughout the war.

† The British 3d Division was a genuine assault force for Normandy and when it went south for Exercise FABIUS it was replaced by a notional British 58th Division in the Highlands.

ders, SHAEF Ops(B) retained control of FORTITUDE NORTH, though in practice implementation by other than special means was effectively delegated to General Thorne. Execution was largely in the hands of the energetic and enthusiastic Colonel Roderick Macleod of the British Army. The methods used were the familiar ones. Training exercises and movements of forces were simulated by dummy wireless traffic, including two phony naval forces, "Force V" for Narvik and "Force W" for Stavanger, simulated by "C.L.H." units. The RAF set out some dummy airplanes and produced some fake wireless traffic to suggest the move of four medium bomber squadrons to eastern Scotland. The Royal Navy conducted a carrier-borne reconnaissance of Narvik in late April (Operation VERITAS), during which it was planned to violate Swedish airspace, though the weather prevented adding this touch.

FORTITUDE NORTH gave Harris's team valuable practice for the more intricate FORTITUDE SOUTH. In early March, Eldredge went up to Edinburgh as the only American in a SHAEF delegation reviewing plans with Scottish Command. He was taken to dine on haggis, and like many before and since, decided "once was enough." From there he flew to Northern Ireland to join Beck and Ingersoll in briefing Major General Wade H. Haislip, the commander of XV Corps, on its role in FORTITUDE NORTH. "This fellow was very bright," recalled Eldredge. Eldredge explained that the corps was to go on radio silence till further notice, and "he was not to tell his staff anything about this. He understood." Eldredge returned with a "fine fat chicken and two dozen eggs" for Diana—precious commodities in wartime Britain. Ingersoll went back to Scotland to work with Macleod.

Haislip subsequently, on April 13 and again on May 10–12, had a more exalted visitor in the person of General Thorne himself. To convince any German agents in Eire who happened to be following events in Ulster that XV Corps really was part of British Fourth Army, Thorne, accompanied by Ingersoll, flew over to Belfast and inspected the corps as if he were their army commander. "He was there for three days," remembered Macleod, "delighted with his visit, and on his return said that the troops were as good as the Guards, a high compliment from one who was himself a Guardsman."

GARBO and BRUTUS were the mainstay for special means in the north as in the south. GARBO worked in particular through his Venezuelan

AGENT THREE in Scotland, aided by THREE's subagent 3(3), the Greek seaman. THREE came to London to help GARBO at the beginning of May, leaving only the rather shaky 3(3) on the scene. BRUTUS worked through his general high-level contacts and a notional visit to Polish troops in Scotland. He passed much of the basic order of battle of Fourth Army to the Germans on his return in mid-April. The American 55th Division in Iceland was identified, and linked to the British VII Corps, by COBWEB and BEETLE, young Norwegian double agents run out of Iceland by MI6, with the further help of a message from PAT J rated by Fremde Heere West as a "believable Abwehr report."

Joint Security Control made another modest contribution to FORTI-TUDE NORTH (and to TURPITUDE in the Mediterranean, described in the next chapter) by having manuals of the Scandinavian and Balkan countries prepared for printing and distribution if needed.

Closely related to FORTITUDE NORTH was GRAFFHAM, a démarche towards Sweden designed to persuade the Germans that the Allies were inducing the active cooperation of the Swedes in projected operations against Norway. In February, the British minister in Stockholm asked the Swedish government for permission to take weather observations and install air navigation equipment in Sweden, to which in early April were added further requests covering such matters as the right to repair aircraft forced down in Sweden and conduct reconnaissance flights over that country. This was all bolstered by conspicuous traveling back and forth between Stockholm and London by British diplomats and high-ranking RAF officers, and by rumors that after the invasion of Norway Allied bombers based in Sweden would join the POINTBLANK offensive; to which, soon after D-Day, was added a joint demand by the United States, Britain, and the Soviet Union for assurance that German troops would not be allowed to transit Sweden from Finland if the Allies landed in Norway. Bevan was in regular contact with Joint Security Control with respect to the plan, and Newman Smith brought the State Department into the picture at an early stage.

The joint demand was not strictly part of GRAFFHAM but of ROYAL FLUSH, a plan devised in April to put diplomatic pressure on Sweden, Turkey, and Spain around the time of D-Day to increase German fears that they might be pushed into the Allied camp. Its Turkish and Spanish phases will be discussed in the next chapter.

FORTITUDE NORTH succeeded in convincing the Germans that Fourth Army existed and planned to do something about Norway, but it did not induce them to do anything they would not have done anyway. Fremde Heere West never believed that anything more than diversionary attacks on Norway was planned or indeed was feasible. There were twelve German divisions in Norway, and they should be able to handle any such threat. They were kept there, and not transferred to France or any more seriously threatened front; but, given Hitler's constant fretting over Norway and Denmark, FORTITUDE NORTH could hardly claim credit for that.

A subsidiary operation, of which not much was expected, was IRONSIDE, a notional invasion of the Bordeaux area. Designed to keep German forces in southwestern France away from Normandy, its original "story" was that on D plus 10 two divisions and a brigade would land on both sides of the Garonne estuary at Royan, the Le Verdon–Soulac-les-Bains area, and Arcachon, building up to a force of ten divisions by D plus 12, opening the port of Bordeaux, and advancing up the Garonne valley to link up with the notional invasion of the south of France (VENDETTA) described in the next chapter.

At Bevan's request O'Connor asked Smith if he could put over by special means the presence of a substantial American force ready for embarkation on the East Coast at the appropriate time, together with such devices as issuing maps of the Bordeaux area. Smith thought that a direct assault from the United States was implausible but that a strong followup force to an assault from Britain was feasible. He proposed to select three or four divisions already under alert for the OVERLORD buildup and scheduled for appropriate sailing dates; tell the division commanders and the convoy captains they were being routed to Bordeaux, changing the captains' orders when at sea; and follow up with low-level leaks within the units and appropriate double agent releases. OPD agreed with this plan; the Navy refused to go along, but that would not greatly matter.

After some debate, the "story" as finally adopted involved a two-division assault from Britain, with six followup divisions direct from the United States; the threat to be fully established by May 29 and maintained until June 28. The genuine 26th, 94th, 95th, and 104th Infantry, and 10th and 11th Armored, divisions were designated as the follow-up task force, to be notionally commanded by Lieutenant General Lloyd Fredendall (Patton's predecessor as commander of II Corps in Tunisia, who had been sent home after the Kasserine Pass debacle). For once, O'Connor told Bevan that Joint Security Control "deserves a good mark" for its work on IRONSIDE.

IRONSIDE was implemented entirely by special means. TATE reported on May 23 that his friend MARY had returned from Washington, having worked on plans for an independent expeditionary force from the United States, and four days later that the force consisted of six divisions commanded by Fredendall. On May 29, BRONX dispatched to Lisbon a telegram asking for fifty pounds immediately, which she needed for her dentist; in the simple code the Germans had given her, this meant that she had definite news that a landing was to be made in the Bay of Biscay in one week. She followed this up with a letter saying that a drunken officer had told her about an airborne attack on the U-boat base at Bordeaux, followed by an invasion; and had told her next day that it had been put off for a month. GARBO advised on June 5 that his AGENT 7(6) in South Wales said there was an American assault division in Liverpool, destined to attack the French southern Atlantic coast in conjunction with a large army direct from the States; GARBO said he himself was skeptical about this. In line with suggestions from Bevan, by messages sent June 2, June 4, June 10, and June 20, RUDLOFF reported that WASCH had discovered that a force (tentatively identified as the divisions selected for Fredendall's notional command) was being diverted for a special operation; the troops were followup troops with no amphibious assault training but were specially trained in river crossing and bridging, and the strictest security measures were being imposed on the staging of the force. When an MI6 agent's transmitter in France known to be under German control asked where escaping Allied POWs should go, MI6 replied that they should gather near Bordeaux.

All this had no effect on the Germans. Though in late January and February ULTRA and MAGIC had shown some German nervousness over possible Allied operations in the Biscay region—Ambassador Oshima reported to Tokyo that on January 22 Hitler had speculated that rather than crossing the Channel the Allies might land in the less heavily defended area around Bordeaux, and the Luftwaffe and the German Navy carried out anti-invasion exercises in that region in February—and the Abwehr did show interest in mid-June, German situation reports dismissed the possibility of Allied activity in the Bay of Biscay as "cover operations of small caliber," "small and insignificant diversionary undertakings." Admiral Abe, the Japanese naval attaché, reported as late as July 3 that Jodl said that the Germans were prepared for landings in the vicinity of Bordeaux, "although there is little probability of such landings."

A bit more effectual, and much more fun, was an escapade called COP-PERHEAD (originally TELESCOPE).

On January 4, while Dudley Clarke, inveterate filmgoer, was visiting Mark Clark's recently opened Fifth Army headquarters at Caserta south of Naples, he saw *Five Graves to Cairo,* a Billy Wilder thriller about the North African war featuring Erich von Stroheim as an implausible but immensely entertaining Rommel. An actor looking remarkably like Montgomery made a brief appearance as a British officer. Inspiration struck. If on the eve of NEPTUNE the Germans thought Montgomery was visiting the Mediterranean they should assume that no cross-Channel operation was imminent. Why not palm the actor off as Montgomery and at the crucial time display him conspicuously at Gibraltar for the benefit of the German agents who surveyed the airport continuously through a telescope located on Spanish soil? Clarke was in London in February, and there on February 23 he finished up a plan along these lines; which in its final version was aimed at aiding VENDETTA, the notional threat to the south of France discussed in the next chapter, as well as at lowering German vigilance on the Channel.

Inquiry proved that Billy Wilder's actor was too tall to pass as Montgomery. A search turned up another suitable actor, who then broke his leg in a motor accident. Finally a dead ringer for Monty was found, Lieutenant Clifton James of the Royal Army Pay Corps. Montgomery cooperated fully; James spent some time with him (supposedly as a journalist) studying his mannerisms. At the last minute it transpired that James had never ridden in an airplane. Suppose an airsick "Monty" should lurch from the plane at Gibraltar? Bevan told Wheatley to get him up for a test flight right away. To everyone's relief, he passed.

On May 26, "Monty" took off from London with a brigadier and an aide in a special plane. They landed at Gibraltar next day, where Harry

Gummer had laid everything on. The Governor and the General greeted one another as old friends; the General and his party breakfasted at Government House; an "accidental" encounter with a known German agent who was visiting a British official went off as planned (the agent rushed off to Spanish soil to put in a telephone call); the agents with the telescope got a good look. Three hours later, a waiting crowd at the Algiers airport saw Monty get off the plane and roar off to meet Jumbo Wilson in the latter's own car with a motorcycle escort. GAOL, the notional airport employee, promptly reported the General's arrival to his control. In Wilson's quarters Monty became once again Lieutenant James, and was whisked away to an isolated villa.

Next day he was quietly taken back to the airport by an American "A" Force driver, where Rex Hamer restored to him his own uniform and papers and put him on a plane for Cairo. There Terence Kenyon met him—rather shaken up and needing a few drinks after the strain of his performance—and put him up in the "A" Force flat until Monty's presence in France was officially disclosed. Betty Crichton helped Kenyon look after him. "A very nice man," she recalled, who "always got a bad press. He was under terrible pressure and strain, and coming out of that part was very difficult for him." Meanwhile, "A" Force special means channels kept the General in North Africa for several more days and then lost track of him.

COPPERHEAD had thus gone off smoothly. (Even the Treasury cooperated; Lieutenant James got a full general's pay for the days he notionally held that rank.) It stirred up considerable interest at the Abwehr; the double agents got many inquiries as to Montgomery's whereabouts. At higher levels it did not affect German plans along the Channel—Rundstedt may not even have been told about Montgomery's travels—but it did have the desired effect in the south of France, Fremde Heere West suggesting that Montgomery's visit might point towards operations against that region in addition to the main invasion.

One odd project with no visible result was called PREMIUM. In February, MULLET, the businessman with connections to an Abwehr contact in Lisbon, sent word that he had gone to work for the Fire Office Committee, an association of British insurance companies possessing extensive records about buildings throughout the world, and that a government entity was asking for copious information about Norway (Narvik and the south) and about Belgium and northern France. Another project, given no codename, consisted of a couple of letters written by SNIPER, a Belgian Air Force sergeant and double agent, advising that he had been given training in airfield construction so he could recruit laborers for such work at a later date in Belgium.

Bevan and Baumer's grueling visit to present BODYGUARD in Moscow proved not to have been in vain. Deane reported on May 19 that the Soviets were threatening seaborne attacks on northern Scandinavia by concentrating ships, troops, and equipment in the Kola Inlet, conducting joint army-navy training, stepping up interservice radio traffic, and reconnoitering the threatened coast, plus troop lectures and map issuance; Soviet special means told the enemy that Soviet officers had flown to Scotland to coordinate Allied-Soviet offensives in the north, and through various channels it was suggested that the summer offensive would not begin until July; while the threat to the Black Sea coast was effectuated by the unexpectedly swift Soviet advance into Rumania itself. There were appropriate leakages of deceptive information in the controlled Soviet press, some of which "were just as startling to Russia's allies as they were to the Germans," said Deane after the war. "Only those who knew of Plan BODYGUARD realized that these revelations were for deceptive purposes."

"It must be said that those phases of the plan requiring Russian cooperation were faithfully carried out," Deane said in his postwar report on his mission. And he later wrote: "It was one of those cooperative ventures that promised results of mutual advantage and in which the Russian part of the operation could be carried out without British or American assistance. When these conditions existed, the Russians were usually cooperative."

(The British were evidently more cynical about Soviet cooperation. General Burrows, Deane's British opposite number, seems to have concluded that it was just a ruse to get hold of British intelligence about northern Scandinavia.)

At the same time as Exercise FABIUS, the double agent system suffered a shock when the Germans kidnapped TRICYCLE's key Abwehr contact, Johann Jebsen, codenamed ARTIST, in Lisbon and spirited him back to Oranienburg concentration camp, from which he would never emerge.

ARTIST had known TRICYCLE from the latter's university days in Germany; it was he who originally recruited TRICYCLE for the Abwehr. In 1943 the Abwehr had sent him to Lisbon to assess TRICYCLE's reliability, and it was with ARTIST that TRICYCLE worked out the plan for an escape route from Switzerland that produced FREAK, among others. While managing the Madrid end of this escape route, ARTIST got in touch with MI6, saying he was in trouble with the Gestapo. He gave them the latest news about Canaris's troubles with the Sicherheitsdienst, and told them about Abwehr agents in England, including GARBO. When these agents continued to send information after he had fingered them, he must have realized that they were under Allied control.

ARTIST disappeared from Lisbon on April 28, just as QUICKSILVER was getting under way. ULTRA revealed that in a joint Abwehr-SD operation (codenamed DORA) he had been kidnapped and taken back to Germany in the false bottom of a trunk. It had to be assumed that under interrogation he would blow TRICYCLE and possibly GARBO and others. TRICYCLE was shut down. Allied intelligence held its collective breath about GARBO and the rest. But nothing disastrous happened. After the war it transpired that the Germans had kidnapped ARTIST so as to forestall any possible defection by him to evade punishment for suspected illegal currency transactions—in order to prevent his unmasking TRICYCLE as a German spy! Under interrogation he apparently never gave anything away about the double agents. Nevertheless, from this point on it was deemed too risky for the double agents to make any direct reference to the Pas-de-Calais as a possible Allied target prior to D-Day, for that would be a certain pointer to Normandy if the Germans had begun to suspect them.

The ARTIST affair helped cure the one weakness in Strangeways's plan, the suggestion that if before NEPTUNE D-Day the Germans should seem to think an attack was coming soon, "the imminence of an attack will be confirmed by special means, but the area of the main attack will remain the Pas-de-Calais." The uncertainty created by the ARTIST fiasco strongly reinforced the consideration that such a "confirmation" could well undermine the credibility of the double agents when the landing actually occurred in Normandy. Far better to focus on keeping German reinforcements away from Normandy after D-Day rather than trying to lure them towards the north before D-Day. On this ground, when at the end of May Montgomery's headquarters grew nervous over the shift of some German divisions closer to Normandy, Hesketh successfully resisted a suggestion that the double agents specifically say that the recent destruction of the Seine bridges by Allied air attack was to prevent moving reinforcements northwards from Normandy. But, he said, "After D-Day we can go absolutely all out."

The glimpses into the German mind afforded by ULTRA and MAGIC all confirmed that FORTITUDE SOUTH, and in particular QUICKSILVER, were on the right track. That the Germans expected that the cross-Channel invasion would be the main Allied effort in 1944 was plain from a number of decrypts. In a mid-December message to Tokyo, Ambassador Oshima reported that Ribbentrop had told him that the likeliest area for a landing was Belgium or the narrows of the Channel, and had agreed with Oshima's suggestion that they might land first in Normandy or Brittany and see how things went. Decrypts of March 10 and April 29 showed that Fremde Heere West had heard that Patton was in England and commanding one or two armies. Admiral Abe reported to Tokyo on May 8 and again on May 13 that the Germans expected the main landing to be centered on Boulogne with the weight east of that town, plus diversionary landings in the Le Havre-Cherbourg area, the Dutch coast, and Denmark. Also encouraging was a decrypt of May 20 from the Japanese military attaché reporting that the Germans were associating FUSAG with Patton's return to Britain. There was a period in May when nerves were frayed by decrypted messages seeming to show that the Germans were focused on a landing in the NEPTUNE area, but by the time the assault loomed the general assessment was that the Germans were uncertain where the blow would fall.

On June 1 came the most dramatic of all these decrypts. The afternoon of May 27, after lunching with Ribbentrop, Oshima had visited Hitler at the Berghof, the Führer's eyrie high above Berchtesgaden, and the next day radioed to his government a verbatim account of his conversation, which was read in Washington as quickly as in Tokyo. As edited for smooth reading by the intelligence people, it ran in pertinent part thus:

"What is your feeling about the second front?" asked Oshima. (He had himself toured the German defenses in the West some months before.)

Hitler replied: "I believe that sooner or later an invasion of Europe will be attempted. I understand that the enemy has already assembled about eighty divisions in the British Isles. Of that force a mere eight divisions are composed of first-class fighting men with experience in actual warfare."

"Does your Excellency believe that those Anglo-American forces are fully prepared to invade?"

"Yes."

"I wonder what ideas you have as to how the second front will be carried out."

"Well," said Hitler, "judging from relatively clear portents, I think that diversionary actions [Oshima used the German word *Ablenkungs-operationen*] will take place in a number of places—against Norway, Denmark, the southern part of western France, and the French Mediterranean coast. After that—when they have established bridgeheads in Normandy and Brittany and have sized up their prospects—they will then come forward with an all-out second front across the Straits of Dover. We ourselves would like nothing better than to strike one great blow as soon as possible. But that will not be feasible if the enemy does what I anticipate; their men will be dispersed. In that event, we intend to finish off the enemy's troops at the several bridgeheads. The number of German troops in the west still amounts to about sixty divisions."

When they received this from the American MAGIC cryptographers, the British codebreakers sent it to SHAEF with top priority—the first diplomatic decrypt to be so treated. To the operations people "it gave ground for some uneasiness," in Hesketh's words, since it allowed pessimists to point to the Führer's expectation of a first landing in or near the real NEPTUNE target area. But the deceivers rejoiced. "When they received this decrypt within a week of D-Day," as the official historian wrote, "Colonel Bevan and his associates could breathe a quiet *Nunc Dimittis.*" "It gave us the first definite assurance that the Germans greatly overestimated our strength," said Hesketh. And it made it crystal clear that in the mind of the enemy supreme warlord, a landing in Normandy would be merely the preliminary bout before the main event.

The first great object of FORTITUDE SOUTH had been accomplished. Now the job would be to "go absolutely all out" after the landing to keep the enemy's eyes focused on the Pas-de-Calais.

The deceivers would have been even happier if they could have seen the whole picture on the other side of the hill.

By the beginning of 1944, Hitler was in worse shape, physically and psychologically, than the Allies realized. The calamity of Stalingrad in early 1943 had been followed in May by the surrender of the German army in Africa; in the summer the Red Army had smashed the last German offensive in the east at the great tank battle of Kursk, and gone on during the rest of the year and into early 1944 to clear the Ukraine and the Crimea of the invader and break the siege of Leningrad. POINTBLANK had smitten the industrial heartland of the Ruhr during the spring and summer of 1943, burned Hamburg in July and a succession of other cities as the year wore on, and wrought destruction on Berlin itself from November to March. So by that spring of 1944 Hitler had to face the fact that of all his grandiose dreams, only the destruction of European Jewry had any prospect of coming true. He could at best only hope that the Western Allies would lose heart if he could drive back their invasion, pound England with the cruise and ballistic missiles that would go on line in mid-1944, and turn the tide in the air and at sea with his new jet aircraft and snorkel submarines; and then fight Stalin to a standstill and a negotiated peace.

In late March 1944, General Hans von Salmuth, commander of the German Fifteenth Army along the Pas-de-Calais, a tough Prussian of the old breed and a veteran of hard fighting in Russia, saw the Führer at a conference of Western Front commanders and was shocked at the sight of "an old, stooping man with an unhealthy, puffy face," who "looked downright worn out, weary—I would even say ill." By early May, Hitler was suffering from stomach spasms and a left leg that shook uncontrollably, and was dosing himself with an extraordinary quantity and variety of medicines. He spent much of the time in the Berghof, its legendary view of the Alps obscured now by the screens of choking smoke that were thrown up several times a day when Allied bombers were near.

Keitel remained head of OKW, and Jodl head of the operations staff. By this time that staff had acquired its own intelligence officer, Colonel

The Success of FORTITUDE.
As of May 15, 1944, Allied forces in England were in reality located generally in
the western part of the country (left). A German intelligence map captured in Italy

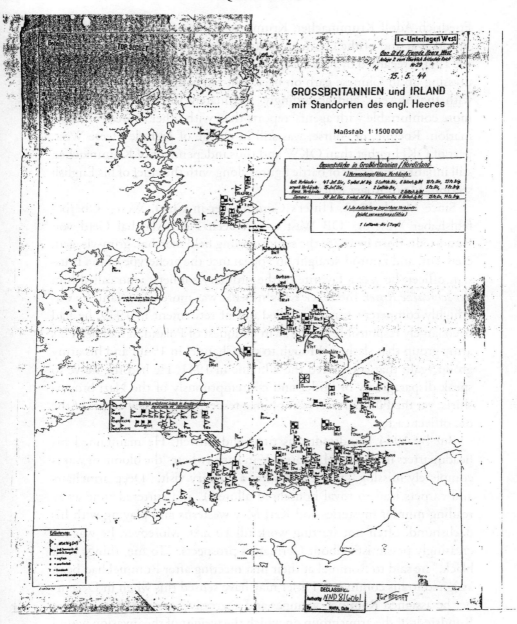

(right) confirmed that the Germans, consistent with FORTITUDE, believed them
to be concentrated instead in the southeast opposite the Pas-de-Calais.

(Electrostatic reproductions of the originals, from the U.S. National Archives.)

Friedrich-Adolf Krummacher. Krummacher was a veteran of the First World War and of the German military mission to Chiang Kai-shek in the 1930s, regarded by his colleagues as *stinkfaul,* bone-lazy, and given to dressing up in Chinese garb on occasion. He had been a liaison officer with the Abwehr, and accordingly, fortunately for the deceivers, was more comfortable with agents' reports than with other sources of information. Roenne, of course, was still in charge of Fremde Heere West, part of OKH rather than OKW and thus independent of Krummacher, still inflating the order of battle figures along with the chief of his English section, Major Roger Michael.

Since March 1942, Hitler's Commander-in-Chief, West—Oberbefehlshaber West or "OB West"—had been Field Marshal Gerd von Rundstedt. Born in 1875, the senior serving field marshal, grim-visaged, crew-cut, and ramrod-straight, in appearance the archetype of the Prussian officer, he was a favorite of the ordinary soldiers, who called him *der schwarze Ritter,* the Black Knight; what was more important, he was a highly competent general. Called out of retirement in 1939, he had been largely responsible for the swift conquest of Poland; it had been his army group that slashed through to the Channel in 1940, and his army group again that conquered the Ukraine in 1941. He had the nerve to speak disparagingly of Hitler and contemptuously of the Nazi brownshirts, yet the Führer reserved for him a respect he gave few others of the old officer caste.

But by 1944 Rundstedt was tired and unwell. He maintained his headquarters in the Hotel Georges V in Paris (where "the bloom of war is completely missing," wrote Jodl after a January visit. "Deep armchairs and carpets lead to royal household allures"), and frittered time away reading murder mysteries and Karl May westerns and playing with his dachshund, often not starting work till 10 A.M. Moreover, he was increasingly pessimistic about Germany's prospects. "To me, things look black," he said to Rommel at their first meeting after Rommel had been assigned to his command. Hitler may have sensed this when in December he assigned Rommel, younger and aggressive, to take charge under Rundstedt of the army group on which the weight of the invasion would certainly fall. Rundstedt's chief of staff was General Günther Blumentritt; his intelligence chief was Colonel Wilhelm Meyer-Detring, a big man, smooth and imposing.

Called Army Group B,* Rommel's command was composed of two armies, the Seventh under General Friedrich Dollmann, a tall, sixty-two-year-old officer from a long line of civil servants, a man of little energy, poorly suited to command the force that would bear the brunt of the Allied attack; and the Fifteenth under Salmuth. The Seventh held France north of the Loire from Brittany to Normandy west of Rouen; the Fifteenth on its right held the northern front up through Belgium. From March onwards, Rommel's chief of staff was General Hans Speidel, a fellow Swabian—Rommel and Speidel would chatter away to each other in their home dialect—who had won the Knight's Cross on the Eastern Front, a bespectacled and polished intellectual who brought to headquarters a comparatively relaxed outlook. He brought something else, too: He was deep in the plot against Hitler.

Rommel's intelligence chief was heavy-set Colonel Anton Staubwasser, mild in manner and straightforward in conduct, a veteran of three and a half years in Fremde Heere West and an expert on the British army; but he had no intelligence sources of his own and was wholly dependent on information from Roenne and Meyer-Detring, "although it was left to [us]," he lamented after the war, "to decide up to what degree the numerous, often very unreliable details, could be accepted as true." He had reason to doubt them. He had still been with Roenne when the latter started inflating his order of battle figures. Staubwasser ventured to remonstrate, and Roenne had told him to go along or ship out. He went along.

Rommel took charge of his new command at the beginning of December, and after whirlwind tours of inspection began to infuse his own legendary energy and enthusiasm into a lethargic command whose soldiers had shown more interest in the black markets and brothels of Paris than in steeling themselves for the Allied onslaught. By March he had established his headquarters in the elegant chateau of the duc de la Rochefoucauld at La Roche-Guyon, in a loop of the Seine forty miles west of Paris, halfway between the capital and the coast. That same month, Rundstedt moved his command post from the Georges V to St. Germain-en-Laye, just west of Paris.

German belief in the Pas-de-Calais as the main Allied target ran far

---

* Under Rundstedt also was Army Group G under General Johannes Blaskowitz, formed in the late spring and composed of Nineteenth Army on the Mediterranean coast and First Army holding the Atlantic coast and inland.

deeper than the Allied deceivers realized; and, what was perhaps equally important, so did German belief that there would be diversionary landings as well.

"We always took the most dangerous area to be the one from the Seine towards the North-East," Rundstedt told Allied interrogators after the war, "because if you think ahead, strategically, to go through Belgium in the direction of the Ruhr, that was the most dangerous thrust." From the outset Rommel thought that sector, the Fifteenth Army's sector, was the likeliest target, and soon after taking over he made a tour of it in a blaze of publicity. He seems never to have wavered from this view. "Rommel's conjectures as to the presumable front of the invasion led him to the opinion that the enemy landing was to be expected above all in the area between Calais and the Somme River," the great armored warfare pioneer General Heinz Guderian—admittedly no admirer of Rommel's plans for defense of the West—told the Allies after the war, adding: "His belief in this theory was so strong that he brushed aside every contention that a landing would also be possible at some other location than the one expected by him."

(Some postwar suggestions to the contrary may have been motivated in part by a desire to seem wise after the event; moreover, Rommel was not above playing on Hitler's "instincts" about a Normandy landing to try to get more control over the central reserve. One German officer shrewdly observed to the American historians in 1947 that "all the diverging views on General Rommel's opinion, have resulted from the General having stressed the possibility of a major landing in *all* sectors, as an incentive to all forces.")

An added reason for regarding the Pas-de-Calais as the logical Allied target was that the launching sites for the V-weapons were there, and it was hoped that they would prove so destructive that the Allies would have to try to overrun their bases as soon as possible. Hitler, in his Führer Directive No. 51 of November 3, had proclaimed that he had decided to reinforce the defenses of the West, "particularly those places from which the long-range bombardment of England will begin. For it is here that the enemy must and will attack, and it is here—unless all indications are misleading—that the decisive battle against the landing forces will be fought." Rommel echoed this view in his first report to Hitler at the end of December.

Helping to deflect Rommel and Rundstedt from Normandy was the counsel of their naval advisors, who claimed that the Allies could not land on the NEPTUNE beaches. The Navy, said Rundstedt, "always told us that, just in the place where the landing came, it was unlikely that one would be made, because they always maintained that a landing *could* only be made at high tide, and there are those reefs, those rocks below the water at high tide. Nobody could land there, said the Navy, as their craft would be wrecked. Instead the invaders landed at *low* tide and were able to use the rocks as cover against the fire from land." (Vice Admiral Friedrich Ruge, the naval liaison officer to Rommel's army group, seems to have forgotten about this in debriefings after the war.) On this ground, Rommel in February rejected Luftwaffe arguments that the invasion would come in Normandy.

Hitler, however, began listening to his intuition, and his intuition began more and more to speak of Normandy and Brittany. At his March 4 main daily war conference he announced that those provinces were especially threatened, and repeated at a March 20 meeting with generals from the West that the invasion would come there. On March 23 he said the same thing to Antonescu, the Rumanian dictator. By May he was having Jodl nag Rundstedt and Rommel continually on the subject. On May 2, Jodl's deputy telephoned Rundstedt's headquarters to say that the Führer considered that besides the Channel front, the two peninsulas of Normandy and Brittany were primarily threatened, and that they should be reinforced because "A partial success by the enemy in the two peninsulas would inevitably at once tie down very strong forces of OB West." Four days later Hitler had Jodl telephone Blumentritt, Rundstedt's chief of staff, to say that he "attached particular importance to Normandy" and that "the Cotentin peninsula would be the first target of the enemy"; two days after that, Jodl telephoned Rundstedt himself to say that Hitler expected the invasion in mid-May with a major effort against Normandy. This was preaching to the choir, however; on April 29 Rundstedt had warned Rommel that Normandy and Brittany, "which are especially open to encirclement and the formation of bridgeheads, and which possess ports, constitute as before a special attraction for the enemy," and on May 15 his situation report expressed the view that a bridgehead on the Cotentin would be a natural first step, which would "perhaps be the enemy's prerequisite condition for a subsequent descent on the Channel coast between Calais and Cherbourg."

This focus on Normandy was always, however, in terms of a preliminary or diversionary attack (except for momentary exasperated outbursts by Hitler, as when, early in April when the ban on travel to the coast and other security precautions began in England, he fumed: "We've no real way of finding out what they are up to over there," adding, "I am for bringing all our strength in here," tapping the Normandy coastline on the map). From the outset Hitler had thought in terms of multiple diversionary landings. His Führer Directive No. 51 had warned that "Holding and diversionary attacks are to be expected on other fronts." "There is no longer any doubt that the invasion in the west will come in the spring," he had mused at a situation conference on December 20. "It would be good if we could know from the start: Where is a diversion and where is the real main attack?" It will be recalled that in January he was even musing to Oshima about the Bay of Biscay. By May, when Jodl was nagging the field commanders, multiple diversionary attacks were uppermost in Hitler's mind. On May 3, he said that Germany must reckon "first of all with individual operations for the creation of several bridgeheads. Above all strong bridgeheads will be attempted in Brittany and the Cherbourg peninsula." And this was the view that he expressed to Oshima more than three weeks later.

The theory of multiple landings was made plausible by the wild overestimates of Allied strength in Britain emanating from Fremde Heere West. At the beginning of 1944 Roenne had estimated fifty-five divisions in England when there were really thirty-seven including garrison troops and training divisions. By May 15 he was showing seventy-seven divisions and nineteen independent brigades, 50 percent greater than the real strength. By May 31 he was up to seventy-nine divisions with independent units equivalent to ten more, and sealift for nearly sixteen, when there were actually some fifty-two divisions. His picture of the structure of these forces moved concurrently into the focus that QUICKSILVER wished to impose. For a long time he envisaged three "groups of forces," one in Scotland, one in central England, and one in the south; an idea apparently derived from the uncontrolled agent JOSEPHINE. In late April, on the eve of the opening of FORTITUDE SOUTH, on the basis mainly of wireless intercepts and analysis of Allied air force organization, he reckoned that Eisenhower commanded FUSAG (American First and Ninth

Armies) under Bradley,* and 21 Army Group (Canadian First Army and a British one tentatively identified as the Fifth) under Montgomery. Then over the next six weeks, largely through GARBO and BRUTUS, he was brought around to the QUICKSILVER order of battle.

At 9:50 P.M. June 5, the very eve of D-Day, when the airborne forces were already aloft and the great armada was afloat, Rundstedt's situation report deemed the "front-line point of main attack still remaining the probable offensive line, extension to the north coast of Brittany inclusive of Brest not impossible," while "Concentration of enemy air action on the coastal defense between Dunkirk and Dieppe . . . might seem to indicate the *Schwerpunkt* of projected landings on a large scale. . . ." Back in Germany, Roenne had finished up a situation report for distribution next day according to which the forces in Britain were organized into 21 Army Group under Montgomery (three armies in south and southwest of England), and FUSAG under Bradley or Patton (Canadian First Army, thirteen divisions, south of the Thames, and American Third Army, twelve divisions, between the Thames and the Wash).

So, as the Führer turned in after midnight that night, this was the picture of the world in his mind and those of his chief commanders. Few, if any, deception schemes have been more spectacularly put over.

Harris and Ingersoll were determined to see D-Day at first hand. For Harris it was something a regular soldier should not miss. Ingersoll wanted to be in at the finish of Nazism, but his motivation was more complex. He was, as has been said, a world-class hypochondriac. He had developed a spot on his back that he was convinced was life-threatening. He did not dare show it to the doctors; better to die with his boots on, he thought, than be sent home for a lingering death. (In fact, the spot proved to be harmless.) So he set out for Utah Beach to carry out with Harris a tactical deception called TROUTFLY.

TROUTFLY aimed to "reinforce" the 82d Airborne by the notional arrival on D plus 1 of a regimental combat team to be simulated by dummy radio traffic. Ingersoll's assignment was to site the equipment. He went in

---

* The American Ninth Army under General William Simpson was at this point little more than a headquarters, meant to be built up by divisions arriving from the United States and to join the other armies on the Continent at a later date.

with the assault on Utah Beach in a landing craft with the commander of a force of tanks whose job it was to fight through to join the 82d Airborne ten miles inland. He wrote an account of the landing that still ranks as one of the great reports to come out of the Second World War. Harris directed the operation from VII Corps headquarters, which was aboard the attack transport *Bayfield* off Utah Beach. It came a cropper because the equipment and personnel never arrived; the ship carrying them was misdirected to Omaha Beach and did not land till D plus 5. Harris had to roam back and forth in a light craft through unswept waters between *Bayfield*, V Corps headquarters on the amphibious command ship *Ancon* off Omaha Beach, and First Army headquarters on the cruiser *Augusta*. He and Ingersoll joined up with one another on D plus 6 and hitched a ride back to England in a B-17 flown by a classmate of Harris's.

They were perhaps the only deceivers who actually crossed the Channel on D-Day, though elements of "R" Force went in very early, ready to simulate the Guards Armored Division, and Strangeways remembered being in Normandy twenty-four hours ahead of the rest of his people. Being cleared for ULTRA, Eldredge would not have been allowed to go even if he had wanted to; which he did not. "I wasn't as brave as I might have been," he told Ingersoll's biographer. "I didn't want to go and have my ass shot off."

Dennis Wheatley was seconded as a war correspondent for a few days and went down to the British 6th Airborne Division to write an account of their taking off for France.

Clare Beck had landed in North Africa and Sicily and sat this one out; he and his MI5 lady stood outside in the darkness the night of June 5 and listened to the air fleet passing overhead, a steady thunder that went on and on.

Down at Southwick Park some hundreds of officers found themselves suddenly summoned to an open-air assembly. Over their heads the same thunder filled the skies as Montgomery himself mounted to the top of a vehicle, announced that the invasion was beginning, and delivered an address on the certainty of victory that his hearers would never forget. More than fifty years later, Captain John Oakes, who would run double agents for the American OSS on the Continent, remembered it as "the most inspiring speech I ever heard."

Down at Canadian headquarters on the south coast, a group from the

3103d sat in a British trailer from which they had been imitating the American XII Corps and listened to the same seemingly endless roar. T/5 Erwin Lochmueller lit up a pipe. A Canadian nudged him and pointed to the No Smoking sign and to the British colonel sitting behind them. Instead of dousing his pipe, Lochmueller passed his pouch back to the colonel. The colonel lit up his own, asked Lochmueller for one of his radios, looked in a frequency table, and tuned in on the voice nets of the landing forces. "Put it on the loudspeaker," he said; and they did, and listened to the troops going in.

But no one who stayed behind passed a more interesting evening than GARBO and his colleagues. Since early May, Tommy Harris had pushed an idea which, as fully developed, was to let GARBO send the Germans advance word of the Normandy landing just too late for it to do them any good; and then, two days after D-Day, to treat the Germans to a full-dress analysis of Allied intentions, concluding that the NEPTUNE landing was only the first prong of a dual invasion, designed to draw reserves to Normandy and lay open the direct route to Berlin. Nervousness over ARTIST having subsided, and the Oshima message having given heart to the deceivers, it seemed safe to implement this plan. Eisenhower approved letting GARBO send warning of the invasion, but no earlier than 3 A.M. Madrid normally shut down between 11:30 P.M. and 7 A.M., so to have them tuned in at the crucial hour GARBO had told them to be listening at 3 A.M. for a report on a division in Scotland in which they had expressed particular interest.

GARBO, Tommy Harris, Tar Robertson, Roger Fleetwood-Hesketh, and GARBO's radio operator Charles Haines gathered around GARBO's transmitter in a house in North London at that hour, ready to send a message that Agent FOUR had broken camp and reached London with word that the Canadian 3d Division had once more been issued cold rations and vomit bags and moved out. At the appointed 3 A.M., Haines called Madrid. No response. Over and over he tried, four times an hour. Nothing. Finally, Madrid came on the air at 8 A.M. and the message—somewhat improved in the interim—was eventually sent.

Mention should be made of a group of tactical deception operations for D-Day that were included in Strangeways's Joint Commanders' Plan,

codenamed PARADISE ONE through FIVE, BIGDRUM, TAXABLE, GLIM-
MER, and TITANIC I through IV.

The PARADISE series belonged strictly to the post-D-Day period, and
encompassed a series of dummy beach exits, beach areas, and bridges to
decoy German air attacks away from the real ones.

The next three were naval operations. BIGDRUM was a diversionary action by a small force of motor launches, supported by aircraft dropping
WINDOW, and complete with phony radio traffic, to engage German
radar stations in the north of the Cotentin peninsula and distract enemy
batteries in that area. GLIMMER was a similar operation at Boulogne.
TAXABLE was a similar but larger demonstration aimed at the coast just
north of Le Havre to give the appearance of a major landing north of the
Seine in addition to the true landings.

The TITANIC series were dummy paratroop drops, some accompanied
by real Special Air Service troops to add to the authenticity. TITANIC was
selected as the codename by Captain M. R. D. Foot, the intelligence officer of the Special Air Service, with the thought that if the Germans
heard of it they might find it impressive (and would not associate it with
a major disaster, as would the British and Americans).

TITANIC I was a drop of dummy paratroops with PINTAILS and appropriate noisemakers between Dieppe and Le Havre at around 2 A.M. of
D-Day, simulating the dropping of an airborne division north of the
Seine in connection with the "landing" simulated by TAXABLE. Two Special Air Service parties accompanied it, their role being to attack dispatch
riders and lone vehicles, allowing some to escape and spread the word
that this was a real paratroop drop. TITANIC II was a planned dummy
parachute drop simulating a parachute brigade on the coast east of Caen
to deter local reserves from moving towards the real landings; it was canceled because there was simply too much air traffic to squeeze it in. TI-
TANIC III was a dummy parachute drop southwest of Caen at the same
time as the actual drop of the British 6th Airborne Division. TITANIC IV
was a dummy parachute drop at Marigny on the base of the Cotentin
peninsula, accompanied by Special Air Service parties, to sow confusion
and hold up the movement of reserves into the Cotentin.

The results of these efforts were generally gratifying. A German
brigade spent D-Day morning looking for the TITANIC IV party; a June
9 MAGIC decrypt showed that Oshima had been told on June 7 that

"powerful airborne units, amounting to about one division, were dropped in the vicinity of Cherbourg but were wiped out," and Admiral Abe reported that Jodl boasted on July 3 that "those of the enemy who dropped" in this area "immediately received an annihilating blow." As for TITANIC I and TAXABLE, Oshima reported on June 8 (decrypt circulated June 12) that "It is uncertain whether the enemy will try to land in the Calais-Dunkirk area; however, an enemy squadron which had been operating off that coast has now withdrawn," and again on the 11th that several attempts to land near Le Havre had been beaten off. Jodl said on July 3, as reported by Admiral Abe, that "we immediately crush[ed] the enemy's plans to land against the strongly defended area of Le Havre and the north side of the Cotentin Peninsula."

The Germans were totally surprised by the landing—whether or not BODYGUARD and its progeny deserve any credit for it. "That the invasion is actually imminent does not seem to be indicated as yet," reported Rundstedt's situation report the night of June 5 already quoted. Rommel was taking a few days' leave at home for his wife's birthday. His chief of staff was entertaining friends (colleagues in the anti-Hitler resistance). Salmuth, commander of Fifteenth Army, was off on a hunting trip. At 4 A.M., one Private Fuchs of Seventh Army's intelligence staff alerted the duty officer to the fact that front-line units were reporting an attack. *"Machen Sie kein Theater, erinnern Sie sich an Dieppe,"* said the major calmly; Don't make a big production out of it, remember Dieppe.

Now came the supreme test of FORTITUDE SOUTH. "Just keep the Fifteenth Army out of my hair for the first two days," Eisenhower had said to Noel Wild in January. "That's all I ask." The deceivers themselves reckoned that the Germans could not be fooled for more than ten days.* SHAEF G-2 had forecast that by D plus 7, six divisions would have been moved south from Fifteenth Army to reinforce the Normandy battle-

---

* This was consistent with experience. Dudley Clarke observed that a large-scale deception "needs at least three weeks implementation before it can have any appreciable effect," and that once the notional D-Day is past, it "loses in effect until it fails to deceive any longer. Thus a deception may be designed to mislead the enemy over a period of three weeks or longer while it is developing, but once it has reached its peak, cannot be expected to remain effective for more than about 14 days."

front (plus two from southwest France and one from the French Mediterranean coast). In line with Hesketh's earlier words, the deceivers now went "absolutely all out."

BRUTUS opened with a message just after 6 P.M. of D-Day itself. "It is clear that the landing was made only by units of the 21 Army Group. . . . FUSAG, as I reported, was ready for an attack which is capable of being released at any moment, but it is now evident that it will be an independent action." That evening and next day he filled in some specific units for FUSAG, ostensibly after touring East Anglia. On D-Day night GARBO sent the purported text of a psychological warfare directive ordering that there be no speculation about future landings—implying, of course, that there would be such landings. (He had to explain away the fact that Churchill had told the House of Commons that NEPTUNE was "the first of a series of landings.") Next day TATE sent word from his post in Kent that he had seen no departures of troops; that, indeed, more were arriving every day. Meanwhile, fake radio traffic was intensified, as were communications with the Resistance in northern France and Belgium.

In the early hours of June 9 came the boldest move of all. Oshima's message about his May 27 meeting with Hitler confirmed the deceivers in carrying out Tommy Harris's audacious plan for GARBO to send a message two days after the landing, laying out a specific judgment that the purpose of the Normandy operation was to draw reserves to that front to ease the way for the main assault. Audacious, because it is asking for trouble to let a double agent paint so specific a picture; moreover, the message itself as drafted was inordinately long, far longer than any sensible agent would allow himself to remain on the air. But in light of the Oshima interview and the extent to which GARBO now had Madrid eating out of his hand, it was decided to take the chance.

So just after midnight the night of June 8–9, GARBO went on the air with a communication than ran for just over two hours. It recounted that he had called in his agents 7(2), 7(4), and 7(7), from Dover, Brighton, and Harwich respectively, to pool all their information. He listed all the units, real and fake, in the southeast of England, and, as already noted, made his first reference to landing craft in eastern rivers. "From the reports mentioned it is perfectly clear that that the present attack is a large-scale operation but diversionary in character for the purpose of establishing a strong bridgehead in order to draw the maximum of our reserves to the area of operation and to retain them there so as to be able to strike a blow somewhere else with ensured success," logically the Pas-de-Calais. He said that from his source J(5), the secretary in the Cabinet Offices, "I learned yesterday that there were 75 divisions in this country before the present assault commenced. Supposing they should use a maximum of 20 to 25 divisions, they would be left with some 50 divisions with which to attempt a second blow." He concluded in typically baroque GARBO style:

> I trust you will submit urgently all these reports and studies to our High Command since moments may be decisive in these times and before taking a false step, through lack of knowledge of the necessary facts, they should have in their possession all the present information which I transmit with my opinion which is based on the belief that the whole of the present attack is set as a trap for the enemy to make us move all our reserves in a hurried strategical disposition which we would later regret.

That was laying it on with a trowel indeed. But, as will be seen, it was accepted all the way up to Hitler.

On into June, BRUTUS and GARBO kept sending stories about the steady flood of American troops into England and the increasing concentrations in the southeast. On June 12, GARBO reported that his talkative American sergeant had reconfirmed that Patton, not Bradley, was the commander of FUSAG. In the bridgehead itself, one day followed another, and the expected reinforcements from Fifteenth Army did not materialize. The crucial first few days passed by and the lodgment was secure. A week after D-Day, only one division had been shifted to Normandy from Fifteenth Army instead of the six that SHAEF G-2 had forecast.

The month wore on, a heartbreaking month, with the Americans bogged down in the hedgerows and the British seemingly stuck at Caen. The ten days that everybody had thought was the maximum that the deception could be maintained passed, and still Fifteenth Army sat in the Pas-de-Calais. Enough reports were decrypted through ULTRA and MAGIC to assure the happy deceivers that it was still waiting for FUSAG. Thus, a June 22 MAGIC decrypt showed Oshima reporting that "23 divisions commanded by General Patton are being held in readiness to make new landings. This is one reason why Germany has avoided pouring a great number of men into the Normandy area," and after several similar reports he was advising Tokyo as late as July 3 (decrypt circulated July 17) that the Foreign Ministry "believe that Patton's forces will land in the neighborhood of Dieppe." "Every day came—dawn came—same thing again. There it was. Nobody could argue against it," recalled Strangeways fifty years later. "I just thought, Crumbs! This is unbelievable!"

They would have been even happier had they known everything transpiring on the other side. Roenne telephoned Staubwasser on D-Day to say that this was not yet the main invasion, another landing would come in the Fifteenth Army area and no troops would be withdrawn from there. His situation report for that day noted that all identified Allied forces were drawn from 21 Army Group and concluded that "the enemy plans a further large-scale undertaking in the Channel area which may well be directed against a coastal sector in the central Channel area." When Rommel arrived back at his headquarters that night he was told that a further major invasion was expected, as Dover was hidden behind a smoke screen. The next day Rommel told Jodl by telephone that "we must assume that the enemy is going to make another invasion focal point elsewhere."

Rundstedt himself, surprised that the Allies had not committed more of the huge host believed to be in England, echoed the expectation that "a landing operation at least as strong, if not stronger, will shortly take place at another part of the coast." Though Jodl seems briefly to have expressed doubts about whether there would be a second invasion—perhaps to buck Rommel up—Roenne's report of June 9 again stressed that the fact that none of the units standing by in southeastern England had been identified "strengthens the supposition that the strong Anglo-American forces which are still available are being held back for further plans." Nevertheless, by that time almost all the German armored reserve had been released to Rommel so as to clear the Normandy bridgehead before the second landing. The last major unit, the 1st SS Panzer Division Leibstandarte Adolf Hitler in Belgium, began to move on June 9.

That same day there was a flurry of excitement at OKW over the possibility of an imminent landing in Belgium, brought about initially by a seeming misinterpretation by the Abwehr of messages to the Belgian Resistance (these may have been part of the deceivers' "absolutely all out" efforts), followed by a fraudulent tidbit supposedly from the freelance agent JOSEPHINE claiming that his sources expected a "second main attack across the Channel," crowned by a summary of GARBO's long message that came in late that night. The last two were shown to Hitler himself. As a cumulative result, the planned move of the Leibstandarte from Belgium to Normandy was countermanded.* The next day, Rommel wrote to his wife: "The invasion is quite likely to start at other places, too, soon. There's simply no answer to it." (That same day, Admiral Dönitz said to his staff: "The invasion has succeeded. The Second Front is now a fact.")

Roenne's situation report for June 11 expressed doubt that Belgium would be the target and instead thought the coast between the Seine and the Somme more likely. At the afternoon situation conference at OKW next day, Keitel and Jodl agreed that the best strategy was to block any other landings—probably on the Dieppe-Boulogne or Calais-Scheldt coasts—and then come back to mop up Normandy. That day the first V-1s were launched on England, and Roenne felt that this would make the Pas-de-Calais an even more desirable objective for the Allies. And that day Dr. Goebbels seems to have decided that the German public needed an excuse for why the Allies had not been driven into the sea; for Sertorius, a prominent commentator, announced that reserves were being held back because other landings were to be expected. By June 15, Roenne had run the total strength of FUSAG up to twenty-six divisions—all those the deceivers wanted included plus sixteen more, including six imaginary ones. And in his situation report of June 24 he was still predicting an attack in the Seine-Somme sector.

Some dissent began to be felt. By mid-month Meyer-Detring at OB West, as well as the Luftwaffe, had noticed that all the best Allied combat units had already showed up in Normandy, and they were beginning to think that perhaps some weakening of Fifteenth Army would be safe. On June 17, Hitler himself made a flying visit to meet Rundstedt and Rommel at Soissons and echoed that view: "The enemy have committed all their battle-experienced divisions to Normandy, which suggests they

* Extravagant claims have been made with respect to this incident. As the official historian has commented, too much should not be made of it, for the division would have had a hard journey under Allied air attack and Resistance harassment (and in fact it joined the battle a week later); still, it did answer those who had doubted that the Germans would move a single division merely on the strength of agent reports.

now have their hands full." This did not persuade Rommel; two days later, and again on June 22, he visited Fifteenth Army's sector, told one of his tank commanders that he expected the next invasion there, and ordered reserves moved closer to the coast.

It did not persuade Jodl either. In his July 3 disquisition reported by Admiral Abe, he said that it was "obvious that Patton's army group (eighteen infantry divisions, six armored divisions, five airborne divisions) is being made ready at London and in southern England for the next landing," which would be in the Fifteenth Army sector; and, since all but two of the enemy's seasoned divisions were in Normandy, and German defenses were strong, "we have every confidence in defeating them." Indeed, *twenty-two* German divisions were being held back in Fifteenth Army's sector.

So, against all expectation, the deception had held for a month, and showed no sign yet of failing. But now the deceivers began to pay the price of success. Units of the artificial FUSAG were beginning to appear in the bridgehead, and Patton would soon be unveiled as commander of the American Third Army and Bradley as commander of the real FUSAG. Could the deception nevertheless be continued?

With Strangeways busy with tactical deception on the Continent, Wild wanted to resume direct control of FORTITUDE SOUTH, and in mid-June Bedell Smith approved this, effective at an appropriate future date. When news of this reached Billy Harris he protested vigorously to Strangeways. Pinning enemy divisions in the Pas-de-Calais was "still an integral part of the battle in Normandy," and it was "unwise to relieve General Montgomery's staff at this time." Until the American army group should take the field, "American cover and deception operations can only be sensibly coordinated with the total objectives of the Field Commander under the direction of General Montgomery's authorized representative." But this was of no avail; Ops (B) was to take over on July 20.

Meanwhile, in mid-June, without Strangeways's participation, Ops (B) was already concocting a follow-on to FORTITUDE SOUTH. Its proposal was based on the assumption that "the bulk of German reserves will be committed to NEPTUNE" by D plus 30. Its object would be "to reduce the rate and weight of reinforcement of enemy forces" for as long as possible. Its "story" would be that NEPTUNE had met less opposition than expected, so Eisenhower had decided to defeat, rather than contain, the Germans in Normandy. He had therefore moved FUSAG forces into the bridgehead, and was forming in England for the Pas-de-Calais a new American 2d Army Group (SUSAG) of fifteen divisions organized into two armies.

The new SUSAG and its elements would be entirely notional. Harris and Ingersoll expressed to Strangeways "grave misgivings over the proposal to activate a large number of wholly notional United States divisions," pointing out that the manpower shortage in the United States was common knowledge. "Building a new FORTITUDE (SOUTH) on a wholly fictitious order of battle means, to Special Plans Branch, a serious overplaying of a winning hand." But this protest was likewise unavailing. Bevan was already cabling Washington urgently for new notional units in addition to those already allocated to the United Kingdom, receiving from Smith and Goldbranson a new army group (SUSAG), two new armies (Twelfth and Fourteenth*), two new corps (XXXIII and XXXVIII), and a new airborne division (18th Airborne), plus transferring some units that had been allocated to the Mediterranean but never employed by "A" Force.

Even more serious was the assumption that the bulk of German resources would soon be committed to Normandy, for that was transparently not the case. Nor was it plausible that Normandy was going better than expected. The whole point was that unless another force was constituted in eastern England, the Germans would assume that the Pas-de-Calais threat no longer existed, and reinforce Normandy from Fifteenth Army. So at a meeting at his headquarters on June 26, Strangeways rewrote this plan as he had rewritten the original. The stated object of the revised plan would now be to *continue* to contain maximum enemy forces in the Pas-de-Calais for as long as possible. Its "story" would be

---

* Use of an even-numbered army could have raised German suspicions, since all the American armies in Europe carried odd numbers while those in the Pacific carried even ones. Though the Germans could not have known it, this in fact did not reflect conscious policy. Indeed, Ninth Army was originally Eighth Army, the number being changed at Eisenhower's request to avoid trespassing on the name of the great British Eighth Army.

that Normandy had *not* gone as well as hoped, so that Ike had had to reinforce it from FUSAG units to the point that he had decided to form an American army group under FUSAG in Normandy, and reorganize the remaining units in England as SUSAG.

The new notional order of battle was sketched out by June 29. SUSAG was never used; instead, Strangeways offered the further suggestion that FUSAG be retained as the Pas-de-Calais force and Bradley's army group be redesignated; it became 12th Army Group. Only one new American army proved necessary, so the notional Twelfth Army was never used either. (That was just as well, for its use might have led to confusion with 12th Army Group, whose new number would, in Ike's words, "obviate confusion resulting if a number identical with one of the armies in this theater were to be selected.")

There remained the problem that Patton would soon appear in Normandy as an army commander. Bradley vetoed the suggestion that Patton be kept in England to continue the deception. The SHAEF staff recommended to Eisenhower that he ask Marshall to send him Lieutenant General Lesley J. McNair, Commanding General, Army Ground Forces, or some similarly well-known senior general, for the new FUSAG commander. Ike gladly complied. "I cannot overemphasize the importance of maintaining as long as humanly possible the Allied threat to the Pas-de-Calais area," he cabled to Marshall on July 6, "which has already paid enormous dividends and, with care, will continue to do so." "Whitey" McNair, in many ways Marshall's right-hand man ever since he became Chief of Staff, was an officer of immense prestige, certain to impress the Germans, whose job of training and organizing a great citizen army was now largely completed. Marshall sent him promptly.

How to justify the downgrading of Patton from army group to army command? The "story" for that apparently came from Eisenhower himself. He was now so understandably enthusiastic about FORTITUDE SOUTH that on July 10 he wrote in person a memorandum for Bedell Smith setting forth the basic ideas of FORTITUDE SOUTH II and suggesting "a story that Patton has lost his high command because of displeasure at some of his indiscretions, and that he has been reduced to Army command." This "story" was adopted. Patton, it ran, had protested in insubordinate terms the denuding of his army group to reinforce Bradley, and an angry Ike, fed up with Patton's behavior,* had demoted him to army command under Bradley as a result. Patton, incidentally, was undisturbed by the role in which he was cast. "As far as the deception planning goes, he was the greatest team player we had over there," Harris told a group of officers shortly after the war. "He would do anything you asked him to do in the interest of the overall picture."

Aside from the genuine American Ninth Army, the reconstituted FUSAG was almost entirely notional—unlike the original FUSAG.† Besides Ninth Army, it consisted of the phony American Fourteenth Army, the British Fourth Army (notionally moved down from Scotland), and the British 2d and American 9th and 21st Airborne Divisions. Every single corps and division was a fake except two British divisions (both in reality only limited service units) and a British armored brigade. Ninth Army, a followup force both supposedly and in truth, was in the Bristol area; the rest, except for the airborne divisions, were in the southeast and east of England.

The final "story" of FORTITUDE SOUTH II, aside from the tale of Patton's insubordination and demotion, was that put forward by Strangeways. Eisenhower had had to reinforce 21 Army Group with so many elements from FUSAG that he had formed a new American 12th Army Group. FUSAG would be rebuilt for the Pas-de-Calais assault with forces from the United States, around a core of at least eight divisions. Training and concentration would be completed by July 26; then would follow an assault "by three infantry and three airborne divisions on the beaches exclusive River Somme to inclusive Boulogne," with the British

---

* Patton's troubles for having slapped a soldier in Sicily were well known, and he had landed in hot water again in late April for an indiscreet speech at the opening of a "Welcome Club" for GIs in the English town of Knutsford in Cheshire (close to the real Third Army headquarters), in which he said that it was "the evident destiny of the British and Americans, and of course the Russians, to rule the world."

† By D-Day FUSAG already had two notional units. It will be recalled that the American VIII Corps and its 79th Division had been replaced in Canadian First Army by the American XII Corps and the 35th Division, from Third Army. These were in turn replaced in Third Army by the notional XXXVII Corps and the notional 59th Division. Then shortly after D-Day, when Canadian II Corps was destined for France, it was replaced by the notional British II Corps, moved down from Scotland in stages so as not to disarrange FORTITUDE NORTH unduly (Operation TWEEZER).

55th and 2d Airborne Divisions on the right, the American 59th and 17th Divisions of XXXVII Corps on the center and left, and supported by the American 9th and 21st Airborne Divisions dropped inland; the immediate followup would be the British 58th Division of II Corps and the American 25th Armored Division of XXXIII Corps. The Allies would be in a position to reinforce the new front with divisions brought directly from the States.

The new FUSAG was built up by the same methods as the old: double agent reports, with heavy reliance on BRUTUS and GARBO, and fake wireless transmissions, particularly a series of wireless exercises representing the notional assault divisions training with their associated naval forces. To avoid detailed reporting of the departure for the front of many old FUSAG units in mid-July, BRUTUS and GARBO were taken out of circulation for a few days; BRUTUS by being sent on an assignment to Scotland, GARBO by being arrested and briefly detained for showing suspicious interest in V-bomb damage. (This led the Abwehr to relieve them both of V-bomb reporting responsibilities.) On his return from Scotland, BRUTUS began feeding the "story" on July 19.

ULTRA and MAGIC continued to show the Germans waiting for the second landing. Churchill sent to Eisenhower a July 15 Oshima decrypt with a note expressing his "feeling . . . that a part of the Patton Army Group might do more good by remaining in England than by coming immediately into the OVERLORD bridgehead. . . . Uncertainty is a terror to the Germans. The forces in Britain are a dominant preoccupation of the Huns. The question must be considered whether, once they know where these are going or have gone, they will not feel free to liberate greater forces than those which we now menace simultaneously at so many points."

Churchill's "feeling" was beginning to be shared by the "Huns" as FUSAG divisions began turning up in the bridgehead. Roenne noted in his situation report of July 10 that FUSAG was now "undoubtedly inferior to that of Montgomery," and speculated that "The massed concentration of Montgomery's army group and its reinforcements from Patton's army group suggests that the latter army group has not been given the decisive role," though it was still likely to assault the Seine-Somme coast. The *Pariser Zeitung,* the German newspaper published in Paris during the occupation, carried on the same front page that told of

the July 20 attempt on Hitler's life a story headlined "PATTON"-DIVISIO-
NEN IM BRÜCKEKOPF, gloating that the appearance of units from the
"southeastern invasion army" "shows to what extent the German defense
in Normandy has used up the enemy forces."

Perceiving Rundstedt as defeatist, Hitler had sacked him as OB West
at the beginning of July and brought in Field-Marshal Hans von Kluge,
a tough Prussian with much experience on the Eastern Front. Rommel
left the scene when injured by an Allied air attack on July 17. Before his
death by forced suicide in October, he told his son that he had realized
that it had been "a decisive mistake to leave the German troops in the
Pas-de-Calais." Though Kluge's staff on July 18 saw "no grounds for
changing our opinion" that when the time was ripe "First U.S. Army
Group can be launched against Fifteenth Army," within another week
they were beginning to wonder, and by July 27 they saw it as increasingly
likely that "the enemy are in fact contenting themselves with the landing
in Normandy." Even Roenne began falling into line on July 31, announcing that "a second major landing on the Channel coast no longer
seems to be so probable in view of the situation in Normandy." On August 4, he considered it "unlikely that early new landings of extensive
scope are envisaged." And on August 7, a broadcast from Berlin quoted
German military spokesmen as saying that "Large-scale landing operations by the Allies need no longer be reckoned with."

Jodl said after the war that he had regarded the Pas-de-Calais threat as
over when Third Army units began appearing in the bridgehead, although Hitler continued to believe in a second landing and to hold Fifteenth Army back. But this did not last long. On July 27, OKW began
authorizing the release of Fifteenth Army divisions to Seventh Army—
two that day, another on July 31, another on August 1; by mid-August
there would be only a dozen divisions left north of the Seine.

On July 25 the great Allied breakout offensive, Operation COBRA,
opened with a stunning bombardment of the German left wing by
massed formations of Flying Fortresses. McNair's participation in FORTI-
TUDE SOUTH II ended in tragedy that day, for while watching the attack
he was killed by misdirected bombs. Eisenhower designated Simpson of
Ninth Army as temporary acting commander of FUSAG. To replace
McNair, Marshall promptly dispatched another prestigious senior officer, Lieutenant General John L. DeWitt. Though less well known than

McNair, DeWitt had a reputation for the type of toughness the Germans could be expected to understand, for it was he who had supervised the removal of Japanese-Americans from the West Coast in 1942. DeWitt arrived in England amid suitable publicity on August 6.

On August 1, both 12th Army Group and Third Army were officially activated, and General Patton—the real General Patton, the larger-than-life, utterly non-notional George S. Patton, Jr.—and his great Third Army began the historic breakout to destroy the German Seventh Army and race across France until they ran out of gas. Though the reconstituted FUSAG would continue to be dangled before German eyes and would play some role in deceptions on the Continent in coming months, FORTITUDE SOUTH was essentially over.

And the Germans never tumbled to it. In the summer of 1945, Eldredge and Cuthbert Hesketh searched through German files and interviewed German officers to determine the effect of the great deception. Eldredge talked with Blumentritt, Rundstedt's chief of staff. "He gave Rundstedt's reasoning about the Pas-de-Calais in detail; it was a fantastic exhibition," reported Eldredge to Newman Smith. "I never expected in my life to hear Blumentritt repeating FORTITUDE." *

Given the enormous stakes, it was perhaps the most successful strategic deception of all time. True, it had a lot of help from Roenne's wild overestimate of Allied strength and perhaps from machinations of anti-Hitler officers who may not have been reluctant to see the Western Allies prevail. But all successful deceptions play up to the enemy's preconceptions. And the basis of almost all successful deceptions in the Second World War was false order of battle. Mere gross numbers of divisions would not have carried the weight they carried with commanders like Rommel without the logical structure imposed by the QUICKSILVER con-

---

* Blumentritt seems to have changed his tune. At another interrogation he claimed that Rundstedt was sure by June 7 that Normandy was the actual invasion, that the capture in the opening days of the American VII Corps's field order and the V Corps's operation order confirmed this, but that Hitler continued to fear a second invasion until the beginning of August, and held Fifteenth Army back. This is inconsistent with other evidence of Rundstedt's state of mind and probably reflects the tendency of German officers to blame all their mistakes on the Führer. The captured orders simply dealt with the two corps's operations in Normandy; they would have done nothing either to disprove or prove the proposition that those operations were a large-scale feint rather than the main thrust.

cept of two separate identifiable army groups, seemingly with two separate clearly defined missions. Special means carried out this concept brilliantly; but the concept itself was due to David Strangeways, and he designed it and supervised its implementation. It was he as well who saw that the attack on the Pas-de-Calais must be presented not as a diversion but as the main offensive; it was he who swept aside thoughts of wasting effort on extensive dummies and false troop movements; and he who saw to it that the basic plot of FORTITUDE SOUTH II would keep the plan going.

Just as no one man or group of men can claim credit for the success of OVERLORD, so no one man or group of men can claim credit for the success of FORTITUDE SOUTH. Hesketh's meticulous planning of special means details, and his firm stand against prematurely mentioning the Pas-de-Calais; the work of the case officers, Tommy Harris and Hugh Astor, and the rest of B1A, all played their part. But, bearing in mind the bloodbath that the invaders suffered on Omaha Beach when confronted merely by one division whose presence had not been detected, and the tough fighting in the bridgehead during June and July, it may be doubted whether OVERLORD would have succeeded if the German Fifteenth Army had not been kept idle for those crucial weeks even when allowance is made for the comparatively low quality of most of its divisions and the extreme difficulty that would have been encountered in moving them to Normandy. And much of the credit for keeping it idle must go to the idea of QUICKSILVER. It is not unfair to say, therefore, that David Strangeways—brash, self-confident, impatient, and withal impishly humorous—contributed as much as any one man to the liberation of Western Europe; and for this he deserves to be remembered.

CHAPTER 14

Mediterranean Finale

Eisenhower realized in the autumn of 1943, at the time of the report on deception directed by the Combined Chiefs, that all the Americans with experience under Dudley Clarke—Goldbranson, Train, and the rest—were reserve or National Guard officers. He told Bedell Smith to assign a young regular officer to learn the trade from Clarke so he could carry that know-how forward into the postwar service. For no discernible reason the choice fell on a thirty-year-old quartermaster and ordnance officer, Major Eugene J. Sweeney.

Though born in Chicago, Gene Sweeney was a Michigander by ancestry and upbringing. His Irish grandfather was one of the pioneers who cleared the land in the "Irish Settlement" in southwestern Michigan. His father had been the first white boy born in the onetime Potawatomi Indian country. From an early age young Gene had dreamed of a military career, though his father, a lawyer, was not enthusiastic over the idea; and after a year of civilian college he secured an appointment to West Point. A popular member of the class of 1938 (his nickname in the yearbook was "Love Bug") who by 1943 sported a large mustache to add years and dignity, he had reached Scotland in the summer of 1942 in the first full convoy from the States to Britain, and had been one of the early arrivals in North Africa in November. There he set up shop at an ordnance depot in Oran where vehicles were assembled; from which work he was summoned one day in the late autumn of 1943 to report to Bedell Smith at AFHQ in Algiers.

Major Sweeney presented himself with some trepidation to Ike's formidable chief of staff. Smith looked Sweeney over; apparently satisfied, he told Sweeney what Eisenhower wanted, and directed him to report to "A" Force headquarters, keep his eyes and ears open, learn all about what they did, and be prepared to report fully on it in due course. "And I didn't have any idea what 'A' Force was like, nor what its particular purpose was," remembered Sweeney. "I had some vague idea it was a highly classified operation. That was about it." So he betook himself to the "A" Force villa and reported to Colonel, shortly to be Brigadier, Clarke; and went on the "A" Force roster officially on November 29.

Sweeney was an unpretentious straightforward Midwesterner, as Carl Goldbranson had been. Clarke was correct with him, and though there were a couple of moments of minor friction between Sweeney and "the Brig," as Clarke came to be known to the Americans, Clarke never seems to have undercut him as he had Goldbranson, and indeed came to respect the aptitude for the business that Sweeney showed. But he did not welcome Sweeney into the inner circle as he had welcomed Ekstrom, Fairbanks, and Train. Sweeney was billeted with a French family and ate at the AFHQ officers' mess (along with Bromley-Davenport and others) rather than at the table in Ekstrom's garden. Sweeney did not let himself be troubled by this. "I had this feeling that initially I was on probation and being ordered around. And I went along and did as I was told because that's what I had been ordered to do, to learn what was going on. . . . I wasn't looking for promotion, I wasn't looking for publicity, or anything else. Just to do my job."

Clarke's style was to throw a new man into the work and let him learn

Eugene Sweeney

by doing. After a short time in Algiers to get Sweeney acclimated, Clarke took him to visit Tac HQ in Italy. There Sam Hood treated him to the only downright rudeness he ever encountered in his dealings with the British. At dinner in the mess, Sweeney's Irish background came up. The Irish aren't worth very much, Hood declared. Sweeney suppressed his normal Hibernian reaction—from the day of his arrival in Britain in 1942 he had had drilled into him Ike's rule that you got along with your allies no matter what—and tried to soothe the situation by observing that on his mother's side he was Scotch-Irish. The only people more worthless than the Irish are the Scotch-Irish, replied Hood. "Which was a hell of a way to start a friendship," observed Sweeney succinctly fifty years later. Fortunately, he did not have to deal with Hood's rudeness for very long, spending more time with Max Niven, who knew how to be polite to a guest. He spent Christmas in Italy in the congenial company of the American 3d Division, attending midnight Mass on a stormy night in a village church with its roof half blasted away.

Then he returned to Algiers to get ready for his first assignment: To take charge of the radio deception element of the deception plan for SHINGLE, the Anzio landing.

SHINGLE, finally ordered on December 26 after some uncertainty, was an effort to break the stalemate along Kesselring's Gustav Line running across Italy north of Naples, by landing behind it at the beaches of Anzio, not far from Rome and close to the German supply line along the Appian Way.

As was described in Chapter 9, FAIRLANDS, the deception plan that "A" Force had operated in the autumn, was transformed in November into a larger operation called OAKFIELD. OAKFIELD was originally designed with the dual object of inducing the Germans to withdraw forces from the main line to guard against threatened landings on both sides of the Italian peninsula, and lowering their vigilance in the eastern Mediterranean in aid of a planned Allied attack on Rhodes. When SHINGLE was decided upon, OAKFIELD was enlarged into a major deception in support of that operation.

Under the final OAKFIELD "story" worked out by Clarke after visits to Alexander's headquarters at Bari and Mark Clark's at Caserta (where he saw *Five Graves to Cairo* and began thinking about COPPERHEAD), a landing would be made on the west coast of Italy near Pisa, far to the north of Anzio, by the American 3d Division and a British division, both setting out from Naples, with a followup by two French divisions from Corsica and Algeria; and a landing on the east or Adriatic coast would be made near Rimini by II Polish Corps from the Middle East, along with the American 82d Airborne and British 1st Airborne Divisions from Sicily, and the American 88th Division from North Africa; with the notional British XIV Corps from North Africa as followup. The whole venture would be under Patton's American Seventh Army.

The plan took several real problems neatly into account. The "Pisa assault force" was in fact the force genuinely earmarked for the Anzio landing; the "story" would account for their being assault-loaded at Naples. The Polish corps was in fact leaving Egypt for the Italian front; the "story" would account for their departure. Especially useful was a Royal Navy force built around three battleships and the aircraft carrier *Illustrious,* which was passing through the Mediterranean bound for the Japanese war. (This was the force that gave rise to DUNDEE, the London Controlling Section's plan with the "story" that its ultimate aim was an operation against Timor.) Initially there had been concern lest the appearance of this flotilla in the Mediterranean might alert the Germans to a landing in Italy. Now it was put to use by depicting it as the covering naval force for the Rimini landing. Moreover, the chief implausibility of a Rimini landing was that it was out of range of Allied supporting aviation. The presence of the aircraft carrier cured that.

OAKFIELD got a full-dress implementation. Near the notional landing areas there were reconnaissance flights, beach reconnaissance, and an occasional bombardment. Maps of the putative target areas were widely issued, as were pamphlets, some in French, on preserving works of art. Troops, airplanes, and landing craft were moved into Corsica, along with substantial construction and extensive camouflage activity. Airborne training was expanded in Sicily and all available gliders were concentrated there. The wolf's head insignia of the fictitious British XIV Corps began to be seen from Algiers to Tunis, and invitations to the corps headquarters Christmas party turned up in Spain and Turkey.

Patton's Seventh Army was now a mere skeleton headquarters and he himself was under a cloud for the soldier-slapping incident, his future in jeopardy; but he threw himself cheerfully into the deception. Radio activity depicted an active Seventh Army preparing for great events. Patton did a conspicuous tour of the Mediterranean with his staff—Corsica, then Malta, then Cairo on December 12, where he spent a week touring historic sites, lecturing the officers of the garrison, visiting the naval base at Alexandria, and, most notably, reviewing the Polish troops soon to embark. A photograph of Patton at the side of the Polish commander, General Anders, with Anders sporting the shoulder patch of Seventh Army, found its way to an Istanbul picture magazine.

Henry Hopkinson in faraway Lisbon even made a small contribution. A British embassy officer called on the pro-German counselor of the papal nuncio's delegation in Lisbon saying that he had heard rumors that there might be an Allied landing north of Rome and wondering whether his son, in a POW camp in that area, could not be moved out of danger's way. He was told that the Vatican could do nothing; but it was assumed that the incident was passed on to the Germans.

OAKFIELD gave Sweeney his first practical experience in deception. Filling in for Bobby Rushton, who was taken ill, he presided over the deceptive radio traffic that suggested intensive invasion preparations in Corsica. Meanwhile, on the tactical level Tac HQ implemented Plan CHETTYFORD, designed to support the attack on the Allied left that was intended to link up with the Anzio landing, by causing the Germans to prepare for an attack on the right.

Through it all, of course, CHEESE and the rest of the Mediterranean stable of double agents fed appropriate misinformation.

SHINGLE achieved total surprise. "Well, well, well," Bevan wrote to Clarke on D plus 1, "I suppose it is a bit too early to speak, but first indications are that you have had another roaring success." Unfortunately, there is no reason to believe that the success was owed to "A" Force. Allied command of the air was so total that the Luftwaffe could not carry out any meaningful reconnaissance; the Germans believed that the Allies were fully engaged on land; the double agent reports were generally lost in the background noise of rumor and speculation. Kesselring himself assured Jodl on January 6, and his chief of staff confirmed on January 15, that there was no reason to expect any substantial landings in the near future, while on January 21, the day before the landings, Roenne at Fremde Heere West found no "credible indications" of any deep flank attack by the Allies.

The Allies failed to follow up their initial success at Anzio, and the ever-resourceful Kesselring struck back quickly. (Kesselring always "showed very great skill in extricating himself from the desperate situations into which his faulty Intelligence had led him," wrote Alexander after the war.) The beachhead found itself isolated until late May, when the Allies finally broke through the Gustav Line and linked up with the Anzio force.

Clarke got back to Algiers from Italy on January 6 and found BODY-GUARD waiting. With OAKFIELD now launched, Clarke turned his attention to constructing an overall plan to meet as many of the BODYGUARD requirements as possible. At the outset it was something of a stopgap. "The peculiar circumstances of the Mediterranean Theater," wrote Clarke after the war, "made it a sheer impossibility to have a hiatus of more than a week or two duration between Deception Plans. By this time the natural impetus of the monster which had been nourished there all through the past three years was such that it would very soon run wild if left to itself. It could certainly not be stopped; and the danger of losing its control was so severe that any form of direction was better than none." Codenamed ZEPPELIN, the new plan was drafted in less than a fortnight, was finished in less than a month, and received Jumbo Wilson's approval on February 4.

The stated object of ZEPPELIN was to contain the maximum German forces in Italy and the Balkans, and weaken German forces south of the line from Pisa to Rimini; while doing nothing to attract German attention to the south of France until a final decision should be reached in an ongoing debate whether to go through with ANVIL, the projected landing in that area. It went through a succession of "stages." The "story" of Stage One, which was put into effect promptly, was that at Tehran the great powers had agreed that all three would launch major operations against the Balkans prior to a cross-Channel assault; this was why Wilson, with his long association with that theater (going back to his command of British forces in Greece in 1941) had replaced Eisenhower. At the subsequent Cairo Conference, the "story" ran, it had been agreed that the British (notional) Twelfth Army, including a division of Greeks, would operate against Greece, while the American Seventh Army with II Polish Corps would operate in the Adriatic. And at a further conference in Algiers at Christmastime it had been decided that if the Anzio landing succeeded there would be further landings on both coasts of Italy to support Alexander's drive up the peninsula. With a D-Day of March 23, therefore, Wilson intended to attack Crete, the Peloponnese, and the eastern shore of the Adriatic at Durazzo in Albania, while Alexander planned to land American forces in Istria at Pola, and the RAF and two armored divisions waited in Syria to move into Turkey in May.

Fundamental once more would be false order of battle. To render the ZEPPELIN "stories" plausible, Clarke reckoned that he would need twenty-four divisions over and above those actually fighting in Italy. They would be needed chiefly to threaten Greece and the Balkans, since the forces training for ANVIL could be shown as the "Istria" force, and a real force was waiting in Syria in case the Turks should decide to enter the war.

So on February 6, 1944, the long-running CASCADE became a new and even more ambitious order of battle plan, WANTAGE, with the audacious goal of inducing the Germans to believe that Allied forces in the Mediterranean were 33 percent greater than they really were—and actually 100 percent greater in the Middle East—with thirty-nine divisions outside Italy (eighteen of them notional), six notional corps, and a bogus army. In early June, after NEPTUNE D-Day, a "second edition" revised the order of battle somewhat and extended the plan to include all French forces.

ZEPPELIN in all its successive stages received the most elaborate implementation yet employed. Special means was of course extensively utilized—some 577 messages went through double agents, plus substantial contributions by Arnold and Wolfson in Turkey and Hopkinson in Lisbon. It was to be an all-out effort in support of the decisive battle. "The machine will be run to a standstill," Clarke told the handlers of double agents in a special order of the day. "Once we have entered the month of June all considerations regarding the safety of our channels (outside Italy) are to be subordinated to the demands of the plans on which we are now working and every risk accepted which can further the success of these plans."

There were the customary administrative measures, with issuance of maps and local guides and calls for interpreters. There were reconnaissance flights and raids by air and sea on the putative targets. Dummy gliders were set out in Sicily and real and dummy gliders at Lecce in the heel of Italy. Especially elaborate was the display of dummies and associated visual and W/T deception in Cyrenaica to support the idea of an invasion of Crete and Greece. Beginning in early February, Colonel Jones began depicting the bogus "8th Armored Division" at Tobruk and the equally bogus "4th Airborne Division" nearby. This display, its activity ebbing and flowing with successive notional postponements of the invasion, continued till the threat to Crete was terminated at the end of Stage Three in early May. (As things turned out, this extensive display served only as insurance, for only a handful of German reconnaissance flights reached Cyrenaica, and these were shot down by the RAF.) Dummies had also to be relied on extensively in Syria after mid-April, when negotiations to induce Turkey to join the Allies had broken down and the genuine forces that had been ready to enter Turkey began moving to Italy.

Clarke was in England for Stage One (from February 8 to March 9), coordinating plans with Wild and waiting for Bevan to come back from Moscow to discuss the next stage. (And enjoying the good life as usual— first-class hotels, clubs, restaurants, West End musicals, socializing with the Wheatleys and with the likes of Henry Hopkinson and Jock Whitney and Joan Bright.) But Bevan's Moscow trip dragged on—not till March 14 did Clarke finally met with Bevan, when the latter reached Algiers with Baumer on their way home—and ZEPPELIN moved to its "Stage Two" on March 10.

This stage was a "postponement"* with a few modifications; landings in Crete, Greece, and Albania would not now take place till April 21, and the operations of the American Seventh Army and the Poles were deferred till May 21 when they would coincide with Soviet landings at Varna in Bulgaria, while the British advance through Turkey was deferred till June, when two armored divisions would invade Thrace and march on Salonika.

---

* The Germans were apprised of the successive "postponements" through an operation called DUNGLOE. One of the London double agents told the Germans in mid-March that arrangements had been made for special Yugoslav leaders to be warned of an Allied invasion by broadcast in the BBC's Yugoslav program of the greeting *"Dragi Slusateli,"* meaning "Dear Friends." This would indicate that the invasion was exactly thirty days away. If the date had to be postponed, the special greeting would be repeated again for two consecutive days. The next appearance of the greeting would indicate that the new D-Day was thirty days off. There were, of course, no such special Yugoslav leaders.

At this point, ZEPPELIN threatened to die on the vine. It was the first major deception that "A" Force had mounted for the benefit of another theater, and Clarke found he was not getting the cooperation he needed. He appealed to "Jumbo," who rose to the occasion. "I now regard main role of Middle East—second only to preserving security in the area—as being to simulate greatest possible threat to Balkans from now till end of June," Wilson cabled to the British commander in Cairo, adding that "considerations of economy and administrative convenience should give way to this overriding need." Clarke had no further problems.

In April, Marlene Dietrich came to Algiers to entertain the troops. Clarke, the film buff, wanted to meet her. The "A" Force officers twitted him. Maybe Doug Fairbanks could fix it, they said. (They probably did not realize that Fairbanks and Dietrich had been lovers; such matters were kept quiet in those days.) Whether or not Fairbanks was responsible—her current flame was the French star Jean Gabin, serving in the Free French forces in Algiers—Marlene accepted an invitation to the mess at Ekstrom's.

Marie balked, fearing to be just the servant girl in the presence of the glamorous beauty. Ekstrom talked her around, telling her that she was fresh and young and pretty while Dietrich was just "an old blonde." So Marie put a lily in her dress and served lunch barefooted, and with her "beautiful figure and very, very pretty face" she more than held her own with Marlene Dietrich; and the party was a great success.

Ekstrom had become a captain in March. Clarke, Crichton, and Train held a little ceremony at which Crichton marched forward to his own martial music and presented Ekstrom with his second silver bars, made of tinfoil.

Another matter that April involving Fairbanks was a suggestion by Admiral Hewitt, when he received JCS 256/2/D, the new augmented charter for Joint Security Control, that Fairbanks be appointed Joint Security Control representative in the Mediterranean theater "in order to promote and facilitate efficient liaison between 'A' Force and JSC." This received short shrift from Bissell, who cabled to Wilson that no representative of Joint Security Control was authorized in any theater of operations.

*   *   *

Stage Three of ZEPPELIN began on April 14: "Postponement" of everything till May 21, with the "story" that the Soviets had asked that all Balkan invasions be timed to coincide with their own entry into Bulgaria. Stage Four began on May 9 with a drastic shift in the "story." Now, operations against Crete and Greece had been abandoned (the excuse given was a politically-inspired mutiny among Greek troops in Egypt, which was known to the Germans); the Poles would still land at Durazzo and Twelfth Army at Pola, though now not till June 19, and Thrace would still be invaded from Turkey, though now not till July or August. Two new notional operations were introduced: a minor one against Rhodes, and a major invasion of southern France by Seventh Army with an attached French corps, likewise with a D-day of June 19 (later moved to June 24), which received the separate codename VENDETTA.

Once it was decided that ANVIL would take place in August, it was settled that Clarke's goal in the western Mediterranean would be to hold German reserves in the south of France for twenty-five to thirty days from NEPTUNE D minus 5. As decided during a visit by Clarke to London the first week of May, VENDETTA involved a landing by the American Seventh Army in the area of Sète and Agde on the western French Mediterranean coast—a location chosen to be as far as possible from the ANVIL target area—for a drive inland to seize the Carcassonne Gap and exploit towards Toulouse and Bordeaux. Seventh Army—commanded since March 2 by Lieutenant General Alexander M. Patch, who succeeded Patton when the latter was transferred to Third Army in England—would be composed largely of notional units spearheaded by the notional XXXI Corps, supposedly commanded by Major General Terry Allen, the hard-driving former commander of the 1st Division.

Implementation of VENDETTA was somewhat hampered by thinness of resources. Not only were nearly all its troops imaginary, but the genuine American 91st Division was called for in Italy, and one of the genuine French divisions, the 9th Colonial Infantry, for the occupation of Elba. Most of the genuine landing craft were in England for OVERLORD. There were not even spare genuine antiaircraft guns.

VENDETTA opened in early May with a whirlwind of activity—smoke screens, dummy landing craft and antiaircraft protection, piling up of supplies—at Bône, Ferryville, and Oran, the ports from which the expedition would notionally set out. Maps and photographs of the target area were issued, instruction in basic French was offered to the troops, civil affairs directives were drawn up, and Allied diplomats asked the Spanish authorities to make available at Barcelona facilities for landing nonmilitary supplies and evacuating wounded.

A huge amphibious exercise was held from June 9 to 11, involving sixty naval vessels (including the British carriers *Indomitable* and *Victorious,* passing through on their way to the Pacific); thirteen thousand men and two thousand vehicles of the 91st were actually put afloat at Oran and kept at sea for three days, while the ether was filled with appropriate wireless traffic, heavy bombers ranged far up the Rhone, and fighters struck targets in the Sète area. On June 11 the frontiers between Algeria and Spanish Morocco were closed, and the cipher and diplomatic bag privileges of neutral diplomats were suspended. But with the 91st sailing away for Italy, the 9th French Colonial for Elba, and the British carriers for the Orient, the pretense could not be maintained indefinitely. So, beginning on June 24, the double agents began feeding the "story" that because reconnaissance had determined that the Germans had not moved their forces in southern France to the Normandy bridgehead, Wilson had decided to postpone the attack. The border was reopened on July 6, and on July 14 the Spanish authorities were advised that it seemed unlikely that the Allies would need the requested humanitarian facilities.

While he was in London in May, Clarke continued Sweeney's education by sending him to visit Advanced HQ (East) in Cairo. Sweeney stopped off in Cyrenaica to see the last few days of the big display of dummies there. The Crichtons welcomed him cordially in Cairo, showed him the town, took him to Groppi's for ice cream; he had the run of No. 9 Kasr-el-Nil, and left with a tarboosh for a souvenir, the gift of Betty Crichton.

Next, Sweeney played a key role in VENDETTA. From the end of May he was attached to Patch's staff as the "A" Force representative. Much of the planning and execution was done on the same basis as a real operation, and an effort was made to minimize the number of officers who knew the operation was phony. Clarke worked with the AFHQ planning staff, which would then issue orders to Patch as to actions he was supposed to take; and Sweeney would advise Patch's staff as to how far Seventh Army should go in carrying out these instructions literally.

*   *   *

The last phase of ZEPPELIN in the east likewise had its own codename, TURPITUDE. Its "story" was that, with the landings in Crete and the Peloponnese canceled, Wilson's only operations in that region would be the capture of Salonika by one Greek infantry and two armored divisions passing through Turkey into Thrace, followed by an advance up the Struma valley to link up with the Soviets. Turkey would be entered about June 1, following on an ultimatum to be delivered immediately after the (notional) Soviet landing at Varna in Bulgaria. A fourth infantry division, with air cover from Turkey, would assault Rhodes in August.

TURPITUDE was largely planned by Crichton in Cairo, and conducted in great measure by the headquarters of Ninth Army, the garrison force for Palestine and Syria, and by the RAF. Jones brought his "brigade" of dummies from Cyrenaica to represent the bogus "20th Armored Division." It joined the genuine 31st Indian Armored Division near Aleppo, while an extensive display of real and dummy antiaircraft and aircraft was mounted in the same region. Road-building, stockpiling, the usual phony wireless traffic and a variety of other activity turned northern Syria into "a hive of military activity," in Clarke's words, until the operation was shut down beginning June 26.

As of July 6, with VENDETTA and TURPITUDE both closed out, ZEPPELIN came to an end.

ZEPPELIN had sown its seeds upon fertile ground.

From the beginning of the year the Germans had been nervous about the eastern Mediterranean. New Year's orders to Field Marshal von Weichs, Rundstedt's opposite number for southeast Europe, warned him that Allied emphasis on a cross-Channel assault might even be cover for a main attack on the Balkans. Through January, Roenne at Fremde Heere West expressed nervousness over southeastern Europe. But over the next few months it became clear that Turkey would not enter the war (the Germans knew a great deal about Allied dealings with the Turks, for the British ambassador's valet was a German agent, the famous CICERO), while in March the last phase of the Soviet offensive in the Ukraine and the opening of an offensive by Tito's partisans in Yugoslavia diverted German attention—and persuaded Roenne briefly that if the Allies had not struck at this favorable moment there was little likelihood of major operations in the region. But by April he was concerned again, and in May Jodl's staff, relying on Roenne's order of battle figures, judged that the Allies must be readying an invasion of Greece, Albania, and the Aegean islands, with fourteen divisions or division equivalents at their disposal; a most gratifying reflection of WANTAGE and of ZEPPELIN.

As for the subsidiary operations, though VENDETTA did not take in the Germans as thoroughly as did FORTITUDE SOUTH, the bottom line came out right. The Germans noticed the heightened activity in North Africa, but to the extent they took it seriously—they suspected that a lot of it was indeed deceptive in nature—they thought that the target might be Italy or the French Riviera rather than the Sète region. MAGIC revealed that the Japanese ambassador in Vichy reported to Tokyo on June 15 that the Germans still believed that landings would be attempted both in the Calais area and in the south of France, and on June 16 that the Allies planned to land in the south of France about June 23 or 24. Jodl told interrogators after the war that he had expected landings in the south of France to precede those in Normandy, mainly because of the large concentrations believed to be in North Africa in April and May.

In any case, the goal of the operation was achieved: During the crucial month of June, of the ten German divisions protecting the French Mediterranean coast, only one was sent to Normandy. (This was 2d SS Panzer Division Das Reich, in the course of whose march north there took place the notorious atrocities of Tulle and Oradour-sur-Glâne.)

TURPITUDE assuredly attracted attention in Turkey. Local officials sent panicky reports to Ankara; the chief of the Turkish General Staff enquired about the goings-on across the border, was told they were "special exercises," and courteously asked if Turkish officers could not perhaps observe them; von Papen, the German ambassador, sent home highly nervous reports. OKW sharply discounted Papen's report, but nevertheless expected action in the Balkans. But Weichs himself was much more concerned with Tito than with hypothetical Allied attacks which he deemed improbable at best and which he could deal with with the forces at hand. So TURPITUDE had no net effect on German dispositions in the Balkans. But, as with VENDETTA, all was well that ended well; the Germans did not draw reinforcements from Weichs to send to any other front.

As Wilson cabled to London when it was over, "Plan ZEPPELIN had in general achieved the required objects."

Parallel with ZEPPELIN was a series of operational deceptions mounted by Tac HQ in Italy.

The first was CLAIRVALE, in some respects a successor to CHETTY-FORD. Designed to continue to induce the Germans to hold forces in front of the British Eighth Army while Alexander mounted an effort on the American Fifth Army's front to try to break the German line at Cassino, its "story" was that the Canadians on Eighth Army's right wing would drive up the Adriatic coast road aided by a landing at Pescara. Implemented largely by conspicuous activity at Termoli, the base from which the Pescara landing would notionally be launched, it ran through most of February, by which time Alexander had called off his effort at Cassino for the time being.

Next came NORTHWAYS, a minor W/T operation in March to cover the move of the British Eighth Army's headquarters westwards in support of Alexander's next effort to break the Gustav Line.

Far more substantial was NUNTON, the deception operation in support of Operation DIADEM, Alexander's offensive in May that at last broke through the Gustav Line at Cassino, linked up with the Anzio force, and captured Rome. Its object was to induce the Germans to reduce their resistance on the southern portion of the front, where Alexander intended to attack, and hold their reserves to the north of Rome. Its "story" was that Alexander, in line with historical precedent under which all successful conquerors of Rome had approached from the north, planned to thin out his main front and conduct merely a holding operation there, while employing his main forces in a major amphibious assault north of Rome at Civitavecchia. The notional D-Day was May 15. The real DIADEM offensive was scheduled to open four days before, on the night of May 11; it was hoped that Kesselring would hold back his reserves from Cassino to meet the threat to Civitavecchia.

NUNTON was conducted largely by the maneuvering of real forces, along with appropriate false information through special means. The land forces involved were in reality those being held in reserve to exploit the breakthrough once it came. The movement of divisions to concentrate behind the Cassino front was explained as movement towards embarkation ports. Fifth Army stationed most of its reserves at Naples and conducted vigorous training in that area, with associated radio traffic. Eighth Army conducted combined exercises at Salerno, again with extensive radio traffic. French forces on Corsica contributed to the picture. Civitavecchia received attention from air bombardment and seaborne reconnaissance, including a demonstration by the Beach Jumpers (Operation SPAM).

The official historian justly called NUNTON "after FERDINAND and HUSKY, the greatest triumph of deception in the Mediterranean." By the beginning of May, both Kesselring and Fremde Heere West were carrying a "Naples Group" of four to six divisions on their Allied order of battle. Already for a fortnight OKW had been noting agents' reports of plans for a landing at Civitavecchia, and OKW thought Civitavecchia was the objective of the "Naples Group"; while Kesselring thought landings at Gaeta or Nettuno to enlarge the Anzio beachhead were at least as likely. Either way, Kesselring was certain that when the attack on the Gustav Line opened it would only be a feint to draw his reserves away from the main amphibious assault. So he kept four good divisions in reserve to meet this phantom attack, one each at Civitavecchia, Gaeta, and Nettuno, with a panzer division positioned to move to the aid of any of them; while OKW left the crack Hermann Goering Division in position north of Rome rather than moving it to France as had been planned. After the battle almost the whole of the intelligence files of the German Fourteenth Army were captured, showing complete acceptance of the NUNTON "story."

DIADEM cracked the Gustav Line at last. Rome fell on June 4. Kesselring conducted a masterly withdrawal north, reaching the Gothic Line, running across the peninsula roughly from Pisa via Florence to Rimini, early in August.

On May 23, Alexander sent for Sam Hood to express his gratitude for the work of "A" Force. "Your contribution towards the success of this operation has been enormous," he said.

The next day, Hood left for England to join Noel Wild. Bromley-Davenport succeeded him in command of Tac HQ.

The physical side of tactical deception had its ups and downs in Italy.

Since the Americans had already employed sonic deception equipment, the British Chiefs of Staff lifted their ban on its use in late 1943, and two "Light Scout Car" companies plus a more heavily equipped unit called the "Field Park," fitted out with POPLIN equipment, were sent to Italy in late 1943 and early 1944. The ship carrying one of the companies and the Field Park was torpedoed off the Italian coast with the loss of almost all the equipment. Then the two company commanders stepped on a land mine; one was killed and the other lost a foot. After a new commander came out from England, the equipment was demonstrated for Alexander and a War Office representative. They did not like what they saw; at least not for the Italian front, where most of the fighting was in mountainous terrain already crowded with real vehicles, amid a civil population that would cause security problems. So the sonic deceivers packed up and went home, feeling no doubt that Fate had dealt them a bad hand.

An American sonic unit, the 3133d Signal Service Company (Special), one of the three trained at Pine Camp in 1943, joined the American Fifth Army a year later, in March 1945. It had been destined for the Japanese war but was diverted to Italy at the last minute. It had better luck, being used twice in the final offensive in April, with satisfactory results; once to draw enemy artillery fire and once to simulate the arrival of armored reinforcements.

British visual deceivers had better luck in Italy than their sonic colleagues. Dummy tanks were useful on a local level in the DIADEM offensive (they were not needed on a higher level, since so many real troops were employed in Plan NUNTON), and in the Anzio beachhead. In PENKNIFE, the operation to cover the departure of Canadian troops presently to be described, they imitated a Canadian corps rest area in Italy after the Canadians had left, and dummy tanks were displayed on board the ships that returned from Marseilles after conveying the Canadians to that port. The dummy tank unit conducted various simulations at the end of the campaign in connection with Operation PLAYMATE, presently to be described, some in conjunction with the American 3133d Signal Service Company.

David Strangeways sent a detachment of his "No. 5 Wireless Group" to Italy to simulate the radio traffic of XIV Corps and otherwise aid in W/T deception—the first time "A" Force had had its own special signal unit available for deception, something Clarke had wanted ever since early 1942. Among other activities, it mounted a simulation of the 42d Division undergoing winter training (Plan SNOWSHOE II), covered the advance of the 2d New Zealand Division in April 1945 as part of PLAY-MATE, and simulated the notional British XIV Corps moving to army group reserve in the spring of 1945 (Plan COMPASS).

In mid-January 1944, Clarke had acquired a new secretary, Daphne Llewellyn, the twenty-one-year-old younger sister of the Countess of Ranfurly, Jumbo Wilson's secretary. She had come out to Cairo in late September to work for Fitzroy Maclean of SOE, and after Maclean left for Yugoslavia had been hired by Clarke (whom she had never met) to work for him in Algiers. Jumbo took her along in his plane when he flew to Algiers to take up his new appointment as Supreme Commander in succession to Eisenhower. "She will not have a dull moment," predicted her sister, who was herself a great admirer of Clarke.

And indeed she did not. She found Clarke a demanding taskmaster who checked every syllable with care. He was never an early riser, and it made for a long day to have to open the office at eight o'clock, wait for him to show up at 10:30 or 11:00, take dictation till Clarke departed for lunch, meetings, and perhaps an inspiring visit to the cinema, and work again into the evening after he reappeared around five o'clock. But he was a good and amusing friend and she was very fond of him, and of such others at Algiers as Michael Crichton and Arne Ekstrom.

Daphne Llewellyn made a small contribution of her own to what might be considered a form of order of battle deception. It was still hard to find anything in the way of entertainment in Algiers; there were very few restaurants, and the sameness of the Cercle Interallié soon palled. The French frowned on levity while the homeland was under the heel of the Boche, and would only allow one regimental dance a year to be held per regiment; and since most of the troops had moved on to Italy this allotment was quickly used up. So she suggested that some regiments be invented that could put in for permission to hold dances. This was the kind of thing Dudley Clarke got great joy from, of course, and he cheerfully went along with the game; "Daphne's Dives," he called the dances that followed; the notional regiments would rent a hotel or some other suitable space, a genuine regimental band would be called in, and once a real dance band visiting to entertain the troops was pressed into service.

* * *

June and July 1944 saw some substantial reorganizations of "A" Force. On June 12, Main HQ closed in Cairo and reopened in Algiers alongside Advanced HQ (West), with Clarke himself based there and Rex Hamer in charge of the office. Cairo under Crichton was renamed "Rear HQ," the title formerly borne by the Nairobi station that had been closed in April. George Train continued in charge of Advanced HQ (West) at Algiers; and Advanced HQ (East) at Cairo was under Tony Simonds. Tac HQ, now "Tac HQ (West)," in Italy continued under Bromley-Davenport, Hood having joined Wild at SHAEF. (There was at one time to have been a Tac HQ (East) to support an invasion of the Dodecanese, but this was never needed.) Outstations continued at Gibraltar, Tunis, Naples, Beirut, and Cyprus.

Within a few weeks, AFHQ main headquarters moved from Algiers to the great royal palace of the Kings of the Two Sicilies—a respectable imitation of Versailles itself—at Caserta some seventeen miles north of Naples. "A" Force Main HQ followed suit, opening officially on July 20

(Rex Hamer)

"A" Force (*circa* 1944)

at Caserta in a house at 46 Via Roma. There was a further reorganization. George Train departed for the United States. Clarke wrote to him a generous personal letter; a sendoff quite different from the one his predecessor Goldbranson had received. After debriefing at SHAEF and in Washington at Joint Security Control, where he prepared a useful summary of deception organization in the western Mediterranean, Train returned to civil life in October. Michael Crichton moved from Cairo to Main HQ, while Rex Hamer took his place at Rear HQ in Cairo. (Hamer was promoted to lieutenant-colonel; the first he knew of it was when Clarke threw a pair of lieutenant-colonel's badges on his desk and said "Put them up.") Main HQ maintained outstations at Rome (Harry Gummer moved over from Gibraltar; his 42 Committee and the 60 Committee, chaired in Lisbon by Henry Hopkinson, were closed down; Terence Kenyon later took over from Gummer), Naples (Major Ron Harvie, a newcomer), Bari (Max Niven), Algiers (a Major William Rose), and Tunis (Major Grandguillot; this was mainly for GILBERT's benefit), with Major Druiff as a spare on Cyprus. Mure in Beirut constituted the one outstation reporting to Rear HQ.

Tac HQ, still under Bromley-Davenport and currently at Lake Bolsena, was renamed "No. 1 Tac HQ," with a new all-American "No. 2 Tac HQ" under Gene Sweeney to be activated for the campaign in France. (No. 1 Tac HQ continued in practice to be called simply "Tac HQ.") The two "Advanced HQ," that for the west at Bari and for the east at Cairo, with their outstations, were now devoted entirely to escape and evasion activity, and indeed this responsibility was to be split off from "A" Force in August. At the same time, the 40 Committee, which had supervised double agent traffic from Algiers, moved to Rome, and the channels in Algiers were gradually eliminated.

The completion of ZEPPELIN marked the end of an era that had begun when Victor Jones took over the dummy tank business nearly three years before. After TURPITUDE, Jones and his "24th Armored Brigade" packed up and left the Middle East for England.

Though Caserta remained the nominal site of Main HQ, as a practical matter the new headquarters was soon in Rome; Clarke always had an eye out for civilized comforts, and Caserta was rather isolated. Rome headquarters was in a house in a seedy little street near the Hotel Excelsior and the headquarters of the Allied Commission. True to the tradition of the Kasr-el-Nil, it had been a brothel. Daphne Llewellyn was the only woman working there, and the neighbors would cheer her from their windows when she left the building.

Ekstrom, who had if anything even more of an eye for civilized comforts than did Clarke, managed to requisition the sumptuous apartment of a former Italian cabinet minister, in a posh neighborhood with a view of St. Peter's and with the minister's excellent butler, who could trade rations with the chef at an Italian cardinal's nearby palace. He should perhaps have yielded it up to his superior officer; but, he reasoned, Clarke did not speak a word of Italian while Ekstrom spoke it fluently; surely Clarke would be happier at a hotel. So when the Brigadier was in Rome, Lieutenant Ekstrom would do his best to steer him towards a hotel, until the day came when Ekstrom had to leave for new duties in France.

It was a fiercely hot summer in Italy. In the British officers' messes they would "drink like fish," Ekstrom recalled; he himself suffered with the heat and could drink only water.

In June, Clarke had sent Freeman Thomas to Italy to look into the potential for new special means channels now that the war was definitively moving away from North Africa. As a result, several new double agents joined the Allied stable on the Italian mainland.

Much the highest-grade of these was ARMOUR. He was an intelligent forty-three-year-old aeronautical expert, an anti-Fascist who had nevertheless been engaged by Italian intelligence in 1941 to penetrate the Abwehr, and had succeeded in doing so to such an extent that he received the rank of lieutenant colonel in the German service and was engaged as a staybehind W/T agent when the Germans evacuated Rome. He turned himself in promptly after the fall of Rome and was operated as a double agent until the end of the war. He and his wife—she was also supposedly his cipher clerk—had broad access to the highest levels of Italian society, reaching to Marshal Badoglio himself, and he was able to pass valuable material of an extremely high grade.

PRIMO was a former airline pilot and radio operator, caught with two helpers behind the Allied lines in January. His helpers were sent to prison in Algeria while he operated a transmitter from a flat in Naples from February to September. Managed by the 44 Committee—Max Niven, who after a brief time with Tac HQ moved to Naples to establish the first "A" Force outstation in Italy, was its "A" Force representative—PRIMO was useful for deceptions involving alleged assault forces gathering in Naples harbor.

APPRENTICE was a White Russian, son of a Czarist officer, with Yugoslav loyalties but fiercely anti-Tito, who was parachuted in from Albania in May and promptly turned himself in. He wanted to serve in the Yugoslav Air Force but was persuaded to function as a W/T double agent instead; which he did from June to the end of the war, transmitting from Bari under the 42 Committee, again with Max Niven as "A" Force representative. He was a great success with the Germans, who not only multiplied the allowance they were paying his wife in Yugoslavia, but parachuted in to him four thousand pounds, cognac and cigarettes, and a Christmas pudding.

ADDICT was a young naturalized Italian subject of Hungarian origin, a staybehind in Rome who transmitted medium-grade information from June to the end of the war under a French case officer.

ARBITER was not an individual but the codename given to transmissions from the radio set of a pro-Fascist ring in Rome, used for deception only briefly, from June to September.

AXE was a staybehind in Florence, the twenty-seven-year-old operator of a radio shop, arrested in late August and on the air from October. After a slow buildup he became so well accepted that in January 1945 he was awarded the Iron Cross—the only agent used by "A" Force to achieve that distinction.

LOYAL was a courier who was arrested bringing in an additional transmitter for AXE. It was used briefly in November and December to pass order of battle information, and then shut down to protect AXE.

Over and above these new channels, the old reliable GILBERT kept up his steady flow of information. In January, his second in command, LE DUC, moved to Naples, at first notionally and then in fact; he eventually made his way to Corsica and ultimately back to France. LE MULET made trips to Algiers and GILBERT himself made one to Morocco.

Money was (notionally) running low in the early part of 1944 and GILBERT constantly pressed the Germans for funds. Three times the Abwehr tried to send in a courier with money; three times the plane was lost. Finally, in late July 1944, a Major Falguière, codenamed LE MOCO, was parachuted in with 950,000 francs, two transmitters, and two companions. (A "Moco" is a man from Toulon, or from the south of France generally; Falguière was a native of Sète, a onetime cavalry noncom who had worked as an accountant in French Morocco and whose main motive was probably to rejoin his family there.) LE MOCO missed his objective by ten miles and broke two ribs in his landing. He was picked up by an Italian Fascist truck driver who helped him hide his paraphernalia and took him to Tunis. There he made contact with GILBERT. The truck driver and LE MOCO's two companions were arrested.

Paillole decided that LE MOCO's arrival afforded an opportunity to dispose of LE MULET. LE MULET was given three hundred thousand francs and told that LE MOCO had brought instructions that he was to work from Algiers. LE MULET was ecstatic at the thought of being able to pay for all the girls he wanted without having to listen to GILBERT's remonstrances, but his joy was short-lived. He was arrested at the Algiers airport; and sent thence to Egypt, according to Sweeney's and Ekstrom's final report, though Paillole's memoirs seem to say he was shot as a traitor.

LE MOCO had hoped to rejoin his family in Morocco after completing his assignment as a courier. But GILBERT was favorably impressed with him, and he was eventually told at least part of the truth and used briefly for deception.

Once ZEPPELIN was terminated in early July, with ANVIL scheduled for mid-August, the goal became to reverse field and lower, rather than raise, German concerns over the south of France. Accordingly, WANTAGE went into a "third edition," the themes of which were that French strength was being cut back; a reduced Seventh Army was training for employment in Italy; while forces in the eastern Mediterranean remained at their prior level, together with a notional force to threaten the Adriatic, built around the notional Polish III Corps. Finally, on August 3, shortly before D-Day for DRAGOON (the new name for ANVIL, as of August 1), the French portion of WANTAGE was again revised so as to include all the French forces that the Germans would soon meet with on the Continent.

The ANVIL—or, now, DRAGOON—landing would be conducted by Seventh Army with American and French troops, under the theater command of AFHQ. Once the DRAGOON force linked up with Eisenhower's forces, it would become 6th Army Group under Eisenhower, composed of Patch's Seventh Army and the French First Army under General Jean de Lattre de Tassigny, and commanded by General Jacob Devers; who, it will be recalled, had moved from ETOUSA in London to the Mediterranean when Eisenhower took over for OVERLORD.* Headquarters 6th Army Group was initially activated on Corsica on July 30, disguised as "Advanced Headquarters AFHQ," and Devers was officially Deputy Supreme Allied Commander under Jumbo Wilson.

Effective July 31, after VENDETTA had been wound up, the American personnel on the deception side of "A" Force were organized into a unit formally assigned to Seventh Army on August 7, designated "No. 2 Tac Headquarters 'A' Force," commonly called "No. 2 Tac." They were understandably proud to carry the "A" Force name. No. 2 Tac had four officers and eight enlisted men. Sweeney, a lieutenant colonel from August 17, was the commanding officer. Chunky Dunn was his executive officer, and had charge of such operational matters as arranging the details of dummy parachute drops and of wireless deception conducted by the regular Seventh Army signal units. Arne Ekstrom was the intelligence officer. Bobby Rushton had gone to join Billy Harris, so administration was handled by a newcomer, Lieutenant Robert Rummel. "Bobby" Rummel was a small and personable young officer from Wyandotte, Michigan, nicknamed "Bubble Jug" for reasons nobody could remember. His duties were purely administrative and he was not generally familiar with actual plans. The four worked together smoothly and made a happy team upon the whole (aside from Ekstrom's mild annoyance at having his efficiency report made out by the literal-minded Dunn).

Clarke pressed Ekstrom to stay on with him in Italy. But Ekstrom felt that as an American officer he belonged with the American unit; moreover, he was anxious to return to his beloved France. His main current responsibility was with respect only to one agent (probably ARMOUR); responsibility for that agent was transferred to Henry Hopkinson, who

---

* When Devers came out to the Mediterranean at the turn of the year he brought General Dan Noce with him. General Lowell Rooks moved up to be AFHQ deputy chief of staff and Noce became G-3 of AFHQ. In this capacity Noce learned at first hand of Dudley Clarke's doings and quickly shed any prejudice against strategic deception resulting from WADHAM and APPENDIX Y; as did Devers himself.

in June had suddenly been shifted to Rome from Lisbon with the title of Deputy High Commissioner to Italy. It is a measure of ARMOUR's stature that he was entrusted to so exalted a figure; though admittedly Hopkinson had little else to do.

Sweeney's orders from Dudley Clarke, dated July 15, were simple and to the point. He was relieved of all "A" Force duties except in connection with ANVIL/DRAGOON. FERDINAND, the cover operation for ANVIL/DRAGOON, would be run by Main HQ "A" Force; Sweeney would ensure liaison between Seventh Army and Main HQ "A" Force. From ANVIL/DRAGOON D-Day on, he was to prepare and implement all deception plans to assist the operations of the ANVIL/DRAGOON forces, coordinating with and getting the approval of "A" Force Main HQ until such time as SHAEF should take over the ANVIL/DRAGOON forces from AFHQ.

Sweeney's final orders from AFHQ were issued on August 6, the day before No. 2 Tac was assigned to Seventh Army. No. 2 Tac's mission was "to act in an advisory capacity in all matters related to tactical deception undertaken by Seventh Army in the field," "operating under and directly responsible to the appropriate general staff section." It would have an engineer camouflage company available "when necessary for training, and in the construction of special displays in the implementation of approved plans." AFHQ stressed that it would be desirable that No. 2 Tac's commanding officer be allowed to attend operations conferences, joint planning staff meetings, and joint intelligence committees, and communicate directly with SHAEF Ops (B) on all technical matters. It was always understood, of course, that No. 2 Tac would become part of 6th Army Group headquarters once that was activated.

For the final planning of ANVIL/DRAGOON, Seventh Army headquarters had moved on July 4 from Algeria to Naples. Sweeney was busy at Naples the rest of the month with the planning for the last phase of FERDINAND, covering the landing itself. He worked closely with Douglas Fairbanks, who was responsible for the naval side of the planning, and got to know him well. "A real nice guy," he recalled in later years. "Very down to earth. I liked Doug as a man." Once he brought Fairbanks to call on the Italian family he was billeted with. *"Doo-glahss Fire-bahnk!"* cried his hostess. "She didn't believe it. . . . So I was a success with that family. After that, they thought I was wonderful."

Sweeney had a couple of minor run-ins with Clarke during the planning for DRAGOON. Once he found himself in a bureaucratic squeeze between two staff elements as a result of some organizational ploy that Clarke had under way. He appealed to the Brig. "Don't you see that you are the ham in the sandwich?" said Clarke. Sweeney was not amused. On another occasion, to obtain Seventh Army's comments on FERDINAND, Sweeney followed U.S. Army procedures, submitted it to G-3, and passed on to Clarke the staff's observations and a note that Patch thought it "logical." This evoked a starchy note from Clarke that "This was NOT sent to you to obtain the personal views of some junior officer in G-3, but to obtain the considered opinion of the C.G. as to whether there was anything in the detail of Plan FERDINAND which ran counter to his own plans." Clarke could never seem to understand that in the United States Army, majors did not approach lieutenant generals directly; nor would a sarcastic rebuke of this kind be usual in that service.

Sweeney respected Clarke's skill but did not carry away the affection for him that Ekstrom did. Clarke, on the other hand, obviously thought highly of Sweeney's skills, and he was later to intervene on Sweeney's behalf at a decisive moment.

ANVIL did not receive its definitive go-ahead until June 14. Promptly thereafter, Bevan came out to Algiers for a four-day visit to work with Clarke on preliminary planning. After some negotiation among the interested parties, the first draft of FERDINAND was approved by Wilson on July 4. It went promptly into effect, with the double agents passing appropriate information, though full formal approval of the final version was not completed till July 28.

It was an elaborate plan—almost as elaborate as ZEPPELIN itself had been. Its "story" was that the Allies had revised their Mediterranean strategy in light of the fact that, contrary to expectations, the Germans had not moved forces from southern France and the Balkans to Normandy. The new strategy, decided upon as a result of a recent (genuine) visit to the theater by Generals Marshall and Hap Arnold, entailed concentration of all American, British, and French forces from North Africa onto the Italian front. The American VI Corps (the real ANVIL/DRAGOON assault force) would set out from Naples and land behind Kesselring at Genoa at the end of July; Seventh Army, with the notional American XXXI Corps and the genuine French II Corps, would form a reserve at

Naples to exploit Alexander's anticipated breakthrough in the north. The notional III Polish Corps and British 5th Airborne Division would attack an undetermined Balkan target from southern Italy; there would be continued threats from the Middle East by the notional British Twelfth Army and by the largely notional British Ninth Army threatening to pass through Turkey.

The centerpiece of this plan—to focus German attention on the Gulf of Genoa—played to a known sensitivity on Hitler's part. ULTRA made clear that he was nervous about the Genoa-Leghorn area. In June he had ordered Kesselring to strengthen defenses there, not only with German troops but with four divisions of pro-Mussolini Italians. Fremde Heere West, on the other hand, was convinced that the area of greatest danger was the French coast between Hyères and St. Raphael—precisely the ANVIL/DRAGOON target area. Then in the latter part of July one of the double agents whom Fremde Heere West described as "a source which up to now has proved to be particularly reliable"—probably GILBERT, but perhaps ARMOUR—sent a message that began turning the German staff towards the Führer's view. Secretary of War Stimson, Jumbo Wilson, Alexander, and Patch, the message ran, had recently met and decided "to take advantage of the successes on the Italian front and make Italy the *Schwerpunkt* of operations while postponing operations against the South of France." Fremde Heere West thought at first that this might be "deception propaganda," but by the first week in August the cumulative effect of agents' messages, together with observations of buildups in Naples, Ajaccio, and Oran, had convinced even Fremde Heere West that "a landing in the area of Genoa appears more likely to us than a landing in the South of France." As for OKW, it had swallowed the basic "story" of FERDINAND soon after GILBERT's (or ARMOUR's) message, warning Kesselring to be alert for landings in his rear in the Gulf of Genoa and the head of the Adriatic, and warning Weichs to be alert for a Polish landing in northwestern Greece and Albania.

The double agent buildup of FERDINAND was supplemented by the pattern of Allied bombing along the French and Italian coasts. The actual landing beaches were in the St. Tropez-Cannes area east of Toulon. This coastline had been heavily fortified by the French before the war, so the operations planners deemed heavy preliminary aerial bombardment to be essential. But that would cut against any deception, by drawing

617

German attention to the very area they should be distracted from. It took several days of intense discussion at the end of August to work out a plan acceptable to all concerned.

Meanwhile, French Resistance activity was controlled and supported in such a way as not to compromise the actual assault area or the cover target.

In aid of FERDINAND, Operation IRONSIDE, the notional seizure of Bordeaux and drive up the Garonne that had been somewhat halfheartedly run back in May and June, was revived as IRONSIDE II.

The Abwehr had shown interest in the agent reports connected with IRONSIDE, and at Bevan's request Newman Smith had continued IRONSIDE in a modest way after the scheduled termination date of June 28. RUDLOFF reported on July 10 that WASCH advised that the IRONSIDE units, equipped with extensive engineering equipment, were held in staging areas as followup forces "for another large scale landing"; on July 12 that an engineer unit specializing in port repair had joined the force; and on July 18 that a special ship used for port repair had arrived in New York.

By this time the Mediterranean theater was pleading for a threat to the Biscay coast to supplement FERDINAND. Bevan thought such an effort would not help much. Wild at SHAEF feared an adverse effect on FORTITUDE. The Combined Chiefs agreed. AFHQ continued to plead for it. So Wild came down to see Clarke in early August, and worked up a "story" under which the Allies had decided to strengthen the French Resistance in southwestern France to the point that they might render Bordeaux and Bayonne untenable and a seaborne force from the United States might then occupy those ports to supply the Resistance with heavy equipment. Fredendall was again nominated for the notional command.

The project got only modest implementation by special means, including a message—probably through the FBI's PAT J—that much heavy army bridging equipment, including steel girders, had been coming into the Port of New York. IRONSIDE II had no effect at all; nor could it have had much, for the Germans had practically denuded the Bordeaux area by the time the plan was approved.

The DRAGOON landings were to be carried out on August 15 by the American VI Corps (3d, 36th, and 45th Divisions), setting out from

Naples, supported by an airborne drop, with French divisions following up the next day. The deceivers were delighted when "Axis Sally," the German propaganda broadcaster, welcomed the men of the 3d, 36th, and 45th, saying that she knew they were training in the vicinity of Naples for an assault against Genoa, and promising them a very warm reception.

The Naples convoys sailed through the Straits of Bonifacio between Sardinia and Corsica, rendezvousing off Corsica with convoys from Oran, Palermo, and the Italian "heel." Radio traffic at Naples and in North African ports was padded while the convoys went on radio silence. The fleet then proceeded up the west coast of Corsica, headed towards Genoa. It changed course only at the last minute, some sixteen hours before H-Hour. Then the tactical component of FERDINAND, planned largely by Sweeney and Fairbanks, took over. Their goal was to create the illusion of a "main" effort seemingly directed initially towards Sète-Narbonne and then swinging towards the Baie de la Ciotat between Marseilles and Toulon, with a "diversionary" effort between Cannes and Nice, east of the actual target area.

During the night before D-Day the eastern "diversionary" action was carried out with some shore bombardment and Commando landings. Fairbanks himself commanded an element of the force conducting the fake attacks. The "main" effort was simulated by airplanes dropping WINDOW which would appear on enemy radar as a convoy ten miles long and six miles wide, while the force conducting the feints shifted to the Baie de la Ciotat and mounted a Beach Jumper fake attack during the predawn hours of D-Day. Enemy radar's ability to see the real invasion armada was blocked by shipboard jamming equipment, supplemented by the MANDREL screen protecting the genuine airborne drop. Meanwhile, several hundred dummy parachutists with PINTAILS and rifle fire simulators were dropped behind the fake target area, with a drop of WINDOW to give the radar appearance of a flight of several hundred transport planes; the real airborne forces were hidden by an airborne MANDREL screen supplemented by jamming from Corsica. (One major radar site in the Marseilles-Toulon area was deliberately not bombed so the WINDOW displays of the diversion could be observed by the Germans.) A smaller group demonstrated briefly off Sète before joining the action at the Baie de la Ciotat.

DRAGOON achieved total surprise. The German corps commander on the scene told Allied interrogators that as late as D minus 1 he had been told that Genoa would be attacked the next day. Reserves were sent scrambling to the site of the dummy paratroop drop. On the higher operational level, Blaskowitz, the commander of Army Group G covering the south of France, had always been convinced that his front would receive the attack (and so had German naval intelligence), and from the pattern of air bombing he had judged that it would come east of the Rhone rather than on the old VENDETTA coast. But this did him little good. Even after the landings, the Germans continued for several days to brace for a second assault against Genoa. On the top level, that of OKW and the Führer himself, FERDINAND, as the official historian said, "was quite the most successful of 'A' Force's strategic operations."

In conjunction with FERDINAND, Tac HQ in Italy ran Operation OT-TRINGTON to support Alexander's attack on Kesselring's Gothic Line in August. Alexander originally planned to breach the Gothic Line in the center at Florence, driving towards Bologna, and the "story" of OT-TRINGTON was that he would in fact attack on the Adriatic end in the Rimini sector, in conjunction with the amphibious left hook into the Gulf of Genoa that was basic to FERDINAND.

OTTRINGTON was under way at full steam in early August, with extensive phony W/T traffic and over a hundred dummy tanks and supporting vehicles beefing up the right wing, when Alexander unexpectedly changed his mind and decided to attack on the right wing after all. Bromley-Davenport hastened to Caserta to confer with Clarke. The upshot was a new plan called ULSTER, the first and only full-scale double bluff ever mounted by "A" Force, under which the OTTRINGTON buildup was gradually revealed as a fake in the hope that the Germans would deduce that the main attack would come in the center after all. It was not a pleasant experience and did nothing to allay Clarke's intense distaste for the double bluff; he would not have considered it at all if it had not been generally believed that the war would soon be over. Nor did it succeed in fooling Kesselring, for a copy of the Eighth Army commander's message to his troops fell into the hands of Kesselring's intelligence, showing him that the offensive on the Adriatic sector was the main drive. No matter; he was unable to stem the onslaught. The Canadians broke the Gothic Line and Alexander advanced steadily towards the plains of Northern Italy.

In the south of France, the success of DRAGOON was followed by an extraordinarily swift advance. Toulon and Marseilles fell to the French on August 28, a month sooner than the planners had expected. Patch drove on up the Rhone. Aix-en-Provence fell on August 21, Grenoble the next day, Avignon on August 25. Blaskowitz, one of the German Army's most skillful tacticians, eluded a trap at Montélimar, but he took terrible losses and Patch's northward drive continued.

No. 2 Tac had supervised much of the deception operations for the landing, but was not to join Seventh Army in France until the lodgment was well established. It finally sailed from Naples on August 27—Clarke had said good-bye to Sweeney on August 22—and landed at St. Tropez on August 30. Sweeney caught up with Seventh Army headquarters at Brignoles, and moved with it to Grenoble on September 3. On the way he passed the appalling carnage left behind by Blaskowitz at Montélimar—dead men, dead horses, abandoned equipment, the very asphalt of the roads set afire by American shells. "The poor Jerries, I felt sorry for the bastards, I really did," he remembered. "We had to use a bulldozer just to bulldoze the stuff off the road to get through."

On September 14, Patch achieved a firm linkup with Bradley's right flank. The next day, as planned, 6th Army Group under Devers took command, consisting now of two armies: Patch's American Seventh Army, and the French Army B (renamed French First Army on September 19) under General Jean de Lattre de Tassigny. On September 20, No. 2 Tac was ordered to join 6th Army Group headquarters at Lyons. There, after a shaky start, it would prove to be the most effective of the deception organizations under Eisenhower's command.

Ekstrom did not accompany his colleagues. He had matters to wind up in Rome—presumably mainly the handing over of ARMOUR to Henry Hopkinson. He was also concerned lest he have to leave behind the expensive tailored summer uniforms in his foot locker, for it had been decreed that only one Val-A-Pak per man would be allowed with Sweeney and the others. He learned that Douglas Fairbanks was flying to the south of France, and hitched a ride with Fairbanks, foot locker and all. Once he got to France he found that things had moved so fast that it was hard to catch up with Sweeney; particularly since he had to be careful about mentioning "A" Force at all.

As for Fairbanks, the flight to France marked his departure from the theater. It was the first leg of a journey to London, and then home to the States for a new assignment in Washington. His departure left "A" Force with no naval connection. James Arbuthnott had returned to London in April. His replacement was Lieutenant-Commander the Earl of Antrim ("Ran" to his friends), "a likeable, though rather cynical man," as Rex Hamer recalled. Antrim had remained only till July, and eventually joined Peter Fleming. But at this stage of the war no naval connection was needed.

FERDINAND continued till September 8, mainly to prolong the threat to Genoa as long as possible, both to protect Patch's right flank as he advanced up the Rhone and to draw German troops away from Alexander's right near Rimini. The double agents told the Germans that plans had changed at the last minute because de Gaulle had insisted that a French army must land in France before the fall of Paris, but that the Genoa operation was still on, to be carried out by the (notional) American XXXI Corps from Naples while Alexander planned to advance on Bologna from Florence.

Clarke worked briefly on a followup strategic plan called BRAINTREE, designed to fulfil continuing obligations under BODYGUARD. But BODYGUARD was effectively terminated in late August and formally ended with Soviet concurrence on September 30. So BRAINTREE was stillborn.

In September came the mildly absurd episode of UNDERCUT already described in Chapter 2, in which a pointless deception plan was forced on Clarke in connection with the landings in Greece. It was followed in October and November by the more logical SECOND UNDERCUT, a small-scale classic order of battle plan in which the small British force entering Greece was augmented by a notional division and two more in backup reserve to become the notional III Corps.

So FERDINAND was "A" Force's last hurrah in the field of strategic deception. Except, that is, for the negative task of liquidating WANTAGE.

That proved to be a slow process. "It was with some chagrin," Clarke recorded, "that 'A' Force soon discovered that to liquidate a notional formation without endangering the secret channels required the exercise of as much ingenuity as its original creation." Little by little, units were notionally transferred to India, broken up for replacements, converted into garrison troops, transferred to England, or reconverted into static headquarters. "One and all died hard," noted Clarke, "and a variety of excuses had to be found for the disappearance of familiar names which had been carried in agents' reports for many months—in some cases for years." Under the "Third (Modified) Edition" of WANTAGE, approved in December, the only ones to remain were the bogus British XIV Corps, now consisting of the 42d and 57th Divisions, which was portrayed as a reserve corps trained in mountain warfare, for use when the Allies in Italy should reach the Alps; the notional British 5th Airborne Division, as a theater reserve to maintain a latent threat against any enemy territory within range; and the notional British 34th Division in the Middle East, able to move to Italy or the Adriatic at need.

Since these few notional units were kept alive till the end, WANTAGE was not formally terminated till May 10, 1945, after the German surrender. As Clarke wrote after the war: "Like its predecessor, CASCADE, it had been perhaps the least spectacular of all the 'A' Force plans, but beyond any doubt the two together had been the most successful and the most important of them all. For three and a half years they had formed the firm foundation upon which had been built every one of the major Strategic Deceptions." Its success had been abundantly confirmed by the German order of battle maps that fell into Allied hands with increasing frequency from the fall of Rome onwards. Almost every CASCADE or WANTAGE unit eventually found its way to them. Particularly gratifying was the puzzlement expressed by one German general in postwar interrogation as to why Alexander had not used in the crossing of the Po the bogus 5th Airborne Division, which did not even appear in the German roster till near the very end.

Along with the task of liquidating the order of battle went the task of liquidating the double agent network that had built it up.

After DRAGOON the question presented itself: What to do with GILBERT? His sources would soon all be gone from North Africa; should he simply be shut down? Considerable debate ensued among "A" Force, the French, MI6, and AFHQ. Paillole pleaded for him to be kept going, and Clarke thought the French deserved consideration for the way they had cooperated on the GILBERT case through thick and thin. So Clarke suggested sending him to France, where the French could use him for political purposes and Ops (B) could use him in strategic deception. Wild agreed to take him on.

GILBERT accordingly radioed to the Germans that he had been assigned to the Marseilles area on the staff of the commanding general of the FFI (the French Forces of the Interior, the army of the Resistance) for southern France, taking with him a (notional) wireless operator known as ROGER, notionally recruited by LE MULET. LE MOCO would open up ATLAS II from Algiers at last, while LE MULET and JEAN would (notionally) open up ATLAS III from Oran. GILBERT contacted the Germans from Marseilles on September 26. Major Grandguillot, who had overseen GILBERT for "A" Force from the beginning, was transferred to Cairo.

LE MOCO was told part of the truth. He was astounded to learn that he was working for French intelligence, but he had faith in GILBERT. He opened up ATLAS II on September 9, but it was quickly shut down; LE MOCO's transmitting was too incompetent, and it had been planned in any case to have him close down in late September, pleading that he was being watched. He remained in Algiers under arrest. ATLAS III apparently never amounted to much.

As for CHEESE, a Cairo-based channel would have little further use, and it was decided as early as May to try to move the channel to Greece once that country was occupied. So on July 6—the very day ZEPPELIN was terminated—NICOSSOF radioed to Athens that once the British returned to Greece he expected to be sent to Athens in his notional job as an interpreter, and asked that a W/T set be left behind for him if the Germans evacuated the country. Athens responded favorably. Over the next month NICOSSOF received, in dribs and drabs, elaborate instructions, codenamed by the Germans ODYSSEE, for finding a buried transmitter and one hundred pounds in gold.

In October, Clarke sent with the occupying force for Greece a team to set up CHEESE at his new base: the radio operator who had faithfully posed as NICOSSOF all these years, the former Sergeant, now Lieutenant, Shears; Lieutenant Klingopoulos of SIME, the talented former case officer for THE LEMONS, now to be case officer for CHEESE; and Major P. E. Phillips (a charming and talented artist who later became president of the Royal Watercolor Society) to open the Athens outstation for "A" Force. Clarke joined them on October 18, a day of wild excitement as the British made a full-scale entrance into the capital with Prime Minister Papandreou. Next day, Klingopoulos conducted a search for the buried treasure. But though the instructions proved to be meticulously correct, neither transmitter nor money was to be found. Perhaps the Germans had left in too much of a hurry; perhaps they had changed their mind. Whatever the reason, that was the end of CHEESE. Clarke stayed in Athens till October 24, enjoying tourism and night clubs as well as business; then sailed from the Piraeus to Haifa and flew to Cairo, arriving the evening of October 27.

The Athens venture was not wholly wasted. Two German staybehind W/T agents were found, turned around, codenamed PEDANT and EFFIGY, and operated in support of SECOND UNDERCUT for a few weeks under a 39 Committee chaired by Phillips. Then in mid-November the Athens outstation was shut down and Phillips went back to Cairo.

The great days were over, and Clarke had already begun phasing "A" Force out. Main HQ at Caserta had been shut down in October. Clarke had held a farewell cinema party the evening of October 13, made his last appearance at the AFHQ chief of staff's morning meeting on October 16, and departed on the 18th for Athens. Daphne Llewellyn eventually went to work for the Allied Commission. Tac HQ took over control of all deception in the Mediterranean theater, including the remaining network of special means in Italy and the Middle East; Michael Crichton took command of it on October 16, with outstations at Rome (Terence Kenyon), Bari (Ron Harvie, who had replaced Max Niven there), and Florence (Freeman Thomas) reporting to him. Rear HQ in Cairo would have administrative control, and Clarke joined Rex Hamer there. Two outstations reported to Cairo: Beirut, where Mure was still plugging away, and Phillips in Athens. Bromley-Davenport and Niven, who had been overseas for five years now, returned to England.

From Cairo Clarke wrote a thoughtful letter to Major General Lowell Rooks, thanking him for "all the interest and help you have given to 'A' Force ever since we first started work at AFHQ in March 1943," and thanking him in particular for "the very generous tribute you paid to our work" in a message Rooks had recently sent to 6th Army Group which had saved Sweeney's No. 2 Tac, as will be described in the next chapter. He enclosed a short memorandum which he said might be of help to American deception planning in the war against Japan, summarizing the history of "A" Force and Clarke's cardinal principles of successful deception.

Rooks passed Clarke's letter and memorandum on to General Hull, the current chief of OPD in Washington, "to further the organization of deception work in our Army." Clarke, he said, "has done an excellent job and, in my opinion, his work has been of the utmost importance and value in all of our operations to date. That we were able to secure almost complete surprise in all of our operations except the one at Salerno is a tribute to the plans he has drawn and implemented." Rooks went on to enlarge on Clarke's theories and the means employed. Hull passed Rooks's letter on to Bissell, who suggested sending it to Newman Smith; and Hull asked Rooks for copies of BARCLAY, ZEPPELIN, and VENDETTA.

In November, Sir John Dill died unexpectedly. Jumbo Wilson was appointed as his successor to head the British delegation to the Combined Chiefs in Washington. He left the theater in December. Alexander succeeded him as Mediterranean supreme commander, and made it clear that he wanted theater-wide deception with him at AFHQ headquarters. So as the second stage of the phasedown of "A" Force, Main HQ was reopened at Caserta under Crichton, now a full colonel, with a portion of Tac HQ; he took Freeman Thomas with him. The strictly tactical element of Tac HQ moved to Florence. Lieutenant-Colonel J. D. Elkington, who had replaced Max Niven at Tac HQ the previous January when the latter moved to open the outstations at Naples and Bari, succeeded Crichton as its commander. Outstations were at Bari (Harvie), Florence (Gummer), and Rome (Kenyon). Cairo was divided into "Rear HQ" and the "Cairo Rear Party."

The last months of the war in Italy saw no major activity. In early December, Crichton presided over an operation in aid of an effort to take Bologna, called SHELLAC. It was operated only briefly, largely because the actual Allied plans kept changing. For 1945, after considering and rejecting a deception policy threatening a spring advance towards Ljubljana by British forces introduced through Yugoslavia during the winter,

Alexander and the Combined Chiefs concluded that the best plan would be to discourage creation of a German central reserve, and keep alive a notional Allied strategic reserve of three divisions in Italy and two in the Middle East. The overall codename RUSTIC was applied to measures designed to induce the Germans not to form a general strategic reserve, but this was never a formal plan. And in fact the only actual operation with this goal was OAKLEAF, begun in January; it sought to persuade the Germans that the winter offensive might imminently be resumed.

Then OAKLEAF was put at risk when in February the Canadian I Corps was ordered moved from Italy to the Western Front (Operation GOLDFLAKE). To cover GOLDFLAKE, Crichton designed Plan PENKNIFE, the "story" of which was that the Canadians were being withdrawn for rest and refit prior to becoming army group reserve in Italy. There were problems coordinating this with SHAEF; and the passage of these substantial forces through Marseilles made difficulties for No. 2 Tac in France, for the Germans might well find it suspicious that their putative agents in the 6th Army Group zone had failed to see any sign of them.

The last deception in Italy was PLAYMATE, in support of the spring offensive that ended with the surrender of the German forces in Italy. Seeking to mislead the Germans as to the timing of the offensive, and to persuade them that Bologna would be outflanked from the east rather than attacked from the south, it got off to a late start and could claim no success. No great matter. The Reich was crumbling on all fronts. Three weeks after the spring offensive opened, the Germans in Italy surrendered.

The Cairo Rear HQ formed in the December reorganization was in fact a one-man outstation consisting of Major Grandguillot. Its function was to control the remaining channels in the Middle East in the hope that they might aid Fleming in India by feeding information to the Germans that they might pass on to the Japanese. At the end of November and in early December, Ronald Wingate visited Cairo on his way home from a visit to D Division, bringing Peter Thorne with him, and held a series of discussions with Clarke and Grandguillot as to what "A" Force's channels could do for SEAC. For a few months Grandguillot functioned as in effect an outstation for SEAC, but as things turned out, not much could be done, and the operation was shut down after the German surrender.

The "Cairo Rear Party" consisted of Clarke and Rex Hamer plus a modest staff. From December on, Clarke devoted most of his time to liquidating WANTAGE and working on the final report and history of "A" Force. He left Cairo for the last time on April 23, 1945. The Rear Party was disbanded. Some went to work for D Division in India. Others went home. Rex Hamer reported to the London Controlling Section, but, though Bevan welcomed him, he found that the others were not happy at the prospect of being joined by a lieutenant colonel; he finished the war with Section V of MI6 instead.

When the Germans surrendered in Italy, Michael Crichton suggested that the time had come to close "A" Force down. Alexander had been a devotee of deception since the days of El Alamein. Tac HQ had followed his headquarters through the years from Tunisia to Florence. He gave proof of his affection for "A" Force by directing that its members of all ranks be sent home in a body, to visit their families and then be disbanded.

They all flew to England in late May and dispersed for a period of home leave. Then they gathered in London, and on June 18, 1945, "A" Force formed up for the last time before Brigadier Clarke, the Master of the Game, in the converted hall of the Great Central Hotel in Marylebone. He dismissed them, and "A" Force passed into history and myth, four years and two months from its birth in a converted bathroom in Cairo in the long-ago days when Britain stood alone.

CHAPTER 15

Last Act in Europe

Returning now to northwestern Europe: Two embarrassments marked the last phase of FORTITUDE SOUTH II. First, without consulting Newman Smith, SHAEF designated real American generals as notional commanders: Major Generals John P. Lucas (the former commander at Anzio) for Fourteenth Army, Matthew B. Ridgway for XXXVII Corps, Stuart Cutler for the 21st Airborne Division, and Terry Allen for the 17th Division. Smith, who might well have been using Lucas and Allen, both of whom were in the States, in other plans, warned the British Staff Mission to clear such matters with him through Bevan. Second, and potentially disastrous, was a *New York Times* story on September 1 that gave much of the credit for victory in Normandy to Allied nurturing of German fears over the Pas-de-Calais, and especially over Patton's army. But the Germans seem never to have picked up on it.

For all its ingenuity, FORTITUDE SOUTH II came too late to affect events. But it tapered FORTITUDE off and modulated into the final phase of BODYGUARD.

On August 26, GARBO sent a long letter to Madrid telling the inside story according to his source 4(3), the American sergeant with access to high-level gossip. The Americans, the account ran, had wanted to hog the glory by letting Montgomery get stuck in Normandy and then delivering the knockout blow in the Pas-de-Calais. But Montgomery, "who has the fame of being an astute intriguer," outfoxed them by drawing the Germans to his front, enabling him to delay the Pas-de-Calais operation by drawing on FUSAG for reinforcements. Now he was trying to get the Pas-de-Calais operation canceled entirely so he could acquire for his army group the new American Fourteenth Army. This, GARBO reported, was a particularly ferocious host, filled with "convicts who were released from prisons in the United States to be enrolled in a foreign legion of the

French or Spanish type. It can almost be said that there are brigades composed of gangsters and bloodthirsty men, specially selected to fight against the Japanese, men who are not supposed to take prisoners, but, instead, to administer a cruel justice at their own hands."

At the London level, the main problem after the Normandy breakout was to dispose of this unsavory outfit. The Germans were told in August that it and the (genuine) American Ninth Army—which began appearing in France at that time—were being detached from FUSAG to constitute the SHAEF strategic reserve in France, while the (genuine) First Allied Airborne Army was added to the notional British Fourth Army to reconstitute FUSAG. Fourth Army itself was notionally shifted north to Yorkshire.

In a report that Fremde Heere West adopted almost verbatim, GARBO advised that the new FUSAG could carry out large-scale airborne operations and could occupy areas the Germans might unexpectedly evacuate. In line with this report, FUSAG was utilized in the latter part of September in connection with the Arnhem operation (Operation MARKET-GARDEN), the airborne effort to seize crossings over the lower Rhine, via reports from BRUTUS, accidentally confirmed by fanciful reports from Kraemer's fictitious source JOSEPHINE, that FUSAG was being readied for a long-range attack, perhaps in the Kiel–Bremen region. But with the failure of MARKET-GARDEN, this threat faded away for the time being.

That was the last gasp of BODYGUARD. Already in late August the British Chiefs had proposed that it be canceled and the European and Mediterranean theaters be left to make suitable deception plans for their respective areas. The Joint Chiefs concurred on September 4, the Combined Chiefs on September 14, and the Soviets on September 28.

SHAEF and London were now left with the task of liquidating the inventory of notional units left over from FORTITUDE SOUTH II. Some, like Fourteenth Army, simply vanished. It moved to France in September, was not heard of again, and was "disbanded" in October. Others were notionally broken up to supply replacement troops. But, like the bogus units of "A" Force, they died hard. Even to the end of the war, it proved well-nigh impossible to get Fremde Heere West to erase a unit once it made their list. In late October, it reckoned the British at fifty-seven divisions with a potential of eighty-nine to ninety-four, and in December, when the Allies were actually close to the bottom of their manpower barrel, it thought divisions might be brought over from Britain in response to the Ardennes offensive. As late as a month before the German surrender, the Japanese naval attaché was reporting to Tokyo that the Luftwaffe was expecting a landing in the Heligoland Bight any day.

The end of BODYGUARD saw also the escape of Bill Baumer from being "mired in the deception game." He finished his last tour with the Joint War Plans Committee in September and left OPD officially on October 31. Though Newman Smith, as Baumer noted in his diary, was "trying some of his nefarious schemes on me for a Pacific job," on November 11 he joined Ops (C) of SHAEF G-3, which was concerned with all the odds and ends of planning other than straight operations (Ops (A)) and deception (Ops (B)); though he did on occasion get involved with Wild's activities. With Kehm already at SHAEF in psychological warfare, OPD had lost its two officers most knowledgeable about deception.

Noel Wild and his Ops (B), after they took charge of deception, did not enjoy a happy relationship with either 21 Army Group or 12th Army Group. He and Strangeways, as we know, detested one another. Strangeways would simply disregard Wild. (Once when Wild expostulated, Chandos Temple told him that he could not expect Marquess of Queensberry rules down in the East End.) Wild regarded Harris's Special Plans team simply as amateurs. There was in fact a perceptible tension between Special Plans and both Wild and Strangeways; reflecting a sense that Special Plans, its experience substantially limited to FORTITUDE, simply had not had a thorough training and did not understand all the aspects of the work.

By contrast, Wild appears to have respected and trusted Sweeney's No. 2 Tac. The prim Etonian and the unpretentious Midwesterner had a close relationship and there was never any conflict; reflecting, no doubt, the fact that Sweeney and Ekstrom had trained under Dudley Clarke. No. 2 Tac's plans even *looked* like the plans Wild was used to. Sweeney, in turn, got on well with Wild. "He wasn't as personable as some people," Sweeney recalled, but "I thought he was a pleasant sort of a guy." (Ekstrom, however, disliked Wild—"that little swine"—and had no respect for him.)

* * *

Tension with Wild first emerged over the use of double agents.

ULTRA had made it clear the previous winter that the Germans were organizing a network of staybehind agents. So it was anticipated that double agents would be run on the Continent. There was some sparring in the early spring among MI5, MI6, SHAEF, and 21 Army Group over responsibility for this. The compromise, formalized on April 7, was to set up at 21 Army Group a detachment called "104 SCI Unit" for the sole purpose of running double agents, nominally under Section V of MI6 but staffed by MI5 and commanded by Christopher Harmer. It would report to Niall Macdermot, 21 Army Group's counterintelligence chief, but Strangeways would have authority to decide whether a double agent was to be used for deception, and control all his traffic. It was to organize and train this unit that Harmer had left SHAEF at the beginning of April and thereafter moved down to Montgomery's headquarters.

As for the American side, Wild confirmed in February that for running agents in the field, X-2 of OSS was the most suitable American counterpart entity to Harmer's organization. X-2's "SCI" field units, whose primary function was to provide ULTRA-derived information to the regular counterintelligence authorities, would have the additional duty of controlling double agents "in collaboration with the deception services."

For deception, the most important of the SCI units in France was 62d SCI Unit, initially assigned to Bradley's army group, within which a "Special Case Unit" ("SCU") was established for double agent work, headed by Captain (later Major) John B. Oakes. During the initial phase of OVERLORD, while American forces in Normandy were still formally under 21 Army Group, the 62d would be attached to Harmer's unit. Ultimately, after the fall of Paris, its SCU would be established in the French capital under Oakes as the controlling element for all the X-2 double agent work in France, reporting to SHAEF.

"Johnny" Oakes, a member of the Ochs family that owned the *New York Times* (his branch had changed the spelling during the anti-German fever of the First World War), was a thirty-one-year-old Princeton graduate and Rhodes Scholar who had been a reporter for the *Trenton Times* and the *Washington Post* before joining the Army. Much liked by those who dealt with him, Oakes, two other officers, and four enlisted men were trained by B1A and Harmer during the spring, and in late May Oakes went down to join the "American increment" at Montgomery's headquarters. There for the first time he made contact with Strangeways and Billy Harris, working at that stage principally with the former. Harris took a liking to Oakes; Oakes in turn came to regard Harris as in many respects his boss; and they worked together harmoniously throughout the war.

In 6th Army group territory, Ekstrom managed to assume control over all matters pertaining to potential use of agents for deception within a territory bounded by the Italian border, the Mediterranean, the Rhone, and the northern boundary of 6th Army Group. Cooperating with the American SCIs in both 12th and 6th Army Group territory was the French "TR," the counterespionage system originally set up by Paillole. There were TR posts with territorial jurisdiction, as well as TR liaison officers with the SCI units and with the headquarters of the armies. There were TR case officers for some captured agents and SCI case officers for others. In No. 2 Tac's territory, Ekstrom tried to bring some order out of this by arranging that an SCI and a TR officer be attached to 6th Army Group; this was done on paper but never in practice.

The first double agent opened on the Continent after D-Day was a Spaniard, Juan Frutos, codenamed DRAGOMAN (originally PANCHO). An interpreter for the American Express office in Cherbourg, he had been a German agent since 1936 and had a staybehind assignment to cover that port. He was caught by the Americans on July 8 through disclosures by an Abwehr defector in Lisbon. Harmer crossed to France, concluded that Frutos was double agent material, and brought Oakes over to oversee the case. Oakes brought over his colleague Lieutenant Edward R. Weismiller of the Marines (a fellow Rhodes Scholar and in civil life a Yale instructor in English), to be case officer.

A joint operation of X-2, MI5, and the French desk of Section V, called the "War Room," had been established in London, supposedly to back up the field units with the vast store of information in the London files. It proved to be more active in second-guessing than in backing up—it never sent much data across the Channel—and took throughout the war a father-knows-best attitude to the officers on the Continent that offended them profoundly. This began with DRAGOMAN. The War

Room questioned the field's judgment that he was suited for double work. They insisted that he be brought over to England for questioning. Eventually this was done; DRAGOMAN was passed; and he functioned as an active and effective double agent for the rest of the war (while in real life serving as an interpreter for the U.S. Army in Cherbourg). But, as Harmer wrote in his final report, the matter "left behind a legacy of suspicion and distrust" between the War Room and the field. The War Room, like Wild himself, particularly mistrusted the inexperienced Americans, fearing that they might do something to give away to the Germans the fact that double agents were being run at all.

When he took overall charge of deception in late July, Wild split Ops (B) into two sections, a Forward section at SHAEF forward headquarters to handle overseas plans and operations, and a Main section back at Norfolk House in London for plans and operations in the United Kingdom. Jervis-Read, Barnes, and Moody were the Forward section. Wild himself remained at Main in London, with Sam Hood, the Heskeths, and Corbett, plus Finter and Birchfield attached. This arrangement continued until SHAEF Main moved to Versailles in September, and Barnes, Corbett, and Moody returned to Washington in September and October. Their departure left Ops (B) purely British, with no one familiar with the peculiarities of the American staff system. Wild then took up station at SHAEF Main in Versailles, leaving Hood in charge at SHAEF Rear in London.

Wild made his own contribution to the resentment the War Room had already stirred up, by decreeing formally in early September that captured agents on the Continent were not to be used for deception because they were not suitable for the purpose. The men in the field—especially the Americans, who saw that Wild did not trust them—took this ruling as questioning their judgment, and when added to the behavior of the War Room it left a lingering resentment. Case officers and the deception agencies were constantly frustrated for the rest of the war by restrictions on what they could send or do with their agents. GILBERT was the sole channel that Wild would permit to be used for high-level deception.

Wild made the situation even worse by failing to clear with the army groups the traffic sent by agents under London control; or indeed to share their traffic with the field at all, "although," as Ingersoll later wrote, "occasional informal summaries were made by junketing members of the Twenty Committee . . . on the drift and content of the traffic of UK based double agents." Eventually an informal understanding was reached that material passed by SHAEF or an army group would be cleared with any other army group that might be affected. But this never extended to England. Thus, in October, GARBO sent a message advising that one of his sources had said that the initiative had passed to the Americans while the British and Canadians in the north would have only "limited and unspectacular objectives." Intended to lull the Germans in aid of the Canadian drive to clean out Antwerp, this message might obviously draw German forces to Bradley's front. Yet nobody cleared it with Harris's Special Plans or any other element of Bradley's headquarters.

Harris and Strangeways had a kindred problem in that Wild would not let them use notional units. The bogus British II Corps and American XXXVII Corps were reported to the Germans as having gone to France, and Strangeways used II Corps briefly. Then Wild reversed field and decided that it was too risky to use notional units on the Continent. As late as March 1945 he still would not let Harris use XXXVII Corps and its divisions. On the other hand, he allowed Sweeney to use essentially notional French corps in his operations ACCORDION and LOAD-LINE.

It was consistent with his overall attitude that Wild refused to adopt the "Strategic Addendum" technique which Dudley Clarke had developed to keep all stations of "A" Force current with developments and policy. Gene Sweeney with his "A" Force training was particularly surprised at this, and found that periodic meetings on the Continent were an inadequate substitute.

Wild's stubbornness was even more exasperating in light of the extensive roster of double agents soon built up on the Continent. By the end of the eleven months of the war in northwestern Europe, the French and the Americans between them had some forty-six controlled agents throughout France; while Harmer, working in conjunction with Belgian intelligence, had a number of such agents in Belgium, some of whom were in a position to cover northern France. For the Channel ports there were DRAGOMAN in Cherbourg, SKULL in Granville, and CAMOUFLAGE in

Rouen and Le Havre. In Bordeaux were KEYNOTE, GUN, and LINEAGE, while HEADLAND at Vannes could cover the south coast of Brittany. The Mediterranean coast was well covered, with FOREST, who though located inland on a farm near Draguignan could cover Marseilles, Cannes, and Nice; MONOPLANE in Marseilles along with SCOUNDREL and his sub-agent ARTHUR; plus JEST and FLORIST, run by the French for some months from Toulon and Cannes respectively. Paris had the old master GILBERT, plus KEEL, MULETEER, and later on GAT, while GAOL notionally moved from his job at the Algiers airport to a similar one at Le Bourget, the airport of Paris. In southwestern France were LITIGANT in Montpellier and the Franco-British PEGASUS in Toulouse. In eastern France there were WITCH, LEAGUE, and ATOM; in Brussels there were SNIPER, formerly in London, and FLAME, FRANK, MINT, and DEPUTY; in Antwerp there was PIP; in Bruges there was DERRICK.

In the last weeks of the war THE RHINEMAIDENS were added to the collection. This was a group of five young female Luftwaffe signal operators from the area around München-Gladbach who were recruited by German intelligence and hastily trained as staybehind agents. One of these, a 19-year-old named Anneliese Peters, turned herself and the rest of the group in to the Americans upon their arrival early in March 1945. "The girls were allowed to stay together," in Christopher Harmer's words, "and they agreed to work for the Allies and send messages under control." Peters herself, codenamed LAZY, and Ingeborg Schotes, code-named BLAZE, proved particularly useful. BLAZE was on the air from March 15 and LAZY from March 28. They were valuable for Strange-ways's Operation TRANSCEND, supporting Montgomery's crossing of the Rhine. Indeed, the Germans thought so well of their work that both BLAZE and LAZY were awarded the Iron Cross. (In its death throes the Reich was handing these out freely; WITCH, BLAZE, and LAZY were hardly in the same class with such earlier recipients as GARBO and SILVER.) In the very last days, BLAZE, who spoke English, notionally met an American captain who was not averse to fraternization, and passed on from him information about the overwhelming American strength in the region.

This was late in the war. Except for the work of No. 2 Tac and except for GILBERT, until near the end Wild clung to his refusal to allow this lineup of double agents to be used for deception. Harris and Eldredge, for example, were involved with DRAGOMAN from early on; from the time 12th Army Group was activated, Harris shared with Strangeways authority to approve agents' military foodstuff (it was Eldredge who actually did the work under Harris). They wanted to use DRAGOMAN in a variety of deception plans, all of which Wild vetoed. Wild's only concession was the grudging one that the army groups could make up their own tactical deception plans, and could use double agents to support deception that was already supported by visual means; to report, for example, the presence of a division that was already being represented on the ground by fake insignia and the like.

GILBERT arrived in Marseilles, it will be recalled, in late September, notionally a lieutenant colonel on the staff of a French general. His current French case officer, Captain Delègue, made contact with Ekstrom on September 24. They worked out a program for GILBERT to send the Germans such information as he might reasonably have gathered so soon after his arrival, together with a plan for sending him to Paris, where he could plausibly gather the high-grade material that was his stock in trade. GILBERT stayed in Marseilles longer than expected, until October 17. Old scores were being settled and collaborators were being shot out of hand, and there were plenty of people around who had known the André Latham of old, veteran of the Vichy police and sometime volunteer to fight with the Germans in Russia. But he made the rounds of restaurants and nightclubs, and chose as his drinking crony the colonel of the local FFI, who took great pleasure in announcing nightly, as he and GILBERT leaned against the bar, that GILBERT would be shot at dawn next day.

GILBERT went on the air from Paris on November 10, telling the Germans that he had had trouble getting a radio operator (ROGER was to return to Marseilles; the new operator was called Jo), and that he was now a lieutenant colonel in the Civil Affairs section of SHAEF. Wild assigned Captain Humphrey Hare of the Welsh Guards to Grandguillot's former role as the link between Allied deception and GILBERT's case officer; he was nominally attached to No. 2 Tac. Hare was an elegant esthete, a would-be poet, "a tall, lanky fellow, with a terrible sort of hysterical laugh," recalled Ekstrom. GILBERT proved to be principally a 6th Army Group instrument, the other two army groups using him only in a few instances. At bottom, however, he remained an entirely French case.

GILBERT's notional position at SHAEF enabled him to supply high-level headquarters information comparable to that which BRUTUS had furnished from England. In addition, unlike BRUTUS, at some risk of exposure he was involved in spycatching activity when the Germans wanted to parachute men and money to him. Embarrassment rooted in his collaborationist past came up continually. He was denounced by patriots and by arrested collaborators. To protect his position the French authorities had to delete information from records and indeed in some cases leave individuals at liberty who would otherwise have been arrested. The problem continued even after the German surrender, when Albert Beugras, who had first recruited GILBERT, identified him to Allied interrogators as the prize mole in Eisenhower's headquarters.

Eldredge crossed over to Normandy three or four weeks after D-Day. Until Bradley's American army group was activated on the first of August, he and the rest of the Special Plans team—bivouacked on a former German position, where the unmistakable cheese-and-sausage smell of the enemy was still discernible—continued to be attached to Strangeways's G(R). Eldredge and Strangeways's intelligence officer, Philip Curtis, worked together. They went touring the battle zone in a jeep from time to time, once running into German fire in what seemed to be a deserted village, and once coming upon the site of an armor battle with the wreckage of burned-out tanks from Curtis's old outfit strewn about. "Philip went and looked in a British tank and drew his head back, blanched, and was nearly sick," wrote Eldredge. "A 'brewed-up' tank is hardly a pretty sight."

Ingersoll and Harris came over at about the same time as Eldredge; and they were at last formally initiated into the ULTRA secret, though they had known something about it unofficially. David Strangeways took Ingersoll for a walk in the Normandy woods one day and suggested that if he was to stay in deception there was something he should be briefed on, but once he was, he could not go into combat or risk capture. Ingersoll decided that he had "already acquired a distaste for being shot at personally" and could be more useful doing deception, and agreed. So he was read into ULTRA soon thereafter.

* * *

When 12th Army Group was activated, Harris, Ingersoll, and Eldredge assumed their proper place as the Special Plans Section of Bradley's G-3. Their administrative officer was Bobby Rushton, formerly of "A" Force (by chance, his elder brother had been a Dartmouth classmate of Eldredge's); "he was a nice little guy," Eldredge recalled, "but never brought into our 'holy' discussions." He left for Joint Security Control at the end of 1944. Lieutenant Colonel Martin Bannister, who had been their signals advisor in connection with FORTITUDE, continued to play that role on the Continent. "A very, very bright guy," remembered Eldredge. "You could tell [him] to dream up something and he would take another swig of hooch and come up with another wonderful idea." Major Albert Davis was their engineer advisor, "a most gentlemanly chap," according to Eldredge, "and he was the guy whose job it was to see that Ralph's inflatable gadgets worked—a pretty difficult job."

After the breakout, Special Plans lived in tents near Laval. For working out possible deceptions, Eldredge rigged up in a converted ambulance a map room showing, based on ULTRA, what the Germans were likely to do next. "But the battle became too fluid and the Krauts couldn't keep up with Patton's gyrations—it finally came to be used by our very bright sergeant as a place to sleep." In due course Bradley divided his army group headquarters in two, a rear headquarters at Verdun, EAGLE MAIN, and an advanced headquarters, EAGLE TAC, housed for most of the period from mid-October to April in the headquarters building of the Luxembourg state railway system in Luxembourg city. Harris, Ingersoll, and Eldredge were at EAGLE TAC throughout, except for a few days at one point when Eldredge was sent back to Main. "Billy let go with a 'scream' and I was brought back," said Eldredge. Rushton, Bannister, and Davis stayed at Main. The leading trio made a good team. Eldredge was critical, Ingersoll was innovative, they would argue, and Harris would referee and needle them both.

They endured the same frustrations as did other deceivers caught in the American staff system. "The G staff is too wooden to handle this very delicate operation," wrote Eldredge long after the war. Moreover, the constituent armies handled deception matters differently. Thus, in Simpson's Ninth Army it was handled by G-3 "straight Leavenworth fashion," while in Hodges's First Army it was the province of the able and astute G-2, Colonel Benjamin "Monk" Dickson. And, particularly in the early stages, field commanders in Bradley's army group regarded cover and deception, in Ingersoll's words, as "decoration which could be superimposed upon the real plan at the last moment—as an afterthought, without adequate planning or coordination." In addition, "it was characteristic of field commanders to pass judgment on what would or would not deceive the enemy. This situation was aggravated and made impossibly difficult by the fact that corps and divisional commanders were not indoctrinated and arguments based on [ULTRA] could not be used."

Here again, Sweeney's No. 2 Tac was more fortunate. Their major deceptions required more of de Lattre's First French Army than of Patch's Seventh Army, and de Lattre was an enthusiast, putting deception plans into action with alacrity as soon as they were received. Indeed, he was running haversack ruses on his own even before No. 2 Tac stepped in.

Unlike Strangeways with Montgomery, Special Plans had no direct access to Bradley, and no way of knowing what his thinking might be. Indeed, Bradley seems barely to have been aware that they existed. Perhaps the last straw came a quarter of a century later, when, as Eldredge wrote, "Omar Bradley said to me at the 25th EAGLE TAC reunion (with his inimitable accent), 'You fellas did more than we thought in the war.'" Also unlike Strangeways, Special Plans had no executive authority. Any tactical deception plan it might manage to get through G-3 would then be carried out by 23d Headquarters Special Troops, which was under the separate command of Colonel Harry L. Reeder.

For Harris and Ingersoll, Reeder was a permanent thorn in their side. Ingersoll, who was the begetter of the 23d, had emphasized from the first that a young man with great imagination should be in charge of it. But the unit was formed in the States under the overall aegis of General Courtney Hodges, now commanding First Army, who wanted to help out Reeder, an old friend who had been relieved of another command. "And they put this old fud in command," said Harris, "and he spent the entire time that he was in Europe trying to get my job and trying to frustrate us in anything we did." "Billy was very short with Reeder," said Eldredge. Reeder, who was not permitted to know about ULTRA, the double agents, or the details of W/T traffic deception, thought "that he was just as wise or wiser and of course Billy could not tell him."

The 23d became enamored of what it called "Special Effects": shoulder patches, bumper markings, and such items, designed to fool low-grade German "line-crosser" agents. However useful such devices may have been for Dudley Clarke in the Middle East, they were largely wasted in France, where the German line-crossing operations were feeble and their W/T agents had been rolled up. Not being privy to ULTRA, both Reeder and field commanders at the division and corps level had wildly exaggerated notions of the efficiency of German intelligence, and correspondingly exaggerated notions of what "Special Effects" could accomplish. Special Plans concluded philosophically that at least they "did improve the morale and efficiency of deception troops by making them feel that they were in constant contact with the enemy."

The 23d ran nearly two dozen operations on the Continent, most of them small-scale and few with any detectable result. There were two in Normandy, neither of which had any effect on German intelligence assessments or operations: one called ELEPHANT and a more elaborate one called BRITTANY. Towards the end of August, it employed sonic, radio, and dummy means to exaggerate the armored forces in front of Brest, in hopes of lowering the will of that garrison to resist (Operation BREST); the final report called this "an example of deception for deception's sake without a valid and attainable object in view." BREST was followed by Operation DIJON, designed to distract German attention from Patton's drive on Metz, which never got off the ground. From September to early November, the 23d conducted a succession of modest operations intended to simulate specific units in one position while they actually moved to another (Operations BETTEMBOURG, WILTZ, DALLAS, ELSENBORN, and CASANOVA).

Then there was a period of idleness. Reeder fretted under it, and nagged Harris to be allowed to do something. His troops needed practice, he said. This was to merge into the disaster of the Ardennes Bulge.

It did not do Ops (B) much good to reserve major deception unto itself.

From the breakout from Normandy at the end of July to the failure at Arnhem in September, things moved too fast. After that there was a period when Wild, and to a lesser extent the army groups, suffered from the same problem that bedeviled Fleming in Burma: There was no agreed strategy for the next phase, so nobody could say what the Allies wanted the Germans to do. An unnamed project by Strangeways in early September to sell the "story" that the major Allied drive would be in the north, a similar Ops (B) plan called AVENGER in early October, and a modified version in November with the added touch of a threatened descent on the north German coast by the British Fourth Army, called AVENGER II, all were stillborn for that reason.

Late January and early February 1945 saw DERVISH, a project for dissipating the German war effort on through to the end of the war by threatening a vital center of communications, industry, and control—specifically, the Stendal-Brandenburg-Wittenberg-Magdeburg area—with a large-scale landing of airborne forces able to maintain themselves for thirty days. It too came to nothing, though Joint Security Control activated the notional 18th Airborne Division in the States for use if necessary, and offered to suggest by special means that it was going overseas. Likewise nothing came of CALLBOY, a proposal to pass off the crossing of the Rhine in the Weser sector in February 1945 as a feint to cover a real thrust against Kassel.

On March 6 Wild gave up, and told the army groups they could initiate plans to support their own operations.

In the middle of that month Ops (B) at last got to manage a deception operation—its only one after FORTITUDE—with TAPER, a small tactical deception suggesting an airborne drop to the north of Cologne to cover the actual drop in the vicinity of Wesel supporting Montgomery's crossing of the Rhine. In the same period, Ops (B) prepared a rough sketch of a plan, never adopted, to be called JURISDICTION, designed to reduce German forces north of the Rhine by telling the Germans by special means that a two-pronged attack towards Frankfurt would be mounted, one prong by 12th Army Group from the Remagen bridgehead, and one by 6th Army crossing between Karlsruhe and Mainz.

David Strangeways and his "R" Force suffered from many of the same frustrations as did 12th Army Group Special Plans.

In Normandy he had a few minor tactical deceptions with names like RAINDROP, HOSTAGE, HOSTAGE II, and TROUSERS. After the breakout, he had special assignments similar to what he had done at Tunis: Rushing into Rouen and Brussels ahead of the main force to seize intelligence material. During the autumn and winter, there were additional modest tactical deceptions, with such names as WARHORSE, HOUSEKEEPER, MAINSTAY, INCLINATION, INFATUATION, and HARLEY STREET.

In October, when Wild briefly allowed bogus units on the Continent, 21 Army Group was notionally reinforced by the fictitious British II Corps, one of the FORTITUDE formations, through Operation IMPERIL, put over mainly by GARBO and by wireless and other activity on the ground. In November and December Strangeways tried TROLLEYCAR, an operation designed to induce the Germans to hold back their reserve by threatening an attack by II Corps west of Arnhem, in conjunction with a sea and airborne attack on the north German coast by the British Fourth Army. Though there was no sign that the Germans fell into line with this story, they did grow nervous again about a landing in the Heligoland Bight.

Strangeways was briefly deflected from deception work during the German Ardennes offensive in December, when "R" Force was pressed into a combat role to defend the crossings of the Meuse at Dinant, Namur, and Givet. His next substantial deception effort, in late January, was an operation to cover Montgomery's offensive through the Reichswald. Called TURBINATE II (TURBINATE I having been canceled when the Ardennes offensive forced postponement of Montgomery's attack), its "story" was that Montgomery would attack in northern Holland to overrun the V-weapon launching sites. Put over largely by special means, including the double agent DEPUTY from Brussels with some help from London double agents, it had no more success than TROLLEYCAR had had. Wild vetoed as too risky the use again of the notional British II Corps in this operation. (Sam Hood visited First Allied Airborne Army in connection with TURBINATE II and discovered that they had produced a notional American 15th Airborne Division without consulting anyone. He alerted Wild, and appropriate steps were no doubt promptly taken.)

Strangeways's last effort of any consequence was TRANSCEND, designed to aid Montgomery's crossing of the Rhine in March by suggesting additional crossings both north and south of the real ones. Aided by both British-controlled and American-controlled double agents, this operation appeared to have succeeded; but, as Harmer wrote in his report, German intelligence was now in disruption and the operation "was in many respects a consolation race with all being allowed to join in and win a prize."

It is a measure of the lack of opportunities for deception in North-western Europe that Strangeways was able to spare a detachment of his "No. 5 Wireless Group" for service in Italy.

The story at 6th Army Group was rather different.

The first thing that Sweeney's No. 2 Tac heard on reporting to 6th Army Group headquarters at Lyons on September 20 was: "Don't un-pack, you're going back where you came from." The excuse given was that Devers was opposed to having operational elements attached to his headquarters. It was true that Devers wanted to keep his headquarters "small, lean, and mobile," in the words of one historian, but in fact Sweeney understood that this hostile reception emanated from the G-2, who considered that a special deception unit would trespass on his turf. Whatever the reason, it looked as if No. 2 Tac would be stillborn.

Sweeney cabled to Dudley Clarke for help.

Clarke in turn called for help from General Rooks. Rooks, an enthu-siast for deception, saved the day. On September 23, he sent a cable, nominally to Devers but as a practical matter to his opposite number and former deputy G-3 at AFHQ, now Devers's G-3, Major General Reuben Jenkins. "Hope this will *not* be construed as desire on our part to force an unwanted detachment upon you," he said, but "It just happens that I know that SHAEF are being asked, on behalf of [the Allied Armies in Italy] to operate deception plan in Southern France with object control-ling divisions on Franco-Italian frontier, thereby preventing their inter-vention in Italian battle." So 6th Army Group "may need retain 'A' Force detachment in France for some little additional time." There would be a meeting in Paris to discuss coordination of deception between SHAEF and the Mediterranean; "suggest you send representative to meeting and defer decision on 'A' Force detachment until then." He concluded with a heartfelt recommendation: "I may say that I have been much impressed with the effectiveness of the British deception organization, of which we never had a counterpart. I am anxious for the good of our service to de-velop this activity. I understand the detachment in question is composed entirely of American officers who have been well trained in deception methods."

The meeting referred to by Rooks was held on September 27 at the Hotel Crillon in Paris, with Bevan, Crichton, Sweeney, Wild, Hesketh, and two more Ops (B) officers. It was decided that No. 2 Tac would be brought under SHAEF's direct control, and that Wild would visit Devers's headquarters to outline deception plans and explain the role of No. 2 Tac. Sweeney explained the limited tactical deception resources available to him, and reported that Ekstrom and his counterintelligence colleagues had already rounded up five potential double agents on the Mediterranean coast. It was agreed that as a general policy, No. 2 Tac's goal in the south of 6th Army Group's territory would be to contain as many Axis forces as possible on the Franco-Italian frontier, while in the north it would be to exaggerate the thrust being made towards the Belfort Gap.

Meanwhile, Rooks's message had proved decisive. Jenkins himself had had favorable experience with the work of "A" Force when he was Rooks's deputy, and effective September 25, No. 2 Tac was formally attached to 6th Army Group with Jenkins's G-3. Like anyone trained under Dudley Clarke, Sweeney would have preferred to be directly under Devers and his chief of staff, but he knew that in the United States Army this was not to be. But he found Jenkins a congenial boss, whose sympathetic attitude overcame most of the shortcomings of No. 2 Tac's organizational position. "He was an intelligent man, a thoughtful man," he said of Jenkins years later. "He would listen to you and whatever it was, you felt you'd made your point, he understood what you were driving at." The main remaining problems were two. First, notwithstanding his final orders from AFHQ, Sweeney was not included in chief of staff's meetings or operational conferences, and this hampered his planning. Second, army group G-2 continued to be rather standoffish.

When 6th Army Group came under SHAEF and its components ceased to be under AFHQ, No. 2 Tac likewise came under the general control of Ops (B) rather than of Dudley Clarke. Sweeney had asked Noel Wild for a written directive setting forth the intended duties of No. 2 Tac and advising Devers as to "the overall cover and deception plan of SHAEF." Instead, on October 7 Wild paid 6th Army Group the personal visit discussed at the Paris meeting and never provided Sweeney with a written charter.

While Sweeney was fighting his organizational battles, Ekstrom was independently busy. When he landed in France he could not at first find

No. 2 Tac, and it was not prudent to make inquiries. So he made contact with the TR and the local OSS, and set out to organize the confused counterintelligence situation in southern France—with the French TR, the OSS, and the military authorities all rounding up suspects in a free-for-all atmosphere Ekstrom called "finders-keepers"—and prepare channels for deception. Though he shortly got in touch with Sweeney, he did not actually catch up with him until the beginning of October, at 6th Army Group's new headquarters in the Heritage Hotel at Vittel. By that time, he had established a happy relationship with all the TR and OSS case officers of agents being doubled and was in a position to review and approve all their material. Throughout the rest of the war, Ekstrom essentially managed the traffic of all the double agents operated from southern France. His skill and experience in double agent work, and his perfect rapport with the French, made him the outstanding deception operative in the European theater.

Sweeney did not realize it at the time, but just when he was getting settled in he was at risk of losing Ekstrom, his most important asset. The China-Burma-India Theater had asked for two captains for deception planning, and on October 17 Bissell cabled SHAEF inquiring about Ekstrom's availability; on the 26th, word came back to Washington that Ekstrom's services were still needed in Europe. "Phoned Haygood to dig up another captain," noted Newman Smith drily on his message slip.

Tension between the practitioners at 21 and 12th Army Groups and the War Room in London eased somewhat after the fall of Paris and Brussels. The American OSS X-2's 62d SCI Unit took over a house and two apartments in Paris, and a joint X-2 and French interrogation center was opened in the capital in the autumn, reducing the pressure from the War Room to send agents back to England for interrogation. Similarly, as Montgomery's army group approached Belgium, Harmer managed finally to get permission for his British 104 SCI Unit to start up at once any agents that might be found in Belgium. Reaching Brussels on September 6, 104 SCI Unit soon set up in that city its own interrogation facility and radio facilities in a commandeered apartment house code-named WUTHERING HEIGHTS.

Meanwhile, at Strangeways's suggestion Harmer had set up a "212 Committee"—the name reflected amalgamation of the two northern army groups—to coordinate and control agents' foodstuff and their use for deception. It met first on August 21. Originally scheduled to meet every Monday, alternately at the headquarters of the two army groups, from November it met only fortnightly, in Paris and in 1945 at SHAEF headquarters at Versailles. But with Ops (B)'s veto of using double agents for more than local tactical deception, the 212 Committee served little purpose. For the American deception officers, attending its meetings was more an annoyance than anything else, aggravated by the fact that the British representatives always seemed to outrank them. But at least the meetings afforded an excuse to visit Paris. "All we really had on our agenda," recalled Ingersoll long after, "was getting to know each other, gossiping, and then going out on the town, each on our own."

Paradoxically, American commanders' belief that "Special Effects" would somehow go straight to German intelligence went hand in hand with appalling indifference to everyday security. The same headquarters whose commander saw spies under every bed in some French village, would authorize a hundred-mile shift in position with all vehicles conspicuously marked, road signs at every intersection, and all troops proudly displaying their shoulder patches. Part of the problem was morale; the men griped over having to cut their patches off and then sew them back on.

Most serious of all was indifference to signal security throughout all four of the American armies. Signal security officers at army level were not indoctrinated in ULTRA until the final months, and "simply could not be convinced that our radio operations were leaking information," wrote Ingersoll later. No. 2 Tac made a nuisance of itself on this subject both at army group level and at SHAEF level, but was steadily rebuffed. The military police were the worst offenders. They used powerful transmitters with a 150-mile range to broadcast to their road nets the identity, strength, routes, and destinations of American units on the move, employing a low-grade cipher that the Germans readily broke. "I went through silently cursing the fact that we could not shut up the MP nets which handed our [order of battle] to the Germans on a silver platter," said Eldredge.

The Ardennes breakthrough put the fear of God into the offenders and for a time signal security showed a marked improvement; but this did not last long. Between D-Day and the spring of 1945, when signal security officers were finally indoctrinated, opened their eyes at last to what was going on, and took appropriate steps, some sixty thousand such messages were transmitted in 12th Army Group territory alone.

"In fact," wrote Bevan toward the end of the war, "during this period [after August 1944] the enemy's knowledge, as disclosed by ULTRA, of our dispositions and intentions was so good (owing to our bad security) that organized deception became not only impossible but positively dangerous." Indeed, he warned Bedell Smith that autumn that "SHAEF could not expect any material assistance from deception in crossing the Rhine unless our security materially improved."

Whatever their professional frustrations, the Special Plans trio managed to live well. Being a bachelor with no one to send him packages from home, Eldredge had arranged with a grocer back in Hanover, New Hampshire, to dispatch a parcel of good things once a month. When Cherbourg was captured, a large supply of alcoholic beverages earmarked for the Wehrmacht was captured with it; Eldredge, though but a lowly lieutenant, drew seven bottles of Cointreau from this loot. Even when they were living under tents near Laval, they entertained in style, with Cointreau and cognac liberated from the Wehrmacht and canned shrimp from Eldredge's store of goodies. Life was even better when EAGLE TAC had settled down in Luxembourg. In the evenings in the Hotel Luxembourg there was usually a party with "Ingersoll martinis" made of British issue gin, Wehrmacht vermouth—a whole truckload had been taken from the U-boat pens in Cherbourg—and lemon from the anti-scurvy packet of lemon powder included in every K-ration. On occasion Marlene Dietrich would be there, and "the party always had two or three Red Cross girls to round it off," Eldredge recalled. They and their colleagues dined well, too, although the beef carcasses supplied to the EAGLE TAC officers' mess always had the filet mignon cut out of them for the General. In April, EAGLE TAC moved to a luxurious hotel in Wiesbaden; the mess was in the main dining room, the sideboard covered with a hundred or so bottles of French and German wines, free for the asking and all marked *Reserviert für die Wehrmacht*.

Social life with the Red Cross girls and their friends was an innocent pastime for Harris and Eldredge. Harris was a married family man, while Eldredge had thoughts only for Diana back in London. The unattached

Ingersoll had a different agenda. "He was a vain man, actually," recorded Eldredge, "who thought that all girls would swoon at the sound of his voice and would, of course, leap into bed as soon as he wished." In January 1945, after the crisis of the Bulge had passed, Ingersoll "was getting restless . . . obviously starved for something with skirts. He impressed upon Billy the absolute necessity of his going to Paris—to consult with the SHAEF staff."

Social life at No. 2 Tac was infinitely simpler. No celebrities seem to have visited 6th Army Group; few journalists and writers came to call.* At army group headquarters there were stretches of time with little to do. Wehrmacht champagne was available at a nickel a bottle. One of the Army's ubiquitous PXs had been set up, and shopping was a pastime; "Dunn was always buying the latest field jacket with the latest fur-lined something or other," remembered Ekstrom. In the evenings there might be dances, with WAC officers, nurses, and Red Cross girls.

Ekstrom was anxious about his wife's health, and not in the mood for such frivolity; moreover, he was on the road much of the time. As always, he managed to make himself comfortable. Though only a captain, in Paris he often arranged to stay at the Raphaël, the smart little hotel reserved for field-grade officers. He had to move around France a good bit, and despite his low rank he rated a jeep with driver for his travels; the driver being a large Brooklyn lad known to all as Earthquake McGoon, after a hulking character in the popular *Li'l Abner* comic strip. His friends in French counterintelligence thought it was uncomfortable for him to have to drive in a jeep, what with raw weather and the mistral blowing. So they presented him with a car commandeered from the villa at Nice of the American millionaire Florence Gould; an extremely elegant vehicle, with red leather seats and a convertible top. Ekstrom in his tailored uniforms sitting in his smart car beside Earthquake McGoon at the wheel was a memorable sight.

* To this day, 6th Army Group is the forgotten orphan of the campaign in Europe. Its front was a long way from the direct line to the Ruhr and Berlin—and from Paris, where newsmen and celebrities wanted to be. During the war and ever since, Devers was overshadowed by Bradley, who was built up by the media as "the GIs' general." Eisenhower disliked and denigrated Devers, with no known rational justification; as did Bradley (of course, for he was quick to agree with Eisenhower on most things). Yet the performance of 6th Army Group was extraordinary, and most students of the military art would rate Devers far higher than Bradley as a general.

By contrast with the other two headquarters, there was little or no social life for David Strangeways's people. His headquarters followed Montgomery's tactical headquarters across northern Europe, and Montgomery was notoriously hostile to having women at headquarters or indeed anywhere in the battle zone. And Strangeways was not buried in an American-style paper-pushing staff organization. He commanded active troops, and he was often in the field with them.

Unlike their colleagues of 12th Army Group, Sweeney and Ekstrom were able to mount some substantial genuine strategic deception operations.

Only a few days after landing in France, Sweeney had received a key assignment on arriving at Patch's headquarters in Grenoble. As the invasion force moved northward, its long line of communication back to the Mediterranean was exposed to a sudden descent by Kesselring's forces swooping down from Italy through the Alpine passes. ULTRA confirmed that the Germans had no such intention for the moment, but Patch wanted to do all he could to deter such a movement. Sweeney accordingly drafted a plan for persuading the Germans that there was a substantial Allied buildup on the border, using radio, special means, and administrative measures.

Action on this project, eventually codenamed JESSICA, was deferred while the question was thrashed out whether there would be a No. 2 Tac at all. Noel Wild approved it at the Paris conference in September. By that time, the Italian front also wanted SHAEF to operate a deception plan the object of which would be to cause the Germans to maintain as many troops as possible along the Franco-Italian border and out of the fighting in Italy. This fit in well with JESSICA, and that operation was officially launched on October 3.

During most of its long life—it endured to April 29, 1945—JESSICA had to walk a narrow line: It had to suggest that Allied forces on the Franco-Italian border were so strong as to require the Germans to hold substantial forces opposite them; and had also to deter a German offensive operation, without posing such a threat as to tempt the Germans to launch a preemptive spoiling attack. The problem was compounded by the fact that SHAEF wanted the size of forces joining Devers to be minimized in order to avoid drawing the Germans to his main front. Moreover, the plan had to deal with political sensibilities, in that French troops were forbidden to operate on the Italian side of the frontier, notwithstanding French eagerness to establish French authority in the Val d'Aosta.

The basic method adopted was to report to the Germans the arrival of actual divisions at Marseilles destined to reinforce 6th Army Group, but leading them to believe that they had remained in the southern region of Devers's front; augmented by suggestions that French Resistance forces were being retrained as regular units destined for the Alpine front. Sweeney wanted to upgrade the force on the border to a notional corps. He asked for the bogus American XXXI Corps from VENDETTA and FERDINAND, but Clarke was still using it. So he asked Wild to obtain a new notional corps for JESSICA. But Wild remained adamant against any order of battle deception on the Continent.

For a fortnight after the Ardennes breakthrough in December, the "story" was that Allied forces on the border were being reduced to reinforce the north. When that threat and the subsequent German offensive on Devers's front had faded, the original JESSICA approach was reverted to; additional notional forces were found by doubling for German consumption the actual reconversion of superfluous antiaircraft troops into ground combat units.

At AFHQ's request, in its very last phase in the closing month of the war JESSICA became more aggressive in order to aid the final offensive in Italy. The genuine forces on the border were reorganized into a "Détachement d'Armée des Alpes" under a French general on March 1. JESSICA was upgraded to present an affirmative threat of an Allied offensive across the border, with the equivalent of two additional French divisions added to the notional lineup, active patrolling on the front, and authority to the French to make raids into Italy.

JESSICA was put over chiefly by double agents and by deceptive radio traffic produced by the regular signals organizations of Seventh Army. The primary special means channel was the double agent FOREST, with the double agent MONOPLANE running a distant second.

FOREST was a forty-four-year-old aviator of some distinction, Lucien Herviou by name, who lived on an isolated farm near the Riviera town of Draguignan and had been recruited by the Germans in 1943. Notionally hiding out from the Allies, he had (notionally) to bicycle into Nice to pick up information; enabling Ekstrom to fine-tune his reports and his

responses to German questions with meticulous precision. His case officer was Gordon Merrick of the OSS. Merrick was a Princeton dropout
who had acted in the original production of *The Man Who Came to Dinner* on Broadway and had then been a reporter for the *New York Post*. Es-
thetically inclined, he had established himself in a luxurious villa at
Cannes with a swimming pool, a first-rate cook, and a considerable staff;
he managed FOREST, who lived twenty-five miles away, as it were by remote control. He had no training in double agent work, took great pride
in being a case officer, and at first resisted taking direction from Ekstrom;
but Ekstrom laid down the law and Merrick became his faithful follower.

MONOPLANE was a ship's radioman named Paul Jeannin, living in
Marseilles. He reported the arrival of troops in that port through December, at which point it seemed desirable to withdraw him from JES-
SICA until he returned to that operation during its last phase in April of
1945.

From time to time minor contributions to JESSICA emanated from
GILBERT in Paris, from LITIGANT, an American-controlled agent in Perpignan, and from PEGASUS, a Franco-British controlled agent operating
from Toulouse.

From the inception of JESSICA to the end of the war, the Axis kept
three divisions on guard on the Franco-Italian border, and added a
fourth in the last weeks. JESSICA was thus another of those deception operations that came out right at the bottom line, though it is impossible
to measure how important its contribution might have been to the final
result.

Devers continued his victorious advance during November. His right
wing, de Lattre's French First Army, liberated all of Alsace except for a
pocket around Colmar. They were the first of the Allies to reach the
Rhine; they broke into Belfort on November 20 and liberated Mulhouse
on November 21; on November 26, General Leclerc of the French 2d
Armored Division took the salute of his men and the citizenry in the
Place Kléber in the powerfully symbolic city of Strasbourg.

Devers next planned an offensive with its weight on his left wing,
Patch's Seventh Army, while de Lattre would reduce the Colmar pocket
and close up to the Rhine. Sweeney was directed to produce a deception
plan for this offensive, with the object of containing the maximum num-

ber of enemy forces east of the Rhine and as far south as possible, all the
way to the Swiss border.

He responded with Plan KNIFEDGE. Its "story" was that after his remarkable performance de Lattre had been given the mission of effecting
a breakthrough in the Mulhouse sector, while Patch's attack further
north would be merely diversionary. Implementation would be by W/T
and administrative measures of the usual sort; diplomatic representa-
tions to Switzerland seeking use of a railroad line through Swiss territory
near Basel and asking that measures to prevent the escape of German war
criminals into Switzerland be intensified; propaganda designed to arouse
public opinion in Switzerland and other neutral countries against the
harboring of war criminals; canvassing by the French around Mulhouse
for guides familiar with the northern Swiss-German border and railroad
men familiar with German rail lines along the border; and, of course,
special means. KNIFEDGE got prompt approval; by 6th Army Group on
November 30 and by Noel Wild on December 2. (Wild reported that
Bevan advised that the proposed diplomatic approaches were already
under way, while the proposed propaganda campaign was contrary to
current policy.)

De Lattre himself, as has been said, was an enthusiast for deception,
and French First Army opened KNIFEDGE at once. Real troops were assembled, physical displays were begun, canvassing for guides and railroad men was started, and appropriate wireless and security measures
were instituted. Already before the attack on Belfort his TR officer, at de
Lattre's express direction, had used a line-crosser double agent to plant
on the Germans phony orders suggesting that he would bypass Belfort,
while another double agent had provided similar material to the Abwehr
in Switzerland. For KNIFEDGE, No. 2 Tac prepared a fake document directing that priority in railroad reconstruction be given to lines in the
southern sector, and passed it through First French Army's TR channel
together with some genuine but harmless papers. On a more significant
level, GILBERT from Paris transmitted messages throughout the first half
of December tailored to the KNIFEDGE "story."

The German attack in the Ardennes, followed by their offensive on
Devers's own front, changed Devers's plans drastically, and KNIFEDGE
was essentially terminated by these developments. It appears not to have
had any particular effect on German dispositions.

By the beginning of March, plans for Seventh Army to close up to the Rhine and cross it were definite enough for a revived version of KNIFEDGE, called ACCORDION, to be approved. Its object was to induce the Germans to retain the maximum number of forces east of the Rhine and south of Karlsruhe. Its "story" was that de Lattre would mount an offensive to breach the Rhine and the Siegfried Line close to the Swiss border during the last half of March, after which Patch would open a similar offensive at Strasbourg. In addition to the usual methods of implementation, SHAEF was asked to arrange for diplomatic negotiations seeking permission to use a railroad passing through Swiss territory near the German border, and for heavy bomber operations in the Basel-Karlsruhe-Ulm area. (Wild subsequently advised that Bevan reported that negotiations with Switzerland were currently impossible and that the SHAEF air staff would not commit the requested bombing resources.)

First French Army's resources were now too thin for much physical implementation. Special Plans offered to lend the 23d, and plans for their use were drawn up; but Harris then decided he needed them for one of 12th Army Group's operations. More significant, as usual, was GILBERT's traffic. He reported that de Gaulle was determined that the French would be the first to cross the Rhine, that French forces were being augmented by various notional or near-notional formations—French III and IV Corps, and the "French Expeditionary Corps for the Far East"—and that recruitment within France was heavier than was actually the case. When Seventh Army launched its actual offensive late in the month, GILBERT conveyed to the Germans the excuse that the unexpected seizure of the Remagen bridge and unexpected German weakness in the Saar had caused a change in plan and a postponement of the French attack.

Seventh Army crossed the Rhine on March 26 and First French Army on March 31. Though Sweeney proposed that ACCORDION be continued in revised form with the object of containing the maximum number of German troops in the Black Forest southwest of Stuttgart, events were moving so fast that there was little more need for it, and it was officially terminated on April 18.

Once again, the bottom line came out right. The Germans kept five to seven divisions in the Black Forest area facing only two French divisions along the southern Rhine; though the contribution, if any, of ACCORDION to this result cannot be determined.

When they were driven from the rest of France the Germans held out in various ports on the Atlantic coast, including pockets at Royan and Pointe de Grave on each side of the Garonne estuary to deny the Allies the use of Bordeaux.

As early as mid-November, a plan (Operation INDEPENDENCE) was prepared for clearing the Garonne estuary by French forces commanded by General Edgar de Larminat, under the overall control of 6th Army Group. Larminat intended to make his initial main effort on the north side of the estuary. Sweeney prepared a modest cover plan codenamed LOADLINE, the thrust of which was to exaggerate the force advancing on the south side. Directed to brief Larminat on the plan and offer it to him to use or not as he saw fit, Sweeney spent several days with Larminat at his headquarters at Cognac. Larminat turned the plan into French and accepted it substantially as drafted. But shortly thereafter, INDEPENDENCE was canceled so the forces earmarked for it could be used in reducing the Colmar pocket. It was not to be revived, under the new codename VENERABLE, until April; when Larminat succeeded in clearing the Garonne estuary in five days. For VENERABLE, Larminat designed his own plan MARÉCAGE, a fictitious threat to the Germans holding out at La Rochelle.

Meanwhile, from the turn of the year to the time of VENERABLE, GILBERT fairly steadily passed information designed to exaggerate the strength of French forces both regular and irregular, as well as appropriate misinformation as to the timing of any offensive against the Gironde pockets.

More important than LOADLINE itself was the work Ekstrom undertook in connection with it to bring some supervision and coordination to the disorganized hodgepodge of special means activity in southwestern France. There was a British SCI station in Bordeaux; there was a French TR station in Toulouse with several outstations; there was an American SCI detachment in Perpignan. All were running agents, with foodstuff wholly uncoordinated and largely unsupervised. Ekstrom found a notable lack of discipline, particularly at the British SCI headquarters; indeed, he found conditions at the headquarters villa that could be described only as decadent. He laid down appropriate directives, and installed Humphrey Hare as No. 2 Tac's representative at Marseilles to vet the foodstuff.

In late September, when First Army was planning an attack at Aachen, 12th Army Group recommended at the suggestion of Special Plans that the offensive include a deception operation designed to induce the Germans to maintain at their present level their forces further south, in the Eifel Mountains opposite the quiet sector of the Ardennes held by VIII Corps. By the time the bureaucrats of its G-3 had completed their elaborate panoply of formal written orders and directives, First Army was bogged down in vicious fighting in the Hürtgen Forest. The project shifted to a new object: Not merely to induce the Germans to hold their forces in the Eifel, but to induce them actually to shift as many forces to the front of VIII Corps as possible.

Harris protested that the timing was now too late. Reeder countered that he had a morale problem; he wanted to work his troops, who were frustrated at not having had an opportunity to mount a division-sized deception. So Army Group directed that the operation take place; "without any real hope on anybody's part for success," said Harris after the war, "but primarily to exercise the deception troops." It became known as Operation KOBLENTZ.

KOBLENTZ had to be cleared with SHAEF. On a cold day in late November, Ingersoll and Eldredge drove to Paris for a meeting on future cover plans. Ingersoll, by now a lieutenant colonel, rated accommodations in the Raphaël, and wangled Captain Eldredge in as well. At the meeting, voicing the party line from Army Group, they suggested that since "there seemed to be nothing happening in the long empty front opposite the Eifel where we were just positioning the 106th Infantry Division to get them used to a quiet front with nothing happening and nothing likely to happen," it would be good experience to create a notional force in this relatively empty space. "The Brits were charmed by this operation," Eldredge wrote, "as we were drawing troops in front of us (the Yanks)."

So Colonel Reeder got his wish, and from December 7 to December 14 the 23d worked hard on Operation KOBLENTZ, simulating the arrival of the 75th Division (a genuine unit, but one which in fact was just disembarking at Le Havre) on the "quiet front" in the Ardennes, notionally preparing to drive down the Moselle towards Koblentz. Reeder pulled out all the stops: fake radio nets, sonic deception, reconnaissances by land and air, feigned preparations for a river crossing, shuffling of real and fake artillery units, propaganda leaflets. The net effect was almost nil; Fremde Heere West speculated on December 15 that perhaps a previously unlocated division might be the 75th, but said that confirmation was awaited.

During this operation, around the 10th of December, an old friend of Ingersoll's showed up at EAGLE TAC: Ted Geisel, known professionally as "Dr. Seuss," who had drawn editorial cartoons for *PM*. Now in uniform working on propaganda films for Frank Capra, he had been in Paris to screen a new film for the generals. He wanted to see the front, but not at any great risk. Ingersoll and Eldredge talked it over and concluded that the very quiet 106th Division front would be the safest place to visit. So Geisel and his entourage set out for that sector.

A few days later, the Germans launched their great counteroffensive in the West, the "Battle of the Bulge." Their opening onslaught broke through the VIII Corps front and rolled over the 106th Division, in the "quiet sector" to which Ingersoll had sent Ted Geisel and where the 23d was mounting Operation KOBLENTZ. The 23d "stormed to the rear," firing its only real shots of the war; as for Dr. Seuss, in his own words, "The retreat we beat was accomplished with a speed that will never be beaten." (Ingersoll and friends did not know what had become of him for three months or so, when a letter arrived from Geisel back in Washington— somewhat caustic about Ingersoll's choice of tourist attractions.)

Ingersoll and Eldredge were in Paris for another meeting when the offensive broke. They got back to Luxembourg that night to find headquarters in something approaching panic. Orders went out to destroy or bury secret documents. Ingersoll had the Special Plans files buried in the courtyard, and marked the spot with a wooden cross, a helmet, and the dog tags of Kelly, their driver. "Ingersoll was in high glee," Eldredge recorded. "Kelly did not think it so funny." But Luxembourg was not threatened; the German focus was further north, their objective Antwerp.

At a famous conference held at EAGLE TAC on December 19, Bradley asked Patton how soon he could turn his Third Army north to hit the enemy in the flank; and was surprised to be told that Patton could start at once—for Patton had suspected that the Germans might have something of the sort up their sleeve and had already made plans to shift his front. For some reason, Ingersoll was there. He was presented to Patton. Patton asked what deception could do to help. The best Ingersoll could come up with on the spur of the moment was an effort to lead the Germans to believe that Patton's troops were moving northwards more slowly than was in fact the case. Harris, Ingersoll, and Eldredge put the plan, dubbed Operation ARDENNES, into effect immediately; greatly aided by the fact that for once, Special Plans was authorized to bypass the staff system and coordinate directly with the units involved.

ARDENNES was the one Special Plans operation that was put through in good measure by "special means." During the autumn, Eldredge and Captain Bittner Browne of the 62d SCI Unit, in civilian life a lawyer in Springfield, Ohio, had developed two double agents in the Metz area, WITCH and LEAGUE, Frenchmen working as German W/T agents who had been parachuted into the Verdun-Metz area and were promptly captured. The basic approach of ARDENNES was for WITCH and LEAGUE to report on the divisions that passed through their areas—WITCH was at Verdun originally and LEAGUE at Longwy, moving to Metz on December 27 and January 15 respectively—timing their passage rather later than their real movement. The agents' reports were confirmed by radio simulation of the units in locations further back than their true ones; plus, on occasion, a panoply of "special effects" by the 23d.

ARDENNES was hindered by the usual appallingly lax American radio security; efforts to squelch the leakage of information through the MP radio nets were only partially successful. Of the first two divisions to move north, the 4th Armored and 80th Infantry, the former was picked up promptly by the German "Y" Service, notwithstanding efforts to confuse them by duplicating the real divisions' transmissions by fake ones (Operation KODAK, December 22–23); the latter achieved tactical surprise. An effort to depict the 35th Division as lying in reserve while it was actually moving up was defeated by the MP radio net, which as so often in the past handed the true information to the Germans "on a silver platter." There was more success in depicting the 87th Division (Operation METZ I, December 28–31) as still in Metz when it moved out to the

Bulge. On the whole, Special Plans was moderately satisfied with the results of ARDENNES.

Harris and Eldredge leafed through the intelligence reports and felt that as cover and deception planners they should have been able to help Intelligence spot enemy cover and deception; should have deduced, for example, that the otherwise unaccountable buzzing of the front lines by German airplanes had been designed to drown the noise of tanks moving up. "We felt very guilty—especially Billy and I," wrote Eldredge long after. "I don't think Ingersoll let it bother him."

They may have felt guilty for another reason. Some officers believed that the 23d's pointless activity in KOBLENTZ might have contributed to the failure of most American intelligence officers to interpret correctly the signs of activity on the German side, thinking that it was simply a response to the 23d's playacting. Another confirmation of the K-SHELL rule: Never, ever, do deception just because you can.

It was a not a festive Christmas at EAGLE TAC. There was a dreary party in the mess that was marked by a bizarre incident, when a drunken lieutenant colonel who handled ULTRA traffic—in civil life an Atlanta lawyer—unaccountably smeared a custard pie in Eldredge's face and a fight nearly broke out. But next day Patton's men punched through to the relief of surrounded Bastogne. The tension let up, and within a few days the atmosphere was so cheerful that the French of the Deuxième Bureau in Metz proposed a joint New Year's Eve party with the OSS 62d SCI Unit. Bittner Browne invited Eldredge to come along. They drove from Luxembourg to Metz in a jeep—heavily armed, and driving largely by moonlight—arriving finally at a splendid mansion which the Deuxième Bureau had taken over from an alleged collaborator. They feasted on American turkey and the contents of the collaborator's cellar with "a beauteous and charming collection of handsomely dressed mademoiselles," who turned out to be "all the naughty little French girls who had shacked up with German officers and assorted French collaborators," and they serenaded the neighborhood through open windows with the *Marseillaise,* the *Star-Spangled Banner* "(rather badly)" and such "captured" songs as *Lili Marlene* and the *Horst Wessel Lied.*

Eldredge turned in about 2 A.M. "Shortly, I awoke with a start to gunfire and machine-gun bullets smacking on the outside of the palatial mansion and nearby buildings. Staggering to the window, .45 in hand, I fired two or three at a black shape, an Me-109 with a white cross on its fuselage, which barreled down the street just a little higher than my window. These were the only shots fired by me in anger during the whole damn war!" (It transpired later that the Luftwaffe had mounted a six-hundred-plane air strike to celebrate New Year's Day.)

Sweeney and Ekstrom likewise had a French New Year's Eve. Officers of French First Army had often asked them to visit their *popote* or mess, and they had never done so. When they were formally invited to a New Year's Eve party, they thought it prudent to accept. At that point, Devers's G-2 anticipated a German offensive on 6th Army Group's front on New Year's Day. (They anticipated correctly, and it continued through much of January.) Sweeney checked with General Jenkins. He told them to go, but to be back by six or seven in the morning. Off they went, Ekstrom suffering with a raging sore throat. When they arrived, the party had not begun. Only a few of their hosts were present; apparently the rest were scouring the neighborhood for mademoiselles and *foie gras*. Time passed, and still no party. By midnight Sweeney decided they had to get back. When they went outside they found that their hosts had covered the hood of their jeep with bottles of champagne and fruitcake liberated from the *Wehrmacht*. They stood in the cold and partook freely. It did wonders for Ekstrom's throat.

Special Plans was involved in a minor way with the German secret weapons that winter.

The V-1 flying bombs or "buzz-bombs"—cruise missiles, in today's parlance—had begun falling on London a few days after D-Day. The Germans had asked GARBO and other agents to report where and when they fell. After vigorous debate over the propriety of doing so, the double agents were utilized to report that the missiles were overshooting central London, in order to induce the Germans to shorten the range so the bombs would in fact fall further east in less densely built-up areas. While the records are incomplete, it is tolerably clear that the same policy was followed with the V-2 rocket bombs—ballistic missiles, in today's parlance—that bombarded London from early September to late March.

The policy was repeated when Antwerp too became a target for German missiles after it was liberated.

Liège, in the 12th Army Group zone, was severely battered as well. In December and January Special Plans repeatedly sought clearance from SHAEF to install a double agent in that city to send similar reports. But Wild did not give permission for this till March, by which time the bombardment had slackened; moreover, by that time rear installations of First Army had been set up in the area where bombs would fall if the range was shortened. So, though tentative arrangements were made to put one of Harmer's double agents in the Liège area for that purpose, the project never came to fruition.

The Allies ground away at the Bulge during January 1945. A succession of small tactical deceptions supported particular moves on the 12th Army Group front in January and February, most of them aimed at showing a division in one place as it moved in for an attack somewhere else.* All were minor, with but modest results. Some were supported by the double agent WITCH. The Germans valued his services enough to award him the Iron Cross on February 10. In March, with the Bulge now but a memory and the Allies crossing the Rhine, there were similar operations with names like LOCHINVAR, BOUZONVILLE, and VIERSEN, the last-named in aid of a Ninth Army operation called EXPLOIT, a feint to cover its crossing of the Rhine.

In the last phase of the war an additional American army, the Fifteenth, was activated in Bradley's army group. Designed eventually for the occupation of Germany, its initial duties were to take over the forces besieging Atlantic ports in which the Germans were still holding out, and after the Rhine crossing to relieve Bradley's other armies of patrolling and mopping-up on the left bank of the river. Harris took the opportunity, with SHAEF participation, to build up this relatively feeble force in German eyes as a formidable new host equipped with advanced weaponry. Wild allowed the major double agents—GARBO and GILBERT—to participate, plus DRAGOMAN, CAMOUFLAGE, ASSASSIN, and ATOM; they

---

* For the record, they were Operation METZ II, also called Operation 90TH INFANTRY DIVISION; Operation L'ÉGLISE; Operation FLAXWEILER; Operation STEINSEL; Operation LANDONVILLERS, also called Operation 26TH AND 95TH INFANTRY DIVISIONS; and Operation 11TH PANZER DIVISION, with subsidiary Operations WHIPSAW and MERZIG.

were, after all, reporting nothing that could be disproved by an observer on the ground. This exaggeration of Fifteenth Army's power, it was hoped, would enhance the credibility of a demonstration by it in early April (Operation RUHR POCKET), threatening a crossing on the Rhine side of the pocket created by Ninth and First Armies surrounding the Ruhr while the forces trapped in the pocket were liquidated.

By the time of the regular meeting of the 212 Committee at Versailles on April 6, Strangeways and 21 Army Group had closed out deception and Ops (B) was ready to wind down. Harris, however, had one more shot in the locker for mid-April, suitably entitled Operation OMEGA. Its object was to deter the Germans from moving forces to oppose the American drive to the south, by the "story" that Bradley had been ordered to drive on Berlin. The 3103d simulated a major buildup behind the front of Ninth Army west of Berlin; SHAEF contributed a high-level report by GILBERT; and BLAZE, LAZY, and other lower-level agents controlled by Oakes and Eldredge reported heavy traffic in the appropriate directions. It is doubtful that this operation had any effect on the rapidly disintegrating enemy.

Harris did not give up easily. At the next 212 Committee meeting, on April 20, when things were really shutting down, he proposed to retain LEAGUE at Metz, hoping that through the Germans he could pass misleading information to the Japanese with respect to American redeployment to the Pacific. But it was too late for such things. On May 7, Jodl signed the German surrender at Eisenhower's forward headquarters in a schoolhouse at Rheims, and the European war was at an end. The Germans in Italy had already surrendered effective May 2, and the army group opposing Devers in the south had surrendered on May 5.

In the final phase 6th Army Group's headquarters was at Heidelberg, the old university town, never bombed and miraculously untouched by the war. ("In all the balconies of all the private houses there were geraniums in bloom" remembered Ekstrom. "I had this terrible resentment and hatred for the Germans. I thought, you know, they don't deserve these flowering plants, this kind of bourgeois ease that they seem to be enjoying. I was absolutely livid.") The 212 Committee had held its final meeting on May 4. Afterwards, Noel Wild took a few days off and visited Gene Sweeney in Heidelberg. They were billeted at the old Schloss on the hilltop, and Wild, who was an amateur artist, did some sketches and watercolors of the town.

On Sunday, May 6, Sweeney received a cable ordering him to report forthwith to Washington, passing through Paris and London for debriefings along the way. He was in Paris when word came of the surrender at Rheims. It was a bright, sunny day; Roger Hesketh managed to get hold of some strawberries, and Sweeney, Hesketh, and Hesketh's driver, a charming British girl, had a pleasant party and celebrated with the rest of Paris. That night he took the boat train for London, and was there for the second day of the great V-E Day celebration—finding the city pretty well exhausted from its first day. After debriefing with the London Controlling Section for several days he left for Washington to report to Joint Security Control.

Ekstrom too was in Paris when the war ended. He was anxious to get home to Pam, whose health had declined during the war. But Noel Wild had gotten him assigned to London to work on the final report of "special means" activity for No. 2 Tac. He was on the point of leaving for London when he received an invitation from Gordon Merrick to spend a few days on the Riviera enjoying the victory celebrations there. So he took a break at Merrick's villa. Merrick had also invited Humphrey Hare, who was a close friend; Ekstrom did not share their interests but it was a luxurious way to end the war.

Eldredge celebrated the surrender with champagne from Wehrmacht supplies in a hotel room in Bad Wildungen near Kassel, EAGLE TAC's last wartime home, in company with a group of regular officers. The regulars were subdued, their occupation gone, wondering whether they would be part of the "first team" in the Japanese war. Eldredge himself was overjoyed. The Nazis were defeated and his war was over. He did not yet know that he was scheduled to join the invasion of Japan.

The elaborate machinery that had been erected for deceiving Germany had already begun to be dismantled when the Germans surrendered, and the process ran swiftly thereafter.

The London Controlling Section began cutting back in the autumn of 1944, after BODYGUARD. Bevan successfully proposed to the Chiefs of Staff in early October a reduction of personnel down to four officers from its current ten. James Arbuthnott was the first to go, to return to his tea-planting in Ceylon at the request of the Colonial Office. Dennis Wheatley found himself with little to do but fuss with a stamp collection, and applied for release in August, though he did not actually leave till December. Ronald Wingate went out to SEAC in November on an inspection trip, the upshot of which was his own assignment to that theater.

The action of the Chiefs of Staff in approving the reduction in force included their placing on record their appreciation of the work of the London Controlling Section "and its subsidiary sections in the operational theaters," whose "record of success has been unique," adding its "opinion that, in at least one instance, 'the Section' made a decisive contribution to the success of a major operation, namely, the Allies' return to the Continent in June 1944."

"A" Force, as described earlier, closed up shop and held its last review on June 18.

The double agent system had begun to wither away as early as mid-1944. At the beginning of the year the Twenty Committee had been overseeing some twenty channels, nine of them communicating by wireless; by the end of the year there were but six, four of them W/T. Fewer opportunities for deception, a reevaluation of agents on the part of the RSHA after it took over the Abwehr, failure of agents to be paid, all played a role in this decline. But none of the agents was compromised during this process.

BRUTUS had to be phased out when arrests and trials after the liberation threatened to give too much publicity to the breakup of his network in France back in 1941. TATE, on the other hand, proved useful to the very end. He was a prime channel for distorting the Germans' aim of their V-2s. Late in the war the U-boat menace revived when the introduction of the snorkel enabled them to lie in wait submerged for long periods. Relying on a notional friend in the Royal Navy, TATE reported that a new type of deep-water mine had been developed to counter this threat and gave particulars of the location of alleged minefields. When this information was seemingly confirmed by the actual loss of U-boats (from other causes, but the Germans could not know that), instructions were issued to avoid important areas of the sea, or to traverse them at shallow depths at which, of course, they were vulnerable to surface action.

GARBO learned in July 1944 that he had been awarded the Iron Cross.

But by September the number of Abwehr defectors in neutral countries was increasing, and MI5 was concerned that the Germans must surely assume that the British had been alerted to GARBO; indeed, in August a would-be defector told Section V in Madrid that such a person as GARBO existed and offered to betray him in exchange for asylum. The offer was declined, but it was clearly time to reorganize the GARBO network. GARBO warned Madrid that Spanish friends had alerted him to this betrayal. He himself went into notional retirement on a remote farm in Wales; MRS. GARBO notionally told the British authorities that he had deserted her and gone back to Spain; and thenceforth he exercised only general supervision over the network from a distance, leaving day-to-day management to AGENT THREE. What the network thereby lost in GARBO's prose style it gained in security, for since THREE did not exist and the Germans did not even have a supposed real-life name for him, he was secure against all betrayal. THREE remained in touch with the Germans to the end of the war.

At the end of 1944, GARBO was awarded the British MBE to go with his Iron Cross. After the German surrender he went to Spain and called upon his handlers; there was briefly some thought that he might be used to penetrate Soviet intelligence through Germans recruited by the Soviets, but this came to nothing. He settled in Venezuela as a language teacher for Shell Oil.

On the other side of the Channel, GILBERT last heard from the Germans on May 2. After the surrender there were some weeks of waiting in case he might be contacted, perhaps through Spain. Meanwhile, protests about letting this collaborator strut about Paris in uniform with a senior staff job had mounted to the point that his commission had been withdrawn. The Deuxième Bureau wanted to get him out of France. Some of his British and American colleagues in SHAEF Civil Affairs—who knew nothing about his real activities—came to the rescue by arranging for him a job with UNRRA in Austria. The cynical and duplicitous André Latham might not have seemed ideally suited for humanitarian work in aid of displaced persons, but it conveniently removed GILBERT from the scene.

In July 1944, with the evident success of FORTITUDE, Joint Security Control had asked SHAEF for a report on it, and SHAEF in turn directed Special Plans to prepare such a report. Ingersoll wrote it during the late summer and early autumn. It was an extensive and unsophisticated essay on the history, theory, and practice of deception, including a discussion of special means, with an account of FORTITUDE and a description of each of the tactical operations of the 23d up through WILTZ in August. It was also implicitly a panegyric to Special Plans.

Early in December, Bobby Rushton, who was on his way to report to Joint Security Control, hand-carried three copies to London with instructions to take a copy to Washington after it had been edited by Bevan. (Shortly before, Bedell Smith had agreed that Bevan should be responsible for editing reports on deception in Europe.) Bevan read it and was horrified. He composed a long letter to Bissell in his courtly style. While expressing "profound admiration for the enthusiasm and thoroughness with which this report has been prepared," he pointed out that it included many matters the authors could have known nothing about at first hand, "many most secret matters which have never before appeared on paper in any report," and matters on which the authors "are not really qualified to express opinions."

Particularly distressing were discussions of intelligence and special means. "Details are set down with regard to the working of the British Secret Intelligence Service which have never before been allowed to appear in any such document." Contrary to policy, there were written references to "double agents" rather than to "special means"; the report seemed to believe that deception of the press was authorized; the accounts of FORTITUDE and BODYGUARD were incomplete and limited only to Special Plans' own experience. Perhaps most distressing, Bevan understood that Special Plans was preparing another report dealing with the use of ULTRA in deception. "On security grounds British Deception Staffs have never been permitted to issue a report on this subject and I would be grateful for your views on the matter."

Bissell passed the report and Bevan's letter to Newman Smith, whose view was that the report contained valuable information and would not harm security if dissemination was carefully controlled. He drafted a mollifying letter for Bissell to send to Bevan.

The promised annex dealing with the use of ULTRA in deception did not reach Bevan till early April and did not reach Bissell till the end of the war in Europe. Bevan pointed out to Bissell that this report too had many shortcomings. "My general impression is that this report does not provide a complete analysis of a complicated and intricate subject. This is not altogether surprising since an Army Group's experience of this sort of work is strictly limited." It did not properly consider the use of ULTRA to follow the enemy's assessment of Allied order of battle, omitted many aspects of the use of ULTRA in connection with special means, and had an unbalanced view of the relationship between a deception staff and the intelligence and security staffs. It unwisely discussed an individual double agent (DRAGOMAN).

By now, Bevan was well aware of the skills of No. 2 Tac. "If you are anxious to obtain any further information on this delicate problem there are, apart from 12th Army Group, two American officers at 6th Army Group deception staff who are fully qualified to discuss it. The two officers are Lt-Col Gene Sweeney and Captain Ekstrom; both were trained and had considerable experience in this subject under 'A' Force in the Mediterranean."

Bissell's response was again rather noncommittal. The paper was "interesting and instructive" as long as its limited perspective was kept in mind. As for Sweeney and Ekstrom, "We are familiar with the record of these officers and have them in mind for future deception requirements."

Those future deception requirements were, of course, in the war against Japan; to which we now return.

Hustling the East (III)

While European deception reached its climax in the late spring and early summer of 1944, Newman Smith and his colleagues were mainly focused on the Japanese war: Nursing along the General Directive for Deception Measures Against Japan, mounting WEDLOCK, and trying to be patient with Ormonde Hunter. Looking ahead, there would plainly be fewer and fewer deception opportunities in Europe; yet for the Japanese war no plans were in sight beyond WEDLOCK, and the theater staffs in the Central and Southwest Pacific knew next to nothing of what had been done in Europe and might be done on their fronts. This led directly to the first and only formal training course for deception officers during the Second World War, the "Young Ladies' Seminary."

It originated with a study of major areas of deception as practiced in the American services, noting specific shortcomings, prepared by Carl Goldbranson in February of 1944, soon after Joint Security Control had received its augmented charter. After some vicissitudes, mainly involving the Joint Planners protecting their turf, this eventually resulted in an August 5 message from the Joint Chiefs to the three theater commanders, Nimitz, MacArthur, and Stilwell, offering to make trained three-man teams available to each of their headquarters for sixty to ninety days to help their staffs develop deception, with the option of sending their "senior deception officers" to Washington for training instead. All three preferred to receive team visits rather than send their own people to Washington, and Stilwell asked that the Army member of the team be available for permanent assignment to his theater.

The training course for the teams that were to be sent out ran from September 4 to September 30. Goldbranson supervised most of the curriculum planning; Burris-Meyer organized and operated the course itself. He dubbed it the "Young Ladies' Seminary," and Goldbranson's curriculum packet he called the "Young Ladies' Guide to Truth and Honour." Recognizing that he should not appear to take himself too seriously when dealing with a "student body" composed of professional officers senior to him, Burris-Meyer armed himself with an old-fashioned school bell, which he would ring to call the class to order.

The "student body" itself consisted of only ten officers, selected by their respective services. From the Army came Colonel John F. Holland, Lieutenant Colonels Frederic W. Barnes and Montgomery B. Raymond, and Major Alfred J. F. Moody. From the Navy were Captains Percival McDowell and George F. Mentz, and Commander John B. Fellows, Jr. From the Air Force were Colonels John A. Hilger, Clyde K. Rich, and Willard Van D. Brown. By design, all were regulars.

Both Barnes and Moody, it will be recalled, had joined SHAEF Ops (B) in May. The Young Ladies' Seminary was already under way when SHAEF advised Washington that since "it now appears that there will be little need for further cover and deception planning and implementation" in Europe, Barnes and Moody now were supernumerary. They arrived in time to participate as instructors in the course. (Subsequently, SHAEF suggested that Jack Corbett of Ops (B) was likewise supernumerary, and Smith asked for him in light of Hunter's stated needs; but Captain Corbett did not leave Europe till early October, too late to participate in the Young Ladies' Seminary.)

Another "student" with practical experience was Captain McDowell (known to his friends as "Peter"), for as the officer in charge of New Weapons, Development, and Research in the Readiness Section of COM-INCH's Readiness Division, he had been involved with Fairbanks's Beach Jumper experiments in early 1943 and with Burris-Meyer's sonic experiments. A graduate of the class of 1923 at Annapolis, where he had roomed with the legendary Arleigh Burke, McDowell had become close friends with Burris-Meyer. The school bell with which Burris-Meyer rang in the Young Ladies' Seminary came from the McDowell nursery; at Burris-Meyer's request, McDowell and his wife entertained the "students" to a corned beef and cabbage dinner to celebrate their "graduation"; and he would eventually be godfather to Burris-Meyer's son, who was named for him.

Other "students" brought a variety of backgrounds to the work. "Egg" Mentz, Annapolis 1919, had won the Navy Cross for his work in removing the North Sea mine barrier in 1919, had been torpedoed off Oran, and had commanded a task group in the landing on Sicily. Fellows, a 1931 Naval Academy graduate, had won the Navy Cross for heroism in the Solomons and the Silver Star as a destroyer commander at New Georgia, and was fresh from a course at the Army-Navy Staff College. Holland was a 1925 West Pointer who had been in charge of special training on the Army Ground Forces staff. Brown was a soft-spoken gentleman from Moultrie, Georgia, known to friends facetiously as "Van Demon" Brown, and more commonly as "Van."

The "student" with the most colorful background was Jack Hilger of the Air Force. A thirty-five-year-old Texan and graduate of Texas A & M, he had been Doolittle's deputy and executive officer in the famous raid on Japan in April 1942. Since then he had commanded a bombardment group in the CBI. The Japanese were known to have placed a price on his head, and it must have seemed desirable to higher authority to rotate him to staff duty out of their reach for a time.

(Joint Chiefs of Staff)

Some members of the Young Ladies' Seminary
Left to right: unidentified, unidentified, McDowell, Holland, unidentified, Burris-Meyer, unidentified, Rich, Hilger, Goldbranson, Mentz

The course was conducted in Room 236 of the Public Health Building, home of the Combined and Joint Chiefs themselves. It covered definitions, missions, the organizational position of Joint Security Control, the history of deception in the war hitherto, strategic objectives, Japanese intelligence, theater deception and channels to the enemy, double agent handling, visual, sonic, and signals deception, plans and planning, and staff organization. Reading matter included relevant CCS and JCS directives; a set of special memoranda prepared by Joint Security Control on specific topics, ranging from notes by Douglas Fairbanks on Beach Jumper details and Office of War Information notes on Japanese psychology, to substantial "Notes of Deception in a Theater of Operations"; and an extensive collection of manuals on relevant subjects. There were field trips to the Army Communications Center, the Navy Special Projects School, the Engineer Board at Fort Belvoir, Virginia, the Bureau of Yards and Docks, and the Army Experimental Station at Pine Camp, New York; a visit to the Beach Jumper training base at Ocracoke, North Carolina, was canceled because of a hurricane, so that Navy tactical deception was covered less well than had been hoped. Some four dozen officers were called in as "faculty," including almost everyone who was in the States and had had anything to do with deception strategic or tactical, from General Bissell, Admiral Schuirmann, and Bill Baumer, through all the officers with the Special Section of Joint Security Control, engineering and signals officers, civilian researchers in sonic deception, and Cunningham of the FBI.

Halfway through the program, Burris-Meyer was able to report a gratifying response from the "students"; and at the end of the course, he expressed general satisfaction, although it had become clear that "staff and probably planning experience should be among the qualifications for officers assigned to similar duties," and that in any future courses more time "devoted to a somewhat detailed study of the enemy, and to the special technique of planning deception would be in order."

Smith allocated Mentz, Rich, and Holland as the visiting team for MacArthur's headquarters; McDowell, Barnes, and Hilger for Nimitz; and Raymond, Brown, Fellows, and Moody for CBI. Raymond had originally been destined for the Pacific and Barnes for CBI, but for some reason their positions were reversed at the last minute. Raymond was technically not a member of the visiting team, on the expectation that he, along with Jack Corbett, the former Ops (B) officer, would be assigned permanently to the theater. The teams for MacArthur's and Nimitz's headquarters left Washington at the end of the first week in October; that for India left in mid-October, to be followed soon thereafter by Raymond and Corbett for their permanent assignments.

The teams had uneven success in the three theaters they visited.

Much the most successful was the team of Peter McDowell, Jack Hilger, and Freddie Barnes in the Central Pacific. It spent six weeks, from October 10 to December 19, in the theater, with gratifying results. McDowell designed BLUEBIRD, the next major American strategic deception, to cover the Okinawa landing, and it was forwarded to Washington for approval on November 21. At the beginning of December a "Special Plans Section" for cover and deception was officially created in Nimitz's joint staff, under the Deputy Chief of Staff, with a clear directive that it should when necessary be assisted by representatives from Future Plans, Current Plans, Intelligence, Operations, Communications, Radar and Radar Countermeasures, and Public Relations. Hilger and Barnes were assigned to it for permanent duty in mid-December.

The new Special Plans Section visited with the commanders and staffs of major headquarters. "All of the officers contacted exhibited considerable interest in cover and deception," reported McDowell, "particularly the Tenth Army"—destined for Okinawa—"and Commander, Fifth Fleet." Requests for Beach Jumper and communications deception units followed; and Special Plans drafted a suggested plan for disguising preparations for Operation DETACHMENT, the seizure of Iwo Jima scheduled for February, as a major training exercise. "The Central Pacific Theater is ready for the employment of *strategic* deception in the conduct of operations against the Japanese," reported McDowell on his return, and what was now needed was "a long range comprehensive deception plan" by the Joint Chiefs, together with improved communication and liaison between Joint Security Control and the theater staff.

McDowell was lost to the cause of deception soon after returning to Washington, taking command in February of USS *Catoctin* in the Mediterranean.

* * *

From the Washington vantage point, MacArthur's headquarters would not have appeared likely to be receptive. "The deception policy of the Southwest Pacific has heretofore been one of saying nothing about future plans and allowing the Japs to sweat it out," one seemingly exasperated OPD officer had written in April. But this was hardly fair. MacArthur had used operational deception in classical forms. The Hollandia landing earlier in 1944 had been supported by an elaborate ruse diverting Japanese attention to Wewak and Hansa Bay, including bombing raids, naval bombardment, PT boat patrols, dummy parachutists, leaks and gossip at General Eichelberger's headquarters, reconnaissance flights, leaflet dropping, submarines leaving empty rafts on shore, and assault convoys seeming at first to be headed for Truk or Rabaul. More serious for the visiting team of Mentz, Holland, and Rich were the touchy self-sufficiency, resentment of all outside interference, and reluctance to share any credit for success that marked MacArthur's headquarters throughout the war. They encountered this in full force.

Holland and Rich drafted a recommendation for establishing a deception subsection in MacArthur's G-3. To encourage this, they included a warning, based presumably on concerns that General Bissell had derived from conversations with the British and no doubt expressed at the Young Ladies' Seminary, that the large British deception staffs from Europe and the Middle East "are now unoccupied, and indications are that they will try to infiltrate these personnel into the Pacific Theater," including officers senior to their American opposite numbers. But their memorandum disappeared into the SWPA staff and was not heard from again.

They had better luck with a proposed deception plan for the key operation in MacArthur's invasion of Luzon, being planned for January: a landing in Lingayen Gulf for a drive up the central valley towards Manila (Operation MIKE ONE). The team prepared a classic plan designed to mislead the Japanese as to the date, scale, and plan of the operation, and to prevent heavy concentration of their forces in the target area. Lieutenant General Richard K. Sutherland, MacArthur's imperious chief of staff, and his G-3, Major General S. J. Chamberlin, did not take kindly to outside advice; they deigned to circulate this plan as a "staff study," but in fact the actual MIKE ONE deception was to follow the team's plan—with MacArthur's staff taking the credit. Otherwise, about the only result of the team's visit was that the theater ordered sample pieces of engineer decoy equipment in December, evidently out of curiosity.

The team's bad luck continued after their formal visit. Holland remained in the theater, and was killed in an airplane accident on Leyte in January 1945. Rich remained in the theater for a time. Though Joint Security Control wanted Mentz back in Washington, unknown to them he took command of a "Diversionary Attack Group" including Beach Jumper units, for an operation against Mindoro in the Philippines, and was badly wounded by a kamikaze at the end of December.

The team that visited China-Burma-India essentially threw up their hands. To put their visit into context, it is necessary to review developments in that unhappy theater after Mountbatten took over.

As 1944 wore on and Nimitz's Central Pacific offensive reached the Marianas and MacArthur returned to the Philippines, China played a diminishing role in American thinking. The Marianas offered an impregnable base from which the B-29s could assault Japan. At Tehran, Stalin undertook to join the war against Japan three months after the war with Germany should end. And Roosevelt began at last to lose patience with Chiang Kai-shek. The Western Allies' main interest in the China theater was now that it keep as many Japanese tied down as possible who might otherwise be returned to defend the home islands.

Stilwell's successful campaign in northern Burma did not improve Chiang's view of him, and in October Roosevelt recalled him at the Generalissimo's request. Wedemeyer replaced him, and as of October 24 the old American CBI Theater was split into a China Theater under Wedemeyer and an India-Burma Theater; the latter, now mainly a logistical and air command, was headed by Lieutenant General Daniel Sultan, Stilwell's former deputy, with his headquarters at Delhi.

The Japanese were not idle. Concurrently with the Imphal offensive (but not in coordination with it) they planned, and in June they mounted, Operation ICHIGO, designed to clear the east China airfields, which were both the tactical base for Chennault's Fourteenth Air Force and a potential base for the B-29s. Conducted in phases, ICHIGO ran until January 1945 and was generally successful. But it was their last major offensive in China.

*　　*　　*

We left Ormonde Hunter in April of 1944 happily transferred to Strate-meyer's staff in Calcutta. His happiness was short-lived. His frustrations soon returned and continued through 1944, accompanied by even louder crashes than before from every china shop he entered. He contin-ued unable to get it through his head that Joint Security Control had no operational authority in the theaters. He could not get straight the rela-tive responsibilities of the British and the Americans for deception. He could not seem to understand that Joint Security Control coordinated with the British through the London Controlling Section rather than dealing directly with British commands. He continued unable to get any of his projects accepted, and to take offense at not being told about proj-ects in distant theaters. But his greatest frustration was his trip to China in May and June, the trip for which he had long pleaded.

Security in China was essentially nil. Communication between Japanese-occupied territory and the rest of the country was open, with travel prac-tically at will. Mail was delivered throughout China; the Bank of China operated in both areas, with its own radio network. In light of the very highly classified nature of deception work, Hunter might therefore have been expected to proceed with the utmost circumspection. But the op-posite was the case. He spent nearly a month, from May 26 to June 23, talking about deception with some four dozen people in Chungking, Kunming, and Kweilin. Some were civilians. Some were British. Some were Chinese. There were military and naval attachés, Foreign Service officers engaged in political advice and psychological warfare, public re-lations officers, OSS officers, Treasury civilians, Office of War Informa-tion civilians, officials with the Chinese customs and the Chinese post office, British officers and a British civilian. With them all he discussed available channels to the Japanese and with many he discussed the spe-cific topic of deception.

His trip coincided with the opening phase of the Japanese ICHIGO of-fensive. From Chennault's headquarters in Kunming he passed on to Washington a deception proposal by Chennault, designed to draw off the Japanese by telling them that the United States was luring them into costly and exhausting operations on the mainland when the real attack on the home islands would come from the Pacific. He asked also for three notional air groups, one heavy bomber group and two fighter groups, to augment Chennault's apparent strength against Japanese lines of communication. The bogus groups he apparently received, but the deception plan was quickly rejected because it would attract Japanese attention to the main thrust in the Pacific.

On his return, Hunter compiled a long report. He dolled it up in a fancy cover, and was eager to have it read and admired in Washington. It was in fact composed almost entirely of his own rough notes from his many interviews. Sultan's chief of staff would not allow it to be sent to Washington.

General Bissell visited the theater in late July and early August. Hunter talked with him and came away optimistic, and was pleased that Bissell directed that his report be sent to Washington. There, Goldbranson read it and was appalled. Hunter seemed to have blabbed about deception to everybody in China. It was not for Goldbranson to suggest directly that Hunter be relieved of duty, but he came as close as propriety allowed. "It is felt that Hunter's value to military deception in that theater has been jeopardized by reason of the number and type of individuals whom he has contacted," he told Smith. "It is reasonable to assume that any future contacts by Hunter will apparently identify deception activities in those areas." He "suggested that restrictions of Hunter's activities and travels in the future would not be injurious to deception in CBI and might be helpful."

On the basis of his trip, Hunter proposed in July a deception organization for the theater consisting of thirteen officers spread over India, China, and Burma, a body more than three times as large as the standard theater organization recommended by Joint Security Control—so large, indeed, as to lead Bevan to make worried inquiries to Washington when he got wind of it; and when Washington turned it down in September he responded bitterly that it was "strange that Joint Security Control, in whom all deception is lodged, cannot provide for adequate personnel in the theater."

By now, he wanted to leave. He evidently wrote to Bissell on October 2 asking to be returned to the States and assigned to Joint Security Control. Bissell replied on October 12 to tell him that the future extent of his activities was in abeyance until the Young Ladies' Seminary teams had completed their studies and made their recommendations to the theater commanders. He advised Hunter that there was no vacancy for him on Joint Security Control, but that he, Bissell, could arrange to have him returned to the States for some suitable assignment, turning his duties in the theater over to Raymond.

The summer and autumn of that year 1944 were not happy times for Fleming. Thorne sensed that Fleming was not getting on as well with Bevan as before. And that was in fact the case. Fleming had visited London early in the summer and had found Bevan less cooperative than he had been. Fleming tended to blame this on war-weariness and on the V-bombs; he told Thorne when he returned that their effect on the conduct of business in London was much worse than had been the far more destructive Blitz of 1940–41. (Many others noticed this. "Real fear came much later [than the Blitz]," wrote Joan Bright after the war, "with the V.1 and V.2 flying-bombs and rockets, when we were, so to speak, 'out of training.' ") In September, Fleming was unhappy about deletions Bevan was making in material Fleming sent for the SUNRISE channel. Bevan in turn was unhappy that Fleming was not keeping him informed as to his special means channels or his overall plans.

Relations with Supremo's headquarters were evidently worse than that. It is hard to see any justification for this; it appears to reflect personal animosity on the part of some individuals and resentment that Fleming had not moved to Kandy. Whatever the reason, it came to a head that same September; early in that month, Fleming advised Air Vice Marshal John Whitworth Jones, Mountbatten's Assistant Deputy Chief of Staff, that his recent proposals "went down so badly" as to make him *persona non grata* at HQSACSEA," and he wanted to depart in a month or two. "I feel," he said, "that a more diplomatic type of officer (and one not handicapped by the faded remnants of a reputation for doing things which appeared initially unsound) would serve the [Supreme Allied Commander] better than at the moment I look like being able to do."

Resignation did not materialize, but Whitworth Jones was a bête noire for Fleming. In December he was angry over a "dangerous" message Fleming had sent to London and Washington involving signals security and relations with the American India-Burma Command, which evidently caused a problem with the Americans. "If Fleming did his business, as he should, from Kandy instead of Delhi," Whitworth Jones told Mountbatten, "the whole thing would have been strangled at birth." In January, Whitworth Jones was referring sarcastically to one of D Division's "Progress Reports" as "the sort of stuff that Fleming has been circulating to the world." "You will note," he added, "that I have decided in my own mind that it is probably better to laugh at Peter Fleming than to get angry with him, which I confess was my first involuntary reaction."

Wedemeyer went from Mountbatten's headquarters to China in October, and Thorne noticed a problem of some kind between Fleming and Wedemeyer, with Fleming non grata in Chungking for a while. (It may be relevant that in his memoirs Wedemeyer singled out Whitworth Jones as one of the three British colleagues at Mountbatten's headquarters whom he found "very congenial.")

Bevan decided, with Fleming's and Mountbatten's concurrence, to send Ronald Wingate out to look the situation over.

So matters in the theater were in this state when the visiting team arrived in October; matters in the two theaters, rather, for more confusion was introduced by the splitting of India-Burma from China effective October 24.

Brown, Fellows, and Raymond arrived in Delhi two days after the split, on October 26. Moody arrived ten days after that. Colonel Dean Rusk, now Sultan's Deputy Chief of Staff, was notably hospitable and sympathetic. Hunter joined them and arranged for conferences and briefings. They had some preliminary briefings and exchanged courtesy calls with Fleming and his D Division. Brown visited with the local OSS to explore what that organization might be able to do, without mentioning cover and deception. Moody, after his arrival, briefed a substantial high-level group of officers on FORTITUDE, and there was another meeting with D Division. They then visited Stratemeyer's headquarters in Calcutta and gave a briefing for top officers, and Brown visited XX Bomber Command at Kharagpur. The conclusion was reached, and embodied in a command decision on November 26, that the India-Burma Theater needed no American cover and deception in its mission, and that the current arrangement—one theater deception officer at New Delhi, one deception advisor to Eastern Air Command at Calcutta, and two liaison officers with Southeast Asia Command—was entirely adequate.

The team next visited China, briefing Wedemeyer's senior officers in Chungking and stopping off at Kunming. It was concluded that in China, deception "on a strategic scale" was impossible because of total lack of security, although "the opportunities are limitless for passage to Japanese of information, real and notional, relating to other theaters."

Back in New Delhi, they met on December 18 with Fleming, Rusk, and Ronald Wingate, who was visiting the theater on the inspection trip presently to be described. They left the theater on December 20, and reported in to Joint Security Control on December 29.

Raymond stayed behind, as the single theater deception officer at New Delhi which had been agreed upon, officially assigned to the G-2 division of the India-Burma theater staff as theater deception officer and liaison with D Division, with the title Chief, Special Planning Branch, for United States Forces, India-Burma Theater. "Chip" Grafton was his deputy.* Hunter, no longer the theater deception officer (and now a full colonel), remained in Calcutta with the theater air headquarters.

Hunter could not understand why an inexperienced officer like Raymond was chosen for the theater staff when an experienced officer like Moody was available. (Unknown to Hunter, Moody was not available. Smith had in fact planned to assign him to SEAC, but he had written to Smith that he did not want the job, and Smith had responded that his wishes would be respected.) "Apparently there is not any great shortage of Deception officers," said Hunter resentfully, "since General Bissell has advised me that there is no opening in this work for me in Washington on my return. . . . I had hoped that Joint Security Control and General Arnold might have decided upon a full time Air Officer on Joint Security Control, and that I would be considered for the position." Smith's blood must have run cold at this. He returned a diplomatic answer, but he hastened to bring Van Brown on board as Joint Security Control's air officer after the latter's return to Washington.

---

* Since October, Fleming had been toying with the idea of a "fully inter-service and inter-Allied" D Division, with an American lieutenant colonel as second in command and an American major also assigned to the staff. O'Connor had advised Bevan that Moody, with a promotion, was a possibility for second in command ("Some experience with SHAEF, did not get on too well with Noel possibly not unilateral fault may as result be leery of combined staffs. Considered able and intelligent. . . . Possibly feels our work backwater but might not feel that way if it meant promotion.") But Fleming thought that "Moody hardly carries guns for this"; he wanted Grafton, "an excellent officer," who he seems unaccountably to have thought was a lieutenant colonel. (British officers "have a great respect for and admiration of [Grafton's] abilities," reported Whitworth Jones.) These ideas did not survive Wingate's reorganization.

Raymond's assignment to India-Burma did not last long; in February 1945, deception liaison for Sultan's headquarters was taken over by a major in the G-2 section, while Raymond was transferred to Chungking with the title of theater antiaircraft officer. This was at least in part a cover assignment, for he acted as Wedemeyer's deception officer from that point on. If Grafton went to China with him the files do not show it.

Jack Corbett, the former Ops (B) officer, was assigned to D Division. His first assignment was as D Division's representative with Northern Combat Area Command, the American-Chinese-British-Indian force in northern Burma. When he visited their front in February, however, he found that "the show is about over up that way with a general withdrawal of the enemy quite evident." After some discussion with Fleming over where he could best be used, Corbett was assigned to D Division Tac HQ in March, briefly covered Raymond's job in Delhi in April, and moved in May to D Division headquarters in Kandy.

Hunter finally left the theater in March for a new assignment in Washington. He wrote to Smith that "it is my hope to be able to continue on the work to which I have been devoting my exclusive time for two years"; a thought that could not have been welcome to Smith and Goldbranson. D Division were not sorry to see Hunter go. He had appeared to them to be increasingly suspicious of D Division's work; they found his successors to be more friendly and cooperative.

With respect to deception planning itself, for Fleming and D Division much of 1944, after Mountbatten opened his headquarters at Kandy, had been little different from what had gone before. There was no opportunity for major strategic deception because there was no agreed major strategic plan, though one proposal after another was made and withdrawn. Fleming's Plan UKRIDGE was replaced by, or renamed as, BLANDINGS, which was designed to make the Japanese prepare for a notional air and seaborne assault, codenamed MALICE, to be mounted between April 21 and 24 against the Sandoway-Taungup-Prome area; it was, of course, subsequently "postponed" several times, finally in mid-May.

As for real operations, after more false starts the Combined Chiefs finally, in the autumn, directed Mountbatten to proceed aggressively with CAPITAL, a three-phased reconquest of northern and central Burma (also including ROMULUS, the recapture of the Arakan coast, Akyab, and

Ramree Island). This at last appeared to give Fleming something he could sink his teeth into. To support CAPITAL he developed a deception plan called STULTIFY (originally FLABBERGAST); approved on November 21, it was the overall blueprint for his operations up until it merged in the late spring and summer of 1945 into SCEPTICAL, the final plan to cover Mountbatten's contemplated campaign into Malaya.

The object of STULTIFY was to induce the Japanese to divert substantial forces from the central Burma front to meet a notional amphibious and airborne threat to southwestern Burma. An additional object, since Mountbatten had still not given up on the idea of an invasion of the Andaman Islands or perhaps the Kra Isthmus, was to induce the Japanese to station the major part of their forces in the Dutch East Indies "and other irrelevant parts of his command to meet an amphibious threat."

The "story" of STULTIFY was that Mountbatten planned to recapture Akyab in early February 1945—this was of course the genuine ROMULUS, which was in fact successfully carried out in December and January, covered by a tactical deception plan called REMUS—and then in March or April seize by seaborne and air assault a position running from the lower Irrawaddy at Prome to the coast at Taungup, with subsequent exploitation to Rangoon: essentially the old notional operation MALICE which had been the subject of BLANDINGS the preceding spring. This would be supported by "a cautious but much publicized overland advance"—the genuine CAPITAL—in northern and central Burma by General Slim's British Fourteenth Army and the Northern Combat Area Command; the limits of this, the "story" ran, "are not likely to be ambitious in view of the appalling health and administrative problems which the 1945 monsoon will bring." (The monsoon was to be expected in early May.) Airborne operations might be carried out in either direct or diversionary support of the overland advance. In addition, probably in April 1945, there would be an operation to seize naval and air bases in northwestern Sumatra, for occupying and immediately developing rubber-producing areas to overcome the severe Allied rubber shortage, followed by exploitation to Java.

The codename for this whole series of operations involved in STULTIFY, both in Burma and elsewhere, would be KNOCKOUT. The threat to southeastern Burma received the codename CLAW; WOLF was the codename given to a plan to explain the actual advance into Arakan as designed to obtain airfields to support CLAW; the threat to Sumatra was FANG; finally, whatever notional airborne operation in support of CAPITAL might be decided upon was called TARZAN (a name that had already been used for one of Mountbatten's abortive proposed operations).

General Slim opened CAPITAL in early December. In January he crossed the Irrawaddy north of Mandalay, while on the west Akyab was recaptured and in the north the Northern Combat Area Command and a force from China reached Lashio and reopened the Burma Road at last. In February Slim made one of the extraordinary moves of the war, crossing the Irrawaddy far below Mandalay to cut off the Japanese in central Burma—greatly aided by deception operation CLOAK, presently to be described. Mandalay fell on March 20. Slim then leapfrogged his divisions south toward Rangoon, and that great port, abandoned by the Japanese, fell to a combined sea and air operation on May 3.

D Division was hard put to explain to the Japanese how, in Fleming's words, "SEAC's idea of a cautious overland advance did in fact include a bloody and determined battle for Upper Burma." By January, the objective for CLAW had to be changed to the Moulmein-Bilin area across the bay from Rangoon, notionally designed to cut off the Japanese retreat from southern Burma. TARZAN finally became an airborne threat to the Japanese escape route through the southern Shan States into northern Thailand, coordinated with a tactical deception plan called CONCLAVE which Slim put into effect beginning in March to threaten the same escape route, directing Japanese attention to the east of Slim's main southward thrust, while CLOAK directed them to the west. CONCLAVE was succeeded in April by notional guerrilla operations against the remaining Japanese lines of communication in the Shan States, called CAPTION, and TARZAN by a similar notional airborne threat called ARAMINTA. In April there was BRUTEFORCE, a simulated beach reconnaissance at the mouth of the Salween River near Moulmein, as well as CLEARANCE, a bogus airfield reconnaissance in the Setul area of extreme southwest Thailand. Fleming devised also a deceptive threat to the mouth of the Rangoon River in aid of the capture of that city, called FIREFLAME, but it was abandoned in view of the actual operations.

These operations gave Fleming an opportunity to put to use the extensive false order of battle in the style of Dudley Clarke that he and Thorne had established over the long dull months to keep DIB's double agent network busy.

Operation CLAW, the sea-air threat to Moulmein implicitly bypassing Rangoon, was to be carried out by an imaginary "Force 144" notionally commanded by Lieutenant General Sir James Gammell, Jumbo Wilson's former chief of staff in the Mediterranean. Force 144 consisted of an airborne element based in east Bengal (XXVI Airborne Corps) and an amphibious element based on the east coast of India, both of them bogus and composed of equally bogus divisions. The notional D-Day for CLAW was originally April 15, but daily "postponements" were planned for such excuses as logistical problems, the success of the central Burma offensive, and breaches of security.

For FANG, the notional invasion of Sumatra, D Division originally intended that the overall command should be Twelfth Army, Dudley Clarke's long-running phony headquarters in the Middle East, which had notionally been transferred to India in the liquidation of the WANTAGE order of battle. But London decided to activate Twelfth Army as a genuine headquarters, so the command for FANG became the "Indian Expeditionary Force." By either name, it was composed of two imaginary corps, the amphibious XXXVIII Indian Corps in Ceylon, notionally under Major-General C. E. N. Lomax (well known to the Japanese as the former commander of the very real 26th Indian Division), and XX Indian Corps, notionally commanded for a brief period by Lieutenant-General Sir Gerald Templer and then by Lieutenant-General W. H. A. Bishop. Both corps and all their component units, once again, were imaginary; some of them were WANTAGE divisions supposedly transferred from the Middle East. The original D-Day for FANG was April 30, but this naturally saw various "postponements."

TARZAN, the airborne threat to the Japanese escape route through the Shan States, was to be conducted by the fictitious 3d Airborne Division of the notional XXVI Airborne Corps. For CONCLAVE, the related operation primarily on land, a bogus 51st Independent Tank Brigade was to advance eastward to complete the encirclement of Japanese troops in central Burma, together with an airborne threat by three battalions of the bogus 3d Indian Airborne Division's notional 5th Parachute Brigade.

It was all to little or no avail. Once again, the Japanese simply were not able to put together the pieces that they were being given—even though, as will be seen, SILVER in effect handed them the whole story. "We must regretfully (but provisionally) record our conviction," wrote Fleming after the fall of Rangoon, "that of the delusions from which [the Japanese] suffered only an uninfluential minority were of our own manufacture."

Made possible by Fleming and Thorne's extensive order of battle, and implemented by a meticulous program of tidbits of information through the network of "special means" channels that Fleming and DIB had built up, together with systematic wireless deception showing the various units concentrating and preparing to move out, STULTIFY was conducted with fully as much skill as were FORTITUDE, FERDINAND, and NUNTON on the other side of the world. The professionalism of D Division and DIB in this work in no way suffers when compared with that of their colleagues in the Mediterranean and Europe. Yet FORTITUDE will be remembered forever, and STULTIFY will no doubt remain forgotten. FORTITUDE was a phenomenal success, in a campaign on which hung the fate of the world. STULTIFY had no perceptible effect on a campaign that, though one of the most masterly of the whole war, was but a sideshow. The fault lies not with D Division but in its stars, as it were. With all the shortcomings of the Abwehr, the Germans were able to build a coherent appreciation of the evidence before them and act upon the information they had. Above the tactical, and occasionally the operational, level the Japanese could do neither.

But CAPITAL saw one very great deception success. This was CLOAK, the operational deception in support of the decisive move of the campaign, Slim's crossing of the Irrawaddy in February and his subsequent clearing of central Burma. It will bear comparison with operation BERTRAM at Alamein. To Slim's own Fourteenth Army staff goes much of the credit for it, but D Division helped.

The Irrawaddy, one of the great rivers of the world, flows southerly through Burma from the Himalayas to Rangoon. Slim's overall plan was to cross the river to its east side, defeat the Japanese forces there, and drive down to Rangoon or Moulmein. On the east bank, not far upstream from the bridge at which Fleming had left the fake papers in Operation ERROR three years before, lies Mandalay, the chief city of central Burma. The Japanese saw Mandalay as Slim's initial objective; but in fact his first target was the town of Meiktila, some seventy miles south of Mandalay and well back from the river, which Slim correctly perceived as the nerve center of Japanese supply and communications.

Slim's Fourteenth Army was made up of IV Corps and XXXIII Indian Corps. His plan was to cross the river first to the north of Mandalay with XXXIII Indian Corps, drawing Japanese attention to that direction; then to make a surprise crossing with IV Corps well south of Mandalay and drive for Meiktila, forcing the Japanese to detach large forces to defend their communications. "This should give me," he afterwards wrote, "not only the major battle I desired, but the chance to repeat our old hammer-and-anvil tactics: XXXIII Corps the hammer from the north against the anvil of IV Corps at Meiktila—and the Japanese between." His main drive would thereafter be a thrust by IV Corps southward from Mandalay towards Rangoon, along the axis of the Mandalay-Rangoon railway.

On January 14, elements of IV Corps crossed the river some fifty miles above Mandalay. A month later, the campaign began in earnest, when elements of XXXIII Indian Corps crossed fairly close downstream from Mandalay, thus giving the impression that Slim planned a pincer movement against that city—IV Corps from the north and XXXIII Indian Corps from the south. In fact, the forces in the north were secretly turned over to XXXIII Indian Corps. What the Japanese thought was IV Corps relatively close to Mandalay was a bogus headquarters imitating the wireless transmissions of that formation (Operation STENCIL); the strength of the corps was notionally augmented by the 11th East African Division—a genuine unit with which the Japanese were well acquainted, but which was not now at the front and was being impersonated by an East African brigade.

Meanwhile, in a move reminiscent of Stonewall Jackson at Second Manassas, the genuine IV Corps made a secret wide-sweeping round-about march under radio silence through difficult country far to the west, then swung back to the river and crossed it some eighty miles downstream from Mandalay. Its crossing was aided by CLOAK, a battery of tricks designed to focus Japanese attention on the "IV Corps" site closer to Mandalay and other crossing points, and to suggest a drive towards the oil fields further south. Feints on the ground were supported by diversionary air bombardment and photoreconnaissance; by D Force sonic deception, including simulated noises of unloading stores, concentrating vehicles, starting outboard motors, and the like; by the collecting of boats and dissemination of rumors among river boatmen; and by miscellaneous timber-felling, light-flashing, dust-raising, and activity at a quickly-built airstrip. There were haversack ruses, including a map left by a party of East Africans showing the oil field area to the south as the objective; an air drop to an imaginary agent of a radio, instructions pointing towards one of the fake crossing-points, and a love letter from his girlfriend; and another air drop of deceptive material seemingly from a crashed airplane. On the eve of the actual crossing came a culminating fake airborne attack by PARAGON dummy paratroopers with associated PARAFEXES, PINTAILS, and BICAT STRIPS, plus a drop of supplies to the notional air landing party. And, needless to say, extensive misinformation was fed to the Japanese through special means.

The Japanese reaction was all that could have been hoped for. They prepared for an attack directed towards the oil fields to the south and for a pincer movement on Mandalay, believing until too late that IV Corps's main crossing was only a feint. Soon relieved of that misconception, they fought and lost a desperate battle at Meiktila. Mandalay was in British hands by March 20–21, and the remaining Japanese forces in Central Burma were left to try to escape eastward through the Shan States (threatened by TARZAN, CONCLAVE, CAPTION, and ARAMINTA) or to be shepherded south by Slim's pursuit.

Slim then made his main drive towards Rangoon with IV Corps, following the railway line to Rangoon down the Sittang valley east of the mountain ridge that divides the Sittang from the Irrawaddy. XXXIII Indian Corps advanced west of the ridge and down both sides of the Irrawaddy, and the object of CLOAK now became to induce the Japanese to defend against this in the belief that it was Slim's main drive. To boost its apparent strength—the pretended 11th East African Division was withdrawn at the end of February—the bogus 18th Indian Division had been added to its lineup. And Slim's great campaign culminated with the fall of Rangoon on May 3.

The network of special means channels which DIB and Fleming had built up continued to grow from strength to strength during 1944–45. It was easily the equal of the MI5 network in Britain and the "A" Force network in the Mediterranean; unfairly, by comparison with them it is largely unsung and likely to remain so. It is convenient to follow them through to the end of the war with Japan, even though that gets ahead of the rest of the story.

SILVER, the prize double agent in Fleming's stable, had made another of his journeys through the hills to Kabul in April of 1944, where he found that of his German contacts only Zugenbühler was left. It was evident, however, that German confidence in him was restored. Meanwhile, his old friend Witzel was back in Germany, and had a new project on the drawing boards: Operation WIDO, pursuant to which Witzel would fly into the North West Frontier with three colleagues and a transmitter, and make his way to Delhi to set up a military intelligence operation. SILVER reported that his Committee declined this offer when it was put to them. What was more important, from this time forward the Germans passed all SILVER's military information to the Japanese. On his preceding visit they had put him in touch with the Japanese in Kabul, in the person of a legation official named Inouye (who was in fact head of the Japanese intelligence organization in Afghanistan). Inouye became one of SILVER's staunchest champions, and by the latter part of 1944 SILVER would become almost a monopoly of the Japanese.

Subhas Chandra Bose had by this time set up headquarters in Burma. The existence of the Committee was foisted upon him—not without difficulty, but he trusted SILVER—and it was logical for the Committee to urge upon both the Japanese and Bose that a radio link be set up between the Committee in Delhi and Bose in Burma, with Japanese assistance. Such a link (codenamed RHINO) was arranged on this visit to Kabul. A station was to be opened at Bose's headquarters, to communicate with the Committee's MARY station at Delhi. SILVER was given all necessary material except the keyword of the cipher to be used; that was promised for his next visit. SILVER left Kabul toward the end of May.

His next visit took place in August. Again the Japanese were particularly receptive, presenting him towards the end of his stay with thirty thousand Afghan rupees and two thousand American dollars. On this visit a German-sponsored wireless link to Bose called ELEPHANT was arranged in addition to the Japanese-sponsored RHINO. SILVER was given the key for ELEPHANT, but without its wireless plan; that allegedly was to be communicated over the MARY-TOM link. Zugenbühler told SILVER that deferring the wireless plan reflected concern over the Soviets, but ULTRA revealed that it reflected renewed distrust of SILVER on the part of the Abwehr.

SILVER got back to Delhi in late September. From October 15 on, MARY tried to communicate with RHINO, but no contact could be made. It transpired that the RHINO schedules, ciphers, and instructions had been tampered with—perhaps by the Soviets, perhaps by SILVER himself; the mystery was never solved. Beginning in December and on into February, MARY told the Germans of the problems with RHINO and asked that ELEPHANT be expedited. They gave a vague response and Fleming concluded that ELEPHANT had probably merged into RHINO.

SILVER's next visit to Kabul took place in March and April of 1945. He was in fine fettle, exuberant over the Soviet military successes, and told Fleming he felt fitter than he had been since 1931 (after the execution of his brother); when he met with Fleming to go over the material to be taken to Kabul he "looked well and was wearing a natty suiting and a regrettable tie," as Fleming recorded. He left Delhi with a colleague on March 5, taking with him a lengthy report filled with false order of battle data and notional information on Allied plans in conformity with STULTIFY and its subsidiaries, and their successor, Plan SCEPTICAL; specifically, that the overall Allied strategy was to destroy the Japanese in Burma and secure a foothold in Sumatra to support a joint airborne and seaborne landing in southeastern Burma around Moulmein to cut the Japanese lines of communication, while the Northern Command captured Lashio and advanced on Thailand.

With three copies of this report and of a corresponding political report—a copy each for the Japanese, the Germans, and the Soviets—SILVER reached Kabul late in the month. By this time the Japanese legation was his primary contact. The Japanese themselves were in some trouble because of subversive activities by a minor official named Kawasaki, with whom SILVER had had some dealings, and MAGIC traffic revealed that the Germans had urged Shimada, the Japanese minister, to break off the SILVER case and that Tokyo had told Shimada to tread carefully and avoid contact with SILVER for the time being.

But this did not affect SILVER's reception by Inouye. He managed to spend a good deal of time with the latter, who was "over full of enthusiasm, and greeted SILVER with many salaams," in Fleming's words, and presented him with twenty thousand Afghani rupees and four thousand American dollars, apologizing for the small size of the gift. Both the Germans and the Japanese had rather inane questions and instructions for SILVER. In addition, the Japanese passed on to him seemingly panicky instructions from Bose for the Committee to begin large-scale sabotage activity throughout India. (Fleming and DIB had judged from instructions received by PAWNBROKER, OWL, and DOUBTFUL that this was coming and had already reported to Bose that serious police raids had taken place, which would be the excuse for not complying with these orders once received.)

Of more interest to D Division was the inability to open RHINO. When SILVER brought this up, the best that Inouye could offer was to tell him to keep listening. On April 23—five days after the MARY-TOM link was cut in the collapse of Germany—Zugenbühler gave SILVER the full code and wireless plan for ELEPHANT at last; Fleming reckoned that now in the death throes of the Reich the Abwehr was no longer interested in the security of the ELEPHANT link. Meanwhile, the Soviets had been told of the problem through diplomatic channels. They disclaimed any further interest in SILVER. According to SILVER himself, he attempted without success to see his contact, one Almazov, at the Soviet mission in Kabul but was turned away, and finally destroyed the copy of his report intended for the Soviets; and after SILVER's return the Soviets lodged a diplomatic protest over SILVER's efforts to get in touch with their officials.

SILVER and his colleague left Kabul on April 26, traveling in a *tonga* or two-wheeled buggy to avoid attention. The Afghans had instituted a system of internal passports for travel, and though he and his colleague had evaded the inspection posts coming into Kabul, the guards stopped them at the main post some twelve miles outside the city; he talked his way through, appealing to the officer in charge "to have faith in us as Pathans as he himself was a Pathan. I was all the while talking in a very polite way and my appeal to his Pathan sentiment had an effect and he allowed us to travel," with a bit of *bakshish* to the officer's men. He was back in Delhi in mid-May. Fleming wanted to send him back as soon as possible, but SILVER announced that he was feeling unfit and wanted to go fishing.

Notwithstanding the material furnished by Zugenbühler, ELEPHANT and MARY never made connection. Presumably coincidentally, on that same April 23 RHINO at last made connection with MARY, using the same schedules SILVER had brought back in October 1944. Through the summer until the Japanese surrender, sporadic but never wholly satisfactory contact was kept up. With Slim's advance and the fall of Rangoon, Bose was on the run; initially to Bangkok, where neither the Japanese nor the Thais wanted him; he sought then to set up operations in Japanese-controlled China.

In connection with ZIPPER, the invasion of Malaya to be described in more detail in Chapter 18, a role for SILVER was planned similar to the just-too-late warnings that GARBO had given for NEPTUNE and GILBERT may have given for HUSKY. He would arrive in Kabul on September 15 with news of ZIPPER—which would have opened a few days before. He got his briefing on August 14 and was flown to Peshawar to make his way to Kabul as usual. But Japan surrendered the next day, and SILVER was recalled. Bose was now dead—he was killed in a plane crash on Formosa on his way to Japan—and the SILVER case was allowed to die a painless natural death.

BACKHAND, the ex-Indian Army officer who had been dropped by parachute north of Rangoon in February 1944 with orders to turn himself in to the Japanese and function as a triple-cross agent, was accepted by the Japanese after some hesitation. In June, he joined the propaganda organization at Bose's headquarters, giving broadcast talks. In them he did his best to convey to the British information as to conditions in the Indian National Army and especially as to friction between the Japanese and the INA. Bose eventually became suspicious and had BACKHAND thrown into a concentration camp, from which he was liberated not long after Rangoon was recaptured.

The BRASS case continued satisfactorily. In an effort to perform order of battle deception of their own, the Japanese opened a notional subsource for BRASS, supposedly in the Meiktila area, to whom the British gave the name McTAVISH. But the complexities of such an operation were too much for them and they finally wrote off the new notional agent and reported his death in an air raid. Fleming tried to induce them to send a courier to make contact with the British forces in the Arakan, and he dropped by parachute an additional wireless transmitter in the hope that

they would bring yet another notional agent on the air; but in their justified lack of self-confidence the Japanese rose to neither bait. More nearly
successful was an audacious project called RASSENDYLL, designed to get
them to bring on the air a notional agent in Tokyo itself. A copy of
Kingsley's *The Water Babies* containing appropriate codes, schedules, and
intelligence data was dropped to the BRASS party with directions to mail
it to Japan. This caused considerable excitement to Fleming's old friend
the BRASS case officer, but he dared do nothing about it and BRASS radioed that the book had not been found.

The Japanese appeared to continue to value the PURPLE WHALES chan-
nel. Through it was sold a memorandum from the manager of the
Burmah-Shell Oil Company in Bombay to an opposite number in
Chungking indicating that Mountbatten planned to bypass southern
Burma in the cold weather of 1944 and attack Sumatra instead. Through
it went also the artist's sketchbook already mentioned, in aid of order of
battle deception. There was a purported personal letter from General
Carton de Wiart, the British military representative in Chungking, to
Chiang explaining why notional operations had been canceled. The Japanese were offered an item called JONAH, purportedly the negative of a
photograph of a map in SEAC's war room, but it evidently was never delivered. In late 1944, the channel cooled off and never came back to life.

Meanwhile, early in 1944 Cheng K'ai-Min, through whose good offices the PURPLE WHALES channel had been operated, had surfaced as
Chiang Kai-shek's chief of military intelligence. He continued to afford
what Fleming after the war called "a noteworthy degree of cooperation."
In February of 1945 he sent a young Chinese naval officer, one Lieutenant Pei (dressed in a lieutenant commander's uniform, presumably
for reasons of "face") to Delhi as liaison to Fleming. Pei remained until
the Japanese surrender, "amiable but fairly useless," Thorne recalled.
Fleming recorded that Pei was "abnormally incurious and his equable
temperament made it easy to deny him access to any Top Secret information."

In May 1944 Cheng told D Division that he had wireless access to the
Japanese through an agent in Macao. When Operation STULTIFY was
getting under way in late 1944, Fleming suggested to Cheng that a regular link to the Japanese be set up on that channel, feeding information

notionally acquired by an imaginary Chinese National Airways pilot on the Calcutta-Chungking run (FREEMAN), who would notionally buy information from a Chinese intermediary in Calcutta (WILLIS), who in turn bought it from a British staff officer in Calcutta in dire financial straits; FREEMAN would then sell it in Chungking to HARDY, a Chinese in Chungking who was in W/T communication with the Japanese in Macao. After some delay, Cheng not only agreed but, as Fleming later found out, upgraded the notional source, by making him a traitorous official at Chungking with access to the dispatches of Chiang Kai-shek's representative at Mountbatten's headquarters.

This channel, called INK, was finally opened early in 1945. It worked not to the Japanese military but to the offices of the Greater East Asia Ministry in Macao, which was believed to be more likely to assimilate subtle hints. (The operator of the Japanese set at the Macao end was a Chinese, second in command of the Japanese intelligence organization in Macao, and himself working for Cheng.) High-level traffic over INK included a considerable amount of order of battle data, news concerning VIPs, and various aspects of STULTIFY and, later, SCEPTICAL. Being located in China, INK should have been controlled by the Americans, as Newman Smith noted with some asperity; but he reluctantly furnished some innocuous material for it through Raymond in Chungking.

INK shut down in early August because the Japanese were moving their Macao intelligence office to Canton, and by the time of the Japanese surrender Cheng had advised that he considered the channel closed for good.

"Our Chinese channels were of definite value," recorded Fleming after the war, "and we saw enough of Chinese aptitude to maneuvers of this kind to be thankful that we were pitting our wits against a much younger Asiatic race."

In April of 1944, a group of eight Indian agents dispatched by Subhas Chandra Bose were landed on the Malabar coast by a Japanese submarine. Their assignment was to gather information in southern India and radio it to a naval station at Port Blair in the Andaman Islands, whence it was relayed to Penang and thence to Singapore. They had orders to split up into three parties (the whole group accordingly received the codename HAT TRICK), one party being destined for Madras city, another for Trichinopoly 250 miles away, and the third for Rameswaram, on the Indian end of the chain of sandbanks between Ceylon and the mainland. The Trichinopoly party was caught soon after the landing, with too much publicity for them to be of any use for deception. The other two groups were arrested more quietly, and were turned into special means channels. The Madras party was dubbed DOUBTFUL, and the Rameswaram party was called AUDREY.

DOUBTFUL, on the air from Madras from June 15, developed into a highly useful channel, becoming even more useful when he was given a notional branch operation at Vizagapatam, five hundred miles northeast of Madras. He helped build up the dummy order of battle, and played a major role in operations STULTIFY, SCEPTICAL, and SLIPPERY. Study of his traffic, together with ULTRA intercepts, confirmed that his information was rated highly by Bose and by Bose's lieutenant known to DIB as SWAMI.

AUDREY was moved at the outset from Rameswaram to Colombo, the main British naval base on Ceylon, where she would be better placed to provide naval intelligence. (One of DOUBTFUL's first tasks was to pass on a cover story to justify AUDREY's move.) She did not go on the air till August 22, and never developed into more than a low-grade channel; even so, she passed on a good deal of misinformation calculated to give the Japanese an erroneous idea of the strength of the British East Indies Fleet. She played a minor role in promoting the STULTIFY threat to Sumatra and the SCEPTICAL threat to the Straits of Malacca. She was still being built up when the Japanese surrendered.

HICCOUGHS and ANGEL ONE, the imaginary espionage and sabotage parties in Burma and Thailand respectively, continued on their way. HICCOUGHS finally acquired a concrete deception role in helping induce the Japanese to keep forces in lower and southwestern Burma that could have been usefully employed in opposing Slim's offensive. HICCOUGHS messages kept the Japanese coastal defenses "in a continual state of jitters," said Fleming, and "even isolated HICCOUGHS messages from time to time generated a state of alarm out of all proportion to their apparent import." Early in 1945, a new torment for the Japanese was devised

when HICCOUGHS traffic began suggesting that the head of the HIC-
COUGHS organization was in touch with an imaginary subversive organization in the Japanese Army known as the Suzuki Kikan.

HICCOUGHS' "cousin" in Thailand, ANGEL ONE, continued to "patrol
no man's land" until the last three months of the war. Having for its first
year been limited to Bangkok, in 1944 it was allowed to take over another notional organization with substantial funds and efficient communication covering Jumbhorn (DICKENS), Kanchanburi (TROLLOPE, with
a subsource CATO), and Chiengrai (THACKERAY); the leader, ANGEL ONE
himself, and his deputy, FOSSIL, remained in Bangkok. ANGEL ONE was
finally put to concrete deceptive use beginning in May 1945, to assist in
putting over two components of Plan SCEPTICAL, the airborne threat to
Bangkok and the amphibious threat to the Kra Isthmus, to be described
in Chapter 18.

The BULL'S EYE channel in Turkey began drying up in the summer of
1944, when Turkey broke off relations with Germany. Low-grade rumors could be passed through journalists and members of the British
colony in Turkey, but that was hardly a satisfactory substitute. The Japanese took over German intelligence operations; then Turkish-Japanese
relations were broken in January 1945. The Spanish ambassador to
Turkey agreed to accept nonmilitary telegrams from Japanese embassies
in Madrid and elsewhere, and there was some hope that BULL'S EYE material could be passed through this channel, but this did not work out.
After Spain broke diplomatic relations with Japan in April 1945, BULL'S
EYE was shut down.

On the other side of Europe from BULL'S EYE there was another channel
controlled directly by MI6 but used by D Division through Bevan,
known as SUNRISE. SUNRISE was a European assistant to Onodera, the
Japanese military attaché in Stockholm, who was supposed to be in indirect touch with a notional agent in London through a member of a
Swedish intelligence service. The notional agent gathered information
through such imaginary entities in London as the "East India Sports
Club" and the "China and Pacific War Investigation Committee." It was
high-grade stuff, liberally spiced with reports on lofty individuals and
inter-Allied disputes. Southeast Asia material for SUNRISE was cabled by

D Division to the London Controlling Section, where it was cleared, modified as appropriate, and sent on to Stockholm.

The human SUNRISE left Stockholm in 1944, but messages notionally from him (or from the imaginary agent) continued to reach Onodera. In May of 1944, use of the channel was offered to Newman Smith through Cunningham of the FBI. SUNRISE would have a fictitious source in Washington; foodstuff would be passed by the FBI (after coordination with Joint Security Control) to FBI liaison in London and thence to the British. With luck, the Japanese would eventually provide a radio transmitter to the notional source in Washington, furnishing a direct link. Smith thought the project seemed unlikely but that it could not hurt to try. Though American material was passed over the channel, no transmitter ever appears to have been furnished.

SUNRISE information was highly regarded both by the Japanese and by the Germans, to whom Onodera often passed it. No doubt it contributed to Onodera's promotion to major general and to his receipt from the Germans of the Verdienst Kreuz with Swords. Unknown to the Axis, of course, Allied codebreakers were picking SUNRISE information up as it went out to Japan; to prevent confusion it was labeled "Su intelligence," and Allied intelligence was warned that it was Allied-controlled.

Fleming put SUNRISE to extensive use; he had been asked by Bevan to do so, in 1943 and again in 1944. In early 1945 Bevan asked him to hold back, saying that the channel could not cope with the volume of material D Division was sending for it. SUNRISE was shut down in early July of 1945 for unspecified political reasons, to Fleming's disappointment, as he had been counting on it for SCEPTICAL.

In the summer of 1944, D Division inherited from MI5 in London one of GARBO's notional subagents, GLEAM by name. GLEAM was THERESA JARDINE, GARBO's Agent 7(3), the supposed mistress of Agent 7(4), an Indian member of the Aryan World Movement, until she was called up for service in the Women's Royal Navy in February of 1944. After her notional Wren training she was notionally posted to SACSEA. GARBO reported all this to the Germans, who directed that she be taught the use of secret ink and sent her three hundred pounds to cover expenses, together with directions to gather information about Allied land, sea, and air units in India. She notionally arrived at SACSEA headquarters at

Kandy in early August of 1944, assigned to the office of the Deputy Chief of Staff for Information and Civil Affairs.

GLEAM got on intimate terms with five different well-placed notional officers—a Royal Navy commander on the planning staff (EDWARD), a wing commander in the RAF (JACK), a commander in the U.S. Navy (DAVE), a British staff colonel (BILL), and a Dutch officer (ANDRIES)—and picked up a great deal of high-level information. Her material was sent in the form of a cover letter to a girlfriend, with secret communications overwritten in high-grade invisible ink. The texts were in fact composed and typed in Delhi and sent by fast bag to London, where they were written on appropriate notepaper and sent to the appropriate cover address in Lisbon.

Her reports disclosed a life that was one gay whirl, and, in her words, "full of surprises." In one week in January 1945, for instance, she lunched with JACK, dined with DAVE, played golf with BILL, and was given "a lovely squishy tea with cream cakes" by ANDRIES. Not long after, she represented her department in a beauty contest for which she wore a blue-and-yellow satin two-piece bathing suit, but lost to a Dutch girl, having eaten too much lunch. A subsequent letter was written from bed, to which she had retired with a tummyache probably resulting from a cocktail she and DAVE had mixed, "a heavenly one which looked like soap-suds and tasted like chocolate blancmange." ("More nuts!!" Newman Smith endorsed on his copy of this account.)

It took some six weeks for each letter to get from Delhi to Lisbon; most of the information was then passed on to the Japanese, taking another six weeks. Notwithstanding the delays, GLEAM was a useful channel. Twelve letters were passed over it between August 1944 and February 1945, ten of which were acknowledged by the Germans; a German message in December 1944 described her product as "stupendous." Then, in January 1945, ULTRA showed Berlin expressing surprise that certain material from JAVELINE (GLEAM's German codename) was the same as certain SILVER material. It transpired that D Division had passed the same information through both. MI5 concluded that this posed too much of a risk of blowing GARBO, and GLEAM was shut down. She had a notional automobile accident in which her notional bottle of secret ink was smashed and she was injured. But Germany collapsed before that could be put through.

* * *

Fleming's other legacy from MI5 in England, FATHER, the Belgian test pilot, got his new radio on the air at the beginning of August, and quickly made contact with Germany. The sender was not in fact FATHER but an operator impersonating him, and FATHER himself did not last long after that. He was temperamental, and in October it was decided to ship him back to Europe. His German control was told that he was returning to Belgium now that it had been liberated, and that he had recruited, to take over his espionage activity and his wireless set, a disaffected Indian courier with Strategic Air Force headquarters in Calcutta whose duties carried him to many parts of India. During November and December, this notional successor—codenamed RAJAH (sometimes spelled RADJA or RADJAH), and in reality portrayed by a sergeant in the Calcutta police—was gradually built up, and on January 3, 1945, RAJAH took over the case entirely. The channel itself was henceforth referred to as DUCK, the name by which the transmitter had been known before it came into FATHER's possession.

DUCK developed into a valuable channel carrying a good deal of high-grade information, at least some of which the Germans passed on to the Japanese. D Division tried to keep DUCK going beyond the German collapse, by having RAJAH ask his control to pass details of his ciphers and frequencies to Subhas Chandra Bose. But nothing came of this, and DUCK expired with the fall of Germany.

Then there were the BATS channels: OWL, in Calcutta, and MARMALADE, in Assam.

MARMALADE never did develop. At the end of 1944, with Slim's offensive getting under way, Fleming decided that moving MARMALADE to Imphal might upgrade his information and stir up some Japanese interest. But this did not help, and he was finally shut down in July 1945.

OWL, by contrast, steadily grew, keeping up a stream of information as to order of battle and other matters of interest. He was useful in STULTIFY, and on January 12, 1945, he kicked off CLOAK by a message that the British would cross the Irrawaddy at Chauk; this produced a special congratulatory signal from his control. After the fall of Rangoon he managed to transfer himself there, and in July and August he was used in aid of SCEPTICAL, to be described in Chapter 18. Alone among

D Division's channels, Owl got a farewell from his control after the Japanese surrender: "Thanks for the trouble," was the message on August 17. "Good-bye."

Pawnbroker, Bose's W/T agent at Calcutta with an outstation at Bombay, developed into a channel as valuable as Owl. From March 1944, he sent information regularly, including high-level material notionally obtained from a genuine organization called the Bengal Volunteer Group. The police broke up that group in the autumn, but early in 1945 he was notionally linked with another genuine organization called the All India Youth League, in which by July he had established such high-grade notional sources as an Indian Air Force officer and an Indian Army staff officer. From March 1945 to the Japanese surrender he was a major contributor both to military deception and to DIB's political counterintelligence, and rated highly by Japanese intelligence. He played a leading role in Sceptical.

The end of October 1944 brought a new channel of the Brass type. On the 31st of that month, SOE—known in those parts as Force 136—sent a party of four Malays codenamed Oatmeal (their wireless codename was Violin) on a sabotage and intelligence mission to the state of Kelantan in Malaya. Oatmeal was composed of a leader, Captain (later General Tan Sri) Ibrahim bin Ismail; a radio operator (there were supposed to be two, but the second operator got too airsick to be left off), and a contact man.

Violin went on the air on November 13. "Have you met Miriam?" asked their control in Colombo. "Yes, I've met Miriam," came the response—the sign that they had been caught. (If all was clear the response should have been "Two Scotsmen left here two days ago.") They had indeed been betrayed and arrested, beaten up, and promised that their lives would be spared if they acted as double agents. Moreover, the Japanese had seized Ibrahim's signal plan, including the security checks. With extraordinary coolness, Ibrahim persuaded the Japanese that the security checks had been written in reverse for additional security, so that the "Miriam" response meant that all was well and the "Scotsman" response was the danger warning. SOE turned the case over to Fleming.

For Oatmeal to be useful for deception its intelligence role had to be refocused on military and naval matters to the exclusion of political affairs. So inquiries directed to OATMEAL concentrated increasingly on military questions; then at the end of the year OATMEAL was asked whether it could relocate to the neighborhood of Penang on the other side of the Malay peninsula, a better vantage point for military intelligence. OATMEAL complied "with implausible speed," as Fleming put it, moving during January 1945 to Wellesley Province. There they were installed in a house north of Taiping, dressed as enlisted men in the local volunteer force; and there they received and responded to a steady flow of questions about troop movements, airfields, defense installations, and such.

The Japanese handling of OATMEAL, "though by no means brilliant," said Fleming, "was more methodical, imaginative and enterprising than any of their work that we came across in connection with other channels." A notable example came at the end of April 1945, when as part of a systematic deception plan they used OATMEAL to report the arrival of five divisions of reinforcements from Japan, complete with plausible details such as codenames, troop movements, and locations and movements of high-ranking officers. Unfortunately for them, from ULTRA intercepts D Division knew what the systematic deception plan was.

OATMEAL was run till the end of the war. The Japanese not only never suspected the true state of affairs; they actually conferred a decoration upon Ibrahim for his services.

In December 1944, three parties of radio-equipped Bose agents were landed from Japanese submarines on the east coast of India and two were promptly caught. One party, given the codename TRAVEL, was landed on the Orissa coast, and another, dubbed TROTTER, further south near the mouth of the Cauvery River. A third, called DUFF, was landed at the same time as TROTTER. It was the only known party of enemy agents entering India that was never caught, but apparently they never established contact with Bose.

The TRAVEL party consisted of a group of two agents, dubbed DUPLEX, and another group of three, dubbed TRUNK, who had received special training at Penang. One of the DUPLEX party was killed on arrival; the other went on a hunger strike and refused to talk, and was sent for trial. The TRUNK party talked freely and was worked as a double under the codename TRAVEL. Their set went on the air in April from Howrah, in the Calcutta suburbs. Communication was spotty, but from the fall of Rangoon to mid-July TRAVEL managed a reasonably steady flow of middle-grade material in aid of SCEPTICAL.

TROTTER consisted of three Madrassis; one was a radio operator who was to settle in Madras, gather intelligence there, and communicate with their control, while the other two were to proceed to Trincomalee and Colombo respectively, in Ceylon, and send reports in secret ink to the operator in Madras. Initial contact was established and good naval misinformation was sent; but the case never worked out, mainly because of the incompetence of the operators at the control end.

The last channel of any significance opened by Fleming, in January 1945, took his special means system full circle to its beginnings with HICCOUGHS. Called COUGHDROP, it too was an imaginary party operating behind Japanese lines to which messages were sent that the Japanese were meant to read.

COUGHDROP was part of FANG, the element of STULTIFY designed to make the Japanese anticipate a landing on Sumatra. It was a notional four-man party composed of an English leader and radio operator, a Dutch officer with experience in Sumatra, and two native couriers, which supposedly landed in northern Sumatra at the end of December 1944 with the mission of collecting military intelligence and reconnoitering suitable landing beaches and sites for airfields. In January, COUGHDROP's control went on the air and went through an act as if trying to establish contact: calling and listening, calling and listening. Eventually he pretended that contact had been established, and thenceforth, as Fleming put it, "the operator at control played his part at the key in much the same way as an actor conducting a stage telephone conversation, acknowledging receipt of traffic, changing to receiving frequencies, giving signal strength, asking for repeats of faulty groups and complaining of interference or atmospheric conditions." Realism was added by housekeeping messages, references to inbound messages, and the occasional repeat back of purported inbound messages to make sure that they had been received correctly—and to enable the Japanese to decipher them.

To make sure the Japanese picked up COUGHDROP, its control transmitted on BRASS's frequency (inbound traffic was notionally on a different frequency) and usually within thirty minutes of Brass's transmissions twice a week; sometimes, indeed, COUGHDROP transmissions were allowed to overlap BRASS's schedule. To make sure that even the densest Japanese cryptographer would be able to decipher the traffic, COUGHDROP's control used a simple substitution cipher with a keyword containing the same number of letters as BRASS's keyword, and on three occasions three sets of messages of the same length were sent, each of which began with the same three letters.

Unfortunately, after the fall of Rangoon it transpired that COUGHDROP had failed because of the fidelity of the Karen agent who acted as the BRASS operator. From associates of his who were liberated at Rangoon it was learned that the first time a COUGHDROP transmission had overlapped a BRASS schedule he had realized that the frequency was being used to support another Allied party and had loyally concealed this from the Japanese. There was obviously nothing more to be expected from COUGHDROP, though for safety's sake it continued to be run on a reduced scale till the end of the war.

Two new channels, ACCOST and POINTER, opened up towards the very end of the war but were not put to use. ACCOST was a party of six Annamese whom MI6 dropped by parachute some thirty-five miles north of Saigon in July 1945. They were soon captured by the Japanese. MI6 turned them over to D Division for deception along the lines of OATMEAL, but the war was over too soon for them to be of any help. POINTER was a group of former MI6 agents, all Karens, whom the Japanese had captured and imprisoned in Rangoon. When the Kempei Tai hurriedly left Rangoon in late April, the POINTER party were released and given the same transmitter that had been used by BRASS's control, with instructions to act as staybehind agents for the Japanese. They of course promptly turned themselves in to the British, and were put on the air as a deception channel on May 20; but it soon transpired that Japanese suspicions made it unpromising for deception purposes, and POINTER saw little further traffic. The case was closed at the end of July.

At the Christmas season of 1944, Fleming had a pleasant reunion with his brother Ian, who had come out to review the intelligence infrastructure of the British Pacific fleet and managed to come up from Colombo to Delhi for a visit.

Another visitor was Ronald Wingate, who arrived in the theater in November on his inspection trip, and stayed for several weeks studying the deception needs and problems of the command, meeting also briefly with the American Young Ladies' Seminary team and stopping off on the way back in Cairo with Peter Thorne to discuss with Dudley Clarke possible help from "A" Force's Middle East channels. On his recommendation, it was decided to split off a small Policy and Plans section of D Division, to function in Kandy as part of the SACSEA headquarters staff, with Fleming continuing as head of the operating bulk of the unit (officially to be renamed "Force 456," though it continued to be called D Division in practice).

With his savior faire, tactfulness, and broad experience, Wingate was the natural choice for the Kandy assignment, bringing his London Controlling Section background to SEAC's headquarters planning, and acting as a buffer between Fleming and those who might share Whitworth Jones's animosity. He arrived back from England to take over on March 25, with the official title "Head of D Division." (Lady Jane Pleydell-Bouverie remembered that when Wingate, who was "terribly keen on making his pennies on the Stock Exchange," was sent to SEAC, "the thing that really worried him was that he couldn't ring his stockbroker up every day.")

In an unrelated change, Fleming's old colleague Bill Cawthorn, who had in time become a major-general, was succeeded in April 1945 as Director of Military Intelligence India by Major-General A. K. Ferguson.

Soon after the fall of Rangoon, an "Advance HQ D Division" was opened in that city under one Major J. M. Howson.

In March 1945, "Ran" Antrim—Lieutenant-Commander the Earl of Antrim RNVR, Dudley Clarke's last naval person—arrived to take over D Division naval duties from Robertson MacDonald. He had very little to do. Once Fleming found him killing time by flipping through the pages of a girlie magazine. "Really, Peter," he said to Fleming's shocked reaction, "you sound almost like a bishop." Fleming always said thereafter that if he ever wrote his autobiography it would be called *Almost a Bishop*.

Meanwhile, in October 1944 the two physical deception units, 303d

Indian Armoured Brigade and 4th and 5th Light Scout Car Companies, were reorganized and consolidated into a combined British and Indian unit called "D Force," commanded by Lieutenant-Colonel Turnbull, with its headquarters at Barasat near Calcutta.

D Division also maintained a Tactical Headquarters under Frank Wilson, with officers attached to field commanders down to corps level.

Paralleling D Force, No. 1 Naval Scout Unit, a naval tactical deception team similar to the American Beach Jumpers, joined the theater early in 1945 attached to the Royal Navy East Indies Fleet.

So with a revised organization that freed Fleming to supervise implementation in the field while Wingate could bring his London experience to strategic planning and a calming presence to Kandy headquarters, a varied menu of special means channels, a full complement of tactical deception units, and the experience of STULTIFY, D Division was ready for the big operations that would come in cooperation with the Americans as the Allies closed in upon Japan.

CHAPTER 17

BLUEBIRD

The autumn and winter of 1944–45 saw the maturing of American deception. After successfully launching the Young Ladies Seminary, Newman Smith at last secured formal authority for Joint Security Control to participate in planning; he at last got the Navy interested; he opened new special means channels directly to the Japanese through neutral countries; and he coordinated the implementation of BLUEBIRD, the most elaborate American strategic deception of the war.

With Bill Baumer leaving the Joint War Plans Committee in September 1944 and departing for Europe at the end of October, and Baron Kehm already gone, there was no longer anyone with a background in deception on any of the regular planning committees. It seemed a good time to get for Joint Security Control the official planning authority that Smith had long craved. A November 17 memorandum over the signatures of General Bissell and Admiral Thebaud, Admiral Schuirmann's successor, pointed out that "As a practical matter, it is impossible under existing conditions for Joint Security Control to prepare a deception plan that would be in consonance with operational plans." As a result, the Joint Planners issued on November 30 an invitation to Joint Security Control "to collaborate with the Joint War Plans Committee in the development of deception objectives and preparation of deception plans for consideration by the Joint Staff Planners." On December 4, Smith and Goldbranson for the Army, and Captain Jeffrey Metzel and Lieutenant Commander Robert F. Nelson for the Navy, were formally designated to collaborate with the Joint War Plans Committee, with an air member to be made available subsequently; on March 21, 1945, Van Brown was appointed for air.

*   *   *

Smith's second accomplishment that winter was to get the Navy seriously interested in large-scale deception at last. Perhaps he did too well from his Army perspective, for the Navy threatened to take bureaucratic control of the whole enterprise.

The Navy's interest had already picked up a bit—perhaps in part because Admiral Bieri, who had been rather obstructive in 1943, departed as Assistant Chief of Staff (Plans) ("F-1") in May of 1944. In the summer of 1944 an additional reserve officer, Lieutenant Commander (subsequently Commander) Robert F. Nelson, was assigned to Smith's section, taking on responsibilities in connection with special means. (Pete Schrup, it would appear, left the section at some time thereafter; his handwriting appears on documents in the file in early 1945 but not subsequently.)

Forty-seven years old in 1944 and a 1918 graduate of Annapolis, Bob Nelson hailed from Moultrie, Georgia. Popular and successful at the Naval Academy, where he had held high midshipman rank, been lightweight wrestling champion, and assistant editor of the class yearbook, he had resigned from the Navy to enter the business world in 1922. In 1925 he was one of the organizers of the Glassine Paper Company in Pennsylvania, and was its president when in the summer of 1944 he rejoined the Navy as a reserve commander. Both George Dyer, the original Navy member of Joint Security Control, and Ray Thurber, soon to have a similar role, had been his Annapolis classmates.

Navy interest may also have been stirred by the return of Douglas Fairbanks to the Washington scene, joining the COMINCH staff at the beginning of October. Bevan asked that he be routed through London to visit LCS; which he did, finding as did others that the V-2s were "really fearsome." (This was evidently another of Bevan's efforts to lobby for reform of Joint Security Control.) Settling in with Mary Lee and their two little girls in a handsome house on Massachusetts Avenue, he wrote for Newman Smith soon after his arrival, with a copy to Admiral Schuirmann, a long memorandum on deception organization based on his own experience. He recommended essentially the British model, with a Bevan-like chief having access to the highest levels, heading an organization, under Operations rather than Intelligence, with both planning and executive responsibilities; which, "in the light of present war trends, should have a strong majority of naval personnel."

At this point, Admiral Schuirmann was reassigned to duty in Europe and Rear Admiral L. Hewlett Thebaud succeeded him as head of naval intelligence and Navy member of Joint Security Control. A fifty-four-year-old Annapolis graduate of the class of 1913, Thebaud was a seagoing officer whose only intelligence experience had been as an assistant naval attaché in Europe for two years in the 1930s. He had commanded a cruiser at Sicily and Salerno in 1943, and a cruiser division in the Pacific at the time of the Battle of the Philippine Sea in the summer of 1944.

Smith knew that the Special Section of Joint Security Control would not be taken seriously by the naval staff, in Washington or the Pacific, until a regular officer of captain's rank could be brought on board. Smith turned to the captain who had shown over the years the most interest in deception work: Jeffrey Metzel of the Readiness Section of the COMINCH Readiness Division, who had followed deception from Fairbanks's early Beach Jumper days. Smith had kept in touch with the Readiness Section; in October, for example, he had sent Goldbranson and Nelson to discuss with the Section the possibility of decoy fleet operations into the Yellow Sea and against the Kuriles. Smith asked that Metzel be given an additional assignment as Senior Naval Member of the Special Section until a captain could be obtained for permanent duty; pointing out, as Metzel reported to Thebaud, that "an experienced Naval partner" was needed to get cover and deception fully functioning in the Pacific, and to strengthen the Special Section "enough to compare favorably with its British counterpart, which had evidenced its desire to assume leadership in cover and deception matters in the naval war against Japan." (Raising this specter could always be relied on to attract the Navy's attention.) Metzel received his assignment on December 5, and served until late February, temporarily joined by a certain Lieutenant Commander R. W. Brandt on December 15.

The officer destined to be the permanent Senior Naval Member of the Special Section reported for duty on January 30. He was Captain Harry R. Thurber, Annapolis 1918, known to his friends as "Ray," a forty-nine-year-old native of Washington State. He had been Halsey's operations officer in the South Pacific in 1942–44, and had commanded the cruiser *Honolulu* at the Marianas, Peliliu, and Leyte; at Leyte she had taken a Japanese torpedo but Thurber brought her home safely. He had no intelligence background, and indeed his only experience outside the conventional Navy career had been in Navy public relations. Events would suggest that he brought more energy than imagination to the job.

At the end of his temporary tour of duty three weeks later, Metzel produced a proposal for radically restructuring Joint Security Control. Effective deception against Japan "can save many thousands of American lives and shorten the war," he said in an "urgent interim report" of February 19 to Admiral Thebaud, "but it will require an adequate deception organization" with adequate directives. "The small staff of the Special Section are able and zealous, and they are striving to advance the introduction and exploitation of deception. They are making progress, but slowly and under great handicaps. . . . Deception in theaters of U.S. strategic responsibility is being planned and implemented largely by virtue of the Special Section getting each step accomplished as a favor, without benefit of directives."

Metzel accordingly recommended to Thebaud a new organization. Defining "cover" as false information given to our own side to prevent leakage of information, and "deception" as "everything involved in feeding to the enemy false or misleading information or information intended to cause him to draw false conclusions," the new organization would leave "cover" with the direct security functions of the Executive Section of Joint Security Control, and transfer the Special Section to a new agency for "deception." The new agency would be denominated "Joint War Information Control," and for cover purposes would have overt responsibility for coordinating propaganda and other psychological warfare with military plans. It would have broad authority to conduct studies, train and assemble a backlog of officers assignable to commanders for temporary duty, administer implementation of deception by other than originating theater commanders, coordinate with allies, study enemy reactions, develop techniques, and initiate deception staff studies. It should be well staffed, for "the stakes are so high that it would be well worth while to over-staff cover and deception many fold than take any risk whatever of under-staffing it." Likewise important was the establishment of an adequate Navy deception section in COMINCH headquarters. "These steps are both urgent," said Metzel, "but neither need wait for the other."

Thebaud passed Metzel's proposal to Bissell with the recommendation that they originate the proposal with the Joint Chiefs. But OPD, while pleased that "The Navy has finally taken an interest," was not receptive, it would appear, to the drastic changes that Metzel had in mind, and the proposal did not go forward.

Meanwhile, Thurber's initial task had been to prepare a study of the organization and requirements of the Navy deception section. Submitted on February 22, his paper was a typical new-broom study. It found what was already known: That "there was a lack of interest and understanding among responsible officers in the Navy Department and on the staffs afloat," and that the naval side of the Special Section needed to be strengthened. True to form—for open-armed cooperation with the other services has never been the Navy's style—Thurber recommended a separate Special Section for the Navy, adequately staffed (fourteen officers and enlisted Wave secretaries), functioning under its own roof away from the Army people in the Pentagon. It should report directly to the COMINCH chief of staff "on the level of command to that allotted to the Division Heads," with office space "near the COMINCH Plans Division to facilitate liaison."

Thurber's recommendations were approved with two major exceptions: Headquarters COMINCH was no more willing to have a Dudley Clarke style stand-alone deception organization "on the level of command to that allotted to the Division Heads" than any other American staff had been. The new "Navy Special Section" was formally inaugurated on February 28 as part of COMINCH Combat Intelligence under Admiral Thebaud, with the designation "F-28." Nor was it as large as Thurber had envisioned; nevertheless, it was a substantial section of eight, soon nine, officers plus enlisted support.

Thurber himself was its chief. (Metzel returned full time to the Readiness Section.) Nelson and Burris-Meyer were brought over from Newman Smith's section, Nelson with nominal responsibility for plans, Burris-Meyer for operations. A reserve captain, F. H. Creech, was the intelligence officer, with a lieutenant to assist him; he had previously been in charge of Navy censorship. Commander J. Q. Holsopple, who had often advised Joint Security Control on communications deception, received a dual assignment to F-28 as well as his basic assignment to the Operations Division; a full-time lieutenant as communications officer assisted him. A Wave lieutenant, assisted by two, later four, enlisted

Waves, handled administration. Except for Creech, the new section had offices in the seventh wing of Main Navy on Constitution Avenue; Creech was with Newman Smith in the Pentagon, as the Navy's intelligence representative for special means matters.

Surprisingly, the one Navy officer with extensive real-life experience in deception was not brought into F-28. Douglas Fairbanks found himself assigned to the "Post War Naval Planning and Sea Frontiers Section" of the Plans Division of COMINCH, with his own designation "F-1421." This section had nothing to do with deception and Fairbanks's assignment to it may have been merely nominal. He himself said he had "no terribly demanding duties," and he continued writing memoranda on the subject of deception for Rear Admiral D. B. Duncan, Bieri's successor as King's "F-1" or Assistant Chief of Staff (Plans), and for Duncan's successor from July 3, Rear Admiral M. B. Gardner. He floated various ideas of his own for deception projects, none of which got very far.

Late in July, he was ordered to go out to his old friend Mountbatten's theater to help with tactical deception for ZIPPER, the invasion of Malaya; and thence to Okinawa by way of Chungking for work in connection with the invasion of Japan. But the war ended before he could leave.

Newman Smith may well have felt that he should have been careful what he wished for when he urged Captain Metzel to persuade the Navy to take a greater interest in deception. After two years of uphill work as the chief deception officer, he was suddenly left as merely the head of an "Army Special Section" that was outnumbered by the Navy, composed, for the time being, of himself, Goldbranson (who would depart for MacArthur's headquarters in May), Van Brown, who joined in March, Bobby Rushton, who joined in early 1945, and Lieutenant Dorothy Stewart; important technical help was also received from Major Jack G. Hines of the Army's Signal Security Agency.

Another blow soon followed, when Thurber was made a full member of Joint Security Control itself. General Hap Arnold wanted Joint Security Control's charter amended to add an Air Staff officer as a full member. The Navy was always on guard against being outnumbered by the Air Force's being treated as a separate service, so Admiral King would approve the suggestion only if another Navy officer was added as well. The

Chiefs approved this on March 13, and the revised charter was promulgated as JCS 234/5/D.

The Army representatives were then Bissell as the G-2 and Major General James P. Hodges, the A-2 or chief Air Force intelligence officer. The Navy designated Thebaud and Thurber. The whole affair was merely symbolic, as far as the Air Force was concerned; Hodges was never active, nor was his successor as of June 12, Major General Elwood R. "Pete" Quesada, the legendary tactical air commander in Eisenhower's theater. The net effect was simply to give Newman Smith's Navy opposite number, Thurber, greater authority on paper than Smith had. Bissell, however, protected Smith's position by ensuring that formal papers from Joint Security Control were signed only by the general or flag officers; he and Thebaud signed as "Senior" Army and Navy Members. (He was an air officer himself, be it remembered.) And by July, Smith was again referred to in the documents as "Chief, Special Section," Joint Security Control.

F-28 tended towards busywork and generating paper. Thurber set out to prepare a document that would summarize the essentials of cover and deception. He had to wrestle with the concepts. Fairbanks tried to help, but could not get very far. Drafts were passed back and forth between Thurber and Newman Smith during March and April. Smith forwarded the final product, substantially as revised by him and Goldbranson, to Bissell on April 18. Describing it as "a sort of 'codification' of many previous JCS and JSC papers," he advised Bissell: "While it is not considered this paper is very vital, Capt. Thurber feels very strongly about it. It can do no harm." After the usual staff massaging, the Chiefs approved it on May 22, and it was circulated to the theaters as JCS 498/10.

Notwithstanding Smith's faint praise, JCS 498/10 was a valuable paper (and remains one; its definitions are those quoted in Chapter 2). It put into one place for the first time a set of definitions and a list of examples of deception techniques, and a codification and amplification of existing Combined Chiefs and Joint Chiefs directives on deception machinery. It outlasted the war as a basic document, known, at least in Navy circles, as "The Gouge."

Smith's Army Special Section was not outnumbered by the Navy for long. Soon after the end in Europe, Harris, Ingersoll, Eldredge, and

Sweeney joined Smith's office. "Our organization has certainly grown since you left here," wrote Brown to Goldbranson in August, just before the war ended, "and is in my opinion asserting itself more every day." Smith put Sweeney in particular charge of monitoring special means. Harris became essentially Smith's deputy, though his intended ultimate destination was MacArthur's headquarters, relieving Goldbranson.

Ingersoll closeted himself in a basement room in the Pentagon dictating the final report of 12th Army Group Special Plans Branch. In later years he remembered it as his masterpiece, "a comprehensive analysis of the principles of and contemporary practice of military deception"; unfortunately, the reality does not live up to this billing. The three prepared a short "Informal Report to Joint Security Control" as well. It was relatively brief, and, like the report that had so much distressed Bevan, it tended to generalize from the very limited perspective of its authors and with no knowledge of the realities Smith had struggled with for two years. It offered recommendations for deception organization for the Pacific. It put in a pitch for Eldredge to return to Europe (and Diana), warning that the British were already researching deception results and suggesting that an American officer should participate to safeguard American interests. It noted of their particular bête noire, Reeder of the 23d, that owing to "lack of flexibility and receptivity to new ideas in [his] planning," it was "felt that he is not suited for this type of work." ("Cuts Reeder's throat," someone wrote by Harris's name on the cover page.) Some kind of difference of opinion developed with respect to these reports, and they were not distributed further, with the exception of a separate annex on air deception.

Ingersoll was gone by July 1 and probably early in June. "With the Third Reich's collapse, I was through with playing soldier," he wrote long after, and he did not want to go to the Pacific. And he wanted to rush into print with the first tell-all book about the European war. In view of his age and accumulated service points, he was able to get himself discharged promptly.

Eldredge, too, in his own words, "did not give a damn about the war with Japan." But he could not get out as easily as Ingersoll. He went on leave back in Hanover as soon as the Informal Report was completed, worked briefly on Japanese invasion planning, and returned to Europe in July and August, working with Cuthbert Hesketh to debrief German officers and examine the German records, not only in Germany but in Copenhagen and Oslo, to evaluate the effectiveness of deception, passing the results to Wild back in London. He did not realize that Harris planned to take him along to the Pacific. But the Japanese surrender relieved him of that. "I thereupon 'stumbled,' " he wrote years later, "and was unable to leave London where I had a delightful time with [Diana]. Billy . . . was furious—*tant pis*—but he got even by keeping me in the army until December, as I worked on 'lessons learned' from G-3, Special Plans, 12th Army Group."

Over on the Navy side, the one thoroughly experienced member of Thurber's group saw a change in his life a few weeks after F-28 was up and running. On April 12, Hal Burris-Meyer and Anita Mersfelder—the bride was a lawyer in an era when there were few women in the profession—were married in the chapel at the Naval Academy in Annapolis. Bob Nelson and his wife Sylvia drove the couple back to Union Station in Washington and saw them off for a honeymoon in Virginia. It must have been the ideal honeymoon; they did not learn for a week that Roosevelt had died the afternoon they were married.

The third of Newman Smith's major initiatives that autumn of 1944 was to accelerate development of special means channels to the Japanese, not through the FBI but through the American military attachés in Argentina, Portugal, and Switzerland.

Still neutral in the war, Argentina had been ruled since a coup d'état in June 1943 by a military junto heavily influenced by Axis sympathizers. Disclosure of an egregious example of German espionage and subversion triggered a break in diplomatic relations with Germany and Japan in late January of 1944. A partial roundup of Axis agents followed, but the internal security forces did not have their hearts in it, and some German channels and a clandestine transmitter still remained. Army officers unhappy with the breach with the Axis conducted what amounted to another coup in February, and by July one of their number, Colonel Juan Perón, had emerged as the most significant figure in the government.

During the next months Perón was more interested in consolidating his domestic position than in faraway wars in which Argentina had little interest. Axis diplomats, no longer formally accredited but unable to go home, moved freely about the country. Eventually, Perón and the Yankees made a deal: The United States would recognize the regime and favor the admission of Argentina to the new United Nations organization to be formed, if Argentina would declare war on Germany and Japan; which she did on March 27, 1945.

Already in mid-1943 Smith had brought up special means in Argentina with Major J. C. King, who was in charge of special intelligence matters in the military attaché's office in Buenos Aires, during a visit of the latter to Washington, and had received assurances from the attaché himself, Brigadier General John W. Lang, that channels were available. As will be recalled, this connection was put briefly to use in February 1944 in connection with BODYGUARD. After some initial irritation over the informal ways of Joint Security Control, Lang joined in the game with a will, suggesting phony documents in PURPLE WHALES style.

Lang was a witty and articulate Mississippian of the class of 1907 at West Point, sixty years old in 1944. A successful infantry officer's career had been punctuated by several special assignments in the Spanish-speaking world, beginning in the Philippines before the First World War, including a term as military attaché in Madrid during that conflict, duty as chief of the American military mission to Colombia from 1939 to 1941, and service as the military attaché in Buenos Aires from February 1941. His deputy for such matters, Major (subsequently Lieutenant Colonel) King, had in civil life been head of South American operations for Johnson & Johnson, the American medical-supply company. Lang and Newman Smith built up a cordial relationship, enhanced by a personal visit to Washington by Lang in February 1945.

Lang was transferred to Mexico City in June 1945. The last few months of special means work in Buenos Aires was overseen by his successor, Brigadier General Arthur R. Harris. Harris was a fifty-four-year-old Nebraska native from the West Point class of 1914, who had seen service in Europe in the First World War, had ridden with the American polo team in the 1920 Olympics, and after a successful career as an artilleryman had headed the Latin America section of G-2 in Washington in 1939–40; as a result of which General Strong had plucked him from troop duty not long after Pearl Harbor and sent him to Mexico as military attaché. Newman Smith discussed special means with Harris before he left Washington for Argentina.

In September 1944, Smith began a program of regular foodstuff for

Argentina, with emphasis on the war with Japan. Cautioning that the use of phony documents "must be reserved for very special and specific purposes on rather rare occasions," he expressed the hope that Lang could develop channels to the Japanese. Lang responded with enthusiasm; a fortnight after receiving Smith's letter he was already cabling back to Smith: "Fishing fine send more bait." (As between Smith and Lang, "fish food" thereafter became the customary term for special means foodstuff. Their correspondence made frequent references to "this piscatorial adventure," "rising to a Royal Coachman," "intimate acquaintance with the eddies and shoals," and similar pleasantries.) The passing of fish food could be a slow process. The weekly supply could take two weeks or more by courier. (For security reasons, Bissell did not want material that was not urgent to go by cable.) Then, at the Buenos Aires end, King had to meet in secrecy with the go-betweens, and at each meeting the next one would be scheduled; it was seldom possible to arrange an unscheduled meeting on short notice.

Lang and Harris worked closely with the FBI's Special Intelligence Service—indeed, Lang had an attractive daughter who went dancing with Ken Crosby of the SIS Argentine station—but kept their own channels to themselves. Sometimes there were problems, as when, in May 1945, an eager new FBI man came out from Washington with word that faint radio communication had been heard going from Argentina to Chile and thence to Japan. Lang knew that the Japanese agent Konomi Miyamoto operated a clandestine short-wave radio near Buenos Aires, and indeed Lang relied on it for Miyamoto to pass his fish food back to Japan. When the FBI offered to tail Miyamoto, Lang fobbed them off, saying that his people were doing it; and he promptly got in touch with Newman Smith and asked that someone tip off J. Edgar Hoover not to let his people "do something which might jeopardize the transmission of ichthyological fodder." Smith spoke with Bissell, and Bissell spoke with Hoover, and there were no more problems.

Lang had lines to three of the principal figures in Japanese intelligence in Argentina. One was Rear Admiral Katsumi Yukishita, the Japanese naval attaché since October 1941 (he had held that post before, in 1930). Yukishita had had a distinguished career in the Japanese Navy— he commanded the battleship *Nagano* in the 1930s, and had headed the Kure Munitions Depot—and was in charge of Japanese maritime espionage in Argentina. He had spent a brief period under house arrest in February 1944, just after the breach of diplomatic relations, but was now at liberty and accessible. A second was that same Konomi Miyamoto who ran the clandestine transmitter. The leading Japanese agent outside the official delegation, Miyamoto was a multilingual world traveler who had lived in the United States, Germany, and Brazil, and was currently employed in Argentina by the important Japanese newspaper *Asahi* (and gave lessons in Japanese on the side). He was believed to head a network of some twenty agents throughout Argentina, and to have been the chief Japanese contact man with German espionage in that country. He had used courier channels to Spain as well as his transmitter. A third key agent, to whom Lang later developed a channel, was Saburo Suzuki, Argentina correspondent for the *Mainichi* newspaper of Tokyo.

Lang opened lines to Yukishita and Miyamoto very quickly. His channel to Yukishita was a former Japanese Army officer codenamed PIN, who had friendly relations with the Admiral. The channel to Miyamoto was RUPERT, originally a British contact, an "Aryan" German who had been approached by Miyamoto to work for Japan. (A third initial possible channel, who apparently did not pan out, was a Ukrainian who had been successfully engaged in counterespionage against the Japanese since 1940.)

PIN was a doubtful character who used a false Japanese name, and spoke no English and little Spanish. He had a home in the resort district outside Córdoba and business interests in Buenos Aires, where he dabbled in real estate and picked up some money from gambling. Although he had been the first to identify Suzuki and Miyamoto as important Japanese agents, and until the spring of 1945 Lang considered him "our best forward passer" of deceptive material, Lang warned Smith that "We have always handled PIN as though he were a double agent, and consider his main value to be that of a channel *to* the Japs."

Certainly there were times when it seemed that PIN was being used by the Japanese to pass misinformation to the Americans. In January 1945, for example, PIN was told that the Japanese were producing a large new tank and a new long-range fighter plane; and shortly after MacArthur landed on Luzon, Lang cabled that PIN had been told that one hundred seventy thousand Japanese were waiting near Manila to defeat the Americans and capture MacArthur. In February PIN was pushing a specific landing site for an invasion of Japan. In March, just at the height of efforts to persuade the Japanese that Formosa rather than Okinawa was the next objective, he was warning that Formosa was heavily defended but no longer of strategic importance. Lang passed such information on for what it was worth. "PIN got this from Admiral Yukishita," he said of one such item. "It sounds screwy but here it is."

PIN fed information to Yukishita steadily from October 1944 onwards. Yukishita expressed gratitude, paying PIN one hundred pesos when he got confirmation of PIN's first piece of information ("Radar-equipped night fighters are now operating in the Southwest Pacific"), and making several payments thereafter. He particularly wanted information on landing craft, weapons and their production, and manpower.

In early January, PIN claimed that the Japanese were desperately searching for a trusted agent with good cover to be sent to the United States. Smith was enthusiastic. "If we can possibly find some means of equipping one of these fellows with a radio and bringing him here under proper control, it would be a most useful acquisition," he told Lang. Two months later, as Lang reported it, PIN "dramatically offered to go as an agent in our service to Japan, even though such a mission were almost certain suicide. While he talked about the great deeds he might accomplish, he worked himself into an emotional state where only a sharp interruption and change of subject prevented him from bursting into tears. There is no doubt about his willingness to go, but for whom would he be working, ourselves or the Japs?"

By then, however, the Argentine declaration of war had given Admiral Yukishita more urgent things to think about. On April 19 he and his colleagues were packed off to internment in the Eden Hotel, a resort near La Falda in the Córdoba hills. It was perhaps not the wisest place to put them, for its owner was a pro-Nazi German named Walter Eichhorn. Eichhorn had an *estancia* called El Cuadro high in the mountains behind the hotel an hour's horseback climb away, fenced and guarded by dogs and by a blond German gatekeeper who warned visitors off. A member of the British mission had seen a radio transmitter in his home before the war.

PIN tried various ways to get in touch with Yukishita. He reported that initially the internees arranged contact with the outside through a local Japanese laundry. But this failed because the Argentine police examined all the washing. PIN then got in touch with Tokujiro Furuta, an ex-correspondent and branch manager for a Tokyo newspaper, who was active in espionage and allegedly had a telephone connection into the Eden from his nearby chalet. Furuta told PIN to work with Yoichi Koko, who was also active in intelligence, but PIN doubted this would come to much; and Furuta himself was arrested in mid-May. Eventually, the internees opened clandestine communication with the outside world through Tadanao Yamazaki, a grocery store owner in the nearby town of Coaquín. Twice a week, Yamazaki's truck brought fish and vegetables to the hotel; and the driver brought in and took out messages as well.

But by June, police surveillance of the Eden had relaxed, people moved relatively freely in and out, and local rumors talked of a clandestine transmitter at Eichhorn's *estancia*. King met PIN in the hills near La Falda and PIN reported on a four-hour visit with Yukishita and the Japanese military attaché, in which they compared the position of Japan to that of Britain in 1940 except that Japan was far stronger; only on the western shore along the Sea of Japan, they said, could beachheads be established, but any fleet that attempted to pass through Tsushima Straits would be destroyed. They directed him to work under a certain Kurajiro Ishikawa, an agricultural expert living in Buenos Aires, to whom he would be introduced by Miyamoto. They also advised that a man named Fushimi in Córdoba city had a radio transmitter.

PIN went to Miyamoto, who introduced him to Ishikawa—and who also received some foodstuff from him and told him that all his reports had been transmitted to Japan. Ishikawa lectured PIN on the need for security and the virtues of patriotic loyalty, and appeared to be looking him over for a possible role as the liaison man between the internees in the Eden and the Japanese net in Buenos Aires. PIN, who had not been interested in fish food since Yukishita was interned in late April, began to want it again.

Late in July, PIN reported to King that he had been summoned before a committee of three leading Japanese—including the president of the Japanese Association of Argentina, and Miyamoto himself—and told that he had been under suspicion for more than a year because of his undetermined source of income and his unknown source for information about the war; that, indeed, some Japanese wanted to liquidate him.

(Yukishita later told PIN that he had not grown suspicious until after he was interned; but then, concerned over PIN's unusually accurate reporting, he had sent instructions to have him shadowed.) But nothing unfavorable had been turned up, and PIN was now to be initiated into the inner circle and engaged on more confidential work. King and Harris were still not sure of PIN, but this did not seem important; whatever the truth, they reckoned that they would now get more information on what was going on in the innermost Japanese circles.

Soon afterwards, PIN got his first assignment from the Miyamoto group: To get in touch with some recently arrived passenger from the United States who was not American or British, so as to inquire as to conditions in the United States; to ascertain how many carriers were operating in the Pacific, and how many were under construction; and to find out how many B-29s were in the Pacific and what their construction rate was. Harris looked for a traveler fitting PIN's requirements and asked Smith for fish food to meet the other requests. Additional requests followed soon after. But the sudden termination of hostilities in mid-August put an abrupt end to these efforts.

The second of Lang's channels, RUPERT, the "Aryan" German, was a pipeline to Miyamoto. He fed Miyamoto steadily from October 1944 onwards, with time out in the summer of 1945 for an extended illness. At the end of February—a few weeks after PIN reported that the Japanese were looking for a good agent to go to the States—RUPERT approached Miyamoto for an assignment with a salary attached, and Miyamoto asked RUPERT if he could travel to neighboring countries. When RUPERT replied in the affirmative, Miyamoto said he would see what he could do. But nothing further seems to have come of this. The channel functioned smoothly, though shortly after the diplomats were interned Miyamoto told RUPERT that getting information out was much slower than before because "the usual channel" had been cut off.

Soon after embarking on the program with PIN and RUPERT, Lang grew concerned that some of the items to be passed were of such a nature as to make it difficult for them to explain their sources if asked. He decided to add a channel with a mercenary approach, so that, as Lang put it, "if he makes a bold approach with something to sell, . . . he most naturally will insist on concealing his source for fear the Japs would cut him out of the syndicate by going direct." Late in November, Lang accordingly opened a third channel, codenamed GUSTAVO or GUS. GUS was a Polish newspaperman who had worked well for the Americans for several years, was known for his anti-Russian sentiments, had worked for a short time in 1943 as a translator for the Japanese Domei news agency, had a wide circle of friends, and could reasonably develop a notional contact leading back to high circles in Washington. He knew Saburo Suzuki, the *Mainichi* correspondent and Japanese agent, and could offer the latter his information for a price, part payment on delivery and the balance when the information was checked.

GUS reestablished contact with Suzuki in December and offered two items to him in January. Though relatively innocuous, they were enough for Suzuki to take him to call upon one Hoshi, his secretary and assistant in intelligence work. GUS told them he had two good sources of American information. One was interviews with recent plane arrivals from the United States. An even better source, he told them, was a weekly airmail letter from his friend Hessell Tiltman, the Washington correspondent for the London *Daily Sketch*. (Tiltman was not notional, though presumably he did not know his name was being taken in vain. He was a prominent British journalist, formerly the Far East correspondent for the *Daily Express*, with eight years' residence in Tokyo; GUS had known him since they met in Warsaw in 1939, and had seen him on a daily basis in New York in the spring of 1940.)

GUS passed a steady diet of fish food to Suzuki. The two became more intimate and often dined together. At breakfast in Suzuki's apartment late in February, Suzuki suddenly asked GUS if he thought he could arrange to go to the United States. Lang read that as a favorable signal and for the next couple of weeks passed the bulk of his material through GUS to improve GUS's position further. But, as with the contemporaneous approaches to PIN and RUPERT, nothing further happened. A substantial flow of material continued to pass through the channel. Suzuki lost interest during the anxious weeks leading up to the Argentine declaration of war—he was concerned lest Japanese civilians be interned, and spoke of hiding out at an *estancia* in San Luís province—but when it became clear that only the diplomats would be rounded up he relaxed, and asked GUS to be on the alert for more naval items.

Late in April, Smith cabled to Lang the breakoff story for BLUEBIRD, presently to be described. Lang passed it to Suzuki through GUS, since PIN was out of touch with Yukishima. This item seems to have confirmed GUS's status with Suzuki. The latter was so pleased that he offered GUS one hundred pesos, but GUS refused the money, saying that he wanted a steady job and not piece work. Suzuki complimented GUS on his getting so many interesting items, and asked him to expand his contacts in the American Embassy.

With Yukishita interned and the PIN channel now more difficult, Lang concluded that GUS should become his "principal bait feeder." He instructed GUS to enhance the seeming importance of his sources by telling Suzuki that some of his information came from off-the-record background statements at high officials' press conferences (presumably passed along by Tiltman). Suzuki spoke highly of GUS's naval material and asked if he could develop an equally good source for aviation items. GUS said he would try, but Lang preferred to continue feeding aviation material through RUPERT and PIN and not risk raising Japanese suspicions by duplicating coverage.

In the latter part of May, Suzuki told GUS what an able and valuable agent he had been, and raised again the idea that he might go to the States. That "would be a most gratifying achievement," wrote Newman Smith. But after discussing it with his superior, Suzuki told GUS that funds were too low for such a venture; indeed, he was having to dispense with Hoshi's services. He apologized for not being able to offer GUS a regular salary. From June on, Suzuki ceased to be talkative about the situation in Japan and avoided efforts by GUS to draw him out; from which GUS inferred a decline in Suzuki's faith in Japanese victory.

Early in July, Smith sent an item of fish food that reflected either a happy coincidence or a risky venture on Smith's part. It was a veiled adumbration of the atomic bomb. "In privately discussing reports that Japan was preparing numerous suicide weapons for use in the event of an invasion of the homeland," it ran, "a Staff Officer stated that Japan might never have the opportunity to use any such weapons, as there was a more than even chance that once the United States had completed their circle of bases there were some surprises in the form of new weapons in store for Japan. He stated this could easily result in Japan asking the U.S. to walk into the mainland without fighting their way in." Smith was indeed privy to the secret of the atomic bomb before Hiroshima—the day before the bomb was dropped, he told Gene Sweeney that something was going to happen today or tomorrow that was going to stop the war—but whether he had been briefed on the matter much before that is unknown. Moreover, as of the time this item of fish food went out, the first bomb was not even to be tested in New Mexico for another two weeks. Perhaps Smith—and Bissell, who would have had to clear an item of this nature if it really referred to the atomic bomb—reckoned that if the bomb was used the item would be regarded as a phenomenal "scoop" confirming that the channel had exceptional sources, and that otherwise it would simply be filed and forgotten. Or perhaps the item was simply a flight of fancy that took on immense significance by sheer chance.

In any case, the item went to GUS, who passed it on to Suzuki on July 17. (If Burris-Meyer's account four years later was correct, GUS embroidered it somewhat, telling Suzuki that the Japanese might expect an attack, the nature and severity of which had no parallel in human experience.) Suzuki's eyes "glared with hate," GUS reported. He was seemingly more and more convinced that Japan's position was hopeless, and showed less interest in fish food than before; but he and GUS continued to be good friends; and Harris advised Smith that after Hiroshima, Suzuki showed "a new respect for GUS" in light of his warning.

After the surrender there was still a Japanese government, though under foreign occupation. Channels such as PIN, RUPERT, and GUS thus could not be abruptly shut down, as the German channels had been after V-E Day. Some material was passed through all three agents during the autumn after the Japanese surrender, but by mid-October Smith was ready to close out the channels in Argentina, while the channels themselves showed little interest. In mid-November, King paid PIN off through the end of the year and suggested that he return home to Córdoba; settled with RUPERT for his services; and redirected GUS towards channels that were opening up towards central and eastern Europe. As far as Joint Security Control was concerned, the Argentine channels were officially shut down at the end of November.

Besides Lang's channels in Buenos Aires, the British had one across the River Plate in Montevideo. His codename was LODGE. As was mentioned in Chapter 11, he was established pursuant to a May 1944 exception to the Strong-Bevan agreement of January 1943, which put Latin America off limits to British special means. He was in place by October 1944, and got into W/T contact with Germany at the beginning of November. He was managed by a local case officer and his traffic—mostly chickenfeed—was coordinated with Newman Smith through British Security Coordination in New York and O'Connor in Washington. He had two notional sources, ALOSI (possibly a typist's error for ALOIS), a Uruguayan clerk in the American embassy in Montevideo, and GOMEZ, who supposedly worked at the American air base in Uruguay. He never played any significant role in deception—partly, in the British view, because the U.S. Navy would not cooperate in providing chickenfeed—and was shut down with the fall of Germany.

Off and on from the autumn of 1944 to the end of the European war, Smith and the FBI tried to lure the Japanese, through RUDLOFF and the Germans, into opening a channel direct from the United States to Tokyo. In October and again in November, RUDLOFF told Hamburg that WASCH's West Coast friend was a radio "ham" with a transmitter, needed money, and would, for pay, open up W/T communication with the Japanese directly. He kept pressing the Germans on this right up to April 1945, but the Germans kept fobbing him off and nothing ever came of it.

With the channels through Buenos Aires in uncertain status after the Argentine declaration of war, and channels via Germany about to shut down, Smith took steps in April 1945 to open lines to the Japanese through the military attachés in Switzerland, Portugal, and Sweden: Brigadier General Barnwell R. Legge in Bern, Colonel Robert A. Solborg in Lisbon, and Brigadier General Alfred A. Kessler in Stockholm. (This was, of course, a clear violation of the Strong-Bevan agreement of January 1943. The British were not told about it. O'Connor picked up the Stockholm inquiry on what he called a "bootleg" basis, and reported it to Bevan, who was understandably annoyed.)

Legge in Bern reported that he had an agent, described by him as "100% trustworthy," who was an intimate acquaintance of a certain Dr. Hach. Hach was a German by birth but apparently no longer a German citizen, and was both strongly anti-Nazi and strongly pro-Japanese. He had lived for a long time in Japan, had been closely associated with the Japanese Navy as an economic advisor, and had been awarded the highest Japanese decoration. An intimate friend of his, Commander Fujimura of the Japanese navy, was in Bern and was in immediate radio contact with the Japanese Navy Ministry. Material could be passed to Hach within twenty-four to forty-eight hours of receipt.

("Hach" was in fact clearly Dr. Friedrich Wilhelm Hack, an old Japan hand who had in the 1930s been involved in German arms sales to Japan and had been an unofficial go-between in discussions that led up to the Anti-Comintern Pact of 1936. Jailed for his outspoken hostility to Nazi racialism, he had been freed through Japanese intercession and ended up in Switzerland as a purchasing agent for the Japanese Navy. There he became a source of information for the OSS station headed by Allen Dulles, with whom Legge worked closely on intelligence matters; and he and Fujimura, inspired by the known role of the Dulles team in arranging the German surrender in Italy, had unsuccessfully sought to act as intermediaries with respect to possible surrender terms for Japan.)

From early May on, Smith furnished Legge with a steady flow of weekly material similar to that being passed to Argentina, substantially all of which reached Fujimura by way of Legge's intermediary and Dr. Hack. Hack himself—he acquired the codename HANS, and Hack, Fujimura, or the intermediary may also have been known as GENTLEMAN JOE—appears to have been an unwitting tool. More than once, Fujimura expressed extreme gratitude to Hack for the information he was getting.

After the Japanese surrender, Smith wanted to keep the channel alive at a reduced level; but early in September Legge learned to his dismay that the OSS station in Switzerland had approached Fujimura to go to Japan and work for that agency. In mid-October Smith wrote the channel off.

Solborg in Lisbon responded promptly: He had three channels ready and able, and added a fourth thereafter.

The first, LESTER, was a Hungarian, the former press attaché to the Hungarian legation in Lisbon, who, said Solborg, was "rather sympathetically inclined toward the Japs on the far-fetched theory of racial similarity." The Hungarian legation were close to their Japanese colleagues—the Hungarian military attaché had been a regular source for his Japanese opposite number, his material receiving the special designation " 'S' Intelligence" when sent to Tokyo—and LESTER was in close contact with the Counselor of the Japanese legation. "Their conversations deal with demography, ethnology, internal economy and related sciences," reported Solborg. "Information of long-range political nature can be planted there."

Solborg's second channel, HENRY, was a Turkish woman, the wife of a German-Jewish refugee; she had been the mistress of the Japanese military attaché and was now close to the Japanese naval attaché. "Spot items relative to military and naval operations can be conveyed through that channel," said Solborg.

The third channel, JIMMIE (always so spelled by Solborg; written as JIMMY in Washington), was described by Solborg as "Director" of the much-feared Portuguese secret police, the Policia Vigilancia e Defesa do Estado or PVDE. (He may thus have been Captain Agostinho Lourenço himself, the head of the PVDE, described by the British as "an extremely energetic and efficient officer, in whom far greater trust is placed than is suggested by his rank . . . smart in appearance, but blunt in manner.") "A clever man of independent financial means, blindly devoted to Salazar and the totalitarian regime," wrote Solborg. "In contact with the Jap Minister. Political information of a nature tending to safeguard Portuguese interests in the Far East by maintaining a fine political balance between the Allies and Japan in that zone can be passed through this medium."

In the latter part of June, Solborg added a fourth channel, KELLY, "a Portuguese Army officer who has access to the highest government circles as well as those of the local Japanese mission."

As with Lang and Legge, over the late spring and summer of 1945 a steady flow of items passed from Newman Smith to Solborg for placement in his channels; a weekly list, plus occasional special pieces by cable. Though Solborg thought the material first sent to him was too commonplace to go very far towards building up credibility and esteem, the system ran smoothly, despite an occasional hitch when the local OSS and FBI unwittingly stumbled through Solborg's territory. LESTER was Solborg's main channel, and, as he reported to Smith, "the whole process was carried out unostentatiously without ever giving the contact or source any inkling of premeditated planning."

Shortly after the end of the war, the channels were closed off; which was all the easier, since LESTER was to leave Lisbon shortly for an indefinite stay in northern Portugal.

General Kessler, the attaché in Stockholm, was able to produce nothing. He reported that he had no suitable agent, though there was one prospect whom he might be able to develop. The local OSS had an agent that might be useful, but that would entail bringing that organization into the picture; which Bissell and Smith were unwilling to do. Nothing further was done about Sweden; which was just as well, since Kessler's prospect turned out to have been unsuitable.

With the broadening of special means channels, foodstuff supply was put on a regular basis. Nelson and Goldbranson initially, and Gene Sweeney in the last months in the summer of 1945, were in charge of preparing a weekly list of items to be passed to the enemy, each serially numbered. Special items, also serially numbered, might be sent outside the regular weekly series. Dorothy Stewart took over from Pete Schrup the job of meticulously recording each item, the channel and date through which it was released, and the deception plan with which it was associated. It has to be said that some of this material was not particularly sophisticated, baldly stating supposed American plans rather than throwing out hints from which the customers could draw their own conclusions. It also has to be said that the practice of giving identical or substantially identical material to two or more different channels was unwise in the extreme. Given the clumsiness of Japanese intelligence, perhaps this did not matter. But an intelligence service with a modicum of professionalism would have spotted this at once and the entire family of double agents would have been blown; and it is surprising that sophisticated operatives like Smith, Goldbranson, and Sweeney could have committed such an elementary mistake.

Special means items were routinely passed to General (as he now was) Carter Clarke, in charge of ULTRA for the Army, so that intelligence specialists monitoring ULTRA and MAGIC could report on anything that appeared in the Japanese traffic. On the British side, such feedback reports were a routine part of D Division's weekly summaries, and arrangements were made in March through O'Connor for exchange of such material between the British and the Americans.

As early as February 1944, Smith had been in touch with the FBI asking them to look out for a possible double agent channel to Japan. The Bureau ran, or tried to open, a few new channels to the Japanese in the last months of the war, without much success.

In January 1945, the FBI alerted Joint Security Control to the fact that ANTHONY (also referred to as CAMCASE), a double agent of Russian origin, had left Bilbao and was due in New York in February to spend three months in the United States gathering aviation intelligence. The FBI hoped that if ANTHONY could be accredited with valuable information, some notional contacts could be established who could be put in direct communication with the Japanese in Madrid. But if anything came of this it does not appear in the surviving available files. Spanish-Japanese relations were broken in April, so the prospects for such channels were limited in any case.

As was mentioned in Chapter 4, José María Aladren, codenamed ASPIRIN, one of Alcázar's people, had been run by the FBI as a double agent under their SPANIP program from October of 1942 to March of 1944, and returned to Washington in February 1945 as a correspondent for a Spanish news agency, equipped to send reports in secret ink, by radio, by coded messages in his newspaper stories, and possibly through a crewman on a Spanish ship acting as a courier. By mid-March the FBI was calling upon Joint Security Control for material to get the channel started. It was sending material by mid-April, through an FBI-operated station in Waldorf, Maryland; but contact did not last beyond May 8. Again, the April breach in Spanish-Japanese relations may have played a role.

As was also mentioned in Chapter 4, in March of 1945 a scheme was worked up under which OUTCAST, a British double agent in the United Kingdom, would pass information to Onodera in Stockholm, purporting to come from a White Russian friend from the United States, codenamed JAM, temporarily visiting Britain. JAM would then return to the

States with a direct connection to Onodera. Nothing seems to have come of this project, so far as the available records disclose.

Also in early March, MOONSTONE, a Swiss Quaker teacher on his way to the United States looking for a job, who had been trained by the Germans in Spain for maritime intelligence, agreed to act as a double agent once he should reach the States and set up radio connection. A number of messages were cleared for him but the available records do not disclose that any were sent; no doubt the European war had ended by the time he would have been of use.

Mention should be made of a curious German agent called ONYX, who reached the States from Spain at the end of 1944. More interesting than his extensive questionnaire was the account his German case officer had given him of alleged secret weapon research by the Germans, including "atom smashing experiments" employing "short radio and sound waves" which if successful would "put the force of a 4000 pound bomb into a 2 inch projectile"; a shell that would kill everyone over a five-kilometer radius; and a long-range flamethrower covering a large area. He was considered for use as a double agent but evidently was never employed and left the country in March.

Ormonde Hunter came crashing into the Washington china shop that spring. He was assigned to the planning section of the Air Staff on his return to the States, and reviewed deception plans on behalf of the Air Force. He seemed, if anything, to understand deception less than he had two years before. When the paper that became BROADAXE, the final overall plan for Japanese deception, crossed his desk he proposed deleting most of the specific deceptions and adding two of his own devising. One was "To promote internal disorder and civil war by fomenting discord between deposed and acting statesmen, between political factions, and between the Army, Navy, and Air Force." The other was "To promote peace movements by the Japanese by deceptively meeting any advance and by holding out encouragement that we are ready to grant them favorable terms" while in fact "strengthening . . . our military position until we are able to enforce complete obedience to our will and unconditional surrender." Smith vigorously blue-penciled these proposals, and the OPD action officer noted that the first proposed new item was "not *military* deception and has no place in a deception paper," while the second was "not acceptable for obvious reasons"; adding that "no effort was made on the part of AAF planners to coordinate their efforts with JSC." (It must be admitted that Hunter's second project had something in common with an abortive scheme to reduce the Italian will to resist by holding out prospects of reasonable peace terms which Bevan floated in early 1943.)

Little attention seems to have been paid to Hunter in the planning process thereafter; though he was still at it when the staff study of deception for CORONET, the final invasion of Japan, was under review at the beginning of August, producing alternative recommendations which Thurber noted "did not take into clear consideration" the existing plans, and proposing an alternate "story" whose objective, said Thurber, "is not stated and is not understood."

In June, Hunter surfaced as chairman of an ad hoc committee of the Joint Planning Staff formed to address ways to improve psychological warfare. At its first meeting he dismayed those in the know by pressing them to talk about Joint Security Control's deception work. All who had been at the meeting were warned to disregard in the national interest the connection between Joint Security Control and deception and not to mention it to anyone else; and a memorandum was drafted advising the chief air planner that "subject officer is considered by members of Joint Security Control to have been indiscreet to a point of bordering on a requirement that disciplinary action be recommended against him." What transpired further is not presently known.

The new deception machinery made possible the most elaborately executed American deception of the war: BLUEBIRD, designed to aid the invasion of Okinawa by focusing Japanese attention on Formosa and South China.

The summer and autumn of 1944 had seen the great debate of the Pacific war. After Nimitz's Central Pacific offensive cleared the Marianas, where should American power be focused? A landing on Formosa and the South China coast, favored by the Navy? Liberation of the Philippines, particularly Luzon, favored by MacArthur? Invasion of the home islands via the Bonins and Ryukyus? Not till October 3 was it settled that MacArthur, after landing on Leyte in the central Philippines later that month, would invade Luzon, with his main landing in Lingayen Gulf (Operation MIKE ONE); and that after supporting that operation Nimitz would land in the Bonins (or, strictly speaking, the Volcanoes) at Iwo Jima (Operation DETACHMENT), and in the Ryukyus at Okinawa (Operation ICEBERG). Though the Formosa-South China option was technically only shelved, as a practical matter it was generally recognized that it was rejected.

This decision was thus brand-new when the Young Ladies' Seminary teams left Washington. It enabled Holland and Rich, of the team sent to MacArthur's headquarters, to design a proposed operational deception to cover the MIKE ONE landing. And it was particularly useful for Peter McDowell and his team. The postponed Formosa operation was a ready-made basis for a strategic deception to cover Nimitz's true offensive against Iwo and Okinawa. On his visit to Nimitz's headquarters McDowell worked out the original design for such a deception, eventually codenamed BLUEBIRD.

For the liberation of Luzon, MacArthur planned to land on Mindoro in the central Philippines in early December (Operation LOVE THREE or L-3), where airfields would be constructed to support operations on Luzon; and on Luzon itself at Lingayen Gulf in the middle of the month (Operation MIKE ONE or M-1). The associated deception plan drafted by Holland and Rich was approved at the end of November, with minor modifications and a "story" and tabulation of actions to be taken prepared by Newman Smith.

As thus approved—it never received a separate codename—its stated objects included causing the dispersion or dislocation of Japanese air and naval forces, causing delay of movement of Japanese reinforcements from outside the Philippines, and providing background and support for a long-term deception. Its "story" was that determined Japanese resistance on Leyte had upset the American schedule for reconquest of the Philippines, but that the Americans also believed that the defense of Leyte had thrown the Japanese off balance in the Formosa area. Consideration was accordingly being given to leaving MacArthur to maintain pressure in the central Philippines with only limited means, and making the next main effort in Formosa (at Christmastime, when the Japanese would presumably assume that the Americans were taking it easy). A detailed schedule of special means implementation was provided, dovetailing with planned air strikes, radio deception, and BAMBINO. A breakoff "postponement" and prolongation "story" for the post-MIKE ONE period were to be devised later, probably using logistical difficulties and the Leyte campaign as excuses.

A message went out to the theaters to this effect on November 30. The next day, MacArthur reported that he had had to defer LOVE THREE on Mindoro till December 15 and MIKE ONE on Luzon till January 9, thus throwing off the proposed Christmas theme. More significantly for deception, McDowell's draft cover plan for the Okinawa operation arrived in Washington.

McDowell's Plan BLUEBIRD—it is convenient to call it by its eventual codename, though that was not assigned until January 12—stated no explicit object in terms of desired Japanese action; by its own terms it was designed to "convince the enemy of our intention of invading the Island of Formosa and the south coast of China between Formosa and Hainan Island in the spring of 1945, and to continue this threat through the summer of 1945, if, and as, necessary." The implicit object, of course, was to induce the Japanese to deploy forces to meet such a threat rather than to meet the actual planned operations against Iwo Jima and Okinawa.

As approved by the Joint Chiefs on December 30, its "story" was that the Formosa-South China operation, with an initial target date of April 1 (referred to as "Cast Day"), had been decided on to relieve pressure on China, gain bases on or near the Asian mainland for further operations and blockade of Japan, seize the China coast from Hong Kong to Amoy to open supply routes to China, and isolate Japanese forces in southeast Asia and the East Indies. After MacArthur launched MIKE ONE, Cast Day would be postponed to April 20, giving as the reasons reverses suffered in Philippine operations and logistical difficulties. Freddie Barnes flew back to Washington from Nimitz's headquarters the week of January 18 to go over implementation plans.

The most extensive and elaborate all-American deception yet mounted, BLUEBIRD was implemented by three major means: deceptive communications, operations and administrative arrangements, and special means. And two separate deception operations, VALENTINE and STURGEON, supported the BLUEBIRD "story."

*   *   *

Communications deception for BLUEBIRD proper, as opposed to the MIKE ONE cover, did not begin until February. Meanwhile, ineptitude on such matters in the field, particularly at MacArthur's headquarters, led to some embarrassing moments.

There were clear signs of trouble ahead when Hilger met on December 17 with Major General Chamberlin, MacArthur's G-3, and with the theater chief signal officer, Major General S. B. Akin, at Tacloban in the Philippines to try to interest them in supporting the plan that became BLUEBIRD. He ran into a stone wall, particularly with Akin—an arrogant officer known to those who worked for him as "S.O.B. Akin," who ran a separate empire in his theater and held Washington, not to mention other theaters, at arm's length. Akin announced that long-range deception would not work against the Japanese. He offered no help, and took offense at what he conceived to be an effort to control his communications. In turn, Chamberlin's plans officer seemed only interested in whether giving the appearance of landing on the China coast without actually doing so would lower MacArthur's image in Chinese eyes.

Shortly thereafter, in the latter part of December, Akin and his people demonstrated an exceptional level of incompetence in signals security and deception. In aid of the MIKE ONE deception, it was decided to give the impression of increased radio traffic between MacArthur's theater and the China Theater by means of dummy traffic out of Delhi, purportedly from Chungking. This dummy traffic was prepared in so amateurish a manner as to be obviously fake; so obviously fake that the British were at once aware of it, and Mountbatten asked angrily why deception was being undertaken in his theater without clearance with him. (Akin had an extensive background in signals interception and cryptography and ran MacArthur's ULTRA activity, and should have known better than this.) Newman Smith prepared a message from Joint Security Control, concurred in by the appropriate experts, directing that this activity be "discontinued immediately" and suggesting that the theaters request that experienced deception officers be sent to them.

Hardly was this out of the way when another problem surfaced with Akin's people. General Carter Clarke, who had had his own run-ins with Akin, consulted Smith upon ascertaining that the Japanese were deriving order of battle information concerning MacArthur's command from intercepted radio communications; apparently an unauthorized effort at order of battle deception. Smith responded that he had suspected as much; moreover, "the methods of controlling traffic for deception purposes in the Southwest Pacific, as you know, have been very amateurish and dangerous both from a point of view of cryptographic security as well as possible compromise of deception activities." He again recommended that a trained signal security officer be sent to the theater. This was finally done.

Communications deception for BLUEBIRD proper began in February. Traffic was substantially increased among Nimitz's advance headquarters on Guam; MacArthur's headquarters on Leyte; Seventh Fleet, which was associated with MacArthur's operations; and Fifth Fleet, which was currently in notional status.*

To monitor the effectiveness of communications deception, the Navy's signal intercept station at Wahiawa, Hawaii, regularly reported their own conclusions as to American fleet activities, drawn from their observation of American radio traffic. Under Major Hines, the Army's Signal Security Agency similarly monitored intertheater traffic. The only serious snag encountered was that although there were as many American submarines in the notional target area of the South China Sea as in the true target area around Okinawa, the latter group were more active in radio transmissions; and the Japanese were known to stress submarine traffic as pointing to areas of intended operations.

Actual operations in January through March were tailored as much as possible to fit the BLUEBIRD story. During January, Halsey's carrier strike

---

* During the last two years of the war, command of the major naval force in the Central Pacific alternated between Admirals Halsey and Spruance; while one, with his staff, was in command, the other and his staff were planning the next operation. Under Halsey the force was called "Third Fleet"; under Spruance it was "Fifth Fleet." Spruance commanded during the Gilberts, Marshalls, and Marianas operations; Halsey then took over and commanded during the major operations in the Philippines; Spruance returned on January 26, 1945, and commanded during the Iwo Jima and Okinawa operations; Halsey took over again on May 27 and it was Third Fleet to the end of the war. Spruance would in turn have taken over for OLYMPIC, the invasion of Kyushu in November; Halsey would then have commanded for CORONET, the invasion of the Tokyo plain in the spring of 1946.

patterns emphasized Formosa and points south, while daily patrol and reconnaissance flights from the Philippines were flown over Formosa, South China, and French Indochina. In mid-February, search and reconnaissance reporting of Japanese activity in the South China Sea and around Formosa was increased. Land-based bombers from China and the Philippines flew one thousand sorties over Formosa and the China coast in January, rising to three thousand in March.

On-site reconnaissance was particularly calculated to attract Japanese attention. During February and March, the "U.S. Naval Group China," an irregular force on the China coast commanded by the legendary Captain Milton E. "Mary" Miles, with OSS assistance, conducted a survey of the area from Hainan Strait to Amoy, covering the part north of Hong Kong first, giving priority to landing beaches and to beach defenses of Swatow and Bias Bay.

Special means implementation of BLUEBIRD saw the most elaborate such effort yet undertaken by the Americans. Key elements, few in number but directly in point, plus supporting items, were fed directly to the Japanese through Buenos Aires, with supporting items to the Germans in the hope that they would be passed on; all surrounded by a protective coating of chickenfeed. PIN in Buenos Aires was advised on November 30 that there was a debate between the Army and the Navy over whether to invade Formosa. PEASANT had already reported to the Germans on November 25 that the WAVE "has been unavailable a great many evenings due to extra work caused by strong Japanese resistance in Philippines. She says plan of Navy relative to Japanese campaign are coming to the fore again, which also means more work for her." He reported on November 28 that he had "overheard discussion in Statler last night between two Army officers, one of whom said Navy's plan against Japan is better than Army's plan." RUDLOFF reported on December 9 that WASCH's friend the West Coast supply officer "states emphatically that conflicting views among high officials as to next move against Japan" had prompted a press statement by Admiral Leahy that Roosevelt did not countermand Joint Chiefs decisions.

A PEASANT report the same day advised that published Army casualties included none of the heavy losses on Leyte, and on December 23 RUPERT in Buenos Aires reported to Miyamoto that the tough fight the

Japanese were putting up on Leyte had come as a surprise to many high officials in Washington. (Lang reported to Smith that Miyamoto responded that "the Japs had not yet begun their real fight, that that would come soon.") After Christmas, PEASANT reported that the newspapers were featuring successes but were writing only short accounts of MacArthur's small advances and the retreats in Europe. PEASANT advised on January 6 that the WAVE said that Navy officers were putting forth their strongest efforts to maintain their own operations even if it meant slowing MacArthur down. On January 11, PEASANT furnished Army casualty figures including the heavy ones on Leyte.

There were scattered messages to encourage the thought that the Americans would be seeking additional bases for the B-29s; on December 2 PEASANT reported that he had learned that the new goal set for the B-29 program was to double their production within the next twelve weeks; PIN received the same information on November 30; and on December 19 RUDLOFF reported that WASCH told him that high officials were pleased that increased B-29 production would provide larger groups for strikes against Tokyo when the Navy reached the China coast.*

The theme of logistical difficulties was encouraged with such items as reports of a shutdown at Ford, passed to the Germans by RUDLOFF on December 5, and to Miyamoto in Buenos Aires by RUPERT on December 23; a report by PEASANT of a telephone strike affecting the war effort; and a report by RUDLOFF that WASCH had told him that railroad freight shipments had been rerouted to the East Coast because of the German breakthrough in the Ardennes.

Miscellaneous items included telling PIN on December 29 that the first and principal phase of a photographic reconnaissance of Formosa, the China coast, and the home islands of Japan had been completed. On the same day, RUPERT was given a story to pass on that "the rubber model

---

* A subsidiary topic, presumably looking ahead to a theme that the Americans would rely on the strategic air effort to break Japan's will without an invasion, was that the Americans were working on a flying bomb similar to the German V-1, called the "JB-2." RUDLOFF reported this on December 20. GUS reported on January 20 that such missiles had been tested and seen to rise several thousand feet before being lost to sight. PEASANT reported on January 25 that they had been tested, were in production, and might be tried first against the Japanese.

project" was proceeding well and that recently "rubber models of the most improved type have been flown to Pearl Harbor, depicting in great detail the islands of the Formosa area and the adjacent China coast," while RUDLOFF sent the Germans a similar message on December 9. PEASANT reported to the Germans on December 5 that a large group of Chinese naval officers were observed at the Army-Navy football game and that the WAVE said that after training they would command Chinese naval units; and another PEASANT report on December 12 said that many naval officers were being transferred from New York to San Francisco.

A "related means" item was the invitation, through the Chinese embassy, of a Formosa-born Chinese naval officer, a certain Lieutenant Liu, to attend a two-day conference of the Joint Intelligence Staff just after Christmas, it being expected that word of this would promptly leak back from Chinese circles to the Japanese. This tended to confirm a report, passed to Miyamoto by RUPERT on December 20, that Pacific Command had asked for the loan of Chinese naval officers training in the United States who were especially familiar with Formosa and the China coast. (Perhaps confirming the leakage through the Chinese, Miyamoto told RUPERT he had already heard this on Radio Tokyo.)

"Postponement" items went through shortly before the MIKE ONE landing on January 9. RUDLOFF reported on January 6 that WASCH's friend had told him that the Army supply people believed that heavy emergency supply demands for the Ardennes battle had caused a several weeks' postponement of the expected mid-January assault against Formosa. On January 8, PEASANT reported that his WAVE's boss had left for conferences in San Francisco because of a sudden change in Pacific plans. Smith had cabled to Lang on January 5 asking him to feed word to PIN, as early as practicable but no later than January 7, that the State Department had postponed preparations for certain unspecified actions to take place after the occupation of Formosa, which had been scheduled for the beginning of February, and that this was probably due to reverses in Europe and the urgent need for supplies there. Because of delays in transmission PIN did not deliver this to Yukishita until the afternoon of January 8; Lang advised Smith that the Admiral "first expressed surprise, then appeared most interested, finally nodded his head as if he were admitting its probability."

*　　*　　*

LOVE THREE, MacArthur's landing on Mindoro, took place on December 15. The Japanese offered virtually no opposition on land, but their kamikazes attacked savagely from the air. As has been mentioned, Captain George Mentz, Young Ladies' Seminary alumnus and head of the deception team that visited MacArthur's headquarters, had taken command of a "Diversionary Attack Group," including Beach Jumper units, that was being moved to Mindoro with an eye to future operations. He was severely wounded in a kamikaze attack on December 30, while the commander of the Beach Jumper unit in his group was killed. The devastation wrought on Mentz's diversionary group meant that no Beach Jumper diversions were attempted in connection with the MIKE ONE landing at Lingayen in January. That was unimportant, for the landings were virtually unopposed except by the kamikazes; Yamashita, the Japanese commander, planned to withdraw to the interior and fight a holding action.

MacArthur's plan for Luzon included two followup landings some three weeks after MIKE ONE: MIKE SEVEN, a landing on January 29 just north of Bataan to cut off possible retreat into that peninsula and support the right of his main drive down the central valley, and MIKE SIX, a landing on January 31 just south of Manila Bay. Some miles east of the MIKE SIX beaches, opposite Mindoro, was Tayabas Bay, and in that area were two Japanese divisions plus additional troops, all well situated to move towards Manila relatively easily. On the night of January 22–23, Mentz's task group, commanded now by Commander A. Vernon Jannotta, simulated a landing force approaching Tayabas Bay, employing scripted radio conversations, corner reflectors and other radar deceptive devices, and a drop of dummy paratroopers further inland. A similar feint was mounted on the night of January 30–31, this time without the dummy parachutists but with a complicated fire support and smoke screen plan. While the operations did not go off wholly without a hitch—the dummy paratroops fell off course, though they did succeed in scaring the guards at the civilian internment camp at Los Baños—they produced a satisfactory result. Japanese forces in the area were thoroughly alerted, destroyed their classified papers in the expectation of immediate action, and did not attempt to interfere with the MIKE SIX landing.

It was a satisfying finale, for this was the last action of the Beach Jumpers during the war.

*   *   *

Japanese expectations for Formosa and South China were teased along until late March, when the imminence of an Okinawa operation was unmistakably apparent. RUDLOFF reported on January 17 that "WASCH's friend" advised that the rate of delivery of materiel to the Soviets might slow down owing to the Navy's demands for possible China coast operations. PIN was told on February 14 that "high officials" were pleased that air and naval bases were available in the Philippines "to cover Nimitz's main assault against Swatow as soon as logistics permit the movement."

Like all good deception plans, BLUEBIRD worked to reinforce the enemy's known preconceptions, for Japanese interest in Formosa and the South China coast was manifest throughout this period.

On February 1, O'Connor reported that the Japanese had radioed to their PURPLE WHALES link to "Find out whether Americans are going to land in South China. We must have early information"; and that Bevan hoped that an interim reply could be given, followed by a circumstantial phony document once plans were firm. But no final operational plan was approved in time to take advantage of this. O'Connor suggested that the channel stall, and an offer was made through the INK radio link to provide intelligence from India meanwhile; which, as will be seen, did not satisfy the Japanese.

A major appraisal of the China situation on February 5 by Wedemeyer's headquarters noted that MacArthur's success in the Philippines was causing the Japanese increased concern about the China coast. The April 15 weekly intelligence circular of the Army General Staff in Tokyo, read promptly by MAGIC, reported that a visit to Washington for conferences early in March by Nimitz, Halsey, Wedemeyer, and the American ambassador to Chungking "points strongly to a landing on the China coast some time in June or later."

To enhance agents' credibility, information was fed from time to time ahead of its public release. Advance notice of the March Washington conference was passed by RUDLOFF on February 13 with further details on February 22 and March 2, and by RUPERT to Miyamoto on February 16. RUPERT was authorized to tell Miyamoto on February 15—four days before the landing on Iwo Jima—that there would be an attack on the Bonins prior to the end of February to protect Nimitz's headquarters on Guam; similarly, RUDLOFF advised the Germans on the same day that

WASCH's friend had informed him that before March 1 there would be an assault operation north of Saipan to protect Nimitz's right flank prior to his drive to the China coast.

As has been mentioned, two minor deception operations in the winter and spring of 1944–45 supplemented BLUEBIRD: VALENTINE and STURGEON.

It will be recalled that HUSBAND, the successor to WEDLOCK in the Aleutians, had been terminated as of October 31, and special means told the Japanese on November 6 that swiftly-moving events in the Central and Southwest Pacific had delayed plans for action in the North Pacific. Aleutian deception could not be simply shut down without compromising the whole operation, so radio deception was maintained to indicate the continued presence of three divisions at Adak, Amchitka, and Attu, with appropriate logistical and administrative support. This operation was codenamed BAMBINO.

As part of the MIKE ONE cover plan, traffic between Nimitz's theater and the Alaskan Command was increased, with a peak on January 2. In support of BLUEBIRD, BAMBINO was then modified by VALENTINE, the "story" of which was that the notional 119th and 141st Divisions were being redeployed from Amchitka and Adak to augment the forces available in the Pacific for a Formosa–China coast operation. Planned in January and executed from February 13 to April 3, the fictitious redeployment was accomplished by radio simulation of daily departures of vessels, associated escort and patrol activity, and arrivals at Seattle and San Francisco. Towards the end of the period, radio traffic simulated the redeployment of the notional 108th Division on Attu, sending a combat team to Amchitka and another to Adak. At the conclusion of the operation, traffic between Alaska and Washington had dropped to the pre-WEDLOCK levels of the spring of 1944. Minor special means support was also accorded to VALENTINE. On April 24, PAT J reported seeing the insignia of three different divisions in Grand Central Station, and on May 1 he identified one of these as the 119th Division. (But by then nobody in Hamburg was paying attention, much less passing information on to the Japanese.)

Related to VALENTINE was a minor cover operation called LOADSTAR, operated from March 1945 to the late summer. Certain Navy and Army ships were being transferred to the USSR under the Lend-Lease program (Operation MILEPOST, and subsequently FOGHORN and ENDRUN), sailing to Russian Pacific ports. LOADSTAR covered this with stories spread in American ports and via special means to imply that only normal Lend-Lease shipments were involved and that those vessels that passed westwards through the Panama Canal were bound for the South Pacific.

Closely allied to LOADSTAR was an informal directive from the Chiefs in mid-February to prepare cover plans for contingency plans under consideration for activities in the North Pacific when the Soviets should enter the Japanese war, including the sending of survey parties into Kamchatka and eastern Siberia, the establishment and defense of American air bases there, and the movement of Soviet troops to the Far East. The war ended before anything along these lines could be done.

Somewhat more elaborate was Operation STURGEON. It will be recalled that in late January 1945 the B-29s of XX Bomber Command left China and moved to India, thence to the Marianas in late February and March for operations against the Japanese homeland. STURGEON was designed to conceal the latter part of this move from the Japanese, so as to continue the apparent focus of the B-29s of XX Bomber Command on China and adjacent territory. It was implemented from January to June, primarily by communications deception and special means.

As initially requested by Nimitz's staff on January 21, the plan consisted only of the suggested "story" that logistical difficulties plus the need for heavy air strikes on the South China coast, Indochina coast, and northern Burma, called for redeployment of the bombers to India or the Philippines. Fleming suggested that XX Bomber Command should notionally remain in SEAC until the monsoon began in June, both to aid his own Plan STULTIFY, and in light of the fact that the Japanese would certainly spot the move. He took it upon himself to give the plan the codename STURGEON. It took some time to clear this augmented plan with all interested parties, and not till March 8 was Joint Security Control able to dispatch to all commands concerned, with a copy for O'Connor, a message confirming that XX Bomber Command would remain notionally in Mountbatten's theater until June 15.

Special means implementation of the move from China to India went forward through RUPERT in Buenos Aires, RUDLOFF in New York, SILVER via the MARY-TOM link, DUCK, OWL, PAWNBROKER, and GLEAM. Implementation was plagued by misunderstandings, with particular confusion at D Division, where Fleming seems not to have briefed Thorne and others as to the object of the plan.

Communications implementation began on March 15; the bombers began flying out a week later, and the heaviest movement took place from April 20 to May 6. SEAC did not have enough personnel to participate in the communications side of the deception. Then, some weeks into the program, MacArthur's signal staff blundered again, issuing confusing instructions which if followed would have given the operation away, since dummy traffic would have been addressed in one manner and genuine traffic in another.

Except that SEAC was never able to participate in the communications phase, all these matters were eventually cleared up. "Workers here all happy," cabled the Army in Hawaii on June 10. "This is our last message under Plan STURGEON. . . . See you all of a sudden." The plan closed formally on June 20, and Newman Smith must have been glad to see it go.

On March 21, with D-Day on Okinawa only nine days off, Joint Security Control asked Nimitz's headquarters for suggestions for a breakoff or postponement "story" for BLUEBIRD. On March 26, Nimitz's headquarters proposed a "story" that neither Okinawa nor Iwo had originally been a major target. Forces from Europe would have been used to seize Okinawa simultaneously with the Formosa operation if the war in Europe had ended as expected by January 1, but it had not; and Japanese resistance in the Philippines had tied down forces there longer than expected. Soon, however, the Americans would be in a strong position to carry out the attack on Formosa and the South China coast. Meanwhile, Nimitz had decided he needed to cover his right flank as he approached China. It was suggested that implementation begin after April 20, the earlier BLUEBIRD Cast Day, and a new Cast Day of July 15 was suggested, with Miles resuming his survey about June 1. The two VALENTINE divisions should be shown as retained in the States for the time being, since Nimitz did not have adequate special signal service units to support fictitious divisions.

Joint Security Control responded on April 3 that the background of this "story" had already begun to be fed prior to the Okinawa landing, and recommended that it be continued without a target date through the summer of 1945 unless a change in the cover target should be made when the next actual objective was determined. The two VALENTINE divisions would be reported as held in the United States as requested; Miles's survey would terminate as planned and could be resumed later. Nimitz's headquarters concurred.

Implementation of this "story" had already begun, with a March 23 message from RUDLOFF to the Germans that because of the heavy fighting on Iwo the American high command might have decided to take and man additional covering airfields in the Ryukyus before the Formosa-China coast operation. Two days later, RUDLOFF advised that a new Tenth Army had been formed in Hawaii, with General Simon B. Buckner, formerly of Alaskan Command, in charge. (Buckner and Tenth Army were in fact the commander and command for the Okinawa operation.) The "story" was passed in greater detail as April wore on. On April 21, GUS gave Suzuki the item already described, to the effect that the determined defense of Iwo plus the diversion of troops to Europe had led the American command to decide that it was essential to obtain a strong base in the Ryukyus to cover the right flank of the westward movement; this was the occasion on which Suzuki was so pleased that he offered GUS money. ("In an emotional voice," Lang reported, "Jap commented that Yankees intend to attack Formosa which is stronger than last fall when attack was threatened.") On the same day, RUDLOFF transmitted a report "on high authority" that it was the need to put an end to harassing aircraft operations from Iwo Jima and Okinawa that brought on those subsidiary operations and delayed the drive to the China coast; and on April 30 he advised that a letter from a friend in San Francisco said that a general and his aide had been in his office last week and had remarked that if the European war had ended at the New Year the Okinawa landings would have been made at about the same time as the landing on Formosa and the South China coast.

BLUEBIRD was a good, thorough plan; Peter McDowell well deserved the commendation for drafting it and starting it on its way that Bissell wrote and sent to Thurber. But, as with most deception operations, it is hard to evaluate the extent to which it aided the Okinawa campaign. Few if any specific BLUEBIRD items were directly confirmed by ULTRA as having passed through the Japanese intelligence system. Comments from the Japanese in Argentina did suggest that the "story" that logistical difficulties would delay American action in the Pacific had been believed.

The Japanese had some 100,000 men on Okinawa to oppose the American invasion, including labor and service troops and local conscripts—rather more than American intelligence estimated. They would have had still more if the veteran 9th Division had not left Okinawa for Formosa in December. This move was not designed to guard against a Formosa threat, however, but to reinforce the Philippines. The 9th Division had not been replaced—again, not because of belief that the Americans would attack elsewhere, but simply owing to shortage of shipping. It is nevertheless true that an all-out effort might have put even more troops on Okinawa in 1944, and that none was made. In the last months before the Okinawa landing the Japanese managed to increase their ground strength on Formosa from 85,000 men to 240,000, and in South China from 67,000 to 161,000. Even with the shipping shortage, it should have been possible to move some of this augmented force to Okinawa.

No such effort was made, because at the end of the year and into 1945, the Japanese high command was uncertain whether the next move would be against Formosa and/or the South China coast. Fundamentally this reflected a dispute between the Army, which placed its bets on Formosa, and the Navy, which preferred Okinawa. Thus, Lieutenant General Seizo Arisue, the chief of Army intelligence, conceded after the war that he had "thought the next attack would come in the Hong Kong area, or possibly in Formosa, and Okinawa as a third possibility." Rear Admiral Tomioka, chief Navy planner for Imperial General Headquarters, told interrogators that the Navy "thought you would attack Okinawa in late March 1945," and his assistant, Commander Miyazaki, confirmed that "The Army considered very strongly the possibility of a U.S. landing attempt on Formosa. The Navy did not agree. We considered that your forces would not take Formosa and thereby spill unnecessary blood." Admiral Mitsumasa Yonai, the Navy Minister, agreed, adding that he "felt that the Army's political influence did have some effect on the Navy's operations." Not until the end of February, after the Iwo landing, did a Japanese consensus begin to develop that Formosa would be bypassed and Okinawa would be the next target. By then it was too late.

It would be pleasing to think that BLUEBIRD was responsible for the Army's insistence on the Formosa option, but there is no evidence of this. The differences between the Army and the Navy probably reflected simply each service's putting priority on the area for which it had primary responsibility. During the January–March period, Japanese intelligence received, as usual, a cacaphony of reports from various sources. A high proportion of these pointed towards Formosa and the China coast, and the Army would have been able to point to such reports as supporting their position. What contribution BLUEBIRD may have made to this cannot be determined. Probably little or none, given the volume and variety of the reports. Indeed, in view of the contretemps that dogged it notwithstanding its elaborate implementation—communications security problems at MacArthur's headquarters, confusion in SEAC, and the inept transmittal of substantially identical messages over multiple "special means" channels—the Americans were lucky that BLUEBIRD did not do permanent damage to their deception effort. But, emerging unscathed from these perils, at a minimum it was, like all deception operations, a valuable insurance policy.

After the Okinawa invasion, BLUEBIRD was tapered off, but the Formosa-South China threat was continued at a reduced level pending approval and implementation of new theater plans.

In keeping with the postponement "story," attention continued to be focused on the China coast, with accompanying reasons why further offensive operations would be delayed. Late in April, Smith sent to Lang for earliest possible use a "story" that Roosevelt's death might cause further postponements in the Pacific; GUS passed it to Suzuki, who observed that it confirmed his own estimate. Early in May, GUS reported that there were delays in shipments to the Pacific coast, and that Carlos P. Romulo of the Philippines had been heard to say that delay in defeating the enemy in his homeland had tied up American forces and prevented their use in other contemplated assaults. Similar items on these general themes moved steadily through the Argentine channels through June and July. Emphasis continued for a time to be laid on the South China coast; consistently with this theme, when Nimitz's headquarters learned late in April that Miles had been directed to do a survey of Shantung in North China, this was suspended at their request. Later, when it appeared that the focus would be moved further north, references to a China landing became more ambiguous.

Appropriate items moved through the new channels in Portugal and Switzerland as well during the summer—thus, KELLY handled a report that an extensive hydrographic study of the China coast had been prepared, and LESTER handled a report that MacArthur wanted Patton to command an army in China—but they appear to have been employed more for chickenfeed than for major items in support of BLUEBIRD.

Meanwhile, D Division reported a curious situation in the Portuguese territory of Macao on the South China coast. The British intercepted a message to Lisbon from the Portuguese governor reporting that the British consul had said that "Allied headquarters in the interior of China" had told him that there would be an Allied landing near Macao in April in conjunction with a landing in Indochina and another on the east coast of China. Nobody knew where the consul got the idea. D Division's inquiries to London and Chungking failed to shed any light on the matter, and the consul could not be asked directly because his cipher was believed to be insecure. O'Connor brought it to Smith's attention, but Joint Security Control was equally mystified. The explanation remains unknown.

In early May, apparent radio linkages between the Central Pacific, Southwest Pacific, and China were increased, and soon afterwards, at Nimitz's request through Joint Security Control, MacArthur's theater began a month-long series of photoreconnaissance missions over Amoy, Swatow, and Hong Kong.

It will be recalled that in February a Japanese request for information about a landing in South China had been fobbed off by offering intelligence about India through INK. This did not satisfy the Japanese, and in April INK was told that they were not particularly interested in India but required all possible information about American plans for landing in China. This led to a round of communications among Fleming, Bevan, O'Connor, Smith, and Raymond in Chungking, in which American primacy in the China Theater was maintained (Raymond pointed out that he had equal access to the INK channel), and a couple of possible responses suggested by D Division were rejected in Chungking as illogical. Ultimately, INK simply reported that he "might have important news regarding landings on the China coast in the near future," and later in the month INK reported that an officer from Manila had joined Wede-meyer's planning staff. The overall debate between D Division and Ray-mond eventually merged into the development later in the summer of the BODYGUARD of the Pacific—an overall deception plan, called BROAD-AXE, for the climactic phase of the Japanese war.

CHAPTER 18

Last Round in Asia

When the Formosa question was settled at the beginning of October 1944, it was possible at last to consider an overall deception strategy for the approach to Japan comparable to BODYGUARD in Europe.

Smith and Goldbranson sketched out in late October a rough draft of an update of CCS 284/10/D, the General Directive for Deception in the War against Japan, and sent it out to the Young Ladies' Seminary teams near the end of that month for their information and comments. (It was in connection with this draft that Goldbranson and Nelson paid their call on the Readiness Section to discuss the possibility of sending decoy fleets into the Yellow Sea and against the Kuriles.) Shortly thereafter, Joint Security Control's participation in planning was assured; and the Joint Staff Planners, no doubt on Smith's inspiration, formally directed the Joint War Plans Committee late in November to revise and update CCS 284/10/D.

But for months nothing was done. First, work was held in abeyance pending the outcome of the summit conferences at Malta and Yalta at the end of January and beginning of February. But thereafter time continued to pass without action, despite some nudging by the British, both in the Combined Chiefs and from Mountbatten's headquarters. This was because Admiral King and the Navy were not yet convinced that an actual invasion of Japan would be necessary, believing that victory might be achieved with pressure and tightening of the blockade. Eventually, however, at the end of April, King signified that he did not object to preliminary planning for an invasion, even though he hoped it would prove unnecessary.

Meanwhile, on April 13 the Joint War Plans Committee and Joint Security Control had reported a proposed overall plan for deception in the war against Japan. The Navy planners held it up until King gave his approval for preliminary invasion planning. A modified version then went to the Chiefs. Even then, Admiral Leahy held it up, on the ground that major items in the "story" might still be things the Allies would wish to do, and he did not want to approve a cover plan that would turn out to be the real plan after all. Douglas Fairbanks pointed out to Admiral Duncan that deception was flexible enough to take care of any such change in plans, and Thurber pointed out that nothing was set in stone until the theaters presented actual implementing measures for approval. General Hull of OPD asked General Marshall to appeal to Leahy on the same grounds. To no avail.

Even after King had given the green light and the Planners submitted to the Chiefs in mid-May a proposed directive for OLYMPIC, the landing on Kyushu projected for November 1945, its approval was held up by disputes between the Army and the Navy as to who would control the landings. Not until May 25 was that settled; on the same day, Leahy went along with the proposed deception policy and it received the Chiefs' approval. It went forward to the British; based on an analysis by Bevan (who, for once, was favorably impressed: "At first glance," he cabled to Wingate, "it seems [a] great improvement on previous efforts") they proposed minor amendments to bring it in line with Mountbatten's current plans; the Americans readily concurred, which was enough for Newman Smith to send an advance summary to the theaters; and on June 15 the plan was finally officially adopted by the Combined Chiefs. Issued next day as CCS 284/16/D, it had already received the codename BROADAXE.

BROADAXE was recognizably derived from Newman Smith's draft of the previous October. Its stated objects were to induce the Japanese to redeploy no ground forces from the mainland to the home islands; to disperse their ground forces at home, with emphasis on the defense of northern Honshu and Hokkaido (since the real invasion would come at the other end of the archipelago); to commit all available sea and air forces against Allied naval and air strikes on the home islands so they could be destroyed before the invasion; and to misinterpret Allied intentions concerning operations in Borneo, Malaya, China, and the Kuriles. The "story" would be that the Allies understood that an invasion of Japan would be costly, could not be undertaken till after a long period of naval blockade and saturation bombardment, and would itself probably not terminate the war since the Japanese government would plan to withdraw to the mainland and continue the struggle. Accordingly, they had decided that Japan must be encircled prior to invasion, with the establishment of additional bomber bases close to Japan, and a buildup in the Aleutians for operations against Hokkaido. There would be landings on the China coast as early as possible to help keep China in the war. Political pressures required the liberation of British, French, and colonial territories as soon as possible.

The notional Allied strategy would thus be:

(1) The seizure of additional bases for saturation air bombardment.

(2) An early amphibious assault on Formosa in late summer of 1945 preliminary to landings on the China coast.

(3) A build-up in the Aleutians in preparation for operations against Hokkaido during the early fall of 1945 to gain additional nearby air and sea bases.

(4) Accelerated supply to China to enable the Generalissimo to intensify his activities in the interior of China with the objective of drawing Japanese strength away from areas of Allied landings on the China coast.

(5) An amphibious assault from the Philippines against French Indo-China in the fall of 1945, coordinated with an overland assault by Chinese forces from the north and in conjunction with a British attack in Siam.

(6) An assault on Sumatra from India in the late fall of 1945.

(7) An advance into the Yellow Sea in the winter of 1945–46 to secure bases for air and inland operations.

Nimitz and MacArthur would be responsible for generating the threat to Formosa and the advance into the Yellow Sea. Nimitz in collaboration with the Alaskan Department would be responsible for the notional Aleutians buildup. MacArthur in collaboration with Mountbatten and Wedemeyer would be responsible for the threatened Indochina landing. Mountbatten in collaboration with Sultan would be responsible for the assault on Sumatra. Sultan in collaboration with Wedemeyer and Mountbatten would be responsible for the apparent acceleration of supply to China, and Wedemeyer would be responsible for intensifying activities in the interior of China to divert Japanese forces and aid the IndoChina project. Twentieth Air Force (the B-29s based in the Marianas) would operate, so far as possible, to support the impressions created by the theater deceptions.

BROADAXE was thus the BODYGUARD of the Pacific in a real sense, for its basic approach was similar to that of Bevan's masterpiece: Distract the enemy's attention in as many directions as possible; induce him to keep forces elsewhere than the real objective; persuade him that the invasion would not come until after a long preparatory bombardment and blockade. It had been, of course, an unconscionably long time aborning—seven months from Newman Smith's original proposal till its final adoption. The MIKE ONE deception, BLUEBIRD, HUSBAND, BAMBINO, and VALENTINE had come and gone except for transitional preservation of BLUEBIRD, as had STULTIFY and CLOAK.

BROADAXE had only two months to run until the Japanese surrender, though nobody could have known this. And some alterations were required even in this short period: The Formosa target would be changed to the Shanghai area; the idea of accelerated supplies to China would be dropped. Still, it was a noteworthy product, this BODYGUARD of the Pacific; taken together with PASTEL and the staff studies presently to be described, it was clear evidence of how far Newman Smith had brought American deception.

In his February report, Metzel had recommended that "as a temporary measure" his proposed "Joint War Information Control" would "initiate deception staff studies and furnish them to the Joint Staff Planners, the Joint War Plans Committee, and theater and area commanders concerned. This is to be done as a service. It is to be made clear that recipients are in no way obligated to incorporate these studies in their plans." In line with this, and in line with Thurber's and F-28's proclivity for generating paper, Smith and Thurber and their colleagues peppered the Joint Staff Planners with papers outlining ideas for each of the component deceptions of BROADAXE.

On May 9 they put forward a staff study for a Formosa deception, suggesting a notional amphibious landing on the west coast of Formosa followed by complete reduction and occupation of that island, seizure of Amoy and occupation of the immediately adjacent Chinese coastal areas, and occupation of the Pescadores.

On May 24 they submitted a staff study for a notional invasion of Hokkaido in the autumn of 1945.

On June 6 came a study for a notional threat to French Indochina, with a revised version three days later.

On June 9 came a staff study for the fictitious advance into the Yellow Sea in the winter of 1945–46; it contemplated a gradual breakoff of the Formosa deception and a transfer of the threat northwards, with an initial notional operation against the Chusan Archipelago, ostensibly to establish air and naval bases to increase the air and sea blockade and to mount an advance into the Yellow Sea (Spruance had proposed an actual Chusan operation, codenamed LONGTOM, which had been the subject of contingency planning at the beginning of the year); followed by a threat to operate against Lienyun to cut the north-south railroad.

They even trespassed on British territory by drafting such a study for a notional threat to Sumatra, but this went forward only very informally.

Each study included a list of possible methods of implementation. They were of the familiar kind—carefully targeted reconnaissance, suggestive air strikes and mining operations, administrative measures, drops of supplies to imaginary agents, W/T deception, special means, and so on. One suspects that Newman Smith used these studies as a device for educating Thurber and his novices.

The Joint Staff Planners deferred all these until final action on the overall policy. They grew a bit impatient at the degree of repetition from one study to another. "We may try the patience of overworked planning staffs in the theaters by sending out papers under different titles which contain about the same material," commented "Abe" Lincoln of OPD. Nevertheless, after making some amendments of their own the Planners sent the studies out to the theater staffs on June 17, the day after BROAD-AXE went out, "for such use as they may deem appropriate, in the thought that it may contain information of value to current deception planning."

In the middle of May, just as the spring's work was clearing through the Joint Chiefs, Newman Smith lost his strong right arm with the departure of Carl Goldbranson for MacArthur's headquarters.

On April 3, the Joint Chiefs had changed the theater designations in the Pacific, establishing MacArthur as Commander in Chief of all Army forces in the Pacific ("CINCAFPAC"), matching Nimitz as "CINC-PAC" commanding all Navy forces there. (MacArthur also continued as "CINCSWPA," the Southwest Pacific theater commander, an Allied command including Australian troops, and Nimitz also continued as "CINCPOA," the Pacific Ocean Areas theater commander.) The new CINCAFPAC headquarters was activated on April 6 in Manila, to which MacArthur had moved his headquarters the month before.

Though the invasion of Japan had not yet been formally decided on, planning for it was certain to proceed. As the definitive historical study has abundantly confirmed, "the estimate that American casualties [in an invasion of Japan] could pass the million mark was set in the summer of 1944 and was never changed." MacArthur's chief of intelligence, basing his calculations on experience in Okinawa, reckoned that "two to two and a half Japanese divisions [could] exact approximately 40,000 casualties on land." Such a calculation would yield by late July an estimate of up to 280,000 casualties during the push to the stop line in Kyushu alone. Faced with the prospect of enormous casualties, and having before them the example of what deception had done as a force ratio multiplier for the Normandy landings, MacArthur's staff at last got seriously interested in the subject. On April 22 they asked for a trained deception officer for their planning group.

Newman Smith reviewed the options. Al Moody was cooling his heels in the officer pool and anxious for assignment but Smith did not deem him qualified; presumably Moody's declining the appointment in India in the autumn of 1944 had something to do with Smith's opinion of him. After discussion with Lincoln and others in OPD, a certain Lieutenant Colonel A. C. Goodwin from the Strategy Section, who had worked on deception planning, was tapped. But Goodwin came to Smith and cried—actually shed tears—at the prospect of being sent overseas. He said he was in love with a girl, was awaiting a divorce from his wife, and was afraid he might not get the divorce and that the girl would not wait for him. It seems unlikely that this evoked much sympathy from a man whose own "girl" had waited for *him* to get a divorce, and who had been sent overseas for two years less than a month after marrying her. But regardless of sympathy, Smith concluded that Goodwin's mental state was such that he would not do a good job if he got the assignment.

The only qualified man available was Goldbranson. Goldbranson was anxious not to go. He had seventeen months overseas already; was in love with his wife of nearly twenty-five years, had a son with thirty-two months overseas and the Purple Heart, a seventeen-year-old daughter about to finish high school, a twelve-year-old son under doctor's care. But he was a good soldier and would accept the assignment and do a first-rate job if ordered. Smith discussed the problem with Bissell on April 25. Bissell decided that there was no one else and Goldbranson must be offered to the theater.

Goldbranson pointed out to Smith another consideration. If he spent two more years in uniform he would be forty-seven when he returned to the Union Pacific after seven years' absence, competing with much younger men firmly established in their positions. Smith saw Bissell again on April 26. The decision is made, said Bissell. No further need for discussion. He would be willing to write a letter to the railroad on Goldbranson's behalf. Goldbranson gave it one more respectful try, by memorandum and in person. Bissell was sympathetic but remained firm. He undertook to provide an understudy for Goldbranson when redeployment from the European theater would permit, so he could get home as soon as possible; and he wrote to the president of the Union Pacific asking that the fact that Goldbranson was "being retained longer in the service of his country will not prejudice his future position in your organization."

Goldbranson left Washington for Manila in mid-May. Shortly before his departure, O'Connor passed on a thoughtful message from Johnny Bevan offering "warmest congratulations" and "kindest regards and best of luck from us all." "We are all delighted to hear this good news," said Bevan, "and feel convinced that you and Wingate will in close collaboration completely confound the enemy." (O'Connor had suggested this, advising Bevan and Wingate that Goldbranson "retains highest respect and liking for yourselves as well as for Dudley Clarke. He is well aware of difficulty we have had here and has done his best to help behind the scenes. . . . He was always only second man and he did what he could.")

Goldbranson had an easy trip out to Manila and received a cordial welcome. "There are some inherent obstacles but a favorable change in the general attitude here is noted," he wrote to Bissell.

Beginning in May with a cover and deception appendix to a CINCPAC staff study for OLYMPIC, on which Douglas Fairbanks commented for Admiral Duncan, and continuing with conferences at Nimitz's advance headquarters on Guam and at Manila, Goldbranson, Hilger, and Barnes, plus several signal officers and senior planners with CINCAFPAC, worked out an outline deception plan for OLYMPIC, codenamed PASTEL, plus a proposed transitional "story" from BLUEBIRD to PASTEL, and Hilger and Barnes left for Washington with it on June 14. They remained in the capital until well into July while it worked its way through the staff system. It was a relatively slow process, causing Goldbranson out in Manila to fret over the delay. Not till June 29, after conferences between Hilger and Barnes and Joint Security Control, principally on June 27,* did it go forward, with Joint Security Control's recommended changes to comply with BROADAXE; after the usual staff committee consideration the Joint Chiefs approved it on July 10.

As befitted the immensity of OLYMPIC—an initial assault of nine divisions, half again as great as D-Day at Normandy—PASTEL was the most elaborate of American deceptions, both the plan and the anticipated execution; more elaborate than BLUEBIRD. As approved by the Chiefs, its stated object was to minimize, to the fullest extent possible, Japanese opposition in the OLYMPIC area by causing them to retain their large ground forces in China, disperse their air and naval effort, lower their vigilance in the OLYMPIC area, reduce the weight and rate of reinforcement, and expend part of their available effort on fortifications in areas other than the target area. Considerations recognized as likely to affect the Japanese estimate of the situation included the need to preserve lines of communication on the continent of Asia; the attrition of Japanese forces; awareness that the Americans must perceive that even after invasion, the government might be transferred to the mainland to continue resistance; the need to retain large forces in Manchuria in the event that the Soviet Union (which had recently terminated its nonaggression pact with Japan) might enter the war; and the obvious utility to the Ameri-

---

* It must have come as a surprise to Smith and Thurber to learn at this conference that Hilger and Barnes were not cleared for ULTRA, that most essential tool of the deceiver's trade. They could do no more than promise to send regular summaries of Japanese reactions to deceptive efforts, since only theater commanders could decide who on their staff should have ULTRA access.

cans of bases on the mainland—for continued restriction of movement in the Yellow Sea and the Sea of Japan, neutralization of home island defenses, and destruction of Japanese industry, as bases for operations against mainland forces if the Japanese government should flee there or if such forces should refuse to honor a capitulation by Tokyo, and to provide token forces if the Soviets should enter Manchuria.

The Chusan-Shanghai area would supply such bases, additional airfields and naval anchorages, and a supply route to Chinese forces. Beaches were available, and the August–March period was suitable for landing operations. (BROADAXE had contemplated a Formosa threat rather than Chusan-Shanghai; as Goldbranson noted in explaining the initial outline to Smith, it was believed that by now the Japanese were aware that Formosa had been bypassed.)

Beginning thirty days before "X-Day," the target date for OLYMPIC, it would probably become apparent to the Japanese that there would be early landings in Japan, and Kyushu should be apparent as the objective beginning X minus 10. Until then, the Japanese could not neglect Shikoku, with its location and its airfields and possible airfield sites; and November–February appeared to offer the most favorable conditions for operations against it.

Accordingly, for maximum effect in inducing the Japanese to pin down forces in China, a threat of landing in the Chusan-Shanghai area with a target date of X minus 30 should be developed. Then about X minus 55 the Imperial General Staff should be made aware that the assault had been postponed (but not abandoned), and should be persuaded instead that there was a specific threat against Shikoku with a target date of X plus 55. (With an X-Day of November 1, therefore, the Chusan-Shanghai landing should be threatened for October 1, and the Japanese should be convinced about September 7 that that attack had been postponed and that an additional landing was planned in Shikoku around Christmastime.)

The first-phase "story" was that the Joint Chiefs had decided that "faced with the severe casualties and delays of the Okinawa campaign, the necessary release of long-service veterans of the Pacific and the logistical difficulties of redeployment from Europe," no invasion of Japan could be undertaken before the autumn of 1946. (The "story" was carefully phrased to make clear, in consonance with the Navy's views, that the need for, or wisdom of, invading the home islands was still not a foregone conclusion.) Meanwhile, to increase strategic bombardment, accelerate delivery of supplies to build up Chinese strength, and be positioned for possible removal of the Japanese government to the mainland and possible Soviet entry into Manchuria, the Chiefs had "decided to develop air and naval bases in the Ryukyus and secure lodgments in the Chusan-Shanghai area about X minus 30 preliminary to a further advance into the Yellow Sea area. Ground forces available for this operation will total sixteen combat divisions and auxiliary troops." In addition, they had directed a six-division buildup in the Aleutians to enable amphibious operations against the Kuriles in late 1945 preparatory to landings on Hokkaido in 1946.

Then, by about X minus 55, special means would have implemented the transitional "story" from BLUEBIRD and BROADAXE that in light of "the deterioration of the Japanese position in China, the rapidly increasing effectiveness of the Chinese forces and the unexpected speed of the redeployment of forces from Europe, coupled with the tremendous destruction in Japan as a result of our all-out bombing effort," MacArthur and Nimitz believed that the best use for forces originally earmarked for followup operations in China after the Chusan-Shanghai landings would be to divert about eight divisions to land in Shikoku about X plus 55 for the development of air bases on that island to support future operations, "increase the effectiveness of our air bombardment and extend to a maximum degree the blockade of Japan."

The plan concluded with a general allocation of responsibility and the requirement that an "Appendix 'A' " be prepared setting forth specific responsibilities and timing.

The addition of the threat to Hokkaido from the Aleutians reflected both BROADAXE and the renewed views of Douglas Fairbanks; but an argument by him that Shantung should be substituted for Chusan-Shanghai, and Hokkaido for Shikoku, was not accepted. A transitional "story" for BLUEBIRD, to the effect that on MacArthur's recommendation the Formosa operation had been indefinitely deferred, was the subject of discussion and agreement but was deemed inappropriate to be included in PASTEL itself.

On July 10, the theaters were formally advised by the Chiefs that PASTEL had been approved, with the suggestion of a conference of theater representatives to prepare the annex on responsibility for, and timing of, implementation.

PASTEL was not the only subject of conferences in those weeks in Washington. In late June, Johnny Bevan came to town for his first visit since the meetings with Strong two and a half years before, and his last before relinquishing the reins of the London Controlling Section.

Bevan continued to be discouraged over—and ignorant of—the state of American deception. Bissell was playing it close to his chest. From the late summer of 1944 on, he was wary of a British effort to take over deception in the Japanese war once the Normandy breakout left them with little more to do in Europe. It was natural therefore for him to hold at arm's length what were no more than genuine offers of help.

A memorandum by Bevan in August 1944 lamented the fact that the Americans had no single body comparable to the London Controlling Section; that Joint Security Control, as far as he could see, was interested mainly in tactical deception, that O'Connor had been told that "they do not at present employ any secret deception channels to the Japanese," and there were no theater staffs in the Far East like Ops (B) and "A" Force. "If we all put our heads together," he said, "I believe we could give the Japanese the information which is good for them! In any case I should like to try and am most anxious to do everything I can possibly do to help, but unless adequate American staffs are established in Washington and at various theater headquarters, and unless secret deception channels are developed and utilized, I fear that we may not achieve very much." (As he wrote this, of course, the Young Ladies' Seminary was gearing up, and Smith had his channels in Buenos Aires, though he would not really begin using them till the next month.)

Ismay passed Bevan's memorandum on to Bedell Smith, wondering whether he or Eisenhower could do anything. Bedell Smith forwarded it to General Handy at OPD, noting that "our experience from Africa through Italy and into France has made me a complete convert to the transcendent value of an experienced deception organization." At Hull's suggestion, Handy responded somewhat evasively that the situation in the Pacific was quite different from Europe, though improved coordination and control might be needed.

Bevan could not have been cheered by seeing the 12th Army Group report about which he wrote to Bissell so eloquently in December; nor by the annex to it dealing with special means which so distressed him in April. The long delay in formulating BROADAXE did not improve his outlook. When in January 1945 a visit by him to Washington was considered, he advised Ismay that having "recently received unofficial indication that Washington have no strategic deception role in view for SEAC at present, there would not appear to be much point in my visiting Washington in [the] near future." In connection with efforts to keep alive the BULL'S EYE channel in Turkey, he had received a discouragingly noncommittal response from Joint Security Control to an offer to try to continue developing channels to the Japanese in the Middle East. In late March, when SEAC, tired of waiting for the overall policy directive, proposed to send a D Division officer to MacArthur's headquarters to coordinate deception planning directly, Bevan again expressed disappointment—

> that the Americans have up to the present not shown much interest in strategical deception for the war against Japan. So far they have responded neither to offers of assistance, particularly with respect to implementation by special means, nor to an invitation to send a representative to the London Controlling Section for purposes of co-ordination. Time is admittedly running very short and continued delay will further reduce the possibility of obtaining appreciable dividends from strategic deception in the Far East.

By that time he was looking ahead to returning to civil life. "Strategical deception in Europe is no longer practicable," he said in a memorandum to Hollis, "while in the Far East American cooperation and interest in strategical deception remains conspicuous by its absence"; and he proposed that he vacate the position of Controlling Officer to return to his firm in the City, which needed him badly, "by, say, 1st September, 1945."

On May 19, Jumbo Wilson wrote to Marshall and King saying that O'Connor was soon to leave, and the British had to decide whether to replace him; while Bevan was retiring, and his organization would be cut back. Did the Americans want or need British help in deception against Japan? Did they want to use British special means channels in that connection? "Some of our people," he said, "have gained the impression that in making these various offers of assistance, as we have done from time to time on the staff level, it has been felt that we are trying to butt in on an affair which is not our business. I should very much like to dispel any such ideas, if indeed they exist. . . . [I]f it would be any use to you we should be glad to ask Colonel Bevan to come out here and discuss with your experts before he leaves the service." Marshall responded on May 22, in a letter provided by Bissell, that a visit by Bevan would be desirable.

(This came just at the time Admiral Leahy was holding up BROADAXE, and it inspired Hull of OPD to try playing the British card with the Admiral. "This guy from London is coming over here, . . . and if we don't have a deception plan at that time, I wouldn't be surprised that the British come up with one," said Hull to General McFarland, the Secretary of the Joint Chiefs. "They believe in it and have done a damned fine job on those things in the past. . . . I think we should either get the damned thing out or revamp it so it can be put out before those blighties get into the picture." "Abe" Lincoln likewise pleaded unsuccessfully with the Navy planners that "if we do not get out a plan soon, we may find that the British will initiate action.")

Leahy had finally approved, and BROADAXE had been adopted, when Bevan ultimately arrived in Washington in response to Marshall's invitation, bringing with him Captain F. M. Gilbertson, formerly with D Division, who had joined the London Controlling Section earlier in the year. Bevan was in Washington by June 22; the principal meeting was held on that same June 27 when the final version of PASTEL was drawn up, with Newman Smith evidently acting on behalf of both his section and Thurber's.

"To avoid a show-down with JSC" over breach of the Strong agreement, as Bevan put it to his colleagues back home, the Strong agreement was replaced by an agreement that in Europe Joint Security Control had channels to the Japanese in two locations as of that date (meaning, presumably, Portugal and Switzerland), while London Controlling Section (and SEAC) had SUNRISE in Europe, PURPLE WHALES (presumably including INK) and SILVER controlled by SEAC; and that no changes or additions would be made to this list of channels without further agreement; although theater commanders could develop channels within their own theaters, and Joint Security Control and the London Controlling Section were free to look into the possibility of new channels (though none could be used without clearing it with Washington). Bevan interpreted this as allowing SEAC to use for its own purposes channels not accepted for use by Joint Security Control.

Though Bevan gave Joint Security Control a full account of his channels (including a warning that SUNRISE might be lost, though he did not realize how soon that would happen), the Americans did not reciprocate. Bevan evidently came away with the impression that ASPIRIN, the Spanish journalist in Washington, was in Lisbon, and that there was only one channel of unspecified grade in Buenos Aires.

Agreement was also reached on a personnel arrangement that might have been of considerable interest had the war lasted. O'Connor was soon to be relieved by an RAF group captain, E. M. Jones. Joint Security Control invited the London Controlling Section to provide an additional British officer with experience in deception to work under Jones, relieve him of detailed liaison work with the Special Section, and sit in most of the day with either Smith's or Thurber's people and assist in the planning and implementation of deception. "He should be a man of considerable imagination," said the agreement, "and well trained for this type of work." (Bevan and Crichton had Dudley Clarke's man Harry Gummer in mind. Given Gummer's past hostility to the Americans in "A" Force, it may be doubted that he would have been a success; but the war ended before his name was ever proposed to Washington.)

From Washington, Bevan went by train to New York, to return on the *Queen Elizabeth* departing July 4, and presumably to look in on British Security Coordination. (Douglas Fairbanks traveled to New York with him. The trip began dramatically, for Fairbanks reached the station at the last minute and rode part way to Baltimore hanging on the outside of the train.) Group Captain Jones arrived a few weeks later. O'Connor took him to New York and Ottawa from July 24 to 29, presumably to introduce him to British Security Coordination and to the Mounties and other appropriate Canadian authorities, and he formally took over from O'Connor on July 30. The war ended a fortnight later.

The closing of SUNRISE left the British with no channel to the Japanese in Europe. Early in August the London Controlling Section proposed to Section V the opening of a channel to be called MOHICAN, by sending to the Japanese mission in Bern a letter purporting to come from a pro-Irish woman whose husband was a prisoner in Japanese hands and whose brother worked for the War Office, offering to sell information from her brother for money and the promise of good treatment for her husband. But the war ended before that idea got any further.

That was substantially Bevan's last service in his accustomed role, except for the transitional period to hand over his post to his successor. Earlier in July he had arranged with Hollis that he should be demobilized on August 1, and that Michael Crichton should become Controlling Officer. Crichton himself was due for demobilization in the autumn, but was willing to remain till the beginning of 1946. In fact, with Ismay and Hollis away at Potsdam, and some high-level problems to be solved, the handover from Bevan to Crichton seems not to have taken place before September 1. Bevan thanked O'Connor for his services. "It has been a terribly uphill task," he said. "Nobody could possibly have done it better."

And so, as it turned out, not until the final curtain had come down did Johnny Bevan pass from the stage on which he had played such a central role since the dark days of 1942. However dim his view of the state of American deception, it is to be hoped that by that time he was at least a bit more encouraged about inter-Allied cooperation on deception in the Japanese war. For that, Ronald Wingate deserves much credit.

With the adoption of BROADAXE and PASTEL, plans for deception in the Japanese war began to converge, as did the real operations that were closing in on Japan. In Southeast Asia, Mountbatten was readying Operation ZIPPER, the invasion of Malaya, to be followed by MAILFIST, the liberation of Singapore. In China, Wedemeyer and Chiang were preparing an operation, ultimately called CARBONADO, to drive to the South China coast and open the ports of Canton and Hong Kong. In the Southwest Pacific, MacArthur's Australians were recovering key points in Borneo, while the slow eradication of Japanese resistance in the Philippines went on and he and Nimitz readied OLYMPIC and its followup, CORONET, an assault on Tokyo itself.

Each of these had its deception plan within the overall BROADAXE framework. Mountbatten had SCEPTICAL and its component SLIPPERY, aided by CONSCIOUS. Wedemeyer had ICEMAN. MacArthur had CON-

SCIOUS to aid Mountbatten, and he and Nimitz had PASTEL. Most significantly, for the first time all four theaters worked in concert.

First in time had been SCEPTICAL, SEAC's successor plan to STULTIFY, to cover ZIPPER.

Under ZIPPER, Mountbatten planned to land in Malaya at Port Swettenham (originally in late August, changed on June 1 to September 9) and at Port Dickson three days later, securing a bridgehead and airfields, and advancing to the Straits of Johore, making amphibious hooks down the west coast of Malaya to cut off retreating Japanese forces.

SCEPTICAL, the ZIPPER cover plan, as cabled in summary form to the British Chiefs on April 21 with a copy to O'Connor for the Americans, had as its object to induce the Japanese not to reinforce Malaya with their forces in Indochina, in Thailand north of the Kra Isthmus, and in Sumatra and Java. Its "story" was that in view of the extensive forces now available, the Combined Chiefs had agreed to add operations against Thailand at the same time and on the same scale as those against Singapore. After the capture of Rangoon, Mountbatten would advance in force down the Moulmein-Bangkok railroad while an airborne assault would seize Bangkok with the help of the Thai underground, all linked with Chinese and American operations against northern and southern Indochina respectively. Seaborne operations would cut the Kra Isthmus and establish air and naval bases in southern Sumatra, looking towards final capture of Singapore by amphibious assault through the Straits of Malacca, probably timed to coincide with an assault on Penang from Phuket. Further details, said Mountbatten, could only be worked out after a visit by his planners to MacArthur's headquarters.

In London, Bevan concluded that the proposal was sound except that, at Foreign Office request, references to the Thai underground should be deleted, since "this might lead the Japanese to take action against elements in Siam who may be of assistance to our operations at a later date." In Washington, Goldbranson got an "unofficial preview" from O'Connor, discussed it briefly with several colleagues, and reported to Smith that all agreed that it would not materially conflict with the general directive that was to become BROADAXE. It passed smoothly through the staffing process and was approved by the Combined Chiefs on May 20.

For direct coverage of the initial ZIPPER landings, the notional operation against the Kra Isthmus was developed into its own subsidiary operation, SLIPPERY, and it was decided to maintain that threat for some weeks after the genuine ZIPPER assault. By the end of July, the fully developed "story" of SCEPTICAL, including SLIPPERY, was that amphibious operations would be impossible during the monsoon, but that on October 15 the (genuine) XV Indian Corps would land in the Victoria Point area, to cut the isthmus and isolate Bangkok from reinforcements from Malaya, and on November 1 the notional XXVI Airborne Corps would assault the Bangkok area, supported by a simultaneous notional amphibious landing (CONSCIOUS) by MacArthur on the southern tip of French Indochina, and by a (genuine) holding attack on the Moulmein area by the (genuine) IV Corps. In the south, the Japanese would be led to expect an attack by the notional Force 144 on the Straits of Malacca planned for October 1.

In early August, however, Mountbatten would notionally receive a fresh directive—information as to which would not reach the Japanese until after the genuine ZIPPER opened on September 9—directing an attack on Singapore at the earliest possible date. He would accordingly "detach" the genuine XXXIV Indian Corps (which would not have previously been mentioned, and which would make the actual ZIPPER attack) to seize a bridgehead in southern Malaya before advancing on Singapore—which of course would be the genuine ZIPPER. Notional preparations would continue for the Kra operation, but in early October the Japanese would be informed that that operation had been canceled and XV Indian Corps was being transferred to the ZIPPER front, because it had become evident both that XXXIV Indian Corps could not capture Singapore unaided and that ZIPPER would effectively delay any reinforcements being sent from Malaya to Bangkok.

D Division merged the old STULTIFY into SCEPTICAL by degrees after April 21, and by late May was implementing by special means the notional movement from Rangoon towards Bangkok and the threat towards Sumatra, assuring the Americans that this would be done only gradually until detailed coordination could be achieved with MacArthur and Wedemeyer.

ICEMAN came in to Washington from Chungking in the latter part of May.

After the ICHIGO offensive ran down, the Japanese in China were on the defensive. Wedemeyer's overall strategy then was to launch in the latter part of 1945 with American-trained Chinese divisions a drive to the south China coast, opening Hong Kong and Canton and freeing the Chinese at last from dependence on the trickle of supplies over the mountains from Burma. This was Operation BETA (subsequently renamed RASHNESS and then again renamed CARBONADO).

It was obvious that the offensive would come on the southern front because of the limited road net of China and the concentration of American-trained divisions there. Moreover, information percolated freely from Chiang Kai-shek's staff into Japanese hands. So Raymond in Chungking adopted an imaginative course. Two outline plans, ICEMAN and RASHNESS, were prepared and given to the Chinese Military Council for the preparation of detailed plans. ICEMAN was a drive eastwards towards Changsha with a diversionary effort in the south. RASHNESS was a drive in the south with a diversionary effort eastwards towards Changsha. No one but Chiang himself, his chief of staff General Chien Ta Cheng, and the Minister of War, General Chen Cheng, was told that ICEMAN was a cover plan and RASHNESS the real one; the Chinese staffs working independently on the two plans did not know a cover plan existed. From time to time, hints would be given to the Japanese that ICEMAN had been selected as the offensive operation.

Smith looked the idea over, discussed it with OPD, and sent it to Thurber with the observation that "no harm can be done and some good *may* be accomplished." Fairbanks and Nelson had much the same reaction, the latter with minor comments which Thurber passed back to Smith.

As events turned out, ICEMAN proved to have been unnecessary. The Japanese began retreating in the south even before RASHNESS (now CARBONADO) could be launched; the offensive essentially began with what was to have been its second phase, and the war ended before it could get very far.

CONSCIOUS, the threat of a landing in southern Indochina by MacArthur to support Mountbatten's threatened Bangkok operation, was the fruit of cordial intertheater cooperation such as had not yet been seen in the Japanese war; reflecting the arrival on the scene of the kindly and tactful Sir Ronald Wingate.

Wingate, it will be recalled, had come out to Kandy in March to be in charge of D Division headquarters planning, while Jack Corbett, on Newman Smith's recommendation, had been promoted to major—though the promotion did not come through till June—and assigned to SEAC. After SCEPTICAL was approved, Wingate arranged for himself and Corbett to visit Raymond in Chungking to discuss liaison as to SCEPTI-CAL and ascertain whether the China Theater needed any help from SEAC with their own plans. Leaving Kandy on June 22, they stopped off in Calcutta to talk with Peter Fleming and Frank Wilson; thence off to Chungking on June 24, running into heavy weather over the Hump—they had to go up to eighteen thousand feet without oxygen, getting "green and sick," recorded Wingate, then "stooging around over the air-field at Kunming for 1-½ hours before we could let down on instruments through the pea soup." On to Chungking next day, flying down the gorges of the Yangtze and landing "a few inches from the water's edge, rushing towards the cliff, hoping the brakes would work." There Raymond "conducted us through the various back streets to Freddy Bishop's quarters at the British Military Mission." (Lieutenant-Colonel F. G. Bishop was now the "Control Section, China," D Division's representative with the British military mission, replacing Major Pierson.) Wingate stayed with General Hayes of the mission; Corbett was billeted in the American compound.

Wingate, Raymond, Bishop, and Corbett spent two productive days together on June 26 and 27. They reviewed BROADAXE in detail, which Raymond had not seen, and agreed on two major points as to which it needed adjustment.

First, MacArthur's notional Indochina landing should not, as BROAD-AXE seemed to contemplate, be targeted on the northern part of that country and coordinated with an overland attack from China. It should rather be portrayed as being pointed at the southern tip, south of Saigon, and as scheduled for early November simultaneously with the SCEPTICAL attack on Bangkok, with the objective of securing the approaches to the Gulf of Siam and establishing fighter cover and convoy protection supporting the Bangkok operation. A genuine blocking action, directed towards northern Indochina from China in support of Wedemeyer's offensive, was already planned; this could be portrayed as intended to support MacArthur's move in the south by containing Japanese forces in the Hanoi area.

Second, the component of BROADAXE that contemplated a notional "accelerated supply to China" and "intensification of activities in the interior of China" should probably be eliminated, since the Japanese were well aware that supply limitations would preclude any operations over and above CARBONADO. As an alternative, Raymond undertook to explore with Joint Security Control the stationing in India of two fictitious American parachute regiments under China Theater's control, which notionally could be freely used in China but would not be dependent on supply over the Hump until actually employed.

"All parties," said the short minutes of the conference, probably prepared by Raymond, "were extremely pleased with the value of this conference which had quickly permitted complete agreement on many confusing matters." It was agreed that Joint Security Control would be urged to arrange for more such intertheater conferences.

The flight back to India was more comfortable than the flight over, for Wingate and Corbett hitched a ride with General Carton de Wiart, head of the British military mission. The general's plane had "armchairs and whisky," recorded Wingate, "and we sailed over the Hump as if it had not been there."

In mid-July, Mountbatten visited MacArthur in Manila to coordinate their respective offensives. Wingate was among the small staff that Mountbatten brought with him, and he, Goldbranson and other officers from MacArthur's staff, plus Raymond from China, met from July 12 to 14 to coordinate their work. Wingate took along an ambitious agenda for discussion, and it was a gratifying session; the Americans agreed to much of what he sought. ("Col. Wingate was quite elated over the whole deal," wrote Corbett to Raymond after Wingate's return.).

Among other things, Plan CONSCIOUS emerged. Its object was to contain Japanese forces in Thailand north of the Kra Isthmus and in French Indochina for ZIPPER, and to draw them away from the flank of CARBONADO. Its "story" was that Mountbatten and MacArthur had agreed that when the monsoon was over, about November 1, Mountbatten would attack Bangkok and MacArthur would support him by a landing in the Saigon area to secure air bases. Subsequently, radio deception and special means would support the further "story" that "through limitation of forces and the necessity for commitments in the Indochina venture," MacArthur's participation in the notional operation against Sumatra under SCEPTICAL would be "limited to air support from South Borneo air fields now in process of establishment." (Australian troops were mopping up the Balikpapan area in South Borneo at that very time and heavy bomber operations from Balikpapan were expected to begin in mid-August.) It was further agreed that Corbett would come to Manila in mid-August for a month's stay as D Division liaison for the implementation of CONSCIOUS, "in the role of 'guest conductor,' " as Corbett put it in a letter to Raymond.

CONSCIOUS was forwarded to the Joint Chiefs, and appears to have met with staff approval, but evidently got no further owing to the end of the war.

Peter Fleming was not at Manila; he was on his way to London and the war would be over when he returned. He appears to have taken little part in military planning after the coming of Ronald Wingate. His last year was marked by two special projects, codenamed GREMLIN and MIMSY, and the concocting of two last phony documents of the PURPLE WHALES type. The more important of these reflected a project codenamed BASSINGTON. None of these projects came to fruition.

GREMLIN was a plan modeled on HICCOUGHS, to feed the Japanese spy mania by sending radio messages, perhaps later even dropping supplies, to a notional spy network in metropolitan Japan itself. Fleming had adumbrated this idea as far back as early 1944. Bevan (who by his own account had been "not particularly keen" on HICCOUGHS when it first opened, though by 1944 he had to admit that it seemed to be having some success), warned him not to extend it to the Japanese homeland "without the full approval of Washington." Fleming became more and more enamored of the idea; O'Connor approached Bissell in Washington for approval, and Fleming himself approached Bissell when he visited the theater; Fleming even suggested the Americans run the project and send an officer out to study HICCOUGHS; all to no avail. "I think it would be a mistake to flog GREMLIN problem at present juncture," Bevan told

Fleming. (These approaches evoked some of Fleming's complaints about Smith and Joint Security Control described in Chapter 11.)

Fleming would not give up. In February 1945 he sent to Bevan a full history of HICCOUGHS with a plea that its success be regarded in considering GREMLIN. By May he had persuaded Wingate, Bevan, and MI6 that GREMLIN could be done by the SEAC psychological warfare branch "provided it is used for 'jitters' and is not used for strategic deception," in Wingate's words. When Bevan went to Washington late that June it was agreed that he would not bring GREMLIN up, probably to leave Wingate a free hand in trying to sell it as support for the invasion of Japan at the Manila conference; but Wingate had no luck there. The war ended before more could be done.

In May 1945 Fleming devised a project called MIMSY. Modeled on the LLAMA affair in the Mediterranean, it proposed to blow a double agent in order to establish communication with the Japanese general staff looking toward surrender. Bevan pointed out that the circumstances were completely different from those surrounding the LLAMA case, and he and Wingate suggested that MIMSY be reduced to having the possibility available as a way to contact the Japanese Southern Army if its commanders wanted to leave Burma. Mountbatten referred the idea to the Chiefs of Staff; they referred it to Bevan; but the war ended four days later.

The more important of Fleming's 1945 PURPLE WHALES ideas reflected a project codenamed BASSINGTON.

Somehow, in all the attention paid in recent years to the crimes of Nazi Germany, the monstrous behavior of the Japanese in the years before 1945 has tended to be overlooked. No one knows how many perished under the heel of that evil empire—something on the order of seventeen million, by one careful reckoning, not to mention the hundreds of thousands of slave laborers seized from the occupied lands. The savage treatment of Allied prisoners of war—from the Bataan Death March in 1942 on through the horrors of slave labor on the Burma-Siam railroad, the dousing of American prisoners on Palawan with gasoline and setting them afire, beheadings, forced starvation, and the cutting-up alive of downed B-29 crewmen at a medical school—was but one element of this mass of atrocities, but it was of course the one of most immediate interest to the Allied military. By the winter of 1944–45 enough was known of these horrors to raise concern that all prisoners might be massacred before Japan should give up.

Fleming constructed a fake document for passing to the Japanese that might do something to ease the situation. Purporting to be "Progress Report No. 7 for the Period 1st October to 31st December 1944" of the "United Nations War Crimes (Far East) Commission," with an attached "Statement of Policy" regarding publication of information regarding Japanese mistreatment of prisoners, it was a typically gleeful Fleming pastiche, with such realistic details as proposed revisions to a supposed separate list of specific incidents, and references to amendments proposed by representatives of various countries and to V-bomb damage to "the Commission's former premises at No. 27 Grosvenor Square." Its thrust was that the Allies were in effect compiling a black book that would be the basis for war crimes trials and would discredit Japan as a respectable member of the family of nations for years to come. Fleming's theory, derived in part from his reading of the INK channel, was that there were civilian elements of the Japanese government, in the Greater East Asia Ministry, the Foreign Ministry, and elsewhere, who disapproved of the mistreatment of Allied prisoners and whose hand would be strengthened if this document reached them.

Mountbatten approved it and it went to London. Bevan sent it to O'Connor and O'Connor gave it to Smith. Smith reviewed it with Rushton and with Goldbranson. He did not like what he saw. Its object was "laudable" but it was "chiefly concerned with diplomatic deception and properly is a function of appropriate propaganda agencies." He discussed it with appropriate General Staff experts and with Bissell. He passed it to Thurber, who felt even more strongly. So Smith told O'Connor that he and Thurber agreed that no action should be taken "without approval of the appropriate civil and military authorities on the combined level," adding that in their opinion this was not a matter for Joint Security Control anyway.

Meanwhile, in London Bevan approved it. The British Chiefs agreed, saying it was "well worth trying," and "in any case, could do no harm." Even more to the point, Churchill, to whom Ismay insisted it be submitted, noted on his copy: "So proceed." So back across the Atlantic it went, this time as a British proposal for Combined Chiefs action. The American Joint Staff Planners concluded, rather like the British Chiefs, that it "can do little if any harm," but suggested a number of amendments. Approved by the Joint Chiefs and the Combined Chiefs as amended, the final version, reprinted by the Foreign Office, reached India in the second week of August. But the war was over before anything could be done about it.

A similar fate befell a copy of minutes of a notional meeting at Mountbatten's headquarters implementing SCEPTICAL and SLIPPERY, which Fleming hoped to plant on the Japanese.

These two efforts were Fleming's last hurrah with D Division. During the summer he had to face a serious potential shortage of help owing to demobilization of long-service officers consequent on the end of the European war. At the end of June, he inserted a rather plaintive "Note on Manpower" in his "Weekly Progress Report" asking recipients to pass on to Bevan the names of officers they might know who would be good at the work and saying that he would be available to interview candidates in London at the end of July. "As in most branches of the military profession," he said, "the basic qualifications are commonsense and (on account of the climate) a strong constitution. Imagination, knowledge of staff duties and some experience of Asiatics are all useful assets."

Several of the surplus staff from "A" Force joined D Division that spring and summer. Tending with some justice to regard themselves as the keepers of the flame, they were not all happy with what they found. Lyndon Laight, for example, an ATS officer who had been with "A" Force since July of 1943 and joined D Division in early June, considered Fleming a "playboy" by contrast with the tight organization she had known. MI6 in India also acquired a recruit from their Middle Eastern operations with experience working with "A" Force, in the person of Wing Commander Michael Ionides, who had been involved in some of the channels to the Abwehr under Dudley Clarke.

Fleming left for England on July 15, shortly after Wingate had left for Manila, having been asked to come by Bevan for a debriefing on the Washington meeting and to meet Crichton. It was the end of his war, for he did not return until the last day of August. The fighting had stopped two weeks before.

A minor but successful tactical deception was mounted in July when Halsey's Third Fleet struck the Japanese home islands. After hitting the

Tokyo area, Halsey's plan was to turn north and strike relatively untouched steel plants and other facilities in northern Honshu and Hokkaido, outside B-29 range. The cruiser *Tucson*, with specially trained communications personnel on board, including radio operators from Halsey's staff, peeled off from the rest of the force on the afternoon of July 10, after the Tokyo strike, and headed south, imitating the usual transmissions of fleet headquarters, including not only Morse but voice transmissions from a prepared script dealing with such matters as stationing of pickets, assignment of night fighter duty, and radar reports. Meanwhile, submarines stationed south of Kyushu radioed reports of the type associated with a carrier strike, and carrier-type airplanes flying from land bases on Okinawa struck Kyushu. A large part of Japanese mobile air strength was diverted southwards, and when Halsey struck at northern Japan on July 14 and 15, he encountered no kamikazes or other air opposition, having evidently achieved total surprise.

After the Manila conference, Newman Smith, recognizing the concerns raised by Raymond over attempting to persuade the Japanese that any China operations other than CARBONADO were feasible at this time, authorized Raymond to do the best he could to suggest interest in the Chusan-Shanghai area looking towards a future junction with Allied forces. As to the related idea of establishing two notional parachute regiments in India at China Theater's disposal, Corbett drafted a "story" to be planted on the Japanese by D Division's special means, under which the regiments would form a special air task force stationed at the former XX Bomber Command base at Kharagpur as a "flying block" available for missions deep into Japanese-held territory in China, after which it would withdraw into one of the islands of resistance in China for subsequent evacuation and reemployment. After CARBONADO, the task force would probably be used to threaten Sumatra. At Raymond's request, two notional parachute regiments, the 613th and 619th, were made available in early August. Then the war ended.

Of all this package of deception plans, only SCEPTICAL received much implementation before the war came to its sudden end. A constant stream of special means messages sought to put over the general BROAD-AXE story. PASTEL got detailed preparations, and sufficient modification to be renamed PASTEL TWO. ICEMAN was overtaken by events, and CONSCIOUS never got started.

SCEPTICAL was implemented by special means furnishing order of battle information, including the steady movement of units into position, together with rumors as to Mountbatten's intentions. The PAWNBROKER channel again carried much of the burden. Its communications were interrupted for a month after Rangoon fell, while Subhas Chandra Bose moved his headquarters to Bangkok. After communication was reestablished in late May, PAWNBROKER, with the failure of SUNRISE and RHINO, became the crucial channel. In July, the disposition of Japanese troops seemed progressively unfavorable, and SEAC called for urgent implementation of the SLIPPERY threat to the Kra Isthmus. By that time, the "All India Youth League" had notionally acquired high-grade sources in the person of an Indian Air Force officer and an Indian Army staff officer, and PAWNBROKER was able to provide what seemed to be very hot news. In the last weeks, ULTRA steadily confirmed the high rating that both Bose and the Japanese placed on PAWNBROKER, and confirmed too that the Japanese were fully persuaded that Mountbatten intended to attack Bangkok with the notional XXVI Airborne Corps and to land on the Kra Isthmus.

As almost always, it is hard to discern what role deception may have played in the faulty Japanese appraisal of Mountbatten's intentions. They did not expect the initial attacks to come in the ZIPPER target area, believing rather that Mountbatten would first secure air bases on Phuket and northern Sumatra, make a major landing around Penang (some two hundred miles north of the ZIPPER target area) in November or December, move down the coast after a month, and only then make a major landing in the ZIPPER target area around Port Swettenham and Port Dickson.

As for PASTEL, preliminary work began even before it was approved. Commander Holsopple outlined communications deception needs for Thurber on June 23. Someone—perhaps Thurber—floated on July 4 the notion that attention might be drawn to the Shanghai area in August if the Chinese government could be induced to make a diplomatic inquiry of the governments with concession rights in the Shanghai International Settlement as to their views on the status of those rights; this got nowhere. After conferences of the communications officers from the theater and fleet staffs plus Major Hines for Joint Security Control, Gold-branson was in a position by July 31 to advise Raymond that on X minus 90 (about August 1, under current planning for OLYMPIC), and continu-ing to X minus 50 with a possible extension thereafter, the radio decep-tion plan linking various headquarters in China with various commands in the Pacific would begin, pointing to preparations for an assault on the Chusan-Shanghai area on X minus 31.

Late in July, a colorful means of focusing Japanese attention on Shanghai presented itself. The Chinese underground in that city was under American control. Miles proposed that his group begin around August 1 to train fifteen hundred key Shanghai personnel for intelli-gence, sabotage of Japanese installations, and protection of the port against Japanese sabotage. Joint Security Control advised Goldbranson that pirate activity in the East China Sea, presumably also controlled by Miles, could help the cause as well.

Meanwhile, detailed planning had gone ahead full speed at Manila. On July 30, MacArthur's G-3 issued a "Staff Study" for PASTEL in its final form. It was now dubbed PASTEL TWO, in view of modifications to the original plan. December 1, rather than Christmastime, was now the no-tional target date for Shikoku; this reflected debate among Goldbranson, Hilger, and Smith both as to the size of the total notional Chusan-Shikoku force to be suggested, and as to the date. Tactical airborne diver-sions the night before the actual landings, in the form of drops of dummy paratroops, PINTAILS, and BUNSEN BURNERS, were added to the plan; and to foreshadow this, a notional airborne corps would be intro-duced into Okinawa beginning about August 20. Detailed responsibili-ties, including special means and related activities, were spelled out and charted.

The staff study was followed on August 5 by formal "Operations In-structions" laying out in great detail the timing of each action and the re-sponsibilities for taking it. Back in Washington, F-28 prepared a draft staff study for a naval tactical deception plan, including a feint towards Shikoku beginning X minus 15 by forces heavily protected against kamikazes, with appropriate Beach Jumper and similar activities, culmi-nating in a simulated landing operation on X minus 5; plus a further feint against northwestern Kyushu, well away from the actual OLYMPIC landing points, from X minus 7 to X plus 15.

On July 31, Goldbranson asked Joint Security Control for unit designations for a notional airborne corps and airborne division for PASTEL Two, together with a supply of one thousand shoulder patches for the latter. He was allocated XXXV Airborne Corps and the 18th Airborne Division. (The genuine 11th Airborne, in the Philippines, was notionally the other division in the corps.) To promote the idea of airborne action, on August 8 Goldbranson asked Joint Security Control for factual or special means implementation of a statement by Major General Matthew B. Ridgway, the notable commander of XVIII Airborne Corps in Europe, to the effect that he was soon coming to the Pacific. Sweeney recommended that it would seem less like a "plant" if it took the form of a War Department press release; but the war ended before anything could be done.

Sweeney and Nelson worked up a proposed master schedule for special means from August through the end of the year, including items requested by Goldbranson.

Throughout the summer, Smith had provided a steady flow of BROADAXE items to Buenos Aires, Lisbon, and Bern. Aside from the usual true chickenfeed—much of it minor news items not yet generally released, and information such as aircraft performance drawn from specialized publications—a significant theme was delay or postponement of forthcoming operations: such as logistical problems with the redeployment from Europe, and a report that General Marshall had told the President that Japan could not be invaded until military and industrial objectives had been destroyed and a full air and sea blockade had been effected.

A general theme, like the similar BODYGUARD theme in Europe, was that the air offensive might render the invasion unnecessary. Just after Hiroshima, on August 8, Group Captain Jones came in with a message from Bevan that an opportunity had arisen to pass to the Japanese minister to Switzerland, via a Swiss industrialist (a "medium grade" channel), the proposition that Allied military opinion favored the theory that with the atomic bomb no invasion would be required. Sweeney had planned to use this story through one of his channels, and did not want that channel tied in with an unproved "medium grade" one. So Smith vetoed the idea.

Focus on China continued with such items as a report that the Coast and Geodetic Survey had completed a study of the China coast, a report that volunteers were being sought to go to China to help train Chinese troops, reports of a new American-equipped Chinese Sixth Army, and a report that a group of Chinese-speaking censorship officers was being assembled in California. There were reports that Patton was being considered for an army command in China.* The northern threat was supported by such items as a report that two infantry divisions stationed in the Southern states had been issued Arctic clothing.

Not much more than that was ever done about the northern threat. On July 5, preliminary investigations began with respect to radio links between Twentieth Air Force and the Aleutians. Work proceeded on communications links between the Aleutians and the other theaters, and OPD produced an idea for moving to the Aleutians actual paratroop units returned from Europe in aid of PASTEL. But representatives of the Alaskan Department never attended the communications planning conferences, and nothing comparable to WEDLOCK ever got under way in the north. It was recognized that too much emphasis on the north might detract from the China threat. Moreover, some elements in the Navy and the Air Force still nursed the hope that there might really be such an operation; KEELBLOCKS, it was called.

While not itself part of PASTEL, the possible need for tactical deception units was not neglected. OPD advised MacArthur early in August that the trusty 3103d Signal Service Battalion was earmarked for the Pacific in October; a new 4030th Signal Service Company with half of the 3103d's capability would be ready in January 1946; the 23d Headquarters Special Troops, reorganized and reequipped, would be ready in February and could be tailored to fit MacArthur's needs; and the 3133d Signal Service Company (sonic deception) returning from Europe, could be made available if desired.

---

* There was truth in this. Marshall offered Wedemeyer three of the best army commanders in Europe: Patton, Simpson, and Truscott. Wedemeyer planned to put Patton in charge of a drive into North China towards Peiping (as Peking was then called) and Tientsin, and Truscott in charge of a drive down the Yangtze towards Shanghai, making Simpson his deputy theater commander. Simpson actually visited China in late July, and Raymond passed advance word to the Japanese that he was rumored to be taking an important assignment there.

On August 7, the go-ahead signal went out to begin communications deception under PASTEL on August 10.

During this orgy of preparation for OLYMPIC, the planners had to look ahead to CORONET, the landing in the Tokyo Bay area scheduled for the spring of 1946. At the end of July the Joint War Plans Committee and Joint Security Control produced a staff study of a cover and deception plan for that operation, the capstone of their series of staff studies, which went forward to the Planners on July 26.

It was an elaborate and detailed 37-page document, with extensive analysis and proposed implementation methods and assignments of responsibility. The basic "story" was that redeployment and retraining of troops from Europe would not make a Honshu landing possible before early 1947, and in the meantime additional blockade bases to supplement the position obtained by OLYMPIC on Kyushu would be secured on the mainland, on Shikoku, and on Hokkaido. Threats would continue against the Yellow Sea area (Shanghai-Shantung), shifting the emphasis to Korea some 120 days before the CORONET target date ("Y minus 120"; "Y-Day" being assumed to be 120 days after OLYMPIC "X-Day," or March 1, 1946), with a notional target date of Y plus 60 for a landing there in the Fusan vicinity; a notional target date of Y plus 30 for the Shikoku assault, and Y plus 90 for Hokkaido and northern Honshu. Detailed measures for implementation were suggested for each notional operation. The study even provided a suggested tactical deception plan for the CORONET landings.

Reviewing the study for OPD, that same Colonel Goodwin who had shed tears rather than go out to Manila suggested that Korea was an illogical notional target. Billy Harris looked at Goodwin's first draft and was blunt as to its "basic misunderstanding of cover and deception capabilities" and its second-guessing of Joint Security Control's ability to "judge whether or not we can get the Jap to believe it." ("I don't believe he read the plan," he said.) Smith concurred that Korea was not very logical, but said that the Navy would insist on keeping it in; Goodwin reported that Ormonde Hunter agreed, but wanted to rewrite the whole thing to adopt a totally different approach. (This was the proposal by Hunter that evoked Thurber's scathing remarks already quoted.) Goodwin's superior judiciously decided to leave well enough alone; it was just a study, and any final plan would take actualities into account; moreover, the Japanese were known to be worried about Korea, and "we should plan to keep them concerned as long as we can."

The study passed muster with the Planners, and went out to the theaters on August 10. (On that same date, Goldbranson submitted within the CINCAFPAC staff some CORONET cover ideas of his own.) By the time it arrived, of course, it was merely a curiosity.

Any hopes that PASTEL TWO would prove to be the FORTITUDE of the Pacific would have been dashed if OLYMPIC had taken place.

MAGIC and ULTRA gave at least some hope that the Japanese might be led to believe that the next blow would fall in China. Throughout the summer, a steady flow of intercepted messages predicted landings up and down the China coast and at various times. Diplomatic intelligence was particularly focused on China. "Perhaps the topic that has most absorbed the energies of the diplomatic intelligence system in recent months," said a comprehensive American study completed just at the end of the war, "is the imminence of American landings on the mainland of Asia. Reports from Shanghai alone have mentioned a dozen different places along the China coast, including nearly all logical places for a landing, as the sites of anticipated American landings. The Greater East Asia office at Canton opened up a special espionage net in December 1944 to collect information on this subject, in response to requests received by the Greater East Asia Ministry from the Army." Scattered lower-level feedback tended in the same direction. Thus, PIN reported in late July, after he was, at least allegedly, cleared of suspicion and upgraded by Miyamoto and Ishikawa, that his spymasters had told him that the Japanese command did not believe that there was enough space on the current American bases to accumulate the men and materiel needed, so they were expected to strike either against additional islands near Japan or to establish bases on the continent of Asia.*

So the China coast remained a logical notional target for deception operations covering OLYMPIC, on the basis of intelligence—and because there was no other real option.

---

* But, of course, if PIN was really working for the Japanese that is just what they would tell him to say.

But MAGIC and ULTRA also disclosed a good deal of traffic to and from Tokyo predicting an invasion of Japan, revealing a steady buildup of forces in Kyushu in May and thereafter and vigorous preparations for defense of the homeland. And in reality there was little doubt at the top in Japan as to where the Americans would attack. Imperial General Headquarters seems to have discounted a China landing after Okinawa. The Navy again made an accurate appraisal of American intentions, though it thought the blow might come sooner than the actual target date of November. "I thought you would have to land on the home islands," said Admiral Tomioka of the Naval General Staff to postwar interrogators. "Once you took Okinawa, we concluded that your first homeland assault must be against Kyushu or Shikoku. Your land based [fighters] did not have enough range to cover Tokyo area landings." General Masakazu Kawabe, effectively the supreme commander of the Japanese Air Force, told postwar interrogators that Imperial General Headquarters estimated two possible moves after Okinawa: "(1) if Russia did not enter the war, we thought you would land in Kyushu in October or November 1945; (2) if Russia did enter the war we thought you would land on the Kanto Plain in the Fall or Winter." The Air Force, according to Kawabe, retained also some expectations for China: "Air General Headquarters estimated that the U.S. might establish B-29 bases on the continent near Shanghai."

On June 8, the generals and admirals told the Emperor that Kyushu or Shikoku might be invaded any time after June 20. All through the summer the Japanese worked frantically to turn Kyushu into one huge fortress, and to prepare for such a bloodbath as would have dwarfed anything Americans had ever seen.

So BROADAXE, and PASTEL on the strategic level, had no chance of success, though the Americans could not know that.

But, of course, the bloodbath did not happen.

On August 14, after two atomic bombs and the entry of the Soviets into the war, the Japanese announced their acceptance of Allied terms.

On August 15, Nimitz's headquarters canceled all outstanding directives requiring communication deception.

And on August 16, the message went out from MacArthur's headquarters:

Deception plan PASTEL TWO is indefinitely suspended effective immediately and all action initiated thereunder should be terminated.

So it was all over; suddenly and abruptly; not, as in Europe, with a grand finale, but as if the play had been stopped and the house lights put up just as the last act curtain was about to be raised.

But it is not likely that any of the players minded.

Epilogue

O what a tangled web we weave
When first we practise to deceive!
But when we've practised quite a while
How vastly we improve our style!

When the war suddenly ended, the Hesketh brothers and Eldredge had already spent the summer searching through German files and interrogating German officers, focusing almost entirely on FORTITUDE SOUTH and its results. Their inquiry did not concentrate so much on whether it had been successful—that was already known, though they turned up a few items, such as the fact that Hitler had read the Abwehr's report of GARBO's long post-D-Day message, that encouraged them to believe that perhaps it had been still more successful than they had thought. More important was *why* it had been successful; and here they soon concluded that the double agents had had far more influence than the elaborate effort at signals deception. This view has stood the test of time.

In Southeast Asia, Peter Fleming toured some liberated areas and interrogated commanders and principal staff officers of the Japanese Southern Army and its "Area Armies" during October. It was plain that they had bought almost all the false information that had been passed on. On that basis, wrote Fleming later, "it would be easy to base sweeping claims for the result of our strategic deception. . . . But such claims would in fact be dishonest and invalid, for, so arbitrary were the workings of the Japanese military machine and so low, within that machine, was the status of their incompetent Intelligence that the dividends which deception paid were only in minor instances truly strategic."

No other followup with the Japanese proved possible. On three separate occasions, Joint Security Control sought permission from

779

MacArthur to send Goldbranson and Corbett to Japan to explore the degree of success of American deception operations. Permission was never granted. (One is inclined to see in this the hand of MacArthur's intelligence chief, Major General Charles A. Willoughby; temperamental and explosive, Willoughby was near-paranoid about any outside interference in his domain.) The fairest conclusion is the one drawn by Burris-Meyer in the unfinished official history of American deception:

> The very absence of this validation, however, should teach us two things; first, to be cautious in assuming that it was our fault that the enemy did as he did, and second, to exploit to the utmost the likelihood that the enemy will do something we want him to do anyway.

What, then, was the final profit and loss of all this nearly five years' effort, from Dudley Clarke's arrival in Cairo to the sudden end in the Pacific? Did it work? Where and when? Did it matter? Did guile really add much to valor? What had been learned of value for the future?

On the strategic level, there was a clear payoff in Europe and particularly in the Mediterranean. Playing up to the Führer's obsession with the Balkans, in BARCLAY and other operations, kept forces tied down in southeastern Europe that could have been used to good effect elsewhere, and contributed particularly to the success of HUSKY. Underpinning this was Dudley Clarke's long and patient building up of the false order of battle and the phantom host that it conjured in the German mind. Less tangible, perhaps, was the repeated playing to Hitler's nervousness over Norway; but the fact remains that three times as many Germans were kept there as would have been needed merely for occupation duties.

Not much can be claimed for strategic deception in the Pacific, though WEDLOCK may have helped keep more forces in the north than the Japanese would otherwise have maintained. BLUEBIRD and related operations had no great effect on Japanese dispositions; indeed, given their lack of mobility by 1945, they could not greatly change their dispositions in any case. They were ready and waiting for the invasion of their homeland. Their strategy was a simple one, as Okinawa had already demonstrated: To stand and fight to the last man (and woman, and child) until the Americans decided that the game was not worth the appalling butcher's bill. PASTEL would have had little or no effect on this.

Southeast Asia saw no dramatic results from strategic deception, but it must always be remembered that until late 1944 the Allies had no real strategy there anyway. The charge sometimes leveled at Peter Fleming, that he became too much immersed in the details of haversack ruses and similar fun and games, is not wholly unfair; but that was only to be expected when a man of such energy and imagination was given so little to do. Moreover, it overlooks the patient and meticulous building up of false order of battle, which Fleming, Peter Thorne, and DIB conducted with a skill that Dudley Clarke would have respected. Not that the Japanese did much with it; but, as Fleming said: "It remains true, however, that the consequences which would have arisen had the Japanese General Staff NOT been deceived about our strength and striking power in South East Asia might have been serious."

On the major operational level, deception could claim some striking successes throughout the war. Strangeways's FORTITUDE SOUTH was extraordinary, one of the great deception payoffs in history. Sam Hood's contemporaneous NUNTON in Italy was equally successful, on a less apocalyptic scale, and has been unjustly forgotten. BARCLAY, including MINCEMEAT, and FERDINAND were major successes on the operational level as on the strategic level. CLOAK, covering Slim's Irrawaddy crossing, was in this league. BERTRAM at Alamein, operational deception shading into tactical, was a classic. MacArthur's deceptions to cover the Hollandia operation were masterly.

It must always be remembered, too, that the resources devoted to deception, while they may seem substantial in an account devoted to that subject alone, were trivial in terms of the Allied war effort. The results of FORTITUDE SOUTH alone would have justified a thousandfold greater expenditure of resources. Someone once asked Dudley Clarke what his operations were worth; and after a moment's thought he answered in terms of the number of corps and divisions that a small expenditure had "bought" in German eyes.

So, to sum up: Did deception help the Allied cause? Yes, sometimes; and when it most mattered. Did it ever harm the Allied cause? No. On balance, would the Allied cause have been better off or worse off without deception? Clearly worse off; very possibly much worse off. What did it cost? Next to nothing, in wartime terms. What utility did it have in the long run? A great deal, if memories did not grow short; for by trial and error the Allies had learned what deception could and could not do, how to structure it, how to guide its implementation, what tools could be used and how to use each of them and harmonize one with the other; and the principles they had developed could be generalized to new situations and new technologies. They had practised to deceive, and they had not merely improved their style but perfected it.

That is a pretty favorable bottom line.

In the autumn of 1945, General Bissell canvassed major U.S. Army commanders for their views on the success of cover and deception and their thoughts for peacetime. As might have been expected, the European and Mediterranean commanders were enthusiastic, while Mac-Arthur and his subordinates were relatively lukewarm. The longest, most thorough, and most thoughtful response came from Devers, drafted by Reuben Jenkins. He had come a long way since the days when he looked at WADHAM and did not like what he saw. He offered critiques of operations in his theater, urged a long-range program at the highest level modeled on the British structure, and at theater and army group level urged the Dudley Clarke principle that the deceivers should be responsible directly to the chief of staff. Dan Noce likewise offered a thoughtful and characteristically blunt assessment. To the topmost commanders—he listed them—he gave high marks for their interest in the subject. But "there were many officers in subordinate positions both in the staff and command who through lack of knowledge had no confidence in these operations, or were too dumb to understand them or would not take the time to understand them, or for other reasons, were constant stumbling blocks with regard to cover and deception." The shortest and bluntest response came, predictably, from Stilwell. "I feel that, in general, too much attention has been paid to deception plans, at least where I served," said Vinegar Joe, "and that unless there is some physical evidence of intentions shown, no enemy commander who knows his job will be thrown off balance by them. However, my experience in this line has been limited." By contrast, Wedemeyer's response was thoughtful, pointing out among other things that commanders should be indoctrinated in advance about deception, and that results cannot be obtained overnight.

* * *

Another matter to be addressed at the close of the war was what should be done with the accumulated knowledge and information with respect to strategic deception. Should security classification be reduced or lifted? How much could safely be told?

It was early decided by both the British and the Americans that the very fact that the Allies had engaged in deception at all (other than on a tactical level) should remain Top Secret. In part this was because it was so closely intertwined with ULTRA, and ULTRA was to remain a closely held secret until the early 1970s. Mainly, though, it was so as not to alert potential future enemies to the fact that deception might be practiced in a conflict with them. (As was noted in Chapter 2, this is not necessarily desirable. An enemy who knows you practice deception may be more likely to think your real operation is a deception and respond accordingly. But the American and British Chiefs came down on the other side of this issue.)

This decision made it even more important that the archival history of deception during the war be written by actual participants. The Combined Chiefs decreed that the history of deception in the war with Germany and Italy should be prepared by the British, with American review and approval. This was set in train as early as October 1944, when the British Chiefs directed Bevan to prepare a history of the London Controlling Section. He presumably completed it before he left at the beginning of September 1945. Dudley Clarke was already at work in Cairo on a history of "A" Force before he left for England in the spring of 1945. The budget gremlins in London could not see why they should have to pay a brigadier to do such work; could not a lieutenant-colonel do just as well? Bevan got Hollis to intercede, and Hollis took care of it. Clarke made it his magnum opus. He worked on it until the end of September 1946. Long and comprehensive but readable, it is a remarkable document.

Clarke's work was supplemented by the reports prepared by Gene Sweeney, with, in his words, "the able and most valuable assistance of Arne Ekstrom," covering FERDINAND and the work of No. 2 Tac on the 6th Army Group front after it separated from the parent "A" Force. Unlike Ingersoll's reports for 12th Army Group, they are unpretentious but thorough and careful, and make a serious contribution to history.

Roger Fleetwood-Hesketh prepared a history of deception in northwestern Europe: FORTITUDE and its predecessors, and activities on the 21 Army Group and 12th Army Group fronts. It was remarkably detailed, and was a work of which he was very proud.

Peter Fleming completed in early 1947 a relatively brief account of deception in Southeast Asia.

Ewen Montagu compiled more than one account of naval deception. One of them was obtained by the Americans—not in a wholly gentlemanly fashion. In September 1945 Montagu completed a detailed account of "Deception on Naval Matters," including ULTRA material, with an extensive description of the double agents. He lent it on a personal basis to Commander William C. Ladd of the U.S. Navy, to read and return but not to copy. Ladd brought it to his boss, Commodore Tully Shelley, the American naval attaché in London. Sensing its great value, they decided to copy it anyway. No enlisted men were available with ULTRA clearance, so Ladd, a slow hunt-and-peck typist, and Lieutenant W. A. Sprague, a faster and more skilled worker, typed it out themselves. Montagu walked in on Sprague in the act; Sprague innocently said he was making a few notes. Sprague typed all night and returned the original to Montagu, saying he had destroyed his notes. Shelley then forwarded the copy, with a handwritten cover note, to Commodore (soon to be Rear Admiral) Thomas B. Inglis, who had replaced Thebaud as Director of Naval Intelligence on September 6.

In a related field, John Masterman wrote for the archives a history of the double agent system in England and the work of the Twenty Committee.

Drawing on these and on the files of the London Controlling Section, Ronald Wingate then prepared the comprehensive history of Allied deception in the war with Germany and Italy which the Combined Chiefs had directed. Completed in April 1947, it is an urbane, literate, and readable work, as befits its author. It does not deal with the personal relationships among the players, but an official report cannot be expected to do that. Nor, understandably, does it deal fully with the American side. Since it was intended as a combined product, interested Americans came to London in June 1947 for a conference to sign off on it, including Harris, Sweeney, and Baumer. Newman Smith did not attend, but he was recalled to active duty for a few days in March and April for the purpose of reviewing it.

Two official reports, one British and one American, dealt with counterintelligence and the running of double agents on the Continent. Christopher Harmer prepared one dealing with such activities in 21 Army Group's sector. The work of OSS X-2 in France after the invasion was covered in a full account prepared by Johnny Oakes and two of his colleagues, Edward Weismiller and Eugene Waith.

No combined history of Allied deception in the war with Japan was ever prepared. The nearest approach to it was made in 1949, when Burris-Meyer presided over the pulling together of a skeleton history of "United States Military Deception in World War II"—informally entitled, in good Burris-Meyer fashion, *Cake Before Breakfast*—but this never passed beyond the very roughest partial draft stage. As has been noted, Newman Smith prepared, before he left, an informal history of the Special Section, Joint Security Control, which seems now to have been lost. So far as is presently known, the FBI prepared no account of its wartime double agent operations.

In the 1980s, the Naval Postgraduate School commissioned the American scholar Katherine L. Herbig to prepare an account of American strategic deception in the Pacific war. Completed in 1985, this relatively brief but very valuable work was declassified in 1987 but has never received wide circulation, though it was the basis of a published article by Herbig and a colleague.

Security was strictly maintained during the early years. There was soulsearching over how much to allude to deception, particularly FORTITUDE, in the final reports of the theater commanders for publication. There was even serious concern about teaching anything about strategic deception at the war colleges. In March of 1946, Montgomery called on Ike personally to urge him to ensure that the Secretary of War let nothing slip about FORTITUDE during a planned eulogy to Patton.

Little by little, cracks in the shield began to appear. General Deane published in 1947 a book on his mission to Moscow that called BODYGUARD by name, described it correctly, and identified Bevan and Baumer. That same year, Gordon Merrick of the OSS, the FOREST case officer to whom Ekstrom had had to read the riot act, published a novel called *The Strumpet Wind*, which gave away some of the use of double agents; a character called "Mercanton" was unmistakably FOREST for those in the know. And again that year, Edward Weismiller began a novel, *The Serpent Sleeping,* based on his running of DRAGOMAN. (He did not complete it until 1961, and it was published in 1962.)

In 1950, Duff Cooper published a novel based on MINCEMEAT called *Operation Heartbreak.* Dismayed by this, and further dismayed by an another proposed book on the subject, the authorities assigned Ewen Montagu to prepare an account from the official records; this was published in 1953 as *The Man Who Never Was* (starring Ewen Montagu), and was subsequently made into a film of the same name. Clifton James published in 1954 an account of his COPPERHEAD adventure called *I Was Monty's Double;* it too reached the silver screen. The same year, the double agent ZIGZAG published a memoir of his adventures called *The Eddie Chapman Story.* The double agent TREASURE published a memoir in 1968 called *Secret Service Rendered.*

Meanwhile, in 1952 the Australian journalist Chester Wilmot published *The Struggle for Europe,* the first comprehensive history of the invasion and its aftermath. It identified FORTITUDE by name and gave a reasonably accurate account of it, except for the double agent implementation. Churchill's memoirs in the 1950s touched on FORTITUDE and MINCEMEAT.

Sir Ronald Wingate published his memoirs in 1959, describing FORTITUDE and touching in very general terms on the work done in Storey's Gate and in the Far East.

The dam broke at the beginning of the 1970s. In 1971 there appeared a book by Sefton Delmer called *The Counterfeit Spy.* Delmer, a journalist fluent in German who during the war had run a "black propaganda" operation beaming broadcasts to Germany purporting to come from dissident Wehrmacht officers clandestinely operating inside the Reich, wrote a detailed account of FORTITUDE with the abundant help of Noel Wild (who came through as very much the central figure except for GARBO himself). Aside from changing the names, and often the nationalities, of many figures, and aside from imaginary conversations and other journalistic flourishes, it was a thorough and highly readable job, with some inaccuracies. It was also copped wholesale from Hesketh's report.

Hesketh was enraged. He had labored for many months on his report and it was his pride and joy. It was plain that Wild had given Delmer access to Hesketh's report; he claimed simply to have talked with Delmer, but Hesketh counted nearly two hundred items taken straight from his text. Hesketh threatened suit for plagiarism. The government intervened, claiming copyright for the Crown. The upshot was a second edition in which Delmer gave Hesketh credit. "I had to be content with the lesser evil of re-editing Delmer's work and removing some of the grosser distortions and inaccuracies," with the help of Bevan, Dudley Clarke, and another colleague, wrote Hesketh to Eldredge.

Hesketh was also given permission to publish the original report. He tried hard to do so, and Eldredge helped flog it round the American publishers. But it was far too detailed to be thought appealing to the general public, especially after Delmer's book. Bootleg copies circulated among intelligence buffs, and it was finally published in 1999 with an introduction by "Nigel West," the nom de plume of Rupert Allason, the authority on British intelligence, who had earlier tracked down GARBO and helped publish his memoirs.

In that same year 1971, Masterman, who had returned to Oxford after the war, agreed with Yale University Press to publish his report of the Twenty Committee's work. The American writer Ladislas Farago, who had just completed a book on the Third Reich's espionage called *The Game of the Foxes,* somehow got access to a prepublication copy, as well as to Delmer's manuscript; mortified to find that the German intelligence coups about which he wrote were mostly British deception coups, he managed a last-minute rewrite before publication. The next year, Masterman's report was published under the title *The Double-Cross System.*

TRICYCLE published an imaginative memoir in 1974, called *Spy Counter-Spy.*

In the early 1970s, most of the files of the Joint Chiefs of Staff for the Second World War period were declassified and made publicly accessible in the National Archives in Washington; but not those of Joint Security Control. For some reason, a relatively small number of files of the former Army Special Section slipped through; in which were the 12th Army Group Special Plans "interim report" of 1944 and "informal report" of 1945. One or two books made use of these files.

Buried in the Joint Chiefs files—and in the files of SHAEF* and OPD, made available at the same time—was a great deal more information about deception than merely the 12th Army Group files and related material. But researchers had to know what they were looking for; and in many cases what they were looking at. The same was true of the F-28 files, which were transferred to the Naval Operational Archives in the Washington Navy Yard in 1960 and declassified on a spotty basis thereafter; of the files of AFHQ; and of sundry British operational files that were declassified in the 1970s. In 1979, a British historian, Charles Cruickshank, published the first serious effort at a history of deception, using mainly the British material of this nature. A workmanlike job, it was necessarily incomplete and in some ways misleading. This was particularly true because the crucial Cabinet Office files, including the histories by Clarke, Wingate, and Fleming, were still classified.

In 1974, Duff Hart-Davis published a biography of Peter Fleming that gave a description of the work of D Division. In 1980, Dennis Wheatley's memoir of life in the London Controlling Section was posthumously published.

Meanwhile, the wraps were at last lifted from ULTRA in 1974. Its relation to deception was explored in several books that appeared thereafter, by such authors as Ronald Lewin; Ewen Montagu came quickly back into the field with a slender book largely based on a report he had compiled at the end of the war. A comprehensive multivolume history of *British Intelligence in the Second World War, Its Influence on Strategy and Operations,* which in fact was overwhelmingly a history of ULTRA and its employment, had been prepared by Professor F. H. Hinsley of Cambridge with the help of others, and began to see publication in 1979.

Hinsley's work was part of the British series of official histories of the Second World War, prepared, with an eye ultimately to general publication, at a higher level than the after-action reports of the participants— which, after all, was what Clarke's, Wingate's, Hesketh's, and Fleming's

---

* At the end of the war it was decided that the original SHAEF files would go to the Americans and the original AFHQ files to the British, with microfilm copies of each given to the other. Missing from the SHAEF files in Washington, however, are the records of Ops (B). In contravention of the understanding, these are in the British Public Record Office; they were presumably taken to London because of the agreement that the British would write the combined history.

accounts really were. A similar history of strategic deception was appropriate; and in 1968 Peter Fleming volunteered himself for the job. The wheels of Whitehall turn slowly, and not until 1971 was Fleming finally appointed to the task. Barely had he begun preliminary work when death overtook him. His successor was a retired journalist, a former assistant editor of *The Times* of London and a close friend of Fleming's, named Oliver Woods. Woods too died suddenly, in late 1972. The assignment then went where it should always have gone, to a professional historian, and one of the very best: Michael Howard of Oxford, Professor Sir Michael Howard, as he would later be; the most distinguished military historian of his generation, who had already worked on the grand strategy of the war for the official history series (and was himself also a friend of Fleming's).

Howard's history, as was to be expected, was thorough, balanced, judicious, and literate. It put myths to rest, notably placing FORTITUDE in context, cutting back some of the more extravagant claims of success in light of German intelligence records, and putting the London Controlling Section and the Twenty Committee into proper perspective. It for the first time gave due credit to the work of Dudley Clarke and to the fundamental importance of the false order of battle. Finished at the end of 1978, it should have been published in 1980. But that was just the time that MI5 and MI6 were rocked with the unmasking of Anthony Blunt and other scandals. When the book was laid on Mrs. Thatcher's desk for final approval of its publication she declared that she had had enough of disclosures about secret service work, that any more of that would just cause more trouble. So for another ten years it lay buried in Whitehall. Howard himself was not cleared to have a copy of his own work.

Meanwhile, unofficial accounts continued to appear, some more balanced and accurate than others. Masterman published his memoirs in 1975, telling the story of his own publication and identifying sundry participants in deception. David Mure, who had played a relatively minor role with "A" Force in the Middle East and had gone into the carpet business after the war, published in 1977 the first account of some of "A" Force's work, a book called *Practise to Deceive*, following it in 1980 with a book, partly about Clarke and partly about a variety of related matters, called *Master of Deception*, and in 1984 with a novel called *The*

Last Temptation. Though Mure's books were valuable in respect of the limited matters within his own experience, that experience was narrow and he offered judgments about things he knew nothing about at first hand. Nevertheless, he developed a reputation as an expert, and in 1986 was invited to the States to address the CIA and the Army War College on the subject of deception.

Rupert Allason tracked down GARBO in Venezuela, was instrumental in bringing him back to England to receive his MBE from Prince Philip himself during the observances of the fortieth anniversary of D-Day, and helped him publish his memoirs in 1985.

Howard's work finally appeared in 1990 as Volume 5 of the official history of British intelligence in the Second World War. At the same time there was published as Volume 4 a history of counterintelligence and security by Hinsley and C. A. G. Simkins, which had likewise been held up by Mrs. Thatcher. It formed an important companion volume to Howard, giving fuller details not only of the double agents controlled from Britain, as Masterman had done, but of activities abroad.

It was generally assumed that these official publications would be the last word, and that the underlying documents would never be declassified. Surprisingly, beginning in the middle 1990s many were; including Wingate's, Clarke's, and Fleming's accounts, many important files of the Cabinet Office (including most of the files of the London Controlling Section), and even MI5 files on double agents. They were essential to the preparation of the present book.

Much is still classified. More will be released over time. (And no doubt that will show up errors, and fill gaps, in the present account.)

Anyone who has worked in them both knows how much tidier and better organized British archives are than American. A large and indiscriminate body of files of Joint Security Control, especially from the pre-1945 period, and of the Army Special Section after the Navy's F-28 was organized, lay for years gathering dust somewhere deep in the Pentagon. Katherine Herbig dug them out for her study in the 1980s; they then continued to gather dust until they were found for the present author through the kindness of Brigadier General David A. Armstrong, Director for Joint History, Office of the Chairman, Joint Chiefs of Staff, and duly declassified.

It seems unlikely that further material on American deception will be discovered. The FBI retains copious files on the double agents it managed; but the Bureau has a way of releasing, years after a request under the Freedom of Information Act, masses of paper covered almost solidly with deletions, and life will probably always remain too short for anyone to pry these loose.

They were a varied lot, the deceivers; and their postwar careers were just as varied.

Dudley Clarke was made Companion of the Bath in the victory honors list of June 1945; one of the two CBs awarded to a deceiver, the other having gone to Johnny Bevan. He retired from the Army soon after the war. He worked on the staff of the Conservative Party, and served on the board of Securicor, Ltd., the leading British private security firm. He kept up with some of the "A" Force alumni; with the Ekstroms—he would bring Pam Ekstrom thoughtful little presents; with Dennis Wheatley—Eldredge remembered him getting cheerfully "squiffed" on Wheatley's hospitality from time to time; with Crichton. He organized an "A" Force Association—he, Crichton, and Sweeney were the forming committee—and kept up a careful card index of members' addresses for many years. When the Association was wound up in 1957, Ken Jones, Maunsell's deputy in the Cairo days, who worked for Ronson after the war, had cigarette lighters made for them all with an "A" Force logo on it.

Clarke lived in Mayfair, not quite able on an Army pension to support the style of life to which he aspired. He published in 1948 a book of memoirs, *Seven Assignments,* about his services during the war up until he joined Wavell; he wrote another, about his years in the Army before 1939, but it was never published. In 1953 he approached the authorities for permission to do a book on deception to be called *The Secret War,* but got nowhere. He hoped someday to publish his "A" Force history, but that was not to be. He died in 1974. Noel Wild delivered the eulogy at a memorial service at Chelsea Old Church, and *The Times* published a supplemental obituary by Rex Hamer.

Johnny Bevan had been made a Companion of the Bath in the New Year's honors for 1945. "I have never known a CB better deserved than yours," wrote Pug Ismay. "I wish it could have been something even bigger," wrote Joe Hollis. "A fully fitting recompense for all the magnificent work you have done," wrote Tar Robertson. After leaving the Army

Bevan continued his distinguished career in the City, but remained on call to be of what help he could to the cause of deception. In 1947, for example, he interceded personally with Eisenhower, then Marshall's successor as Chief of Staff of the Army, in hopes of reviving the wartime collaboration on strategic deception. Annually for many years he presided over a dinner at Brooks's for the old deceivers and their colleagues. He died in 1978.

David Strangeways continued in the Army; first in occupation duty in Germany, then in charge of the Visual Interservice Training and Research Establishment (VISTRE) in England, and in another training assignment. Accustomed to speaking freely and to the latitude Montgomery had accorded him, he did not always endear himself to higher authority. He served with the British military mission to Greece and from 1952 to 1955 commanded a battalion in Malaya. He was slated to command the task force responsible for nuclear testing on Christmas Island in 1957, but he had conscientious scruples about nuclear weapons, resigned from the Army, and after theological training was ordained in the Church of England. He served as a parish vicar in Dorset and Wiltshire until 1971, was chaplain of the British embassy in Stockholm, and became chancellor and senior canon of St. Paul's Anglican cathedral in Malta. He retired to Suffolk in 1981 and died in 1998.

Michael Crichton had received the OBE in 1944. He returned to the City. Betty adjusted readily to English life, and they lived happily ever after. Crichton died in 1970; Betty was still living when this book was being written.

Dennis Wheatley returned to his novel-writing and to the life of a *bon viveur* which he had in fact never left. He died in 1977.

Sir Ronald Wingate received the OBE for his work at Southeast Asia Command. He succeeded Crichton as Controlling Officer in early 1946. His chief duty was the production of his comprehensive history. He left that post in early 1947 to become the No. 2 member of the British reparations commission in Belgium. In 1959 he published a charming memoir of his varied career, called *Not in the Limelight*.

Peter Fleming likewise received the OBE for his work; the same decoration, he said wrily, that his grandmother had received in 1918 for running the family house in London as a hospital for wounded officers. He received also the Chinese Order of the Cloud and Banner. He returned to the life of a country squire, journalist, and author; presided over his estates, shot large numbers of game birds, and regularly contributed pieces to the *Spectator* under the name "Strix" and anonymous Fourth Leaders to the *The Times*. He published no fewer than ten books, including the first account of Operation SEALION, the German invasion of 1940 that never happened, and an account of the siege of Peking in the Boxer Rebellion that formed a basis for a star-studded film. In August of 1971, he had just set out on the official history of deception when he died of a sudden heart attack—fittingly, while out shooting in Scotland.

Peter Thorne had a distinguished postwar career in Parliament, serving from 1948 to 1982 successively as Assistant Sergeant-at-Arms, Deputy Sergeant-at-Arms, and Sergeant-at-Arms of the House of Commons. Knighted in 1981, he was still living when this book was compiled.

Tar Robertson continued with MI5 and, among other things, presided over the "Inter-Service Communications Intelligence Committee," the immediate postwar equivalent of the Twenty Committee. He died in 1994.

Roger Fleetwood-Hesketh became MP for Southport in 1952. An amateur architect, he devoted much of his time to his ancestral country house, Meols Hall in Lancashire. He died in 1987.

Ewen Montagu returned to the law and became a sometimes controversial judge in the lower levels of the English judicial system. He died in 1985.

Rex Hamer returned to the educational world. After the war he was a housemaster at Stowe, and then headmaster of various schools in the colonies. He married Lyndon Laight, the former ATS officer with "A" Force who subsequently joined D Division. Clarke was godfather to their daughter. He was still living when this book was prepared.

Christopher Harmer returned to his solicitor's practice in Birmingham. He kept up with his old colleagues in many ways; and acted as Masterman's solicitor in the dispute over publishing his report on the double agents. He died in the late 1990s.

Noel Wild had received the OBE in 1943. He remained involved in deception, and apparently in other security matters, until the late 1960s. He died in 1995.

Lieutenant-Colonel Jervis-Read, who had headed the Ops (B) physical deception section, went on to a brilliant career, reaching the rank of full general, Quartermaster-General of the Army, and Army Aide-de-Camp to the Queen.

Newman Smith was given a handsome accolade by "Abe" Lincoln of OPD in October 1945, receiving the Legion of Merit for his work. He retired from active duty in January 1946 and turned the Army Special Section over to Billy Harris. Too old to return to commercial banking, he took charge of a division of the Veterans' Administration loan guarantee service. He continued as a reserve colonel in the Army until retirement in 1949, commanding the Inspector-Instructor Group of the Military District of Washington and lecturing on military intelligence and internal security to college ROTC units. In 1949, at the request of the Secretary of Commerce, he surveyed the security requirements of that department, subsequently becoming its security control officer. He was reappointed a reserve colonel in March of 1953 and may have had some brief duties in connection with the Korean War.

Zelda Fitzgerald had died tragically in a fire at her sanitarium in 1948. Scott's and her daughter Scottie had married a young Washington lawyer. "Aunt Rosalind's finest hour came when we moved to Georgetown in 1955," Scottie wrote later; "she threw herself into the decorating of the 30th Street house, with all the enthusiasm of a frustrated decorator." In November of that year, Smith retired from the Commerce Department, and at the end of the year he and Rosalind moved back to Montgomery, "to be near Grandma, who by now was in her nineties," according to Scottie.

By contrast with his earnest demeanor with his Pentagon colleagues, the Colonel was fondly regarded as a humorous and witty fixture of the Montgomery community. Every morning he could be seen in a jaunty beret walking his dog down Perry Street, where they had settled in at No. 1339. He was a member of a select fifteen-member group founded in 1916 called the Unity Club, at one meeting of which he presented a paper on his old sport of fencing. He died in Montgomery of heart failure on July 28, 1964. Rosalind survived him by fifteen years and died in 1979.

Clayton Bissell served as military attaché in London from 1946 to 1948, and then at the Air Force headquarters in Wiesbaden. He retired in 1950 and died in 1972.

Bill Baumer served as a military adviser at Potsdam and several other international conferences in 1945–46, and then as deputy chief of staff of the U.S. Army Constabulary in occupied Germany; and thereafter in Washington in charge of the Armed Forces Radio Network. He left active duty in 1950 to enter the business world, though remaining in the Army Reserves, in which he reached major general's rank. His business career was eminently successful, as an officer and director of major corporations and a mutual fund. He died in 1989.

Baron Kehm spent five postwar years on the faculty of the Command and General Staff School at Fort Leavenworth, and then served for three years as Army attaché in Dublin. He performed several special assignments for the State Department after his retirement in 1954. He died in 1979.

Carl Goldbranson returned to civil life and to the Union Pacific, but only for a few months; perhaps his forebodings had been right. He returned to the service, this time in the Air Force, and was evidently again engaged in deception activity for the rest of his career. By December of 1948 he was a full colonel, assigned at that time to the "Special Plans Section" of the Strategy Branch, Headquarters U.S. Air Force. He died in 1957.

Gene Sweeney was awarded the OBE by the British. Continuing his military career, he remained with the Special Section for a time, after Harris took it over. Then the Army gave him the opportunity to train in civilian business management with the Chrysler Corporation; Michael Crichton once looked him up there when Crichton happened to be visiting Detroit. He may have been seconded to the CIA for one or more tours. After he retired from the Army a full colonel in 1960, he attended seminary in Rome and in 1965 was ordained a Catholic priest. He died in 2002.

When Arne Ekstrom returned home, holder of the American Legion of Merit and the French Legion of Honor—whose ribbon he proudly wore in his buttonhole for the rest of his life—he soon concluded that it would be desirable to take the family out of New York for a time. So he accepted a (still classified) position with the CIA in Paris, and the Ekstroms lived there from 1946 to 1955; lived very well, as one could do even in Paris in those days of the almighty dollar, renting the house of Prince Poniatowski, with a large garden, trees twice as high as the house, butlers and chauffeurs and gardeners.

When they returned to New York, Ekstrom, still with the CIA, commuted to Washington until 1958 or 1959, at which point he retired from the Agency. He invested in a new art gallery, and found himself unexpectedly having to manage it single-handed when the original manager suddenly decamped. Though he had no experience in the business, it flourished, as everything Ekstrom touched seemed to do; and Cordier & Ekstrom took its place as one of the leading galleries of New York. He died in 1996, urbane, witty, articulate, and charming to the end.

Billy Harris, who received the French and Belgian Croix de Guerre, stayed in charge of the Army Special Section and its subsequent incarnations until late 1948. By that time he had spent over four years in this odd business—"mired in the deception game," as Bill Baumer had put it—and further promotion looked bleak. But he managed at last to return to his real trade as an artilleryman, commanding a battalion in Japan. It crossed over into Korea early in that conflict, and supported the 7th Cavalry; in September 1950, Harris succeeded to command of that regiment and led it on a 120-mile push through armed resistance in twenty-one hours—the longest sustained advance against an armed enemy in the history of the United States Army. Back on the right professional track after this feat, he had a successful career and ended up a major general. He retired in San Antonio in 1966 and died in 1986.

Ralph Ingersoll, newly and happily married soon after leaving the Army, rushed into print with a book called *Top Secret*. It was shrilly anti-British in tone; its thrust was that the Limeys had done little or nothing to help win the war, Eisenhower had been their patsy, while our gallant Russian allies were to be trusted. One might also have concluded from it that Omar Bradley was not only a military genius but the only worthwhile Allied top commander. It was briefly a best-seller but was soon justly forgotten.

He returned to *PM*, but not for long. There was turmoil on the staff between Communist and anti-Communist factions. He was accused by the anti-Communists of catering too much to the Communist side. Then, against Ingersoll's most basic belief, the owner decided the paper should accept advertising. So Ingersoll quit. One misfortune trod on another's heels: a failed novel, the death of his wife, an unsuccessful second marriage, another failed novel. But with the help of an admirer he rebuilt his life as owner of a successful chain of small-town newspapers.

He married once more, very happily. He kept up with Went Eldredge and to a lesser extent with Billy Harris; with his old boss General Noce, and with Clare Beck (whose daughter was for a time married to one of Ingersoll's sons). He published a partial autobiography and worked on a continuation, never published. He died in 1985.

Went Eldredge returned to the Dartmouth faculty. Diana crossed the ocean and they were married. He kept up with Ingersoll, and with Roger Fleetwood-Hesketh. He carved out a distinguished academic career in city planning, and in later years gave lectures—to classified audiences—about his experiences in deception. In the 1970s, while delivering one of these, he suffered a stroke that curtailed his activities thereafter. He died in 1991.

Jack Corbett served briefly as headquarters commandant for the American India-Burma theater after the Japanese surrender. He remained in the Army, and was in the Army Special Section in the immediate postwar years, making with Packard of the Navy a noteworthy visit to London in 1947, presently to be described.

Hal Burris-Meyer remained in the Navy, still working on deception matters, until the early 1950s. Briefly with the Muzak Corporation, he was soon tempted back into government service and joined the CIA. He died in 1984.

Ray Thurber left the deception business in the autumn of 1947 as a rear admiral. He served in various posts, both staff and line—his last seagoing duty was as commander of a battleship division—none of them, apparently, having any relation to deception. He retired in 1953 with a "tombstone promotion" to vice admiral, and died in 1967.

Bob Nelson left active duty in November 1945, with the Chinese Medal of Merit. The next month he sold the Glassine Paper Company, and spent some time with the Office of Price Administration until it closed; he was then an executive with major paper companies until retirement in 1958. He died at his Florida home in 1963.

Douglas Fairbanks returned to Hollywood and made a few films after the war, but gradually faded from the screen. He remained active in the Naval Reserve, reaching the rank of captain. He published two volumes of memoirs, and died in 2000.

Jeffrey Metzel retired from the Navy in 1949 with a "tombstone promotion" to rear admiral, and served as a consulting engineer to the

Owens-Corning Fiberglas Corporation. He died in a bizarre fashion in 1952, stabbing himself in the abdomen with a Japanese sword and leaping to his death from the third floor of his home in Chevy Chase, Maryland.

Pete Schrup returned to his old agency, the Federal Trade Commission, where he had a successful career as a staff lawyer and subsequently as an administrative law judge. He died in 1989.

Johnny Oakes returned to journalism, now at the *New York Times;* from 1949 to 1961 he was a member of the editorial board, and from 1961 to 1976 was editor of the editorial page. He died in 2001.

Mark Felt, who handled the PEASANT case for the FBI, rose eventually to be associate director of the Bureau. Fined for alleged conspiracy to violate the civil rights of friends and relatives of terrorists—this was perceived in many quarters as a politically-motivated prosecution by the Carter administration—he was pardoned by Reagan and retired to California. Persistent rumors that he was "Deep Throat" of Watergate fame were finally confirmed in 2005. He was still living when this book was being written.

Ormonde Hunter had been a square peg in a round hole; it was unfortunate both for him and for the country, for in some role other than the one Fate assigned him he could have made a great and satisfying contribution. He returned to Savannah and to a law practice and civic service that were infinitely more distinguished and successful than his career in the deception business had been. He died in 1989.

C. W. Grafton returned to Louisville, to the practice of law, and to writing. He published another successful mystery and a novel about college football; but his law practice absorbed his writing career. He died in 1982. His daughter Sue followed in his footsteps as a mystery writer; more prolifically, and even more successfully.

Gordon Merrick lived after the war in France, Greece, Normandy, and Ceylon. He wrote nothing dealing with intelligence after *The Strumpet Wind.* In 1971 he published a novel called *The Lord Won't Mind.* A few years before, it would have been deemed homosexual pornography, but times had changed. It was the first of a trilogy on the same theme. He died in 1988.

* * *

It is no part of this book to tell the story, if there is one, of strategic deception in the Cold War years. It is doubtful that enough material has been declassified to tell it. The present author has been able to discern a few bits and pieces that may be of interest.

In the immediate postwar years, both the British and the Americans desired to keep the capability alive, in terms both of organization and of know-how. As to organization, the debate on the American side continued as to whether it belonged in operations and plans or in intelligence; and as to whether there should be a Joint Security Control or its equivalent, or whether the work should be done by the regular staff. The latter view won, together with assignment to operations and plans; and by 1951 deception planning was under the Joint Strategic Plans Group of the Joint Staff.

Organizationally, things were more interesting on the British side. After a period of uncertainty, the London Controlling Section continued as a three-man body, with one officer from each of the services. By 1947, the equivalent of the wartime organization was in effect on a sub rosa basis—almost an unconstitutional one, it is tempting to say. Joe Hollis, who at the end of 1946 had succeeded Ismay as Military Deputy Secretary to the Cabinet and Chief of Staff to the Minister of Defense, put together the "Hollis Committee," which functioned as the equivalent of the old W Board, though its emphasis was on operations rather than intelligence. Composed of the three Chiefs of Staff, the chiefs of MI5 and MI6, a Foreign Office representative, and the Controlling Officer, with the Navy member of the London Controlling Section as its secretary, it was a sub rosa subcommittee of the Committee of Imperial Defense, and the fact that it existed at all was known to extremely few; for the lawful Committee included ministers with avowed or suspected Communist affiliations. The old Twenty Committee, whose name had become too well known, was reincarnated as the "Inter-Service Communications Intelligence Committee," chaired by Tar Robertson and with members from MI5, MI6, the service intelligence divisions, the Home Office, and the London Controlling Section, with the function of coordinating the intelligence implementation of deception plans and the counterintelligence efforts of the various intelligence agencies.

In 1947 and 1948, the British made a number of efforts to reestablish combined deception activity with the Americans, all of which were met noncommittally. In 1947, Jack Corbett, now with the Army's successor to the old OPD, and Commander Wyman H. "Packy" Packard, an assistant plans officer in the old F-28 and now with its successor organization, were sent to London to explore what the British were up to. They learned of the existence of the Hollis Committee. Two of the three members of the current London Controlling Section (Wing Commander Pat Saunders, RAF, and Lieutenant John H. Harvey-Jones, RN*) they found hospitable and cooperative; the third was Noel Wild, representing MI5 and MI6, and presumably the Army. They brought back amusing thumbnail sketches of the old hands they met. Wild they found "a slippery customer who is by nature anti-American, at least American officers have always had trouble dealing with him. Although cordial, and socially friendly, he is still reluctant to play ball with us and can be counted to run away from all business transactions, or short change you in those he is forced into. His word is not considered reliable. He remained absent during the entire visit, after an initial contact." "Various other personalities [who] are ever present behind the scenes at London Controlling Section" whom they also met included Bevan ("Very cooperative to Americans"); Dudley Clarke, recently retired but "occasionally borrows office space at LCS" ("Not very cooperative to Americans"); Wingate, who "frequently pops in, borrows secretarial staff" and whom Corbett of course already knew ("Extremely cooperative to Americans"); Roger Fleetwood-Hesketh, now assigned to M15 ("friendly, clever, psychologically abnormal, takes all cues from Col. Wild"); and Tar Robertson of MI5 ("extremely cooperative and pro-American. . . . More than superiors").

At least in the United States, substantive understanding of deception above the tactical level seems to have declined—among the military, at least—after the 1950s, perhaps reflecting a loss of institutional memory as the relatively few veteran operatives passed from the scene. In 1947, Eisenhower, as Chief of Staff, had decreed that "no major operations should be undertaken without planning and executing appropriate deceptive measures." The 1948 contingency plan for war with the Soviet

---

* Harvey-Jones, a young submarine officer, subsequently left the Navy for the business world, and in 1982 reached the pinnacle of British industry as Chairman of ICI, the great chemical concern.

Union (FROLIC), which accepted that the Soviets would overrun continental Europe at the outset, contained such an annex, with the "story" that the Americans would hold on in Western Europe.* Again, in 1951 CINCPAC prepared such an annex to its current contingency plan for war with the Soviet Union (Joint Plan SCHOOLYARD). Evidently called BRIDEGROOM, the deception plan contemplated a threat against the Kuriles (now Soviet-occupied) and Kamchatka, including reinforcement of the Pacific by notional forces; and a threat against the Shanghai area "in the event it becomes necessary to employ Chinese Nationalist forces to halt Communist aggression in Southeast Asia."

On the operational level, MacArthur's brilliantly successful landing at Inchon in 1950 was covered by a deception operation threatening a landing at Kunsan, much further to the south, which included false briefings of the troops involved, diversionary operations off Kunsan, and use of double agents.

On the highest strategic level, both the British and the Americans saw from the outset that the most potentially valuable line of deception with the Soviets was to mislead them as to the state of Western research and development, and in particular to induce them to fritter their resources in directions known to the West to be unproductive. In 1948 the codenames THUMBTACK, CLASSROOM, and BOYHOOD, respectively, were assigned to an unspecified "cover and deception plan in support of U.S. military position and policies in current European situation," to another unspecified "cover and deception plan to misguide and waste scientific effort of USSR in biological warfare field," and to "cover and deception implementation during peacetime in support of emergency war plans."

The wartime Allies had practised to deceive, and they had not merely improved their style but perfected it. But memories grew short, and the skills faded away. Overall, it would appear that most of the lessons of the Second World War were forgotten within twenty years, at least by the Americans.

* Noting that, because "available intelligence indicates that the matters discussed in conferences, in which Western European nations (except the United Kingdom and Canada) take part, become the knowledge of the Soviets," the planners, perhaps remembering ICEMAN, recommended that "conferences with foreign planners should follow a substitute, or false, FROLIC implementing the deception annex to the real one."

In 1947, Eisenhower, as Chief of Staff, had warned that "As time goes on individuals familiar with these means of warfare"—meaning deception and psychological warfare—"are likely to become progressively less available in the Regular Army and there is danger that these two means may in the future not be considered adequately in our planning." He accordingly directed that Plans and Operations, the successor to OPD, "maintain the potential effectiveness of these arts." For a few years after the war, knowledgeable officers—Goldbranson, Burris-Meyer, Harris, Thurber, Jack Deane, son of General Deane of the Joint Chiefs and Moscow, who worked for Harris after the war—would lecture at the war colleges on wartime lessons learned. Commander Holsopple left behind a short but thorough treatise on the art of communications deception. But institutional memory faded. In 1950, the Joint Chiefs were already working on a project, evidently called HOARDING, for "improving" American capability to employ cover and deception in support of current war plans.

Given Eisenhower's enthusiasm for the subject, it might have been expected that during his administration, the height of the Cold War, efforts at strategic deception at a high level would be made. If they were, we do not know about them. The Kennedy administration's national security team would, one thinks, have been charmed by such trickery. But by then, memory appears to have been at a low ebb, and to have remained so through the 1960s.

Double agent operations run for penetration purposes during this era sometimes included deceptive material. For example, from 1959 to 1980 the FBI ran an operation called SHOCKER, in which an Army sergeant allowed the GRU, the Soviet military intelligence organization, to believe that they had recruited him as an agent. Among the foodstuff provided for him to pass to the Soviets was material designed to suggest that a line of research on nerve gas which American scientists had determined to be hopeless was in fact productive. But this appears to have been merely an ad hoc move and not a systematic strategic deception.

In the Vietnam conflict an organization known as Military Assistance Command Vietnam Special Operations Group ("MACVSOG" or more usually "SOG") ran various operations that were more psychological warfare than deception, and would have made Dudley Clarke shake his head. The most elaborate of these was Project HUMIDOR, designed to

convince the Hanoi leadership that there was a serious opposition movement in North Vietnam. Its centerpiece was a notional underground resistance organization called the "Sacred Sword of the Patriots League."
The "League" had membership cards, a clandestine radio, leaflets and
gift kits; and in its palmiest days SOG operated on a South Vietnamese
island a dummy village to which kidnapped North Vietnamese were
brought and told that it was in League-controlled territory in North
Vietnam.

On the other side of the hill, all or nearly all agents sent into North
Vietnam under American auspices appear to have been doubled, though
it took an embarrassingly long time for the Americans to realize it. In
1968, after finally tumbling to this fact, SOG opened a program to extend the HUMIDOR myth, called FORAE. Three of its subprojects, BOR-
DEN, URGENCY, and OODLES, rose above ordinary psychological warfare.
BORDEN was a program to recruit North Vietnamese prisoners of war as
agents to return home with the assignment to get in touch with the notional resistance movement in North Vietnam, or at a minimum to
spread the belief among their fellow prisoners that there was such a
movement. URGENCY was a project to frame hard-core North Vietnamese prisoners as American agents and allow them to fall back into
North Vietnamese hands. OODLES was an operation consisting of radio
messages and supply drops to nonexistent underground teams in North
Vietnam, reminiscent of Peter Fleming's HICCOUGHS, ANGEL ONE and
GREMLIN. The FORAE program was shut down on Washington's orders in
late 1968 and 1969.

Neither HUMIDOR nor FORAE nor anything else mounted by SOG—
or by any other American agency in Vietnam above the tactical level—
seems to have been a true deception operation designed with the
CAMILLA principle in mind that you start with what you want the enemy
to *do*. The only thing besides FORAE that came close to anything Dudley
Clarke and the rest did in the Second World War was a 1968 SOG project called ELDEST SON in which, just as in the COPPERS SCHEME of 1941,
captured enemy ammunition was doctored so as to blow up when fired.
And, like the COPPERS SCHEME, this hardly qualified as deception.

Then in the early 1970s a few officers in the Pentagon seem to have got
an inkling as to what they had been missing. In June 1971, the opera-

tions division of the Joint Staff sponsored a conference on deception and brought Baumer, Harris, and Eldredge in to talk about their recollections. Ingersoll was invited but could not attend; he was glad he had not done so, when he learned from Eldredge that the old-timers were excused after they had said their say, leaving current matters to be discussed by the younger folk; like a professor lecturing in an occupied country, he said. From that time on, occasional lectures and conferences were arranged for various groups, by Eldredge in particular. In 1982, William Casey, Reagan's first Director of Central Intelligence, who had been in London with OSS and knew something at first hand about FORTITUDE—Hesketh had sent him a copy of his history some years before—organized a "Lessons Learned" conference attended by Baumer, Harris, Eldredge, Douglas Fairbanks, Hesketh, and R. V. Jones (who had advised British intelligence from the science and engineering viewpoint during the war).

This renewal of interest was a good thing; but it is noteworthy that the emphasis was on the limited perspective of those who had participated in FORTITUDE. The long-range perspective that Dudley Clarke and Johnny Bevan had brought to the work was missing; especially any experience of the slow, patient order of battle work that had underpinned everything. It is strong evidence of institutional amnesia that Sweeney and Ekstrom—Bevan's choices as the ablest of the American deceivers, trainees of Dudley Clarke, with more real high-level experience than any other living Americans—were forgotten. Still, the early 1980s seems to have marked a new awareness at the top of the American government of the possibilities of strategic deception; and almost certainly, Casey played a key role in this.

Within a limited specialist group, academic study accompanied this revival of interest. In the 1980s appeared several scholarly works by Barton Whaley, Michael Handel, Katherine Herbig, Donald C. Daniel, and others; historical articles on particular deceptions were published in journals devoted to intelligence; and in 1985, Katherine Herbig prepared her study for the Naval Postgraduate School. Except for Herbig's valuable work, all relied heavily if not entirely on the limited declassified information then available. Some writers engaged in verbose theoretical analysis and high-sounding taxonomy of little or no practical use.

Equally fancy "research" by contractors such as RAND followed suit in the 1990s.

But at least the basic lesson seems to have been learned. It had clearly been learned by 1991, when Schwartzkopf's successful offensive in the Gulf War against the Iraqi right was supported by a carefully planned deception operation to induce the Iraqis to prepare for an offensive against their left instead. So far as is presently known, this was conducted as a classical feint, involving extensive actual operations on the threatened flank together with sonic deception and an enhanced volume of radio traffic.

In more recent years there has been much talk about "information warfare" and "perception management," with deception regarded as an element of this activity. In line with this thinking, in 1996 the Joint Chiefs of Staff published a document entitled "Joint Doctrine for Military Deception" (revised in 2006 as "Military Deception"). It is written in the gobbledygook of the late twentieth-century military bureaucracy, and relies heavily on the pretentious theoretical analysis of the preceding decade; it is hard to imagine that it could be of much practical use. Nevertheless, most of the basic elements as Dudley Clarke had developed them by 1942 can be discerned within this near-impenetrable thicket by the knowing eye; though it still contemplates that deception planning will be buried deep within the operations staff. A tangled web this "joint doctrine" may be, but it is at least a sign that to practise to deceive has not once again been forgotten.

In the future as in the past, much will depend on the willingness of senior commanders, and civilian leaders, to embrace the concept of deception and give the deceivers their full support; to follow the example of Wavell, Eisenhower, Patton, Montgomery, and Alexander, rather than that of Bradley and Stilwell. If they do, the spirits of Dudley Clarke, Johnny Bevan, Newman Smith, Peter Fleming, and their extraordinary menagerie of colleagues will look down from whatever very special Valhalla they inhabit, and smile.

Addendum

In a mere thirty-seven pages, Newman Smith's long-lost November 1945 "Informal Memorandum of the Origin, Development, and Activities of the Special (Deception) Section, Joint Security Control, 1942-1945" not merely "presents the hazards which beset, and the barriers which impeded, the progress of the concept of military deception and tells how they were overcome to gain acceptance for deception as a military instrument," in the words of Burris-Meyer's unfinished *Cake Before Breakfast*; more significantly, it points up Smith's appreciation of the superiority of British deception organization and doctrine, and his awareness of the reasons for American shortcomings—including stubborn confusion of deception with psychological warfare; failure to assimilate the British experience; divorce of implementation from planning, and of deception planning from operational planning; assignment of too few officers, especially too few regulars; focus on deception as only a theater responsibility; and the indifference of some theater commanders, notably in the Pacific. (All reconfirming, incidentally, how wrong were British misapprehensions of Smith's understanding of these matters.)

Equally noteworthy is Smith's perception of American needs for the future—needs that began being met only relatively recently. Warning that "unless every effort is made to study, evaluate and assimilate the mass of material and experience gained by the handful of officers of the United States Army for this purpose, the situation with respect to deception operations at the beginning of any future war will be as hopeless and ineffective as it was at the beginning of World War II," he urged the "development of deception doctrine and technique, and their incorporation in a Top Secret Cover and Deception Manual"; the preparation of deception training courses for staff officers; a permanent joint cover and deception organization; and study of the potential for peacetime deception. A far-reaching program—and one which, as described in the Epilogue, appears now to have been largely adopted, but only after nearly half a century.

In short, the long-lost "Informal Memorandum" abundantly confirms Newman Smith's standing as the godfather of modern American military deception.

APPENDIX I

Allied Deception Operations

(Selected genuine operations included are in *italics*)

A-R   ("Anti-Rommel"). Plan, April-May 1941, to make Germans believe British planned landing behind Rommel between Tripoli and Benghazi, and that Wavell was reinforced by additional armor (including notional new air-conditioned tanks to enable attack on Tripoli during hot months of May and June).

ABAFT   Cover plan for a projected (but later called off) combined operation in Red Sea for landing in Eritrea, February 1941.

ABEAM   Plan to lead Italians to believe there were airborne troops ("1 S.A.S. Brigade") in Middle East, January–July 1941.

ACCOLADE   Genuine plan (abandoned) to invade Rhodes, September 1943.

ACCORDION   Notional attack by 6th Army Group to breach Rhine-Siegfried Line, March 1945, designed to contain maximum number of German forces east of Rhine and south of Karlsruhe in support of actual Seventh Army attack and Rhine crossing.

ACCUMULATOR   Simulated diversionary attack against Coutances-Granville area on west coast of Cotentin peninsula, June 12–13, 1944, to hold German troops on that side of Cotentin.

ACROBAT   Plan for pursuit of Rommel to Tripoli if Axis should be routed as result of CRUSADER, December 1941. Cover plan: ADVOCATE.

ADORATION   Complete radio silence on every nonoperational radio circuit in the UK from *NEPTUNE* H-Hour until July 6, 1944.

ADVOCATE   (Originally called XMAS PLAN II.) Cover plan for *ACROBAT,* December 1941. "Story": No further advance would take place, but Auchinleck would build a strong defensive flank in Libya and transfer forces to Persia and Iraq to reinforce Russians in Caucasus and encourage Turks.

ALACRITY Planned occupation of Azores, October 1943, not carried out. Cover plan: DUMMER.

ALKEY London Controlling Section cover plan for visit of King George to Mediterranean, June 1943.

ANAGRAM Plan to cover move of Australian forces from Middle East to Far East, January 1942, by depicting it as concentration of Australian forces in Egypt for invasion of Crete.

ANAKIM Seaborne attack on Burma, authorized at Casablanca for autumn 1943; canceled for lack of resources.

ANVIL Original name for southern France landing, 1944. Renamed *DRAGOON*.

APPENDIX Y COSSAC cover and deception plan for *NEPTUNE*. Originally called TORRENT. Replaced by MESPOT/FORTITUDE.

ARAMINTA Part of STULTIFY, 1944–45. Notional airborne operations against remaining Japanese lines of communication in the Shan States. Related: FANG, CLAW, CAPTION, WOLF.

ARCADIA Washington Conference, December 1941–January 1942.

ARDENNES Operation, December 16–30, 1944, designed to cause Germans to delay reinforcing southern shoulder of the Ardennes bulge by representing Third Army's advance northwards as slower than it actually was. Related operation: METZ I.

ARGONAUT Yalta Conference, February 1945.

ASSASSIN Plan proposed by Ewen Montagu to Twenty Committee in July 1942 to discredit a certain Tome, Icelandic vice consul at Vigo who had recruited agents for the Germans, by passing word that he had betrayed recruited agents to the British. Only partially implemented, with no result. Related plan: SPANNER.

AVALANCHE Salerno landing, September 1943.

AVENGER Cover plan, October 1944, for possible thrust into north Germany by 21 Army Group, by giving impression that main attack would be on south Germany across upper Rhine by U.S. Third Army and 6th Army Group, thus drawing Germans to the south; stillborn because there was no decision as to which sector would have priority in real operations.

AVENGER II   Cover plan, November 1944, adding to AVENGER a notional threat to ports of north Germany; stillborn for same reason as AVENGER.

AXTELL   Plan (never implemented) for notional increase in number of antisubmarine vessels in area of Mozambique Channel. Revised version: a plan to entice the U-boat pack to the area south of Cape Town, August–September 1943. Neither version was ever implemented, as *HUSKY* had opened the Mediterranean route to the Middle East and India.

BALDERDASH   Projected operation, April–May 1942, to lure German naval forces to battle by reporting through the transmitter of a seized German espionage ship and through double agent COBWEB in Iceland that the British Home Fleet had sailed to Iceland. Canceled.

BAMBINO   Continuation of Kuriles threat after HUSBAND, beginning November 1944; tapering off the WEDLOCK series. Related operations: WEDLOCK, HUSBAND, VALENTINE.

BARCLAY   Overall Mediterranean deception plan for 1943, approved April 1943, amended May 20. British Twelfth Army (notional) in Egypt would invade Peloponnese and Crete, Turkey would hopefully come into the war, then substantial forces would be moved through Turkey to link with Russians in Bulgaria and Thrace. Diversionary attacks on north and south France; British Eighth Army to land in south France, French forces to exploit up Rhone; simultaneously Patton (U.S. Seventh Army) to attack Corsica and Sardinia. Sicily and Italy to be bypassed. Original D-Days May 25 for Greece and Crete, June 5 for others; on May 20 to be postponed to June 25 and July 3; on June 20 another month's postponement. (*HUSKY* actually planned for first half July.) Subsidiary operation: WATERFALL. Related operations: LEYBURN, MINCEMEAT.

BARDSTOWN   Extension of ERASMUS, October 1943. In light of submarine menace in Indian Ocean, Axis told by special means that troops moving by convoy from Africa to India are going by way of Nile. Related: ERASMUS.

BARONESS   A fake message from Foreign Secretary Anthony Eden to British ambassador in Chungking shortly before TORCH and referring to that operation, passed to the Japanese through PURPLE WHALES in November 1942 shortly after TORCH.

BASSINGTON Plan by Peter Fleming toward close of war to pass to Japanese a forged report concerning war crimes, designed to secure improved treatment for Allied prisoners. Overtaken by Japanese surrender.

BASTION Plan, February–March 1942, to make Rommel believe he was running into a trap at the Gazala Line, by swelling apparent armored reserves with dummies, beefing up Tobruk defenses.

BAYTOWN Crossing of Straits of Messina, September 1943.

BERTRAM Deception plan for El Alamein. Subsidiaries: BRIAN, CANWELL, DIAMOND, MARTELLO, MELTING POT, MUNASSIB, MURRAYFIELD. Associated strategic plan: TREATMENT.

BETA Overland operation against Canton and Hong Kong, autumn 1945. Name changed to *RASHNESS,* then to *CARBONADO.* Related plan: ICEMAN.

BETTEMBOURG An operation of 23d Headquarters Special Troops, September 15–22, 1944, to simulate 6th Armored Division north of Metz to shore up a weak point in the line.

BIGDRUM One of the *NEPTUNE* D-Day naval diversions: Small force of motor launches to engage German radar stations on the north of the Cotentin Peninsula and distract enemy batteries in that area. Related diversions: TAXABLE, GLIMMER.

BIJOU London Controlling Section plan, December 1942–December 1943, to make Axis believe aircraft carrier *Indefatigable* (which was not in fact completed till end of 1943) was launched in the Clyde and sailed to the Far East.

BLANDINGS Tactical extension of UKRIDGE. To deter Japanese in Burma from reinforcing areas of British interest by threatening a notional air and seaborne assault codenamed MALICE, to be mounted between April 21 and 24, 1944 against the Sandoway-Taungup-Prome area by four divisions, including a notional airborne division, to cut Japanese line of communication to Akyab, establish bridgehead on lower Irrawaddy, and secure base for further coast-hopping; diversionary landings on Cheduba and Ramree Islands.

BLARNEY STONE London Controlling Section 1943–44 plan (never accomplished) for a POW to escape with important fake papers via Eire. *Cf.* IDES.

BLAST Original name of WADHAM.

BLUE BOOT PLAN A Twenty Committee idea, January 1941, to tell the Germans that all British troops will have some special identifying mark in event of invasion, to induce the Germans to use it while the British in fact would not. One suggestion, which gave the project its name, was to tell them that British soldiers would have their right boots painted blue. Blue paint would be issued to the troops to add conviction to the story. Not carried out.

BLUEBIRD   Deception plan for Okinawa operation: Notional invasion of Formosa and southeast China coast, spring 1945.

BOARDMAN   Plan July–September 1943 to weaken enemy forces in Italy, especially central and southern, and contain maximum enemy forces in Mediterranean, especially southern Balkans and particularly Greece. "Story": Landings in Sardinia and Corsica, then heel of Italy, early September, followed by landing in Peloponnese and then in either southern France or northwest Italy.

BODEGA   A fictitious vast underground depot supposedly built in the Chislehurst caves, in which GARBO's AGENT FOUR worked.

BODYGUARD   Overall deception plan for Europe and Mediterranean, 1944. "Story": Invasion impossible till late summer, heavy reliance on POINTBLANK to bring Germany down. Multiple threats to Norway, Balkans, France other than cross-Channel, with associated diplomatic initiatives. Related plans: FORTITUDE NORTH, FORTITUDE SOUTH, GRAFFHAM, IRONSIDE, ROYAL FLUSH, TURPITUDE, VENDETTA, ZEPPELIN.

BOODLE   Recapture of Attu, 1943. Renamed COTTAGE.

BOOTHBY   Deception plan for BAYTOWN (crossing of Straits of Messina): Notional attack on Crotone ("ball" of "foot" of Italy) by forces which were actually to carry out BAYTOWN.

BOUZONVILLE   A 23d Headquarters Special Troops operation, March 11–13, 1945, simulating a buildup in the Bouzonville-Saarlautern area by the 80th Division, to cover a real attack between Trier and Saarburg.

BRAINTREE   Intended follow-on plan to FERDINAND, to continue the Mediterranean theatre's obligations under BODYGUARD. Never implemented, since BODYGUARD went into abeyance late August 1944; a much-modified FERDINAND continued till September instead.

BREST   Tactical deception by 23d Headquarters Special Troops designed to exaggerate armored forces in front of Brest, August 20–27, 1944.

BRIAN   Dummy petrol, food, and ammunition dumps and associated administrative camps, to suggest attack in south at Alamein, October 1942. Subsidiary of BERTRAM.

BRIMSTONE Plan to invade Sardinia after the clearing of North Africa, considered and rejected at the Casablanca Conference. Related: CANUTE, GARGOYLE, JIGSAW.

BRITTANY Operation of 23d Headquarters Special Troops during August 9–12 1944 to simulate movement of regimental combat teams from 35th, 80th, and 90th Infantry Divisions and 2d Armored Division into Brittany to distract from German counterattack at Mortain.

BROADARROW (Sometimes written as "BROADBARROW.") Plan, January 1943, to discourage U-boat operations in eastern Mediterranean by exaggerating size of antisubmarine fleet.

BROADAXE Overall plan for Pacific deception 1945. Object: Induce Japanese not to move troops to home islands, deploy troops on home islands to meet threats away from planned attacks on Kyushu and Honshu. "Story": No invasion of Japan until further bases seized; operations to be conducted against Formosa, China coast, Thailand, French Indochina, Sumatra, Kuriles and Hokkaido, Yellow Sea. Related operations: PASTEL, PASTEL TWO, CONSCIOUS.

BROADSTONE Operation October–November 1943 to cover British X Corps first attack on Gustav Line by threatening landing behind German flank.

BROADSWORD Second name of STARKEY.

BROCK Blowing up of installations in Hampshire to enhance the credibility of MUTT and JEFF as saboteurs, 1942.

BRUNETTE Abortive project, January–February 1943, to enhance in German naval eyes the risk to shipping of reinforcing and supplying Tunisia in hopes Navy would persuade Army to cut losses in Tunisia.

BRUTEFORCE A simulated beach reconnaissance at the mouth of the Salween River, night of March 29–30, 1945.

BUCCANEER Plan for amphibious attack on Andaman Islands; finally canceled end 1943.

BUNBURY Sabotage of a power station at Bury St. Edmunds to enhance credibility of MUTT and JEFF, 1943.

BUXOM Cover plan for *BOODLE/COTTAGE* (recapture of Attu), 1943. Troops were led to believe they were training for Central Pacific.

CADBOLL   Part of FORTITUDE NORTH, to portray the false order of battle, preparations for operations overseas, and concentration of British 3d and 52d Divisions in the Clyde in preparation for embarkation. Canceled and replaced by CADBOLL II.

CADBOLL II   An element of FORTITUDE: Shift of British II Corps to Cambridge, where it could be available either for Norway or for the Pas-de-Calais. Followed by TWEEZER.

CALLBOY   Attempt to pass off crossing of Rhine in Weser sector February 1945 as a feint to cover a real thrust against Kassel.

CAMILLA   Deception plan (December 1940) for Wavell's attack on Italian East Africa, 1941, to induce Italians to believe that British Somaliland was objective.

CANUTE   Part of London Controlling Section cover plan for the canceled operation *BRIMSTONE*, the invasion of Sardinia after the clearing of North Africa: Plan to direct threat against Norway or Northern France, to be managed by LCS. Related: GARGOYLE, JIGSAW.

CANWELL   Wireless deception plan at Alamein, October 1943, in conjunction with BERTRAM to conceal movement of X Corps to staging area.

CAPITAL   Slim's offensive to clear Japanese from Burma, 1944–45.

CAPTION   April–May 1945 sequel to CONCLAVE. Notional guerrilla operations against remaining Japanese lines of communication in the Shan States. Related: FANG, CLAW, ARAMINTA, WOLF.

CARBONADO   Overland operation against Canton and Hong Kong, autumn 1945. (Originally *BETA*, then *RASHNESS*.) Related plan: ICEMAN.

CARROT I   Signal exercise simulating an amphibious exercise of naval Force "V" with British 52d Division, April 12, 1944. Part of FORTITUDE NORTH.

CARROT II   Signal exercise simulating an amphibious exercise of naval Force "V" with British 52d Division, April 15, 1944, suggesting landing in a fjord region. Part of FORTITUDE NORTH.

CARNEGIE   Deception plan early September 1943 to increase notionally strength of forces at Salerno and Taranto.

CARTER PATERSON   December 1942 effort to plant a fake minefield chart on the Germans through TATE and set up a channel through which he could pass documents to them; only partially successful.

Casanova Operation, November 4–9, 1944, in which elements of 23d Head-quarters Special Troops simulated the 90th Division 11 miles from its actual point of attack on Metz.

Cascade First comprehensive order of battle deception plan for whole of Middle East theatre, beginning March 1942, frequently updated; edition of July 1942 little changed till replaced by "Cascade (1943)" in March 1943; widened several times in 1943; replaced by Wantage February 1944.

Catspaw London Controlling Section 1943–44 plan (never accomplished) to plant fake documents on the Swedish air attaché in a taxi.

Cattle Bogus beach reconnaissance of the Bireuen area in northeastern Sumatra, April 1945.

Cavendish Amphibious exercise planned to be held in the Channel in October 1942; canceled. Part of Overthrow.

Cent (Originally Dryshod.) Part of Quicksilver II: a notional combined exercise in Studland area by 4th Brigade of 2d Canadian Division assisted by naval Force "G," April 16, 1944. Related operation: Dollar.

Chatter A suggestion, not adopted, by Ewen Montagu to support the Bodyguard "postponement" theme in the spring of 1944 by telling the Germans that the western Allies were seeking another summit conference to induce Stalin to agree that the 1944 invasion would be limited to Norway.

Chesterfield Operation December 1943 to cover Fifth Army attack in Italy by threatening landing behind German line at Gaeta.

Chettyford Tactical aspect of Oakfield, covering pre-Anzio regrouping of 15th Army Group in Italy, January 1944. "Story": Fifth Army on left will stay on defensive, detaching divisions for Pisa landings, while Eighth Army on right will advance to link up with Rimini landing.

Christmas Cake London Controlling Section 1943–44 plan (never carried out) to allow a POW to escape by stealing an airplane in which there would be forged compromising documents.

Clairvale A notional drive up the Adriatic coast with an outflanking landing at Pescara, January–February 1944, designed to deter Germans from moving forces from Adriatic flank to Anzio and Cassino.

Claw   Threat to Moulmein (with implicit bypassing of Rangoon) 1945, part of Stultify.

Clearance   Bogus airfield reconnaissance in the Setul area, extreme southwest Thailand, April 1945.

Cloak   Deception operation for IV Corps's crossing of Irrawaddy in Operation *Capital,* 1945. Related operations: Conclave, Pippin, Stencil.

Cockade   Umbrella name for Starkey, Wadham, and Tindall, 1943. Threats to French Channel coast, Brittany, and Norway, to bring on air battle and pin down maximum German forces in the West.

Colfax   Operation, early September 1943, to contain Hermann Goering and 15th Panzer Divisions and prevent their moving south to Salerno, by creating threat of landing in Gulf of Gaeta north of Naples.

Collect   Deception plan in support of *Crusader* offensive, 1941, to lull Germans into relaxing their guard by repeated reports that offensive imminent, followed by postponements. Related: Tripoli Plan.

Compass   Wireless deception in Italy, March–April 1945, to indicate move of notional British XIV Corps to army group reserve.

Compass   Wavell's attack in the Western Desert, November 1940.

Conclave   Operation, March 1945, to threaten Japanese escape route through the Shan States, linked with Tarzan threat, directing Japanese attention to the east of main thrust south from Meiktila, while Cloak directed them to west. To be conducted by an ad hoc "scratch corps" composed of the genuine 20th Indian Division and the bogus 51st Indian Tank Brigade.

Conrad   Operations against Phuket, 1945.

Conscious   Notional attack by MacArthur on Cochin-China in support of Sceptical and Iceman, autumn 1945.

Copperhead   (Originally Telescope.) Visit to Mediterranean by Montgomery's double, end May 1944, designed to make Germans believe no cross-Channel operation imminent.

Coppers   Planting of doctored explosive ammunition on Axis troops, July 1941.

CORDITE   Projected invasion of Rhodes, 1941.

CORDITE COVER PLAN   Cover plan for projected invasion of Rhodes (*CORDITE*), March 1941. Italians to be made to believe objective was Scarpanto seven days after real Rhodes date. Canceled by German offensive in Libya.

CORKSCREW   Seizure of Pantelleria, June 1943.

CORONET   Planned invasion of Honshu at Kanto Plain (Tokyo area), March 1946.

COTTAGE   Recapture of Attu, 1943. Originally *BOODLE*.

COWPER   Tactical plan covering unsuccessful attack on Tunis late April 1943 (COWPER I) and successful attack early May (COWPER II), both by British First Army. "Story": Eighth Army would launch final assault from south in early May.

CRUSADER   Auchinleck's offensive in Cyrenaica, late 1941.

CRYSTAL   Plan to make arrival of M-3 Grant tanks as much of a surprise to the Germans as possible, spring 1942, by deceiving them as to when they would be ready (notional ammunition troubles) and which units were receiving them.

CULVERIN   Possible attack on Sumatra, abandoned spring 1944.

CYPRUS DEFENSE   Plan to exaggerate strength of Cyprus garrison after fall of Crete, June 1941.

DALLAS   Operation of 23d Headquarters Special Troops November 2–10, 1944, to cover a change of position of XX Corps artillery during the attack on Metz.

DANTE   A tactical deception plan by British Twelfth Army in Burma, July 1945, to support planned operations (rendered unnecessary by termination of hostilities).

DEFRAUD   Bogus beach reconnaissance on the Nicobar Islands, April 1945, part of a series to cover the compromise of beach reconnaissances in connection with *CONRAD*.

DERRICK   Tactical deception plan in support of *HUSKY*, July 1943. Aimed at containing enemy forces in west of Sicily as long as possible. Related: FRACTURE.

DERVISH   Abortive project, January–February 1945, for dissipating German war effort by threatening a vital center of communications, industry, and control; specifically, threatening Stendal-Brandenburg-Wittenberg-Magdeburg area with a large-scale landing of airborne forces able to maintain themselves for thirty days.

DETACHMENT   Seizure of Iwo Jima, 1945.

DIADEM   Alexander's offensive in Italy, May 1944.

DIAMOND   Subsidiary of BERTRAM. Dummy water pipeline ostensibly for troops on south front at Alamein, October 1943.

DIJON   Effort, August 25–30 1944, to distract German attention away from Patton's drive on Metz by indicating that it might be diverted southwards towards Dijon or northwards towards Belgium.

DOLLAR   (Originally DRYSHOD.) Part of QUICKSILVER II: a notional combined exercise in Studland area by 4th Brigade of 2d Canadian Division assisted by naval Force "G," April 25, 1944. Related operation: CENT.

DOMESDAY   Original name of STARKEY.

DOWNFALL   Conquest of Japan. Related operations: *OLYMPIC, CORONET,* PASTEL, PASTEL Two.

DRACULA   Seaborne assault on Rangoon, approved by Combined Chiefs at *OCTA-GON* conference, August 1944.

DRAGOON   Revised name for southern France landings, August 1944 (originally *ANVIL*).

DRINK   Operation to divert German U-boat effort from the South to the North Atlantic, and from the Atlantic to the Mediterranean, winter 1942–43.

DRYSHOD   Original name of CENT.

DUMMER   Cover plan for *LIFEBELT,* planned seizure of Azores, spring 1943, and subsequently for *ALACRITY* (peaceful occupation, October 1943). Troops were told they were bound for Corsica and the Mediterranean.

DUNDAS   London Controlling Section operation starting at beginning of 1943 to build up enlarged notional order of battle for British forces in the United Kingdom. Associated operation for American order of battle: LARKHALL.

DUNDEE   London Controlling Section plan November 1943, implemented January 1944, to deceive Axis as to destination of strong force of battleships and carriers sent through Mediterranean to Far East. Designed to induce Japanese to reinforce Timor area and discourage reinforcement of Bay of Bengal area, by suggesting operation against Timor (as quid pro quo for use of Azores bases). Integrated with OAK-FIELD for passage of ships through Mediterranean.

DUNGLOE   Purported radio messages to notional Yugoslav resistance giving notional D-Days for notional invasion of Balkans under ZEPPELIN.

DUNTON   Plan to support ZEPPELIN by indicating direct sailing of an American division in two big ships lying idle, the *Île de France* and the *Nieuw Amsterdam.*

DYNAMITE   Contingency plan for recapture of Stavanger, 1941–42.

E.S.   Small operation, June 1942, to alert the Germans to a possible operation against southern Norway, to distract attention from a convoy to Murmansk.

ELEPHANT   Tactical deception by 23d Headquarters Special Troops to cover move from reserve of 2d Armored Division near Cerisy-le-Forêt in early July 1944.

11TH PANZER DIVISION   A tactical operation designed to keep 11th Panzer Division south of Trier during American offensive further north, February 1–15, 1945. Related operation: WHIPSAW.

ELSENBORN   An operation, November 3–12, 1944, in which elements of 23d Headquarters Special Troops simulated the 9th Division remaining in barracks at Elsenborn while it moved to the Hürtgen Forest offensive.

ENVY   Proposed abandonment of glider on a sandbank in the Irrawaddy near Prome, containing documents and equipment to give impression that airborne forces would soon attack Taungu-Sandoway-Prome area, March–April 1944. Presumably part of BLANDINGS. Cancelled.

ERASMUS   Combined Army/Navy order of battle deception plan, February–August 1943, to cover transfer of troops from Africa to India, to exaggerate size of transfer and to deter Japanese submarines from interfering. Related: BARDSTOWN.

ERROR   Leaving of phony documents suggesting formidable buildup of British troops in India, for Japanese to find at retreat over Irrawaddy, April 1942.

EUREKA   Tehran Conference, November 1943.

EXPLOIT   Deception plan for *FLASHPOINT,* Ninth Army's crossing of the Rhine, March 21–April 1, 1945. Simulated buildup for crossing at Uerdingen in XIII Corps sector, with actual crossing further north in XVI Corps sector. Related operation: VIERSEN.

EUPHRATES PLAN   Plan, July 1941, to deter any German advance into Turkey by making Germans believe that substantial armor was being left on northern borders of Syria and Iraq.

EXPORTER   Invasion of Syria, June 1941.

EXPORTER COVER PLAN   Cover plan for invasion of Syria, June 1941 *(EXPORTER)*. Notionally, De Gaulle leaving Egypt in disgust because Wavell would not attack Syria.

FABIUS   Genuine amphibious exercise out of Southampton, early May 1944.

FABRIC   Strategic/tactical plan, May–June 1942, to cover Auchinleck's planned June 1942 offensive by persuading Germans that Eighth Army planned to stay on defensive till September, and to make it appear that when Auchinleck did attack it would be on his right rather than (as planned) his left. Prematurely ended by Rommel's late May offensive. Related plan: MAIDEN.

FAIRLANDS   Follow-on to BOARDMAN, 1943. Notional attack on Rhodes and Crete (later changed to Western Greece) from Middle East; together with continued threat to Leghorn region, to prevent transfer of forces in that area to main German line south of Rome.

FALSE ARMISTICE PROJECT   Project for fake proclamation of armistice by King of Italy, July 1943. Vetoed in London.

FANG   Part of STULTIFY, 1944–45. Notional seaborne operation against Sumatra. Related: CAPTION, ARAMINTA, CLAW, WOLF.

FANTO   A July 1943 London Controlling Section proposal in aid of *HUSKY*, to send through a double agent in Britain information pointing towards an invasion of Sicily some days after the actual planned D-Day. Apparently not implemented.

FATHEAD   Dropping of the body of a Bengali Hindu with accoutrements proper to an agent, with a notional Burmese companion, and two containers, behind Japanese lines, November 1943. Aim: To provide Japanese commander in Akyab with a wireless set, cipher, and questionnaire, and hopefully open new BRASS type channel. No result; known that Japanese recovered corpse and searched vigorously for his notional companion.

FERDINAND   Deception plan for *DRAGOON*. "Story": South of France landings abandoned, landings to be made at Rimini and at Genoa (U.S. VI Corps), French II Corps and notional U.S. XXXI Corps in reserve for Alexander to exploit any breach in Gothic Line; British (notional) Twelfth Army and (near-notional) Ninth Army in Middle East in reserve to exploit any weakening of Germans in Balkans or Russian landing in Bulgaria. Related operations: OTTRINGTON, IRONSIDE II, BRAINTREE.

FERNBANK   Plan, September 1943 continuing into 1944, to induce U-boats to stay within depth suitable for depth charges, by passing "story" through special means that Royal Navy had developed rocket-propelled depth charge which was stronger, more accurate, would go deeper, but were ineffective at shallower depths.

FILM STAR   Supplementary operation to ROSEBUD. Second run-up to VIOLA III. Simulated notional 59th Division conducting dry run for VIOLA III; a combined signal exercise on assault force/divisional level.

FIREFLAME   Plan for deception operations in area of Rangoon River mouth to assist in capture of Rangoon, 1945.

FLABBERGAST   A D Division plan operated for only one month, September–October 1944, aimed at supporting a plan to retake central Burma in conjunction with a seaborne assault on Rangoon, by suggesting that the Allied main thrust would be towards Malaya by way of Sumatra. Replaced by STULTIFY when the actual plan was changed by eliminating the Rangoon landing and otherwise.

FLASHPOINT   U.S. Ninth Army crossing of the Rhine, March 21–April 1, 1945.

FLATON   Brief operation, August 1944, to protect certain sources that had been useful in Sicily by belittling through special means the efficiency of the Allied intelligence service in Sicily.

FLAXWEILER   A sonic deception by 23d Headquarters Special Troops to support a river crossing demonstration across the Moselle at Flaxweiler, Luxembourg, the night of January 17–18, 1945.

FLINTLOCK   Operations against the Marshall Islands, 1944.

FLOUNDERS   Follow-on to PURPLE WHALES.

FORAGER   Invasion of Marianas, 1944.

FORFAR   Part of STARKEY: Series of 14 commando raids on French Channel coast (only eight carried out) July–August 1943, called FORFAR BEER, FORFAR EASY, etc., to suggest that information was being sought in preparation for an invasion.

FORTITUDE   *NEPTUNE* deception plan.
  FORTITUDE NORTH: Notional invasion of Norway.
  FORTITUDE SOUTH: Notional invasion across Pas-de-Calais. Subsidiaries: QUICKSILVER I through VI, BIGDRUM, TAXABLE, GLIMMER, TITANIC I through IV, PARADISE ONE through FIVE. Related: GRAFFHAM, IRONSIDE, SKYE. Continuation: FORTITUDE SOUTH II.

FOYNES   Cover to conceal from Germans weakening of Allied forces in Italy resulting from redeployment of divisions and landing craft to England for *OVERLORD,* 1944, by largely notional transfer of troops to Mediterranean. Related: GOTHAM.

FRACTURE   U.S. naval diversionary operation to west of *HUSKY* assault beaches, in support of DERRICK.

FRANK   Supplementary operation to ROSEBUD. Simulated notional 59th Division conducting 48-hour command post exercise in Felixstowe area.

GALVANIC   Operations against Gilbert Islands, 1943–44.

GARFIELD   Operation, October 1943, to divert German attention from attack that crossed the Volturno, by threatening landing behind German right flank.

GARGOYLE   Part of London Controlling Section cover plan for *BRIMSTONE,* the canceled 1943 invasion of Sardinia: Plan to direct threat against Crete, to be managed by "A" Force (corresponding to original Stage A of WAREHOUSE). Related: CANUTE, JIGSAW.

GIBRALTAR COVER PLAN   (Given no specific codename.) Part of cover for *TORCH,* making Axis believe that accumulation of shipping and stores at Gibraltar for *TORCH* was to support expedition to relieve Malta.

GILMERTON   London Controlling Section 1943–44 plan to brief General von Cramer, a POW to be repatriated, as an unconscious agent as to Allied intentions to destroy completely all German industry by bombing. Abandoned because of opposition by Clement Attlee, the Deputy Prime Minister.

GLENANNE   A London Controlling Section project to support IRONSIDE II by having British ambassador ask Spanish Foreign Ministry whether Spanish government has heard anything about German intention to cross into Spain, impressing Spanish with the importance of the question to the British. Not clear whether ever implemented.

GLENDON   Notional convoys from Canada to India, 1944. Canceled.

GLIMMER   Air-sea operation at *NEPTUNE* D-Day simulating large assault convoy approaching Boulogne coast.

GLOSSOP   Fleming plan, December 1943, to cover operations to be carried out from India in spring of 1944. Aborted for lack of clear strategic decisions for Southeast Asia.

GOLDLEAF-HERITAGE   Cover plan for invasion of Madagascar (*IRONCLAD*), April–May 1942, portraying the invasion force as destined for operation against Dodecanese.

GOTHAM   Plan in support of FOYNES, December 1943, by appearing to be shipping landing craft into Mediterranean: dummy craft on decks of freighters passing Gibraltar. Failed because of weather and lightness of inflatable craft.

GOWRIE   A London Controlling Section scheme, never implemented, to contribute to deception of Germans as to timing of *NEPTUNE* D-Day. "Special means" at D-minus 30 would tell the Germans D-Day was D plus 45. At D minus 10 the British minister in Dublin would demand that the German embassy be closed because there is information that vital military information has been transmitted to Berlin. If the Irish refuse, threaten sanctions and ban shipping to Northern Ireland and travel between Northern Ireland and Eire.

GRAFFHAM   Effort spring 1944 to make Germans believe Allies enlisting cooperation of Sweden in connection with contemplated operations against northern Norway. Related operations: FORTITUDE NORTH, ROYAL FLUSH.

GRANDIOSE   Cover plan for Churchill's visit to Middle East, August 1942: Bogus itinerary passed through channels. Anticipated further development terminated in August on orders from London Controlling Section because *TORCH* upcoming.

GREMLIN   A project by Fleming in 1944–45, modeled on HICCOUGHS, to radio messages to a notional spy network in Japanese home islands. Never implemented.

GRIPFIX   Effort to get German garrison at Halfaya to surrender by dropping in a forged order from Rommel to do so, December 1941.

GUY FAWKES   Sabotage by setting off a bomb at a food dump at Wealdstone, November 1941, to enhance credibility of MUTT and JEFF.

H.D.F. ("HEADACHE FOR DER FUEHRER.") Attempt to upset morale and efficiency of 8th Panzer Regiment, April 1942, by passing into German hands a forged letter notionally from former commander (an anti-Nazi) to current commander hinting at anti-Nazi plots.

HAIRCUT   Supplementary operation to ROSEBUD. Simulated notional 293d Infantry Regiment with notional 467th Field Artillery Battalion of notional 17th Division practicing amphibious assault followed by ground exercise in Studland Bay area.

HALLUCINATE   An "R" Force tactical operation, August 1944.

HARDBOILED   Notional attack on Norway, spring 1942.

HARDIHOOD   Military aid to Turkey, 1943.

HARDMAN   Operation October–November 1943 to try to induce Germans to abandon line of the River Sangro by threatening seaborne and airborne landings near Pescara.

HARLEQUIN   Army training exercise loading troops on South Coast of England, run in conjunction with STARKEY, 1943.

HARLEY STREET   An "R" Force tactical operation: Notional reconnaissance near Nijmegen, October 1944.

HENGEIST   Operation, April 1943, to make Germans believe that attack of British 6th Armored Division at Fondouk was directed further south, towards Sfax.

HERITAGE   See GOLDLEAF-HERITAGE.

HICCOUGHS   Phony messages broadcast to Burma, 1942–45, on All-India Radio in a breakable cipher, suggesting existence of network of British agents in Burma and showing topics of purported interest to Allied intelligence.

HONEYSUCKLE   Supplementary operation to ROSEBUD. Third run-up to VIOLA I. Combined signal exercise on assault force/divisional level of notional British 55th Division making amphibious assault with naval Force "M" in Hayling Island area.

HOSTAGE   Tactical operation to help American attack from Gafsa towards Gabès to try to hold the panzer divisions on their front, in aid of Eighth Army's attack on Mareth Line, March 1943. "Story": U.S. II Corps will attack from Maknassy towards Mahares (well north of Gabes).

HOSTAGE   Diversionary deception operation of "R" Force to support Operation *EPSOM*, attack on Caen June 23–25, 1944.

HOSTAGE II   Diversionary deception operation of "R" Force to support Operation *GOODWOOD,* offensive in Caen area, July 1944.

HOTSTUFF   Radio implementation of QUICKFIRE. Cover for U.S. task force for *TORCH,* indicating by radio deception passage of U.S. task force from Norfolk to Syria via Trinidad, Recife, Cape Town and Aden.

HOUSEKEEPER   "R" Force radio representation, October 1944, of notional 76th Division moving into area behind where actual 49th Division moved out.

HUSBAND   Continuation of Kuriles threat after WEDLOCK. Began June 1944. Related operations: WEDLOCK, BAMBINO, *STALEMATE,* VALENTINE.

HUSKY   Invasion of Sicily, July 1943.

IAGO   Project, never implemented, to frame a suitable U-boat captain as a British agent by sending him secret ink messages and tipping off the Gestapo through an MI6 agent known to have been "blown."

ICEBERG   Seizure of Okinawa, 1945.

ICEMAN   Cover plan for *CARBONADO.* Alternative plan for operations against Hong Kong in autumn 1945, designed for security, in that the Chinese staff studied both plans without knowing which was the real one. Related: CONSCIOUS.

IDES   London Controlling Section 1943–44 plan (never accomplished) to have an airplane containing a mailbag of documents, including false documents, stolen. *Cf.* BLARNEY STONE.

IMPERIL   Special means, wireless, and other activity, October–November 1944, to increase apparent strength of 21 Army Group by representing the notional British II Corps as present in France.

INCLINATION   An "R" Force tactical operation: Radio representation of notional 27th Armored Brigade under Canadian 3d Division concentrating north of Maldegen, October 1944.

INDEPENDENCE   Plan for clearing the Garonne estuary by French troops, November 1944; canceled; replaced by *VERITABLE,* April 1945.

INFATUATION   Deception plan for Canadian operation against Walcheren, October–November 1944.

IRAN COVER PLAN   Rumors and leakages that a division was moving from India to Palestine, July–August 1941, as cover for occupation of Iran.

IRONCLAD   Invasion of Madagascar, March 1942.

IRONSIDE   Notional attack to take Bordeaux in conjunction with *NEPTUNE.*

IRONSIDE II   Notional seizure of Bayonne and Bordeaux by force direct from USA supported by French Resistance, July 1944.

JAEL   October 1943 plan to convince Germans cross-Channel invasion deferred until Germany weakened by bombing and operations in Balkans (main thrust) and Scandinavia. Forerunner of BODYGUARD.

JANTZEN   Genuine training exercise in South Wales, July 1943, embarking and disembarking assault troops on division scale, passed off as part of WADHAM.

JESSICA   Operation to pin down German troops on Franco-Italian frontier, while discouraging enemy debouchment in rear of 6th Army Group, 1944–45.

JIGSAW   Cover for *BRIMSTONE* by telling troops from UK and USA they were destined to reinforce North Africa. Related: CANUTE, GARGOYLE.

JITTERBUG   Supplementary operation to ROSEBUD. Simulated notional 17th Division conducting dry run for VIOLA II.

JONAH   Project to send through PURPLE WHALES a negative of a photograph of the Bombay Assault Training Area map in the SEAC war room showing the units of the bogus XX Indian Corps, late 1944. Not carried out.

JUPITER   Actual plan for attack on Norway, 1942.

JURISDICTION   A plan drafted at SHAEF Ops (B) in March 1945 and never adopted, designed to reduce German forces north of the Rhine by telling the Germans by special means that a two-pronged attack towards Frankfurt would be mounted, one prong by 12th Army Group from the Remagen bridgehead, and one by 6th Army taking the Saar, crossing between Karlsruhe and Mainz, and advancing on Frankfurt.

K-SHELL   Rumor, spread in Middle East in January 1941, that the British had a new type of shell, working by concussion.

KENNECOTT   Plan to make the Axis believe, once the *TORCH* convoys were discovered, that their objective was Sicily and the toe of Italy together with relief of Malta. Related plan: TOWNSMAN.

KINKAJOU Open aspect of proposed cover plan for *ANAKIM*, February 1943: Press manipulation and public statements suggesting offensive along lines of actual *ANAKIM* operation. Related: WALLABY.

KNIFEDGE Notional attack by First French Army in Mulhouse-Vesoul sector to contain maximum German forces east of the Rhine and south of Lauterbourg, December 1944.

KNOCKOUT Overall name for the series of notional SEAC operations involved in STULTIFY.

KOBLENTZ Operation, December 6–14, 1944, in which 23d Headquarters Special Troops simulated the 75th Division as part of a feint down the Moselle valley towards Koblentz; terminated by the Ardennes counteroffensive.

KODAK Operation, December 22–23, 1944, in which radio transmissions of the 80th Infantry and 4th Armored Divisions were simulated by 23d Headquarters Special Troops so as to confuse the Germans as to which were the real signals from those units.

LANDONVILLERS A 23d Headquarters Special Troops operation, January 28–February 2, 1945, imitating the 95th Division at Landonvillers near Metz, while the real division was replaced by the 26th Division.

LANGTOFT London Controlling Section operation from January 1943 to conceal Allied reduction in production of chemical weapons, discourage Germans from starting chemical warfare, and give them no excuse to relax their costly preparations for defense against chemical warfare.

LARKHALL London Controlling Section operation starting at beginning of 1943 to build up enlarged notional order of battle for American forces in UK. Associated operation for British order of battle: DUNDAS.

LEAPYEAR Fourteenth Air Force plan, early 1945, for controlling volume of high-priority traffic when a major operation was taking place. Never put into effect and rescinded June 1945.

LEEHOLME A London Controlling Section idea, never implemented, for a Norwegian officer, supposedly a liaison in Moscow, to send to his wife in Norway false messages in clear for the Germans to intercept.

LEEK I Signal exercise simulating entire British 52d Division in landing exercises with naval Force "V," May 4, 1944. Part of FORTITUDE NORTH.

LEEK II Signal exercise simulating full-scale landing exercise of British 52d Division and naval Force "V," May 10–11, 1944. Part of FORTITUDE NORTH.

L'ÉGLISE A 23d Headquarters Special Troops operation, January 10–13, 1945, to show the 4th Armored Division as moving to reserve near L'Église, Belgium, when it was actually moving to a new attack position east of Luxembourg.

LEMSFORD London Controlling Section project, October 1943, to induce Bulgaria to leave the Axis and/or bring Bulgarian troops home, and induce Germans to reinforce Bulgarian defenses, by passing information to Germans of planned heavy air raids, dropping of saboteurs, and eventual dropping of airborne forces at Sofia. Not carried out.

LEYBURN Miscellaneous plan in support of BARCLAY and STARKEY, May–August 1943, showing, by various means, interest in preserving works of art in notional target areas.

LIFEBELT Planned seizure of the Azores, spring 1943. Cover plan: DUMMER.

LIGHTFOOT Offensive beginning at El Alamein, 1942.

LOADER Cover plan for the King's visit to North Africa, June 1943.

LOADLINE Cover plan for Operation *INDEPENDENCE,* December 1944, to contain maximum number of enemy troops in Gironde South during operations against Gironde North, December 1944.

LOADSTAR Plan, April 1945, to cover *MILEPOST, FOGHORN,* and *ENDRUN,* transfer of Lend-Lease vessels to Soviets, by concealing location and movement of transferred vessels (West Coast and Alaska across North Pacific) and indicating South Pacific destination for vessels entering Pacific from Panama Canal. Plan coordinated with VALENTINE.

LOBSTER Decoy airfield to divert German attacks from the real airfield at Djedjelli near Bougie in Algeria prior to *HUSKY,* 1943.

LOCHIEL Tactical plan in support of Eighth Army's attack on Mareth Line, February 1943, telling Germans there would be no attack till French and Americans took Cairouan, with diversionary landing in Gulf of Gabès.

LOCHIEL II Tactical plan, follow-on to LOCHIEL but discarded in favor of HOSTAGE. Similar to LOCHIEL but would have featured a drive on Tunis by British First Army from Teboursouk.

LOCHINVAR   A 23d Headquarters Special Troops operation, March 1–11, 1945, near Saarlautern, designed to cover the withdrawal of the 94th Division into reserve.

LONGTOM   Operation proposed by Spruance in autumn 1944 for implementation July 1945, for seizure of a base on the China coast at Ningpo peninsula–Chusan Archipelago area (south of mouth of Yangtze).

MACHIAVELLI   Passing to Germans of phony charts of minefields on the east coast of England by TRICYCLE, January 1941.

MAIDEN   Plan to make Germans believe Auchinleck intended to visit London at scheduled time of opening of anticipated June 1942 offensive. Related plan: FABRIC.

MAINSTAY   An "R" Force signal exercise for the notional 7th Division, August 1944.

MAILFIST   Liberation of Singapore, planned for 1945.

MALICE   Notional codename for notional airborne and seaborne operation in Plan BLANDINGS, notionally postponed in April 1944.

MANDIBLES   Projected invasion of Dodecanese, spring 1941. Elements: *ARMATURE, CORDITE.*

MANNA   Actual operation for British force to occupy Athens on heels of retreating Germans, September 1944. Cover plans: UNDERCUT, SECOND UNDERCUT.

MALINGER   A D Division operation, January–March 1944. Part of UKRIDGE, simulating an airborne threat to the Mandalay area. Related operation: UKRIDGE.

MARCHIONESS   A fake message from Foreign Secretary Anthony Eden to British ambassador in Chungking, making prognostications about Italy, passed to the Japanese through PURPLE WHALES after the fall of Tunis in 1943.

MARÉCAGE   Fictitious threat to La Rochelle to cover Operation *VERITABLE,* clearing of Garonne estuary, April 1945.

MARTELLO   Subsidiary of BERTRAM. Concealment of move of British X Corps at Alamein by dummies and SUNSHIELDS. Associated operation: MURRAYFIELD.

MELTING POT   Subsidiary of BERTRAM. Concentration of dummy armor on left at Alamein covering movement of 10th Armored Division.

MERCURY   The "Southern Railway Plan": Leakage through controlled agent TATE of plans for troop movements to southeastern embarkation points, June 8–14, 1944.

MERZIG   Sonic element of second part of Operation 11TH PANZER DIVISION, February 13–15, 1945. 23d Headquarters Special Troops sonic display, simulating a concentration of armor some 15 miles southwest of Merzig, Germany.

MESPOT   Original name of FORTITUDE.

METZ I   A small effort by 23d Headquarters Special Troops to depict the 87th Division as still in Metz when it moved out to the Ardennes, December 28–31, 1944; part of ARDENNES.

METZ II   A small effort by 23d Headquarters Special Troops, January 6–9, 1945, to show the 90th Division as in reserve in Metz while it moved to the Ardennes.

MIDAS   Not strictly a deception scheme. A 1941 operation whereby twenty thousand pounds was transferred from the Germans through an innocent third party to TATE, supposedly to pay agents in England.

MIMSY   A May 1945 proposal by Fleming, modeled on the LLAMA affair in the Mediterranean, to afford the Japanese command a channel for communicating with the Allies through a blown double agent. Never implemented.

MINCEMEAT   Floating of a body bearing fake documents suggesting invasion of Sardinia and Greece onto Spanish shore in aid of *HUSKY,* 1943.

MINOTAUR   Planting through the Spanish embassy in London of a fake War Cabinet paper on the defense of open spaces against an airborne invasion, November 1941.

MULLINER   A D Division plan, summer 1944, to induce Japanese to contain ground and air forces outside Burma by threatening operations against the Andamans and Nicobars, with various suggestions (including a prediction by an Indian holy man) that operations during the monsoon season were contemplated. Approved by Mountbatten only in much reduced form from the original, and received no concentrated attention.

MUNASSIB   Subsidiary of BERTRAM. Double bluff at Alamein: Dummy guns in south, secretly replaced by real guns before battle, to encourage German view that main attack in south.

MURRAYFIELD Subsidiary of BERTRAM. Replacement of moved vehicles in MARTELLO at Alamein.

MUSTACHE Supplementary operation to ROSEBUD. First run-up to VIOLA III. Simulated regimental combat team of notional 59th Division practicing amphibious assault with naval Force "F" followed by ground exercise, in Felixstowe area.

NAN Supplementary operation to ROSEBUD. Simulated notional 17th Division conducting 48-hour command post exercise in Studland Bay–Wareham area.

NEPTUNE Landing in Normandy, June 1944.

NETTLE A D Division operation, March–April 1945, portraying actual move of 15th Indian Division from Madras to Ramree as an amphibious threat aimed at Bassein west of Rangoon. Cut back or terminated because of inadequate W/T simulation by Army signals troops.

NEWSPAPER London Controlling Section plan, never implemented, to smuggle into Lisbon a specially printed copy of some innocuous daily paper following a ban on export abroad of such papers.

NORTHWAYS Tactical plan in Italy, March 1944, mainly a security plan to cover move of Eighth Army headquarters from Adriatic coast to left center of the line near Cantalupo.

NUNTON Deception plan for *DIADEM*, April–May 1944. "Story": Alexander would conduct only a holding operation on the Gustav Line, main attack would be a landing on Civitavecchia coast north of Rome. Related operation: SPAM.

OAKFIELD Follow-on to FAIRLANDS, November 1943, revised January 1944 to cover *SHINGLE*. Threatened landings by Patton with U.S. Seventh Army at Pisa on west, by Anders with Polish Corps and British troops at Rimini on east, meeting at Bologna to seal off Germans in Italy. Tactical aspect: CHETTYFORD.

OAKLEAF Tactical plan in Italy, January 1945, to give impression that 15th Army Group's winter offensive had not been called off but was imminent, so Germans would not withdraw forces from line to form large central reserve.

OCTAGON Quebec Conference, September 1944.

OLYMPIC Planned invasion of Kyushu, November 1945.

Appendix I

Omega   Operation, April 17–25, 1945, to deter movement of German forces to South Germany by representing that forces freed by liquidation of the Ruhr Pocket would be used to drive for Berlin.

Omnibus   Double agent messages suggesting preparation for an attack on Norway, July 1941.

Onion I   Signal exercise simulating an amphibious exercise of naval Force "V" with British 52d Division, April 20, 1944. Part of Fortitude North.

Onion II   Signal exercise simulating an assault brigade of British 52d Division in landing exercises with naval Force "V," April 24, 1944. Part of Fortitude North.

Ottrington   Deception plan for assault on Gothic Line in Italy, August 1944, dovetailed with Ferdinand. Notional landing at Genoa to outflank Gothic Line, with coordinated attack on right towards Rimini. Terminated early August when Alexander decided to make real main attack on right. Follow-on operation: Ulster.

Overlord   Invasion of Western Europe and defeat of Germany.

Overthrow   Threat of operation to be mounted from England in October 1942 to seize and retain a bridgehead in the Pas-de-Calais, with the object of deterring Germans from reinforcing Russian front or Mediterranean; part of deception for *Torch*. Originally called Passover, then Steppingstone. Related: Cavendish, Kennecott, Pender, Quickfire, Solo, Sweater, Townsman.

Padlock   Original name of Quickfire.

Paprika   A Twenty Committee plan in 1942 to cause friction in the German higher echelons in Belgium by sending messages from which it might be inferred that certain high officials were plotting to make contact with the British to open peace negotiations. Never implemented because of disagreements as to its political wisdom.

Paradise   Overall code word for dummy lighting of *Neptune* beaches and beach exits to decoy German bombing, beginning night of D plus 1.
    Paradise One: Dummy beach exits east of Asnelles-sur-Mer.
    Paradise Two: Dummy beach exit near Langrun.
    Paradise Three: Dummy beach between Arromanches and Port-en-Bessin.
    Paradise Four: Dummy beach between rivers Orne and Dives.
    Paradise Five: Dummy bridges in areas of bridges at Benouville and Ranville.

Passover   Name originally proposed by Bevan for Overthrow; rejected. Replaced by Steppingstone, also rejected.

Appendix I

PASTEL   The original deception plan for Operation *OLYMPIC,* the invasion of Kyushu scheduled for November 1945. Replaced by PASTEL Two to add notional airborne operations.

PASTEL TWO   Revision of PASTEL, final deception plan for *OLYMPIC,* the invasion of Kyushu, November 1945. "Story": Landings planned in Chusan-Shanghai area for *OLYMPIC* D minus 30, preparatory to further advance into Yellow Sea area; then beginning D minus 55, "story" changed to amphibious and airborne seizure of Shikoku to set up airfields for air coverage of future operations against the home islands.

PATENT   Plan to ease pressure on supply convoys to Malta by threats to invade Dodecanese (subsequently Crete), June–July 1942. Merged thereafter into RAYON.

PENDER   Part of Gibraltar cover in *TORCH.* PENDER I: rumor that Eisenhower recalled to Washington for important conferences and could not be in Gibraltar. PENDER II: Rumor that Cunningham in Gibraltar en route for Lagos, ultimately to take up appointment as C-in-C Far Eastern fleet.

PENKNIFE   Plan, February–March 1945, to conceal withdrawal of various Canadian and British forces from Italian front to Northwest Europe.

PEPPER   A 1941 plan to discredit the German consul-general in Lisbon on the basis of information from the double agent CELERY. Result unknown.

PIPPIN   Deceptive signal operation to cover relief of 5th Indian Division and 11th East African Division by 2nd Division and 28th East African Brigade at crossing of the Chindwin in 1944. Related operations: CLOAK, STENCIL.

PLAN A.B.   Planting of a fake listing of divisional insignia through the Spanish embassy in London, August 1941.

PLAN I   Project, January–March 1941, to test whether the Germans had sufficient faith in the double agents to base an operational action on their reports, by informing them through TATE that a hidden ammunition dump had been found in the New Forest. Actual dummy dump was built, but no reconnaissance or other action by the Germans resulted.

PLAN IV   A 1941 Twenty Committee operation: Passing through the Spanish embassy of a file of phony air raid damage reports designed to lead the Germans to concentrate their bombing on well-defended air bases rather than on cities.

PLAYBOY-COMPASS   (Originally PLAYBOY, renamed COMPASS.) Project, February–August 1943, to use special means channels to lure a U-boat alongside a boat with trick keel loaded with explosives which could be slipped and magnetically attached to U-boat, subsequently sinking it. Many experiments but never implemented.

PLAYMATE   Plan, March 1945, to cover 15th Army Group spring offensive. "Story": Bologna to be outflanked to the east, including small amphibious operation, rather than attacked from the south.

PLUNDER   21 Army Group crossing of Rhine, March 1945.

POINTBLANK   The Combined Bomber Offensive, the Anglo-American strategic bombing campaign against Germany.

POUND   Commando raid on Ushant designed to lend plausibility to WADHAM.

PREMIUM   Special means "story" about interest of allied force in insurance records in Norway and Pas-de-Calais, March–April 1944.

QUADRANT   Quebec conference, August 1943.

QUAKER   Radio cover plan for British Second and Canadian First Armies as they prepared for D-Day. (Not part of FORTITUDE.) Purpose: Eliminate radio silence before D-Day, cover final moves of assault corps to embarkation points, avoid prolonged radio silence if invasion postponed. U.S. counterpart: WILLIAMS.

QUICKFIRE   (Originally PADLOCK.) September 1942 cover plan for TORCH forces from U.S.: Troops were told that they were an expeditionary force to occupy Syria and Cyprus. Associated plan: SWEATER.

QUICKSILVER   Overall codename for the various operations under FORTITUDE SOUTH that implemented the threat to the Pas-de-Calais area either before or after D-Day.
    QUICKSILVER I: The basic "story" for FORTITUDE SOUTH: FUSAG (Ninth Air Force associated), in southeast of England, to land in Pas-de-Calais after German reserves were committed to Normandy.
    QUICKSILVER II: Radio deception plan to shows FUSAG concentrated in southeast. Included CENT and DOLLAR.
    QUICKSILVER III: Display of dummy landing craft, including associated simulated wireless traffic and signing of roads and special areas.
    QUICKSILVER IV: Air plan for QUICKSILVER: Consisted of FORTITUDE air plan plus bombing of Pas-de-Calais beach area and tactical railway bombing immediately before D-Day.
    QUICKSILVER V: Increased activity around Dover (giving impression of extra tunneling, additional wireless stations), to suggest embarkation preparations.

QUICKSILVER VI: Night lighting to simulate activity at night where dummy landing craft were situated.

RAINDROP   "R" Force operation June 12, 1944, to draw German attention away from an attack by 51st (Highland) Division by suggesting an attack eastwards to link up with possible landings at Le Havre.

RAMSHORN   Fleming's July–December 1943 operation to divert Japanese forces away from planned Allied offensive in northern Burma and against Akyab and Ramree, by threatening Andamans and Sumatra.

RASHNESS   Overland operation against Canton and Hong Kong, autumn 1945. Formerly *BETA;* name changed again to *CARBONADO,* May 1945. Related plan: ICEMAN.

RASPUTIN   Plan proposed by Fleming in March 1943, to convey to Japanese that War Cabinet had decided to employ gas warfare against Japan when expedient after the elimination of Germany, to create uneasiness at high levels and divert Japanese shipping space, trained personnel, etc. Rejected by Bevan because it cut across LANGTOFT and might induce the Germans and Japanese to initiate gas warfare.

RASSENDYLL   A copy of *The Water Babies* in which were concealed codes, schedules, and intelligence data, was dropped to the BRASS party, directing them to mail it back to Japan; hoped Japanese would be induced to establish a notional subsource in Tokyo, but they did not.

RAYON   Notional invasion of Crete, August–September 1942, to cover Malta convoy and prevent further reinforcement of Rommel from Crete.

RED BAR   A series of Royal Navy messages, early in the war, in a cipher a copy of which had been sold to the Italians.

REMUS   Tactical deception plan in support of *ROMULUS,* December 1944.

RHINO   Plan for radio communication between Subhas Chandra Bose and the All India Revolutionary Committee, 1944.

RISSOLE   London Controlling Section 1943–44 double-bluff plan (never accomplished); a reverse MINCEMEAT: Planting of a body with obviously fake documents pointing to the real objective.

RITZ   London Controlling Section project (never accomplished) to have a British official leave false documents in an unlocked bag in the Ritz Hotel, Madrid, where it had been learned that all luggage was searched by German agents.

ROMULUS   Recapture of Arakan coast, Akyab, and Ramree, 1944–45.

ROOFTREE   A D Division operation, May 1944, to support a carrier strike against Sabang by suggesting to the Japanese that the target was Port Blair.

ROSCOR   A D Division operation, November 1943–January 1944, using W/T deception to build up the apparent sea and air strength of the British Eastern Fleet so as to discourage the Japanese fleet at Singapore from sortieing into the Bay of Bengal.

ROSEBUD   Signal deception supporting FORTITUDE SOUTH II, simulating FUSAG from July 21, 1944. Originally scheduled for fifty days but terminated August 17. Supplementary operations: VANITY I, VANITY II, HONEYSUCKLE, VIOLA I, HAIRCUT, JITTERBUG, VIOLA II, MUSTACHE, FILM STAR, VIOLA III, FRANK, NAN.

ROUBLE   Notional command post exercise for notional U.S. tactical air command held July 26–28, 1944 linking it with notional U.S. Fourteenth Army in England and its notional corps and divisions.

ROYAL FLUSH   Diplomatic demarches on neutrals in support of BODYGUARD: Sweden, Turkey, Spain, suggesting operations in Norway, through Turkey, south of France. Related operation: GRAFFHAM.

RUHR POCKET   Operation, March 28–April 9, 1945, to induce Germans to hold troops along the Rhine in the Ruhr Pocket by threatening crossing in the vicinity of Cologne by U.S. Fifteenth Army.

RUSTIC   Deceptive measures in 1945 designed to further overall policy of preventing diversion of German forces from Italy and the Balkans to the West and preventing formation of general strategic reserve in the south. Not a formal plan.

RUTHLESS   Proposed operation, autumn 1940, to capture a German air-sea rescue boat by crashing a captured Heinkel in the Channel. Not carried out.

"S" FORCE   (Not a deception operation.) Dash into Tunis and Bizerte by Tac H.Q. "A" Force, augmented, to seize intelligence materials and facilitate entry of Allied intelligence and military government personnel.

SCEPTICAL   April 1945 deception plan to cover *ZIPPER*, landing on Malaya coast. Notional assaults against Thailand (overland and airborne against Bangkok) and Singapore (establish bases on Sumatra and Java, then attack on Singapore through Straits of Malacca). Related operations: CONSCIOUS, SLIPPERY.

SECOND UNDERCUT Deception plan for *MANNA*, October–November 1944, aimed at exaggerating size of British force moving into Greece.

SEE SAW Signal deception June 1–3, 1944, supplementing QUICKSILVER I. Represented U.S. 28th Division conducting three-day amphibious assault exercises with Royal Navy force.

SENTINEL Tactical plan with dummies and other devices, to make the Nile Delta appear more strongly defended than it was, July–September 1942.

SEXTANT Cairo Conference, November–December 1943.

SHELLAC Plan, December 1944, to cover Fifth Army assault on Bologna. Mainly a simulated concentration on right flank near Rimini.

SHINGLE Anzio landing, January 1944.

SHOTGUN London Controlling Section 1943–44 plan (never accomplished) to pass to Germans a forged document of a technical nature on the state of equipment of the British Army, exaggerating its manpower and war potential.

SKYE Radio deception component of FORTITUDE NORTH with respect to British Fourth Army. SKYE I was Fourth Army headquarters, SKYE II was the British II Corps, SKYE III was the American XV Corps, SKYE IV was the British VII Corps.

SLIPPERY Notional attack on Kra Isthmus to cover accumulation of landing craft for *ZIPPER* (landing on Malaya coast). Absorbed into SCEPTICAL.

SLY-BOB Dummy submarine project, Gulf of Suez, February 1943.

SNOWSHOE II Wireless deception (winter 1944–45), to indicate two notional British divisions undergoing mountain training in Rieti area.

SOCIETY Cover plan for movement of British 5th Division from India to Iran, September 1942.

SOLO I Notional landings at Trondheim and Narvik, autumn 1942; part of *TORCH* cover.

SOLO II Cover plan for *TORCH* assault forces from Britain; troops were told they were headed for Middle East but would take Dakar first. See also SWEATER.

SPAM   A suggestion by Ewen Montagu and Noel Wild, not adopted, to aid FORTI-
TUDE SOUTH by dropping a corpse in the Channel carrying a notebook suggesting
that NEPTUNE was just an exercise.

SPAM   Demonstration near Civitavecchia, May 24, 1944, in aid of NUNTON.

SPANNER   Plan proposed by Montagu to Twenty Committee in July 1942, when
vigorous British protests to Spain about German intelligence activity in the Strait of
Gibraltar were producing results, to cover the fact that ULTRA was the source of
British information, by having a double agent report that a Foreign Office official
had said that a German agent in the Straits had been bribed to betray the activity.
Not implemented. Related plan: ASSASSIN.

STAB   Original name for SWEATER.

STALEMATE   Invasion of the Palaus, 1944.

STARKEY   (Originally DOMESDAY, then BROADSWORD.) Part of COCKADE: Notional attack by fourteen British and Canadian divisions to establish a bridgehead
on either side of Boulogne between September 8 and 14, 1943. Related operation:
LEYBURN.

STEINSEL   A 23d Headquarters Special Troops operation, January 27–29, 1945,
imitating the radio of the 4th Division around Steinsel, Luxembourg, while the real
division moved north towards Houffalize.

STENCH   A project, October 1941, to misdirect German poison gas research by
passing to the Germans a fake British gas mask employing crystals of a new kind and
seemingly protecting against a gas that would affect the ears. The Germans showed
little interest in the offer of such a mask.

STENCIL   Deceptive signal operation, part of CLOAK, simulating HQ IV Corps remaining in previous position to cover right hook down Gangaw Valley, January
1945. Related operations: CLOAK, PIPPIN.

STEPPINGSTONE   Name proposed by Bevan for OVERTHROW after rejection of
PASSOVER.

STIFF   Twenty Committee project, 1941, to drop a W/T set, instructions, and ciphers, and perhaps a dead body, for Germans to find, in hopes they would play back
the set. Never implemented.

STULTIFY   (Replaced FLABBERGAST.) Deception plan for SACSEA 1944–45 strategy. "Story": Mountbatten planned to recapture Akyab in early February 1945, and

then in March or April seize by seaborne and air assault a position running from the lower Irrawaddy at Prome to the coast at Taungup, with subsequent exploitation to Rangoon. This would be supported by "a cautious but much publicized overland advance" (the genuine CAPITAL) in northern and central Burma, and followed in April by an operation to seize naval and air bases in northwestern Sumatra, for occupying and immediately developing rubber-producing areas, followed by exploitation to Java. Subsidiaries (collectively called KNOCKOUT): FANG, CLAW, CAPTION, ARAMINTA, WOLF. Merged into SCEPTICAL.

STURGEON   Cover plan, January–June 1945, for transfer of XX Bomber Command from China to the Pacific beginning January 1945. "Story": B-29s were being moved from China to India, perhaps ultimately to the Philippines, where they would have better logistical support, would be used to strike south China and Southeast Asia targets.

SWEATER   (Originally STAB.) Cover for TORCH: TORCH troops from U.S. were told that objective was tropical training in Haiti.

SYMBOL   Casablanca Conference, January 1943.

TAPER   Notional airborne landing north of Cologne, March 1945, to cover VARSITY.

TARZAN   Airborne threat to route through southern Shan States into north Thailand, 1945. (Name originally assigned to genuine proposed overland advance in North Burma, 1944.)

TAXABLE   One of the NEPTUNE naval diversions, simulating large assault convoy approaching Cap d'Antifer-Fécamp coast. Linked with TITANIC I.

TELESCOPE   Original name of COPPERHEAD.

10TH ARMORED DIVISION   Plan to make the enemy believe Wavell had an additional armored division, February 1941. Merged into A-R PLAN in March.

TERMINAL   Potsdam Conference, July–August 1945.

THERAPY   Misleading of Germans as to British night fighter devices, from January 1942.

TINDALL   (Originally UPSHOT.) Part of COCKADE: notional attack on Norway, four divisions from the sea and an airborne division to seize Stavanger; activated July 1943, stood down end of August, allegedly to release forces for STARKEY, reactivated after STARKEY as a threat in being till onset of winter (TINDALL II).

TITANIC   Overall code word for the deceptive airborne operations under FORTI-
TUDE in direct support of *NEPTUNE* landings.

TITANIC I: D-Day dummy parachutist and SAS drop at Yvetot, thirty miles
southwest of Dieppe, simulating dropping of an airborne division north of the
Seine to retain German forces north of Seine and draw northwards reserves
south of Seine.

TITANIC II: D-Day dummy parachute drop east of Dives River to delay local reserves immediately east of Dives from moving westwards. Canceled.

TITANIC III: D-Day dummy parachute drop southwest of Caen at same time as
dropping of British 6th Airborne Division.

TITANIC IV: D-Day dummy parachute drop at Marigny on base of Cotentin at
same time as U.S. 101st Airborne drop, to draw westwards German counterattack forces in area of St. Lô.

TORCH   Invasion of French North Africa, November 1942.

TORRENT   Original name for COSSAC cover and deception plan for *NEPTUNE;*
subsequently APPENDIX Y.

TOWNSMAN   "A" Force share in KENNECOTT, supporting notional objective of
TORCH convoys as Sicily and the toe of Italy, plus diversionary attack on Crete.

TRANSATLANTIC TELEPHONE PLAN   London Controlling Section plan, never implemented, to fake conversations over the transatlantic telephone.

TRANSCEND   21 Army Group cover operation, March 1945, to cover *PLUNDER* by
making Germans defend against notional crossings of Rhine at Emmerich and in
Düsseldorf area.

TREATMENT   Strategic deception plan for Alamein, 1942. "Story": British planned
to invade Crete, but meanwhile Eighth Army would attack at Alamein, to improve
its position, during moonless period around November 6 (actually planned full-
scale attack in full moon October 23). Associated tactical plan: BERTRAM.

TRIDENT   Washington Conference, May 1943.

TRIPOLI PLAN   Continuation of threat to Tripoli begun in PLAN A-R, in support of
COLLECT, July 1941.

TROJAN HORSE   Original name for MINCEMEAT.

TROLLEYCAR   (Subsequently TROLLEYCAR II.) 21 Army Group operation November–December 1944 to induce Germans to hold back reserve, mainly in Holland,

to counter notional attack by First Canadian Army west of Arnhem in conjunction with sea and airborne attack from U.K.

TROUSERS   An "R" Force tactical deception operation to cover Canadian First Army's attack on the Falaise pocket, August 1944, by inducing the Germans to retain their armor on the right flank.

TROUTFLY   Tactical support of 82d Airborne at Ste. Mère Église on NEPTUNE D plus 1, by simulation through dummy radio traffic of a regimental combat team of the 9th Division. Failed because ship carrying troops and equipment was misdirected.

TURBINATE   21 Army Group operation, January 1945, to cover VERITABLE by making Germans believe 21 Army Group to undertake operation against Northern Holland (Utrecht) to relieve London from V-bomb attacks.

TURPITUDE   Eastern Mediterranean element of ZEPPELIN Stage 4 implementation, complement to VENDETTA, May–June 1944. Notional landing at Salonika and thrust up Struma valley to link with Russian forces to land at Varna; preparatory facilities to be set up in Turkey and preliminary assault on Rhodes. Related: VENDETTA, ZEPPELIN, DUNGLOE.

TWEEZER   Plan issued June 8, 1944, for substituting II British Corps in FUSAG to replace Canadian II Corps and U.S. XX Corps actually sent to Normandy.

TWELVEBORE   Operation of uncertain nature in 21 Army Group area, November 1944.

UKRIDGE   A D Division operation, January–March 1944, to deter Japanese reinforcement of Burma by maintaining the illusion that amphibious operations were to be mounted outside Burma (Andamans, Ramree, northern Sumatra), and focus Japanese attention on central Burma by simulating an airborne threat to the Mandalay area (Operation MALINGER). Related operation: MALINGER.

ULSTER   Double bluff, August 1944, growing out of OTTRINGTON, after genuine attack was set for Adriatic side rather than center.

UNDERCUT   Deception plan for MANNA, September 1944, aimed at concealing objective of MANNA from Greeks themselves. Follow-on: SECOND UNDERCUT.

UNNAMED PLAN   Cover plan for BATTLEAXE offensive in Western Desert, June 1941. Abortive.

UNNAMED PLAN   London Controlling Section plan, never implemented, to lose compromising documents in Lisbon and Madrid.

APPENDIX I

UNNAMED PLAN   London Controlling Section plan to make enquiries through the Fire Offices Committee, asking for information regarding certain areas in Europe. Not originally implemented but later became PREMIUM.

UNNAMED PLAN   London Controlling Section plan, February 1943, to reduce the Italian will to resist by holding out prospects of reasonable peace terms. Disapproved by the Foreign Office.

UNNAMED PLAN   Plan to lead Italians to believe Wavell weakening his forces in Libya to provide an expeditionary force for Greece, end 1940.

UNNAMED PLAN   London Controlling Section proposal, March 1943, to reduce output from Ploesti oil fields and discourage air defense of the fields by leading Germans to believe that there would be no air raids on them, while issuing propaganda to the oil workers threatening such raids; not adopted.

UNNAMED PLAN   Unnamed cover plan for *VAULT*, a proposed quick seizure of the Azores by a small force in June 1943. Troops to be told they were proceeding to the Mediterranean to take over garrison duties on Pantelleria. Operation VAULT never took place.

UPSHOT   Original name of TINDALL.

VALENTINE   Final, modest early 1945 Kuriles deception, beginning December 23, 1944. Terminated WEDLOCK. Related operations: WEDLOCK, BAMBINO, HUSBAND, BLUEBIRD.

VAN DYKE   Signal deception June 16–20, 1944, supplementing QUICKSILVER I. Represented joining up of air support parties of XIX Tactical Air Command with Third Army.

VANITY I   Supplementary operation to ROSEBUD. First run-up to VIOLA I. Simulated assault brigade of notional British 55th Division making amphibious assault with naval Force "M" in Hayling Island area.

VANITY II   Supplementary operation to ROSEBUD. Second run-up to VIOLA I. Simulated another assault brigade of notional British 55th Division making amphibious assault with naval Force "M" in Hayling Island area.

VARSITY   Allied airborne drop in vicinity of Wesel to accompany assault across the Rhine.

VENDETTA   Western Mediterranean element of ZEPPELIN Stage 4 implementation, complement to TURPITUDE, May–July 1944. Notional assault on Narbonne

841

region by U.S. Seventh Army, exploiting inland to Toulouse; one real, one notional French corps, one notional U.S. corps; eight notional, four real divisions. Related plans: IRONSIDE II, TURPITUDE, ZEPPELIN.

VERITABLE 21 Army Group offensive through Reichswald, February 1945.

VERITAS Carrier-borne air reconnaissance operation off Narvik, April 26, 1944, in aid of FORTITUDE NORTH.

VIERSEN A 23d Headquarters Special Troops operation, March 18–24, 1945, in connection with EXPLOIT, the deception plan for Ninth Army's crossing of the Rhine, simulating the 30th and 79th Divisions in assembly areas in XIII Corps zone. Related operation: EXPLOIT.

VIOLA I Supplementary operation to ROSEBUD. Simulated notional British 55th Division making amphibious assault with naval Force "M" followed by a ground exercise in Hayling Island area, showing lateral links relating the force to the task force commander. Preliminary run-ups: VANITY I, VANITY II, HONEYSUCKLE.

VIOLA II Supplementary operation to ROSEBUD. Simulated notional U.S. 17th Division making amphibious assault with naval Force "N" followed by a ground exercise in Studland Bay area, showing lateral links relating the force to the task force commander. Preliminary run-ups: HAIRCUT, JITTERBUG.

VIOLA III Supplementary operation to ROSEBUD. Simulated notional U.S. 59th Division practicing amphibious assault followed by ground exercise in Felixstowe area, showing lateral links that established position of naval Force "F" in task force for FORTITUDE SOUTH II landing. Preliminary run-ups: MUSTACHE, FILM STAR.

WADHAM (Originally BLAST.) Part of COCKADE: Notional landing September 22, 1943 (when STARKEY would have drawn in German reserves) by a U.S. corps of five divisions (largely notional) from England to capture Brest, followed by U.S. VII Corps of seven divisions direct from U.S. to exploit; to be notionally called off and replaced by TINDALL.

WALLABY Secret aspect of proposed cover plan for *ANAKIM*, February 1943: Special means and misleading activity to persuade Japanese that Allies planned offensive against Sumatra and Straits of Malacca and that public discussion (KINKAJOU) was a cover operation. Related: KINKAJOU.

WANTAGE Enlargement of notional order of battle in Mediterranean, February 1944–May 1945, to induce Germans to believe that there were sufficient reserves in both eastern and western Mediterranean to undertake large-scale operations against southern Europe.

WAR OF NERVES Rumor program, February 1941, designed to worry the enemy and make him dissipate his forces in the Mediterranean.

WAREHOUSE Post-Alamein notional threat to Greece ("Stage 'A'," notional landing on Crete scheduled for January 1943, as first step), to contain Axis forces in Balkans and Aegean, November 1942. "WAREHOUSE (1943)" drafted March 1943, but replaced by BARCLAY after *BRIMSTONE* cancellation and *HUSKY* decision. Related plan: WITHSTAND.

WARHORSE An "R" Force radio deception operation to attract German attention to the Nijmegen area, November–December 1944, canceled when the Germans broke the dikes and flooded the terrain.

WATERFALL Massive display of dummies in Cyrenaica to give the impression of large force poised to invade Crete and the Peloponnese in support of BARCLAY. Related operation: BARCLAY.

WEDLOCK Notional threat to Kuriles to aid Marianas invasion, spring–summer 1944. Related operations: BAMBINO, *FORAGER,* HUSBAND, *STALEMATE,* VALENTINE.

WETSHOD Original name of DOLLAR.

WHIPSAW Sonic and visual element of first part of Operation 11TH PANZER DIVISION, February 1–4, 1945. 23d Headquarters Special Troops presented sonic deception representing three tank battalions near Grevenmacher and Wormeldingen east of Luxembourg, while two dummy field artillery batteries were displayed around Saarlautern.

WHITWOOD Minor plan, November 1943–January 1944, to cover purpose of antiaircraft units and ground crews gathered on Turkish-Syrian border in hopes that Turkey would allow Allies to use airfields in southwestern Turkey.

WILDERNESS I Wireless simulation of an amphibious assault landing exercise by the Canadian 1st Infantry Division south of Salerno April 1–2, 1944, in aid of NUNTON.

WILDERNESS II Wireless simulation of an amphibious assault landing exercise by the U.S. 36th Division under Canadian I Corps in the Gulf of Pozzuoli April 6–7, 1944, in aid of NUNTON.

WILLIAMS   Radio cover plan for U.S. First Army as it prepared for NEPTUNE D-Day. (Not part of FORTITUDE.) Simulated First Army from May 9, when real First Army went off air, to June 6 (radio silence throughout U. K. May 27–30). British counterpart: QUAKER.

WILTZ   An operation of 23d Headquarters Special Troops, October 4–10, 1944, simulating 5th Armored Division, to cover its movement northward from Luxembourg to Malmédy.

WINDSCREEN   Plan (three different plans, as situation changed) to try to jockey Rommel out of Agheila and then Buerat positions, in January 1943.

WITHSTAND   From January 1943, spreading "story" that Allies so concerned over possible invasion of Turkey that they planned preemptive attack against Dodecanese via Crete. Follow-on and intensification of WAREHOUSE.

WOLF   Subsidiary plan to STULTIFY, to correlate the factual advance in Arakan with the KNOCKOUT threat against Southwest Burma, December 1944.

WORKHOUSE   Plan, January–August 1943, to make Axis believe the Straits of Hormuz had been mined.

WYANDOTTE   Idea, December 1942, for plan never implemented, to make Japanese believe Q-ships being operated on sea-lanes between Australia and South Africa.

XMAS PLAN II   Original name for ADVOCATE.

ZEPPELIN   Overall Mediterranean plan, February–July 1944, for keeping German forces in south of France and Balkans from reinforcing Normandy. Complex, five stages. Elements: VENDETTA, TURPITUDE. Related plan: DUNGLOE.

APPENDIX II

Special Means Channels

Note: Some double agent channels are included that were used primarily for penetration rather than deception purposes. A few captured agents who for one reason or another never became successful double agents, as well as some proposed channels that did not materialize, are also included.

(UNIDENTIFIED) An "A" Force channel in Istanbul to a Central European German-language newspaper (perhaps the *Pester Lloyd*).

ABDEL A source for PESSIMIST Y *(q.v.)*, a notional Damascus businessman with political connections.

ACCOST A party of six Annamese dropped by MI6 some thirty-five miles north of Saigon in July 1945; captured by the Japanese and turned over to D Division for deception. The Japanese surrender forestalled its use.

ADDICT An "A" Force (French-controlled) channel. Hungarian, naturalized Italian. A staybehind in Rome who transmitted medium-grade information to the Germans from June 1944 to the end of the war.

AGENT J1 A subagent of GARBO *(q.v.)*. A notional KLM airline official, supposedly the courier who arranged for GARBO's messages to be carried from Britain to Lisbon.

AGENT J2 A source for GARBO *(q.v.)*. A notional talkative RAF officer employed at Fighter Command headquarters. Originally, "J2" designated a notional KLM pilot who was supposedly a backup courier; he was soon dropped.

AGENT J3 A source for GARBO *(q.v.)*. A notional highly-placed official in the Spanish section of the British Ministry of Information, who believed GARBO to be an anti-Axis Spanish Loyalist refugee, spoke freely with him, often shared sensitive information, and eventually procured part-time employment in the Ministry for him.

AGENT J4   A source for GARBO *(q.v.)*. A notional left-wing employee of the censorship department of the Ministry of Information.

AGENT J5   A source for GARBO *(q.v.)*. A notional secretary in the "Ministry of War," a bit in love with GARBO.

AGENT ONE   Known as CARVALHO. A subagent of GARBO *(q.v.)*. A notional Portuguese traveling salesman based at Newport in Wales, positioned to observe traffic on the Bristol Channel and to make contact with Welsh dissidents.

AGENT TWO   Known as GERBERS. A subagent of GARBO *(q.v.)*. A notional British subject of German-Swiss descent, a businessman in Bootle, positioned to observe traffic in the Mersey. Notionally died of cancer in the autumn of 1942 before he could discover the buildup of shipping for TORCH. His work carried on thereafter by AGENT 2(I), his notional widow. See AGENT 2(I); GARBO; GERBERS, MRS.; WIDOW, THE.

AGENT 2(I)   Known as MRS. GERBERS or THE WIDOW. A subagent of GARBO *(q.v.)*. Notional widow of GARBO's AGENT TWO *(q.v.)*. Carried on her husband's work after his notional death and became in effect GARBO's personal assistant.

AGENT THREE   A subagent of GARBO *(q.v.)*. A notional rich Venezuelan student in Glasgow, positioned to observe shipping in the Clyde. Efficient and able; became GARBO's second in command and took over the GARBO network after GARBO went into hiding in 1944. Maintained a network of three subagents or sources, AGENTS 3(I) through 3(3) *(qq.v.)*. Brother of AGENT FIVE, MOONBEAM *(q.v.)*.

AGENT 3(I)   A source for GARBO's AGENT THREE *(q.v.)*. A notional noncommissioned officer in the RAF.

AGENT 3(2)   A source for GARBO's AGENT THREE *(q.v.)*. A notional talkative lieutenant in the British 49th Division.

AGENT 3(3)   A subagent of GARBO's AGENT THREE *(q.v.)*. A notional Greek sailor in Scotland, a Communist deserter from the British Merchant Navy who believed that in working for GARBO he was working for the Soviets.

AGENT FOUR   Known as FRED. A subagent of GARBO *(q.v.)*. A notional Gibraltarian. Originally a waiter in Soho; then directed by the Ministry of Labor to work, first, in an imaginary underground munitions depot in the Chislehurst caves, then in military canteens, and then in a sealed military camp near Southampton, out of which he broke to give GARBO warning, just too late, of the launching of the Normandy invasion.

AGENT 4(1)   GARBO's notional radio operator. Recruited by FRED, GARBO's AGENT FOUR *(q.v.)*. A notional Spanish Republican friend of FRED with access to an amateur radio transmitter, who thought the enciphered messages he sent were being directed to an anti-Franco underground in Spain. Operated from March 1943. Known to the Germans as ALMURA.

AGENT 4(2)   A source for FRED, GARBO's AGENT FOUR *(q.v.)*. A notional guard in the notional subterranean munitions depot in the Chislehurst caves.

AGENT 4(3)   A source for FRED, GARBO's AGENT FOUR *(q.v.)*. A notional sergeant in the American Services of Supply.

AGENT FIVE   A subagent of GARBO *(q.v.)*. Notional brother of AGENT THREE *(q.v.)*. Supposedly moved to Toronto as a traveling salesman in 1943, where he was codenamed MOONBEAM (FBI codename GLOCASE). Reported initially by secret writing; acquired a radio operator in Montreal in early 1945. See GARBO; MOONBEAM; AGENT 5(1).

AGENT 5(1)   Known as CON. A subagent of MOONBEAM, GARBO's AGENT FIVE *(q.v.)*, during his period in Canada; an imaginary cousin of MOONBEAM, living in Buffalo.

AGENT SIX   Known as DICK. A subagent of GARBO *(q.v.)*. A notional anti-British South African whose linguistic skills gave him employment in the War Office and had good contacts in various government departments. Killed in North Africa in July 1943.

AGENT SEVEN   Known as STANLEY. A subagent of GARBO *(q.v.)*. A notional seaman in Swansea, an anti-English Welsh nationalist. Head of a substantial network, AGENTS 7(1) through 7(7) *(qq.v.)*, codenamed DAGOBERT by the Germans.

AGENT 7(1)   A source for STANLEY, GARBO's AGENT SEVEN *(q.v.)*. A notional soldier in the British 9th Armored Division.

AGENT 7(2)   Known as DAVID. A subagent of STANLEY, GARBO's AGENT SEVEN *(q.v.)*. A notional retired seaman operating in the Dover area. Supposedly the leader of the Aryan World Order, an imaginary subversive movement in Britain. Known to the Germans as DONNY.

AGENT 7(3)   Known as GLEAM. An imaginary subagent of STANLEY, GARBO's AGENT SEVEN *(q.v.)*. Also known as THERESA JARDINE. Known to the Germans as JAVELINE. An MI5 and subsequently D Division channel. The mistress of AGENT

7(4) *(q.v.)*, RAGS, in Britain; called to active duty as a Wren and posted in 1944 to Mountbatten's headquarters in Ceylon, whence she passed information back to the GARBO network, ultimately intended for the Japanese, by letters in secret ink. Discontinued (through a notional automobile accident) as the result of German suspicions over the similarity between her information and that of SILVER *(q.v.)*.

AGENT 7(4) Known as RAGS. A subagent of STANLEY, GARBO's AGENT SEVEN *(q.v.)*. A notional Indian poet and member of the Aryan World Order; GLEAM's lover. Operated in the Brighton area. Known to the Germans as DICK.

AGENT 7(5) A subagent of STANLEY, GARBO's AGENT SEVEN *(q.v.)*. A notional Brother in the Aryan World Order in Swansea; a relative of DAVID.

AGENT 7(6) A subagent of STANLEY, GARBO's AGENT SEVEN *(q.v.)*. A notional Welsh Brother in the Aryan World Order in Swansea.

AGENT 7(7) A subagent of STANLEY, GARBO's AGENT SEVEN *(q.v.)*. The notional Treasurer of the Aryan World Order. Operated in the Harwich area. Known to the Germans as DORICK.

ALBERT Real name Blondeau. French. A former petty officer in the French navy; GILBERT's radio operator in North Africa until he was removed in the autumn of 1943 lest he be recognized by LE MULET when the latter joined GILBERT. See GILBERT; LE MULET.

ALBERTO A source for BROMO *(q.v.)*; a notional military policeman of Spanish descent.

ALERT A source for CRUDE *(q.v.)*, 1942–44. An imaginary civilian orderly at an area headquarters in Northern Syria who hated the British because his mother had been raped by an English sergeant-major in the First World War.

ALOIS See ALOSI.

ALOSI (Possibly a typist's error for ALOIS.) A notional Uruguayan clerk in the American embassy in Montevideo, a source for LODGE *(q.v.)*.

ALMURA German name for GARBO's AGENT 4(1) *(q.v.)*.

ANDRIES A notional Dutch liaison officer, a source for GLEAM *(q.v.)* at Mountbatten's headquarters.

ANDROS A British Admiralty channel. An agent of the British naval attaché in the Spanish navy or Ministry of Marine; called BLIND *(q.v.)* when used as a special means channel.

ANGEL ONE   A D Division channel. An imaginary group of spies and saboteurs in Thailand, headquartered in Bangkok and controlled from India, to whom daily messages were broadcast over All-India Radio. Opened in the spring of 1943 and augmented in May 1944 by taking over an imaginary existing operation with agents in Jumbhorn (DICKENS), Kanchanburi (TROLLOPE), and Chiengrai (THACKERAY) *(qq.v.)*. Employed in support of SCEPTICAL in 1945. See FOSSIL.

ANTHONY   An FBI channel; FBI file name CAMCASE. Of Russian origin. An agent of the Madrid Abwehr, sent to the United States in February 1945 to gather aviation intelligence.

ANTOINE   A SHAEF channel. A double agent run from Cannes by the French for a short time in 1944.

APOSTLE   An "A" Force channel. A French-run channel, operating the transmitter of a staybehind SD agent in Rome after its fall.

APPRENTICE   An "A" Force channel. Russian, naturalized Yugoslav. Son of a Czarist officer, with Yugoslav loyalties but fiercely opposed to Tito and his Communists, who was parachuted into Italy from Albania in May 1944 and promptly turned himself in. He wanted to serve in the Yugoslav Air Force but was persuaded to act as a double agent, notionally teaching Serbian to Allied officers and reporting their indiscretions to the Germans by W/T. Functioned from June 1944 to the end of the war, transmitting from Bari.

ARABEL   (Sometimes spelled ARABAL.) German name for the GARBO network.

ARBITER   An "A" Force channel. Not an individual but the codename given to transmissions from the radio set of a pro-Fascist ring in Rome, used for deception from June to September 1944.

ARMOUR   An "A" Force channel. Italian. A staybehind agent for the Germans when they evacuated Rome in 1944. In reality an anti-Fascist aeronautical expert who had penetrated the Abwehr for Italian intelligence and was commissioned a lieutenant colonel in the German service. He turned himself in promptly after the fall of Rome and was operated as a double agent until the end of the war. He and his wife, who was also supposedly his cipher clerk, had broad access to the highest levels of Italian society, reaching to Marshal Badoglio himself, and he was able to pass valuable material of an extremely high grade.

ARTHUR   An "A" Force channel, made available by the French and used for deception, March to October 1943. A rich Spanish Jew in Oran, a financier on a fairly large scale, with some shady financial dealings with the Spanish consul at Oudjda in

French Morocco, headquarters of the American Fifth Army. Passed along, by letter, miscellaneous observations picked up more or less at random. Moved in high circles, and could plausibly pass high-grade information.

ARTHUR   A SHAEF channel. A French-controlled agent in Marseilles, 1944–45; operated through SCOUNDREL *(q.v.)*.

ARTIST   Real name Johann Jebsen. German. Not a deception channel but an Abwehr officer. Recruiter and chief contact of TRICYCLE. An Allied sympathizer who agreed through TRICYCLE to work with the British from September 1943. Arrested and sent to Oranienburg concentration camp in April 1944 under suspicion of illegal currency transactions; died there.

ASPIRIN   An FBI channel. Spanish. Real name José María Aladren. A journalist in Washington, an agent of the Spanish freelance Angel Alcázar de Velasco, recruited by the FBI as a double agent in October 1942. Sent back to Spain in March 1944 and arranged to work for the Germans and the Japanese jointly. Returned to Washington in February 1945 as a correspondent for a Spanish news agency, equipped to send reports in secret ink, by radio, by coded messages in his newspaper stories, and possibly through a crewman on a Spanish ship acting as a courier.

ASSASSIN   A SHAEF channel. Real name Auguste Mauboussin. French. A German staybehind agent in Rouen, run by the French from December 1944. French codename: MARTIN.

ATLAS I   Codename for GILBERT's principal transmitter, in Tunis, in the original Abwehr plan for GILBERT when he was sent to North Africa.

ATLAS II   Codename for an outlying transmitter in Algiers, in the original Abwehr plan for GILBERT when he was sent to North Africa. Eventually opened by LE MOCO *(q.v.)* in September 1944 when GILBERT went to France, but soon shut down owing to LE MOCO's incompetence.

ATLAS III   Codename for an outlying transmitter in Oran, in the original Abwehr plan for GILBERT when he was sent to North Africa. Eventually opened up, notionally by LE MULET with JEAN as operator, in September 1944 when GILBERT went to France, but apparently did not accomplish much. See GILBERT; JEAN; LE MULET.

ATOM   A SHAEF channel. Real name Paul Julien Michel. French. American-run case from February 1945, notionally at Longwy.

AUDREY   A D Division channel. A portion of the HAT TRICK party, assigned originally to operate from Rameswaram; moved notionally to Ceylon, from which naval information was furnished by W/T to the Japanese until the end of the war. See HAT TRICK; DOUBTFUL.

AXE   An "A" Force channel. Italian. A staybehind in Florence, the twenty-seven-year-old operator of a radio shop arrested in late August 1944 and on the air from October. After a slow buildup he began to be accepted by the Germans to such an extent that in January 1945 he was awarded the Iron Cross.

B.G.M.   ("Blonde Gun Moll.") An "A" Force term for the woman who played MISANTHROPE *(q.v.)* in connection with efforts to pay CHEESE *(q.v.)*.

BACKHAND   A D Division channel. Real name Mohammed Zahiruddin. An Indian Army officer, cashiered in 1940 for anti-British activities, who changed his views and offered his services to the British. Parachuted into Japanese-held Burma in February 1944 with orders to give himself up and function as a triple-cross agent. Assigned by Subhas Chandra Bose in June 1944 to make propaganda broadcasts for Bose's organization. In these he sought to convey information to the British; Bose became suspicious and had him thrown into a concentration camp, from which he was liberated after the recapture of Rangoon in 1945.

BALLOON   An MI5 channel. Real name Dickie Metcalfe. British. Supposedly a cashiered British officer engaged in the arms business; notionally a source for TRICYCLE *(q.v.)*. Operated from Britain, May 1941 to November 1943.

BARONESS   An MI6 and "A" Force channel. A Russian (naturalized Swedish) businessman long resident in Istanbul, who offered his services to MI6 when the Germans recruited him as a possible agent in Syria and Turkey, then in India, then as a staybehind in Turkey. Nothing came of these plans and he moved to Argentina in 1944.

BASILE   See PESSIMIST Z.

BASKET   An MI5 channel. Real name Lenihan. Irish. Parachuted into Eire near Dublin in July 1941, with two radio sets, a good secret ink, and over four hundred pounds cash. Went to Northern Ireland and surrendered himself. Main mission, to send weather reports from Sligo, could not be allowed. Efforts were made to run him as a letter-writer.

BATES   One of PEASANT's notional sources, a notional aviator friend of Shell Oil's Washington manager. See PEASANT.

BATS   A group of seven parties of three Indian agents each, equipped with W/T transmitters, parachuted by the Japanese into Assam in April 1943. Two of the parties were used as D Division deception channels under the codenames OWL and MARMALADE *(qq.v.)*.

BEETLE   An MI6 channel. Real name Petur Thomsen. Icelander. Landed in Iceland by U-boat with a wireless set and a barometer September 1943. A channel for sending weather information and in connection with naval deception.

BENEDICT   German name for the network of GARBO's AGENT THREE *(q.v.)*.

BIG LEMON   An "A" Force channel, leader of THE LEMONS *(q.v.)*. A middle-aged Istanbul Greek, imprisoned when LITTLE LEMON turned the party in, and depicted to the Germans as being at large, providing information, and active in the movement for union of Cyprus with Greece. See THE LEMONS; LITTLE LEMON.

BILL   A notional British colonel on Mountbatten's staff, a source for GLEAM *(q.v.)* at Mountbatten's headquarters.

BISCUIT   An MI5 channel. Real name Sam McCarthy. British. An MI5 informant before he was brought into controlled agent work; a reformed drug smuggler, petty thief, and confidence man. Made a colleague of SNOW when the latter's German control asked that he recruit a subagent. After a number of adventures, terminated with the termination of the SNOW case in March 1941. See SNOW.

BLACKGUARD   A SIME penetration agent and MI6 channel. Iranian. Former propaganda broadcaster for Radio Berlin. Abwehr agent in Istanbul recruiting agents for Iran, Egypt, and India, who volunteered to work for MI6 in the autumn of 1943. Procured the arrest of KISS *(q.v.)* in 1943 and assisted in getting a radio transmitter to FATHER *(q.v.)* in India in 1944.

BLAZE   A SHAEF (21 and 12th Army Groups) channel in the last weeks of the war in Europe. Real name Ingeborg Schotes. German. One of THE RHINEMAIDENS *(q.v.)*, one of whom, LAZY *(q.v.)*, turned herself and the rest of the group in to the Americans upon their arrival early in March 1945; BLAZE and LAZY were used for deception thereafter, transmitting from Rheydt.

BLIND   A British Admiralty channel. Codename employed for ANDROS *(q.v.)* in his capacity as a channel to the German Sicherheitsdienst in Madrid.

BOILER   A SHAEF channel. Real name Menard. French. French-controlled Sicherheitsdienst agent in Paris after the liberation; never made contact with the Germans and disappeared during the course of an attempted French double agent operation.

BOOTLE   An MI5 and Deuxième Bureau channel. Real name Fressay. French. Wrote letters in secret ink between December 1943 and February 1944.

BOVRIL   Original codename for GARBO *(q.v.)*.

BOWSPRIT   An apparently abortive OSS X-2 channel from Italy after the surrender of that country. Real name Ercole Pugliesi. An Italian SIA officer in Sicily with a W/T set who gave himself up after the surrender of Italy.

BRASS   A D Division channel. A three-man team of Karens dropped by MI6 near Rangoon in November 1942. It was evident from their W/T messages that they had been captured and the Japanese were attempting to use the channel for deception. Transmissions to them were thereafter used for purposes of deceiving and confusing the Japanese until after the fall of Rangoon in 1945.

BRISTLE   Apparently a channel, probably diplomatic, to the Italians from May 1942. Used in connection with STARKEY and BOARDMAN. No further information.

BROMO   An FBI channel. Sometimes referred to as LITTLE JOE. Real name José Laradogoitia. Spanish (Basque). Lived in Idaho in the 1930s; deported in 1941 after being convicted of passing bad checks. Recruited in Spain by the Abwehr; sent to Rio as a letter writer in 1942 and to the United States in 1943. Alerted by the British, the FBI made a double agent of him, situated in New York. Communicated by secret ink with Madrid until March 1944, when he apparently acquired an operator and began sending by radio. Acquired at least two notional subsources, ALBERTO and LUÍS *(qq.v.)*. His material was low-grade, mainly information about equipment and shipping.

BRONX   An MI5 channel. Real name Elvira Chaudoir. Peruvian. Daughter of a diplomat; recruited by the Germans in Paris, October 1942. Operated by MI5 as a letter writer from Britain till end of the war.

BROUWER   A British-run channel, operating the transmitter of a staybehind SD agent in Rome after its fall.

BRUTUS   An MI5 channel. Real name Roman Garby-Czerniawski. Also known as Armand Walenty. Polish Air Force officer. Operated by MI5 from Britain from October 1942 to January 1945. Had headed a secret anti-German organization in France in 1940–41, captured by the Germans, believed by them to have been turned, and allowed to "escape" in July 1942 with orders to create a Polish fifth column in England. Promptly turned himself in and used as a double agent after some hesitation. Notionally assigned to high headquarters in Britain and furnished high-level information, particularly in aid of FORTITUDE in 1944.

BULL'S EYE   A multiple channel from D Division to the Japanese via "A" Force and thence through the British military attaché in Ankara. Initially, from April 1943, a Japanese journalist, but he was arrested by the Turks. The next channel was via the Polish military attaché, and after the Italian surrender via the Italian military attaché. Began drying up in the summer of 1944 when Turkey broke off relations with Germany; there was some hope that material could be passed through the Spanish ambassador to Turkey, but the channel was finally shut down after Spain broke diplomatic relations with Japan in April 1945.

BYZANCE   An "A" Force channel controlled by the French. Arrived in Algiers when the landings in Southern France were already under way in August 1944; worked by W/T to Berlin for two or three weeks.

CAMCASE   See ANTHONY.

CAMOUFLAGE   A SHAEF channel. Real name André Schurmann, alias Henri. French. A ship's radio operator recruited by the Abwehr in November 1941 and thereafter a major recruiter at Le Havre. Run as a double agent by the French from December 1944. French codename SERRE.

CAPRICORN   An "A" Force channel. An MI6 agent in the Peloponnese, working back to Cairo by W/T, made available to "A" Force for deceptive purposes when MI6 concluded in late 1942 that he was operating under Italian control. From mid-February 1943 on, he was sent questionnaires about Axis defenses and units in the Peloponnese to help the successive deceptions aimed at creating a threat in that area. After Italy changed sides the Italians who had run the case told the British that they had had an impression that perhaps the Allies had tumbled to the fact that they were running CAPRICORN, but that all his traffic was sent to the Italian general staff nevertheless.

CARELESS   An MI5 channel. A Polish airman, operated from Britain from July 1941. Difficult to control; imprisoned and discontinued in January 1943.

CARROT   An MI5 channel. Luxembourger. A promising case in Britain in June 1942, designed to discover whether Vichy was trying to penetrate organizations in Britain. Failed to develop owing in part to unspecified complications over his dealings with the Deuxième Bureau, and discontinued in December 1942.

CARVALHO   GARBO'S AGENT ONE (q.v.).

CATO   A D Division channel. A notional source for ANGEL ONE's notional agent TROLLOPE, located at Banpong, Thailand, forty miles west of Bangkok. See ANGEL ONE; TROLLOPE.

CATO A German codename for GARBO *(q.v.)*.

CELERY An MI5 channel. Real name Walter Dicketts. British. Recruited to accompany SNOW to Lisbon in early 1941 and proceed to Germany to attempt to penetrate the German intelligence services. Visited Germany and returned; substantially discontinued with the discontinuance of SNOW, though he made one more abortive visit to Lisbon before final discontinuance in August 1941. See SNOW.

CHARLES German name for SKULL *(q.v.)*.

CHARLIE An MI5 channel. Real name Kiener. A loyal British subject born in England but with a German father, recruited by the Germans in 1938 through threats to his brother in Germany. A photographer, made part of the SNOW team to work with microdots. Discontinued with termination of the SNOW case. See SNOW.

CHEESE An "A" Force channel. ROBERTO to the Germans. One of the great deception channels of the war. The original CHEESE was Renato Levi, a British national of Italian Jewish ancestry, recruited by MI6 before the war, and subsequently by the Abwehr and the Italian SIM, and sent by them to Cairo in February 1941. There notionally acquired a W/T operator, PAUL NICOSSOF, supposedly a Syrian of Slavic extraction. Sent back to Italy by the British in the spring of 1941 to report these developments, and there jailed for black market activities. The CHEESE channel was continued by NICOSSOF, communicating regularly with the Abwehr until February 1945. See GAULEITER OF MANNHEIM; LAMBERT; MISANTHROPE; NICOSSOF; STEPHAN.

CHER BÉBÉ An "A" Force channel. A double agent made available to "A" Force for deception by the French. A Spanish mechanic. Owed his codename to a currently popular song in Algerian music halls. Recruited by the Germans and reported to them through the Spanish consul in Oran. Spent more than a year in jail before agreeing to work for the French; then served as a deception channel for some months from May 1943.

CHOPIN BRUTUS's fictitious wireless operator (in reality an MI5 operator), described by BRUTUS to the Germans as an elderly retired Polish Air Force officer whose family had perished in Russia and who worked for idealistic motives. See BRUTUS.

CLIFFORD, ALEXANDER An "A" Force channel. British. The genuine special correspondent of the *Daily Mail* in Cairo, utilized in May 1942 in aid of Operation FABRIC to publish a story implying that no British offensive was possible during the *Khamseen* season.

COBWEB   An MI6 channel. Real name Ib Arnason Riis. Danish-born Icelander. Landed in Iceland by U-boat April 1941; operated by radio until the end of the war.

COCAINE   An "A" Force channel, made available by the French and used for deception from July 1943. Transmitted by W/T daily from Algiers to the Germans in Melilla till he was shut down in September 1944.

COCONUT   A July 1943 project to open a channel for passing deceptive information through an agent in Istanbul who would claim to the Germans to be a "post box" handling messages between Cairo and British intelligence in Athens and willing to betray the British. Nothing further known.

COCASE   See TOM X.

COG   A SHAEF channel. Real name Michel Lamour. French. Sent into Bordeaux by the Germans in March 1945, captured and run by the French.

CON   See AGENT 5(I).

CORAL   A November 1942 proposal to set up a deception channel to the Italians in Turkey through a former Italian agent in Palestine. Apparently not adopted.

COSSACK   A SHAEF channel. Turkish, naturalized French citizen. A staybehind agent captured near Lille in November 1944 and run by the French under British overall supervision.

COSTA   See PESSIMIST X.

COUGHDROP   A D Division channel. A notional four-man party of agents supposedly landed in northern Sumatra in December 1944, to whom regular W/T messages were addressed in hopes of making the Japanese believe that the Allies intended to invade Sumatra. Operated to the end of the war with no known result.

CRUDE   Also known as MARQUIS. An "A" Force channel. A Turkish businessman resident in Iraq who ran a notional spy ring, communicating with the Abwehr in Turkey by secret ink letters and occasional personal visits. Operated from January 1943 until well into 1944. Notional source: ALERT (q.v.).

CRUZ   A British Admiralty channel. Night watchman in the offices of British naval attaché's office in Lisbon, passing documents supposedly retrieved from wastebaskets.

CUPID   An "A" Force channel, made available by the French. A German Jew. An attractive and intelligent young woman who ran a bar in Casablanca. She wrote a letter in secret ink once a week to the Germans in Barcelona, but her information was low-grade and there was not much success in elevating it. Efforts to use her for deception began in March 1943 and were given up by June.

DAGOBERT   German name for GARBO's STANLEY (AGENT SEVEN) network. See GARBO; AGENT SEVEN.

DAVE CLOSE   A notional commander in the U.S. Navy, a source for GLEAM *(q.v.)* at Mountbatten's headquarters. See GLEAM.

DAVID   See AGENT 7(2).

DAVIL   An "A" Force channel, made available by the French. French. Working for French intelligence, he got himself recruited by the Abwehr in Madrid and sent to Casablanca with a W/T set, working back to Hamburg. Employed at the Casablanca air base, from January to June 1944 he sent misleading aviation information and order of battle and shipping data.

DAVIT   A SHAEF channel. A French-controlled agent in Paris after the liberation, who succeeded in making W/T contact with the Germans but never succeeded in transmitting messages.

DE CONNINCK   (His real name.) A SHAEF (21 Army Group) channel. Belgian. An Abwehr staybehind in Torhout; put on the air after the liberation of that town in autumn 1944 but no contact made.

DELPIÈRE, JOSEPH   See DUTEIL.

DENSEL   See DENZIL.

DENZIL   An "A" Force channel. A notional RAF technician supposedly preparing Turkish airfields for British use; a source for GALA *(q.v.)* in the spring of 1944.

DEPUTY   A SHAEF (21 Army Group) channel. Belgian. A ship's radio operator who was stranded in a British port at the fall of Belgium in 1940 and recruited by SOE with orders to jump ship and return to Belgium. Upon his return in 1942 he was recruited by the Abwehr and was left in Brussels as a staybehind agent in 1944, reporting to the Allies upon the liberation of that city. Utilized to the end of the war, particularly in connection with false reports as to the fall of V-2s in Antwerp. See DOMINANT.

DERRICK   A SHAEF (21 Army Group) channel. Belgian. A hydrographer on the Ostend coast and an agent for Belgian intelligence from 1940. Recruited by the

Germans as a staybehind and reported to the Allies upon their arrival. Operated from Bruges to the end of the war; chiefly valuable as a penetration agent and for cryptanalysis.

DESIRE   A SHAEF channel. Real name Alfred Gabas, alias André Dumont. French. Arrested in Cherbourg when he attempted to contact DRAGOMAN (*q.v.*) in August 1944, and imprisoned thereafter. Considered too dangerous to run and never used.

DICK   See AGENT SIX.

DICK   German name for GARBO's AGENT 7(4) (RAGS) (*q.v.*).

DICKENS   Notional agent of ANGEL ONE (*q.v.*) at Jumbhorn, Thailand, on the Kra Isthmus.

DITCH   A SHAEF channel. A French agent who would have been run from Rennes in 1944–45 except for the shortage of personnel to run double agent operations.

DINOS   See RIO.

DOLEFUL   Also known as DOMINO. A SIME penetration agent and "A" Force channel. Turkish. A Turkish secret service agent posing as a Wagon-Lit attendant on the Taurus Express between Istanbul and Baghdad who had served the German Army in 1918; made available to the British through the good offices of the Turkish secret services. Passed information from Baghdad to a former German officer in Turkey. Used from November 1942 but never fully trusted, and utilized only for low-level information. Attempts to transfer him to the service of the Japanese towards the end of the war were frustrated when in January 1945 Turkey expelled Japanese representatives.

DOMINANT   A SHAEF (21 Army Group) channel. Belgian. A former Communist seamen's union leader in Antwerp, a genuine contact for DEPUTY (*q.v.*), utilized as the notional source for false information transmitted by DEPUTY concerning V-2 damage in Antwerp. Died in a (genuine) gas accident at the end of December 1944.

DOMINO   See DOLEFUL.

DONNY   German name for GARBO's AGENT 7(2) (*q.v.*) See GARBO.

DORICK   German name for GARBO's AGENT 7(7) (*q.v.*).

DOUBTFUL   A D Division channel. A portion of the HAT TRICK party *(q.v.)*, operating from Madras with a branch in Vizagatapam. Provided W/T traffic to the Japanese from June 1944 to the end of the war, particularly in support of order of battle, STULTIFY, and SCEPTICAL.

DRAGOMAN   A SHAEF (12th Army Group) channel. Real name Juan Frutos. Spanish. Originally codenamed PANCHO. An interpreter for the American Express office in Cherbourg, a staybehind caught soon after the Americans took that port in 1944, who had been a German agent since 1936. Became an interpreter for the U.S. Army and utilized as a double agent under American control to the end of the war.

DRAGONFLY   An MI5 channel. Real name Mr. George. A German born in England who left Germany for Holland just before the outbreak of war in 1939; recruited by the Germans and reported this on coming to England in 1940. Operated from 1941 to November 1943 as a W/T agent from England. Finally discontinued in January 1944 over failure of Germans to pay him.

DREADNOUGHT   An MI5 channel. Real name Ivo Popov. Yugoslav. A doctor; older brother of TRICYCLE *(q.v.)*. Helped select trusted Yugoslavs to be smuggled out of Yugoslavia to Britain as purported Abwehr agents. Escaped to Italy in 1944.

DUCK   A D Division channel. Originally, the codename for the W/T transmitter sent to FATHER in India, which first went on the air to the Germans in August 1944; subsequently the name for the former FATHER channel via this radio when it was taken over by RAJAH from January 1945 to the end of the war, with FATHER's departure from the theater. See FATHER; RAJAH.

DUTEIL   A real name, not a codename. Also known as Joseph Delpière. French. A former pimp who served on the Russian front with Jacques Doriot, chief of the collaborationist Parti Populaire Français, and had been secretary to Beugras, head of the party's special intelligence branch. The saboteur of GILBERT's group when dispatched to North Africa in 1943, assigned also to kill GILBERT if he thought he was double-crossing the operation. When GILBERT turned himself in and became a double agent, Duteil joined the French Army, planning to desert and make his way back to France through Spain. Eventually arrested and executed.

ECCLESIASTIC   An MI6 channel. A double agent through whom letters were sent to the Germans in 1944 designed to make them suspect OSTRO's loyalty. See OSTRO.

EDITH   A British Admiralty channel. English seaman used by British in 1942 as a channel to Germans in Lisbon.

ÉDOUARD   A French-run "A" Force channel from North Africa. French. A member of the RAM trio *(q.v.)*. A sergeant in the intelligence office of the French XIX Corps. See NORBERT; RAM.

EDWARD   A notional Royal Navy commander on Mountbatten's planning staff, a source for GLEAM *(q.v.)* at Mountbatten's headquarters.

EFFIGY   An "A" Force channel. A German staybehind W/T agent in Athens after the British occupation in October 1944, utilized for a few weeks in support of Operation SECOND UNDERCUT.

EL GITANO   An "A" Force channel controlled by the French. Spanish (Catalan). A hairdresser, smuggler, and pimp in Oran, recruited by the Germans in December 1942 and subsequently by the Italians in June 1943. Utilized to pass low-level shipping information and deceptive material about the French Army, on a much reduced scale after the Italian surrender till finally shut down in February 1944.

ELEPHANT   A D Division channel. German-sponsored wireless link between the station of the notional All India Revolutionary Committee in Delhi (MARY) and Subhas Chandra Bose, arranged between SILVER and the German legation in Kabul in August 1944. Successful connection was never achieved. See SILVER; MARY.

ELIAS   A notional part-time booking clerk in the Haifa railroad station, a source for THE PESSIMISTS *(q.v.)*.

EMILE   Another name for CHEESE *(q.v.)*.

ENTWHISTLE   An "A" Force channel; the codename for SIME's organization for spreading rumors.

FALCON   A SHAEF channel. Real name Georges Gaspari. French (Corsican). Dropped by parachute near Dijon and captured, February 1945. Run from Lyons jointly by the French and Americans until the end of the war.

FAN   A SHAEF (21 Army Group) channel. French. Recruited as an agent by the French fascist party; left as a low-level staybehind agent in Normandy, arrested and attempted to be run as a double agent, but contact was never made.

FANNY   An "A" Force channel. A French-educated Syrian. A triple agent in Damascus, known by the Germans to be working under control; information appearing to contradict desired deception stories was accordingly passed through him.

FARCE   A SHAEF channel. An agent caught by the French at Lyons; never developed into a double agent.

FATHER   An MI5 and D Division channel. Real name Henri Arents. Belgian. A distinguished pilot who got in touch with the Germans as a device for escaping from occupied Belgium. They assigned him the task of traveling to the United States and sending reports by W/T and secret ink. Made his way to Portugal; could not get an American visa, so went to England, arriving June 1941. Received queries and instructions by radio and replied by secret writing. Employed in connection with TORCH cover plans. Received technical questions about aviation which were hard for a skilled aviator to evade, so in June 1943 was posted to India. Sent information from there by secret writing until provided with a transmitter (DUCK), which went on the air in August 1944. Temperamental, and sent home in October 1944; DUCK continued to be operated till the end of the war by a notional disaffected Indian courier, RAJAH. See DUCK; RAJAH.

FICKLE   Planned as a SHAEF channel but not used. Real name Pierre Laurent. French. Parachuted into Allied territory in February 1945 along with FISH *(q.v.)*, with instructions to set up a W/T operation from Lyons. Promptly captured; never used as a channel despite various plans to do so.

FIDELIO   German name for FOREST *(q.v.)*.

FIDO   An MI5 channel. Real name Grosjean. A French air force pilot arriving in England in July 1943 with the assignment to send intelligence and steal an airplane and fly it back to Germany. Discontinued February 1944 owing to suspicions and poor communications.

FIREMAN   (Also known as MADELEINE.) A British Admiralty channel. A Russian-born seaman of Belgian parentage, used briefly by the British in 1942 as a double agent to the Germans in Lisbon.

FISH   A SHAEF channel. Parachuted into Allied territory in Febrary 1945 along with FICKLE *(q.v.)*, with assignment to set up a W/T network centered at Rheims. Surrendered on landing and run from Troyes by the French and an American case officer.

FLAME   A SHAEF (21 Army Group and Belgian Deuxième Direction) channel. Belgian. A staybehind who turned himself in to the British at Ghent in late September and transmitted from Brussels and later from Rumst, from November 1944 to the end of the war. A low-grade case yielding no results.

FLESHPOTS, THE   Real names John Eppler and Peter Sandstetter. Germans who made their way to Cairo across the desert to Asyut and by train to Cairo in July 1942. Not used for deception because of publicity given their capture.

FLORIST   A SHAEF channel. Real name Roger Hardouin. French. A French-controlled agent operating from Cannes.

FOREST   A SHAEF (6th Army Group) channel. Real name Lucien Guillaume Herviou. Known to the Germans as LUC. French. A forty-four-year-old aviator of some distinction, recruited by the Germans in 1943. Lived on an isolated farm near the Riviera town of Draguignan, notionally hiding out from the Allies; notionally bicycled in to Nice to pick up information which he passed on to the Germans by W/T. A major channel for 6th Army Group in 1944–45.

FORGE   A SHAEF (21 Army Group and Belgian Deuxième Direction) channel. Belgian. A Sicherheitsdienst agent sent with a radio operator through the lines in South Holland in November 1944. Efforts to establish W/T contact with the Germans were unavailing.

FOSSIL   ANGEL ONE's deputy, in Bangkok. See ANGEL ONE.

FRAIL   A SHAEF channel. Real name Jacques Michel. French. Parachuted into France in Haute-Savoie with a W/T set in April 1945 and promptly arrested. Not used as a channel because of newspaper publicity given to his capture.

FRANK   A SHAEF (21 Army Group) channel. Belgian. A former Belgian Army radio operator who joined the pro-German Rexist Party in 1942 to spy on it for the Resistance and was recruited by the Sicherheitsdienst as a staybehind. He gave himself up to the British on their arrival in Brussels and was run from Brussels, subsequently from Cenval, as a W/T channel from mid-September 1944 to the end of the war.

FREAK   Real name Marquis Frano de Bona. Yugoslav. Operated from London by MI5. A Yugoslav officer who passed out by escape route from Switzerland set up by TRICYCLE and DREADNOUGHT, after training in W/T and secret ink. Reached England December 1943, and became the W/T operator for TRICYCLE's network. See TRICYCLE.

FRED   GARBO's AGENT FOUR *(q.v.)*.

FREDDIE   A channel to the French intelligence service in French Indochina who would transmit to the Japanese. Reported by Fleming in March 1943; nothing further known.

FREELANCE   A SHAEF potential channel, not used. Real name Georges Quantin. French. Surrendered to the Americans at Lyons in April 1945, having been directed by the Germans to contact certain other agents; never became a double agent.

FRENCH AIRMAN   An "A" Force channel. A notional French airman who passed the "story" of Operation COWPER by W/T from Algiers in April 1943.

FRITZCHEN   German name for ZIGZAG *(q.v.)*.

G.W.   An MI5 channel. Real name Gwilym Williams. Welsh. A retired police inspector, purportedly SNOW's contact in the Welsh Nationalist Party and a potential saboteur, 1939; subsequently sent information to the Germans through Spanish diplomatic channels, 1940–41. Discontinued in February 1942 when his Spanish contact was arrested. See SNOW.

GABBIE   Hungarian. Codename for one of a group of genuine chorus girls who worked the cabarets of Nicosia in Cyprus and were notional sources for LITTLE LEMON *(q.v.)*. See HELGA; MARIA; MARKI; SWING-TIT; TRUDI.

GALA   An "A" Force channel. Real name Anna Agiraki. Greek. Mistress of the chief of the Italian SIM in Athens; she grew difficult to the extent of attracting the attention of the OVRA, and was made part of the QUICKSILVER *(q.v.)* team to get rid of her. Notionally operated as a high-level prostitute in Beirut (in reality imprisoned in Palestine), gathering information from Allied officers for QUICKSILVER. See DENZIL; PAPADOPOULOS; STEVEN; TAKIS; TOLLUS.

GANDER   An MI5 channel. Real name Hans Reysen. German. Parachuted into England in October 1940; operated as a W/T agent for approximately one month.

GANTZ   Original name of KLEIN *(q.v.)*.

GAOL   An "A" Force, and subsequently a SHAEF, channel. A former wireless operator for the French air force, recruited by the Gestapo in Paris and parachuted into Algeria in August 1943; turned himself in immediately. Notionally employed at the Algiers airport, transmitting to the Germans in Dijon. Shut down in North Africa at the end of August 1944 by notionally transferring him to France, where from October he was notionally employed at Le Bourget airport, Paris, and was run from there by the French.

GARBO   (Originally BOVRIL; ARABEL to the Germans.) An MI5 channel. One of the great deception double agent channels of the war. Real name Juan Pujol García. Spanish. After unsuccessfully offering in Spain to work for British intelligence, from October 1941 he transmitted to the Abwehr in Madrid a stream of letters purportedly sent from Britain (he was actually in Portugal) conveying wholly fanciful information. Brought to England by MI6 in April 1942 and run by MI5 thereafter; as a letter writer and after March 1943 by W/T as well. Developed an extensive and ramified network of notional subagents and sources. Employed especially in connection with TORCH and in COCKADE, FORTITUDE, and the V-1 and V-2 campaigns. Notionally went into hiding in the latter part of 1944 and his work continued by his notional deputy, AGENT THREE, to the end of the war. Received both the MBE from the British and the Iron Cross from the Germans. See AGENT ONE through AGENT 7(7) and AGENT J1 through AGENT J5; ALMURA; ARABEL; BENEDICT; CARVALHO; CATO; DICK; DONNY; DORICK; FRED; GERBERS; GERBERS, MRS.; GLEAM; GLOCASE; MOONBEAM; STANLEY; WIDOW, THE.

GAT   A SHAEF channel. A French-controlled agent who parachuted into Allied-held territory and was operated from Paris from March 1945; never an important case.

GAULEITER OF MANNHEIM   Supposedly an "A" Force channel, a source for CHEESE. According to various postwar sources, a German Jew who parachuted into Palestine and turned himself in, claiming to be a Jewish Abwehr recruit, and notionally became a waiter in a British officers' mess and a source for CHEESE up through the CRUSADER offensive of late 1941. Very probably the original of these accounts is Ernst Paul Fackenheim, a winner of the Iron Cross in the First World War, trained by the Abwehr in 1940 to be dropped into England but not dropped because of adverse flying conditions. Then thrown into Dachau as a Jew; released by agreeing to serve the fatherland and to get good treatment for his aged mother and his motherless son. Parachuted into Palestine in October 1941, tasked by the Abwehr to claim Jewish refugee status, report on troop movements for six or seven weeks, then escape. Offered to be a double agent, but believed by the British to intend to remain loyal to the Germans as a triple agent, and tried for espionage. See CHEESE.

GELATINE   An MI5 channel. Real name Friedle Gaertner. Austrian. A subagent of TRICYCLE. Supposedly a lively lady with many friends in the military. Operated from England from May 1941 to May 1945, sending political intelligence.

GENTLEMAN JOE   See HANS.

GERARD   Formerly PALIMPSEST. A D Division channel. No further information.

GERBERS   GARBO's AGENT TWO (q.v.).

GERBERS, MRS.   GARBO's AGENT 2(1) (q.v.).

GESQUIÈRE   (His real name.) A SHAEF (21 Army Group) channel. Belgian. An Abwehr staybehind in Belgium; put on the air in autumn 1944 but no contact made.

Appendix II

Gilbert A French-controlled "A" Force, subsequently SHAEF, channel. One of the great deception double agent channels of the war. Real name André Latham. French. A 1914 St. Cyr graduate with service in the French Army in both wars, sent by the Abwehr to Tunis shortly before the German surrender in North Africa to head a staybehind network. Turned himself in when the Allies entered Tunis, and was a major W/T channel to the Germans throughout the war, operating from North Africa until the autumn of 1944, when he was moved to Marseilles and then to SHAEF. See Albert; Atlas I; Atlas II; Atlas III; Duteil; Jean; Le Duc; Le Moco; Le Mulet.

Giraffe An MI5 channel. Real name Georges Graf. Czech. Had served in the French Army and was recruited by Sweetie (*q.v.*). Reached England in September 1940 with Spanehl (*q.v.*). Wrote a few letters and received some money, but the Germans considered the material scanty and he was discontinued and allowed to serve in the French forces in the Middle East.

Glass A SHAEF channel. A French-run agent at Bayonne who could never make contact with the Germans.

Gleam An MI5 and subsequently D Division channel, Garbo's Agent 7(3), Theresa Jardine (*q.v.*).

Glocase An MI5 and FBI channel. FBI name for Moonbeam (*q.v.*).

Godsend Nationality unknown. An intermediary dispatched from Egypt to Istanbul through whom Nicossof (*q.v.*) was enabled to receive a payment by the Germans at the beginning of 1944.

Godstone A channel arranged by Bevan during his visit to Moscow in 1944. A high-level contact of the British ambassador, through whom information could be passed on to the Japanese. Apparently given only limited use, if any, because it required high-grade material which Fleming could not supply because of lack of definite plans in Southeast Asia.

Gomez A notional employee at the American air base in Uruguay, a source for Lodge (*q.v.*).

Guinea An "A" Force channel. Real name James Ponsonby. A forty-two-year-old Englishman, commercial attaché to the British Consul-General in Tangier. Joined SOE in July 1941. In July 1943, MI6 in Tangier hatched a plot to establish him as a German agent by appearing to be desperately in need of money and volunteering his services to the Germans. Notionally appointed "Coordinator of the War Effort" at Tangier, giving him notional access to high-grade information. Furnished deception information to the Germans, sometimes at personal risk, from the invasion of Italy until the spring of 1944, when he was withdrawn to England for his own safety. Awarded the MBE for his services.

GUN   A SHAEF channel. Real name Guy Godet-la-Loi, alias Garcia. French. Deposited behind French lines on the Gironde in March 1945 and arrested in Bordeaux soon after. Run by the French as a double agent in counterintelligence operations.

GUS   Also called GUSTAVO. A Joint Security Control channel through the U.S. military attaché in Buenos Aires. Polish. A journalist who had worked for a short time in 1943 as a translator for the Japanese Domei news agency in Buenos Aires and was a channel to Saburo Suzuki, a Japanese agent who was the Buenos Aires correspondent for the *Mainichi* newspaper. From December 1944, fed Suzuki information supposedly derived from interviews with recent plane arrivals from the United States and from a weekly airmail letter from his friend HESSELL TILTMAN *(q.v.)* in Washington.

GUSTAVO   See GUS.

HAMLET   An MI6 channel. Real name Dr. Koestler. An Austrian Jew and former cavalry officer, in business in Lisbon. Associated with the Abwehr until April 1944, in touch with anti-Hitler elements in Germany. A contact of MULLET *(q.v.)* from late 1941.

HANS   A Joint Security Control channel through the U.S. military attaché in Bern. Real name Dr. Friedrich Wilhelm Hack. German by birth but stateless in 1945, in Switzerland under a Nansen passport. Anti-Nazi and pro-Japanese; had lived for a long time in Japan and had been economic advisor to the Japanese Navy. Utilized from the spring of 1945, through an intermediary, as an unwitting channel of information to Commander Fujimura of the Japanese Navy in Bern, who was in immediate radio contact with the Navy Ministry in Tokyo. Hack (or Fujimura, or perhaps the intermediary) may also have been known as GENTLEMAN JOE.

HARI SINGH   A notional Sikh taxi driver in Calcutta, a source for OWL *(q.v.)*.

HARRY   French. A (genuine) former warrant officer in the French Army, a friend of and source for ARTHUR *(q.v.)*.

HAT TRICK   An eight-man group of Indian agents landed on the Malabar coast by a Japanese submarine in April 1944 and divided into three parties, allotted to the Trichinipoly, Madras, and Ramaswaram areas respectively. The arrest of the Trichinopoly party received too much publicity for them to be used as double agents. The

Madras party became DOUBTFUL *(q.v.)* and the Ramaswaram party became AUDREY *(q.v.)*, both D Division channels.

HEADLAND   A SHAEF channel. A French-controlled agent operating from Vannes from February 1945.

HEIR   A SHAEF channel. A French-controlled agent in Paris whose W/T set was unearthed at Gerardmer; made contact with the Germans but never succeeded in producing anything of value.

HEKTOR   One of the fictitious agents in England from whom Karl Heinz Kraemer, an Abwehr officer attached under diplomatic cover to the German embassy in Stockholm, claimed to obtain high-grade information for purported intelligence reports which Kraemer concocted and sent to his superiors in the Abwehr. See JOSEPHINE.

HELGA   Austrian. Codename for one of a group of genuine chorus girls who worked the cabarets of Nicosia in Cyprus and were notional sources for LITTLE LEMON *(q.v.)*. See GABBIE; MARIA; MARKI; SWING-TIT; TRUDI.

HENRY   A Joint Security Control channel through the U.S. military attaché in Lisbon. Turkish. Wife of a German-Jewish refugee in Lisbon, former mistress of the Japanese military attaché and in 1945 close to the Japanese naval attaché. Employed as an unwitting channel to the Japanese from the spring of 1945 to the close of the war.

HERMAN   A subagent of PAT J *(q.v.)*, a notional seaman first class of German extraction and sympathies in the Philadelphia Navy Yard.

HICCOUGHS   A D Division channel. An imaginary network of staybehind agents in Burma, to whom instructions in an easily-broken cipher were broadcast twice daily over All-India Radio from July 1942 to the end of the Japanese war.

HITTITE   A potential MI5 channel that did not materialize. Recruited by the German consul in New Orleans and told he would be approached by a German contact if he ordered late breakfast at the Regent Palace Hotel in London before the end of November 1940, and to go to the Cumberland Hotel if no approach had been made after three days. Supposed contact never appeared and was apparently never identified.

HOLTZ   A subagent of PAT J *(q.v.)*, a notional seaman second class in the United States Navy of German extraction and sympathies.

HONKY TONK An "A" Force contact. Real name Madame Holstein, born Su Yang. Indochinese. *Poule de luxe* of Beirut and sometime Abwehr agent; never consciously used as a deception channel.

HOST A SHAEF channel. Real name Pierre Eugene Schmidt. French (Alsatian). A former interpreter for the Sicherheitsdienst, given espionage training and left, along with HOSTESS *(q.v.)* as a staybehind in Alsace, where they gave themselves up when the Americans arrived in early December 1944. Operated, never satisfactorily, as a W/T channel from Strasbourg intermittently till April 1945.

HOSTESS A SHAEF channel. Real name Elizabeth Steulet. French (Alsatian). Given espionage training and left, along with HOST *(q.v.)* as a staybehind in Alsace, where they gave themselves up when the Americans arrived in early December 1944. Never satisfactory; a project to employ her as a triple-cross in conjunction with HOST never came to fruition.

HULSMANN A SHAEF channel. An agent who for a time was to be played by the French as a triple-cross from Calais, but the project died.

HUMBLE A source for SMOOTH *(q.v.)*. A notional keeper of a fruit and vegetable shop in Aleppo, supposedly supplying produce to military camps and messes in the area.

IDOL Real name Jacques Moglia. Argentine. A courier sent to LEGION *(q.v.)* in March 1945. Not used as a deception channel.

IMPULSE A SHAEF channel. A German agent captured in Amiens and opened from Paris under Franco-British control; Germans, suspicious, told him to go off the air.

INFAMOUS An Armenian carpet merchant of Turkish nationality who had been working for the Abwehr in Istanbul and offered his services to MI6 (continuing to pass unauthorized false information to the Germans). Employed as an intermediary in an unsuccessful effort by the Germans to get payment to NICOSSOF *(q.v.)* in 1943 and a successful effort to pay QUICKSILVER *(q.v.)* in 1944.

INK A D Division channel. A W/T channel to the Japanese in Macao set up for D Division's use by General Cheng K'ai-min, Chiang Kai-shek's director of military intelligence, in early 1945. Notionally an official in Chungking with access to the dispatches of General Feng Yi, Chiang's representative at Mountbatten's headquarters.

JACK See PESSIMIST Y.

Jack A notional wing commander in the RAF, a source for GLEAM *(q.v.)* at Mountbatten's headquarters.

Jacqueline A SHAEF channel. A French double agent at Bordeaux who gave some promise but never materialized as a long-term double agent case.

Jam A notional White Russian in the United States, supposedly a friend of the British double agent OUTCAST *(q.v.)*. Part of a March 1945 scheme to create a direct connection for the United States to Onodera, the Japanese military attaché in Stockholm. OUTCAST would pass information to Onodera, purporting to come from Jam, temporarily in Britain. Jam would then return to the States with a direct connection to Onodera. The scheme was apparently never carried through.

Jardine, Theresa See AGENT 7(3) and GLEAM.

Javeline German name for GLEAM *(q.v.)*.

Jean Notional backup wireless operator for GILBERT *(q.v.)*, replacing the original genuine backup operator, one Falcon, who was put into the French Army when GILBERT came over to the Allies. Notionally opened up the ATLAS III *(q.v.)* station from Oran with LE MULET in 1944 after GILBERT went to France.

Jeep A British agent in the Gibraltar area as of August 1944; no further information.

Jeff An MI5 channel. Real name Tor Glad. Norwegian. A fluent linguist. MUTT and JEFF were originally primarily sabotage agents; the Germans had to be persuaded to use them as spies. Arrived in Britain in April 1941 with a W/T transmitter and other equipment in a boat from a German seaplane, and promptly turned themselves in. JEFF was unjustly judged unreliable and spent the war imprisoned on the Isle of Man, Stafford Gaol, and Dartmoor. MUTT and JEFF notionally carried out Plan GUY FAWKES in November 1941, Plan BROCK in 1942, and Plan BUNBURY in 1943, all acts of sabotage. Employed in connection with TORCH cover plans in November 1942, and with notional threats to Norway in April, August, and October of 1943. Their perceived value to the Germans appeared to fluctuate. By the beginning of 1945 the Germans had lost faith in them. See MUTT.

Jest A French-controlled W/T double agent in Toulon, 1944–45, taken off the air in February 1945 when ULTRA suggested that he had alerted the Germans that he was under control.

Jewel An "A" Force channel, French-run. French. An agriculturalist (an expert on grasses) recruited by the Germans before the war and sent to North Africa in

1941. Run as a W/T channel from Casablanca from early 1943 to December of that year, and thereafter until September 1944 from Algiers.

JIMMIE   Also spelled JIMMY. A Joint Security Control channel through the U.S. military attaché in Lisbon. Portuguese. Director of the Portuguese secret police, the Policia Vigilancia e Defesa do Estado or PVDE; perhaps Captain Agostinho Lourenço himself, the head of the PVDE; described by the military attaché as a clever man of independent financial means, blindly devoted to Salazar and the totalitarian regime. Utilized in 1945 as an unwitting channel for information to the Japanese minister to Portugal.

JIMMY   See JIMMIE.

JO   Notional wireless operator for GILBERT (q.v.) when the latter moved to Paris in November 1944.

JOB   An agent captured in November 1943 as a result of DRAGONFLY's communications.

JOHNNY   German codename for SNOW (q.v.).

JOSEF   A British Admiralty channel. A seaman of Russian origin, uncertain nationality. A trained Soviet agent used by the British from August 1942 to December 1944 in an effort to open a channel to Japanese intelligence in Lisbon.

JOSEPHINE   One of the fictitious agents in England from whom Karl Heinz Kraemer, an Abwehr officer attached under diplomatic cover to the German embassy in Stockholm, claimed to obtain high-grade information for purported intelligence reports which Kraemer concocted and sent to his superiors in the Abwehr. See HEKTOR.

JOY   A SHAEF channel. A French-run channel in Marseilles from mid-September 1944 to the end of the war.

JURY   A SHAEF channel. A French-controlled agent in Paris; failed to make contact with the Germans.

JUSTICE   General codename for deceptive material intended to be leaked by British naval attachés as gossip, rumors, or indiscretions. Messages to attachés containing such information would begin "Justice for Germany," "Justice for Italy," etc.

KANSHI RAM   A notional Indian Navy sailor, a source for OWL (q.v.).

KARL   A notional friend of, and source for, RZ 282 *(q.v.)*, having some connection with Singapore and the Malay Peninsula.

KEEL   A SHAEF channel run by OSS. Real name Jean Carrère. French. A Sicherheitsdienst agent left in Paris as a staybehind who turned himself in when the Allies entered Paris. Run as a W/T double agent from early September 1944 until the end of the month, when he was kidnapped by the French Resistance in revenge for his activities for the Sicherheitsdienst since 1940, and was not seen again. Run as a notional thereafter until April 1945.

KELLY   A Joint Security Control channel through the U.S. military attaché in Lisbon. Portuguese. A Portuguese Army officer with access to the highest government circles as well as those of the local Japanese mission. An unwitting channel through whom information was passed to the Japanese through the spring and summer of 1945.

KEYNOTE   A SHAEF channel, French-run. Real name Louis Valette. French codename VERDIER. A staybehind in Bordeaux, arrested in early November 1944 and run to the end of the war.

KHALIL   A notional launderer at British Ninth Army headquarters; a source for QUICKSILVER *(q.v.)*.

KISS   An "A" Force or MI6 channel. Iranian. A student sent on an Abwehr mission to Tehran; arrested at the end of 1943 through BLACKGUARD *(q.v.)* and used from March 1944 to March 1945 to pass to the Germans deceptive information about Soviet troop movements and the political situation.

KLEIN   (Originally GANTZ.) One of PEASANT's sources, a notional employee of Chrysler who furnished production information. See PEASANT.

KNOCK   A notional traveler in surgical goods throughout the Middle East; a source of information from Iran and Iraq for HUMBLE *(q.v.)*.

KYRIAKIDES   A notional businessman in Palestine, traveling frequently to Cairo; many Greek royalist connections and an important source for QUICKSILVER *(q.v.)*, particularly as to Greek affairs.

LAMBERT   An early name for CHEESE *(q.v.)*.

LAZY   A SHAEF (21 and 12th Army Groups) channel in the last weeks of the war in Europe. Real name Anneliese Peters. German. One of THE RHINEMAIDENS *(q.v.)*. A 19-year-old former Luftwaffe signal operator trained as a staybehind in the

München-Gladbach area, who turned herself and the rest of the group in to the Americans upon their arrival early in March 1945; she and BLAZE *(q.v.)* were used for deception thereafter, transmitting from Rheydt.

LE DUC   GILBERT's *(q.v.)* second in command. Real name Captain Dutey-Marisse. French. Turned himself in with GILBERT in Tunis in 1943; subsequently notionally, and eventually in reality, moved to Sicily, Italy, Corsica, and France as a source for GILBERT.

LE MOCO   An "A" Force channel. Real name Major Falguière. French. Parachuted into North Africa in July 1944 with money and equipment for GILBERT *(q.v.)*. Briefly opened the ATLAS II *(q.v.)* transmitter from Algiers after GILBERT went to France in September 1944.

LE MULET   Part of GILBERT's *(q.v.)* group. Real name Charmain, alias Caron. French. Parachuted into North Africa in October 1943 with funds and a new cipher for GILBERT, and an assignment to set up a network for the French fascist political party. Allowed to function until LE MOCO's *(q.v.)* arrival in July 1944, when he was arrested and disposed of.

LE PETIT   A French-run "A" Force channel. Czech. An interpreter for the Americans on the Oran docks, employed by the Spanish secret service. Used from April 1943 to August 1944 to pass high-grade material to the Germans via the Spanish vice-consul at Oudjda.

LEAGUE   A SHAEF (12th Army Group) channel. Real name Paul Collignon, alias Gobin. French. A staybehind who turned himself in to the Americans when they entered his village near the Franco-German border in late November 1944. Run as a W/T double agent from St. Avold from early December 1944 and from Metz from mid-January 1945.

LEGION   A SHAEF channel jointly run by the French and the Americans. Real name Robert Pennors. French. Dropped by parachute near Lyons on December 6, 1944, and arrested by the French five days later. Run as a double agent, primarily for penetration purposes, from early January 1945. The Germans suspected that he had been turned and he was never of great value.

LEMON   Apparently a (female) MI5 channel to the Swiss, and perhaps to others, in London.

LEMONS, THE   An "A" Force channel (BIG LEMON and LITTLE LEMON). Greeks. BIG LEMON, the leader, and LITTLE LEMON, his wireless operator, were dispatched by the Abwehr in the spring of 1943 to join a supposed espionage group in Beirut but rowed ashore in Cyprus instead, claiming to be refugees. LITTLE LEMON turned them in; BIG LEMON was imprisoned while LITTLE LEMON was set up as a deception channel, passing information to the Germans from June 1943 to September 1944. See BIG LEMON; LITTLE LEMON.

LESTER   A Joint Security Control channel through the U.S. military attaché in Lisbon. Hungarian. Former press attaché to the Hungarian legation in Lisbon; sympathetically inclined toward the Japanese on a theory of racial similarity between Japanese and Hungarians, and close to the counselor of the Japanese legation. An unwitting channel through whom information was passed to the Japanese through the spring and summer of 1945.

LIBERATORS, THE   An "A" Force channel operated from Baghdad. A minor channel in the 1942–43 period, with a wireless set that never worked properly. No further information.

LILOU   Also known as MISCHIEF or QUACK. A British W/T agent (MISCHIEF) in Sicily who radioed in February 1943 that he was under control and that any message signed LILOU would be genuine. Judged by the British to be under control in his LILOU capacity as well, and deception information was accordingly passed to him. This judgment proved to have been correct; the original LILOU message was genuine but it was monitored by the Italians, MISCHIEF was shot, and the channel worked by an Italian operator imitating MISCHIEF's fist.

LINEAGE   A French-controlled agent in Bordeaux in early 1945; not a success.

LIPSTICK   An MI5 channel. Real name Josef Terradellas. Spanish (Catalan). A Catalan separatist who persuaded the Germans to send him to England as a trained letter-writing agent and promptly reported this to the British in Madrid. Operated from November 1942 to December 1944; but was more interested in Catalan separatist activities than in double agent work.

LITIGANT   A SHAEF channel run by OSS. Real name Mohammed Charrad. Tunisian. An Arab nationalist and fanatical hater of France, formerly active in Arab propaganda; installed as a staybehind at Sète in the summer of 1944 and arrested in late November. Operated as a W/T channel from December 1944 to April 1945.

LITTLE LEMON   An "A" Force channel, wireless operator of THE LEMONS (q.v.). Greek. Son of an Athens restaurateur. Turned the party in soon after arrival on Cyprus, and operated as a deception channel until September 1944. Depicted to the Germans as a ladies' man with key sources among the chorus girls of the local cabarets. See THE LEMONS; GABBIE; HELGA; MARIA; MARKI; SWING-TIT; TRUDI.

LLAMA An "A" Force channel. Libyan. A staybehind left by the Italians to operate a transmitter from the hills behind Tripoli after Montgomery occupied Tripolitania. Employed from March 1943 to send false order of battle and troop movement information. Proposed as a secure channel to the Italian military at the time of the surrender negotiations, but this was vetoed by London.

LODGE A British (MI6) double agent in Montevideo, transmitting material approved by Joint Security Control. Presumably Uruguayan. Worked by W/T and secret ink to Hamburg from November 1944, with two notional sources, ALOSI and GOMEZ *(qq.v.)*.

LOYAL An "A" Force channel. Italian. A courier arrested bringing in a transmitter for AXE *(q.v.)*. The transmitter was used briefly in November and December 1944 to pass order of battle information, and then shut down to protect AXE.

LUC German name for FOREST *(q.v.)*.

LUÍS A notional source for BROMO *(q.v.)*, a notional employee of the War Production Board.

LUIZOS An "A" Force channel. A notional Greek Cypriot in the British Army, a source for THE PESSIMISTS *(q.v.)*.

LYNCH An MI5 channel to the Spanish military attaché in London.

MADELEINE See FIREMAN.

MAGNET An informant for the Sicherheitsdienst in Belgium, arrested through information provided by FRANK *(q.v.)*, and subsequently a notional source for the latter.

MANAGER One of PEASANT's notional sources, the unnamed manager of Shell Oil's Washington office. See PEASANT.

MARCHAND A French-run SHAEF channel. Presumably French. Located in Grenoble. Of considerable utility, but little further is known.

MARGERY A British Admiralty channel. British subject; a seaman used in 1942 as a contact with German intelligence in Lisbon.

MARIA Rumanian. Codename for one of a group of genuine chorus girls who worked the cabarets of Nicosia in Cyprus and were notional sources for LITTLE LEMON *(q.v.)*. See GABBIE; HELGA; MARKI; SWING-TIT; TRUDI.

MARIE NICOSSOF's notional girlfriend. See MISANTHROPE.

MARKI Bulgarian. Codename for one of a group of genuine chorus girls who worked the cabarets of Nicosia in Cyprus and were notional sources for LITTLE LEMON (q.v.). See GABBIE; HELGA; MARIA; SWING-TIT; TRUDI.

MARMALADE A D Division channel. Real names Lal Khan (leader of the party), Ganga Bahadur, and Abdullah Khan. One of the BATS parties (q.v.), put on the air from Dibrugarh in May 1943; moved to Imphal in January 1945. Never high-grade or satisfactory, with no visible results, but operated to the end of the war.

MARTIN A SHAEF channel. Operated briefly as an unofficial French double agent at Beziers. (Not to be confused with ASSASSIN (q.v.), whose French codename was MARTIN.)

MARTIN French codename for ASSASSIN (q.v.).

MARQUIS See CRUDE.

MARY A D Division channel. A wireless station of SILVER's notional All-India Revolutionary Committee set up at Delhi in July 1943. It communicated initially with a station at the German legation in Kabul called OLIVER; and soon thereafter also with a station at Burg, near Magdeburg, called TOM (qq.v.), with which a link was maintained to the fall of Germany. In 1944 it was arranged that MARY would communicate also with a Japanese-sponsored station to be set up at Subhas Chandra Bose's headquarters in Burma called RHINO, and subsequently as well with a German-sponsored station at Bose's headquarters called ELEPHANT (qq.v.). Fragmentary communication was achieved with RHINO in 1945 but the MARY-ELEPHANT link never opened.

MARY A notional girlfriend and occasional source for TATE (q.v.). Supposedly an Admiralty cipher clerk; temporarily transferred to the U.S. Navy mission in London, late 1943 to spring 1944.

MASK A SHAEF channel. A French-controlled agent in Paris, whose W/T set was recovered by the American SCI near Cherbourg; failed to make contact with the Germans.

McTAVISH Name assigned by the British to a notional agent of the BRASS group (q.v.) situated in the Meiktila area, created by the Japanese in November 1944 probably to give the impression of a stop-line in Central Burma further north than the real one.

MEADOW   A SHAEF (21 Army Group) channel. Belgian. A Flemish nationalist who had been a double agent for Belgian intelligence, penetrating the Abwehr, since 1939, and was left by the Germans as a staybehind in Brussels with the assignment to organize a network of Flemish subagents reporting military information. But he had no radio operator and despite considerable effort in view of his great potential, it was never possible to activate him as a channel to the Germans.

MECHANIC   A SHAEF (21 Army Group) channel, never successfully opened. A Belgian merchant captain, interned in Germany after his ship was torpedoed, who (after consulting his senior officers and managing to notify London) accepted training for an assignment in North or South America as an Abwehr agent in return for being released. When summoned to Germany on the advance of the Allies in August 1944 he refused to go and turned himself in to the Allies on their arrival. Was to have been run as a double agent channel with VAN MELDERT *(q.v.)* as his radio operator, but the latter never made contact.

MERCY   An informant for the Sicherheitsdienst in Belgium, arrested through information provided by FRANK *(q.v.)*, and subsequently a notional reserve W/T operator for the latter.

METEOR   Real name Eugn Sostaric. A Yugoslav officer trained by the Germans to serve as a triple-cross agent in Britain. Worked with TRICYCLE *(q.v.)* from November 1943 to May 1944.

MICASE   See MIKE.

MIKE   (FBI file name MICASE.) An FBI channel. Real name Letsch. Dutch. A motion picture executive sent as a courier to the United States with money for PAT J *(q.v.)*. Turned himself in to the American embassy in Madrid and arrived in New York in November 1944. Remained in the United States, cooperating with the FBI, to the end of the war.

MIMI   German codename for PESSIMIST Y *(q.v.)*.

MINARET   British codename for RUDLOFF *(q.v.)*.

MINT   A SHAEF (21 Army Group and Belgian Deuxième Direction) channel. Belgian. An Abwehr agent, son of the former Belgian consul in Zurich; parachuted south of Brussels in mid-March 1945. Operated as a penetration channel only.

MISCHIEF   See LILOU.

MISANTHROPE   NICOSSOF's notional girlfriend MARIE. Notionally a clever young Greek of good education who hated the British and had a wide circle of acquaintances in Cairo military circles; she not only picked up information for Nicossof but over time learned to encipher his messages and operate his transmitter when he was absent. See Cheese; Marie; B.G.M.

Model   A SHAEF (21 Army Group) channel. Belgian. A Sicherheitsdienst stay-behind agent in Belgium who did nothing to carry out his assignment but in fact joined the Canadian Army and fought bravely for the Allied side. When detected he was sent through the lines to penetrate the Sicherheitsdienst; though he was well received by the Germans and promised the Iron Cross, the war ended before they could give him a mission.

Mohican   An August 1945 London Controlling Section proposed channel, in which a notional Irish-sympathizing wife of a British POW in Japanese hands (a genuine one known to be dead would be selected) would write to the Japanese embassy in Bern offering to pass them information to be supplied by her brother in the War Office, in exchange for money and the promise of good treatment for her husband. Never implemented because of the end of the war.

Monarch   An "A" Force channel. A minor channel in the Middle East in late 1943 or 1944; nothing further known.

Monarch   A French-controlled channel at Arcochon, apparently briefly in contact with the Germans in Spain just after the end of the war; nothing further known.

Monkey   A channel to Sweden; nothing further known.

Monoplane   A SHAEF (6th Army Group) channel. Real name Paul Jeannin. French. Prior codenames Jacques and Twit; German codename Normandie. Former radio operator on the French liner *Normandie;* embittered against the United States by having been interned with the ship after the fall of France. Recruited by the Abwehr and installed in Marseilles by March 1943. Began transmitting shortly after the Allied landing in southern France, August 1944. Arrested in October and operated as a W/T channel thereafter until April 1945. A major channel for Operation Jessica.

Moon   A channel in Istanbul. Nothing further known.

Moonbeam   Garbo's Agent Five *(q.v.);* FBI codename Glocase *(q.v.).*

Moonstone   An FBI channel. Swiss. A Quaker teacher on his way to the United States looking for a job in early 1945, who had been trained by the Germans in Spain for maritime intelligence and agreed to act as a double agent once he reached the States and set up radio connection in May. The end of the European war presumably forestalled any use of the channel.

MORON   A December 1942 Middle East project for passing deceptive material by handling W/T links with "studied incompetence." Nothing further known.

MORTAL   A SHAEF channel. A French double agent at Marseilles who gave some promise but never materialized as a long-term double agent case.

MOSELLE   An "A" Force channel. A party of Spanish and Sardinian saboteurs that the British infiltrated into Sardinia in 1943, communicating by wireless with Algiers. It was soon evident that they had been captured and were working under control, so they were turned over to "A" Force for use as a deception channel. Employed for keeping alive Axis expectations of an attack on Sardinia; discontinued when the Axis evacuated Sardinia early in September 1943.

MULETEER   A SHAEF channel. Real name Marcel Altazin. French. A staybehind in Paris. Arrested in early October 1944 and operated until April 1945, primarily for penetration.

MULLET   Real name Thornton. British, born and lived in Belgium. Operated as a source for HAMLET *(q.v.)* from late 1941 to May 1944.

MUTT   Real name John Moe. Norwegian. See under JEFF.

MYRIAM   A SHAEF (6th Army Group) channel. A female line-crosser run by the French in the 6th Army Group area and employed in connection with Operation ACCORDION.

NETTLE   Real name Goldschmidt. German. A talent spotter and recruiter for the Abwehr in Spain who actually worked for MI6.

NEVI   A subagent of RUDLOFF. A notional civilian employee of the Navy Department in Washington. See RUDLOFF; OSTEN.

NICOSSOF, PAUL   An imaginary Syrian of Slavic extraction, originally the notional W/T operator for the CHEESE channel, becoming himself the CHEESE channel after Levi's departure. See CHEESE.

NORBERT   A French-run "A" Force channel from North Africa. French. Leader of the RAM trio *(q.v.)*. An extremely intelligent flight sergeant in the French air force with years of experience as a double agent. See ÉDOUARD; RAM.

NORMANDIE   German name for MONOPLANE *(q.v.)*.

OAF   A SHAEF channel. A French-run agent at Lille who could never make contact with the Germans.

OATMEAL   A D Division channel. A party of Malays landed by SOE in Malaya at the end of October 1944, captured and turned by the Japanese, and turned over to D Division as a deception channel; employed to the end of the war.

ODIOUS   An "A" Force channel. Real name Max Brandli. German; a Swiss citizen; a watch salesman in Istanbul. Recruited by MI6 in 1942; eventually confessed that he continued to work for the Abwehr. Never apparently used for deception.

OFFICER   A subsource for RUDLOFF; a notional friend of RUDLOFF's friend WASCH in the Army Service Forces in San Francisco. See RUDLOFF; WASCH.

OLIVER   An "A" Force channel. A fictitious South African field security policeman who supposedly landed in North Africa in November 1942, and sent order of battle information by mail to a friend in London who was supposedly a German agent. Operated from January to July 1943.

OLIVER   A D Division channel. The German wireless station in Kabul to which MARY worked. See MARY; SILVER.

OMELETTE   A July 1943 "A" Force project for passing deceptive information by messages in a readily breakable cipher on normal communications channels. Nothing further known.

ONYX   An FBI channel, apparently never used. Probably Spanish. Reached the United States from Spain at the end of 1944 with an extensive questionnaire from the Abwehr. Considered for use as a double agent but evidently was never employed and left the country in March 1945.

OPTIMIST   An "A" Force channel. Real name Gulzar Ahmed. Indian. Utilized to leak false information concerning the movement of the 5th Indian Division in 1943.

ORLANDO   According to one source, may have been a German name for CHEESE *(q.v.)*.

OSTEN   A notional subsource of RUDLOFF's notional source NEVI. See RUDLOFF; NEVI.

OSTRICH   A SHAEF channel. A double agent, apparently in the Cherbourg area.

OSTRO   Real name Paul Fidrmuc. Czech. Not a "special means" channel; a source
of uncontrolled misinformation to German intelligence. A former Abwehr agent
who settled in Lisbon from 1940 and throughout the war furnished bogus information to the Abwehr from purported agents abroad. See ECCLESIASTIC.

OTTO   A subagent of PAT J *(q.v.);* a notional laborer in the Brooklyn Navy Yard of
German extraction and sympathies.

OUTCAST   An MI5 or MI6 channel. A British double agent in the United Kingdom who passed information to Onodera, the Japanese military attaché in Stockholm. Possibly the notional source for SUNRISE *(q.v.).* See SUNRISE; JAM.

OWL   A D Division channel. Real name Adjudya Das. One of the BATS parties
(q.v.), put on the air from Calcutta in May 1943 and operated to the end of the war.
Not high-grade but increasingly valuable, with a number of notional sources in-
cluding HARI SINGH, KANSHI RAM, and the RADIO BABU *(qq.v.).* Transferred to
Rangoon after its fall and transmitted from there in July–August 1945. Useful in
STULTIFY, CLOAK, SCEPTICAL, and SLIPPERY.

PALIMPSEST   See GERARD.

PANCHO   See DRAGOMAN.

PANDORA   Anonymous letters written to the German minister in Dublin, supposedly from an Irish hater of the English, into which deception items were occasionally inserted.

PAPADOPOULOS   A notional dockyard superintendent in Beirut; a source for GALA
(q.v.).

PAT J   An FBI channel. Real name Alfred Meiler. Dutch. A diamond cutter by
trade; a German agent in the First World War. Assigned to make his way to the
United States in 1942; turned himself in to the Dutch embassy in Madrid, passed to
MI6 and in turn to the Americans. Established by the FBI as a diamond cutter in
New York, and from February 1943 to the end of the European war notionally
transmitted regularly to Germany from a station on Long Island; transmitter actu-
ally operated by the FBI. Subagents HOLTZ, HERMAN, and OTTO *(qq.v.);* see also
MIKE.

PAWNBROKER   A D Division channel. A party of eight agents reporting to Subhas
Chandra Bose, landed by Japanese submarine on the Kathiawar coast in December

1943. Captured, and thereafter notionally established with W/T stations in Bombay and Calcutta, the latter communicating with Bose in Rangoon and subsequently in Bangkok. In communication with Bose and the Japanese to the end of the war. Became a high-grade channel, important in CLAW, SCEPTICAL, and SLIPPERY.

PEANUT   A notional source for HICCOUGHS *(q.v.)* in Penang, who indiscreetly discussed naval affairs.

PEASANT   An FBI channel. Real name Helmut Siegfried Goldschmidt. A Dutch Jew. An adventurer recruited by the Abwehr to go to the Western Hemisphere. Turned himself in to the British in Lisbon in 1943 and was taken to England. A compulsive womanizer and otherwise difficult to handle; kept under surveillance in England and notionally reached the United States, where he was notionally employed by Shell Oil in Washington. Communicated (actually the FBI) by wireless with Germany from August 1944. Sources BATES, KLEIN, MANAGER, ROBERTS, SAUNDERS, WAVE *(qq.v.)*.

PEASANT   An unsuccessful French-run channel in Paris; nothing further known.

PEDANT   An "A" Force channel. A German staybehind W/T agent in Athens after the British occupation in October 1944, utilized for a few weeks in support of Operation SECOND UNDERCUT.

PEGASUS   A SHAEF channel. A Franco-British controlled agent operating from Toulouse.

PENNANT   A French-controlled Sicherheitsdienst agent in Paris; made brief W/T contact with the Germans which could not be maintained.

PEPPERMINT   An MI5 channel. Real name Jose Brugada Wood. Spanish. An assistant press attaché in the Spanish embassy in London, recruited as a German agent by Alcázar de Velasco and employed to send controlled data via the Spanish diplomatic bag from December 1941 to April 1943.

PESSIMISTS, THE   An "A" Force channel. A three-man party, speaking several languages and well trained by the Abwehr, assigned to establish themselves in Damascus in the autumn of 1942 and report to the Abwehr in Athens. Landed near Tripoli in Syria in a small boat from a refugee schooner in October 1942 and promptly captured. A deception channel from November 1942. See PESSIMIST X, PESSIMIST Y, PESSIMIST Z.

PESSIMIST X   Also called COSTA. An "A" Force channel, one of THE PESSIMISTS *(q.v.)*. A young Italian, half Swiss in ancestry. Originally the leader of the PESSIMIST party. Imprisoned after capture. Notionally traveled round Syria gathering information for PESSIMIST Y to send, and from time to time, at the request of the Abwehr in Athens, wrote letters in secret ink to sundry Axis contacts.

PESSIMIST Y   Also called JACK (MIMI to the Germans, short for Demetrios). An "A" Force channel, one of THE PESSIMISTS *(q.v.)*. An Alexandrian Greek. A professional singer, trained as the wireless operator of the party. Middle-aged. Had been in touch with MI6 in Athens before going to work for the Abwehr. In contact with the Abwehr in Athens from November 1942, working from the same house as QUICK-SILVER *(q.v.)*, transmitting information notionally gathered by PESSIMIST X and PESSIMIST Z.

PESSIMIST Z   Also called BASILE. An "A" Force channel, one of THE PESSIMISTS *(q.v.)*. An Alexandrian Greek, a thug with a drug-smuggling conviction on his record. Imprisoned after capture. Notionally established a link with a smuggler operating into Palestine, enabling him to furnish PESSIMIST Y with information supposedly from that country; also notionally got a job with a cross-desert transportation company, which enabled him notionally to visit Iraq frequently and bring back items of information for PESSIMIST Y.

PIET   A notional disaffected South African confidential clerk to the chief of the operations staff at GHQ Middle East, a habitual complainer dogged by money troubles and problems with women; supposedly a source for CHEESE *(q.v.)* in 1941.

PIG   An "A" Force channel. A German agent in Istanbul who had been told to infiltrate himself into the British service. Utilized in BARCLAY in 1943, when he was given a packet of money to take to CAPRICORN *(q.v.)*, which included pound notes surcharged for use by Allied forces in Greece.

PILGRIM   A SHAEF channel. A French double agent based in Mulhouse, run from February 1945.

PIN   A Joint Security Control channel through the U.S. military attaché in Buenos Aires, 1944–45. A former Japanese army officer who had friendly relations with Rear Admiral Katsumi Yukishita, the Japanese naval attaché in Buenos Aires, and subsequently had connections with Konomi Miyamoto, the leading Japanese agent in Argentina.

PIP   A SHAEF (21 Army Group and Belgian Deuxième Direction) channel. Belgian. The radio operator of a party of three Flemings (PIP, SQUEAK, and WILFRED) dropped by parachute into Belgium near Malines in early February 1945 under the auspices of Skorzeny's SS Jagdverbände, with orders to conduct espionage, sabotage, and assassination. Run as a penetration channel. See SQUEAK; WILFRED.

PITT   A SHAEF channel. An American agent in the Stuttgart vicinity, caught and turned by the Germans and run as a triple-cross by the Allies late in the war.

PLATO   A British Admiralty channel. A Greek seaman used briefly by the British in 1942 as a double agent to the Germans in Lisbon.

POINTER   A D Division channel. A group of ex-MI6 agents, Karens, captured by the Japanese and then left as staybehind agents in Rangoon and turned by D Division shortly after the liberation of that city, late spring and summer 1945.

PORTUGUESE CHANNELS   A group of 16 "A" Force channels to the Germans in Portugal established August–September 1941. Generally unsuccessful.

PRIMO   An "A" Force channel. Italian. A former airline pilot and radio operator who was caught with two helpers behind the Allied lines in January 1944. His helpers were sent to prison in Algeria while he operated a radio set from a flat in Naples from February to September.

PUPPET   Real name Mr. Fanto. British. An associate of HAMLET and a friend of von Falkenhausen, the governor of occupied Belgium; passed information through HAMLET from April 1943 to May 1944. See HAMLET.

PURPLE WHALES   A D Division channel. Possibly the agent known to the Japanese as "GH"; real name possibly Lin Ch'ing-shan. A channel controlled by Chinese intelligence through which false high-level documents originating with D Division were passed to the Japanese.

PYRAMIDS, THE   An "A" Force channel. An Egyptian gang who cooked up on their own the idea of building a transmitter and selling information to the Germans. A gang member gave the plan away and agreed to open up communication himself with a set furnished by SIME, transmitting to Budapest. Never got in satisfactory touch with Budapest; terminated in early 1943.

QUACK   See LILOU.

QUAIL   A SHAEF (12th Army Group) channel. Real name Pierre Schmittbuhl. French; a German agent since 1933. Parachuted near Metz in early February 1945 and run as a W/T channel in Nancy until April.

QUALITY   A SHAEF channel. A German staybehind agent at Kaiserslautern; made contact at the very end of the war but never actually operated as a double agent.

QUAY   A SHAEF (6th Army Group) channel. A French agent in Baden-Baden run as a "quintuple cross" by the French counterintelligence unit attached to 6th Army Group.

QUEASY   A SHAEF channel. A French double agent who obtained a W/T set and was to be run as a double agent at Giromagny, but the project never developed.

QUICKSILVER   An "A" Force channel. Real name George Liossis. Greek; an officer in the Greek air force and nephew of General Liossis, a high-ranking Greek officer in Cairo. Probably in touch with British intelligence before offering himself to the Abwehr. After training by the Germans, he was dispatched by boat from Athens in August 1942 with two companions, RIO and GALA (qq.v.), with orders to land in Syria and transmit from somewhere near Beirut. Intercepted by the Royal Navy, and promptly gave himself up; a W/T channel for "A" Force for two years from October 1942, operating from the same house as PESSIMIST Y (q.v.). See KHALIL; KYRIAKIDES.

RA 506   A notional agent of HICCOUGHS in Akyab.

RADIO BABU   A notional Calcutta Bengali with a radio shop and a brother in Ranchi; a source for OWL (q.v.).

RADJA   See RAJAH.

RADJAH   See RAJAH.

RAJAH   (Also spelled RADJA and RADJAH.) A D Division channel. A notional disaffected Indian courier for Strategic Air Force headquarters in Calcutta (role played by a sergeant in the Calcutta police), supposedly recruited by FATHER to take over his duties when he himself should be returned to Belgium. Took over the DUCK transmitter in January 1945. See FATHER; DUCK.

RAGS   GARBO's AGENT 7(4) (q.v.).

RAINBOW   An MI5 channel. Real name Pierce. A British subject educated and employed in Germany till 1938. Returned to Britain, reported an effort to recruit him for espionage after becoming aware of espionage by an acquaintance. Allowed to be "recruited" as a double agent in early 1940. Initially a pianist in a dance band, subsequently employed in a factory. Utilized to send commercial and industrial information and as a channel to other agents in Britain. Discontinued June 1943 owing to weakening of German interest.

RAM   A French-run "A" Force channel from North Africa. French. A sergeant in the French Army, recruited by the Germans in Paris and sent to Algiers in early

1942, working in fact for French intelligence. Notionally employed in the communications branch of the French Army headquarters in Algiers. Radioed regularly to the Germans in Paris from March to September 1943. Not himself particularly bright and obtained only low-grade material on port activities in Algiers; but he was the W/T operator for two others, NORBERT and ÉDOUARD *(qq.v.)*.

REP   A source for RUDLOFF *(q.v.)*. A notional engineer at the Republic Aviation aircraft factory on Long Island and at the Brooklyn Navy Yard; brother of WASCH *(q.v.)*.

RG 273   A notional agent of HICCOUGHS *(q.v.)* in Taungup, subsequently in Sandoway.

RHINEMAIDENS, THE   A SHAEF (21 and 12th Army Groups) channel in the last weeks of the war in Europe. A group of five young (ages 18 to 21) female Luftwaffe signal operators from the area around München-Gladbach who had been released from active duty when the airfield there became nonoperational in September 1944, and were subsequently recruited and hastily trained to act as staybehind agents to send reports on troop movements, unit insignia observed, and the like. One of these (LAZY, *q.v.*), turned herself and the rest of the group in to the Americans upon their arrival early in March 1945, and she and BLAZE *(q.v.)* were used for deception thereafter.

RHINO   A D Division channel. A Japanese-sponsored wireless station at Subhas Chandra Bose's headquarters to communicate with the MARY station of SILVER's notional All-India Revolutionary Committee at Delhi. Achieved only fragmentary communication with MARY from the spring of 1945. See MARY; SILVER.

RI 919   Notional agent of HICCOUGHS *(q.v.)* in the Andaman Islands.

RIO   Real name Bonzos. Greek. A thug who had worked for the Gestapo in Greece, fobbed off by them on the Abwehr and made part of the QUICKSILVER party. Factually imprisoned in Palestine but notionally a soldier in the Greek Army in the Middle East and thereafter a sailor in the Greek Navy, notionally furnishing information to QUICKSILVER in both capacities.

RK 459   A member of the HICCOUGHS *(q.v.)* network; moved from Prome to Yenanyaung in February 1945.

RO 914   Notional agent of HICCOUGHS *(q.v.)* in Rangoon.

ROBERTO   German codename for CHEESE *(q.v.)*.

ROBERTS One of PEASANT's sources, a notional employee of the War Production Board. See PEASANT.

ROGER Notional wireless operator for GILBERT *(q.v.)* when the latter moved to the south of France in the autumn of 1944. Notionally replaced by Jo *(q.v.)* when GILBERT moved to Paris.

ROMNEY A November 1942 proposal to set up a notional smuggling ring on the Syria-Turkey frontier as a deception channel. Apparently not adopted.

ROSE, MR. Codename for Renato Levi (CHEESE) after his return to Egypt in 1944.

ROSIE A SHAEF channel. A triple-cross agent in San Remo who was worked for a time by the OSS in Nice in 1944–45.

ROVER An MI5 channel. A Polish naval officer recruited by the Abwehr as a prisoner of war and forced laborer; reached England in May 1944 and his transmitter was operated by MI5 to the end of the war.

RP 790 Notional agent of HICCOUGHS *(q.v.)* in Bhamo.

RQ 318 Notional agent of HICCOUGHS *(q.v.)* in Shwebo.

RR 545 HICCOUGHS collective designation for all agents. See HICCOUGHS.

RUBY An "A" Force channel. A low-grade W/T channel run from North Africa from December 1943 to February 1944.

RUDLOFF An FBI channel. Real name Jorge José Mosquera. Argentine. British codename MINARET. A leather merchant in Buenos Aires, Montevideo, and Germany, who spent some years in Hamburg in the 1930s and undertook to work for the Abwehr in order to be allowed to take his profits out of Germany. Sent to Montevideo in 1941, and there got in touch with the United States embassy. Brought to New York in November 1941 and set up by the FBI as a trading company. Communicated with Hamburg by radio from February 1942, via a station on Long Island operated by the FBI, and continued transmitting several times a week until the fall of Germany. Acquired a substantial network of notional subagents, including REP, WASCH, NEVI, OSTEN, OFFICER, and UNNAMED FRIEND *(qq.v.)*.

RUPERT A Joint Security Control channel through the U.S. military attaché in Buenos Aires, 1944–45. A German through whom information was fed to Konomi Miyamoto, the leading Japanese agent in Argentina.

RY 327   The notional head of the HICCOUGHS *(q.v.)* network.

RZ 282   Notional female agent of HICCOUGHS *(q.v.)* in Rangoon.

SANDRU   A notional porter at a Damascus hotel; a source for THE PESSIMISTS *(q.v.)*

SAUNDERS   One of PEASANT's sources, a notional Air Force captain. See PEASANT.

SAVAGES, THE   An "A" Force channel. Greek and Greek Cypriot. A pair of newly-weds and their best man, who landed on Cyprus in a caïque from the Piraeus in July 1943 with money and a transmitter. Their principal motivation had been to get out of Greece, but they unwisely delayed in admitting their real nature to the British. See SAVAGE I; SAVAGE II; SAVAGE III.

SAVAGE I   An "A" Force channel. Greek Cypriot. The leader of the SAVAGES party and best man at the wedding of the others. A law student and former soldier in the Greek Army. A wireless operator; was moved from Cyprus to Cairo, where he was supposedly employed in the Allied Liaison Branch at GHQ Middle East, and trans-mitted from there from August 1943 to the end of the war.

SAVAGE II   Greek Cypriot. A physician resident in Athens, recruited by the Abwehr in December 1942; he in turn recruited SAVAGE I in January 1943. Married SAVAGE III shortly before their venture in espionage. Was kept in a detention camp for the duration of the war.

SAVAGE III   Greek. A physician who married SAVAGE II shortly before their venture in espionage. She likewise spent the rest of the war in a detention camp.

SCOOT (SKOOT)   Original codename for TRICYCLE *(q.v.)*. Sometimes appears as SCOUT.

SCOUNDREL   A SHAEF channel. A French-controlled agent in Marseilles from February 1945.

SCRUFFY   Real name Alphonse Timmerman. A Belgian seaman who reached Britain in September 1941 as a secret ink letter writer. Operated by MI5 in a delib-erately clumsy fashion to try to convince the Germans that the British did not run double agents skillfully. Abandoned when the Germans appeared unable to recog-nize that he was under control.

SHACKLE   A SHAEF channel. Codename for two agents parachuted near Chartres in January 1945. French-controlled; gave promise but failed to materialize as a dou-ble agent operation.

SHEPHERD A SHAEF channel. A Frenchman who arrived in Britain in March 1944 equipped with secret ink for reporting to the Germans. Not a valuable channel; MI5 allowed NETTLE *(q.v.)* to recruit him in the hope that this would enhance NETTLE's credibility with the Germans.

SILENT A SHAEF channel. A trio of agents named Durandeau, Themez, and Dubois. French. Parachuted into France February 1945; a French-run case notionally located at Lille.

SILVER A D Division channel. One of the great deception double agent channels of the war. Real name Bhagat Ram Talwar. An up-country Hindu in his early thirties, able to pass as a Moslem tribesman of the northwestern hill country. An Indian nationalist whose brother had been hanged for assassinating a British official, SILVER was a Communist whose primary loyalty was to the Soviet Union. Played in 1941 a major role in the escape of the Indian nationalist Subhas Chandra Bose from British custody and his flight to Afghanistan and thence to Germany. Recruited by the Germans in Kabul as an Axis agent (actually a Soviet double agent); arrested by the British in 1942 and his true allegiance revealed to the British by the Soviets, who consented to his acting as a double agent for the British. Notionally head of an imaginary anti-British All India Revolutionary Committee with members throughout India. Made five journeys to Kabul from 1943 to 1945 to report to the Germans, and, later, the Japanese, and from July 1943 to the end of the German war communicated directly with Kabul and with Germany through a wireless station in Delhi codenamed MARY *(q.v.)*. A high-quality channel through which much information of a very high grade was passed. Awarded the Iron Cross by the Germans.

SINK A French-controlled agent operating from Montpellier from March 1945.

SKULL A SHAEF channel run by the OSS. Real name Jean Senouque. French. German codename CHARLES. A staybehind agent to cover the western base of the Cherbourg peninsula. Arrested August 1944 and used as a W/T double agent from late November 1944 to the end of the war.

SLAVE An "A" Force channel. Egyptian. A young journalist recruited by the Germans in the Balkans in 1941. The British laid hands on him in Sofia and he agreed to bring a W/T set to Egypt and transmit at British direction. When he got to Egypt he was told that the transmitter was useless and that the project was dropped. The transmitter, which in fact was in perfectly good working order, was taken from him and run by SIME. Carried some "A" Force items in the period May to July 1942, but it had some technical problems and once CHEESE was rehabilitated SLAVE was allowed to lapse.

SMOOTH   An "A" Force channel. A Turkish customs official living in Antioch, believed by the Germans to be in their service with a network of agents but in fact employed by the British. The channel through which information from HUMBLE *(q.v.)* was passed.

SNARK, THE   An MI5 channel. Real name Mihailovic. Yugoslav. A female domestic servant who arrived in Britain in July 1941 equipped with secret ink. Utilized to report food prices and living conditions until discontinued in March 1943 for lack of German interest.

SNIPER   An MI5 and subsequently SHAEF (21 Army Group) channel. Belgian air force pilot who reached Britain in November 1943 with assignment to report on technical matters and antisubmarine warfare. Reported by secret ink from Britain; returned to Belgium after the liberation, was provided by the Germans with a buried W/T set, and operated by 21 Army Group from early January 1945 to the end of the war.

SNOW   Real name Arthur Owens. Welsh. The first of the double agents. German codename JOHNNY. In the 1930s, traveled between England and Germany on business and would report information to the British Admiralty and subsequently to MI6. Recruited by the Germans in 1936 and functioned as a German agent to the outbreak of war, being furnished with a W/T set in January 1939, which was operated from Wormwood Scrubs prison in his brief internment in September 1939. Made several direct contacts with the Germans in Rotterdam and Antwerp in 1939–40 and Lisbon in 1941; brought G. W., BISCUIT, CELERY, and CHARLIE *(qq.v.)* into his network, and was the initial contact for SUMMER *(q.v.);* his dealings gave early insight into German methods and ciphers. Discontinued in March 1941 after a visit to Lisbon at which he may or may not have disclosed his role to the Germans.

SPANEHL   An MI5 channel. Real name Ivan Spaneil. A Czech who had served in the French Army and was recruited by SWEETIE *(q.v.)*. Reached England with GIRAFFE *(q.v.)* in September 1940; case soon discontinued as unproductive and he was allowed to serve in the French forces in the Middle East.

SPANIP   General codeword for FBI activity targeted against Spanish cooperation with Axis espionage (from "Spanish" plus "Nipponese").

SPRINGBOK   An MI6, subsequently MI5 and OSS, channel. Real name Hans Christian von Kitze. German. He had lived in South Africa and was recruited by the Abwehr to go back via South America. Turned himself in at Rio and on MI6's advice persuaded the Abwehr to let him go to Canada instead. Taken over by MI5 at the end of 1942 and attempted to be revived with OSS; a failure owing to U.S. State Department incompetence, and shut down in August 1943.

SQUEAK   A SHAEF (21 Army Group and Belgian Deuxième Direction) channel. Belgian. One of a party of three Flemings (PIP, SQUEAK, and WILFRED) dropped by parachute into Belgium near Malines in early February 1945 under the auspices of Skorzeny's SS Jagdverbände, with orders to conduct espionage, sabotage, and assassination. PIP was run as a penetration channel. See PIP; WILFRED.

SRAIEB   An Arab saboteur and agent of the Parti Populaire Français who was briefly involved with GILBERT *(q.v.)* in 1943.

STANDARD   A SHAEF channel. A Sicherheitsdienst double agent supposed to operate in Rouen but never able to make contact.

STANLEY   GARBO'S AGENT SEVEN *(q.v.)*.

STEPHAN   Real name said to be Klein. Said to be Austrian. According to some sources, one of the original components of the CHEESE channel *(q.v.)*, an agent in Cairo working back to the Abwehr in Athens, who turned himself in together with a radio transmitter. But the declassified records do not confirm this story.

STEPHEN   See STEVEN.

STEVEN   Also spelled STEPHEN. A notional naval liaison officer from Alexandria, a source for GALA *(q.v.)*.

SUMMER   An MI5 channel. Real name Gøsta Carol. Swedish. Parachuted into England in early September 1940 and promptly captured. Operated by W/T to Germany; notionally operated from Cambridge area until January 1941, when discontinued because notionally under suspicion. In fact discontinued after an attempt to escape.

SUNRISE   A channel controlled by MI6 but used extensively by D Division. A genuine European assistant to Onodera, the Japanese military attaché in Stockholm, who was supposed to be in indirect touch with a notional agent in London (possibly OUTCAST *(q.v.)*) through a member of a Swedish intelligence service with worldwide ramifications. The notional agent gathered high-grade information with respect to Southeast Asia through such imaginary entities in London as the "East India Sports Club" and the "China and Pacific War Investigation Committee." The original SUNRISE left Stockholm in 1944, but messages notionally from him (or from the imaginary agent) continued to reach Onodera at least until the end of the European war.

SUNSET   A German intelligence officer in Lisbon who was an informant for the United States. Not used as a channel, although the FBI suggested such use in the summer of 1945 at the prompting of the OSS, when SUNSET claimed he had been approached by a Japanese consular official and asked whether he had anyone in the United States with a radio transmitter who could secure certain information. Joint Security Control vetoed such use on security grounds and because of perceived unreliability of SUNSET.

SWAMI   British codename for Subhas Chandra Bose's lieutenant; not himself a channel.

SWEET WILLIAM   An MI5 channel. Real name William Jackson. British. An employee of the Spanish embassy in London, recruited by Alcázar de Velasco and used by MI5 to send controlled data via the Spanish diplomatic bag from August 1941 to August 1942.

SWEETIE   Real name Wiesner. French. A double agent for MI6 in Lisbon, specially employed by the Germans, and encouraged by MI6, to recruit Czech agents for work in Britain. See GIRAFFE; SPANEHL.

SWING-TIT   Rumanian. Codename for one of a group of genuine chorus girls who worked the cabarets of Nicosia in Cyprus and were notional sources for LITTLE LEMON (q.v.). See GABBIE; HELGA; MARIA; MARKI; TRUDI.

TAKIS   A notional Greek submarine officer, a source for GALA (q.v.)

TATE   An MI5 channel. Real name Wulf Schmidt. Danish (born German in Schleswig-Holstein). Trained with SUMMER, parachuted into England in September 1940, promptly captured and put on the air from October 1940. Held the "long-distance record" for a W/T double agent, communicating regularly with Hamburg until May 1945. Allowed to work as a photographer; notionally working on a farm. Played significant roles in FORTITUDE SOUTH and in naval deception. Awarded the Iron Cross.

TAZY   German name for MULETEER (q.v.).

TEAPOT   An MI6 channel. German. Manager of a food co-op near Hamburg with branches across Germany and in Istanbul. Recruited by British in Istanbul in late 1943. Discovered to be a German plant, but continued as a triple-cross. Communication by radio between MI6 and his residence near Hamburg.

THACKERAY   Notional agent of ANGEL ONE (q.v.) at Chiengrai, Thailand.

TILTMAN, HESSELL   A genuine individual who was a notional source for GUS (q.v.). Washington correspondent for the London *Daily Sketch* and former Far East correspondent for the *Daily Express,* with eight years' residence in Tokyo. One of the notional sources for GUS in Buenos Aires was a weekly airmail letter from Tiltman; GUS had known Tiltman since they met in Warsaw in 1939, and had seen him on a daily basis in New York in the spring of 1940.

TOLLUS   Notionally an informant for GALA *(q.v.);* her distortion of the name of a genuine brigadier (George Tolhurst).

TOM   D Division codename for German wireless station at Burg, near Magdeburg, as a link with MARY, SILVER's station in Delhi. See MARY; SILVER.

TOM X   An FBI channel (FBI file name COCASE). French. Real name Dieudonné Costes. A noted French aviator who was pressured into undertaking an espionage mission to the United States for the Germans in 1942; turned himself in to the Americans in Spain and operated under FBI control in New York. Some of his traffic aided BODYGUARD in early 1944; instrumental in the capture of Paul Cavaillez, a genuine German agent, in 1945.

TRAVEL   A D Division channel. A three-man party reporting to Subhas Chandra Bose, landed by Japanese submarine on the Orissa coast in December 1944. In W/T communication with Bose's headquarters from February 1945, initially from Amritsar and subsequently from Howrah, with a notional agent in Vizagapatam.

TREASURE   An MI5 channel. Real name Nathalie ("Lily") Sergueiev. French citizen of Russian origin, intelligent and temperamental. Reached Britain in August 1943; communicated by secret writing and received messages by wireless; traveled to Lisbon in March 1944 and received a transmitter, which was operated by MI5 because she was too difficult a personality to be trusted to do so. After the liberation of Paris she moved to France in the women's army service, but continued notionally to operate from Britain until terminated for lack of further utility in December 1944. Useful in connection with FORTITUDE SOUTH and in keeping up with the French underground.

TRICYCLE   An MI6 channel. Originally SCOOT or SKOOT. Real name Dusko Popov. A Yugoslav commercial lawyer of good family, educated in France and Germany. Recruited by the Abwehr in 1940; reported this to the British in Belgrade; reached England under MI6's auspices in December 1940. Became one of the Germans' most trusted agents, recruiting subagents and traveling to Lisbon twice in early 1941 and meeting with the Abwehr directly. Lived in London, and later in the United States, the life of a wealthy and dissolute playboy. Sent by the Abwehr to the United States to August 1941 to open a network there. Controlled unsatisfactorily by the FBI. Visited Rio de Janeiro for the Abwehr in the autumn to set up a Brazilian network. In W/T communication with Germany from February 1942. Returned to London in August 1942. Set up with the Germans an escape route out of Switzerland for interned Yugoslav officers; one of these, FREAK, became his radio operator. Shut down in May 1944 when Jebsen (ARTIST) *(q.v.)*, his key Abwehr contact, was arrested and imprisoned.

TROLLOPE   Notional agent of ANGEL ONE *(q.v.)* at Kanchanburi, Thailand, eighty miles west of Bangkok.

TROTTER   A D Division channel. A party of three Madrassi agents of Subhas Chandra Bose, landed by Japanese submarine near the mouth of the Cauvery River in December 1944. Went on the air from Madras in April 1945, but was never a satisfactory channel owing to Japanese ineptness.

TRUDI   Austrian. Codename for one of a group of genuine chorus girls who worked the cabarets of Nicosia in Cyprus and were notional sources for LITTLE LEMON *(q.v.)*. See GABBIE; HELGA; MARIA; MARKI; SWING-TIT.

TURBOT   An MI5 diplomatic channel.

TURKISH CHANNELS   A group of ten "A" Force channels to the Germans in Turkey (nine in Istanbul and one in Ankara) established April–May 1941. They and subsequent Turkish channels were useful throughout the war.

TWERP, THE   Notional member of the PAWNBROKER *(q.v.)* party transmitting from Bombay.

TWIST   An MI6 and "A" Force channel. Turkish. An employee of the Italian consulate in Istanbul who was variously recruited by MI6, the Abwehr, and the Sicherheitsdienst. Used to transmit deception information about Allied plans in the Balkans in the summer of 1944. Shut down after Turkey expelled his German contacts in August 1944; some ineffectual efforts to make him a Japanese agent thereafter.

TWIT   See MONOPLANE.

UMBERTI   An OSS channel in Italy from May 1944. Real name Cesidio Seri. Italian. A sergeant in the Italian air force, trained by the Germans in Rome and put ashore at Mondragone, fifty miles north of Naples, with three other agents in May 1944. Employed as a triple agent to build up the credibility of a double agent run by the British.

UNCLE   The actual uncle of PESSIMIST Y *(q.v.)*, from Zamalek; a source of notional information for him.

UNNAMED FRIEND   An unnamed friend of RUDLOFF's source WASCH, who notionally wanted to set up a W/T operation to the Japanese from the West Coast in 1945. See RUDLOFF; WASCH.

VAN MELDERT   (His real name.) A SHAEF (21 Army Group) channel. Belgian. A hotelkeeper in a village between Antwerp and the Dutch frontier; an Abwehr staybehind in whose hotel MECHANIC *(q.v.)* had been lodged during his training. Gave himself up when his village was liberated, but no radio contact was ever established.

VERDIER   French codename for KEYNOTE *(q.v.)*.

VIOLET   A French/OSS channel in France. Real name Maurice van Dest. French. Arrested in Paris in September 1944, briefly and unsuccessfully attempted as a W/T agent in the last weeks of the war.

WAIT   A notional doctor, a source for KNOCK *(q.v.)*.

WASCH   A source for RUDLOFF *(q.v.)*. A notional civilian with the War Department in Washington, the brother of REP *(q.v.)*.

WASHOUT   An MI6 channel. Real name Ernesto Simoes. Portuguese. Son of night watchman at British Embassy in Lisbon. Sent to England in hopes the case would develop, but he turned out to be unsuited for the work by temperament and ability.

WATCHDOG   An MI5 and Royal Canadian Mounted Police channel. Real name Waldemar Janowsky. German. Landed by U-boat in Canada in November 1942 or January 1943 with a W/T set. Run as a double agent by the RCMP and MI5's Canadian representative until discontinued in the summer of 1943 owing to lack of security among Canadian press and politicians.

WAVE   One of PEASANT's sources. An otherwise unnamed notional Wave, secretary to a high Navy official, supposedly met by PEASANT at a cocktail party and a source of high-level information thereafter. See PEASANT.

WEASEL   An MI5 channel. Belgian. A doctor, experienced and intelligent, who had received secret writing and W/T training from the Germans and fell into British hands in May 1940. There was hope for him but his secret letters received no reply and MI5 suspected that he had managed to insert into them a warning that he was under control.

WHISKERS   A French-run "A" Force channel. An exiled Spanish officer in Algeria whom the Spanish vice-consul believed to be his loyal agent; the vice-consul passed his material to the Abwehr in Spanish Morocco. Utilized from April to August 1943; in July 1943 appointed head of the Spanish secret service in Algeria. A high-grade channel until he was suddenly cut off when the vice-consul was removed from his post for black marketeering.

WHISKY   Not a channel but an abortive plan in April 1945 to put THE SAVAGES in direct W/T communication with the Japanese.

WIDOW, THE   See GERBERS, MRS.

WILFRED   A SHAEF (21 Army Group and Belgian Deuxième Direction) channel. Belgian. One of a party of three Flemings (PIP, SQUEAK, and WILFRED) dropped by parachute into Belgium near Malines in early February 1945 under the auspices of Skorzeny's SS Jagdverbände, with orders to conduct espionage, sabotage, and assassination. PIP was run as a penetration channel. See PIP; SQUEAK.

WIT   A source for KNOCK *(q.v.)*, a notional acquaintance of his with business in the Persian Gulf dock areas.

WITCH   A SHAEF (12th Army Group) channel. Real name Henri Gaillard. French. Dropped by parachute near Etain in late October 1944. Operated by W/T from the vicinity of Verdun and subsequently Metz to the end of the war. Notionally employed at the American Red Cross Club in Metz from late February 1945. Utilized particularly in covering the northward shift of Third Army after the Ardennes offensive. Awarded the Iron Cross.

WIZARD   French. A courier sent in to WITCH and LEAGUE *(qq.v.)* in February 1945, bringing equipment, money, and a new cipher. Allowed to disappear after reporting safe arrival.

WORM, THE   Real name Stefan Zeis. Yugoslav. Reached England in September 1943, and was able to confirm to TRICYCLE the reliability of DREADNOUGHT and ARTIST *(qq.v.)*.

ZIGZAG   An MI5 channel. Real name Eddie Chapman. Known to the Germans as FRITZCHEN. British. Had done prison time in the Channel Islands for safecracking. Volunteered to work for the Germans to return to Britain. Well trained by the Abwehr. Parachuted into England in December 1942 with assignment to sabotage the de Haviland aviation works, report certain information by W/T, and return. Notionally carried out these assignments and returned to occupied France via Lisbon. Parachuted back into England in June 1944 with assignment to send information by W/T; moderately useful but grew talkative about his work and was discontinued in November 1944.

ZULU   An "A" Force channel. Iraqi. The head of the Iraqi police in Mosul. A member of the pro-German Iraqi "National Liberation Column" or "Ratl"; through DOLEFUL *(q.v.)* he procured a transmitter for its use in December 1942 but it was never satisfactorily operated.

APPENDIX III

The Phantom Armies

This is as complete a list of the notional units employed in Allied strategic military deception in the Second World War as could be compiled from the available declassified records.

Component elements of a division sometimes changed, and those given should be considered representative only.

UNITED STATES UNITS

1st Army Group ("FUSAG")

Insignia: A black Roman numeral "I" on a blue pentagon with a double border, white inside red. The insignia was designed by the headquarters itself, and approved by the Quartermaster General after some grousing that "numerals are prohibited in accordance with War Department policy and the placing of black on blue violates the law of visibility." As described by its originators, "This design combines the use of the National colors with the representation of the First U.S. Army Group by the Roman numeral one."

A FORTITUDE SOUTH and FORTITUDE SOUTH II unit. (Originally Bradley's genuine army group, renamed 12th Army Group, after which FUSAG became notional.) Composed initially (FORTITUDE SOUTH) of Canadian First and U.S. Third Armies; subsequently (FORTITUDE SOUTH II) of U.S. Fourteenth and Ninth and British Fourth Armies; plus army troops. Commanded successively by Patton, McNair, and DeWitt. Notional headquarters at Wentworth near Ascot. Scheduled to land in the Pas-de-Calais and drive on Antwerp and Brussels. In August 1944, Fourteenth and Ninth Armies were detached to SHAEF reserve, and First Allied Airborne Army—genuine, but with additional notional components—was added to FUSAG. FUSAG was employed to pose a threat of airborne operations in the Kiel-Bremen area at the time of the Arnhem operation in September. It was disbanded in October.

2d Army Group

This unit, sometimes called SUSAG, was assigned to SHAEF for FORTITUDE SOUTH II to replace FUSAG when that should be activated in Normandy, but was never used because of the change of number of the real FUSAG to 12th Army Group.

No insignia designed.

Twelfth Army

Allocated to SHAEF but never used. No insignia.

Fourteenth Army

Insignia: A white letter "A" on a red acorn standing in its cup. According to the Quartermaster General, "The acorn is a symbol of strength and the letter 'A' is the initial letter of its organization. The colors are those of the Army distinguishing flag."

A FORTITUDE SOUTH II unit. Part of FUSAG. Composed of XXXIII and XXXVII Corps, 9th and 21st Airborne Divisions, and army troops. Commanded by General John P. Lucas. Landed in Liverpool in May and June 1944 (headquarters at Mobberly in Cheshire) and moved to East Anglia between July 11 and 21, with headquarters at Little Waltham in Essex. Scheduled to form the center and left of the Pas-de-Calais invasion, plus airborne drops inland. Detached from FUSAG and placed in SHAEF strategic reserve in August. Moved to Southampton-Brighton area, headquarters at Fareham near Portsmouth, in late August. Moved to France in September, and withdrawn from FUSAG to SHAEF command. GARBO reported to the Germans that in its ranks were "many convicts who were released from prisons in the United States to be enrolled in a foreign legion of the French or Spanish type. It can almost be said that there are brigades composed of gangsters and bloodthirsty men, specially selected to fight against the Japanese, men who are not supposed to take prisoners, but, instead, to administer a cruel justice at their own hands." Not heard of again after it moved to France. Disbanded in October. GARBO reported in March 1945 that it had been disbanded and most of its troops used to replace the heavy American casualties in the Ardennes.

IX Amphibious Corps

Notional amphibious corps associated with Ninth Fleet, established in Alaska in 1944 for WEDLOCK. Composed of 108th, 119th, 130th, 141st, and 157th Infantry Divisions plus supporting services. No insignia.

XVII Airborne Corps

Added briefly to the notional order of battle (in early September 1944) but never communicated to the Germans, this notional corps was formed in September for the Arnhem cover operation and consisted of the notional 9th and 21st Airborne

Divisions and the genuine 17th Airborne Division (still in England at that time). Originally a notional corps of the genuine First Allied Airborne Army, after Arnhem it was transferred to the notional Fourth British Army and then disbanded. No insignia.

XXX Corps

Originally allocated to the United Kingdom; allotted to SHAEF for FORTITUDE SOUTH II but never used. No insignia.

XXXI Corps

Insignia: Three white arrow tails pointing inwards on a blue triangle truncated at all three vertices. According to the Quartermaster General, "The design is blue and white the colors of the corps and the three arrow tails give the impression of all meeting at one point making a perfect score, conveying the impression of the number of the corps."

A VENDETTA unit, notionally commanded by Major General Terry Allen. Part of Seventh Army. Notionally arrived in North Africa from the United States in spring 1944. Composed of genuine 91st Infantry Division and French 1st Armored Division, plus notional 7th and 8th Algerian Infantry Divisions. Headquarters at Mostaganem near Algiers. In VENDETTA, assigned, with attached British 42d Division, to seize the Montpellier Gap and establish a defensive flank in the area Montpellier-Clermont l'Hérault.

Subsequently also used in FERDINAND, the deception plan to cover the Southern France landing in August of 1944. Headquarters at Naples. In an early version of FERDINAND, was an element of Seventh Army designated as a strategic reserve to exploit any breach in the Gothic Line. In its final version, ordered to prepare to mount an amphibious assault on Genoa. Disbanded in autumn 1944.

XXXIII Corps

Insignia: A hexagonal annulet with its parallel sides horizontal, with a arrowhead emanating from each corner, upon a disk divided vertically blue and white, with the colors counterchanged (*i.e.,* the right half white and blue, and the left half blue and white). No official description of the significance of the insignia, but obviously based on three-and-three arrowheads emanating from the annulet.

A FORTITUDE SOUTH II unit. Part of Fourteenth Army. Composed originally of 11th and 48th Infantry and 25th Armored Divisions, plus corps troops. Arrived in England in June, headquarters at Marbury in Cheshire. Moved to Bury St. Edmunds in Suffolk in July. Scheduled for second-wave followup and support for Pas-de-Calais landing. Moved to Romsey near Southampton in August. The 25th Armored Division was transferred out to XXXVII Corps in September; 17th Infantry Division transferred in from XXXVII Corps in first week of October. Disbanded in October and its divisions used as sources for replacements.

XXXV Airborne Corps

Composed of the genuine 11th Airborne and notional 18th Airborne Divisions. Notionally to be stationed on Okinawa from September of 1945 to lend plausibility to diversionary dummy airborne drops scheduled as part of PASTEL TWO, the cover and deception operation in conjunction with Operation OLYMPIC, the invasion of Kyushu planned for November 1945. Headquarters notionally to arrive on Okinawa about September 1, 1945. No insignia adopted.

XXXVII Corps

A FORTITUDE SOUTH II unit. Part of Fourteenth Army. Arrived in England in May–June, with headquarters at Great Baddow, Chelmsford, Essex. Composed of 17th and 59th Infantry Divisions plus corps troops. Originally assigned to Third Army; transferred to Fourteenth Army in mid-July. Scheduled to be in the first wave on the beach at the Pas-de-Calais. Moved to Worthing in Sussex in late August, headquarters at Goring. The 25th Armored was transferred in from XXXIII Corps in September; 17th Infantry transferred out to XXXIII Corps in the first week of October. Withdrawn from FUSAG in September. Left Southampton, apparently for France, with 59th Infantry and 25th Armored Divisions at end of September and never heard of again, though the abortive Plan AVENGER in October 1944 proposed using it as a notional increment to 12th Army Group. No insignia adopted.

XXXVIII Corps

Allocated to SHAEF for FORTITUDE SOUTH II but never used. No insignia.

XXXIX Corps

Originally allocated to the United Kingdom; allotted to SHAEF for FORTITUDE SOUTH II but never used. No insignia.

5th Motorized Division

Allocated to North Africa; could not be changed to 5th Infantry Division when motorized divisions were abolished, since there was a genuine 5th Infantry Division; never used.

Appendix III

6th Airborne Division

Insignia: A blue shield with a white chief invected of three, with white bands pile-wise at the junction of the invects; above it an Airborne tab, the letters AIRBORNE in yellow on black. According to the Quartermaster General, "The shield is blue for the sky and the white chief and bands convey the impression of an open parachute, representing flight in the skies."

A VENDETTA unit. Part of Seventh Army. Headquarters in Sicily. Arrived from the United States in May 1944. Composed of 517th Regimental Combat Team (517th Parachute Infantry Regiment, 596th Airborne Engineer Company, 460th Field Artillery Battalion), 1st Battalion 551st Parachute Infantry Regiment, and 550th Airborne Battalion. To be dropped astride communications at Paulhan to block the roads from the north. Disbanded in Italy in July 1944.

9th Airborne Division

Insignia: A circle of nine white clouds surrounding a blue field bearing a vertical yellow lightning flash. According to the Quartermaster General, "The nine clouds represent the number of the division and the lightning flash represents the striking power coming from the skies and out of the clouds."

A FORTITUDE SOUTH II unit. Date of arrival in England not recorded. Allotted to SHAEF by Bevan in April 1944. Part of Fourteenth Army. Composed of 196th and 199th Infantry Glider Regiments and 523d Infantry Parachute Regiment, plus divisional troops. Headquarters at Leicester. Scheduled to drop inland at Pas-de-Calais. Came under direct command of FUSAG in mid-August, transferred to First Allied Airborne Army, and used as part of the airborne threat to the Kiel-Bremen area in September. Assigned to (notional) XVII Airborne Corps; disposed of in November by telling the Germans that it and the notional 21st Airborne had been merged to form the (genuine) 13th Airborne Division (which was soon in fact to arrive in France).

10th Infantry (Light) Division

Allotted to the notional Middle East order of battle in 1942. Deception was kept so super-secret that Army Ground Forces did not know of this, and duly activated the genuine 10th (Light) Division; fortunately it was not too late to deactivate the notional one, and AGF was thereafter brought into the deception picture. In late April 1945, enemy intelligence was told by a double agent that men wearing the insignia of the 10th Light Division had been seen in New York. By that time the real unit had been renamed "10th (Mountain) Division" in November 1944.

11th Infantry Division

Insignia: A disk segmented in twelve pieces alternating white and blue except for the twelfth, which is black. According to the Quartermaster General, "The division of the clock has been used as the basis for the 11th Division shoulder sleeve insignia, the 12th hour being blacked out."

Originally the 11th Motorized Division, allotted to the United Kingdom. Converted to an infantry division for FORTITUDE SOUTH II. Part of XXXIII Corps. Composed of 178th, 352d, and 392d Infantry Regiments, plus division troops. Arrived in England in June, "well trained," BRUTUS reported to German intelligence. Originally occupied the area round Northwich, Cheshire, with headquarters at Delamere House. To Bury St. Edmunds in Suffolk in mid-July (the former area of the genuine 4th Armored Division). Scheduled to be in reserve at Pas-de-Calais. Moved to Winchester in August, and to Abergavenny, South Wales in October; personnel used for replacements.

Note that there was a genuine 11th Airborne Division, which served in the Pacific.

13th Airborne Division

Allocated to the Middle East but never used. Subsequently a genuine airborne division, notionally formed from the notional 9th and 21st Airborne Divisions. No insignia as a notional division.

14th Infantry Division

Insignia: A yellow saltire on a blue square with chamfered corners. According to the Quartermaster General, "The blue is for the Infantry and the saltire produces the effect of a Roman numeral while the long sides of the square produce the number of the division."

Originally the 14th Motorized Division, allocated to the United Kingdom. Converted to an infantry division for FORTITUDE SOUTH II but never used.

15th Infantry Division

Originally the 15th Motorized Division, allocated to the United Kingdom. Converted to an infantry division and allocated to SHAEF for FORTITUDE SOUTH II but never used. No insignia.

15th Armored Division

Insignia: All American armored divisions employed the same insignia except for the number; see under 25th Armored Division.

Allocated to North Africa and then to SHAEF for FORTITUDE SOUTH II but never used.

16th Infantry Division

Allocated to Middle East notional order of battle in 1942 but never used. No insignia apparently prepared.

17th Infantry Division

Insignia: A white saltire on a septfoil, the top and bottom portions blue and the left and right portions red. According to the Quartermaster General, "The saltire or Saint Andrew's cross produces the effect of a Roman numeral 10 which placed on the septfoil, a seven-lobed figure, produces the number of the division. The background is blue and red, the colors of the Infantry division distinguishing flag."

Originally the 17th Motorized Division, allocated to the United Kingdom. Converted to an infantry division and allotted to SHAEF for FORTITUDE SOUTH II. Part of XXXVII Corps and subsequently XXXIII Corps. Composed of 293d, 336th, and 375th Infantry Regiments and division troops. Arrived in England in June 1944, headquarters at Birmingham. Moved to Hatfield Peverel in Essex in July. Associated with notional naval "Force N," with which it notionally trained at Southampton and Studland. Scheduled to be in the first wave at the Pas-de-Calais. Moved to Brighton-Burgess Hill on the south coast in August. Transferred to XXXIII Corps in early October. Moved to South Wales in October; personnel used for replacements. Disbanded early 1945.

18th Airborne Division

Insignia: A blue shield with white clouds in the upper and right portions with the head of a yellow battleaxe issuing from the clouds on the right. According to the Quartermaster General, "The insignia is representative of an Airborne organization, illustrating armed might sailing out of the clouds."

Composed of 566th Parachute Infantry Regiment and 567th and 570th Glider Infantry Regiments plus division troops. Originally allocated to SHAEF for use in Europe but never directly used there. Reactivated in the States in March 1945, and special means suggested to the Germans that it was being shipped overseas; this was in support of Plan DERVISH, a project for luring German forces away from the Western Front by feigning a major airborne landing deep in central Germany. It was notionally to be stationed on Okinawa in September of 1945 as part of the notional XXXV Airborne Corps, for use in the invasion of Japan. Quartering parties were notionally to arrive on Okinawa about August 15, 1945, and the division itself about September 1.

19th Infantry Division

Allocated to SHAEF for FORTITUDE SOUTH II but never used. Composed of 572d, 573d, and 578th Infantry Regiments and division troops. No insignia.

21st Infantry Division

Originally the 21st Motorized Division, allocated to the United Kingdom. Converted to an infantry division and allotted to SHAEF for FORTITUDE SOUTH II. Never used; presumably replaced by 21st Airborne. No insignia.

21st Airborne Division

Insignia: On a blue disk, a white cloud emitting a yellow lighting flash vertically, and above it another white cloud emitting two yellow lightning flashes chevronwise. According to the Quartermaster General, "The insignia is blue for Infantry, the lightning flashes allude to the number of the division."

A FORTITUDE SOUTH II unit. Composed of 277th and 278th Glider Infantry Regiments and 521st Parachute Infantry Regiment, plus divisional troops. Arrival in England not recorded. Part of Fourteenth Army. Headquarters at Fulbeck in Lincolnshire. Scheduled to drop inland at Pas-de-Calais. Came under direct command of FUSAG in August; transferred to First Allied Airborne Army, and used as part of the airborne threat to the Kiel-Bremen area in September. Assigned to (notional) XVII Airborne Corps; disposed of in November by telling the Germans that it and the notional 9th Airborne had been merged to form the (genuine) 13th Airborne Division (which was soon in fact to arrive in France).

22d Infantry Division

Insignia: A gold scorpion in the attitude of striking, on a red disk. According to the Quartermaster General, "The symbol of the Gold Scorpion Division's insignia represents the striking and stinging power inflicted by this division upon its opponents in battle."

Notionally sent to the Middle East to reinforce the British in September–November 1942. Specific number requested by Dudley Clarke on his visit to Washington, September 1942.

23d Infantry Division

Allocated to Middle East notional order of battle in 1942 but never used. Subsequently allotted to SHAEF but again never used. No insignia apparently prepared.

APPENDIX III

25th Armored Division

All armored divisions wore (and wear) the same insignia except for the divisional number. It is based on the insignia of the old Tank Corps from World War I, a triangle divided in three portions colored blue in the lower left, yellow above, and red in the lower right (the colors of the infantry, cavalry, and artillery), with the tracks of a tank, the barrel of a cannon, and a red lightning flash imposed upon the center and the number of the division in black upon the yellow upper portion.

A FORTITUDE SOUTH II unit. Part of XXXIII Corps and subsequently XXXVII Corps. Activated at Pine Camp, New York, in 1941. Composed of 72d, 73d, and 74th Tank Battalions, and 498th, 499th, and 500th Armored Infantry Battalions, plus divisional troops. Arrived in England in June 1944, headquarters at Wincanton in Somerset. Moved to East Dereham in Norfolk in July. Scheduled for second wave at Pas-de-Calais. Moved to Tidworth in Hampshire in August. Came under XXXVII Corps in September. Sailed from Southampton, apparently for France, at the beginning of October and not heard of again.

39th Armored Division

Allocated to North Africa and then to SHAEF for FORTITUDE SOUTH II but never used.

46th Infantry Division

Insignia: An olive drab disk, thereon a yellow square with its long axis vertical, thereon a blue six-pointed star. According to the Quartermaster General, "The insignia consisting of the square of four sides and the star of six points designates the number of the division. The blue is for Infantry."

Reported to the Germans as having arrived in the United Kingdom in May and June 1943 as part of LARKHALL, an operation designed to convey the impression of greater American strength in Britain than was the case. The June edition of *National Geographic,* featuring illustrations of divisional insignia, showed no 46th Division (or other notional divisions); Bevan therefore sought permission to substitute a real division but this was denied by Joint Security Control. Held in readiness for use in FOYNES to be sent to Italy to replace divisions transferred to England, but this proved unnecessary. Allocated to SHAEF, but not used in FORTITUDE, and allowed to fade away.

A genuine National Guard division after the war (with different insignia).

47th Infantry Division

Allocated originally to North Africa and then to SHAEF for FORTITUDE SOUTH II, but never used. No insignia.

48th Infantry Division

Insignia: A four-pointed star, point up, divided into eight portions alternately white and red. According to the Quartermaster General, "The form of pointed cross alludes to the number '4,' while the blue and white alternated alludes to the number '8.' "

A FORTITUDE SOUTH II unit. Part of XXXIII Corps. Formed in 1942 at Camp Clatsop, Oregon; participated in maneuvers on the Olympic Peninsula and at the Desert Training Center; guard duty on the Alcan Highway. Composed of 80th, 95th, and 146th Infantry Regiments, plus divisional troops. Arrived in England in June 1944, "well trained," BRUTUS reported to German intelligence. Headquarters originally at Newcastle-Under-Lyme in Staffordshire. Assigned to XXXIII Corps in early July. Moved to Woodbridge in Suffolk in July. Scheduled to be in reserve at Pas-de-Calais. Moved to Brockenhurst in Hampshire in August. Commenced air landing training in August; personnel attached to 21st Airborne for training. Moved to Newbury area in Berkshire at the end of September to continue airborne training. Continued airborne training in autumn of 1944; came under British Base Section when XXXIII Corps was disbanded in October. Moved to Dundee in Scotland at beginning of December and continued airborne training. Disbanded January 1945; cadre for special troops sent to the United States, rest used for replacements.

(A genuine National Guard division after the war, with different insignia.)

50th Infantry Division

Insignia: A shield bearing a monogram of the Greek capital letters delta and gamma; no colors assigned. According to the Quartermaster General, "The device in the center of the shield is the Attic (Greek) symbol for the number 50. It is composed of the symbol for $5 = \Gamma$ with the symbol for $10 = \Delta$, i.e., $5 \times 10 = 50$."

Was allocated to North Africa. Employment unknown. May never have been used.

55th Infantry Division

Insignia: A gold pentagon inside a larger gold pentagon, the interior space and the space between them blue. No official interpretation of the insignia, but the two pentagons obviously suggest the two 5s of the number 55.

A FORTITUDE NORTH unit. Notionally stationed in Iceland from August–October 1943 as replacement for the (genuine) 5th Division, which was (genuinely) moved to Britain. Accepted by German intelligence by February 1, 1944. Under notional British VII Corps, March 1944. Composed of 78th, 83d, and 96th Infantry Regiments plus division troops. Joined by notional 7th, 9th, and 10th Ranger Battalions in April of 1944. The "story" of FORTITUDE NORTH contemplated that the

55th Division and the associated Ranger battalions would be the followup force on the Narvik front after the initial assault by the British 52d (Lowland) Division. Was never put to actual use, being deemed too far from Norway to be a plausible part of a Scandinavian operation. Ceased to be under British VII Corps in July. Held along with the genuine British 48th Division and the notional British 90th Division as a force prepared to occupy southern Norway immediately if there were any sign of a German withdrawal. Left Iceland by stages in winter of 1944–45. Fully withdrawn from Iceland March 1945, and not heard of again.

59th Infantry Division

Insignia: A white rattlesnake coiled to strike, on a blue disk with the ground under the rattlesnake yellow. No official interpretation of the insignia (it was designed not by the Quartermaster General but by the American FORTITUDE deception team), but obviously suggestive of the "Don't Tread on Me" rattlesnake flags of the Revolutionary War.

A FORTITUDE SOUTH II unit. Activated at Fort Custer, Michigan, in 1942, and participated in maneuvers in Tennessee and Minnesota and at the Desert Training Center before arriving in England in May 1944, with headquarters at Harwich. Composed of 94th, 139th, and 171st Infantry Regiments plus divisional troops. Originally used to replace the genuine 35th Infantry Division in Third Army. Part of XXXVII Corps. Headquarters moved to Ipswich. Associated with notional naval "Force F" at Harwich. Conducted three amphibious exercises in July. Scheduled to be in the first wave at the Pas-de-Calais. Moved to Rowlands Castle in Hampshire in August. Sailed from Southampton, apparently for France, at the end of September and not heard of again.

108th Infantry Division

Insignia: A yellow mace with an eight-spiked ball, on a scarlet oval edged in yellow. According to the Quartermaster General, "The mace insignia is described as a 'skull buster' indicative of the power and crushing effect that this division anticipates displaying against its aggressors."

Composed of 433d, 434th, and 436th Infantry Regiments, 636th, 637th, 638th, and 642d Field Artillery Battalions, plus appropriate supporting units. Used in WEDLOCK and subsequent Kuriles deceptions. One of the assault divisions for the notional attack on Paramushiro and Shumushu, June 1944. Headquarters at Camp Earle, Attu.

Appendix III

109th Infantry Division

Allocated to the Middle East but never used; originally the 19th Infantry Division, changed to 109th in July 1944.

No insignia.

112th Infantry Division

Allocated to SHAEF but never used. No insignia.

119th Infantry Division

Insignia: Red flames edged in yellow on a black disk. According to the Quartermaster General, "This insignia is symbolic of the ability, tenacity and courage of the organization in overcoming opposition with the speed and all consuming qualities of a furious fire."

Composed of 488th, 489th, and 491st Infantry Regiments, 639th, 640th, 641st, and 649th Field Artillery Battalions, plus appropriate supporting units. Used in WEDLOCK and subsequent Kuriles deceptions. Headquarters on Amchitka. As part of VALENTINE, notionally redeployed to Seattle in February–April 1945 for rehabilitation and training prior to planned movement to Central Pacific in summer 1945, probably for notional operations on the China coast. In mid-April 1945, enemy intelligence was told by a double agent that men wearing the insignia of the 119th had been seen in Grand Central Station on April 20 and again the next week.

125th Infantry Division

Allocated to Middle East notional order of battle in 1942. No insignia apparently prepared.

130th Infantry Division

Insignia: A white flying serpent embowed, on a blue disk. According to the Quartermaster General, "The blue is for infantry while the flying serpent is representative of the swift striking power of the division."

Composed of 492d, 493d, and 494th Infantry Regiments, 642d, 643d, 644th, and 650th Field Artillery Battalions, plus appropriate supporting units. Used in WEDLOCK and subsequent Kuriles deceptions. One of the assault divisions for the notional attack on Paramushiro and Shumushu, June 1944. Headquarters at Fort Mears (Dutch Harbor).

135th Airborne Division

Insignia: A black widow spider on a yellow disk; above it an Airborne tab, the letters AIRBORNE in yellow on black. According to the Quartermaster General, "The insignia represents the Airborne fighter striking from above with venomous stinging power."

Consisted of 551st Parachute Infantry Regiment and 524th and 525th Glider Infantry Regiments. Allocated for use by "A" Force in May 1944; its component 551st Parachute Infantry Regiment was "due to arrive in Naples shortly." May never have been used otherwise.

141st Infantry Division

Insignia: A white thistle blossom on a blue disc. According to the Quartermaster General, "The blue is for Infantry while the thistle, which is the type of plant which has protecting spines which ward off those approaching it, represents the punishment inflicted by the organization on aggressors."

Composed of 495th, 496th, and 497th Infantry Regiments, 645th, 646th, 647th, and 651st Field Artillery Battalions, plus appropriate supporting units. Used in WEDLOCK and subsequent Kuriles deceptions. One of the assault divisions for the notional attack on Paramushiro and Shumushu, June 1944. Headquarters at Fort Greeley, Kodiak. As part of VALENTINE, notionally redeployed to San Francisco in February–April 1945 for rehabilitation and training prior to planned movement to Central Pacific in summer 1945, probably for notional operations on the China coast.

157th Infantry Division

Insignia: A yellow heraldic tiger with red tongue and claws on a blue disk. According to the Quartermaster General, "The blue is for Infantry while the heraldic tiger is noted for its savageness and bloodthirstiness, wonderful in strength and most swift in flight, as it were an arrow."

Composed of 557th, 558th, and 565th Infantry Regiments, 944th, 946th, 952d, and 956th Field Artillery Battalions, plus appropriate supporting units. Used in WEDLOCK and subsequent Kuriles deceptions. Probably the reserve division for the notional attack on Paramushiro and Shumushu, June 1944. Headquarters on Adak. In the first week of May 1945, a double agent fed enemy intelligence the story that a patch consisting of a "blue circle with a yellow dog or lion in the center" had been seen in New York.

9th Armored Regiment

A separate notional unit assigned to Middle East order of battle in 1942. Notionally introduced element by element into Egypt in June–November 1942, as part of CAS-

CADE. Notionally part of GHQ reserve in Cairo (where in fact a number of American tank technicians were stationed to service the American tanks used in the Western Desert). No insignia known.

Actually an element of the genuine 20th Armored Division, which served in Europe in the last months of the war.

613th Parachute Regiment

Made available to China Theater as one of two notional United States parachute regiments to be reported as arriving in India about November 1, 1945. To be stationed at the former XX Bomber Command base at Kharagpur, forming with the 619th Parachute Regiment a "flying block" available for missions deep into Japanese-held territory in China, after which it would withdraw into one of the islands of resistance in China for subsequent evacuation and reemployment. No insignia.

619th Parachute Regiment

Made available to China Theater as one of two notional United States parachute regiments to be reported as arriving in India about November 1, 1945. To be stationed at the former XX Bomber Command base at Kharagpur, forming with the 613th Parachute Regiment a "flying block" available for missions deep into Japanese-held territory in China, after which it would withdraw into one of the islands of resistance in China for subsequent evacuation and reemployment. No insignia.

7th Ranger Battalion

Insignia: The usual Ranger Battalion shoulder patch without numerical designation.

Designated for use in Iceland in connection with FORTITUDE NORTH, March 1944, as part of British VII Corps.

8th Ranger Battalion

Insignia: The usual Ranger Battalion shoulder patch without numerical designation.

Designated for use in Iceland in connection with FORTITUDE NORTH, March 1944, but never used.

9th Ranger Battalion

Insignia: The usual Ranger Battalion shoulder patch without numerical designation.

Designated for use in Iceland in connection with FORTITUDE NORTH, March 1944, as part of British VII Corps.

10th Ranger Battalion

Insignia: The usual Ranger Battalion shoulder patch without numerical designation.

Designated for use in Iceland in connection with FORTITUDE NORTH, March 1944, as part of British VII Corps.

11th Ranger Battalion

Insignia: The usual Ranger Battalion shoulder patch without numerical designation.

Designated for use in Iceland in connection with FORTITUDE NORTH, March 1944, but never used.

Ninth Fleet

Notional fleet established in North Pacific in 1944 for WEDLOCK. No insignia.

Task Force 23

A notional task force imitated by radio deception under Operation HOTSTUFF, supposedly sailing from Norfolk, Virginia, for the eastern Mediterranean via the South Atlantic, the Indian Ocean, and the Red Sea, to cover the actual Task Force 34 carrying Patton's force to the landings in French Morocco, November 1942.

Task Force 69

Notional task force to convoy VII Corps from the East Coast to invade Brittany as part of WADHAM, 1943. Commanded by Rear Admiral Alan G. Kirk.

VIII Tactical Air Command

This unit was obtained for SHAEF at its request (in July 1944), but was never used. Notionally commanded by Brigadier General Jesse Auten. No insignia.

20th Troop Carrier Group

Designation of XX Bomber Command for local deception purposes in India, early 1944. The B-29 was referred to as the "C-75-A." No insignia adopted.

66th Heavy Bombardment Group

A notional bomber group (a genuine group previously disbanded) designated June 1944 to augment the apparent strength of Chennault's Fourteenth Air Force in resisting the Japanese Ichigo offensive. No new insignia adopted.

550th Fighter Group

A notional fighter group designated June 1944 to augment the apparent strength of Chennault's Fourteenth Air Force in resisting the Japanese Ichigo offensive. No insignia adopted.

551st Fighter Group

A notional fighter group designated June 1944 to augment the apparent strength of Chennault's Fourteenth Air Force in resisting the Japanese Ichigo offensive. No insignia adopted.

BRITISH UNITS

Fourth Army

Sign: A rectangle divided into three horizontal stripes of red, black, and red, with a design like the figure 8 in yellow, the bottom part of the lower circle being missing; this being the medieval "4" upon the normal Army colors of red, black, red.

Scottish Command disguised as a field army, employed to threaten Norway in Tindall and Fortitude North. Brought down to Heathfield, Sussex, for Fortitude South II as a component of FUSAG, consisting of British II and VII Corps and the American 35th Division, plus army troops. Then moved to Essex, with headquarters at Colchester, for FUSAG's threat to the Kiel-Bremen area in connection with the Arnhem operation. Moved to Yorkshire in early December, and employed as a threat to Holland in Operation Trolleycar, in November–December 1944. Disbanded February 1945.

Sixth Army

Eastern Command disguised as a field army. Situated at Luton in East Anglia, designed to represent an army in being in 1943 capable of invading anywhere on the coast of northeastern Europe. Not used thereafter, but remained in the order of battle held by the Germans till the end of the war.

Twelfth Army

Sign: A trained seal balancing on its nose a terrestrial globe showing the Eastern Hemisphere, black on a white background. Noel Wild took credit for the sign, symbolizing an amphibious creature treating the whole world as its own.

A CASCADE May 1943 unit to represent an army in being in Egypt and Palestine as a standing threat to the Balkans. Included at one time or another III British Corps (largely notional), XVI British Corps (notional) and III Polish Corps (notional); originally also XIV British Corps (notional). Role initially played by "Force 545," the headquarters planning Montgomery's part in the Sicilian campaign; subsequently by the Air Defense Headquarters of Egypt.

Subsequently assigned to India-Burma, and intended by D Division to be the force for FANG, the notional invasion of Sumatra in 1945, to be composed of the notional XX Indian Corps and XXXVIII Indian Corps; but this idea was abandoned when a genuine Twelfth Army with a different sign (formerly Burma Army) was activated in Burma in May 1945.

II Corps

Sign: A red leaping salmon upon three wavy blue bands against a white background, all in an oblong red border.

Originally a genuine corps in France in 1940, commanded by Sir Alan Brooke; hence the divisional sign of a salmon leaping up a brook. Was being disbanded in early 1944 when selected to be one of the two corps of the notional British Fourth Army for FORTITUDE NORTH. Headquarters at Stirling in Scotland, with under command the genuine British 3d Infantry Division (shortly replaced by the notional British 58th Infantry Division), the genuine British 55th Division in Northern Ireland, and the genuine 113th Independent Infantry Brigade in the Orkneys. II Corps was to attack Stavanger, with the 3d (later the 58th) Division and supporting Commandos and paratroops seizing the airfields, the 55th Division joining as followup; the genuine U.S. XV Corps from Northern Ireland would augment the force, which would advance on Oslo.

Transferred to FUSAG in early June and moved to Lincolnshire; restored to Fourth Army when that formation joined FUSAG for FORTITUDE SOUTH II, headquarters now at Tunbridge Wells in Kent, with under command the British 55th and 58th Divisions and the British 35th Armored Brigade. Transferred to France in late September, consisting of the essentially notional 55th Division, the genuine 79th Armored Division, and the essentially notional 76th Division; also apparently at times the genuine 59th Division, disbanded but notionally kept alive. Notionally part of Canadian First Army in TROLLEYCAR II in November 1944.

Appendix III

III Corps

Sign: A green fig leaf (stem upwards) on a white square. Adopted early in the war as a reference to its commander, General Adam.

A Cascade May 1943 unit. Originally a genuine corps, serving in France in 1940 and evacuated at Dunkirk. "Embarked" for Iran in 1943; became part of bogus Twelfth Army. Subsequently in Syria, Egypt, Italy, and Greece. Notionally to land in the heel of Italy as part of Boardman. Notionally to seize Corfu as part of Fairlands. In Greece in 1944, became Headquarters Land Forces, Greece, notionally composed of the bogus 34th and 5th Airborne Divisions under Second Undercut.

VII Corps

Sign: A scallop shell on a blue ground.

Originally a genuine corps formed in 1940 to defend against a German invasion. Disbanded on Christmas Day, 1940. Notionally reactivated as a unit of British Fourth Army in Fortitude North with headquarters at Dundee. Composed of the genuine British 52d (Lowland) Division at Dundee, the notional U.S. 55th Division in Iceland, a Norwegian brigade, and three notional American ranger battalions in Iceland, plus corps troops.

Moved south with Fourth Army for Fortitude South II, with headquarters at Folkestone in Kent and under command the British 61st and 80th Divisions and 5th Armored Division, the latter two notional and the 61st a genuine but low-establishment formation.

Moved to East Anglia in September, to Yorkshire in December, and disbanded in January 1945.

XIV Corps

Sign: A black wolf's head with a lolling red tongue on a white square.

A Cascade December 1943 unit, formed for Oakfield, the deception plan in support of the Anzio landing. Notionally arrived in North Africa in December 1943. Composed of notional 40th, 42d, and 57th Divisions. Represented in Algeria and Tunisia by various administrative establishments; Christmas cards were distributed, and invitations to a "XIV Corps Staff" Christmas party found their way to Axis agents in Spain and Turkey. Notionally to be employed in landings in the Leghorn-Pisa area under Oakfield, winter 1943–44. Notionally placed under American Seventh Army for Zeppelin, early 1944. Notional British 5th Airborne Division added; 40th Division and 5th Airborne were notionally in Sicily by April 1944. In Western Mediterranean in May 1944. In the winter of 1944–45, XIV Corps, its insignia adopted by the Mountain Warfare Training Center in Italy, was portrayed as a reserve corps located in the Terni area, now consisting of the 42d and

914

57th Divisions, trained in mountain warfare for use when the Allies in Italy would reach the Alps. Notionally placed under command of 15th Army Group in the spring of 1945, and simulated army group reserve in April 1945 in support of the spring offensive.

XVI Corps

Sign: A phoenix rising from red flames and bearing a flaming torch in its mouth, on a white ground.

Part of notional Twelfth Army, notionally formed in Egypt in November 1943. Included 8th Armored Division, 15th Motorized Division, and 34th Division. Notionally to seize Cephalonia and Zante as part of FAIRLANDS.

XVIII Corps

The original notional corps on Cyprus in the CYPRUS DEFENSE PLAN of 1941. Subsequently revised to XXV Corps. Sign unknown.

XIX Airborne Corps

Sign unknown. A notional corps of the genuine First Allied Airborne Army in the autumn of 1944, with under command the notional British 2d Airborne Division and the genuine British 6th Airborne Division. Headquarters at Salisbury. Disposed of in December by telling the Germans it had only been an administrative unit.

XXI Corps

A unit of British Fourth Army in TINDALL. Sign unknown.

XXV Corps

Sign: On a yellow rectangular background, a red lion passant gardant, Richard Coeur-de-Lion's badge and the emblem of Cyprus; originally the badge of Headquarters British Troops in Cyprus, taken over for XXV Corps.

Originally XVIII Corps. Created as part of the CYPRUS DEFENSE PLAN in 1941, when the genuine 50th Division joined the notional 7th Division on Cyprus. Subsequently part of CASCADE, composed of notional 7th Division and notional 1st Indian Division. In WANTAGE, composed only of 7th Mixed Division after amalgamation of 1st Indian and 7th Divisions into 7th Mixed Division.

Appendix III

XXVI Airborne Corps

Sign: Mercury with winged feet. Part of Force 144 for Operation CLAW, with headquarters at Comilla, composed of 9th Indian Airborne Division, 32d Division (Air Transit), and 89th Armored Brigade Group in that area and 10th African Airborne Division at Chittagong. Commanded by Lieutenant-General the Hon. H. K. M. Kindersley. Concentrating in East Bengal in February–April 1945. Reassigned to the genuine Twelfth Army in June 1945, destined for airborne attack on Bangkok as part of SCEPTICAL.

2d Airborne Division

Sign: None known; presumably the standard representation of Bellerophon astride a Pegasus in pale blue on a dark maroon background, worn by all British airborne troops.

A FORTITUDE SOUTH unit. Part of British II Corps, located at Grantham and Skegness in Lincolnshire. Composed of 11th and 12th Parachute Brigade and 13th Air Landing Brigade plus division troops. Placed under direct command of FUSAG for FORTITUDE SOUTH II. Part of the airborne force threatening airborne operations in the Kiel-Bremen area at the time of the Arnhem operation. Disbanded to provide replacements for the genuine 1st and 6th Airborne Divisions, December 1945.

3d Airborne Division

Sign: A white winged scimitar. (Evidently the fact that all airborne troops wore the blue Bellerophon and Pegasus was overlooked when the 3d, 4th, and 5th Airborne Divisions were assigned their own insignia.)

Moved from Agartala to Sylhet/Sylchar area in February–March 1945 in preparation for Operation CLAW. Training for possible long-range penetration missions in March. Three battalions of its 5th Parachute Brigade were to participate in support of Operation CONCLAVE, March 1945.

4th Airborne Division

Sign: An open white parachute with black wings on a blue background.

A CASCADE March 1943 division. Composed after April 1943 of 6th (Gurkha) Parachute Brigade (6th Battalion, 6th Gurkha Regiment; 160th and 161st Parachute Regiments Army Air Corps (Gurkha), 1st Special Air Service Brigade (1st Special Air Service Regiment, Detached Special Air Service Brigade, 6th Special Group), and 7th Airborne Brigade (Greek Squadrons Special Forces, French Parachute Battalion). Formed in January 1943 from airborne troops in the Middle East and units from the United Kingdom that arrived in December 1942 and January 1943; 6th Brigade arrived from India in April 1943 to replace 4th Brigade, which was transferred to North Africa. Located in Palestine. Role played by the 11th Parachute Battalion. Notionally destined for Crete and Greece in 1943 as part of BARCLAY. In ZEPPELIN, 1944, located in Cyrenaica as part of Twelfth Army, notionally destined for Crete.

5th Airborne Division

Sign: a light blue bolt of lightning, formed by five zigzags on a dark red square; discontinued in 1945 when it was realized that all British airborne forces wore the same sign of a blue Pegasus.

A component of British XIV Corps in Sicily in late 1943, activated as part of FOYNES as a "replacement" for units transferred to Britain for OVERLORD. Notionally arrived in the Mediterranean from the United Kingdom in January 1944. Built up around the genuine 2d Parachute Brigade of the genuine 1st Airborne Division, which was left behind in North Africa when the division was sent to England for OVERLORD. Composed of genuine 2d Parachute Brigade, 8th Parachute Brigade (8th Highland Light Infantry, 8th Cameronians, 12th Black Watch), and 9th Air Landing Brigade (4th Rifle Battalion, 5th Royal Ulster Rifles, 15th King's Royal Rifle Corps). In Western Mediterranean in May 1944. As part of FERDINAND, together with the notional Polish III Corps, would notionally launch an attack in late summer 1944 across the Adriatic against an undetermined Balkan target.

In SECOND UNDERCUT in Greece in late 1944, notionally standing by in Sicily as part of III Corps to join the notional 34th Division if necessary.

From December 1944 to the end of the war it was maintained as a notional small theater reserve of airborne troops so as to sustain a latent threat to any part of German-held territory within range. Represented during that period by 313 Company R.A.S.C. in Sicily, while the airborne staff at AFHQ was called "HQ 5th Airborne Division (Plans)." After a number of enemy questionnaires concerning its location were directed to the double agents, followed by the parachuting of German agents into Sicily, it was decided that security of the "division" was in jeopardy. So headquarters of the division and its 9th Brigade were notionally moved to Rome, where they were represented by the genuine British 2d Parachute Brigade, the only real airborne troops in Italy, while its 8th Brigade notionally moved to Malta for training and to receive reinforcements from England.

5th Armored Division

Sign: None known. A FORTITUDE SOUTH II unit, under British VII Corps of British Fourth Army, with headquarters at Newmarket. Composed of 37th Armored Brigade and 43d Infantry Brigade, plus division troops. Moved to Yorkshire with VII Corps in November. Disbanded January 1945.

7th Division

Sign: A black cat on a white background. Originated with the CYPRUS DEFENSE PLAN of June 1941. Arrived in Egypt from the United Kingdom in 1941, moved to Cyprus in July 1941. Role originally played by the command on Cyprus, subsequently, as part of CASCADE, adopted by one of the defense sectors on Cyprus. Originally consisted of the 21st, 22d, and 23d Infantry Brigades, plus divisional troops, four squadrons of tanks, and antiaircraft and miscellaneous other units. Initially in CASCADE composed of three notional brigades, the Northern, Southern, and 41st Indian Brigades. In Cyprus in May 1944. Became 7th Mixed Division on amalgamation with 1st Indian Division, May 1943; composed from February 1944 of 2d and 41st Indian Infantry Brigades and 1st Cyprus Brigade.

8th Division

Originally a genuine division formed in Palestine in 1938 from two infantry brigades, with its sign a red cross on a blue shield. Disbanded in early 1940.

"Revived" as an original CASCADE division in 1942, notionally in Syria as part of the British Ninth Army. Sign: Crossed battleaxes. Role played by the former Desert Brigade in northern Syria. Composed of 18th, 19th, and 40th Brigades, all bogus.

8th Armored Division

Sign: The word "GO" in black letters on a green disk in a black square, representing a traffic light; the division's motto was No Stop; No Caution; Go On.

A CASCADE March 1943 division. Originally a genuine division organized in England in 1940, sent out to Egypt in the spring of 1942, broken up shortly thereafter and subsequently represented by notional means. Simulated by the dummy tank brigade in the Tobruk area until June 1943 in support of WATERFALL, being notionally destined to land in the Peloponnese; and again in the autumn of 1943 in support of BOARDMAN and FAIRLANDS. Notionally destined to land in Crete in the spring of 1944 as part of ZEPPELIN, as an element of XVI Corps, Twelfth Army; role once more played in Cyrenaica by the dummy tank brigade.

9th Armored Division

Sign: The face of a giant panda (said to have been adopted as a pun on "panzer").

Originally a genuine Home Forces division, disbanded in 1944. Included by the Germans in their order of battle for FUSAG and accordingly maintained as a notional division and allotted to 21 Army Group for use in deception. Composed of 28th Armored Brigade and 7th Infantry Brigade plus division troops.

Expected in 1943 to play a major part in deception under APPENDIX Y, and for this reason GARBO was given a notional AGENT 7(I), a soldier in this division. With the development of FORTITUDE, the division played only an accidental role and AGENT 7(I) was never a significant source for the GARBO network.

12th Mixed Division

Sign: A yellow giraffe on black.

There was a genuine 12th Division, a second-line Territorial division formed in 1939. It fought in France in 1940, and was subsequently disbanded when the home forces were reorganized.

"Revived" as an original CASCADE division, 1942. Notionally part of the internal garrison of Egypt in 1942, having come overland from East Africa during the autumn of 1941. Role played by the 1st Sudan Defense Force Brigade; notionally included the 2d Sudan Defense Force Brigade (still in the Sudan) and an imaginary 182d Infantry Brigade. In Libya in May 1944, composed of 1st and 2d Sudan Defense Brigades and 38th Indian Infantry Brigade.

15th Armored Division

Sign: a white unicorn on a black ground.

An original CASCADE division (July 1942). Notionally held in GHQ reserve in Cairo in the spring of 1942. Composed of the 7th Armored Brigade (bogus), the 35th Armored Brigade (bogus), and the 200th Guards Motor Brigade (genuine). Role taken by the vehicles of the Royal Armored Corps Base Depot in Cairo; role of the 7th Armored Brigade taken by the "A" Force Depot. Converted to 15th Motorized Division in November 1943.

15th Motorized Division

Sign: A white unicorn on black ground.

Composed of 33d and 77th Motorized Brigades. Notionally arrived in Middle East from the United Kingdom and April 1942 as the 15th Armored Division, converted to 15th Motorized Division in November 1943. As 15th Motorized Division, in Eastern Mediterranean, part of XVI Corps, Twelfth Army, in May 1944. Moved to Southeast Asia Command, and in early 1945 undergoing jungle training at Belgaum, destined to replace the notional 21st Indian Division in the notional XX Indian Corps which was to attack Sumatra.

20th Armored Division

Sign: A tortoise, orange on black.

A CASCADE March 1943 division. Composed of 87th Armored Brigade (Cheshire Yeomanry, 3d R.G.H., 4th Northants Yeomanry) and 88th Motorized Brigade (4th Sikhs, 4th P.P. Rifles, Alwar, 88th Motorized Brigade Transportation

Company). Supposedly arrived in Palestine from the United Kingdom in mid-August 1942. Located in Palestine, role played by the Army Tank School. In Eastern Mediterranean, part of Ninth Army, in May 1944. Under Plan TURPITUDE, poised on the Turkish border near Aleppo.

21st Division

Sign: A lion rampant on a disk.

Part of XX Indian Corps, at Ranchi, until moved to Madras early 1945 (replaced by 15th Motorized Division) as part of Force 144 for Operation CLAW. Transferred to IV Corps of Fourteenth Army in May 1945.

23d Division

Originally a genuine division, whose sign was a white Tudor rose on a blue (sometimes green) background. Fought with the B.E.F. in France in 1940, evacuated and subsequently disbanded.

"Revived" as an original CASCADE division (July 1942). Sign: the colors of the Royal Marines (red, yellow, and blue) on a foul anchor, a sign already in use at that time by the Royal Marine Mobile Naval Base Defense Organization, which played the role of the division in 1942.

Notionally held in reserve for the defense of Egypt in 1942. Composed of one Royal Marine brigade, the 84th Infantry Brigade, and the 38th Indian Infantry Brigade; all fictitious, although some real battalions were included in them. Removed from order of battle with the CASCADE revision of May 1943, when the Marines moved to the Far East.

26th Armored Division

Sign: A front view of a white elephant's head on a blue square.

Components included 28th Hussars and 29th Lancers. At Ranchi in early March 1945.

32d Division (Air Transit)

Sign: A front view of a white elephant's head on a green square.

Training with transport aircraft under Central Command, India, March 1945. Part of XXVI Airborne Corps, July 1945.

33d Division

Sign: A yellow hammer on a black circle.

Notionally reached Ceylon in Autumn 1942 and amphibiously trained there.

Transferred to Egypt January–March 1944 under WANTAGE. Part of Twelfth Army in May 1944. Composed of 97th, 98th, and 109th Infantry Brigades. Returned to Ceylon as an element of XXXVIII Indian Corps for Operation FANG, 1945.

34th Division

Sign: A checkerboard, four by four, top left square white.

A CASCADE May 1943 division, located in Algiers, landed from the United Kingdom March 25, 1943. Transferred to the Middle East in January 1944, pursuant to WANTAGE, and part of XVI Corps, Twelfth Army, by May 1944. Composed of 83d Armored Brigade and 39th and 81st Infantry Brigades. Represented in SECOND UNDERCUT as part of "III Corps," occupying Greece in late 1944. In Plan PENKNIFE, intended to be notionally transferred to Italy to replace the genuine British 1st Division, which was being transferred to the Middle East.

40th Division

Sign: A brown acorn on a white diamond. The sign was made up on the scene in Sicily when the "division" was activated in 1943. It was designed to be reminiscent of the 40th Division in the First World War, whose sign was a bantam cock on which was superimposed a white diamond bearing an oak leaf and an acorn, this being an augmentation granted by GHQ to recognize the division's role in the capture of Bourlon Wood.

A CASCADE December 1943 division, activated as part of FOYNES as a replacement for units transferred to Britain for OVERLORD. Part of notional XIV Corps. Composed of three notional brigades, whose roles were played by the 30th battalions of the Royal Norfolk Regiment, the Somerset Light Infantry, and the Green Howards, respectively; with the battalion commanding officers flying brigadier's pennants and the battalion adjutants signing correspondence as "Brigade-Majors." The "brigades" were the 119th, 120th, and 121st Infantry Brigades. In Tunis in February 1944; in Sicily in May 1944.

42d Division

Sign: A small white diamond inside a small red diamond.

Originally the genuine 42d (East Lancashire) Division, which fought in Belgium in 1940, was evacuated at Dunkirk, was converted to an armored division in the United Kingdom, and was disbanded in 1943. A CASCADE December 1943 division, activated as part of FOYNES as a "replacement" for units transferred to Britain for OVERLORD. Part of notional XIV Corps. Composed of 133d, 142d, and 149th Infantry Brigades. In Western Mediterranean in May 1944. In VENDETTA, attached to notional American XXXI Corps for its mission to seize the Montpellier Gap and establish a defensive flank in the area Montpellier-Clermont l'Hérault. From December 1944 to the end of the war it was a component of the notional XIV Corps, notionally in the Terni area of Italy, represented by the genuine Armored Reinforcement Regiment.

55th Division

Sign: The red rose of Lancashire.

A training division in Home Forces, essentially disbanded by late in the war; notionally kept alive as a component of II Corps.

57th Division

Sign: The letter "D" on its side in white and red on a black background.

A CASCADE December 1943 division, activated as part of FOYNES as a "replacement" for units transferred to Britain for OVERLORD. Notionally a revived First World War division. Arrived in North Africa from the United Kingdom in September 1943. Part of notional XIV Corps. Composed of 170th, 171st, and 172d Infantry Brigades. In Western Mediterranean in May 1944. From December 1944 to the end of the war it was a component of the notional XIV Corps, notionally in the Terni area of Italy, represented by the genuine R.A.C. Training Depot. Notionally standing by to reinforce notional III Corps in Greece as part of SECOND UNDERCUT, October 1944.

58th Division

Sign: A stag's head full face on a black square, intended to suggest a connection with the Scottish Highlands.

An element of British II Corps for FORTITUDE NORTH. Created to substitute for the genuine British 3d Division as the assault division for the attack on Stavanger, when the 3d was designated to participate in Exercise FABIUS. Number selected because ULTRA showed that the Germans believed there was a 58th Division, which they had earlier located in the Windsor area. Composed of 173d, 174th, and 175th Infantry Brigades, plus division troops. First appeared at Aberlour, south of Inverness; engaged in mountain training at Aviemore, Kincraig and Dufftown till mid-April, then moved south towards Glasgow, conducting a series of mountain exercises as it went. Conducted an exercise in embarkation and debarkation at Greenock in late April, then continued training east of Glasgow. Moved to Lincolnshire in July; and to Kent for FORTITUDE SOUTH II in June, headquarters at Gravesend. Moved to East Anglia in September, to Yorkshire in November, back south to Hertfordshire in early 1945 and disbanded in April.

59th Division

Sign: Pit-head gear in red against a black slagheap on a blue background.

A genuine division, broken up after the Normandy campaign but notionally kept alive as a component of II Corps.

64th Division

Sign: Apparently a castle flanked by dolphins, with a trident.

Part of notional XX Indian Corps in early 1945, training for FANG.

68th Division

Sign: A red diamond on a blue background.

Formerly the genuine 61st Division, a "Lower Establishment" division, briefly earmarked for service in France but stood down to become merely a Home Forces training and drafting division. Notionally continued as a fighting division for use as part of FOYNES as a replacement for units transferred to Britain for OVERLORD, but never actually used.

70th Division

Sign: A red four-pointed star on a white square.

Originally the genuine 6th Division, which withstood the siege of Tobruk and then was sent to India in the winter of 1941–42, renumbered the 70th Division but retaining its sign. The 70th Division was subsequently broken up after distinguished service, but its existence was apparently notionally continued for deception purposes. Notionally disbanded June 1945.

76th Division

Sign: A red sailing ship (a Norfolk wherry) on a black square.

Originally a genuine home service reserve division, renumbered as the 47th Division. Designation and sign retained for deception purposes and assigned to 21 Army Group. Composed of 213th, 25th, and 215th Infantry Brigades plus division troops. Notionally involved in Operation TROLLEYCAR, notional attack west of Arnhem, in November 1944, as part of notional II Corps.

77th Division

Sign: A white arm emerging from blue waves holding aloft a red sword, on a black background; representing King Arthur's Excalibur.

Originally a genuine home service holding division, renumbered as the 45th Division. Designation and sign retained for deception purposes; held in reserve and apparently never used. Composed of 103d, 203d, and 209th Infantry Brigades plus division troops.

80th Division

Sign: A red ocean liner steaming across a light blue sea, emitting light blue smoke from her funnel, on a yellow background with light blue border.

Originally a genuine home service reserve division, renumbered as the 38th Division. Designation and sign retained for deception purposes. Composed of 50th, 211th, and 208th Infantry Brigades plus division troops. Assigned to British VII Corps of British Fourth Army for FORTITUDE SOUTH II, moving from Lancashire to new headquarters at Canterbury. Assigned to British II Corps, remaining at Canterbury, in September, and reverted to VII Corps and transferred to Suffolk in October. Moved to Yorkshire with Fourth Army in December, thence south to St. Neots in East Anglia and was disbanded in April.

90th Division

Sign, if any, unknown. Composed of 72d, 73d, and 74th Infantry Brigades plus division troops. In autumn 1944, held along with the genuine British 48th Division and the notional American 55th Division as a force prepared to occupy southern Norway immediately if there were any sign of a German withdrawal.

10th Armored Brigade

A CASCADE 1942 formation. Sign, if any, unknown. Part of the Cyprus garrison in CASCADE 1942.

27th Armored Brigade

Sign, if any, unknown. Represented as concentrating north of Maldegen under Canadian 3d Division, Operation INCLINATION, October 1944.

33d Armored Brigade

A CASCADE 1942 formation, composed of the dummy tank regiments in the desert. Subsequently a genuine 33d Armored Brigade formed part of the armored force of Montgomery's 21 Army Group on the Continent in 1944–45.

42d Tank Brigade

Sign: A triangle inside a circle inside a square. One of D Division's notional units; nothing further known.

Appendix III

1st Special Air Service Brigade

Originally created for ABEAM in early 1941. Continued as part of CASCADE in 1942, represented by a dummy glider unit, "K" Detachment. In CASCADE March 1943, part of the notional 4th Airborne Division in Palestine. No sign known.

7th Air Landing Brigade

A CASCADE formation, notionally to land in Tripoli in Rommel's rear in the final version of WINDSCREEN, December 1943. No sign known.

103d Special Service Brigade

Sign: A spiked club on a square. One of D Division's notional units; nothing further known.

140th Special Service (Commando) Brigade

Sign: A black mask. In Ceylon in early 1945.

North Eastern Task Force

A bogus naval task force created in connection with FORTITUDE SOUTH and FORTITUDE SOUTH II, as well as to "pad" the call sign book for Operation NEPTUNE in case a copy should fall into enemy hands. Consisted of Force F, Force M, and Force N.

Force F

A bogus naval force at Sheerness and Harwich (headquarters at Sheerness) simulated by W/T traffic in May and June in conjunction with BIGBOB displays on the East Coast. In early June 1944 simulated traffic of amphibious training with the American 28th Division (SEE SAW), then in July and August with the American 59th Division (FILM STAR, FRANK, MUSTACHE, VIOLA III), as part of the buildup for FORTITUDE SOUTH II.

Force M

A bogus naval force at Portsmouth and Newhaven (headquarters at Portsmouth), notionally composed largely of landing craft returned from the Normandy landings, simulated by W/T traffic in July and August 1944 conducting amphibious exercises with the British 55th Division at Hayling Island (VANITY I, VANITY II, HONEYSUCKLE, VIOLA I) as part of the buildup for FORTITUDE SOUTH II.

925

APPENDIX III

Force N

A bogus naval force at Southampton, notionally composed largely of landing craft returned from the Normandy landings, simulated by W/T traffic in July and August 1944 conducting amphibious exercises with the American 17th Division at Studland (HAIRCUT, JITTERBUG, VIOLA II) as part of the buildup for FORTITUDE SOUTH II.

Force V

A bogus naval force in the Clyde for the assault on the Narvik area in FORTITUDE NORTH, simulated by W/T traffic in April and May 1944 conducting amphibious exercises with the British 52d Division (CARROT I, CARROT II, ONION I, ONION II, LEEK I, LEEK II).

Force W

A bogus naval force in the Clyde for the landing in the Stavanger area in FORTITUDE NORTH, simulated by W/T traffic in April and May 1944 conducting embarkation and disembarkation exercises with the notional British 58th Division.

INDIAN UNITS

Indian Expeditionary Force

Sign: A sword (Army), a trident (Navy), and wings (RAF); a combined operations sign similar to that of the Commandos.

The initial army-level force for Operation FANG, the invasion of Sumatra as part of STULTIFY. Composed of XX Indian Corps at Poona and XXXVIII Indian Corps in Ceylon. Apparently replaced in June 1945 by Force 144, with the same corps under command and a similar mission as part of SCEPTICAL. This name was originally assigned to the genuine unit formed in March 1943 that was renamed XXXIII Indian Corps in October of that year.

Force 144

Sign unknown unless when it took over the notional Sumatra operation in mid-1945 it adopted the sign of its predecessor, the "Indian Expeditionary Force."

Also known as Task Force 144. Initially, the army-level force for Operation CLAW, the threat to Moulmein as part of STULTIFY in early 1945. Headquarters at Barrackpore. Composed of British XXVI Airborne Corps in eastern Bengal, and a seaborne element on the east coast of India (British 21st Division at Madras, 21st Indian Division and the Frontier Armored Division at Vizagapatam, and 15th

Indian Division at Madras and later at Ramree Island). Commanded by Lieutenant-General Sir James Gammell, replaced in May 1945 by Major-General H. L. Davies.

In June 1945, placed in charge of notional operations against Sumatra and the Straits of Malacca in connection with SCEPTICAL, with under command XX Indian Corps and XXXVIII Indian Corps, apparently replacing the "Indian Expeditionary Force" in that role.

XX Indian Corps

Sign: A Roman numeral "XX" on a square.

Briefly commanded by Lieutenant-General Sir Gerald Templer in early 1945 and then by Lieutenant-General W. H. A. Bishop. In early 1945 part of Indian Expeditionary Force, headquarters at Poona, engaged in training for FANG, the notional seaborne expedition against Sumatra. Composed of 21st Division, 24th Indian Division, 64th Division, 83d West African Division, and a commando brigade, all notional, with 21st Division to be replaced by notional 15th Motorized Division from Egypt after completion of its jungle training. Assigned in June 1945 to Force 144.

XXXVIII Indian Corps

Sign unknown. Notional amphibious corps in Ceylon, April 1945, allotted to Twelfth Army (when that was still notional) for Operation FANG, the notional seaborne operation against Sumatra. Comprised of 34th Indian Division, 33d Division, 13th East African Division and 140th Commando Brigade. Commanded by Major-General Lomax. Reassigned to Force 144 in June 1945.

Airborne Raiding Task Force

A bogus entity added to D Division order of battle in July 1945. Sign unknown.

1st Indian Division

Sign: A black buck.

A CASCADE division from 1942; part of the notional XXV Corps defending Cyprus, along with the notional 7th Division. Composed of the 2d 10th, and 20th Indian Brigades, "a mixture of the real and false." Removed from order of battle in May 1943 revision of CASCADE, Cyprus no longer being threatened by the enemy.

2d Indian Division

Sign: A yellow hornet on a black rectangular background.

A CASCADE division from 1942, notionally in Iraq with the British Tenth Army.

Role played by No. 2 Lines of Communication Area troops in southern Iraq and southwestern Iran. The formation sign is said to have been adopted because of the prevalence of hornets in the area at the time. Composed of 34th, 39th, and 40th Indian Infantry Brigades. In Persia-Iraq Command in May 1944 for internal security in Iraq.

6th Indian Division

Sign: The head of a Deccan tiger in black and yellow on a square black background.

A genuine division formed in 1941 and engaged in protecting the supply routes to the USSR in Iraq until disbanded in late 1944. Notionally brought to Karachi, rehabilitated, and held in reserve.

9th Indian Airborne Division

Sign: Blue winged horse on a red ground.

Composed of 6th (Gurkha) Parachute Brigade and two other brigades. Part of XXVI Airborne Corps, moved from Secunderabad to Comilla early 1945.

12th Indian Division

Sign: The lion of Persia (a maned lion passant gardant holding a scimitar) outlined in blue on a yellow background.

A CASCADE March 1943 division. Formed in November 1942 from brigades formerly in 2d Indian Division; composed in 1944 of 31st, 34th, 39th, and 60th Indian Infantry Brigades. The 34th Brigade came from Sudan in February 1942, 39th formed in Iraq in October 1941. In Persia-Iraq Command in May 1944, for internal security in Iran; role played by line of communication troops in similar fashion to "2d Indian Division."

15th Indian Division

Sign: A black head wearing a fez in profile on a white ground.

In Persia-Iraq Command in May 1944. In Madras in early 1945, moved to Ramree Island end March as part of Force 144 threatening Moulmein area as part of Operation CLAW. Transferred to IV Corps of Fourteenth Army in May 1945.

16th Indian Division

Sign: A white mahseer on a blue background.

Composed of 8th Madras Division Headquarters Battalion and 68th, 70th, and 76th Indian Infantry Brigades. Raised in Northwest India in winter 1942–43; combined a long period of training with an operational role of frontier defense reserve;

arrived in Iraq piecemeal in spring 1944. In Persia-Iraq Command in May 1944, for internal security in Iraq. Notionally brought to India in 1944–45 and after rehabilitation was reformed at Quetta to go into reserve.

18th Indian Division

Sign: An eagle swooping towards its prey.

Raised in Quetta early 1942. Moved to Shillong in the spring of 1944 and went forward into the Burma valley as Fourteenth Army reserve during the Japanese invasion of Manipur. Withdrawn to Ramchi to reorganize, July 1944. Allotted to Fourteenth Army to participate in CLOAK and moved from Ranchi to Imphal via Calcutta, March 1945. In CLOAK, notionally augmented Slim's secondary drive down the Irrawaddy. Joined IV Corps in early April and withdrew to Monywa as Fourteenth Army reserve in late April.

21st Indian Division

Sign: A pouncing panther.

A buildup division for Operation CLAW, leaving notional XX Indian Corps to replace 15th Indian Division at Madras early 1945. At Vizagapatam as part of Force 144 for Operation CLAW.

32d Indian Armored Division

Sign: A black rhinoceros head on a red square.

Moving from Ranchi to Ahmednagar via Calcutta in early 1945.

34th Indian Division

Sign: A blue triangle.

In Ceylon spring 1945. Doing amphibious training March–April 1945. Part of XXXVIII Indian Corps for FANG.

Frontier Armored Division

Sign: A flying eagle.

Moved from Jubbulpore to Vizagapatam early 1945. Part of Force 144. Part of it receiving jungle training under Central Command early 1945.

Nepalese Division

Sign: A pouncing tiger.

Reported to the Japanese in February 1945 as guarding the India-China pipeline.

Appendix III

1st Burma Division

Sign: A white pagoda on a black circle.

Included British regiments of Inniskillings and Lancashire Fusiliers. Training in Imphal region, Silchar-Sylhet area, in February–March 1945, for possible long-range penetration missions, and allocated to Tarzan. Disbanded July 1945.

51st Indian Tank Brigade

A notional element which was to advance eastward to complete the encirclement of Japanese troops in central Burma in Operation Conclave, Match 1945. No known sign.

NEW ZEALAND UNITS

3d New Zealand Division

Sign: A kiwi.

An original Cascade unit in 1942; soon changed to 6th New Zealand Division, no doubt because a genuine 3d New Zealand Division was being held in readiness in New Zealand to defend against the Japanese.

6th New Zealand Division

Sign: A kiwi.

An original Cascade division, 1942 (having briefly been designated "3d New Zealand Division"). Notionally part of the internal garrison of Egypt in 1942. Role played by the headquarters of the New Zealand Expeditionary Force at Na'adi Camp, near Cairo, whose name was officially changed to "6th New Zealand Division." Training units, reinforcements, and lines of communication units were renamed for its components, the 9th, 10th, and 11th New Zealand Infantry Brigades, plus divisional troops.

In North Africa in May 1944. Offered for use by MacArthur as notionally transferred from the Mediterranean, October 1944.

SOUTH AFRICAN UNITS

7th South African Division

Sign: An ox wagon silhouette on dark green and tangerine.

A Cascade March 1943 division. Units arrived in Egypt from South Africa in May and June of 1942; division originally formed as part of the Delta defense force in the emergency of July 1942. Stationed in Egypt; role played by the Union of

South Africa Defense Force Base in Egypt in similar fashion to the "6th New Zealand Division." Composed of 5th and two unnumbered South African Brigade Groups.

AFRICAN COLONIAL UNITS

10th African Airborne Division

Sign: A buffalo's head.
    Part of XVI Airborne Corps. Moved from Gwalior to Chittagong early 1945.

13th East African Division

Sign: A buffalo's head.
    In reality the 21st East African Brigade, in Ceylon from March 1942, but represented to the Japanese as the 13th East African Division. Still at Hambantota in Ceylon as of early 1945. Part of XXXVIII Indian Corps for Operation FANG.

83d West African Division

Sign: A palm tree.
    Part of XX Indian Corps early 1945, training to invade Sumatra or Java.

FRENCH UNITS

III Corps

No known insignia.
    A genuine corps in early 1945 but not up to strength or activated, and essentially notional. Employed for notional strengthening of French First Army in connection with Operation ACCORDION, notional crossing of the Rhine by French forces, spring 1945. Composed of French 1st, 10th, 36th, and 2d Armored Divisions. Notionally to be commanded by General Jacques Leclerc.

IV Corps

No known insignia.
    A corps which in early 1945 had not yet been formed, though there were plans for it. Employed for notional strengthening of French First Army in connection with Operation ACCORDION, notional crossing of the Rhine by French forces, spring 1945. Composition not specified other than French 27th Alpine Division.

French Expeditionary Corps for the Far East ("FEFEO")

No known insignia.

A corps with nominal existence in 1944–45, with elements scattered from France to Algiers and Ceylon, containing only one unequipped division equivalent stationed in the south and southwest of France. Offered by the French for service against the Japanese in the French colonies in the Far East, but declined by the Combined Chiefs of Staff. Composed of 1st and 2d Colonial Far East Divisions and nondivisional elements. Notionally expanded to an active corps in support of Operations KNIFEDGE and ACCORDION.

7th Algerian Infantry Division

No known insignia.

Formed in North Africa after the Tunisian campaign. In North Africa in May 1944, under American XXXI Corps, Seventh Army.

8th Algerian Infantry Division

No known insignia.

Formed in North Africa after the Tunisian campaign. In North Africa in May 1944, under American XXXI Corps, Seventh Army.

3d Armored Division

No known insignia.

Formed in North Africa after the Tunisian campaign. In garrison in North Africa in May 1944.

10th Colonial Infantry Division

No known insignia.

Formed in North Africa after the Tunisian campaign. In North Africa in May 1944, part of "2d French Expeditionary Corps," the name by which de Lattre's French Army B (subsequently French First Army) was known during the WANTAGE period.

GREEK UNITS

1st Greek Division

Sign: Head of Athena in white on a blue square.

A CASCADE May 1943 division. Formed in the Middle East in 1942, reconstituted 1944. In eastern Mediterranean, part of Ninth Army, in May 1944. Composed of 1st, 2d, and 3d Greek Brigade Groups.

POLISH UNITS

III Polish Corps

Sign: Three red towers on a blue ground (for personnel) or a white ground (for vehicles).

A CASCADE May 1943 unit, originally part of Twelfth Army. Formed in Palestine in winter 1943–44. As part of FERDINAND, together with the notional British 5th Airborne Division, would notionally launch an attack in late summer 1944 from southern Italy against an undetermined Balkan target. Composed of 2d Polish Armored Division and 7th and 8th Polish Infantry Divisions.

2d Polish Armored Division

Sign: A white lion rampant holding a mailed arm and sword on a blue and red diagonal ground.

A CASCADE May 1943 division, originally notionally in Palestine, in Italy by May 1944. Formed in Spring 1943 from elements of the disbanded 6th Polish Division added to 2d Polish Army Tank Brigade; moved from Persia and Iraq Command in August 1943. Composed of 4th Carpathian Lancers, 2d Polish Armored Brigade (4th, 5th, and 6th Polish Armored Regiments), and 5th Polish Motorized Brigade (7th Polish Field Regiment, 7th Polish Antitank Regiment, and 16th, 17th, and 18th Polish Motorized Battalions).

7th Polish Division

Sign: A red griffin rampant on a white background, borrowed from the arms of the province of Pomurce.

A CASCADE March 1943 division. Notionally arrived from Russia during summer of 1942 and moved from PAIC in July 1943. Role played by Polish base and training units in Palestine. Composed of 7th Polish Reconnaissance Regiment, 7th Polish Machine Gun Battalion, 7th Polish Rifle Brigade (22d, 23d, and 24th Polish Rifle Battalions), and 8th Polish Rifle Brigade (19th, 20th, and 21st Polish Rifle Battalions).

8th Polish Division

Part of III Polish Corps, Twelfth Army, in eastern Mediterranean in May 1944. Insignia unknown.

APPENDIX IV

Maps

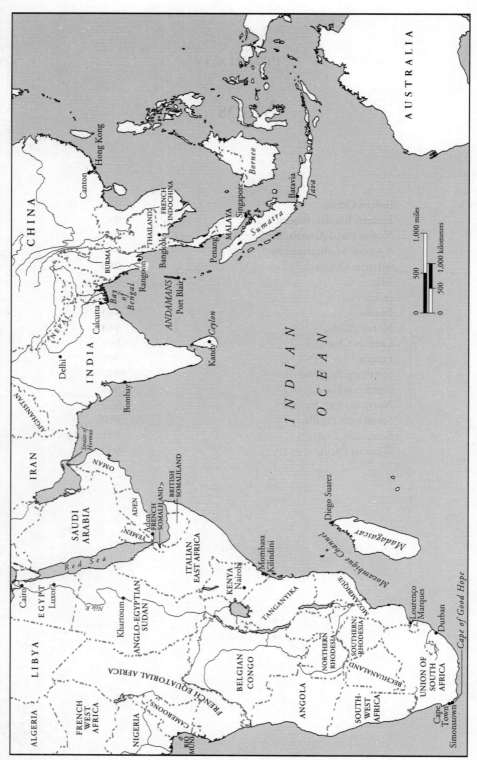

Indian Ocean

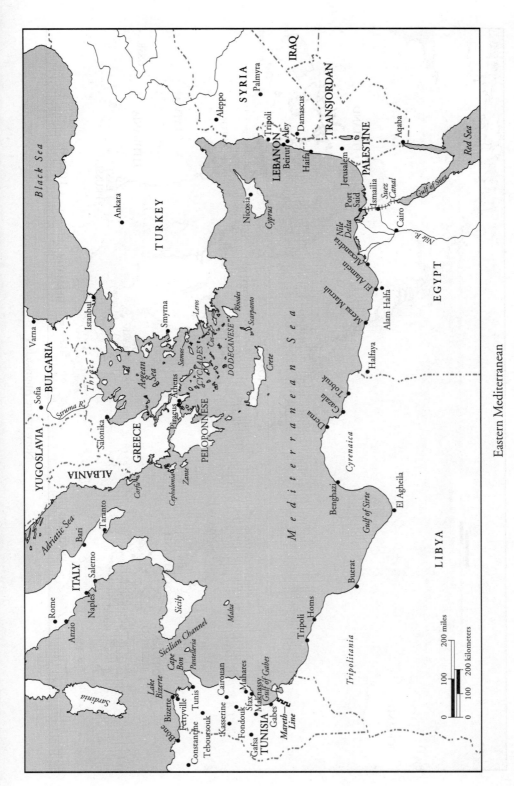

Eastern Mediterranean

937

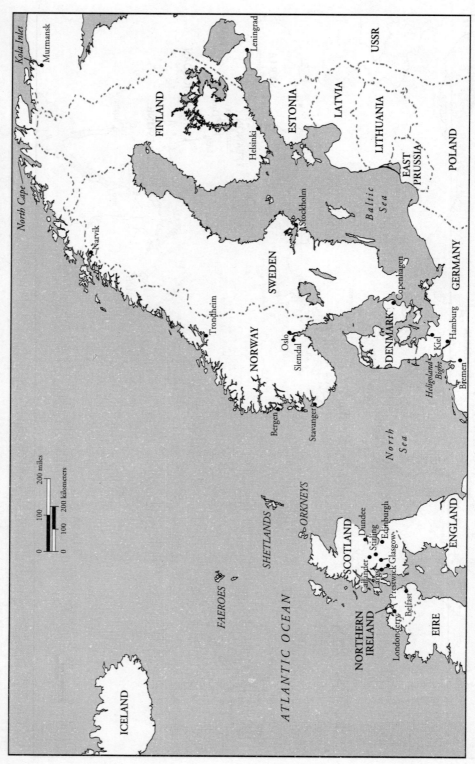

Norway and Approaches

Western Mediterranean

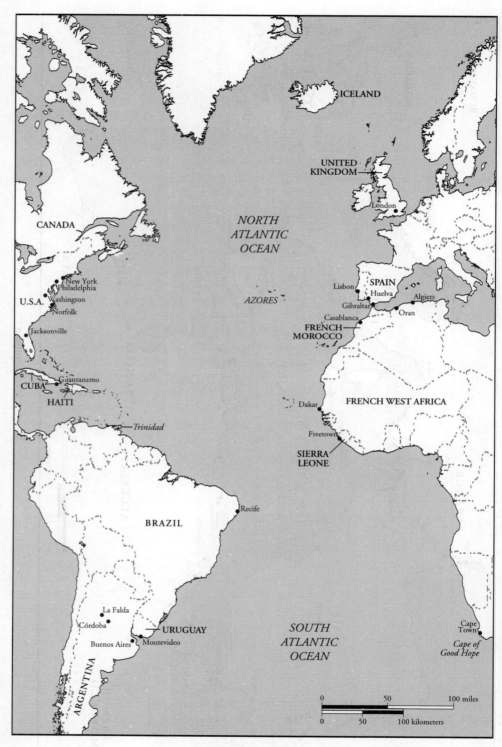

Atlantic Ocean

940

Italy

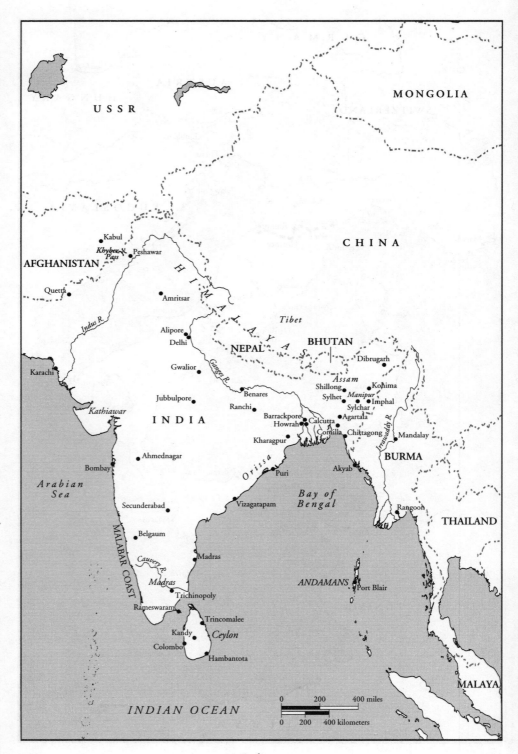

India

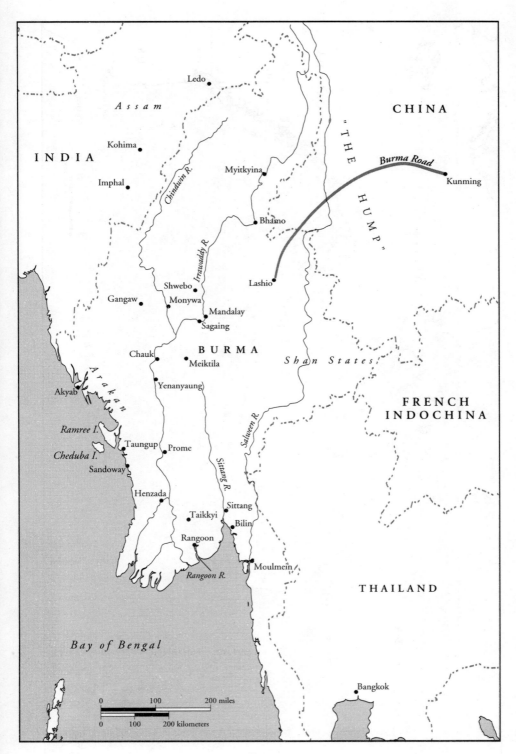

Burma

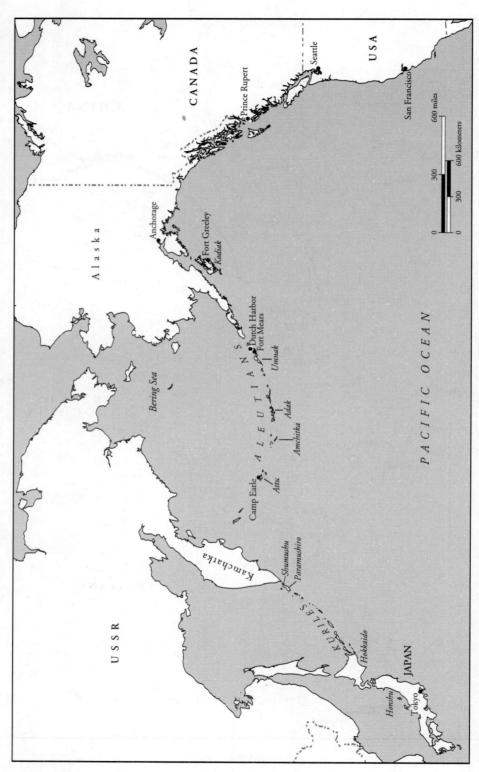

N rth Pacific

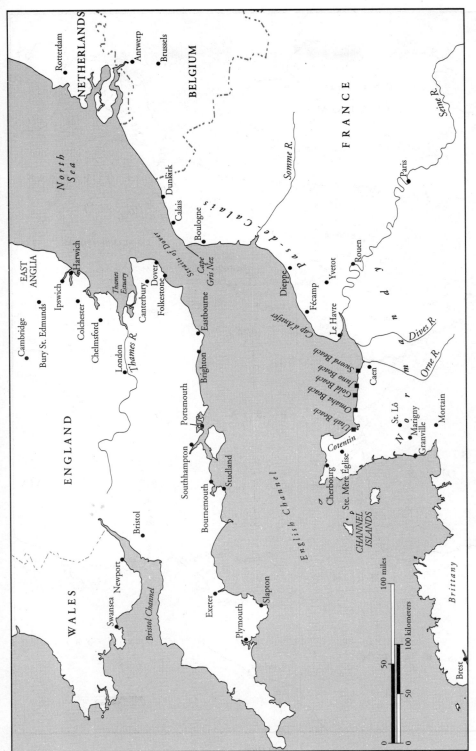

English Channel

France

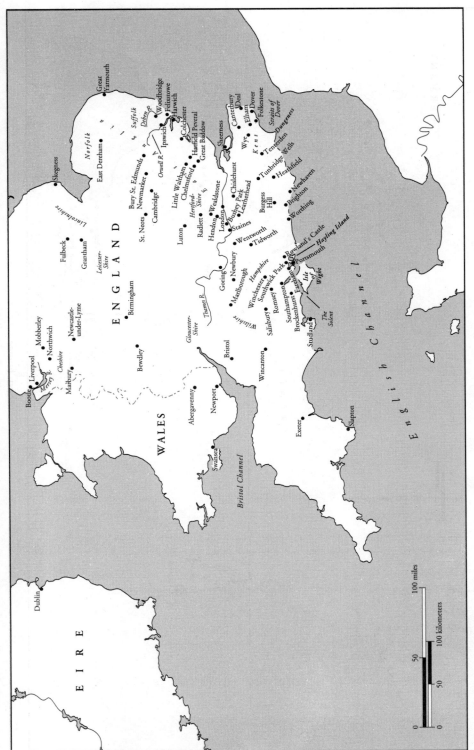

Southern England

947

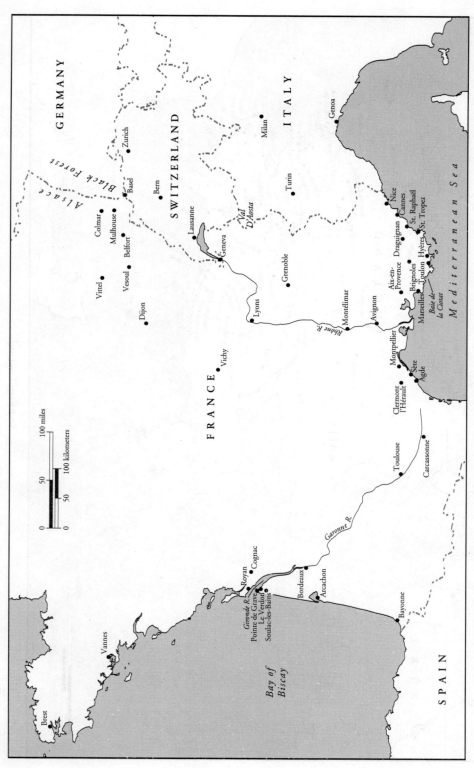

Southern France

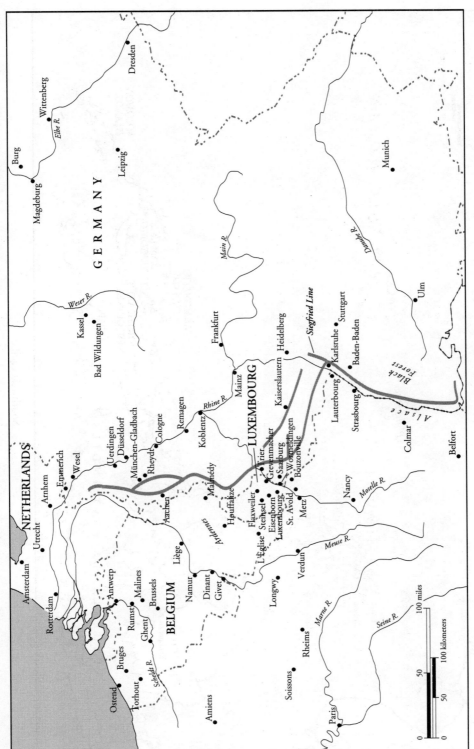

Northwestern Europe

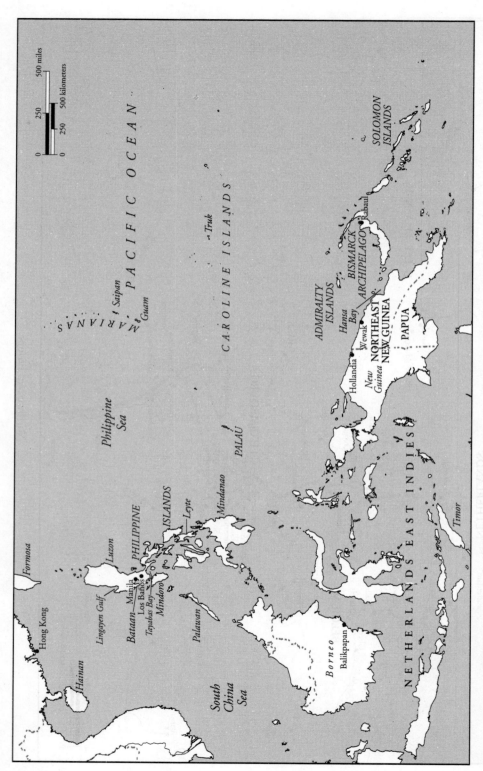

Southwest Pacific

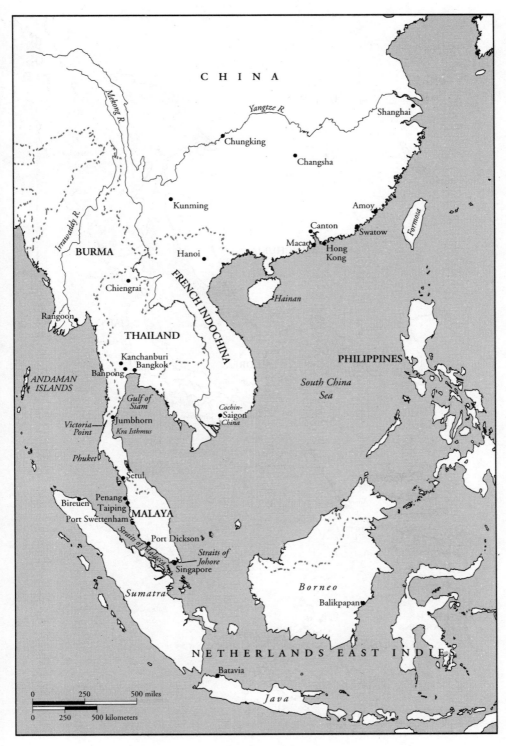

Southeast Asia

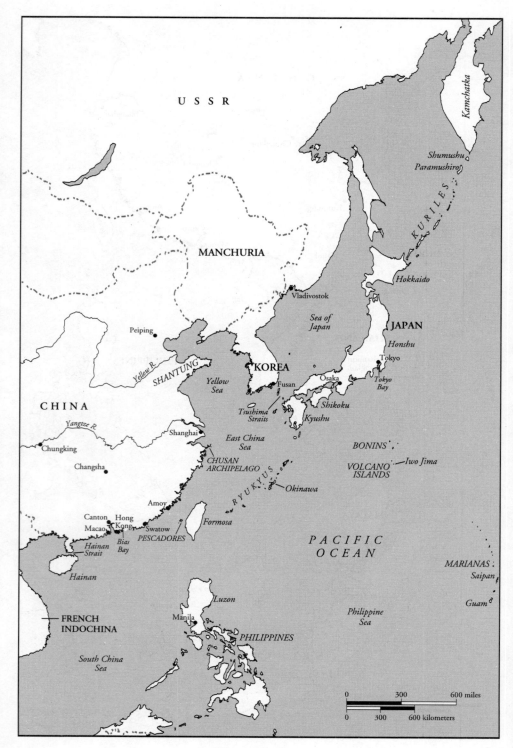

Western Pacific

BIBLIOGRAPHY

United States National Archives, Washington and College Park.

RG 165 ABC File ABC 313 (12 July 44) Sec. 1-A—Deception History.
RG 165 ABC File ABC 334 (7 August 1943).
RG 165 ABC File ABC 381 (22 Jan 43).
RG 165 ABC File ABC 381 (22 Jan 43) Sec. 1.
RG 165 ABC File ABC 381 (4-8-43) Sec. 1-A—Deception policy, organization, etc.
RG 165 ABC File ABC 381 (4-8-43) Sec. 1-B—Deception policy, organization, etc.
RG 165 ABC File ABC 381 (7-20-42) Origins of JSC.
RG 165 ABC File ABC 381 (7-25-42) Secs 1 and 1½ Torch C&D.
RG 165 ABC File ABC 381 (8-15-42) JSC.
RG 165 ABC File ABC 381 Japan 15 Apr 43 Sec. 1-A.
RG 165 ABC File ABC 381 Japan 15 Apr 43 Sec. 1-B.
RG 165 ABC File ABC 381 Japan 15 Apr 43 Sec. 3—Pacific.
RG 165 ABC File ABC 381 Japan 15 Apr 43 Sec. 4—Forager.
RG 165 ABC File ABC 381 Japan 15 Apr 43 Sec. 5—Pastel.
RG 165 ABC File ABC 381 United Nations (1-23-42) Sec. 3-B—JSC charter.
RG 165 ABC File ABC 384 China (15 Dec 43) Sec. 1-B—Iceman.
RG 165 ABC File ABC 385 Europe (23 Sep 43) Sec. 2.
RG 165 ABC File ABC 385 Europe (23 Sep 43) Sec. 4-B.
RG 165 ABC File ABC 470.2 (12 July 43).
RG 200 Box 40 (Sutherland Papers) Folder War Dept Nos. 701–800 4/2/44–8/6/44, re 1944 "General Directive for Deception Measures Against Japan."
RG 218 (1948–50) Folder CCS 385 (1-23-48) Sec. 1 Re Improving Cover and Deception Capability (1950).
RG 218 (1948–50) Folder CCS 385 (1-23-48) Sec. 3 Instruction in Use of Communications in Cover Plans by Army and Navy Staff College.
RG 218 Box 211 Folder CCS 334 (4-30-42) Joint Security Committee.
RG 218 Box 211 Folder CCS 334 (8-4-42) Joint Security Control (Sec. 1).
RG 218 Box 211 Folder CCS 334 JSC (8-4-42) Sec. 2.
RG 218 Box 211 Folder CCS 334 JSC (8-4-42) Sec. 3.
RG 218 Box 211 Folder CCS 334 Joint Security Control (Sec. 2).
RG 218 Box 211 Folder CCS 334 Joint Security Control (Sec. 3).
RG 218 Box 261 Folder CCS 353 (10-1-42) Amphibious Training Operations in the Caribbean Area.
RG 218 Box 262 Folder CCS 354.2 (10-30-42) Report of Observation of Joint Army and Navy Tactical Exercise.
RG 218 Box 290 Folder CCS 380.1 (10-22-43) Security Precautions.
RG 218 Box 360 Folder CCS 385 (11-16-44) Plans for Operation "M-1."
RG 218 Box 360 Folder CCS 385 (5-19-44) Plan "Royal Flush."

RG 218 Box 360 Folder CCS 385 (9-22-43) Necessity of Coordinating Communications With Deception or Cover Plans.

RG 218 Box 361 Folder CCS 385 (6-25-43) Planning for Deception for Cross-Channel Operation (Sec. 1).

RG 218 Box 361 Folder CCS 385 (6-25-43) Deception Plans for Overlord (Sec. 2).

RG 218 Box 362 Folder CCS 385 (4-8-43) (2) Cover and Deception Planning (Sec. 1).

RG 218 Box 362 Folder CCS 385 (4-8-43) (2) Cover and Deception Planning.

RG 218 Box 363 Folder CCS 385 (12-31-42) Deception Policy.

RG 218 Box 363 Folder CCS 385 (2-10-43) Deception Plan for "Boodle" and "Cottage."

RG 218 Box 363 Folder CCS 385 (2-10-43) Deception Policy (Germany and Italy).

RG 218 Box 364 Folder CCS Plans for Strategic Deception.

RG 218 Box 370 CCS 385.2 (10-17-45) Release of Information on Rubber Deception Ships.

RG 218 Box 372 Folder CCS 385.2 (4-2-44) Technical Information Concerning the Existing Means of Tactical Deception.

RG 218 Box 45 Folder CCS 300 (1-8-43) (Sec. 1).

RG 218 Box 51 Folder CCS 311 (3-5-42) Policy of the U.S. Army Governing Radio Deception and Radio Interference.

RG 218 Box 652 Folder CCS 385 Japan (7-2-45) Plan Pastel.

RG 218 Box 663 Folder CCS 385 Marshall Islands (1-6-44) Propaganda and Deception Plans for Flintlock.

RG 218 Box 688 Folder CCS 385 Pacific Ocean Areas (12-1-44) Cover Plans for Detachment, Iceberg, and Subsequent Operations in the Pacific Ocean Areas During 1945.

RG 218 Box 688 Folder CCS 385 Pacific Ocean Areas (4-5-44) Propaganda Plan for Forager.

RG 218 Box 688 Folder CCS 385 Pacific Ocean Areas (6-9-44) Deception and Propaganda Plans for Stalemate.

RG 218 Box 693 Folder CCS 385 Pacific Theater (4-1-43) Deceptive Policy in the Pacific Theater (Sec. 1).

RG 218 Box 693 Folder CCS 385 Pacific Theater (4-1-43) Deceptive Policy in the Pacific Theater (Sec. 2).

RG 218 Box 694 Folder CCS 385 Pacific Theater (4-1-43) Deceptive Policy in the Pacific Theater (Sec. 3).

RG 218 Box 694 Folder CCS 385 Pacific Theater (4-1-43) Deceptive Policy in the Pacific Theater (Sec. 4).

RG 218 Box 694 Folder CCS 385 Pacific Theater (4-1-43) Deceptive Policy in the Pacific Theater (Sec. 5).

RG 218 Box 713 Folder CCS 385 Southeast Asia (5-2-45) Cover Plans for SEA.

RG 218 Folder CCS 313.6 (7-11-44) Sec. 3, Receipt and Disposal of Records of Combined Operations.

RG 313 Entry 1 Box 69 Folder 381 Bodyguard.

RG 319 Entry 101 Box 1 Folder 1 Cover & Deception Synopsis of History.

RG 319 Entry 101 Box 1 Folder 2 Cover & Deception Report ETO Exhibit 1 Basic Directive, SHAEF.

RG 319 Entry 101 Box 1 Folder 3 Cover & Deception Report ETO Exhibit 2, Basic Directive, AC/S, G-3, FUSAG.

RG 319 Entry 101 Box 1 Folder 4 Cover & Deception Report ETO Exhibit 3, Cover & Deception, Definition & Procedure.

RG 319 Entry 101 Box 1 Folder 5 Cover & Deception Report ETO Exhibit 4, Use of Intelligence in Cover & Deception.

RG 319 Entry 101 Box 1 Folder 6 Cover & Deception Report ETO Exhibit 5, Cover & Deception, Recommended Organization.

RG 319 Entry 101 Box 1 Folder 7 Cover & Deception Report ETO Exhibit 5, Operations in Support of Neptune (A) Plan Fortitude.

RG 319 Entry 101 Box 1 Folder 8 Cover & Deception Report ETO Exhibit 6, Operations in Support of Neptune (B) Fortitude North.

RG 319 Entry 101 Box 1 Folder 9 Cover & Deception Report ETO Exhibit 6, Operations in Support of Neptune (C) Fortitude South I.

RG 319 Entry 101 Box 1 Folder 10 Cover & Deception Report ETO Exhibit 6, Operations in Support of Neptune (D) Fortitude South II.

RG 319 Entry 101 Box 1 Folder 11 Cover & Deception Report ETO Exhibit 6, Operations in Support of Neptune (E) Diversionary Plan, Summary and Results.

RG 319 Entry 101 Box 1 Folder 12 Cover & Deception Report ETO Exhibit 6, Operations in Support of Neptune (F) Results.

RG 319 Entry 101 Box 1 Folder 13 Cover & Deception Report ETO Exhibit 7, Technical Signal Report on Fortitude.

RG 319 Entry 101 Box 1 Folder 14 Cover & Deception Report ETO Exhibit 8, Tactical Operation A.

RG 319 Entry 101 Box 1 Folder 15 Cover & Deception Report ETO Exhibit 8, Tactical Operation B.

RG 319 Entry 101 Box 2 Folder 16 Cover & Deception Report ETO Exhibit 8, Tactical Operation C.

RG 319 Entry 101 Box 2 Folder 17 Cover & Deception Report ETO Exhibit 8, Tactical Operation D.

RG 319 Entry 101 Box 2 Folder 18 Cover & Deception Report ETO Exhibit 8, Tactical Operation E.

RG 319 Entry 101 Box 2 Folder 19 Cover & Deception Report ETO Exhibit 9, Report on Pneumatic Dummy Equipment.

RG 319 Entry 101 Box 2 Folder 20 Cover & Deception Report ETO Exhibit 10, Present Policy, Cover & Deception.

RG 319 Entry 101 Box 2 Folder 21 Cover & Deception Report ETO Exhibit 11, Extract From New York Times, 1 September 1944.

RG 319 Entry 101 Box 2 Folder 27 Informal Report to Joint Security Control by Special Plans Branch, G-3, 12th Army Group (basic report) 5/25/45.

RG 319 Entry 101 Box 2 Folder 28 Informal Report to Joint Security Control by Special Plans Branch, G-3, 12th Army Group (Appendix 1, Enemy Reactions to Fortitude April–June 1944) 5/25/45.

RG 319 Entry 101 Box 2 Folder 30 Informal Report to Joint Security Control by Special Plans Branch, G-3, 12th Army Group (Appendix 3, Cover and Deception in Air Force Operations, ETO) 5/25/45.

RG 319 Entry 101 Box 3 Folder 15 Deception in UK—Larkhill and Cockade.

RG 319 Entry 101 Box 3 Folder 64 Report of Special Signal Operations in the United Kingdom 3/14/45.

RG 319 Entry 101 Box 3 Folder 68, Strategic Deception Conference, Machinery for Deception Against Japan.

RG 319 Entry 101 Box 4 Folder 70 Anvil Plans (#1).

RG 319 Entry 101 Box 4 Folder 71 Wantage Order of Battle Plan.

RG 319 Entry 101 Box 4 Folder 72 Ferdinand.

RG 319 Entry 101 Box 4 Folder 73 Vendetta.

RG 319 Entry 101 Box 4 Folder 75 Enemy Agent Summary.

RG 319 Entry 101 Box 4 Folder 76 Report on Special Operations.

RG 319 Entry 101 Box 4 Folder 78, Operation "Anvil."

RG 319 Entry 101 Box 4 Folder 79, Report on the Activities of No. 2 Tac H.Q. A Force (Sixth Army Group) 31 July 1944–5 May 1945, Part I.

RG 319 Entry 101 Box 4 Folder 80, Report on the Activities of No. 2 Tac H.Q. A Force Sixth Army Group, Part II.

RG 319 Entry 101 Box 4 Folder 81 Report on the Activities of No. 2 Tac H.Q. A Force (Sixth Army Group) Part III, Operation Loadline, Southern France.

RG 319 Entry 101 Box 4 Folder 82, Report on the Activities of No. 2 Tac H.Q. A Force (Sixth Army Group) Part IV, Conclusions.

RG 319 Entry 101 Box 5 Folder 87, Normandy, Cobra and Mortain, de Gersdorff interview.

RG 319 Entry 101 Box 6 Folder 77, Projects.

RG 319 Entry 101 Box 6 Folder 89, RAF history of visual deception.

RG 319 Entry 101 Box 6 Folder 91 (unnamed).

RG 319 Entry 101 Box 6 Folder 92, VISTRE.

RG 319 Entry 101 Box 7 Folder 1, Pastel Two.

RG 319 Entry 101 Box 7 Folder 2, Cover Plan Pastel.

RG 319 Entry 101 Box 7 Folder 3, Barclay.

RG 319 Entry 101 Box 7 Folder 4, Operation Zipper and Cover and Deception Plan Conscious.

RG 319 Entry 101 Box 7 Folder 5, map of Pacific Theatre summer 1945.

RG 319 Entry 101 Box 7 Folder 7, Zeppelin.

RG 319 IRR Files Box 35 X1.10.72.06 Peter J. W. De Bye, Folder Volume 1 of 2—Folder 1 of 1.

RG 319 IRR Files Box 35 X1.10.72.06 Peter J. W. De Bye, Folder Volume 2 of 2—Folder 1 of 3.

RG 319 IRR Files Box 35 X1.10.72.06 Peter J. W. De Bye, Folder Volume 2 of 2—Folder 2 of 3.

RG 319 IRR Files Box 35 X1.10.72.06 Peter J. W. De Bye, Folder Volume 2 of 2—Folder 3 of 3.

RG 331 Entry 1 Box 11 Folder Plan Fortitude (SHAEF G-2 files).

RG 331 Entry 1 Box 45 Folder Organization and Personnel G-3 SHAEF.

RG 331 Entry 1 Box 64 Folder 370.5/3 Movement of 23d Headquarters and Headquarters Detachment Special Troops.

RG 331 Entry 1 Box 69 Folder 381 Avenger.

RG 331 Entry 1 Box 73 Folder 381 Fortitude Vol. 1.

RG 331 Entry 1 Box 73 Folder 381 Fortitude Vol. 2.

RG 331 Entry 199 Box 66 Folder Fortitude (fat folder of that name).

RG 331 Entry 199 Box 66 Folder Fortitude (thin folder of that name).

RG 331 Entry 3 Box 122 Folder COSSAC (43) Papers II Nos. 23–70.

RG 331 Entry 3 Box 122 Folder Minutes of COSSAC Staff Conferences.

RG 331 Entry 3 Box 122 Folder (loose) List of Papers—Primary Papers of COSSAC Serial No. 1.

RG 331 Entry 3 Box 122 Folder COSSAC (43) Papers I Nos. 1–22.

RG 331 Entry 3 Box 123 Folder COSSAC (43) Papers III Nos. 71–99.

RG 331 Entry 3 Box 123 Folder COSSAC PSOs' Meetings.

RG 331 SHAEF G-3 General Records Subject File 1942–45 Combating the Guerrilla to Daily Reports, Folder Cover Plan—Overlord (Bull Papers).

RG 457 Box 72 SRDJ decrypts.

United States Navy, Naval Operational Archives, Washington Navy Yard.
(All From Strategic Plans Division, Series XXII).

Box 529, Folder Agents (1945–1946).

Box 530, Folder C&D (Cover and Deception) in World War II (General).

Box 530, Folder C&D Lecture for War College.

Box 530, Folder CEAs (1943, 1948).
Box 530, Folder Combined Planning, Cover and Deception.
Box 530, Folder Communications Security.
Box 530, Folder Lectures on C&D.
Box 531, Folder Deception Methods Techniques and Devices.
Box 531, Folder Deception Staff Studies Pacific.
Box 531, Folder Documentation Relating to the Recording and Evaluation of Cover and Deception Plans.
Box 531, Folder Eisenhower's Summary (1944).
Box 531, Folder Notes on Deception in a Theater of Operations.
Box 531, Folder OP-30C-Out-622 and OP-30C-Out-623.
Box 532, Folder Douglas Fairbanks.
Box 532, Folder F-28 Dispatch File.
Box 533, Folder C&D Formulation of Plan.
Box 533, Folder Germany Deception.
Box 535, Folder India-Burma.
Box 535, Folder Instructional Material, Plans.
Box 535, Folder Japan Deception.
Box 535, Folder Japanese Intelligence Analysis of WW II.
Box 535, Folder Japanese Intelligence in Europe.
Box 535, Folder Japanese Intelligence Organization.
Box 536, Folder Japanese Intelligence Services.
Box 537, Folder M-1.
Box 537, Folder Misc C&D, Psychological Warfare.
Box 537, Folder N.I.D. (British Naval Intelligence Division) Admiralty Reports on Naval Deception.
Box 539, Folder OSS.
Box 543, Folder Radio Deception.
Box 543, Folder Russia—Deception.
Box 544, Folder SEAC.
Box 544, Folder Special Means—A.
Box 544, Folder Special Means—B.
Box 544, Folder Special Means—D.
Box 544, Folder Special Means—General Measures.
Box 544, Folder Special Means—Record.
Box 544, Folder Special Means (unlabeled; first of two).
Box 544, Folder Special Means (unlabeled; second of two).
Box 545, Folder Presentation—Strategic C&D, National War College.
Box 545, Folder SWPA.
Box 546, Folder Tactical C&D—Navy Department Organization and Fleets Letter.
Box 546, Folder Tactical C&D Directives.
Box 546, Folder Tactical Deception—Ruses.
Box 546, Folder Tactical Deception, RCM.
Box 546, Folder USS Tucson Operations 10–11 July.
Box 548, Folder Bassington.
Box 548, Folder Bluebird—Dispatches.
Box 548, Folder Bluebird (first of two of that name).
Box 548, Folder Bluebird (second of two of that name).
Box 548, Folder Intelligence—Pastel, Bluebird, Broadaxe.
Box 548, Folder Sample Plans.
Box 549, Folder Bodyguard.
Box 549, Folder Broadaxe.

Box 549, Folder Broadaxe—Dispatches.
Box 549, Folder Broadaxe—Staff Studies.
Box 549, Folder Carbonado.
Box 549, Folder Conscious.
Box 549, Folder Coronet (Staff Studies).
Box 549, Folder Fortitude.
Box 549, Folder Plan—Coronet.
Box 550, Folder Husky.
Box 550, Folder Iceberg (first of two of that name).
Box 550, Folder Iceberg (second of two of that name).
Box 550, Folder Iceman.
Box 550, Folder Longtom.
Box 550, Folder Mike One.
Box 550, Folder Milepost.
Box 550, Folder Milepost Dispatches.
Box 550, Folder Olympic.
Box 550, Folder Pastel.
Box 550, Folder Pastel (unlabeled; first of two).
Box 550, Folder Pastel (unlabeled; second of two).
Box 550, Folder Pastel Two (first of two of that name).
Box 550, Folder Pastel Two (second of two of that name).
Box 550, Folder Plan—Iceman.
Box 550, Folder Plan—Olympic.
Box 551, Folder Penknife.
Box 551, Folder Sceptical.
Box 551, Folder Sturgeon.
Box 551, Folder Torch.
Box 551, Folder Wedlock.
Box 551, Folder Wedlock—Dispatches.
Box 551, Folder Wedlock [Bluebird] Section III.
Box 551, Folder Wedlock OP-03-C-Out 204.
Box 551, Folder Wedlock Sections I–II.
Box 551, Folder Wedlock Section III.
Box 551, Folder Wedlock Section IV Folder No. 1—Enemy Reaction.
Box 551, Folder Wedlock Section IV Folder No. 2—Enemy Reaction.
Box 551, Folder Wedlock Section V—Preliminary Material for Critique.
Box 551, Folder Wedlock Section V—Summary and Evaluation.

Public Record Office (National Archives), Kew

ADM 223/794, Reports of Lieutenant-Commander Montagu on naval deception matters.
ADM 1/25230, Major Martin (Mincemeat).
CAB 21/1422, Cabinet Office Registered Files, Inter-Service Security Board.
CAB 79, War Cabinet, Chiefs of Staff Committee, Minutes of Meetings 1939 to 1946.
CAB 81/76 through 81/86, War Cabinet, Chiefs of Staff Committees and Sub-Committees Minutes and Papers, Combined Chiefs of Staff.
CAB 119/66, Joint Planning Staff, Strategic deception cover plans for operations.
CAB 119/67, Joint Planning Staff, Strategic deception policy and machinery.
CAB 121/105 through 121/110, Cabinet Office Special Secret Information Centre, A/Policy/Deception.
CAB 154/1 through 154/112, War Cabinet and Cabinet Office: London Controlling Section Correspondence and Papers.

HW 41/159, Records of the GC&CS, Miscellaneous Papers on Deception Far East.

KV 2/467, Records of the Security Service, Peasant.

KV 2/1133, Records of the Security Service, Cheese.

KV 2/1163, Records of the Security Service, Ernst Paul Fackenheim.

KV 4/63 through 4/69, Records of the Security Service, Formation and minutes of the Twenty Committee in connection with Double Cross agents.

KV 4/100, Records of the Security Service, Functions and Organization of CI War Room 1944–45.

KV 4/101, Records of the Security Service, Minutes of the 212 Committee in connection with traffic for special agents in Europe.

KV 4/185 through 196, Records of the Security Service, Diaries of Guy Liddell.

KV 4/197, Records of the Security Service, SIME.

PREM 3/117, Records of the Prime Minister, Deception.

WO 106/5921, Operation Mincemeat: Copies of documents made available to the press.

WO 171/3831 War Diaries, 21 Army Group, G(R) Main including Camouflage Pool, June 1944–December 1944.

WO 171/3832 War Diaries, 21 Army Group, Headquarters "R" Force Main, June 1944–December 1944.

WO 171/3868 War Diaries, 21 Army Group, G(R) Main including Camouflage Pool, January 1945–March 1945.

WO 171/3869 War Diaries, 21 Army Group, Headquarters "R" Force Main, January 1945–April 1945.

WO 203/20, Deception 14th Army.

WO 203/21, Deception Post Rangoon Operations.

WO 203/3313, D Division Policy Etc.

WO 203/3314, Deception Division: Operational Planning and Reports.

WO 203/5746, Plan Mulliner.

WO 203/5769A, Plan Mimsy.

WO 204/1561, Allied Force Headquarters, Deception plans: Policy and operations planning April 1943–February 1944.

WO 204/1562, Allied Force Headquarters, Deception plans: Policy and operations planning February 1944–August 1944.

WO 204/1563, Allied Force Headquarters, Deception plans: Policy and operations planning August 1944–May 1945.

WO 208/3163, Directorate of Military Intelligence, Operation Mincemeat: German reaction to deception plan.

WO 219/2204A, SHAEF G-3 Ops (B), Plan Mespot: Comments by British Chiefs of Staff; Cover operations for Overlord.

WO 219/2205, SHAEF G-3 Ops (B), Plan Mespot: Chiefs of Staff directives.

WO 219/2206, SHAEF G-3 Ops (B), Essential long-term measures within the framework of Plan Bodyguard.

WO 219/2207, SHAEF G-3 Ops (B), Wireless deception and security.

WO 219/2208, SHAEF G-3 Ops (B), Combined Signal Board; Signal Security Committee.

WO 219/2209, SHAEF G-3 Ops (B), Movement and administrative preparations.

WO 219/2210, SHAEF G-3 Ops (B), Concealment and display of camps: Report by GHQ.

WO 219/2211, SHAEF G-3 Ops (B), Camouflage and visual misdirection policy.

WO 219/2212, SHAEF G-3 Ops (B), Camouflage and visual misdirection policy: Agenda and minutes.

WO 219/2213, SHAEF G-3 Ops (B), Assistance by resistance groups.

WO 219/2214, SHAEF G-3 Ops (B), Naval plan: Liaison with the Admiralty.

WO 219/2215, SHAEF G-3 Ops (B), Operation Bigbobs: Policy, reports.

WO 219/2216, SHAEF G-3 Ops (B), Exercise Fabius: Plans and appreciation.

WO 219/2217, SHAEF G-3 Ops (B), Deliberate release to the enemy of commanders' names for forces under 1 U.S. Army Group.

WO 219/2218, SHAEF G-3 Ops (B), Plan Ironside: Outline operational policy.

WO 219/2219, SHAEF G-3 Ops (B), Plan Copperhead: Preliminary arrangements.

WO 219/2220, SHAEF G-3 Ops (B), Plans for Scandinavia.

WO 219/2221, SHAEF G-3 Ops (B), Exercise Skye: Instructions.

WO 219/2222, SHAEF G-3 Ops (B), North 2 Scandinavia.

WO 219/2223, SHAEF G-3 Ops (B), South: (Pas de Calais).

WO 219/2224, SHAEF G-3 Ops (B), South 2: Preparation Planning.

WO 219/2225, SHAEF G-3 Ops (B), South 2 and North 2: Order of battle.

WO 219/2226, SHAEF G-3 Ops (B), Redesignation of First U.S. Army Group and employment of Generals L. J. McNair and J. L. DeWitt.

WO 219/2227, SHAEF G-3 Ops (B), Wireless plan and signal instructions.

WO 219/2228, SHAEF G-3 Ops (B), Wireless silence and activity.

WO 219/2229, SHAEF G-3 Ops (B), Exercise Rosebud: Move of 4 Army.

WO 219/2230, SHAEF G-3 Ops (B), Exercise Rosebud: Move of 4 Army: General instructions.

WO 219/2231, SHAEF G-3 Ops (B), Air plan and bombing programme.

WO 219/2232, SHAEF G-3 Ops (B), Decoy lighting: East and southeast England.

WO 219/2233, SHAEF G-3 Ops (B), Release of resources.

WO 219/2234, SHAEF G-3 Ops (B), Restriction of fishing.

WO 219/2235, SHAEF G-3 Ops (B), Code words and meanings.

WO 219/2236, SHAEF G-3 Ops (B), Wireless deception and restrictions. R Force wireless activity.

WO 219/2237, SHAEF G-3 Ops (B), Mulberries: Camouflage of Phoenix components and artificial ports.

WO 219/2238, SHAEF G-3 Ops (B), Security restrictions in the U.K. after D-Day.

WO 219/2239, SHAEF G-3 Ops (B), Use of police wireless: Security.

WO 219/2240, SHAEF G-3 Ops (B), Security ban on visitors to restricted coastal areas.

WO 219/2241, SHAEF G-3 Ops (B), Leave and travel restrictions.

WO 219/2242, SHAEF G-3 Ops (B), Postal arrangements: Censorship, directive.

WO 219/2243, SHAEF G-3 Ops (B) (number not used).

WO 219/2244, SHAEF G-3 Ops (B), Organization of 12 Reserve Unit.

WO 219/2245, SHAEF G-3 Ops (B), Organization of 13 Reserve Unit: 24 Armoured Brigade.

WO 219/2246, SHAEF G-3 Ops (B), Cover and Deception Policy.

WO 219/2247, SHAEF G-3 Ops (B), 21 Army Group.

WO 219/2248, SHAEF G-3 Ops (B), 12 U.S. Army Group.

WO 219/2249, SHAEF G-3 Ops (B), 6 U.S. Army Group.

WO 219/2250, SHAEF G-3 Ops (B), Plan Ironside 2.

WO 219/2251, SHAEF G-3 Ops (B), Plan Avenger.

WO 219/2252, SHAEF G-3 Ops (B), Plan to Lower Enemy Morale.

WO 219/2253, SHAEF G-3 Ops (B), Plan Dervish: Submission to the Directors of Plans.

WO 219/2254, SHAEF G-3 Ops (B), Plan Callboy.

WO 219/2255, SHAEF G-3 Ops (B), Plan Jurisdiction.

WO 219/2256, SHAEF G-3 Ops (B), Plans Taper 1 and 2.

WO 219/2257, SHAEF G-3 Ops (B), Transfer of 1 Canadian Corps from Italy to N.W. Europe.

WO 219/2258, SHAEF G-3 Ops (B), Operation Overlord: Security measures.

BIBLIOGRAPHY

Records of Joint Security Control, in Records of the Joint Staff,
The Pentagon, Washington.

Acco binder "1."
Acco binder "3" "Portugal."
Acco binder "4" "South America."
Acco binder "5" "File 'C'."
Acco binder "6" "File 'D'."
Acco binder "7" "File 'A'."
Acco binder "8" "File 'B'."
Acco binder "9" "201–400 File 'B'."
Acco binder "10" "B File."
Acco binder "11" "Index."
Acco binder "12" "File X."
Acco binder "13" "File Z—Salesmen's Qualifications."
Acco binder "14."
Acco binder "15."
Acco binder "16."
Acco binder "18."
Acco binder Deception Policy 1943–1944 Europe.
Acco binder Deception Policy Pacific Areas.
Acco binder Deceptive Warfare Operational Planning Handbook.
Acco binder Report on Operations in Northwestern Europe.
Acco binders with final reports of No. 2 Tac HQ.
Acco binder Trident Conclusions.
Folder #6 LEN Harris/Ingersoll/Harris/Eldredge.
Folder #8—Hist Decep in West Europe.
Folder #10 H.S.A.
Folder 10 Operation Husky.
Folder 11—Tactical and Strategic Deception—Capt. Hershey.
Folder 13–8 Communications Deception—Simulative.
Folder 13-12/5 Communication Decep: RPTS-India Burma.
Folder 13-12/6 Communications Deception Reports: Europe, Africa.
Folder 13-16/8 Radio Deception Manual (Prot. Sec. Br.).
Folder 13-19/2 Notes on W/T Security.
Folder 13-25/3 Bluebird (Radio Decep. Plan of Iceberg).
Folder 13-25/3 Wedlock.
Folder 13-25/7 Dragoon.
Folder 15 Amphibious Training Manuals.
Folder 21 Radio Deception by ASF.
Folder 26—Sonic Deception.
Folder 000—To 312.1(1) China-Burma-India Theater.
Folder 000—To 312.1(2) China-Burma-India Theater.
Folder 100.00 Radio Cover and Deception.
Folder 370.2 Bodyguard.
Folder 370.2 Broadaxe.
Folder 370.2 Cockade.
Folder 370.2 Detachment.
Folder 370.2 Flounders.
Folder 370.2 Forager.
Folder 370.2 Fortitude.
Folder 370.2 Graffham.

Folder 370.2 Ironside.
Folder 370.2 Mediterranean.
Folder 370.2 Operations and Results.
Folder 370.2 Operations and Results, Subfolder "Informal Report to Joint Security Control."
Folder 370.2 Overlord.
Folder 370.2 Pacific Theater.
Folder 370.2 Pastel.
Folder 370.2 Royal Flush.
Folder 370.2 Sturgeon.
Folder 370.2 Sweater.
Folder 370.2 TAC, No. 2.
Folder 370.2 Wadham.
Folder 370.2 Wadham—Captured Documents, Jan 1944 through May 1944.
Folder 370.2 Wadham—Captured Documents, June 1944.
Folder 370.2 Wadham—Captured Documents, June 1944 Continued.
Folder 370.2 Wadham—Captured Documents, Jul 1944 through Spring 1945.
Folder 370.2 Wadham, Continued.
Folder 500.00 Security Classification Studies—Ana.
Folder Barclay.
Folder Bluebird.
Folder C&D Instructions—Course Outlines (WWII—1944, 1945).
Folder Colonel Bevan Use of Special Intelligence in Cover and Deception.
Folder Communications Security Organization.
Folder Conscious.
Folder D Division Weekly Implementation Series.
Folder Deception—SE Asia—WWII Historical D. Div—SACSEA—16 Dec 44–15 Mar 45.
Folder Deception—SE Asia—WWII Historical D. Div—SACSEA—24 Mar 45–4 Jun 45.
Folder Deception—SE Asia—WWII Historical D. Div—SACSEA—30 Dec 44–23 Mar 45.
Folder Deception—SE Asia—WWII Historical D. Div.—SACSEA—5 Jun–10 Sep 45.
Folder Deception—SE Asia—WWII Historical D. Div.—SACSEA—5 Jun–10 Sep 45 (second folder).
Folder Deception Units.
Folder Deceptive Equipment—Navigation—BJU's—Resume—Aug 44.
Folder Dervish.
Folder European Theater 000.
Folder German Deception; Nazi Deception Navy; ASF Radio Deception.
Folder German Efforts WWII.
Folder German Espionage in Argentina.
Folder German Estimate of Allied OB.
Folder Germany.
Folder Great Britain.
Folder Hanover File (Materials to be forwarded to CRREC).
Folder History—Miscellaneous.
Folder History of Traffic Control Section of Protective Branch.
Folder Japan.
Folder Japanese Deception Methods.
Folder Japanese Estimates of Allied OB.
Folder Japanese Interception of 14th Air Force Messages.
Folder JCS To Handle Deception.
Folder Joint Security Control Memos 1–9.
Folder Letters From World War II Commanders Re Value of Cover and Deception.

Folder Loose copy of "Action Report Western Naval Task Force The Sicilian Campaign."
Folder M-1.
Folder Miscellaneous Deception Plans—WWII.
Folder Olympic.
Folder Operation Wedlock.
Folder Oplan "Bluebird"—Pacific WWII.
Folder Peter Fleming.
Folder Plan for Deception Section for China.
Folder Portugal.
Folder REF 403.00—Operation Reports 23d Headquarters.
Folder South America.
Folder Southeast Asia.
Folder Sweden.
Folder Switzerland.
Folder The Japanese Intelligence System.
Folder Torch.
Folder Traffic Sent.
Folder Wedlock—Subfolder "370.2 Wedlock—Adm Impl."
Folder Wedlock—Subfolder "370.2 Wedlock—Impl Phys."
Folder Wedlock—Subfolder "370.2 Wedlock—Plan."
Folder Wedlock—Subfolder "JT COMCENTER 1 May–31 Dec 44."
Folder Wedlock—Subfolder "Operation Wedlock."
Loose document, "12th Army Group Report Copy 1."
Loose document, "Reports on the Wireless Deception and Security Measures Taken by the Three Services in Connection With Operation Neptune July 1944" with appendices and notes.
Loose document, "Cake Before Breakfast" and attached memorandum.
Loose document, "Report of Signal Division SHAEF in Operation Overlord."
Loose document, "US and British Order of Battle and German Intelligence."
Loose document, COSSAC July 1943 "Operation Overlord Report and Appreciation."
Loose document, COSSAC, Operation Overlord—Maps.
Loose documents on German deception measures.
Loose envelope, "Retain as Photos of Examples of Tactical C&D Equipment."
Loose envelopes, unlabeled (two) [Maps and planning materials for Wadham].
Loose file binder "Re: Rudloff Case Received Messages."
Loose file binder "Re: Rudloff Case Sent Messages."
Loose pack of 3 x 5 cards [Tabulation of deception operations].
Ring binder "Task Force Security."
Unlabeled folder [CBI Theater material].
Unlabeled folder [Navy material].
Unlabeled folder [Radio deception in ZEPPELIN and ACCUMULATOR].

Other Archival Material

1st Joint Air Course, Functions and Set-Up of the Army Staff, 4 Oct 43, USAFHRA 505.89-38, in United States Air Force Historical Research Center, Maxwell AFB, Alabama.
Baumer, William H., Jr., Papers, Hoover Institution; Diary, privately held, copy in author's possession.
Bevan, John H., Papers, privately held, some copies in author's possession.
Burris-Meyer, Harold, Papers, privately held, some copies in author's possession.
Campbell, Dr. J., *Operation Starkey 1943—"a piece of harmless playacting"?*, in Mure papers, Imperial War Museum.

Churchill Archives, Churchill College, Cambridge.

Clarke, Dudley, *A Quarter of My Century,* Imperial War Museum.

Clarke, Dudley, Papers, Imperial War Museum.

Eldredge, H. Wentworth, Papers, privately held, copies in author's possession.

Evans, *Organization and Functioning of Navy Staffs,* Naval War College, Newport, Rhode Island, 31 May 1945, copy in United States Air Force Historical Research Center, Maxwell AFB, Alabama.

Fleming, Peter, Papers, Reading University Library.

Goldbranson, Carl, Papers, privately held, copies in author's possession.

[Harmer, Christopher], *Report on Controlled Enemy Agents in 21 Army Group Area During Period July 1944 to May 1945,* in Norman Holmes Pearson Papers, Wooden File, Box 3, Folder "XX," Beinecke Library, Yale University.

Harris, Hunter, Jr., U.S. Air Force Oral History Interview No. K239.0512-811, United States Air Force Historical Research Center, Maxwell AFB, Alabama.

Harris, William A., Lecture, Command and General Staff College, Fort Leavenworth, Kansas [1947].

Harris, William A., Papers, United States Army Military History Institute, United States Military Academy Library, and privately held, some copies in author's possession.

Hunter, E. Ormonde, Diary, in author's possession.

Hunter, E. O., Collection, Georgia Historical Society.

Ingersoll, Ralph, papers, in Boston University Library.

Ingersoll, Ralph, *Time Out for a War,* copy in author's possession.

Johnson to Files, Subject: Dr. Kayl—German Espionage Agent, 20 January 1947, in files of the Federal Bureau of Investigation, copy in author's possession.

Kehm, Harold D., Papers, privately held, copies in author's possession.

Macleod, Roderick, Papers, Imperial War Museum.

"Magic" Summaries and Diplomatic Summaries, United States Army Military History Institute, Carlisle, Pennsylvania.

Maunsell, R. J., Papers, Imperial War Museum.

Mure, David, Papers, Imperial War Museum.

Nevins, Arthur S., Papers, United States Army Military History Institute, Carlisle, Pennsylvania.

Oakes, John B., Weismiller, Edward R., and Waith, Eugene M., *A History of OSS/X-2 Operation of Controlled Enemy Agents in France and Germany, 1944–1945, Together with a Study of the Theory and Practice of CEA Operation. Volume I. Study and History; Volume II, Appendices* (Strategic Services Unit, War Department, Washington, 1946), in files of Central Intelligence Agency, Langley, Virginia; copy in author's possession.

Pearson, Norman Holmes, Papers, Beinecke Library, Yale University.

Plan Bodyguard, USAFHRA 119.04-8; Plan Cockade, USAFHRA 505.61-15; Plan Fortitude, USAFHRA 505-61-21, in Air Force Historical Agency, Maxwell AFB, Alabama.

Praun, German Radio Intelligence, OCMH MS P-038 Abr, U.S. Army Center for Military History, Washington.

Smith, Bedell, Papers, Eisenhower Library, Abilene, Kansas.

Smith, Newman, personnel file, in files of United States Department of Veterans Affairs, copy in author's possession.

Smith, Newman, Papers, privately held, some copies in author's possession.

Spanip files, in files of the Federal Bureau of Investigation, copies of some documents in author's possession.

Teeple, Captain David S., Information Obtained by the German Intelligence Service Relative to Allied Atomic Research, 23 October 1945, in files of the Federal Bureau of Investigation, copy in author's possession.

Train, George F., Papers, privately held, some copies in author's possession.

Unit Files, United States Army Institute of Heraldry, Arlington, Virginia.

Verbatim Transcript of Stratagem Conference, Pentagon, 21 June 1971, Sponsored by DOCSA, J3, JCS, Joint Staff, Contract Support by SURC, in files of the Joint Staff, United States Department of Defense.

York Associates, Transcripts of Interviews With David Strangeways and Hugh Astor, 1994.

Books

Alexander of Tunis, Field-Marshal Earl, *The Alexander Memoirs* (Cassell, London, 1962).

Andrew, Christopher, *Her Majesty's Secret Service* (paperback ed., Penguin Books, New York, 1987).

Appleman, Roy E., Burns, James M., Gugeler, Russell A., and Stevens, John, *Okinawa: The Last Battle* (Historical Division, Department of the Army, Washington, 1948).

Arnold, H. H., *Global Missions* (Harper & Bros., New York, 1949).

Arnold, Ralph, *A Very Quiet War* (Rupert Hart-Davis, London, 1962).

Astley, Joan Bright, *The Inner Circle* (Little, Brown, Boston, 1971).

Bhattacharyya, S. N., *Subhas Chandra Bose in Self-Exile: His Finest Hour* (Metropolitan Book Co., Delhi, 1975).

Bidwell, Bruce W., *History of the Military Intelligence Division, Department of the Army General Staff: 1775–1941* (University Publications of America, Frederick, Maryland, 1986).

Boyd, Carl, *Hitler's Japanese Confidant* (University Press of Kansas, Lawrence, 1993).

Bradley, Omar, and Blair, Clay, *A General's Life* (Simon & Schuster, New York, 1983).

Bristow, Desmond, with Bristow, Bill, *A Game of Moles* (Little, Brown, London, 1993).

Browne, Joseph Edward, *Deception and the Mediterranean Campaigns of 1943–1944* (U.S. Army War College, Carlisle Barracks, 1986).

Bruccoli, Matthew J. (ed.), *F. Scott Fitzgerald: A Life in Letters* (Scribners, New York, 1994).

Bruccoli, Matthew J., Duggan, Margaret M., and Walker, Susan, *Correspondence of F. Scott Fitzgerald* (Random House, New York, 1981).

Bruccoli, Matthew J., *Some Sort of Epic Grandeur* (Harcourt Brace Jovanovich, New York, 1981).

Bryan, C. D. B., *The National Geographic Society: 100 Years of Adventure and Discovery* (H. N. Abrams, New York, 1987).

Bryant, Arthur, *The Turn of the Tide* (Doubleday, Garden City, 1957).

Bullock, David L., *Allenby's War: the Palestine-Arabian Campaigns, 1916–1918* (Blandford Press, New York, 1988).

Butcher, Harry C., *My Three Years With Eisenhower* (Simon & Schuster, New York, 1946).

Butow, Robert J. C., *Japan's Decision To Surrender* (Stanford University Press, Palo Alto, 1954).

Calvert, Michael, *Fighting Mad* (Jarrolds, London, 1964).

[Carboni, Giacomo], *Memorie Segrete 1935–1948* (Parenti Editore, Florence, n.d.).

Carboni, Giacomo, *Più Che Il Dovere* (Daneri in Via Margutta, Rome, 1952).

Clark, George Rogers, *The Conquest of the Illinois* (Quaife ed., R. R. Donnelly, Chicago, 1920).

Clarke, Dudley, *Seven Assignments* (Jonathan Cape, London, 1948).

Clarke, Jeffrey J., and Smith, Robert Ross, *Riviera to the Rhine* (Center for Military History, Washington, 1993).

Clarke, T. E. B., *This Is Where I Came In* (Michael Joseph, London, 1974).

Cline, Ray S., *Washington Command Post: The Operations Division* (Office of the Chief of Military History, Washington, 1951).

Cole, Howard N., *Heraldry in War: Formation Badges 1939–1945,* (3d ed., Wellington Press, Gale & Polden Ltd., Aldershot, 1950).

Coll, Blanche D., Keith, Jean E., and Rosenthal, Herbert A., *The Corps of Engineers: Troops and Equipment* (Government Printing Office, Washington, 1958).

Collier, Richard, *The Sands of Dunkirk* (Dutton, New York, 1961).

Collins, Frederick C., *The FBI in Peace and War* (G. P. Putnam's Sons, New York, 1943).

Colville, John, *The Fringes of Power* (Norton, New York, 1986).

Colyton, Henry, *Occasion, Chance and Change* (Michael Russell, London, n.d.).

Connelly, T. L., *Army of the Heartland: The Army of Tennessee, 1861–1862* (Louisiana State University Press, Baton Rouge, 1967).

Cooper, Artemis, *Cairo in the War: 1939–1945* (Hamish Hamilton, London, 1989).

Copeland, Miles, *Beyond Cloak and Dagger: Inside the CIA* (Paperback ed., Pinnacle Books, New York, 1975).

Corr, Gerard H., *The War of the Springing Tigers* (Osprey Publishing Ltd., London, 1975).

Craven, Wesley Frank, and Cate, James Lea, *The Army Air Forces in World War II,* vol. 4 (University of Chicago Press, Chicago, 1950).

Cross, Colin, *The Fascists in Britain* (St. Martin's Press, New York, 1961).

Cruickshank, Charles, *Deception in World War II* (Oxford University Press, Oxford and New York, 1979).

Cruickshank, Charles, *SOE in the Far East* (Oxford University Press, Oxford and New York, 1983).

Crute, Joseph H., Jr., *Units of the Confederate States Army* (Derwent Books, Midlothian, Virginia, 1987).

Cullum, George W., *Biographical Register of the Officers and Graduates of the United States Military Academy at West Point, New York, Since Its Establishment in 1802,* and supplements (Houghton Mifflin, Boston, various dates).

D'Este, Carlo, *Patton: A Genius for War* (HarperCollins, New York, 1995).

Davie, Maurice R., Donaldson, Norman V., English, Philip H., and Wayland, Elton S. (eds.), *History of the Class of 1915, Yale College, Volume III* (Yale University Press, New Haven, 1952).

Davis, Vernon E., *The History of the Joint Chiefs of Staff in World War II: Organizational Development* (Historical Division, Joint Secretariat, Joint Chiefs of Staff 1972) (Draft 1996–97).

De Risio, Carlo, *Generali, Servizi Segreti e Fascismo* (Arnoldo Mondadori, Milan, 1978).

Deane, John R., *The Strange Alliance* (New York, Viking, 1947).

Delmer, Sefton, *The Counterfeit Spy* (Harper & Row, New York, 1971).

DeLoach, Cartha, *Hoover's FBI* (Henry Regnery, Washington, 1995).

Deutsch, Harold C., *The Conspiracy Against Hitler* (University of Minnesota Press, Minneapolis, 1968).

Dictionary of National Biography, with Supplements (Oxford University Press, Oxford and New York, various dates).

Dobinson, Colin, *Fields of Deception: Britain's Bombing Decoys of World War II* (Methuen, London, 2000).

Drea, Edward, *MacArthur's Ultra* (University Press of Kansas, Lawrence, 1992).

Dufty, William, *Sugar Blues* (Warner Books, New York, 1975).

Dunlop, Richard, *Behind Japanese Lines With the OSS in Burma* (Rand McNally, Chicago, 1979).

Dwyer, John B., *Seaborne Deception* (Praeger, New York, 1992).

Eichelberger, Robert L., *Our Jungle Road to Tokyo* (Viking, New York, 1950).

Eisenhower, Dwight D., *At Ease: Stories I Tell to Friends* (Doubleday, New York, 1967).

Eisenhower, John S. D., *The Bitter Woods* (paperback ed., Ace Books, New York, 1970).

Ely, Albert, Jr. (ed.), *History of the Class of 1915* (Yale University Press, New Haven, 1915).

Ely, Albert, Jr. (ed.), *History of the Class of 1915, Yale College, Volume II* (Yale University Press, New Haven, 1930).

Fairbanks, Douglas, Jr., *A Hell of a War* (St. Martin's, New York, 1993).

Farago, Ladislas, *Patton, Ordeal and Triumph* (paperback ed., Dell, New York, 1965).

Farago, Ladislas, *The Game of the Foxes* (David McKay, New York, 1971).

Felt, Mark, *The FBI Pyramid* (G. P. Putnam's Sons, New York, 1979).

Fishel, Edwin C., *The Secret War for the Union* (Houghton Mifflin, Boston, 1996).

Fitzgerald, F. Scott, *Stories of F. Scott Fitzgerald, Edited by Malcolm Cowley* (Scribners, New York, 1951).

Fitzgerald, Zelda Sayre, *Save Me the Waltz* (Paperback ed., Signet Books, New York, 1968).

Foot, M. R. D., and Langley, J. M., *MI9* (Little, Brown, Boston, 1980).

Foot, M. R. D., *SOE in France* (HMSO, London, 1966).

Foot, M. R. D., *SOE: An Outline History of the Special Operations Executive 1940–46* (Revised ed., University Publications of America, Frederick, 1986).

Fox, John P., *Germany and the Far Eastern Crisis, 1931–1938: A Study in Diplomacy and Ideology* (Oxford University Press, Oxford, 1982).

Frank, Richard B., *Downfall* (Random House, New York, 1999)

Freeman, Douglas Southall, *Lee's Lieutenants* (Scribners, New York, 1942–44).

Freeman, Douglas Southall, *The South to Posterity* (Scribners, New York, 1939).

Fuchida, Mitsuo, and Okumiya, Masatake, *Midway, The Battle That Doomed Japan* (Ballentine, New York, 1955).

Furer, Julius Augustus, *Administration of the Navy Department in World War II* (Government Printing Office, Washington, 1959).

Gilbert, Martin, *Winston S. Churchill, Volume IV, 1916–1922, The Stricken World* (Houghton Mifflin, Boston, 1975).

Gill, Brendan, *Here at The New Yorker* (Random House, New York, 1975).

Glines, Carroll V., *Doolitle's Tokyo Raiders* (Van Nostrand, Princeton, 1964).

Graham, Dominick, and Bidwell, Shelford, *Tug of War: The Battle foe Italy: 1943–45* (New York, St. Martin's, 1986).

Graham, Sheilah, *The Real F. Scott Fitzgerald 35 Years Later* (Grosset & Dunlap, New York, 1976).

Harries, Meirion, and Harries, Susie, *Soldiers of the Sun* (Random House, New York, 1991).

[Harris, Tomás], *Garbo: The Spy Who Saved D-Day* (Public Record Office, Kew, 2000).

Hart-Davis, Duff, *Peter Fleming: A Biography* (Jonathan Cape, London, 1974).

Harvey-Jones, John, *Getting It Together* (Heinemann, London, 1991).

Harvison, C. W., *The Horsemen* (Macmillan, London, 1967).

Hayes, Grace Person, *The History of the Joint Chiefs of Staff in World War II: The War Against Japan* (Naval Institute Press, Annapolis, 1982).

Headquarters 12th Army Group (Final After Action Report), n.p. n.d., USAMHI.

Henderson, G. F. R., *Stonewall Jackson and the American Civil War* (1 vol. ed., Longmans, New York, 1936).

Henry, Robert Selph, *"First With the Most" Forrest* (Bobbs-Merrill, New York and Indianapolis, 1944).

Herbig, Katherine L., *American Strategic Deception in the Pacific War 1942–1945*, unpublished work, copy in author's possession (Naval Postgraduate School, Monterey, 1985).

Herzstein, Robert E., *Henry R. Luce* (Macmillan, New York, 1994).

Hesketh, R. F., *Fortitude: A History of Strategic Deception in North Western Europe, April, 1943 to May, 1945* (printed page proofs, n.p., n.d.; published as *Fortitude: The D-Day Deception Campaign; Introduction by Nigel West* (Overlook Press, Woodstock, 2000).

[Hill, Tom, and others], *British Security Coordination: The Secret History of British Intelligence in the Americas 1940–45* (Nigel West ed., St Ermin's Press, London, 1998).

Hinsley, F. H., *British Intelligence in the Second World War* (Volumes 1–3, Cambridge University Press, New York, 1977–88).

Hinsley, F. H., and Simkins, C. A. G., *Security and Counterintelligence (British Intelligence in the Second World War,* Volume 4) (Cambridge University Press, New York, 1990).

Hoopes, Roy, *Ralph Ingersoll* (Atheneum, New York, 1985).

Howard, Sir Michael, *Strategic Deception (British Intelligence in the Second World War,* Volume 5) (Cambridge University Press, New York, 1990).

Howe, George F., *Northwest Africa: Seizing the Initiative in the West* (Office of the Chief of Military History, Washington, 1957).

Imperial War Museum, *The Cabinet War Rooms* (Imperial War Museum, London, n.d.).

Irons, Peter H., *Justice at War* (Oxford University Press, New York, 1983).

Irving, David, *Goebbels, Mastermind of the Third Reich* (Focal Point, London, 1996).

Irving, David, *Hitler's War* (Hodder & Stoughton, London, 1977).

Irving, David, *The Trail of the Fox* (paperback ed., Avon, New York, 1978).

James, D. Clayton, *The Years of MacArthur,* vol. 2 (Houghton Mifflin, Boston, 1975).

Kahn, David, *Hitler's Spies* (Macmillan, New York, 1978).

Kahn, David, *Seizing the Enigma* (Houghton Mifflin, Boston, 1991).

Kahn, David, *The Codebreakers* (Macmillan, New York, 1967).

Keegan, John, *The Second World War* (New York, Viking, 1990).

Keegan, John, *Waffen SS* (New York, Ballantine, 1970).

Kipling, Rudyard, *Rudyard Kipling's Verse, Definitive Edition* (Doubleday Doran, New York, 1945).

Kirkpatrick, Ivone, *Mussolini, Study of a Demagogue* (Odhams, London, 1964).

Kronman, Mark, *The Deceptive Practices of the 23rd Headquarters, Special Troops During World War II,* Tactical Operations Analysis Office, Interim Note No. T-111 (Aberdeen Proving Ground, Maryland, January 1978).

Lanahan, Eleanor, *Scottie the Daughter of . . .* (HarperCollins, New York, 1995).

Lanahan, Eleanor, *Zelda, An Illustrated Life* (Harry N. Abrams, New York, 1996).

Lawrence, T. E., *Seven Pillars of Wisdom* (Doubleday Doran, New York, 1935).

Lewin, Ronald, *The Chief* (Farrar Straus Giroux, New York, 1980).

Liddell Hart, B. H., ed., *The Rommel Papers* (Harcourt Brace, New York, 1953).

Lorillard, Lieutenant-Colonel, *Ruses de Guerre et Contre-Ruses* (Charles-Lavauzelle & Cie, Paris, 1935).

Lycett, Andrew, *Ian Fleming* (Weidenfeld & Nicolson, London, 1995).

Magan, William: *Middle East Approaches: Experience and Travels of an Intelligence Officer 1939–1948* (Michael Russell, Wilby, 2001).

Maheu, Robert, and Hack, Richard, *Next to Hughes* (HarperCollins, New York, 1992).

Malory, Sir Thomas, *The Works of Sir Thomas Malory,* ed. Eugene Vinaver (Oxford University Press, Oxford, 1954).

Manhattan Directory, December 1941–June 1942 (New York Telephone Company, New York, 1941).

Marshall, George C. (Blond, Larry I., and Blond, Joellen K., eds.), *George C. Marshall Interviews and Reminiscences for Forrest C. Pogue* (George C. Marshall Research Center, Lexington, 1991).

Maskelyne, Jasper, *Magic—Top Secret* (Stanley Paul, London, 1949).

Masterman, J. C., *On the Chariot Wheel* (Oxford University Press, London, 1975).

Masterman, J.C., *The Double-Cross System in the War of 1939 to 1945* (Yale University Press, New Haven, 1972) (reprinted Pimlico, London, 1995).

Matloff, Maurice, and Snell, Edward M., *Strategic Planning for Coalition Warfare 1941–1942* (Office of the Chief of Military History, Washington, 1953).

Matloff, Maurice, *Strategic Planning for Coalition Warfare 1943–1944* (Office of the Chief of Military History, Washington, 1953).

Mayer, Martin, *Madison Avenue U.S.A.* (paperback ed. Pocket Books, New York, 1959).

McMillan, Malcolm Cook, *Constitutional Development in Alabama: A Study in Politics, the Negro, and Sectionalism* (University of North Carolina Press, Chapel Hill, 1955).

Milford, Nancy, *Zelda* (Harper & Row, New York, 1970).

Milkman, Paul, *PM: A New Deal in Journalism* (Rutgers University Press, New Brunswick and London, 1997).

Mitchell, Samuel W., Jr., *Hitler's Legions* (Stein & Day, New York, 1985).

Mizener, Arthur, *The Far Side of Paradise* (Houghton Mifflin, Boston, 1951).

Montagu, Ewen, *Beyond Top Secret U* (Peter Davies, London, 1977).

Montagu, Ewen, *The Man Who Never Was* (revised paperback ed., Oxford University Press, Oxford and New York, 1996).

Moore, A. B., *History of Alabama* (Alabama Book Store, Tuscaloosa, 1951).

Morgan, Lieutenant-General Sir Frederick, *Overture to Overlord* (Doubleday, New York, 1950).

Morgan, Judith and Neil, *Dr. Seuss and Mr. Geisel* (Random House, New York, 1995).

Morison, Samuel Eliot, *Aleutians, Gilberts and Marshalls* (Little, Brown, Boston, 1951).

Morison, Samuel Eliot, *The Liberation of the Philippines* (Little, Brown, Boston, 1959).

Morison, Samuel Eliot, *New Guinea and the Marianas* (Little, Brown, Boston, 1953).

Morison, Samuel Eliot, *Sicily, Salerno, Anzio* (Little, Brown, Boston, 1954).

Morison, Samuel Eliot, *The Invasion of France and Germany* (Little, Brown, Boston, 1957).

Morison, Samuel Eliot, *The Rising Sun in the Pacific* (Little, Brown, Boston, 1948).

Morison, Samuel Eliot, *Victory in the Pacific, 1945* (Little, Brown, Boston, 1960).

Muggeridge, Malcolm, *The Infernal Grove* (William Morrow, New York, 1973).

Mure, David, *Master of Deception* (William Kimber, London, 1980).

Mure, David, *Practise To Deceive* (William Kimber, London, 1977).

Naftali, Timothy J., *X-2 and the Apprenticeship of American Counterespionage, 1942–1944* (UMI Dissertation Services, Ann Arbor, 1993).

Niven, David, *The Moon's a Balloon* (Putnam's, New York, 1972).

Packard, Wyman H., *A Century of U.S. Naval Intelligence* (Department of the Navy, Washington, 1996).

Page, Joseph A., *Perón* (Random House, New York, 1983).

Paillole, Paul, and Minella, Alain-Gilles, *L'Homme des Services Secrets* (Éditions Juillard, Paris, 1995).

Petersen, Neal H. (ed.), *From Hitler's Doorstep: The Wartime Intelligence Reports of Allen Dulles, 1942–1945* (Pennsylvania State University Press, University Park, Pennsylvania, 1996).

Pillon, Giorgio, *Spie per l'Italia* (I Libri del NO, Rome, 1968).

Pogue, Forrest C, *The Supreme Command* (Office of the Chief of Military History, Washington, 1954).

Popov, Dusko, *Spy Counter-Spy* (Grosset & Dunlap, New York, 1974).

Potter, E. B., *Nimitz* (Naval Institute Press, Annapolis, 1976).

Powers, Thomas, *Heisenberg's War* (Knopf, New York, 1993).

Preston, Dickson J., *Talbot County: A History* (Tidewater, Centreville, Maryland, 1986).

Pujol, Juan, with Nigel West [Rupert Allason], *Operation Garbo* (Random House, New York, 1985).

Ranfurly, Countess of, *To War With Whitaker* (Heinemann, London, 1994, paperback ed., Mandarin, London, 1995).

Register of Rhodes Scholars, 1903–1945 (Oxford University Press, London and New York, 1950).

Riebling, Mark, *Wedge: The Secret War Between the FBI and CIA* (Knopf, New York, 1994).

Riva, Maria, *Marlene Dietrich* (Knopf, New York, 1993).

Robertson, Sir William, *From Private to Field-Marshal* (Houghton Mifflin, Boston, 1921).

Rogers, James Grafton, *Wartime Washington: The Secret OSS Journal of James Grafton Rogers 1942–43,* ed. Thomas F. Troy (University Publications of America, Frederick, Maryland, 1987).

Rogers, William W., Ward, Robert David, Atkins, Leah Rawls, and Flynt, Wayne, *Alabama: The History of a Deep South State* (University of Alabama Press, Tuscaloosa, 1994).

Romanus, Charles F., and Sunderland, Riley, *Stilwell's Command Problems* (Office of the Chief of Military History, Washington, 1956).

Romanus, Charles F., and Sunderland, Riley, *Stilwell's Mission to China* (Office of the Chief of Military History, Washington, 1953).

Romanus, Charles F., and Sunderland, Riley, *Time Runs Out in CBI* (Office of the Chief of Military History, Washington, 1959).

Rout, Leslie B., Jr., and Bratzel, John F., *The Shadow War* (University Publications of America, Frederick, 1986).

Rusk, Dean, *As I Saw It* (Norton, New York, 1990).

Salmaggi, Cesare, and Pallavisini, Alfredo, *2194 Days of War* (Gallery Books, New York, 1977).

Sears, Steven W., *To the Gates of Richmond* (Ticknor & Fields, New York, 1992).

Shirer, William L., *The Rise and Fall of the Third Reich: A History of Nazi Germany* (Simon & Schuster, New York, 1960).

Shultz, Richard H., Jr., *The Secret War Against Hanoi* (HarperCollins, New York, 1999).

Slim, Field-Marshal the Viscount, *Defeat Into Victory* (David McKay, New York, 1961).

Spector, Ronald H., *Eagle Against the Sun* (Free Press, New York, 1985).

Stafford, David, *Churchill and Secret Service* (Overlook Press, Woodstock and New York, 1999).

Stanton, Shelby L., *Order of Battle U.S. Army, World War II* (Presidio Press, Novato, California, 1984).

Staubwasser, Colonel Anton, *Army Group B Intelligence Estimate,* OCMH MS B-675, B-782.

Sun Tzu, *The Art of War,* tr. Ralph D. Sawyer (Westview Press, Boulder, 1984).

Swanberg, W. A., *Luce and His Empire* (Scribners, New York, 1972).

Talwar, Bhagat Ram, *The Talwars of Pathan Land and Subhas Chandra's Great Escape* (People's Publishing House, New Delhi, 1976).

Theoharis, Athan G., and Cox, John Stuart, *The Boss: J. Edgar Hoover and the Great American Inquisition* (Temple University Press, Philadelphia, 1988).

Tuchman, Barbara W., *Stilwell and the American Experience in China* (Macmillan, New York, 1971).

United States Navy Department, Office of Naval History, *Glossary of U.S. Naval Code Words* (Navexos P-474) (2d ed. Revised March 1948) (Government Printing Office, Washington, 1948).

Ure, Percy Neville, *Justinian and His Age* (Penguin Books, Harmandsworth, 1951).

Waller, John H., *The Unseen War in Europe* (Random House, New York, 1996).

War Department, U.S., *Battle Participation of Organizations of the American Expeditionary Forces in France, Belgium and Italy 1917–1918* (Government Printing Office, Washington, 1930).

Warner, Philip, *Auchinleck, the Lonely Soldier* (paperback ed., Sphere Books, London, 1982).

Wavell, Sir Archibald, *Allenby* (Oxford University Press, New York, 1941).

Wedemeyer, Albert C., *Wedemeyer Reports!* (Henry Holt & Co., New York, 1958).

Weigley, Russell, *Eisenhower's Lieutenants* (Indiana University Press, Bloomington, 1981).

West, Nigel [Rupert Allason], *MI5* (Stein and Day, New York, 1981).

West, Nigel [Rupert Allason], *MI6* (Random House, New York, 1983).

West, Nigel [Rupert Allason], *Seven Spies Who Changed the World* (Secker & Warburg, London, 1991).

West, Nigel [Rupert Allason], and Tsarev, Oleg, *The Crown Jewels* (HarperCollins, London, 1998).

Wheatley, Dennis, *Stranger Than Fiction* (Paperback edition, Arrow Books, London, 1965; Hutchinson, London, 1959).
Wheatley, Dennis, *The Deception Planners* (Hutchinson, London, 1980).
Whitehead, Don, *The FBI Story* (Random House, New York, 1956).
Whitehill, W. M., *United States Naval Administration in World War II: Commander in Chief, United States Fleet Headquarters* (Division of Naval History, Washington, 1946).
Williams, Mary H., *Chronology 1941–1945* (Office of the Chief of Military History, Washington, 1960).
Wintle, A. D., *The Last Englishman* (Michael Joseph, London, 1968).
Wise, David, *Cassidy's War* (Random House, New York, 2000).
Woodburn Kirby, S., Roberts, M. R., Wards, G. T., and Desoer, N. L., *The War Against Japan, Volume V, The Surrender of Japan* (HMSO, London, 1969).
Works Progress Administration, *Washington, City and Capital* (Government Printing Office, Washington, 1937).
Yardley, H. O., *The American Black Chamber* (Bobbs-Merrill, Indianapolis, 1931).
Yu, Maochun, *OSS in China* (Yale University Press, New Haven, 1995).
Ziegler, Philip, *Mountbatten* (Knopf, New York, 1985).

Articles

"A Classmate," "Obituary, Harold David Kehm," in *Assembly,* September 1960, p. 123.
Baumer, Natalie B., "Mission to Moscow and the Normandy Invasion," in *Assembly,* September 1994, p. 22.
Bonham, Francis G., "Deception in War," in *The Infantry Journal,* Vol. 41 (July–August 1934) p. 272.
Bruce-Briggs, B., "Another Ride on Tricycle," in *Intelligence and National Security,* Vol. 7 No. 2, p. 77.
Campbell, John P., "Some Pieces in the Ostro Puzzle," in *Intelligence and National Security,* Vol. 11 No. 2 (April 1996) p. 245.
Carr, Caleb, "The Black Knight," *MHQ,* Vol. 6 no. 3 (Spring 1994) pp. 90–97.
Constable, Trevor J., "Bagnold's Bluff," in *The Journal for Historical Review,* Vol. 18 No. 2 (March/April 1999) p. 6.
Dierdorff, R. A., "Obituary of Robert F. Nelson," in Class Notes, Class of 1919, in *Shipmate,* August 1963, p. 29.
Dihm, Lieutenant General Friedrich, "MS B-259," in *World War II German Military Studies, Vol. 12 Part V, The Western Theater* (New York and London, Garland Publishing, Inc., 1979).
Dovey, H. O., "Maunsell and Mure," in *Intelligence and National Security,* Vol. 8 No. 1 (January 1993) p. 165.
Dovey, H. O., "The Eighth Assignment, 1941–1942," in *Intelligence and National Security,* Vol. 11 No. 4 (October 1996) p. 672.
Dovey, H. O., "The Eighth Assignment, 1943–1945," in *Intelligence and National Security,* Vol. 12 No. 2 (April 1997) p. 69.
Dovey, H. O., "The False Going Map at Alam Halfa," in *Intelligence and National Security,* Vol. 4 No. 1 (January 1989) p. 165.
Dovey, H. O., "The Intelligence War in Turkey," in *Intelligence and National Security,* Vol. 9 No. 1 (January 1994) p. 59.
Dovey, H. O., "Operation Condor," in *Intelligence and National Security,* Vol. 4 No. 2 (April 1989) p. 357.
Eldredge, H. Wentworth, "Biggest Hoax of the War," in *Air Power History,* Vol. 37 No. 3 (Fall 1990) p. 15.
Eldridge, Justin L. C., "The Blarney Stone and the Rhine: 23rd Headquarters, Special Troops

and the Rhine River Crossing, March 1945," in *Intelligence and National Security,* Vol. 7 No. 3 (July 1992) p. 211.

Gersdorff, Major General Freiherr von, "General Remarks With Regard to the German Defense in the Invasion," MS No. B-122, in *World War II German Military Studies, Vol. 12 Part V, The Western Theater* (New York and London, Garland Publishing, Inc., 1979).

Giangreco, D. M., "Casualty Projections for the U.S. Invasion of Japan, 1946: Planning and Policy Implications," in *The Journal of Military History,* Vol. 61 No. 3 (July 1997) p. 521.

Guderian, Colonel General Heinz, "The Interrelationship Between the Eastern and Western Front," May 1948, MS No. T-42, in *World War II German Military Studies, Vol. 21 Part VIII, Diplomacy, Strategy, and Military Theory, Continued* (New York and London, Garland Publishing, Inc., 1979).

Huber, Thomas M., "Deceiving the Enemy in Operation Desert Storm," Spiller, Roger J. (ed.), in *Combined Arms in Battle Since 1939,* (Fort Leavenworth, U.S. Army Command and General Staff Press, 1992) p. 59.

Hunt, David, "Remarks on 'A German Perspective on Allied Deception Operations,' " in *Intelligence and National Security,* Vol. 3 No. 1 (January 1988) p. 190.

Hoover, J. Edgar, "The Spy Who Double-Crossed Hitler," in *The American Magazine,* May 1946, p. 23.

Imboden, John D., "Stonewall Jackson in the Shenandoah," in *Battles and Leaders of the Civil War,* Vol. 2 (The Century Company, New York, 1888) p. 297.

Jones, Robert F., "The Kipkoror Chronicles," *MHQ,* Vol. 3 No. 3 (Spring 1991) pp. 38–47, reprinted in Cowley (ed.), *Experience of War* (Norton, New York, 1992) pp. 253–266.

Knox, Macgregor, "Fascist Italy Assesses Its Enemies," in *Knowing One's Enemies: Intelligence Assessment Before the Two World Wars* (Ernest R. May, ed.) (Princeton University Press, Princeton, 1984), pp. 347–72.

Martin, Bernd, "Die deutsch-japanischen Beziehungen während des Dritten Reichs," Funke, Manfred (ed.), in *Hitler, Deutschland und die Mächte: Materialen zur Aussenpolitik des Dritten Reichs* (Droste Verlag, Düsseldorf, 1976) pp. 454–70.

McCausland, Jeffrey, "The Gulf Conflict: A Military Analysis," Adelphi Paper No. 282 (International Institute for Strategic Studies and Brassey's, London, 1993).

McKay, C. G., "MI5 on Ostro: A New Document From the Archives," in *Intelligence and National Security,* Vol. 12 No. 3 (July 1997) p. 158.

Morgan, Roger, "The Man Who Almost Is," in *After the Battle,* No. 54 (1986) p. 1.

Morgan, Roger, "The Second World War's Best Kept Secret Revealed," in *After the Battle,* No. 94 (1996) p. 31.

Naftali, Timothy J., "De Gaulle's Pique and the Allied Counterespionage Triangle in World War II," in *In the Name of Intelligence,* Peake and Halpern (eds.) (NIBC Press, Washington, 1994) pp. 379–410.

Park, Edwards, "A Phantom Division Played a Role in Germany's Defeat," in *Smithsonian,* April 1985.

Paschall, Rod, "Deception for St. Mihiel, 1918," in *Intelligence and National Security,* Vol. 5 No. 3 (July 1990) p. 158.

Perras, Galen Roger, "We Have Opened the Door to Tokyo: United States Plans To Seize the Kurile Islands, 1943–1945," in *The Journal of Military History,* Vol. 61 No. 1 (January 1997) p. 65.

Ruge, Vice Admiral Friedrich, "Rommel's Measures To Counter the Invasion," 31 Apr 46, MS No. A-982, in *World War II German Military Studies, Vol. 12 Part V, The Western Theater* (New York and London, Garland Publishing, Inc., 1979).

Ruge, Vice Admiral Friedrich, "The Trail of the Fox: A Comment," in *Military Affairs,* Vol. 43 No. 12 (October 1979) p. 158.

Sheffy, Yigal, "Institutionalized Deception and Perception Reinforcement: Allenby's Campaigns in Palestine," in *Intelligence and Military Operations,* Vol. 5 No. 2 (April 1990) p. 173.

Smyth, Denis, "Screening 'Torch': Allied Counter-Intelligence and the Spanish Threat to the Secrecy of the Allied Invasion of French North Africa in November, 1942," in *Intelligence and National Security,* Vol. 4 No. 2 (April 1989) p. 335.

Sondern, Frederick, Jr., and Coster, Donald Q., "But We Expected You at Dakar," in *American Legion Magazine,* Vol. 40 (August 1946) p. 26.

Speidel, Lieutenant General Hans, "Ideas and Views of Gefldm Rommel, Commander of Army Group B, on Defense and Operations in the West in 1944," 11 Mar 47, MS No. B-720, in *World War II German Military Studies, Vol. 12 Part V, The Western Theater* (New York and London, Garland Publishing, Inc., 1979).

Sullivan, Brian R., "A Deal With the Devil: Italian Military Intelligence under the Fascist Regime, 1922–1943" (unpublished).

Thorne, Lieutenant-Colonel Sir Peter, "Hitler and the Gheluvelt Article," in *Guards Magazine,* August 1987, p. 106.

Troy, Thomas F., Letter to the Editor, 12 June 1992, in *Intelligence and National Security,* Vol. 7 No. 4 (October 1992), p. 516.

Troy, Thomas F., "The British Assault on J. Edgar Hoover: The Tricycle Case," in *International Journal of Intelligence and Counterintelligence,* Vol. 3 No. 2 (Summer 1989), p. 169.

Miscellaneous Printed Material

Joint Chiefs of Staff, *Joint Doctrine for Military Deception,* Joint Pub 3–58, 31 May 1996.

REFERENCES

In most of the cited files from the British Public Record Office, individual pages (or "folios") have been Bates-numbered. A citation such as "PRO CAB 119/67/310" therefore normally refers to Bates-numbered page 310 in piece 67 in Cabinet Office records series 119. For files which have not been Bates-numbered, the third number refers to a handwritten document number, such as "PRO 154/26/9A."

Exceptions: Dudley Clarke's six-volume final history of "A" Force (CAB 154/1 through 154/6, London Controlling Section, Clarke, Dudley, "A" Force Permanent Record File, Narrative War Diary), is cited by a Roman numeral reflecting the volume, plus the original (not the Bates) page number. Thus, "II/26" refers to original page 26 in CAB 154/2. Similarly, Sir Ronald Wingate's two-volume history (CAB 154/100 and 154/101, London Controlling Section, Wingate, Sir Ronald, Historical Record of Deception in the War Against Germany and Italy), is cited as "Wingate" plus the original page number (the two volumes are paginated consecutively). And Peter Fleming's account of his work in India and Burma (CAB 154/99, London Controlling Section, Fleming, Peter, A Report on Strategic Deception as Practised in the War Against Japan From the India and South East Asia Commands) is cited as "Fleming" plus the original page number.

Documents in the Public Record Office are subject to Crown copyright. Quotations from them are made pursuant to the blanket waiver granted in HMSO Guidance Note 3.

Material from the United States National Archives and Records Administration is cited "NARA" with the appropriate record group number, entry number where applicable, folder name, and document name.

Records from the Naval Operational Archives at the Washington Navy Yard are all from Strategic Plans Division, Series XXII, and are cited "NOA" plus box number, document name, and folder name. Thus, "NOA 535" refers to "Naval Operational Archives, Washington Navy Yard, Strategic Plans Division, Series XXII, Box 535."

Files of Joint Security Control are cited "JSC" plus folder name.

Citations to Hesketh, *Fortitude: A History of Strategic Deception in North Western Europe, April, 1943 to May, 1945,* are not to the published version but to the original page proofs with corrections in Hesketh's hand, copies of which have circulated in what might be called samizdat form. There is a copy in the USAMHI library.

Other short references to works listed in the Bibliography should be self-evident.

Introduction

xi *Dolus:* Aeneid ii.390.

Prologue, 1862–1940

1 Jackson in 1862: Henderson 299–302, 333, 365. "Always mystify": Imboden 297. Joshua: Joshua 8:3–22. Tao: Sun Tzu 168. Ancient deceptions: Ure 75–76; Wingate 244–45. Marlborough, Frederick, Napoleon: Wingate 244–50.

2 Jackson's tight security: Henderson 196, 298, 319, 333–34; Freeman, Lee's 1:482–83. Cavalry screen: Henderson 319. Secret marches: Henderson 221, 319–20. False rumors: Henderson 299. Feigned retreats: Henderson 419–20. Deserters: Sears 182. "He exercised": Dictionary of National Biography. Visited Virginia: Freeman, South 160–61. "He who compels": Henderson 206.

3 "Both were masters": Henderson 708. "Always an ardent advocate": Robertson 106. Henderson's ruses: Robertson 106–109.

4 Meinertzhagen: Jones. "His intelligence": Sheffy 175–81.

4 Turkish expectations: Sheffy 182. Allenby's measures: Sheffy 188, 191–92. Haversack Ruse: Sheffy 189–90; Bullock 68; Wavell 202.

5 Allenby before Megiddo: Sheffy 210–11; Bullock 122–24.

6 One step further: Sheffy 227–29. "Deceptions which": Lawrence 537.

6 Spreading word, administrative measures, radio traffic: Howard 33.

7 Hints to press: Maunsell 17–20. Races, party: Cooper 57; Maunsell 18. "A special section": I/2. "recognized an original": Clarke, Seven 7.

Chapter 1: The Master of the Game

9 Clarke family: Clarke, Quarter 7; Clarke, This 18–19, 47–48; author's telephone interviews with Mrs. David Greenwood, daughter of T. E. B. Clarke, 1995–97.

10 Running battle: Simonds MS memorandum n.d., in Mure Papers, Box 2.

10 Clarke's early career: Clarke, Quarter; Mure, Master 40–58; unsigned, undated biographical summary in Clarke Papers.

10 Clarke in Germany: Clarke, Quarter 503–508.

11 Restaurant against wall: Clarke, Quarter 580. "Two at least": Clarke, Quarter 602.

11 Clarke early in WWII: Clarke, Seven.

11 "Would you like . . . as good as fixed": Clarke, Seven 75.

11 Eire mission: "Calais and Dublin" in Clarke Papers. Commandos: Clarke, Seven 206, 218–19. Niven brothers: Clarke, Seven 214–16; Niven 241; Astley 54. Dunkirk raid: Clarke, Seven 225–39.

12 "Merry-eyed potato": Author's telephone interview with Sir Edgar Williams, 24 Oct 92.

12 Guinea pig: Author's telephone interview with Lady Maynard (Daphne Llewellyn), 10 Apr 96. "Sharp little man": Muggeridge 179. Blinking: Thynne to Mure, 27 Nov [78?], in Mure Papers, Box 2. "Gently booming": Mure, Practice 41. Neat, hair, pipe: Author's telephone interview with Mrs. David Greenwood, 31 Dec 95. "A man of few words": Mure, Practice 249. "Beneath his bland": Ranfurly 214. "Quiet chuckle": Wheatley 97. "An excellent raconteur": Wheatley 19–20. "Truly legendary": Wheatley 97.

13 Garrison rations: Simonds MS memorandum, n.d., in Mure Papers, Box 2. "To flout": Simonds to Mure, 22 Jun 80, in Mure Papers, Box 2. Bronco: Unsigned, undated biographical summary in Clarke Papers. Cinema: Author's telephone interview with Lady Maynard (Daphne Llewellyn), 10 Apr 96. "He did not suffer fools": Address by Colonel Noel Wild, Chelsea Old Church, Memorial Service for Brigadier Dudley Wrangel Clarke, C.B., C.B.E., May 30th 1974, in Clarke Papers.

13 Detested children: Author's telephone interview with Mrs. David Greenwood, 31 Dec 95. Little gifts, feminine sensibility: Author's interviews with Arne Ekstrom, March and May 1993; Wheatley 97. "She had a distinctive Slav beauty": Clarke, Quarter 246. Nina: Clarke, Quarter 246. "A romance": Clarke, Quarter 434. "Treacle tarts": Hamer to Mure, 21 Feb 81, in Mure Papers, Box 2. Women, "Dudley's Duchesses": Author's interviews with Arne Ekstrom, January 4, March, May 1993. Mayfair house: Clarke, Seven 19; Clarke, Quarter 601. Shepheard's: Author's telephone interview with Rex Hamer, 19 Jun 97.

14 "He had such a fantastically": Thynne to Mure, 27 Feb 78, in Mure Papers, Box 2. "At any time": Thynne to Mure, 9 May 78, in Mure Papers, Box 2. "He was certainly": Mure, Practice 249. House would collapse: Clarke, Seven 258–59. Guise of American: Simonds to Mure, n.d., in Mure Papers, Box 2.

15 Journey to Cairo: Clarke, Seven 260–61. Plus-fours: Simonds to Mure, n.d., in Mure Papers, Box 2. "Behind an inarticulate": Howard 32. Inarticulateness: Cooper, 43, 61; Clarke, Seven 261.

16 Egypt: Cooper 45, 57–59, 101, 133–34. Axis influence: Maunsell 7, 13. Enemy agents: Cooper 57, 123. Security tight: Maunsell 14–16, 22. Cairo: Cooper 4–5, 24–26, 29, 34, 305; Hart-Davis 263. British Army: Cooper 36, 125, 128–29.

18 Gabardine Swine, Short Range Shepheard's Group: Mure, Master 103. Groppi's Light Horse: Hamer to Mure, 21 Feb 81, in Mure Papers, Box 2. "Personal Intelligence Officer": I/6; Wingate 63. MI9 until August 1944: I/6.

19 Clerical staff: West, MI5 286; I/22. Rice pudding: Author's interview with Arne Ekstrom, 25 Mar 93. No clear understanding: Shearer to Mure, 21 Jul 77, in Mure Papers, Box 2.

19 SIME: Hinsley & Simkins 150–51; Maunsell 6. Maunsell and SIME: West, MI5 286; Maunsell 4, 35, 37–38, 44–49; Dovey, Maunsell 61–62.

20 Turkish relations: Hinsley & Simkins 151–52; Maunsell 9. Some success: I/1; Maunsell 18.

20 Sansom: Cooper 202. Russell: Cooper 32, 305. Foreign Office: Cooper 101.

21 CAMILLA: I/2–4. Its unfortunate result: Clarke, Some Notes on the Organization of Deception in the United States Forces, 30 Oct 44, in NARA RG 165 ABC File ABC 381 (4-8-43) Sec. 1-A—Deception policy, organization, etc. Reading Italian messages: Howard 35.

22 ABEAM: I/7–8.

23 K-SHELL: I/8–11. TENTH ARMORED DIVISION PLAN: I/11.

24 Jaundice: I/11–12. ABAFT: I/12. WAR OF NERVES PLAN: I/12. CORDITE COVER PLAN: I/15.

25 DOLPHIN: I/16–17. BYNG BOYS: I/18.

25 A-R: I/19–28.

27 Clarke's HQ: I/14.

27 "A" Force: Wingate 63; I/22–23. Ladies: Mure, Practice 81; Mure, Master 80.

28 Jones description: Mure, Master 96. Social: Hamer to author, 2 Jun 97. Put in charge: I/48. Wavell dummies: I/34.

28 Royal Tank Regiments: I/34–35, I/74. "By combining": I/52.

29 Maskelyne: Maskelyne *passim*. POW escape: NOA 532 Folder Douglas Fairbanks. Titters the Taster: Maunsell 5.

29 "A" Force Technical Unit: I/78–80; PRO CAB 154/9 through 154/12. Clarke to Turkey: I/25.

30 Wolfson: Lycett 147; Dovey, Maunsell 73. "A long and profitable": I/25. Arnold: West, MI6 126; Hinsley 1:198. "An equally": I/25. Turkish channels: I/25, I/49. Tensions rising: I/152; Hinsley 1:423–24.

31 Conspicuous: Cooper 75. EXPORTER COVER PLAN: I/27–29. Catroux: Cooper 51.

31 Resistance light: I/27. Wavell and Catroux: Hinsley 1:422–23.

31 Cyprus threat: Hinsley 1:424. CYPRUS DEFENSE PLAN: I/31–35. Druiff: I/48.

33 Churchill: Barnett 67–69. Cover for BATTLEAXE: I/30.

33 GSI(d): I/36–37, II/19. COPPERS SCHEME: I/37–39. Wintle: I/53–54.

35 EUPHRATES PLAN: I/44–46. IRAN COVER PLAN: I/47.

35 COLLECT: I/41–49. TRIPOLI PLAN: I/42.

36 To Lisbon: I/49. Fortnums: Andrew 457.

36 Different names: I/62. Beginnings of CHEESE: Report on Cheese, PRO KV 2/1133/3;

The Case of Cheese @ Lambert, PRO KV 2/1133/71a; Memorandum by Kenyon-Jones, The Origin of Cheese, in Clarke Papers; Howard 36; Wingate 30–31; Hinsley & Simkins 166. Robertson: Maunsell 26. Simpson: West, MI6 126; Hamer to author, 31 Jul 97. Novelist: II/74.

39 CHEESE and postponement: I/60, I/71.
40 STEPHAN and GAULEITER: Dovey, Maunsell; Mure, Master; Mure, Practice; West, MI6 190–91. Fackenheim: PRO KV 2/1163.
40 "Derek William Carter": Passport and travel orders in Clarke Papers. "As a spy center": I/49. "Germans, Portuguese," disappointing: I/50.
41 "Aimed more": I/56–57. Clarke in London: I/56. Twenty Committee: Minutes, PRO KV 4/64/88. October 7 meeting: Minutes of COS (41) 344th Meeting, 7 Oct 41, PRO CAB 119/67/310. "Considerable ingenuity": Strategic Deception, JP (41) 819, 8 Oct 41, PRO CAB 119/67/307–309. October 9 meeting: I/57; Howard 23; Minutes of COS (41) 348th Meeting, 9 Oct 41, PRO CAB 119/67/306.
42 "Who alone": Clarke to Fleming, 1 Jun 81, in Clarke Papers ("Wavell" corrected to "Auchinleck"). Chiefs and Stanley: L. C. H. [ollis] to Howkins, 10 Oct 41, PRO CAB 119/67/305; Howard 23; I/57. COLLECT: I/60–61.
43 "Wrangal Craker": Crown Jewels 308–309. "I am afraid": Liddell diary 23 Oct 41, PRO KV 4/188. Liddell met Clarke: Liddell diary 29 Sep 41, 3 Oct 41, PRO KV 4/188. Ordered back, torpedoed; cabled London; "in explanation"; "if he considers": Dill to Prime Minister, CIGS/PM/BM/194, 31 Oct 41, Churchill Archives CHAR 20/25/42. Churchill agreed: Notation on Dill to Prime Minister, CIGS/PM/BM/194, 31 Oct 41, Churchill Archives CHAR 20/25/42; A. B. to Nutting, 2 Nov 41, Churchill Archives CHAR 20/25/43. "Mentally stable"; "we can reasonably"; "a foolhardy": Dill to Prime Minister, CIGS/PM/194, 18 Nov 41, Churchill Archives CHAR 20/25/44. Definite purpose: Dill to Prime Minister, CIGS/PM/194, 21 Nov 41, Churchill Archives CHAR 20/25/46. "The case is being shrouded": West, Crown Jewels 308–309. Whispered around: Wheatley 97. Rumors: Wild to Mure, 30 Jan 80, in Mure Papers, Box 2. Vigorously denying: Wild to Mure, 30 Jan 80, in Mure Papers, Box 2; Author's interviews with Arne Ekstrom, March and May 1993; author's telephone interview with Mrs. David Greenwood, 31 Dec 95.
44 Sonic deception: I/87.
44 GRIPFIX: I/88–90.
45 ADVOCATE: I/82–86; Howard 25. "There is so much"; Liddell diary 30 Oct 41, PRO KV 4/188.
47 Use of press; "A very ticklish": I/85–86; Howard 25.
47 Bagnold affair: II/8–9, 20. "The closely": II/20. "A frank": II/22.
48 BASTION: II/9–16. "A neat leak": II/14. "We learnt": II/16. "Led to a good deal": II/18.
50 End of Bagnold affair: II/20–23. Bagnold career: Constable. Promotion and reorganization: II/26–28.
50 "The organ": I/92.

Chapter 2: The Art of Deception

52 Uther and Igraine: Malory 2–5.
53 "Planned measures for disguising"; "for revealing": JCS 498/10, 14 May 45, in NARA RG 218 Box 694 Folder CCS 385 Pacific Theater (4-1-43) Deceptive Policy in the Pacific Theater (Sec. 4).
53 Insistence of Navy: Goldbranson to Smith, Subject: Captain Thurber's Memorandum for F-01 (Rough Draft), 14 Mar 45, in JSC folder Japan. "Is to cause the enemy": JCS 498/10, 14 May 45, in NARA RG 218 Box 694 Folder CCS 385 Pacific Theater (4-1-43) Deceptive Policy in the Pacific Theater (Sec. 4).

54 "It is often possible": Mao Tse-tung, Protracted War, quoted Delmer 23.

54 Morgan suggestion: Morgan to Bevan, 16 Jul 1943, in NARA RG 331 Entry 1 Box 73 Folder 381 Fortitude Vol. 1. Bevan reply, "my channels": Bevan to Morgan, 20 Jul 1943, *ibid.*

55 Stretch effect: Wingate 291.

56 Clyde error: Hesketh 99 n. 7. "The target": JCS 498/10, 14 May 45, in NARA RG 218 Box 694 Folder CCS 385 Pacific Theater (4-1-43) Deceptive Policy in the Pacific Theater (Sec. 4).

57 "Psychologically the deceivers": Wingate 9.

57 Norway: Wingate 8, 291. Foreign Office: Wingate 200. Amphibious warfare: Wingate 292.

58 Fighter cover: Wingate 292. "A far more potent": Wingate 9. Focus on own vulnerability: Wingate 292.

58 "If these two": Wingate 10. Fourteen divisions: Howard 146–47.

60 Risk smaller: Wingate 7–8.

60 Unnecessarily cautious: Wingate 7–8, 91, 141–42. Differences in theaters: Wingate 7.

60 COCKADE: Wingate 7. Japanese limitations: Report on Execution and Planning of Bluebird, 26 May 1945, p. 1, in JSC folder Hanover File (Materials to be forwarded to CRREC). For successful deception: Wingate 11. "It is impossible": Fleming 5.

61 "Once you have started": Wingate 132. Theobald: Kahn, Codebreakers 571.

62 Three-tiered system: Schuirmann and Bissell to addressees, Subject: TOP SECRET Control Procedure, 29 Feb 44, JSC/C15 Serial 2992, in NARA RG 165 ABC File ABC 381 (8-15-42)—JSC. SCU/SLU: Hinsley 1:571–72. "Link": Eldredge to Hesketh, 20 Dec 71, Eldredge papers. Eyes-only communications: Author's interview with Eugene Sweeney.

63 C.C. Procedure: I/58.

63 C.C. List: I/58. "Tactical Deception": II/51.

64 Assimilated over time: Wingate 12.

65 "The lesson": IV/129. Eisenhower decreed: Eisenhower to Director, Plans and Operations Division, WDGS, 19 Jun 47, in National Archives Collection, Xerox 2102, Marshall Memorial Library.

66 "Both sides": II/23. "The real answer": II/23; see also Clarke and Bevan, Suggested Deception Organization in Peace, First Revise, 5 Jun 45, PRO CAB 121/105/56–69; Unsigned memorandum, Deception Organization in Peace, 14 Sep 45, PRO CAB 121/105/35–39. "They had 'invented' ": II/23–24. "They are the real users": II/24. Field representatives: II/24. "Sometimes appeared": II/43; Wingate 92.

67 SHAEF G-3: NARA RG 331 Entry 1 Box 45 Folder Organization and Personnel G-3 SHAEF. U.S. Navy staff system: Evans.

67 British staff system: Notes on British Staff System, 17 Feb 43, in Nevins papers, folder Codenames, British procedure, D-Day memos, USAMHI; *1st Joint Air Course, Functions and Set-Up of the Army Staff, 4 Oct 43,* USAFHRA 505.89-38; Graham & Bidwell 243–44.

68 Clarke and Strong: Rooks to Hull, 3 Nov 44, in NARA RG 165 ABC File ABC 381 (4-8-43) Sec. 1-A—Deception policy, organization, etc.

69 "Used to open his heart"; "I would make a signal": Author's telephone interview with David Strangeways, 26 Aug 92.

69 "Went down to tell": Author's interview with Arne Ekstrom, 25 Mar 93.

70 10th Division: National War College Lecture, Lieutenant Colonel John R. Deane, Jr., 1 March 1950, in NOA 546, Folder Tactical Deception—Ruses, p. 15. Seen in New York: "Implementation of Bluebird Special Means for Week Ending 28 April 1945," in NOA 548, Folder Bluebird; File Log, Item A 150, 27 Apr 45, JSC Acco Binder "7 File A." "In this vast": Wingate 4.

70 But it was concluded: Wingate 129–30.
72 Diffusion of effort: Wingate 136. "Got themselves involved": Wingate 137.
72 "Truth deserves": Extract from 4th Eureka meeting, Tehran, Iran, 30 Nov 43, in NOA 543, Folder Russia—Deception. Tennyson: The Grandmother, Stanza 8. Western landing the basis: Wingate 183.
73 Real Spitfires as fakes: III/86. Double bluffs dangerous: Wingate 310–12.
73 "We should never": Wingate 312. August 1944: I/45. "Poor man": I/45. "Get out": Wingate 11. "Breakoff": *E.g.,* Joint Security Control to CINCPOA, 141846, 14 Apr 45, in JSC folder Bluebird.
75 Relying on POINTBLANK: Wingate 231. BLUEBIRD breakoff "story": CINCPOA to COMINCH (JSC), 260031, 26 Mar 45, NOA 548, Folder Bluebird. "You don't take": Young & Stamp 48.
75 Risk compromise: Wingate 265. Hard to induce Germans to remove units: Wingate 206–207. Liquidation after FORTITUDE SOUTH: Hesketh 156–58. "A" Force liquidation: IV/41; Wingate 265–66.
77 "All the business of war": Duke of Wellington, Croker Papers, 1885, Vol. III, p. 276 (quoted as tagline to Foreword, Freya Stark, Alexander's Path, p. xi (Overlook Press, Woodstock, New York, 1988). Clarke's poster: Thynne to Mure 27 Nov [78?], Mure Papers (Box 2).
78 Watson's watch: Doyle, *The Sign of Four,* Ch. 1. Price of oranges: Copeland 46.
78 Buying drachmas: III/83. "A" Force Instructions and Strategic Addendum: II/45–49.
79 "In it there appeared": II/48. "To control both the thoughts . . . the essentials": II/48–49.
80 More than one hundred paragraphs: II/48.
81 Quaker guns: Sears 17, 355. Low-level oblique photography: Wingate 19. "A distressing struggle": I/45.
81 Tank blew off: I/45. Dummy landing craft: Wingate 316, 333–34; IV/5. SUNSHIELD, CANNIBAL, associated equipment, imitation bridges, observation posts, smoke and flash simulators: NARA RG 218 Box 372 Folder CCS 385.2 (4-2-44) (Technical Information Concerning the Existing Means of Tactical Deception) 22–23, 34–36.
82 PARAGONS: Fleming 13. SAINTS: II/88. Composition: NARA RG 218 Box 372 Folder CCS 385.2 (4-2-44) (Technical Information Concerning the Existing Means of Tactical Deception) 24. Fairbanks model: Facilities for Operational Deception, Cominch P-005, p. 56, in NOA 537, Folder Misc C&D, Psychological Warfare. "A" Force self-destroying model: II/88. PD Packs: Beach Jumper Tactical Deception, 1 Dec 43, pp. 1–2; Fairbanks to The Admiral, Subject: U.S. Navy Devices Employed in Tactical Deception, 25 Jul 44, p. 2, both in NOA 531 Folder Deception Methods Techniques and Devices; Joint Security Control Information Bulletin No. 3, 1 Jun 45, JSC/J14 Serial 5122, p. 2, in NARA RG 165 ABC File ABC 381 Japan 15 Apr 43 Sec. 1-B. Intermingling dummies: Sweeney to G-3, Seventh Army, 29 Jul 44, in NARA RG 319 Entry 101 Box 4 Folder 78, Operation Anvil. "Could only have been conceived": Morison, Invasion 249. PINTAILS: Fleming 13; III/107; NARA RG 218 Box 372 Folder CCS 385.2 (4-2-44) (Technical Information Concerning the Existing Means of Tactical Deception), p. 24.
82 GOONEY BIRDS: Dwyer 20; Fairbanks to The Admiral, Subject: U.S. Navy Devices Employed in Tactical Deception, 25 Jul 44, p. 8, in NOA 531, Folder Deception Methods Techniques and Devices. Dummy aircraft: NARA RG 218 Box 372 Folder CCS 385.2 (4-2-44) (Technical Information Concerning the Existing Means of Tactical Deception), p. 22; Facilities for Operational Deception, Cominch P-005, p. 24, NOA 537, Folder Misc C&D, Psychological Warfare. "Q" Sets: NARA RG 218 Box 372 Folder CCS 385.2 (4-2-44) (Technical Information Concerning the Existing Means of Tacti-

cal Deception), p. 25. "C.T.D.": Wingate 17. Tankers disguised: NARA RG 218 Box 372 Folder CCS 385.2 (4-2-44) (Technical Information Concerning the Existing Means of Tactical Deception), p. 17.

83   Flame and smoke floats, WATER SNOWFLAKES, SNOWFLAKES: NARA RG 218 Box 372 Folder CCS 385.2 (4-2-44) (Technical Information Concerning the Existing Means of Tactical Deception), p. 19. AQUASKITS and AQUATAILS: Fleming 13. Drone boats: Fairbanks to The Admiral, Subject: U.S. Navy Devices Employed in Tactical Deception, 25 Jul 44, p. 8, NOA Box 531, Folder Deception Methods Techniques and Devices. Dummy landing craft: II/87–88; Hesketh 188; NARA RG 218 Box 372 Folder CCS 385.2 (4-2-44) (Technical Information Concerning the Existing Means of Tactical Deception), p. 32.

84   Fleet Tenders, *Centurion:* PRO ADM 223/794/430–431. U.S. Navy disguises, "Swiss Navy": Facilities for Operational Deception, Cominch P-005, pp. 39–48, NOA 537, Folder Misc C&D, Psychological Warfare.

84   FILBERTS: Facilities for Operational Deception, Cominch P-005, pp. 17–18, in NOA, 537, Folder Misc C&D, Psychological Warfare. British usage: Cole 1.

85   British use of WWI units; British concocted own; role-playing; "In order factually": Train to Davy, ADV"A"F/36/1, 12 Feb 44, PRO WO 204/1561/50–51; Cole 1.

86   HEATER: Facilities for Operational Deception, Cominch P-005, p. 66, NOA 537, Folder Misc C&D, Psychological Warfare. POPLIN; employment and characteristics of sound systems: NARA RG 218 Box 372 Folder CCS 385.2 (4-2-44) (Technical Information Concerning the Existing Means of Tactical Deception), pp. 27–29; IV/26. Dropping with paratroops: Young & Stamp 78.

87   POLLY, AIR HEATER, WATER HEATER, CANARY HEAD, DUCK HEAD, BUNSEN BURNER: Burris-Meyer papers; Fairbanks to The Admiral, Subject: U.S. Navy Devices Employed in Tactical Deception, 25 Jul 44, p. 8; Facilities for Operational Deception, Cominch P-005, pp. 1–2, 72–75, 86–87, NOA 537, Folder Misc C&D, Psychological Warfare.

87   Proficiency of Beach Jumpers, Light Scout Car units: Facilities for Operational Deception, Cominch P-005, p. 71, NOA 537, Folder Misc C&D, Psychological Warfare; NARA RG 218 Box 372 Folder CCS 385.2 (4-2-44) (Technical Information Concerning the Existing Means of Tactical Deception), p. 27; IV/26–28. Bomb simulators, PARAFEXES: NARA RG 218 Box 372 Folder CCS 385.2 (Technical Information Concerning the Existing Means of Tactical Deception), pp. 23, 24, 27, 29. Bicat devices: Operations of 303 Ind. Bde. During Winter 1943–44, No. 6036/D Division, 2 Jun 44, in JSC unlabeled folder; O'Brien 172.

88   Platoon level firefight: Fleming 13. TITANIC: Young & Stamp 78–80. Diesel fumes, gas: NARA RG 218 Box 372 Folder CCS 385.2 (4-2-44) (Technical Information Concerning the Existing Means of Tactical Deception), p. 30. Cordite: Young & Stamp 78.

88   Devices for radar jamming and spoofing: NARA RG 218 Box 372 Folder CCS 385.2 (4-2-44) (Technical Information Concerning the Existing Means of Tactical Deception), pp. 10–14; Facilities for Operational Deception, Cominch P-005, p. 4, 8, 9, 14–15, NOA 537, Folder Misc C&D, Psychological Warfare; Glenn, *The BJs,* p. 5, NOA 532, Folder Douglas Fairbanks.

90   Japanese fists at Midway: Potter 95–97.

90   Radio traffic analysis: NARA RG 218 Box 372 Folder CCS 385.2 (4-2-44) (Technical Information Concerning the Existing Means of Tactical Deception), p. 1. American priority levels: "History of Radio Deception" undated, p. 7; "When a deception unit reaches . . ." undated, p. 1, both in JSC folder 100.00 Radio Cover and Deception. German identification of raids: NARA RG 218 Box 372 Folder CCS 385.2 (4-2-44)

(Technical Information Concerning the Existing Means of Tactical Deception), p. 6. Japanese intercepts at Rabaul: Barasch to McCartney, Uses by the Japanese of Traffic Analysis To Predict B-29 Raids, 27 January 1945, in JSC folder 13-25/3 Bluebird; Project No. 1471-A, The Japanese Intelligence System, p. 26, in JSC folder The Japanese Intelligence System.

91   Manipulative and imitative: Army Service Forces, Office of the Chief Signal Officer, Radio Deception, in JSC folder 21 Radio Deception by ASF, pp. 5, 7–8.

91   Fake messages and dummy traffic: Galli and Maher to Sheetz, 6 Oct 43; Instructions for Cryptographic Office, both in JSC folder 13-12/6 Communications Deception Reports: Europe, Africa; Army Service Forces, Office of the Chief Signal Officer, Radio Deception, in JSC folder 21 Radio Deception by ASF, pp. 7–8, 11–12.

93   Spoof vans: Army Service Forces, Office of the Chief Signal Officer, Radio Deception, in JSC folder 21 Radio Deception by ASF, p. 8; NARA RG 218 Box 372 Folder CCS 385.2 (4-2-44) (Technical Information Concerning the Existing Means of Tactical Deception), p. 8. Beach Jumpers comparably equipped: Facilities for Operational Deception, Cominch P-005, p. 11, in NOA 537, Folder Misc C&D, Psychological Warfare. Imitative deception: "Imitative Deception," undated, in JSC folder 100.00 Radio Cover and Deception; Army Service Forces, Office of the Chief Signal Officer, Radio Deception, pp. 13–14, in JSC folder 21 Radio Deception by ASF; Quiz and Answers to Quiz, in JSC folder 100.00 Radio Cover and Deception. Jamming: "Imitative Deception" undated, in JSC folder 100.00 Radio Cover and Deception; NARA RG 218 Box 372 Folder CCS 385.2 (4-2-44) (Technical Information Concerning the Existing Means of Tactical Deception), pp. 12–13. Analyzing own traffic: "American Traffic Analysis" undated, in JSC folder 100.00 Radio Cover and Deception.

94   Wahiawa: E.g., CINCPOA Advance to CINCPAC Pearl, No. NCR 7223, DTG 160636, 16 Mar 45, in JSC loose pack of radiograms in JSC folder Bluebird. Tight control over radio deception: Army Service Forces, Office of the Chief Signal Officer, Radio Deception, p. 14, in JSC folder 21 Radio Deception by ASF; Protective Security Branch, Radio Deception, April 1, 1943, pp. 1–2, in JSC folder Hanover File (Materials to be forwarded to CRREC); NARA RG 218 Box 372 Folder CCS 385.2 (4-2-44) (Technical Information Concerning the Existing Means of Tactical Deception), p. 9.

95   False going map: Dovey, False Going; Liddell diary, 18 Feb 44, PRO KV 4/193.

95   Rumormongering before TREATMENT: II/110.

95   Press policy: Bevan to Strong, 31 Dec 43; Strong to Bevan, 21 Jan 44; Wingate to Bissell, CO/530, 28 Feb 44; Bissell to Wingate, JSC/J1 Serial 316, 6 Mar 44, all in JSC folder 370.2 Bodyguard; J. K. W[oolnough] to Roberts, Re: JCS 806, 15 Apr 44, in NARA RG 165 ABC File ABC 381 Japan 15 Apr 43 Sec. 4—FORAGER. WEDLOCK: Herbig 48. Useful mistake: IV/15. "A certain Central European": II/54.

96   Pester Lloyd: see IV/15. Uses made of it: II/54–55, II/111, III/75, IV/15. Turkish newsman: III/75. FABRIC: II/59.

97   Resistance and COCKADE: Wingate 151.

97   "I know half": Mayer 259. 70,000 men in Kuriles: Herbig 54. MacArthur refused: Cake Before Breakfast, p. 12, JSC loose document. Captured maps: Wingate 92.

Chapter 3: The Customers

99   Liss and FHW: Hinsley 1:10; Kahn, Spies 405, 420–21. Zossen headquarters: Kahn, Spies 50, 486. 1940 performance: Kahn, Spies 423. Broadened: Kahn, Spies 423, 462–65. North Africa: Kahn, Spies 424, 465. Liss after FHW: Kahn, Spies 424.

100 Roenne: Kahn, Spies 424; Delmer 139–40, 240. "An intellectual": Irving, Fox 434–35. Roenne's surveys: Kahn, Spies 427. Inflating order of battle: Kahn, Spies 496.

101 Devout: Delmer 240. "Deception can never": Howard 52. Bürkelein: Kahn, Spies 427.

REFERENCES

101 Communications intelligence: Kahn, Spies 172–87, 198–212. "Was a thoroughly corrupt": Maunsell 27.
102 Abwehr origins: Hinsley & Simkins 295. Canaris: Rout & Bratzel 1–3; Hinsley & Simkins 295–96.
103 Move to Zossen: Kahn, Spies 238; Hinsley & Simkins 297.
103 Abt I, Pieckenbrock: Kahn, Spies 236–37; Hinsley & Simkins 296. Hansen: Howard 48. Abt II and Lahousen, Abt III and Bentevegni: Hinsley & Simkins 296. Abt Z and Oster: Hinsley & Simkins 296; chart at Rout & Bratzel 4. V-Männer: Rout & Bratzel 7.
104 Abt IIIF, Abwehr headquarters: Hinsley & Simkins 297. Asts: Hinsley & Simkins 297–98; Rout & Bratzel 5; Kahn, Spies 240–41. Aussenstellen, KOs: Hinsley & Simkins 297–98, Howard 46.
104 Brandenburg units: Hinsley & Simkins 298–99. Inefficient organization: Rout & Bratzel 5; Hinsley & Simkins 297–98; Kahn, Spies 239–43. "The whole organization": Maunsell 27. Easy prey: Maunsell 18. "It is probably fair": Report on the Activities of No. 2 Tac H.Q. A Force (Sixth Army Group) 31 July 1944–5 May 1945, Part I, p. 68, in NARA RG 319 Entry 101 Box 4 Folder 79.
105 More tangible corruption: Howard 49. Canaris agonizing: Kahn, Spies 234–36. Franco: Waller 155–56. Oster and Dochnanyi ringleaders: Hinsley & Simkins 296, 302. Pieckenbrock and Hansen anti-Nazi: Howard 48.
106 Spanish and Portuguese facilities: Hinsley & Simkins 159–60, 301. Latin American radio: Rout & Bratzel 9.
106 Norden bombsight: Kahn, Spies 328–31. Roundup, SNOW, offensive: Hinsley & Simkins 41, 87–98. Abwehr in Middle East: Howard 47; Hinsley & Simkins 163–64. Sealed off: Maunsell 21, 28.
107 RSHA: Hinsley & Simkins 299.
107 Ten Commandments: Hinsley & Simkins 300. Schellenberg: Kahn, Spies 256–58; Hinsley & Simkins 300. Abwehr failures: Hinsley & Simkins 301–302.
108 Downfall of Canaris and others, Schellenberg takeover: Kahn, Spies 269–71; Hinsley & Simkins 302–303.
109 Saiko Senso Shodo Kaigi; Imperial General Staff; Gunji Sangi In: ONI Monograph, Japanese Intelligence Services, May 1947, in NOA 535, Folder Japanese Intelligence Organization, pp. 1–2.
110 Army Department II: ONI Monograph, Japanese Intelligence Services, May 1947, in NOA 535, Folder Japanese Intelligence Organization, p. 23. Air intelligence section: Same, p. 29. Department I prestige: Same, pp. 19–20. "There was a feeling": A Review of Japanese Intelligence, 18 April 1946, in NOA 535, Folder Japanese Intelligence Analysis of WW II, p. 8. Army officers: Same, p. 9.
110 Naval intelligence: ONI Monograph, Japanese Intelligence Services, May 1947, in NOA 535, Folder Japanese Intelligence Organization, pp. 6–11, 13, 15–16.
110 Foreign Affairs Ministry: ONI Monograph, Japanese Intelligence Services, May 1947, in NOA 535, Folder Japanese Intelligence Organization, p. 32. Greater East Asia Ministry: Same, pp. 33, 73–74; Project No. 1471-A, The Japanese Intelligence System, 4 September 1945, pp. 8, 74, in JSC folder The Japanese Intelligence System. Techniques stressed: ONI Monograph, Japanese Intelligence Services, May 1947, in NOA 535, Folder Japanese Intelligence Organization, p. 62.
111 Focus on spies; Tonegi Kikan: Project No. 1471-A, The Japanese Intelligence System, 4 September 1945, pp. 2, 81, 92–106, in JSC folder The Japanese Intelligence System. Commingling of propaganda, journalism: ONI Monograph, Japanese Intelligence Services, May 1947, pp. 7, 50, in NOA 535, Folder Japanese Intelligence Organization. Rumors and fantasies: Project No. 1471-A, The Japanese Intelligence System, 4 September 1945, pp. 77, 87, 89–91, in JSC folder The Japanese Intelligence System.

Cryptanalysis, success against Chinese: Project No. 1471-A, The Japanese Intelligence System, 4 September 1945, p. 16, in JSC folder The Japanese Intelligence System; D Division Weekly News Letter No. 37 30 Dec 44–5 Jan 45 Part II paragraph 25; D Division Weekly News Letter No. 40 20 Jan–26 Jan 45, Part I paragraph 8, both in JSC folder Deception—SE Asia—WWII Historical D. Div—SACSEA—30 Dec 44–23 Mar 45. Traffic analysis: ONI Monograph, Japanese Intelligence Services, May 1947, p. 12, in NOA 535, Folder Japanese Intelligence Organization; Project No. 1471-A, The Japanese Intelligence System, 4 September 1945, pp. 7, 8, in JSC folder The Japanese Intelligence System.

111 Special operations units: ONI Monograph, Japanese Intelligence Services, May 1947, p. 24, in NOA 535, Folder Japanese Intelligence Organization. Kempei Tai: ONI Monograph, Japanese Intelligence Services, May 1947, pp. 26–29, in NOA 535, Folder Japanese Intelligence Organization. Opium dens: Harries 244. Collapse after Pearl Harbor: ONI Monograph, Japanese Intelligence Services, May 1947, p. 41, in NOA 535, Folder Japanese Intelligence Organization.

112 Western Hemisphere: ONI Monograph, Japanese Intelligence Services, May 1947, pp. 45–47, in NOA 535, Folder Japanese Intelligence Organization; Japanese Intelligence Services in Latin America and Europe, 28 January 1947, pp. 1–9, in NOA 535, Folder Japanese Intelligence Organization. Focus on Europe: ONI Monograph, Japanese Intelligence Services, May 1947, pp. 49–50, in NOA 535, Folder Japanese Intelligence Organization; Project No. 1471-A, The Japanese Intelligence System, 4 September 1945, pp. 17, 160–61, in JSC folder The Japanese Intelligence System.

112 Onodera generally: OP-16-B-7, A Summary of Recent Development[s] in Japanese Espionage in the Neutral States of Europe, pp. 1–2, 7 (Sweden), in NOA 535, Folder Japanese Intelligence in Europe; Japanese Intelligence Services in Latin America and Europe, 28 January 1947, pp. 14–16, in NOA 535, Folder Japanese Intelligence Organization; Strategic Services Unit, War Department, Japanese Wartime Intelligence Activities in Northern Europe, in NOA 535, Folder Japanese Intelligence Organization.

113 Take over Oslo Ast: Central Intelligence Group, Japanese Collaboration With the German Intelligence Service, n.d. [1946?], in NOA 535, Folder Japanese Intelligence Organization. Kraemer: Strategic Services Unit, War Department, Japanese Wartime Intelligence Activities in Northern Europe, p. 28, in NOA 535, Folder Japanese Intelligence Organization. Maasing, Grip: Project No. 1471-A, The Japanese Intelligence System, in JSC folder The Japanese Intelligence System, pp. 155–56.

113 SUNRISE: Fleming 30–31; Project No. 1471-A, The Japanese Intelligence System, in JSC folder The Japanese Intelligence System, p. 156. "There can be little doubt": Japanese Intelligence Services in Latin America and Europe, 28 January 1947, p. 16, in NOA 535, Folder Japanese Intelligence Organization.

114 Second-rate personnel: A Review of Japanese Intelligence, 18 April 1946, p. 8, in NOA 535, Folder Japanese Intelligence Analysis of WW II. Rarely graded, little exchange, so-called summaries: Project No. 1471-A, The Japanese Intelligence System, in JSC folder The Japanese Intelligence System, pp. 65–68, 78, 162.

114 Poor training: ONI Monograph, Japanese Intelligence Services, May 1947, pp. 78–79, in NOA 535, Folder Japanese Intelligence Organization. Field and signal security: Project No. 1471-A, The Japanese Intelligence System, in JSC folder The Japanese Intelligence System, pp. 59–61, 67–68. U.S. radio deception: Report on Execution of Plan Bluebird, 26 May 1945, p. 1, in JSC folder Hanover File (Materials to be forwarded to CRREC). "Only the most rudimentary": Fleming 8.

115 "The Japanese have been": Project No. 1471-A, The Japanese Intelligence System, in JSC folder The Japanese Intelligence System, p. 69. February 1945 summary: Project No. 1471-A, The Japanese Intelligence System, in JSC folder The Japanese Intelligence System, pp. 58, 69; Japanese Estimate of Allied Order of Battle, n.d., in JSC folder Jap-

anese Estimates of Allied OB. Predictions: Project No. 1471-A, The Japanese Intelligence System, in JSC folder The Japanese Intelligence System, pp. 71–72.

115 Fleming rated: Fleming 2. "Could in no"; "crudity": Fleming 3. "The credulity"; "it was clear": Fleming 6. "They are rather": DMI/2681 "Cover Plan," 3 Feb 43, NARA RG 319 Entry 101 Box 3 Folder 68 "Anglo-American Meeting (May-June '43)" (signed Cawthorn but internal evidence suggests drafted by Fleming). "While . . . there was never": Fleming 7.

116 Unintelligent and unreliable: Fleming 8. Inadequately trained and supported: Fleming 71–72.

117 Impractical: Fleming 8. Multilevel: Fleming 9. Locations: Fleming App. B. "Scarcely credible": Fleming 9. Frequencies: Fleming 70. Equipment: Fleming 9, 70.

117 Communication incompetence: Fleming 9, 16, 18–20, 69–71. Doubling: Fleming 26, 28.

118 "Which resulted": Fleming 6–7.

119 No coordination: De Risio 16. France and Yugoslavia: Knox 351–52. Chiefs of SIS and SIA: De Risio 15. Overestimated French: Knox 352. SIM structure, chiefs: De Risio 15, 20; Sullivan 12, 25; Knox 349, 351.

119 Dapper: De Risio facing p. 142. Careerist: De Risio 58. Best in Europe: Carboni, Più 32. Germany report: De Risio 58–59.

120 Carboni order of battle inflation: De Risio 59–61. Airpower inflation: MacGregor Knox to author, 20 May 97. Ingratiating: Kirkpatrick 380. Amè: Pillon 11–12 note 1; De Risio 69, facing p. 142. Badoglio rude: De Risio 73.

120 Greek adventure: Kirkpatrick 456–58, De Risio 72. SIM beforehand: De Risio 72; MacGregor Knox to author, 20 May 1997.

121 SIM: Knox 349–50, 356. King of Yugoslavia: Sullivan 15. Spanish Civil War: Sullivan 26. Ethiopia: Sullivan 17. Hoax photographs: Sullivan 26.

122 Detachment and reorganization; "One of the secrets"; "It is unnecessary": Report on the Organization and Activities of the Italian CS Service—Jan 1941–Sept 1943—by Lt. Col. Giulio Fettarappa-Sandri, in Norman Holmes Pearson Papers, folder Other Systems, Wooden File, Box 1. Fettarappa-Sandri: Pillon 97 note 1. SIM and British agents: Liddell diary, 23 Aug 44, PRO KV 4/194; C. M. Woods to author, January 31, 1994.

122 SIM cryptography: Kahn, Codebreakers 468–73; Knox 350. Stealing codebooks: Knox 350; Naftali, X-2 413-14 note 61; Sullivan 8. British bluff: Knox 369; Sullivan 18. Mail opening: Knox 350. Espionage lagged: Knox 350. Roatta and Canaris: Sullivan 22–25. Amè and Canaris: De Risio 203–204.

123 "We also hope": De Risio 213–16; Waller 314–15. "As for any other": Report on the Organization and Activities of the Italian CS Service—Jan 1941–Sept 1943—by Lt. Col. Giulio Fettarappa-Sandri, in Norman Holmes Pearson Papers, folder Other Systems, Wooden File, Box 1. Germans unaware: C. M. Woods to author, January 31, 1994.

123 Same quarry: C. M. Woods to author, January 31, 1994. Carboni returns: De Risio 223–24. Turkey agents: Dovey, Turkey 80. Insufficiently skeptical: Knox 350–51.

124 "Made far more": Liddell diary, 18 Feb 44, PRO KV 4/193. "Was one of several cases": III/44. "Showed us": III/48.

Chapter 4: Most Secret Sources and Special Means

125 Three essentials: Stratagem Conference 28. TRIANGLE: Dovey, Condor 361 and note 26.
126 Enigma: Kahn, Seizing 88–89.
126 Abwehr hand ciphers: Hinsley & Simkins 44. Wave of agents: Hinsley & Simkins 88. ISOS; main group: Hinsley & Simkins 89; Hinsley 1:120 note*. Combined Bureau

Middle East: Hinsley 2:20; Howard 31. Knox breakthrough, ISK: Hinsley & Simkins 159.

127 GGG: Hinsley 2:668. ISTUN: Howard 15 note†. ISOS generically: Hinsley & Simkins 44 note‡, 108 note†. "Ice": Maunsell 26; Hamer to Mure, 21 Feb 81, in Mure Papers, Box 2. January 1944: Hinsley 2:20. SD: Hinsley & Simkins 132–33 and note, 182 note*.

127 Loss and recovery of ISOS: Hinsley & Simkins 266. Volume: Hinsley & Simkins 182. Uses of ISOS: Hinsley & Simkins 182–83. Sharing with MI5 and Twenty Committee: Howard 20, Masterman to Menzies, B.1.A./JCM, 5 Jun 42, PRO KV 4/65/169A; Menzies to Masterman, 11 Jun 42, PRO KV 4/65/171A.

128 MAGIC: Kahn, Codebreakers 21–24. "More Nazi": Shirer 872. Oshima: Boyd.

129 British censorship, FBI sharing: Naftali, X-2 326; Hinsley & Simkins 143–44, 158–59, 185–86, 201–202; Whitehead 193–94.

130 DESPOT: E.g., G(R) Main War Diary January 1945, App. J-7, PRO WO 171/3868; G(R) Main War Diary March 1945, 26 Mar 45, PRO WO 171/3868. Purists: Oakes 1:5 note; Copeland 198 note; Paillole, Services 11. CEAs: Oakes 1:5, 1:340. Adopted by SHAEF: Report on the Activities of No. 2 Tac H.Q. A Force (Sixth Army Group) 31 July 1944–5 May 1945, Part I, p. 61, in NARA RG 319 Entry 101 Box 4 Folder 79.

130 W, W2: Paillole, Services 11.

130 MI5: Howard 16; Hinsley & Simkins 4, 7–8, 141. Able men: Masterman, Chariot 219. B Division: Hinsley & Simkins 10, 70, 178. B1, B1A: Hinsley & Simkins 178–79. Kell, Petrie: Masterman, Chariot 217, Hinsley & Simkins 65, Andrew 478.

131 "A rock": Masterman, Chariot 218. "One of the best": Andrew 479. Reorganization: Hinsley & Simkins 69–70. Liddell: Masterman, Chariot 218. "A big haughty fellow": Bristow 43–44. "A perfect officer type": Author's interview with Sir Michael Howard, 11 May 92. Natural leader: Masterman, Chariot 219. Marriott: Author's interview with Sir Michael Howard, 11 May 92. Masterman: Dictionary of National Biography.

132 Masterman surprised: Masterman, Chariot 211. St. James's: Masterman, Chariot 220. "The Force": Harvison xi. WATCHDOG: Harvison 105–116; PRO ADM 223/794/340, 356–358; Masterman 121, 144; Hinsley 201, 228; Pujol 113–14.

133 Size of Section V: Hinsley & Simkins 10, 180. Turf battle: Hinsley & Simkins 132–37.

133 Cowgill: Naftali, X-2 198–201; West, MI6 63. "A born": Naftali, X-2 226. Embarrassed by not in uniform: Whitehead 219, 343 note 5. Political hacks: DeLoach 14–16.

134 Number doubled: Whitehead 185; but see Naftali, X-2 56 note 39. Detestation of OSS: E.g., Ladd to Director, Re: Spanip, 25 Sep 43, in FBI Spanip file.

135 Half of the building: Felt 28. Hard-driving: Felt 30. Ladd: Whitehead 345 note 9. "Took the heat": Felt 30. Espionage Section: Felt 30–31; author's telephone interview with Mark Felt, 19 Jan 96. Tamm: Author's telephone interview with Richard D. Fletcher, 19 Jan 96.

135 SIS: Whitehead 212. OSS in North Africa: Naftali, X-2 212–13, 217–18.

136 OSS with Section V: Naftali, X-2 212–13, 217–18, 347–49. 104 SCI: Naftali, X-2 609–11; Harmer 1, 3. Indoctrination of Americans: Oakes 66–67; Naftali, X-2 616–17. Three SCI units: Naftali, X-2 619–20.

137 "One of those people": Bristow 109. TR: Oakes 80; Naftali, De Gaulle's 385. Paillole to North Africa: Naftali, De Gaulle's 388, 391; Naftali, X-2 263–66. Rounding up: Hinsley & Simkins 262.

137 Paillole 1944–45: Naftali, De Gaulle's 393, 395–96, 399, 404.

138 Long Island saboteurs: Whitehead 199–205, 346 note 6. Churchill asked: Liddell diary, 7 Oct 40, PRO KV 4/187. MI5 limited: Hinsley & Simkins 96–98. Jakobs: Hinsley & Simkins 92, 97 note*.

140 TREASURE and BRUTUS after ISOS loss: Hinsley & Simkins 182, 274–75. SIME and U-boats: Maunsell 27.

140   $366,125: Whitehead 345 note 8. Paillole visits England: Naftali, De Gaulle's 384 and notes 12–16, 405 and note 22; Masterman 36.

141   Couriers: Whitehead 195, 395 note 8.

142   The *Manuel Calvo:* Author's telephone interview with Ray Wannall, 19 Dec 96. Japanese attaché: Masterman 93–94.

143   German radio sets: Hinsley & Simkins 311. GARBO: Hinsley & Simkins 313. Secret ink in WWI: Yardley 55–89.

143   Sophisticated secret ink: Hinsley & Simkins 226. Poisons: Hinsley & Simkins 92 note*. Pyramidon: Felt 32. Hollow tooth: Hinsley & Simkins 341. Matches: Hinsley & Simkins 126, Whitehead 345. RAINBOW alerted: Hinsley & Simkins 43; Whitehead 195–96 is probably the same. Actual use: Hinsley & Simkins 103–104; Whitehead 196.

144   PAT J peculiarity: Hoover 120. MOONBEAM handwriting sample: Hinsley & Simkins 228 note‡. PAT J's idioms: Hoover 120. MOONBEAM cheapskate: Pujol 113.

144   *Chars allemands:* Mure, Practise 51. TREASURE: Hinsley & Simkins 241.

145   GARBO's network: Harris, *passim;* Pujol x–xi; Howard 231–32.

146   Tonic: Report on the Activities of No. 2 Tac H.Q. "A" Force (Sixth Army Group) Part IV, Conclusions, p. 23, in NARA RG 319 Entry 101 Box 4 Folder 82.

146   General to Middle East: II/47.

147   "Trying to control": Hinsley & Simkins 128. Four new double agents: Hinsley & Simkins 91.

147   Stress the strength: Howard 6; Hinsley & Simkins 98. Genesis of W Committee and Twenty Committee: Howard 6–7; Masterman 61; Hinsley & Simkins 98. "U" Section, Twenty Club: PRO ADM 223/794/7.

148   Masterman embarrassed: Masterman, Chariot 223. Over the objection: Liddell diary, 8 Jan 41, PRO KV 4/187. Meetings: PRO ADM 223/794/7; Hinsley & Simkins 100. Vote only once: Masterman, Chariot 223. W Board meetings: Howard 10. SIME Special Section: III/59; Hinsley & Simkins 153.

149   The 30, etc., Committees: III/59–62.

150   "Altogether the 30-Committees": III/62.

150   "A dull sort": Hinsley & Simkins 128, Howard 12. Controlled every agent: Hinsley & Simkins 128; Howard 20–21.

151   *Intoxication:* Naftali, De Gaulle's 384–85. "The German General Staff": Howard 12. MACHIAVELLI and PLAN IV: Masterman 83–84; Howard 17.

151   MI5 reiterated: Hinsley & Simkins 101. Montagu: Liddell diary, 23 Oct 41, 27 Oct 41, 29 Oct 41, PRO KV 4/188. "Not to be informed": Minutes, 30 Oct 41, PRO KV 4/64/101; Hinsley & Simkins 101–102; Howard 23. Disappointed, stayed out: Liddell diary, 29 Oct 41, 30 Oct 41, PRO KV 4/188. "Exactly what machinery": Liddell diary, 20 Oct 41, PRO KV 4/188.

152   "Men who seldom"; "of filling": Howard 12. "In MI5 double agents": Howard 20–21.

153   "Not, as it had been": Masterman 55. "Reference sheets"; "Reports received by the Germans": Memorandum, NID 12 [Montagu], Machinery of Deception Outside N.I.D—W Board, Twenty Committee & London Controlling Section, 18 Sep 45, PRO ADM 223/794/13. Bevan proposed, declined: Liddell diary, 25 Aug 42, PRO KV 4/190; Howard 29.

154   Rumrich case: Hinsley & Simkins 142; Whitehead 162–63. BSC: Hinsley & Simkins 143.

154   Sebold case: Kahn, Spies 333; Hinsley & Simkins 157; Whitehead 168–69. JOE K: Hinsley & Simkins 157–59; Whitehead 193–94.

155   TRICYCLE: Hinsley & Simkins 95; Masterman 55–56, 79; Howard 17; Whitehead 196, 344 note 5; Troy, British. Pearl Harbor: Masterman 79–80, 196–98; Popov; Troy, British; Naftali, X-2 182–83; Morison, Rising Sun 43; Riebling 472–76; Hill 388–93.

156   Two ladies: Hill 357. Two subagents: West, Seven 13.

156 Eight letters: Naftali, X-2 63 note 7. No point: Naftali, X-2 42.

157 Montagu and Cowgill to New York: Montagu, Beyond 77–84. Abwehr suspicious: Hinsley & Simkins 124. Debt: Troy, British 188.

157 TRICYCLE's return: Hinsley & Simkins 125–26. Donovan Hoover's boss: Theoharis & Cox 187–88. Stephenson's link: Hinsley & Simkins 147. MI5 unhappy: Hinsley & Simkins 131 note*. Liaison: Hinsley & Simkins 147–48.

158 Mills visits: Hinsley & Simkins 186–87; Pujol 113. Improved relations: Hinsley & Simkins 202. MOONBEAM: Howard 223; Hinsley & Simkins 228; Pujol 113–14.

159 "It is believed": Hinsley & Simkins 229.

159 Felt, Levy: Author's telephone interview with Richard D. Fletcher, 19 Jan 96.

160 PEASANT: Unsigned memorandum, "Re: [Deleted] Espionage—G," 2 Sep 44, in JSC brown Acco binder labeled "14," p. 23. San Diego bars: Unsigned memorandum, "Re: Pat J. Case," 16 Mar 45, in JSC brown Acco binder labeled "14," p. 15.

161 Schellenberg: Hinsley & Simkins 279.

161 OSTRO: Campbell. OSTRO and D-Day: Hinsley & Simkins 199–200. V-1s: Howard 178–80. Efforts to discredit: Montagu, Twenty Committee, 30 Nov 44, PRO ADM 223/794/70; Hinsley & Simkins 279.

161 Kraemer: Hinsley & Simkins 200–201, 278. Invasion for June: Hinsley & Simkins 258. 1st SS Panzer: Howard 188–89; Hesketh 102. Arnhem: Hinsley & Simkins 278–79. Inaccurate American information: Project No. 1471-A, The Japanese Intelligence System, in JSC folder The Japanese Intelligence System, p. 155.

162 Del Pozo: Hinsley & Simkins 94–95; Masterman 92. POGO: Liddell diary 10 Oct 40, PRO KV 4/187. Calvo: Masterman 113. Other Spanish channels: Hinsley & Simkins 104–105; Masterman 99, 113.

163 Calvo arrest and results: Hinsley & Simkins 107–110. "By far the best": Masterman 113. Alcázar and Japanese: Hinsley & Simkins 108 note*; Project No. 1471-A, The Japanese Intelligence System, in JSC folder The Japanese Intelligence System, p. 151. Toh Reports: Smyth 347–350; ADM 223/794/349–55. SPANIP: Memorandum, [Deleted] to [Deleted], undated, FBI File 65-46383-2143, in FBI Spanip file; Report, 30 Jan 43, in FBI File No. 65-46383, Spanip Section 8; OP-16-B-7, A Summary of Recent Development[s] in Japanese Espionage in the Neutral States of Europe, p. 1 (Spain), in NOA 535, Folder Japanese Intelligence in Europe; Foxworth to Director, 28 Oct 42, in FBI File No. 65-46383, Spanip Section 1; report dated 5 Nov 42, p. 38, in FBI File No. 65-46383, Spanip Section 2.

164 Alcázar leaves Spain: Hinsley & Simkins 206; OP-16-B-7, A Summary of Recent Development[s] in Japanese Espionage in the Neutral States of Europe, p. 1 (Spain), in NOA 535, Folder Japanese Intelligence in Europe. Aladren: Project No. 1471-A, The Japanese Intelligence System, 4 September 1945, p. 151; unsigned memorandum, 13 Mar 45, Double Agent Aspirin, in JSC brown Acco binder labeled "14," pp. 52–53.

164 Abwehr Lisbon station: Hinsley & Simkins 200. Pacheco: Project No. 1471-A, The Japanese Intelligence System, 4 September 1945, p. 154. Fülop: Project No. 1471-A, The Japanese Intelligence System, 4 September 1945, p. 151; Hinsley & Simkins 278.

Chapter 5: London Control

166 Stanley: DNB; Colville 767. "Possessed what": DNB 1941–1950, p. 826. YMCA Conservatives: Cross 21, 40.

167 FOPS: Davis 2:1118–19. In that capacity: Wingate 49, 52.

167 War Cabinet; Defense Committee: Davis 1:89–91. Secretariat: Cline 99–100; Cabinet War Rooms 6, 18; Davis 1:91; Wheatley 38. Hollis: DNB.

168 Stanley memorandum: Stanley, Strategic Deception, 31 Oct 41, Annex to COS (41) 245 (O), 31 Oct 41, PRO CAB 119/67/294–296. Approval: Minutes, COS (41)

374th Meeting, 3 Nov 41, PRO CAB 119/67/288. Halahan: Montagu, Beyond 48; Wheatley 20.

169 Lumby: Wheatley 16, 20, 26–27, 32, 70.

169 Wheatley papers: Wheatley, Stranger 10–13, 149. Wheatley recruitment: Wheatley, Stranger 411; Wheatley 21.

170 Wheatley description; Wheatley 40; Young & Stamp 177. Wheatley biography: DNB. Grammar and spelling: Young & Stamp 177. Spy novels: Wheatley, Stranger 19–20. Family in MI5: Wheatley, Stranger 19–20; West, MI5 123, 137.

171 Invasion paper: Wheatley, Stranger 20–39. Chatsworth Court: Wheatley, Stranger 13. GELATINE: West, MI5 198.

171 Friends: Wheatley, Stranger 411. Induction, basic training, reports for duty: Wheatley, Stranger 411–13; Wheatley 26. New Public Offices: Cabinet War Rooms 3.

172 Office; "acres": Wheatley 22, 26, 66–67. War Rooms: Cabinet War Rooms; Wheatley 24; CWR Annexe: e-mails of Phil Reed and Neil Cooke to author, August 2003. "Storey's Gate": Author's telephone interview with Phil Reed, Curator of the Cabinet War Rooms, 8 Apr 97. Mr. Rance: Cabinet War Rooms 7; Wheatley 67; Young & Stamp 172–73. JPS mess: Wheatley 24, 28.

173 Wheatley's first days: Wheatley 27–31. No general directive: Wingate 49, 56. Nothing to do: Wheatley 28, 32–33.

173 Secretary to section: Wingate 49, 55; Wheatley 33, 42. Thinkpieces: Wheatley 34–35.

174 Stavanger project: Wheatley 36; Wingate 83–84. "The fate": Wingate 83. Pecking out: Wheatley 36–37. "A pleasant and efficient": Wheatley 37.

175 Plan as real operation plus rumors: Wingate 83–84. HARDBOILED; "bloody fool": Wheatley 37–38.

175 Training, supplies: Wheatley 39; Wingate 83. Meetings: Wheatley 38–39. Directives: Wingate 83. Foot-dragging: Wheatley 40–41.

176 No physical implementation: Wheatley 45; Wingate 84–85. Double agents: Masterman 86; Wingate 84; PRO KV 4/197/12, p. 2.

176 Detectable results, German moves: Wingate 84–85; Wheatley 55. Inactivity, fancy uniform: Wheatley 40–41.

177 Wheatley's papers: Wheatley 41–42, 44–45.

177 GOLDLEAF-HERITAGE: Wheatley 46–47; Wingate 86; II/60–61.

178 "Hazy": Wingate 87. Success, useful: Wingate 87, II/60.

178 "Putting up suggestions": Wingate 49. Major from ISSB: II/60; Wingate 86; Wheatley 29. Wheatley papers: Wheatley 49, 50–52, 55–56. Staff moved: Wheatley 52.

179 Lumby transferred; "this deadly": Wheatley 53; Davidson to Ismay, 11 May 42, PRO CAB 121/105/531.

179 Lumby killed: Wheatley 81; Muggeridge 186. Stanley on leave, asks release: Wheatley 53–54; Howard 25. Wheatley alone, more makework: Wheatley 53–56.

180 Stanley memorandum; "a civilian"; "a soldier": Machinery for Strategic Deception, Note by the Controller, 2 Jun 42, CAB PRO 119/67/206–209. Godfrey: Andrew 341, 455.

180 Built up: Andrew 455–56. Godfrey's papers: D. N. I. [Godfrey], Ruses de Guerre, 29 Oct 39; Unsigned, undated memorandum Notes on A/S Warfare; untitled memorandum, D. N. I. [Godfrey], 2 Nov 39; Godfrey, Special Intelligence—Deception, Ruses de Guerre, Passing on False Information, Etc., 8 Nov 39; all attached to Godfrey, Ruses de Guerre, 13 Nov 39, PRO ADM 223/794/216–32. Visits to USA: Lycett 142–44. Montagu: Montagu, Beyond 16–21, 46.

181 "The chief difficulty": Machinery for Strategic Deception, Note by the Controller, 2 Jun 42, CAB PRO 119/67/206–209. "I have always had": Wavell to Churchill, 12461/C, 21 May 1942, CAB 121/105/525–26, PRO PREM 3/117/69; quoted Howard 25–26. Took hint: Howard 26. Bevan appointed: Minutes, COS (42) 157th

Meeting, 21 May 42, PRO CAB 121/105/527. Already designated: Davidson to Ismay, 11 May 42, PRO CAB 121/105/531. Joined June 1: Wheatley 57. Bevan: Who Was Who, 1971–1980. "A rather frail-looking": Wheatley 58.

183   "His most notable": Wheatley 59. Countryman, outdoorsman: Wheatley 59–60. Children: Letter, Julian Bevan to Jennifer Bevan Lowther, 14 Feb 97, original in possession of Julian Bevan, copy in author's possession. Essentially unmilitary: Author's interview with Sir Michael Howard, 11 May 92. "He disliked": Wheatley 227, also 88, 132. Friends, Brooke: Young & Stamp 177; Astley 165; Wheatley 60. Baumer: Stratagem Conference 35. Churchill's interest: Young & Stamp 174. All three together: PRO PREM 3/117/49. "So English": Stratagem Conference 29.

184   Schooldays: Letter, Julian Bevan to Jennifer Bevan Lowther, 14 Feb 97, original in possession of Julian Bevan, copy in author's possession. "When things were looking": Wheatley 58. Western Front: Wheatley 58–59. "I had great fun": Bevan to Eldredge, 9 Apr 70, p. 3, Eldredge papers. January 1918: Letter, Jennifer Bevan Lowther to author, 28 Feb 97.

184   Assessment of German intentions: Wheatley 60. Churchill in France February 1918: Gilbert 66–71. Kept in Army: Letter, Jennifer Bevan Lowther to author, 28 Feb 97. Would have liked to remain: Wheatley 60. "Bless my soul"; "We lost": Stratagem Conference 114. Denmark: Letter, Julian Bevan to Jennifer Bevan Lowther, 14 Feb 97, original in possession of Julian Bevan, copy in author's possession.

185   Partner in 1925; "the very special": Letter, Eric Faulkner to Bevan, 17 Apr 74, original in possession of Julian Bevan, copy in author's possession. "To God and history": Stratagem Conference 29; see also letter, W. A. Harris to Bevan, 24 Dec 74, original in possession of Julian Bevan, copy in author's possession. "You will no doubt": Letter, G. O. Allen to Bevan, 2 Jan 45, original in possession of Julian Bevan, copy in author's possession. MI5 officer: Wheatley 60–61.

186   Haversack ruse: Hart-Davis 230. Home Guard: Wheatley 42, 60–61. "Has very special": Stanley to Ismay, 20 May 42, PRO CAB 121/105/529.

186   Fleming: Wheatley 36. Chiefs directed Planners: Minutes, COS (42) 160th Meeting, 25 May 42, PRO CAB 121/105/525. Unimaginative suggestions: Strategic Deception—Machinery, JP (42) 545 (S), 2d Preliminary Draft, 29 May 42, PRO CAB 119/67/224–244. "A kind of unofficial": Hollis to Simpson, 29 May 45, PRO CAB 121/105/84. Sharp departure: Strategic Deception—Machinery, JP (42) 545 (S) (Draft), 1 Jun 42, PRO CAB 119/67/224–234.

187   Stanley memorandum; "greater experience"; "I am sure"; "Controlling Officer"; "I had, in fact"; "we should": Machinery for Strategic Deception, Note by the Controller, 2 Jun 42, CAB PRO 119/67/206–209.

187   Fourth version; "We are satisfied"; "The Controlling Section": Strategic Deception—Machinery, JP (42) 619, 18 Jun 42, PRO CAB 121/105/516–522; another copy, PRO CAB 119/67/169–175; Howard 26. Drafting directive; "two-thirds"; "Dennis, we": Wheatley 64–65. "It was cock-eyed": Bevan to Eldredge, 9 Apr 70, p. 5, Eldredge papers.

188   Chiefs' approval: Minutes, COS (42) 184th Meeting, 20 Jun 42, PRO CAB 121/105/506; amendments marked on JP (42) 619, Strategic Deception—Machinery, 18 Jun 42, PRO CAB 121/105/516–522. London Controlling Section, LCS: Young & Stamp 174.

189   Text of directive: Howard 243; Wingate 54; COS (42) 180 (0), Strategic Deception, PRO CAB 121/105/508.

189   "And, from that point": Wheatley 65. Formally appointed: Minutes, COS (42) 228th Meeting, 5 Aug 42, PRO CAB 121/105/487. Office, files: Wheatley 63.

190   "Staff duties, "deception bible": Wheatley 35, 62–63. Sweden proposal: CAB 119/67/137–151, 163–168, 177.

190 Deception part of operations: Howard 27. Functioning of LCS: Wingate 50, 56.

191 Move to "Mr. Rance's room"; its appearance: Wheatley 24; Young & Stamp 173; Astley 59. Furnishings: Wheatley 67. Increased establishment: Wingate 55. Wheatley bumped up: Wheatley 78.

192 No dividing line: Wingate 55. ULTRA, double agents: Wheatley 188. Complex task: Howard 28; COS (42) 208 (0), 20 Jul 1942, CAB 121/105/503–504. TORCH ratified: Howard 55. July 28 directions: Howard 28. "Positively ate": Wheatley 111. "Had Johnny": Wheatley 57.

192 Wheatley with senior officers: Wheatley 61–62, 90. Hungaria: Wheatley 137, 143, 225.

193 "Trotted out": Baumer, But General ch. II, Baumer papers. Bevan at home: Wheatley 61–62. "Dennis, we can't": Wheatley 78. Petavel; "with a mass": Wheatley 78–79. Pipe: Young & Stamp 178, 180.

194 Petavel handled office work: Wheatley 79. Intelligence subsection; "Enormous charts": Wheatley 129. "Sticking pins"; Twenty Committee: Wheatley 148. Order of battle: Wheatley 129. Wingate of the Sudan: Encyclopedia Britannica, 14th ed., *sub nom.* Wingate, Sir Francis Reginald.

194 Wingate: Wheatley 79–80, 111, 137; Young & Stamp 177. "Bland likeable"; "a rod of steel": Baumer, But General, ch. II, Baumer papers. "Charming and clever": Wheatley 110. "Encyclopedic"; "amusing cynicism"; "a vast experience"; "He used at times": Wheatley 82.

195 Left in charge: Wheatley 16. Deputy Controlling Officer: Wingate 55. Montagu assured: Wheatley 88. Admiralty hunted: Howkins to Bevan, 26 Aug 42, 42/932, PRO CAB 119/67/112. Arbuthnott; "in a very high"; good man in a conference: Wheatley 89–91.

195 Arbuthnott briefing: Wheatley 89. Arbuthnott and Wheatley: Wheatley 91. Amicably with Montagu: Memorandum, NID 12 [Montagu], Machinery of Deception Outside N.I.D.—W Board, Twenty Committee & London Controlling Section, 18 Sep 45, PRO ADM 223/794/15. Arrangement of "schoolroom": Wheatley 92.

196 Obtaining new space; its configuration: Wheatley 127–28. (Lady Jane Bethell mentions only four rooms. Young & Stamp 173. The current reconstruction of the floor plan by the Cabinet War Rooms differs in some respects.)

197 Wheatley's decor: Wheatley 128–29; Young & Stamp 177. JPS mess: Wheatley 136–37. March 1943 enlargement: Wingate 55. Gordon Clark: Wheatley 129, 133. Morley: Wheatley 129–30; Young & Stamp 178; "A strange cross": Wheatley 129–30. With Goldbranson: Goldbranson, Summary of Conversation With Colonel Dudley W. Clarke, Commander of "A" Force, Monday, 28 June, 1943, Goldbranson papers, p. 1.

198 Morley abroad: Wheatley 130–31; Wingate 58. Hoare: Wingate 55; Wheatley 132. "An elderly": Wheatley 132.

198 "He treated me": Wheatley 132. Andrade: Wheatley 132–33. Floating roadway: Memorandum, NID 12 [Montagu], Machinery of Deception Outside N.I.D.—W Board, Twenty Committee & London Controlling Section, 18 Sep 45, PRO ADM 223/794/147–48. Support staff: Wheatley 128; Young & Stamp 174.

199 Lady Jane: Young & Stamp 172, 174–76. Happy shop: Wheatley 90; Young & Stamp 176. Bevan modest: Wheatley 58. Overconscientious: Wheatley 164. "Whenever a crisis": Wheatley 111. "Very kind": Young & Stamp 177. "That smile": Wheatley 59. Loved the work: Stratagem Conference 105.

200 Not rigid: Wheatley 90, 131; Young & Stamp 176, 178. Night work; "unannounced strike": Wheatley 92–93, 110, 162.

200 "Obsessed by security": Wheatley 64. "Security is a fetish": Baumer, Report on Visit to London and Algiers Planning Sections, 10 May 43, p. 3, in Hoover Institution,

Baumer Papers, Box 10, Folder 2. Petavel and wife: Young & Stamp 175. "If any of us": Wheatley 110–111.

201 ISSB meetings: Wingate 57. Committees: Wingate 57–58; Wheatley 154 (Wheatley says three committees); Memorandum, NID 12 [Montagu], Machinery of Deception Outside N.I.D.—W Board, Twenty Committee & London Controlling Section, 18 Sep 45, PRO ADM 223/794/1; Baumer, Report on Visit to London and Algiers Planning Sections, 10 May 42, p. 3, in Hoover Institution, Baumer Papers, Box 10, Folder 2. (Ladd, Deception Through Section V, 19 Oct 43, in NOA 530, Folder Combined Planning, Cover and Deception, seems to misapprehend the TWIST Committee somewhat.) TWIST: Hesketh 15. TORY: See Annex B to Suggested Cooperation by Joint Security Control, 24 Apr 43, in JSC folder 370.2 Operations and Results. RACKET: PRO CAB 154/42; Wingate 59.

201 TWIST and RACKET abolished, TORY retained: Hesketh 15–17; Hesketh to Casey 1–2; minutes 24 Feb 44, PRO KV 4/67/351. Mock telegram: Memorandum, NID 12 [Montagu], Machinery of Deception Outside N.I.D—W Board, Twenty Committee & London Controlling Section, 18 Sep 45, PRO ADM 223/794/16. PWE liaison: Wingate 58; Wheatley 135.

202 Foreign Office, MEW liaison: Wingate 58. Mutt and Jeff: Wheatley 157. Reading assignments: Wheatley 133.

203 Staff meetings: Baumer, Report on Visit to London and Algiers Planning Sections, 10 May 42, p. 2, in Hoover Institution, Baumer Papers, Box 10, Folder 2.

204 Many drafts: Young & Stamp 176. Same brainstorming: Wingate 59. Two resolutions of commendation: Wheatley 221–22. "If in some future": Letter, Wheatley to Bevan, 29 Dec 1944, original in possession of Julian Bevan, copy in author's possession.

204 Godfrey transferred: Montagu, Beyond 164. "Is almost completely ignorant"; "He went to America"; "often insisted": Lieut. Comdr. R.N.V.R. [Montagu] to Hutchison et al., 1 Mar 43, PRO ADM 223/794/104–105V. "Considerable friction"; "Lost much of his original": Memorandum, NID 12 [Montagu], Machinery of Deception Outside N.I.D—W Board, Twenty Committee & London Controlling Section, 18 Sep 45, PRO ADM 223/794/13.

206 Complaining after war: Memorandum, NID 12 [Montagu], Machinery of Deception Outside N.I.D—W Board, Twenty Committee & London Controlling Section, 18 Sep 45, PRO ADM 223/794/13.

207 "Personal" file: Wheatley 63–64. ARABEL: Hinsley & Simkins 113; Pujol 97–98.

207 Windermere: Harris 58; Pujol 69. "Drunken orgies"; "a liter"; Brighton; pounds etc.: Harris 58–59; Pujol 81–82. Malta convoy: Harris 41; Pujol 84. Action through BALLOON: Hinsley & Simkins 113 note*. Jarvis: Pujol 62.

208 Bringing Pujol to England: Harris 66–67; Hinsley & Simkins 113–14; Howard 19. Arrival: Harris 67. (Arrived April 25, per Pujol 92 and Bristow 38. This may be date of arrival in London.) "Warm brown": Bristow 38.

209 Pujol's life: Harris 42–43; Pujol 22, 28–38. January 1941 approach: Harris 43; Pujol 56. Taken on by Germans: Harris 46–50; Pujol 46–55.

209 To Portugal: Harris 51; Pujol 93. Unsuccessful efforts with British: Harris 52–53, 61; Hinsley & Simkins 112. Nearly 40 messages: Howard 19.

210 Contact through Americans: Harris 62–66; Hinsley & Simkins 113; Pujol 90. To England: Pujol 88; Bristow 33, 38. Debriefing: Bristow 38–42. BOVRIL became GARBO: Pujol 94. Harris: Pujol 79–80. Bristow: Bristow 1–7. Meetings with Twenty Committee: Bristow 43–44. Evidence incontrovertible: Hinsley & Simkins 114. Bristow transferred: Bristow 45. Family smuggled out: Harris 80–82; Hinsley & Simkins 114; Pujol 92–93.

211 "I do not wish": Howard 116. "This unfolding": Howard 241. Number of messages: Howard 231.

212 Small office; restaurants: Pujol 93. "We tried to report": Howard 231.

212 Refugee, freelance: Hinsley & Simkins 115. Building of network: Hinsley & Simkins 115, 225–27; Howard 232–33; Pujol 94, 97, 100–114.

213 Radio link: Hinsley & Simkins 115–16, 225–27, Pujol 102.

214 Never again; "Became entirely dependent": Howard 231.

214 BRUTUS: Hinsley & Simkins 117–19.

Chapter 6: The Turning of the Tide

215 Establishment touch: Rex Hamer to author, 2 Jun 97, p. 3. Reorganization: II/56. Druiff to Beirut: II/116. Substantial private army: II/28. Small and dapper: Author's interview with Arne Ekstrom, 27 May 93. Prim-looking: Mure, Master, facing p. 96. "A slim": Eldredge to Dwyer, 28 Sep 89, Eldredge papers. Morocco, Palestine: Mure, Master 48, 88.

216 Wild up to recruitment by Clarke: Mure papers, Box 1, small folder; Mure, Master 88.

217 McCreery: Mure, Master 91. Reports for duty: Mure, Master 91; A" Force card index, H. N. H. Wild, Clarke papers. Eulogy: Copy in Clarke papers. Detested: Author's telephone interview with David Strangeways, 26 Aug 92; Author's interview with Arne Ekstrom, 4 Jan 1993; Thynne to Mure, 27 Feb 1978, in Mure Papers, Box 2. "Fancy for his own": Thynne to Mure, 27 Feb 78, in Mure Papers, Box 2. Thynne: "A" Force card index, O. St. M. Thynne, Clarke papers; Mure papers; Mure, Master 106; Thynne to Mure, 27 Feb 78, in Mure Papers, Box 2. "Having appreciated": Thynne to Mure, 27 Feb 78, in Mure Papers, Box 2.

217 Bromley-Davenport a "nice guy": Author's interview with Eugene Sweeney, January 1993; Author's telephone interview with Alice Crichton, 12 Oct 95. "A very over-dressed man"; "Of high color": Mure, Practice 136. "Decent enough": Hamer to Mure, 21 Feb 81, in Mure Papers, Box 2. "Pathological horror"; "charming, if eccentric": Mure, Practice 136. Detail work: Mure, Practice 139.

218 De Salis: "A" Force card index, P. F. de Salis, Clarke papers. "Highest authority": Cooper 126. Chairmanship: Colyton 187–88. Mrs. Hopkinson with "A" Force: Colyton 188; "A" Force card index, Mrs. A. Hopkinson, Clarke papers. To Lisbon: Colyton 216–18.

219 Mrs. Hopkinson: Colyton 26. "Alice had": Colyton 61–62. Fast friends: Author's telephone interview with Alice Crichton, 12 Oct 95. Lasseter: Letter, Rex Hamer to author, 2 Jun 97, p. 4. Duchess of Split: Author's telephone interview with Alice Crichton, 20 Nov 93; Author's interviews with Eugene Sweeney and Arne Ekstrom. Cuba Mason: Author's interviews with Eugene Sweeney and Arne Ekstrom. Ekstrom recalled that "the Duchess of Split" was Clarke's humorous name for Cuba Mason rather than for Margery Hopkins.

219 Crichton's background: Mure, Master 103. Anxious to get back: Author's telephone interview with Alice Crichton, 20 Nov 93. Simple cabaret: Letter, Hamer to Mure, 21 Feb 81, p. 3, in Mure Papers, Box 2. Betty-to-You: Cooper 123.

220 Alice Sims; recruiting Crichton: Author's telephone interview with Alice Crichton, 12 Oct 95. "An individual": Mure, Practice 27.

221 Shearer and de Guingand: Warner 158. Northern flank, Persia and Iraq Command: II/89, II/116.

221 Advanced headquarters "X" Force; staffing: II/116.

222 Dummy tank activity: II/62–65. "74th Armored brigade": II/83–84, 90. 303d Indian: II/89–90.

222 Prototype DRYBOB and WETBOB; SAINTS tested: II/87–88. Sonic deception: II/84–87.

222 CRYSTAL: II/29–31.

223 H.D.F.; "this admittedly": II/33; PRO KV 4/197/2A.

223 Planned counteroffensive: Warner 160, 172. FABRIC: II/50–58. MAIDEN: II/59. "In no way": II/55.

224 More frayed: II/53–54, 56–58. Lessons: II/58–59.

225 CASCADE: II/35–42. White unicorn, yellow hornet: II/38–39; Cole 114. History, components: II/38–39. Book of Reference: II/39.

226 Later editions: II/41–42; III/58. German overestimates: II/43–44. "The amount of inconvenience": II/40.

227 The Flap: Cooper 192–97. "Today I drive": Cooper 196.

227 "A" Force in the Flap: II/68–72. Stealing license plates: Mure, Master 103.

229 Crichton marriage: Author's telephone interview with Alice Crichton, 29 Apr 97; Alice Crichton to author, n.d. (11 Nov 96). Formal wedding: Clarke diary 9 Jul 42, Clarke papers; author's telephone interview with Alice Crichton, 12 Oct 95.

229 Revival of CHEESE: II/73–76. "Gradually building up": II/74. "Be very active": II/73.

230 Control to Athens: Hinsley & Simkins 166. Fellers: Irving, Fox 184, 234; Kahn, Codebreakers 476; Hinsley 2:31, 2:331, 2:361, 2:389, 2:640. Seebohm: Howard 65; Hinsley 2:298, 2:404.

231 July 4 ISOS: Hinsley & Simkins 166. Committee, Thynne: Mure, Master 106. Traffic last half of year: III/75–76. "Nicossof's a Russian name": Mure, Master 64 (name changed to "Fornikov").

231 Funding CHEESE: II/91–98. MISANTHROPE: Hinsley & Simkins 166. Hamada: Hinsley & Simkins 165. "The cries were raised": II/92.

232 "An attractive Cretan"; "B.G.M.": II/75. Pushed off roof: Mure, Master 107.

232 "Dressed in white": II/92–95.

233 "Do you think"; "The money": II/92–95.

234 SLAVE; THE PYRAMIDS, FLESHPOTS: II/119; Dovey, Condor. QUICKSILVER: II/120–21. Liossis: Handwritten memorandum, n.d., in Mure Papers (Box 2). In touch early: Mure, Practise 37–38. Bonzos: West, MI6 193.

235 Gestapo happy: Mure, Practise 38. Agaraki: West, MI6 193. SIM: Mure, Practise 39. Intercepted; radio contact: Hinsley & Simkins 165; II/120–21. "Military and naval": II/120. GALA: Mure, Practice 39.

235 "Acquired quite a reputation"; "Became also": II/120. Villa: Mure, Practise 117. Africans: Mure, Practice 117–18; Mure, Master 111. "Glowering concentration": Mure, Practise 119. Escapes: Mure, Practise 118. THE PESSIMISTS: II/121–23. PESSIMIST Y: Mure, Practise 63.

236 "They were altogether": II/121. Alerted by ULTRA: Hinsley & Simkins 165. "Q and P Hall": Mure, Practise 256. "Was a fat": Mure, Practise 63. Permanent smile, treacherous: Mure, Practise 63, 118–19. Location, radio masts: Mure, Practise 117–18. Wills: Mure, Practise 78; West, MI6 194.

237 DOLEFUL/DOMINO: I/117–18; Hinsley & Simkins 165. PATENT: II/66–67, 77.

238 RAYON: II/77–79.

238 CHEESE leading role: II/78. Convoy loss: Salmaggi & Pallavisini 278. Admiral's satisfaction: II/79. SENTINEL: II/80–82.

239 GRANDIOSE, GRANDIOSE II: II/99–100.

240 Delaying Rommel: II/79. Williams haversack ruse: Liddell diary, 18 Feb 44, PRO KV 4/193; Dovey, False Going 165–68. Learning of TORCH: II/102.

240 Montgomery briefing: Clarke diary 14 Sep 42, Clarke papers; II/108. TREATMENT approved: II/108. BERTRAM, Richardson: II/109; Young & Stamp 68–74.

241 TREATMENT, BERTRAM stories: II/108–109. Implementation divided: II/109.

242 American traveler; rumors, administrative arrangements; CHEESE prediction, gratifying request: II/111.

242 Implementation of BERTRAM and subsidiaries: II/109–113.

243 DIAMOND: Cruickshank, Deception 30; Young & Stamp 70–71. BRIAN: Cruickshank,

Deception 30; Young & Stamp 73. Munassib: Cruickshank, Deception 30–31; Young & Stamp 74.

243 Martello: Cruickshank, Deception 31; Young & Stamp 73. Murrayfield: Cruickshank, Deception 31–32; Young & Stamp 73.

243 Melting Pot: II/84, 113; Cruickshank, Deception 32; Young & Stamp 73. Canwell: II/114. "Serious": Howard 66. Pajamas: II/114. Stumme: Howard 66–67. Thoma: Cruickshank, Deception 33.

244 Rommel shifts: Howard 67. Buggush operation: II/86. Demonstration: II/89. McCreery: Mure, Master 91.

Chapter 7: Enter the Yanks

246 Vincennes: Clark, Conquest 132–33. St. Michaels: Preston 166. Manassas, Harrison's: Sears 17, 355. Magruder: Sears 37, 216–17, 247. "Reinforcements": Connelly 1:177. Forrest: Henry 156–57. Myer: Fishel 347–49, 566–67. Backfired: Fishel 480–83, 567.

246 Belfort Ruse: Paschall. Fox Conner: Paschall 174; Eisenhower, At Ease 183–92. 1934 article: Bonham.

247 Midway: Herbig 15–16, 18–19; Fuchida & Okumiya 200. Air Force separate: Cline 68. Arcadia, CCS: Cline 68, 100; Davis 1:13, 1:137, 1:176–77.

247 JCS: Cline 98–99; Davis 1:281–83. Reorganized committee structure, planning procedures: Davis 2:1132, 2:1138–40, 2:1149, 2:1173, 2:1178, 2:12024–26; Cline 239–42.

248 OPD, S&P: Cline 106–107, 135, 363–70. Navy planning: Whitehill 57–58, 62 and note 4.

249 Offices: WPA Washington 860–61, 863, 885. Decision for Torch: Matloff & Snell 176, 185–87, 236, 277–84.

250 Bevan directed: Howard 28. "Security Committee": NARA RG 218 Box 211 Folder CCS 334 (4-30-42) Joint Security Committee. Strong: Biographical release, USAMHI.

250 "No fool": Liddell diary, 6 Sep 43, KV 4/192. "A very good man"; "able but pretty impossible": Rogers, Wartime 180. "A sound, solid": Rogers, Wartime 17 note 16. "The only affirmatively": Rogers, Wartime 17, 90. Strong in touch: Wingate 60. Visits London: Kroner to Chief of Staff, 4 Aug 42, p. 2, in JSC folder Torch. Upshot: Kroner to Chief of Staff, 4 Aug 42, p. 2; W. B. S. [Bedell Smith] to Handy, 3 Aug 42; Marshall to CG ETOUSA, 4 Aug 42; Eisenhower to AGWAR 7 Aug 42; Kroner to Secretary, JCS, 11 Aug 42; Wedemeyer to Strong with enclosure, 17 Aug 42; Kroner for Record, 18 Aug 42, all in JSC folder Torch; Minutes, JCS 29th Meeting, 18 Aug 42, in NARA RG 218 Box 211 Folder CCS 334 (8-4-42) Joint Security Control (Sec. 1); Davis 2:9034–38. Dyer: Biographical release, Naval Operational Archives.

251 Dyer Navy responsibility: Dwyer 5. Fiske: Kroner to CG Army Ground Forces et al., 14 Aug 42, in JSC folder Torch. Rome: John Slonaker, USAMHI to author, by telephone. "An appalling": Fiske to Strong, 19 Aug 42, in JSC folder Torch. "The Chief": Marshall to USFOR London (Personal Strong to Eisenhower) 19 Aug 42, in JSC folder Torch. "Clothed with": Cooke to King, 21 Aug 42, in NARA RG 218 Box 211 Folder CCS 334 (8-4-42) Joint Security Control (Sec. 1). Chiefs' new decision: JCS 79/4/D, 26 Aug 42, in same.

252 "Regulating and coordinating": JCS 79/4/D, 26 Aug 42, in same; Smith to Strong et al., 23 Aug 42, with attachment, in JSC folder Torch. Strong and Dyer designated: Marshall to Secretary JCS, 26 Aug 42; Cominch to Dyer, 26 Aug 42, both in NARA RG 218 Box 211 Folder CCS 334 (8-4-42) Joint Security Control (Sec. 1); Davis 2:9041 and n. 57. Leahy wrote: Davis 2:9041–42 and n. 58; letters in NARA RG 218 Box 211 Folder CCS 334 (8-4-42) Joint Security Control (Sec. 1). "Peter": Bevan to

Cook, C/O/43/264, 29 Apr 43, in JSC folder 370.2 Operations and Results. Strong notifies: Strong to Cook, 3 Sep 42, in NARA RG 218 Box 211 Folder CCS 334 (8-4-42) Joint Security Control (Sec. 1). Pierce designated: Pierce to JSC, 26 Dec 42, in JSC folder 370.2 Flounders; Strong to CG Army Ground Forces *et al.*, 24 Aug 42, in JSC folder Torch. "A most intelligent": Wheatley 95; see Wingate 98.

252 "Attend the portions": Handy to Strong, Subject: British and U.S. Deception Policy, 10 Oct 42, Marshall Library, Reel 374 Item 5606. "Cover Officer": Notes on Conference—20 Oct 42, Goldbranson papers. Goldbranson: Goldbranson papers, *passim*.

254 "Both competent": Wheatley 95. "A pretty solid": Author's interview with Arne Ekstrom, March 1993. "He didn't take long": Author's telephone interview with Desmond Bristow, 28 Jan 94.

254 AFHQ summer 1942: Cline 376. Limey S.O.B.: Author's interviews with Eugene Sweeney. U.S. versus British committees: Davis 2:10036–40. "The whistle": Eldredge, Stratagem Conference 60; Eldredge autobiography, ch. 14, "The Battle of the Bulge, Christmas, 1944," p. 2, Eldredge papers.

255 "They treated us": Stratagem Conference 115.

255 Council Bluffs: Author's telephone interview with Rex Hamer, 19 Jun 97. British proposal: CCS 98, 20 Jul 42, in NARA RG 218 Box 362 Folder CCS 385 (4-8-43) (2) Cover and Deception Planning (Sec. 1). "Decided objections"; referred back: Minutes JCS 26th Meeting, 28 Jul 42, in same.

256 BBC; "an interesting": Baumer, But General, ch. 2, Baumer papers. Ornithologist; casual uniform: Wheatley 70–71. Clarke visit to USA: II/104; Clarke diary, 20 Sep 42-8 Oct 42, Clarke papers.

257 Clarke memorandum: Memorandum, 29 Sep 42, in JSC folder 370.2 Sweater. Attendees at gathering: Wheatley 95–96; Wingate 99.

257 "For days": Wheatley 96. October 15: Clarke diary 15 Oct 42, Clarke papers; Wheatley 98. Wheatley oversleeps: Wheatley 96–97. Division of responsibility: Howard 71. Clarke in London: Clarke diary 9 Oct 42–1 Nov 42, Clarke papers.

258 Meeting with Goldbranson: Notes on Conference—20 Oct 42, Goldbranson papers. Clarke and Fleming to Egypt: Clarke diary, 1 Nov 42–3 Nov 42, Clarke papers.

259 "It must be assumed": Marshall to USFOR London (Personal Strong to Eisenhower) 19 Aug 42, in JSC folder Torch. Cooke: Cooke to King, 21 Aug 42, in NARA RG 218 Box 211 Folder CCS 334 (8-4-42) Joint Security Control (Sec. 1). Codename: Plan for Special Operation, 14 Sep 42; Appendix Afirm, 10 Sep 42, both in NARA RG 218 Box 364 Folder CCS Plans for Strategic Deception. "The target date": Bryant 415; Kahn, Spies 471–72. Churchill memo: Bryant 415. Hedda Hopper: Clipping, "Hollywood Does Its Bit," 2 Oct 42, in JSC Ring Binder "Task Force Security." Officer arrested: Fitzpatrick to Bissell, 2 Oct 42, in same. "All equipped": Black to Chief of Staff, Task Force "A," 16 Oct 42, in same. Patton reception: Conduct of Officers, in same.

260 "The American staff": Resume of Torch and Associated Plans, OP-30C-OUT-214, p. 8, in JSC folder Conscious. Carrier: Carpenter to ACS G-2, 27 Oct 42, in JSC Ring Binder "Task Force Security." Wife in code: Conduct of Officers in same. Sweater: Resume of Torch and Associated Plans, OP-30C-OUT-214, p. 8, in JSC folder Conscious. Stab, Padlock: Appendix "A," 14 Sep 42, in NARA RG 218 Box 693 Folder CCS 385 Pacific Theater (4-1-43) Deceptive Policy in the Pacific Theater (Sec. 2). Quickfire, Hotstuff: "Deception Operations for Torch," 25 Jan 46, in JSC folder Conscious. Australia vetoed: General Report, "Torch" Cover and Deception Plans, COS(42)416(O) 26 Nov 42, p. 3, PRO CAB 81/76/18.

260 Sweater implementation: Resume of Torch and Associated Plans, OP-30C-OUT-214, pp. 8–13, in JSC folder Conscious. Haiti: Leahy to Hull, 29 Sep 42; Hull to American Legation, Port-au-Prince, 29 Sep 42; White to Secretary of State, No. 384, 30 Sep 42; Cominch to Grannis, 2 Oct 42, all in JSC folder 370.2 Sweater.

260 QUICKFIRE implementation: Joint Operation Plan No. 1, 21 Oct 42; CCS [Draft] Combined Directive for Preparation of War Plan "Quickfire," 19 Sep 42, both in JSC folder 370.2 Sweater. Actually considered: Matloff 249, 279, 290. Dodecanese: Eisenhower to AGWAR, No. 3187, 6 Oct 42, in JSC folder 370.2 Sweater; Resume of TORCH and Associated Plans, OP-30C-OUT-214, p. 17, in JSC folder Conscious. British channels: Howard 59.

261 HOTSTUFF implementation: Radio Deception Plan for Operation "Torch"; "Operation 'Torch' W/T Security Measures," both in JSC folder 13-12/6 Communications Deception Reports: Europe, Africa; Resume of TORCH and Associated Plans, OP-30C-OUT-214; Map "Task Force 34 Sailed 23 Oct 1942," both in JSC folder Conscious.

262 Dakar suggested: Howard 60. Risky: General Report, "Torch" Cover and Deception Plans, COS (42)416(O) 26 Nov 42, pp. 2–3, PRO CAB 81/76/17-18. German propaganda: Wingate 107–108; Cruickshank, Deception 42–43. Diplomatic speculation: Howard 60. Maintained track: Resume of TORCH and Associated Plans, OP-30C-OUT-214, p. 14 note*, in JSC folder Conscious. OSS: Sondern & Coster.

262 OWI attempted deception: Bowes-Lyon to Lockhart, 10 Aug 42; Marshall to Smith, 13 Aug 42, both in NARA RG 218 Box 362 Folder CCS 385 (4-8-43) (2) Cover and Deception Planning (Sec. 1). GIBRALTAR COVER PLAN: Wingate 104. Strangeways: Wheatley 18; Young & Stamp II/102.

263 "A somewhat dicey": Young & Stamp 35. KENNECOTT and TOWNSMAN: II/103–104; Wingate 107.

264 SOLO I: Wingate 103; Operation "Torch" Wireless Security Measures, Dec 1942, pp. 10–12, in JSC folder Operation "Torch" W/T Security Measures; Cruickshank, Deception 41–42. Lascar seamen: Dyer to Elmer Davis, 14 Sep 42, in JSC folder 370.2 Flounders; Howard 59. PASSOVER, STEPPINGSTONE: Cruickshank 35–36; PRO CAB 121/105/493. OVERTHROW: Wingate 100–101.

265 CAVENDISH: Wingate 100; Cruickshank, Deception 37–38. AFLAME: COS(47) (416) (O), "Torch" Deception and Cover Plans, 26 Nov. 42, Annex p. 2, PRO CAB 81/76/17; COS (47) (416) (O), "Torch" Deception and Cover Plans, 26 Nov. 42, Appendix I, p. 8, PRO CAB 81/76/23. SOLO II: Howard 58; Cruickshank, Deception 44.

265 KENNECOTT: Wingate 108–109.

266 TOWNSMAN: II/105–106. "The American"; "The American raids": II/104–105. PENDER I: Wingate 105. PENDER II: Wingate 106.

267 "On the aerodrome": II/106. Dummy drop: II/105. Sighting convoy: Wireless Security Measures, Dec 1942, p. 8, in JSC folder Operation "Torch" W/T Security Measures. Axis swamped: Howard 61–62; Kahn, Spies 470–78.

267 "Pushing at an open door": Howard 59. "Zone of destiny": Kahn, Spies 466. "The constant threat": Howard 59. Orkneys: Wingate 103. Narvik; full alert: Howard 59. January dangerous: Wingate 103–104. None withdrawn: Wingate 103. Rundstedt's sitrep: Wingate 101. "Feeling the way"; "dress rehearsal"; "every possibility": Wingate 101, 380; Howard 57–58.

268 "His opinion": Wingate 381; Howard 58. "On the basis": Howard 57. "Enemy attack possible"; "are more occupied": Wingate 381. Canterbury raids: Wingate 102. "There will be no": Wingate 382. Weather impossible: Wingate 102.

269 Dispatch from Stockholm: Wingate 105–106. Rome radio: Wingate 109. Never spotted: Kahn, Spies 477. Axis reconnaissance: Wingate 109. Madrid attaché: Howard 62. Embassy reported: Kahn, Spies 478. "Sicily, Sardinia": Kahn, Spies 477.

269 "Agents' reports": Wingate 383. Submarines, bombers: Wingate 109. One torpedoing: II/106. "When the African": Wingate 110. Caught napping: Wingate 109; II/107. Hitler to Munich: Kahn, Spies 478.

270 First token units: II/107. Agents used: Masterman 109–110; Howard 62. Timetable: Howard 61.

271 GARBO starred; Gerbers; Nov. 19: Pujol 98 (misstated as "59 years."); Liverpool *Daily Post,* 24 Nov 42. Clipping to Lisbon: Pujol 98.

271 "Your last reports": Howard 63. Formal action: Wingate 110. "So the next time"; "literally a day and night": Bevan to his mother, 22 Nov 42, in Bevan papers. Enlarge staff: Torch Cover and Deception Plans, Recommendations in Light of Experience Gained, LCS(42) (8), 15 Dec 42, pp. 2, Appendix, PRO CAB 81/76/4, 6. Operations and intelligence branches: Torch Cover and Deception Plans, Lessons Learnt, LCS(47) (7), 7 Dec 42, p. 6, PRO CAB 81/76/12.

272 "Must be undertaken": Torch Cover and Deception Plans, Lessons Learnt, LCS(47) (7), 7 Dec 42, p. 3, PRO CAB 81/76/9. Improved rumor-spreading: Torch Cover and Deception Plans, Lessons Learnt, LCS(47) (7), 7 Dec 42, PRO CAB 81/76/7–14. "Duplication and crossing": Torch Cover and Deception Plans, Lessons Learnt, LCS(47) (7), 7 Dec 42, p. 5, PRO CAB 81/76/11. "Apparently one of the major": Hollis to Ismay, 7 Dec 42, PRO CAB 121/105/474. Trip approved: Minutes COS(42) 339th Meeting, 8 Dec 42, PRO CAB 121/105/473. Proposed overall policy: L.C.S.(42)(P)(1), Second Draft, 6 Dec 42; Strong and Dyer to JCS through Joint Staff Planners, 31 Dec 42, both in JSC Loose Acco Binder "Deception Policy 1943–1944 Europe."

273 December 26: Goldbranson to Adjutant General 26 Dec 42, Goldbranson papers. Authority to veto: CS 79/5, 24 Dec 42. Admiral Train: Cominch to Vice CNO, 2 Dec 42, in NARA RG 218 Box 211 Folder CCS 334 (8-4-42) Joint Security Control (Sec. 1); Davis 2:12001–12002. "Plodding, goodnatured": Baumer diary, p. 56, 17 June 43.

273 Attaché in Bangkok: John Slonaker, USAMHI, to author. Internment: Baumer diary, p. 56, 17 Jun 43. December 31 meeting: Minutes of a Meeting Held at U.S. War Department . . . 31 Dec 1942, in JSC folder Great Britain. Abbott: Goldbranson to Smith, undated, in JSC folder 370.2 Wadham. Nondeception work: Joint Security Control Organization Chart, 10 Mar 44, in NARA RG 218 Box 211 Folder CCS 334 JSC (8-4-42) Sec. 2.

275 "Emphasized the fact": Minutes of a Meeting Held at U.S. War Department . . . 31 Dec 1942, in JSC folder Great Britain. Policy go forward: Memorandum for Record, undated [31 Dec 42], in JSC folder Torch. PAT J: Hoover 120. "Established American channels"; "In other words": Pierce Memorandum for Record, Subject: South American Channels, 2 Jan 43, in JSC folder South America. "For instance": Minutes of a Meeting—General Strong & Colonel Bevan, 5th January 1943, in JSC folder Great Britain. Gone forward: Strong and Dyer to JCS through Joint Staff Planners, 31 Dec 42, attachment p. 4, in JSC Loose Acco Binder "Deception Policy 1943–1944 Europe." Navy downplay: TWX Cominch to JSC (Dyer to Pierce) 1 Jan 43, in same. "That the policy"; "deferred": Minutes, JPS 53d Meeting, 6 Jan 43, in NARA RG 218 Box 363 Folder CCS 385 (12-31-42) Deception Policy.

276 Two bogus units: National War College Lecture, Lieutenant Colonel John R. Deane, Jr., 1 Mar 50, in NOA 546, Folder Tactical Deception—Ruses, pp. 4–5. Goldbranson with OPD: Minutes of a Meeting Held at U.S. War Department . . . 31 Dec 1942, in JSC folder Great Britain. Table of units: Table dated 22 Jun 1943, row 3 column D item (v), in JSC folder European Theater 000; Goldbranson to Baumer, 7 May 43, paragraph 5, in JSC folder 000—To 312.1(1) China-Burma-India Theater.

277 Final White House meeting: Cline 216. Baumer early life: Baumer, But General, ch. 3, Baumer papers.

278 "I was at West Point": Baumer, But General, ch. 4, Baumer papers.

279 "His face fell a foot": Baumer diary, p. 57, 10 July 43. "Get me every son of a bitch": Baumer, But General, ch. 1, Baumer papers. Patton out October 23: D'Este, Patton 425. Baumer and Patton: Baumer diary, p. 36, 28 Jan 43; Baumer, But General, ch. 1, Baumer papers. "Were walking over": Baumer, But General, ch. 1, Baumer papers.

279   "Seven-day-a-week": Baumer, But General, ch. 1, Baumer papers; Baumer diary p. 42, 15 Mar 43. "Always kind"; "a nice gentleman"; "what it took"; nature of work: Baumer diary p. 66, 27 Aug 43; Baumer, But General, chs. 1–2, Baumer papers. "A real blitz": Baumer diary p. 36, 28 Jan 43; Baumer, But General, ch. 2, Baumer papers.

280   "Immediately after": Baumer diary, p. 37, 28 Jan 43. Kehm early life: Cullum No. 6982; obituary, Assembly, September 1980. "Colloquial-speaking": Baumer, But General, ch. 2, Baumer papers.

281   Jackson's health: Baumer diary p. 59, 10 Aug 43; Baumer, But General, ch. 2, Baumer papers. Smith description: Newman Smith VA records, medical; Military Record and Report of Separation, n.d. (1946), Newman Smith VA file; enlistment form, Newman Smith VA file; Discharge Certificate, 5 Jun 1909, Newman Smith papers. "Miniature Pershing": Author's telephone interview with Will Hill Tankersley, 28 Mar 94.

282   Lumber boomtown: Moore, 521–23; Rogers 301; Newman Smith résumé, Newman Smith papers. Enlistment: Enlistment form, Newman Smith VA file. Service: Military Record; Newman Smith résumé, both in Newman Smith papers. "Excellent"; "honest and faithful": Discharge certificate, 5 Jun 1909, Newman Smith papers. Howard College: Newman Smith résumé, Newman Smith papers; Officer's and Warrant Officer's Qualification Card Copy, WD AGO Form 66-4, n.d. [1946?], Newman Smith personnel file. Read law: Newman Smith résumé, Newman Smith papers. Mildred Beasley: VA Form 21-534; 1916 application for enlistment, both in Newman Smith VA file. Business college: Separation Qualification Record, Newman Smith VA records. Bank: Newman Smith VA file, undated Statement of Applicant [1916?]. Cashier for women's department: Undated clipping, "Captain Newman Smith Weds Miss Rosalind Sayre"; undated clipping, "Major and Mrs. Newman Smith, After Stay in Europe, Are Home for Visit," both in Newman Smith papers. Own age: Lanahan 45.

283   Newspaper work, film scenario: Undated clipping, "Captain Newman Smith Weds Miss Rosalind Sayre"; undated clipping, "Major and Mrs. Newman Smith, After Stay in Europe, Are Home for Visit," both in Newman Smith papers. Captain: Officer's, Warrant Officer's, and Flight Officer's Qualification Record, WD AGO Form 66, 26 Apr 1949, in Newman Smith personnel file. Old Fourth: Crute 5–6. Machine gun company: Undated clipping, "Maj. Newman Smith Back After Varied Service Overseas," in Newman Smith papers. With Pershing: Newman Smith résumé, Newman Smith papers. Divorce final: VA Form 21-534, Newman Smith VA file. To Mineola: Undated clipping, "Maj. Newman Smith Back After Varied Service Overseas," in Newman Smith papers.

284   Married: Certificate of Marriage, Newman Smith papers. Shipped out: Alabama Adjutant General's certificate, 23 Nov 49, in Newman Smith personnel file. Hotel bill: Bill from Hotel McAlpin, Broadway at 34th St., 2 Nov 17, Newman Smith papers. "Capitán": Rosalind to Aunt Nina, 6 Feb 20, in Newman Smith papers. "My sweetheart": E.g., Hotel McAlpin envelope "For My Sweetheart," in Newman Smith papers; diary entry, 29 Nov 42, in Newman Smith papers. "A protected"; "always felt"; "marveling once again": Lanahan 210. Reached France: Diary entries on sheets numbered 77–78, Newman Smith papers. Rainbow Division: Battle Participation sub nom. 42d Division.

284   Battalion command: Newman Smith résumé, Newman Smith papers. "You deserve": MS diary note on page (not necessarily actual date of writing) for May 31, 1918, in Newman Smith papers. Major: Officer's, Warrant Officer's, and Flight Officer's Qualification Record, WD AGO Form 66, 26 Apr 1949, in Newman Smith personnel file. "I have lived centuries": Undated clipping, "Capt. Julian Strassburger Loses Life in France, Says Letter to Local People," in Newman Smith papers. Gassed: Undated clipping, "Maj. Newman Smith Back After Varied Service Overseas," in Newman Smith papers; Newman Smith résumé, Newman Smith papers. Battle stars: Newman Smith

résumé, Newman Smith papers. G-3: Officer's, Warrant Officer's, and Flight Officer's Qualification Record, WD AGO Form 66, 26 Apr 1949, in Newman Smith personnel file. Relief work: Undated clipping, no headline, Newman Smith papers. Hoover aides: Author's telephone interview with George Nash, 23 Dec 96. Old Fourth homecoming; Milford 47. Mustering out: Alabama Adjutant General's certificate, 23 Nov 49; Officer's, Warrant Officer's, and Flight Officer's Qualification Record, WD AGO Form 66, 26 Apr 1949, both in Newman Smith personnel file.

284   In New York: Officer's and Warrant Officer's Qualification Card Copy, WD AGO Form 66-4, n.d. [1946?], Newman Smith personnel file. Constantinople: Draft statement, Newman Smith papers; Author's telephone interview with Will Hill Tankersley, 28 Mar 94. Brussels: *C.I.T. Topics* "Brussels—Headquarters of Commercial Investment Trust Societe Anonyme"; draft statement, both in Newman Smith papers; Smith to VA, 27 Jul 33; Smith to Veterans Bureau, 21 Nov 33; Smith to McCoy, 22 Jan 34 (misdated 1933), all in Newman Smith VA file. Borrowing: Smith to Treasury Department, 29 Oct 24; Smith to Veterans Bureau, 21 Nov 33; Smith to Veterans Bureau, 7 Jul 35; Stirling to Smith, 31 Jul 42, all in Newman Smith VA file; Scott to Zelda, 6 Oct 39, quoted Lanahan 107. Fencing, polo, golf: Officer's and Warrant Officer's Qualification Card Copy, WD AGO Form 66-4, n.d. [1946?]; Officer's, Warrant Officer's, and Flight Officer's Qualification Record, WD AGO Form 66, 26 Apr 1949, both in Newman Smith personnel file; Newman Smith résumé, FIE membership card, both in Newman Smith papers. Barclay de Tolly: Author's telephone interview with Will Hill Tankersley, 28 Mar 94.

285   Zelda and Scott: Bruccoli 87, 131; Milford 62. In Europe: Lanahan, Zelda 122–23. Smiths witnessed: Author's telephone interview with Will Hill Tankersley, 28 Mar 94. Zelda breakdown: Lanahan 41; Milford 160–62. Rosalind blamed: Bruccoli 296; Bruccoli & Duggan 515. "The mad world": Bruccoli & Duggan 236.

286   Adopt: Lanahan 45. "Tearing around Europe": Fitzgerald 397. Borrowed from Fitzgerald: Bruccoli, Letters 413. Back to USA: Smith to Veterans Bureau, 21 Nov 33; Smith to VA, 30 Dec 33; Smith to McCoy, 22 Jan 34 (misdated 1933); Smith to Veterans Bureau, 20 Feb 35, all in Newman Smith VA file; Officer's and Warrant Officer's Qualification Card Copy, WD AGO Form 66-4, n.d. [1946?]; Oath form 7 Aug 35, both in Newman Smith personnel file; Draft statement; undated clipping, evidently from a CIT house publication; Newman Smith résumé, all in Newman Smith papers; Milford 324; Bruccoli & Duggan 515. Last years of Fitzgerald: Graham 67; Bruccoli 481; Bruccoli & Duggan 522–23. Funeral: Bruccoli 490; Milford 350; Mizener 299. "I can easily": Lanahan 45. Reserve commission: Newman Smith résumé, Newman Smith papers; Officer's and Warrant Officer's Qualification Card Copy, WD AGO Form 66-4, n.d. [1946?], Newman Smith personnel file. Eyes and ears: Author's telephone interview with Will Hill Tankersley, 28 Mar 94. "Available immediately": Newman Smith résumé, Newman Smith papers. Took a while: Rosalind to Smith, 29 Nov 42, in Newman Smith papers. To Washington: Diary pages dated 29 and 30 Nov 1942, in Newman Smith papers. Sworn in; operations officer: Officer's and Warrant Officer's Qualification Card Copy, WD AGO Form 66-4, n.d. [1946?], Newman Smith personnel file.

Chapter 8: Hustling the East (I)

289   "At the end": Chapter Headings, The Naulahka, ch. 5, Kipling 537. "It is impossible": Fleming 5.

290   Wavell's command: Lewin 147. "A" Force arrangements: I/56, I/76; Fleming 1. General: I/76.

290   Request for Fleming: Hart-Davis 258; CinC Far East to War Office personal for CIGS,

No. 024682, Fleming papers, folder 5/1 War Miscellaneous Folder. Fleming early life: Hart-Davis 13, 18, 38–63, 70–79, 83–124, 151–184, 199–205.

292 Fleming appearance: Hart-Davis 255; Wheatley 98; Astley 76. "Motor-tires": Hart-Davis 52. Pipe: Wheatley 98. Sense of smell: Hart-Davis 15. Handsome: Hart-Davis 51. "Modern Elizabethan": Hart-Davis 126.

292 "A four-square": Astley 76. Brothers: Hart-Davis 237; Kate Grimond to author, 15 Oct 96.

293 Early war service: Hart-Davis 222–57.

293 Cairo: Hart-Davis 258–63. Arrival India: Hart-Davis 262–64. "Found the long": Hart-Davis 264.

293 Cawthorn: Hart-Davis 268, 278. "A one-horse show": Wheatley 42. Office: Hart-Davis 269. "Have a look round": Hart-Davis 264–65. GOLDLEAF-HERITAGE: Wheatley 48.

294 Planning ERROR: Hart-Davis 266. Bright letter; "veiled and slangy": Hart-Davis 266; Fleming 3. ERROR material: Lewin 197.

295 Carrying out of ERROR: "Everyone seemed"; "had too much"; "a few wretched"; "Most sympathetic"; "Gurkhas doing"; "It flounced"; "bumping over"; "Typewriters, stationery"; "Dear G"; "Any harm": Fleming's contemporary notes, in Fleming papers folder 4/4; Error Notebook and Report on Operation Error, May 6, 1942, in Fleming papers, papers, folder 5/1.

297 "Might help": Lewin 198. 1944 report: Hart-Davis 269; PRO CAB 154/41/5A.

297 "Far above": Fleming 3. "The situation": Fleming 2. "What we want": Fleming 33. Forged Tanaka Memorial: Proposal for a Second Edition of the "Tanaka Memorial," undated document, in Fleming papers, 5/1 War Miscellaneous Folder; Hart-Davis 216.

298 "Local equivalent": "Report on Purple Whales," para. 30(b), in Fleming papers, papers, folder 5/1. "The verbatim medium": Fleming 33. "This method of approach": "Report on Purple Whales," para. 5, Fleming papers, folder 5/1. Initial thought: Howard 205. Wavell decision: "Report on Purple Whales," para. 16, Fleming papers, folder 5/1.

298 Churchill's signature: Fleming to ISSB, 7 May 42, PRO CAB 121/105/537. "Rather grandiose"; "better a tamer": Stanley to Hollis, "Purple Whales," 8 May 42, PRO CAB 121/105/533. "That London"; a bit better: "Report on Purple Whales," paras. 12, 15, Fleming papers, folder 5/1. Communication might cease: "Report on Purple Whales," para. 12, Fleming papers, folder 5/1. Kabul visit: Hart-Davis 270; "Report on Purple Whales," para. 13, Fleming papers, folder 5/1; see Hunter to JSC, 11 Oct 43, in JSC folder 000—To 312.1(1) China-Burma-India Theater. Bruce report: "Report on Purple Whales," para. 14, Fleming papers, folder 5/1.

299 "At odd moments": "Report on Purple Whales," para. 18, Fleming papers, folder 5/1.

299 Text of purported transcript: "Verbatim Record . . . ," Fleming papers folder 5/1. (Summary in Appendix "A" to [Cook] to Strong, 5 Dec 42, in JSC Folder Misc Deception Plans—WWII.)

299 Bruce note: Hartley to Bruce, D.O. No. 1064/8, 17 Jun 42, in Fleming papers, folder 5/1. "Approximate[d] in genre": "Report on Purple Whales," para. 18, Fleming papers, folder 5/1.

300 Stilwell quotes: "Verbatim Record . . . ," pp. 5, 6, 9, Fleming papers, folder 5/1. Air raid on England: "Report on Purple Whales," paras. 19–22, Fleming papers, folder 5/1.

300 To China; passing of document: "Report on Purple Whales," paras. 23–32, Fleming papers, folder 5/1.

300 Payment: Fleming 33, 35; "Report on Purple Whales," handwritten endorsement, Fleming papers, folder 5/1; [Cook] to Strong, 5 Dec 42, in JSC folder Misc Deception Plans—WWII. Cheng K'ai-min: Fleming 9–10, 35. Total of six: Chart, Principal Plans . . . , PRO WO 203/3313.

301   GH: Project No. 1471-A, The Japanese Intelligence System, pp. 102–103, in JSC folder The Japanese Intelligence System. (PRO CAB 154/66 suggests that GH was the INK channel, opened later.) Soviet threat idea: Stanley to Howkins, 14 Jan 42, PRO CAB 119/67/273. Rumor-mill themes: JP (42) 459 (S) (Preliminary Draft), Report by the Joint Planning Staff, 28 Apr 42, PRO CAB 119/67/259–66; Howkins to Directors of Plans, 42/455, Japan—Strategic Deception, 2 May 42, PRO CAB 119/67/257–58. Sketchy plan: JP (42) 477, Strategic Deception, 5 May 42, PRO CAB 119/67/252/55. Enlargement of shop: Hart-Davis 275; Fleming papers, folder 5/1, No. 45257/VI/ G.S.S.D.6, 2 Jun 42.

301   Better accommodations: Hart-Davis 274; Thorne to author, 27 Apr 97. One to two: Fleming 38. Thorne: Hart-Davis 278. Sleep: Arnold, Quiet 97. His assignments; Thorne to author, 17 Jan 97.

302   Bicat: Hart-Davis 111, 281. Ralli: Thorne to author, 12 Jan 97. Wilson: Hart-Davis 281. Johnson: Thorne to author, 27 Apr 97. MacDonald: PRO CAB 119/66/184. "Never really"; Antrim: Thorne to author, 27 Apr 97.

303   Horder: Obituary, Lord Horder, The Times, 7 July 97; Thorne to author, 12 Jan 97. "The most literate": Thorne to author, 27 Apr 97. Clerical: Hart-Davis 279; Thorne to author, 12 Jan 97. DIB: Andrew 277. Short on personnel and equipment: Liddell diary, 15 Jun 43, PRO KV 4/191. DIB officers: Thorne to author, 27 Apr 97. Trevor-Roper: Thorne to author, 12 Jan 97. Marriott visit: Hinsley & Simkins 232. "Quarrels": Liddell diary, 15 Jun 43, PRO KV 4/191. Lecture: Liddell diary, 4 Jun 43, 15 Jun 43, PRO KV 4/191. "Finally succeeded": Liddell diary, 19 Aug 43, PRO KV 4/191. SILVER case officers: Magan 84. "An irresponsible": PRO KV 4/197/5A, p. 2. Fretting: Bevan to Wingate, L.C.S. (45)I/C.42, 14 May 45, PRO CAB 154/37/67A. Thorne day-to-day; DIB and police; MI5 and MI6: Thorne to author, 27 Apr 97.

304   "Extremely comfortable": Arnold, Quiet 94. Comfortable life: Hart-Davis 270; Arnold, Quiet 94; Thorne to author, 17 Nov 96.

305   HICCOUGHS (full summary): Head of D Division to London Controlling Section, 3 Feb 45, Subject: Deceptive Broadcasts, in JSC folder Deception—SE Asia—WWII Historical D. Div—SACSEA—16 Dec 44–15 Mar 45.

305   "By trading": Fleming 37. Breakable cipher: O'Connor to Bissell HMOC/992, May 1, 1944, in JSC folder Great Britain; Fleming 37.

306   "We were often reduced"; effort of faith: Fleming 38. "Only the Japanese": Fleming 38.

306   Goebbels: Irving, Goebbels 333, 340; Hinsley & Simkins 67 note*. London and Cairo: Hart-Davis 275–77; Clarke diary 1–3 Nov 42, Clarke papers. "Vaguely known"; "having as he put it": Muggeridge 179.

307   Bose: Fleming's own draft for his possible history of deception, untitled draft found in Fleming papers, folder 14/1, Deception History. See also entry for Bose in Current History, 1944.

308   Escape to Germany: Bhattacharyya 6–13, 41–42; Liddell diary, 4 Aug 42, 4 Nov 42, PRO KV 4/190.

308   Kischen Chand: Thorne to author, 27 Apr 97. SILVER's appearance, character: Talwar 128; photographs in Talwar; Fleming 44; Thorne to author, 27 Apr 97. Chopped tiger whiskers: Hart-Davis 280. Pushtu; Ramat Khan; could make his way: Fleming 44, 47. Kabul intelligence center: Fleming 50. All under control: Project No. 1471-A, The Japanese Intelligence System, pp. 144–45, in JSC folder The Japanese Intelligence System. German staff: Peter Geissler, Auswärtiges Amt, Bonn, to author, 14 Jun 96. Recruited: Talwar 142–45; Fleming 45; Liddell diary, 4 Nov 42, PRO KV 4/190. Brother: Bhattacharyya 9. "As a kind": Howard 209. Soviets agree to double agent; "They seem to have": Liddell diary, 4 Aug 42, 4 Nov 42, PRO KV 4/190; Hinsley & Simkins 232–33. DIB and SILVER: Fleming 44. Journeys to Kabul: Talwar 186–90, 192–98, 198–216, 217–23; Fleming 45–53. All India Revolutionary committee: Howard 209. January

1943 report: Howard 209. (Thorne believed that this report must have been taken by SILVER on a later trip to Kabul, since he recalled explaining a point in it to SILVER during Fleming's absence in May 1943. Thorne to author, 27 Apr 97. But Howard had access to the contemporary documents. Perhaps the meeting referred to by Thorne took place in connection with a later report in some other absence of Fleming's.) Germans more interested in insurgency and sabotage; wireless link: Howard 209. April 2: Talwar 183.

311 ULTRA and MAGIC: *E.g.*, Appendix A to Part II of "D Division Weekly Progress Report No. 3:7–13 Apr 45," in JSC folder Deception—SE Asia—WWII Historical D. Div—SACSEA—24 Mar 45–4 Jun 45. BRASS: Fleming 25; Howard 207. Tiger hunt: Hart-Davis 277–78. Field-marshal: Lewin 208. Lieutenant-colonel: as of 6 Mar 42, according to Fleming Papers 15/6 Service Record, p. 6; but he was still signing as Major on his PURPLE WHALES report in July.

312 "Doing this ungentlemanly": Hart-Davis 277. Arakan offensive: Hart-Davis 288–89. KINKAJOU-WALLABY: Wavell to COS, 17 Feb 43, 38748/COS, NARA RG 319 Entry 101 Box 3 Folder 68 "Anglo-American Meeting (May–June '43)." "Put across": same, para. 3.

312 Wedemeyer: Wedemeyer 196–97; Arnold, Global 407–409. "This ambitious": Howard 212. Bevan/Wingate reaction, response: "Anakim" Cover Plan, LCS (43) 3 (Revised Final), 26 Mar 43, PRO CAB 119/66/160; Le Mesurier to Wingate, 43/401, 28 Mar 43, PRO CAB 119/66/168–69; COS to C-in-C India, OZ 893, 29 Mar 43, paras. 5(a), 7, NARA RG 319 Entry 101 Box 3 Folder 68 "Anglo-American Meeting (May–June '43)." ANAKIM abandoned: Matloff 142; Hayes 399. RASPUTIN: PRO CAB 119/66/145–50.

313 Wavell proposes conference: Wavell to COS, 17 Feb 43, 38748/COS, para. 1, NARA RG 319 Entry 101 Box 3 Folder 68 "Anglo-American Meeting (May–June '43)." London concurs: COS to C-in-C India, OZ 893, 29 Mar 43, para. 8, in same folder. "Agreed that considerable": Smith to Deane, 14 July 1943, in JSC loose Acco binder Deception Policy Pacific Areas.

314 Air organization: Craven & Cate 4:414, 4:454. FLOUNDERS; "Our representative"; acknowledgement: Cook to Kroner, 29 Aug 1942; Kroner to Cook, 3 Sep 1942, both in NARA RG 218 Box 363 Folder CCS 385 (12-31-42) Deception Policy. October, November, December: Marshall to CGUSAFCBI Branch HQ New Delhi, 24 Oct 42, No. 101, CM-OUT-08245; New Delhi to AGWAR, 15 Nov 42, No. Ammdel 193, CM-IN-6888, both in JSC folder 370.2 Flounders; [Cook] to Strong, 5 Dec 42, in JSC folder Misc Deception Plans—WWII. Strong told them: PRO CAB 154/73. "Countering organization"; no American units: Cook to ISSB, 2 Dec 42, PRO CAB 154/73.

314 Pentagon meeting: Minutes of a Meeting Held at U.S. War Department . . . 31 Dec 1942, in JSC folder Great Britain. American material would be furnished: Blizzard to Wedemeyer, 5 Apr 43, in JSC folder 370.2 Flounders. Hunter: Clippings filed in Biography file, E. O. Hunter, and in E. O. Hunter collection, No. 1342, Items 2147, 2150, 2152, Georgia Historical Society. "Hormones": Author's telephone interview with Malcolm McLean, 15 Apr 93. Bon vivant: Baumer diary p. 56, 17 Jun 43, Baumer papers.

315 Hunter's first weeks: Hunter diary.

316 Fleming in Cairo: Clarke diary, 4–5 May 43, Clarke papers. Marriott: PRO CAB 119/66/184; Thorne to author, 27 Apr 97; Hinsley & Simkins 232. Fleming in England: Hart-Davis 285. *Queen Mary:* Hart-Davis 195. Wavell: Lewin 217–20. Carey, Smith-Hutton: Packard 205; Whitehill 103, 144–48. "A small group": Minutes JCS 66th meeting, 16 March 43, in NARA RG 218 Box 211 Folder CCS 334 Joint Security Control (Sec. 2). Room 2E816: Hunter diary 17 May 43. Executive officers: See Cominch to VCNO, 2 Dec 42, in NARA RG 218 Box 211 Folder CCS 334 (8-4-42) Joint Security Control (Sec. 1). Smith shared: Hunter diary 19 May 43.

318 Dyer to Med: Dyer biographical sheets, Naval Operational Archives. Full-time job: Minutes JCS 65th Meeting, 9 Mar 43, in NARA RG 218 Box 211 Folder CCS 334 (8-4-42) Joint Security Control (Sec. 1). Grosskopf: Morison, Victory 40; Rogers 17, 73; Packard 205; Directive, COMINCH to Grosskopf, 25 Mar 43, in NARA RG 218 Box 211 Folder CCS 334 Joint Security Control (Sec. 2). F-24: Whitehill 157. JSC original charter: JSC 79/4/D, 26 Aug 42, paras. 2a(1), (2,) and (4), 3(a) and (b), in NARA RG 218 Box 211 Folder CCS 334 (8-4-42) Joint Security Control (Sec. 1). QUICKFIRE planning: Deane to Leahy *et al.*, 22 Sep 42, in NARA RG 218 Box 364 Folder CCS Plans for Strategic Deception.

318 Rebuffed as to planning: Minutes, JCS 60th Meeting, 2 Feb 43; Deane to Strong and Train, 2 Feb 43; JPS 128, 9 Feb 43; Minutes JPS 59th Meeting, 10 Feb 43; Minutes JCS 62d Meeting, 16 Feb 43, all in NARA RG 218 Box 363 Folder CCS 385 (2-10-43) Deception Plan for "Boodle" and "Cottage." BUXOM: Morison, Aleutians 38. "In a complete muddle"; "come to what I hope"; "fed up"; "whilst Kehm": Cook to Bevan, LS/39, 4 Feb 43, PRO CAB 154/73/49A. Productive exchange: Strong to Wedemeyer, Subject: Deception Planning and Execution, 5 Jun 43, in NARA RG 165 ABC File ABC 381 (4-8-43) Sec. 1-A—Deception policy, organization, etc.; Baumer, Report on Visit to London and Algiers Planning Sections, 10 May 42, in Hoover Institution, Baumer Papers, Box 10, Folder 2; Maddocks to Secretariat, JCS, 4 Feb 43, Enclosure "B" to Paper No. 17, Committee on JCS. War Planning Agencies, 10 Feb 43, in NARA RG 218 Box 45 Folder CCS 300 (1-8-43) (Sec. 1); Kehm to Wedemeyer, Subject: Revision of JCS 256/1/D, no date; Kehm to Chief, Strategy & Policy Section, Subject: General Strong's Memorandum on Deception Planning and Execution, no date; Train to Wedemeyer, 8 Jun 43, in NARA RG 165 ABC File ABC 381 (4-8-43) Sec. 1-A—Deception policy, organization, etc.; K[ehm] to Colonel Blizzard, undated buck slip attached to Train to Wedemeyer, 8 Jun 43, all in NARA RG 165 ABC File ABC 381 (4-8-43) Sec. 1-A—Deception policy, organization, etc.

319 April 15 decision: Marshall to Secretariat, JCS, 8 Apr 43; Deane to Leahy *et al.*, 14 Apr 43; Minutes JCS 74th Meeting, 13 Apr 43; JCS 256/1/D, Cover and Deception Planning, 15 Apr 43, all in NARA RG 218 Box 362 Folder CCS 385 (4-8-43) (2) Cover and Deception Planning (Sec. 1). Draft 1943 policy withdrawn, replaced: Minutes JPS 59th Meeting, 10 Feb 43, in NARA RG 218 Box 363 Folder CCS 385 (12-31-42) Deception Policy; CCS 184, Deception Policy 1943, 3 Mar 43, in NARA RG 218 Box 363 Folder CCS 385 (2-10-43) Deception Policy (Germany and Italy). JCS direction: Royal to Secretaries, JSP, Subject: Deception Policy 1943, 3 Mar 43, in NARA RG 218 Box 363 Folder CCS 385 (2-10-43) Deception Policy (Germany and Italy).

320 British proposal considered, adopted: McFarland to Grosskopf *et al.*, 5 Mar 43; JPS 139/1, 9 Mar 43; JPS 139/1, 9 Mar 43, para. 10, pp. 8–9; Minutes JPS 65th Meeting, 20 Mar 43; CCS 184/3/D, all in NARA RG 218 Box 363 Folder CCS 385 (2-10-43) Deception Policy (Germany and Italy). Broad policy for Pacific: McFarland to Cooke *et al.*, Subject: Deception Policy, Pacific Theater, 1943, 1 Apr 43; Minutes JPS 68th Meeting, 7 Apr 43; JPS 162, 14 Apr 43; Minutes JPS 70th Meeting, 21 Apr 43; Wedemeyer to Joint Staff Planners, 10 May 43; Cooke to Wedemeyer, 11 May 43; Wedemeyer to Cooke, 14 May 43; Minutes JSP 75th Meeting, 19 May 43, all in NARA RG 218 Box 693 Folder CCS 385 Pacific Theater (4-1-43) Deceptive Policy in the Pacific Theater (Sec. 1).

320 Baumer report; "certain cohesiveness"; "particularly at a loss": Baumer, Report on Visit to London and Algiers Planning Sections, 10 May 42, in Hoover Institution, Baumer Papers, Box 10, Folder 2. Hunter reports: Hunter diary 16 May 43. Ray: Biographical sheet, Rear Admiral Herbert J. Ray, Jr., Naval Historical Center. "A quite capable man"; "I liked": Baumer diary p. 56, 10 Jul 43.

321 "Complete accord": Hunter diary 17 May 43.

321 Meetings: Hunter diary; Baumer diary; Minutes of a Meeting Held at the U.S. War Department, 1430—25 May 1943; Minutes of the Third Meeting, 1430—27 May 1943, both in JSC loose Acco binder Deception Policy Pacific Areas; JP (42) 545 (S) (2d Preliminary Draft), Strategic Deception—Machinery, 29 May 42, PRO CAB 119/67/224–30; Machinery for Deception of Japan: Views of the British Representatives at Washington Cover Plan Conference, 25 May 43; Memorandum for the Joint U.S. Staff Planners, Subject: Machinery for the Deception of Japan, 28 May 43; Enclosure A to Memorandum For the Joint Chiefs of Staff Through the Joint Staff Planners, Subject: Final Report of the Washington Anglo-American Cover and Deception Conference, 2 Jun 43, all in NARA RG 319 Entry 101 Box 3 Folder 68, Strategic Deception Conference, Machinery for Deception Against Japan.

322 Recycled paper: JPS 162/1, 28 May 43, in NARA RG 218 Box 693 Folder CCS 385 Pacific Theater (4-1-43) Deceptive Policy in the Pacific Theater (Sec. 1); Enclosure B to Memorandum For the Joint Chiefs of Staff Through the Joint Staff Planners, Subject: Final Report of the Washington Anglo-American Cover and Deception Conference, 2 Jun 43, in NARA RG 319 Entry 101 Box 3 Folder 68, Strategic Deception Conference, Machinery for Deception Against Japan.

322 Recommended general plan: General Deception Plan for the War Against Japan, Enclosure C to Memorandum For the Joint Chiefs of Staff Through the Joint Staff Planners, Subject: Final Report of the Washington Anglo-American Cover and Deception Conference, 2 Jun 43, in NARA RG 319 Entry 101 Box 3 Folder 68, Strategic Deception Conference, Machinery for Deception Against Japan. "An attractive pair": Astley 96; Hart-Davis 285. Kehm and Baumer warning: Notes on JPS 77th Meeting, 2 Jun 43, in Hoover Institution, Baumer Papers, Box 2.

323 "The British lost": Jackson to Strong, 4 Jun 43, in JSC loose Acco binder Deception Policy Pacific Areas. Smith replaces Jackson: McFarland to Jackson and Smith, 12 Jun 43, in NARA RG 218 Box 693 Folder CCS 385 Pacific Theater (4-1-43) Deceptive Policy in the Pacific Theater (Sec. 1). Wrestling match: Hayes 418–427. "In June": Baumer diary 59. "The Navy wouldn't": Stratagem Conference 41. Back and forth: Minutes JSP 78th Meeting, 9 Jun 43; McFarland to Grosskopf et al., with enclosure, 11 Jun 43; McFarland to Secretariat, JWPC, 11 Jun 43; Barber to Secretary JCS, 15 Jun 43; Miller and Barber to Secretary, JSP, 17 Jun 43; JPS 162/2 and Enclosure B, 19 Jun 43; Minutes JSP 82d Meeting, 23 Jun 43; JPS 162/3, 9 Jul 43; all in NARA RG 218 Box 693 Folder CCS 385 Pacific Theater (4-1-43) Deceptive Policy in the Pacific Theater (Sec. 1).

323 Deane's smooth management: Redman to Deane, 5 Jul 43; Smith to Deane, 14 Jul 43; Minutes JPS 85th Meeting, 14 Jul 43; McFarland to Secretariat, JCS, 16 Jul 43; all in NARA RG 218 Box 693 Folder CCS 385 Pacific Theater (4-1-43) Deceptive Policy in the Pacific Theater (Sec. 1). "That the work": Supplementary Minutes CCS 103d Meeting, 23 Jul 43; CCS 284, 20 Jul 43, both in NARA RG 218 Box 693 Folder CCS 385 Pacific Theater (4-1-43) Deceptive Policy in the Pacific Theater (Sec. 1). Review and minor changes: J.S.M. Washington to War Cabinet Offices, JSM 1091, 21 Jul 43, PRO CAB 121/105/286–88; COS (43) 419 (O), Deception Measures Against Japan, PRO CAB 121/119/283–84; Minutes COS (43) 176th Meeting (O), 30 Jul 43, PRO CAB 121/105/282; COS to Joint Staff Mission, 30 Jul 43; CCS 284/1, 30 Jul 43; CCS 284/2, 5 Aug 43; Minutes, JCS 100th Meeting, 6 Aug 43, all in NARA RG 218 Box 693 Folder CCS 385 Pacific Theater (4-1-43) Deceptive Policy in the Pacific Theater (Sec. 1). Adopted: Supplementary Minutes, CCS 105th Meeting, 6 Aug 43; CCS 284/3/D, 6 Aug 43, both in NARA RG 218 Box 693 Folder CCS 385 Pacific Theater (4-1-43) Deceptive Policy in the Pacific Theater (Sec. 1). "I have the impression"; "a short account": Minutes COS (43) 176th Meeting (O), 30 Jul 43, PRO CAB 121/105/282.

324 "General overall"; "provide for the implementation"; coordinating; indicate parts; direct communication; "provide for continuity": CCS 284/3/D, Deception Measures Against Japan, 6 Aug 43, paras. 1*a*(4), 1*a*(5), 1*b*(2) (a) and (b), 1*c*(1), (3), (5), all in NARA RG 218 Box 693 Folder CCS 385 Pacific Theater (4-1-43) Deceptive Policy in the Pacific Theater (Sec. 1)). With the understanding: Royal to Secretary Joint Security Control *et al.*, Subject: Deception Measures Against Japan, 7 Aug 43, in NARA RG 218 Box 693 Folder CCS 385 Pacific Theater (4-1-43) Deceptive Policy in the Pacific Theater (Sec. 1). JCS 234/3/D: JCS 234/3/D, 27 Apr 44, in NARA RG 218 Box 211 Folder CCS 334 Joint Security Control (Sec. 2).

325 "Prepare an overall": Deane to JSP, 11 Aug 43, in NARA RG 218 Box 693 Folder CCS 385 Pacific Theater (4-1-43) Deceptive Policy in the Pacific Theater (Sec. 1). Navy sidestep: Minutes JCS 106th Meeting, 17 Aug 43; Deane to JSP, 17 Aug 43; McFarland to Secretariat, JCS, 18 Aug 43; Minutes JCS 107th Meeting, 18 Aug 43; Minutes CCS 111th Meeting, 18 Aug 43, all in NARA RG 218 Box 693 Folder CCS 385 Pacific Theater (4-1-43) Deceptive Policy in the Pacific Theater (Sec. 1). Settled: Hayes 428–32. "Deception came up again": Baumer diary 65. O'Connor replaces: Wingate 134. "Tim": O'Connor to Bevan, 16 Jan 45, in Bevan papers. Financial background: Stratagem Conference 37. Petavel's assistant: C/O/(43)/358, Bevan to D. M. O., Establishment of London Controlling Section, 5 Sep 43, PRO CAB 121/105/252.

325 "General . . . The basic": JCS 498, 13 Sep 43, in NARA RG 218 Box 693 Folder CCS 385 Pacific Theater (4-1-43) Deceptive Policy in the Pacific Theater (Sec. 1). Appendices: Same, Apps. A and B. "In view of the fact": Supplementary Minutes JCS 104th Meeting, 14 Sep 43, in NARA RG 218 Box 693 Folder CCS 385 Pacific Theater (4-1-43) Deceptive Policy in the Pacific Theater (Sec. 1). "Would not carry": Supplementary Minutes CCS 119th Meeting, 17 Sep 43, in NARA RG 218 Box 693 Folder CCS 385 Pacific Theater (4-1-43) Deceptive Policy in the Pacific Theater (Sec. 1). Approved: CCS 284/5/D, 17 Sep 43, in NARA RG 218 Box 693 Folder CCS 385 Pacific Theater (4-1-43) Deceptive Policy in the Pacific Theater (Sec. 1). "An unusual"; "On this occasion": Bevan to Hollis, C/O/385, 28 Sep 43, PRO CAB 121/105/246.

Chapter 9: The Soft Underbelly

328 "Not only the most": III/183. WAREHOUSE: II/125. CANUTE, GARGOYLE, JESSICA: II/12; Bevan to Howkins with attachment, C/O/240, 5 Dec 42, PRO CAB 119/66/242–46. Bevan directed: L. C. H. to Bevan, Deception Plans for 1943, 29 Jan 43, PRO CAB 121/105/433. Approved 4 Feb: Minutes, COS 16th Meeting (O), COS(43)44(O), 4 Feb 43, PRO CAB 119/67/52–53. "It is now in our interest": Cavendish-Bentinck to Bevan, D/C/8, 2 Feb 43, PRO CAB 119/67/71–72. "And if the Russians": Notation on W. A. H. to Buzzard et al., 43/171, PRO CAB 119/67/65.

329 General policy for 1943: CCS 184/3, Deception Policy 1943, 3 Apr 43, in NARA RG 218 Box 363 Folder CCS 385 (2-10-43) Deception Policy (Germany and Italy). Its adoption: CCS 184, Deception Policy 1943, 3 Mar 43; Royal to Secretaries, JSP, Subject: Deception Policy 1943, 3 Mar 43; McFarland to Grosskopf *et al.*, 5 Mar 43; JPS 139/1, Deception Policy 1943, Report by the Joint Staff Planners, 9 Mar 43; CCS 184/1, Deception Policy 1943, Memorandum by the United States Chiefs of Staff, 22 Mar 43; Notes on CCS 78th Meeting, 2 Apr 43; CCS 184/2, Deception Policy 1943, Memorandum From the Representatives of the British Chiefs of Staff, 1 Apr 43; Minutes CCS 79th Meeting, 2 Apr 43, all in NARA RG 218 Box 363 Folder CCS 385 (2-10-43) Deception Policy (Germany and Italy); Minutes COS 63d (O) Meeting, COS(43) (165(O), 1 Apr 43, PRO CAB 119/67/37.

330 WITHSTAND: III/26–28; Howkins to Buzzard *et al.* with attachment, 43/269, 26 Feb 43, PRO CAB 119/66/53–60; II/124–25, III/28–29. "At the present": II/124–25.

331 Divisions; III/28. WAREHOUSE (1943): III/58. Goldbranson: II/126. "We all did": Goldbranson, Summary of Conversation With Colonel Dudley W. Clarke, Commander of "A" Force, Monday, 28 June 1943, Goldbranson papers, p. 1.

331 Operators etc.: II/132.

332 Hamer: Hamer to author, 2 Jun 97. Expansion of premises: II/134. Inexperience: II/126. "Virtually no": III/30; Browne 59, citing unidentified communication from London to Alexander, 1 Feb 43, and Alexander to Nye, 2 Feb 43, both in PRO AIR 20/4535. Mure: II/117–18; Mure to Thynne, 9 Apr 77, in Mure papers, Box 2; Hamer to author, 2 Jun 97.

333 Strangeways: Young & Stamp 33–35; Obituary, *The Times,* 5 Aug 98; Collier 230; author's telephone interviews with Strangeways, Philip Curtis, Sir Edgar Williams, Christopher Harmer. "If you can't": Young & Stamp 36. "He was a small": Wheatley 85. "A nice man": Author's telephone interview with Sir Edgar Williams, 24 Oct 92. Wild loathed: Marginal note by Mure on Hamer to Mure, 21 Feb 81, in Mure papers, Box 2. "Thought Noel": Author's telephone interview with Philip Curtis, 21 Oct 92. "Pain in the neck": Author's telephone interview with David Strangeways, 26 Aug 92. "Both these": Fairbanks to Thurber, Subject: Deception, 6 Feb 45, p. 2, in NOA 532, Folder Douglas Fairbanks.

335 Temple, Fillingham, setting up shop: II/30–31. LOCHIEL: II/33–35. WINDSCREEN: II/128–31. HENGEIST: III/35–36; PRO WO 204/1561/379–80.

335 Final assault: Howe 598–99. COWPER, COWPER II: III/36–38. "S" Force: III/38–39. "This is a line": III/40. Planning for Sicily: III/39.

336 Dunn, Rushton: Author's interview with Eugene Sweeney, December 1992–January 1993. Lightweight: Stratagem Conference 142. Goldbranson in Cairo: Goldbranson, Summary of Conversation With Colonel Dudley W. Clarke, Commander of "A" Force, Monday, 28 June, 1943, Goldbranson papers, pp. 1–2; III/54. Got on with Wild: Wild to Mrs. Goldbranson, 20 Aug 57, in Goldbranson papers.

337 New organization: III/49–53. Thomas: Author's interview with Arne Ekstrom, March 1993; Hamer to Mure, 21 Feb 81, in Mure papers, Box 2; author's telephone interview with Desmond Bristow, 28 Jan 94. Mockler-Ferryman replaced: Howe 489.

337 Goldbranson while waiting: Goldbranson, Summary of Conversation With Colonel Dudley W. Clarke, Commander of "A" Force, Monday, 28 June, 1943, Goldbranson papers, p. 2. Carl Jr.: C. E. Goldbranson, Jr., to author, n.d. (c. June 1994); Goldbranson to Leafy, 7 Mar 43, Goldbranson papers. St. Georges: Author's telephone interview with Lady Maynard (Daphne Llewellyn), 10 Apr 96; Picture in History of AFHQ, in Eisenhower Library, Bedell Smith Papers, Box 20, Folder History of AFHQ Section 2; Howe 405. "A tremendous": Howe 405; Bristow 105. "What particularly": Muggeridge 192.

338 Nothing to do: Baumer diary p. 52, 18 May 43. Picked up: Muggeridge 192. Cercle Interallié, "in the end": Author's interview with Arne Ekstrom, March 1993. Clarke, Bevan to Algiers: III/52–53; Clarke diary 15 Mar 43, Clarke papers. One room, garage: III/53. First draft: III/68.

339 No independent position; "He still doesn't like it": Rooks to Hull, 3 Nov 44, in NARA RG 165 ABC File ABC 381 (4-8-43) Sec. 1-A—Deception policy, organization, etc. G-2, G-3: Rooks to Hull, 3 Nov 44, in NARA RG 165 ABC File ABC 381 (4-8-43) Sec. 1-A—Deception policy, organization, etc.; Clarke to Whiteley, 19 Jul 43, cited Browne 48; Report on the Activities of No. 2 Tac H.Q. A Force (Sixth Army Group) Part IV, Conclusions, p. 10, in NARA RG 319 Entry 101 Box 4 Folder 82; Rooks to Clarke, 3 Aug 43 (G-3 AFHQ file: 74/1), quoted in History of AFHQ Part Two, Section 2, p. 297, in Eisenhower Library, Bedell Smith Papers, Box 20, Folder History of AFHQ Part Two, Section 2. New quarters: III/53; Author's interview with Eugene J. Sweeney, December 1992–January 1993; Author's telephone interview with Lady Maynard (Daphne Llewellyn), 10 Apr 96.

339 Moved in: Clarke diary 28 Mar 43, Clarke papers. Clarke's activity, back to Cairo: Clarke diary 22–24 Mar 43, 11 Apr 43, 12–16 Apr 43, Clarke papers.
340 Goldbranson: Goldbranson, Summary of Conversation With Colonel Dudley W. Clarke, Commander of "A" Force, Monday, 28 June, 1943, Goldbranson papers.
340 Made fun: Author's interview with Arne Ekstrom, 4 Jan 93.
341 Contretemps with OSS, Strong: Clarke to Smith, ADV. "A"F/23/1, 7 Apr 43; Smith to AGWAR for Donovan, No. 2515, 8 Apr 43; Strong to Smith, No. 5648, 9 Apr 43; Smith to Strong, No. 3047, 10 Apr 43; Smith to Donovan, unnumbered, 10 Apr 43 Strong to Smith, No. 5776, 12 Apr 43; Notation on Strong to Smith, No. 5776, 12 Apr 43, all in Eisenhower Library, Bedell Smith Papers, Box 13, Folder Eyes Only Cables (incoming and outgoing) (February 15–April 30, 1943). OSS: Naftali, X-2 248–78; Naftali, De Gaulle's 387–401. "These people": Author's interview with Arne Ekstrom, March 1993.
341 Bristow transfers: Bristow 46, 102, 149. Meets with Clarke: Clarke diary 20 Mar 43, Clarke papers. Ekstrom in airplane: Author's interview with Arne Ekstrom, 4 Jan 93.
342 Ekstrom early life: Author's interviews with Arne Ekstrom, 4 Jan 93 and March 1993; Obituary, Parmenia Migel Ekstrom, *New York Times,* 16 Nov 89; *New York Times,* 31 Dec 33, p. 24, col. 1; Separation Qualification Record; Military Record and Report of Separation Certificate of Service, Arne H. Ekstrom, copies in author's possession. "We had a passionate": Author's interview with Arne Ekstrom, 4 Jan 93.
343 Meeting with Clarke: Clarke diary, 9 Apr 43, 10 Apr 43, Clarke papers. Initial time with "A" Force: Author's interview with Arne Ekstrom, March 1993. Oran station: II/120. 41 Committee: Bristow 122. "Ekstrom Team": Progress Report of Plan "Barclay" as of Midnight June 6th 1943. p. 1, in NARA RG 319 Entry 101 Box 7 Folder 3 "Barclay." " 'A' Force was to reap": III/120.
344 No fresh water: Author's interview with Arne Ekstrom, 4 Jan 93. "Lives up": Baumer diary 52. Barclay: Author's interview with Arne Ekstrom, March 1993. Called "Bobby Lloyd" in Bristow 120–23. Oran billet, Marie: Author's interviews with Arne Ekstrom, 4 Jan and March 1993. "If I had to get": Author's interview with Arne Ekstrom, March 1993.
345 Ekstrom to Algiers; mess: Same.
345 Fairbanks earlier life, Navy career: Fairbanks 59–83, 95–103, 130–43, 156. Knew Mountbatten: Ziegler 71. With Mountbatten, training: Fairbanks 145–48, 159–62.
347 "A shy and soft-spoken": Fairbanks 170. Hewitt: Glenn, *The BJs,* p. 9, in NOA 532, Folder Douglas Fairbanks; Morison, Sicily 14–15. Readiness Section: Whitehill 119–20. Sponsoring study: Sonic Deception: The Reproduction, Transmission, and Perception of Deceptive Sounds, OSRD Report No. 4094, p. 2, in JSC folder 26—Sonic Deception.
347 Burris-Meyer: Author's telephone interviews with Anita Burris-Meyer, 4 Apr 94 and 19 Feb 95; Burris-Meyer papers. "An amusing": Eldredge to Dwyer, 17 Aug 89, Eldredge papers. Terrifying sounds, beer bottles: Dwyer 16.
348 Sonic bomb: Burwell to Coordinator of Research, Subject: Meeting of Section C-5 of the National Defense Research Committee on July 9, 1942, 11 Jul 42, pp. 2–4, in NOA 531, Folder Deception Methods Techniques and Devices. Battle of Sandy Hook: Sonic Deception: The Reproduction, Transmission, and Perception of Deceptive Sounds, OSRD Report No. 4094, p. 3, in JSC folder 26—Sonic Deception.
349 Origin of name: Dwyer 16. HEATER: Sonic Deception: The Reproduction, Transmission, and Perception of Deceptive Sounds, OSRD Report No. 4094, p. 3, in JSC folder 26—Sonic Deception. Burris-Meyer commissioned: Dwyer 17–18. Dummy paratrooper: Fairbanks to McDowell, Subject: Project #191, Tests of, 11 Mar 43; CO, VC-22 to Fairbanks, Subject: Dummy Paratroop Experiments, Report on, 10 Mar 43;

Clendenen, Memorandum, 8 Mar 43; Fairbanks, Memorandum, 11 Mar 43; all in NOA 531, Folder Deception Methods Techniques and Devices. Dummy landing craft: Fairbanks to McDowell, Subject: Project #192, Tests of, 16 Mar 43, in NOA 531, Folder Deception Methods Techniques and Devices." Other projects: Fairbanks to McDowell, Subject: Special Operations, New List of Projects for, 12 Mar 43, in NOA 531, Folder Deception Methods Techniques and Devices.

350   Trials on Lake Bizerte: Dwyer 27. Fairbanks follows: Fairbanks 174–75. Commander removed: Fairbanks 189. Andrews: Dwyer 32. "An experienced": Fairbanks 189. Fairbanks assignment: Memorandum for the Admiral, Subject: Special Operations—Summation Report on (January 1943 to September 1944), 1 Sep 44, p. 2, in NOA 532, Folder Douglas Fairbanks; Dwyer 32–33; Fairbanks 190. "Allied Naval Liaison": III/24–25, 53; Clarke to Commander-in-Chief Mediterranean, undated; Hewitt to C-in-C, Allied Forces, Mediterranean Theater, Serial 00185, 19 Feb 44, PRO WO 204/1562/125. Burris-Meyer, Crichton: Crichton to G-3 (Org.), ADV "A"F/22/11, 26 Aug 43, in NOA 531, Folder Deception Methods Techniques and Devices.

351   Baumer to London: Baumer diary 42, 47, 49. "A delightful fellow": Wheatley 143. "Of them all"; "Arranged several"; Wheatley would never: Baumer diary 50. Sightseeing: Wheatley 143; Baumer diary 50. "From such an officer": Wheatley 143. 45th Division: Hull to COMGENNATO, 26 Apr 43, with Memorandum for Record; Hull to COMGENNATO, 3 May 43, with Memorandum for Record, both in NARA RG 319 Entry 101 Box 7 Folder 3 "Barclay." To Oran: Baumer diary 50–52. "As usual": Baumer diary 52.

352   "Old OPD": Baumer diary 52. Meeting with Goldbranson: Goldbranson to Baumer, 7 May 43, in JSC folder 000—To 312.1(1) China-Burma-India Theater. Set of notes: Military Deception, n.d., in same JSC folder. "Drink vin rouge": "To a dinner": Baumer diary 53. Baumer flight home: Baumer diary 53–54.

352   Clarke in Algiers: Clarke diary 18 Jun 43, 3 Jul 43. Clarke and Goldbranson: Goldbranson, Summary of Conversation With Colonel Dudley W. Clarke, Commander of "A" Force, Monday, 28 June, 1943; Goldbranson to Roderick, 28 Jun 43, both in Goldbranson papers.

353   "Colonel Goldbranson well known": AGWAR, Strong to NAGAP, No. 1843, 6 Jul 43; "Dudley Clarke wishes": NAGAP to AGWAR, Strong, No. W-4492, 9 Jul 43; "War establishment": Goldbranson to Roderick, 28 Jun 43; all in Goldbranson papers.

353   Goldbranson return: Debarkation endorsement, travel orders, 2 Jul 43; Goldbranson to Quartermaster, Subject: Commissary Privileges, 27 Jul 43, both in Goldbranson papers. Bevan letter: Goldbranson to Bevan, 17 Aug 43, Goldbranson papers. Train background: History of [Yale] Class of 1915 (1930), 227–28; History of [Yale] Class of 1915 (1952), 227–28; Train papers. Knew Crichton: Author's interview with Arne Ekstrom, March 1993. Joins "A" Force: Train to Joint Security Control, Subject: Deception in the Western Mediterranean, in JSC folder 11—Tactical and Strategic Deception—Capt. Hershey, 2 Aug 44. Commander: III/180. Appearance: Susan Train to author, 28 May 93. "George Train, the American": Hamer to Mure, 21 Feb 81, in Mure papers, Box 2.

354   ARTHUR, EL GITANO, LE PETIT, CHER BÉBÉ, CUPID: Plan To Build Up Special Agents for Coming Operations, ADV "A"F/21/3c, 4 Jul 43, Goldbranson papers; III/121–22.

356   40 Committee: III/117; Goldbranson, Summary of Conversation With Colonel Dudley W. Clarke, Commander of "A" Force, Monday, 28 June, 1943, Goldbranson papers, p. 3; Bristow 119–120. RAM, NORBERT, EDOUARD, WHISKERS: Plan To Build Up Special Agents for Coming Operations, ADV "A"F/21/3c, 4 Jul 43, Goldbranson papers; III/117–18.

356   COCAINE: III/116. GAOL: III/118. RUBY, BYZANCE, OLIVER: III/119.

357   MOSELLE: III/145–46. LLAMA: Wingate 166–67.

358   Déricourt: Hinsley & Simkins 197; Foot, France 290–305.
358   GILBERT narrative: Report on the Activities of No. 2 Tac H.Q. A Force Sixth Army Group, Part II, pp. 111–13, 138, 141–44, 149, 154–55, in NARA RG 319 Entry 101 Box 4 Folder 80.
358   Reile: Kahn, Spies 504. Falcon: Paillole, Services 484.
359   Paillole breaks rule: Paillole, Services 487. Duteil: Paillole, Services 486–87.
360   From June 10: Paillole, Services 487. 43 Committee, Grandguillot: III/125; Hamer to author, 2 Jun 97. Confident tone: Plan To Build Up Special Agents for Coming Operations, ADV "A"F/21/3c, 4 Jul 43, para. 2e, Goldbranson papers.
360   "The absence"; the greatest prima donna"; "stood in awe": Report on the Activities of No. 2 Tac H.Q. A Force Sixth Army Group, Part II, pp. 112–13, in NARA RG 319 Entry 101 Box 4 Folder 80. Germain: Bristow 109. "Consortium d'intelligence": Report on the Activities of No. 2 Tac H.Q. A Force Sixth Army Group, Part II, pp. 154–55, in NARA RG 319 Entry 101 Box 4 Folder 80.
361   THE LEMONS: II/136–38. Tall and blond: Mure, Practice 137; Mure, Master 113. "A boastful": III/136. "But for an endless": III/137. Klingopoulos: Mure, Practise 136.
362   Chorus girls: Mure, Practise 138. THE SAVAGES: II/138–40.
362   "Produced a fiction": III/139. SAVAGE I: Hinsley & Simkins 229. "Biggest 'scoop' ": Notes on D/A activities in the Middle East, PRO KV 2/197/1, p. 7. WHISKY: D Division Weekly Progress Report No. 1 24–30 Mar 45, p. 12, in JSC folder Deception—SE Asia—WWII Historical D. Div—SACSEA—24 Mar 45–4 Jun 45.
363   Getting NICOSSOF paid: II/96–98. Cohen: PRO KV 2/1133/80a. Levi return: PRO KV 2/1133.
363   "Exercise the very greatest": III/115. Less useful: III/116. "Occupied Enemy": II/96. Name merited: Hinsley & Simkins 231.
365   CAPRICORN: III/42–44.
365   LILOU: III/45–48.
366   "The peak": III/114. Odds against: III/64–67, 80.
366   BARCLAY: III/68–75, 78–85, 90, 93–96, 98, 102–103,
367   ULTRA revealed: Howard 90. "Be careful": Summary of Plan Barclay, undated, attached to Hollis to CIGS et al., 29 Apr 43, PRO CAB 121/105/327; III/68.
367   Goldbranson bootlegged: Goldbranson, Summary of Conversation With Colonel Dudley W. Clarke, Commander of "A" Force, Monday, 28 June 1943, Goldbranson papers, p. 3. Bevan tidbits: Cruickshank, Deception 58, citing PRO AIR 20/964 (14.7.1943), PRO WO 199/84, 5B, 16A, PRO ADM 179/272 (9.2.1943). WATERFALL: III/77–82, 86. "Twelfth Army": III/82–83.
368   Tunis bookshops: III/83; Progress Report of Plan "Barclay" as of Midnight June 6, 1943. p. 1, in NARA RG 319 Entry 101 Box 7 Folder 3 "Barclay."
369   ANIMALS: Foot, SOE 235. Montgomery leave: III/85; Galveston to Crichton, Mideast I/68424, 19 Jan 43, in Eisenhower Library, Bedell Smith Papers, Box 5 Folder Cable Log (incoming) June 16–30, 1943 (1), p. 213. Wireless security: Wingate 72; III/81–82. "Their contribution": II/82.
369   Air disappointing: III/78–80, 94–96.
370   Air appendix: III/95–96. "Who really triggered": Robertson to Mure, 24 Apr 79, Mure papers (Box 2). Equal credit: Bevan to Lamplough, 21 Aug 43, PRO CAB 154/67/48–54.
370   Cholmondeley: Howard 89; Morgan, Almost 4. "A most extraordinary": Robertson to Mure, 24 Apr 79, Mure papers (Box 2). TROJAN HORSE: Untitled undated account of MINCEMEAT, PRO CAB 154/112/1; PRO CAB 154/67/242. "Normal peacetime"; "Whilst this courier": Plan Trojan Horse, undated, PRO CAB 154/67/242.
371   Spilsbury: Morgan, Almost 4. Drowning not necessary: Montagu, Man 24–25. Purchase: Morgan, Almost 4; Howard 89. Michael: Morgan, Secret 31. Three months:

Operation Mincemeat, Twenty Committee, 4 Feb 43, p. 4, PRO CAB 154/67/235.

371 "If the real"; "A different"; "This is made": Operation Mincemeat, Twenty Committee, 4 Feb 43, p. 4, PRO CAB 154/67/232–36. Assignments: Masterman 137 note *; Operation Mincemeat, Twenty Committee, 4 Feb 43, PRO CAB 154/67/231.

372 Fine-tuned: Cholmondeley, Operation Mincemeat, Meeting held in Major Robertson's room at 16.30 hours on 10.2.43, 11 Feb 43, PRO CAB 154/67/229–30. Army officers: Untitled undated account of MINCEMEAT, PRO CAB 154/112/1.

373 "Thought that the contents": [Bevan] to Robertson, Operation Mincemeat, C/O/43/66, 12 Feb 43, PRO CAB 154/67/227–28. Montagu suggested: Montagu to Robertson, 16 Feb 43, PRO CAB 154/67/226.

373 Clarke favored; Bevan return; Planners: Corrected text for "Paragraph 13," enclosure to Bevan to Montagu, 3 Jun 45, PRO CAB 154/67/23. Huelva; submarine; Jewell; airtight: Untitled, undated account, p. 3, PRO CAB 154/112/3–4. Container made: Montagu to Bevan, Mincemeat, 26 Mar 43, PRO CAB 154/67/218–19; untitled, undated account, p. 3, PRO CAB 154/112/5.

373 Montagu revisions, drafted: 7B, undated, PRO CAB 154/67/220; 8D, undated, PRO CAB 154/67/217. Marine officer; Untitled, undated account, p. 3, PRO CAB 154/112/6; see Montagu, Man 51–56.

374 Items on body: Personal Documents and Articles in Pockets, PRO WO 106/5921/13; photograph, PRO WO 106/5921/24; see Montagu, Man 82–84. Snapshot: PRO WO 106/5921/21. Leigh; writing of letters: Found: The Fiancée Who Never Was, *Daily Telegraph,* 29 Oct 96, p. 1. "Bill darling": Pam to [Bill], 18 [Apr 43], PRO WO 106/5921/20; Montagu, Man 73. Fleming's secretary: Monypenny Is Shaken But Not Deterred by Trial, *The Times,* 10 May 97. Letters, other items: PRO WO 106/5921/13, 15–18, 22, 25–28; see Montagu, Man 68–78. "Your cousin Priscilla": PRO WO 106/5921/26; see Montagu, Man 75. Reed: Found: The Fiancée Who Never Was, *Daily Telegraph,* 29 Oct 96, p. 1; see Montagu, Man 55. Three id. cards: Untitled, undated account, p. 3, PRO CAB 154/112/7. Commandos book: Mountbatten to Eisenhower, 22 Apr 43, PRO WO 106/5921/28; see Montagu, Man 60–61.

375 "Let me have him back": Mountbatten to Cunningham, 21 Apr 43, PRO WO 106/5921/23; see Montagu, Man 58. Excuse for briefcase: Untitled, undated account, p. 3, PRO CAB 154/112/8. Nye letter drafts: Galveston to Chaucer, 112077, 1 Apr 43, PRO CAB 154/67/207; AFHQ to War Office, 112000, 2 Apr 43, PRO CAB 154/67/205; Galveston to Chaucer, 112134, 2 Apr 43, PRO CAB 154/67/197; Draft Letter From V.C.I.G.S. to General Alexander, 23 Apr 43, PRO CAB 154/67/198; Draft Letter From V.C.I.G.S. to General Alexander, 23 Apr 43, PRO CAB 154/67/199; Minutes COS (43) 68th Meeting (O), 7 Apr 43, PRO CAB 154/67/194–95; Bevan to V.C.I.G.S., C/O/43/208, 8 Apr 43, and Annexes, PRO CAB 154/67/165–69; Bevan to V.C.I.G.S., C/O/43/215, 10 Apr 43, PRO CAB 154/67/156–57.

376 Nye version: Howard 245–46; Bevan to Hollis, C/O/43/218, 12 Apr 43, PRO CAB 154/67/141–43. COS approval with revision: Bevan to V.C.I.G.S., C/O/43/225, 13 Apr 43, PRO CAB 154/67/137; Minutes COS (43) 76th Meeting (O), 14 Apr 43, PRO CAB 154/67/123–24. "Is Alexander": Montagu to Robertson, 16 Feb 43, PRO CAB 154/67/226. "But what is wrong": Enclosure to Bevan to Hollis, C/O/43/218, 12 Apr 43, PRO CAB 154/67/143; see Montagu, Man 50. "Now I hope": Nye to Bevan, 14 Apr 43, PRO CAB 154/67/93–94. "I was instructed": Bevan memorandum, Mincemeat, undated, PRO CAB 154/67/63.

377 Bedell Smith approval: Freedom to Air Ministry, 4589, [16 Apr 43], PRO CAB 154/67/119. "Handle with care": Untitled, undated account, p. 3, PRO CAB 154/112/9. Delivery of corpse: Untitled, undated account, PRO CAB 154/112/7–14. "The body was": Untitled, undated account, p. 4, PRO CAB 154/112/10.

377 Effects returned: Untitled, undated account, p. 6, PRO CAB 154/112/12. Casualty list: Morgan, Almost 21; untitled, undated account, p. 5, PRO CAB 154/112/11; Montagu, Man 106. Card, tombstone: PRO WO 106/5921/33–34. ANDROS report: Untitled, undated account, pp. 7–8, PRO CAB 154/112/13–14.

378 ULTRA report: Untitled, undated account, p. 5, PRO CAB 154/112/11; quoted in full, Browne 134–35 n. 7, citing USAMHI Ultra Reel 127, 5 to 15 May 43, ML 1955; Browne 81 n. 19, citing PRO DEFE 3/815, ML 1955.

378 BARCLAY success: Howard 90–91. Swallowed whole: Wingate 385–86, 388–92 "The circumstances": Wingate 386. "To have priority": Howard 91; untitled, undated account, p. 9, PRO CAB 154/112/15. "The discovered Anglo-Saxon order"; Mussolini; "You can forget": Hunt 190–91. Goebbels: Irving, Goebbels 433. Keitel dispatch: Wingate 396–97. Double: III/110.

379 Rommel, Roenne: Howard 91–92. Hitler's suspicion: Howard 92. General recognition: III/110; Hinsley 3/1, 78–79; Howard 93; Browne 67–68. Italian skill: Browne 115–17 and n. 44; Cruickshank, Deception 60, citing PRO WO 204/758 (24.6.1943, 1.7.1943, 12.8.1943).

380 DERRICK: III/105–108. Fake move: Subject: Recce for Move of HQ Force 141 to Oran, 28 Jun 43, Goldbranson papers; Browne 117; TROOPERS 92452 SIGS 3 to AFHQ, 25 Jan 43, in Eisenhower Library, Bedell Smith Papers, Box 5 Folder Cable Log (incoming) June 16–30, 1943 (3), p. 260; Bromley-Davenport, Subject: Move of HQ Force 141 to Oran, 29 Jun 43, in Goldbranson papers. LOBSTER: Summary Report of Results of Plan "Lobster," n.d., in JSC folder 370.2 Operations and Results; AADC Djidjelli, Standing Operation Instruction No. 17, 3 Jun 43; Special Report— Raid of Night—18 Jun 43, Ref DJ/INTEL/59, in Goldbranson papers. Bizerte anchorage: III/106.

380 Unimpressive: Morison, Sicily 20 n. 16; Dwyer 28–32. Baxter's plane: III/107–108. FALSE ARMISTICE: III/108–109; Cruickshank, Deception 57–58, citing PRO AIR 20/4534.

381 Bolstering double agents; "Most important": Plan To Build Up Special Agents for Coming Operations, ADV "A"F/21/3c, 4 Jul 43; Thomas, Resume of Plan To Build Up Special Agents for Coming Operations, ADV "A"F/21/3c, 8 Jul 43; Time Programme for Ram-Norbert, all in Goldbranson papers. "I feel I must": Crichton to Clarke Adv "A"F/21/3(c), 8 Jul 43, in Goldbranson papers.

382 DERRICK successful: III/111–12. Honorable peace project: Wingate 165; Bevan to Howkins, C/O/43/98 with attachment, 21 Feb 43; note on R. Le M. to Buzzard et al., 43/243, 21 Feb 43; Bevan to Howkins, C/O/43/112, 25 Feb 43, PRO CAB 119/66/208–213.

382 LLAMA project: Wilson to Eisenhower, No. CIC/82, 8 Aug 43, in Eisenhower Library, Bedell Smith Papers, Box 17, Folder Capitulation of Italy (July-September 1943) (3), p. 73; Wingate 166–67. BAYTOWN: III/141–45.

383 GUINEA: III/146–48; Wingate 36–38; Howard 94; author's telephone interview with Katya Doolittle Coon, 17 Mar 96.

384 Reorganization: III/154. Gummer: Author's interview with Arne Ekstrom, March 1993; Author's interview with Eugene Sweeney, January 1993. Strangeways to Sicily; FLATON: III/154. BOOTHBY: II/149, 1566–56.

385 GILBERT reports: Paillole, Services 410–11. Beach Jumpers: Morison 300–301; Dwyer 36–42; Fairbanks 191–200. Strangeways in Italy; III/157–58.

386 "A prospect": Graham & Bidwell 78–81, 104. Headquarters Bari: IV/24. Bromley-Davenport: III/159.

386 Short-term deceptions: III/170–72, 176–77.

387 FAIRLANDS: III/164–67. WHITWOOD: III/176.

387 OAKFIELD: IV/6–8.

388 LE DUC to Sicily: Paillole, Services 490; Report on the Activities of No. 2 Tac H. Q. A Force Sixth Army Group, Part II, p. 148 note *, in NARA RG 319 Entry 101 Box 4 Folder 80. LE MULET: Report on the Activities of No. 2 Tac H. Q. A Force Sixth Army Group, Part II, pp. 144–46, in NARA RG 319 Entry 101 Box 4 Folder 80; Paillole, Services 488–89 (says arrived November 25). "Rhine maidens": Muggeridge 205. "Between two": Paillole, Services 494.

389 WORKHOUSE, BIJOU, WYANDOTTE: III/2–7; Howard 225. Finter: III/2. "Charming, but": Young & Stamp 178. Temperamental: Hamer to author, 2 Jun 97.

390 Six naval deceptions: III/8–9, 14–15; Howard 227.

392 Nairobi operations: III/19–24. Huff: Hamer to author, 2 Jun 97. Arbuthnott: III/17, 180. Impressed: Hamer to author, 2 Jun 97. Finter to OVERLORD: III/24.

393 Giving thought: Clarke, The Future of the "A" Force Organization, 12 Sep 43, in AFHQ Records, cited and discussed in Browne 148–49 and 165–66 n. 23. Bevan and Clarke attended: III/178. Reallocation, Nairobi, personnel shifts: III/174, 179–82. Lisbon; Colyton 226–27; III/175.

394 One-star flag: Author's interview with Eugene J. Sweeney, December 1992–January 1993.

Chapter 10: Hustling the East (II)

395 Grand strategy: Romanus & Sunderland, Mission 388–89; Romanus & Sunderland, Command 18. SAUCY: Outline Cover Plan for Operation Saucy, 28 May 1943, in JSC folder Misc Deception Plans—WWII; Romanus & Sunderland, Command 52; Romanus & Sunderland, Mission 332; Matloff 141–42, 234–36; Tuchman 372.

396 SEAC: Romanus & Sunderland, Mission 364, 380–81; Romanus & Sunderland, Command 6, 49.

397 Wavell, Fleming, Auchinleck: Lewin 220, 242–51; Hart-Davis 286; Warner 226. Bissell, Stratemeyer: Romanus & Sunderland, Mission 285, 346; Romanus & Sunderland, Command 7, 84. Variety of plans: Romanus & Sunderland, Command 50–51, 66, 71, 76, 81; Slim 184–85.

397 SEXTANT: Romanus & Sunderland, Command 55. 1944 operations: Slim 185–86, 213, 254–95; Romanus & Sunderland, Command 159, 166–68, 172–75, 192–96, 256. XX Bomber Command: Romanus & Sunderland, Command 15–17, 369–70.

398 Famine: Romanus & Sunderland, Command 12. Gandhi: Lewin 189. SAUCY cover: Hunter diary, 28 May 43; Outline Cover Plan for Operation Saucy, 28 May 1943, in JSC folder Misc Deception Plans—WWII.

399 "Materially assisted": Same. "India can": JPS 162/2, 19 Jun 43, in NARA RG 218 Box 693 Folder CCS 385 Pacific Theater (4-1-43) Deceptive Policy in the Pacific Theater (Sec. 1). RAMSHORN: Hunter to Joint Security Control, GSOPD, Subject: Plan C.B.I. Theater, 2 Jul 43; Hunter to Jackson, Subject: Plan C.B.I. Theater, 15 July 43, both in JSC folder 000—To 312.1(1) China-Burma-India Theater; Cover Plan for India Command Operations 1943–1944, Plan "Ramshorn," LCS (43) 12, JIC (I), No. DMI/6876, 12 Jul 43, PRO CAB 119/66/44–48. "The main difficulty": Report on Cover Plan "Ramshorn," LCS (43), 2 Aug 43, PRO CAB 121/105/263–68.

399 Chiefs' directions: Minutes, COS (43) 121st Meeting, COS (43) 441 (0), 6 Aug 43, PRO CAB 121/105/261; Air Ministry to Armindia, OZ 2326, 6 Aug 43, PRO CAB 121/105/260. "The illusion": Chart, "Principal Plans," PRO WO 203/3313.

399 DUNDEE: Outline Plan "Dundee," LCS (43) 15 (Final Draft), 31 Oct 43, in JSC folder 000—To 312.1(1) China-Burma-India Theater; Plan Dundee, undated; Smith to Strong, Subject: Plan DUNDEE—Prepared by London Controlling Section, 23 Nov 43, both in JSC folder 000—To 312.1(1) China-Burma-India Theater. GLOSSOP:

Wingate to Jacob, C/O/385, 5 Dec 43, PRO CAB 121/105/231; Jebb to Price, un-dated, PRO CAB 121/105/228; Price to Wingate 9 Dec 43 and reply, 10 Dec 43, PRO CAB 121/105/143; Note, C.R. P[rice], 22 Dec 43, PRO CAB 121/105/144; Price to Bevan, 20 Mar 44, PRO CAB 121/105/197.

401 "Just as infantry": Fleming 40. Angel One: Fleming 40–41.

401 Bull's Eye: Fleming 29. "Jump four": Fleming 30.

402 Bats, Owl, Marmalade: Fleming 20–22; Hart-Davis 280; Hunter to Joint Security Control, GSOPD, Subject: Plan C.B.I. Theater, 2 Jul 43 and 9 Jul 43, in JSC folder 000—To 312.1(1) China-Burma-India Theater. Hari Singh, "guide, philosopher": D Division Weekly Progress Report No. 10 29 May–4 Jun 45, p. 8, in JSC folder Decep-tion—SE Asia—WWII Historical D. Div—SACSEA—24 Mar 45–4 Jun 45.

403 Purple Whales: Fleming 9, 35. Eden telegrams: PRO CAB 154/74. Fake cipher: Cawthorn "Cover Plan" 9 Feb 43 (Kinkajou-Wallaby memo), p. 3. Quadrant re-port; Wallace idea: O'Connor to Bissell, HMOC/990, 1 May 44, in JSC folder Great Britain. "An operational matter": Bevan to O'Connor, 23 Sep 43, PRO CAB 154/74/96.

403 Battle carrier; "Who's that blighter": PRO ADM 223/794/130–32. "A personal letter": Fleming 35. Carrier battleship: File Log, Item 237, in JSC brown Acco binder labeled "9 201-400 File 'B' "; in JSC brown Acco binder labeled "1," entry for "Code A-X-19 27 Jan.," Item 2; see also Goldbranson to Creech, 26 Apr 45, in JSC brown Acco binder "6 File D."

404 Silver: Talwar 184, 186–90; Peter Geissler, Auswärtiges Amt, Bonn, to author, 14 Jun 96; Fleming. Revision of Fleming's charter: III/179. "By nature a dull": Fleming App. 2.

405 Hiccoughs: Fleming 38–40; O'Connor to Bissell HMOC/992, May 1, 1944, in JSC folder Great Britain.

406 Brass: Fleming 25–27. "Once, in a lean": Fleming 26. Seruta: D Division Weekly Progress Report No. 12 12–18 Jun 45, in JSC folder Deception—SE Asia—WWII Historical D. Div. SACSEA 5 Jun–10 Sep 45. "Our spy"; "our special source": Fleming 26.

407 Map case: Young & Stamp 224; O'Brien 47–49. Carrier pigeons: Young & Stamp 225; O'Brien 218–19; Hunter to Smith, Subject: Channels, 15 Jan 44; Hunter to Smith, Subject: Homing Pigeon Experiments, 1 Feb 44, both in JSC folder 000—To 312.1(1) China-Burma-India Theater. Sacks of mail: Young & Stamp 224–25; O'Brien 275–76. "If any pilot": O'Brien 320.

408 Abortive SOE: Young & Stamp 220–21.

408 Fathead: Fleming 59–60; O'Brien 138–43, 151–52; Young & Stamp 222–23; Hart-Davis 282. "Two hours": Fleming 59. "And, like the Ancient": Young & Stamp 222. "A project too boldly": Fleming 10. "The European": O'Brien 153.

410 Father: Fleming 24; Hinsley & Simkins 233.

411 Bose to Japan: Bhattacharyya 70–72, 77–81, 102, 110.

412 Spy schools; "His friends": Untitled draft in Fleming papers, folder 14/1, Deception History.

412 Pawnbroker party: D Division Weekly News Letter No. 37, Part II paragraph 10, in JSC folder Deception—SE Asia—WWII Historical D. Div—SACSEA—30 Dec 44–23 Mar 45; Fleming 16; Hinsley & Simkins 233.

412 Puri party: Fleming 71; D Division Weekly News Letter No. 39 13 Jan–19 Jan 45, Part I Paragraph 11, in JSC folder Deception—SE Asia—WWII Historical D. Div—SAC-SEA—30 Dec 44–23 Mar 45. Backhand: D Division Weekly Progress Report No. 13 19–25 Jun 45, in JSC folder Deception—SE Asia—WWII Historical D. Div. SAC-SEA 5 Jun–10 Sep 45; Fleming 60; D Division Weekly Progress Report No. 7 5–14 May 45, in JSC folder Deception—SE Asia—WWII Historical D. Div—SACSEA—24 Mar 45–4 Jun 45; O'Brien 19–21.

413 Sketchbook: Fleming 10, 35; Hart-Davis 281. "This was much": Fleming 10. Posters: Young & Stamp 226–27; O'Brien 220–21. Japanese overestimate: Fleming 7; Howard 214 note*. JSC skeptical: Unsigned, undated Draft Analysis of D Division's Letter Dated 5 August 1944 Subject: Japanese Intelligence Capabilities, in JSC folder Great Britain.

414 303d: Hunter to Joint Security Control, Subject: Deception Field Units, 12 Jul 43, in JSC folder 000—To 312.1(1) China-Burma-India Theater; Hunter to Smith, Subject: Eastern Army (British) Deception Staff, 23 Jan 44; Operations of 303 Ind. Bde. During Winter 1943–44; No. 6036/D Division, 2 Jun 44, both in JSC unlabeled folder [CBI].

414 Light Scout Cars: Fleming 12, 14.

414 Hunter's return: Hunter to Joint Security Control, GSOPD, Subject: Plan C.B.I. Theater, 2 Jul 43, in JSC folder 000—To 312.1(1) China-Burma-India Theater; Hunter diary 23 Jun 43. Merrill: Tuchman 272; Romanus & Sunderland, Mission 347. "Exploded. There is not": Hunter diary 28 Jun 43. "Under the circumstances": Hunter to Joint Security Control, GSOPD, Subject: Plan C.B.I. Theater, 2 Jul 43, in JSC folder 000—To 312.1(1) China-Burma-India Theater.

415 "The General Staff": Hunter diary 26 May 43. JSC in OPD: *E.g.,* Hunter to Joint Security Control, GSOPD, Subject: Plan C.B.I. Theater, 2 Jul 43; Hunter to Jackson, Subject: Channels, 9 Aug 43, both in JSC folder 000—To 312.1(1) China-Burma-India Theater. Carey: Hunter to Smith and Carey, Subject: B-29s, 29 Dec 43; Hunter to Carey, 14 Jan 44, both in same. General Staff/JCS: Hunter to Smith, Subject: Organization, 30 Jan 44; Smith to Hunter, Subject: Organization, 10 Feb 44, JSC/LB Serial 242, both in JSC unlabeled folder [CBI]. LCS: *E.g.,* Hunter diary 29 Jul 43; Hunter to Smith, Subject: Organization, 9 Feb 44, in JSC unlabeled folder [CBI].

415 Submission to British Chiefs: Hunter to Smith, Subject: Directive on Deception Planning, 15 Mar 44; Smith to Hunter, WAR 15186, 28 Mar 44, CM-OUT-15186 (28 Mar 1944), both in JSC folder 000—To 312.1(1) China-Burma-India Theater. Sometimes Baumer: *E.g.,* Hunter to Baumer, Subject: Air Corps Strategy, 17 Feb 44, in JSC folder 000—To 312.1(1) China-Burma-India Theater. Five afternoons: Hunter to Carey, 14 Jan 44, in JSC folder 000—To 312.1(1) China-Burma-India Theater. Arnold: Hunter to Smith, Subject: Air Deception Officer, 2 May 44, in JSC unlabeled folder [CBI]. "I was delighted": Hunter to Bissell, 14 Feb 44, in JSC unlabeled folder [CBI]. "Offered his complete": Hunter diary 31 May–2 Jun 43.

416 Control struggle: Yu 111–12. Japanese money: Hunter to Joint Security Control, OPD, Subject: Plan C.B.I. Theater, 2 Jul 43 and 9 Jul 43, in JSC folder 000—To 312.1(1) China-Burma-India Theater. Eifler: Dunlop 69–70; Yu 25–26, 113; Tuchman 340. Hunter to Baumer, Subject: Progress Report, 7 Oct 43, in JSC folder 000—To 312.1(1) China-Burma-India Theater. Heppner: Yu 108–114; Hunter diary 12 Aug 43. HEATER; "Deception of a strategic": C. E. G[oldbranson], Memorandum for Record, Subject: Conference With OSS re Junior Heaters, 24 Mar 44; Smith to Hunter, Subject: Organization, 28 Mar 44, JSC/J3A6 Serial 392, both in JSC unlabeled folder [CBI].

417 Lorenço Marques, Engert, bombarded, Kweilin, Chang: Hunter to Allen, Subject: Channels, 14 Jul 43; Hunter to JSC, 11 Oct 43; Enclosure to Hunter to Jackson, Subject: Wallaby, 14 Jul 43; Hunter to Jackson, Subject: Channels, 9 Aug 43; Hunter to Ferris, Subject: U.S. Channel, 14 Dec 43, all in JSC folder 000—To 312.1(1) China-Burma-India Theater. "A professional": Hunter to Smith, Subject: Channel— Afghanistan, 31 Dec 43, in same. Pigeons: Hunter to Smith, Subject: Channels, 15 Jan 44; Hunter to Smith, Subject: Homing Pigeon Experiments, 1 Feb 44; Smith to Hunter, Subject: Channels, 26 Jan 44, JSC/D1 Serial 189, all in JSC folder 000—To 312.1(1) China-Burma-India Theater.

417 Pamphlet, additional ideas: Hunter to Carey, 14 Jan 44; Hunter to Smith, Subject: Notes on Deception, 5 Feb 44, both in JSC folder 000—To 312.1(1) China-Burma-India Theater. Estimate; "the cover plan": Hunter to Baumer, Subject: Progress Report, 7 Oct 43; Hunter, Estimate of the Situation, 28 Aug 43, both in JSC folder 000—To 312.1(1) China-Burma-India Theater. "Extremely dangerous"; "would seem": Smith to Hunter, 7 Feb 44, JSC/D1 Serial 224; Smith to Hunter, Subject: Censor, 28 Dec 43, JSC/D1 Serial 1298, both in JSC folder 000—To 312.1(1) China-Burma-India Theater. Censor: Hunter to Smith, Subject: Censor, 15 Dec 43, in same. Argentina: Hunter to Smith, Subject: Channel Argentina, 15 Nov 43; Smith to Hunter, Subject: Channel Argentina, 2 Dec 43, JSC/D1 Serial 1206, both in same. *Gripsholm:* Hunter to Smith, Subject: Exchange Nationals, 21 Oct 43; Smith to Hunter, Subject: Exchange Nationals, 2 Nov 43, JSC/C10/C14 Serial 1113, both in same.

418 Set of JSC documents: Smith to Hunter, Subject: Deception and Cover—Information, Devices, etc., 23 Oct 43, JSC/D1 Serial 1076, with enclosure, in same. Packet of ideas: Hunter diary 13 Jul 43; Hunter to Jackson, Subject: Plan C.B.I. Theater, with attachments, 15 July 43, in same. "Are we going": Hunter diary 22 Jul 43. "Act with GSI(d)": Ferris to Cawthorn, Subject: Plans, 24 Jul 43, in JSC folder 000—To 312.1(1) China-Burma-India Theater.

419 Stilwell in India: Romanus & Sunderland, Mission 347; Tuchman 381, 388. "Thought it all": Hunter diary 23 Jul 43. Get appointment: Hunter diary 31 Jul 43. Bissell; "And when you got here": Hunter diary 29 Jul 43. Stilwell interview; "Very much impressed": Hunter diary 31 Jul 43. Why RAMSHORN approved: Hunter to Jackson, Subject: Plan for CBI Theater, 31 Jul 43, in JSC folder 000—To 312.1(1) China-Burma-India Theater.

420 Fleming to China: Hunter diary 7 Sep 43; Hunter to Ferris, Subject: Special Planning, 14 Sep 43, in JSC unlabeled folder [CBI]; Fleming to Bevan, 27 Sep 43, PRO CAB 154/66/2C; Hart-Davis 287. Chennault: Hunter, Trip to Kunming, 8 Nov 43, in JSC folder 000—To 312.1(1) China-Burma-India Theater. Pierson: Fleming 5; Hunter to Hearn, Subject: Major S. C. F. Pierson, 29 Feb 44, in JSC unlabeled folder [CBI]. Stratemeyer and Davidson: Hunter diary 4 Sep 43. "Told him of Fleming's": Hunter diary 7 Sep 43.

420 Visit to Eifler: Hunter to Baumer, Subject: Progress Report, 7 Oct 43, in JSC folder 000—To 312.1(1) China-Burma-India Theater. Grafton: Hunter to Smith, Subject: Personnel, 10 Mar 44, in JSC unlabeled folder [CBI]; author's telephone interview with Mikell Grafton, 2 Dec 97. Proposed reorganization, trip to Kunming: Hunter to Commanding General, U.S. Army Forces, China-Burma-India, 4 Feb 44; Hunter to Smith, Subject: Organization, 26 Nov 43; Hunter to Bissell and Smith, No. AG 505, 26 Jan 44, CM-IN-17419 (17 Jan 44), all in JSC unlabeled folder [CBI]; Hunter, Trip to Kunming, 8 Nov 43; Hunter to Smith, Subject: Trip to China, 17 Nov 43; Hunter to Smith, 18 Nov 43, all in JSC folder 000—To 312.1(1) China-Burma-India Theater. "More deceptive minded": Hunter diary 18 May 43.

421 "A tendency"; a lot of what the Japs": Enclosure to Hunter to Carey, 14 Jan 44, section 7, in JSC folder 000—To 312.1(1) China-Burma-India Theater. "What was done": Hunter to Fleming, 7 Jan 44, in JSC unlabeled folder [CBI]. Chickenfeed: Hunter to Smith, Subject: Channels—Preservation of, 30 Jan 44; Smith to Hunter, Subject: Channels—Preservation of, 10 Feb 44, JSC/LB Serial 941; Smith to Hunter, Subject: Channels, 25 May 44, JSC/J8 Serial 593; Smith to Hunter, Subject: Channels, 16 Jun 44, JSC/J8 Serial 667, all in JSC folder 000—To 312.1(1) China-Burma-India Theater. "One of the duties": Hunter to Chennault, Subject: Discord Enemies, 18 Oct 43, in same.

421 Doctored sitrep: Headquarters Tenth Air Force (Rear Echelon), Daily Intelligence Extracts for 11 October 1943; Hunter, Trip to Kunming, 8 Nov 43; Hunter to Ferris,

Subject: Organization of Deception Section, 14 Nov 43; Hunter to Smith, Subject: Japanese-German Discord, 23 Nov 43, all in JSC folder 000—To 312.1(1) China-Burma-India Theater. "Enough fuel": Hunter to Chennault, Subject: Discord Enemies, 18 Oct 43, in same. "Well launched": Hunter to Ferris, Subject: Organization of Deception Section, 14 Nov 43, in same.

422 B-29 snafu: Hunter to Smith and Carey, Subject: B-29s, 29 Dec 43, in JSC folder 000—To 312.1(1) China-Burma-India Theater; Hunter to Ferris, Subject: Twilight, 29 Dec 43, enclosure to Hunter to Smith, Subject: Twilight, in JSC unlabeled folder [CBI]; Smith to Hunter, 4 Mar 44, JSC/J1 Serial 313; Westlake to Arnold, Subject: Progress Report of B-29 Project, 15 Feb 44 ("this could cause"); Westlake to Roberts, No. 6394, 17 Feb 44, CM-OUT-7707 (18 Feb 44) ("treatment of both"); Smith to Schuirmann and Bissell, Subject: Preparation, Coordination, and Implementation of Deception Plans, 26 Feb 44; Schuirmann and Bissell to Commanding General, Army Air Forces, and Director, Bureau of Public Relations, Subject: Planning and Implementation of Cover and Deception Measures, 28 Feb 44 ("that in the future"), all in JSC folder 000—To 312.1(1) China-Burma-India Theater.

424 Hunter and B-29s: Hunter to Smith, No. AMMDEL AG 992, 16 Feb 44, CM-IN-11117 (16 Feb 44); Smith to Hunter, No. 6338, 16 Feb 44, CM-OUT-6766 (16 Feb 44) ("more appropriate"); Hunter to Bissell, Nos. AM342 AMMDEL, AG1073 AM-MDEL, 20 Feb 44, CM-IN-14204 (20 Feb 44); Bissell to Hunter, No. 6497, 22 Feb 44, CM-OUT-9357 (22 Feb 44); Hunter to Smith, Subject: B-29, 25 Feb 44 ("I am of the opinion"); Smith to Hunter, 4 Mar 44, JSC/J1 Serial 313 ("Please be assured"); Hunter to Smith, Subject: Deception Policy, 5 Mar 44 ("I would be very much"); Smith to Hunter, Subject: Deception Policy, 20 Mar 44, JSC/J1F2 Serial 355 ("An overall"); Hunter to Smith, Subject: Deception B-29s, 6 Apr 44 ("If my method"), all in JSC folder 000—To 312.1(1) China-Burma-India Theater.

425 "More or less promised"; Fleming: Thorne to author, 12 Jan 97.

425 GSI(d) enlarged: DMI/2629/GSI(d), 14 Oct 43; Goldbranson to Smith, 29 Oct 43 ("A definite attempt"); DMI/6937, 25 Nov 43, enclosure to Hunter to Smith, Subject: Organization, 26 Nov 43; Hunter to Baumer, Subject: SEAC Activities, 27 Nov 43, all in JSC unlabeled folder [CBI]; Hunter to Smith, Subject: Trip to China, 17 Nov 43, in JSC folder 000—To 312.1(1) China-Burma-India Theater.

425 Limbo: Hunter diary, 1 Nov 43, 9 Nov 43; Hunter to Smith, Subject: Organization, 26 Nov 43 ("Whether General"); Smith to Hunter, Subject: Organization, 22 Dec 43, JSC/D1 Serial 1278, both in JSC unlabeled folder [CBI].

426 Kehm trip: Kehm to Assistant Chief of Staff, OPD, Subject: Report on Foreign Duty [20 Jan 44], in Kehm papers. "Central authoritative": same, para. 8. Internee letters: Hunter to Smith, Subject: Internees, 20 Dec 43; Smith to Hunter, Subject: Internees, 30 Dec 43, JSC/D1 Serial 1302; Smith to Hunter, Subject: Internees, 26 Jan 44, JSC/F5 Serial 190, all in JSC folder 000—To 312.1(1) China-Burma-India Theater.

427 January 1944 reorganization: Fleming 3–4; Hunter to Smith, Subject: Organization, 7 Jan 44 ("Expressed the definite"); Hunter to Smith, Subject: Organization, 30 Jan 44; Hunter to Smith, Subject: Organization of Special Planning Section, 7 Feb 44, all in JSC unlabeled folder [CBI]. Hunter transfer: Hunter to Commanding General, United States Army Forces, China-Burma-India, Subject: Assignment of Lt. Col. Edward O. Hunter, Air Corps; Hunter to Commanding General, United States Army Forces, China-Burma-India, Subject: Transfer of Special Planning Section G-3, C.B.I., 4 Feb 44; Hunter to Smith, Subject: Organization of Special Planning Section, 7 Feb 44 ("Definitely decided"); Goldbranson to Smith, 27 Jan 44; Smith to Hunter, No. 6072, 4 Feb 44, CM-OUT-1571 (4 Feb 44); Smith to Hunter, Subject: Deception Officer for Assignment to CBI, 17 Feb 44; Hunter to Smith, Subject: Organization, 8 Mar 44; Schwab to Commanding General, Tenth Air Force, *et al.*, Subject: Deception

Planning and Implementation, 21 March 44; Barrett to Gates and Clark, No. W 853, 24 Mar 44, CM-IN-17563 (25 Mar 44); Hunter to Smith, Subject: Organization and Personnel, 27 Mar 44 ("Already I feel"); Smith to Hunter, No. WAR 22711, 12 Apr 44, CM-OUT-22711 (13 Apr 44), all in JSC unlabeled folder [CBI].

427 Sickman: Hunter to Smith, Subject: Personnel, 7 Mar 44; Hunter to Smith, Subject: Organization and Personnel, 27 Mar 44, both in JSC unlabeled folder [CBI].

427 Fleming to front: Thorne to author, 12 Jan 97; Hart-Davis 289–300.

428 BLANDINGS, UKRIDGE: Extract From Weekly Letter No. 5 of 27 March, in JSC folder 370.2 Operations and Results; Chart, Principal Plans . . . , PRO WO 203/3313; Grafton to Hunter and Fleming, Subject: Report on Tour, 21 Apr 44, enclosure to Hunter to Smith, Subject: Trip—Lieut. Cornelius W. Grafton, 29 Apr 44, in JSC folder 000—To 312.1(1) China-Burma-India Theater ("Tell Colonel Fleming"; "I'm afraid I didn't"). "Their complete confidence": Hunter to Smith, Subject: Personnel, 10 Mar 44, in JSC unlabeled folder [CBI]. Cochrane: Calvert 151; Romanus & Sunderland, Mission 366.

429 Move to Kandy: Ziegler 279; Thorne to author, 27 Apr 97. "Lavish array": same. Full colonel: Service Record p. 6, in Fleming papers 15/6 Service Record. Over his own protest: Fleming to ASCOS (A), DMI/3085/GSI(d), 12 Jan 44, PRO WO 3313/23A. Lavish scale; "A perpetual"; de Wiart: Ziegler 279. Lax security: Cary to Smith, Subject: Security in India, 5 Sep 44, in JSC unlabeled folder [CBI]. Hunter to Calcutta: Hunter to Smith, Subject: Deception B-29s, 6 Apr 44, in JSC file 000—To 312.1(1) China-Burma-India Theater.

Chapter 11: American Deception Grows Up

431 Schrup: Author's telephone interview with Mrs. E. P. Schrup, 12 Apr 94. "Presents the hazards": Cake Before Breakfast p. 15, JSC loose document.

432 Officers formally: AG 210.61 General Staff (10 Nov 43) PO-A-A, 12 Nov 43, in Goldbranson papers. Special Section: Joint Security Control Organization Chart, 10 Mar 44, in NARA RG 218 Box 211 Folder CCS 334 (8-4-42) Joint Security Control (Sec. 2). Room 2B656: Volkel to Peck, undated [Aug 44], JSC/G1 Serial 895, in NARA RG 218 Box 694 Folder CCS 385 Pacific Theater (4-1-43) Deceptive Policy in the Pacific Theater (Sec. 3); buck slip, 2 Jul 45, in JSC folder Portugal. Full colonel: Officer's, Warrant Officer's, and Flight Officer's Qualification Record, WD AGO Form 66, 26 Apr 1949, in Newman Smith personnel file. Officer backgrounds and assignments: Smith to Deception Section, Subject: Allocation of Duties, 10 Feb 44; List, Advance HQ "A" Force U.S.A. Personnel, 12 Jul 43 (Stewart), both in Goldbranson papers; Smith to Bentley, 10 May 44, in JSC folder 000—To 312.1(1) China-Burma-India Theater; Bentley, Operational Deception, 29 Sep 44, p. 8, in NOA 530, Folder Combined Planning, Cover and Deception; Creighton and Cockrell to Chief Signal Officer, 3 Jul 44, JSC/J1E1 Serial 705; Creighton and Cockrell to CNO, 3 Jul 44, JSC/J1E1 Serial 707; C. E. G[oldbranson] to Record, Subject: Communications Plan "WEDLOCK"; all in JSC folder Wedlock—subfolder 370.2 Wedlock—Impl Phys; Whitehill 221 (Grosskopf). "Lack of adequate": Smith to Bentley, 10 May 44, in JSC folder 000—To 312.1(1) China-Burma-India Theater.

433 "With Joint Security Control": Cake Before Breakfast, p. 19, JSC loose papers. Kehm and Baumer: Baumer, But General ch. 6; Cullum No. 6932 (Kehm), No. 9734 (Baumer). Put Smith on: Author's telephone interview with Anita Burris-Meyer, 4 Apr 94. "Everyone regarded him": Eldredge to Dwyer, 17 Aug 89, Eldredge papers.

434 "Absolutely no"; breakfast; Schrupping: Author's telephone interview with Anita Burris-Meyer, 19 Feb 95. Smith apartment: Officer's, Warrant Officer's, and Flight Officer's Qualification Record, WD AGO Form 66, 26 Apr 1949, in Newman Smith

personnel file. Bissell succeeds: AGWAR to AFHQ, Ref. No. 95, in Eisenhower Library, Bedell Smith Papers, Box 21, Folder Cable Log—In (Feb 44) (2).

434 Bissell: Author's telephone interview with Anita Burris-Meyer, 19 Feb 95; Clayton Bissell Oral History, Air Force Historical Research Center, No. K 239.052-987, 22 Feb 66, pp. 4, 10–14; Bissell biographical release, USAMHI.

435 BODYGUARD briefing: Notation in Smith's handwriting on Smith to Bissell, 9 Feb 44, in JSC folder 370.2 Bodyguard. "There has been a general increase": Smith to Hunter, Subject: Organization, 10 Feb 44, JSC/1B, Serial 242, in JSC unlabeled folder [CBI].

435 "On account of the possible": O'Connor to Bissell, n.d., HMOC/990, in JSC folder Great Britain. Schuirmann: Biographical release, Naval Operational Archives, Washington Navy Yard. Greater interest: General Chart Plan "Wedlock," under 9 Nov 43, in Cake Before Breakfast p. 36, JSC loose papers. O'Connor not clearing: Hunter to Smith, Subject: Channels—Preservation of, 30 Jan 44; Goldbranson to Smith, 10 Feb 44, with endorsement, both in JSC file 000—To 312.1(1) China-Burma-India Theater; Goldbranson to Smith, Subject: Channels, 22 Mar 44; Smith to Record, Subject: Use of Japanese channels in U.K. and neutral countries by London Controlling Section, memo from Colonel O'Connor dated 29 February 1944, 13 Apr 44, both in JSC folder Great Britain.

436 Bevan agreement: Bissell to O'Connor, Subject: Coordination of Deception Planning, 8 May 44, JSC/J8 Serial 543; Smith to Bissell, Subject: Coordination of Deception, 18 May 44, JSC/J8 Serial 586; Bissell to O'Connor, 18 May 44, JSC/J8 Serial 587. Suspicious: Smith to Bissell, Comment No. 2, 23 Mar 45, in JSC folder Japanese Estimates of Allied OB. Ability hampered: Bissell to O'Connor, Subject: Purple Whales Channel, Operation Hiccoughs, 8 May 44, JSC/J8 Serial 538; Bissell to Cowgill, Subject: Establishment of Double Agent in South America, 10 May 44, JSC/J8 Serial 550, both in JSC folder Great Britain. British clearance: Smith to Record, Subject: Use of Double Agents in Canada by British, 10 Feb 44; Entry for 10 May 44 in tabulation J/NS/3a, DA's British; Cowgill to Smith, No. 6973, 10 May 44, all in JSC folder Great Britain.

437 Section V misunderstanding: Bouverie to Bissell, Subject: Establishment of Double Agent in South America, 24 May 44; Unsigned memorandum British Use of Double Agents in South America, 7 Jun 44, with endorsement in Newman Smith's handwriting, 8 Jun 44; Smith handwritten notes 28 Jul 44; Schuirmann and Bissell to Stephenson, Subject: Establishment of Double Agent in South America, 16 Aug 44, JSC/J5 Serial 861; Stephenson to Bissell, 3 Oct 44, all in JSC folder Great Britain. LODGE material: Smith to O'Connor, Subject: Montevideo Double Agent, 8 Feb 45, JSC/J5 Serial 242, in JSC folder Great Britain. New master list: Lieutenant Colonel John R. Deane, Jr., National War College Lecture, Order of Battle, 1 March 1950, pp. 15–16, in NOA 546, Folder Tactical Deception—Ruses; Hesketh 120. Identify division only: Smith to Commanding General, Alaskan Department, Subject: Plan "WEDLOCK"—Order of Battle, 10 May 44, JSC/J1E1 Serial 549, in JSC file Wedlock—subfolder "370.2 Wedlock—Impl Phys."

437 Baumer to Heraldic Section: Stratagem Conference 23; J. K. W[oolnough] to Roberts, Re: JCS 806, 15 Apr 44, in NARA RG 165 ABC File ABC 381 Japan 15 Apr 43 Sec. 4—Forager; Notes of David Kahn telephone interview with Baumer, 10 Sep 76, copy in author's possession; Lieutenant Colonel John R. Deane, Jr., National War College Lecture, Order of Battle, 1 March 1950, p. 16, in NOA 546, Folder Tactical Deception—Ruses.

438 National Geographic: Table dated 22 Jun 1943 [Bratby], rows 4–5 columns B and D, in JSC folder European Theater 000; Bevan letter 9 Apr 45, both in JSC folder European Theater 000; Stratagem Conference 22; Lieutenant Colonel John R. Deane, Jr., National War College Lecture, Order of Battle, 1 March 1950, p. 17, in NOA 546,

Folder Tactical Deception—Ruses. Rattlesnake patch: Lieutenant Colonel John R. Deane, Jr., National War College Lecture, Order of Battle, 1 March 1950, p. 16, in NOA 546, Folder Tactical Deception—Ruses. "Which has the insignia": Hesketh 236, serial no. 41.

438 Tactical deception: Cake Before Breakfast pp. 32, 37, 40, 42, JSC loose papers. 23d Headquarters: Appendix to Enclosure "A" to Cover and Deception Planning—Proposed Amendment to JCS 256/1/D, 28 Jan 44, in NARA RG 218 Box 362 Folder CCS 385 (4-8-43) (2) Cover and Deception Planning (Sec. 1); Stratagem Conference 22; Cover & Deception Synopsis of History, p. 3, in NARA RG 319 Entry 101 Box 1 Folder 1 "Cover & Deception Synopsis of History"; Kronman 1, 20. Commercial artists: Dahl to Ingersoll, 12 Jun 46, in Ingersoll papers. Sonic units: Cake Before Breakfast pp. 33, 39, JSC loose papers; Kronman 13–15. 3103d: Cake Before Breakfast p. 31, JSC loose papers.

440 Bringing Navy and Army Engineers together: Appendix to Enclosure "A" to Cover and Deception Planning—Proposed Amendment to JCS 256/1/D, 28 Jan 44, in NARA RG 218 Box 362 Folder CCS 385 (4-8-43) (2) Cover and Deception Planning (Sec. 1). "So far this war": Report on Cover Plans, COS (43) 706 (O), 13 Nov 43, PRO CAB 121/105/236. CCS request for report: Minutes COS (43) 280th Meeting, 16 Nov 43, PRO CAB 121/105/235; Air Ministry to JSM, Washington, OZ 3757, 16 Nov 43, PRO CAB 121/105/234; Planning and Implementation of Deception, CCS 434, 18 Nov 43; Planning and Implementation of Deception, CCS 434/1, 22 Nov 43, both in NARA RG 218 Box 362 Folder CCS 385 (4-8-43)(2) Cover and Deception Planning (Sec. 1).

441 AFHQ report; "While 'A' Force"; "Due to the extent": Cover and Deception Plans Period November 1942–November 1943, CCS 434/2, 6 Jan 44 (also at PRO CAB 121/105/216–224); Algiers to AGWAR, W/9516/25103, NAF 581, 7 Jan 44, CM-IN-4473 (7 Jan 44), both in NARA RG 218 Box 362 Folder CCS 385 (4-8-43) (2) Cover and Deception Planning (Sec. 1).

441 JSC re AFHQ report: Cover and Deception Planning—Proposed Amendment to JCS 256/1/D, JCS 683, 28 Jan 44, in NARA RG 218 Box 362 Folder CCS 385 (4-8-43) (2) Cover and Deception Planning (Sec. 1). Amended JSC charter: Cover and Deception Planning—Proposed Amendment to JCS 256/1/D, JCS 683/1, 1 Feb 44; Joint Chiefs of Staff Directive Cover and Deception Planning, JCS 256/2/D, 1 Feb 44; Ulio to Commanding Generals et al., AG 381 (5 Feb 44) OB-S-E, 7 Feb 44; Graves to Secretary, British Joint Staff Mission, Subject: Cover and Deception Planning, 4 Feb 44; Redman to Royal, Cover and Deception Planning, 40/R, 14 Feb 44, all in NARA RG 218 Box 362 Folder CCS 385 (4-8-43) (2) Cover and Deception Planning (Sec. 1). Fairbanks: Fairbanks 214, 218, 220; Fairbanks to Metzel, 21 Dec 43, in JSC folder Deceptive Equipment—Navigation—BJU's—Résumé—Aug 44. Joint Communications: Myers to Secretary, Joint Security Control, Subject: Coordination of Communications With Deception (Cover) Plans, 7 Feb 44; Smith to Myers, Subject: Coordination of Communications With Deception (Cover) Plans, 8 Feb 44, both in NARA RG 218 Box 360 Folder CCS 385 (9-22-43) Necessity of Coordinating Communications With Deception or Cover Plans.

442 Bevan summary: Minutes, COS (44) 11th Meeting, 13 Jan 44, PRO CAB 121/105/214; Cover and Deception Plans, COS (44) 132 (0), 6 Feb 44, PRO CAB 121/105/204–210; Minutes, COS (44) 44th Meeting, 10 Feb 44, PRO CAB 121/105/203. To Japanese through Germans: Cake Before Breakfast pp. 60, 63, JSC loose papers.

442 RUDLOFF history; "Nice-looking tall": Unsigned, undated document, "R.," pp. 26–28, in JSC brown Acco binder labeled "14"; Hill 396–97. A.3778: Farago, Game 654. Real name: Hill; Undated, unsigned chart in Schrup's handwriting, in NOA 544, Folder Special Means—A. ND98: Whitehead 196. "Max Fritz Ernst Rudloff"; La Plata: Con-

roy to Director, FBI, Re: Max Fritz Ernst Rudloff, with aliases, et al., 24 Nov 43, in JSC brown Acco binder labeled "18."

444 Early messages: Messages 1–4, 20 Feb 42, 12–14, 13 Mar 42, 15–18, 17 Mar 42, 57–59, 24 Apr 42, undated tabulation of Rudloff messages in JSC brown Acco binder labeled "14." FBI transmitter: Undated hand-drawn map with notes in Schrup's handwriting, in JSC brown Acco binder labeled "14."

444 Three to four a week: Undated chart in JSC brown Acco binder labeled "14."

445 Further activities, big spender, ladies' man: Undated document, "R.," in JSC brown Acco binder labeled "14"; Hill 397–98. RUDLOFF's sources: Undated tabulation of Rudloff messages; Unsigned memorandum Re: Rudloff Case, B-205, 23 Oct 44, both in same. Levy: Author's telephone interview with Richard G. Fletcher, 19 Jan 95.

446 Tokyo raid foodstuff; "were trying to pass off": Liddell diary, 4 Jun 42, PRO KV 4/190. British got own back: Hill 397. Liddell-Tamm-Ladd agreement: Liddell diary, 18 Jun 42, PRO KV 4/190. Hoover loathed: Liddell diary, 25 Aug 42, 4 Sep 42, 15 Sep 42, 28 Sep 42, 22 Dec 42, PRO KV 4/190; 23 Aug 44, PRO KV 4/194. GC&CS: PRO CAB 154/41/157A.

446 PAT J history; "expressionless": Undated memorandum re: PAT J, in JSC Acco Binder labeled "14"; list of agents in JSC Folder Portugal; Hoover; Hinsley & Simkins 228–29; Hill 395–96. Meiler: Teeple 2; Powers 345. Köhler: Farago, Game 655; Kahn, Spies 494. Hübner, lecture: Teeple 2. Messages: File Log, in JSC brown Acco binder labeled "7 File A"; File Log; unsigned, undated memorandum "Summary PAT J Messages," both in JSC brown Acco binder labeled "8 File B"; File Log, in JSC brown Acco binder labeled "9 201–400 File 'B' "; File Log, in JSC brown Acco binder labeled "10 B File." "The informant": Unsigned, undated memorandum endorsed "Pat J 10/20/43" in Newman Smith's handwriting, in JSC brown Acco binder labeled "18." "Uncle will protect": Hoover 122.

449 MIKE: Hoover ("forthright"); Unsigned memorandum, Re: MICASE, 26 Mar 45; Unsigned, undated memorandum, "Comments: B-404"; Unsigned memorandum, Re: MICASE, 18 May 45, B-521, all in JSC brown Acco binder labeled "10 B File"; Author's telephone interview with Ray Wannall, 22 Jan 98. Stockmann: Teeple 2. Letsch real name; "Had trustworthy": Unsigned, undated memorandum, "PJ—Outgoing," Message No. 202, 21 Nov 44, in JSC brown Acco binder labeled "D."

450 Farago claims: Farago, Game 653–57.

450 De Bye: Undated, unsigned chart in Schrup's handwriting, in NOA 544, Folder Special Means—A; Powers 45, 64–66; Teeple 3–4; Farago, Game 231; Hoover to Bissell, Attention: Colonel Newman Smith, 16 Nov 44, in JSC folder Traffic Sent; Unsigned memorandum, Military Intelligence Division, Subject: Debye, Peter Joseph William, 15 Sep 43; Unsigned, undated memorandum of telephone call with FBI, stamped "Received Back MIS Record Section Sep 10 1943"; Undated Memorandum From CIA, Subject: DeBye, Prof. Peter Joseph William, receipt-stamped 26 Sep 51, all in NARA RG 319 IRR Files Box 35 X1.10.72.06 Peter J. W. De Bye, Folder "Volume 2 of 2—Folder 3 of 3"; FBI Report, Peter Josef William Debye, File No. NY 116-12889, 23 May 50, in same, "Folder 2 of 3". Sent worthless information: Teeple 3–4; Hoover 122. No disloyalty, cleared: Quinto to CG Second Army, Subject: Debye, Peter Joseph William, 22 Aug 56, in NARA RG 319 IRR Files Box 35 X1.10.72.06 Peter J. W. De Bye, Folder "Volume 2 of 2—Folder 3 of 3."

451 Van de Grint: Undated, unsigned chart in Schrup's handwriting, in NOA 544, Folder Special Means—A. "V.G.": Unsigned, undated memorandum, "PJ—Outgoing," Message 237, 22 Dec 44, in JSC brown Acco binder labeled "D."

451 BROMO history: Undated memorandum, Re: [Deleted], in JSC brown Acco binder labeled "File X"; Unsigned memorandum, Re: [Deleted], 15 Mar 45, both in JSC brown Acco binder labeled "14. Real name: Undated, unsigned chart in Schrup's handwriting,

in NOA 544, Folder Special Means—A; File "B" Log, in JSC brown Acco binder labeled "8 File B." Spencer: Author's telephone interview with Ray Wannall, 19 Dec 96. Transmissions: File "B" Log, in JSC brown Acco binder labeled "8 File B"; Undated chart in JSC brown Acco binder labeled "14." Subsources: Unsigned memoranda, B-410, 27 Mar 45, and B-422, 31 Mar 45, in JSC brown Acco binder labeled "10 B File."

452 PEASANT history: Unsigned, undated memorandum "G. Peasant Case"; Unsigned memorandum, Re: [Deleted] Espionage—G, 2 Sep 44; Unsigned memorandum, Re: Peasant Case, 3 Aug 44, all in JSC brown Acco binder labeled "14"; Unsigned, undated memoranda re letters sent 7 Jun 43 and 14 Jun 43, in subfile "P Sent Traffic," in JSC file Traffic Sent; KV 2/467; Felt 36. Not rated very highly: "Poor rating," per undated, unsigned chart in Schrup's handwriting, in NOA 544, Folder Special Means—A. Real name: Cited memoranda in JSC brown Acco binder labeled "14"; PRO KV 2/467. Not without some concern: PRO KV 2/467/179A.

454 "Because attitude": Unsigned, undated memorandum re letter sent 13 Nov 43, in subfile "P Sent Traffic," in JSC file Traffic Sent. Money, communication: Unsigned, undated memoranda re cables sent 15 Apr 44 and 22 Jul 44, in subfile "P Sent Traffic"; Unsigned, undated memorandum Message No. 1 received 19 Jul 44, in subfile "P Received Traffic," all in JSC file Traffic Sent. On the air: Messages Nos. 1 sent 26 Aug 44 and 2 sent 1 Sep 44, in subfile "P Sent Traffic"; Message No. 2 received 1 Sep 44, in subfile "P Received Traffic," all in JSC file Traffic Sent. Felt alerted: Unsigned memorandum, Re: Peasant Case, 3 Aug 44, in JSC brown Acco binder labeled "14." "Inasmuch as time": Unsigned memorandum, Re: [Deleted] Espionage—G, 2 Sep 44, in JSC brown Acco binder labeled "14." "Have good contact": Messages Nos. 5, 6, 7 sent 16 Sep 44, in subfile "P Sent Traffic," in JSC file Traffic Sent. R-4360: Message (unnumbered) sent 23 Sep 44, in subfile "P Sent Traffic." "Secretary to a high": Message (unnumbered) sent 28 Sep 44, in subfile "P Sent Traffic," in JSC file Traffic Sent.

455 Willow Run: Message No. 17 sent 5 Oct 44, B-146, in subfile "P Sent Traffic," in JSC file Traffic Sent. PEASANT's informants: Messages Nos. 14 sent 30 Sep 44, 30 sent 26 Oct 44, 33 sent 30 Oct 44, all in subfile "P Sent Traffic," in JSC file Traffic Sent; Unsigned memorandum, Re: Peasant Case, 4 Oct 44, in JSC brown Acco binder labeled "14"; unsigned memorandum, Re: Peasant Case, B-204, 19 Oct 44, in JSC brown Acco binder labeled "9 201-400 File 'B.' " Japanese real customer: Felt 37. Nags for money: Messages Nos. 4 sent 12 Sep 44, 5, 6, 7 sent 19 Sep 44, 45 sent 11 Nov 44, 46 sent 14 Nov 44, No. 101 sent 27 Jan 45, No. 102 sent 10 Feb 44, No. 110 sent 10 Mar 45, No. 113 sent 17 Mar 45, Nos. 115 and 116 sent 22 Mar 45, in subfile "P Sent Traffic"; Message (unnumbered) received 10 Apr 45, No. 42 received 28 Apr 45, in subfile "P Received Traffic," all in JSC file Traffic Sent. Loose tradecraft: Messages No. 8 sent 19 Sep 44, No. 8 received 23 Sep 44, (unnumbered) sent 26 Sep 44, No. 33 sent 30 Oct 44, in subfile "P Sent Traffic"; No. 18 received 31 Oct 44, in subfile "P Received Traffic," all in JSC file Traffic Sent. Failing recognize: Message No. 51 sent 25 Nov 44, in subfile "P Sent Traffic"; Message No. 23 received 28 Nov 44 ("since they are"), in subfile "P Received Traffic," both in JSC file Traffic Sent.

455 Spanish: Message No. 15 received 28 Oct 44, in subfile "P Received Traffic"; Message No. 35 sent 31 Oct 44, in subfile "P Sent Traffic"; Message No. 19 received 3 Nov 44, in subfile "P Sent Traffic"; Message No. 93 sent 20 Jan 45, in subfile "P Sent Traffic"; Message No. 33 received 23 Jan 45 and notation thereon, in subfile "P Received Traffic," all in JSC file Traffic Sent. Engine production: Felt 36–37; Messages Nos. 40 and 41 sent 7 Nov 44, in subfile "P Sent Traffic"; Message No. 24 received 6 Dec 44 ("unglaubwürdig"), in subfile "P Received Traffic"; Message No. 62 sent 19 Dec 44 ("my friend"), in subfile "P Sent Traffic," all in JSC file Traffic Sent; Unsigned memorandum Re: Peasant Case, 8 Dec 44 ("Your suggestion"), in JSC brown Acco binder la-

beled "9 201-400 File 'B.' " "Hearty Christmas": Message No. 28 received 21 Dec 44, in subfile "P Received Traffic," in JSC file Traffic Sent.

456 Colepaugh and Gimpel: Kahn, Spies 3–26; Whitehead 205–206. "Did they have money": Message No. 76 sent 5 Jan 45, in subfile "P Sent Traffic," in JSC file Traffic Sent. "No reason to worry": Message No. 31 received 6 Jan 45, in subfile "P Received Traffic," in JSC file Traffic Sent.

456 Costes history: Noted French Flier Poses as Nazi Agent To Trap Spy, *New York Times,* 26 May 45, p. 1; Former French Flier Is Sentenced Here, *New York Times,* 26 Sep 45; Unsigned memorandum, Re: Cocase, 7 Mar 44, in JSC brown Acco binder labeled "File X"; Maheu 29–32; Ladd to Tamm, Re: Spanip and Cocase, 19 Jun 43, in FBI Spanip file. "If you will give us": Newman Smith endorsement, 14 Feb 44, on unsigned, undated memoranda, Re: Cocase, Number 2/2, B-32, in JSC brown Acco binder labeled "8 File B." Provoke reply: Unsigned, undated memorandum, Re: Cocase, Number 2/7, in JSC brown Acco binder labeled "8 File B." Series of messages: Unsigned, undated memorandum, Summary COCASE Messages, in same. "Although he might": Unsigned memorandum, Re: Cocase, 7 Mar 44, in JSC brown Acco binder labeled "File X."

457 Entrapping Cavaillez: Noted French Flier Poses as Nazi Agent To Trap Spy, *New York Times,* 26 May 45, p. 1; Maheu 32–35; Smith to Record, Subject: [Deleted], 21 Jun 44, in JSC folder Traffic Sent. "A" "B" "C" and "D" messages: JSC brown Acco binder labeled "11 Index."

458 "It is preferable": Smith to Record, 12 Jul 44, in JSC brown Acco binder labeled "7 File A." Tracking deception items: JSC brown Acco binder labeled "5" "File 'C' "; Summaries of Implementation by Special Means, in JSC folder Great Britain. British material: Clarke to Bissell, Subject: HMOC/2407, with endorsements; Bissell to O'Connor, Subject: Exchange of Deception Material, 13 Mar 45, in JSC folder Great Britain; O'Connor to Clarke, HMOC/2805, 10 Apr 45, in JSC folder D Division Weekly Implementation Series; Bissell to O'Connor, Subject: Exchange of Deception Material, 13 Mar 45, in JSC folder Great Britain.

459 Continuity of deception: Strong to Deane, Subject: Deception Measures Against Japan—CCS/284/3/D and CCS/284/5/D, 21 Sep 43; JPS 162/7/D, Continuity of Deception Measures Against Japan, 24 Sep 43; JWPC 96/D, Directive to White Team, Continuity of Deception Measures Against Japan, 25 Sep 43; McFarland to Senior Member, Joint Security Control, Subject: Deception Measures Against Japan, 9 Oct 43, all in NARA RG 218 Box 693 Folder CCS 385 Pacific Theater (4-1-43) Deceptive Policy in the Pacific Theater (Sec. 1). Sending material to theaters: Baumer to Goldbranson, Subject: Implementation of CCS 284/5/D, 17 September, 24 Sep 43, in Goldbranson papers; Bissell and Schuirmann to Commander-in-Chief, Pacific Ocean Areas, et al., Subject: Deception Measures Against Japan in 1943–44, 4 Oct 43; Smith to Strong, Subject: Suggestions to Theater Commanders on Deception, 6 Oct 43; Schuirmann to Strong, 1 Nov 43; Bissell and Schuirmann to Commander-in-Chief, Pacific Ocean Areas, et al., Subject: Deception Measures Against Japan in 1943–44, 4 Oct 43, all in JSC loose Acco binder, "Deception Policy Pacific Areas."

459 Revision of CCS 254/5/D: Baumer to [blank], Subject: Revision of CCS 284/5/D, 6 Jan 44 ("An additional reason"), in JSC loose Acco binder, "Deception Policy Pacific Areas"; General Directive for Deception Measures Against Japan in 1943–1944, JIS 20/M, 26 Jan 44; General Directive for Deception Measures Against Japan in 1943–1944, JIS 20/1, 29 Jan 44, both in NARA RG 218 Box 693 Folder CCS 385 Pacific Theater (4-1-43) Deceptive Policy in the Pacific Theater (Sec. 2); Kehm to Strategy Section OPD *et al.,* Subject: Revision of CCS 284/5/D, undated; Smith to Kehm, Subject: Revision of CCS/284/5/D, 4 Feb 44, both in JSC loose Acco binder, "Deception Policy Pacific Areas"; General Directive for Deception Measures Against Japan, JWPC

190, 15 Feb 44; General Directive for Deception Measures Against Japan, JCS 490/1, 17 Feb 44; General Directive for Deception Measures Against Japan, CCS 284/6, 20 Feb 44, all in NARA RG 218 Box 693 Folder CCS 385 Pacific Theater (4-1-43) Deceptive Policy in the Pacific Theater (Sec. 2); JSM Washington to War Cabinet Offices, London, ZO 76, 30 Mar 44, PRO CAB 121/105/196; JSM Washington to War Cabinet Offices, London, ZO 88, 12 Apr 44, PRO CAB 121/105/195; JSM Washington to WCO London, JSM 1483, 3 Feb 44, PRO CAB 121/105/213; Minutes, COS (44) 34th Meeting (O), 4 Feb 44, PRO CAB 121/105/212; Air Ministry to Britman, Washington, OZ 633, 4 Feb 44, PRO CAB 121/105/211; Wingate to Carver, CO/385, 12 Apr 44, PRO CAB 121/105/193; Air Ministry to JSM Washington, OZ 1949, 14 Apr 44, PRO CAB 121/105/192; General Directive for Deception Measures Against Japan, CCS 284/7, 19 Apr 44, in NARA RG 218 Box 693 Folder CCS 385 Pacific Theater (4-1-43) Deceptive Policy in the Pacific Theater (Sec. 2); General Directive for Deception Measures Against Japan, JPS 372/3, 27 Apr 44; Minutes JPS 150th Meeting, 3 May 44; General Directive for Deception Measures Against Japan, JCS 498/2, 3 May 44; General Directive for Deception Measures Against Japan, CCS 284/8, 6 May 44; General Directive for Deception Measures Against Japan, CCS 284/9, 25 May 44; General Directive for Deception Measures Against Japan, CCS 284/10/D, 26 May 44, all in NARA RG 218 Box 693 Folder CCS 385 Pacific Theater (4-1-43) Deceptive Policy in the Pacific Theater (Sec. 3).

460 "On a typical day": Cake Before Breakfast, p. 10, JSC loose papers.

461 No Kuriles before 1945: Perras 75–80; Hayes 482–86.

461 "A gradual buildup": Outline Plan for Attaining First Objective in Deception Against Japan, 30 Sep 43, in JSC loose Acco binder "Deception Policy Pacific Areas." Basic idea: Cake Before Breakfast, p. 49, JSC loose papers; Herbig 33. Papers, directive to commands in Alaska: J. H. C. for Record, 23 Oct 43; Handy to Commanding General, Western Defense Command, OPD 381 (23 Oct 43), 23 Oct 43, both in NARA RG 218 Box 693 Folder CCS 385 Pacific Theater (4-1-43) Deceptive Policy in the Pacific Theater (Sec. 2); OPD 381 Sec. (3 Nov 43), cited in General Chart Plan "Wedlock," under 22 Oct 43, 26 Oct 43, 3 Nov 43, cited Cake Before Breakfast. p. 36, JSC loose papers.

462 RUDLOFF item: Whitehead 197–98; undated, unsigned memorandum, A-1, initialed "Passed 11/3/43" in Newman Smith's handwriting, in JSC brown Acco binder labeled "7" "File 'A.' " Buckner interview: Clipping from unidentified newspaper, Gen. Buckner Urges Land Occupation of Japanese Mainland, 2 Nov 43; Nigashi to ACofS G-2, Subject: Radio Intercept, 22 May 44, attached to Collins to Smith, 22 May 44, both in JSC folder Wedlock, subfolder "370.2 Wedlock—Adm Impl." AD-JAPAN-44: Deceptive Measures Against Japan, Alaskan Department, JPS 372 Appendix, 9 Jan 44, in NARA RG 218 Box 693 Folder CCS 385 Pacific Theater (4-1-43) Deceptive Policy in the Pacific Theater (Sec. 2).

462 JCS approval: Deceptive Measures Against Japan, Alaskan Department, JWPC 176/1, 19 Jan 44; Deceptive Measures Against Japan, Alaskan Department, JPS 372/1, 12 Feb 44, both in NARA RG 218 Box 693 Folder CCS 385 Pacific Theater (4-1-43) Deceptive Policy in the Pacific Theater (Sec. 2); CINCPAC-CINCPOA to COMINCH, Serial 00012, 3 Feb 44; Smith to Record, Subject: Deception Measures Against Japan, Alaskan Department Plan AD-Japan-44, 11 Feb 44, both in JSC folder Japan; Deceptive Measures Against Japan, Alaskan Department, JCS 705, 13 Feb 44; Graves to Aide to COMINCH et al., Subject: Deceptive Measures Against Japan, Alaskan Department 17 Feb 44, both in NARA RG 218 Box 693 Folder CCS 385 Pacific Theater (4-1-43) Deceptive Policy in the Pacific Theater (Sec. 2). OPD transmittal: Handy to Commanding General, Alaskan Department, OPD 381 Sec. (17 Feb 44), 17 Feb 44, in NARA RG 218 Box 693 Folder CCS 385 Pacific Theater (4-1-43) Deceptive Policy

in the Pacific Theater (Sec. 2). Smith and Bissell: Notation in Smith's handwriting on unsigned, undated Memorandum for General Bissell, Subject: Brief of Deception Plan, Alaskan Department (JCS 705), in JSC folder Wedlock—subfolder "370.2 Wedlock—Plan." Deceptive radio: Herbig 42; C. E. G[oldbranson] to Record, Subject: "Wedlock" Plan, 26 Feb 44, in JSC folder Wedlock—subfolder "370.2 Wedlock—Plan"; Smith to CG Alaskan Department, No. 7056, 15 Mar 44, CM-OUT-6572 (16 Mar 44), in untitled, unsigned, undated account of Operation WEDLOCK, p. 38½, in NOA 551, Folder Wedlock Section III.

463 Alaska Command planning: General Chart Plan "Wedlock," under 12 Feb 44, in Cake Before Breakfast. p. 37, JSC loose papers; Verbeck to Assistant Chief of Staff, G-2, Subject: Plan "Wedlock," 18 Mar 44, in JSC folder Wedlock—subfolder "370.2 Wedlock—Plan."

464 Nimitz participation: Smith to Verbeck, Subject: Implementation Plan "Wedlock," 13 Mar 44, JSC/J1E1 Serial 336; Hq Als Dept Ft Richardson to WAR, No. A 3876, 12 Mar 44, CM-IN-8271 (12 Mar 44), both in JSC folder Wedlock, subfolder "370.2 Wedlock—Adm Impl"; HQ Als Dept to WAR, No. A3856, 11 Mar 44, CM-IN-7510 (11 Mar 44), in NARA RG 218 Box 693 Folder CCS 385 Pacific Theater (4-1-43) Deceptive Policy in the Pacific Theater (Sec. 2); CINCPAC Serial 00029, Subject: Propaganda Plan for "Forager," 22 Mar 44; Deception Plans for "Forager," JPS 433/1, 13 Apr 44; Deception Plans for "Forager," JCS 806/1, 16 Apr 44; Supplemental Minutes, JCS 159th Meeting, 18 Apr 44; Memorandum of informal action, Deception Plan for "Forager," JCS 806/2, all in NARA RG 218 Box 688 Folder CCS 385 Pacific Ocean Areas (4-5-44) Propaganda Plan for Forager.

464 San Francisco meeting, plan: Buckner, Redman, and Holsopple, Statement of Principles Agreed Upon in Conference at San Francisco 23 March 1944 Regarding Plan Wedlock; CINCPOA to COMINCH, 300544Z, NCR 7525-S, 31 Mar 44; Smith to Commanding General, Alaskan Department, 31 Mar 44, CM-OUT-16880 (31 Mar 44), all in JSC folder Wedlock—subfolder "370.2 Wedlock—Plan." Collins: General Chart Plan "Wedlock," under 2 Apr 44, in Cake Before Breakfast. p. 37, JSC loose papers; Smith to Verbeck, Subject: Implementation Plan "Wedlock," 13 Mar 44, JSC/J1E1 Serial 336; Smith to Verbeck, Subject: Plan "Wedlock," 12 Apr 44, JSC/J1E1 Serial 436, both in JSC folder Wedlock, subfolder "370.2 Wedlock—Adm Impl." Bogus divisions: Smith to Commanding General, Alaskan Department, Subject: Plan "WEDLOCK"—Order of Battle, 10 May 44, JSC/J1E1 Serial 549; Smith to CG, Alaskan Department, Subject: Plan "Wedlock," 1 Jun 44, JSC/J1E1 Serial 601, both in JSC file Wedlock—subfolder "370.2 Wedlock—Impl Phys." Canadian troops: Smith to CG, Alaskan Department, Subject: Plan "Wedlock," JSC/J1E1 Serial 549, 10 May 44, with enclosures, in JSC folder Wedlock—subfolder "370.2 Wedlock—Impl Phys"; Collins to Smith, 22 May 44, in JSC folder Wedlock—subfolder "370.2 Wedlock—Adm Impl."

465 Air and sea strikes: Herbig 48. Special means: Message A-17, 22 Mar 44 (Rudloff 3/22); Message A-27A, 10 Apr 44 (Fleming channel); Message A-20, 30 Mar 44 (invasion craft), all in unsigned memorandum Wedlock Series Special Means, 5 Apr 46, p. 1, in NOA 551, Folder Wedlock OP-03-C-Out 204. Ike complaint: Message 650, 2 May 44, in undated tabulation of Rudloff Wasch messages, in JSC brown Acco binder labeled "14" (slightly different dates in other files).

465 Seattle tonnage; fifty carriers; floating docks; halftrack gear; Stabinol; Canadians; Arctic clothing: General Chart Plan "Wedlock," Cake Before Breakfast pp. 38–39, JSC loose papers (slightly different dates for some items in other files). Navy at Adak; West Coast bars: Messages A-47A and A-48, 20 May 44, in unsigned memorandum Wedlock Series Special Means, 3 Apr 46, p. 5, in NOA 551, Folder Wedlock OP-03-C-Out 204. Press speculation: Unsigned, undated draft Resume and Evaluation of

AD-JAPAN 1944 . . . , in NOA 551, Folder Wedlock. Patches: Smith to Commanding General, Alaskan Command, Subject: Plan "Wedlock," 24 Jun 44, JSC/J1E1 Serial 6105, and associated documents, in JSC folder Wedlock, subfolder "370.2 Wedlock—Adm Impl."

466 Soviets: Herbig 47 (list to Moscow); untitled, unsigned, undated account of Operation WEDLOCK, p. 56-1/2, in NOA 551, Folder Wedlock Section III (Adak). Halsey: CINCPOA to COMINCH 13 May 44, cited in unsigned, undated memorandum, Selected Briefed Dispatches from NorPac and OP-20-K Files . . . , in NOA 551, Folder Wedlock (Nimitz suggestion); Smith to Grosskopf, Subject: Cable 130422 From CINCPOA to COMINCH, 17 May 44, JSC/J1E1 Serial 580 (mention suppressed); C. E. G[oldbranson] to Smith, 8 Apr 44 ("going to a new"), both in JSC folder Wedlock, subfolder "370.2 Wedlock—Adm Impl."

467 Communications deception: Joint Communications Center, APO 980, Report of Operations From 1 May 1944 Thru 31 Dec 44, in JSC folder Wedlock—subfolder "JT COMCENTER 1 May–31 Dec 44" (Task Group Nan; joint center; stations; port traffic); Undated, unsigned memorandum, Principles Agreed Upon in Conference at San Francisco, 23 March 1944, Regarding Plan Wedlock, Appendix A (Navy Dept.), in JSC file Wedlock—subfolder "370.2 Wedlock—Impl Phys" (special crypto); Herbig 43, 45, 49 (call signs; Sicily; discontinuance). Operation launching, breaking off: Herbig 45; Joint Communications Center, APO 980, Report of Operations From 1 May 1944 Thru 31 Dec 44, in JSC folder Wedlock—subfolder "JT COMCENTER 1 May—31 Dec 44"; General Chart Plan "Wedlock," in Cake Before Breakfast, JSC loose papers; Untitled, unsigned, undated account of Operation WEDLOCK, in NOA 551, Folder Wedlock Section III; CINCPOA to COMINCH and COMNORPAC, 282205Z [28 May 44]; CINCPOA to JANCOMCEN Adak, 022050Z [2 June 44], in JSC folder Wedlock—subfolder "370.2 Wedlock—Impl Phys"; Collins to Smith, Subject: Plan Wedlock, 25 Jun 44, in JSC folder Wedlock—subfolder 370.2 Wedlock—Plan.

468 STALEMATE proposal approved: CINCPAC to COMINCH, Subject: Deception and Propaganda Plans for "Stalemate," 9 Jun 44, Serial 00072; Deception and Propaganda Plans for "Stalemate," JPS 479/1, 20 Jun 44; Deception and Propaganda Plans for "Stalemate," JCS 915, 22 Jun 44; Memorandum of informal action, Deception and Propaganda Plans for "Stalemate," JCS 915, 25 Jun 44, all in NARA RG 218 Box 688 Folder CCS 385 Pacific Ocean Areas (6-9-44) Deception and Propaganda Plans for Stalemate.

468 HUSBAND planning, approval: Unsigned, undated memorandum on planning and execution of deceptive measures in the Alaskan-Aleutian theater in 1944–45, p. 15, in JSC folder Operation Wedlock; Joint Security Control to CINCSWPA et al., WARX 56822, 27 Jun 44, CM-OUT-56822 (27 Jun 44); CG Alaskan Department to War Department et al., 30 Jun 44, CM-IN-44 (1 Jul 44); CINCSWPA to War Department et al., No. CX 14423, 3 Jul 44, CM-IN-2475 (4 Jul 44); Joint Security Control to CINCPOA et al., WARX 60049, 4 Jul 44, CM-OUT-60049 (4 Jul 44); Joint Security Control to CINCPOA et al., WARX 60049, 4 Jul 44, CM-OUT-60049 (4 Jul 44), all in NARA RG 218 Box 688 Folder CCS 385 Pacific Ocean Areas (6-9-44) Deception and Propaganda Plans for Stalemate; various messages in JSC file Wedlock—subfolder "370.2 Wedlock—Impl Phys"; Undated Report of Conference Regarding Future Employment of Plan Wedlock, enclosure to CINCPAC and CINCPOA to Commanding General, Alaskan Department, et al., Subject: Report of Conference Regarding Future Employment of Plan Wedlock, in JSC folder Wedlock—subfolder "370.2 Wedlock—Plan"; Report of Conference Regarding Plan "Wedlock," JCS 705/1, 17 Aug 44; Report of Conference Regarding Plan "Wedlock," JPS 372/5/D, 18 Aug 44; Donnelly to Duncan et al., Subject: Report of Conference Regarding Plan Wedlock, 20 Aug 44; Re-

port of Conference Regarding Plan "Wedlock," JCS 705/2, 24 Aug 44; Buck slip, undated, all in NARA RG 218 Box 694 Folder CCS 385 Pacific Theater (4-1-43) Deceptive Policy in the Pacific Theater (Sec. 3).

469 HUSBAND implementation: Collins to Smith, Subject: Plan Wedlock, 21 Jul 44; Collins to Smith, Subject: Plan Husband, 4 Sep 44, both in JSC folder Wedlock, subfolder "370.2 Wedlock—Adm Impl"; Message 062338, 6 Sep 44, in unsigned, undated tabulation Selected Briefed Dispatches from NorPac and Op-20-K files . . . , in NOA 551, Folder Wedlock; Unsigned, undated memorandum on planning and execution of deceptive measures in the Alaskan-Aleutian theater in 1944–45, pp. 14–15, in JSC folder Operation Wedlock; Collins to Smith, Subject: Plan Wedlock, 25 Jun 44 (CUB FILBERTS); Collins to Smith, Subject: Plan Wedlock, 21 Jul 44 (BIGBOBS in storms), both in JSC folder Wedlock—subfolder "370.2 Wedlock—Plan"; Untitled, unsigned, undated account of Operation WEDLOCK, p. 92½, in NOA 551, Folder Wedlock Section III; unsigned, undated draft Resume and Evaluation of AD-JAPAN 1944, . . . in NOA 551, Folder Wedlock; Cake Before Breakfast, pp. 37, 41, 46, JSC loose documents.

469 Special means: E. P. S[chrup] to Smith, Subject: Messages designed to confirm presence of five (fictional) divisions in the Aleutians as indicated by Plan "Wedlock," 21 Jun 44; draft message Smith to Collins, 5 Aug 44, sent 7 Aug 44, both in JSC folder Wedlock, subfolder "370.2 Wedlock—Adm Impl." Message 73, 10 Sep 44 (submarine); Message 118, 21 Sep 44 (language instruction); Message A-X-5 (delayed plans), all in File Log, JSC brown Acco binder labeled "7" "File 'A.' "

470 Technical weaknesses: Unsigned memorandum, Evaluation by OP-20K of Wedlock Communications, October 1945, in NOA 551, Folder Wedlock.

470 Special means: Unsigned memorandum, Evaluation Report on Operation Wedlock, Deception Implementation, 17 Apr 46, p. 4, in NOA 551, Folder Wedlock OP-03-C-Out 204. Absorbed much; desired appreciation: Unsigned memorandum, Evaluation Report on Operation Wedlock, Deception Implementation, 17 Apr 46, pp. 1–2, 4, in NOA 551, Folder Wedlock OP-03-C-Out 204. 400,000 men, 700 planes: Tabulation and charts attached to Nelson to Smith, Subject: Japanese Estimate of US Strength—Accuracy Check on, 15 Sep 45, F-28 Out No. 162, in NOA 551, Folder Wedlock OP-03-C-Out 204. Powerful naval force; Naval General Staff: Untitled, unsigned, undated account of Operation WEDLOCK, p. 174, in NOA 551, Folder Wedlock Section V—Preliminary Material for Critique. Vice Chief: Unsigned, undated Evaluation of Army Phase of Operation "Wedlock" (April–July 1944), Based on Enemy Analysis of the Operation, p. 10, in JSC folder Wedlock—subfolder "Operation Wedlock." "Will one of these": Associated Press dispatch, Japs Believe Yank Offensive Is Coming by Way of Aleutians, attached to Collins to Smith, 22 May 44, in JSC folder Wedlock, subfolder "370.2 Wedlock—Adm Impl." "The communication traffic"; special means; press releases: Unsigned memorandum, Evaluation Report on Operation Wedlock, Deception Implementation, 17 Apr 46, pp. 1–4, in NOA 551, Folder Wedlock OP-03-C-Out 204. Main American thrust: Morison, New Guinea 145; Ritchie to Chief, Strategy & Policy Group, Subject: Comments on JPS 433/1 and JCS 806, 15 Apr 44, in NARA RG 165 ABC File ABC 381 Japan 15 Apr 43 Sec. 4—FORAGER; Herbig 53, 55, 167. Japanese reinforcements: Untitled, unsigned, undated account of Operation WEDLOCK, pp. 139, 153, 164, in NOA 551, Folder Wedlock Section IV Folder No. 2—Enemy Reaction. Number of troops: Herbig 164–65 and note 17, citing 60,000 for 14 Jun 44 according to researches in the war diary of Army Imperial Headquarters by Professor Ichiki, Professor of Military History, National Defense College, Tokyo. According to Major Shimada, Military Affairs Section, Army GHQ, postwar interview by USSB, quoted in A Review of Japanese Intelligence, 18 Apr 1946, p. 10, in NOA 535, Folder Japanese Intelligence Analysis of WW II, the total number of ground

troops in the Kuriles rose from 14,000 in 1943 to 41,000 in 1944, and in Hokkaido
from 20,600 to 34,000. Tabulation and charts attached to Nelson to Smith, Subject:
Japanese Estimate of US Strength—Accuracy Check on, 15 Sep 45, F-28 Out No. 162,
in NOA 551, Folder Wedlock OP-03-C-Out 204 shows 25,000 Japanese army troops
in January 1944 and 70,000 ending June 1944. Air reinforcements: A Review of Japanese Intelligence, 18 Apr 1946, p. 10, in NOA 535, Folder Japanese Intelligence
Analysis of WW II; untitled, unsigned, undated account of Operation WEDLOCK, p.
145, in NOA 551, Folder Wedlock Section IV Folder No. 2—Enemy Reaction; tabu-
lation and charts attached to Nelson to Smith, Subject: Japanese Estimate of US
Strength—Accuracy Check on, 15 Sep 45, F-28 Out No. 162, in NOA 551, Folder
Wedlock OP-03-C-Out 204. Subsequent reduction: Unsigned, undated Analysis of
the planning and execution of cover and deception measures for the Alaskan-Aleutian
Island theater during the period from 1 November 1943 to 3 April 1945 entailing
Wedlock, Husband, Bambino, and Valentine, p. 19, in NOA 551, Folder Wedlock
OP-03-C-Out 204. Subsequent cutback: A Review of Japanese Intelligence, 18 Apr
1946, p. 10, in NOA 535, Folder Japanese Intelligence Analysis of WW II. SHO plan:
Spector 421–22. Northern anchor: Herbig 164–65.

472 Nursed and nudged: Schedule of Events by Which Plan Wedlock Will Be Implemented, 22 May 44, in JSC folder Wedlock—subfolder "370.2 Wedlock—Impl
Phys"; Chart "Wedlock" "Forager," in JSC folder 370.2 Forager"; notation by C. E. G.
to Schrup on Collins to Smith, Subject: Plan Wedlock, 25 Jun 44, in JSC folder Wedlock—subfolder "370.2 Wedlock—Plan"; chart headed "Wedlock" (Schrup's handwriting), in JSC file Wedlock—subfolder "370.2 Wedlock—Impl Phys."

473 "He knows very well": O'Connor to Bevan, HMOC/1010, 6 May 44, PRO CAB
154/75/119. "The nakedness of the land": O'Connor to Bevan HMOC/1015, 9 May
44, PRO CAB 154/37/15A. Even if Bissell consent: O'Connor to Bevan, No. 900, 27
Jul 44, PRO CAB 154/37/20A. "Feeble bromide": O'Connor to Bevan, No. 573, 8 Jan
45, PRO CAB 154/36/41B. "G2 have no expectation": O'Connor to Bevan, GOR
198, 27 Mar 45, PRO HW 41/159. "They speak of [their] channels": O'Connor to
Bevan, OC/281, 2 Feb 44, PRO CAB 154/45/31A. "Certain personnel": Bevan to
Fleming, 8 Sep 44, PRO CAB 154/26/15A. Any British recommendation: O'Connor
to Bevan, 11 Mar [sic; Apr?] 45, PRO CAB 154/36/50A. "Our friends here"; "in an
amateurish way": O'Connor to Bevan, HMOC 3064, 10 May 45, PRO 154/40.
"Overruled on all counts": O'Connor to Bevan, No. GOR 269, 7 May 45, PRO HW
41/159. "How hopeless": Bevan to O'Connor, C/O.430, 23 Sep 43, PRO CAB 154/
75/96. "Washington simply": Bevan to Fleming, 8 Sep 44, PRO CAB 154/26/15A.
"If, therefore, J.S.C.": Bevan to O'Connor, COO/385, 7 Apr 45, PRO CAB 154/36/
47A. "Signs of the Americans": Bevan to O'Connor, CO/385, 3 Apr 45, PRO CAB
154/36/46A. "Since the Americans": Bevan to O'Connor, LCS (45) I/D.5, 18 Apr 45,
PRO CAB 154/40/17A. "Judging by local talent": Fleming to Bevan No. 128, 31 Aug
43, PRO CAB 154/75/81. "For your information": Fleming to Horder, No. 527, 24
Sep 44, PRO WO 203/3314. "A body in whose capacity": Fleming to Bevan, O'Connor, and Horder, No. IND 5186, 25 Mar 45, PRO HW 41/159. Strong keeping matters to himself: O'Connor to Bevan, OC/429, 19 Apr 44, PRO CAB 154/75/110.
"Told me that deception needed": Liddell Diary, 6 Sep 43, PRO KV 4/192. "A recipe
for inaction": Howard 216. Approach to Strong: Bevan to Fleming, OZ 2651, 3 Sep
43, PRO CAB 154/75/83. Approach to Kehm: Bevan to O'Connor, C/O/465, 18 Sep
43, PRO CAB 154/75/93; Bevan to O'Connor, C/O.430, 23 Sep 43, PRO CAB 154/
75/96. Approach to Bissell: O'Connor to Bevan, HMOC/1010, 6 May 44, PRO CAB
154/37/14A; O'Connor to Bevan HMOC/1015, 9 May 44, PRO CAB 154/37/15A.
Approach to Fairbanks: Bevan to Fleming, L/99, 5 Sep 44, PRO HW 41/159; O'Connor to Bevan, HMOC/1915, 7 Dec 44, PRO CAB 154/75/26A. Approach to

Thurber: Bevan to O'Connor, Fortnightly Letter No. 37, CO/426/1, 20 Jul 45, PRO CAB 154/36/25A. Navy cooperate if asked: O'Connor to Bevan, HMOC/849, 22 Mar 44, PRO CAB 154/36/10A. Bissell ambitious; "far more concerned": Petavel to Bevan, 21 Apr 45, PRO CAB 154/36/52A.

Chapter 12: BODYGUARD

477 Morgan, COSSAC: Dictionary of National Biography 1961–1970; Morgan, Overture 208–10; Marshall to Eisenhower, No. 5585, 23 Dec 43; Smith to Eisenhower, No. W-9154, 2 Jan 44, both in Bedell Smith Papers, Box 19, Folder COSSAC (Chief of Staff Supreme Allied Command) Staff Organization Eyes Only Cable File (1), Eisenhower Library. Formal directive; "an elaborate": Combined Chiefs of Staff, Organization of Command, Control, Planning and Training for Cross-Channel Operations, Directive to the Chief of Staff to the Supreme Commander (Designate), CCS 169/8/D, 23 Apr 43, para. 5a, in NARA RG 165 ABC File ABC 381 (22 Jan 43) Sec. 1, COSSAC directives etc.; Howard 75; History of COSSAC, p. 13, in Bedell Smith Papers, Box 33, Folder History of COSSAC (Chief of Staff to Supreme Allied Commander), 1943–44 (May 1944), Eisenhower Library.
478 "Camouflage and pretense": Howard 74. Ops (A) and (B): Hesketh 15–16. Jervis-Read: Howard 75. Did not foresee: Hesketh 15–16.
478 "A charming": Wheatley 189. Seconded: Memorandum, Hesketh to Casey, attached to USAMHI copy of Hesketh, p. 1. Only planning authority: Minutes, COS 86th (O) Meeting, COS(43)208(O), 27 Apr 43, PRO CAB 119/67/34. Hollis suggestion: Howard 74; Hesketh 3. Outline scheme: Deception Plan for 1943, COSSAC/00/512/Ops, 29 Apr 43, in NARA RG 331 Entry 3 Box 122 Folder COSSAC (43) Papers I Nos. 1–22. UPSHOT: LCS Summary of Information No. 2, 30 May 43, in JSC folder 370.2 Operations and Results. "There has been an appalling": Howard 154. TINDALL: Howard 81–82.
479 STARKEY: Wingate 150. WADHAM: Operation "Wadham," Appreciation and Outline Plan, COSSAC (43) 24, 15 Jun 43, para. 2a, in JSC folder 370.2 Wadham.
479 Narvik: Wingate 151. Montagu: Deception on Naval Matters, p. 16, PRO ADM 223/794/143. Resistance forces: Wingate 150–51. Reporters: Wingate 150; Kehm to Joint Security Control, Subject: Comments on Psychological Warfare Plan for COCK-ADE, 20 Aug 43, in JSC folder European Theater 000.
480 LARKHALL: Wingate 139. April 23 memo: Suggested Co-Operation by Joint Security Control in the Implementation of Deception Policy 1943 Part I—Germany and Italy, 23 Apr 43, in NARA RG 319 Entry 101 Box 3 Folder 15 "Deception in UK—Larkhall and Cockade." OPD, Strong: Wedemeyer to Hull, Subject: Deception Measures, 7 May 43; Hull to Joint Security Control, Subject: Deception Measures, 7 May 1943, both in NARA RG 319 Entry 101 Box 3 Folder 15 "Deception in UK—Larkhall and Cockade"; Table dated 22 Jun 43, rows 1 and 2 column B, in JSC folder European Theater 000.
481 LARKHALL implementation: Scenario for Operation "Wadham" (as of July 10 1943), in JSC folder 370.2 Wadham; Smith to Strong, Subject: Implementation—LARKHALL, 14 Jul 43, in JSC folder European Theater 000; Wingate 139–40. DUNDAS implementation: Wingate 140–42, FHW estimate: Howard 76. Jodl estimate: Wingate 143–44, 405–406.
482 Abbott: Goldbranson to Smith, 20 Aug 43, in JSC folder 370.2 Wadham; Cook to Strong, DS/41, 8 Mar 43, in JSC folder 370.2 Operations and Results.
483 Ingersoll early life: Hoopes 18, 31–61. At the New Yorker: Hoopes 61–79; Gill 46. At Time Inc.: Hoopes 29, 79–80, 88–118, 123–24, 152–53, 163–65, 168–69, 174–76,

183–87, 222–23; Swanberg 38–41, 148, 158; Herzstein 58, 98. At *PM:* Hoopes 210, 219–21, 232–33, 241–53, 253 ("perhaps the most celebrated"), 272; Milkman 59 ("egocentric hysteria"). Enlisted: Hoopes 260–63; Milkman 58–59.

484 Noce: Noce biographical release, USAMHI; obituary, *Washington Post,* 20 Feb 76. Training manuals: Ingersoll memorandum, undated, c. 1965, Re: 7 Army Manuals Herewith, in Ingersoll papers, box unknown. PA officer: Hoopes 265–69.

484 To London: Ingersoll undated memorandum, "Dorchester Hotel Bill," in Ingersoll papers, box marked "Ingersoll, R. 5-19-80 PM First Class 6R." Book: Hoopes 270–74. Hit by cab: Hoopes 274–75. Treated loftily: Hoopes 276.

485 "Hard-driving": Herzstein 58. "Insulting"; "manipulative": Hoopes 129. "Capricious": Swanberg 145. "Erratic": Hoopes 195. "Conceited egoist": Hoopes 185. Hypochondriac: Swanberg 121, 123. "Ingersoll was the trickiest": Hoopes 278. Memo-writer: Hoopes 129. "He blew": Hoopes 185. "Annoying habit": Hoopes 184.

486 RATTLE: Wingate 156; Morgan, Overture 138. TINDALL implementation: Howard 81–82; Air Ministry to B.A.D. Washington, 10 Aug 43; Scenario for Operation Wadham, p. 3, Item 12, both in JSC folder 370.2 Wadham.

486 STARKEY implementation: Morgan, Overture 93–98; Howard 78–79; Wingate 151–52. HARLEQUIN: Wingate 152; Howard 78. FORFAR raids: Cruickshank, Deception 68; Young & Stamp 75–76. Letters, leave stopped; PRO CAB 154/20. "Explosion that shattered": Morgan, Overture 99. GARBO: Pujol 104; Howard 78. Other double agents: Masterman 118–19; Howard 76. Pigeons: PRO 154/35. "It was an inspiring": Morgan, Overture 100. German reaction: Howard 80. "A German coast-artillery"; troops disappointed: Morgan, Overture 100–101. "It appears": Howard 78.

489 WADHAM planning: Wingate 154; Strong to Devers, Subject: Implementation of Operation "Wadham," 12 Aug 43; unsigned, unaddressed, undated text of message (V Corps); Smith to Strong, Subject: Attached Letter From E. T. O, 27 Aug 43; Operation "Wadham," Appreciation and Outline Plan, COSSAC (43) 24, 15 Jun 43; Noce to COSSAC, Subject: Implementation of Deception Scheme "Wadham," AG-381, 3ETO-CW-3-(11/7/43), 11 Jul 43; Woodruff to Commanding General, ETOUSA, Subject: Operation Wadham as Pertains to Follow-Up Force, 20 Jul 43, all in JSC folder 370.2 Wadham; Andrews to Marshall, 26 Apr 43; Devers to Marshall, 19 May 43, both in JSC folder 370.2 Operations and Results; Kehm to Handy, Subject: COCKADE and WADHAM, 26 Jul 43, in JSC folder 370.2 Cockade.

489 WADHAM implementation: Wingate 154–56; Cruickshank, Deception 82; Edwards to Cusick, Subject: Directive for trip to the United States, 26 Jul 43; Goldbranson to Smith, 20 Aug 43; Smith to Strong, Subject: Comment on Implementation of Operation "Wadham" in U.S., 30 Aug 43; Cort to Kehm, Subject: Points Relative to Wadham Still Undecided, 27 Jul 43; Edwards to Commanding General VII Corps, Subject: Directive for Establishing Troop Basis, 20 Jul 43; Commanding General VII Corps to G-2 MIS Washington, 30 Jul 43 ("the European fortress"); Noce to COSSAC, Subject: Implementation of Deception Scheme "Wadham," AG-381, 3ETO-CW-3-(11/7/43), 11 Jul 43; Scenario for Operation "Wadham" (as of July 10 1943), all in JSC folder 370.2 Wadham; Kehm to Christiansen, Subject: Designation of Units, 26 Jul 43, in JSC folder 370.2 Cockade; "Should the U.S.": Kehm to Blizzard, Subject, Deception Planning in the European Theater, 2 Jul 43, in NARA RG 319 Entry 101 Box 3 Folder 15 "Deception in UK—Larkhall and Cockade."

490 Special means: Strong to Devers, Subject: Implementation of Operation "Wadham," 12 Aug 43, in JSC folder 370.2 Wadham; Tabulation, Larkhall, undated, in JSC folder European Theater 000; Messages 175–77, 29 Jul 43; 229–32, 3 Sep 43; 233–38, 8 Sep 43, in undated tabulation of Rudloff messages, p. 1, in JSC brown Acco binder labeled "14." Navy participation: Goldbranson to Smith, item 2(d), 20 Aug 43; Sherburne to Chief of Strategy Section, OPD, WD, Subject: Termination of VII Corps Participation

in WADHAM, 10 Sep 43; Hull to Bieri, Subject: Implementation of Wadham, 10 Aug 43; Cominch to CinC United States Atlantic Fleet, Subject: Task Force 69—Directive for the Composition, Assembly, Loading, Overseas Movement, and Disposition of, FF1/A16-3, Serial 001671, 17 Aug 43; Kehm, Conclusions in Respect to Wadham Reached 22 August 1943; Smith to Strong, Subject: Comment on implementation of operation "WADHAM" in U.S., 30 Aug 43 ("done too late"), all in JSC folder 370.2 Wadham; "A mounting wave"; "the explanation": Wingate 154.

491   Goldbranson critique, V Corps: Goldbranson to Smith, 20 Aug 43; Woodruff to Commanding General, OPD, Subject: Control of Deceptive Planning, 24 Sep 43; Sherburne to Chief of Strategy Section, OPD, Subject: Termination of VII Corps Participation in WADHAM, 10 Sep 43, all in JSC folder 370.2 Wadham. TINDALL II: Wingate 151.

492   "The operational reactions": Note on Enemy Reaction to "Cockade," M.I.14/15/590/43, 1 Oct 43, p. 4, PRO CAB 154/61/6A. COCKADE results: Howard 79 ("the *Schwerpunkt*"), 81–83; Wingate 403–404; Morgan, Overture 100. "It never took much"; "all the evidence": Howard 82. Hesketh procedure: Hesketh to Casey 3. "Your activity and that": Howard 233–34.

494   Not Pas-de-Calais: Morgan, Overture 142; Operation "Overlord" Report and Appreciation With Appendices, July 1943, COS (43) 416 (O), Appendix C, copy in Nevins papers, Box "COSSAC/SHAEF: Overlord Plans 1943–44" ("the most heavily defended"; "in spite of").

494   "Our first survey": Wingate 196. "The sole favorable": Wingate 197. "To induce the enemy to weaken"; "to contain"; "to induce the enemy to dispose": Deception Policy in Relation to Overlord, First Thoughts, 13 Jul 43, in NARA RG 331 Entry 1 Box 73 Folder 381 Fortitude Vol. 1.

495   DUMMER: Wingate 162–63. "More despairing"; "resulted in a good deal": Wingate 199. Hard thinkers: Goldbranson draft memorandum to General Strong, 27 Aug 43; Smith to Strong, Subject: Comment on Colonel Bevan's "First Thoughts" in Relation to "OVERLORD," 30 Aug 43; Kehm to Joint Security Control, Subject: Comments on "First Thoughts," 21 Aug 43, all in JSC folder 370.2 Overlord.

496   JAEL: Deception Policy for the War Against Germany—Plan Jael, L. C. S. (43)(P)(5) (First Draft), 22 Sep 43, PRO CAB 119/66/10–18; Bevan to Cornwall Jones, 22 Sep 43, PRO CAB 119/66/19; Wingate 199–200.

497   COSSAC reaction: Barker to Secretary COS Committee, Subject: Plan "JAEL," COSSAC (43) 65 (Final), in NARA RG 331 Entry 3 Box 122 Folder COSSAC (43) Papers I Nos. 1–22. Clarke visit: Clarke diary 26 Sep 43 to 28 Oct 43, Clarke papers. COSSAC concurs: Wingate 199. United States reactions: "generally sound": Goldbranson to Smith, 8 Oct 43, in JSC folder European Theater 000; Unsigned memorandum initialed WHB, Notes on CPS [blank] Meeting, October 1943, Deception Policy for the War Against Germany—Plan Jael; Bieri to Kuter and Wedemeyer, Subject: Deception Policy for the War Against Germany—Plan Jael, 9 Oct 43, with endorsement Kuter to Roberts, 13 Oct 43, both in NARA RG 165 ABC File ABC 385 Europe (23 Sep 43) Sec. 2. Bevan and colleagues: Howard 105. Chiefs defer: Minutes, COS 260th (O) Meeting, COS(43)651(O), 26 Oct 43, PRO CAB 119/66/6.

498   Stopgap; LARKHALL revised: Wingate 202.

498   FOYNES: Plan "Foynes," LCS (43)16 FINAL (Revised), 9 Nov 43, PRO CAB 81/77/22–23; Wingate 205–206; IV/5–6. Straits, GOTHAM: IV/5; Wingate 316, 333–34. Double bluff: Minutes, COS 269th (O) Meeting, COS(43)680(O), 4 Nov 43, PRO CAB 119/66/32–33.

499   Roenne: Wingate 206–207; Howard 113–14 ("no indication"; "had no information"; "a masterly grasp"). Paternity message: Praun 73.

500   LANGTOFT: Wingate 158–59; Deception—Notes on Various Incidents and Matters,

PRO ADM 223/794/340. LEYBURN: Wingate 163. SHOTGUN: Wingate 209. GILMERTON, BLARNEY STONE, CHRISTMAS CAKE, CATSPAW, RISSOLE: PRO CAB 154/42; Wingate 209. PREMIUM: Hesketh 40, 55, 92 note.

501   Personnel: Wheatley 160. "Devers watched": Casey, The Grand Deception or Confusion to the Enemy, pp. 5–6, in Mure papers, Box 3.

502   Ingersoll self-credit: Hoopes 278. "These conclusions": Operation "Overlord" Report and Appreciation With Appendices, July 1943, COS (43)416(O), Appendix C, copy in Nevins papers, Box "COSSAC/SHAEF: Overlord Plans 1943–44," USAMHI.

502   TORRENT/APPENDIX Y: COSSAC/00/6/3/Ops, Subject: Operation "TORRENT," 6 Sep 43; Operation "Torrent," Appreciation, 16 Sep 43, COSSAC (45) 39 (Second Draft for Head Planners); COSSAC/23FK/INT. 7 Oct 43; various other drafts, all in NARA RG 331 Entry 1 Box 73 Folder 381 Fortitude Vol. 1; COSSAC (43) 28, 20 Nov 43, in NARA RG 331 Entry 3 Box 122 Folder COSSAC (43) Papers I Nos. 1–22; Hesketh 8; Howard 105–106; Campbell 27.

502   Harris history: Hunter Harris oral history 2–3, 6, 51; Harris biographical sketch, USAMHI; Cullum No. 9608; Noce to The Adjutant General, 5 Aug 48, Harris papers; Dufty 82. Negative report: Howard 107. "Putting a hooped": Hoopes 277. Ingersoll idea: Hoopes 277; Bull to Noce, Subject: Reaction to "Torrent," 16 Nov 43, in NARA RG 331 SHAEF G-3 General Records Subject File 1942–45 "Combatting the Guerrilla" to "Daily Reports," Folder Cover Plan—Overlord (Bull Papers).

504   "To weaken"; "the most important"; "might well be": Clarke to Jervis Read, C/O/348/28, 28 Oct 43, PRO WO 219/2204A/204–206. "Long-term preparations": Operation "Overlord"—Cover Plan, Note by Controlling Officer, 11 Nov 43, COS(43)701(O), in NARA RG 331 Entry 1 Box 73 Folder 381 Fortitude Vol. 1. Held up with three exceptions: Minutes, COS 292d (O) Meeting, COS(43)735(O), 30 Nov 43, PRO WO 219/2204A/265; Hesketh 8; Wingate 248; Howard 106. Harris and Ingersoll to FUSAG: Cullum No. 9605.

505   "Truth deserves"; "This is what we call": Extract from 4th Eureka meeting, Tehran, Iran, 30 Nov 43, in NOA 543, Folder Russia—Deception. Bevan at SEXTANT: Clarke diary 27 Nov 43, 8 Dec 43, Clarke papers; Wingate 230. Directive: Howard 107.

505   "Haggard"; "it had become"; "lazy and indifferent": Wheatley 162–64.

506   "Drafted, redrafted"; "it is difficult": Gordon Clark to Mure, 6 Nov 80, Mure papers (Box 2).

506   Approval by British: Minutes COS (43) 315th Meeting (O), 24 Dec 43, COS (43) 779 (O); Minutes COS (43) 318th Meeting (O), 27 Dec 43, COS (43) 779 (O) (Revise), both in NARA RG 313 Entry 1 Box 69 Folder 381 Bodyguard; Wingate 230. Twenty Committee; "in vacuo": Minutes, 6 Jan 44, PRO KV 4/67/337. Approval by Americans: JPS 366/1, 10 Jan 44 ("unable to propose"; "we feel"); Minutes JPS 121st Meeting, 12 Jan 44; JCS 661, 13 Jan 44; Minutes JCS 142d Meeting, 18 Jan 44; CCS 459/1, 18 Jan 44, all in NARA RG 218 Box 361 Folder CCS 385 (6-25-43) Planning for Deception for Cross-Channel Operation (Sec. 1). CCS approval: CCS 459/2, 20 Jan 1944, in same. To Ike: Memorandum for the Supreme Commander, Allied Expeditionary Force, Subject: Overall Deception Policy for War Against Germany, 22 Jan 44, in NARA RG 165 ABC File ABC 385 Europe (23 Sep 43) Sec. 2.

507   Specifics of BODYGUARD: Text in Howard 247–53.

509   Oshima fifty divisions: Hinsley 3 part 2 p. 45.

511   Baumer to London: Baumer diary 92. Noce, Harris, Ingersoll: Baumer diary 97, 102. "A good man": Baumer diary 73. "He must have bought": Baumer diary 98. Wild, mess: Baumer diary 99–100.

511   "London is the Paris": Baumer diary 103. Summons to Moscow: AGWAR to ETOUSA, Ref. No. R-8356, 20 Jan 44, in NARA RG 313 Entry 1 Box 69 Folder 381

Bodyguard; JCS to Deane, No. 191, CM-OUT-7929, 20 Jan 44, in NARA RG 165 ABC File ABC 385 Europe (23 Sep 43) Sec. 2. "Would only mire"; Roberts; "he laughed": Baumer diary 104.

512 British Chiefs, Ismay: Baumer to Roberts, Subject: Planning Report for Week Ending on Jan. 22, 23 Jan 43, p. 3, Baumer papers; Redman to Royal, 18 Jan 44; McFarland to Redman, 19 Jan 44, both in NARA RG 218 Box 361 Folder CCS 385 (6-25-43) Planning for Deception for Cross-Channel Operation (Sec. 1). Waiting for visa: Baumer diary 102–103. "Tore a huge hole": Baumer diary 101. Ismay: Baumer diary 104.

513 To Moscow: Baumer diary 104–109.

513 Harriman: Baumer diary 109. Deane: Baumer diary 110. "It took": Deane 147.

514 Waiting: Baumer diary 110–13. "Living in hope": Cruickshank, Deception 115. "The days at the office"; "in the first tier": Baumer diary 111.

514 First meeting: Baumer diary 115; Report by Colonel Bevan and Lieutenant-Colonel Baumer Regarding their Visit to Moscow, LCS (44)10, 2 Apr 44, PRO CAB 81/78/46. "A red-headed"; "out of his depth"; "a long-winded": Baumer diary 115. "He started every"; "a mental giant": Deane 147.

515 More waiting; "poor emaciated": Baumer diary 115. Second meeting: Report by Colonel Bevan and Lieutenant-Colonel Baumer Regarding their Visit to Moscow, LCS (44)10, 2 Apr 44, PRO CAB 81/78/47; Moscow to AGWAR, No. 205, 11 Feb 44, CM-IN-9039 (13 Feb 44); Moscow to AGWAR, No. 206, 11 Feb 44, CM-IN-9068 (13 Feb 44); CS to Deane, No. 279, 13 Feb 44, CM-OUT-5701 (13 Feb 44); JCS to Deane, No. 279, 13 Feb 44, CM-OUT-5701 (13 Feb 44), all in JSC folder 370.2 Bodyguard; Bodyguard—Comments on, COS(44)(154)(O), 14 Feb 44, LCS 44(5), PRO CAB 81/78/68–70.

516 Third meeting: Baumer diary 116; Moscow to AGWAR, No. 217, 16 Feb 44, CM-IN-12176 (18 Feb 44); Moscow to AGWAR, No. 218, 16 Feb 44, CM-IN-12231 (18 Feb 44), both in JSC folder 370.2 Bodyguard; Report by Colonel Bevan and Lieutenant-Colonel Baumer Regarding their Visit to Moscow, LCS (44)10, 2 Apr 44, PRO CAB 81/78/47. Clarke: Bodyguard—Comments on Telegram No. 325 Dated 17th February, 1944 from Colonel Bevan, Moscow, LCS (44)6, 17 Feb 44, PRO CAB 81/78/65–67. More waiting: Baumer diary 116–17. "When Stalin got back": Report by Colonel Bevan and Lieutenant-Colonel Baumer Regarding their Visit to Moscow, LCS (44)10, 2 Apr 44, p. 5, PRO CAB 81/78/48.

516 Reception: Baumer diary 117–18; Baumer, But General ch. 5 ("Every morning"). "About as sumptuous"; "when at midnight"; "when I saw"; Baumer diary 117–18.

517 Another copy: Baumer diary 119; Report by Colonel Bevan and Lieutenant-Colonel Baumer Regarding their Visit to Moscow, LCS (44)10, 2 Apr 44, p. 5, PRO CAB 81/78/48. 1:30 A.M. meeting: Baumer diary 119; Moscow to WAR, No. 278, 5 Mar 44, CM-IN-3468 (5 Mar 44), in NARA RG 218 Box 361 Folder CCS 385 (6-25-43) Planning for Deception for Cross-Channel Operation (Sec. 1); Baumer diary 120 ("we didn't know"). Protocol agreement: Moscow to AGWAR, No. 283, 6 Mar 44, CM-IN-4713 (6 Mar 44), in NARA RG 218 Box 361 Folder CCS 385 (6-25-43) Planning for Deception for Cross-Channel Operation (Sec. 1); Report by Colonel Bevan and Lieutenant-Colonel Baumer Regarding their Visit to Moscow, LCS (44)10, 2 Apr 44, pp. 5–6, PRO CAB 81/78/48–49; Baumer diary 120; Baumer, But General ch. 5 ("when we said").

518 To Cairo: Baumer diary 121–26; Clarke diary 12 Mar 44, Clarke papers; Extract from Weekly Letter No. 3, 14 Mar 44, in JSC folder 370.2 Bodyguard. "One of those": Baumer diary 123. "Got into your eyes"; "excellent improvisation": Baumer diary 124. To Algiers: Baumer diary 125–26. "Royal welcome": Baumer diary 125.

519 To London: Extract from Weekly Letter No. 5, 27 Mar 44, in JSC folder 370.2 Bodyguard.

519 "Was working in a nebulous"; "was interested in Russia": Baumer diary 127. "Titular boss": Ingersoll to Bowman, 17 Aug 79, Ingersoll papers, box "8-11-86 5 of 5." Colonel's eagles: Baumer diary 128. Socializing: Baumer diary 128–30.

519 Broke out caviar: Baumer diary 129. Wedemeyer: Baumer diary 130; Wedemeyer 262–63. Hunter Harris, Bevan: Baumer diary 131–33. "Everything there": Baumer diary 93. Handy would not: Cline 306–307; Baumer diary 93, 127. Home by air: Baumer diary 93, 133. Others making colonel: Baumer diary 94.

Chapter 13: QUICKSILVER

521 Five-division assault: Smith to Eisenhower, No. W-9389, 5 Jan 44, in Bedell Smith papers, Box 19, folder COSSAC (Chief of Staff Supreme Allied Command) Staff Organization, Eyes Only Cable File (2); Minutes of Meeting, 21 Jan 44, in NARA RG 331 Entry 3 Box 122 Folder Minutes of COSSAC Staff Conferences.

522 Strangeways to England: III/80. G(R) name: Hesketh 18; Young & Stamp 41.

523 Reporting, working with Williams: Author's telephone interview with Philip Curtis, 21 Oct 92; Young & Stamp 50. Temple: Young & Stamp 43 ("probably the biggest"); York Strangeways interview p. 12 ("drippy, lanky"). "A wonderful chap": Author's telephone interview with Philip Curtis, 21 Oct 92. Curtis: Young & Stamp 49; author's telephone interview with Philip Curtis, 21 Oct 92.

523 Informally assigned: Noce to The Adjutant General, Subject: Lieutenant Colonel William A. Harris, O-18976, 5 Aug 48, Harris papers. "Special Plans Section"; promotions; orders: Smith to Commanding General, FUSAG, Subject: Cover and Deception, SHAEF/18209/1/Ops, 26 Mar 44, in NARA RG 319 Entry 101 Box 1 Folder 2 "Cover & Deception Report ETO Exhibit 1 Basic Directive, SHAEF"; Kibler to Harris, Subject: Cover and Deception, 321 (G-3), 7 Apr 44, in NARA RG 319 Entry 101 Box 1 Folder 3 "Cover & Deception Report ETO Exhibit 2, Basic Directive, AC/S, G-3, FUSAG." Part of G(R): Cover & Deception Synopsis of History, p. 5, in NARA RG 319 Entry 101 Box 1 Folder 1 "Cover & Deception Synopsis of History." "But we are not": Eldredge to Harrison, 31 Aug 88; Eldredge to Dwyer, 17 Aug 89, Eldredge papers.

524 Beck: "Informal Report to Joint Security Control by Special Plans Branch, G-3, 12th Army Group" 25 May 45, Part IV, p.1, in NARA RG 319 Entry 101 Box 2 Folder 27; Author's telephone interview with Mrs. Clarence Beck, 12 Oct 98; AGWAR to ETOUSA, Ref. No. R-8124, 15 Jan 44; ETOUSA to AGWAR, Ref. No. W-10097, 27 Jan 44; Beck to Reeder, Ref. No. E 20256, SMC OUT 381, 24 Mar 44, all in NARA RG 331 Entry 1 Box 64 Folder 370.5/3 Movement of 23d Headquarters and Headquarters Detachment Special Troops.

524 Fast friends: Author's telephone interview with Mrs. Clarence Beck, 13 Oct 98; Ingersoll to Rippert, 1 Jul 46; Ingersoll to Went [Eldredge], 11 Oct 54; Ingersoll to Woodworth, 19 Feb 64, all in Ingersoll papers, box unmarked. Seaman: "Informal Report to Joint Security Control by Special Plans Branch, G-3, 12th Army Group" 25 May 45, Part IV, p.1, in NARA RG 319 Entry 101 Box 2 Folder 27.

524 Ingersoll's female friend: Hoopes 281.

525 Ingersoll to USA: Murphy to Ingersoll, 11 May 58, Ingersoll papers, box unmarked; Ingersoll to Harris, Ref No. W-22171, 12 Apr 44, SMC IN 868, in NARA RG 331 Entry 1 Box 64 Folder 370.5/3 Movement of 23d Headquarters and Headquarters Detachment Special Troops.

525 Eldredge history: Eldredge autobiography, Eldredge papers; Eldredge to Ingersoll with attached curriculum vitae, 12 Mar 64; Eldredge to Ingersoll, 28 Sep 69, both in Ingersoll papers, box unlabeled. "Some bright American"; "wanted to be doing something"; "God he was ugly": Eldredge autobiography, ch. 11, Eldredge papers.

527 "Johnnie tried to feed me"; "I thought, no drink": Eldredge autobiography ch. 11, El-

dredge papers. "A gentleman": Author's telephone interview with David Strangeways, 26 Aug 92. "A ragamuffin": Eldredge autobiography, ch. 11, Eldredge papers.

528 Wild to be sized up: Bull to Barker, Subject: Cover and Deception, 20 Nov 43, in NARA RG 331 SHAEF G-3 General Records Subject File 1942–45 "Combatting the Guerrilla" to "Daily Reports," Folder Cover Plan—Overload (Bull Papers). Wild to England: III/180. Bedell Smith: Delmer 36–37. Hesketh: Wild to Mure, 30 Jan 80, Mure papers (Box 2). Lash and Brown: Tentative Assignments, NARA RG 331 Entry 1 Box 45 Folder Organization and Personnel G-3 SHAEF. Barnes, Moody, Corbett: CG SHAEF to War Department, Nr. S-50175, 5 Apr 44, CM-IN-3820 (6 Apr 44), in NARA RG 319 Entry 101 Box 6 Folder 91; Barnes, Cullum 9984; Moody, Cullum 12240. Moody horse arm: Coll 3. Hood: III/180; "A" Force card index, S. B. D. Hood, Clarke papers. Rushton: SHAEF to AFHQ, No. S-50853, 28 Apr 44, SMC-OUT-661 (28 Apr 44), in NARA RG 331 Entry 1 Box 45 Folder Organization and Personnel G-3 SHAEF; Cover & Deception Synopsis of History, p. 1, in NARA RG 319 Entry 101 Box 1 Folder 1. Harmer: Hesketh to Casey 4; Hesketh introduction 4; Harmer to author, 14 Apr 93.

529 Phyllis White: Hesketh to Casey 4. Cuthbert: Hesketh introduction 3–4. Taciturn: Author's interview with Arne Ekstrom, 4 Jan 93. Wild learns of Twenty Committee: Delmer 40. "Had suggested": Minutes 10 Feb 44, PRO KV 4/67/347. "Williams does not think": Liddell diary, 18 Feb 44, PRO KV 4/193. Twist abolished: Hesketh 16–17. Racket abolished: Minutes, 24 Feb 44, PRO KV 4/67/351. Wild full member: Delmer 40, Wheatley 188.

530 Desk in LCS: Delmer 40. Ops (B) deal directly: Hesketh 20–21. LCS central: Hesketh 17.

530 "Noel Wild was useless": Author's telephone interview with Christopher Harmer, 19 Dec 92. Macdermot: Author's interview with John B. Oakes, 29 Dec 98 ("tough-minded and brilliant"); Harmer to author, 31 Jan 93 ("The great thing"). "Anyone looking back": Harmer to author, 31 Jan 93.

530 "Appendix 'Y' had a certain": Jervis-Read to G-3, COSSAC/18.200/Ops, 30 Dec 43, PRO WO 219/2204A/8.

531 Appendix Y approach: Hesketh 4–8, 45. January meetings and drafts: Jervis-Read to multiple addressees, COSSAC/18216/Ops, 3 Jan 44, PRO WO 219/2204A/10–15; Jervis-Read to multiple addressees, COSSAC/18216/Ops, 3 Jan 44, PRO WO 219/2204A/10–15; McLean to Head Planners, COSSAC/18216/Ops, 11 Jan 44, PRO WO 219/2204A/16–21; Hesketh 12; copy of Appendix Y in NARA RG 331 Entry 1 Box 73 Folder 381 Fortitude Vol. 1.

532 Two British generals: Hesketh 11–12. "Unofficially, I believe": McLean to DC of S G-3, COSSAC/18216/Ops, 13 Jan 44, PRO WO/2204A/22. "I was not a much": Author's telephone interview with David Strangeways, 21 Jul 92.

532 "Much disliked": Author's telephone interview with Christopher Harmer, 19 Dec 92. "He was a 'snotty' ": Eldredge autobiography, 11, Eldredge papers. De Guingand letter: De Guingand to West, 21 A Op/25/COS, 25 Jan 44, PRO WO 219/2204A/33; another copy at PRO WO 219/2205/5.

533 Final draft approved: Smith to British Chiefs of Staff, Subject: Plan "Mespot," 30 Jan 44, PRO WO 219/2204A/36; another copy in NARA RG 331 Box 11 Folder Plan Fortitude (SHAEF G-2 files); Jervis-Read to AC of S G-3, Subject: Plan "Mespot," SHAEF/18216/Ops, 29 Jan 44, PRO WO 219/2204A/35; Plan "Mespot," SHQ AEF (44)(13), PRO WO 219/2204A/37–44; McLean to AC of S G-3, Subject: Plan "Fortitude," SHAEF/18216/Ops, 18 Feb 44, PRO WO 219/2204A/47. Stories A and B: Plan Fortitude, SHAEF 44 (13), paras. 23, 28, printed in Hesketh 184–85.

534 "We should, therefore": Plan Fortitude, SHAEF 44 (13), para. 14, printed in Hesketh 184. "Fully developed": Hesketh 12. New codename: Hesketh 10 note 10; Cruick-

shank, Deception 99 note*, citing WO 106/4165/82B; Jervis-Read to AC of S G-3, Subject: New code word for "MESPOT," SHAEF/17213/Ops, 10 Feb 44, PRO WO 219/2204A/46.

535 Wild's plan; "with four additional"; "Where were these four": Hesketh 46.

535 "Spread themselves about": Hesketh 41–42. Strangeways opposed: Hesketh 42–43. Cut camouflage staff: Young & Stamp 41–42. "We have worked on the principle": West to de Guingand, SHAEF/18216/Ops, 1 Feb 44, PRO WO 219/2204A/45; another copy at PRO WO 219/2204A/5.

536 "Directing the threat"; "making preparations": Smith to Ramsay *et al.*, Subject: Plan "Fortitude" (Camouflage policy and administrative preparations) SHAEF/18216/Ops, Draft,—Feb 44, PRO WO 219/2205/6–7. Successive drafts: SHAEF/18216/Ops (First Draft), 8 Feb 44, PRO WO 219/2205/9–12; SHAEF/18216/Ops (Second Draft), 16 Feb 44, PRO WO 219/2205/15–18.

537 Final version; "subject to adjustment": SHAEF 44 (21), 26 Feb 44, Hesketh 44, 186–88. February 23 meeting: Hesketh 43. SHAEF overrules: Hesketh 44; Smith to Headquarters 21 Army Group, Subject: Plan "Fortitude" (Camouflage Policy and Administrative Preparations), 4 Mar 44, NARA RG 331 Entry 1 Box 73 Folder 381 Fortitude Vol. 1. "Arrange direct"; coolly adjusted: Hesketh 44.

537 Hesketh and Harmer: Author's telephone interview with Christopher Harmer, 19 Dec 92.

538 Launched early April: Hesketh 51. "The story on which": Hesketh 189.

539 Germans had discovered: Hinsley 3 part 2:48–49.

539 Germans to believe: Hesketh 190. QUICKSILVER II: Hesketh 203–204.

540 QUICKSILVER III: Hesketh 191. QUICKSILVER V: Hesketh 190. QUICKSILVER VI, QUICKSILVER IV: Hesketh 191–92.

541 "Their best general": Irving, Hitler's War 683–84.

541 Patton to command FUSAG: Author's telephone interview with Christopher Harmer, 19 Dec 92. No. 3 Wireless Group: Hesketh 19; Young & Stamp 42, 98–99. "C.L.H." units: Hesketh 19. 3103d: McKenney to Gross, 25 Jun 85, in author's possession; Hesketh 19; Nickel to Dwyer, 7 Feb 90, in author's possession; Author's telephone interview with former T/5 Erwin C. Lochmueller, June 1992.

542 Work of 3103d: Report of Special Signal Operations in the United Kingdom, in NARA RG 319 Entry 101 Box 3 Folder 64; Cover & Deception Report ETO Exhibit 7, Technical Signal Report on Fortitude, in NARA RG 319 Entry 101 Box 1 Folder 13; Cover & Deception Synopsis of History, in NARA RG 319 Entry 101 Box 3 Folder 1.

543 Signalmen: Author's telephone interview with Colonel John R. Nickel, 21 Sep 92; Author's telephone interview with T/5 Erwin C. Lochmueller, June 1992.

543 Net opens: Hesketh 48, 89; 3103d Signal Service Battalion Operations Quicksilver and Williams, Appendix A, in NARA RG 319 Entry 101 Box 3 Folder 64 "Report of Special Signal Operations in the United Kingdom." Germans not listening: Wingate 91, 141; Eldredge to Smith, 25 Jul 45, in JSC folder Deception Units. Field Security Reports: Hesketh 23.

544 TRICYCLE, FREAK: Masterman 139–40; Hinsley & Simkins 222–23, 238. BRUTUS, CHOPIN: Masterman 142; York Astor interview p. 8; Hinsley & Simkins 239; Hesketh 27, 63; Author's telephone interview with Christopher Harmer, 19 Dec 92. TATE: West, Seven 33, 56; Hesketh 92, 111 note 23. TREASURE: Hinsley & Simkins 241. Coast ban settled March 10: Hinsley & Simkins 254.

546 GARBO deployment: Howard 233; Hesketh 26, 49, 66–68; Pujol 112. GARBO busiest: Hinsley & Simkins 239. Traceable messages: Hesketh Introduction 5. PANDORA: Hesketh 29.

547 Special means methodology: Hesketh 193–204; Hesketh introduction 4; Hesketh to Casey 4.

548 Cuthbert courier: Hesketh introduction 3, 197.

548 "General Montgomery has asked": Hoopes 287.

548 Sample move of VIII Corps: Hesketh 50–51, 65 note 15, 108, 202, 233 Messages 12 and 15, 234 Message 23, 235 Message 32; 3103d Signal Service Battalion Operations Quicksilver and Williams, Appendix A, in NARA RG 319 Entry 101 Box 3 Folder 64 "Report of Special Signal Operations in the United Kingdom"; Cover & Deception Report ETO Exhibit 7, Technical Signal Report on Fortitude, Appendix A, in NARA RG 319 Entry 101 Box 1 Folder 13; Ingersoll to Harris, Subject: Alteration in Fortitude South, 28 May 44, in NARA RG 331 Entry 199 Box 66 Folder Fortitude (fat folder of two by that name).

549 TRICYCLE in Lisbon: West, Seven 29; Hinsley & Simkins 238; Hesketh to Casey 5; Howard 121 ("particularly valuable").

550 BRUTUS summaries: Hesketh 63, 235 Messages 27 and 32 ("FUSAG gives").

550 Dummy landing craft: Hesketh 20, 48, 60, 102; RAF history of visual deception, ch. XV, in NARA RG 319 Entry 101 Box 6 Folder 89; Cruickshank, Deception 182–84.

550 QUICKSILVER V, VI: Cover Plan Fortitude, RAF Visual Deception, in JSC folder 370.2 Fortitude; Cruickshank, Deception 184–85.

551 QUICKSILVER IV: Hesketh 61.

551 Radio silence: Cover & Deception Synopsis of History, p. 7, in NARA RG 319 Entry 101 Box 1 Folder 1.

551 VAN DYKE, DRYSHOD, WETSHOD, SEE SAW: Hesketh 60–61, 190; Cover & Deception Report ETO Exhibit 7, Technical Signal Report on Fortitude, pp. 6–9, in NARA RG 319 Entry 101 Box 1 Folder 13; Report: Operation Neptune—Radio Deception, Appendix "C," p. 2, in JSC loose item "Report . . . Neptune July 1944"; Report of Special Signal Operations in the United Kingdom, pp. 170–202, in NARA RG 319 Entry 101 Box 3 Folder 64; Cover and Deception Synopsis of History, p. 8, in NARA RG 319 Entry 101 Box 1 Folder 1 "Cover & Deception Synopsis of History." TIGER: Author's telephone interview with David Strangeways, 26 Aug 92.

552 GARBO and FABIUS; "You should give": Hesketh 68–69. End of "postponement": Hesketh 68–69.

553 Postponement special means: Howard 122; Hesketh 30, 78; 60 Committee: III/175; Clarke diary 18 Feb 44, 22 Feb 44, Clarke papers. "I went on the wagon": Colyton 231–32.

554 Bevan summit proposal: O'Connor to Joint Security Control, HMOC/991, 1 May 44, in JSC folder 370.2 Overlord. Theme dropped: Hesketh 31. "That only preparations": Howard 122. Germans clogged: Kahn, Spies 507–509. "Enemy diplomatic"; "Numerous reports": Hesketh 78.

554 Suggested topics: O'Connor to JSC, HMOC/661, 21 Jan 44; O'Connor to LCS, HMOC/734, 11 Feb 44; Chart, undated, all in JSC folder 370.2 Bodyguard. Close coordination: E.g., O'Connor to Joint Security Control, HMOC/843, 20 Mar 44, in JSC folder 370.2 Bodyguard. RUDLOFF messages: Messages A-6, A-13, A-11, A-15, A-22, A-33, A-34, in JSC brown Acco binder labeled "15."

555 BROMO message: File Log, in JSC brown Acco binder labeled "8 File B." B-29 to England: Craig to Cook, Subject: Special Instructions for B-29 Destined for United Kingdom, 28 Feb 44, in JSC brown Acco binder labeled "16." Leaflets: Smith to Hunter, Subject: B-29, 22 May 44, in JSC folder 000—To 312.1(1) China-Burma-India Theater. Clarke leak: Extract From Weekly Letter No. 9 from LCS dated 27 April 1944; Undated, unsigned Notes on Part II Weekly Letter No. 9 from LCS, 27 April 44, both in JSC folder 370.2 Operations and Results. Bissell/Bevan: Bissell to O'Connor, Subject: Coordination of Deception Planning, JSC/J8 Serial 543, 8 May 44; O'Connor to Bissell through Smith, HMOC/1032, 15 May 44; Smith to Bissell, Subject: Coordination of Deception, JSC/J8 Serial 586, 18 May 44; Bissell to O'Connor., JSC/J8, 18

May 44; Bevan to O'Connor, LCS (44) I/D.3, 26 May 44; Bevan to Clarke, Subject: Coordination of Implementation as Between the British and Americans, LCS (44) I/D.3, 29 May 44, all in JSC folder Great Britain.

556 Argentina: Smith to Lang, JSC/LB Serial 261, 2 Feb 44; King to Smith, 11 Apr 44; Lang to Smith, 1 Jul 44, all in JSC folder South America. "Bulgy": Delmer 18. Hitler: Thorne to author, 12 Jan 97; Thorne, Hitler.

556 FORTITUDE NORTH: 10, 33–39; Howard 115; PRO WO 219/2220 passim; Macleod, The Story of the Fourth Army, in Macleod papers; "Cover & Deception Report ETO Exhibit 6, Operations in Support of Neptune (B) Fortitude North," in NARA RG 319 Entry 101 Box 1 Folder 8; Cover Plan Fortitude, RAF Visual Deception, in JSC folder 370.2 Fortitude; Report on Naval Assistance to Plan "Fortitude," undated, PRO WO 219/2214/6.

557 Eldredge and Ingersoll: Eldredge autobiography, Ch. 11, Eldredge papers ("once was enough") ("This fellow") ("he was not to tell") ("fine fat"); ETOUSA to CG XV Corps, Ref. No. 7210, 8 Mar 44, SMC-OUT-95, PRO WO 219/2220/743–76; Macleod to Jervis-Read, 13 Mar 44, PRO WO 219/2220/10. Thorne at XV Corps: Jervis-Read to Wild, SHAEF 18216/1/3, 3 Apr 44, PRO WO 219/2220/30; Jervis-Read to Hern, 18216/1/Ops, 6 May 44, PRO WO 219/2220/138. "He was there": Macleod p. 5.

557 Special means: Hesketh 29–30, 38–39; Masterman 114; Kahn, Spies 494 ("believable Abwehr"). Scandinavian manuals: Smith to Record, Subject: Manuals on Balkan and Scandinavian Countries, 13 Apr 44, in JSC brown Acco binder labeled "16."

558 GRAFFHAM, ROYAL FLUSH: Wingate 242–46; Howard 117–19; Smith to Record, Subject: Plan BODYGUARD (GRAFFHAM), 9 Mar 44, in JSC folder 370.2 Graffham; and that folder passim. FORTITUDE NORTH results: Howard 117.

559 IRONSIDE: Operation "Ironside" Outline Plan Approved by SHAEF 15 May 1944; Unsigned memorandum, SHAEF/18237/Ops, 8 May 44; Bevan to O'Connor, 1 May 44; O'Connor to JSC, HMOC/994, Reference: Plan BODYGUARD, 2 May 44; O'Connor to Bevan OC/467, 5 May 44; Unsigned, undated memorandum, Implementation of Plan "Bodyguard"; O'Connor to Bevan, OC/467, 5 May 44; Bevan to O'Connor, OC/489, 10 May 44; Bevan to O'Connor, L/107, 8 May 44; Undated memorandum Plan IRONSIDE; Unsigned, undated Implementation Chart, Plan "IRONSIDE," all in JSC folder 370.2 Ironside. "Deserves a good mark"; O'Connor to Bevan, 24 May 44, PRO CAB 154/57/23.

560 Special means: Hesketh 53–54; Messages A-54, A-56, A-57, A-58, in JSC brown Acco binder labeled "15"; items 59, 60, 62, 65 in undated document, "R.," in JSC brown Acco binder labeled "14"; PRO CAB 154/57/85.

560 No effect: Hinsley & Simkins 250; Re: Rudloff Case, 15 Jun 44, in JSC Acco binder "7 File A"; Hesketh 94 ("cover operations"), 99 ("small and insignificant"); Wingate 418; Howard 190–91 ("although there is little").

561 COPPERHEAD: IV/85–90; Clarke diary 4 Jan 44, Clarke papers; III/8; NARA RG 331 Box 11 Folder Plan Fortitude (SHAEF G-2 files); Hesketh 62, 93; Wheatley 192; Hamer to Mure, 21 Feb 81, Mure papers (Box 2); Author's telephone interview with Alice Crichton, 20 Nov 93 ("a very nice man"); Howard 126; James.

562 PREMIUM: Hesketh 40; Hinsley & Simkins 241. SNIPER: Hesketh 55.

563 Soviets: United States Military Mission to Moscow to War Department, No. 577, 19 May 44, CM-IN-14952 (20 May 44), in JSC folder 370.2 Bodyguard. "Were just as startling": Deane 148. "It must be said": Report of United States Military Mission to Moscow, unpaginated, in JSC folder 370.2 Bodyguard. "It was one": Deane 149. Burrows: Stafford 284.

564 ARTIST: Hinsley & Simkins 95, 222–24; Hesketh 57–58.

565 "After D-Day": Hesketh 58.

565 German thinking: Hinsley 3 part 2:45–49, 56–60, 62–64. Oshima message ("What is

your feeling"): Original unedited text, SRDJ 59973, in NARA RG 457 Box 72; "Magic" Summary No. 798, 1 Jun 44, USAMHI. First diplomatic decrypt: Hinsley 3 part 2:61. "It gave ground"; "it gave us": Hesketh 95 note 4. "When they received": Howard 131.

567 Salmuth: Irving, Fox 379. "An old, stooping": Irving, Hitler's War 614. Stomach spasms, leg, medicines, Berghof: Irving, Hitler's War 624–25. Krummacher: Hesketh 101 and note 15; Delmer 13–14; Kahn, Spies 395. Roenne, Michael: Irving, Fox 434.

570 Rundstedt: Irving, Fox 378 ("To me"); Carr 90–94. "The bloom": Irving, Fox 383. Meyer-Dietring: Kahn, Spies 487.

571 Army Group G: Kahn, Spies 487. Dollmann: Irving, Fox 388, 471. Speidel: Irving, Fox 406, 416–18. Staubwasser: Kahn, Spies 487; Irving, Fox 408; Staubwasser B-675, p. 7 ("although it was left to us"); Delmer 145–46.

571 Rommel in charge: Irving, Fox 379–84, 398. Rundstedt move: Kahn, Spies 487. "We always took": Report From Captured Personnel and Material Branch, B-823, 6 Aug 45, p. 2, in JSC folder 370.2 Operations and Results. Rommel tour: Irving, Fox 379. No great admirer: Irving, Fox 411–12. "Rommel's conjectures": Guderian 26.

572 Postwar suggestions: Ruge 5; Dihm 35; Hesketh 96–97. "All the diverging": Brandenberger, in Staubwasser B-675, p. 1. "Particularly those places": Howard 108.

572 First report: Rommel Papers 453. "Always told us": Report From Captured Personnel and Material Branch, B-823, 6 Aug 45, p. 2, in JSC folder 370.2 Operations and Results; see also Gersdorff p. 4. Ruge: Ruge 5. Rejected Luftwaffe: Irving, Fox 392. March 4: Irving, Fox 396; Irving, Hitler's War 614. March 20: Irving, Fox 420; Irving, Hitler's War 614. Antonescu: Irving, Hitler's War 625.

573 "A partial success": Howard 130. "Attached particular importance": Kahn, Spies 502; Howard 130. "The Cotentin": Irving, Hitler's War 626. Jodl telephoned: Irving, Hitler's War 883. "Which are especially"; "perhaps be the enemy's": Howard 130–31. "We've no real": Irving, Hitler's War 625; Irving, Fox 408–409.

574 "Holding and diversionary": Howard 108. "There is no longer": Kahn, Spies 479. "First of all": Kahn 502. Roenne estimates: Hesketh 89, 146; Howard 115; Kahn, Spies 496.

575 Structure: Delmer 157; Hesketh 85–86. "Front-line point": Howard 132.

575 Roenne sitrep: Howard 131. Führer turns in: Irving, Hitler's War 634.

575 Ingersoll hypochondria: Hoopes 288. Ingersoll and Harris at invasion; TROUT-FLY: Hoopes 287–98; "Cover & Deception Synopsis of History," pp. 9–10, in NARA RG 319 Entry 101 Box 1 Folder 1; Weigley 74, 78; Noce to The Adjutant General, Subject: Lieutenant Colonel William A. Harris, O-18976, 5 Aug 48, in Harris papers.

576 "R" Force: "Cover & Deception Synopsis of History," p. 9, in NARA RG 319 Entry 101 Box 1 Folder 1. Strangeways: Young & Stamp 46. "I wasn't as brave": Hoopes 287. Wheatley: Wheatley 194–211. Beck: Author's telephone interview with Mrs. Clarence Beck, 12 Oct 98.

576 "The most inspiring": Author's interview with John B. Oakes, 29 Dec 98. 3103d; "Put it on the loudspeaker": Author's telephone interview with former T/5 Erwin C. Lochmueller, June 1992.

577 GARBO: Pujol 135, 138, 142; Hesketh 55–56, 60–61, 70–71.

578 PARADISE series: Operation Overlord, Cover and Diversionary Plans NJC/00/261/33, Part III, 18 May 44, Decoy Lighting of Neptune Beaches and Beach Exits, in Folder 505-61-21, Plan Fortitude folder, AFHRA. BIGDRUM, GLIMMER, TAXABLE: Wingate 260; Cruickshank, Deception 199; Young & Stamp 169; Operation Overlord, Cover and Diversionary Plans NJC/00/261/33, Part II, Diversionary Plans, in Folder 505-61-21, Plan Fortitude folder, AFHRA. TITANIC: Young & Stamp 77; Author's conversation with Foot, 9 Jan 94 (name); Wingate 260–61; Cruickshank, Deception 197–98;

Young & Stamp 77–79; Operation Overlord, Cover and Diversionary Plans NJC/00/261/33, Part II, Diversionary Plans, in Folder 505-61-21, Plan Fortitude folder, AFHRA.

578 German brigade: Young & Stamp 80. "Powerful airborne": MAGIC Decrypt No. 806, 10 Jun 44, USAMHI. "Those of the enemy"; "It is uncertain": "Magic" Summary No. 809, p. 3, 12 Jun 44, USAMHI. Several attempts: "Magic" Summary No. 810, p. 2, 13 Jun 44, USAMHI. "We immediately": Wingate 261, 416.

579 "That the invasion": Howard 132. Rommel leave: Irving, Fox 441. Chief of staff: Irving, Fox 435–36. Salmuth: Irving, Hitler's War 883. *"Machen Sie kein"*: Eldredge autobiography, 11, Eldredge papers. "Just keep": Delmer 167. Not more than 10 days: Harmer to author, 14 Apr 93. "Needs at least": Memorandum Deception, P/149 (1st Draft), 7 Apr 44, PRO WO 204/1561/63.

579 G-2 forecast: Hesketh 115. "It is clear": Hesketh 98; see also Hesketh 74. Specific units: Hesketh 64 note 9. GARBO: Hesketh 72–73. TATE: Hesketh 74. Intensified: Howard 188.

580 GARBO long message; "From the reports"; "I learned yesterday"; "I trust": Hesketh 95 note 4, 103.

581 On into June: Howard 189–90. June 12: Hesketh 90. No German moves: Hesketh 110, 115; Hinsley 3 part 2:182 note†. "23 divisions": "Magic" Summary No. 819, p. 2, 22 Jun 44, USAMHI.

582 "Believe that": "Magic" Diplomatic Summary No. 844, p. 3, 17 Jul 44, USAMHI. "Every day came": York Strangeways interview p. 24. "I just thought": Young & Stamp 46. Roenne telephoned: Kahn, Spies 514–15; Irving, Fox 443. "The enemy plans": Howard 185–86; Hesketh 90, 98. Dover smoke: Irving, Fox 445. "We must assume": Irving, Fox 446–47. Rundstedt surprised: Hesketh 104 note 30. "A landing operation at least": Hesketh 100.

582 Jodl doubts: Irving, Fox 450. "Strengthens the supposition": Howard 189. 1st SS Panzer: Howard 187–88; Hesketh 101–104 ("second main attack"). Too much should not be made: Howard 188–89.

583 "The invasion is quite"; "The invasion has succeeded": Irving, Fox 454. June 11 sitrep: Howard 189. Keitel and Jodl agreed; first V-1s: Kahn, Spies 517. Sertorius: Hesketh 110. Strength of FUSAG: Hesketh 90. June 24 sitrep: Hesketh 115.

583 Best units in Normandy: Irving, Fox 461. "The enemy have committed"; Rommel not persuaded: Irving, Fox 465–66. "Obvious that Patton's": Wingate 417–18; Howard 190–91. Twenty-two divisions: Howard 191.

584 Smith approves: Smith handwritten notation on Jervis-Read to Deputy G-3, 15 Jun 44, PRO WO 219/2223/45. "Still an integral": Harris to Chief of Section, G(R) 21 Army Group, Subject: Reaction to SHAEF Paper on the Subject "Fortitude (South) Outline Plan for Special Means, Post D-Day Period," 22 Jun 44, in NARA RG 331 Entry 199 Box 66 Folder Fortitude (fat folder of two by that name). July 20: Wild to War Office, Subject: Plan Fortitude South II, SHAEF/18208/Ops(B), 20 Jul 44, PRO WO 219/2215/63. Ops (B) followon; "The bulk of German": Plan "Fortitude II," SHAEF 18236/3/Ops(B) (Second Draft), 22 Jun 44, PRO WO 219/2224/12–25; Plan "Fortitude II," SHAEF 18236/3/Ops(B) (Third Draft), 23 Jun 44, PRO WO 219/2224/26–36 (another copy in NARA RG 331 Entry 199 Box 66 Folder Fortitude (fat folder of two by that name).

585 "Grave misgivings": Harris to Chief of Section, G(R) 21 Army Group, Subject: Reaction to SHAEF Paper on the Subject "Fortitude (South) Outline Plan for Special Means, Post D-Day Period," 22 Jun 44, in NARA RG 331 Entry 199 Box 66 Folder Fortitude (fat folder of two by that name). Eighth Army: Eisenhower to Marshall, No. E-27881, 15 May 44, in Bedell Smith papers, Box 24, folder Cable Log—Out (May 1944) (1). Bevan request: L/12 cipher 21 June, 22 Jun 44, in JSC folder European The-

ater 000; Extract From Weekly Letter No. 17 of 5 July 1944, in JSC folder 370.2 Operations and Results.

585 Strangeways rewrote: Hesketh 118 note 1; G(R) War Diary 26 Jun 44, PRO WO 171/3831. Revised plan: Plan "Fortitude/South II," SHAEF/18236/3/Ops(B) (Fourth Draft), 26 Jun 44, PRO WO 219/2224/39–40; Plan "Fortitude/South II," SHAEF/18236/3/Ops(B) (Fifth Draft), 26 Jun 44, PRO WO 219/2224/45–46. "To obviate confusion": Hesketh 121. Bradley vetoed: William A. Harris, Lecture on Deception, Command and General Staff College, c. 1947, C&GSS Library, No. U167.5.D37 H37, Lecture II, p. 7.

586 "I cannot overemphasize": Eisenhower to Marshall, Eyes Only, No. S-55125, 6 Jul 44, in NARA RG 331, Entry 199, Box 66, Folder Fortitude (fat one of two of that name). (Original hand-corrected staff draft in Arthur S. Nevins Papers, USAMHI, Box "COSSAC and SHAEF Overlord Plans 1943–45," Folder "7/43–7/44"). McNair sent: Marshall to Eisenhower, Ref. No. W-61630, 7 Jul 44, in NARA RG 331 Entry 199 Box 66 Folder Fortitude (fat folder of two by that name). "A story that Patton": Hesketh 122. "The evident destiny": D'Este, Patton 586–88. "As far as deception planning goes": William A. Harris, Lecture on Deception, Command and General Staff College, c. 1947, C&GSS Library, No. U167.5.D37 H37, Lecture II, p. 8.

587 Units of reconstituted FUSAG: Hesketh 33, 68 note 1, 107–109, 119.

587 Final "story": SHAEF/18250/Ops(B), July 19, 1944, Appendix "B," in NARA RG 331 Entry 1 Box 73 Folder 381 Fortitude Vol. 1; printed at Hesketh 208. Wireless exercises: JSC files loose item "Report . . . Neptune July 1944," Appendices B and C. BRUTUS and GARBO: Hesketh 125–26, 130. ULTRA and MAGIC: Hinsley 3 part 2:200–103, 215–16. "Feeling . . . that": Hinsley 3 part 2:216 and note *.

588 "Undoubtedly inferior": Hesketh 127–28. *Pariser Zeitung:* Wingate 234–35; Hesketh 130. Kluge: Irving, Fox 482. "A decisive mistake": Rommel Papers 478. "No grounds"; beginning to wonder; "the enemy are in fact"; "a second major": Howard 193. "Unlikely that early": Hesketh 134. "Large-scale landing": Hesketh 137. Jodl: Hesketh 130. On July 27: Howard 193. By mid-August: Hesketh 136–37.

589 Simpson: Hesketh 131. Japanese-Americans: Irons. Arrival: Hesketh 132. "He gave Rundstedt's": Eldredge to Smith, 14 Jul 45, in JSC folder Deception Units.

590 Changed tune: Paul to Gruenther, Subject: Associated Press release re U.S. Field Orders Captured by Germans, 25 Feb 49, in JSC folder History—Miscellaneous.

591 Low quality, difficulty moving: Bickell.

Chapter 14: Mediterranean Finale

592 Sweeney assignment, background, early days with "A" Force; "And I didn't"; "I had this feeling"; "Which was a hell": Author's interview with Eugene J. Sweeney, December 1992–January 1993. November 29: III/180.

594 OAKFIELD: IV/6–12.

596 Patton tour: Farago, Patton 351–52. Picture: IV/12. Hopkinson: Colyton 227.

596 Corsica deceptive radio; CHETTYFORD: IV/19–21, 23–24. "Well, well": IV/15. "Credible indications": Howard 139–41.

597 "Showed very great": Alexander 125. Clarke and BODYGUARD: Clarke diary 6 Jan 44, Clarke papers; IV/30. "The peculiar circumstances": IV/31.

597 ZEPPELIN: IV/31–34.

598 Need twenty-four divisions: IV/32. WANTAGE: PRO WO 204/1562 *passim;* IV/38–40.

598 ZEPPELIN implementation: IV/36–37, 43–48, 57–61, 94; NARA RG 319 Entry 101 Box 7 Folder 7 "Zeppelin" *passim.* "The machine": IV/94.

599 Clarke in England: Clarke diary 8 Feb 44–9 Mar 44, Clarke papers; IV/37.

599 DUNGLOE: IV/48. "I now regard": Wilson to Paget, Ref. No. F 27345, MC OUT 779, 4 Apr 44, PRO WO 204/1562/226. No further problems: IV.47.

600 Dietrich: Riva 540–44; Clarke diary, 12 Apr 44, Clarke papers; Author's interview with Eugene J. Sweeney, December 1992–January 1993; Author's interview with Arne Ekstrom, 25 Mar 93 ("beautiful figure"). Ekstrom a captain: Author's interview with Arne Ekstrom, undated 1993.

600 JCS representative: Proposed Dispatch to Joint Chiefs of Staff, undated, PRO WO 204/1562/126 ("In order to promote"); Clarke, endorsement on informal routing slip, 9 Apr 44, PRO WO 204/1562/194; Chief of Staff to CinC Mediterranean with enclosure, Subject: Joint Security Control Representative in the Mediterranean, 18 Apr 44, PRO WO 204/1562/138–39; AGWAR to Wilson, Ref. No. W-32453, 5 May 44, MC IN 2995 (5 May 44), PRO WO 204/1562/106.

601 ZEPPELIN Stages 3 and 4: IV/57, 66–68, 73.

601 VENDETTA: IV/43–44, 63–70, 75–77; Howard 126–27, 148–50; NARA RG 319 Entry 101 Box 4 Folder 73 "Vendetta," *passim;* Wilson to Clarke, Ref. No. 60910, 7 Mar 44, Bedell Smith papers, Eisenhower Library Box 21, Folder Cable Log––In (March 1944) (2); Unsigned, undated memorandum in JSC folder 370.2 Ironside (Terry Allen).

602 Sweeney to Libya, Cairo: Author's interview with Eugene J. Sweeney, December 1992–January 1993.

602 Attached to Patch's staff: IV/72, 75.

603 TURPITUDE: IV/80–84.

603 "A hive": IV/83. Warning to Weichs; Roenne: Howard 142–43. CICERO: Hinsley & Simkins 213–15. Little likelihood; Howard 145.

604 Must be readying: Howard 146. German reaction to VENDETTA: Howard 151–52. MAGIC revealed: "MAGIC" Summary No. 816, 19 Jun 44, pp. 1–2, USAMHI. Jodl: Wingate 286. Das Reich: Mitchell 442; Keegan, Waffen SS 117.

604 TURPITUDE reactions: IV/84–85; Wingate 289; Howard 152–54. "Plan ZEPPELIN had": IV/96. CLAIRVALE: IV/21–23.

605 NORTHWAYS: Wingate 304; IV/23–24.

605 NUNTON: IV/50–52. "After FERDINAND": Howard 161. German reactions: IV/53–55; Howard 161–62; Graham & Bidwell 265–66.

606 "Your contribution": IV/55. Hood to England; Bromley-Davenport: III/180; IV/55. Sonic fiasco: II/87; IV/26–28.

607 3133d: Cake Before Breakfast p. 33, JSC loose papers; Goldbranson to record, 6 Feb 45; Goldbranson to Smith, 7 Feb 45, in Goldbranson papers. Dummy tanks: IV/55–56, V/20–21.

607 No. 5 Wireless Group: II/5; V/5–6, 22–23.

608 Daphne Llewellyn: Ranfurly 192–93, 211–12; Author's telephone interview with Lady Maynard (Daphne Llewellyn), 10 Apr 96. "She will not have": Ranfurly 214.

609 June–July reorganizations: III/175; IV/91–92, 119, 121–27; Clarke diary 18 Oct 44 (supplemental pages), Clarke papers; Report on the Activities of No. 2 Tac H.Q. "A" Force (Sixth Army Group) 31 July 1944–5 May 1945, Part I, p. 14, in NARA RG 319 Entry 101 Box 4 Folder 79. Train departure, sendoff: IV/121–22; Clarke to Train, 16 Jul 44; Orders, 15 Jul 44; Certificate of Service, 15 Oct 44, all in Train papers; Train to Joint Security Control, Subject: Deception in the Western Mediterranean, in JSC folder 11—Tactical and Strategic Deception—Captain Hershey, 2 Aug 44. "Put them up": Hamer to author, 2 Jun 97.

610 Jones departure: IV/84.

610 Rome: Author's telephone interview with Lady Maynard (Daphne Llewellyn), 10 Apr 96; Author's interview with Arne Ekstrom, 25 Mar 93.

611 Freeman Thomas agent hunt: IV/98. ARMOUR: IV/100–101.

611 PRIMO: IV/98–99. Niven: III/174. APPRENTICE: IV/99–100.

612 ADDICT: IV/100. ARBITER: IV/100. AXE: IV/101. LOYAL: IV/101–102.

612 GILBERT; LE MOCO: III/133; Paillole, Services 492–95; Report on the Activities of No. 2 Tac H.Q. A Force Sixth Army Group, Part II, p. 149, in NARA RG 319 Entry 101 Box 4 Folder 80.

613 WANTAGE revisions: IV/40.

614 "Advanced HQ AFHQ": Operation Anvil, p. 3, in NARA RG 319 Entry 101 Box 4 Folder 78. No. 2 Tac: Report on the Activities of No. 2 Tac H.Q. A Force (Sixth Army Group) 31 July 1944–5 May 1945, Part I, p. 7, 12, in NARA RG 319 Entry 101 Box 4 Folder 79; Sweeney to Wild, TAC2/KFE/109, 1 Dec 44, PRO WO 219/2249/41; Author's interview with Eugene Sweeney, December 1992–January 1993; Author's interview with Arne Ekstrom, 27 May 93.

614 Ekstrom: Author's interview with Arne Ekstrom, 27 May 93. Hopkinson: Colyton 235–36.

615 Sweeney's orders: Report on the Activities of No. 2 Tac H.Q. A Force (Sixth Army Group) 31 July 1944–5 May 1945, Part I, pp. 12–14, 17, in NARA RG 319 Entry 101 Box 4 Folder 79. Planning FERDINAND: Report on the Activities of No. 2 Tac H.Q. A Force (Sixth Army Group) 31 July 1944–5 May 1945, Part I, p. 14, in NARA RG 319 Entry 101 Box 4 Folder 79; Anvil Plans (#1), in NARA RG 319 Entry 101 Box 4 Folder 70. "A real nice guy"; "Doo-glahss"; "Don't you see": Author's interview with Eugene J. Sweeney, December 1992–January 1993.

616 "Logical"; "This was NOT": Callahan to Guthrie, Notes on "Ferdinand," with endorsement, 27 Jul 44; JEB (?) to Sweeney, undated; Clarke to Sweeney, 31 Jul 44, all in NARA RG 319 Entry 101 Box 4 Folder 72 Ferdinand. ANVIL go-ahead; Bevan in Algiers: IV/109–110.

616 FERDINAND: Plan "Ferdinand" (Approved Version), PRO 204/1562/10–23; NARA RG 319 Entry 101 Box 4 Folder 72 Ferdinand; IV/111–12.

617 Hitler nervous about Genoa-Leghorn; FHW; "A source"; "a landing"; OKW: Howard 157–58. Bombing plan: IV/114–17; Operation Anvil, pp. 10–16, in NARA RG 319 Entry 101 Box 4 Folder 78; Morison, Invasion 243–44.

618 Resistance: Operation Anvil, p. 3, in NARA RG 319 Entry 101 Box 4 Folder 78. IRONSIDE continued: O'Connor to Bevan OC/41, 17 Jun 44, PRO CAB 154/57/38; Bevan to O'Connor, L/179, 27 Jun 44, PRO CAB 154/57/39; O'Connor to Bevan, No. 812, 1 Jul 44, PRO CAB 154/57/40; Bevan to O'Connor, L/191, 3 Jul 44, PRO CAB 154/57/41; From London, 28 June 1944, L/179, 27 Jun 44, in JSC folder 370.2 Ironside; Messages A-63 ("for another large scale landing"), A-64, A-65 in JSC brown Acco binder labeled "15." Plea for Biscay threat: COS to AFHQ, 15 Jul 44, CM-IN-13971 (17 Jul 44); SHAEF to AMSSO, NO. 55814, 19 Jul 44, CM-IN-16557 (20 Jul 44); Chiefs of Staff to SACMED, No. COSMED 158, 26 Jul 44, CM-IN-22168 (27 Jul 44), all in JSC folder 370.2 Ironside.

618 IRONSIDE II: IV/112; Plan "Ironside II," date illegible; L/225 cipher 4 August, both in JSC folder 370.2 Ironside; File—Log, in JSC Acco binder "7 File A"; cf. unsigned, undated memorandum, Implementation During Week Ending 19 August 1944, in JSC folder 370.2 Ironside; Hesketh 138. DRAGOON: Weigley 222–28. Axis Sally: "Operation Anvil," p. 2, in NARA RG 319 Entry 101 Box 4 Folder 78.

619 FERDINAND implementation: "Operation Anvil," passim, in NARA RG 319 Entry 101 Box 4 Folder 78; Morison, Invasion 245–50; IV/114; Fairbanks 236–42; Dwyer 65–75.

619 Total surprise: "Operation Anvil," p. 8, in NARA RG 319 Entry 101 Box 4 Folder 78; Joint Security Control Information Bulletin No. 3, 1 Jun 45, JSC/J14 Serial 5122, p. 2, in NARA RG 165 ABC File ABC 381 Japan 15 Apr 43 Sec. 1-B. Blaskowitz: Howard 159; Morison, Invasion 244. "Was quite the most": Howard 159.

References

620 OTTRINGTON; ULSTER: IV/104–107; Howard 163; Graham & Bidwell 347–48, 352.

621 No. 2 Tac to France: IV/118; Clarke diary 22 Aug 44, Clarke papers; Report on the Activities of No. 2 Tac H.Q. A Force (Sixth Army Group) 31 July 1944–5 May 1945, Part I, p. 15–16, in NARA RG 319 Entry 101 Box 4 Folder 79. "The poor Jerries": Author's interview with Eugene J. Sweeney, December 1992–January 1993.

621 Ekstrom to France: Author's interview with Arne Ekstrom, March 19, 1993. Fairbanks: Fairbanks 244–45. "A likeable": Hamer to author, 2 Jun 97. Antrim until July: III/24. FERDINAND conclusion: IV/119–20.

622 BRAINTREE: IV/119–20. SECOND UNDERCUT: Clarke to AC of S G-3, Main "A"F/44/68, 1 Oct 44, PRO WO 204/1563/35–37; IV/130–31. "It was with some chagrin": IV/41. Liquidation of units: Wingate 265–66. "One and all died hard": IV/41; Wingate 265.

623 WANTAGE Third (Modified) Edition: IV/108; V/4. WANTAGE terminated: IV/40–41; PRO WO 204/1563/3. "Like its predecessor": IV/42. German maps: PRO WO 204/1562/60–62. General's puzzlement: IV/41–42.

624 GILBERT: Report on the Activities of No. 2 Tac H.Q. A Force Sixth Army Group, Part II, pp. 105, 147–152, in NARA RG 319 Entry 101 Box 4 Folder 80; Paillole, Services 495; III/134.

624 CHEESE: IV/136–38. Phillips: Hamer to author, 2 Jun 97. Clarke in Athens: Clarke diary 18 Oct 44 (supplemental pages), 24 Oct 44–27 Oct 44, Clarke papers.

625 PEDANT and EFFIGY, 39 Committee: IV/138. Caserta: Clarke diary 13 Oct 44, 16 Oct 44, 18 Oct 44, Clarke papers; Ranfurly 306; Author's telephone interview with Lady Maynard (Daphne Llewellyn), 10 Apr 96. Tac HQ, outstations: IV/134.

625 Rear HQ: IV/133; "A" Force card index, Clarke papers; Author's telephone interview with Rex Hamer, 19 Jun 97. "All the interest"; enclosure: Clarke to Rooks, RHQ"A"F/10/21, 20 Oct 44 (sic; should probably be 30 Oct 44); Clarke, Some Notes on the Organization of Deception in the United States Forces, RHQ "A"F/10/21, 30 Oct 44, both in NARA RG 165 ABC File ABC 381 (4-8-43) Sec. 1-A—Deception policy, organization, etc. "To further the organization": Rooks to Hull, 3 Nov 44, in NARA RG 165 ABC File ABC 381 (4-8-43) Sec. 1-A—Deception policy, organization, etc.

626 Hull, Bissell: Buck slip on Rooks to Hull, 3 Nov 44; A. C. G. to Record, 30 Dec 44, both in NARA RG 165 ABC File ABC 381 (4-8-43) Sec. 1-A—Deception policy, organization, etc. Main reopened; reorganization: IV/135, V/3, V/24. SHELLAC: IV/134–35; V/1–2.

626 Plan for 1945: AFHQ to Chiefs of Staff, No. FX 7546, MEDCOS 220, 18 Dec 44, CM-IN-18687 (19 Dec 44), in NARA RG 165 ABC File ABC 381 (4-8-43) Sec. 1-A—Deception policy, organization, etc.; V/8, V/14; CAB 121/105/166–180. RUSTIC: V/8; Crichton to AC of S G-3, HQ"A"F/22/1, 10 Jan 45, WO PRO 204/24–25. OAKLEAF: V/9–10. PENKNIFE: V/13–16; Oakes 298. PLAYMATE: V/17–19.

627 Wingate: IV/139; Clarke diary 29 Nov 44, Clarke papers. Grandguillot: IV/139–40. Rear Party: Author's telephone interview with Rex Hamer, 19 Jun 97; IV/140.

628 Home leave, dismissal: V/24.

Chapter 15: Last Act in Europe

629 SHAEF generals: L/231 9 Aug 44; SHAEF to War Department, No. S 56342, 27 Jul 44, CM-IN-22846 (27 Jul 44); Smith to Godley, Subject: L/231, 9 August, Your OC/78, JSC/J1C2c, Serial 870, 18 Aug 44, all in JSC folder 370.2 Ironside. New York Times story: McFarland to Cornwall-Jones, Subject: Cover and Deception Planning, 4 Sep 44, in NARA RG 218 Box 361 Folder CCS 385 (6-25-43) Deception Plans for Overlord (Sec. 2).

REFERENCES

629 GARBO; "who has the fame"; "convicts": Hesketh 140, quoting GARBO Letter No. 26, August 26, 1944; Howard 194.

630 New FUSAG; MARKET-GARDEN: Howard 195–97

630 BODYGUARD canceled: Memorandum by the Representative of the British Chiefs of Staff, 26 Aug 44, CCS 495/5; Plan "Bodyguard," Report by the Joint Staff Planners, JCS 661/4, 30 Aug 44; Minute of Informal Action, 4 Sep 44; Plan "Bodyguard," CCS 459/9; United States Military Mission Moscow to War Department, No. MX 21177, 28 Sep 44, CM-IN-26704 (29 Sep 44), all in NARA RG 218 Box 361 Folder CCS 385 (6-25-43) Deception Plans for Overlord (Sec. 2). Liquidating notional units; FHW: Hesketh 139, 156; Howard 199–200; Notional U.S. Formations Employed in the Western European Theater During 1944, SHAEF/24134/SM/Ops, 11 Jul 45, in JSC folder European Theater 000; Weigley 570–71; 12th Army Group Report, Copy 2, p. 44, in JSC folder Operation Wedlock. Baumer: Cullum No. 9734; Baumer diary, p. 144, 28 Aug 44 ("trying some"); Stratagem Conference 11 (occasional involvement).

631 Wild: Mure, Master 123; Author's interview with Eugene J. Sweeney, December 1992–January 1993 ("He wasn't as personable"); Author's interviews with Arne Ekstrom, 4 Jan 93, 27 May 93 ("that little").

632 104 SCI: Oakes 65 note*, 67; Harmer 2.

632 X-2; "in collaboration": Oakes 61–62; Report on the Activities of No. 2 Tac H.Q. A Force (Sixth Army Group) 31 July 1944–5 May 1945, Part I, pp. 60–61, in NARA RG 319 Entry 101 Box 4 Folder 79. 62d SCI, SCU: Oakes 64; Harmer 3.

633 Oakes: Register of Rhodes Scholars; Author's interview with Arne Ekstrom, 4 Jan 93; Oakes 66 note**, 67; Harmer 3; Author's interview with John B. Oakes, 28 Dec 98. Ekstrom: Report on the Activities of No. 2 Tac H.Q. A Force (Sixth Army Group) 31 July 1944–5 May 1945, Part I, pp. 61–63, in NARA RG 319 Entry 101 Box 4 Folder 79.

633 DRAGOMAN: Oakes 70, 93–94; Harmer 8–9, 42–43. Weissmiller: Register of Rhodes Scholars; Author's interview with John B. Oakes, 28 Dec 98. War Room: Oakes 67–70, 93–154. "Left behind": Harmer 9.

634 Ops (B) organization: Wild to Distribution, SHAEF/18214/1/Ops(B), 24 Jul 44, NARA RG 331 Entry 1 Box 45 Folder Organization and Personnel G-3 SHAEF. Americans return: SHAEF Forward to War Department, No. 14492, 11 Sep 44, CM-IN-10096 (11 Sep 44); Handy to SHAEF, No. WARX 29124, 11 Sep 44, CM-OUT-29124 (12 Sep 44); Marshall to Eisenhower, No. WAR 34145, 20 Sep 44, CM-OUT-34145 (21 Sep 44); HQComZ to War Department, No. EX 51906, 3 Oct 44, CM-IN-3949 (5 Oct 44), all in JSC folder 000—To 312.1(2) China-Burma-India Theater; Report on the Activities of No. 2 Tac H.Q. A Force (Sixth Army Group) Part IV, Conclusions, p. 12, in NARA RG 319 Entry 101 Box 4 Folder 82. Wild decree; resentment; frustration; GILBERT: Oakes 85–88, 242; Harmer 34, 67–70; Report on the Activities of No. 2 Tac H.Q. A Force (Sixth Army Group) 31 July 1944–5 May 1945, Part I, pp. 20, 63, in NARA RG 319 Entry 101 Box 4 Folder 79.

635 "Although occasional"; informal understanding; "limited and unspectacular": 12th Army Group Report, Copy 2, pp. 19–20, 30, in JSC folder Operation Wedlock. Notional units: Oakes 80; Hesketh 155–56; PRO WO 219/2247/2–4; 12th Army Group Report, Copy 2, p. 75, in JSC folder Operation Wedlock; Report on the Activities of No. 2 Tac H.Q. A Force Sixth Army Group, Part II, pp. 96–98, 121, in NARA RG 319 Entry 101 Box 4 Folder 80.

635 No Strategic Addendum: Report on the Activities of No. 2 Tac H.Q. A Force (Sixth Army Group) 31 July 1944–5 May 1945, Part I, pp. 19–20, in NARA RG 319 Entry 101 Box 4 Folder 79.

635 Roster of double agents: Oakes *passim*. THE RHINEMAIDENS: Harmer 20 ("the girls

were allowed"), 40; Oakes 277–78; 12th Army Group Report, Copy 2, p. 84, in JSC folder Operation Wedlock.

637 Authority to approve: Oakes 78–79; Eldredge to Herbig, 3 Sep 84, Eldredge papers. DRAGOMAN vetoes: Oakes 124. Concession: Oakes 84.

637 GILBERT: Wingate 35 (Delègue); NARA RG 319 Entry 101 Box 4 Folder 80, "Report on the Activities of No. 2 Tac H.Q. A Force Sixth Army Group, Part II," pp. 105–109, 125, 149, 151–54. Hare: "A" Force card index, Clarke papers; III/125; IV/122; Author's interview with Arne Ekstrom, 27 May 93 ("a tall"). Beugras: Naftali, De Gaulle's 382.

638 Eldredge to France: Eldredge autobiography, Ch. 2; Eldredge to Clendennin, 16 Jul 90, both in Eldredge papers. Eldredge and Curtis: "R" Force War Diary 26 Jul 44, PRO WO 171/3832; Curtis to author, 29 Jun 93. "Philip went": Eldredge autobiography, ch. 4 p. 4, Eldredge papers. Ingersoll and Harris: Eldredge to Dwyer, 11 Sep 89, Eldredge papers; Ingersoll, Time Out 2–4 ("already acquired").

639 Special Plans organization: Cover and Deception, Synopsis of History, p. 10, in NARA RG 319 Entry 101 Box 1 Folder 1. "He was a nice"; "A very, very bright"; "a most gentlemanly": Eldredge to Dwyer, 11 Sep 89, Eldredge papers. "You could tell": Stratagem Conference 88. Near Laval: Eldredge autobiography, ch. 12, pp. 4–5, Eldredge papers. Map room: Cover & Deception Report ETO, Special Exhibit, "Use of Ultra in Cover and Deception," undated, pp. 2–3, in JSC folder Colonel Bevan Use of Special Intelligence in Cover and Deception. "But the battle": Eldredge to Harrison, 31 Aug 88, Eldredge papers.

640 Bradley headquarters: Weigley 497, 499. "Billy let go"; stayed at Main: Eldredge to Dwyer, 11 Sep 89, Eldredge papers. Team: Stratagem Conference 16. "The G staff": Eldredge to Harrison, 31 Aug 88, Eldredge papers; see 12th Army Group Report, Copy 2, p. 19, in JSC folder Operation Wedlock. "Straight Leavenworth": Stratagem Conference 39. Dickson: Eisenhower, Bitter Woods 208. "Decoration which could"; "it was characteristic": 12th Army Group Report, Copy 2, pp. 19, 25, in JSC folder Operation Wedlock.

640 No access to Bradley: Stratagem Conference 27, 34. "Omar Bradley said": Eldredge to Herbig, 25 Jun 83, Eldredge papers.

640 "And they put this old fud": Stratagem Conference 27. "Billy was very": Eldredge to Dwyer, 11 Sep 89, Eldredge papers. "Special Effects"; "did improve": 12th Army Group Report, Copy 2, pp. 25–26, in JSC folder Operation Wedlock.

641 23d operations: 12th Army Group Report, Copy 2, *passim,* in JSC folder Operation Wedlock ("An example of deception": p. 19); Kronman 20–29; Park.

641 No agreed strategy: Informal Report to Joint Security Control by Special Plans Branch, G-3, 12th Army Group, 25 May 45, in NARA RG 319 Entry 101 Box 2 Folder 27.

642 AVENGER, AVENGER II: Wingate 344; Howard 198; SHAEF/19018/Ops(B) GCT/370.28-205/Ops(B), 6 Oct 44, in JSC folder European Theater 000; Report on the Activities of No. 2 Tac H.Q. A Force Sixth Army Group, Part II, pp. 6–7, in NARA RG 319 Entry 101 Box 4 Folder 80. DERVISH: PRO WO 219/2253/11–15, 20–30; Wingate 344; Howard 198. 18th Airborne: Smith to O'Connor, JSC/J1C2, Serial 382, 14 Mar 45, in JSC folder Dervish. CALLBOY: PRO WO 219/2254; Howard 198. Wild gave up: Wingate 344; Report on the Activities of No. 2 Tac H.Q. A Force Sixth Army Group, Part II, pp. 83–84, in NARA RG 319 Entry 101 Box 4 Folder 80. TAPER: Wingate 351; Howard 198.

642 JURISDICTION: PRO WO 219/2255. Strangeways special assignments: "R" Force War Diary September 1944, PRO WO 171/3832; Young & Stamp 51–52. Strangeways tactical deceptions: "R" Force and G(R) war diaries, *passim,* PRO WO 171/3831–32. IMPERIL: PRO WO 219/2247/2. TROLLEYCAR: PRO WO 219/2247/2, 38–54; Wingate 351; Harmer 33; Hesketh 154, 159–61; Howard 199.

REFERENCES

642 Combat role: "R" Force War Diary 20–21 Dec 44, PRO WO 171/3832. TURBINATE, TURBINATE II: PRO WO 219/2247/3–4, 55–64; Wingate 351; Harmer 33. 15th Airborne: Hood to Wild, 8 Feb 45, PRO WO 219/2247/63. TRANSCEND: PRO WO 219/2247/65–69.

643 "Was in many respects": Harmer 33. No. 5 Wireless Group: V/5–6. "Don't unpack": Author's interview with Eugene J. Sweeney, December 1992–January 1993; Report on the Activities of No. 2 Tac H.Q. A Force (Sixth Army Group) 31 July 1944–5 May 1945, Part I, p. 17, in NARA RG 319 Entry 101 Box 4 Folder 79. "Small, lean": Weigley 345. Cabled to Clarke: Report on the Activities of No. 2 Tac H.Q. A Force (Sixth Army Group) 31 July 1944–5 May 1945, Part I, p. 103, in NARA RG 319 Entry 101 Box 4 Folder 79. Rooks: Clarke to Rooks, 20 Oct 44; Rooks to Hull, 3 Nov 44, both in NARA RG 165 ABC File ABC 381 (4-8-43) Sec. 1-A—Deception policy, organization, etc.; Jenkins biographical release, USAMHI.

644 "Hope this will *not*": Report on the Activities of No. 2 Tac H.Q. A Force (Sixth Army Group) 31 July 1944–5 May 1945, Part I, p. 17, in NARA RG 319 Entry 101 Box 4 Folder 79. Crillon meeting: Minutes of Meeting Held Hotel Crillon, 27th September 1944, at 1000 hours, 27 Sep 44, PRO WO 219/2246/103–104; Clarke diary 26 Sep 44, Clarke papers.

645 Jenkins; No. 2 Tac to G-3: Report on the Activities of No. 2 Tac H.Q. A Force (Sixth Army Group) 31 July 1944–5 May 1945, Part I, p. 10–13, 28, in NARA RG 319 Entry 101 Box 4 Folder 79. "He was an intelligent": Author's interview with Eugene J. Sweeney, December 1992–January 1993. Directive; "the overall"; Wild visit: Report on the Activities of No. 2 Tac H.Q. A Force (Sixth Army Group) 31 July 1944–5 May 1945, Part I, p. 18, in NARA RG 319 Entry 101 Box 4 Folder 79.

646 Ekstrom: Report on the Activities of No. 2 Tac H.Q. A Force (Sixth Army Group) 31 July 1944–5 May 1945, Part I, pp. 16, 60, 61, 63–64, 66–67, in NARA RG 319 Entry 101 Box 4 Folder 79. Ekstrom to CBI: CG US Army Forces CBI to War Department, No. CRA 15171, 28 Sep 44, CM-IN-27005 (29 Sep 44); Bissell to SHAEF, WAR 47800, 17 Oct 44, CM-OUT 47800 (17 Oct 44); Handwritten slip, [Schrup] to Smith, with endorsement, 26 Oct 44 ("Phoned Haygood"), all in JSC folder 000—To 312.1(2) China-Burma-India Theater.

646 X-2: Oakes 68–69, 197, 199, 210. 104 SCI: Harmer 10–11. 212 Committee: PRO KV 4/101; Harmer 6; Oakes 79, 82, 89; Harmer to author, 6 Nov 92.

647 Americans: Stratagem Conference 69. "All we really": Ingersoll, Time Out, ch. XI p. 16. Indifference to security; "simply could not": 12th Army Group Report, Copy 2, pp. 26–27, in JSC folder Operation Wedlock. No. 2 Tac rebuffed: Report on the Activities of No. 2 Tac H.Q. A Force Sixth Army Group, Part II, pp. 43–46, in NARA RG 319 Entry 101 Box 4 Folder 80.

647 "I went through": Eldredge to Herbig, 25 Jan 80, Eldredge papers. Improvement, not for long: Report on the Activities of No. 2 Tac H.Q. A Force Sixth Army Group Part II, pp. 43–46, in NARA RG 319 Entry 101 Box 4 Folder 80. Sixty thousand: 12th Army Group Report, Copy 2, pp. 26–27, in JSC folder Operation Wedlock. "In fact, during": Bevan to Bissell, LCS (45)I/D.11, 13 Apr 45, p. 2, in JSC folder Colonel Bevan Use of Special Intelligence in Cover and Deception.

648 Eldredge's goodies; "Ingersoll martinis"; "the party always"; beef; Wiesbaden; Diana: Eldredge autobiography, ch. 12, pp. 1, 2, 4–5, Eldredge papers.

649 "He was a vain"; "was getting restless": Eldredge autobiography, ch. 15, pp. 1–2, Eldredge papers. Bradley: Weigley 580; Bradley & Blair 210, 390. Wehrmacht champagne: Author's interview with Eugene J. Sweeney, December 1992–January 1993. "Dunn was always": Author's interview with Arne Ekstrom, 27 May 93.

649 Ekstrom at Raphaël, car, driver: Author's interview with Arne Ekstrom, 27 May 93.

650 JESSICA: Report on the Activities of No. 2 Tac H.Q. A Force (Sixth Army Group) 31

July 1944–5 May 1945, Part I, pp. 15–16, 26–28, 39–46, 50–54, 83–89, in NARA RG 319 Entry 101 Box 4 Folder 79. "Détachement d'Armée des Alpes": Williams 419.

651 Deceptive signaling: Myers to Chief, Security Division, Subject: Report of Conference with Captain Gilden, 26 Nov 45, p. 4, in JSC folder 13–12/6 Communications Deception Reports: Europe, Africa.

651 FOREST: Oakes 231–40; Report on the Activities of No. 2 Tac H.Q. A Force (Sixth Army Group) 31 July 1944–5 May 1945, Part I, pp. 68–80, 96, in NARA RG 319 Entry 101 Box 4 Folder 79. Merrick: Obituary, *New York Times,* 23 Apr 88; Author's interview with Arne Ekstrom, 27 May 93.

652 MONOPLANE, GILBERT, LITIGANT, PEGASUS: Report on the Activities of No. 2 Tac H.Q. A Force (Sixth Army Group) 31 July 1944–5 May 1945, Part I, pp. 40, 68–75, 84, 85, in NARA RG 319 Entry 101 Box 4 Folder 79. Axis divisions: Report on the Activities of No. 2 Tac H.Q. A Force (Sixth Army Group) 31 July 1944–5 May 1945, Part I, pp. 100–109, in NARA RG 319 Entry 101 Box 4 Folder 79.

653 KNIFEDGE: Report on the Activities of No. 2 Tac H.Q. A Force Sixth Army Group, Part II, pp. 9, 11, 14–15, 25–29, 58–62, 110–13, in NARA RG 319 Entry 101 Box 4 Folder 80; Paillole, Services 497–99.

654 ACCORDION: Report on the Activities of No. 2 Tac H.Q. A Force Sixth Army Group Part II, pp. 49a–56, 64–67, 78–79, 82, 87, 91, 119, 121–23, 129, 133–34, 157, in NARA RG 319 Entry 101 Box 4 Folder 80.

655 LOADLINE: Report on the Activities of No. 2 Tac H.Q. A Force (Sixth Army Group) Part III, Operation Loadline, Southern France, pp. 8–10, 15–19, 29, 31–33, in NARA RG 319 Entry 101 Box 4 Folder 81. MARÉCAGE: Larminat to Commanding General 6th Army Group with enclosure, No. 1178 EM/3, 31 Mar 45, PRO WO 219/2249/115–22. GILBERT: Report on the Activities of No. 2 Tac H.Q. A Force Sixth Army Group, Part II, pp. 117–19, 124, 131, in NARA RG 319 Entry 101 Box 4 Folder 80; Report on the Activities of No. 2 Tac H.Q. A Force (Sixth Army Group). Part III, Operation Loadline, Southern France, pp. 46–49, in NARA RG 319 Entry 101 Box 4 Folder 81.

655 Ekstrom coordination: Author's interview with Arne Ekstrom, 27 May 93; Oakes 243; Report on the Activities of No. 2 Tac H.Q. A Force (Sixth Army Group) Part III, Operation Loadline, Southern France, pp. 40–43, in NARA RG 319 Entry 101 Box 4 Folder 81.

656 KOBLENTZ: "Without any real hope": William A. Harris, Lecture on Deception, Command and General Staff College, c. 1947, C&GSS Library, No. U167.5.D37 H37, Lecture III, p. 17. "There seemed to be"; "The Brits": Eldredge autobiography, "The Battle of the Bulge, Christmas, 1944," Eldredge papers, pp. 1–3. FHW: 12th Army Group Report, Copy 2, pp. 41–42, in JSC folder Operation Wedlock.

657 Geisel; "stormed to the rear"; caustic letter: Eldredge autobiography, "The Battle of the Bulge, Christmas, 1944," Eldredge papers, pp. 8–9; "The retreat we beat": Morgan & Morgan 111–13. "Ingersoll was in high": Eldredge autobiography, "The Battle of the Bulge, Christmas, 1944," Eldredge papers, pp. 4–5.

658 Patton plans: Weigley 498–99. ARDENNES: 12th Army Group Report, Copy 2, pp. 42–43, in JSC folder Operation Wedlock; Eldredge autobiography, "The Battle of the Bulge, Christmas, 1944," p. 5; Eldredge to Herbig, 3 Sep 84, both in Eldredge papers. WITCH and LEAGUE: 12th Army Group Report, Copy 2, p. 39, in JSC folder Operation Wedlock; Oakes 251–70. KODAK, METZ I: 12th Army Group Report, Copy 2, pp. 43, 45–49, in JSC folder Operation Wedlock; Kronman 31–32. Buzzing: Stratagem Conference 93–95. "We felt very guilty": Eldredge to Harrison, 24 Sep 89, Eldredge papers.

659 Response to playacting: Author's interview with Eugene J. Sweeney, December 1992–January 1993. Christmas at EAGLE TAC; "a beauteous"; ("rather badly)";

"Shortly, I awoke": Eldredge autobiography, "The Battle of the Bulge, Christmas, 1944," pp. 5–8, Eldredge papers. Six-hundred-plane raid: Weigley 573.

660 German offensive: Weigley 552–56. Sweeney and Ekstrom's Christmas: Author's interview with Arne Ekstrom, 27 May 93; Author's interview with Eugene J. Sweeney, December 1992–January 1993. Reporting V-1s and V-2s in London: Howard 170–77, 181–83. Antwerp: Harmer 33, 55. Liège: 12th Army Group Report, Copy 2, p. 59, in JSC folder Operation Wedlock.

661 Small tactical deceptions; Iron Cross: Kronman 32–41; 12th Army Group Report, Copy 2, pp. 60–77, in JSC folder Operation Wedlock.

661 Fifteenth Army: Weigley 668. Exaggeration: 12th Army Group Report, Copy 2, pp. 78–80, in JSC folder Operation Wedlock; Oakes 143, 190, 274, A358. Ops (B) winding down: Oakes 89. OMEGA: 12th Army Group Report, Copy 2, pp. 81–84, in JSC folder Operation Wedlock.

662 Harris's last shot: Oakes 89. "In all the balconies": Author's interview with Arne Ekstrom, 27 May 93. 212 Committee: Harmer 7. Sweeney at end: Author's interview with Eugene J. Sweeney, December 1992–January 1993.

662 Ekstrom at end: Author's interview with Arne Ekstrom, 27 May 93. Eldredge at end: Eldredge autobiography, " 'Rigours' of War Department, July 1944–September 1945 Normandie to Germany," Eldredge papers, p. 6.

663 Winding down LCS: Note by Controlling Officer, COS (44)201, 6 Oct 44, PRO CAB 121/105/182-182A; Extract from COS (44) 336th Meeting, 11 Oct 44, CAB 121/105/181; Wheatley 220–24. "And its subsidiary sections": Copy of a Minute (COS 1716/4) dated 11th October, 1944, from Secretary, Chiefs of Staff Committee to London Controlling Officer, Annex III to COS (44) 336th Meeting Min Y, CAB 121/105/180; Wheatley 221.

664 Winding down double agents: Masterman 168; Montagu, Twenty Committee, 9 Nov 44, PRO ADM 223/794/66. BRUTUS: Masterman 173. TATE: Howard 182, 228; Masterman 183–84.

664 GARBO: Pujol 159–62, 169–71, 183–84; Howard 236.

665 GILBERT: Report on the Activities of No. 2 Tac H.Q. A Force Sixth Army Group, Part II, p. 154, in NARA RG 319 Entry 101 Box 4 Folder 80.

666 Ingersoll report: Bissell to Bevan, 8 Jan 95, JSC/J3A4, Serial 131, p. 1; Rushton to Bissell, 16 Dec 44, both in NARA RG 319 Entry 101 Box 1 Folder 1 Cover & Deception Synopsis of History. "Profound admiration": Bevan to Bissell, CO/560, 11 Dec 44, in NARA RG 319 Entry 101 Box 1 Folder 1 Cover & Deception Synopsis of History. Smith/Bissell response: Buck slip, Smith to Bissell, 5 Jan 45; Buck slip, Bissell to Smith, 7 Jan 45; Bissell to Bevan, JSC/J3A4 Serial 131, 8 Jan 45, all in NARA RG 319 Entry 101 Box 1 Folder 1 "Cover & Deception Synopsis of History."

666 ULTRA annex: Cover & Deception Report ETO, Special Exhibit, "Use of Ultra in Cover and Deception," undated; Taylor to Bevan, 4 Apr 45; Clarke to Smith, 7 May 45, all in JSC folder Colonel Bevan Use of Special Intelligence in Cover and Deception. "My general impression": Bevan to Bissell, LCS (45) I/D.11, 13 Apr 45, in JSC folder Colonel Bevan Use of Special Intelligence in Cover and Deception. "Interesting and instructive": Bissell to Bevan, JSC/J/NS/2e, Serial 531, 8 May 45, in JSC folder Colonel Bevan Use of Special Intelligence in Cover and Deception.

Chapter 16: Hustling the East (III)

668 Knew next to nothing: Cake Before Breakfast, p. 56, JSC loose papers. Goldbranson study: C. E. G[oldbranson] to Smith, 19 Feb 44, in JSC folder Communications Security Organization. Vicissitudes: Draft Memorandum by Joint Security Control, With Concurrence of Joint Staff Planners, undated, attached to Peck to McFarland, Subject:

Deception Planning and Information, 17 May 44, in NARA RG 218 Box 362 Folder CCS 385 (4-8-43)(2) Cover and Deception Planning (Sec. 1); Draft, Volkel to Secretary JCS, with attachment, JCS/J1 Serial 6114, 26 Jun 44; buck slip, date illegible; Deception in the War Against Japan, JCS 498/3, 27 Jun 44; Peck to Joint Staff Planners, Subject: Deception in the War Against Japan, 28 Jun 44; Deception in the War Against Japan, JPS 481/D, 28 Jun 44; Donnelly to Joint War Plans Committee, Subject: Deception in the War Against Japan, 30 Jun 44; Deception in the War Against Japan, JWPC 190/3/D, 30 Jun 44; Memorandum for the Secretary, Joint Staff Planners, Subject: Deception in the War Against Japan, JWPC 190/4, 5 Jul 44; Memorandum to Holders of JCS 498/3, Dated 27 June 1944, 7 Jul 44, all in NARA RG 218 Box 694 Folder CCS 385 Pacific Theater (4-1-43) Deceptive Policy in the Pacific Theater (Sec. 3). JCS to commanders: Draft attached to Stockhausen to Secretary, Joint Chiefs of Staff, Subject: Deception in the War Against Japan, 20 Jul 44; Deception in the War Against Japan, JCS 498/4, 21 Jul 44; McFarland to Joint Staff Planners, Subject: Deception in the War Against Japan, 23 Jul 44; Deception in the War Against Japan, JPS 372/4/D, 23 Jul 44; Donnelly to Joint War Plans Committee, 24 Jul 44; Deception in the War Against Japan, JWPC 190/6; Deception in the War Against Japan, Report by the Joint Staff Planners, JCS 498/5, 1 Aug 44; approved by JCS informal action as JCS 493/6, 5 Aug 44, all in NARA RG 218 Box 694 Folder CCS 385 Pacific Theater (4-1-43) Deceptive Policy in the Pacific Theater (Sec. 3).

668 Responses: Rear Echelon GHQ, SWPA, to War Department, No. C 15847, 7 Aug 44, CM-IN-6250 (7 Aug 44); CINCPOA to COMINCH, No. 100498, NCR 2883, 10 Aug 44; CGUSACBI to War Department, No. CRA 10418, 10 Aug 44, CM-IN-8791 (10 Aug 44), all in NARA RG 218 Box 694 Folder CCS 385 Pacific Theater (4-1-43) Deceptive Policy in the Pacific Theater (Sec. 3). Planning and organizing course; school bell: Cake Before Breakfast, p. 56, JSC loose papers; Burris-Meyer papers; Author's telephone interview with Anita Burris-Meyer, 22 May 94. Students: List in JSC folder Communications Security Organization; Cake Before Breakfast, p. 57, JSC loose papers. "It now appears": SHAEF Forward to War Department, No. 14492, 11 Sep 44, CM-IN-10096 (11 Sep 44); War Department to SHAEF, WARX 29124, 11 Sep 44, CM-OUT-29124 (12 Sep 44), both in JSC folder 000—To 312.1(2) China-Burma-India Theater.

669 Corbett: Marshall to Eisenhower, No. WAR 34145, 20 Sep 44, CM-OUT-34145 (21 Sep 44); HQ COM Z ETO to War Department, No. EX 51906, 3 Oct 44, CM-IN-3949 (5 Oct 44), both in JSC folder 000—To 312.1(2) China-Burma-India Theater. McDowell: Author's telephone interview with Anita Burris-Meyer, 19 Feb 95; Whitehill 213; Fairbanks to McDowell, Subject: Project #192, Report on Trials of, 16 Mar 43, in NOA 531, Folder Deception Methods Techniques and Devices. Mentz: Author's telephone interview with Anita Burris-Meyer, 19 Feb 95; Mentz biographical release, Naval Operational Archives, Washington Navy Yard. Fellows: Fellows biographical release, Naval Operational Archives, Washington Navy Yard.

670 Holland: Cullum no. 7819; obituary, Assembly, April 1945, pp. 10–11. Brown: Author's telephone interview with Anita Burris-Meyer, 16 Feb 95; Bentley to Brown, 30 Jan 47, in NOA 548, Folder Bluebird. Hilger: Glines 42, 46, 66, 264, 266, 420; author's telephone interview with Anita Burris-Meyer, 16 Feb 95.

671 YLS course: JSC folder C&D Instructions—Course Outlines (WWII—1944, 1945); JSC folder Communications Security Organization; JSC folder Joint Security Control Memos 1–9; JSC folder Deception Units; Cake Before Breakfast, p. 56, JSC loose papers; Burris-Meyer to Volkel, Subject: Conference Room Requirements, 26 Aug 44; Volkel to Peck, JSC/G1 Serial 895, n.d., both in NARA RG 218 Box 694 Folder CCS 385 Pacific Theater (4-1-43) Deceptive Policy in the Pacific Theater (Sec. 3). "Staff and probably planning": Burris-Meyer to Smith, Subject: J.S.C. Indoctrination Course,

Report on, 17 Oct 44, in JSC folder C&D Instructions—Course Outlines (WWII—1944, 1945).

671 Teams: Undated memoranda in Smith's handwriting; undated list, all in JSC folder 000—To 312.1(1) China-Burma-India Theater.

672 Departures: Bissell to CGUSAFCBI, No. WAR 40565, 2 Oct 44, CM-OUT-40565 (3 Oct 44); Bissell to CINCPOA, No. WAR 40566, 3 Oct 44, CM-OUT-40566 (3 Oct 44); Bissell to CINCSWPA, No. WAR 40567, 3 Oct 44, CM-OUT-40567 (3 Oct 44), all in JSC folder 000—To 312.1(2) China-Burma-India Theater. BLUEBIRD designed; Special Plans Section: McDowell to Joint Security Control, Subject: Cover and Deception, Report of Progress, Central Pacific, 29 Dec 44, with Enclosure A, in JSC folder German Efforts WWII. Hilger and Barnes assigned: Glines 420; CINCPOA to COMINCH, 7 Dec 44, CM-IN-7785 (8 Dec 44); Bissell to CGUSAFPOA, No. WAR 76636, 13 Dec 44, CM-OUT-76636 (13 Dec 44); TAG to CGUSAFPOA, No. WAR 78642, 16 Dec 44, CM-OUT-78642 (17 Dec 42); CINCPOA to COMGENPOA, 23 Dec 44, CM-IN-24414 (26 Dec 44), all in JSC folder 000—To 312.1(2) China-Burma-India Theater. "All of the officers contacted"; "The Central Pacific Theater": Enclosure E to McDowell to Joint Security Control, Subject: Cover and Deception, Report of Progress, Central Pacific, 29 Dec 44, in JSC folder German Efforts WWII.

672 McDowell lost: McDowell biographical release, Naval Operational Archives, Washington Navy Yard. "The deception policy": Ritchie to Chief, Strategy & Policy Group, Subject: Comments on JPS 433/1 and JCS 806, 15 Apr 44, in NARA RG 165 ABC File ABC 381 Japan 15 Apr 43 Sec. 4—Forager. Hollandia: Spector 286–87, citing Morison New Guinea 66, Eichelberger 102; Drea 116–18, citing MacArthur history 1:144–45, and GHQ, Far East Command, MIS, GS, "A Brief History of the G-2 Section, GHQ, SWPA, and Affiliated Units" 25–28, Plate 9, and accompanying text.

673 Conversation with British: Handwritten notes, unsigned, undated, in Bissell's handwriting, in JSC folder 000—To 312.1(2) China-Burma-India Theater. "Are now unoccupied": Memorandum for the Chief of Staff, 27 Oct 44, Enclosure F to McDowell to Joint Security Control, Subject: Cover and Deception, Report of Progress, Central Pacific, 29 Dec 44, in JSC folder German Efforts WWII.

673 Staff study: Staff Study, Cover and Deception, Operation "Mike-One," Musketeer Operations, 15 Nov 44, in NOA 550, Folder Mike One. Actual plan: GHQ SWPA Hollandia to War Department, 16 Nov 44, CM-IN-15326 (16 Nov 44), in NOA 537, Folder M-1. Sample pieces: Cake Before Breakfast p. 43, JSC loose papers. Holland: Bissell to CINCSWPA, No. WAR 77716, 15 Dec 44, CM-OUT-77716 (15 Dec 44), in JSC folder 000—To 312.1(2) China-Burma-India Theater; Cullum no. 7819; obituary, Assembly, April 1945, pp. 10–11. Rich: Smith to CINCSWPA Attention Rich, Subject: Implementation of CULTURE, JSC/J5 Serial 161, 13 Jan 45, in JSC folder Olympic. Mentz: Dwyer 83–84; Morison, Liberation 44–46; Bissell to CINCSWPA, No. WAR 86691, 4 Jan 45, CM-OUT-86691 (Jan 45); GHQSWPA to War Department, No. X 56062, 8 Jan 45, CM-IN-7430 (9 Jan 45), both in JSC folder 000—To 312.1(2) China-Burma-India Theater.

674 ICHIGO: Romanus & Sunderland, Command 316; Romanus & Sunderland, Time 179. Hunter problems: Hunter to Sultan, Subject: Organization, 16 Apr 44; Hunter to Smith, Subject: Deception Section CBI, 9 May 44; Hunter to Smith, Subject: Organization, 13 May 44; Smith to Hunter, No. WAR 57183, 27 Jun 44, CM-OUT-57183 (28 Jun 44); Smith to Hunter, Subject: Theater Deception Staff, JSC/J3A6 Serial 709, 3 Jul 44; Hunter to Smith, Subject: Organization CBI Deception Section, 1 Jul 44 [misdated 1 Apr 44]; Hunter to Smith, Subject: Organization 19 Jul 44, all in JSC unlabeled folder [CBI]; CGUSAFCBI to War Department, No. CRAX 7285, 3 Jul 44, CM-IN-[illegible] (7 Jul 44); Handy to CGUSAFCBI, No. WARX 61924, 5 Jul 44,

CM-OUT-61942 (7 Jul 44); Milani to Joint Security Control, Subject: Deception Activities in India Burma Theater, 8 Jan 45; Hunter to Smith, Subject: Channel, 18 Oct 44, all in JSC folder 000—To 312.1(1) China-Burma-India Theater.

675   Security in China: C. E. G. [oldbranson] to Smith, Subject: Plan for Deception Section for China Prepared by Lt. Col. E. O. Hunter, 28 Aug 44, para. 3g, in JSC folder 000—To 312.1(1) China-Burma-India Theater.

675   Hunter in China; his report: Plan for Deception Section for China, 25 Jun 44, in JSC folder Plan for Deception Section for China. Chennault proposal: Same, page headed "Kunming"; GG 14th Air Force to War Department, No. CAKX 3670, 18 Jun 44, CM-IN-14879 (18 Jun 44); GG 14th Air Force to War Department, 20 Jun 44, CM-IN-16334 (20 Jun 44); Smith to Hunter, No. WAR 53919, 21 Jun 44, CM-OUT-53919 (21 Jun 44), all in JSC unlabeled folder [CBI]; CG USAFCBI to War Department, No. CRAX 7284, 3 Jul 44, CM-IN-2220 (3 Jul 44); Donnelly to Joint Security Control, Subject: Deception Plan in Central China, 4 Jul 44; Bissell to CGAAF India-Burma Sector, No. WARX-60630, 5 Jul 44, CM-OUT-60630 (5 Jul 44), all in JSC folder 000—To 312.1(2) China-Burma-India Theater. Sultan's chief not allow: Hunter to Smith, 18 Aug 44, in JSC unlabeled folder [CBI]. Bissell: Hunter to Smith, 28 Jul 44, in JSC unlabeled folder [CBI]; Shaw to Chief of Staff, Subject: Deception Operations, 15 Aug 44, in JSC folder 000—To 312.1(1) China-Burma-India Theater. "It is felt": C. E. G. [oldbranson] to Smith, Subject: Plan for Deception Section for China Prepared by Lieutenant Colonel E. O. Hunter, 28 Aug 44, in JSC folder 000—To 312.1(1) China-Burma-India Theater.

676   Hunter proposal: Hunter to Smith, Subject: Organization CBI Deception Section, 1 Jul 44 [misdated 1 Apr 44]; Smith to Hunter, Subject: Theater Deception Staff, JSC/J3A6 Serial 709, 3 Jul 44, both in JSC unlabeled folder [CBI]; CGUSAFCBI to War Department, No. CRAX 13990, 16 Sep 44, CM-IN-15741 (17 Sep 44); Hunter to Smith, 18 Sep 44; Shaw to Chief of Staff, Subject: Deception Operations, 15 Aug 44, all in JSC folder 000—To 312.1(1) China-Burma-India Theater. Bevan: Copy of Letter from Colonel Bevan dated 23rd August 1944, in JSC unlabeled folder [CBI]. Turned down: Smith to Bissell, Subject: Deception Personnel, CBI, 19 Sep 44; Handy to CGUSAFCBI, No. WAR 33977, 20 Sep 44, CM-OUT-33977 (21 Sep 44), both in JSC folder 000—To 312.1(2) China-Burma-India Theater. "Strange that Joint": Hunter to Smith, 1 Oct 44, in JSC unlabeled folder [CBI]. Hunter and Bissell: Bissell to Hunter, 12 Oct 44, in JSC unlabeled folder [CBI].

677   Thorne sensed; V-bombs: Thorne to author, 27 Apr 97. "Real fear": Astley 63. SUNRISE deletions: Fleming to Bevan, 25 Aug 44, PRO CAB 154/26/13A; Bevan to Fleming, 1 Sep 44, PRO CAB 154/26/16A; Bevan to Fleming, 8 Sep 44, PRO CAB 154/26/15A; Fleming to Bevan, 23 Sep 44, PRO CAB 154/36/9A. Not keeping informed: Bevan to Fleming, LCS(44)I/C1(d), 11 Oct 44, PRO CAB 154/36/11A; Wingate to Bevan, 9 Oct 44, PRO CAB 154/36/10A; Bevan to Fleming, LCS(44)I/C1(d), 11 Oct 44, PRO CAB 154/36/11A. "Went down"; "persona non grata": "I feel": Fleming to Whitworth Jones, D.O. No. 400/D.Div., 6 Sep 44, PRO WO 203/3314. "Dangerous"; "If Fleming did": Whitworth Jones to SAC, 28 Dec 44, PRO WO 203/3314. "The sort of stuff"; "You will note": Whitworth Jones to Chief of Staff, 6 Feb 45, WO 203/3314. Wedemeyer: Thorne to author, 27 Apr 97. "Very congenial": Wedemeyer 270. Send Wingate: Fleming to Bevan, 23 Sep 44, PRO CAB 154/36/9A; Bevan to Fleming, LCS(44)I/C1(d), 11 Oct 44, PRO CAB 154/36/11A; Ismay to Pownall, 24 Nov 44, PRO WO 203/3313.

678   Team in India: Report on Deception in INDIA and CHINA Theaters, 30 Dec 44, with appendices, in NOA 535, Folder India-Burma; Hunter to Smith, 27 Nov 44, in JSC unlabeled folder [CBI]; Brown to Rusk, Subject: Assignment of Special Plans Officer to SEAC, JSC/J3A6 Serial 133, 8 Jan 45; Raymond to Brown, Subject: Assignment of

Special Plans Officer to SEAC, 18 Jan 45, both in JSC folder 000—To 312.1(2) China-Burma-India Theater.

678 Team in China; "on a strategic scale"; "the opportunities": Report on Deception in INDIA and CHINA Theaters, 30 Dec 44, with appendices, in NOA 535, Folder India-Burma; Brown to Smith, Subject: Deception personnel for Fourteenth Air Force, 18 Dec 44, in JSC unlabeled folder [CBI]. New Delhi and return: Report on Deception in INDIA and CHINA Theaters, 30 Dec 44, with appendices, in NOA 535, Folder India-Burma; Milani to Joint Security Control, Subject: Deception Activities in India Burma Theater, 8 Jan 45, in JSC folder 000—To 312.1(2) China-Burma-India Theater; Smith to Raymond, JSC/J3A6 Serial 1292, 29 Dec 44, in JSC unlabeled folder [CBI]. "Fully inter-service": ADCOS(A) to DCOS, 25 Oct 44, PRO WO 203/3313/68A. "Some experience": Bevan to Fleming, No. GCCS 8386, 15 Dec 44, PRO HW 41/159. "Moody hardly carries"; "an excellent officer": Fleming to O'Connor, IND 3158, 18 Dec 44, PRO HW 41/159. "Have a great respect": Whitworth Jones to D/SAC, 11 Dec 44, PRO WO 203/3314. Moody not available: Moody to Smith, 17 Nov 44; Smith to Moody, 28 Nov 44, both in JSC unlabeled folder [CBI].

679 "Apparently there is not"; Smith reply: Hunter to Smith, 27 Nov 44; Smith to Hunter, undated, both in JSC unlabeled folder [CBI]. Raymond: CGUSAFIB to War Department, No. CRA 3514, 10 Feb 45, CM-IN-10570 (11 Feb 45), in JSC folder 000—To 312.1(2) China-Burma-India Theater; D Division Weekly News Letter No. 42 3 Feb–9 Feb 45 Part I, in JSC folder Deception—SE Asia—WWII Historical D. Div—SACSEA—30 Dec 44–23 Mar 45. Corbett: Milani to Joint Security Control, Subject: Deception Activities in India Burma Theater, 8 Jan 45; CGUSAFIB to War Department, CRA 23947, 29 Dec 44, CM-IN-28412 (29 Dec 44), both in JSC folder 000—To 312.1(2) China-Burma-India Theater; D Division Weekly News Letters No. 41, 27 Jan–2 Feb 45, Part I, and No. 44, 17–23 Feb 45, Part I, both in JSC folder Deception—SE Asia—WWII Historical D. Div—SACSEA—30 Dec 44–23 Mar 45; Unsigned [Corbett] to Fleming, 20 Feb 45; Unsigned [Corbett] to Fleming, 5 Mar 45, both in JSC folder Southeast Asia; D Division Weekly Progress Report No. 1 24–30 Mar 45; No. 5 21–27 Apr 45, Part I; No. 7 5–14 May 45, all in JSC folder Deception—SE Asia—WWII Historical D. Div—SACSEA—24 Mar 45–4 Jun 45. Hunter departure; "it is my hope": Hunter to JSC, in JSC folder Misc Deception Plans—WWII; Hunter to Smith, 18 Jan 45, in JSC folder 000—To 312.1(2) China-Burma-India Theater; Thorne to author, 27 Apr 97.

680 BLANDINGS: Extract from LCS Weekly Letter No. 5, 27 Mar 44, in JSC folder 370.2 Operations and Results; Head of D Division to London Controlling Section, 3 Feb 45, Subject: Deceptive Broadcasts, pp. 11-A, 12-A, in JSC folder Deception—SE Asia—WWII Historical D. Div—SACSEA—16 Dec 44–15 Mar 45. CAPITAL: Romanus & Sunderland, Time 85–87.

681 STULTIFY: Plan STULTIFY, Report by D Division, 21 Nov 44, SAC(44)13(O) Revised, in JSC folder Misc Deception Plans—WWII; D Division Progress Report No. 1 16–31 Dec 44, in JSC folder Deception—SE Asia—WWII Historical D. Div—SACSEA—16 Dec 44–15 Mar 45.

681 Codenames: Plan STULTIFY, Report by D Division, 21 Nov 44, SAC(44)13(O) Revised, in JSC folder Misc Deception Plans—WWII; D Division Progress Report No. 1 16–31 Dec 44; No. 2 1–15 Jan 45; No. 3 16–31 Jan 45, all in JSC folder Deception—SE Asia—WWII Historical D. Div—SACSEA—16 Dec 44–15 Mar 45.

682 "SEAC's idea of a cautious"; CLAW objective: D Division Progress Report No. 3 16–31 Jan 45, in JSC folder Deception—SE Asia—WWII Historical D. Div—SACSEA—16 Dec 44–15 Mar 45. CLAW, TARZAN, CONCLAVE, CAPTION, ARAMINTA, BRUTEFORCE, CLEARANCE, FIREFLAME: Amendment No. 1 to Fourteenth Army Operation "CON-

CLAVE," 10066/G(O)1, 22 Mar 45; COS ALFSEA to Headquarters Fourteenth Army, Subject: Deception Plan Conclave, with attachments, 10 Apr 45; Control Section, D Division, 12 ABPO, to Warde-Aldam *et al.*, Subject: Deception and Cover, 16 Feb 45; Unsigned draft memorandum to Tenth Air Force, 221 Group, Strategic Air Force, Subject: "Conclave," all in JSC folder Southeast Asia; D Division Weekly Progress Report No. 1 24–30 Mar 45; No. 2 31 Mar–6 Apr 45; No. 3 7–13 Apr 45; No. 4 14–21 Apr 45; No. 5 21–27 Apr 45; No. 8, 15–21 May 1945, all in JSC folder Deception—SE Asia—WWII Historical D. Div—SACSEA—24 Mar 45–4 Jun 45; D Division Weekly News Letter No. 43 10 Feb–16 Feb 45; No. 44 17 Feb–23 Feb 45, both in JSC folder Deception—SE Asia—WWII Historical D. Div—SACSEA—30 Dec 44–23 Mar 45.

683 Units for CLAW: D Division Weekly News Letter No. 40 20 Jan–26 Jan 45; No. 42 3 Feb–9 Feb 45; No. 47 10 Mar–16 Mar 45, p. 2; No. 48 10 Mar–16 Mar 45, p. 2, all in JSC folder Deception—SE Asia—WWII Historical D. Div—SACSEA—30 Dec 44–23 Mar 45; D Division Weekly Progress Report No. 2 31 Mar–6 Apr 45, in JSC folder Deception—SE Asia—WWII Historical D. Div—SACSEA—24 Mar 45–4 Jun 45. Units for FANG: D Division Weekly Progress Report No. 1 24–30 Mar 45, para. 14, in JSC folder Deception—SE Asia—WWII Historical D. Div—SACSEA—24 Mar 45–4 Jun 45; D Division Weekly News Letter No. 43 10 Feb–16 Feb 45; No. 44 17 Feb–23 Feb 45, both in JSC folder Deception—SE Asia—WWII Historical D. Div—SACSEA—30 Dec 44–23 Mar 45. Units for TARZAN: D Division Weekly News Letter No. 48 10 Mar–16 Mar 45, in JSC folder Deception—SE Asia—WWII Historical D. Div—SACSEA—30 Dec 44–23 Mar 45; Amendment No. 1 to Fourteenth Army Operation "CONCLAVE," 10066/G(O)1, 22 Mar 45, in JSC folder Southeast Asia; D Division Weekly Progress Report No. 1 24–30 Mar 45, in JSC folder Deception—SE Asia—WWII Historical D. Div—SACSEA—24 Mar 45–4 Jun 45. "We must regretfully": D Division Weekly Progress Report No. 6 28 Apr–4 May 45, in JSC folder Deception—SE Asia—WWII Historical D. Div—SACSEA—24 Mar 45–4 Jun 45.

685 Slim's plan; "This should give me": Slim 327–28.

685 STENCIL: Report on Signal Deception Measures Adopted During Operations of the Fourteenth Army, Dec 44–Mar 45, pp. 2–8, in JSC folder 3-12/5 Communication Decep: RPTS-India Burma.

686 CLOAK: Slim 328–29; Howard 219; O'Brien 170–83; Young & Stamp 228–29; Corbett to Hunter, 10 Mar 45, p. 1, attached to Hunter to Joint Security Control 10 Mar 45, in JSC folder Misc Deception Plans WWII; D Division Weekly Progress Report No. 1 24–30 Mar 45, in JSC folder Deception—SE Asia—WWII Historical D. Div—SACSEA—24 Mar 45–4 Jun 45; D Division Weekly News Letter No. 43 10 Feb–16 Feb 45, in JSC folder Deception—SE Asia—WWII Historical D. Div—SACSEA—30 Dec 44–23 Mar 45.

686 Japanese reaction, culmination: D Division Progress Report No. 5 15–28 Feb 45, para. 1B; No. 6 1–15 Mar 45, para. 1B, in JSC folder Deception—SE Asia—WWII Historical D. Div—SACSEA—16 Dec 44–15 Mar 45; D Division Weekly News Letter No. 46 3 Mar–9 Mar 45, in JSC folder Deception—SE Asia—WWII Historical D. Div—SACSEA—30 Dec 44–23 Mar 45; D Division Weekly Progress Report No. 1 24–30 Mar 45, in JSC folder Deception—SE Asia—WWII Historical D. Div—SACSEA—24 Mar 45–4 Jun 45; D Division Weekly News Letter No. 45 24 Feb–2 Mar 45, in JSC folder Deception—SE Asia—WWII Historical D. Div—SACSEA—30 Dec 44–23 Mar 45; Fleming 67; Howard 219; Slim 404–405.

687 SILVER April–May 1944: Talwar 189, 192–98; Peter Geissler, Auswärtiges Amt, Bonn, to author, June 14, 1996. Inouye: Liddell diary, 15 Sep 44, PRO KV 4/190.

687 SILVER August–September 1944: Talwar 198–216; Appendix "A" to D Division

Weekly Progress Report No. 8, 25 May 45, para. 5, in JSC Folder Deception—SE Asia—WWII Historical D. Div—SACSEA—24 Mar 45–4 Jun 45.

688 MARY-RHINO-ELEPHANT: D Division Weekly News Letter No. 45 24 Feb–2 Mar 45, para. 9(b)(i), in JSC folder Deception—SE Asia—WWII Historical D. Div—SACSEA—30 Dec 44–23 Mar 45.

688 SILVER March–April 1945: Talwar 211, 216–23 ("to have faith"); D Division Weekly News Letter No. 41 27 Jan–2 Feb 45 p. 5 ("looked well"); No. 44 17 Feb–23 Feb 45 para. 12; No. 46 3 Mar–9 Mar 45, Part II para. 12 and Appendix; No. 48 17 Mar–23 Mar 45 p. 7 paras. 14–15, all in JSC folder Deception—SE Asia—WWII Historical D. Div—SACSEA—30 Dec 44–23 Mar 45; D Division Weekly Progress Report No. 5 21–27 Apr 45, p. 10; No. 8 25 May 1945, Appendix "A" paras. 5 ("over full"), 8–11, 13–15, 20, 23–27; No. 9 22–28 May 45 p. 9, all in JSC folder Deception—SE Asia— WWII Historical D. Div—SACSEA—24 Mar 45–4 Jun 45.

690 End of Bose and SILVER: D Division Weekly Progress Report No. 15 3–9 Jul 45 p. 3; No. 19 31 Jul–6 Aug 45 p. 2, in JSC folder Deception–SE Asia–WWII Historical D. Div. SACSEA 5 Jun–10 Sep 45.

690 BACKHAND: D Division Weekly Progress Report No. 7 5–14 May 45, in JSC folder Deception—SE Asia—WWII Historical D. Div—SACSEA—24 Mar 45–4 Jun 45; D Division Weekly Progress Report No. 13 19–25 Jun 45, in JSC folder Deception—SE Asia—WWII Historical D. Div. SACSEA 5 Jun–10 Sep 45.

690 BRASS: Fleming 26–27; Head of D Division to Tactical H.Q. D Division, 13 Jan 45, in JSC folder Japanese Deception Methods; D Division Weekly Progress Report No. 12 12–18 Jun 45 p. 2, in JSC folder Deception—SE Asia—WWII Historical D. Div. SACSEA 5 Jun–10 Sep 45.

691 PURPLE WHALES: Fleming 35; Fleming to ADCOS(A), 122/D.Div., 10 Nov 44, PRO CAB 154/75/19A; D Division Weekly News Letter No. 39 13 Jan–19 Jan 45 para. 24; No. 41 27 Jan–2 Feb 45, both in JSC folder Deception—SE Asia—WWII Historical D. Div—SACSEA—30 Dec 44–23 Mar 45; see Project No. 1471-A, The Japanese Intelligence System, p. 103, in JSC folder The Japanese Intelligence System. "A noteworthy degree": Fleming 34. Lt. Pei: D Division Weekly News Letter No. 44, 17–23 Feb 45, Part I, in JSC folder Deception—SE Asia—WWII Historical D. Div—SACSEA—30 Dec 44–23 Mar 45; Fleming 10. "Amiable but fairly useless": Thorne to author, 27 Apr 97. "Abnormally incurious": Fleming 34.

692 INK: Fleming 34–35; D Division Weekly Progress Report No. 1 24–30 Mar 45 p. 17; No. 6 28 Apr–4 May 45 p. 14; No. 8 15–21 May 45 p. 10; No. 9 22–28 May 45 p. 10, all in JSC folder Deception—SE Asia—WWII Historical D. Div—SACSEA—24 Mar 45–4 Jun 45; D Division Weekly Progress Report No. 11 5–11 Jun 45 p. 9; No. 17 17–23 Jul 45; No. 19 31 Jul–6 Aug 45 p. 9; No. 21 14–20 Aug 45 p. 6, all in JSC folder Deception—SE Asia—WWII Historical D. Div. SACSEA 5 Jun–10 Sep 45; Fleming 34–35; D Division Weekly Progress Report No. 1 24–30 Mar 45 p. 17; No. 6 28 Apr–4 May 45 p. 14; No. 8 15–21 May 45 p. 10; No. 9 22–28 May 45 p. 10, all in JSC folder Deception—SE Asia—WWII Historical D. Div—SACSEA—24 Mar 45–4 Jun 45; D Division Weekly Progress Report No. 11 5–11 Jun 45 p. 9; No. 17 17–23 Jul 45; No. 19 31 Jul–6 Aug 45 p. 9; No. 21 14–20 Aug 45 p. 6, all in JSC folder Deception— SE Asia—WWII Historical D. Div. SACSEA 5 Jun–10 Sep 45. "Our Chinese channels": Fleming 35.

693 DOUBTFUL, AUDREY: Fleming 19–20.

693 HICCOUGHS; "In a continual state": Fleming 40. Suzuki Kikan: D Division Weekly News Letter No. 42 3 Feb–9 Feb 45 Part II para. 34, in JSC folder Deception—SE Asia—WWII Historical D. Div—SACSEA—30 Dec 44–23 Mar 45.

694 ANGEL ONE: Fleming 40–41; D Division Weekly News Letter No. 39 13 Jan–19 Jan

45 Part II para. 27; No. 47 10 Mar–16 Mar 45 Part II para. 16(b)(ii), in JSC folder Deception—SE Asia—WWII Historical D. Div—SACSEA—30 Dec 44–23 Mar 45.

694 Bull's Eye: Fleming 30; Project No. 1471-A, The Japanese Intelligence System, 4 Sep 45, p. 152, in JSC folder The Japanese Intelligence System.

694 Sunrise: PRO ADM 223/794/157; Smith to Record, Subject: Japanese Channel, 10 May 44, in JSC folder 000—To 312.1(1) China-Burma-India Theater; Smith to Record, Subject: Stockholm Channel (Japanese), 12 Jul 44, in JSC folder Great Britain; Project No. 1471-A, The Japanese Intelligence System, 4 Sep 45, p. 156 ("Su Intelligence"), in JSC folder The Japanese Intelligence System; PRO CAB 154/36; PRO CAB 154/26/9A; LCS to Fleming, 21 Feb 45, PRO CAB 154/26/42A; D Division Weekly News Letter No. 45 24 Feb–2 Mar 45, p. 7, in JSC folder Deception—SE Asia—WWII Historical D. Div—SACSEA—30 Dec 44–23 Mar 45; D Division Weekly Progress Report No. 15 3–9 Jul 45 p. 7; No. 19 31 Jul–6 Aug 45 p. 8, both in JSC folder Deception—SE Asia—WWII Historical D. Div. SACSEA 5 Jun–10 Sep 45; LCS to Fleming and Wingate, 8 Jul 45, PRO CAB 154/26/49A.

695 Gleam: Fleming 23–24; Hinsley & Simkins 234; D Division Weekly News Letter No. 39 13 Jan–19 Jan 45, Part II para. 25 (names); No. 40 20 Jan–26 Jan 45 p. 6 ("lovely squishy tea"); No. 43 10 Feb–16 Feb 45; No. 45 24 Feb–2 Mar 45 p. 8, Part II para. 14 ("a heavenly"; "More nuts") ("stupendous"), all in JSC folder Deception—SE Asia—WWII Historical D. Div—SACSEA—30 Dec 44–23 Mar 45; D Division Progress Report No. 6 1–15 Mar 45, in JSC folder Deception—SE Asia—WWII Historical D. Div—SACSEA—16 Dec 44–15 Mar 45.

697 Father/Duck: Fleming 24–25; Hinsley & Simkins 234; Notes on D/A Activities in the Middle East, PRO KV 4/197, p. 10; D Division Weekly News Letter No. 37 30 Dec 44–5 Jan 45; No. 39 13 Jan–19 Jan 45, p. 2, both in JSC folder Deception—SE Asia—WWII Historical D. Div—SACSEA—30 Dec 44–23 Mar 45.

697 Marmalade; Owl; "Thanks for the trouble": Fleming 21–22. Pawnbroker: Fleming 16–17.

698 Oatmeal: Cruickshank, SOE 204, 205 and note, 206; Fleming 27–29. "With implausible speed": Fleming 28. "Though by no means": Fleming 29.

699 Travel, Trotter: Fleming 17–19, 71; D Division Weekly Progress Report No. 7 5–14 May 45, pp. 6–7, in JSC folder Deception—SE Asia—WWII Historical D. Div—SACSEA—24 Mar 45–4 Jun 45; D Division Weekly News Letter No. 37 30 Dec 44–5 Jan 45 Part II para. 25; No. 40 20 Jan–26 Jan 45, Part II para. 27, both in JSC folder Deception—SE Asia—WWII Historical D. Div—SACSEA—30 Dec 44–23 Mar 45.

700 Coughdrop: Fleming 42–43. "The operator": Fleming 42.

701 Accost: D Division Weekly Progress Report No. 20 7–13 Aug 45, pp. 8–9, JSC folder Deception—SE Asia—WWII Historical D. Div. SACSEA 5 Jun–10 Sep 45. Pointer: D Division Weekly Progress Report No. 9 22–28 May 45 p. 4, in JSC folder Deception—SE Asia—WWII Historical D. Div—SACSEA—24 Mar 45–4 Jun 45; D Division Weekly Progress Report No. 12 12–18 Jun 45, pp. 8–9; No. 19 31 Jul–6 Aug 45 p. 7, both in JSC folder Deception—SE Asia—WWII Historical D. Div. SACSEA 5 Jun-10 Sep 45.

701 Ian: Lycett 155. Wingate reorganization: Fleming 4; D Division Weekly News Letter No. 38, 6 Jan–12 Jan 45, in JSC folder Deception—SE Asia—WWII Historical D. Div—SACSEA—30 Dec 44–23 Mar 45. Wingate arrival: D Division Weekly Progress Report No. 1, 24–30 Mar 45, in JSC folder Deception—SE Asia—WWII Historical D. Div—SACSEA—24 Mar 45–4 Jun 45. "Terribly keen": Young & Stamp 177. Ferguson: D Division Weekly Progress Report No. 4 14–21 Apr 45, cover page, in JSC folder Deception—SE Asia—WWII Historical D. Div—SACSEA—24 Mar 45–4 Jun 45. Advance HQ D Division: D Division Weekly Progress Report No. 9 22–28 May 45, cover page and p. 5, in JSC folder Deception—SE Asia—WWII Historical D.

Div—SACSEA—24 Mar 45–4 Jun 45. Antrim: D Division Weekly News Letter No. 46 3 Mar–9 Mar 45; No. 47 10 Mar–16 Mar 45, both in JSC folder Deception—SE Asia—WWII Historical D. Div—SACSEA—30 Dec 44–23 Mar 45. "Really, Peter": Hart-Davis 303.

703 D Force: Wingate 21; Fleming 5, 11–12, 20–21. No. 1 Naval Scout Unit: Fleming 14.

Chapter 17: BLUEBIRD

704 "As a practical matter": Thebaud and Bissell to Joint Staff Planners, 17 Nov 44, JSC/J1A Serial 1187, in NARA RG 165 ABC File ABC 381 Japan 15 Apr 43 Sec. 1-B. As a result: JPS 560/M, 18 Nov 44; JPS 560/1, 27 Nov 44, both in NARA RG 165 ABC File ABC 381 Japan 15 Apr 43 Sec. 1-B; Minutes, JPS 181st Meeting, 29 Oct 44, in NARA RG 218 Box 694 Folder CCS 385 Pacific Theater (4-1-43) Deceptive Policy in the Pacific Theater (Sec. 3). "To collaborate": JPS 560/2, 30 Nov 44, in NARA RG 165 ABC File ABC 381 Japan 15 Apr 43 Sec. 1-B.

705 Appointments: Thebaud and Bissell to Joint Staff Planners, Subject: Collaboration with Joint War Plans Committee by Deception Section, Joint Security Control, 4 Dec 44, JSC/J6 Serial 1209, in NARA RG 218 Box 694 Folder CCS 385 Pacific Theater (4-1-43) Deceptive Policy in the Pacific Theater (Sec. 3); Thebaud and Bissell to Joint Staff Planners, 21 Mar 45, JSC/J6 Serial 3318, in NARA RG 165 ABC File ABC 381 (8-15-42)—JSC. Bieri: Whitehill 62. Nelson: 1919 *Lucky Bag*, entry for Robert Franklin Nelson; obituary in *Shipmate*, August 1963 p. 29. Fairbanks to COMINCH via London: Fairbanks to The Admiral, Subject: Special Operations—Summation report on (January 1943 to September 1944), 1 Sep 44, in JSC folder Deceptive Equipment—Navigation—BJU's—Résumé—Aug 44; O'Connor to Schuirmann, 6 Sep 44, HMOC/1476, in NOA Box 532, Folder Douglas Fairbanks; Fairbanks 245 ("really fearsome"). Bevan lobbying: Bevan to Fleming, 2/99, 5 Sep 44, PRO HW 41/159. "In the light of": Fairbanks to Smith, Subject: Deceptive Warfare and Special Operations—Comments on, undated, in NOA 532, Folder Douglas Fairbanks. Thebaud: Thebaud biographical release, Navy Operational Archives.

706 Metzel: Whitehill 119–20. Decoy fleet: Goldbranson to Record, 30 Oct 44, in JSC folder Japan. "An experienced Naval partner": Metzel to Thebaud, Subject: Organization and Directives for Cover and Deception, 19 Feb 45, in JSC folder JCS To Handle Deception. Metzel, Brandt assigned: Edwards to Secretariat, Joint Chiefs of Staff, Subject: U.S. Navy Assignment to Joint Security Control, undated, Serial 9347 (Metzel), Serial 9521 (Brandt), in NARA RG 218 Box 694 Folder CCS 385 Pacific Theater (4-1-43) Deceptive Policy in the Pacific Theater (Sec. 3).

706 Thurber: Unsigned memorandum Summary of Activities by the U.S. Navy Cover and Deception Organization in World War II, 10 Jul 46, p. 13, in NOA 530, Folder C&D (Cover and Deception) in World War II (General); Morison, New Guinea vii; Thurber biographical release, Navy Operational Archives, Washington Navy Yard.

707 "Can save many thousands"; "The small staff"; "everything involved"; "the stakes are so high"; "These steps are both"; proposed organization: Metzel to Thebaud, Subject: Organization and Directives for Cover and Deception, 19 Feb 45; Draft Memorandum for the Joint Chiefs of Staff, Subject: Responsibility for Cover and Deception, 19 Feb 45; Draft Directive, Joint War Information Control, 19 Feb 45; Draft Enclosure (A), Staff Study of Directives Governing Cover and Deception, 19 Feb 45; Draft Appendix (A), Staff Study of Directives Governing Cover and Deception, Discussion, 19 Feb 45; Draft Directive, Joint Security Control, 19 Feb 45; Draft Directive, Cover and Deception, 19 Feb 45, all in JSC folder JCS To Handle Deception.

707 Thebaud recommendation: Thebaud to Bissell, Subject: Responsibility for Cover and

Deception, 20 Feb 45, in JSC folder JCS To Handle Deception. "The Navy has finally": T. D. R. to Craig, 6 Feb 45, in NARA RG 165 ABC File ABC 381 (4-8-43) Sec. 1-A—Deception policy, organization, etc. Thurber recommendation; "there was a lack"; "on the level of command": Unsigned memorandum Summary of Activities by the U.S. Navy Cover and Deception Organization in World War II, 10 Jul 46, in NOA 530, Folder C&D (Cover and Deception) in World War II (General).

708 F-28 organization, personnel: Whitehill 232; Packard 302–303; Whitehill 223 (Creech); Joint Security Control Information Bulletin No. 3, 1 Jun 45, JSC/J14 Serial 5122, p. 3, in NARA RG 165 ABC File ABC 381 Japan 15 Apr 43 Sec. 1-B; unsigned memorandum Summary of Activities by the U.S. Navy Cover and Deception Organization in World War II, 10 Jul 46, pp. 18–19, in NOA 530, Folder C&D (Cover and Deception) in World War II (General).

709 Fairbanks as F-1421: Fairbanks 246 ("no terribly demanding duties"). Fairbanks ideas: Fairbanks 246–251; Fairbanks to Thurber, Subject: Deception, 6 Feb 45; Fairbanks to Duncan, Subject: Accelerating Japanese Surrender by Deceptive Measures, Proposals for, 17 May 45; Fairbanks to F-4, Subject: Decoy Vessels, Proposals for, 12 Jun 45, all in NOA 532, Folder Douglas Fairbanks; Fairbanks, Coordination of United States National Information for Use in Support of Overall Deception Against Japan, Draft II, undated; Draft "X," undated, both in JSC folder Japan. Ordered to SEA, Okinawa: Fairbanks 254–57. Hines: E.g., Hines to Smith, 20 Feb 45, in JSC folder Wedlock—subfolder 370.2 Wedlock—Adm Impl.

709 Thurber full JSC member: Memorandum by Commanding General Army Air Forces, Revision of Charter of Joint Security Control, JCS 234/4, 11 Jan 45; King to Joint Chiefs of Staff, Subject: Revision of Charter of Joint Security Control, 7 Mar 45; Giles to Secretary, Joint Chiefs of Staff, Subject: Revision of Charter of Joint Security Control, 8 Mar 45; Revision of Charter of Joint Security Control, informal action, JCS 234/4; Charter, Joint Security Control 13 Mar 45, JCS 234/5/D; all in NARA RG 218 Box 211 Folder CCS 334 Joint Security Control (Sec. 2); Dunlop to Commanding Generals, 24 Mar 45, AG 334 (23 Mar 45) OB-S-E, in NARA RG 165 ABC File ABC 381 (8-15-42)—JSC; JCS Info Memo 379, 21 Mar 45; JCS Info Memo 411, 12 Jun 45, both in NARA RG 165 ABC File ABC 381 (8-15-42)—JSC. "Chief, Special Section": Buck slip, 2 Jul 45, in JSC folder Portugal.

710 JCS 498/10: Fairbanks to Thurber, Subject: Cover and Deception Definitions, 7 Apr 45, in NOA 532, Folder Douglas Fairbanks; Smith to Bissell, Subject: Report by Joint Security Control and Implementation of JCS Directive 498/6, 18 Apr 45, in JSC folder Oplan "Bluebird"—Pacific WWII; Goldbranson to Smith, Subject: Captain Thurber's Memorandum for F-01 (Rough Draft); Smith to Thurber, Subject: Your Rough Draft Inclosure to Memorandum for F-1, 15 Mar 45; Undated draft, Deception in the War Against Japan; C. E. G[oldbranson] to Smith, handwritten notes, 29 Mar 45; Undated draft Enclosure A, Subject: Cover and Deception in the War Against Japan; Undated, unsigned handwritten memorandum [Goldbranson]; Draft NS/dms 7 Apr 45; Unsigned, undated draft, 2. DEFINITIONS; Draft, Cover and Deception in the War Against Japan, 11 Apr 45; C. E. G[oldbranson] to Smith, 12 Apr 45; Smith to Thurber, Subject: Report on JCS 498/5, and Proposed Enclosure, 13 Apr 45, all in JSC folder Japan; Draft, Cover and Deception in the War Against Japan, 14 Apr 45; Cover and deception in the War Against Japan, S&P:TDR:rak:(1), 10 May 45; Routing Information Form and attachments, 24 May 45; Handwritten note, undated, all in NARA RG 165 ABC File ABC 381 Japan 15 Apr 43 Sec. 1-B; JCS 498/10, 14 May 45; Cover and Deception in the War Against Japan, 22 May 45, in NARA RG 218 Box 694 Folder CCS 385 Pacific Theater (4-1-43) Deceptive Policy in the Pacific Theater (Sec. 4). The Gouge: Undated Draft Lecture on Cover and Deception for High Command and Staff Schools, OP-30C-OUT-305, in NOA 530, Folder Lectures on C&D.

710 Harris *et al.* to JSC: Joint Security Control Information Bulletin No. 3, 1 Jun 45, JSC/J14 Serial 5122, pp. 2–3, in NARA RG 165 ABC File ABC 381 Japan 15 Apr 43 Sec. 1-B. "Our organization has certainly": Brown to Goldbranson, 10 Aug 45, in Goldbranson papers.

711 "A comprehensive analysis": Ingersoll to Harris, 16 Sep 70, in Ingersoll papers. Does not live up: 12th Army Group Report, Copy 2, in JSC folder Operation Wedlock. Interim Report: Informal Report to Joint Security Control by Special Plans Branch, G-3, 12th Army Group, 25 May 45, with appendices, in NARA RG 319 Entry 101 Box 2 Folders 27–29. "Cuts Reeder's throat": Same, cover page. Not distributed further: Smith to Eldredge, JSC/J1Z Serial 821, 8 Aug 45, in JSC folder Deception Units. Ingersoll gone: Still listed in Joint Security Control Information Bulletin No. 3, 1 Jun 45, JSC/J14 Serial 5122, in NARA RG 165 ABC File ABC 381 Japan 15 Apr 43 Sec. 1-B. "With the Third Reich's": Ingersoll to Bowman, 17 Aug 79, in Ingersoll papers, unlabeled box.

711 Age and points: Hoopes 309. "Did not give a damn": Eldredge to Herbig, 3 Sep 84, Eldredge papers. Eldredge to Hanover, Europe: Eldredge to Smith, 30 May [45]; Smith to Eldredge, 1 Jun 45, in NARA RG 319 Entry 101 Box 6 Folder 91; Eldredge to Hesketh, 20 Dec 71; Eldredge to Herbig, 3 Sep 84, Eldredge papers. "I thereupon 'stumbled' ": Eldredge to Dwyer, 17 Aug 89, in Eldredge papers. Burris-Meyer marriage: Author's telephone interview with Anita Burris-Meyer, 19 Feb 95.

712 Argentina: Page 50–78; German Espionage in Argentina, 1943–1944, 14 Jul 44, in JSC folder German Espionage in Argentina. Moved freely: Lapping to Joint Security Control, Subject: Japanese in Argentina, 8 Aug 45, in JSC folder South America.

713 Smith with Lang and King: Harris to Smith, 29 Aug 45; Lang to Smith, 14 Jul 43; Smith to King, 28 Jul 43, JSC/LB Serial 261; Smith to Lang, 16 Mar 44, JSC/J8 Serial 347; Smith to King, 18 Feb 44, JSC/LB Serial 261, 18 Feb 44; King to Smith, 4 Mar 44; King to Smith, 11 Apr 44; Lang to Chief, MIS, War Department, Subject: Proper Procedure, 4 Mar 44; Smith to Lang, 16 Mar 44, JSC/J8 Serial 347; Lang to Smith, 31 Mar 44; Lang to Smith, 31 Mar 44, all in JSC folder South America. Lang: Biographical release, John W. Lang, USAMHI. King: Harris to Smith, 5 Oct 45, in JSC folder South America. Personal visit: Lang to Smith, 31 Jan 45; Smith to Lang, 24 Feb 45, JSC/J5 Serial 2119, both in JSC folder South America.

713 Harris: Cullum 524; Smith to Lang, 4 Jun 45, JSC/J5 Serial 608, in JSC folder South America. "Must be reserved": Smith to Lang, 25 Sep 44, JSC/J8 Serial 9992, with enclosed list of foodstuff A-X-1, in JSC folder South America. "Fishing fine": US Military Attaché, Buenos Aires, to War Department, No. 380, 1 Nov 44, CM-OUT-950 (1 Nov 44), in JSC folder South America. "This piscatorial adventure:" Lang to Smith, 7 Apr 45; "Rising to a Royal Coachman": Lang to Smith, 21 Apr 45; "Intimate acquaintance with the eddies and shoals": Smith to Lang, 27 Mar 45, JSC/J5 Serial 3140, all in JSC folder South America.

714 Not by cable: Smith to Lang, 16 Jan 45, JSC/J5 Serial 173, in JSC folder South America. At Buenos Aires end: Lang to Smith, 21 Dec 44, in JSC folder South America. SIS: Author's interview with Kenneth M. Crosby, 20 Mar 97. "Do something which might jeopardize": Lang to Smith, 11 May 45, in JSC folder South America. No more problems: Smith to Lang, 23 May 45, JSC/J5 Serial 587; Smith to Lang, 30 May 45, JSC/J5 Serial 5123, both in JSC folder South America.

714 Yukishita, Miyamoto, Suzuki: Lapping to Joint Security Control, Subject: Japanese in Argentina, 8 Aug 45, in JSC folder South America. Miyamoto: Harris to Smith, 5 Sep 45, in JSC folder South America. PIN: Cake Before Breakfast, p. 65, JSC loose papers; Lang to Smith, 13 Oct 44; 30 Oct 44; 1 Nov 44; 23 Nov 44, all in JSC folder South America. RUPERT: Cake Before Breakfast, p. 65, JSC loose papers; Lang to Smith, 13 Oct 44, in JSC folder South America. Ukrainian: Lang to Smith, 14 Jul 43, 13 Oct 44, both in JSC folder South America.

715 PIN's character: Harris to Smith, 24 Jul 44; Lang to Smith, 21 Dec 44, 7 Apr 45, all in JSC folder South America. First to identify: Lang to Smith, 3 Mar 45, in JSC folder South America. "Our best forward passer": Lang to Smith, 21 Dec 44, in JSC folder South America. "We have always handled": Lang to Smith, 12 Jan 45, in JSC folder South America. Tank and fighter: Lang to Smith, 12 Jan 45, in JSC folder South America. 170,000 waiting: US Military Attaché Buenos Aires to War Department, No. 395, 16 Jan 45, CM-IN-15460 (17 Jan 45), in JSC folder South America. Landing site: Davis to Smith, 21 Feb 45, in JSC folder South America. Warning Formosa: CG US Military Attaché Buenos Aires to War Department, No. 413, 22 Mar 45, CM-IN-23763 (23 Mar 45); Lang to Smith, 23 Mar 45; US Military Attaché, Buenos Aires to War Department, No. 417, 2 Apr 45, CM-IN-1852 (3 Apr 45), all in JSC folder South America. "PIN got this": US Military Attaché Buenos Aires to War Department, No. 395, 16 Jan 45, CM-IN-15460 (17 Jan 45), in JSC folder South America.

716 100 pesos: Lang to Smith, 2 Dec 44, in JSC folder South America. "Radar-equipped": A-X-1, enclosure to Smith to Lang, 25 Sep 44, Item 6, JSC/J8 Serial 9992, in JSC folder South America. Several payments: Lang to Smith, 29 Dec 44, in JSC folder South America. Landing craft: Lang to Smith, 30 Oct 44, in JSC folder South America. Weapons, manpower: Lang to Smith, 1 Nov 44; Lang to Smith, 23 Nov 44; Lang to Smith, 2 Dec 44, all in JSC folder South America. "If we can": Smith to Lang, 20 Mar 45, JSC/J5 Serial 3108, in JSC folder South America. "Dramatically offered": Lang to Smith, 3 Mar 45, in JSC folder South America. Eden Hotel: Lang to Smith, 21 Apr 45, in JSC folder South America.

716 Eichhorn: Lang to Smith, 27 Jun 45; Davis to Smith, 15 June 45, both in JSC folder South America. PIN's efforts to reach Yukishita: Lang to Smith, 24 Apr 45, 16 May 45, 23 May 45, all in JSC folder South America. Furuta: Harris to Smith, 5 Sep 45. Koko; Yamazaki: Lapping to Joint Security Control, Subject: Japanese in Argentina, 8 Aug 45, in JSC folder South America. Relaxed: Davis to Smith, 15 Jun 45, 27 Jun 45, both in JSC folder South America. Four-hour conversation: Davis to Smith, 27 Jun 45, in JSC folder South America.

717 Ishikawa; PIN interested again: Harris to Smith, 10 Jul 45, in JSC folder South America. Summoned before committee: King to Smith, 18 Oct 45; Harris to Smith, 24 Jul 45, both in JSC folder South America.

718 First assignment, further requests: Harris to Smith, 28 Jul 45, 1 Aug 45, both in JSC folder South America.

718 RUPERT: Full summary of fish food in JSC brown Acco binder labeled "1"; Harris to Smith, 11 Aug 45; Lang to Smith, 3 Mar 45; Lang to Smith, 16 May 45, all in JSC folder South America. "If he makes a bold": Lang to Smith, 23 Nov 44, in JSC folder South America. Gus: Cake Before Breakfast, p. 65, JSC loose papers; Lang to Smith, 23 Nov 44; Lang to Smith, 29 Dec 44; unsigned memorandum 4 Sep 45; Lang to Smith, 3 Mar 45; Lang to Smith, 7 Apr 45; Lang to Smith, 21 Apr 45; Joint Security Control to Lang, No. WAR 70642, 20 Apr 45, CM-OUT-70642 (Apr 45); US Military Attaché Buenos Aires to War Department, No. 433, 23 Apr 45, CM-IN-21655 (23 Apr 45); US Military Attaché Buenos Aires to War Department, No. 435, 23 Apr 45, CM-IN-22046 (24 Apr 45) ("principal bait feeder"); Lang to Smith, 1 May 45; Lang to Smith, 16 May 45; Smith to Lang, 4 Jun 45, JSC/J5 Serial 608 ("would be a most gratifying"); Lang to Smith, 23 May 45; Harris to Smith, 10 Jul 45, all in JSC folder South America.

720 "In privately discussing reports": A-X-45 Item 2, enclosure to Smith to Davis, J/NS/5b Serial 705, 3 Jul 45, in JSC folder South America. Told Sweeney: Author's interview with Eugene Sweeney, January 1993. To Suzuki; "glared with hate": Harris to Smith, 1 Aug 45, in JSC folder South America. No parallel: Cake Before Breakfast, p. 59, JSC loose papers. "New respect": Harris to Smith, 29 Aug 45, in JSC folder South America.

721 Closing down Argentine channels: JSC brown Acco binder "4 South America"; Smith to Harris, 18 Oct 45, J/NS/5b Serial 1074 King to Smith, 26 Oct 45; King to Smith, 13 Nov 45; King to Smith, 13 Nov 45; Smith to Harris, 26 Nov 45, J/NS/54 Serial 1171; Gibbons to Smith, 30 Nov 45; E. J. S[weeney] to Harris and Smith, 12 Dec 45, all in JSC folder South America.

721 LODGE: Hill 593–95; O'Connor to Joint Security Control, 11 Oct 44, HMOC/1628, in JSC folder Great Britain; O'Connor to Joint Security Control, 8 Nov 44, HMOC/1769; O'Connor to Joint Security Control, 21 Nov 44, HMOC/1832, both in JSC folder South America.

722 WASCH's friend: RUDLOFF messages no. 189, 28 Oct 44; no. 224, 15 Nov 44; nos. 303–304, 23 Dec 44; nos. 392–93, 1 Feb 45; no. B-424, 8 Apr 45, in JSC brown Acco binder labeled "14." "Bootleg"; Bevan annoyed: O'Connor to Bevan, No. GOR 163, 17 Apr 45, PRO CAB 154/36/51A; Bevan to O'Connor, CO 426/1, 21 Apr 45, PRO CAB 154/36/53A. Switzerland: Bissell to Legge, 11 Apr 45; Legge to Bissell, 29 May 45; Legge to Smith, 3 Aug 45; Legge to Smith, 1 Oct 45; all in JSC folder Switzerland.

723 Hack: Fox 20, 179–83, 190, 196, 211; Martin 460; Petersen 5, 379–80, 523–24; Butow 103–109. Use of HANS, writeoff: Unsigned, undated memorandum, Information for Implementation Book, in JSC brown Acco binder "6 File D"; Legge to Smith, 3 Aug 45; Legge to Smith, 10 Sep 45; Smith to Legge, 15 Oct 45, J/NS/5a Serial 1075, all in JSC folder Switzerland; Cake Before Breakfast, p. 65, JSC loose papers.

723 Lisbon: Bissell to Solborg, 11 Apr 45, J/NS Serial 441. LESTER: Cake Before Breakfast, p. 65 (former press attaché), JSC loose papers; Solborg to Bissell, 24 Apr 45 ("rather sympathetically"; "Their conversations"), in JSC folder Portugal; Project No. 1471-A, The Japanese Intelligence System, 4 Sep 45, p. 153, in JSC folder The Japanese Intelligence System (" 'S' intelligence"). HENRY: Cake Before Breakfast, p. 66, JSC loose papers; Solborg to Bissell, 24 Apr 45 ("spot items"), in JSC folder Portugal.

724 JIMMIE: Cake Before Breakfast, p. 66, JSC loose papers; Solborg to Bissell, 24 Apr 45 ("A clever man"), in JSC folder Portugal. "An extremely energetic": Pujol 86.

724 KELLY; "a Portuguese Army": Solborg to Smith, 20 Jun 45, in JSC folder Portugal. "The whole process": Solborg to Smith, 23 Aug 45, in JSC folder Portugal. Material to Lisbon: Solborg to Smith, 6 Jun 45; Unsigned memorandum, Re: Sunset Case, 23 Jun 45; JSC to Military Attaché Lisbon, No. WAR 24859, 30 Jun 45, CM-OUT-24859 (June 45); U.S. Military Attaché, Lisbon, to War Department, No. 259, 2 Jul 45, CM-IN-1424 (2 Jul 45); E. S. [weeney] to Smith, 29 Aug 45; Smith to Solborg, 30 Aug 45, J/NS/5c Serial 8130; Sweeney to Record, 30 Aug 45; Solborg to Smith, 23 Aug 45 ("The whole process"), all in JSC folder Portugal.

725 Lisbon closed out: Smith to Fuqua, 26 Sep 45, J/NS/5a Serial 984; Solborg to Smith, 23 Aug 45; Fuqua to Smith, 27 Aug 45, all in JSC folder Portugal. Stockholm: Bissell to Kessler, 20 Apr 45, J/NS Serial 489; Kessler to Bissell, 14 May 45; Bissell to Kessler, 28 May 45, JSC/J5 Serial 5111; Kessler to Smith, 20 Sep 45; Smith to Kessler, 5 Oct 45, J/NS/5d Serial 1047, all in JSC folder Sweden.

725 Weekly list: Cake Before Breakfast, p. 61, JSC loose papers. Meticulous record: JSC brown Acco binder labeled "1."

725 ULTRA and MAGIC monitoring: O'Connor to Bissell, HMOC/2417, 24 Feb 45; Clarke to Smith, 2 Mar 45, endorsed Smith to Clarke, 8 Mar 45, approved Bissell 10 Mar 45, concurred Thurber 13 Mar 45; Bissell to O'Connor, Subject: Exchange of Deception Information, 13 Mar 45; O'Connor to Bissell, HMOC/2592, 14 Mar 45; Bissell to O'Connor, Subject: Exchange of Deception Information (ULTRA), Your HMOC/2592, 14 March 45, 17 Mar 45, all in JSC folder Great Britain. Asking FBI: Hoover to Legal Attaché, Lisbon, Re: Japanese Activities Espionage—J, Spanip Espionage—S, 14 Feb 44; Hoover to SAC Washington, re Spanip, 22 Feb 44; both in FBI Spanip file.

726 ANTHONY/CAMCASE: Unsigned memorandum, 23 Jan 45, Re: Camcase, in JSC brown Acco binder labeled "14," pp. 44–45; unsigned memorandum, 13 Nov 44, Re: Double Agents, in JSC folder Traffic Sent. ASPIRIN: Unsigned memorandum, 13 Mar 45, Double Agent Aspirin, in JSC brown Acco binder labeled "14"; File Log, items for "Spanip," in JSC brown Acco binder labeled "10" "B File"; [Deleted] to Ladd, Re: Spanip Case Espionage G and J, 2 Jul 45, in FBI Spanip file. JAM: Unsigned memorandum, 2 Mar 45, Re: Jam Case, in JSC brown Acco binder labeled "14."

727 MOONSTONE: Cake Before Breakfast, p. 62, JSC loose papers; chart, p. 1, and unsigned memorandum, Re: Moonstone, in JSC brown Acco binder labeled "14"; File Log, in JSC brown Acco binder labeled "10" "B File." ONYX: Cake Before Breakfast, p. 62, JSC loose files; unsigned memorandum, Re: Onyx Case, 28 Dec 44 (secret weapon research); chart, both in JSC brown Acco binder labeled "14."

727 "To promote internal disorder"; "To promote peace movements": Hunter, Deception Measures Against Japan (JPS 372/7); blue-pencil: unsigned Proposed Changes to JPS 372/7, 19 Apr 45; "not *military* deception": unsigned Analysis of Proposed AAF Changes, [26 Apr 45], all in NARA RG 165 ABC File ABC 381 Japan 15 Apr 43 Sec. 1-B. Italy scheme: Deception Plan for Italy, 20 Feb 43, PRO CAB 119/66/211–13. "Did not take into clear"; "is not stated": Thurber to F-1, Subject: AFAEP Notes on J.W.P.C. 190/16. ("Staff Study of Cover and Deception Objectives for CORONET"), F-28-OUT-94, with enclosure A, 7 Aug 45, in JSC folder Misc Deception Plans—WWII.

728 Hunter at meeting: Martin to Thurber, Subject: Ad Hoc Committee of the JPS on Psychological Warfare, 29 Jun 45; unsigned draft Memorandum for the Assistant Chief of Staff, Air Plans, Army Air Forces, Subject: Colonel E. O. Hunter, A.C., 7 Jul 45 ("subject officer"), both in NOA 532, Folder Douglas Fairbanks.

728 Debate settled: Hayes 603–24; Spector 418–20. MIKE ONE deception planning: GHQ SWPA Hollandia to War Department, No. CX 52283, 16 Nov 44, CM-IN-15326 (16 Nov 44), in NARA RG 218 Box 360 Folder CCS 385 (11-16-44) Plans for Operation "M-1"; CINCPOA to CINCSWPA, No. NCR 7265, 22 Nov 44, CM-IN-21934 (22 Nov 44), in JSC folder Olympic; GHQ SWPA Hollandia to War Department, No. CX 52782, 23 Nov 44, CM-IN-22748 (23 Nov 44); Peck to McFarland, 22 Nov 44, in NARA RG 218 Box 360 Folder CCS 385 (11-16-44) Plans for Operation "M-1"; Peck to Joint Security Control, Subject: Proposed Cover Plan for Forthcoming Operation, 22 Nov 44; JPS 565/D, 23 Nov 44; JPS 565/1, 24 Nov 44; Minutes JPS 181st Meeting, 29 Nov 44; Donnelly to Joint Security Control, Subject: Proposed Action by Joint Security Control in Support of Theater Cover Plan for Operation M-1, 29 Nov 44, with enclosure, all in NARA RG 218 Box 360 Folder CCS 385 (11-16-44) Plans for Operation "M-1."

729 Approved plan for M-1: Enclosure to Donnelly to Joint Security Control, Subject: Proposed Action by Joint Security Control in Support of Theater Cover Plan for Operation M-1, 29 Nov 44, in NARA RG 218 Box 360 Folder CCS 385 (11-16-44) Plans for Operation "M-1." Theaters advised, operations postponed: Joint Security Control to CINCSWPA and CINCPOA, No. WARX 70546, 30 Nov 44, CM-OUT-70546 (30 Nov 44), in NARA RG 218 Box 360 Folder CCS 385 (11-16-44) Plans for Operation "M-1"; CINCPOA to COMINCH, NCR 2610, [5] Dec 44, CM-IN-4368 (5 Dec 44), in JSC folder Olympic; Hayes 357.

730 BLUEBIRD: Outline of Plan, in Nimitz to Cominch, Subject: Cover Plans for Detachment, Iceberg, and Subsequent Operations in the Pacific Ocean Areas During 1945, A1/A8 Serial 0001011, 21 Nov 44; King to Secretary Joint Chiefs of Staff, Subject: Cover Plans for Detachment, Iceberg, and Subsequent Operations in the Pacific Ocean Areas During 1945, FF1/A1/A8 Serial 003476, 1 Dec 44; JCS 1184, 3 Dec 44; JPS 573/D, 3 Dec 44; JPS 573/1, Report by the Joint Staff Planners, 14 Dec 44; Minutes

JPS 183d Meeting, 20 Dec 44; JCS 1184/1, 22 Dec 44; Graves to Aide to Cominch, Subject: Cover Plans for "Detachment," "Iceberg," and Subsequent Operations in the Pacific Ocean Areas During 1945, 30 Dec 44, all in NARA RG 218 Box 688 Folder CCS 385 Pacific Ocean Areas (12-1-44) Cover Plans for Detachment, Iceberg, and Subsequent Operations in the Pacific Ocean Areas During 1945; Unsigned, undated Operational Report—Bluebird, in NOA 548, Folder Bluebird (2d folder); Joint Security Control to CINCSWPA et al., No. WARX 84948, 30 Dec 44, CM-OUT-84948 (31 Dec 44); CINCPOA to COMINCH, DTG 051026, No. NCR 6958, 5 Jan 45; Joint Security Control to CINCPOA, DTG 121807, 12 Jan 45; Joint Security Control to CINCSWPA, No. WAR 90933, 12 Jan 45, CM-OUT-90933 (Jan 45), all in JSC folder Bluebird. Codename BLUEBIRD: Joint Security Control to CINCPOA, 12 Jan 45, in JSC folder Bluebird. Barnes to Washington: CINCPOA to COMINCH, DTG 150502, No. NCR 7383, 15 Jan 45; Joint Security Control to CINCPOA (Advance), DTG 311603, No. NCR 8119, 30 Jan 45, both in JSC folder Bluebird.

731 Akin: Drea 18, 26, 28–31. Meeting with Akin and Chamberlin: Hilger to CINCPAC and CINCPOA, Subject: Conference Between Colonel Hilger and Representatives of General Headquarters, Southwest Pacific Area, at Tacloban—Report of, 17 Dec 44, Enclosure G to McDowell to Joint Security Control, Subject: Cover and Deception, Report of Progress, Central Pacific, 29 Dec 44, in JSC folder German Efforts WWII.

731 Dummy traffic contretemps: US Forces India Burma Theater to SSA, No. DN 5, 2 Jan 45, AS CM-IN 354; US Forces India Burma Theater to SSA, No. DN 5, 2 Jan 45, AS CM-IN 354; Smith to Hines, 5 Jan 45; Draft message to CINCSWPA, 5 Jan 45 ("discontinued immediately"); Corderman to CINC GHQ SWPA Rear Echelon Brisbane, 6 Jan 45, all in JSC folder Olympic.

732 Unauthorized deception: Drea 30 (Clarke and Akin); Clarke to Smith, Subject: Reports From SWPA; Smith to Clarke, Subject: Reports From SWPA; Corderman to Clarke, Subject: Reports From SWPA, 17 Jan 45 ("the methods of controlling"), all in JSC folder Olympic. Halsey, Spruance alternating: Morison, Victory 20, 169, 356–57. Communications deception for BLUEBIRD: Unsigned, undated Operational Report—Bluebird, p. 5–7, in NOA 548, Folder Bluebird; CINCPOA Advance to CINCPAC Pearl, No. NCR 7223, DTG 160636, 16 Mar 45, in JSC folder Bluebird; CINCPAC Adv HQ to CINCPAC Pearl HQ, DTG 250809 April, 25 Apr 45, in NOA 548, Folder Bluebird—Dispatches.

732 Actual air operations tailored: Unsigned, undated Operational Report—Bluebird, pp. 5, 11–12, in NOA 548, Folder Bluebird (2d folder). NavGroup China activity: CINCPOA to COMINCH, No. NCR 6958, 5 Jan 45, in JSC folder Bluebird; JCS to Commanding General U.S. Forces China Theater, No. WARX 28598, 27 Jan 45 (CM-OUT-28598 Jan 45); CG, U.S. Army Forces, China Theater, to War Department, No. CFB 32624, 7 Feb 45, CM-IN-7633 (8 Feb 45; CG, U.S. Army Forces, China Theater, to War Department, No. CFBX 33529, 27 Feb 45, CM-IN-28164 (27 Feb 45); Gross to Joint Chiefs of Staff, Subject: Letter of Transmittal, 312.2 (Engr), 27 Mar 45; Malloy to Joint Chiefs of Staff, Subject: Letter of Transmittal, 312.2 (Engr), 16 May 45, all in NARA RG 218 Box 688 Folder CCS 385 Pacific Ocean Areas (12-1-44) Cover Plans for Detachment, Iceberg, and Subsequent Operations in the Pacific Ocean Areas During 1945; CINCPOA Adv Hq to ComNavGroup China, DTG 100531, 10 Feb 45; COMINCH to ComNavGroup China, DTG 110045, 11 Feb 45; COMINCH to ComNavGroup China, DTG 210205 February, 21 Feb 45, all in NOA 548, Folder Bluebird—Dispatches.

733 PIN re debate: Smith to Lang, No. WAR 66688, 22 Nov 44, CM-OUT-66688; Lang to Smith, 2 Dec 44, both in JSC folder South America. "Has been unavailable": Unsigned, undated memorandum re message sent 25 Nov 44, in subfile "P Sent Traffic,"

in JSC file Traffic Sent. "Overheard discussion": Unsigned, undated memorandum re message sent 28 Nov 44, in subfile "P Sent Traffic," in JSC file Traffic Sent. "States emphatically": Message A-96, sent 9 Dec 44, in unsigned, undated memorandum "Bluebird Pertinent Messages," in JSC brown Acco binder labeled "15." Casualties: Unsigned, undated memorandum re message sent 2 Jan 45, in subfile "P Sent Traffic," in JSC file Traffic Sent. "The Japs had not yet": Smith to Lang, 30 Nov 44, JSC/J5 Serial 11155; Lang to Smith, 29 Dec 44, both in JSC folder South America. PEASANT re successes: Unsigned, undated memorandum re message sent 5 Dec 44, in subfile "P Sent Traffic," in JSC file Traffic Sent. PEASANT re strongest efforts: Unsigned, undated memorandum re message sent 6 Jan 45, in subfile "P Sent Traffic," in JSC file Traffic Sent.

733 PEASANT casualty figures: Unsigned, undated memorandum re message sent 11 Jan 45, in subfile "P Sent Traffic," in JSC file Traffic Sent. B-29s: Unsigned, undated memorandum re message sent 5 Dec 44, in subfile "P Sent Traffic," in JSC file Traffic Sent (PEASANT); Smith to Lang, No. WAR 69365, 28 Nov 44, CM-OUT-69365; Lang to Smith, 2 Dec 44, both in JSC folder South America (PIN); Message A-107, sent 19 Dec 44, in unsigned, undated memorandum "Bluebird Pertinent Messages," in JSC brown Acco binder labeled "15" (RUDLOFF). Flying bomb: Message B-231, sent 20 Dec 44, in unsigned, undated memorandum "Bluebird Pertinent Messages," in JSC brown Acco binder labeled "15" (RUDLOFF); A-X-15 Item 1, through Gus 20 Jan 45, in JSC brown Acco binder labeled "1" (GUS); unsigned, undated memorandum re message sent 25 Jan 45, in subfile "P Sent Traffic," in JSC file Traffic Sent (PEASANT). Logistical difficulties: Message A-100, sent 5 Dec 44; A-118, sent 6 Jan 45, both in unsigned, undated memorandum "Bluebird Pertinent Messages," in JSC brown Acco binder labeled "15" (RUDLOFF); Smith to Lang, 30 Nov 44, JSC/J5 Serial 11155; Lang to Smith, 29 Dec 44 both in JSC folder South America (RUPERT); unsigned, undated memorandum re message sent 9 Dec 44, in subfile "P Sent Traffic," in JSC file Traffic Sent; Message, in unsigned, undated memorandum "Bluebird Pertinent Messages" in JSC brown Acco binder labeled "15" (PEASANT).

734 Miscellaneous: Smith to Lang, 30 Nov 44, JSC/J5 Serial 11155; Lang to Smith, 29 Dec 44, both in JSC folder South America (PIN, RUPERT); Message A-106, sent 9 Dec 44, in unsigned, undated memorandum "Bluebird Pertinent Messages," in JSC brown Acco binder labeled "15" (RUDLOFF); Unsigned, undated memorandum re message sent 12 Dec 44, in subfile "P Sent Traffic," in JSC file Traffic Sent (PEASANT). Lt. Liu: Appendix "A," Implementation of CULTURE, in JSC folder Olympic. Chinese officers: Smith to Lang, No. WAR 69365, 28 Nov 44, CM-OUT-69365; Lang to Smith, 29 Dec 44, both in JSC folder South America.

735 Postponement: Message A-118, sent 6 Jan 45, in unsigned, undated memorandum "Bluebird Pertinent Messages," in JSC brown Acco binder labeled "15" (RUDLOFF); Unsigned, undated memorandum re message sent 8 Jan 45, in subfile "P Sent Traffic," in JSC file Traffic Sent (PEASANT); Smith to Lang, No. WAR 87247, 5 Jan 45, CM-OUT-87247 (Jan 45) (PIN); Lang to Smith, 12 Jan 45 ("first expressed surprise"), both in JSC folder South America. LOVE THREE: Spector 518. Kamikazes, Mentz: Morison, Liberation 44–46; Dwyer 83–84; GHQSWPA to War Department, No. X 56062, 8 Jan 45, CM-IN-7430 (9 Jan 45); Joint Security Control to CINCSWPA, No. WAR-86691, 4 Jan 45, CM-OUT-86691 (Jan 45), in JSC folder 000—To 312.1(2) China-Burma-India Theater; CTF 77 to COMINCH, No. NCR 8045DTG 160101/Z, 16 Jan 45, in JSC folder Bluebird.

736 No BJs at Lingayen: CTF 77 to COMINCH, No. NCR 8045DTG 160101/Z, 16 Jan 45, in JSC folder Bluebird; Spector 520 (Yamashita). MIKE SIX, SEVEN: Morison, Liberation 185–93.

736 Tayabas Bay: Dwyer 86–89; GHQ SWPA to War Department, No. C 10906, 24 Feb

45, CM-IN-24861 (24 Feb 45), in JSC folder M-1; Thurber, Operational Deception—Deception Evaluation Report Number Two, 7 Mar 47, pp. 11–15, in NOA 535, Folder Instructional Material, Plans. RUDLOFF re slowdown: Message B-234, sent 17 Jan 45, in unsigned, undated memorandum "Bluebird Pertinent Messages," in JSC brown Acco binder labeled "15." PIN re officials: Smith to Davis, No. WAR 37636, 14 Feb 45, CM-OUT-37636 (Feb 45); Davis to Smith, 21 Feb 45, both in JSC folder South America. "Find out whether": O'Connor to Thebaud and Bissell, 1 Feb 45, HMOC/2240; To London [O'Connor to Bevan], 21 Feb 45, both in JSC folder Great Britain; D Division Progress Report No. 3 16–31 Jan 45, Appendix A p. 13, in JSC folder Deception—SE Asia—WWII Historical D. Div—SACSEA—16 Dec 44–15 Mar 45; O'Connor to Bissell, 24 Apr 45, HMOC/2917, in JSC folder Japan.

737 Major appraisal: CGUSAFCT to War Department, No. CFBX 32476, 5 Feb 45, CM-IN-5121 (6 Feb 45), in JSC folder 000—To 312.1(2) China-Burma-India Theater. "Points strongly": Extract From Magic Diplomatic Summary No. 1119 of 18 April 45, attached to O'Connor to Bissell, 24 Apr 45, HMOC/2917, in JSC folder Japan. RUDLOFF advance notice: Message A-132, sent 13 Feb 45; A-136, sent 22 Feb 45; A-140, sent 2 Mar 45, all in unsigned, undated memorandum "Bluebird Pertinent Messages," in JSC brown Acco binder labeled "15." RUPERT re Bonins: Smith to Davis, No. WAR 37636, 14 Feb 45, CM-OUT-37636 (Feb 45); Davis to Smith, 21 Feb 45, both in JSC folder South America. RUDLOFF re assault: Message A-137, sent 2 Mar 45, in unsigned, undated memorandum "Bluebird Pertinent Messages," in JSC brown Acco binder labeled "15."

738 BAMBINO, VALENTINE: Unsigned, undated memorandum on planning and execution of deceptive measures in the Alaskan-Aleutian theater in 1944–45, pp. 16–18, in JSC folder Operation Wedlock; Hines to Smith, 10 Apr 45, in JSC folder Wedlock—subfolder 370.2 Wedlock—Adm Impl. PAT J: Messages A-147, A-150, File Log, in JSC brown Acco binder labeled "7" "File 'A.' "

738 LOADSTAR: JCS 1255/3, 20 Mar 45; approved by the Joint Chiefs, McFarland to Joint Security Control, Subject: Cover and Deception Plan for MILEPOST, 26 Mar 45; Joint Security Control, Detailed Implementation—Plan "Loadstar," undated, all in JSC folder 370.2 Pacific Theater; Cake Before Breakfast, p. 75, JSC loose papers. Soviet contingencies: Memorandum Report for Captain Thurber, undated, in NOA 530, Folder Combined Planning, Cover and Deception.

739 STURGEON planning: Unsigned memorandum, p. 3, Operation Sturgeon, 17 Jan 46, in JSC folder Bluebird; CINCPOA to COMINCH, No. NCR 3654, DTG 210155, 21 Jan 45, in JSC folder 370.2 Sturgeon; Smith to London Controlling Section, Subject: Implementation in S.E.A.C., Request for, 23 Jan 45, JSC/J5 Serial 198; O'Connor to Joint Security Control, Reference Your JSC/J5 Serial 198 of 23 Jan Move of XX Bomber Command, 1 Feb 45, HMOC/2241; O'Connor to Joint Security Control Subject: Implementation in SEAC, Request for, 2 Feb 45; Joint Security Control to CINCPOA Adv HQ et al., No. WARX 49890, 8 Mar 45, CM-OUT-49890 (Mar 45); Smith to O'Connor, Subject: D Division Weekly Newsletter No. 42, 9 Mar 45, JSC/J11 Serial 348, all in JSC folder 370.2 Sturgeon. RUPERT: Smith to Lang, No. WAR 25599, 23 Jan 45, CM-OUT-25599 (Jan 45); Smith to Lang, No. WAR 26030, 24 Jan 45, CM-OUT-6030 (Jan 45); Lang to Smith, 27 Jan 45; all in JSC folder South America. RUDLOFF: Message A-125, sent 27 Jan 45, in unsigned, undated memorandum Bluebird Pertinent Messages, in JSC brown Acco binder labeled 15. SILVER: D Division Weekly News Letter No. 41 27 Jan–2 Feb 45, Part I, pp. 1, 4, in JSC folder Deception—SE Asia—WWII Historical D. Div—SACSEA—30 Dec 44–23 Mar 45. DUCK: D Division Progress Report No. 4 15–28 Feb 45, Appendix A p. 7; OWL: D Division Progress Report No. 4 1–14 Feb 45, Appendix A p. 1; PAWNBROKER: D Division Progress Report No. 4 15–28 Feb 45, Appendix A pp. 1–2; GLEAM: D Division

Progress Report No. 4 15–28 Feb 45, Appendix A p. 9, all in JSC folder Deception—SE Asia—WWII Historical D. Div—SACSEA—16 Dec 44–15 Mar 45.

740 Misunderstandings: D Division Weekly News Letter No. 47 10 Mar–16 Mar 45, p. 6, in JSC folder Deception—SE Asia—WWII Historical D. Div—SACSEA—30 Dec 44–23 Mar 45; Smith to O'Connor, Subject: Exchange of Deception Information, 27 Mar 45, JSC/J11 Serial 3137; Smith to Thurber, Subject: Lieutenant Martin's Recommendation re Plan STURGEON, 19 Mar 45, JSC/J11 Serial 3101; CGUSAFIB to War Department, No. CRA 5063, 28 Feb 45, CM-IN-29265 (28 Feb 45); Marshall to Sultan, No. WAR 48588, 4 Mar 45, CM-OUT-48588 (Mar 45), all in JSC folder 370.2 Sturgeon. Not briefed: D Division Progress Report No. 3 16–31 Jan 45, p. 1, in JSC folder Deception—SE Asia—WWII Historical D. Div—SACSEA—16 Dec 44–15 Mar 45; O'Connor to Joint Security Control, Subject: Plan STURGEON, 31 Mar 45, HMOC/2736, enclosing Thorne to King, Subject: Plan STURGEON, 22 Mar 45; Smith to O'Connor, Subject: Plan STURGEON, 23 Apr 45, all in JSC folder 370.2 Sturgeon. Communications implementation: Hines to Smith, Subject: Summary of Radio Deception Activities Under Plan STURGEON for the Period 15 March–21 March 1945, 21 Mar 45; Hines to Smith, Subject: Movement of Planes of 58th Wing, 20th Bomber Command, 19 Apr 45; Hines to Smith, Subject: Special Traffic Analysis Report on Plan STURGEON, 25 Apr 45; Telecon Msg No W-2011, COMGENBOMCOM XX to COMGENAF TWENTY, Subject: Telecon Note, TOD 240437 YFK, [24 Mar 45]; Smith to O'Connor, Subject: Implementation of Plan STURGEON, 27 Mar 45, JSC/J11 Serial 3138, all in JSC folder 370.2 Sturgeon. MacArthur blunder: GHQ SWPA to War Department et al., DTG 10/0940Z, 10 Apr 45, CM-IN-8812 (10 Apr 45); Joint Security Control to CINCSWPA et al., No. WARX 66466, 11 Apr 45, CM-OUT-66466 (Apr 45); CG 20th Bomber Command to War Department et al., No. 1794, 16 Apr 45, CM-IN-14877 (16 Apr 45), all in JSC folder 370.2 Sturgeon. SEAC unable: Joint Security Control Information Bulletin No. 3. 1 Jun 45, JSC/J14 Serial 5122, p. 1, in NARA RG 165 ABC File ABC 381 Japan 15 Apr 43 Sec. 1-B. "Workers here all happy": CGUSARPOA to War Department et al., DTG 100147Z, 10 Jun 45, CM-IN-9423 (10 Jun 45), in JSC folder 370.2 Sturgeon. Closed: Marshall to Sultan, No. WARX 79792, 9 May 45, CM-OUT-79792 (May 45); Joint Security Control to CINCAFPAC et al., No. WARX 81067, 11 May 45, CM-OUT-81067 (May 45), both in JSC folder 370.2 Sturgeon.

740 Breakoff story: COMINCH for JSC to CINCPOA Adv, DTG 211646, 21 Mar 45; CINCPOA to COMINCH (JSC), DTG 260031, 26 Mar 45; COMINCH (for JSC) to CINCPOA Adv, DTG 032057, 3 Apr 45, all in NOA 548, Folder Bluebird—Dispatches; CINCPOA Adv HQ to COMINCH (JSC), DTG 042250, 4 Apr 45, in NOA 550, Folder Iceberg.

741 Special means: RUDLOFF: Message A-140, sent 23 Mar 45 (Iwo); Message A-141, sent 25 Mar 45 (Tenth Army), in tabulation of RUDLOFF messages, in JSC brown Acco binder labeled "14." GUS: Smith to Lang, No. WAR 70642, 20 Apr 45, CM-OUT-70642 (Apr 45); Lang to Smith, No. 433, 23 Apr 45, CM-IN-21655 (23 Apr 45)("In an emotional"), both in JSC folder South America. RUDLOFF: Message A-146, sent 21 Apr 45 ("on high authority"); Message A-141, sent 25 Mar 45 (general and aide), both in tabulation of Rudloff messages, in JSC brown Acco binder labeled "14."

741 McDowell commendation: Draft, pencil notation dated 23 Apr 45, in NOA 550, Folder Iceberg (2d folder).

742 Logistical believed: Lang to Smith 2 Jun 45; Harris to Smith, 1 Aug 45, both in JSC folder South America. 100,000 men; 9th Division: Appleman 85, 89, 103, 483–85. Formosa/South China strength: Unsigned, undated Operational Report—Bluebird, p. 14, in NOA 548, Folder Bluebird (2d folder). Uncertain at year end: Appleman 95.

742 Arisue, Tomioka, Miyazaki, Yonai: Navy Section, Joint Security Control, A Review

of Japanese Intelligence, 18 Apr 46, pp. 11–12 (Yonai), 13 (Arisue), 14 (Tomioka, Miyazaki), in NOA 535, Folder Japanese Intelligence Analysis of WW II. Consensus: Appleman 96. Great portion of reports: Unsigned memorandum, Japanese Appreciation of Allied Intentions 1 Jan–1 Mar 1945, 24 Aug 45, in JSC folder Japanese Appreciation of Allied Intentions.

743 Reduced level: Joint Security Control Information Bulletin No. 3. 1 Jun 45, JSC/J14 Serial 5122, p. 1, in NARA RG 165 ABC File ABC 381 Japan 15 Apr 43 Sec. 1-B. Roosevelt death: Smith to Lang, No. WAR 72322, 24 April 45, CM-OUT-72322; Lang to Smith, No. 439, 27 Apr 45, CM-IN-26109 (28 Apr 45), both in JSC folder South America. Shipment, Romulo: A-X-35, 2 May 45, Items 1 and 16; Lang to Smith, 16 May 45, both in JSC folder South America.

743 Shantung survey: COMINCH and CNO to ComNavGroup China, DTG 261650, 26 Mar 45, in NOA 550, Folder Pastel; COMINCH to ComNavGroup China, No. NCR 41803, DTG 261921, 27 Apr 45; CINCPOA Adv HQ to Joint Security Control, No. NCR 811, DTG 250710, 25 Apr 45, both in JSC folder Bluebird. KELLY: A-167 item 2, sent 19 Jun 45; LESTER: A-176 item 1, sent 14 Jul 45, both in unsigned, undated tabulation Messages Recorded, in JSC brown Acco binder labeled "1." Macao puzzle: D Division Weekly Progress Report No. 5 21–27 Apr 45, p. 5; No. 6 28 Apr–4 May 45, p. 2, both in JSC folder Deception—SE Asia—WWII Historical D. Div—SACSEA—24 Mar 45–4 Jun 45.

744 Apparent linkages; photoreconnaissance: CINCPAC Adv HQ to CINCSWPA, No. NCR 7267, DTG 121213, 3 May 45; CINCPOA Adv HQ to Joint Security Control, No. NCR 1901, DTG 080206 May, 8 May 45; Joint Security Control to CINCAFPAC and CINCPOA, No. WARX 79655, 9 May 45, CM-OUT-79655 (May 45), all in JSC folder Bluebird; CINCSWPA Manila to War Department, No. CX 16135, 13 May 45, CM-IN-12086, in NOA 548, Folder Bluebird—Dispatches. All possible information: D Division Weekly Progress Report No. 3 7–13 Apr 45, p. 13; "Might have important news": No. 8, 15–21 May 1945, p. 10; Officer from Manila: No. 9 22–28 May 45, p. 10, all in JSC Folder Deception—SE Asia—WWII Historical D. Div—SACSEA—24 Mar 45–4 Jun 45. Debate: O'Connor to Bissell, 24 Apr 45, HMOC/2917; Bissell to O'Connor, Subject: Most Secret Channels to the Japanese through China, Your HMOC/2917, 26 Apr 45, JSC/J/NS/2C Serial 4115, both in JSC folder Japan; CGUSARCT to War Department, No. CFB 36008, 20 Apr 45, CM-IN-18750 (20 Apr 45); Joint Security Control to CGUSARCT, No. WAR 70641, 20 Apr 45, CM-OUT-70641 (April 45); Joint Security Control to CGUSAFCT, No. WAR 73365, 26 Apr 45, CM-OUT-18750 (Apr 45); CGUSARCT to War Department, No. CFB 36383, 27 Apr 45, CM-IN-25730 (27 Apr 45); CGUSARCT to War Department, No. CFB 36354, 27 Apr 45, CM-IN-25544 (27 Apr 45); Joint Security Control to CGUSAFCT, No. WAR 74884, 28 Apr 45, CM-OUT-74884 (Apr 45); CGUSARCT to War Department, No. CFB 36665, 2 May 45, CM-IN-1484 (2 May 45); Joint Security Control to CGUSAFCT, No. WAR 76399, 2 May 45, CM-OUT-76399 (May 45); CGUSARCT to War Department, No. CFB 37714, 20 May 45, CM-IN-18885 (20 May 45); Joint Security Control to CGUSAFCT, No. WAR 85530, 2 May 45, CM-OUT-85530 (May 45); CGUSARCT to War Department, No. CFB 38824, 6 June 45, CM-IN-5486 (6 Jun 45); R. V. R[ushton] to Smith, 7 Jun 45; Joint Security Control to CGUSAFCT, No. WAR 13786, 8 Jun 45, CM-OUT-13786 (Jun 45); Joint Security Control to CGUSAFCT, No. WAR 13787, 8 Jun 45, CM-OUT-13787 (Jun 45); all in JSC folder 000—To 312.1(2) China-Burma-India Theater; Fleming to O'Connor, Wingate, Bevan, IND 6217, 4 May 45, PRO CAB 154/75/151; Bevan to O'Connor, CO/426/1, 4 May 45, PRO CAB 154/75/153B. Compare INK report, D Division Weekly Progress Report No. 14 26 Jun–2 Jul 45, p. 11, in JSC folder Deception—SE Asia—WWII Historical D. Div. SACSEA 5 Jun–10 Sep 45.

Chapter 18: Last Round in Asia

746 Draft update: Draft General Directive for Deception Measures Against Japan in 1944–1945, undated, in JSC folder Japan. Sent out: Smith to McDowell, Subject: Implementation, 27 Oct 44, JSC/J5 Serial 10103; Smith to Mentz, Subject: Implementation, 27 Oct 44, JSC/J5 Serial 10104; Smith to Brown, Subject: Implementation, 27 Oct 44, JSC/J5 Serial 10105, all in JSC folder Japan. Decoy fleets: Goldbranson to Record, 30 Oct 44, in JSC folder Japan. Formally directed: JWPC 190/7/D, 28 Nov 44, in NARA RG 165 ABC File ABC 381 Japan 15 Apr 43 Sec. 1-B. Pending Malta/Yalta: J. C. C. to Colonel Wood, Subject: Deception Plans in Force, 24 Jan 45, para. 4, in NARA RG 165 ABC File ABC 381 Japan 15 Apr 43 Sec. 1-B. Nudging: Minutes, CCS 189th Meeting, 16 Mar 45; CPS 156/D, 17 Mar 45, both in NARA RG 218 Box 694 Folder CCS 385 Pacific Theater (4-1-43) Deceptive Policy in the Pacific Theater (Sec. 3); SACSEA to British Chiefs of Staff, 28 Mar 45, SEACOS 345; CCS 284/12, 7 Apr 45, in NARA RG 218 Box 694 Folder CCS 385 Pacific Theater (4-1-43) Deceptive Policy in the Pacific Theater (Sec. 4); Minutes, COS (45) 88th Meeting, 5 Apr 45, PRO CAB 121/105/142; AMSSO to JSM Washington, COS (W) 738, 5 Apr 45, PRO CAB 121/105/140; Deception Policy in the Far East SEACOS 345, Note by the Controlling Officer, LCS (45)1, 3 Apr 45, PRO CAB 121/105/144–45; JPS 372/8, 17 Apr 45; Minutes, JPS 198th Meeting, 18 Apr 45; Donnelly to Joint Chiefs of Staff, Subject: General Directive for Deception Measures Against Japan; JCS to MacArthur, WAR.72645, 24 Apr 45, CM-OUT-72645 (Apr 45); GHQSWPA to War Department, No. 14938, 27 Apr 45, CM-IN-25326 (27 Apr 45), all in NARA RG 218 Box 694 Folder CCS 385 Pacific Theater (4-1-43) Deceptive Policy in the Pacific Theater (Sec. 4).

746 King not object: Hayes 704. Proposed plan: JPS 372/7, 13 Apr 45, in NARA RG 218 Box 694 Folder CCS 385 Pacific Theater (4-1-43) Deceptive Policy in the Pacific Theater (Sec. 4). Navy holdup: Minutes JPS 198th Meeting, 18 Apr 45; Minutes JPS 199th Meeting, 25 Apr 45, both in NARA RG 218 Box 694 Folder CCS 385 Pacific Theater (4-1-43) Deceptive Policy in the Pacific Theater (Sec. 4); Thurber to F-1, Subject: JPS 372/7, undated, in Naval Operational Archives, Washington Navy Yard, Strategic Plans Division, Series XXII, Box 549, Folder Broadaxe. Leahy holdup: J. E. H[ull] to Marshall, Subject: General Directive for Deception Measures Against Japan (JCS 498/8); Leahy to Joint Chiefs of Staff, Subject: J.C.S. 498/9, 8 May 45, both in NARA RG 218 Box 694 Folder CCS 385 Pacific Theater (4-1-43) Deceptive Policy in the Pacific Theater (Sec. 4); Record of telephone conversation between General Hull and General McFarland, 23 May 45, National Archives Collection, Xerox 2272, Marshall Memorial Library. Fairbanks, Thurber: Fairbanks to Duncan, Subject: Memorandum by the Chief of Staff (Admiral Leahy) Regarding Directive for Deception Measures against Japan (JCS 498/8), 12 May 45, in NOA 549, Folder Broadaxe—Staff Studies; Thurber to F-1, Subject: J.C.S. 498/9, Memo by the Chief of Staff to the Commander in Chief of the Army and Navy, 10 May 45, in NOA 549, Folder Broadaxe. Hull: J. E. H[ull] to Marshall, Subject: General Directive for Deception Measures Against Japan (JCS 498/8), 24 May 45, in JSC folder Japan. Control disputes: Hayes 704–706. Broadaxe finally approved: Supplement to CD-73 Dated 28 May 1945, 28 May 45, in NARA RG 218 Box 694 Folder CCS 385 Pacific Theater (4-1-43) Deceptive Policy in the Pacific Theater (Sec. 4); CCS 284/14, 25 May 45, in NARA RG 165 ABC File ABC 381 Japan 15 Apr 43 Sec. 1-B; "At first glance": Bevan to Wingate, No. GCCS 0420, 29 May 45, PRO HW 41/159. Deception Directive—War Against Japan, LCS (45) 6 (Final), 5 Jun 45, PRO CAB 121/105/75–79; Minutes COS (45) 145th Meeting, 6 Jun 45, PRO CAB 121/105/74; AMSSO to JSM Washington, COS (W) 945, 6 Jun 45, PRO CAB 121/105/72–73; CCS 284/15, 7 Jun 45; Minutes JPS

205th Meeting, 7 Jun 45; Donnell to Joint Chiefs, 7 Jun 45, both in NARA RG 218 Box 694 Folder CCS 385 Pacific Theater (4-1-43) Deceptive Policy in the Pacific Theater (Sec. 4); Joint Security Control to CINCPOA et al., No. WARX 14392, 8 Jun 45, CM-OUT-14392 (Jun 45), in JSC folder Bluebird; Memorandum of informal action, 15 Jun 45, in NARA RG 218 Box 694 Folder CCS 385 Pacific Theater (4-1-43) Deceptive Policy in the Pacific Theater (Sec. 4); General Directive for Deception Measures Against Japan, CCS 284/16/D, 16 Jun 45, in NARA RG 218 Box 694 Folder CCS 385 Pacific Theater (4-1-43) Deceptive Policy in the Pacific Theater (Sec. 5). Codename: JSC/A6-3, 21 May 45, in JSC folder 370.2 Broadaxe.

749 "Initiate deception staff studies": Metzel to Thebaud, Subject: Organization and Directives for Cover and Deception, 19 Feb 45, in JSC folder JCS To Handle Deception.

749 Studies: JWPC 190/12, 9 May 45 (Formosa); JWPC 190/13, 24 May 45 (Hokkaido); JWPC 190/14/M, 6 Jun 45; JWPC 190/14/M (Revised), 2 Jun 45 (French Indochina); JWPC 190/15/M, 6 Jun 45 (Yellow Sea), all in NARA RG 218 Box 694 Folder CCS 385 Pacific Theater (4-1-43) Deceptive Policy in the Pacific Theater (Sec. 4); Nelson to Thurber 28 May 45 with attached draft, in NOA 549, Folder Broadaxe—Staff Studies (Sumatra). LONGTOM: Hayes 702. Deferred: Undated memorandum signed T. D. R., in NARA RG 165 ABC File ABC 381 Japan 15 Apr 43 Sec. 1-B; Minutes JPS 202d Meeting, 16 May 45; Minutes JPS 204th Meeting, 30 May 45, both in NARA RG 218 Box 694 Folder CCS 385 Pacific Theater (4-1-43) Deceptive Policy in the Pacific Theater (Sec. 4); Minutes JPS 206th Meeting 13 Jun 45, in NARA RG 218 Box 694 Folder CCS 385 Pacific Theater (4-1-43) Deceptive Policy in the Pacific Theater (Sec. 5) ("We may try"). Sent out: Minutes JPS 205th Meeting 7 Jun 45, in NARA RG 218 Box 694 Folder CCS 385 Pacific Theater (4-1-43) Deceptive Policy in the Pacific Theater (Sec. 4); Minutes JPS 206th Meeting 13 Jun 45, Items 2 and 3 ("for such use"); Donnelly to various theater commanders, 17 Jun 45, in NARA RG 218 Box 694 Folder CCS 385 Pacific Theater (4-1-43) Deceptive Policy in the Pacific Theater (Sec. 5).

751 New theaters: Williams 472. Manila: Williams 478; James 2:657. "The estimate": Giangreco 580. "Two to two and a half": Giangreco 549. 280,000: Giangreco 577–78.

751 Officer requested; Goldbranson assignment: Goldbranson to Smith, 27 Apr 45; Goldbranson to Bissell, 28 Apr 45; Bissell to Cockrell, 1 May 45; Bissell to Jeffers, 12 May 45 ("being retained"); Movement Orders, 5 May 45, all in Goldbranson papers. "Warmest congratulations": Bevan to Goldbranson, 12 May 45, Goldbranson papers. "Retains highest": O'Connor to Bevan and Wingate, No. GOR, 8 May 45, PRO HW 41/159.

752 "There are some inherent": Goldbranson to Bissell, 11 Jun 45, Goldbranson papers. Fairbanks: Fairbanks to Duncan, Subject: Comments on CinCPac Joint Staff Study "OLYMPIC," Cover and Deception Appendix, 2 Jun 45, in NOA 532, Folder Douglas Fairbanks. Preparation of PASTEL: T. D. R[oberts] to Smith, 20 Jun 45, in JSC folder Olympic; CINCAFPAC to War Department, No. C 18160, 10 Jun 45, CM-IN-9211; CINCAFPAC to War Department, No. C 18203, 10 Jun 45, CM-IN-9212; Larr to Record with enclosures, 13 Jun 45, all in NOA 550, Folder Pastel; Martin to Thurber, Subject: Ad Hoc Committee of the JPS on Psychological Warfare, 29 Jun 45, in NOA Folder Douglas Fairbanks. Fret over delay: CINCAFPAC to War Department, No. CX 22986, 3 Jul 45, CM-IN-2233, in JSC folder 370.2 Pastel. JSC conferences: JCS 1410, Report by the Joint Staff Planners, 6 Jul 45, para. 3, in NARA RG 218 Box 652 Folder CCS 385 Japan (7-2-45) Plan Pastel. See also unsigned, undated memorandum headed Discussion, [25 Jun 45]; unsigned, undated memorandum headed The Problem; Suggested Revision, 27 Jun 45; Suggested Revision, 27 Jun 45; Suggested Revision; all in NOA 550, Folder Pastel (second unlabeled folder). ULTRA clearance: Memorandum for the record, 9 Jul 45, in JSC folder 370.2 Pastel. Adoption: Memorandum for Joint

War Plans Committee, Subject: Cover Plan PASTEL, JSC Serial 6117, 29 Jun 45, in NOA 550, Folder Pastel; JWPC 383; Minutes JPS 210th Meeting, 4 Jul 45; JCS 1410, Plan "Pastel," 6 Jul 45; Decision on JCS 1410, Plan "Pastel," 10 Jul 45, all in NARA RG 218 Box 652 Folder CCS 385 Japan (7-2-45) Plan Pastel.

753 Details of PASTEL: JCS 1410, Appendix "B," Plan "Pastel," in NARA RG 218 Box 652 Folder CCS 385 Japan (7-2-45) Plan Pastel.

754 Japanese aware Formosa bypassed: Joint Security Control to CINCAFPAC *et al.*, No. WARX 11443, 4 Jun 45, CM-OUT-11443 (Jun 45); CINCAFPAC to War Department, No. CX 17832, 6 Jun 45, CM-IN-5154 (6 Jun 45), both in JSC folder Bluebird.

755 Navy re foregone conclusion: Frank 146–48, 212.

755 Hokkaido, Fairbanks: Larson to Duncan, 22 Jun 45; Fairbanks to F-1, Subject: Brief of and Comments on "Over-all Cover and Deception Plan for the OLYMPIC Operation"—as Prepared by CinCPac, in Collaboration With CinCAFPac, 22 Jun 45, both in NOA 550, Folder Pastel. Transitional "story": Original version, Transitional Story, Bluebird to ———, 13 Jun 45; suggested revision, 27 Jun 45; Memorandum for Joint War Plans Committee, Subject: Cover Plan PASTEL, JSC Serial 6117, 29 Jun 45, paragraph 9, all in NOA 550, Folder Pastel. Theaters advised: JCS to CINCAFPAC et al., No. WARX 29914, 10 Jul 45, in NARA RG 218 Box 652 Folder CCS 385 Japan (7-2-45) Plan Pastel.

756 "They do not at present": Minute by the Controlling Officer, 26 Aug 44, National Archives Collection, Verifax 4098, Marshall Memorial Library.

756 Ismay: Ismay to Smith, 29 Aug 44, National Archives Collection, Verifax 4098, Marshall Memorial Library; Bevan to Fleming, L/99, 5 Sep 44, PRO HW 41/159. "Our experience"; Hull, Handy: Smith to Handy, 31 Aug 44; J. E. H[ull] to Handy, 4 Sep 44; Handy to Smith, 5 Sep 44, all in Reel 123, Item 3094, Marshall Memorial Library. "Recently received unofficial": Foreign Office to Tolstoy, undated (but filed with material from January 1945), PRO CAB 121/105/165. Noncommittal response: O'Connor to Thebaud and Bissell, HMOC/2037, 28 Dec 44; Thebaud and Bissell to O'Connor, 4 Jan 46, both in JSC folder Great Britain.

757 SEAC send officer: SACSEA to British Chiefs of Staff, 28 Mar 45, SEACOS 345; CCS 284/12. 7 Apr 45, both in NARA RG 218 Box 694 Folder CCS 385 Pacific Theater (4-1-43) Deceptive Policy in the Pacific Theater (Sec. 4); Minutes, COS (45) 88th Meeting, 5 Apr 45, PRO CAB 121/105/142; AMSSO to JSM Washington, COS (W) 738, 5 Apr 45, PRO CAB 121/105/140. "That the Americans have": Deception Policy in the Far East SEACOS 345, Note by the Controlling Officer, LCS (45)1, 3 Apr 45, PRO CAB 121/105/144–45. "Strategical deception in Europe": Bevan to Hollis, CO/358, 26 Mar 45, PRO CAB 121/105/150.

758 "Some of our people": Wilson to Marshall, 19 May 45, National Archives Collection, Xerox 2272, Marshall Memorial Library. Marshall response: Bissell to Marshall, 21 May 45, Reel 123, Item 3138, Marshall Memorial Library; Marshall to Wilson, 2 May 45; Record of telephone conversation between General Hull and General McFarland, 23 May 45 ("This guy from London"), both in Xerox 2272, National Archives Collection, Marshall Memorial Library. "If we do not": Minutes JPS 203d Meeting, 23 May 45, in NARA RG 218 Box 694 Folder CCS 385 Pacific Theater (4-1-43) Deceptive Policy in the Pacific Theater (Sec. 4).

758 Bevan visit: Deception Directive—War Against Japan, LCS (45) 6 (Final), 5 Jun 45, para. 2 last subparagraph, PRO CAB 121/105/75; Joint Security Control to CGUSAFCT, No. WAR 20799, 22 Jun 45, CM-OUT-20799 (Jun 45), in NOA 532, Folder F-28 Dispatch File; Summary of Activities by the U.S. Navy Cover and Deception Organization in World War II, 10 Jul 46, pp. 23–24, in NOA 530, Folder C&D (Cover and Deception) in World War II (General); Agreement in Committee by Joint Security Control and London Controlling Section, [27 Jun 45], in NOA 544, Folder Special Means—A. (Note

that latter document says that other details of agreement are in Appendix "B" of the Report of Committee of Joint Security Control and London Controlling Section, JSC Serial 6106 of 27 June 45. This has not been found.) "To avoid a show-down"; new agreement: Minute of a Meeting . . . , 19 Jul 45, PRO CAB 154/36/24A. Sunrise closing, Aspirin, Buenos Aires: Bevan to O'Connor, 20 Jul 45, PRO CAB 154/26/49A. "He should be a man": Note of Decision Reached in Meeting Between Joint Security Control (J.S.C.) and the London Controlling Section (L.C.S.), 27 Jun 45, in JSC folder Great Britain. Gummer: Bevan to O'Connor, LCS(44) I/C 42, 2 Aug 45, PRO CAB 154/36/26A; Jones to Bevan, EMJ 41 16 Aug 45, PRO CAB 154/36/27A. Bevan to New York: Bevan to Wingate, Fleming, LCS, No. GOR 410, 27 Jun 45, PRO HW 41/159; Fairbanks 256.

759   Jones: O'Connor to Smith and Thurber, HMOC 3652, 21 Jul 45, in JSC folder Great Britain. Mohican: PRO CAB 154/19. Bevan departure; Crichton: Hollis to CIGS *et al.*, COS 878/5, 5 June 45, PRO CAB 121/105/80; D Division Weekly Progress Report No. 24 4–10 Sep 45, distribution list, in JSC folder Deception—SE Asia—WWII Historical D. Div. SACSEA 5 Jun–10 Sep 45; Bevan to O'Connor, 20 Jul 45, PRO CAB 154/36/25A. "Terribly uphill": Bevan to O'Connor, LCS(44) I/C 42, 2 Aug 45, PRO CAB 154/36/26A.

760   Zipper: Williams 543; CGUSFIB to War Department, NO. CRA 18330, 3 Jul 45, CM-IN-2248 (3 Jul 45), in JSC folder 370.2 Pacific Theater.

761   Sceptical: Mountbatten to British Chiefs of Staff, SEACOS 370, 21 Apr 45, PRO CAB 121/105/106–107. "This might lead": Deception Policy in the Far East, SEACOS 370, LCS(45)3, 28 Apr 45, PRO CAB 121/105/101–102. "Unofficial preview"; not conflict: O'Connor to Joint Security Control, HMOC/2961, 28 Apr 45; Goldbranson to Smith, 26 Apr 45, both in JSC folder Misc Deception Plans—WWII. Approval: Minutes COS (45) 112th Meeting, 30 Apr 45, PRO CAB 121/105/100; AMSSO to JSM Washington, COS(W) 821, 30 Apr 45, PRO CAB 121/105/98–99; Chiefs of Staff to SACSEA, No. COSSEA 271, 20 May 45, CM-IN-20131, in JSC folder 370.2 Pacific Theater.

762   Fully developed: D Division Weekly Progress Report No. 12 12–18 Jun 45, p. 1; No. 18 24–30 Jul 45, Part I para. 3, both in JSC folder Deception—SE Asia—WWII Historical D. Div. SACSEA 5 Jun–10 Sep 45. Stultify merging: Fleming 17. Special means implementation: Cornwall-Jones to Joint Security Control, Subject: Plan SCEPTICAL, HMOC/3202; O'Connor to Joint Security Control, Subject: Plan SCEPTICAL, HMOC/3226, 29 May 45, both in JSC folder Misc Deception Plans—WWII.

762   Iceman: Shaw to Joint Security Control, Subject: Outline Plan, Operation Iceman, 12 May 45; Thurber to F-1, Subject: Outline Plan for Operation "Iceman," 31 May 45 ("no harm can be done"), both in NOA 550, Folder Plan—Iceman; Fairbanks to F-10, Subject: Outline Plan for Operation "Iceman," 1 Jun 45; R. F. N[elson] to Thurber, Subject: Plan ICEMAN, undated; Thurber to Smith, Subject: Outline Plan, Operation Iceman, 4 Jun 45, all in NOA 550, Folder Iceman. Carbonado: Romanus & Sunderland, Time 354–55.

764   Corbett promotions: Corbett to Rusk, 13 Jun 45 (Captain); Minutes, 26 Jun 45, (Major), both in JSC folder Southeast Asia. Wingate and Corbett to Chungking: D Division Weekly Progress Report No. 11 5–11 Jun 45, in JSC folder Deception—SE Asia—WWII Historical D. Div. SACSEA 5 Jun–10 Sep 45; Handwritten memorandum (Wingate), Tour Notes: Visit to Chungking, undated, in JSC folder Southeast Asia ("stooging around"; "conducted us"); Wingate, Limelight 213 ("green and sick"; "a few inches").

764   Chungking meeting; "All parties": Minutes of a Meeting To Discuss Plans of Coordinating the Deception Activities of China Theater and Southeast Asia Command, 30

Jun 45, in JSC folder Southeast Asia. "Armchairs and whisky": Wingate, Limelight 213. Mountbatten visit: James 2:662; Ziegler 296; Wingate, Limelight 215. Manila conference: Corbett to [Raymond], ———Jul 45; Raymond to Corbett, 10 Aug 45; Brief by Head of D Division, Suggested Points To Be Made to General MacArthur, unsigned [Wingate], undated; Draft Agenda for Manila, unsigned, undated; Corbett to [Raymond], ———Jul 45 ("Colonel Wingate was quite"), all in JSC folder Southeast Asia.

765 Conscious: Cover and Deception Plan "Conscious," in Russell, Memo for Record, undated, in NARA RG 319 Entry 101 Box 7 Folder 4, Operation Zipper and Cover and Deception Plan Conscious. Balikpapan: James 2:756. "In the role of": Corbett to [Raymond], ———Jul 45, in JSC folder Southeast Asia. Conscious forwarded, staffed: Roberts to Chief, Joint Security Control, Subject: Plan "CONSCIOUS," 24 Jul 45; Draft Appendix, undated, both in JSC folder Conscious; JWPC 190/17, 30 Jul 45, in NARA RG 165 ABC File ABC 381 Japan 15 Apr 43 Sec. 1-B; CCS 899, 10 Aug 45, in NARA RG 218 Box 694 Folder CCS 385 Pacific Theater (4-1-43) Deceptive Policy in the Pacific Theater (Sec. 5).

766 "Not particularly keen"; "without the full approval": Bevan to Menzies, CO/436, 14 Apr 44, PRO CAB 154/37/9A. Approaches to Bissell, Smith reaction, nothing expected: O'Connor to Bevan, HMOC/1010, 6 May 44, PRO CAB 154/37/14A; Purple Whales Channel and Operation Hiccoughs, signed Bissell, n.d., PRO CAB 154/37/15B; O'Connor to Bevan, HMOC/1015, 9 May 44, PRO CAB 154/37/15A. Americans run project: Fleming to Bevan, GRO.235, 28 Aug 44, PRO HW 41/159. "I think it would be": Bevan to Fleming, L/99, 5 Sep 44, PRO HW 41/159. Full history: Head of D Division to London Controlling Section, 3 Feb 45, Subject: Deceptive Broadcasts, in JSC folder Deception—SE Asia—WWII Historical D. Div—SACSEA—16 Dec 44–15 Mar 45. Psychological warfare; "provided it is used": Wingate to Bevan, 103/D.Div., 8 May 45, PRO CAB 154/37/68A; Cohen to Bevan, CX.69651/700/V.X., 12 May 45, PRO CAB 154/37/66A; Bevan to Wingate, L.C.S. (45) I/C.42, 14 May 45, PRO CAB 154/37/67A; D Division Weekly Progress Report No. 12 12–18 Jun 45, p. 3, in JSC folder Deception—SE Asia—WWII Historical D. Div. SACSEA 5 Jun–10 Sep 45. Not bring up in Washington: 116/D Div, 31 May 45, PRO CAB 154/37/69A. Wingate sell in Manila: Fleming to Wingate, 116/D Div, 31 May 45; Wingate to Fleming, B/203/8/D.Div,—July 45, both in PRO CAB 154/37. Mimsy: PRO CAB 154/39; PRO WO 203/5769A; Bevan to Wingate, No. GCCS 03951, 26 May 45, PRO HW 41/159.

767 Seventeen million: Frank 162–63. Bassington: Fleming 35; Plan Bassington, attached to O'Connor to Joint Security Control, HMOC/2460, 2 Mar 45, in JSC folder 370.2 Operations and Results; other copies, PRO CAB 121/105/130–37, and CCS 843, 25 Apr 45, in NOA 548, Folder Bassington. Smith review: Rushton to Smith, Subject: Summarization of Plan Bassington, 8 Mar 45; C. E. G[oldbranson] to Smith, Subject: Plan BASSINGTON, HMOC/2460, 8 Mar 45; Smith to O'Connor, Subject: Plan "BASSINGTON"—HMOC/2460, JSC/J10 Serial 349, 14 Mar 45 ("laudable"; "chiefly concerned"); Smith to Bissell, 12 Mar 45, Subject: Plan "BASSINGTON"—HMOC/2460, 2 March 1945, 12 Mar 45; Smith to Thurber, Subject: Plan "BASSINGTON," JSC/J10 Serial 3116, 21 Mar 45, and Thurber endorsement, 21 Mar 45, all in JSC folder 370.2 Operations and Results. "Without approval": Smith to O'Connor, JSC/J10 Serial 3129, 23 Mar 45, in NOA 548, Folder Bassington. British review: Bevan to Hollis, CO/567, 6 Apr 45, PRO CAB 121/105/125–26; Minutes COS (45) 92d Meeting, PRO CAN 121/105/122 ("well worth"); Notation, 15 Apr 45, on Ismay to Prime Minister, 9 Apr 45, PRO CAB 121/105/120–21 ("So proceed"). Back across: CCS 843, 25 Apr 45; JCS 1365, 24 May 45 ("can do little"), both in NOA 548, Folder Bassington; D Division Weekly Progress Report No. 20 7–13 Aug 45, para. 3, in JSC folder Deception—SE Asia—WWII Historical D. Div. SACSEA 5 Jun–10 Sep 45.

769 Minutes: Fleming 36. "As in most branches": D Division Weekly Progress Report No. 13 19–25 Jun 45, in JSC folder Deception—SE Asia—WWII Historical D. Div. SACSEA 5 Jun–10 Sep 45. Laight: "A" Force card index, entry Laight; D Division Weekly Progress Report No. 11 5–11 Jun 45, in JSC folder Deception—SE Asia—WWII Historical D. Div. SACSEA 5 Jun–10 Sep 45; Author's telephone interview with Rex Hamer, 19 Jun 97 ("playboy").

769 Ionides: Mure, Master 112; D Division Weekly Progress Report No. 11 5–11 Jun 45, p. 1, in JSC folder Deception—SE Asia—WWII Historical D. Div. SACSEA 5 Jun–10 Sep 45. Fleming in England: Bevan to Wingate and Fleming, No. GOR 388, 26 Jun 45, PRO HW 41/159; Bevan to Wingate, No. GOR 413, 27 Jun 45, PRO HW 41/159; COS to SACSEA, 28 Jun 45, PRO WO 203/3313/104A; D Division Weekly Progress Report No. 16 10–16 Jul 45; No. 23 28 Aug–3 Sep 45, in JSC folder Deception—SE Asia—WWII Historical D. Div. SACSEA 5 Jun–10 Sep 45. *Tucson:* Tactical Deception in Support of Third Fleet Raids on Japan, 10–14 July 1945, Enclosure (A) to ONO Serial 0974P34 of 18 Dec 1946, in NOA 531, Folder Deception Staff Studies Pacific; extensive details in NOA 546, Folder USS Tucson Operations 10–11 July; Morison, Victory 309–316.

770 Chusan-Shanghai interest: CGUSFCT to War Department, No. CFB 1710, 26 Jul 45, CM-IN-26269 (26 Jul 45); Smith to Weckerling, 27 Jul 45; Joint Security Control to CGUSFCT, No. WAR 39468, 27 Jul 45, CM-OUT-39468 (Jul 45); CGUSFCT to War Department *et al.,* No. CFBX 2167, 30 Jul 45, CM-IN-31215 (30 Jul 45); Joint Security Control to CGUSFCT, No. WAR 45475, 7 Aug 45 (Aug 45), all in JSC folder 370.2 Broadaxe. Parachute regiments: Deception Plan: To Represent the Presence in India of Two Notional U.S. Parachute Regiments for Operations in China, undated, evidently an attachment to Memorandum to Chief of Staff, Subject: Deception Planning, unsigned [Corbett], 19 Jul 45, both in JSC folder Southeast Asia; CGUSFCT to War Department, No. CFB 2870, 6 Aug 45, CM-IN-6129 (6 Aug 45); CGUSFCT to War Department, No. CFB 2377, 1 Aug 45, CM-IN-645 (1 Aug 45; Joint Security Control to CGUSFCT, No. WAR 43968, 3 Aug 45, CM-OUT-43968 (Aug 45), all in JSC folder 000—To 312.1(2) China-Burma-India Theater.

770 Special means implementation of SCEPTICAL: Fleming 17.

771 Japanese expectations: Woodburn Kirby 72. Holsopple: CGUSFCT to War Department, No. CFB 2377, 1 Aug 45, CM-IN-645 (1 Aug 45); Joint Security Control to CGUSFCT, No. WAR 43968, 3 Aug 45, CM-OUT-43968 (Aug 45), both in JSC folder 000—To 312.1(2) China-Burma-India Theater. Concession rights: Unsigned, undated memorandum to F-00, dated 4 Jul 45, by hand, in NOA 550, Folder Pastel. Radio deception plan: CINCAFPAC to War Department, No. CX 26961, 20 Jul 45, CM-IN-20267 (20 Jul 45); CINCAFPAC to CGUSFCT, No. CX 28361, 26 Jul 45, CM-IN-26166 (26 Jul 45), both in JSC folder 370.2 Pastel; CINCAFPAC to CINCPAC Adv HQ, No. CX 24942, DTG 121159, 12 Jul 45; CINCPOA Adv HQ to CINCAFPAC, DTG 140233, 14 Jul 45; complete copy of plan, all in NOA 550, Folder Pastel.

772 Miles, Shanghai: Joint Security Control to COMUSAFPAC, No. WARX 38701, 26 Jul 45, CM-OUT-38701 (Jul 45) in JSC folder 370.2 Pastel. Staff study, PASTEL Two: Johnson to Chief, S&P Group, Subject: Brief of Cover and Deception Plan "PASTEL-TWO," 9 Aug 45, in NARA RG 319 Entry 101 Box 7 Folder Pastel Two; GHQ USAFPAC, Staff Study "Pastel-Two," Cover and Deception Olympic Operations, 30 Jul 45, in NOA 550, Folder Pastel Two. Debate over size, date: Joint Security Control to CINCAFPAC, No. WARX 41505, 31 Jul 45, CM-OUT-41505 (Jul 45); CINCAFPAC to War Department, No. CX 29108, 28 Jul 45, CM-IN-28969 (28 Jul 45), both in JSC folder Olympic; CINCPAC Adv HQ to COMINCH, No. NCR 4393, DTG 022301, 2 Jul 45, in JSC folder 370.2 Broadaxe. Operations Instructions: GHQ US-

AFPAC, Operations Instructions Number 3, in NOA 550, Folder Pastel Two. Tactical draft study: Draft Outline of a Staff Study for Tactical Cover and Deception Plan for Olympic, 3 Aug 45, F-28-OUT-92, in NOA 550, Folder Pastel Two.

773 Airborne corps: CINCAFPAC to War Department, No. CX 29841, 31 Jul 45, CM-IN-32110 (31 Jul 45); CINCAFPAC to War Department, No. CX 30582, 3 Aug 45, CM-IN-2997 (3 Aug 45), both in NOA 532, Folder F-28 Dispatch File; CINCAFPAC to War Department, No. CX 31629, 8 Aug 45, CM-IN-7461 (8 Aug 45); Sweeney to Harris, 9 Aug 45; Rushton to Harris, Subject: Cable CM-IN-7461, 9 Aug 45; Smith to Goodwin, Subject: Cable COM-IN 7461, 10 Aug 45, all in JSC folder 370.2 Pastel. Master schedule: Unsigned, undated list in JSC folder 370.2 Broadaxe; Marshall to Chief of Staff, Subject: Special Means Implementation, PASTEL-TWO, 7 Aug 45, with enclosure, in NARA RG 165 ABC File ABC 381 Japan 15 Apr 43 Sec. 5—Pastel. Logistical problems: A-X-45 Item 3, through PIN 16 Jul 45, in JSC brown Acco binder labeled "1."

773 Marshall to President: A-X-47 Item 2, through Gus 7 Aug 45, in JSC brown Acco binder labeled "1." Swiss industrialist: From London, 8 Aug 45, GOR 595 from GCCS 09857; Sweeney to Harris and Smith, 9 Aug 45; Smith to Bevan, 11 Aug 45, all in JSC folder Japan. Special means messages: Week ending 9 Jun 45, Item 7, through KELLY 19 Jun 45 (coast study); A-X-42 Item 10, through Gus 6 Jul 45 (volunteers); A-161 Item 1, through LESTER 5 Jun 45 (Sixth Army); A-X-47 Item 6, through PIN 9 Aug 45 (censorship officers); A-X-43 Item 4, through Gus 23 Jul 45; A-176 Item 1, through LESTER 14 Jul 45 (Patton); A-X-47 Item 4, through RUPERT 10 Aug 45 (Arctic clothing), all in JSC brown Acco binder labeled "1." Wedemeyer commanders: Wedemeyer 331–32. Simpson visit: CGUSAFCT to War Department, No. CFB 2067, 29 Jul 45, CM-IN-29926 (29 Jul 45); Joint Security Control to CGUSFCT, No. WAR 40813, 30 Jul 45, CM-OUT-40813 (Jul 45); Joint Security Control to CGUSFCT, No. WAR 42059, 1 Aug 45, CM-OUT-42059 (Aug 45), all in JSC folder 370.2 Broadaxe.

774 Tactical units: Hull to CINCAFPAC, DTG 071715Z, 7 Aug 45, in NOA 550, Folder Pastel.

775 Begin: CINCARPAC to War Department et al., No. X 31397, 7 Aug 45, CM-IN-6746 (7 Aug 45), in JSC folder 370.2 Pastel. CORONET staff study: JWPC 190/16, 26 Jul 45, in NARA RG 218 Box 694 Folder CCS 385 Pacific Theater (4-1-43) Deceptive Policy in the Pacific Theater (Sec. 5).

775 Goodwin: Goodwin to Roberts, Subject: Comments on JWPC 190/16 (Staff Study of Cover and Deception Objectives for "CORONET"), 4 Aug 45, in NARA RG 165 ABC File ABC 381 Japan 15 Apr 43 Sec. 1-B. "Basic misunderstanding": Harris to Smith, Subject: Comments on Colonel Goodwin's Recommendations Concerning Cover and Deception for CORONET, 6 Aug 45, in JSC folder 370.2 Pacific Theater. "We should plan to keep": Memorandum, T. D. R[oberts], Staff Study of Cover and Deception Objectives for Coronet (JWPC 190/16), in NARA RG 165 ABC File ABC 381 Japan 15 Apr 43 Sec. 1-B. Passes Planners, to theaters: Minutes JPS 213th Meeting, 8 Aug 45; Donnelly to CINCAFPAC, Subject: Staff Study of Cover and Deception Objectives for CORONET, 10 Aug 45, and similar cover letters, all in NARA RG 218 Box 694 Folder CCS 385 Pacific Theater (4-1-43) Deceptive Policy in the Pacific Theater (Sec. 5). Goldbranson: Goldbranson to Russell, Subject: Strategic and Tactical Deception, Operation "CORONET," 10 Aug 45, in Goldbranson papers.

776 Steady flow: Tab A, undated, in JSC folder Japanese Appreciation of Allied Intentions. (Probably Tab A to Japanese Predictions of Allied Landings, in same folder.) "Perhaps the topic": Project No. 1471-A, The Japanese Intelligence System, 4 Sep 45, pp. 79–80, in JSC folder The Japanese Intelligence System. PIN report: Harris to Smith, 1 Aug 45, in JSC folder South America.

777 Steady buildup: Drea 208–22; Frank 211–13. Vigorous preparations: Unsigned, un-
     dated memorandum, Japanese Predictions of Allied Invasion of Japanese Homeland—
     10 August 1945, in JSC folder Japanese Appreciation of Allied Intentions. "I thought
     you would have to land"; "(1): If Russia"; "Air General Headquarters": Navy Section,
     Joint Security Control, A Review of Japanese Intelligence, 18 Apr 46, pp. 11–12, in
     NOA 535, Folder Japanese Intelligence Analysis of WW II. Kawabe: Frank 84. Told
     Emperor: Drea 211.
777 All through summer: Frank 164–96. Directives canceled: CINCPAC Adv HQ to
     COMNORPAC et al., DTG 150018, 15 Aug 45, in NOA 550, Folder Pastel (second
     unlabeled folder). "Deception Plan Pastel Two": CINCAFPAC to War Department et
     al., No. CX 34124, 16 Aug 45, CM-IN-16189 (16 Aug 45), in NARA RG 218 Box
     652 Folder CCS 385 Japan (7-2-45) Plan Pastel.

                                        Epilogue

779 O what a tangled: J. R. Pope, A Word of Encouragement, Punch, 1958. Double agents
     more than signals: Eldredge to Smith, 14 Jul 45, 25 Jul 45, 3 Aug 45, in JSC folder De-
     ception Units. Fleming: Hart-Davis 302. "It would be easy": Fleming 3.
780 No permission: Cake Before Breakfast, p. 12, JSC loose papers. Willoughby: Drea
     16–17. "The very absence": Cake Before Breakfast, pp. 12–13, JSC loose papers. Lev-
     eled at Fleming: Alan Stripp in introduction to 1996 edition of The Man Who Never
     Was, p. 5.
781 "It remains true": Fleming 3.
782 Bissell canvassed: Bissell to MacArthur, JIZ Serial 1037, 9 Oct 45, and similar, in JSC
     folder Letters From World War II Commanders Re Value of Cover and Deception.
782 Letters from commanders: Devers to Bissell, 5 Nov 45 (Devers); Noce to Bissell, JSC-
     SSP-108, 15 Oct 45 ("there were many") (Noce); Stilwell to Bissell, Subject: Your
     JIZ—Serial 1028, 5 Nov 45 ("I feel that, in general") (Stilwell); Wedemeyer to Assis-
     tant Chief of Staff G-2, Subject: Deception, 25 Nov 45 (Wedemeyer).
783 Remain Top Secret: CCS 281/8, approved by informal action 20 Nov 45, in NARA
     RG 218 Box 362 Folder CCS 385 (4-8-43) (2) Cover and Deception Planning. His-
     tory, Germany and Italy: CCS 701/23, 17 Dec 45, in NARA RG 218 Folder CCS
     313.6 (7-11-44) Sec. 3, Receipt and Disposal of Records of Combined Operations.
     Bevan directed: Military Deception Present and Future Position, COS (45) 214 (0), 27
     Mar 45, para. 3, PRO CAB 121/105/151.
783 Hollis took care: Bevan to Hollis, CO/358, 26 Mar 45, PRO CAB 121/105/150;
     Bevan to Secretary, COS Committee, CO/358, 11 Apr 45, PRO CAB 121/105/116;
     Hollis to Simpson, 29 May 45, PRO CAB 121/105/84. Till end Sep 46: Theobald to
     Dorothy Clarke, 11 Dec 78, in Clarke papers. Clarke report: London Controlling Sec-
     tion, Clarke, Dudley, "A" Force Permanent Record File, Narrative War Diary, PRO
     CAB 154/1 through 154/6. Sweeney report: Report on the Activities of No. 2 Tac
     H.Q. A Force (Sixth Army Group), 31 July 1944–5 May 1945, Parts I-IV and Conclu-
     sions, NARA RG 319 Entry 101 Box 4 Folders 79–83. "The able and most valuable":
     Sweeney notes on earlier draft. Hesketh history: Hesketh. Fleming report: Fleming.
784 Copying Montagu paper: Ladd and Sprague to Inglis, Subject: Naval Deception,
     N.I.D., Admiralty Study on, 11 Sep 45; Shelley to Inglis, 11 Sep 45, both in Admiralty
     Study on Naval Deception, in NOA 537, Folder N.I.D. (British Naval Intelligence Di-
     vision) Admiralty Reports on Naval Deception. Ladd, Inglis: Packard 26. Masterman
     report: Masterman. Wingate completed April 1947: Wingate to Bevan, 11 Mar 47,
     Bevan papers. Wingate history: Wingate. June 1947 conference: WDGPO to Hq
     EUCOM, No. WAR 98147, 14 May 47, CM-OUT-98147 (15 May 47), in JSC folder
     Great Britain. Smith recalled: Recalled to active duty 31 Mar–4 Apr 1947, Officer's,

Warrant Officer's, and Flight Officer's Qualification Record, WD AGO Form 66, 26 Apr 1949, in Newman Smith personnel file.

785 Harmer report: Harmer. Oakes report: Oakes. Patton eulogy: Cornwall-Jones to Wingate, No. ZO 900, 12 Mar 46, PRO CAB 121/105/3. Deane: Deane. Merrick: Morrow, New York, 1947. Weismiller: The American Oxonian, vol. 89 no. 2 (Spring 2002) at 227–28. ZIGZAG; TREASURE: Nigel West in Masterman 1996 reprint xiii note *. Wingate: Wingate, Limelight. Delmer: Delmer.

786 Hesketh enraged; "I had to be content": Hesketh to Eldredge, 10 Dec 71; Hesketh to Eldredge, 27 Nov 72, both in Eldredge papers.

787 TRICYCLE: Popov.

788 Cruickshank: Cruickshank, Deception. Slender book: Montagu, Beyond.

788 Fleming: Fleming to Clarke, 24 May 71, in Clarke papers. Woods: Bevan to Clarke, 11 Aug 72; Woods to Clarke, 22 Aug 72, both in Clarke papers; author's telephone interview with Sir Michael Howard, 14 Mar 00.

789 Howard: Howard to Bevan, 13 Jul 78, Bevan papers; Howard to Eldredge, 17 Jun 85, in Eldredge papers; author's telephone interview with Sir Michael Howard, 14 Mar 00.

790 Mure to USA: Mure, unsigned, undated, unaddressed draft letter, in Mure papers, Box 2.

791 Clarke: Eldredge to Dwyer, 11 Sep 89, in Eldredge papers; Clarke papers passim; Wild to Mure, 1 Jul 78, in Mure papers (Box 2); author's telephone interview with Ann M. Greenwood, 31 Dec 95; book proposal, 18 Jul 53, in Clarke papers; funeral program in Clarke papers; The Times, 14 May 74. Bevan: "I have never known": Ismay to Bevan, 4 Jan 45; "I wish": Hollis to Bevan, 1 Jan 45; "A fully fitting": Robertson to Bevan, 4 Jan 45, all in Bevan papers.

791 Bevan obituary: Daily Telegraph, 5 Dec 78. Strangeways: Author's telephone interview with Sir Edgar Williams, 24 Oct 92; Stratagem Conference 90; obituary, The Times, 5 Aug 98. Crichton: Author's telephone interview with Alice Crichton, 20 Nov 93.

792 Fleming: Hart-Davis 303 and passim.

793 Thorne: Entry in Who's Who. Robertson: Unsigned Memorandum for Record, 9 Sep 47, paragraph 3b(3), in NOA 530, Folder Combined Planning, Cover and Deception. Hesketh: Introduction to Hesketh 1999 edition, pp. xii–xiii. Montagu: Obituary, New York Times, 21 Jul 85. Hamer: Author's telephone interview with Rex Hamer, 19 Jun 97. Harmer: Author's telephone interview with Christopher Harmer, 19 Dec 92.

793 Wild: Eldredge to Ingersoll, 30 Dec 71, in Ingersoll papers; author's telephone interview with Alice Crichton, 12 Oct 95. Jervis-Read: Author's conversations with Colonel P. N. Trustram Eve. Smith: Accolade: Lincoln to Assistant Chief of Staff, G-2, 4 Oct 45, in NARA RG 165 ABC File ABC 381 (4-8-43) Sec. 1-B—Deception policy, organization, etc. Reserve, Interior, Commerce: Military Record and Report of Separation, n.d. (1945), Newman Smith VA file; Officer's, Warrant Officer's, and Flight Officer's Qualification Record, WD AGO Form 66, 26 Apr 1949; Weeks to Smith, 5 Oct 55, in Newman Smith personnel file; Biographical Sketch, 15 Sep 53, in Newman Smith papers. "Aunt Rosalind's finest"; "to be near": Lanahan 210.

794 In Montgomery: Author's telephone interview with Will Hill Tankersley, 28 Mar 94; VA Form 8-526, 24 Jun 57; William U. Cawthon, MD, 10 Aug 64, both in VA file. Bissell: Obituary, Washington Post, 30 Dec 72. Baumer: Obituary, Westfield, New Jersey Leader, 23 Feb 89.

795 Kehm: Obituary, Assembly, September 1980, pp. 123–34. Goldbranson: Author's telephone interview with Carl E. Goldbranson, Jr., 2 Mar 00; Goldbranson to Campbell, Subject: Ruses and Stratagems—Historical Examples, 14 Dec 48, in NOA 530, Folder C&D Lecture for War College; Wild to Mrs. Goldbranson, 20 Aug 57, in Goldbran-

son papers. Sweeney: Author's interview with Eugene Sweeney, December 1992–January 1993.

795 Ekstrom: Author's interviews with Arne Ekstrom, 1993; author's conversations with Nicolas Ekstrom. Harris: Cullum No. 9608; Biographical sketch, Major General William A. Harris, USAMHI. Ingersoll: Hoopes 395.

797 Eldredge: Eldredge papers, *passim*. Corbett: D Division Weekly Progress Report No. 22 21–27 Aug 45, in JSC folder Deception—SE Asia—WWII Historical D. Div. SACSEA 5 Jun–10 Sep 45; Draft Memorandum, Subject: Implementation of CCS 701/23, in JSC folder #6 LEN Harris/Ingersoll/Harris/Eldredge. Burris-Meyer: Author's telephone interview with Anita Burris-Meyer, 19 Feb 95. Thurber: Thurber biographical release, Naval Operational Archives. Nelson: Obituary, *Shipmate*, August 1963. Metzel: Admiral Stabs Self, Leaps to His Death, *Washington Post*, 27 Jul 52. Schrup: Author's telephone interview with Mrs. E. P. Schrup, 12 Apr 94.

798 Hunter: Obituary, *Savannah Evening Post*, 1 Mar 89. Merrick: Obituary, *New York Times*, 23 Apr 88.

799 Ops/plans or intelligence: Cabell, Record of Conversations re Joint Security Control, 6 Oct 47, in NARA RG 218 Box 211 Folder CCS 334 Joint Security Control (Sec. 3). JSC or regular staff: Whitehill 37; NOA, separate item "Strategic Plans Division, Series XXII: OP-607 Support Plans Branch and Predecessor Offices, ca. 1942–1961, chiefly 1942–1952"; JCS Memorandum for Information No. 549, 25 Aug 47, in NARA RG 218 Box 211 Folder CCS 334 Joint Security Control (Sec. 3). By 1951: Note by the Secretaries, 16 Mar 51, in NARA RG 218 Box 362 Folder CCS 385 (4-8-43) (2) Cover and Deception Planning. Period of uncertainty: Clarke and Bevan, Suggested Deception Organization in Peace, First Revise, 5 Jun 45, PRO CAB 121/105/56-69; L. C. H[ollis] to Clarke, 11 Sep 45, PRO CAB 121/105/48–49; Deception Organization in Peace, Note by the Secretary, COS (45) 564 (O), 8 Sep 45, PRO CAB 121/105/52–55; Minutes COS (45) 220th Meeting, 11 Sep 45, PRO CAB 121/105/50–51; Unsigned memorandum, Deception Organization in Peace, 14 Sep 45, PRO CAB 121/105/ 35–39; Minutes COS (45) 233d Meeting, 25 Sep 45, PRO CAB 121/105/33; Hollis to Menzies, COS 204/6, 19 Feb 46, PRO CAB 121/105/4–5. Hollis: Dictionary of National Biography 1961–70.

799 Hollis Committee; Inter-Service Communications Intelligence Committee: PRO CAB 81/80; Unsigned Memorandum for Record, 9 Sep 47; C.V.R.S. to Norstad, Subject: Cover and Deception, 20 Oct 47, both in NOA 530, Folder Combined Planning, Cover and Deception.

799 Efforts to reestablish: Unsigned memorandum, Possible British Deception Approach During Your Visit to U.K., Col. Harris/SPB/72269, 22 Jun 48 and Tab "I"; Glover to Sherman, Subject: British Peacetime Cover and Deception Organization, 22 Sep 47; C.V.R.S. to Harris, 17 Oct 47, all in NOA 530, Folder Combined Planning, Cover and Deception.

800 Corbett and Packard; "a slippery," etc.: Unsigned Memorandum for Record, 9 Sep 47, pars. 3b(2) (a)-(d), in NOA 530, Folder Combined Planning, Cover and Deception. "No major operations": Eisenhower to Director, Plans and Operations Division, WDGS, 19 Jun 47, in JSC folder Letters From World War II Commanders Re Value of Cover and Deception.

801 FROLIC; "available intelligence": Conferences With Foreign Planners, Second Draft, 25 Mar 48, in NOA 530, Folder Combined Planning, Cover and Deception. BRIDE- GROOM; "in the event": Operation Plan CINCPAC No. 12–51, last change date 27 Aug 52, in NOA 535, Folder Instructional Material, Plans. Inchon: Unsigned, un- dated lecture [Captain Frank M. Adamson, USN?], in NOA 537, Folder Misc C&D, Psychological Warfare.

801   Fritter resources: Unsigned Memorandum for Record, 9 Sep 47, para. 5b, in NOA 530, Folder Combined Planning, Cover and Deception. THUMBTACK, CLASSROOM, BOYHOOD: File No. JSC/A6-3, Serial 873, 18 Aug 48, in NARA RG 319 Entry 101 Box 4 Folder 93. "As time goes on": Eisenhower to Director, Plans and Operations Division, WDGS, 19 Jun 47, in JSC folder Letters From World War II Commanders Re Value of Cover and Deception.

802   Holsopple treatise: Holsopple to Thurber with attached Notes for Deception Planners—Re Communications, 29 Aug 45, in NOA 543, Folder Radio Deception. HOARDING: NARA RG 218 (1948–50) Folder CCS 385 (1-23-48) Sec. 1, Re Improving Cover and Deception Capability (1950). SHOCKER: Wise, Cassidy's War.

802   HUMIDOR: Shultz 139–57. All agents doubled: Shultz 90. FORAE: Shultz 114–20, 124–27.

803   ELDEST SON: Shultz 158–60. June 1971 conference: Stratagem Conference. Occupied country: Ingersoll to Eldredge, 18 Nov 71, in Ingersoll papers, unmarked box.

804   Sent to Casey: Handwritten notation on Hesketh to Casey 1. Lessons Learned conference: Eldredge, Hoax.

805   Schwartzkopf: McCausland; Huber, Desert Storm 59. Joint Doctrine: Joint Chiefs of Staff, Joint Doctrine for Military Deception, Joint Pub 3-58, 31 May 1996.

Appendix I: Operations

A-R: I/19–21, 24, 25, 42; Howard 35. ABAFT: I/12. ABEAM: Wingate 87–88; I/7–10, 24, 28. *ACCOLADE:* Wingate 212; III/167. ACCORDION: Wingate 34, 349–50; No. 2 Tac Report. ACCUMULATOR: G(R) War Diary June 1944 App. J; 21AGp/00/286/1/G(R), PRO WO 171/3831; Appendix F in JSC loose item "Report . . . Neptune July 1944"; Cruickshank, Deception 200–01; Cruickshank, Deception 200–01; Young & Stamp 46. ADORATION: NARA RG 319 Entry 101 Box 1 Folder 1 "Cover & Deception Synopsis of History," p. 7. ADVOCATE: Wingate 88–89; I/82–86, II/8; Howard 24–25, 39n. *ALACRITY:* Wingate 167–68. ALKEY: III/124. ANAGRAM: Wingate 89; II/2–3, 8. *ANAKIM:* Howard 211–12. APPENDIX Y: Wingate 248. ARAMINTA: D Division Weekly Progress Report No. 8, 15–21 May 1945, in JSC Folder Deception—SE Asia—WWII Historical D. Div—SACSEA—24 Mar 45–4 Jun 45. ARDENNES: 12th Army Group Report, in JSC folder Wedlock, pp. 42–49. ASSASSIN: PRO ADM 223/794/338–39. AVENGER: Wingate 344, 346; Howard 198. AVENGER II: Wingate 344; Howard 198; No. 2 Tac Report Part II 7. AXTELL: Wingate 332–33; III/22–24. BALDERDASH: PRO ADM 223/794/289–90, 305–306. BAMBINO: Herbig 279–81. BARCLAY: Wingate 73, 76, 173–83; III/58, 62–105, 109–14, 183; Howard 85–98; Cruickshank, Deception 51–54; CAB 119/66/81–133. BARDSTOWN: Wingate 331–32; III/21–22. BARONESS: PRO CAB 154/74. BASSINGTON: Fleming 35. BASTION: Wingate 89; II/8–18, 22, 26, 50; Howard 40. BERTRAM: Wingate 116–121; II/108–115; Howard 65–67; Cruickshank, Deception 26–27; Young & Stamp 73–74, 149–50. BETTEMBOURG: Kronman 26–27; 12th Army Group Report, in JSC folder Wedlock, pp. 22–24. BIGDRUM: Wingate 260. BIJOU: Wingate 316–18; III/4–6, 10; Mure, Master 145–48; PRO ADM 223/794/132–34. BLANDINGS: Chart, Principal Plans . . . , PRO WO 203/3313; 3/27/44 LCS Weekly Letter No. 5, in JSC folder 370.2 Operations and Results. BLARNEY STONE: PRO CAB 154/42; Wingate 209. BLAST: LCS Summary of Information No. 2, in JSC folder 370.2 Operations and Results. BLUE BOOT PLAN: Liddell diary 23 Jan 41, PRO KV 4/187. BLUEBIRD: Herbig 281, 285, 287; Graves to Aide to Cominch, Subject: Cover Plans . . . , 30 Dec 44, in NARA RG 218 Box 688 Folder CCS 385 Pacific Ocean Areas (12-1-44) Cover Plans for Detachment, Iceberg, and Subsequent Operations in the Pacific Ocean Areas During 1945. BOARDMAN: Wingate 73, 189–92, 211; III/44, 141–50, 161; Howard 92–94; PRO CAB 119/66/78–80. BODEGA: Hinsley & Simkins 225. BODYGUARD: Wingate 230–38, 291–92; Howard *passim.* BOOTHBY: Wingate 73, 190, 192–93; III/148–49, 155–56;

REFERENCES

Howard 94. Bouzonville: Kronman 37–38; 12th Army Group Report, in JSC folder Wedlock, p. 77. Braintree: IV/118. Brest: Kronman 25–26; 12th Army Group Report, in JSC folder Wedlock, pp. 18–19. Brian: Cruickshank, Deception 30; Young & Stamp 73. Brittany: Kronman 22–23; 12th Army Group Report, in JSC folder Wedlock, p. 13. Broadarrow: Wingate 322–323; III/7–9. Broadaxe: General Directive for Deception Measures Against Japan, CCS 84/16/D, 16 Jun 45, in NARA RG 218 Box 694 Folder CCS 385 Pacific Theater (4-1-43) Deceptive Policy in the Pacific Theater (Sec. 5). Broadstone: Wingate 22–23; III/171–72. Broadsword: LCS Summary of Information No. 2, in JSC folder 370.2 Operations and Results. Brock: Masterman 126. Brunette: Wingate 323–25; III/9–12. Bruteforce: D Division Weekly Progress Reports Nos. 1, 2, 3, in JSC folder Deception—SE Asia—WWII Historical D. Div—SACSEA—24 Mar 45–4 Jun 45. Bunbury: Masterman 133. Buxom: Attachment to Strong and Train to Secretariat, JCS, 6 Feb 43, in JSC folder 370.2 Operations and Results. Cadboll: Hesketh memo, PRO WO 219/2221/2. Cadboll II: Hesketh memo, PRO WO 219/2221/2. Callboy: Howard 198. Camillia: I/3–5, 92, II/4; Howard 34–35. Canute: II/126; CAB 119/66/243. Canwell: Wingate 120; II/114. Caption: D Division Weekly Progress Report No. 8, 15–21 May 1945, in JSC Folder Deception—SE Asia—WWII Historical D. Div—SACSEA—24 Mar 45–4 Jun 45. Carrot I: Appendix "B" in JSC Loose Item Report . . . Neptune July 1944. I: Appendix "B" in JSC Loose Item Report . . . Neptune July 1944. Carrot II: Appendix "B" in JSC Loose Item Report . . . Neptune July 1944. Carnegie: Wingate 73, 193–94; III/156–57. Carter Paterson: PRO ADM 223/794/146–47. Casanova: Kronman 29; 12th Army Group Report, in JSC folder Wedlock, pp. 32–38. Cascade: Wingate 90–93, 263; II/34–44, 41–42, 52, 117, 125, 137, III/58, IV/2, 4, 37; Howard 42–44, 64, 93, 96, 137, 150; Cruickshank, Deception 146, 156. Catspaw: PRO CAB 154/42; Wingate 209. Cattle: D Division Weekly Progress Report No. 5 21–27 Apr 45, in JSC folder Deception—SE Asia—WWII Historical D. Div—SACSEA—24 Mar 45–4 Jun 45. Cavendish: Wingate 100; Howard 57. Chatter: PRO CAB 154/42. Cent: Hesketh 60, 190; Appendix "C" in JSC loose item Report . . . Neptune July 1944, p. 2. Chesterfield: Wingate 223; Report . . . Neptune July 1944, p. 2. Chesterfield: Wingate 223, 300; III/172–73. Chettyford: Wingate 219, 300–302; IV/13, 19–21; Howard 138–39, 159–60. Christmas Cake: Wingate 209; PRO CAB 154/42. Clairvale: Wingate 302–304; IV/1, 21–23. Claw: Fleming 17. Clearance: D Division Weekly Progress Report Nos. 3, 5, in JSC folder Deception—SE Asia—WWII Historical D. Div—SACSEA—24 Mar 45–4 Jun 45. Cloak: Howard 205, 218–19; Fleming 66–68; Young & Stamp 227–29; Chart, Principal Plans . . . , PRO WO 203/3313; "Subject Deception Plan Conclave," with attachment, in JSC folder Southeast Asia. Cockade: Wingate 146–50; Howard 71–83, 109, 112, 233–34; Cruickshank, Deception 61, 86, 215, 221. Colfax: Wingate 73, 193–94; III/158–59. Collect: Wingate 88; I/40–42, 48, 49, 60–62, 92. Compass: Clarke V/22–23. Conclave: Subject Deception Plan Conclave with attachments; From Control Section with attachments; Amendment No. 1 to 14th Army Operation "Conclave" with attachments, all in JSC folder Southeast Asia; D Division Weekly News Letter No. 48, Weekly Progress Report Nos. 1, 2, 4, all in JSC folder Deception—SE Asia—WWII Historical D. Div—SACSEA—30 Dec 44–23 Mar 45; O'Brien 306. Conscious: Cover and Deception Plan "Conscious," 13 July 1945, pp. 3–4, in NARA RG 319 Entry 101 Box 7 Folder 4, Operation 13 July 1945, pp. 3–4, in NARA RG 319 Entry 101 Box 7 Folder 4, Operation Zipper and Cover and Deception Plan Conscious. Copperhead: Wingate 259; IV/69, 85–90; Howard 125–27, 151; Cruickshank, Deception 96–98. Coppers: I/37–39, 48, II/19. Cordite Cover Plan: I/15–16, 22, 23. Cowper: Wingate·171; III/36–37. Crystal: II/29–31. Cyprus Defense: Wingate 87–88; I/31–33, 92. Dallas: Kronman 28; 12th Army Group Report, in JSC folder Wedlock, pp. 32–36. Dante: D Division Weekly Progress Report No. 19, in JSC folder Deception—SE Asia—WWII Historical D. Div. SACSEA 5 Jun–10 Sep 45. Defraud: D Division Weekly Progress Report No. 5 21–27 Apr 45, in JSC folder Deception—SE Asia—WWII Historical D. Div—SACSEA—

24 Mar 45–4 Jun 45. DERRICK: Wingate 73, 183–88; III/68, 105–108, 154. DERVISH: Wingate 344, Howard 198, Plan "Dervish" in JSC folder Dervish. DIAMOND: Cruickshank, Deception 30. DIJON: 12th Army Group Report, in JSC folder Wedlock, pp. 20–21. DOLLAR: Hesketh 60, 190; Appendix "C" in JSC loose item Report . . . Neptune July 1944 p. 2. DOMESDAY: LCS Summary of item Report . . . Neptune July 1944 p. 2. DOMESDAY: LCS Summary of Information No. 2, in JSC folder 370.2 Operations and Results. DRINK: CAB 119/66/207. DRYSHOD: Hesketh 190. DUMMER: Wingate 162–63; PRO CAB 81/77/63–68. DUNDAS: Wingate 140–45, 250; Howard 75–76. DUNDEE: Wingate 215, 316, 319–20; III/177, IV/1–5; Smith to Strong. 23 Nov 43, in folder 000—To 312.1(1) China-Burma-India Theater; CAB 119/66/2–5, 81/77/24–25. DUNGLOE: Wingate 273–74; IV/48–49. DUNTON: PRO CAB 154/56; Plan BODYGUARD report on position at 5/17/44, in JSC folder 370.2 Operations and Results. E.S.: PRO ADM 223/794/123, 291, 306–311. ELEPHANT: Kronman 20–24; 12th Army Group Report, in JSC folder Wedlock, p. 12. 11TH PANZER DIVISION: 12th Army Group Report, in JSC folder Wedlock, pp. 67–73. ELSENBORN: Kronman 28–29. ENVY: Enclosure, Hunter to Smith, 29 Apr 44, in JSC folder 000—To 312.1(1) China-Burma-India Theater. ERASMUS: Wingate 329–31; III/19–21. ERROR: Howard 203–204; Fleming papers. EXPLOIT: Eldredge papers. EUPHRATES PLAN: I/44–46, 48. EXPORTER COVER PLAN: I/29–30. *FABIUS:* Hesketh 60. FABRIC: Wingate 89–90; II/50–59, 62–66. FAIRLANDS: Wingate 73, 210–13; III/161, 164–69; Howard 96–97, 138. FALSE ARMISTICE PROJECT: Wingate 167–68; III/108–109. FANG: D Division Weekly Progress Report No. 8, 15–21 May 1945, in JSC Folder Deception—SE Asia—WWII Historical D. Div—SACSEA—24 Mar 45–4 Jun 45. FANTO: PRO CAB 154/34. FATHEAD: Fleming 59–60. FERDINAND: Wingate 254, 285, 293–99; IV/78, 103, 109–120; Howard 155–59; Cruickshank, Deception 96, 165–68; No. 2 Tac Report. FERNBANK: Wingate 327–29; III/14–15; Howard 226–27; PRO ADM 223/794/135–36, 208–211. FILM STAR: Appendix "D" p. 6 in JSC loose item "Report . . . Neptune July 1944." FIREFLAME: D Division Weekly News Letter No. 44, 17 Feb–23 Feb 1945, in JSC folder Deception—SE Asia—WWII Historical D. Div—SACSEA—30 Dec 44–23 Mar 45. FLABBERGAST: Chart, Principal Plans . . . , PRO CAB WO 203/3313; O'Connor to JSC, HMOC/1931, with enclosure, in JSC folder Misc Deception Plans—WWII. FLATON: III/154. FLAXWEILER: Kronman 34; 12th Army Group Report, in JSC folder Wedlock, pp. 62–63. FLOUNDERS: Cook to Hayes, 29 Aug 1942, NARA RG 218 Box 363 Folder CCS 385 (12-31-42) Deception Policy. FORFAR: Cruickshank, Deception 68; Young & Stamp 75–76. FORTITUDE: Hesketh *passim.* FOYNES: Wingate 74, 202, 203–208; III/177, IV/1–5; Howard 112–15, 136, 151; PRO CAB 119/66/1–38. FRACTURE: III/106; Action Report Western Naval Task Force The Sicilian Campaign, in JSC loose papers. FRANK: Cover & Deception Report ETO Exhibit 7, Technical Signal Report on Fortitude, in NARA RG 319 Entry 101 Box 1 Folder 13 pp. 14–15. GARFIELD: Wingate 222; III/170. GARGOYLE: II/126; PRO CAB 119/66/243. GIBRALTAR COVER PLAN: Wingate 104–105. GILMERTON: Wingate 209; PRO CAB 154/42; PRO CAB 81/77/4–5; PRO ADM 223/794/13–14. GLENANNE: PRO CAB 154/58/51A. GLENDON: 3/19/44 LCS Weekly Letter No. 4, 19 Mar 44, in JSC folder 370.2 Operations and Results. GLIMMER: Wingate 260; Cruickshank, Deception 199; Young & Stamp 169. GLOSSOP: PRO CAB 121/105/197, 226, 228, 231. GOLDLEAF-HERITAGE: Wingate 85–87; II/60–61. GOTHAM: Wingate 316, 333–34; IV/5. GOWRIE: PRO CAB 154/42; buck slip N. S[mith] to Record and attached buck slips, in JSC folder European Theater 000. GRAFFHAM: Wingate 242–46, 252; Howard 117–19, 126; Cruickshank, Deception 123, 125–26, 219; Hesketh 40–41, 82–83. GRANDIOSE: Wingate 115; II/99–100, 105, 124, 138; Howard 64. GREMLIN: D Division Weekly Progress Report No. 13 19–25 Jun 45, p. 3, in JSC folder Deception—SE Asia—WWII Historical D. Div. SACSEA 5 Jun–10 Sep 45; Head of D Division to London Controlling Section, 3 Feb 45, Subject: Deceptive Broadcasts, in JSC folder Deception—SE Asia—WWII Historical D. Div—SACSEA—16 Dec 44–15 Mar 45. GRIPFIX: I/88–92. GUY FAWKES: Masterman 88. H.D.F.:

II/32–33. HAIRCUT: Cover & Deception Report ETO Exhibit 7, Technical Signal Report on Fortitude, in NARA RG 319 Entry 101 Box 1 Folder 13 p. 14; Appendix "C" p. 7, in JSC loose item "Report . . . Neptune July 1944." HALLUCINATE: Young & Stamp 46; "R" Force War Diary 18 Aug 44, PRO WO 171/3832. HARDBOILED: Wingate 83–85; Howard 23–24, 59. HARDMAN: Wingate 222; III/171. HARLEQUIN: Wingate 152, 156; Cruickshank, Deception 61–62. HARLEY STREET: "R" Force War Diary October 1944 App. J-10A, PRO WO 171/3832; Young & Stamp 46. HENGEIST: Wingate 171; III/35–36. HICCOUGHS: Fleming 36–38. (Erroneously described in Young & Stamp 221–23.) HONEYSUCKLE: Appendix "C" p. 7, in JSC loose item "Report . . . Neptune July 1944.". HOSTAGE: Wingate 170–71; III/34–35. HOSTAGE: Young & Stamp 92–93. HOSTAGE II: G(R) War Diary July 1944, App. J, 21AGp/00/286/1/G(R), PRO WO 171/3831; "R" Force War Diary July 1944 App. I, PRO WO 171/3832. HOTSTUFF: Resume of Torch and Associated Plans, in JSC folder Conscious. HOUSEKEEPER: "R" Force War Diary 25–26 Oct 44 and App. J-11, PRO WO 171/3832. HUSBAND: Herbig 279–81. IAGO: PRO ADM 223/794/145–46; Montagu, Beyond 130–31. ICEMAN: Shaw to Joint Security Control, Subject: Outline Plan, Operation Iceman, 12 May 45, in NOA 550, Folder Plan—Iceman. IMPERIL: G(R) War Diary October 1944, App. J-3, 9 Oct 44, PRO WAR 171/3831; "R" Force War Diary 28 Sep 44, October 1944 passim, 5 Nov 44, App. D; "R" Force War Diary 5 Nov 44, PRO WO 171/3832. INCLINATION: "R" Force War Diary 18–19 Oct 44 and App. J-12, J-12A, PRO WO 171/3832. INFATUATE: "R" Force War Diary 31 Oct 44, PRO WO 171/3832. IRAN COVER PLAN: I/47. IRONSIDE: Wingate 254; Howard 125; Hesketh 53–54, 99. IRONSIDE II: Wingate 295; IV/112; Howard 156, 191; Hesketh 138. JAEL: Wingate 199–201, 230; Howard 104–105, 113; PRO CAB 119/66/6–19. JESSICA: Wingate 34, 346–48; Howard 198; No. 2 Tac Report. JIGSAW: Attachment Cook to Col. Pierce, 6 Jan 43, in JSC folder 370.2 Operations and Results; PRO CAB 119/66/243, 227. JITTERBUG: Cover & Deception Report ETO Exhibit 7, Technical Signal Report on Fortitude, in NARA RG 319 Entry 101 Box 1 Folder 13 p. 14; Appendix "C" p. 7, in JSC loose item "Report . . . Neptune July 1944." JONAH: D Division Weekly News Letter No. 41 27 Jan-2 Feb 45, in JSC folder Deception—SE Asia—WWII Historical D. Div—SACSEA—30 Dec 44–23 Mar 45. JURISDICTION: PRO WO 219/2255. K-SHELL: I/11. KENNECOTT: Wingate 98, 106–110; Howard 59–63. KINKAJOU: Wavell to COS, 17 Feb 43, 38748/COS, NARA RG 319 Entry 101 Box 3 Folder 68 Anglo-American Meeting (May-June '43); Howard 211–12. KNIFEDGE: Wingate 348; Howard 198; No. 2 Tac Report Part II, pp. 13–14, NARA RG 319 Entry 101 Box 4 Folder 80. KNOCKOUT: D Division Weekly Progress Report No. 1, 16–31 Dec 1944, in folder Deception—SE Asia—WWII Historical D. Div—SACSEA—30 Dec 44-23 Mar 45. KOBLENTZ: Kronman 30–31; 12th Army Group Report, in JSC folder Wedlock, pp. 40–41. KODAK: Kronman 31–32; 12th Army Group Report, in JSC folder Wedlock, pp. 43, 45–46. LANDONVILLERS: Kronman 35; 12th Army Group Report, in JSC folder Wedlock, pp. 65–66. LANGTOFT: Wingate 158–60, 209; PRO ADM 223/794/340; Chemical Warfare Plan "Langtoft," in JSC folder 370.2 Operations and Results. LARKHALL: Wingate 138–40, 143, 202; Howard 75–76. LEAPYEAR: JSC folder Bluebird. LEEK I: Appendix "B," in JSC loose item "Report . . . Neptune July 1944," p. 4. LEEK II: Appendix "B," in JSC loose item "Report . . . Neptune July 1944," p. 4. L'ÉGLISE: Kronman 33–34. LEMSFORD: PRO CAB 81/77/26. LEYBURN: Wingate 163. LOADER: Wingate 161. LOADLINE: Wingate 34; No. 2 Tac Report Part III, in NARA RG 319 Entry 101 Box 4 Folder 81, p. 29. LOADSTAR: Cake Before Breakfast, in JSC loose papers. LOBSTER: Summary Report of Results of Plan Lobster; Display Lighting . . . , both in JSC folder 370.2 Operations and Results. LOCHIEL: Wingate 170–71; III/32–34. LOCHIEL II: III/34. LOCHINVAR: Kronman 36–37; 12th Army Group Report, in JSC folder Wedlock, p. 76. MACHIAVELLI: Masterman 83; Howard 17. MAIDEN: II/59. MAINSTAY: "R" Force War Diary 27 Aug 44, PRO WO 171/3832. MALICE: LCS Weekly Letter No. 5, 27 Mar 44, in JSC folder 370.2 Operations and Results; Head of D Division to London Controlling Section, 3 Feb 45, Subject: Deceptive Broadcasts, pp. 11-A, 12-A, in JSC folder Deception—SE Asia—WWII Historical D. Div—SACSEA—16 Dec 44–15 Mar 45. Malinger: Chart, Principal Plans . . . , PRO WO 203/3313. Marchioness: PRO CAB 154/74. Marécage: Larminat to Commanding General 6th Army Group with enclosure, No. 1178 EM/3, 31 Mar 45, PRO WO 219/2249/115–22. Martello: Cruickshank, Deception 31; Young & Stamp 73. Melting Pot: II/84, 113; Cruickshank, Deception 32; Young & Stamp 73. Mercury: Hesketh 110–111. Merzig: Kronman 36; 12th Army Group Report, in JSC folder Wedlock, pp. 71–73. Metz I: Kronman 32; 12th Army Group Report, in JSC folder Wedlock, pp. 48–49. Metz II: Kronman 32–33; 12th Army Group Report, in JSC folder Wedlock, pp. 60–61. Midas: West, Seven 14; PRO ADM 223/794/9. Mimsy: PRO CAB 154/39. Minotaur: PRO KV 4/197/12. Mincemeat: Wingate 177–79; III/68, 75–78, 99; Howard 89–92, 206, 245–46; Montagu, Man. Mulliner: PRO WO 203/5746; Chart, Principal Plans . . . PRO WO 203/3313. Munassib: Cruickshank, Deception 30–31; Young & Stamp 74. Murrayfield: Cruickshank, Deception 31–32; Young & Stamp 73. Mustache: Cover & Deception Report ETO Exhibit 7, Technical Signal Report on Fortitude, in NARA RG 319 Entry 101 Box 1 Folder 13 pp. 14–15; Appendix C, p. 6, in JSC loose item "Report . . . Neptune July 1944." Nan: Cover & Deception Report ETO Exhibit 7, Technical Signal Report on Fortitude, in NARA RG 319 Entry 101 Box 1 Folder 13 pp. 14–15; Appendix C, p. 6, in JSC loose item "Report . . . Neptune July 1944." Nettle: PRO WO 203/5759. Newspaper Plan: PRO CAB 154/42; Wingate 209. Northways: Wingate 304; IV/23–24. Nunton: Wingate 304–308; IV/50–55; Howard 161–62. Oakfield: Wingate 213–21, 270, 300; III/169, 171, 177, IV/5, 6–15, 31; Howard 97, 138; PRO CAB 119/66/20–31. Oakleaf: Clarke V/9–10, 14, 17, 20. Omega: 12th Army Group Report, in JSC folder Wedlock, pp. 81–84. Omnibus: Masterman 86; PRO ADM 223/794/335; PRO KV 4/197/12, p. 1a. Onion I: Appendix "B," in JSC loose item Report . . . Neptune July 1944 p. 2. Onion II: Appendix "B," in JSC loose item Report . . . Neptune July 1944, p 3. Ottrington: Wingate 308–10; IV/103–105, 117; Howard 156, 162–63. Overthrow: Wingate 98, 100–103, 110; Howard 57–58, 61–63 Padlock: Appendix "A," 14 Sep 42, in NARA RG 218 Box 693 Folder CCS 385 Pacific Theater (4-1-43) Deceptive Policy in the Pacific Theater (Sec. 2). Paprika: Masterman 84–85, 125. Paradise: AFHRA Folder 505-61-21. Passover: Cruickshank 35–36; PRO CAB 121/105/493. Pastel, Pastel Two: GHQ USAFPAC, Staff Study "Pastel-Two," Cover and Deception Olympic Operations, 30 Jul 45, in NOA 550, Folder Pastel Two. Patent: Wingate 111–12; II/66–67, 77. Pender I: Wingate 98, 105–106. Pender II: Wingate 106. Penknife: Clarke V/12–16, 20–21, 22. Pepper: Masterman 84. Pippin: Report on Signal Deception Measures Adopted During Operations of the Fourteenth Army Dec 44–Mar 45, in JSC folder 3-12/5 Communication Decep: RPTS-India Burma. Plan A.B.: PRO KV 4/197/12. Plan I: PRO ADM 223/794/149–51; Masterman 82–83. Plan IV: Masterman 83–84, 93; Howard 17. Playboy-Compass: Wingate 326–27; III/14–15. Playmate: Clarke V/17–19, 22. Pound: Cruickshank, Deception 84. Premium: Masterman 159–60; Hesketh 40, 55, 92n. Purple Whales: Wingate 128; Howard 205–206; Fleming 33–36. Quaker: Wingate 259. Quickfire: Deception Operations for Torch, 25 Jan 46, in JSC folder Conscious. Quicksilver I through VI: AFHRA Folder 505-61-21. Raindrop: G(R) War Diary June 1944 App. J, 21AGp/00/286/G(R), 7 Jun 44, PRO WAR 171/3831; Young & Stamp 46. Ramshorn: Howard 214; PRO CAB 121/105/260–279; PRO CAB 119/66/39–53; Chart, Principal Plans . . . , PRO CAB WO 203/3313. Rasputin: PRO CAB 119/66/145–50. Rassendyll: Fleming 26. Rayon: Clarke V/22. Red Bar: Deception on Naval Matters, in NOA 537, Folder N.I.D. (British Naval Intelligence Division) Admiralty Reports on Naval Deception. Remus: D Division Weekly Progress Report No. 1, 16–31 Dec 1944, in Folder Deception—SE Asia—WWII Historical D. Div—SACSEA—30 Dec 44–23 Mar 45. Rhino: Fleming. Rissole: PRO CAB 154/42; Wingate 209. Rosebud: Technical Signal Report on Fortitude; Report of Special Signal Operations in the United Kingdom, in NARA RG 319 Entry 101 Box 1 Folders 13, 64. Ritz: PRO CAB 154/42. Rooftree, Roscor: Chart, Principal . . . , PRO WO 203/3313. Rou-

BLE: Technical Signal Report on Fortitude, in NARA RG 319 Entry 101 Box 1 Folder 13. ROYAL FLUSH: Wingate 239–42, 283; IV/68, 69–70, 74; Howard 126, 151–52. RUHR POCKET: 12th Army Group Report, in JSC folder Wedlock, pp. 78–80. RUSTIC: Clarke V/8. RUTHLESS: Cruickshank, Deception 13–14. "S" FORCE: III/34, 38–39, 41. SCEPTICAL: Chart, Principal Plans . . . , PRO WO 203/3313, PRO CAB 121/105/106–107. SECOND UNDERCUT: Wingate 314–15; IV/130–31. SEE SAW: Technical Signal Report on Fortitude; Report of Special Signal Operations in the United Kingdom, in NARA RG 319 Entry 101 Box 1 Folders 13, 64. SENTINEL: Wingate 113–15; II/80–82. SHELLAC: Wingate 312; IV/134. SHOTGUN: PRO CAB 154/68; Wingate 209. SKYE: Hesketh memo, PRO WO 219/2221/2; Macleod to Jervis-Read, 12 Mar 44, PRO WO 219/2221/6; Cruickshank, Deception 104–105. SLIPPERY: D Division Weekly Progress Report Nos. 12, 19, in JSC folder Deception—SE Asia—WWII Historical D. Div. SACSEA 5 Jun–10 Sep 45. SLY-BOB: Wingate 325–26; III/12–14. SNOWSHOE II: Clarke V/22. SOCIETY: Wingate 115–16; II/101. SOLO I: Wingate 98, 103–104; Howard 59–63. SOLO II: Wingate 99; Howard 58–63. SPAN-NER: PRO ADM 223/794/338–39. SPAM (Channel): PRO CAB 154/42. SPAM (Italy): Navy memo. STAB: Appendix "A," 14 Sep 42, in NARA RG 218 Box 693 Folder CCS 385 Pacific Theater (4-1-43) Deceptive Policy in the Pacific Theater (Sec. 2). STARKEY: Wingate 146, 148, 149–53, 339–40; Howard 75, 81–82, 103, 112, 233–34. STEINSEL: Kronman 34–35; 12th Army Group Report, in JSC folder Wedlock, p. 64. STENCH: Minutes 2 Oct 41, PRO KV 4/64/88; Liddell diary 4 Oct 41, PRO KV 4/188; PRO KV 4/197/12 p. 3. STENCIL: Report on Signal Deception Measures Adopted During Operations of the Fourteenth Army Dec 44–Mar 45, in JSC folder 3-12/5 Communication Decep: RPTS-India Burma. STEP-PINGSTONE: Cruickshank, Deception 36. STIFF: PRO ADM 223/794/337–38, 435; Master-man 83. STULTIFY: Plan STULTIFY, Report by D Division, 21 Nov 44, SAC (44) 13 (O) Revised, in JSC folder Misc Deception Plans—WWII; D Division Progress Report No. 1 16–31 Dec 44, in JSC folder Deception—SE Asia—WWII Historical D. Div—SACSEA—16 Dec 44–15 Mar 45; D Division Weekly Progress Report No. 8, 15–21 May 1945, in Folder Deception—SE Asia—WWII Historical D. Div—SACSEA—24 Mar 45–4 Jun 45; Fleming 17; Chart, Principal Plans . . . , PRO WO 203/33/3. STURGEON: Unsigned memo-randum, p. 3, Operation Sturgeon, 17 Jan 46, in JSC folder Bluebird. SWEATER: Wingate 60, 99; Resume of TORCH and Associated Plans, OP-30C-OUT-214, p. 8, in JSC folder Con-scious. TAPER: Wingate 351; Howard 198. TARZAN: D Division Weekly Progress Report No. 1, 24–30 March 1945, in JSC Folder Deception—SE Asia—WWII Historical D. Div—SACSEA—24 Mar 45–4 Jun 45. TAXABLE: Wingate 260; AFHRA Folder 505-61-21; Cruickshank, Deception 199; Young & Stamp 169. THERAPY: PRO KV 4/197/12. TELE-SCOPE: IV/85–90. 10TH ARMORED DIVISION: Wingate 87–88; I/11. TINDALL: Wingate 146, 148, 151–53, 156, 339–40; Howard 75, 77, 81–82, 103, 112; Cruickshank, Deception 74–81, 104; PRO ADM 223/794/143. TITANIC I through IV: Wingate 260–61; AFHRA Folder 505-61-21; Young & Stamp 77–82; Cruickshank, Deception 197–98. TORRENT: COSSAC/00/6/3/Ops, Subject: Operation "TORRENT," 6 Sep 43, in NARA RG 331 Entry 1 Box 73 Folder 381 Fortitude Vol. 1. TOWNSMAN: II/102–107, 124. TRANSATLANTIC TELEPHONE PLAN: PRO CAB 154/42; Wingate 204. TRANSCEND: Wingate 351; Harmer 14, 20, 32, 33. TREATMENT: Wingate 116, 121; II/108–115, 116, 124; Howard 65–67. TRIPOLI PLAN: I/42–43. TROJAN HORSE: PRO CAB 154/67/242. TROLLEYCAR: Wingate 351; Hes-keth 154, 159–61; Harmer 14, 32–33; Montgomery to SHAEF, 21AGp/00/27569/G(R), 5 Nov 44, G(R) War Diary November 1944 App. J-2; PRO WAR 171/3831. TROUSERS: "R" Force War Diary August 1944 App. J-2, PRO WO 171/3832; Young & Stamp 46. TROUTFLY: 12th Army Group Report, in JSC folder Wedlock. TURBINATE: Wingate 351; Harmer 14, 32, 33; PRO WO 219/2247/3–4, 55–64. TURPITUDE: Wingate 279, 286–90; IV/69, 80–85; Howard 152–55; Naftali 613. TWEEZER: Hesketh 107–110. TWELVEBORE: G(R) War Diary 23 Nov 44, PRO WAR 171/3831. UKRIDGE: Chart, Principal Plans . . . , PRO WO 203/3313; Extract From Weekly Letter No. 5 of 27 March, in JSC folder 370.2 Operations and Results; Grafton to Hunter and Fleming, Subject: Report on Tour, 21 Apr 44, enclosure to Hunter to Smith, Subject: Trip—Lieut. Cornelius W. Grafton, 29 Apr 44, in JSC folder 000—To 312.1(1) China-Burma-India Theater. ULSTER: Wingate 310–12; IV/105–108; Cruickshank, Deception 202–204. UNDERCUT: Wingate 313–14; IV/128–30. UNNAMED PLAN (BATTLEAXE): I/30. UNNAMED LCS PLANS (documents, Fire Offices) Wingate 209. UNNAMED PLAN (Libya): I/1. UNNAMED LCS PLAN (Ploesti): PRO CAB 119/66/195–200. UNNAMED PLAN *(VAULT)*: PRO CAB 119/66/76–77. UPSHOT: LCS Summary of Information No. 2, in JSC folder 370.2 Operations and Results. VALENTINE: Herbig 281. VAN DYKE: Technical Signal Report on Fortitude; Report of Special Signal Operations in the United Kingdom, in NARA RG 319 Entry 101 Box 1 Folders 13, 64. VANITY I: Appendix "C" p. 7, in JSC loose item "Report . . . Nepture July 1944." VANITY II: Appendix "C," p. 7, in JSC loose item "Report . . . Nepture July 1944." VENDETTA: NARA RG 319 Entry 101 Box 4 Folder 73 "Vendetta"; Wingate 241–42, 279–86; IV/68, 71–80; Howard 148–52, 155–56; Hesketh 52–53. VERITAS: Hesketh 36. VIERSEN: Kronman 38–41. VIOLA I: Cover & Deception Report ETO Exhibit 7, Technical Signal Report on Fortitude, in NARA RG 319 Entry 101 Box 1 Folder 13 p. 14; Appendix "C" p. 7, in JSC loose item "Report . . . Nepture July 1944." VIOLA II: Cover & Deception Report ETO Exhibit 7, Technical Signal Report on Fortitude, in NARA RG 319 Entry 101 Box 1 Folder 13 p. 14; Appendix "C" p. 7, in JSC loose item Report . . . Nepture July 1944. VIOLA III: Cover & Deception Report ETO Exhibit 7, Technical Signal Report on Fortitude, in NARA RG 319 Entry 101 Box 1 Folder 13 p. 14; Appendix "C" p. 6, in JSC loose item Report . . . Nepture July 1944. WADHAM: Operation "Wadham," Appreciation and Outline Plan, COSSAC (43) 24, 15 Jun 43, in-JSC folder 370.2 Wadham; Wingate 146, 148, 149–53, 339–40; Howard 75, 81–82, 103. WALLABY: Wavell to COS, 17 Feb 43, 38748/COS, in NARA RG 319 Entry 101 Box 3 Folder 68 Anglo-American Meeting (May–June '43); Howard 211–12; PRO CAB 119/66/160–194. WANTAGE: Wingate 263–66; II/42, 44, IV/4, 32, 38–42, 67, 108, V/4–7; Howard 137–38, 150. WAR OF NERVES: I/12. WAREHOUSE: Wingate 122, 174; II/100, 124–27, 138, III/27, 57–58, 63–64; Howard 85–86. WARHORSE: G(R) War Diary 28 Nov 44, 2 Dec 44, PRO WAR 171/3831. WATERFALL: Wingate 180–81; III/68, 87–89; Howard 88. WEDLOCK: Deceptive Measures Against Japan, Alaskan Department, JCS 705, 13 Feb 44, in NARA RG 218 Box 693 Folder CCS 385 Pacific Theater (4-1-43) Deceptive Policy in the Pacific Theater (Sec. 2). WETSHOD: Hesketh 190. WHIPSAW: Kronman 35–36; 12th Army Group Report, in JSC folder Wedlock, 68–71. WHITWOOD: Wingate 224; III/176. WILDERNESS: PRO CAS 154/103/21–22. WILLIAMS: Technical Signal Report on Fortitude; Report of Special Signal Operations in the United Kingdom, in NARA RG 319 Entry 101 Box 1 Folders 13, 64. WILTZ: Kronman 27–28; 12th Army Group Report, in JSC folder Wedlock, pp. 28–29. WINDSCREEN: Wingate 123–24, 170; II/126–31, III/33. WITHSTAND: Wingate 173; III/26–29; Howard 85–86; PRO CAB 119/66/53 ff. WOLF: D Division Weekly Progress Report No. 1, 16–31 Dec 1944, in Folder Deception—SE Asia—WWII Historical D. Div—SACSEA—30 Dec 44–23 Mar 45. WORKHOUSE: Wingate 321–22; III/2–3. WYANDOTTE: Wingate 316, 318; III/6–7. XMAS PLAN II: I/82. ZEPPELIN: Wingate 239, 263–73; IV/5, 29–37, 39, 40, 43–48, 57–69, 93–97; Howard 147–55 and *passim;* Cruickshank, Deception 146–55.

Appendix II: Channels

(UNIDENTIFIED): II/54–55, 111, III/75, IV/15. ABDEL: Mure, Practise 146, 149, 253. ACCOST: D Division Weekly Progress Report No. 20 7–13 Aug 45, pp. 8–9, JSC folder Deception—SE Asia—WWII Historical D. Div. SACSEA 5 Jun–10 Sep 45. ADDICT: Wingate 41; IV/100; Naftali 628, 648 n. 103. AGENT J1: Harris 293–94; Howard 231–32. AGENT J2: Harris 295; Howard 232; Hinsley & Simkins 115; Pujol 97. AGENT J3: Harris 296; Howard 232; Hinsley & Simkins 227, 240. AGENT J4: Harris 297; Howard 232. AGENT J5: Harris

298; Howard 232. AGENT ONE: Harris 299–300; Howard 232; Hinsley & Simkins 113, 225–27. AGENT TWO: Harris 301; Howard 232; Hinsley & Simkins 113, 115, 225. AGENT 2(1): Harris 302; Howard 232. AGENT THREE: Harris 303; Howard 232, 233, 234, 236; Hinsley & Simkins 113, 117, 243, 275. AGENT 3(1): Harris 304; Howard 232. AGENT 3(2): Howard 232; Hinsley & Simkins 227, 240. AGENT 3(3): Harris 306–307; Howard 232, 233; Hinsley & Simkins 227. AGENT FOUR: Harris 308–309; Howard 232–33, 234; Hinsley & Simkins 115, 116, 226, 228, 240, 287. AGENT 4(1): Harris 310; Howard 233, Hesketh 73 n. 26. AGENT 4(2): Harris 311; Howard 233. AGENT 4(3): Harris 312; Howard 233; Hinsley & Simkins 227. AGENT FIVE: Harris 313–14; Howard 233. AGENT 5(1): Harris 315; Howard 233; Hinsley & Simkins 228; Pujol 113–14; memo 23 Jan 45 in brown Acco binder labeled "14." AGENT SIX: Harris 316–17; Hinsley & Simkins 115, 117, 225; Howard 233; Pujol 107. AGENT SEVEN: Harris 318–19; Howard 233, 234; Hinsley & Simkins 115, 225; Hesketh 50. AGENT 7(1): Harris 320; Hinsley & Simkins 223. AGENT 7(2): Harris 321; Hesketh 50–51; Hinsley & Simkins 227. AGENT 7(3): Harris 322; Fleming 22; Howard 208, 233. AGENT 7(4): Harris 323; Hinsley & Simkins 227; Howard 233; Hesketh 51. AGENT 7(5): Harris 324; Hinsley & Simkins 227, 239. AGENT 7(6): Harris 325; Hinsley & Simkins 227, 239; Hesketh 51. AGENT 7(7): Harris 326; Hinsley & Simkins 227, 239; Hesketh 51. AL- BERT: Wingate 32–33; III/128, 130, 131, 135; No. 2 Tac Report Part II 140, 142–48. AL- BERTO: JSC Brown Acco Binder "10" "B File." ALERT: Mure, Practise 14, 66, 89, 90, 122, 127, 140, 149, 150, 172, 174, 175–77, 181, 189, 197–98, 215, 220, 237, 234, 253; Mure, Master 112, 114. ALOSI: O'Connor to JSC, HMOC/2013, 24 Dec 44, in JSC Folder South America. ALMURA: Hesketh 37 n. 26. ANDRIES: D Division Weekly News Letter No. 39 13 Jan–19 Jan 45, Part II Paragraph 25, in JSC folder Deception—SE Asia—WWII Historical D. Div—SACSEA—30 Dec 44–23 Mar 45. ANDROS: PRO ADM 223/794/428. ANGEL ONE: Fleming 40–41; Howard 206. ANTHONY: Memo, Re CAMCASE, 23 Jan 45, in JSC Acco Binder Labeled "14." ANTOINE: Oakes 330. APOSTLE: Naftali 628; Naftali 648 n. 103. AP- PRENTICE: Wingate 40–41; IV/99; Howard 136, 141. ARBITER: IV/100; Naftali 627–28. ARMOUR: Wingate 39–40; IV/100–101 Howard 136, 141, 157; Naftali 628. ARTHUR (North Africa): III/120, 122; Plan To Build Up Special Agents for Coming Operations, ADV "A"F/21/3c, 4 Jul 43, Goldbranson papers. ARTHUR (France): Oakes 3?8. ARTIST: Master- man 139–40, 153–54, 178; Hinsley & Simkins 95, 104, 217, 221–24, 235, 238, 241, 264; West, Seven 24–31. ASPIRIN: JSC folder Portugal; JSC Acco Binder Labeled "14." ASSASSIN: Oakes 158, 160, 165, 171, 174, 176, 178, 181, 184, 185, 186, 189, 190, 191, 193, 328; 12th Army Group Report p. 80, in JSC folder Wedlock. ATLAS I, II, III: No. 2 Tac Report Part II, pp. 143, 148, 149, 152; Paillole, Services 495; III/134. ATOM: Oakes 267, 271–75, 276, 300, 329, 330, 333. AUDREY: Fleming 19; Howard 207. AXE: IV/101; Hinsley & Simkins 269n.; Naftali 648 n. 103. B.G.M.: II/75. BACKHAND: D Division Weekly Progress Report Nos. 7, 13, in JSC folder Deception—SE Asia—WWII Historical D. Div. SACSEA 5 Jun–10 Sep 45; Fleming 10. BALLOON: PRO KV 2/1070 through 1083; List of real names in 1995 Masterman reprint; Howard 17, 77, 223, 226, Masterman 78, 79, 86, 96, 109, 138, 140; Hinsley & Simkins 104, 113n., 123–24, 127, 219, 221, 235, 236, 248n.; PRO ADM 223/794/342–43; West, Seven 13–14. BARONESS: Notes on D/A Activities in the Middle East, PRO KV 4/197/1, pp. 14–15; Hinsley & Simkins 230. BASKET: List of real names in 1995 Masterman reprint; Masterman 99–100; Hinsley & Simkins 92, 194–95. BATES: Mes- sage No. 30 sent 26 Oct 44, in subfile "P Sent Traffic," in JSC file Traffic Sent; JSC Folder Portugal. BATS: Fleming 20–22; Howard 207. BEETLE: List of real names in 1995 Masterman reprint; Howard 115–16, 224; Masterman 114, 143; Hinsley & Simkins 193n., 217, 218, 240; PRO ADM 223/794/295–97. BENEDICT: Howard 62, 78, 232, 234, 236. BILL: D Di- vision Weekly News Letter No. 39 13 Jan–19 Jan 45, Part II Paragraph 25, in JSC folder De- ception—SE Asia—WWII Historical D. Div—SACSEA—30 Dec 44–23 Mar 45. BISCUIT: List of real names in 1995 Masterman reprint; Howard 4; Masterman 43–45, 50–51, 92; Hinsley & Simkins 87–88, 102, 103, 322, 323, 325. BLACKGUARD: Hinsley & Simkins 231;

PRO ADM 223/794/362, 363, 367; Notes on D/A Activities in the Middle East, PRO KV 4/197/1, p. 9 Blaze: Harmer 8, 20, 24, 25, 27, 33, 40; Naftali 638 n. 4; Oakes 276–79, 326, 330; 12th Army Group Report p. 84, in JSC folder Wedlock. Blind: PRO ADM 223/794/428. Boiler: Oakes 205, 329. Bootle: Masterman 1995 reprint xix. Bovril: Bristow 38. Bowsprit: Naftali 643 n. 51. Brass: Fleming 25–26; Howard 207–208. Bristle: Minutes 21 May 42, PRO KV 4/65/165; PRO CAB 154/14; PRO CAB 154/63. Bromo: JSC Folder Portugal; JSC Acco Binder Labeled "14"; JSC Brown Acco Binder "10" "B File"; JSC Brown Acco Binder Numbered "12" "File X"; real name in undated, unsigned chart in Schrup's handwriting, in NOA 544, Folder Special Means—A. Bronx: List of real names in 1995 Masterman reprint; Howard 76, 125; Masterman 9, 120, 149, 160–62; Hinsley & Simkins 112, 121, 241, 243, 274; PRO ADM 223/794/363. Brouwer: Naftali 628. Brutus: List of real names in 1995 Masterman reprint; PRO KV 2/72, 2/73; Wingate 43, 250–52, 255–56, 258; Howard 77, 115–17, 121, 131, 168, 172, 174, 186, 189, 192–94, 196–97, 199, 234; Masterman 120–21, 140–42, 148–49, 152, 155, 156, 169, 173; Hinsley & Simkins 112, 117–19, 182, 188, 196, 220, 238–43, 273, 275; Hesketh *passim.* Bull's Eye: Fleming 29–30. Byzance: III/120, 122. Camouflage: Oakes 157–58; Oakes 160, 171, 174, 176, 178, 181, 184, 185, 186, 189, 190, 191, 193, 328; 12th Army Group Report p. 80, in JSC folder Wedlock. Capricorn: Wingate 36; III/42–44, 45. Careless: Masterman 97–98, 109, 143–44; Hinsley & Simkins 93–94, 105, 123, 219, 220. Carrot: Masterman 117. Cato: D Division Weekly News Letter No. 39 13 Jan–19 Jan 45, Part II para. 25, in JSC folder Deception—SE Asia—WWII Historical D. Div—SACSEA—30 Dec 44–23 Mar 45. Cato: Howard 235 footnote. Celery: PRO KV 2/674; List of real names in 1995 Masterman reprint; Masterman 63, 73, 84, 90–92; Hinsley & Simkins 102–103, 223n.; West, Seven 27, 41. Charlie: PRO KV 2/454; List of real names in 1995 Masterman reprint; Howard 4; Masterman 40–41, 43, 92; Hinsley & Simkins 42, 102–103. Cheese: PRO KV 2/1133; List of real names in 1995 Masterman reprint; Wingate 30, 170; I/62–72, 92, II/ 73–76, 78, 82, 91–98, 104–105, 110, 111, 137, III/33, 37, 115–16, 139, IV/136–38; D Division Weekly News Letter No. 44, in JSC folder Deception—SE Asia—WWII Historical D. Div—SACSEA—30 Dec 44–23 Mar 45; Maunsell 25–27; Howard xi, 36–37, 63, 67–68, 84, 136–37; Mure, Practise 23–34, 37, 50–51, 101, 128, 134, 151, 153, 186, 211, 217, 251, 254, 255; Mure, Master 18, 177–78, 92, 105, 108–109, 119–20, 122, 128, 130, 133, 146, 230, 242; Masterman 107; Hinsley & Simkins 165–67, 229, 230. Cher Bébé: III/120, 122. Chopin: Hesketh 27, 63–64, 164. Clifford, Alexander: II/55. Cobweb: PRO KV 2/1137; List of real names in 1995 Masterman reprint; Howard 115–16, 224; Masterman 200; Hinsley & Simkins 112, 121, 218, 240; PRO ADM 223/794/290–95. Cocaine: III/117–18. Cog: Oakes 177, 179, 303, 305. Cossack: Harmer 17, 22, 24, 27, 40–41; Oakes 178, 179, 181, 182, 185, 187, 191, 224, 329, 331. Coughdrop: Fleming 42; Howard 206. Crude: II/118; Notes on D/A Activities in the Middle East, KV 4/197/1, p. 8; Mure, Practise 9, 89, 149–50, 179, 189, 207–208, 220, 227, 254; Mure, Master 114. Cruz: PRO ADM 223/794/394–95. Cupid: III/120, 122. Dagobert: Howard 233–34. Dave Close: D Division Weekly News Letter Nos. 39, 40, in JSC folder Deception—SE Asia—WWII Historical D. Div—SACSEA—30 Dec 44–23 Mar 45. Davil: III/120, 121–22. Davit: Oakes 329. De Conninck: Harmer 15, 24, 25. Denzil: Mure, Practise 201, 226, 254. Deputy: Hinsley & Simkins 383; Harmer 12–14, 20, 22, 23, 24, 25, 27, 29, 31, 32, 33, 41, 49, 53. Derrick: Hinsley & Simkins 382–83; Harmer 14–15, 19, 20, 22, 23, 24, 25, 27, 29, 31, 32, 33, 37, 41–42, 49; Oakes 185. Desire: Oakes 108. Dick: Howard 121, 188. Dickens: D Division Weekly News Letter No. 40, 20 Jan–26 Jan–45, Part II para. 23, in JSC folder Deception—SE Asia—WWII Historical D. Div—SACSEA—30 Dec 44–23 Mar 45. Ditch: Oakes 330. Doleful: II/117–18; Hinsley & Simkins 165, 210, 229, 230; Notes on D/A Activities in the Middle East, PRO KV 4/197/1, p. 9; Mure, Practise 52–53, 117, 123, 188, 205–206, 254. Dominant: Harmer 13–14. Donny: Howard 121, 188. Dorick: Howard 121, 188. Doubtful: Fleming; Howard 207. Dragoman: Oakes 70,

93–154, 157, 161, 162, 165, 166, 168, 169, 170, 171, 172, 174, 177, 178, 181, 184, 185, 189, 190, 191, 192, 223, 298, 320, 321, 326, 327, 328, 330, 331; Hinsley & Simkins 379–80; Harmer 5, 8–10, 24, 27, 29, 31, 42, 67; Naftali 623–24, 628–636, 646 n. 87, 648 nn. 104–126; PRO ADM 223/794/10. DRAGONFLY: List of real names in 1995 Masterman reprint; Masterman 56–57, 72–74, 84, 86, 95, 98, 108, 109, 111, 143, 144; Hinsley & Simkins 104–105, 123, 219–21, 235–36, 244. DREADNOUGHT: PRO KV 2/867 through 2/870; List of real names in 1995 Masterman reprint; Masterman 138–40; Hinsley & Simkins 217, 218, 222–24; West, Seven 8, 20, 24, 29, 32. DUCK: Fleming 24–25. DUTEIL: Wingate 32–33; III/126–27, 129, 130, 135; No. 2 Tac Report Part II 140–42, 148. ECCLESIASTIC: PRO ADM 223/794/70. EDITH: PRO ADM 223/794/389. EDOUARD: Plan To Build Up Special Agents for Coming Operations, ADV "A"F/21/3c, 4 Jul 43, Goldbranson papers. EDWARD: D Division Weekly News Letter No. 39 13 Jan–19 Jan 45, Part II para. 25, in JSC folder Deception—SE Asia—WWII Historical D. Div—SACSEA—30 Dec 44–23 Mar 45. EFFIGY: IV/138. EL GITANO: III/120–21. ELEPHANT: D Division Weekly News Letter No. 42 3 Feb–9 Feb 45 p. 6 para. 28(b), in JSC folder Deception—SE Asia—WWII Historical D. Div—SACSEA—30 Dec 44–23 Mar 45; Fleming. ELIAS: Mure, Practise 216, 254. EMILE: PRO KV 2/1133 ENTWHISTLE: II/48. FALCON: Oakes 264–65, 300–302, 329, 330. FAN: Harmer 10, 24, 25, 27, 43. FANNY: Handwritten memo, n.d., in Mure Papers (Box 2); Mure, Practise 9, 15, 175, 188, 221, 254. FARCE: Oakes 330. FATHER: List of real names in 1995 Masterman reprint; Masterman 74, 98, 99, 109, 144; Fleming 24–25; Howard 208; Masterman 74, 98–99, 109, 144; Hinsley & Simkins 93, 105, 123, 219, 220, 231, 233–34, 236. FICKLE: Oakes 303–308, 322, 329. FIDELIO: No. 2 Tac Report Part I 96. FIDO: List of real names in 1995 Masterman reprint; Masterman 143. FIREMAN: PRO ADM 223/794/389V–390. FISH: Oakes 179, 218, 302–308, 329, 330. FLAME: Harmer 12, 20, 25, 27, 43, 49, 52. FLESHPOTS, THE: II/119. FLORIST: Oakes 231, 293–95, 328. FOREST: No. 2 Tac Report Part I 75–89, 92, 96–98; Oakes 178, 179, 181, 182, 185, 187, 218, 224, 225, 227, 231–40, 243, 293, 326, 328, 330. FORGE: Harmer 18–19, 24, 25, 44. FOSSIL: D Division Weekly News Letter No. 47 10 Mar–16 Mar 45, Part II paragraph 16, in JSC folder Deception—SE Asia—WWII Historical D. Div—SACSEA—30 Dec 44–23 Mar 45 FRAIL: Oakes 301. FRANK: Hinsley & Simkins 383–84; Harmer 11–12, 20, 24, 25, 27, 29, 31, 32, 33, 44–45, 49, 50–51, 67; Oakes 205. FREAK: PRO KV 2/1069; West, Seven; Masterman 139–40, 149, 154. FREDDIE: Fleming to LCS, 17 Mar 43, PRO CAB 154/31. FREELANCE: Oakes 204, 205, 330. FRENCH AIRMAN: III/37. FRITZCHEN: Masterman 203. G.W.: PRO KV 2/468; List of real names in 1995 Masterman reprint; Masterman 40, 42, 57–58, 77, 84, 87, 92–93, 113. GABBIE: Mure, Practise 138, 159, 164, 225, 255. GALA: West, MI5 193; Wingate 31, 322; II/120; Mure, Practise 8, 13, 39, 80–83, 118, 120, 123, 125, 127–31, 133, 147–48, 150, 164, 180, 191, 196, 201, 207, 216, 218–19, 222–23, 226, 255; Howard 68. GANDER: List of real names in 1995 Masterman reprint; Masterman 53; Hinsley & Simkins 91, 96, 97, 325; West, Seven 43, 47, 48. GAOL: III/17, IV/89; Oakes 329, Oakes A284. GARBO: PRO KV 2/39 through 2/42, 2/63 through 2/71; Harris passim; Wingate 18, 41–43, 56, 160, 252, 253, 255, 256, 258; Howard 18–20, 49, 52, 62–63, 77–79, 104, 116–17, 121–22, 121–22, 124–25, 128, 167–70, 172–74, 179, 185–86, 188–92, 194–95, 199, 208, 223, 226, 231–41; Masterman ix, xiii, 16, 28, 109, 114–17, 125, 142–43, 147, 148, 149, 152–57, 168, 173–75, 178; Hinsley & Simkins 112–17, 126, 127, 131, 134, 182, 219, 220, 223, 225–28, 234–36, 238–40, 242–43, 275–76, 278, 287, 312–13; Pujol passim. GAT: Oakes 329, 333. GAULEITER OF MANNHEIM: PRO KV 2/1163; Maunsell to Petrie, 9 Oct 42, PRO KV 4/197; Mure, Master 19, 73–77, 109, 119, 128, 131, 133, 135, 160, 165, 207, 229, 242. GELATINE: List of real names in 1995 Masterman reprint; Howard 17, 62, 76; Masterman 86, 96, 109, 138, 140, 149, 176; PRO ADM 223/794/363; West, Seven 12–13. GENTLEMAN JOÉ: JSC Brown Acco Binder Labeled "6" "File D." GERARD: Fleming to Bevan, 17 Nov. 42, PRO CAB 154/73/21A. GESQUIÈRE: Harmer 15, 21, 24, 25. GILBERT: Naftali, De Gaulle 382; Wingate 32–36; III/126–35, 172; Howard xi, 84, 136, 137, 157; No. 2 Tac

Report Part I 85; No. 2 Tac Report Part II 138–55; No. 2 Tac Report Part III 46–50; Oakes 153 n. *, 329, 330; Thomas, Resume of Plan To Build Up Special Agents for Coming Operations, ADV "A"F/21/3c, 8 Jul 43, in Goldbranson papers; 12th Army Group Report pp. 49, 79, 83, in JSC folder Wedlock; Paillole, Services 485–96. GIRAFFE: List of real names in 1995 Masterman reprint; Masterman 54–55, 59n; West, Seven 10–11. GLASS: Oakes 330. GLO-CASE: JSC Folder Portugal; JSC Acco Binder Labeled "14." GODSEND: II/98. GODSTONE: PRO CAB 154/44. GOMEZ: O'Connor to JSC, HMOC/2013, 24 Dec 44, in JSC Folder South America. GUINEA: PRO CAB 154/21; Wingate 36–38, 192; III/146–48; Howard 94. GUN: Oakes 175, 176, 177, 178, 179, 181, 183, 184, 224, 304, 305, 328. GUS: JSC Acco Binder "File 'Z'—Salesmen's Qualifications"; JSC Folder South America; Cake Before Breakfast, JSC loose papers. HAMLET: PRO KV 2/325; list of real names in 1995 Masterman reprint; Mure, Master 249–50; Masterman 118–20, 159–60, 178. HANS: JSC Acco Binder "File 'Z'—Salesmen's Qualifications"; JSC Folder Switzerland. HARI SINGH: *E.g.,* D Division Weekly News Letter No. 39 13 Jan–19 Jan 45 para. 13 and nearby documents in JSC folder Deception—SE Asia—WWII Historical D. Div—SACSEA—30 Dec 44–23 Mar 45. HARRY: III/120, III/122. HAT TRICK: Fleming 19; Howard 207. HEADLAND: Oakes 148, 181, 239, 240, 245, 246, 322, 329, Maps. HEIR: Oakes 329. HEKTOR: Hesketh 102; Hinsley & Simkins 200–201, 278. HELGA: Mure, Practise 138, 166, 255. HENRY: JSC Acco Binder "File 'Z'—Salesmen's Qualifications"; Solborg to Smith, 24 Apr 45, in JSC folder Portugal. HERMAN: JSC Folder Portugal. HICCOUGHS: Fleming 37–38; Howard 206; Head of D Division to London Controlling Section, 3 Feb 45, Subject: Deceptive Broadcasts, in JSC folder Deception—SE Asia—WWII Historical D. Div—SACSEA—16 Dec 44–15 Mar 45. HITTITE: Masterman 3n.* HOLTZ: JSC Folder Portugal; JSC Brown Acco Binder "10" "B File." HONKY TONK: Mure, Practise 35–36, 119, 151, 255. HOST: Oakes 166, 283–89, 329, 330. HOSTESS: Oakes 166, 283–89, 329, 330. HULSMANN: Oakes 330. HUM-BLE: Mure, Practise 9, 15, 66, 88–91, 96–99, 103–13, 127, 140, 165, 168, 175, 176, 189, 195–96, 205, 227, 235, 236, 255. IDOL: Oakes 299. IMPULSE: Oakes 329. INFAMOUS: Wingate 31; II/96, 121; Hinsley & Simkins 231; Notes on D/A Activities in the Middle East, PRO KV 4/197/1, pp. 12–13. INK: Fleming 34; D Division Weekly Progress Report No. 1 p. 17, in JSC folder Deception—SE Asia—WWII Historical D. Div—SACSEA—24 Mar 45–4 Jun 45. JACK (for GLEAM): D Division Weekly News Letter No. 39 13 Jan–19 Jan 45, Part II paragraph 25, in JSC folder Deception—SE Asia—WWII Historical D. Div—SACSEA—30 Dec 44–23 Mar 45. JACQUELINE: Oakes 330. JAM: Memo re Jam Case, 2 Mar 45, in JSC Acco Binder Labeled "14." JARDINE, THERESA: Hanis 322; Howard 233. JAVE-LINE: D Division Weekly News Letter No. 45 24 Feb–2 Mar 45, Part II para. 14, in JSC folder Deception—SE Asia—WWII Historical D. Div—SACSEA—30 Dec 44–23 Mar 45. JEAN: III/131, 132–35; No. 2 Tac Report Part II 144, 148, 153. JEEP: PRO ADM 223/794/57. JEFF: PRO KV 2/1068; List of real names in 1995 Masterman reprint; Master-man 73, 86, 87, 96–97, 109, 126, 130, 133, 148, 150, 149, 168. Wingate 156; Hinsley & Simkins 92, 105, 123, 219, 223, 235, 236, 312; Mure, Master 265; Howard 62, 77; Young & Stamp 182; PRO ADM 223/794/336. JEST: Oakes 238, 239, 328. JEWEL: III/120. JIM-MIE: JSC Acco Binder "File 'Z'—Salesmen's Qualifications"; Solborg to Smith, 24 Apr 45, in JSC Folder Portugal. JO: No. 2 Tac Report Part II 152. JOB: Masterman 57. JOHNNY: Howard 3. JOSEF: Masterman 120; PRO ADM 223/794/388–389V. JOSEPHINE: Howard 188, 191, 196, 197; Hinsley & Simkins 200, 277, 278. JOY: Oakes maps. JURY: Oakes 329. JUSTICE: PRO ADM 223/794/422, 427–28. KANSHI RAM: *E.g.,* D Division Weekly News Letter No. 39 13 Jan–19 Jan 45 para. 13 and nearby documents in JSC folder Deception—SE Asia—WWII Historical D. Div—SACSEA—30 Dec 44–23 Mar 45. KARL: Head of D Division to London Controlling Section, 3 Feb 45, Subject: Deceptive Broadcasts, p. 17-A, in JSC folder Deception—SE Asia—WWII Historical D. Div—SACSEA—16 Dec 44–15 Mar 45. KEEL: Oakes 197–205, 326, 329, 330. KELLY: JSC Acco Binder "File 'Z'—Salesmen's Qualifica-tions"; Solborg to Smith, 20 Jun 45, in JSC Folder Portugal. KEYNOTE: No. 2 Tac Report Part

REFERENCES

I 92; Oakes 159–160, 165, 171, 172, 173, 175, 176, 177, 178, 179, 181, 183, 184, 189, 191, 193, 223, 224, 303, 304, 322, 328, 330. KHALIL: Mure, Practise 255. KISS: Notes on D/A Activities in the Middle East, PRO KV 4/197/1, pp. 9–10; Hinsley & Simkins 231; PRO CAB 154/38; Liddell diary 3 Mar 45, PRO KV 4/196. KLEIN: Message No. 14 sent 30 Sep 44, in subfile "P Sent Traffic," in JSC file Traffic Sent; unsigned memorandum, Re: Peasant Case, 4 Oct 44, in JSC brown Acco binder labeled "14," p. 21. KNOCK: II/118; Mure, Practise 89, 255. KYRIAKIDES: Mure, Practise 255. LAMBERT: PRO KV 2/1133, LAZY: Harmer 8, 20, 24, 25, 27, 40; Naftali 638 n. 4; Oakes 276–79, 326, 330. LE DUC: Wingate 32; III/128, 130–35; No. 2 Tac Report Part II 140, 142, 144, 148. Wingate 32; III/128, 130–35. LE MOCO: Wingate 33; III/133–35; No. 2 Tac Report Part II 146–48. LE MULET: Wingate 33; III/131–35; No. 2 Tac Report Part II 144–48. LE PETIT: III/120, 122; Thomas, Resume of Plan To Build Up Special Agents for Coming Operations, ADV "A"F/21/3c, 8 Jul 43, in Goldbranson papers. LEAGUE: Oakes 187, 256–70, 271, 276, 319, 322, 323, 329, 330; 12th Army Group Report p. 39, 42, 67–73, in JSC folder Wedlock. LEGION: Oakes 296–300, 302, 329, 330. LEMON: Liddell Diary 10 Feb 44, PRO KV 4/193; PRO CAB 154/14; PRO CAB 154/65. LEMONS, THE: III/136–38; Mure, Practise 8, 79, 82–83, 123, 133, 135–40, 159–64, 179, 202, 215, 234–37, 255–56; Mure, Master 102, 113–14, 208; Howard 68; Hinsley & Simkins 229–30. LESTER: JSC Acco Binder "File 'Z'—Salesmen's Qualifications"; JSC Folder South America; Solborg to Smith, 24 Apr 45, in JSC Folder Portugal. LIBERATORS, THE: II/118. LILOU: Wingate 36–37; III/45–48. LINEAGE: Oakes 181, 328. LIPSTICK: List of real names in 1995 Masterman reprint; Masterman 121, PRO ADM 223/794/9, 248 ff. LITIGANT: No. 2 Tac Report Part I 85, 92; Oakes 148, 168, 178, 179, 182, 224, 241–47, 297, 318, 328, 330. LITTLE LEMON: III/136–38; Mure, Practise 8, 79, 82–83, 123, 133, 135–40, 159–64, 179, 202, 215, 234–37, 255–56; Mure, Master 102, 113–14, 208; Howard 68; Hinsley & Simkins 229–30. LLAMA: Wingate 165–67; III/151–53; Howard 84. LODGE: Hill 393–95; O'Connor to Joint Security Control, 11 Oct 44, HMOC/1628, in JSC folder Great Britain; O'Connor to Joint Security Control, 8 Nov 44, HMOC/1769; O'Connor to Joint Security Control, 21 Nov 44, HMOC/1832, both in JSC folder South America. LOYAL: IV/101–102. LUC: No. 2 Tac Report Part I 96. LUÍS: Unsigned memorandum, B-422, 31 Mar 45, in JSC brown Acco binder labeled "10 B File." LUIZOS: Mure, Practise 82, 223–24, 227, 256. LYNCH: PRO CAB 154/14, PRO CAB 154/63, PRO CAB 154/65. MAGNET: Harmer 11. MANAGER: JSC Folder Portugal. MARCHAND: Oakes 148, 181, 182, 185, 187, 188, 245, 297, 329, 330, 333. MARGERY: PRO ADM 223/794/390. MARIA: Mure, Practise 138, 159, 164, 225, 257. MARIE: II/75, 92–94. MARKI: Mure, Practise 159, 256. MARMALADE: Fleming 21–22; Howard 207; D Division Weekly News Letter No. 39 13 Jan–19 Jan 45 para. 14, in JSC folder Deception—SE Asia—WWII Historical D. Div—SACSEA—30 Dec 44–23 Mar 45. MARTIN (at Beziers): Oakes 330. MARY (radio): Fleming 51. MARY (TATE's friend): West, Seven 57, 60; Howard 223; Hesketh 53. MASK: Oakes 329. MCTAVISH: Buck slip Smith to Bissell w/ attached memo "Head of 'D' Division," in JSC folder Japanese Deception Methods. MEADOW: Harmer 16–17, 18, 23, 24, 25, 45. MECHANIC: Harmer 16. MERCY: Harmer 11. METEOR: List of real names in 1995 Masterman reprint; Howard 227; Masterman 138–39; Hinsley & Simkins 217–18, 222, 224; PRO ADM 223/794/262–72; West, Seven 20–21. MIKE: JSC Folder Portugal; JSC Brown Acco Binder "10" "B File." MINARET: Hill 393–95. MINT: Harmer 15, 19, 20, 25, 27, 31, 45–46. MISANTHROPE: Hinsley & Simkins 166; Notes on D/A Activities in the Middle East, KV4/197/1; p. 5. MODEL: Harmer 17–18, 23, 24, 25, 29, 46. MOHICAN: PRO CAB 154/19. MONARCH: II/118. MONARCH: Oakes 90, maps. MONKEY: PRO CAB 154/63. MONOPLANE: No. 2 Tac Report Part I 66–75, 84, 89–94; No. 2 Tac Report Part I 66–75, 84, 89–94; Oakes 148, 158, 160, 161, 168, 169, 172, 174, 175, 177, 178, 181, 182, 191, 192, 221–228, 234, 239, 240, 243, 246, 303, 308, 321, 322, 326, 328, 330. MOON: PRO CAB 154/24. MOONBEAM: Masterman 173; Hinsley & Simkins 115, 225, 227, 228; PRO ADM 223/794/363; Pujol 113–14. MOONSTONE: JSC Folder Portugal; JSC Acco

Binder Labeled "14." MORTAL: Oakes 330. MOSELLE: III/145–46. MULETEER: Hinsley & Simkins 380–82; Oakes 173, 176, 178, 179, 182, 189, 205–218, 224, 302, 305, 322, 326, 329, 330, 331. MULLET: PRO KV 2/326; List of real names in 1995 Masterman reprint; Howard 77; Masterman 118–19, 149, 159–60; Hinsley & Simkins 121–22, 240–41; Mure, Master 166–67, 231. MUTT: PRO KV 2/1067; List of real names in 1995 Masterman reprint; Young & Stamp 182; Wingate 156; Howard 62, 77, 116; Masterman 73, 86, 87, 96–97, 109, 126, 130, 133, 148, 149, 168; Hinsley & Simkins 92, 101, 105, 123, 219, 221, 235, 236, 312; Mure, Master 235; PRO ADM 223/794/336. MYRIAM: No. 2 Tac Report Part II 106–107. NETTLE: List of real names in 1995 Masterman reprint; Masterman 169–70. NEVI: JSC Folder Portugal; JSC Brown Acco Binder Labeled "11" Index. NICOSSOF, PAUL: Wingate 31; I/71–72, II/73–76, 91–98; Hinsley & Simkins 166–67. NORBERT: Thomas, Resume of Plan To Build Up Special Agents for Coming Operations, ADV "A"F/21/3c, 8 Jul 43, in Goldbranson papers. NORMANDIE: No. 2 Tac Report Part I 91, 93–94. OAF: Oakes 330. OATMEAL: Fleming 27–29; Howard 208; Cruickshank, SOE 204, 205 and note, 206. ODIOUS: Notes on D/A Activities in the Middle East, PRO KV 4/197/1, pp. 13–14; Mure, Practise 212–23, 135, 153, 188, 251, 256; Mure, Master 165, 262; Hinsley & Simkins 231; Dovey, Intelligence War in Turkey 86 n. 56; Dovey, Maunsell and Mure 73–74. OFFICER: JSC Folder Portugal. OLIVER: III/117–19. OLIVER: See MARY. ONYX: JSC Folder Portugal; JSC Acco Binder Labeled "14." OPTIMIST: II/118; Gulzar Ahmed to William Kimber & Co., 18 Oct 80, in Mure Papers Box 2; Mure, Practise 65–72, 99, 256. ORLANDO: Mure, Practise 19–22. OSTEN: JSC Brown Acco Binder Labeled "11" "Index." OSTRICH: Oakes 148. OSTRO: Oakes 142; Howard 177–78, 180; Masterman 4, 151, 180; Hinsley & Simkins 199–200, 235, 256, 279; PRO ADM 223/794/74–75, 297. OTTO: JSC Folder Portugal. OUTCAST: Memo re JAM Case, 2 Mar 45, in JSC Acco Binder Labeled "14." OWL: Hart-Davis 280; Fleming 20–22; D Division Weekly News Letter No. 39 13 Jan–19 Jan 45 para. 13, in JSC folder Deception—SE Asia—WWII Historical D. Div—SACSEA— 30 Dec 44–23 Mar 45; Howard 207. PANCHO: Naftali 646 n. 87. PANDORA: Hesketh 29. PAPADOPOULOS: Mure, Practise 81, 126–27, 133, 148, 151, 180, 186, 191, 227, 257; Mure, Master 208, 215. PAT J: JSC Folder Portugal; JSC Acco Binder Labeled "14"; JSC Brown Acco Binder "10" "B File"; Hinsley & Simkins 228–29; Hill 395–96. PAWNBROKER: Fleming 16–17; Howard 207; Hinsley & Simkins 233. PEANUT: PRO CAB 154/37/34A; Head of D Division to London Controlling Section, 3 Feb 45, Subject: Deceptive Broadcasts, p. 18-A, in JSC folder Deception—SE Asia—WWII Historical D. Div—SACSEA—16 Dec 44–15 Mar 45. PEASANT: JSC Folder Traffic Sent; Folder Portugal; JSC Acco Binder Labeled "14"; PRO KV 2/467. PEASANT (France): Oakes maps. PEDANT: IV/138. PEGASUS: No. 2 Tac Report Part I 85 Part II 104. PENNANT: Oakes 329. PEPPERMINT: List of real names in 1995 Masterman reprint; Masterman 113, 24; PRO ADM 223/794/143–44 also 56. PESSIMISTS, THE: Wingate 32, 327; II/121–22, 137, III/14; Mure, Practise 8, 14–15, 32–33, 51, 63–66, 76–84, 96–101, 117–20, 123–27, 144, 146, 153–56, 165–72, 176–79, 186, 188, 196, 201–203, 215–16, 224–25, 234–35, 238–42, 248, 256; Mure, Master 19, 110, 111–12, 133, 146, 165–66, 208, 215; Howard 68, 165, 229, 230. PIET: I/71–72; Hinsley & Simkins 166. PIG: III/43–44. PILGRIM: Oakes 329, maps. PIN: JSC Acco Binder "File 'Z'—Salesmen's Qualifications"; JSC Folder South America. PIP: Harmer 19, 20, 24, 25, 27, 32, 34, 46–47. PITT: Oakes 330. PLATO: PRO ADM 223/794/389V. POINTER: D Division Weekly Progress Report No. 9 22–28 May 45 p. 4, in JSC folder Deception—SE Asia—WWII Historical D. Div—SACSEA—24 Mar 45–4 Jun 45; D Division Weekly Progress Report No. 12 12–18 Jun 45, pp. 8–9; No. 19 31 Jul–6 Aug 45 p. 7, both in JSC folder Deception—SE Asia— WWII Historical D. Div. SACSEA 5 Jun–10 Sep 45. PORTUGUESE CHANNELS: I/49–50. PRIMO: IV/98–99; Howard 136, 141; Hinsley & Simkins 269 n.; Naftali 612–614, 643 n. 48. PUPPET: PRO KV 2/327; List of real names in 1995 Masterman reprint; Mure, Master 167; Howard 77; Masterman 119, 159; Hinsley & Simkins 122, 218, 240–41. PURPLE WHALES: Fleming 9–10; Project No. 1471-A, The Japanese Intelligence System, pp.

102–103, in JSC folder The Japanese Intelligence System. PYRAMIDS, THE: II/119. QUAIL: Oakes 275. QUALITY, QUAY, QUEASY: Oakes 330. QUICKSILVER: Handwritten memo, n.d., in Mure Papers (Box 2); Wingate 31–32, 317, 322, 327–28; II/119–20, 137, III/6–7, 14, 17; Mure, Practise 14–15, 32–33, 36–39, 41, 63–66, 78, 81, 83–84, 99, 103, 114, 117, 119, 120, 123–25, 131, 237–41, 243, 256; Mure, Master 19, 110–12, 133, 146, 165–66, 208, 215; Howard 68; Hinsley & Simkins 165, 229, 230. RA 506: Head of D Division to London Controlling Section, 3 Feb 45, Subject: Deceptive Broadcasts, p. 5-A, in JSC folder Deception—SE Asia—WWII Historical D. Div—SACSEA—16 Dec 44–15 Mar 45 RADIO BABU: *E.g.*, D Division Weekly News Letter No. 39 13 Jan–19 Jan 45 para. 13 and nearby documents in JSC folder Deception—SE Asia—WWII Historical D. Div—SACSEA—30 Dec 44–23 Mar 45. RAJAH: Fleming 25; Hinsley & Simkins 234; D Division Weekly News Letter No. 37 30 Dec 44–5 Jan 45, in JSC folder Deception—SE Asia—WWII Historical D. Div—SACSEA—30 Dec 44–23 Mar 45. RAINBOW: PRO KV 2/1066; Masterman 46–47, 59n., 78, 93, 95; Hinsley & Simkins 43, 92, 102, 103, 123, 185, 219, 220; West, Seven 3, 30–43, 47, 56. RAM: III/117; Howard 84; Thomas, Resume of Plan To Build Up Special Agents for Coming Operations, ADV "A"F/21/3c, 8 Jul 43, in Goldbranson papers. REP: JSC Folder Portugal. RG 273: Head of D Division to London Controlling Section, 3 Feb 45, Subject: Deceptive Broadcasts, pp. 5-A, 9-A, in JSC folder Deception—SE Asia—WWII Historical D. Div—SACSEA—16 Dec 44–15 Mar 45 RHINEMAIDENS, THE: Harmer 8, 20, 24, 25, 27, 40; Naftali 638 n. 4; Oakes 276–79, 326, 330. RHINO: D Division Weekly News Letter No. 45 24 Feb–2 Mar 45 para. 9(b)(i), in JSC folder Deception—SE Asia—WWII Historical D. Div—SACSEA—30 Dec 44–23 Mar 45; Fleming. RI 919: Head of D Division to London Controlling Section, 3 Feb 45, Subject: Deceptive Broadcasts, p. 19-A, in JSC folder Deception—SE Asia—WWII Historical D. Div—SACSEA—16 Dec 44–15 Mar 45. RIO: West, MI5 193; Wingate 322, 328, 329; II/20, III/17; Howard 68; Mure, Practise 8, 38, 80–83, 108, 127, 181, 188, 191, 216, 227–29, 254. RO 914: Head of D Division to London Controlling Section, 3 Feb 45, Subject: Deceptive Broadcasts, p. 21-A, in JSC folder Deception—SE Asia—WWII Historical D. Div—SACSEA—16 Dec 44–15 Mar 45. ROBERTO: PRO KV 2/1133. ROBERTS: JSC Folder Portugal. ROGER: No. 2 Tac Report Part II 152. ROSIE: Oakes 330. ROVER: Howard 182; Masterman 170–71, 181; Hinsley & Simkins 219, 274; PRO ADM 223/794/362. RP 790: Head of D Division to London Controlling Section, 3 Feb 45, Subject: Deceptive Broadcasts, p. 13-A, in JSC folder Deception—SE Asia—WWII Historical D. Div—SACSEA—16 Dec 44–15 Mar 45. RQ 318: Head of D Division to London Controlling Section, 3 Feb 45, Subject: Deceptive Broadcasts, p. 5-A, in JSC folder Deception—SE Asia—WWII Historical D. Div—SACSEA—16 Dec 44–15 Mar 45. RR 545: Head of D Division to London Controlling Section, 3 Feb 45, Subject: Deceptive Broadcasts, p. 5-A, in JSC folder Deception—SE Asia—WWII Historical D. Div—SACSEA—16 Dec 44–15 Mar 45. RUBY: III/119. RUDLOFF: JSC Folder Traffic Sent; JSC Folder Portugal; JSC Acco Binder Labeled "14." RUPERT: JSC Acco Binder "File 'Z'—Salesmen's Qualifications"; JSC Folder South America; Hill 396–97. RK 459: D Division Weekly News Letter No. 44 p. 8, in JSC folder Deception—SE Asia—WWII Historical D. Div—SACSEA—30 Dec 44–23 Mar 45. RY 327: D Division Weekly News Letter No. 44 p. 8; Head of D Division to London Controlling Section, 3 Feb 45, Subject: Deceptive Broadcasts, p. 5-A, both in JSC folder Deception—SE Asia—WWII Historical D. Div—SACSEA—16 Dec 44–15 Mar 45. RZ 282: Head of D Division to London Controlling Section, 3 Feb 45, Subject: Deceptive Broadcasts, p. 17-A, in JSC folder Deception—SE Asia—WWII Historical D. Div—SACSEA—16 Dec 44–15 Mar 45. SANDRU: Mure, Practise 98, 257. SAUNDERS: Message No. 33 sent 30 Oct 44, in JSC brown Acco binder "9." SAVAGES, THE (I, II, and III): III/138–40; Howard 68; Hinsley & Simkins 229, 230; D. Div. Weekly Implementation Series No. 7, in JSC folder D Division Weekly Implementation Series. SCOOT (SKOOT): PRO ADM 223/794/44; 360. SCOUNDREL: Oakes 328, maps. SCRUFFY: Pujol 65–66; Masterman 99; West, Seven 21–22; Hinsley & Simkins 329. SHACKLE: Oakes

329. SHEPHERD: Masterman 169–70; Hesketh 74. SILENT: Oakes 264–65, 329, 333. SILVER: Fleming; Howard xi, 208–10; Hinsley & Simkins 232–34; Talwar; Hart-Davis. SINK: Oakes 329. SKULL: Oakes 119, 121, 122, 129, 148, 150, 152, 158–94, 221, 224, 269, 283, 303, 308, 320, 322, 326, 329, 330, 331, 334; Hinsley & Simkins 379–80; Harmer 9–10, 27; No. 2 Tac Report Part I 92. SLAVE: II/119. SMOOTH: Notes on D/A Activities in the Middle East, KV 4/197/1, p. 8; Mure, Practise 9, 87–88, 96, 176, 205, 213, 224–25, 236, 257; Mure, Master 112. SNARK, THE: PRO KV 2/669 through 2/673; List of real names in 1995 Masterman reprint; Masterman 77, 78, 99; Hinsley & Simkins 93, 94, 105. SNIPER: PRO KV 2/1138; Masterman 143, 168; Hinsley & Simkins 218–19, 384; Harmer 18, 20, 24, 29, 33, 47–48; PRO ADM 223/794/362–63. SNOW: PRO KV 2/444 through 2/453; List of real names in 1995 Masterman reprint; Mure, Master 154; Howard 3–6, 14; Masterman ix, xiii, 28, 36–47, 49, 50, 57, 59n. 61, 63, 90–93, 114; Hinsley & Simkins 41–44, 87–88, 91, 92, 94, 95, 102–103, 107–108, 280, 311–23, 325–26; PRO ADM 223/794/177; West, Seven 8, 35, 38, 46, 58, 215. SPANEHL: Masterman 54; West, Seven 10. SPANIP: FBI File No. 65-46383, Spanip. SPRINGBOK: PRO KV 2/1134 through 2/1136; Hinsley & Simkins 228; Naftali 294–98, 333 note 22; Hill 390–93. SQUEAK: Harmer 19, 20, 24, 25, 27, 32, 34, 46–47. SRAIEB: Wingate 33. STANDARD: Oakes 330. STEPHAN: List of real names in 1995 Masterman reprint; Mure, Master 19, 75–76, 109, 119–20, 128, 130, 135, 165, 207, 230, 242, 263. STEVEN: Mure, Practise 81, 99, 123, 131, 168, 180, 186, 222, 257; Mure, Master 146. STIFF: PRO KV 4/197/1a. SUMMER: PRO KV 2/60; List of real names in 1995 Masterman reprint; Masterman 50–52; Hinsley & Simkins 91, 96, 97, 102, 221n., 312, 322; West, Seven 34. 35, 38, 43, 60, 61, 217. SUNRISE: Fleming 30–31; PRO ADM 223/794/157. SUNSET: Memo, Re Sunset Case, 23 Jun 45; Nr. 255; WAR 22093; WAR 24859, all in JSC folder Portugal. SWAMI: Fleming 20. SWEET WILLIAM: List of real names in 1995 Masterman reprint; Masterman 78, 99. SWEETIE: Masterman 54; West, Seven 10. SWING-TIT: Mure, Practise 138, 159, 164, 225, 257. TAKIS: Mure, Practise 80, 150, 181, 188, 257. TATE: PRO KV 2/61, 2/62; West, Seven 33–65; Howard 10–11, 14, 62, 121, 125, 168–69, 175n, 182, 223, 226–30; Masterman 17, 52–53, 59n., 73, 74, 78, 83, 85, 93–95, 109, 149, 150, 152, 160–61, 183–85; Hinsley & Simkins 91, 92, 97n., 101, 103, 104, 107–108, 123, 221n., 236, 240, 242, 273–74, 312, 322–23, 331–33; Hesketh 27–28, 30, 41, 49–50, 53, 61, 92, 110–111, 126, 129, 131, 137, 142, 150, 160, 163, 164. TEAPOT: Hesketh 29, 39; PRO ADM 223/794/125, 155–56, 338. THACKERAY: D Division Weekly News Letter No. 41 27 Jan–2 Feb 45, Part II para. 21, in JSC folder Deception—SE Asia—WWII Historical D. Div—SACSEA—30 Dec 44–23 Mar 45. TILTMAN, HESSELL: Cake Before Breakfast 65, in JSC loose papers; unsigned memorandum in JSC folder South America, 4 Sep 45. TOLLUS: Mure, Practise 257. TOM: Fleming. TOM X: Unsigned memorandum, Re: Cocase, 7 Mar 44, in JSC brown Acco binder Numbered "12" "File X"; Maheu 29–32; Ladd to Tamm, Re: Spanip and Cocase, 19 Jun 43, in FBI Spanip file; unsigned, undated memoranda, Re: Cocase, Number 2/2, B-32, and Number 2/7; Summary COCASE Messages, all in JSC brown Acco binder labeled "8 File B." TRAVEL: Fleming 17–18; Howard 207. TREASURE: PRO KV 2/464 through 2/466; List of real names in 1995 Masterman reprint; Howard 175n, 181; Masterman 143, 149, 161, 169; Hinsley & Simkins 182, 218, 241, 274; PRO ADM 223/794/362, 364V; Mure, Master 235, 248, 249. TRICYCLE: KV 2/845 through 2/866; Wingate 160, 250, 251; PRO ADM 223/794/44, 360; Howard 17–18, 49, 62, 77, 121, 148–49, 152–55, 167, 176–77, 223, 226, 228; Masterman 55–56, 59n, 73–74, 79–81, 83, 85–86, 95–96, 120, 138–39, 147–48, 178, 196–98; Hinsley & Simkins 95, 102–104, 107–108, 113, 119n., 122–27, 131, 134, 147, 183, 185, 217, 221–25, 238, 241; West, Seven 7–32; Hill 388–90; Mure, Practise 153, 176, 258; Mure, Master 19, 161–73, 176, 178, 186–90, 195, 209, 228–29, 248. TROLLOPE: D Division Weekly News Letter No. 39 13 Jan–19 Jan 45, Part II para. 25, in JSC folder Deception—SE Asia—WWII Historical D. Div—SACSEA—30 Dec 44–23 Mar 45 TROTTER: Fleming 16–17; Howard 207. TRUDI: Mure, Practise 138, 159, 225, 257. TURBOT: PRO CAB 154/14; PRO CAB 154/65. TURK-

ish Channels: I/25–26. Twerp, The: D Division Weekly News Letter No. 37 30 Dec 44–5 Jan 45, Part II paragraph 10(b), in JSC folder Deception—SE Asia—WWII Historical D. Div—SACSEA—30 Dec 44–23 Mar 45. Twist: Notes on D/A Activities in the Middle East, PRO KV 4/197/1, p. 15; Hinsley & Simkins 230–31. Twit: No. 2 Tac Report Part I 91. Umberti: Naftali 614–16, 643 n. 51, 644 n. 57, 644 n. 61. Uncle: Mure, Practise 258. Unnamed Friend: Unsigned memorandum Re: Rudloff Case, B-205, 23 Oct 44, in JSC brown Acco binder labeled "9 21-400 File 'B.' " Van Meldert: Harmer 15, 16, 24, 25. Verdier: Oakes 160. Violet: Oakes 311–14, 329. Wait: Mure, Practise 89, 97, 258. Wasch: JSC Folder Portugal. Washout: List of real names in 1995 Masterman reprint; Masterman 117–18. Watchdog: PRO ADM 223/794/340, 356–58; List of real names in 1995 Masterman reprint; Masterman 121, 144; Hinsley & Simkins 201, 228; Pujol 113–14; Harvison 105–116. Wave: JSC Folder Portugal. Weasel: PRO KV 2/244; Masterman 117. Whiskers: Wingate 39; III/117–18; Howard 184, 136. Whisky: D Division Weekly Progress Report No. 3, p. 12, in JSC folder Deception—SE Asia—WWII Historical D. Div—SACSEA—24 Mar 45–4 Jun 45. Wilfred: Harmer 19, 20, 24, 25, 27, 32, 34, 46–47. Wit: Mure, Practise 97, 258. Witch: Oakes 251–70, 271, 276, 278, 326, 329, 330, 333; 12th Army Group Report pp. 39, 42, 48, 49, 60, 67–73, in JSC folder Wedlock. Wizard: Oakes 266–67, 270, 271, 300. Worm, The: List of real names in 1995 Masterman reprint; Masterman 139–40; Hinsley & Simkins 217, 224; West, Seven 24–25. Zigzag: PRO KV 2/455 through 2/463; List of real names in 1995 Masterman reprint; Howard 172n, 175n, 227; Masterman xiv, 122–23, 131–32, 171–73; Hinsley & Simkins 111, 112, 120, 183, 194, 196, 219–20, 235, 274–75, 312; PRO ADM 223/794/340–41, 412–14; Chapman. Zulu: II/118; Notes on D/A Activities in the Middle East, PRO KV 4/197/1, p. 12; Dovey, Intelligence War in Turkey 79; Hinsley & Simkins 210.

Appendix III: Fictitious Units

United States Units

1st Army Group ("FUSAG"): Reynolds to ACS G-1, 21 Mar 44; drawing dated 13 Apr 44; Doriot to Military Personnel Division, A.S.F, 13 Mar 44; AG, First Army Group, to AG, Washington, March 7, 1944, 421 (G-1), all in U.S. Army Institute of Heraldry Unit Files, First Army Group; Fortitude Order of Battle, SHAEF/18216/3/Ops (B), 29 Aug 44, in NARA RG 331 Entry 199 Box 66 Folder Fortitude (fat folder of two by that name); Hesketh 141, 148, 157. *2d Army Group:* Bevan letter, Notional U.S. Formations, 9 Apr 45; From London, 22 Jun 44, both in JSC folder European Theater 000. *Twelfth Army:* Bevan letter, Notional U.S. Formations, 15 Mar 45; From London, 22 Jun 44, both in JSC folder European Theater 000; National War College Lecture, Lieutenant Colonel John R. Deane, Jr., 1 March 1950, in NOA XXII, Box 546, Folder Tactical Deception—Ruses, p. 11. *Fourteenth Army:* Doriot to Director, JSC, 3 Aug 44; drawing dated 30 Jun 45, in U.S. Army Institute of Heraldry Unit Files, XXXI Corps; Fortitude Order of Battle, SHAEF/18216/3/Ops (B), in NARA RG 331 Entry 199 Box 66 Folder Fortitude (fat folder of two by that name); From London, 12 July 1944; Notional U.S. Formations Employed in the Western European Theater During 1944, SHAEF/24134/SM/Ops, 11 Jul 45, both in JSC folder European Theater 000; Hesketh 120, 135, 139, 140 (quoting Garbo Letter No. 26, 26 Aug 44), 141, 156; 12th Army Group Report p. 79, in JSC folder Wedlock. (9th and 21st Airborne Divisions formed into XIX Airborne Corps, per Notional U.S. Formations Employed in the Western European Theater During 1944, SHAEF/24134/SM/Ops, 11 Jul 45, in JSC folder European Theater 000. Notation that this was "news" to JSC penciled in on JSC copy; presumably error for XVII Airborne Corps.) *IX Amphibious Corps:* Cake Before Breakfast p. 50, JSC loose papers. *XVII Airborne Corps:* Hesketh 158. *XXX Corps:* Bevan letter, Notional U.S. Formations, 15 Mar 45; From London, 22 Jun 44; notes on Bevan letter 22 Mar 45, all in JSC folder European Theater 000. *XXXI Corps:* Doriot to JSC, 3 Aug 44, SPQRD 421.4, 31st Corps, in U.S. Army Institute of Heraldry Unit Files, XXXI Corps; Unsigned, undated memorandum in JSC folder 370.2 Ironside; PRO WO 204/1562/89; AFHQ Mediterranean Joint Planning Staff P/143 (Final), Annexure B, p. 2 para. 4 and Appendix 1, in NARA RG 319 Entry 1 Box 4 Folder 73; NARA RG 319 Entry 101 Box 4 Folder 72 "Ferdinand" 28 Jul 44); Howard 156; Smith to O'Connor, JSC/J4 Serial 1265, 12 Dec 44, in JSC folder 370.2 Mediterranean; Fortnightly Letter No. 29, 20 Jan 45, in JSC folder 370.2 Overlord. *XXXIII Corps:* Drawing dated 30 Jun 45, in U.S. Army Institute of Heraldry Unit Files, XXXIII Corps; From London, 22 Jun 44; Notional U.S. Formations Employed in the Western European Theater During 1944, SHAEF/24134/SM/Ops, 11 Jul 45, all in JSC folder European Theater 000; Fortitude Order of Battle, SHAEF/18216/3/Ops (B), in NARA RG 331 Entry 199 Box 66 Folder Fortitude (fat folder of two by that name); Hesketh 141, 156. *XXXV Airborne Corps:* CominchAFPac Manila to War Department, July 31, 1945, No. CX 29841; JSC to CominchAFPac Manila, August 2, 1945, No. WAR 42688; CominchAFPac Manila to War Department, August 3, 1945, No. CX 30582; all in NOA 532 Folder F-28 Dispatch File; GHQ USARPAC, Operation Instructions No. 3, 5 Aug 45, in NOA 550 Folder Pastel Two. *XXXVII Corps:* Notional U.S. Formations Employed in the Western European Theater During 1944, SHAEF/24134/SM/Ops, 11 Jul 45; SHAEF/19018/Ops(B) GCT/370.28–205/Ops(B), 6 Oct 44, in JSC folder European Theater 000; Fortitude Order of Battle, SHAEF/18216/3/Ops (B), in NARA RG 331 Entry 199 Box 66 Folder Fortitude (fat folder of two by that name); Hesketh 141, 156. *XXXVIII Corps:* Bevan letter, Notional U.S. Formations, 15 Mar 45; From London, 22 Jun 44; notes on Bevan letter 22 Mar 45, all in JSC folder European Theater 000. *XXXIX Corps:* Bevan letter, Notional U.S. Formations, 15 Mar 45; From London, 22 Jun 44; notes on Bevan letter 22 Mar 45, all in JSC folder European Theater 000. *XXXVIII Corps:* Bevan letter, Notional U.S. Formations, 15 Mar 45; notes on Bevan letter 22 Mar 45, all in JSC folder European Theater 000. *5th Motorized Division:* From London, 22 Jun 44; Notes on Bevan letter 22 Mar 45, both in JSC folder European Theater 000. *6th Airborne Division:* Doriot to Director, JSC, 3 Aug 44; drawing dated 15 July 45, in U.S. Army Institute of Heraldry Unit Files, 6th Airborne Division; AFHQ Mediterranean Joint Planning Staff P/143 (Final), Annexure B, p. 2 para. 4a and Appendix 1, in NARA RG 319 Entry 1 Box 4 Folder 73; PRO WO 204/1562/88; From London, 25 July 1944; O'Connor to JSC, HMOC/1917, 7 Dec 44, both in JSC folder 370.2 Mediterranean. *9th Airborne Division:* Doriot to Director, JSC, 3 Aug 44; drawing dated 30 Jun 45, both in U.S. Army Institute of Heraldry Unit Files, 9th Airborne Division; Notional U.S. Formations Employed in the Western European Theater During 1944, SHAEF/24134/SM/Ops, 11 Jul 45, in JSC folder European Theater 000 (XIX Airborne Corps); Extract From Weekly Letter No. 10, 3 May 44, in folder 370.2 Graffham; Fortitude Order of Battle, SHAEF/18216/3/Ops (B), in NARA RG 331 Entry 199 Box 66 Folder Fortitude (fat folder of two by that name); Hesketh 147–48, 158. *10th Infantry (Light) Division:* National War College Lecture, Lieutenant Colonel John R. Deane, Jr., 1 March 1950, p. 15, in NOA 546 Folder Tactical Deception—Ruses; Implementation of Bluebird Special Means for Week Ending 28 April 1945, in NOA 548 Folder Bluebird; File Log, Item A-150, 27 Apr 45, JSC Acco Binder "7 File A"; Stanton 92. *11th Infantry Division:* Undated, unsigned memorandum; drawing dated 30 Jun 45, both in U.S. Army Institute of Heraldry Unit Files, 11th Division; Units—Required—Now Available—JSC To Furnish; From London, 22 Jun 44; Bevan letter 22 Mar 45; Notional U.S. Formations Employed in the Western European Theater During 1944. SHAEF/24134/SM/Ops, 11 Jul 45, all in JSC folder European Theater 000; Fortitude Order of Battle, SHAEF/18216/3/Ops (B), in NARA RG 331 Entry 199 Box 66 Folder Fortitude (fat folder of two by that name); Hesketh 123, 141, 156. *13th Airborne Division:* Bevan letter, Notional U.S. Formations, 9 Apr 45. *14th Infantry Division:* Doriot to Director JSC, 7 Aug 44, in U.S. Army Institute of Heraldry Unit Files, 14th Infantry Division; Units—Required—Now Available—JSC To Furnish, undated; From London, 22 Jun

44; Bevan letter 22 Mar 45, all in JSC folder European Theater 000. *15th Infantry Division:* Notes on Bevan letter 22 Mar 45; Units—Required—Now Available—JSC To Furnish, undated; From London, 22 Jun 44; Bevan letter, Notional U.S. Formations, 15 Mar 45, all in JSC folder European Theater 000. *15th Armored Division:* From London, 22 Jun 44; Bevan letter, Notional U.S. Formations, 15 Mar 45; Units—Required—Now Available—JSC To Furnish, undated; notes on Bevan letter 22 Mar 45, all in JSC folder European Theater 000. *16th Infantry Division:* National War College Lecture, Lieutenant Colonel John R. Deane, Jr., 1 March 1950, p. 4, in NOA 546 Folder Tactical Deception—Ruses; Bevan letter, Notional U.S. Formations, 9 Apr 45. *17th Infantry Division:* Doriot to Director, JSC, 3 Aug 44; drawing dated 30 Jul 45, both in U.S. Army Institute of Heraldry Unit Files, 17th Infantry Division; Units—Required—Now Available—JSC To Furnish, undated; From London, 22 Jun 44; Bevan letter 22 Mar 45; Notional U.S. Formations Employed in the Western European Theater During 1944, SHAEF/24134/SM/Ops, 11 Jul 45, all in JSC folder European Theater 000; Fortitude Order of Battle, SHAEF/18216/3/Ops (B), in NARA RG 331 Entry 199 Box 66 Folder Fortitude (fat folder of two by that name); Hesketh 121, 141, 156. *18th Airborne Division:* Unsigned, undated memorandum, 18 Airborne Division, with notation "Approved July 5, 1944"; drawing dated 15 Jul 45, both in U.S. Army Institute of Heraldry Unit Files, 18th Airborne Division; Roster of component units, 18th Airborne Division, 27 Jun 44; Bevan letter, Notional U.S. Formations, 15 Mar 45, both in JSC folder European Theater 000; Smith to O'Connor, JSC/J1C2, March 14, 1945; O'Connor to JSC, HMOC/2515 8 Mar 45, both in JSC folder Dervish; CominchAFPac Manila to War Department, July 31, 1945, No. CX 29841; JSC to CominchAFPac Manila, August 2, 1945, No. WAR 42688; CominchAFPac Manila to War Department, August 3, 1945, No. CX 30582, all in NOA 532 Folder F-28 Dispatch File; GHQ USARPAC, Operation Instructions No. 3, 5 Aug 45, in NOA 550 Folder Pastel Two. *19th Infantry Division:* Units—Required—Now Available—JSC To Furnish, undated; From London, 22 Jun 44; Bevan letter, Notional U.S. Formations, 15 Mar 45; Bevan letter 22 Mar 45, all in JSC folder European Theater 000. *21st Infantry Division:* Units—Required—Now Available—JSC To Furnish, undated; From London, 22 Jun 44; Bevan letter, Notional U.S. Formations, 15 Mar 45; Bevan letter 22 Mar 45, all in JSC folder European Theater 000. *21st Airborne Division:* Unsigned, undated memorandum, "21st Airborne Division"; drawing dated 30 Jun 45, both in U.S. Army Institute of Heraldry Unit Files, 21st Airborne Division; Fortitude Order of Battle, SHAEF/18216/3/Ops (B), in NARA RG 331 Entry 199 Box 66 Folder Fortitude (fat folder of two by that name); Roster, 21st Airborne Division, in JSC folder 370.2 Bodyguard; Roster, 21st Airborne Division; Notional U.S. Formations Employed in the Western European Theater During 1944, SHAEF/24134/SM/Ops, 11 Jul 45 (XIX Airborne Corps), both in JSC folder European Theater 000; Hesketh 147–48, 158. *22d Infantry Division:* Doriot to Director, JSC, 3 Aug 44; drawing dated 30 Jun 45, both in U.S. Army Institute of Heraldry Unit Files, 22d Infantry Division; National War College Lecture, Lieutenant Colonel John R. Deane, Jr., 1 March 1950, pp. 4–5, in NOA 546 Folder Tactical Deception—Ruses; Clarke memo 29 Sep 42, in JSC folder 370.2 Sweater. *23d Infantry Division:* National War College Lecture, Lieutenant Colonel John R. Deane, Jr., 1 March 1950, p. 4, in NOA 546 Folder Tactical Deception—Ruses; Bevan letter, Notional U.S. Formations, 9 Apr 45; Smith to Bissell, Comment No. 2, 23 Mar 45, in JSC folder Japanese Estimates of Allied OB. *25th Armored Division:* Notional U.S. Formations Employed in the Western European Theater During 1944, SHAEF/24134/SM/Ops, 11 Jul 45, in JSC folder European Theater 000; Fortitude Order of Battle, SHAEF/18216/3/Ops (B), in NARA RG 331 Entry 199 Box 66 Folder Fortitude (fat folder of two by that name); Hesketh 141, 156. *39th Armored Division:* From London, 22 Jun 44; Bevan letter, Notional U.S. Formations, 15 Mar 45; Units—Required—Now Available—JSC To Furnish, undated; notes on Bevan letter 22 Mar 45, all in JSC folder European Theater 000. *46th Infantry Division:* Doriot to Director JSC, 3 Aug 44; drawing dated 30 Jun 45, both in U.S. Army Institute of Heraldry Unit Files, 46th Infantry

Division; National War College Lecture, Lieutenant Colonel John R. Deane, Jr., 1 March 1950, p. 7, in NOA 546 Folder Tactical Deception—Ruses; Table dated 22 Jun 1943 [Bratby], rows 4–5 columns B and D; Bevan letter, Notional U.S. Formations, 9 Apr 45; O'Connor to Smith, HMOC 3640, 19 Jul 45, all in JSC folder European Theater 000; Smith to Bissell, Comment No. 2, 23 Mar 45, in JSC folder Japanese Estimates of Allied OB; Wingate 205; Howard 75–76. *47th Infantry Division:* From London, 22 Jun 44; Bevan letter, Notional U.S. Formations, 15 Mar 45; Units—Required—Now Available—JSC To Furnish, undated; notes on Bevan letter 22 Mar 45, all in JSC folder European Theater 000. *48th Infantry Division:* Doriot to Director JSC, 3 Aug 44; drawing dated 16 Feb 49, both in U.S. Army Institute of Heraldry Unit Files, 48th Infantry Division; Fortitude Order of Battle, SHAEF/18216/3/Ops (B), in NARA RG 331 Entry 199 Box 66 Folder Fortitude (fat folder of two by that name); From London, 22 Jun 44; Notional U.S. Formations Employed in the Western European Theater During 1944, SHAEF/24134/SM/Ops, 11 Jul 45, both in JSC folder European Theater 000; Hesketh 123, 141, 156. *50th Infantry Division:* Unsigned, undated handwritten memorandum, both in U.S. Army Institute of Heraldry Unit Files, 50th Infantry Division; Bevan letter, Notional U.S. Formations, 9 Apr 45; Smith to O'Connor, JSC/J4 Serial 3149, 29 Mar 45, in JSC folder European Theater 000. *55th Infantry Division:* Doriot to Director JSC, 3 Aug 44; drawing dated 30 Jun 45, both in U.S. Army Institute of Heraldry Unit Files, 55th Division; National War College Lecture, Lieutenant Colonel John R. Deane, Jr., 1 March 1950, p. 8, in NOA 546 Folder Tactical Deception—Ruses; Fortitude Order of Battle, SHAEF/18216/3/Ops (B), in NARA RG 331 Entry 199 Box 66 Folder Fortitude (fat folder of two by that name); Notional U.S. Formations Employed in the Western European Theater During 1944, SHAEF/24134/SM/Ops, 11 Jul 45, in JSC folder European Theater 000; Extract From Letter from L.C.S. Dated 1 February 1944, in JSC folder 370.2 Operations and Results; Hesketh 33–34, 77. *59th Infantry Division:* Drawing dated 31 Aug 49, in U.S. Army Institute of Heraldry Unit Files, 59th Division; National War College Lecture, Lieutenant Colonel John R. Deane, Jr., 1 March 1950, p. 16, in NOA 546 Folder Tactical Deception—Ruses; From London, 22 Jun 44; Notional U.S. Formations Employed in the Western European Theater During 1944, SHAEF/24134/SM/Ops, 11 Jul 45, both in JSC folder European Theater 000; Fortitude Order of Battle, SHAEF/18216/3/Ops (B), in NARA RG 331 Entry 199 Box 66 Folder Fortitude (fat folder of two by that name); Hesketh 108, 121, 141, 156. *108th Infantry Division:* Doriot to Director JSC, 3 Aug 44; drawing dated 15 Jul 45, both in U.S. Army Institute of Heraldry Unit Files, 108th Infantry Division; Smith to CG, Alaskan Department, Subject: Plan "Wedlock," JSC/J1E1 Serial 549, 10 May 44, in JSC folder Wedlock—subfolder 370.2 Wedlock—Adm Impl; subfolder "General Plan—Wedlock—Copy No. 2," in JSC folder Wedlock; Draft "Present Evaluation of AD-Japan 1944," p. 6, in NOA 551 Folder Wedlock. *109th Infantry Division:* Bevan letter, Notional U.S. Formations, 15 Mar 45; Bevan letter, Notional U.S. Formations, 9 Apr 45; HMOC/1307, 19 Jul 44, all in JSC folder European Theater 000. *112th Infantry Division:* Bevan letter, Notional U.S. Formations, 15 Mar 45. *119th Infantry Division:* Doriot to Director JSC, 3 Aug 44; drawing dated 15 Jul 45, both in U.S. Army Institute of Heraldry Unit Files, 119th Infantry Division; Smith to CG, Alaskan Department, Subject: Plan "Wedlock," JSC/J1E1 Serial 549, 10 May 44, in JSC folder Wedlock—subfolder 370.2 Wedlock—Adm Impl; subfolder "General Plan—Wedlock—Copy No. 2," in JSC folder Wedlock; Draft "Present Evaluation of AD-Japan 1944," p. 6, in NOA 551 Folder Wedlock; "Analysis of the Planning and Execution of Cover and Deception Measures . . . ," p. 25, in NOA 551 Folder Wedlock Op-03-C-OUT 204; Anonymous Air War College lecture, March 6, 1947, p. 12, in USAFHRA, Maxwell AFB, Alabama Folder 180.9811-4 47/03/06; Implementation of Bluebird Special Means for Week Ending 21 April 1945, and same for week ending 28 April 1945, in NOA 548 Folder Bluebird. *125th Infantry Division:* National War College Lecture, Lieutenant Colonel John R. Deane, Jr., 1

REFERENCES

March 1950, p. 4, in NOA 546 Folder Tactical Deception—Ruses. *130th Infantry Division:* Doriot to Director JSC, 3 Aug 44; drawing dated 15 Jul 45, both in U.S. Army Institute of Heraldry Unit Files, 130th Infantry Division; Smith to CG, Alaskan Department, Subject: Plan "Wedlock," JSC/J1E1 Serial 549, 10 May 44, in JSC folder Wedlock—subfolder 370.2 Wedlock—Adm Impl; Draft "Present Evaluation of AD-Japan 1944," p. 6, in NOA 551 Folder Wedlock. *135th Airborne Division:* Doriot to Director JSC, 3 Aug 44; drawing dated 15 Jul 45, both in U.S. Army Institute of Heraldry Unit Files, 135th Airborne Division; AMSSO to BAD, 8 May 44, with handwritten notation; Smith to O'Connor, JSC/J4 Serial 545, 9 May 44, both in JSC folder 370.2 Mediterranean. *141st Infantry Division:* Doriot to Director JSC, 3 Aug 44; drawing dated 15 Jul 45, in U.S. Army Institute of Heraldry Unit Files, 141st Infantry Division; Smith to CG, Alaskan Department, Subject: Plan "Wedlock," JSC/J1E1 Serial 549, 10 May 44, in JSC folder Wedlock—subfolder 370.2 Wedlock—Adm Impl; Draft "Present Evaluation of AD-Japan 1944," p. 6, in NOA 551 Folder Wedlock; subfolder "General Plan—Wedlock—Copy No. 2;" in JSC folder Wedlock; but see "Analysis of the Planning and Execution of Cover and Deception Measures . . . ," p. 25, in NOA 551 Folder Wedlock Op-03-C-OUT 204 (Adak); Draft "Present Evaluation of AD-Japan 1944," pp. 6, 21, in NOA 551 Folder Wedlock; CINCPOA to COMINCH (JSC), 260031 March 1945; COMINCH (FOR JSC) to CINCPOA ADV, 032057 April 1945, copies of both in author's possession; Anonymous Air War College lecture, March 6, 1947, p. 12, in US-AFHRA, Maxwell AFB, Alabama, Folder 180.9811-4 47/03/06. *157th Infantry Division:* Doriot to Director JSC, 7 Aug 44; drawing dated 15 Jul 45, both in U.S. Army Institute of Heraldry Unit Files, 157th Infantry Division; Smith to CG, Alaskan Department, Subject: Plan "Wedlock," JSC/J1E1 Serial 601, 1 Jun 44, in JSC folder Wedlock—subfolder "370.2 Wedlock—Adm Impl"; Draft "Present Evaluation of AD-Japan 1944," p. 6, in NOA 551 Folder Wedlock; subfolder "General Plan—Wedlock—Copy No. 2," in JSC folder Wedlock; "Implementation of Bluebird Special Means for Week Ending 7 May 1945," in NOA 548 Folder Bluebird. *9th Armored Regiment:* National War College Lecture, Lieutenant Colonel John R. Deane, Jr., 1 March 1950, pp. 4–5, in NOA 546 Folder Tactical Deception—Ruses; II/41–42. *613th Parachute Regiment:* Minutes of a Meeting To Discuss Plans of Coordinating the Deception Activities of China Theater and South East Asia Command, 30 Jun 45; CGUSFCT to War Department, CFB 2377, 1 Aug 45; R. V. R[ushton] to Smith, 3 Aug 45; CFB 2870, CG US Forces China Theater to War Department, 6 August 45; Bissell to CGUSFCT, WAR 43968, 3 Aug 45, all in JSC folder 000—To 312.1(2) China-Burma-India Theater; Deception Plan: To Represent the Presence in India of Two Notional U.S. Parachute Regiments for Operations in China, undated, evidently an attachment to Memorandum to Chief of Staff, Subject: Deception Planning, unsigned [Corbett], 19 Jul 45, in JSC folder Southeast Asia. *619th Parachute Regiment:* Same as preceding. *7th Ranger Battalion:* Smith to O'Connor, 28 Mar 44, JSC/J4 Serial 383, in JSC file 370.2 Bodyguard. *8th Ranger Battalion:* Smith to O'Connor, JSC/J4 Serial 383, 28 Mar 44; Smith to O'Connor, 18 Mar 44, JSC/J4 Serial 383, both in JSC file 370.2 Bodyguard; Bevan letter, Notional U.S. Formations, 9 Apr 45, in JSC folder European Theater 000. *9th Ranger Battalion:* Same. *10th Ranger Battalion:* Same. *11th Ranger Battalion:* Same. *Ninth Fleet:* Cake Before Breakfast p. 50, in JSC loose papers. *Task Force 23:* Resume of TORCH and Associated Plans, OP-30C-OUT-214, p. 17, in JSC folder Conscious. *Task Force 69:* Cominch to Cinc United States Atlantic Fleet, Subject: Task Force 69—Directive for the Composition, Assembly, Loading, Overseas Movement, and Disposition of, FF1/A16-3, Serial 001671, 17 Aug 43, in JSC folder 370.2 Wadham. *VIII Tactical Air Command:* From London, 13 Jul 44 endorsed C. E. G. 17 Jul 44; Bevan letter, Notional U.S. Formations, 9 Apr 45, both in JSC folder European Theater 000; National War College Lecture, Lieutenant Colonel John R. Deane, Jr., 1 March 1950, p. 11, in NOA 546 Folder Tactical Deception—Ruses. *20th Troop Carrier Group:* Bissell to Hunter, No. 6497, 22 Feb 44, CM-OUT-9357 (22 Feb 44), in JSC file

REFERENCES

000—To 312.1(1) China-Burma-India Theater. *66th Heavy Bombardment Group:* CG Fourteenth Air Force to War Department, 20 Jun 44, CM-IN-16334 (20 Jun 44); undated, unsigned memorandum, both in JSC unlabeled folder [CBI]. *550th Fighter Group:* Same. *551st Fighter Group:* Same.

British Units

Fourth Army: Hesketh 38, 119, 142, 159, 162; Wingate 357; Fortitude Order of Battle, SHAEF/18216/3/Ops (B), in NARA RG 331 Entry 199 Box 66 Folder Fortitude (fat folder of two by that name). *Sixth Army:* Wingate 357; Hesketh 2. *Twelfth Army:* PRO WO 204/1562/83; Mure, Master 94; II/42; III/71; III/144; Wingate 164; Cole 28, 31; D Division Weekly Progress Report Nos. 8, 9, both in JSC folder Deception—SE Asia—WWII Historical D. Div—SACSEA—24 Mar 45–4 Jun 45. *II Corps:* Cole 31; Wingate 357, 425; Hesketh 33–34, 38, 76, 119, 155–56; Harmer 32–33; Fortitude Order of Battle, SHAEF/18216/3/Ops (B), 29 Aug 44; Wild to Controlling Officer *et al.,* SHAEF/18216.3/Ops (B) GCT 322-2 Ops (B), 5 Dec 44, both in NARA RG 331 Entry 199 Box 66 Folder Fortitude (fat folder of two by that name); G(R) War Diary October 1944, App. J-3, 9 Oct 44, PRO WAR 171/3831; "R" Force War Diary 28 Sep 44, October 1944 App. D, 5 Nov 44, PRO WO 171/3832. *III Corps:* Cole 31; III/71, 143, 167; IV/130; Wingate 164. VII Corps: Hesketh 34, 38, 119–20, 142, 161–62; Cole 33; Fortitude Order of Battle, SHAEF/18216/3/Ops (B), 29 Aug 44, in NARA RG 331 Entry 199 Box 66 Folder Fortitude (fat folder of two by that name). *XIV Corps:* PRO WO 204/1562/90; Wingate 218; II/42; IV/11–12, 34, 108; V/6, 16, 18; Extract From Weekly Letter No. 5 of 3/27/44, in JSC folder 370.2 Operations and Results; Allied Order of Battle, Med. Theater on 28 May 1944, in JSC folder 370.2 Mediterranean. *XVI Corps:* PRO WO 204/1562/85; Allied Order of Battle, Med. Theater on 28 May 1944, in JSC folder 370.2 Mediterranean; III/167. *XVIII Corps:* I/32. *XIX Airborne Corps:* Hesketh 158, 162. *XXI Corps:* Wingate 357. *XXV Corps:* Cole 37; I/32; II/36; PRO WO 204/1562/81. *XXVI Airborne Corps:* Sheet of formation insignia, undated, in JSC folder Southeast Asia; D Div. Weekly Implementation Series No. 2 3–9 Mar 1945; No. 3 10/16 Mar 1945, both in JSC folder D Division Weekly Implementation Series; D Division Weekly News Letter No. 41 27 Jan–2 Feb 45; No. 44 17 Feb–23 Feb 45, both in JSC folder Deception—SE Asia—WWII Historical D. Div—SACSEA—30 Dec 44–23 Mar 45; D Division Progress Report No. 3 16–31 Jan 45, Group II(b), in JSC folder Deception—SE Asia—WWII Historical D. Div—SACSEA—16 Dec 44–15 Mar 45; D Division Weekly Progress Report No. 3 7–13 Apr 45 p. 23; No. 5 21–27 Apr 45; No. 11 5–11 Jun 45 para. 5, all in JSC folder Deception—SE Asia—WWII Historical D. Div—SACSEA—24 Mar 45–4 Jun 45; D Division Weekly Progress Report No. 16 10–16 Jul 45, p. 6, in JSC folder Deception—SE Asia—WWII Historical D. Div. SACSEA 5 Jun–10 Sep 45. *2d Airborne Division:* Cole 73; Hesketh 109, 119–20, 147–49, 162; Fortitude Order of Battle, SHAEF/18216/3/Ops (B), 29 Aug 44, in NARA RG 331 Entry 199 Box 66 Folder Fortitude (fat folder of two by that name). *3d Airborne Division:* Sheet of formation insignia, undated; Amendment No. 1 to 14th Army Operation "Conclave" with attachments, both in JSC folder Southeast Asia; D Division Weekly News Letter No. 41 27 Jan–2 Feb 45; No. 44 17 Feb–23 Feb 45; No. 46 3 Mar–9 Mar 45 Appendix, all in JSC folder Deception—SE Asia—WWII Historical D. Div—SACSEA—30 Dec 44–23 Mar 45; D Division Progress Report No. 6 1–15 Mar 45, in JSC folder Deception—SE Asia—WWII Historical D. Div—SACSEA—16 Dec 44–15 Mar 45; D Div. Weekly Implementation Series No. 3 10/16 Mar 1945, in JSC folder D Division Weekly Implementation Series. *4th Airborne Division:* PRO WO 204/1562/83; II/41, 86; IV/58–61; Allied Order of Battle, Med. Theater on 28 May 1944, in JSC folder 370.2 Mediterranean. *5th Airborne Division:* PRO WO 204/1562/88; Extract From Weekly Letter No. 5 of 3/27/44, in JSC folder 370.2 Operations and Results; Allied Order of Battle, Med. Theater on 28 May 1944, in JSC folder 370.2 Mediterranean; IV/111, 130; V/4, 6–7;

1098

Wingate 204–205. *5th Armored Division:* Hesketh 119–20, 160, 162; Fortitude Order of Battle, SHAEF/18216/3/Ops (B), 29 Aug 44, in NARA RG 331 Entry 199 Box 66 Folder Fortitude (fat folder of two by that name). *7th Division:* PRO WO 204/1562/81; II/32, 39; Allied Order of Battle, Med. Theater on 28 May 1944, in JSC folder 370.2 Mediterranean. *8th Division:* Cole 51; II/39. *8th Armored Division:* Cole 43; III/86–87, 161; IV/41, 58–61; Howard 88; Allied Order of Battle, Med. Theater on 28 May 1944, in JSC folder 370.2 Mediterranean. *9th Armored Division:* Cole 43; Hesketh 25 n. 2, 90 n. 31, 112; Fortitude Order of Battle, SHAEF/18216/3/Ops (B), 29 Aug 44, in NARA RG 331 Entry 199 Box 66 Folder Fortitude (fat folder of two by that name). *12th Mixed Division:* PRO WO 204/1562/82; Cole 52; II/36, 38; Allied Order of Battle, Med. Theater on 28 May 1944, in JSC folder 370.2 Mediterranean. *15th Armored Division:* PRO WO 204/1562/85; II/35, 38. *15th Motorized Division:* PRO WO 204/1562/85; Allied Order of Battle, Med. Theater on 28 May 1944, in JSC folder 370.2 Mediterranean; D Division Weekly News Letter No. 46 3 Mar–9 Mar 45 Appendix, in JSC folder Deception—SE Asia—WWII Historical D. Div—SACSEA—30 Dec 44–23 Mar 45; D Division Weekly Progress Report No. 5 21–27 Apr 45 p. 21, in JSC folder Deception—SE Asia—WWII Historical D. Div—SACSEA—24 Mar 45–4 Jun 45. *20th Armored Division:* PRO WO 204/1562/81; II/41; IV/83; Allied Order of Battle, Med. Theater on 28 May 1944, in JSC folder 370.2 Mediterranean. *21st Division:* Sheet of formation insignia, undated, in JSC folder Southeast Asia; D Div. Weekly Implementation Series No. 2 3–9 Mar 1945, in JSC folder D Division Weekly Implementation Series; D Division Weekly News Letter No. 46 3 Mar–9 Mar 45 Appendix; No. 48 10 Mar–16 Mar 45, in JSC folder Deception—SE Asia—WWII Historical D. Div—SACSEA—30 Dec 44–23 Mar 45; D Division Weekly Progress Report No. 8 15–21 May 45 para. 7, in JSC folder Deception—SE Asia—WWII Historical D. Div—SACSEA—24 Mar 45–4 Jun 45. *23d Division:* Cole 54; II/35–39, 42. *26th Armored Division:* Sheet of formation insignia, undated, in JSC folder Southeast Asia; D Division Weekly Progress Report No. 3 7–13 Apr 45 p. 19, in JSC folder Deception—SE Asia—WWII Historical D. Div—SACSEA—24 Mar 45–4 Jun 45; D Division Weekly Progress Report No. 18 24–30 Jul 45 p. 4, No. 19 31 Jul–6 Aug 45 p. 6, both in JSC folder Deception—SE Asia—WWII Historical D. Div. SACSEA 5 Jun–10 Sep 45; D Div. Weekly Implementation Series No. 2 3–9 Mar 1945, No. 4 17/23 Mar 1945, both in JSC folder D Division Weekly Implementation Series. *32d Division (Air Transit):* Sheet of formation insignia, undated, in JSC folder Southeast Asia; D Division Weekly News Letter No. 46 3 Mar–9 Mar 45 p. 4 and Appendix, in JSC folder Deception—SE Asia—WWII Historical D. Div—SACSEA—30 Dec 44–23 Mar 45; D Division Weekly Progress Report No. 16 10–16 Jul 45, in JSC folder Deception—SE Asia—WWII Historical D. Div. SACSEA 5 Jun–10 Sep 45. *33d Division:* PRO WO 204/1562/83; IV/34; Allied Order of Battle, Med. Theater on 28 May 1944, in folder 370.2 Mediterranean; D Division Weekly Progress Report No. 3 7–13 Apr 45, in JSC folder Deception—SE Asia—WWII Historical D. Div—SACSEA—24 Mar 45–4 Jun 45. *34th Division:* PRO WO 204/1562/85; PRO WO 204/1563/36–37; II/34, 42; IV/34, 130; V/15; Allied Order of Battle, Med. Theater on 28 May 1944, in folder 370.2 Mediterranean; Wingate 241. *40th Division:* PRO WO 204/1562/87; Cole 56; Wingate 205; II/42; Extract From Weekly Letter No. 5 of 3/27/44, in JSC folder 370.2 Operations and Results; Train to Davy, ADV"A"F/36/1, 12 Feb 44, PRO WO 204/1561/50; Allied Order of Battle, Med. Theater on 28 May 1944, in JSC folder 370.2 Mediterranean. *42d Division:* PRO WO 204/1562/90; Cole 45, 57; II/42; V/6; Extract From Weekly Letter No. 5 of 3/27/44, in JSC folder 370.2 Operations and Results; Allied Order of Battle, Med. Theater on 28 May 1944, in JSC folder 370.2 Mediterranean; AFHQ Mediterranean Joint Planning Staff P/143 (Final), Annexure B, Appendix 1, in NARA RG 319 Entry 1 Box 4 Folder 73; Wingate 205. *55th Division:* Cole 67. *57th Division:* PRO WO 204/1562/90; V/6; Extract From Weekly Letter No. 5 of 3/27/44, in JSC folder 370.2 Operations and Results; Allied Order of Battle, Med. Theater on 28 May 1944, in folder 370.2 Mediterranean; Wingate 205. *58th Division:* Hesketh 34, 35–36, 38, 77, 82,

107–108, 119, 142, 160, 163; Fortitude Order of Battle, SHAEF/18216/3/Ops (B), 29 Aug 44, in NARA RG 331 Entry 199 Box 66 Folder Fortitude (fat folder of two by that name). *59th Division:* Cole 68–69. *64th Division:* Sheet of formation insignia, undated, in JSC folder Southeast Asia; D Division Weekly News Letter No. 46 3 Mar–9 Mar 45, Appendix, in JSC folder Deception—SE Asia—WWII Historical D. Div—SACSEA—30 Dec 44–23 Mar 45. *68th Division:* Sheet of formation insignia, undated, in JSC folder Southeast Asia; Cole 69; Wingate 205. *70th Division:* Sheet of formation insignia, undated, in JSC folder Southeast Asia; Cole 70; D Division Weekly Progress Report No. 11 5–11 Jun 45, in JSC folder Deception—SE Asia—WWII Historical D. Div. SACSEA 5 Jun–10 Sep 45. *76th Division:* Cole 70; Harmer 33; Hesketh 120; Wild to Controlling Officer *et al.,* SHAEF/18216.3/Ops (B) GCT 322-2 Ops (B), 5 Dec 44; Fortitude Order of Battle, SHAEF/18216/3/Ops (B), 29 Aug 44, both in NARA RG 331 Entry 199 Box 66 Folder Fortitude (fat folder of two by that name). *77th Division:* Cole 71; Hesketh 120; Fortitude Order of Battle, SHAEF/18216/3/Ops (B), 29 Aug 44, in NARA RG 331 Entry 199 Box 66 Folder Fortitude (fat folder of two by that name). *80th Division:* Cole 72; Hesketh 119–20, 157, 160, 163; Fortitude Order of Battle, SHAEF/18216/3/Ops (B), 29 Aug 44, in NARA RG 331 Entry 199 Box 66 Folder Fortitude (fat folder of two by that name). *90th Division:* Hesketh 77 n. 7; Fortitude Order of Battle, SHAEF/18216/3/Ops (B), 29 Aug 44, in NARA RG 331 Entry 199 Box 66 Folder Fortitude (fat folder of two by that name). *10th Armored Brigade:* II/39. *27th Armored Brigade:* "R" Force War Diary 18–19 Oct 44 and App. J-12, J-12A, PRO WO 171/3832. *33d Armored Brigade:* II/39; Cole 198. *42d Tank Brigade:* Sheet of formation insignia, undated, in JSC folder Southeast Asia. *1st Special Air Service Brigade:* II/39, 41. *7th Air Landing Brigade:* II/129. *103d Special Service Brigade:* Sheet of formation insignia, undated, in JSC folder Southeast Asia. *140th Special Service (Commando) Brigade:* Sheet of formation insignia, undated, in JSC folder Southeast Asia; D Division Weekly News Letter No. 46 3 Mar–9 Mar 45, Appendix, in JSC folder Deception—SE Asia—WWII Historical D. Div—SACSEA—30 Dec 44–23 Mar 45. *North Eastern Task Force:* Appendix "C" to JSC loose item Report . . . Neptune July 1944. *Force F:* Hesketh 108; Appendix "C" to JSC loose item Report . . . Neptune July 1944 paras. 11–16. *Force M:* Appendix "C" to JSC loose item "Report . . . Neptune July 1944" para. 17. *Force N:* Appendix "C" to JSC loose item "Report . . . Neptune July 1944" paras. 18–20. *Force V:* Appendix "B" to JSC loose item "Report . . . Neptune July 1944" paras. 6(i)–(iv). *Force W:* Appendix "B" to JSC loose item "Report . . . Neptune July 1944" paras. 6(v)–(vi).

Indian Units

Indian Expeditionary Force: Sheet of formation insignia, undated, in JSC folder Southeast Asia; Cole 5 note*, 113; D Division Weekly News Letter No. 46 3 Mar–9 Mar 45 Appendix paras. 6, 8, in JSC folder Deception—SE Asia—WWII Historical D. Div—SACSEA—30 Dec 44–23 Mar 45; D Division Weekly Progress Report No. 11 5–11 Jun 45 para. 5, in JSC folder Deception—SE Asia—WWII Historical D. Div. SACSEA 5 Jun–10 Sep 45. *Force 144:* D Division Weekly News Letter No. 41 27 Jan–2 Feb 45, in JSC folder Deception—SE Asia—WWII Historical D. Div—SACSEA—30 Dec 44–23 Mar 45; D Division Weekly Progress Report No. 8 15–21 May 45 para. 5; No. 10 29 May–4 Jun 45 para. 5, both in JSC folder Deception—SE Asia—WWII Historical D. Div—SACSEA—24 Mar 45–4 Jun 45; D Division Weekly Progress Report No. 11 5–11 Jun 45 para. 5, in JSC folder Deception—SE Asia—WWII Historical D. Div. SACSEA 5 Jun–10 Sep 45. *XX Indian Corps:* Sheet of formation insignia, undated, in JSC folder Southeast Asia; D Division Weekly News Letter No. 46 3 Mar–9 Mar 45 Appendix; No. 47 10 Mar–16 Mar 45 para. 11, both in JSC folder Deception—SE Asia—WWII Historical D. Div—SACSEA—30 Dec 44–23 Mar 45; D Division Progress Report No. 3 16–31 Jan 45 Group III(b), in JSC folder Deception—SE Asia—WWII Historical D. Div—SACSEA—16 Dec 44–15 Mar 45; D Division Weekly

REFERENCES

Progress Report No. 1 24–30 Mar 45 para. 14, in JSC folder Deception—SE Asia—WWII Historical D. Div—SACSEA—24 Mar 45–4 Jun 45; D Division Weekly Progress Report No. 13 19–25 Jun 45, in JSC folder Deception—SE Asia—WWII Historical D. Div. SACSEA 5 Jun–10 Sep 45; D Division Weekly News Letter No. 46 3 Mar–9 Mar 45 Appendix, in JSC folder Deception—SE Asia—WWII Historical D. Div—SACSEA—30 Dec 44–23 Mar 45. *XXXVIII Indian Corps:* D Division Weekly Progress Report No. 3 7–13 Apr 45; No. 4 14–21 Apr 45 para. 9, both in JSC folder Deception—SE Asia—WWII Historical D. Div—SACSEA—24 Mar 45–4 Jun 45; D Division Weekly Progress Report No. 11 5–11 Jun 45, in JSC folder Deception—SE Asia—WWII Historical D. Div. SACSEA 5 Jun–10 Sep 45. *Airborne Raiding Task Force:* D Division Weekly Progress Report No. 16 10–16 Jul 45, in JSC folder Deception—SE Asia—WWII Historical D. Div. SACSEA 5 Jun–10 Sep 45. *1st Indian Division:* II/36, 39 ("a mixture"), 42. *2d Indian Division:* Cole 114; Mure, Practice 127; II/36, 39; Allied Order of Battle, Med. Theater on 28 May 1944, in JSC folder 370.2 Mediterranean; PRO WO 204/1562/80. *6th Indian Division:* Cole 117; Notional Order of Battle, 15 Feb 45, in JSC folder Southeast Asia. *9th Indian Airborne Division:* Sheet of formation insignia, undated, in JSC folder Southeast Asia; D Div. Weekly Implementation Series No. 1 24 Feb–2 Mar 1945; No. 2 3–9 Mar 1945; No. 3 10/16 Mar 1945, all in JSC folder D Division Weekly Implementation Series; D Division Weekly News Letter No. 44 17 Feb–23 Feb 45; No. 45 24 Feb–2 Mar 45 p. 4; No. 46 3 Mar–9 Mar 45, all in JSC folder Deception—SE Asia—WWII Historical D. Div—SACSEA—30 Dec 44–23 Mar 45; D Division Weekly Progress Report No. 3 7–13 Apr 45 p. 17, in JSC folder Deception—SE Asia—WWII Historical D. Div—SACSEA—24 Mar 45–4 Jun 45. *12th Indian Division:* Cole 121; II/41; PRO WO 204/1562/80; Mure, Practice 127; Allied Order of Battle, Med. Theater on 28 May 1944, in JSC folder 370.2 Mediterranean. *15th Indian Division:* Sheet of formation insignia, undated, in JSC folder Southeast Asia; D Div. Weekly Implementation Series No. 1 24 Feb–2 Mar 1945; No. 9 21–27 April 1945, both in JSC folder D Division Weekly Implementation Series; D Division Weekly News Letter No. 41 27 Jan–2 Feb 45; No. 42 3 Feb–9 Feb 45 Part I paras. 8–9, both in JSC folder Deception—SE Asia—WWII Historical D. Div—SACSEA—30 Dec 44–23 Mar 45; D Division Weekly Progress Report No. 3 7–13 Apr 45 p. 7; No. 5 21–27 Apr 45 pp. 6, 9; No. 8 15–21 May 45 para. 7, all in JSC folder Deception—SE Asia—WWII Historical D. Div—SACSEA—24 Mar 45–4 Jun 45; Allied Order of Battle, Med. Theater on 28 May 1944, in JSC folder 370.2 Mediterranean; PRO WO 204/1562/80. *16th Indian Division:* PRO WO 204/1562/80; Allied Order of Battle, Med. Theater on 28 May 1944, in JSC folder 370.2 Mediterranean; Notional Order of Battle, 15 Feb 45, in JSC folder Southeast Asia. *18th Indian Division:* Sheet of formation insignia, undated, in JSC folder Southeast Asia; D Division Weekly News Letter No. 41 27 Jan–2 Feb 45, in JSC folder Deception—SE Asia—WWII Historical D. Div—SACSEA—30 Dec 44–23 Mar 45; D Division Weekly Progress Report No. 1 24–30 Mar 45; No. 3 7–13 Apr 45 p. 6, both in JSC folder Deception—SE Asia—WWII Historical D. Div—SACSEA—24 Mar 45–4 Jun 45; Fleming 67. *21st Indian Division:* Sheet of formation insignia, undated, in JSC folder Southeast Asia; D Division Weekly News Letter No. 41 27 Jan–2 Feb 45; No. 42 3 Feb–9 Feb 45 Part I paras. 9, 12, both in JSC folder Deception—SE Asia—WWII Historical D. Div—SACSEA—30 Dec 44–23 Mar 45. *32d Indian Armored Division:* D Div. Weekly Implementation Series No. 1 24 Feb–2 Mar 1945; No. 3 10/16 Mar 1945, both in JSC folder D Division Weekly Implementation Series; Sheet of formation insignia, undated, in JSC folder Southeast Asia; D Division Weekly News Letter No. 44 17 Feb–23 Feb 45, in JSC folder Deception—SE Asia—WWII Historical D. Div—SACSEA—30 Dec 44–23 Mar 45. *34th Indian Division:* D Div. Weekly Implementation Series No. 4 17/23 Mar 1945, in JSC folder D Division Weekly Implementation Series; D Division Weekly Progress Report No. 3 7–13 Apr 45 p. 8, in JSC folder Deception—SE Asia—WWII Historical D. Div—SACSEA—24 Mar 45–4 Jun 45. *Frontier Armored Division:* D Division Weekly News Letter No. 41 27 Jan–2 Feb 45 p. 2; No. 46 3 Mar–9 Mar 45 Appendix, both in JSC folder Deception—SE Asia—WWII Historical D. Div—SACSEA—30 Dec 44–23 Mar 45. *Nepalese Division:* Sheet of formation insignia, undated, in JSC folder Southeast Asia; D Division Progress Report No. 4 1–14 Feb 45 Group I, in JSC folder Deception—SE Asia—WWII Historical D. Div—SACSEA—16 Dec 44–15 Mar 45. *1st Burma Division:* Sheet of formation insignia, undated, in JSC folder Southeast Asia; D Division Weekly News Letter No. 44 17 Feb–23 Feb 45 p. 2; No. 46 3 Mar–9 Mar 45 Appendix, both in JSC folder Deception—SE Asia—WWII Historical D. Div—SACSEA—30 Dec 44–23 Mar 45; D Division Weekly Progress Report No. 5 21–27 Apr 45 p. 18, in JSC folder Deception—SE Asia—WWII Historical D. Div—SACSEA—24 Mar 45–4 Jun 45; D Division Weekly Progress Report No. 14 26 Jun–2 Jul 45 p. 17; No. 16 10–16 Jul 45, both in JSC folder Deception—SE Asia—WWII Historical D. Div. SACSEA 5 Jun–10 Sep 45. *51st Indian Tank Brigade:* Amendment No. 1 to 14th Army Operation "Conclave" with attachments, in JSC folder Southeast Asia; D Division Weekly Progress Report No. 1 24–30 Mar 45, in JSC folder Deception—SE Asia—WWII Historical D. Div—SACSEA—24 Mar 45–4 Jun 45.

New Zealand Units

3d New Zealand Division: PRO WO 204/1562/82; II/38; Cole 106. *6th New Zealand Division:* PRO WO 204/1562/82; II/36–39; Cole 107; Allied Order of Battle, Med. Theater on 28 May 1944, in JSC folder 370.2 Mediterranean; Smith to Holland, Subject: Availability of Fictional Division, JSC/J4 Serial 1102, 1 Nov 44, in JSC folder 370.2 Mediterranean.

South African Units

7th South African Division: PRO WO 204/1562/82; II/41; Allied Order of Battle, Med. Theater on 28 May 1944, in JSC folder 370.2 Mediterranean.

African Colonial Units

10th African Airborne Division: Sheet of formation insignia, undated, in JSC folder Southeast Asia; D Division Weekly News Letter No. 41 27 Jan–2 Feb 45; No. 44 17 Feb–23 Feb 45, both in JSC folder Deception—SE Asia—WWII Historical D. Div—SACSEA—30 Dec 44–23 Mar 45. *13th East African Division:* Sheet of formation insignia, undated, in JSC folder Southeast Asia; III/20; Wingate 331; D Division Weekly News Letter No. 46 3 Mar–9 Mar 45, Appendix, in JSC folder Deception—SE Asia—WWII Historical D. Div—SACSEA—30 Dec 44–23 Mar 45; D Division Weekly Progress Report No. 3 7–13 Apr 45, in JSC folder Deception—SE Asia—WWII Historical D. Div—SACSEA—24 Mar 45–4 Jun 45. *83d West African Division:* Sheet of formation insignia, undated, in JSC folder Southeast Asia; D Division Weekly News Letter No. 46 3 Mar–9 Mar 45, Appendix, in JSC folder Deception—SE Asia—WWII Historical D. Div—SACSEA—30 Dec 44–23 Mar 45.

French Units

III Corps: No. 2 Tac Report Part II pp. 97–98, 121, 123, 131, in NARA RG 319 Entry 101 Box 4 Folder 80. *IV Corps:* No. 2 Tac Report Part II pp. 98, 121, 123, in NARA RG 319 Entry 101 Box 4 Folder 80. *French Expeditionary Corps for the Far East ("FEFEO"):* No. 2 Tac Report Part II pp. 96–97, 121, in NARA RG 319 Entry 101 Box 4 Folder 80. *7th Algerian Infantry Division:* PRO WO 204/1562/88–89; Allied Order of Battle, Med. Theater on 28 May 1944, in JSC folder 370.2 Mediterranean. *8th Algerian Infantry Division:* Same. *3d Armored Division:* PRO WO 204/1562/87; Allied Order of Battle, Med. Theater on 28 May 1944, in JSC folder 370.2 Mediterranean. *10th Colonial Infantry Division:* PRO WO 204/1562/91; Allied Order of Battle, Med. Theater on 28 May 1944, in JSC folder 370.2 Mediterranean.

REFERENCES

Greek Units

1st Greek Division: PRO WO 204/1562/81; Hesketh 247a; see Cole 247; II/42; Allied Order of Battle, Med. Theater on 28 May 1944, in JSC folder 370.2 Mediterranean.

Polish Units

III Polish Corps: PRO WO 204/1562/86; Wingate 164; IV/40, 111. *2d Polish Armored Division:* PRO WO 204/1562/86; II/42; Allied Order of Battle, Med. Theater on 28 May 1944, in JSC folder 370.2 Mediterranean. *7th Polish Division:* Cole 253; PRO WO 204/1562/86; II/41. *8th Polish Division:* Allied Order of Battle, Med. Theater on 28 May 1944, in JSC folder 370.2 Mediterranean.

INDEX

Note: Page numbers in *italics* refer to illustrations.